STARCH: CHEMISTRY AND TECHNOLOGY

Third Edition

Food Science and Technology
International Series

Series Editor

Steve L. Taylor
University of Nebraska – Lincoln, USA

Advisory Board

Ken Buckle
The University of New South Wales, Australia

Mary Ellen Camire
University of Maine, USA

Roger Clemens
University of Southern California, USA

Hildegarde Heymann
University of California – Davis, USA

Robert Hutkins
University of Nebraska – Lincoln, USA

Ron S. Jackson
Quebec, Canada

Huub Lelieveld
Bilthoven, The Netherlands

Daryl B. Lund
University of Wisconsin, USA

Connie Weaver
Purdue University, USA

Ron Wrolstad
Oregon State University, USA

A complete list of books in this series appears at the end of this volume

Starch: Chemistry and Technology

Third Edition

Edited by

James BeMiller and Roy Whistler

AMSTERDAM • BOSTON • HEIDELBERG • LONDON • NEW YORK
OXFORD • PARIS • SAN DIEGO • SAN FRANCISCO • SINGAPORE
SYDNEY • TOKYO

Academic Press is an imprint of Elsevier

Academic Press is an imprint of Elsevier
30 Corporate Drive, Suite 400, Burlington, MA 01803, USA
32 Jamestown Road, London NW1 7BY, UK
525 B Street, Suite 1900, San Diego, CA 92101-4495, USA
360 Park Avenue South, New York, NY 10010-1710, USA

Second edition 1984
Third edition 2009

Copyright © 1984, 2009 Elsevier Inc. Apart from Chapter 19 which is in the public domain. All rights reserved

No part of this publication may be reproduced, stored in a retrieval system or transmitted in any form or by any means electronic, mechanical, photocopying, recording or otherwise without the prior written permission of the publisher

Permissions may be sought directly from Elsevier's Science & Technology Rights Department in Oxford, UK: phone (+44) (0) 1865 843830; fax (+44) (0) 1865 853333; email: permissions@elsevier.com. Alternatively visit the Science and Technology Books website at www.elsevierdirect.com/rights for further information

Notice
No responsibility is assumed by the publisher for any injury and/or damage to persons or property as a matter of products liability, negligence or otherwise, or from any use or operation of any methods, products, instructions or ideas contained in the material herein. Because of rapid advances in the medical sciences, in particular, independent verification of diagnoses and drug dosages should be made

Library of Congress Cataloging in Publication Data
A catalog record for this book is available from the Library of Congress

British Library Cataloguing in Publication Data
A catalogue record for this book is available from the British Library

ISBN: 978-0-12-746275-2

For information on all Academic Press publications
visit our website at www.elsevierdirect.com

Typeset by Macmillan Publishing Solutions
(www.macmillansolutions.com)

Working together to grow
libraries in developing countries

www.elsevier.com | www.bookaid.org | www.sabre.org

ELSEVIER BOOK AID International Sabre Foundation

Contents

Preface to the Third Edition xvii
List of Contributors xix

1   History and Future of Starch 1
    I.   History 1
         1. Early History 1
         2. 1500–1900 2
         3. 1900–Present. 4
    II. Development of Specialty Starches 5
         1. Waxy Corn Starch 5
         2. High-amylose Corn Starch. 5
         3. Chemically Modified Starches 6
         4. Other Naturally Modified Corn Starches. 6
    III. Other Products from Starch 6
         1. Sweeteners 6
         2. Ethanol 7
         3. Polyols. 8
         4. Organic Acids 8
         5. Amino Acids 8
    IV. Future of Starch 9
         1. Two New Starches for Industry 9
         2. Present American Companies 9
    V.   References 10

2   Economic Growth and Organization of the US Corn Starch Industry 11
    I.   Introduction 11
    II.   Extent and Directions of Market Growth 11
    III. High-fructose Syrup Consumption 13
    IV.   Fuel Alcohol 15
    V.   Technical Progress 16
    VI.   Plant Location 16
    VII. Industry Organization 16
    VIII. Effects of Corn Price Variability 18
    IX.   International Involvement 19
    X.   Future Industry Prospects 20
    XI.   References 20

3 Genetics and Physiology of Starch Development 23
   I. Introduction 24
   II. Occurrence 25
      1. General Distribution 25
      2. Cytosolic Starch Formation 25
      3. Starch Formed in Plastids 26
   III. Cellular Developmental Gradients 26
   IV. Non-mutant Starch Granule Polysaccharide Composition 28
      1. Polysaccharide Components 28
      2. Species and Cultivar Effects on Granule Composition 30
      3. Developmental Changes in Granule Composition 31
      4. Environmental Effects on Granule Composition 32
   V. Non-mutant Starch Granule and Plastid Morphology 33
      1. Description 33
      2. Species and Cultivar Effects on Granule Morphology 33
      3. Developmental Changes in Average Starch Granule Size 34
      4. Formation and Enlargement of Non-mutant Granules 34
   VI. Polysaccharide Biosynthesis 36
      1. Enzymology 36
      2. Compartmentation and Regulation of Starch Synthesis and
         Degradation in Chloroplasts 37
      3. Compartmentation and Regulation of Starch
         Synthesis in Amyloplasts 40
   VII. Mutant Effects 43
      1. Waxy 44
      2. Amylose-extender 50
      3. Sugary 53
      4. Sugary-2 56
      5. Dull 57
      6. Amylose-extender Waxy 58
      7. Amylose-extender Sugary 59
      8. Amylose-extender Sugary-2 60
      9. Amylose-extender Dull 61
      10. Dull Sugary 61
      11. Dull Sugary-2 62
      12. Dull Waxy 62
      13. Sugary Waxy 63
      14. Sugary-2 Waxy 63
      15. Sugary Sugary-2 64
      16. Amylose-extender Dull Sugary 64
      17. Amylose-extender Dull Sugary-2 65
      18. Amylose-extender Dull Waxy 65
      19. Amylose-extender Sugary Sugary-2 66
      20. Amylose-extender Sugary Waxy 66
      21. Amylose-extender Sugary-2 Waxy 67
      22. Dull Sugary Sugary-2 67
      23. Dull Sugary Waxy 67
      24. Dull Sugary-2 Waxy 68

| | | 25. Sugary Sugary-2 Waxy .... 68 |
| | | 26. Amylose-extender Dull Sugary Waxy .... 68 |

    VIII. Conclusions 69
    IX. References 71

4 Biochemistry and Molecular Biology of Starch Biosynthesis 83

    I. Introduction 84
    II. Starch Synthesis in Plants: Localization 84
        1. Leaf Starch 84
        2. Starch in Storage Tissues 85
    III. Enzyme-catalyzed Reactions of Starch Synthesis in Plants and Algae and Glycogen Synthesis in Cyanobacteria 85
    IV. Properties of the Plant 1,4-α-Glucan-Synthesizing Enzymes 87
        1. ADP-glucose Pyrophosphorylase: Kinetic Properties and Quaternary Structure 87
        2. Relationship Between the Small and Large Subunits: Resurrection of ADPGlc PPase Catalysis in the Large Subunit 91
        3. Phylogenetic Analysis of the Large and Small Subunits 95
        4. Crystal Structure of Potato Tuber ADPGlc PPase 95
        5. Supporting Data for the Physiological Importance of Regulation of ADPGlc PPase 104
        6. Differences in Interaction Between 3PGA and Pi in Different ADPGlc PPases 105
        7. Plant ADPGlc PPases can be Activated by Thioredoxin 107
        8. Characterization of ADPGlc PPases from Different Sources 108
        9. Identification of Important Amino Acid Residues Within the ADPGlc PPases 111
        10. Starch Synthase 114
        11. Branching Enzyme 129
        12. Other Enzymes Involved in Starch Synthesis 136
    V. Abbreviations 138
    VI. References 139

5 Structural Features of Starch Granules I 149

    I. Introduction 149
    II. Granule Architecture 153
        1. An Overview of Granule Structure 153
        2. Molecular Organization of Crystalline Structures 153
        3. Crystalline Ultrastructural Features of Starch 158
        4. The Supramolecular Organization of Starch Granules 160
    III. The Granule Surface 167
        1. Starch Granule Surface and Chemistry and Composition 168
        2. Surface-Specific Chemical Analysis 169
    IV. Granule Surface Imaging 170
        1. Granule Imaging by SEM Methods 170
        2. Principles of AFM 171
        3. Sample Preparation for AFM Imaging of Granular Starch 172
        4. Surface Detail and Inner Granule Structure Revealed by AFM 173
        5. Interpretation of AFM Images with Respect to Granule Structure .. 175

      6. Discussion of Granule Surface Imaging by Scanning Probe Microscopy (SPM)....................................177
      7. Future Prospects of SPM of Starch 179
  V. A Hypothesis of Starch Granule Structure: The Blocklets Concept.....180
  VI. Location and State of Amylose Within Granules....................184
  VII. Surface Pores and Interior Channels of Starch Granules.............186
  VIII. Conclusions 187
  IX. References 188

6 Structural Features of Starch Granules II 193
  I. Introduction...193
  II. General Characteristics of Starch Granules........................194
      1. Granule Shapes, Sizes and Distributions 194
      2. Porous Structures of Starch Granules 195
      3. Shapes of Gelatinized Starch Granules 200
  III. Molecular Compositions of Starch Granules 201
      1. Amylopectin and Amylose............................201
      2. Intermediate Material and Phytoglycogen................202
      3. Lipids and Phospholipids 204
      4. Phosphate Monoesters 205
  IV. Structures of Amylose and Amylopectin 205
      1. Chemical Structure of Amylose 205
      2. Single Helical Structures (V-Complexes) of Amylose 208
      3. Double Helical Structures of Amylose 211
      4. Chemical Structure of Amylopectin......................212
      5. Cluster Models of Amylopectin.........................218
      6. Effects of Growing Temperature and Kernel Maturity on Starch Structures...................................224
  V. Locations of Molecular Components in the Granule 225
  VI. References 227

7 Enzymes and Their Action on Starch...............................237
  I. Introduction...238
  II. Amylases..238
      1. Action of Endo-Acting α-Amylases 238
      2. Action of Exo-Acting β-Amylases.......................244
      3. Amylases Producing Specific Maltodextrin Products.......246
      4. Action of Isoamylases 247
      5. Archaebacterial Amylases 248
      6. Action of Cyclomaltodextrin Glucanosyltransferase 250
  III. Relation of Structure with Action of the Enzymes..................253
      1. Relation of Structure with Action of Endo-Acting α-Amylases......253
      2. Structure and Action of Soybean β-Amylase 257
      3. Structure and Action of Glucoamylases..................257
      4. Specific Amino Acids at the Active-Site Involved in Catalysis and Substrate Binding...........................261
      5. Structure and Function of Domains in Amylolytic Enzymes.......262
  IV. Mechanisms for the Enzymatic Hydrolysis of the Glycosidic Bond 264
  V. Action of Amylases on Insoluble Starch Substrates..................267

      1. Action of α-Amylases on Amylose-V Complexes and
         Retrograded Amylose ... 267
      2. Action of Amylases with Native Starch Granules ... 269
  VI. Inhibitors of Amylase Action ... 272
  VII. Action of Phosphorylase and Starch Lyase ... 276
      1. Plant Phosphorylase ... 276
      2. Starch Lyase ... 277
  VIII. Enzymic Characterization of Starch Molecules ... 278
      1. Determination of the Nature of the Branch Linkage in Starch ... 279
      2. Identification and Structure Determination of Slightly
         Branched Amyloses ... 280
      3. Formation of β-Amylase Limit Dextrins of Amylopectin and
         Determination of their Fine Structure ... 282
  IX. References ... 284

8 Structural Transitions and Related Physical Properties of Starch ... 293
  I. Introduction ... 293
  II. Starch Structure, Properties and Physical Methods of Analysis ... 295
      1. Ordered and Amorphous Structural Domains (See Also
         Chapters 5 and 6) ... 296
      2. Physical Properties of Starch in Water ... 301
  III. State and Phase Transitions ... 310
      3. Glass Transitions of Amorphous Structural Domains ... 311
      4. Annealing and Structural Modifications by Heat–Moisture
         Treatments ... 320
      5. Melting Transitions of Crystallites in Granular Starch ... 323
      6. Gelation and Retrogradation of Starch and its Polymeric
         Components ... 332
      7 Phase Transitions and Other Properties of V-Structures ... 354
  IV. References ... 359

9 Corn and Sorghum Starches: Production ... 373
  I. Introduction ... 374
  II. Structure, Composition and Quality of Grain ... 375
      1. Structure ... 376
      2. Composition ... 381
      3. Grain Quality ... 385
  III. Wet-milling ... 391
      1. Grain Cleaning ... 392
      2. Steeping ... 394
      3. Milling and Fraction Separation ... 408
      4. Starch Processing ... 421
      5. Product Drying, Energy Use and Pollution Control ... 421
      6. Automation ... 423
  IV. The Products ... 423
      1. Starch ... 423
      2. Sweeteners ... 423
      3. Ethanol ... 424
      4. Corn Oil ... 425
      5. Feed Products ... 426

	V. Alternative Fractionation Procedures 427
	VI. Future Directions in Starch Manufacturing...................429
	1. Continued Expansion into Fermentation Products 429
	2. Biosolids as Animal Food 429
	3. Processing of Specific Hybrids..........................430
	4. New Corn Genotypes and Phenotypes via Biotechnology and Genetic Engineering..................................430
	5. Segregation of the Corn Starch Industry 430
	VII. References 431
10	**Wheat Starch: Production, Properties, Modification and Uses 441**
	I. Introduction 442
	II. Production 442
	III. Industrial Processes for Wheat Starch Production 444
	1. Conventional Processes................................446
	2. Hydrocyclone Process (Dough–Batter) 448
	3. High-pressure Disintegration Process 450
	IV. Properties of Wheat Starch and Wheat Starch Amylose and Amylopectin...451
	1. Large Versus Small Granules 452
	2. Fine Structures of Amylose and Amylopectin................457
	3. Partial Waxy and Waxy Wheat Starches 465
	4. High-amylose Wheat Starch 470
	5. A Unique Combination of Properties 471
	V. Modification of Wheat Starch 475
	1. Crosslinking..475
	2. Substitution..478
	3. Dual Derivatization 479
	4. Bleaching, Oxidation and Acid-thinning 480
	VI. Uses of Unmodified and Modified Wheat Starches 481
	1. Role in Baked Products 481
	2. Functionality in Noodles and Pasta.......................485
	3. Other Food Uses 488
	4. Industrial Uses......................................489
	VII. References 491
11	**Potato Starch: Production, Modifications and Uses 511**
	I. History of Potato Processing in The Netherlands 512
	II. Starch Production...514
	1. World Starch Production 514
	2. Potato Starch Production in Europe 514
	III. Structure and Chemical Composition of the Potato................515
	1. Formation and Morphology of the Tuber 515
	2. Anatomy of the Tuber 516
	3. Chemical Composition 518
	4. Differences Between Commercial Starches 519
	5. New Development: The All-amylopectin Potato 521
	IV. Potato Starch Processing 522
	1. Grinding 525

		2. Potato Juice Extraction 525
		3. Fiber Extraction 526
		4. Starch Classification 527
		5. Starch Refinery 529
		6. Sideline Extraction 530
		7. Removal of Water from the Starch 532
		8. Starch Drying and Storage 533
	V.	Potato Protein 534
		1. Environmental Aspects 534
		2. Protein Recovery 535
		3. Properties and Uses 535
	VI.	Utilization 535
		1. Substitution (See Also Chapters 17 and 20) 535
		2. Converted Starches (See Also Chapters 17 and 20) 536
		3. Crosslinked Starches (See Also Chapters 17 and 20) 536
		4. The Preference for Potato Starch in Applications 537
	VII.	Future Aspects of Potato Starch Processing 538
	VIII.	References 538

12 Tapioca/Cassava Starch: Production and Use 541

	I.	Background 541
	II.	Processing 545
	III.	Tapioca Starch 550
	IV.	Modification 555
	V.	Food Applications 556
	VI.	Industrial Applications 563
	VII.	Outlook 564
	VIII.	References 564

13 Rice Starches: Production and Properties 569

	I.	Rice Production and Composition 569
		1. Rice Production 569
		2. Rice Milling and Composition 570
	II.	Uses of Milled Rice and Rice By-products 571
		1. Milled Rice 571
		2. By-products 572
	III.	Preparation of Rice Starch 573
		1. Traditional Method 573
		2. Mechanical Method 574
	IV.	Properties of Rice Starch 574
		1. General Properties Unique to Rice Starch 574
		2. Pasting Properties 575
	V.	Factors Affecting Rice Starch Properties 575
		1. Rice Variety: Common Versus Waxy 575
		2. Protein Content 576
		3. Method of Preparation 576
		4. Modification 577
	VI.	Rice Starch Applications 577
	VII.	References 578

14 Rye Starch 579
    I. Introduction 579
    II. Isolation 580
        1. Industrial 580
        2. Laboratory 580
    III. Modification 582
    IV. Applications 582
    V. Properties 582
        1. Microscopy 582
        2. Composition 583
        3. X-Ray Diffraction Patterns 584
        4. Gelatinization Behavior 584
        5. Retrogradation 584
        6. Amylose–Lipid Complex 584
        7. Swelling Power and Amylose Leaching 584
        8. Rheology 585
        9. Falling Number 586
    VI. References 586

15 Oat Starch 589
    I. Introduction 589
    II. Isolation 589
        1. Industrial 590
        2. Laboratory 590
    III. Modification 591
    IV. Applications 591
    V. Properties of Oat Starch 591
        1. Microscopy 591
        2. Chemical Composition 592
        3. X-Ray Diffraction 594
        4. Gelatinization 594
        5. Retrogradation 595
        6. Swelling Power and Amylose Leaching 596
        7. Rheological Properties 597
    VI. References 598

16 Barley Starch: Production, Properties, Modification and Uses 601
    I. Introduction 601
    II. Barley Grain Structure and Composition 602
    III. Barley Starch 604
        1. Isolation and Purification 604
        2. Chemical Composition of Barley Starch 605
        3. Granule Morphology 607
        4. X-Ray Diffraction and Relative Crystallinity 607
        5. Gelatinization 607
        6. Swelling Factor and Amylose Leaching 610
        7. Enzyme Susceptibility 612
        8. Acid Hydrolysis 613
        9. Pasting Characteristics 615

	10. Retrogradation 618
	11. Freeze–Thaw Stability..619
	12. Chemical Modification 619
	13. Physical Modification 621
IV.	Resistant Barley Starch 621
V.	Production and Uses of Barley Starch..............................623
VI.	Conclusion 625
VII.	References 625

17 Modification of Starches 629

- I. Introduction 629
- II. Cationic Starches 632
    1. Dry or Solvent Cationization 633
    2. Polycationic Starches..634
    3. Amphoteric Starch or Starch-containing Systems 635
    4. Cationic Starches with Covalently-reactive Groups 636
- III. Starch Graft Polymers (See Also Chapter 19) 637
- IV. Oxidation of Starch 638
- V. Starch-based Plastics (See Also Chapter 19)......................640
- VI. Encapsulation/Controlled Release 642
- VII. Physically Modified Starch 644
    1. Granular Cold-Water-Swellable (CWS) and Cold-Water-Soluble Starch (Pregelatinized Granular Starch)........................644
    2. Starch Granule Disruption by Mechanical Force 646
- VIII. Thermal Treatments 646
- IX. Enzyme-catalyzed Modifications...................................647
- X. References 648

18 Starch in the Paper Industry 657

- I. Introduction to the Paper Industry 658
- II. The Papermaking Process..660
- III. Starch Consumption by the Paper Industry 662
- IV. Starches for Use in Papermaking 663
    1. Current Use 663
    2. Recent Trends 665
- V. Application Requirements for Starch 666
    1. Viscosity Specifications 666
    2. Charge Specifications 668
    3. Retrogradation Control 669
    4. Purity Requirements..671
- VI. Dispersion of Starch 672
    1. Delivery to the Paper Mill.....................................672
    2. Suspension in Water 673
    3. Dispersion Under Atmospheric Pressure 674
    4. Dispersion Under Elevated Pressure........................674
    5. Chemical Conversion 676
    6. Enzymic Conversion..677
- VII. Use of Starch in the Papermaking Furnish 681
    1. The Wet End of the Paper Machine 681

        2. Flocculation of Cellulose Fibers and Fines 681
        3. Adsorption of Starch on Cellulose and Pigments 682
        4. Retention of Pigments and Cellulose Fines 683
        5. Sheet Bonding by Starch 684
        6. Wet-end Sizing 685
        7. Starch Selection for Wet-end Use 687
  VIII. Use of Starch for Surface Sizing of Paper 688
        1. The Size Press in the Paper Machine 688
        2. The Water Box at the Calender 693
        3. Spray Application of Starch. 693
        4. Starch Selection for Surface Sizing. 693
  IX. Use of Starch as a Coating Binder 695
        1. The Coater in the Paper Machine 695
        2. Starch Selection for Paper Coating 698
  X. Use of Starch as Adhesive in Paper Conversion. 700
        1. Lamination of Paper 700
        2. The Corrugator for Paperboard. 700
        3. Starch Selection for Use in Corrugation and Lamination. 702
  XI. Use of Starch in Newer Specialty Papers 703
  XII. Environmental Aspects of Starch Use in the Paper Industry 703
  XIII. Starch Analysis in Paper 705
  XIV. References 706

19 Starch in Polymer Compositions 715

  I. Introduction 715
  II. Starch Esters 717
  III. Granular Starch Composites 719
  IV. Starch in Rubber. 724
  V. Starch Graft Copolymers 726
  VI. Thermoplastic Starch Blends 731
  VII. Starch Foams 735
  VIII. References 737

20 Starch Use in Foods 745

  I. Introduction 746
        1. First Enhancement of Starch for Foods 747
        2. Modern Use of Starch in Foods. 747
        3. Development of Crosslinking. 747
        4. Development of Monosubstitution 747
        5. 'Instant' Starches. 748
        6. Improvement of Starch Sources (See Also Chapter 3) 748
  II. Functions of Starch in Food Applications. 748
        1. Starch Structures Relevant to Foods 749
        2. Gelatinization and Pasting 749
        3. Changes During Cooking. 750
  III. Impact of Processing and Storage on Foods
      Containing Cooked Starch 751
        1. Concentration During Cooking. 751
        2. Effects of Time and Temperature 751

- 3. Effects of Shear ... 752
- 4. Comparison of Food Processing Equipment ... 753
- 5. Impact of Processing and Storage ... 754
- 6. Changes that Occur During Cooling, Storage and Distribution ... 754
- 7. Recommended Processing ... 755

IV. Modified Food Starches (See Also Chapter 17) ... 756
- 1. Why Starch is Modified ... 756
- 2. Derivatizations ... 756
- 3. Conversions ... 760
- 4. Oxidation ... 761
- 5. Physical Modifications ... 762
- 6. Native Starch Thickeners ... 767

V. Starch Sources (See Also Chapters 9–16) ... 767
- 1. Dent Corn ... 768
- 2. Waxy Corn ... 768
- 3. High-amylose Corn ... 769
- 4. Tapioca ... 770
- 5. Potato ... 770
- 6. Wheat ... 770
- 7. Sorghum ... 771
- 8. Rice ... 771
- 9. Sago ... 772
- 10. Arrowroot ... 772
- 11. Barley ... 772
- 12. Pea ... 772
- 13. Amaranth ... 773

VI. Applications ... 773
- 1. Canned Foods ... 774
- 2. Hot-filled Foods ... 775
- 3. Frozen Foods ... 775
- 4. Salad Dressings ... 776
- 5. Baby Foods ... 777
- 6. Beverage Emulsions ... 777
- 7. Encapsulation ... 777
- 8. Baked Foods ... 778
- 9. Dry Mix Foods ... 778
- 10. Confections ... 778
- 11. Snacks and Breakfast Cereals ... 779
- 12. Meats ... 780
- 13. Surimi ... 781
- 14. Pet Food ... 781
- 15. Dairy Products ... 781
- 16. Fat Replacers ... 782

VII. Interactions with Other Ingredients ... 783
- 1. pH ... 783
- 2. Salts ... 783
- 3. Sugars ... 784
- 4. Fats and Surfactants ... 784

　　　　5. Proteins 785
　　　　6. Gums/hydrocolloids 786
　　　　7. Volatiles 786
　　　　8. Amylolytic Enzymes 786
　　VIII. Resistant Starch 787
　　IX. References 788

21　Sweeteners from Starch: Production, Properties and Uses 797
　　I. Introduction 797
　　　　1. History 797
　　　　2. Definitions 799
　　　　3. Regulatory Status 800
　　II. Production Methods 800
　　　　1. Maltodextrins 800
　　　　2. Glucose/corn Syrups 802
　　　　3. High-fructose Syrups 808
　　　　4. Crystalline Fructose 813
　　　　5. Crystalline Dextrose and Dextrose Syrups 813
　　　　6. Oligosaccharide Syrups 815
　　III. Composition and Properties of Sweeteners from Starch 817
　　　　1. Carbohydrate Profiles 817
　　　　2. Solids 818
　　　　3. Viscosity 819
　　　　4. Browning Reaction and Color 821
　　　　5. Fermentability 822
　　　　6. Foam Stabilization and Gel Strength 823
　　　　7. Freezing Point Depression 824
　　　　8. Boiling Point Elevation 824
　　　　9. Gelatinization Temperature 824
　　　　10. Humectancy and Hygroscopicity 825
　　　　11. Crystallization 826
　　　　12. Sweetness 827
　　　　13. Selection of Sweeteners 828
　　IV. References 829

22　Cyclodextrins: Properties and Applications 833
　　I. Introduction 833
　　II. Production 835
　　III. Properties 837
　　IV. Toxicity and Metabolism 838
　　V. Modified Cyclodextrins 840
　　　　1. Hydroxyalkylcyclodextrins 840
　　VI. Complex Formation 842
　　VII. Applications 845
　　VIII. References 848

Index 853

Preface to the Third Edition

Work towards production of the third edition of *Starch: Chemistry and Technology* was begun by Professor Roy L. Whistler and myself, but shortly thereafter Professor Whistler was unable to continue with the project. I was pleased to be able to see this edition through to completion.

Many developments have occurred in the world of starch chemistry, genetics, biochemistry, molecular biology and applications since the second edition was published in 1984. This edition, like the previous two editions, covers the isolation processes, properties, functionalities and uses of the most commonly used starches, viz., normal maize/corn, waxy maize, high-amylose maize, cassava (tapioca), potato and wheat starch, with emphases on those aspects of production, properties and uses that are unique to each; but not in single chapters. It also covers those starches that are generally available in only limited or potentially limited amounts, viz., rice (including waxy rice, but not all varieties of rice), sorghum, barley (including waxy barley), oat and rye starches. Chapters on the latter three starches are new to this edition. Not included are other starches that may be isolated from plants that are grown in limited areas and may be localized commercial products. These include amaranth, arrowroot, banana, canna, kuzu, millet, mung bean, pea (smooth and wrinkled), quinoa, sago, sweet potato and taro starch, except that some are mentioned in the chapter on starch use in foods and two are mentioned in the first chapter. Where available, many of these starches are available as flours, rather than pure starch. There has been an interest in small granule starch that can be obtained from cattail roots, dasheen tubers, and the seeds of amaranth, canary grass, catchfly, cow cockle, dropwort, pigweed and quinoa. None of these are covered except as noted above. However, properties and uses of small granule wheat starch are covered in the chapter on wheat starch.

All chapters/subjects that were also in the previous edition have been updated. Chapters have been added on the biochemistry and molecular biology of starch biosynthesis, structural transitions and related physical properties of starch, and cyclodextrins. There are two chapters on the structural features of starch granules that present not only advances in understanding the organization of starch granules, but also advances in understanding the fine structures of amylose and amylopectin, both of which are based on techniques that have been developed since 1984.

The chapter on corn and sorghum starch production not only thoroughly covers advances in understanding and in carrying out the wet-milling process, but also alternative corn kernel fractionation techniques, the relationship of starch production to other products from corn grain and future directions.

The greatly enlarged chapter on wheat starch presents advances in its production, the differences between large and small granules, the fine structures of wheat starch amylose and amylopectin, genetic and chemical modification of wheat starch, and its functionalities and uses, especially in food products.

The past two decades have also seen a considerable enlargement and maturation of the cassava (tapioca) starch industry that is reflected in another larger chapter, which also compares the characteristics of tapioca/cassava starch with those of other starches. The chapter on potato starch has also been considerably updated, especially from a processing standpoint. The latter chapter contains a discussion of all-amylopectin potato starch.

Because consumers have become more mindful of what is in their diet, and because in the European Economic Community chemically-modified starches must be labeled as such, there has developed an interest in starches that have only been heated to achieve the process tolerance and short texture of a lightly-crosslinked starch. Such developments in modifying the properties of starch without chemical derivatization are discussed in two chapters.

Also greatly enlarged and updated is the thorough chapter on the applications of starch products in the paper industry.

James N. BeMiller
West Lafayette, Indiana USA
May 2008

List of Contributors

Karin Autio, VTT Biotechnology and Food Research, VTT, Finland (Chapter 14, 15)
Paul M. Baldwin, Centre de Recherches Agro-Alimentaires, INRA, Nantes, France (Chapter 5)
Sukh D. Bassi, MGP Ingredients Inc., Atchison, Kansas, USA (Chapter 10)
Costas G. Biliaderis, Department of Food Science and Technology, Aristotle University, Thessaloniki, Greece (Chapter 8)
Charles D. Boyer, Department of Horticulture, Oregon State University, Corvallis, Oregon (Chapter 3)
William F. Breuninger, National Starch and Chemical Company, Bridgewater, New Jersey, USA (Chapter 12)
Chung-wai Chiu, National Starch and Chemical Co., Bridgewater, New Jersey (Chapter 17)
Steven R. Eckhoff, Department of Agricultural Engineering University of Illinois, Urbana, Illinois, USA (Chapter 9)
Ann-Charlotte Eliasson, Department of Food Technology, Lund University, Lund, Sweden (Chapter 14, 15)
Paul L. Farris, Department of Agricultural Economics, Purdue University, West Lafayette, Indiana USA (Chapter 2)
Daniel J. Gallant, Centre de Recherches Agro-Alimentaires, INRA, Nantes, France (Chapter 5)
Douglas L. Garwood, Golden Harvest Seeds, Stonington, Illinois (Chapter 3)
Hielko E. Grommers, AVEBE U.A., P.O. Box 15, 9640 AA, Veendam, The Netherlands (Chapter 11)
Allan Hedges, Consultant, Crown Point, Indiana USA (Chapter 22)
Larry Hobbs, (Chapter 21)
Ratnajothi Hoover, Department of Biochemistry, Memorial University of Newfoundland, St. John's, Canada (Chapter 16)
Jay-lin Jane, Department of Food Science and Human Nutrition and the Center for Crops Utilization Research, Iowa State University, Ames, Iowa, USA (Chapter 6)
Gerald D. Lasater, MGP Ingredients Inc., Atchison, Kansas, USA (Chapter 10)
Clodualdo C. Maningat, MGP Ingredients Inc., Atchison, Kansas, USA (Chapter 10)
William R. Mason, Formerly of National Starch and Chemical Co., Bridgewater, New Jersey, USA (Chapter 20)
Hans W. Maurer, Highland, Maryland 20777 (Chapter 18)

Cheryl R. Mitchell, Creative Research Management, Stockton, California, USA (Chapter 13)

Serge Pérez, Centre de Recherches sur les Macromolécules Végétales (affiliated with the Université Joseph Fourier, Grenoble), CNRS, Grenoble, France (Chapter 5)

Kuakoon Piyachomkwan, National Center for Genetic Engineering and Biotechnology, Pathumthani, Thailand (Chapter 12)

Jack Preiss, Department of Biochemistry and Molecular Biology, Michigan State University, East Lansing, Michigan, 48824, USA (Chapter 4)

John F. Robyt, Laboratory of Carbohydrate Chemistry and Enzymology, Department of Biochemistry, Biophysics, and Molecular Biology, Iowa State University, Ames, Iowa, 50011, USA (Chapter 7)

Deborah Schwartz, Corn Refiners Association, Inc., Washington, D.C. (Chapter 1)

Paul A. Seib, Department of Grain Science and Industry, Kansas State University, Manhattan, Kansas, USA (Chapter 10)

Jack C. Shannon, Department of Horticulture, The Pennsylvania State University, University Park, Pennsylvania (Chapter 3)

Daniel Solarek, National Starch and Chemical Co., Bridgewater, New Jersey (Chapter 17)

Klanarong Sriroth, Department of Biotechnology, Kasetsart University, Bangkok, Thailand (Chapter 12)

Do A. van der Krogt, AVEBE U.A., P.O. Box 15, 9640 AA, Veendam, The Netherlands (Chapter 11)

Thava Vasanthan, Department of Agricultural, Food and Nutritional Sciences, University of Alberta, Edmonton, Canada (Chapter 16)

Stanley A. Watson, Ohio Agricultural Research and Development Center The Ohio State University, Wooster, Ohio, USA (Chapter 9)

Roy L. Whistler, Whistler Center for Carbohydrate Research, Purdue University, West Lafayette, Indiana (Chapter 1)

J. L. Willett, Plant Polymer Research, National Center for Agricultural Utilization Research, Agricultural Research Service, US Department of Agriculture, Peoria, Illinois, USA (Chapter 19)

Kyungsoo Woo, MGP Ingredients Inc., Atchison, Kansas, USA (Chapter 10)

History and Future of Starch

Deborah Schwartz[1] and Roy L. Whistler[2]

[1] *Corn Refiners Association, Inc., Washington, D.C.*
[2] *Whistler Center for Carbohydrate Research, Purdue University, West Lafayette, Indiana*

I. History	1
1. Early History	1
2. 1500–1900	2
3. 1900–Present	4
II. Development of Specialty Starches	5
1. Waxy Corn Starch	5
2. High-amylose Corn Starch	5
3. Chemically Modified Starches	6
4. Other Naturally Modified Corn Starches	6
III. Other Products from Starch	6
1. Sweeteners	6
2. Ethanol	7
3. Polyols	8
4. Organic Acids	8
5. Amino Acids	8
IV. Future of Starch	9
1. Two New Starches for Industry	9
2. Present American Companies	9
V. References	10

I. History

1. Early History

Humans and their ancestors have always eaten starchy foods derived from seeds, roots, and tubers. It is fascinating to read the known history of crops and especially to follow the very early agricultural production of grain crops such as barley, rice, wheat and corn, with the latter having become the major source of isolated starch. Trace amounts of rice found in underground excavations along the middle region of the Yangtze River in Hubei and Hunan provinces have been radioactive carbon dated to a medium age of 11 5000 by a team of Japanese and Chinese archaeologists.[1] This predates the previous earliest known site for domestication of barley in China, indicated as 10 000 years ago.[1]

Corn (see Chapter 9), the only important cereal crop indigenous to the Americas, probably originated in Mexico, the oldest record (dating back 7000 years) being found in Mexico's valley of Tehuacan.[2] By 5000 BC, the teosinte plant must have interbred with the original corn plant to give the female inflorescence a degree of specialization that precluded the possibility of natural seed dissemination with the positive requirement that human activity was required for continuing survival. Corn apparently spread rapidly throughout the Americas, as far as the regions that are now Argentina and Canada.

Wheat (see Chapter 10) is the number one food grain consumed by humans, and its production leads all crops, including rice and corn. Wheat is a cool-season crop, but it flourishes in many different agroclimate zones. It is believed to have originated in the fertile crescent of the Middle East, where radiocarbon dating places samples at, or before, 6700 BCE, with wheat grains existing in the Neolithic site of Jarno, Northern Iraq.[3]

The practical use of starch products and, perhaps of starch itself, developed when Egyptians, in the pre-dynastic period, cemented strips of papyrus together with starch adhesive made from wheat. Early documents were lost, but Caius Plinius Secundus, Pliny the Elder, 23–74 AD (who died in the eruption of Vesuvius), described documents made by sizing papyrus with modified wheat starch to produce a smooth surface. The adhesive was made from fine ground wheat flour boiled with diluted vinegar. The paste was spread over papyrus strips, which were then beaten with a mallet. Further strips were lapped over the edges to give a broader sheet. Pliny stated that the 200-year-old sheets which he observed were still in good condition. Pliny also described the use of starch to whiten cloth and to powder hair. Chinese paper documents of about the year 312 are reported to contain starch size.[4] At a later date, Chinese documents were first coated with a high fluidity starch to provide resistance to ink penetration, then covered with powdered starch to provide weight and thickness. Starches from wheat and barley were common at that time.

A procedure for starch production was given in some detail in a Roman treatise by Cato in 184 BCE.[5] Grain was steeped in water for ten days and then pressed. Fresh water was added. Mixing and filtration through linen cloth gave a slurry from which the starch was allowed to settle. It was washed with water and finally dried in the sun.

2. 1500–1900

In the Middle Ages the manufacture of wheat starch became an important industry in Holland, and Dutch starch was considered to be of high quality. An early form of starch modification practiced in this period involved the starch being slightly hydrolyzed by vinegar. At that time, starch found its principal use in the laundry for stiffening fabrics and was considered a luxury suitable for the wealthy. During the mid-1500s, starch was introduced into England during the reign of Queen Elizabeth, who is said to have appointed a special court official for laundry starching. The custom of powdering the hair with starch appears to have become popular in France in the sixteenth century, and by the end of the eighteenth century, the use of starch for this purpose was generally practiced.

In the eighteenth century, more economical sources of starch than wheat were being sought. In 1732, the Sieur de Guife recommended to the French government that potatoes be used to manufacture starch. The potato starch industry in Germany dates from 1765 (see Chapter 11).

The nineteenth century witnessed an enormous expansion of the starch industry, due largely to demands of the textile, color printing and paper industries, and to the discovery that starch can be readily converted into a gum-like product known as dextrin. In the early 1800s, gum substitutes from starch were first made. A textile mill fire in 1821, however, is generally credited with the founding of the British gum industry. After the blaze was extinguished, one of the workmen noticed that some of the starch had been turned brown by the heat and dissolved easily in water to produce a thick, adhesive paste. The roasting of new starch was repeated, and the product was shown to have useful properties. Commercial dextrins were made in Germany in 1860 by an acid process. An American patent for dextrin manufacture that appeared in 1867 incorporated roasting of starch after it had been moistened with acid.

The early 1800s also saw development of the basic technology which would lead to today's starch-derived sweetener industry. The discovery that starch could be transformed into a sweet substance by heating with dilute acid was made in 1811 by the Russian chemist G.S.C. Kirchoff, who was trying to develop a substitute for the gum arabic that was then used as a soluble binder for clay. The first American facility to produce starch syrups was established in 1831. In 1866, production of D-glucose (dextrose) from starch was realized. A number of glucose manufacturing plants were built in Europe in the 1800s. Manufacture of crystalline dextrose began in 1882.

The first American starch plant, a wheat starch production facility, was started by Gilbert in Utica, New York in 1807. It was converted to a corn starch production facility in 1849. Industrial production of corn starch in the United States had begun in 1844, when the Wm. Colgate & Co. starch plant in Jersey City, New Jersey, switched from manufacture of wheat starch to manufacture of corn starch using a process developed by Thomas Kingsford in 1842, in which crude starch was extracted from corn kernels using an alkaline steep. In 1848, Kingsford started his own firm in Oswego, New York. By 1880, this firm had grown to be the largest company of its kind in the world. Other US wheat starch plants began operating in this period, but within a few years all were converted to corn starch plants.

In 1820, the production of potato starch had begun in Hillsborough County, New Hampshire. Potato starch use grew rapidly until 1895, at which time 64 factories were operating. They manufactured 24 million pounds (11 million kg) of starch annually during the production season, which lasted about three months. Rice starch manufacture began in the United States in 1815. However, production did not expand significantly, and the little rice starch used was mainly imported.

By 1880 there were 140 US plants producing corn, wheat, potato and rice starches. By 1900 the number of American starch facilities had decreased to 80, producing 240 million pounds (110 million kg) per year. Although a number of small plants continued to be built they could not compete and, in 1890, a consolidation took place to form the National Starch Manufacturing Company of Kentucky, representing 70% of the corn starch capacity. Although National Starch Manufacturing did not perform well, in the 1890s a number of glucose manufacturers tried to relieve their problems through similar consolidations. In 1897, six of the country's seven glucose factories were consolidated and became known as the Glucose Sugar Refining Company. In 1899, some of the remaining independent firms formed the United Starch Company.

3. 1900–Present

In 1900, the United Starch Company and the National Starch Manufacturing Company joined forces to form the National Starch Company of New Jersey. In 1902, Corn Products Company, representing 80% of the corn starch industry with a daily grind of 65 000 bushels (1800 tons), was formed by union of the National Starch Company of New Jersey, the Glucose Sugar Refining Company, the Illinois Sugar Refining Company, and the Charles Pope Glucose Company. In 1906, Corn Products Company and the National Starch Company merged to become Corn Products Refining Company, with a daily grinding capacity of 140 000 bushels (3900 tons). This was soon reduced to 110 000 bushels (3100 tons), or 74% of the US total. The Corn Products Refining Company is known today as Corn Products International, Inc.

Many of today's US starch companies also have their roots in the early 1900s. In 1906, the Western Glucose Company was incorporated; in 1908, it became the American Maize-Products Company, which was purchased by Cerestar in 1996, then Cargill gained complete control of Cerestar in 2002. The Clinton Sugar Refining Company began as a subsidiary of the National Candy Company in 1906. It underwent a series of ownership and name changes, beginning with the Clinton Corn Syrup Refining Company. The plant in Clinton, Iowa was acquired by Archer Daniels Midland Co. in 1982. The A.E. Staley Manufacturing Company was organized in 1906 and began with corn starch production in Decatur, Illinois. In 1903, the J.C. Hubinger Brothers Company began corn starch production in a factory in Keokuk, Iowa. This firm was purchased by Roquette in 1991 and became Roquette America, Inc. Douglas & Company was organized and began corn starch production in a plant in Cedar Rapids, Iowa in 1903. In 1920, the company was purchased by Penick & Ford, Ltd. It became Penford Products Company in 1998. A facility built by Piel Brothers Starch Company was organized in 1903. Its plant in Indianapolis, Indiana became the core of the starch business of National Starch and Chemical Corporation upon its acquisition by National Adhesives Corporation in 1939 and reorganization as National Starch Products, Inc. A number of other companies, including Union Starch, Huron Milling Company, Keever Starch Company, Anheuser-Busch, and Amstar Corporation operated starch facilities during the period from 1902 through the 1970s, but then either stopped production or sold the facility. A surplus government grain alcohol plant in Muscatine, Iowa was acquired after World War II by the Grain Processing Company and was modified to produce commercial starch in addition to ethanol.

Archer Daniels Midland Company and Cargill, Inc. both entered the starch industry through purchase of plants that were originally built by entrepreneurs in Cedar Rapids, Iowa. The Corn Starch & Syrup Company was acquired by Cargill in 1967 and a substantial interest in Corn Sweeteners was purchased by ADM in 1971. The most recent entry in the US corn starch industry is Minnesota Corn Processors, a farmer-owned cooperative which began its wet-milling operations in Marshall, Minnesota in 1983.

The US corn wet-milling industry is represented by the Corn Refiners Association, Inc., a Washington, DC-based trade association which provides technical, regulatory, legislative and communications support for its members.

II. Development of Specialty Starches

Starch in its native form is a versatile product, and the raw material for production of many modifications, sweeteners and ethanol. Starting in the 1930s, carbohydrate chemists have developed numerous products that have greatly expanded starch use and utility.

1. Waxy Corn Starch

Waxy corn starch, also known as waxy maize starch, consists of only amylopectin molecules, giving this starch different and useful properties (see Chapter 3). This genetic variety of corn was discovered in China in the early 1900s, when corn plants were transferred from the Americas. The starch stains red with iodine, not blue as ordinary starches do. When the corn kernel is cut, the endosperm appears shiny and wax-like, and the corn was termed waxy corn or waxy maize. However, it contains no wax.

Waxy-type corn was brought to the United States in 1909 and remained a curiosity at agricultural experiment stations until World War II cut off the supply of cassava (tapioca) starch from the East Indies. During a search for a replacement, waxy corn starch was found to be a suitable alternative. In the 1940s, geneticists at Iowa Agricultural Experiment Station developed waxy corn into a high-yielding hybrid. After waxy corn was introduced as a contract crop, its starch developed rapidly into a valuable food starch. Although other all-amylopectin starches, such as waxy sorghum and glutinous rice, and now waxy wheat and all amylopectin potato starches, are also composed only of amylopectin molecules, they have not had the industrial acceptance of waxy corn, since corn also supplies quality oil and protein products. Acreage planted to waxy corn in the United States, Canada and Europe has expanded rapidly. An estimated 550 000–600 000 acres (220 000–250 000 hectares) of waxy corn was grown in the United States in 1996.

2. High-amylose Corn Starch

Although the term amylose dates to 1895, it was not until the 1940s that it became associated with the mainly linear chains of starch (see Chapter 3). Before this, little was known about the structure or identity of starch polymers. In 1946, R.L. Whistler, a carbohydrate chemist, and H.H. Kramer, a geneticist, set out to produce a corn modification that would be the opposite of waxy corn, i.e. one in which the starch would be composed only of amylose molecules. Whistler and Kramer were able to increase the amylose content from the 25% normally found in corn to 65%. As high-amylose corn became further developed by other researchers, the amylose content was increased to 85%, with approximately 55% and 70% being common in commercial varieties.

High-amylose starch is used primarily by candy manufacturers who utilize high-strength gels to help give candy shape and integrity. Addition of modified high-amylose starch can enhance the texture of foods such as tomato paste and apple sauce. The ability of amylose starches to form films led to widespread investigation of its use in industrial products, including degradable plastics.

3. Chemically Modified Starches

The performance and quality of starch can be improved through chemical modification (see Chapter 17). Chemical modifications provide processed foods, such as frozen, instant, dehydrated, encapsulated and heat-and-serve products, the appropriate texture, quality and shelf life (see Chapter 21), and improved processing condition tolerance, such as improved heat, shear and acid stability. Modification also allows starches to be used in the paper industry (see Chapter 19) as wet-end additives, sizing agents, coating binders, and adhesives and as textile sizes.

4. Other Naturally Modified Corn Starches

In recent years, developments in corn genetics have suggested that many of the valuable properties of modified starches could be produced through changes in the biosynthesis of starch in the corn plant, rather than through chemical modification. Corn starch companies, in conjunction with corn seed companies and scientists at universities and agricultural experiment stations, have undertaken extensive investigation of such a possibility. In addition to amylose and waxy genes, other genes affect the production of starch. Some of these genes are dull, sugary 1, sugary 2, shrunken 1, shrunken 2, soft starch (horny) and floury 1 (see Chapter 3). These genes affect synthesis of starch (see Chapter 4) and lead to the production of starches with altered structural and functional characteristics. Work has been pursued rapidly over the past ten years to evaluate the starches produced. Some starches evaluated include amylose extender dull, amylose extender sugary 2, dull sugary 2, dull soft starch amylose extender waxy, dull waxy, waxy shrunken 1, waxy floury 1, waxy sugary 2 and sugary 2. Since genes determine the structures of both amylopectin and amylose molecules and their ratio, unique waxy types, intermediate-amylose and high-amylose starches are produced via cross-breeding.

III. Other Products from Starch

1. Sweeteners

Kirchoff's discovery of starch hydrolysis led eventually to today's modern starch sweetener industry. The original starch-derived sweeteners, which were produced by acid-catalyzed hydrolysis of starch and which contained varying amounts of dextrose, other saccharides and polysaccharides, are known as glucose syrups. Glucose syrups of the 1800s and early 1900s were produced in both solid and liquid forms. Solid forms were made by casting and drying liquid products. In the 1920s, Newkirk developed technology needed to fully hydrolyze starch to dextrose (D-glucose), and crystalline dextrose production developed quickly.

Advances in enzyme technology in the 1940s and 1950s (see Chapter 7) enabled precise control of products and the degree and conditions of hydrolysis, greatly expanding the range and utility of glucose syrup products. At the same time, new purification techniques were introduced which permitted production of syrups of high purity.

Isomerases which convert glucose into the sweeter fructose were commercially introduced in the 1960s. Their introduction, coupled with manufacturing technology to immobilize these enzymes, led to the introduction of high-fructose syrup (HFS) in the United States in 1967. Refinements in production processes produced a liquid sweetener that could replace liquid sucrose on a one-to-one basis. At the same time, major upheavals in the world sugar market caused major sugar users to seek such an alternative.

During the late 1970s and early 1980s, numerous US beverage companies began using HFS to replace some of the sucrose in their drinks, and HFS growth far outpaced population growth. In 1984, the corn wet milling industry achieved the goal of capturing the beverage market when all major soft drink bottlers in the US began using HFS for much of their nutritive sweetener needs. Since then, HFS growth has continued to outpace increases in population as per capita annual soft drink consumption grew from around 44 gallons (165 liters) in 1985, to over 50 gallons (190 liters) in 1995 (see Chapter 22).

2. Ethanol

Glucose syrups are easily fermented by yeast to ethanol. While beverage ethanol has been produced from many sources of sugar and starch for countless centuries, large-scale production of fuel-grade ethanol by fermentation is attributed to a demand for combustible motor fuel additives.

Automobile pioneer Henry Ford first advocated the use of alcohol as a fuel in the 1920s as an aid to American farmers. During the 1930s, more than 2000 Midwestern service stations offered gasoline containing between 6% and 12% ethanol made from corn. Because of its high cost and the opening of new oil fields, 'gasohol' disappeared in the 1940s. However, in response to the oil supply disruption of the mid-1970s, ethanol was reintroduced in 1979. US ethanol production grew from a few million gallons in the mid-1970s to about 1.6 billion gallons (6×10^9 liters) in 1996 (see Chapter 2).

Today, most ethanol is made from corn starch. After separation from corn by wet milling, starch slurry is thinned with alpha-amylase and saccharified with amyloglucosidase. The resulting sugar solution is fermented by *Saccharomyces* yeast. Modern US ethanol plants use simultaneous scarification, yeast propagation and fermentation. The major portion of fuel-grade ethanol is now produced by continuous fermentation, which offers the advantages over batch fermentation of lower capital cost for fermenters, improved microbiological control, and ease of automating control of the process. From the 32 pounds (14.5 kg) of starch in a bushel of corn, about 2.5 gallons (9.5 liters) of ethanol is produced.

3. Polyols

Hydrogenation of sugars produces a class of materials known as sugar alcohols or polyols. Major commercial sugar alcohols include mannitol, sorbitol (D-glucitol), malitol, and xylitol and syrups related to these products, with all but xylitol being obtainable from starch by hydrolysis, isomerization in the case of mannitol, and hydrogenation. Sugar alcohols are found naturally in some plants, but commercial extraction is not feasible. Polyols were first discovered by the isolation of 'manna' from the mountain ash tree, and sorbitol was isolated from rowan berries in 1872 by the French chemist Joseph Boussingault.

Polyols are unique among simple carbohydrates in their low ability to be fermented. This characteristic enables them to impart sweetness to foods while exhibiting lower caloric values than other carbohydrates and reducing the formation of dental caries. Polyols are used in a variety of applications in foods, confections, pharmaceuticals and industrial uses. Rising demand for low- and reduced-calorie foods and confections that contribute to a reduction in dental caries has contributed to the growth of these starch-derived products.

4. Organic Acids

Organic acids are found throughout nature. Citric, lactic, malic and gluconic acids have become large-scale food and industrial ingredients. Originally produced from fermentation of sucrose or sugar by-products, they are now mainly produced from fermentation of dextrose. Major new plants were built by US starch producers for organic acid production in the 1980s and 1990s.

Citric acid makes up almost 85% of the total volume of the organic acid market. It was first described in 1784 when isolated from lemon juice. In 1917, it was discovered that certain fungi accumulate citric acid. In 1923, the first US commercial plant was built to produce citric acid by fermentation; citric acid is now used mainly in soft drinks, desserts, jams, jellies, candies, wines and frozen fruits.

Lactic acid, initially produced in 1880, was the first organic acid made industrially by fermentation of a carbohydrate. Nowadays it is made both by fermentation and by chemical synthesis. About 85% of the use of lactic acid is in food and food-related applications, with some use in the making of emulsifying agents and poly(lactic acid).

5. Amino Acids

During the 1980s, advances in fermentation technology allowed the economic production of a number of amino acids from starch hydrolyzates. Examples are lysine, threonine, tryptophan, methionine and cysteine. Starch-derived amino acids are generally used as animal nutrition supplements, enabling animal nutritionists to formulate finished animal feeds tailored to nutrient requirements of individual animals. Feeds supplemented with these products can also reduce feed costs, animal waste and nitrogen pollution.

IV. Future of Starch

1. Two New Starches for Industry

Banana starch is certain to join the group of industrial starches, because it can be obtained from cull bananas discarded by large banana plantations. Banana bunches are cut from trees in plantations and sent to a central processing station, where culls consisting of small or damaged fruit are removed. Such culls represent 25% of the banana crop and 25% of the green banana is starch. The starch can be readily recovered from banana pulp in a four-hour steep at an appropriate pH. Banana starch consists of large (20 μm) granules with properties suitable for a variety of applications. The production costs, essentially of cartage plus that of starch extraction, are expected to give a market price that approaches or equals that of corn starch.

Amaranth has been used for dietary 'greens' and its seeds as storable food grain (see Chapter 17). Its use reached a zenith during the Mayan and Aztec period in Central America. A tithe of 200 000 bushels (9000 m^3) per year was placed on farmers by Montezuma, but production was stopped in that region by the conquistador Cortez in 1519, since he abhorred the pagan use of ground grain mixed with blood for shaping into conformations of animals, birds and human heads, which were then eaten. Amaranth was later grown in the mountains of South and Central America and now is grown in the northern United States. It is often popped and mixed with sugar syrup and sold as candy bars. The flour, mixed in low levels with wheat flour, produces an interesting flavor in bread and pancakes. Amaranth seeds contain about 67% starch, with granules being about 1 μm in diameter. Its characteristics could be useful in foods, and tests have shown that it may have applications as a fat replacer.

2. Present American Companies

The vast majority of starch produced in the United States, either for sale as starch or for conversion to other products, is derived by the wet-milling of corn. A small amount of starch is also produced by isolation from potatoes or extraction from wheat or rice flour. Current US companies involved in starch production are as follows.

Corn Starch Producers
- ADM Corn Processing (a division of Archer Daniels Midland Company) has plants in Decatur, Illinois; Cedar Rapids, Iowa; Clinton, Iowa; and Montezuma, New York.
- Cargill, Incorporated has plants in Blair, Nebraska; Cedar Rapids, Iowa; Dayton, Ohio; Eddyville, Iowa; Hammond, Indiana; Memphis, Tennessee; Decatur, Alabama; Dimmit, Texas; and Wahpeton, North Dakota (ProGold).

- Corn Products International, Inc. has plants in Bedford Park, Illinois; Stockton, California; and Winston-Salem, North Carolina.
- Grain Processing Corporation has plants in Muscatine, Iowa and Washington, Indiana.
- Minnesota Corn Processors has plants in Marshall, Minnesota and Columbus, Nebraska.
- National Starch and Chemical Company (a subsidiary of ICI) has plants in Indianapolis, Indiana and North Kansas City, Missouri.
- Penford Products Co. (a company of Penford Corporation) has a plant in Cedar Rapids, Iowa.
- Roquette America, Inc. has a plant in Keokuk, Iowa.
- Tate & Lyle North America has plants in Decatur, Illinois; Lafayette, Indiana (2); and Loudon, Tennessee.

Wheat Starch Producers
- ADM Arkady, a division of ADM Millings, has a plant in Keokuk, Iowa.
- Heartland Wheat Growers has a plant in Russell, Kansas.
- Manildra Milling Corporation, owned by Honan Holdings, Inc. has plants in Minneapolis, Minnesota and Hamburg, Iowa.
- Midwest Grain Products, Inc. has plants in Atchison, Kansas and Pekin, Illinois.

Potato Starch Producers
- Penford Food Ingredients (a company of Penford Corporation) has plants in Monte Vista, Colorado; Murtaugh, Idaho; Stanfield, Oregon; and Houlton, Maine.
- Tate & Lyle North America has plants in Idaho Falls, Idaho; Richland, Washington; and Plover, Wisconsin.
- Western Polymer Corporation has a plant in Moses Lake, Washington.

V. References

1. Normile D. *Science*. 1997;275:309.
2. Benson GO, Pearce RB. In: Watson SA, Ramstad PE, eds. *Corn Chemistry and Technology*. St. Paul, MN: American Association of Cereal Chemists; 1991 [Chapter 1].
3. Inglett GE. *Wheat, Production and Utilization*. Westport, CT: AVI Publishing; 1974 [Chapter 1].
4. Wiesner L. *Papier-Fabr*. 1911;9:886. Marus Procius Censorius Cato., *De Agriculture*, 184 BCE, Scriptores rei Rustica.

Economic Growth and Organization of the US Corn Starch Industry

Paul L. Farris
Department of Agricultural Economics, Purdue University, West Lafayette, Indiana, USA

I. Introduction	11
II. Extent and Directions of Market Growth	11
III. High-fructose Syrup Consumption	13
IV. Fuel Alcohol	15
V. Technical Progress	16
VI. Plant Location	16
VII. Industry Organization	16
VIII. Effects of Corn Price Variability	18
IX. International Involvement	19
X. Future Industry Prospects	20
XI. References	20

I. Introduction

The US starch industry, also known as the wet corn milling, corn wet-milling, and the corn refining industry, has grown rapidly and starch production has expanded in several other countries. Although people continue to consume some starch directly from starch-bearing plants, either raw or cooked, their demands for commercially produced starch to be added to foods and beverages have increased significantly. Starch use in a broad range of industrial products such as paper, textiles, building materials and alcohol for fuel has also expanded.

II. Extent and Directions of Market Growth

As a consequence of overall market growth, the quantity of corn (including minor quantities of sorghum grain) that was processed by the wet corn milling industry

Table 2.1 US Corn: food, seed and industrial use, 1980–81 to 1999–2000[a,b,c]

Year[b]	HFS[d]	Glucose syrups and dextrose[d]	Starch[d]	Fuel alcohol[d]	Beverage alcohol[d]	Cereals and other products[d]	Seed[d]	Total[d]	US corn production[d]	Total FSI use as % of US production
1980–1981[e]	165	156	151	35	78	54	20	659	6639	9.9
1981–1982[e]	183	160	146	86	86	53	19	733	8119	9.0
1982–1983[e]	214	165	150	140	110	60	15	854	8235	10.4
1983–1984[e]	265	167	161	160	88	70	19	930	4174	22.3
1984–1985[e]	310	167	172	232	84	81	21	1067	7672	13.9
1985–1986[e]	327	169	190	271	83	93	19	1152	8875	13.0
1986–1987[e]	338	171	214	290	95	109	17	1233	8226	15.0
1987–1988	358	173	226	279	85	113	17	1252	7131	17.6
1988–1989	361	182	215	287	117	117	18	1298	4929	26.3
1989–1990	368	193	219	321	129	120	19	1370	7532	18.2
1990–1991	379	200	219	349	135	124	19	1425	7934	18.0
1991–1992	392	210	225	398	161	128	20	1534	7475	20.5
1992–1993	415	214	218	426	136	129	19	1556	9477	16.4
1993–1994	441	219	225	458	110	140	20	1613	6336	25.4
1994–1995	459	224	230	533	100	150	18	1715	10 103	17.0
1995–1996	473	227	226	396	125	161	20	1628	7374	22.1
1996–1997	492	233	238	429	130	172	20	1714	9233	18.6
1997–1998	513	229	246	481	133	182	20	1805	9207	19.6
1998–1999	531	219	240	526	127	184	20	1846	9759	18.9
1999–2000	540	222	251	566	130	185	20	1914	9431	20.3

[a]Source: references 2, 3, 4 and 5
[b]Marketing year beginning 1 September
[c]Amount subjected to wet-milling is that in columns 2–5
[d]10^6 bushels. To convert to 10^6 metric tons, multiply by 0.02541
[e]Crop year began 1 October prior to 1986

increased five-fold between 1972 and 1992, from 262 million bushels (6.66×10^6 metric tons) to 1303 million bushels (33.11×10^6 metric tons) during the 20 year period.[1] This rapid increase took an increasing share of expanding corn production in the United States. The manufacture of wet-milled products, which accounted for ~5% of US corn production in the 1960s and 1970s, averaged close to 20% between 1990 and 1999 (Table 2.1). Yearly percentages fluctuated due mainly to variations in corn production, because wet-milling demands for corn increased quite steadily.

Large increases in demands for two major products, high-fructose corn syrup and fuel alcohol, propelled high industry growth beginning in the 1970s. The market for high-fructose syrup was stimulated by the growing acceptance of corn sweeteners in food and, especially, beverage products. Prices of corn sweeteners were competitive with US raw sugar prices, which were substantially higher than world sugar prices due to government policies (Table 2.2). Production of alcohol for engine fuel also increased greatly in the 1970s, motivated by rising crude oil prices and stimulated by government subsidies, which continued to exist in the 1990s.

Relatively steady increases in the production of standard starch and syrups accompanied the accelerating expansion of high-fructose syrup (HFS) that began in the 1960s, and the rapid increase in production of fuel alcohol in the 1980s and 1990s

Table 2.2 HFS and sugar prices and producer price index for total finished goods, 1981–2000[a]

Year	HFS-42 wholesale list prices, midwest markets, dry weight[b]	HFS-55 wholesale list prices, midwest markets, dry weight[b]	US raw sugar prices, duty fee paid, New York[b]	World raw sugar prices[b,d,c]	World refined sugar prices[b,d]	US producer price index for total finished goods, 1982 = 100
1981	21.47	23.59	19.73	16.93	20.51	96.1
1982	14.30	18.81	19.92	8.42	11.36	100.0
1983	18.64	21.06	22.04	8.49	11.40	101.6
1984	19.94	22.69	21.74	5.18	7.71	103.7
1985	17.75	19.95	20.34	4.04	6.79	104.7
1986	18.07	19.96	20.95	6.05	8.47	103.2
1987	16.50	17.46	21.83	6.71	8.75	105.4
1988	16.47	18.68	22.12	10.17	12.01	108.0
1989	19.24	21.41	22.81	12.79	17.16	113.6
1990	19.69	21.88	23.26	12.55	17.32	119.2
1991	20.93	23.25	21.57	9.04	13.41	121.7
1992	20.70	23.00	21.31	9.09	12.39	123.2
1993	18.83	20.93	21.62	10.03	12.79	124.7
1994	18.77	22.47	22.04	12.13	15.66	125.5
1995	15.63	NA	22.96	13.44	17.99	127.9
1996	14.46	NA	22.40	12.23	16.64	131.3
1997	10.70	NA	21.96	12.06	14.33	131.8
1998	10.58	NA	22.06	9.68	11.59	130.7
1999	11.71	NA	21.16	6.54	9.10	133.0
2000	11.32	NA	19.09	8.51	9.97	138.0

[a]Source: references 6, 7, 8, 9 and 10
[b]Cents/lb.
[c]Contract No. 11 – f.o.b. stowed Caribbean port (including Brazil) spot price
[d]Contract No. 5 London daily price for refined sugar, f.o.b., Europe, spot price

(Table 2.1). These two products, which together accounted for less than one-third of all food, seed and industrial uses of corn in 1980–1981, accounted for almost 60% of these uses in the 1990s. HFS production more than tripled, requiring 165 million bushels (4.219×10^6 metric tons) of corn in 1980–1981 and 540 million bushels (13.72×10^6 metric tons) in 1999–2000. The rise in fuel alcohol production was even more dramatic, expanding from 35 million bushels (0.89×10^6 metric tons) of corn in 1980–1981 to 566 million bushels (14.37×10^6 metric tons) in 1999–2000. Fuel alcohol production alone consumed more than 5% of total US corn production during the ten year period 1990–1991 through 1999–2000.

Numerous edible and industrial products are manufactured by the corn wet-milling industry through further refining. The dollar value of industry product shipments more than doubled (from $3.1 billion to $7.2 billion) between 1982 and 1997 (Table 2.3).

III. High-fructose Syrup Consumption

Until the late 1960s, sweeteners derived from starch accounted for less than 15% of the US total caloric sweetener market. By the mid-1980s, the starch-derived sweetener

Table 2.3 Corn product shipments by the wet corn milling industry, 1982, 1987, 1992 and 1997[a]

Product	1982		1987		1992		1997	
	million dollars	%	million dollars	%	million dollars	%	million dollars	%
Sweeteners	1610.4	51.8	2182.5	49.1	2911.0	45.4	3056.2	42.5
Manufactured starch	655.1	21.1	774.3	17.4	1318.1	20.5	1526.1	21.2
Corn oil	234.9	7.6	613.1	13.8	801.6	12.5	980.4	13.6
By-products	577.7	18.6	845.8	19.0	1363.5	21.3	1585.0	22.1
Not specified by kind	27.6	0.9	30.5	0.7	21.4	0.03	40.7	0.6
Total	3105.7	100.0	4446.2	100.0	6415.5	100.0	7188.4	100.0

[a]Source: reference 1

Table 2.4 Starch-derived sweetener consumption per capita in the United States (including Puerto Rico), 1992–2000[a]

Calendar year	HFS	Sweeteners[b]		Total	Total caloric sweeteners[b]	Starch-derived sweeteners share of total caloric sweeteners, %
		Glucose syrup	Dextrose			
1992	50.6	17.7	3.8	72.0	137.8	52.2
1993	53.3	17.9	3.8	75.0	140.4	53.4
1994	55.0	18.1	3.8	76.9	143.0	53.8
1995	55.2	18.4	3.9	77.5	143.6	54.0
1996	56.5	18.3	3.9	78.7	145.3	54.2
1997	59.2	19.8	3.7	82.7	149.6	55.3
1998	61.4	18.8	3.6	83.9	150.7	55.7
1999	62.8	18.2	3.5	84.5	152.8	55.3
2000	61.6	17.9	3.3	82.9	150.1	55.2

[a]Source: reference 11
[b]10^6 pounds, dry basis. To convert to metric tons, multiply by 453.5

share of total caloric sweeteners had risen to more than 50%, and the upward trend, though slowing, was continuing (Table 2.4). HFS had rapidly replaced most other sweeteners in the nonalcoholic beverage market (Table 2.5). Beverage use accounted for about 75% of HFS production in the mid-1990s.

Production of HFS in other industrialized countries is far lower than that in the United States, but production elsewhere has also been increasing (Table 2.6). Future significant growth is expected in other countries, but if relatively lower world sugar prices prevail abroad, the pace of HFS growth in other countries will likely be slower than past HFS growth in the United States.

Table 2.5 US domestic food and beverage uses of HFS-42 and HFS-55, 1980 and 1995[a]

Domestic use	HFS-42[a]		HFS-55[b]	
	1980	1995	1980	1995
Cereal and bakery products	304	441	4	14
Confectionery, including chewing gum	11	44	1	12
Processed foods	334	657	17	163
Dairy products	110	210	5	27
Multiple and miscellaneous	380	427	38	110
Beverages, mainly soft drinks	372	1310	525	4327
Total	1511	3089	591	4653

[a]Source: references 10 and 12

[b]1000 short tons, dry basis. To convert to metric tons, multiply by 0.9074

Table 2.6 World production of high-fructose syrup in selected countries, selected years[a,b]

	1982	1987	1992	1995
United States	2846	5145	6236	7121
Canada	110	202	250	255
Argentina	40	169	180	220
EU	260	265	286	303
Japan	579	724	761	750
South Korea	69	182	263	250
Taiwan	NA	15	125	180
Others	60	81	133	335
World total	3964	6783	8134	9414

[a]Source: reference 10

[b]1000 metric tons, dry basis

IV. Fuel Alcohol

The search for alternative energy sources, beginning with the crude oil shortage and crisis in the 1970s, led to renewed emphasis on alcohol as an automotive fuel. As a liquid fuel that could be used to help power a large motorized transportation sector, alcohol became a very desirable form of energy. Concurrently, the 1977 Clean Air Act and the phaseout of lead from gasoline provided added stimulus. Federal and state legislative changes were enacted, and subsidies were given to encourage ethanol production for so-called gasohol, which is one part ethanol and nine parts gasoline. Further inducement came from the 1990 Clean Air Act Amendments to reformulate gasoline to meet certain oxygen levels to help control carbon monoxide and ground-level ozone problems. Although there are alternative fuel oxygenators, and some questions about the benefit of ethanol in vehicles using fuel injectors, US legislation continues to specify the use of corn- or grain-based ethanol in gasoline.

This added demand, along with research emphasis in both private and public sectors, produced significant gains in technical efficiency. One study reported that ethanol production from corn went from a process that required 16% more energy than it produced to a net energy surplus of 33%.[13] Research involving applications of biotechnology and chemistry may achieve further gains. Also, corn-processing firms are finding plant operating advantages in co-product initiatives and in balancing seasonal production of alternative products, such as fuel alcohol and HFS. Lower feedstock costs may be realized if efficiencies in producing short rotation woody crops and grass, as well as corn, are realized.

V. Technical Progress

The complex and highly technical process of corn refining requires large economies of scale that, in turn, require substantial capital investment. Also, technological improvements in recent years have greatly reduced the amount of labor needed, so that the number of employees in the US corn wet-milling industry decreased from 12 100 in 1972 to 9200 in 1997, even though the amount of corn processed approximately tripled. Payroll expenses for all employees declined from 42% of value added by manufacture in 1972 to slightly less than 14% in 1997. As a consequence of technological advances and plant scale economies, the industry is dominated by a relatively few large plants. In 1997, 26 establishments with 100 or more employees each accounted for 93.4% of value added by manufacture. The other 25 plants produced the remaining 6.6%.[1]

VI. Plant Location

Corn refining plants tend to be located near sources of raw material. In 1992, nearly 75% of the US corn wet-milling industry value of shipments was from plants located in Iowa, Indiana and Illinois.[1] In the early 1990s, these three states accounted for about 45% of US corn production. Processing plants in the Corn Belt are substantially larger than those in locations away from major corn producing areas.

VII. Industry Organization

Although market growth and technological progress have advanced remarkably in recent decades, the corn refining industry has continued to be dominated by relatively few firms. Some seventy years ago, the industry was judged to be monopolistic and

Table 2.7 Percentage of total value of shipments accounted for by largest companies in the wet corn milling industry. Selected years[a]

Year	Number of companies	Total value of shipments 10^6\$	Share of value of shipments			Primary product specialization %
			Four largest companies %	Eight largest companies %	Twenty largest companies %	
1947	47	460.0	77	95	99+	88
1954	54	458.4	75	93	99	91
1958	53	522.0	73	92	99	91
1963	49	622.4	71	93	99	83
1967	32	751.3	67	89	99+	84
1972	26	832.3	63	86	99+	97
1977	22	2014.8	63	89	99+	93
1982	25	3268.4	74	94	99+	92
1987	32	4788.9	74	94	100	94
1992	28	7045.2	73	93	99+	88
1997	30	8455.2	72	90	99+	84

[a]Source: reference 16

the leading firm, Corn Products Refining Company, was required to divest a portion of its assets. Nevertheless, this company remained the dominant firm for many years, although its 60% market share in 1918 gradually declined to ~45% in 1945.[14,15]

In 1947, the four largest firms in Census Industry 2046 (311221 in 1997), wet corn milling, accounted for 77% of industry value of shipments. The four-firm concentration declined to 63% in 1972, but rose to 74% in 1982 and remained at that level in 1987. It was 73% in 1992 and 72% in 1997 (Table 2.7). Only a dozen or so companies accounted for essentially all US industry output.

Data are available by company for industry capacity of two important sweetener products, HFS-42 and HFS-55 (see Chapter 22) (Table 2.8). The largest firm in both 1987 and 1992 was Archer Daniels Midland, with about one-third of total US HFS industry capacity. When capacities for the next three, Tate and Lyle North America (formerly A.E. Staley Mfg. Co.), Cargill and CPC International are included, the largest four accounted for ~85% of the total capacity for manufacturing HFS. In the early 1990s, Archer Daniels Midland produced an estimated half of the US fuel alcohol.

Corn refining firms have become increasingly diversified, with the expansion of new food products for the consumer market. They have also acquired other lines of business and become more conglomerate in character, along with their expansion through direct investment in other countries. Marion and Kim[17] estimated that about half of the change in four-firm concentration between 1977 and 1988 came from internal growth and the other half from mergers and acquisitions. In 1991, the total sales of the four largest corn sweetener producing firms ranked among the 50 largest US food processing companies.[18] In 1993, three of the four largest in the United States were among 50 of the world's largest food processing firms.

Table 2.8 US production capacity for high-fructose syrup by company, 1987 and 1992a

Company	HFS-42		HFS-55		Total	
	1000 short tons[b] dry weight	%	1000 short tons[b] dry weight	%	1000 short tons[b] dry weight	%
1987						
Archer Daniels Midland	675	36.2	1200	31.1	1875	32.8
American Fructose[c]	128	6.9	348	9.0	476	8.3
Cargill[c]	355	19.0	540	14.0	895	15.7
CPC International	213	11.4	348	9.0	561	9.8
Golden Technologies	18	1.0	38	1.0	56	1.0
Hubinger[c]	120	6.4	130	3.4	250	4.4
A.E. Staley Mfg. Co.[c]	355	19.0	1250	32.4	1605	28.1
Total	1864	100.0	3854	100.0	5718	100.0
1992						
Archer Daniels Midland	816	30.6	1425	33.1	2241	32.2
American Fructosec	284	10.7	350	8.1	634	9.1
Cargill[c]	570	21.4	755	17.6	1325	19.0
CPC International	290	10.9	365	8.5	655	9.4
Golden Technologies	38	1.4	54	1.3	92	1.3
Hubinger[c]	130	4.9	177	4.1	307	4.4
A. E. Staley Mfg. Co.[c]	535	20.1	1175	27.3	1710	24.6
Total	2663	100.0	4301	100.0	6964	100.0

[a]Source: reference 6

[b]To convert to metric tons, multiply by 0.9074

[c]See Chapter 1 for changes

VIII. Effects of Corn Price Variability

Corn is the major cost ingredient used in producing corn refinery products, amounting to 81% of total materials, ingredients, containers and supplies purchased by the industry in 1992.[1] However, because the price of corn is more variable than that of other components, due primarily to corn production variability, the relative cost of corn varies considerably between years. The net cost of corn to US millers, after allowance for by-product credit, was estimated at $1.26 per bushel in 1997, $0.92 in 1992, $0.26 in 1987 and $1.03 in 1982 (Table 2.9). The 56 pounds (25 kg) in a bushel of corn yields 31.5 pounds (14.3 kg) dry weight, of starch (56%), 1.55 pounds (0.703 kg) of corn oil (2.8%), 2.65 pounds (1.20 kg) of corn gluten meal (47%) and 13.5 pounds (0.680 kg) of corn gluten feed (24%). The 6.8 pounds (3.1 kg) of residual is mainly moisture (12%). By-product credit values are based on the amounts of corn oil and corn gluten feeds times their prices.[6]

To illustrate further, although corn accounted for 51% of the corn wet-milling industry value of product shipments in 1997, the percentage was only 32 in 1987 when corn supply was abundant and prices low (Table 2.9). These percentages were 44% in 1982 and 40% in 1992. Because product selling prices fluctuate through a

Table 2.9 Corn price and cost to the wet corn milling industry, 1982, 1987, 1992 and 1997[a]

Year	Corn price[b]	By-product allowance[b]	Net corn cost[b]	Total corn cost[c]	Value of product shipments[c]	Corn cost as % of value of product shipments
1982	2.48	1.45	1.03	1370.8	3105.7	44.1
1987	1.59	1.33	0.26	1435.9	4446.2	32.3
1992	2.33	1.41	0.92	2587.7	6415.5	40.3
1997	2.67	1.41	1.26	3675.0	7188.4	51.1

[a] Source: reference 8
[b] Dollars per bushel
[c] 10^6 dollars

Table 2.10 US high-fructose syrup (HFS) supply and use, 1992–2000[a,b]

Year	Supply[c]		Total	Utilization[c]	
	Domestic production	Imports		Exports	Domestic use
1992	6634	193	6827	100	6727
1993	7097	189	7286	113	7173
1994	7467	137	7605	123	7481
1995	7759	79	7838	104	7733
1996	8157	123	8280	224	8057
1997	8677	116	8793	276	8517
1998	9150	117	9267	388	8879
1999	9412	121	9533	350	9183
2000	9367	121	9488	324	9164

[a] Source: reference 8
[b] Includes Puerto Rico
[c] 1000 short tons, dry basis. To convert to metric tons, multiply by 0.9074

much narrower range than purchase prices for raw grain, industry earnings from corn refining tend to be inversely related to the price of corn.

IX. International Involvement

Exports of a variety of products manufactured by the corn refining industry expanded significantly during the 1980s and 1990s, building on the continuing large foreign sales of corn gluten feed and meal. Additionally, corn oil, starches and sweeteners all posted major gains. Total exports of processed corn wet-milling food products reached 1.375×10^9 in 1992,[19] which amounted to 19.5% of the total industry value of shipments in 1992.[1] Some trade has occurred in HFS (Table 2.10).

Expanded US production and consumption of two major corn refinery products, high-fructose syrup and fuel alcohol, probably contributed positively to the US trade balance by substituting to some extent for imported sugar and oil. US corn refining firms have also expanded their direct investment in foreign facilities, such investment occurring in several countries, and foreign companies have purchased US facilities (see Chapter 1).

X. Future Industry Prospects

The dramatic growth in HFS production and consumption began in the early-1970s and moderated in the late-1980s. Per capita consumption continued to rise slightly through the early 1990s. Additional HFS marketing expansion is expected as the US population increases. Some further penetration also appears probable in selected food uses. Much will depend, however, on sugar import policies and on technological developments that might further lower manufacturing costs or lead to advances in the development of products such as crystalline fructose.

Fuel alcohol production could well expand further if government policies provide incentives for more widespread use as automotive fuel, and if technical developments continue to lower alcohol manufacturing costs. Some efficiency has occurred in overall plant operating costs due to seasonal balancing of the production of HFS and fuel alcohol. Plant capacity used in producing HFS for surging summer soft drink demands can be used during other times to produce fuel alcohol.

Food demands for other corn refinery products are expected to expand with population growth or possibly more rapidly with new product development and technological improvements that would lower unit costs. Similarly, expanded industrial uses can be expected with growth in the US economy and possible new uses and applications made feasible through research.

A continued upward trend in sales abroad in both food and industrial markets appears promising. Several firms are expanding their foreign direct investments in corn refining operations and participating in joint ventures with foreign firms.

The total number of corn refining firms in the United States is not expected to change greatly, although further investment in corn refining capacity can be expected with increased demand for starch for food and industrial uses. Given the large economies of scale and the highly technical nature of the business, capacity expansion is expected to continue to occur, primarily within existing firms rather than through new entrants. The conglomerate character of firms with wet corn milling operations is expected to become more prevalent, although the four-firm concentration may not change greatly.

XI. References

1. US Department of Commerce, Bureau of the Census: *Census of Manufacturers*. Industry Series, for indicated years.
2. US Department of Agriculture. *Feed Situation and Outlook Yearbook*. FDS-1999; April 1999.
3. US Department of Agriculture. *Feed yearbook*. FDS-2000; April 2000.

XI. References

4. US Department of Agriculture. *Feed Outlook*. FDS-0301; March 12, 2001.
5. US Department of Agriculture. *Sugar and Sweetener Situation and Outlook Yearbook*. SSS-228; May 2000.
6. Gray F, Buzzanell P, Moore, W. *US Corn Sweetener Statistical Compendium*. US Department of Agriculture, Statistical Bulletin Number 868; 1993.
7. US Department of Agriculture. *Economic Research Service, Sugar and Sweetener Situation and Outlook Yearbook*. SSSV19N2; June 1994.
8. US Department of Agriculture. *Economic Research Service, Sugar and Sweetener Situation and Outlook Report*. SSS-230; January 2001.
9. US Department of Labor. *Monthly Labor Review*, various issues.
10. Buzzanell P, US Department of Agriculture. Corn Sweeteners: Recent Developments and Future Prospects. Paper presented at Azucar '95 Foro Internacional, Guadalajara, Mexico; October 10, 1995.
11. US Department of Agriculture. *Economic Research Service, Sugar and Sweetener Situation and Outlook Yearbook*. SSS-231; May 2001.
12. US Department of Agriculture. *Economic Research Service, Sugar and Sweetener Situation and Outlook Yearbook*. SSSV21N1; March 1996.
13. Lee H. Ethanol's Evolving Role in the U.S. Automobile Fuel Market. In: *Industrial Uses of Agricultural Materials*: US Department of Agriculture; 1993. [1US-1].
14. Watkins MW. *Industrial Combinations and Public Policy*. Cambridge, UK: Houghton-Mifflin; 1927.
15. Whitney SN. *Antitrust Policies*. Vol. II. New York: The Twentieth Century Fund; 1958.
16. US Department of Commerce, Bureau of the Census. *Concentration Ratios in Manufacturing*, for indicated years.
17. Marion BW, Kim D. *Agribusiness*. 1991;7(5).
18. US Department of Agriculture: Economic Research Service, *Food Marketing Review*, 1994–95. Agricultural Economic Report, No. 743; 1996.
19. US Department of Agriculture: Economic Research Service, *Food Marketing Review*, 1992–93. Agricultural Economic Report No. 678; 1994.

Genetics and Physiology of Starch Development

Jack C. Shannon[1], Douglas L. Garwood[2] and Charles D. Boyer[3]

[1]Department of Horticulture, The Pennsylvania State University, University Park, Pennsylvania
[2]Garwood Properties LLC, Stonington, Illinois
[3]College of Agricultural Services and Technology, California State University Fresno, California

I. Introduction	24
II. Occurrence	25
1. General Distribution	25
2. Cytosolic Starch Formation	25
3. Starch Formed in Plastids	26
III. Cellular Developmental Gradients	26
IV. Non-mutant Starch Granule Polysaccharide Composition	28
1. Polysaccharide Components	28
2. Species and Cultivar Effects on Granule Composition	30
3. Developmental Changes in Granule Composition	31
4. Environmental Effects on Granule Composition	32
V. Non-mutant Starch Granule and Plastid Morphology	33
1. Description	33
2. Species and Cultivar Effects on Granule Morphology	33
3. Developmental Changes in Average Starch Granule Size	34
4. Formation and Enlargement of Non-mutant Granules	34
VI. Polysaccharide Biosynthesis	36
1. Enzymology	36
2. Compartmentation and Regulation of Starch Synthesis and Degradation in Chloroplasts	37
3. Compartmentation and Regulation of Starch Synthesis in Amyloplasts	40
VII. Mutant Effects	43
1. Waxy	44
2. Amylose-extender	50
3. Sugary	53
4. Sugary-2	56
5. Dull	57
6. Amylose-extender Waxy	58

7. Amylose-extender Sugary	59
8. Amylose-extender Sugary-2	60
9. Amylose-extender Dull	61
10. Dull Sugary	61
11. Dull Sugary-2	62
12. Dull Waxy	62
13. Sugary Waxy	63
14. Sugary-2 Waxy	63
15. Sugary Sugary-2	63
16. Amylose-extender Dull Sugary	64
17. Amylose-extender Dull Sugary-2	65
18. Amylose-extender Dull Waxy	65
19. Amylose-extender Sugary Sugary-2	66
20. Amylose-extender Sugary Waxy	66
21. Amylose-extender Sugary-2 Waxy	67
22. Dull Sugary Sugary-2	67
23. Dull Sugary Waxy	67
24. Dull Sugary-2 Waxy	67
25. Sugary Sugary-2 Waxy	68
26. Amylose-extender Dull Sugary Waxy	68
VIII. Conclusions	69
IX. References	71

I. Introduction

Starch, a common constituent of higher plants, is the major form in which carbohydrates are stored. Starch in chloroplasts is transitory and accumulates during the light period and is utilized during the dark. Storage starch accumulates in reserve organs during one phase of the plant's lifecycle and is utilized at another time. Starches from reserve organs of many plants are important in commerce.

The complete pathway of starch synthesis is complex and not completely understood. Although considerable effort has been directed at characterizing the enzymes involved in starch synthesis, the role of these enzymes and other factors in determining subtle variations in starch granule structure and starch fine structure remain largely unknown. Certainly gross starch structure is similar in various species. Variations in granule structure and in starch fine structure are well documented and described elsewhere in this volume. Variation can be associated with plant species, cultivars of a species, the environment in which a cultivar is grown, and genetic mutations.

This chapter first reviews non-mutant starch granule composition and development and then focuses on genetic mutants and how they have been useful in understanding the complexity of polysaccharide biosynthesis and development. Due to the limitations of space, attention is given only to a few of the plant species which are important sources of commercial starch production; the discussion will focus on maize

(*Zea mays* L.), because of the many known endosperm mutants of maize which affect polysaccharide biosynthesis. Although developing maize kernels have been used for many of the investigations of starch biosynthesis, the information gained applies to other species, and these effects are illustrated whenever appropriate. As a result of this approach, it has been necessary to be selective in choosing examples to illustrate general trends in the genetics and physiology of starch development. Apology is given to those whose papers could also have been used to illustrate similar points.

No chapter can adequately cover all aspects of starch development, biosynthesis and genetics. Readers wishing more detailed information should consult other reviews by Boyer,[1] Boyer and Hannah,[2] Boyer and Shannon,[3] Hannah,[4,12] Hannah et al.,[5] Nelson and Pan,[6] Preiss,[7] Preiss et al.,[8] Preiss and Sivak,[9] Smith and Martin,[10] Wang and Headley,[11] Smith et al.,[13] James et al.,[14] Thomlinson and Denyer,[15] Ball and Morell[16] and Chapter 4.

II. Occurrence

1. General Distribution

Starch can be found in all organs of most higher plants.[17,18] Organs and tissues containing starch granules include pollen, leaves, stems, woody tissues, roots, tubers, bulbs, rhizomes, fruits, flowers, and the pericarp, cotyledons, embryo and endosperm of seeds. These organs range in chromosome number from the haploid pollen grain to the triploid endosperm, the main starch-storing tissue of cereal grains.

In addition to higher plants, starch is found in mosses and ferns, and in some protozoa, algae and bacteria.[18] Some algae, namely the Cyanophyceae or bluegreen algae,[19,20] and many bacteria produce a reserve polysaccharide similar to the glycogen found in animals.[18,21] Under growth-restrictive culture conditions, *Chlamydomonas*, a single-celled algae, accumulates polysaccharides with characteristics very similar to those of starch in higher plants.[22] Both starch and a water-soluble polysaccharide, similar to glycogen and termed phytoglycogen, occur in sweet corn and other maize genotypes,[23] as well as related genotypes of sorghum[24] and rice.[25] A glycogen-type polysaccharide also has been reported in the higher plant *Cecropia peltata*.[26] Badenhuizen[18] classified starch-producing species into two groups: plants in which starch is formed in the cytosol of a cell and plants in which starch is formed within plastids.

2. Cytosolic Starch Formation

Starch granules are formed in the protozoa *Polytomella coeca*[21,27] but other species of protozoa produce amylopectin-type polysaccharides, glycogen or laminaran.[21,27] Red algae, Rhodophyceae, produce a granular polysaccharide called Floridean starch on particles outside the chloroplasts. In many of its properties this starch resembles the amylopectin of higher plants, but in other properties it is intermediate between amylopectin and glycogen. Floridean starch contains no amylose.[19,28] Free polysaccharide granules are also produced in the Dinophyceae, but their chemical nature is unknown.[19]

Starch-like substances are produced in several species of bacteria.[18,21] For example, *Escherichia coli* produces a linear glucan.[21,29] *Corynebacterium diphtheriae* produces a starch-like material and *Clostridium butyricum* produces a glucan with some branching.[21] *Neisseria perflava* produces a glucan, intermediate in structure between amylopectin and glycogen;[29] however, more recent work shows that the structure more closely approaches that of glycogen.[30]

3. Starch Formed in Plastids

Starch is formed in chloroplasts of moss, fern and green algae.[18] Chlorophyceae (green algae) starch is similar to that of higher plants, and several species have been used in studies of starch biosynthesis.[19,22,29] In a recent set of studies, Ball et al.[22] used *Chlamydomonas reinhardtii* to study starch biosynthesis. They produced several *Chlamydomonas* mutants which produce starch with characteristics similar to starches produced by maize endosperm mutants.[31-34] The various starch mutations of *Chlamydomonas* will be discussed in Section 3.7. Other classes of algae which produce starch are Prasinophyceae[19,35] and Cryptophyceae.[35,36]

In plastids of higher plants, starch granules are classified as transitory or reserve.[17] Transitory starch granules accumulate for only a short period of time before they are degraded. Starch formed in leaf chloroplasts during the day, which is subsequently hydrolyzed and transported to other plant parts at night in the form of simple sugar, is an example of transitory starch. Transitory starch is also formed in lily (*Lilium longiflorum*) pollen during germination of the pollen grains.[37] A transient form of starch accumulates in heterotrophically grown suspension cultured plant cells shortly after subculture to fresh medium containing sugar, but the starch is metabolized for energy and growth later in the culture cycle. Transitory and reserve starch granules can be differentiated by the fact that transitory starch granules lack the species-specific shape associated with reserve starch granules. Furthermore, when exogenous sugar is supplied, the number, but not the size, of granules in a chloroplast increases, while the reverse occurs in amyloplasts.[17]

Reserve starch is usually formed in amyloplasts, although it is occasionally formed in chloroamyloplasts. These are chloroplasts that have lost their lamellar structure and subsequently start producing fairly large reserve starch granules.[17] Chloroamyloplasts form starch independent of photosynthesis. They have been described in tobacco (*Nicotiana tabacum* L.) leaves, *Aloe* leaves and flowers, central pith of potato (*Solanum tuberosum* L.) fruit, *Pellionia* and *Dieffenbachia* stems, and other tissues.[17,18] Such sources of reserve starch are insignificant, however, when compared to the reserve starch formed in roots, tubers and seeds.

III. Cellular Developmental Gradients

To properly evaluate data relating to reserve starch development and composition, cellular development of tissues in which this starch is formed must be appreciated. Enlarging potato tubers,[38] cotyledons of developing pea seeds[39] and endosperms

III. Cellular Developmental Gradients

of developing maize,[40–44] rice (*Oryza sativa* L.),[45] sorghum (*Sorghum bicolor* L. Moench),[46–48] wheat (*Triticum aestivum* L.),[49,50] rye (*Secale cereale* L.),[51] triticale (X *Triticosecale* Wittmack),[52] and barley (*Hordeum vulgare* L.)[53] kernels are composed of a population of cells of varying physiological ages.

In maize kernels, the basal endosperm cells begin starch biosynthesis late in development and contain small starch granules.[40,43,44,54] Peripheral maize endosperm cells, which are the last to develop, also contain small starch granules.[41,43,44,54] Thus, a major gradient of cell maturity from the basal endosperm to the central endosperm and a minor gradient from the peripheral cells adjacent to the aleurone layer inward exists in *normal* (non-mutant) maize endosperm. A similar cellular developmental gradient occurs in sorghum.[46–48] In barley, starch formation begins at the apex of the grain and around the suture across the central region.[51] Deposition occurs last in the youngest cells near the aleurone layer.[53] Related gradients occur in rice,[45] rye[51] triticale[52] and wheat.[49,55]

Since all endosperm cells are not the same age, the physiologically younger cells may undergo the same developmental changes in starch biosynthesis as older cells, but at a later time in grain or kernel development. Shannon[56] divided 30-day-old *normal* maize kernels into seven endosperm zones and found that the sugar and starch composition of the lower zone corresponds to that found in whole endosperms 8, 10 and 12 days post-pollination, while the carbohydrate composition of upper zones is similar to that in kernels 22–28 days post-pollination. When starch granules from 36-day-old *normal* maize kernels were separated into different size classes, a decline in apparent amylose percentage with decreasing granule size was observed, which reflected the characteristics of unfractionated starch isolated from endosperms earlier in kernel development.[57] Although variations in granule size occur throughout the endosperm, starch granules within a given cell of *normal* maize endosperm are similar in size.[42,54]

The existence of cellular developmental gradients has two important ramifications when studying the genetics and physiology of starch development. First, evaluations of developing tissue using whole tissue homogenates are based on polysaccharides and enzymes isolated from cells of differing physiological age. Thus, such whole tissue data represents only an average stage of cellular development at the date of sampling. Secondly, tissue that does not reach maturity because of environmental or other reasons will differ in composition from fully mature tissue, and variation in starch composition can occur between samples.

As tissues storing reserve starch develop and the cells fill with starch granules, the starch concentration, expressed as a percentage of tissue weight, increases. For example, the starch content of potatoes increases from 5% to 18% of the fresh weight as tuber size increases from 0–1 cm to 10–11 cm.[58] In maize, numerous workers have demonstrated a similar increase, with data reported by Wolf et al.[59] and Earley[60] being typical. At 7–10 days post-pollination, starch comprises less than 10% of kernel weight. This percentage increases to 55–60% by 30–35 days, and then remains fairly constant until maturity. The starch content of barley kernels rises in a sigmoid pattern with time, and 95% is deposited between 11 and 28 days after ear emergence.[61] Similar increases are observed in the reserve starch concentration in other species.[62–65]

IV. Non-mutant Starch Granule Polysaccharide Composition

1. Polysaccharide Components

Non-mutant (*normal*) reserve and transitory starch granules are composed primarily of amylose and amylopectin. Amylose is essentially a linear polymer consisting of (1→4)-linked α-D-glucopyranosyl units. Amylopectin is a branched polymer of α-D-glucopyranosyl units primarily linked by (1→4) bonds with branches resulting from (1→6) linkages. Properties of these two major starch components are summarized in Table 3.1.

To determine the relative amounts of amylose and amylopectin in starch and the properties of these components, starch granules must first be isolated and purified from the plant species to be studied.[66,67] Fractionation of the starch into its components can be achieved through two basic methods, involving either selective leaching of the granules or complete granule dispersion.[66–68] Methods based on granule dispersion are more satisfactory.[68] Fractionation methods have been extensively reviewed.[67–72] Thus, only the basic aspects of these methods needed to establish a framework for discussing the starch composition of different species and genotypes will be presented. Methods for dispersing the granule have included autoclaving in water, solubilization in cold alkali, treatment with liquid ammonia, and solubilization in dimethyl sulfoxide, with the latter method being preferred.[66–72]

Once dispersed, the differential iodine-binding properties of amylose and amylopectin (Table 3.1) can be utilized to estimate the amount of linear polysaccharide present in the starch without fractionating the starch.[73] Amylose can be determined either by measuring the absorbance of the starch–iodine complex (blue value procedure) and relating this absorbance to that obtained for amylose and amylopectin standards[74–80] or by the method of potentiometric iodine titration in which the amount (mg) of iodine bound per 100 mg of polysaccharide is determined and this amount is related to the amount bound by an amylose standard.[67,73,81,82] For non-mutant starches, these procedures give similar results;[73] however, absolute results can vary with both procedures, depending on the iodine-binding properties of the amylose and

Table 3.1 Properties of the Amylose and Amylopectin Components of Starch[a]

Property	Amylose	Amylopectin
General structure	Essentially linear	Branched
Color with iodine	Dark blue	Purple
λ_{max} of iodine complex	~650 nm	~540 nm
Iodine affinity	19–20%	<1%
Average chain length (glucosyl units)	100–10 000	20–30
Degree of polymerization (glucosyl units)	100–10 000	10 000–100 000
Solubility in water	Variable	Soluble
Stability in aqueous solution	Retrogrades	Stable
Conversion to maltose by crystalline β-amylase	~70%	~55%

[a] Adapted from Marshall,[416] Williams,[66] and Radley[417]

amylopectin standards. Lansky et al.,[83] for example, showed that iodine affinities for purified amyloses could range from 18.5% to 20.0%, with some amylose subfractions having iodine affinities of 20.5–20.8%. Furthermore, the amylose content estimated by all procedures based on iodine complex formation should be considered 'apparent amylose,'[84] i.e. occurrence of branched chain components with long external chains results in an overestimation of the amylose content.[73,85] Likewise, the presence of short chain length amylose causes the amylose content to be underestimated,[73] because absorption of the starch–iodine complex is reduced when the average degree of polymerization is less than 100.[86] These limitations should be remembered when amylose percentages are presented.

Dispersed starch can be separated into the amylose and amylopectin components by adding a polar organic substance, such as thymol or 1-butanol, to produce an insoluble amylose complex.[72] This initial precipitate is usually purified by solubilizing the complex and precipitating the amylose again, as above. Amylopectin may be recovered from the initial supernatant by freeze-drying or by precipitation with alcohol.[66–68] Alternatively, the amylopectin component can be removed first from the dispersion by high-speed centrifugation followed by the addition of a polar organic substance to precipitate the amylose from the supernatant.[87] Dispersed starch also has been fractionated using size exclusion column chromatography (SEC).[88–90] All these procedures will permit quantitative estimation of the amount of amylose in the starch.

Amylose and amylopectin preparations isolated following fractionation consist of a population of molecules that vary in their degree of polymerization (Table 3.1). For example, amylose can be subfractionated into a graded series of molecular sizes;[83,91,92] the amylopectin fraction also has a broad distribution of molecular weights.[93,94] In addition to heterogeneity of molecular sizes, amylose also appears to consist of a mixture of both linear and slightly branched chains, the proportions of which may vary with the source of the starch and with the maturity of the source.[67] The laboratory of Hizukuri[95–98] has fractionated amyloses from various botanical sources by size. Generally, three fractions are obtained, with a predominate fraction having a mean degree of polymerization (DP) of 400–800, a second fraction with a mean DP of approximately 1500, and a third fraction with a mean DP of approximately 2500. By labeling the non-reducing ends of these fractions and additional structural characterization, the proportion of linear molecules in these fractions, as well as chain lengths and numbers, was determined.[99] As the size of the fractions increased, the proportion of the linear amylose molecules decreased. Similarly, the number of chains increased with the size of the amylose molecules.

Current models of amylopectin structure are based on the cluster model first proposed by Nikuni[100] and French.[101] In this model, the amylopectin is composed of repeating clusters of similar size, chain numbers and chain lengths. Hizukuri[102,103] expanded on this model from results of the distribution of chain lengths obtained from various debranched amylopectins. When these chains from debranching are chromatographically fractionated a periodic distribution of chain lengths is observed.[104] Depending on the source of the starch, the chain distribution is trimodal or tetramodal and based on intervals of DP 12 to 15. Longer chains are thought to span two or more clusters depending on their length. Amylopectins with long chains have been associated with starches with B-type x-ray crystalline patterns and shorter-chained amylopectins have been associated with starches having A-type patterns.[102]

However, starch polymers cannot be divided sharply into amylose and amylopectin fractions. Rather, the two major fractions blend into each other through intermediate fractions. The presence of intermediate polysaccharides in the starch granule is apparent from the SEC elution profile of *normal* maize starch when compared to the profile of a mixture of purified amylose and amylopectin.[88,90] Based on indirect evidence from iodine affinities, Lansky et al.[83] suggested that 5–7% of normal maize starch consists of material intermediate between the strictly linear and highly branched fractions. Subsequently, several 'non-amylopectin' types of branched polysaccharides have been recovered by various modifications of the previously described fractionation procedures. For example, Erlander et al.[105] recovered a low molecular weight component from the supernatant following amylose precipitation with thymol and removal of amylopectin by centrifugation. The polysaccharide remaining in the supernatant had a β-amylolysis limit and degree of branching similar to that of amylopectin. Perlin[86] obtained an intermediate component following removal of amylopectin by centrifugation and precipitation of amylose with amyl alcohol. The polysaccharide remaining in the supernatant was more highly branched than amylopectin, based on reduced β-amylolysis limits, and was of lower molecular weight. A related highly branched polysaccharide with viscosity similar to amylopectin was recovered from the supernatant following recomplexing the amylose fraction of starch from potato tuber, rubber (*Havea brasiliensis*) seed, barley kernels and oat (*Avena sativa* L.) kernels.[106,107] A 'loosely' branched polysaccharide related to amylopectin, but with greater average chain lengths and higher β-amylolysis limits, was recovered from rye and wheat starches[107] and from *normal* maize starch.[108] 'Hinoat' oat starch was found to contain 26% of an intermediate molecular weight branched starch component following SEC, while wheat starch contained 10% of a similar fraction.[108] Hizukuri et al.[95] concluded that the structures of the branched fraction of amylose are intermediate to true linear amylose and amylopectin. Another polysaccharide reported in small amounts in starch of non-mutant rye,[107] wheat[107] and maize[110] is short chain length amylose. In *normal* maize starch, this linear polysaccharide has an average chain length of 58.[110] Given the polydisperse and polymolecular nature of the two basic fractions of starch, it is not surprising that different methods have yielded various fractions of 'intermediate' structure.

2. Species and Cultivar Effects on Granule Composition

The percentage of amylose in non-mutant reserve starch of higher plants varies, depending on the species and cultivar from which the starch is isolated. Deatherage et al.[111] analyzed starch from 51 species and reported an amylose content of from 11% to 37%. A summary of data from the literature for 23 species indicates a range of from 11% to 35% amylose.[112] Starches of six species of legumes investigated had amylose percentages which varied from 29% to 37%.[113]

Almost as much variation in amylose percentage has been observed among cultivars of a single species. For example, amylose percentage of starch ranges from 20%

to 36% for maize (399 cultivars),[111,114] from 18% to 23% for potatoes (493 cultivars),[115] from 21% to 35% for sorghum (284 cultivars),[111,116,117] from 17% to 29% for wheat (167 cultivars),[111,118] from 11% to 26% for barley (61 cultivars, including 5 genetic lines),[119,120] from 8% to 37% for rice (74 cultivars)[121-124] and from 34% to 37% for eight cultivars of peas (*Pisum sativum* L.).[64,111] Because of the variation in amylose percentage among species and among cultivars within a species, no average amylose percentage will be meaningful for non-mutant starches *per se* or for non-mutant starches of a given species. However, all non-mutant starches have more amylopectin than amylose.

Species and/or cultivar differences are also observed in other starch properties and in the properties of isolated amylose and amylopectin. To illustrate, purified amylose samples have been shown to differ in β-amylolysis limit and average DP.[64,67,124] Purified amylopectin samples have also been shown to differ in β-amylolysis limit, average length of unit chains and viscosity.[64,66,67,124,125] Campbell et al.[121] observed a range of amylose content from 22.5% to 28.1% in 26 maize inbreds selected for maturity, kernel characteristics and pedigree. Starches from these non-mutant genotypes also differed in thermal properties (DSC), paste viscosities and gel strengths.

3. Developmental Changes in Granule Composition

Increased amylose percentages have been observed for various plant species as a function of the age of the tissue from which the starch was isolated. Several investigators[57,59,127-129] reported increased amylose percentages in maize endosperm during kernel development. For example, Tsai et al.[129] reported an amylose increase from 9% to 27% from 8 to 28 days post-pollination. The percentage of amylose in potato starch increased from 12% in 0- to 1-cm tubers, and to 20% in 15- to 16-cm tubers.[58] In starch from cassava (*Manihot utilissima*) roots harvested at various maturities, significant variation in amylose percentage (16–17%) has been observed; however, the net increase in roots from 5 to 9 months of age amounted to only 0.3%.[130] In starch of developing rice grains, amylose increased from 23% to 27% in kernels of cultivar 'IR8' 4 to 39 days post-pollination[131] and from 30% to 37% in kernels of cultivar 'IR28' 3 days post-pollination to maturity, with 41% observed in kernels 7 days post-pollination.[62] Various workers[63,132-135] have reported that the percentage of amylose in wheat starch increases with kernel development; however, the amount of increase varies with the initial sampling date and the cultivar examined. In starch of developing barley kernels, the percentage of amylose increased from 16% to 28% from 9 to 46 days after anthesis,[137] from 13% to 25% and from 14% to 26% for two cultivars during a 12-week period,[138] and from 14% to 22% from 14 to 30 days after ear emergence, with the percentage remaining constant from 30 days until maturity.[61] The amylose concentration in smooth-seeded pea starch increases from 15% in 2- to 6-mm peas to 37% in 11- to 12-mm peas.[64] Developmental differences are also observed in other starch properties, and in the properties of isolated amylose and amylopectin.[58,64,130,131,134-136]

Similar increases in amylose percentage are observed as a function of increasing granule size when granules from a developing tissue at a single stage of development

Table 3.2 Amylose content of starch granules of various size classes isolated from maize endosperm 36 days post-pollination and from intermediate size (5–6 cm) potato tubers

Maize[a]		Potato[b]	
Granule size, μm	Amylose, %	Average granule size, μm	Amylose, %
Unfractionated	25.4	28 (Unfractionated)	17.2
10 to 20	26.4	37	19.5
5 to 10	23.0	28	18.0
Less than 5	20.5	16	16.7
		10	16.0
		7	14.4

[a]Data from Boyer et al.[57]
[b]Data from Geddes et al.[58]

are separated into various size classes (Table 3.2). This increase in amylose percentage with granule size has also been observed in starches from mutant geno-types.[57,139] Since smaller granules have lower amylose percentages similar to those starch granules from younger tissue, the smaller granules are presumably isolated from the physiologically younger cells present in the developing tissue (see Section 3.3). This effect of granule size on percentage amylose is not applicable to the small starch granules found in mature wheat and barley endosperms, since the small and large populations have similar properties.[63,139,140] In barley and wheat, these small granules are formed late in the growth cycle and represent a second discrete population of granules formed in cells already containing the larger starch granules, and thus do not represent granules from physiologically less mature cells.[140,141]

Because the percentage of amylose varies with maturity of the tissue, starches from tissues that do not reach final maturity will be altered in their physicochemical properties from the corresponding mature starch.

4. Environmental Effects on Granule Composition

Growing conditions associated with different locations, years, planting dates, etc. can also affect the polysaccharide composition of non-mutant granules. Location and year of production and environmental conditions affect the percentage of amylose in rice,[121,142] with milled samples of 'IR8' rice ranging from 27% to 33% amylose.[143] The percentage of amylose in 'Selkirk' wheat grown at four locations ranged from 23.5% to 24.7%,[118] and that of 'Katahdin' potatoes grown at three locations ranged from 21% to 24%.[112] The amylose percentage in starch from 30 samples of 'Compana' barley representing different environmental and cultural practices ranged from 19% to 23%.[119] Limited variation was seen for amylose percentage in maize starch from plants grown for three years in each of eight states. Year effects ranged from 26.2% to 26.8% averaged over locations, and location effects ranged from 25.5% to 27.7% averaged over years.[144] Although present, environmental effects are not as large as those associated with cultivars or cultivar maturity.

V. Non-mutant Starch Granule and Plastid Morphology

1. Description

Reserve starch granules in higher plant tissues develop in organelles called amyloplasts.[17] An amyloplast may contain one starch granule or it may contain several granules, depending on the plant species or genetic mutant. When only one granule is produced in an amyloplast, such as in endosperms of wheat, barley and maize, potato tubers, pea embryos, and others, it is called a simple granule.[18] When two or more granules occur in one amyloplast, they form the parts (granula) of one compound granule. Such granules are often rounded at first, but become angular as they pack together within the amyloplast. The granula of the compound granule are separated by a narrow layer of stroma.[18] Examples of species having compound granules include endosperms of rice and oats, cassava and sweet potato (*Ipomoea batatas* L.) roots, sago (*Metroxylon* sp.), and dasheen (*Colocasia esculenta*). In an extreme case of compound granules, amyloplasts in quinoa (*Chenmopodium quinoa*) seeds contain more than 100 granula.[145] Badenhuizen[18] called granules that are initially compound but become united by the deposition of a common surrounding layer of starch, semi-compound granules. Starch granules from the bulb of *Scilla ovatifolia* are semi-compound.[18] Goering et al.[146,147] described the presence of large 'starch chunks' in seeds of *Amaranthus retroflexus* (pigweed). The starch chunks are composed of many small granula cemented together with amorphous starch,[147] and can be considered semi-compound granules.

Wheat, rye and barley produce two types of granules. The first granules produced in the endosperm cells develop into large lenticular granules.[18] However, about two weeks after initiation of the first granules, additional small granules are produced within evaginations of the original amyloplasts, which then separate from the original amyloplasts by constriction.[148] The secondary granules are generally spherical and remain small. Although two basic size classes exist, no abrupt size change occurs, and some intermediate size granules are present.[50,149] In mature barley kernels, the large granules constitute about 90% of the total starch volume, but represent only 12% of the total number of granules.[150] Large starch granules in 17 wheat flours averaged 12.5% of the total granule number, while accounting for 93% of the starch granule weight.[149] Many of the large lenticular granules of wheat and barley have an equatorial groove or furrow.[151,152] Buttrose[151] suggested that starch-synthesizing enzymes may be concentrated within the equatorial groove.

2. Species and Cultivar Effects on Granule Morphology

Size and shape of reserve starch granules are extremely diverse and are species specific.[18] This diversity is illustrated in photographs of starch granules from over 300 species and varieties.[153] Microscopic characteristics of various starches are also summarized by Moss[154] and Kent[155] (see Chapter 23). Scanning electron microscopy (SEM) has been used to reveal the topography of starch granules. Hall and Sayre published SEM pictures of various root and tuber starches,[156] cereal starches[152] and 16 other miscellaneous starches.[157]

Smaller granules are often found in tissues of species producing compound granules such as rice,[152] malanga (*Xanthosoma sagittifolium*)[156] and cowcockle (*Saponaria vaccaria* L.).[157] As noted earlier, the secondary granules in barley, wheat and rye remain small, with most less than 10 μm in diameter.[155] At the other extreme, large granules in potato tubers can exceed 120 μm in diameter.[156] Starch granules from most species are non-uniform in size. The amount of this variation can be seen, for example, by examining granule size distributions for wheat,[158] rye[158] triticale,[158] potato,[58] barley,[61] maize[57,59] and dropwort (*Filipendula vulgaris*).[159]

Differences in average starch granule size in cultivars from a single species have been also reported. For example, average starch granule diameter ranged from 8.2 to 17.5 μm in 12 sorghum cultivars,[116] from 17.8 to 25.6 μm in six triticale cultivars,[158] and from 3.8 to 5.7 μm in 10 rice cultivars.[124]

In addition to having an effect on amylose percentage (Section 3.4), varying environmental conditions also affect average starch granule diameter. Data on average starch granule size for rice cultivars grown in two different seasons[122] and dropwort grown at varying fertility levels[159] illustrate this effect.

In contrast to the species-specific shape and size of reserve starch granules, transitory starch granules in chloroplasts appear similar in all species.[18] In chloroplasts, the assimilatory starch granules are small and disk-shaped.[17]

3. Developmental Changes in Average Starch Granule Size

As tissues storing reserve starch mature, starch content (Section 3.3) and percentage amylose (Section 3.4) increase. Similarly, average starch granule size increases with increasing age of the storage tissues. Such increases have been documented in maize[57–59] and rice[131] endosperm, in potato tubers[58] and in pea cotyledons.[64] This trend does not apply to average starch granule size in barley, wheat and rye, where a second population consisting of a large number of small granules are formed late in development. In kernels of these species, average granule size initially increases; however, as the small granules are formed, average granule size decreases.[61,63] In barley, the maximum average granule diameter of 10.5 μm was observed 16 days after ear emergence.[61]

4. Formation and Enlargement of Non-mutant Granules

Plant cells have several types of plastids, such as proplastids, chloroplasts, chloroamyloplasts, chromoplasts and amyloplasts, depending on the species and tissues. Badenhuizen[17] contended that, although certain plastids do not form starch under natural conditions, all can be induced to form starch by floating tissue pieces on a sugar solution. Although starch can be produced in a variety of plastids when supplied with sugar, chloroplasts and amyloplasts are the primary sites of starch accumulation in nature. Transitory starch is produced in chloroplasts during the day and mobilized at night. During extended light periods, the number of small granules in a chloroplast increases, but granule size remains relatively small.[17] The control of starch synthesis and degradation in chloroplasts is discussed in Section 3.6.

Reserve or storage starch accumulates in specialized leucoplasts called amyloplasts and occasionally in chloroamyloplasts. Amyloplasts are organelles bounded by a double membrane which develop from proplastids. Duvick[160] studied early plastid and starch development in maize endosperm cells. He described small filaments which developed knobs in maize endosperm cells. Then, according to him, starch granules formed within the filament knobs. Based on electron microscope observations,[17,151,161] the filaments observed in living cells by Duvick[160] apparently are proplastids developing into amyloplasts (knobbed filaments). Proplastids and young amyloplasts in fixed sections have very irregular shapes and likely assume various shapes (amoeboid) in the living cell.[161]

The inner membrane of young amyloplasts from barley[151] and maize[161,162] has been shown to be extensively invaginated to form tubuli, stroma lamellae or vesicles. Badenhuizen[162] observed that starch granules are formed in the 'pockets' provided by the lamellar structure. He suggested that these pockets are necessary for initiation of starch granule formation, perhaps by promoting locally elevated concentrations of enzymes and substrates. The inner membrane of chloroplasts contain the specific translocators necessary for transfer of metabolites between the chloroplast stroma and the cytosol.[163,164] It is assumed, based on the similarity between chloroplasts and amyloplasts, that the inner membrane of the amyloplasts also functions in the regulation of metabolite transfer. Thus, the invaginations of the inner membrane noted above would effectively increase the surface area of the membrane, and perhaps allow for more effective substrate transfer into the amyloplasts.[151]

Tandecarz et al.[165] reviewed several years of evidence supporting the conclusion that starch biosynthesis involves a specific initiation event mediated by UDP-Glc:protein transglucosylase (UPTG) (EC 2.4.1.112), a 38 000 molecular weight polypeptide. UPTG, which has an almost absolute requirement for Mn^{2+}, is the active enzyme, and at the same time, the glucosyl acceptor. The resulting glucosylated polypeptide serves as the glucosyl primer required for polymer elongation via the action of starch synthase.

Starch polysaccharide initiation and elongation is assumed to occur in the amyloplast stroma. The stroma (ground substance) of amyloplasts appears homogeneous by electron microscopic examination.[17,161] However, Badenhuizen[17,18,162] observed granular particles in the amyloplast stroma of tissue fixed in potassium permanganate. Although the granular structure observed may have been an artifact caused by the fixation procedure, it did show the presence of material in the stroma that accumulated in amyloplasts and then declined with formation of the starch granule.[18] Accumulation and decline of these particles also occured during starch granule growth.[18] Badenhuizen[18] called these particles coacervate droplets, and suggested that they become attached to the periphery of the starch granule. He concluded[17,18] that starch molecules are produced in the amyloplast stroma and then the completed molecules become part of the growing starch granule. Shannon et al.[166] exposed maize plants to $^{14}CO_2$ and determined the distribution of ^{14}C in the amylose and amylopectin components of starch 1–6 hours later. They found that the specific activity (^{14}C/mg of polysaccharide) of amylose and amylopectin increased at a similar rate, and that the radioactivity was distributed throughout the polysaccharide molecules.

They concluded that these data supported Badenhuizen's[17,18] suggestion that starch molecules are completely synthesized in the amyloplast stroma and are then deposited on the granule surface. Once the polysaccharides are part of the granule, there was no evidence of subsequent conversion of amylose to amylopectin.[166] This is in contrast to conclusions drawn from long-term ^{14}C-labeling studies of wheat starch.[167,168] In results of these studies, amylose appeared to be synthesized first, and then transformed into amylopectin. These differences may be due to the different species used or to the widely different sampling times.

VI. Polysaccharide Biosynthesis

1. Enzymology

Enzymes responsible for the synthesis of transitory starch in leaves and reserve starch in seeds, tubers, etc. are generally considered to be the same in both types of tissues.[7] Chloroplast starch is synthesized and accumulates during the light period when photo-synthetic carbon fixation exceeds assimilate demand by the plant, and it is hydrolyzed at night or at any time when assimilate demand exceeds current carbon fixation. Thus, synthesis and degradation of transitory starch in chloroplasts are finely regulated. In starch storage tissues, starch synthesis is the predominant function of the amyloplast enzymes during tissue development. Thus, it is likely that the activities of amyloplast enzymes may be regulated by mechanisms different from those in chloroplasts. The first phase of starch synthesis is synthesis of the glucosyl primer (initiation) followed by primer extension in phase two.[165] The α-glucan primer initiation phase of starch biosynthesis is less well characterized than the elongation phase. As noted in the previous section, in potato tubers and maize endosperm α-glucan primer synthesis is catalyzed by UPTG, a 38 000 molecular weight polypeptide which itself becomes glucosylated.[165] It is proposed that the glucosylated UPTG functions as the primer for starch synthase, but it is not known whether UPTG glycosylation occurs in the cytosol prior to transfer into the amyloplast stroma or whether it occurs in the amyloplast stroma.

It is generally agreed that reactions catalyzed by ADP-Glc pyrophosphorylase (AGPase) (EC 2.7.7.27), starch synthase (EC 2.4.1.21), and starch-branching enzyme (SBE) (EC 2.4.1.18) are the final three reactions in the starch biosynthetic pathway in both photosynthetic and non-photosynthetic tissues.[7] Ball et al.[169] recently reviewed evidence supporting the function of a debranching enzyme in starch synthesis. According to their model, a more highly branched 'preamylopectin' is selectively debranched to yield an amylopectin-like polymer and short chain malto-oligosaccharides. Maize *su1* mutant endosperm cells are deficient in debranching enzyme,[170,171] and the more highly branched polysaccharide, phytoglycogen, accumulates. Recent studies have confirmed a role for both isoamylase and pullulanase debranching enzymes in starch biosynthesis.[172–174] A potential role for disproportionating enzyme or D-Enzyme (EC 2.4.1.25) in starch biosynthesis has also been proposed.[175] This enzyme is located in amyloplasts and can use maltoheptaoses as a donor for the addition of glucans to the outer chains of amylopectin. During the last decade much progress has been made in characterizing the various starch biosynthetic enzymes and their kinetic and regulatory properties. Results of these studies have been summarized in several comprehensive reviews; see references 1–16 and Chapter 4. This section is directed to discussion of the cellular compartmentation of starch synthesis in photosynthetic and nonphotosynthetic tissues and the *in vivo* regulation of the pathways. Enzymic modifications associated with various genetic mutations are included in the discussion of specific endosperm mutant genotypes (Section 3.7).

2. Compartmentation and Regulation of Starch Synthesis and Degradation in Chloroplasts

Photosynthesis occurs in plastids called chloroplasts. Chloroplasts develop from proplastids and are bounded by a double membrane.[17] The outer membrane is freely permeable to small molecules and it contains a number of enzyme activities, pore protein and phosphoproteins.[163,164] Several of these function in protein transport into the chloroplasts.[163,164] The membrane transporters important in carbohydrate partitioning in photosynthetic tissues are localized in the inner membrane. Thus, this review is restricted to the inner membrane transporters which function in carbohydrate transfer. The reader is referred to the review by Douce and Joyard[163] for more information on the properties of the chloroplast outer membrane. The inner membrane is the functional barrier to the exchange of several metabolites between the chloroplast stroma and the cytosol.[163,164] Flugge and Heldt[164] reviewed the various transporters of the inner membrane. The triose phosphate/inorganic phosphate translocator (Pi translocator) is primarily responsible for transport of the products of carbon reduction, 3-phosphoglycerate (3-PGA) and triose phosphates (G3P, DHAP), out of the chloroplast in exchange for inorganic phosphate (Pi) from the cytosol.[164] The Pi translocator from root plastids and plastids from other nonphotosynthetic tissues are capable of transporting G6P in exchange for Pi or triose phosphates.[176] The chloroplast Pi translocator from leaves of normally growing spinach (*Spinacia oleracea*) plants does not transfer G6P,[163] but Quick et al.[177] reported that a Pi translocator in spinach leaf chloroplasts capable of transporting G6P was induced by incubating the leaf petiole in 50 mM glucose for several days. It was also noted that the mRNA encoding the Pi/triose phosphate translocator was strongly repressed, and they concluded that incubation of the leaf in glucose activated transcription of a gene responsible for synthesis of a Pi/hexose phosphate translocator which was capable of transferring G6P, but not G1P. A similar treatment of potato leaves likewise induced synthesis of a chloroplast Pi/G6P translocator.[177] Transgenic potato[178,179] and tobacco[180] plants expressing antisense RNA to the Pi/triose-P translocator were used to demonstrate the significance of the Pi translocator in the *in vivo* partitioning of carbohydrates between the chloroplast stroma and the cytosol. Antisense potato plants with about a 25% reduction in expression of the Pi translocator showed a large accumulation of starch and a substantial reduction in sucrose synthesis during the day.[178,179] However partial inhibition of sucrose synthesis in the Pi translocator-antisense plants did not lead to a consistent inhibition of growth,[178] apparently because the increased quantity of starch accumulating during the day provided an additional source of carbohydrate for export from the leaf and growth during the night. Among other metabolite translocators in the inner chloroplast membrane are the dicarboxylate and monocarboxylate translocators and an ATP/ADP translocator.[62,164] The reader is referred to the review by Flugge and Heldt[164] for more information on these transporters.

Chloroplasts which produce starch under natural conditions reduce carbon by the Calvin cycle of photosynthesis. The mesophyll cells of C4 plants, such as maize, fix carbon by the C4 pathway of photosynthesis,[181] but under normal growth conditions they produce very little starch.[182,183] Rather, the four-carbon acids produced in the mesophyll cells are transferred to the bundle sheath cells where they are decarboxylated and the resulting carbon dioxide is refixed by Calvin cycle enzymes in the bundle sheath chloroplast.[181] Bundle sheath chloroplasts of C4 plants accumulate starch.[183] However, it should be noted that if assimilate transport from maize leaves is restricted, as in the *sucrose export defective-1* mutant, starch accumulates in chloroplasts of both mesophyll and bundle sheath cells.[184] In the Calvin cycle, the enzyme ribulose bisphosphate carboxylase/oxygenase (Rubisco) catalyzes carboxylation of ribulose 1,5-bisphosphate and the enzyme-bound intermediate undergoes hydrolysis to yield two molecules of 3-PGA.[185] The 3-PGA molecules thus produced by photosynthesis are: (a) used as a substrate to maintain the function of the Calvin cycle; (b) utilized in the production of chloroplast starch; and/or (c) unloaded from the chloroplasts via the Pi translocator discussed above.[163,164] Partitioning of 3-PGA between sucrose and starch is highly regulated.[184] For example, in the starch biosynthesis pathway within the chloroplasts, there are two regulatory enzymes: fructose 1,6-bisphosphatase (EC 3.1.3.11) (F1,6BPase) and AGPase. Activity of plastid F1,6BPase, which functions in both the ribulose 1,5-bisphosphate regeneration pathway and the starch biosynthetic pathway, is light regulated via a ferredoxin-thioredoxin mediated oxidation/reduction system. In the light, it is reduced and active; but in the dark, it is oxidized and inactive.[185] AGPase is allosterically regulated by the ratio of Pi to 3-PGA and triose phosphates with Pi, and the 3-PGA plus triose phosphates functioning as negative and positive effectors, respectively.[7]

Triose phosphates transferred from the chloroplasts are used for sucrose synthesis in the cytosol by the combined activity of the enzymes of gluconeogenesis and sucrose synthesis.[7,186] This pathway is also controlled by two regulatory enzymes, the cytosolic F1,6BPase and sucrose-P synthase (EC 2.4.1.14). Regulatory control of the cytosolic F1,6BPase differs from that of the plastid isozyme. Activity of the cytosolic F1,6BPase is regulated by the quantity of the metabolite fructose 2,6-bisphosphate (F2,6BP) and AMP.[187] As quantities of F2,6BP and AMP increase, cytolosic F1,6BPase activity declines. Synthesis and breakdown of the regulatory metabolite, F2,6BP, is also allosterically regulated.[187] For example, Pi and 3-PGA (and triose phosphates) are positive and negative effectors of F6P 2-kinase, the enzyme responsible for synthesis of F2,6BP, and these effectors have the opposite effect on F2,6BPase, the enzyme which hydrolyzes F2,6BP. Thus, a relatively small change in the ratio of Pi to 3-PGA and triose phosphates affects the synthesis of F2,6BP, which in turn regulates the activity of F1,6BPase.

Sucrose-P synthase (SPS) activity in spinach is regulated at three levels, protein quantity, protein phosphorylation/dephosphorylation, and allosteric control.[188,189]

SPS protein quantity varies with tissue (leaf) development. SPS phosphorylation in the dark and dephosphorylation in the light provides control by modifying the sensitivity of SPS to allosteric regulation by G6P and Pi, positive and negative effector metabolites, respectively.[188,189] Only phosphorylated SPS is responsive to allosteric regulation by G6P and Pi. It should be noted that the precise regulatory mechanisms of SPS vary depending on the species.[189] *In vivo* control of the sucrose biosynthetic pathway was shown to be shared between the two regulatory enzymes.[190] Given the regulatory controls just reviewed, it follows that during periods of active photosynthesis, energy production in the form of NADPH and ATP is high, which results in reduction of CO_2 to form 3-PGA and a decline in chloroplastic Pi (due to the synthesis of ATP). The resulting increase in positive effectors (3-PGA and triose phosphates) and decline in the negative effector (Pi) of AGPase yields increased ADP-Glc synthesis, resulting in enhanced starch biosynthesis.[7] At the same time, the increased 3-PGA and triose phosphates and reduced Pi in the chloroplasts enhances transfer of 3-PGA and triose phosphates to the cytosol in exchange for Pi from the cytosol. Reduced Pi and increased 3-PGA and triose phosphates in the cytosol stimulates the F2,6BPase-mediated decline in F2,6BP, thus activating cytosolic F1,6BPase.[187] Increased production of hexose Ps coupled with reduced Pi in the cytosol activates the phosphorylated SPS, leading to increased sucrose synthesis. If sucrose synthesis exceeds the capacity for transport or utilization, sucrose and Pi accumulate, resulting in increased F2,6BP, which in turn inhibits cytosolic F1,6BPase. F1,6BP accumulates and Pi declines, so that more 3-PGA and triose phosphates remain in the chloroplasts and starch accumulates.[7,187] In the dark, photosynthetic production of 3-PGA ceases, the level of ADP-Glc declines, and starch synthesis ceases.[7] Also in the dark, Pi in chloroplasts increases by 30–50%, and the pH of the chloroplasts' stroma declines. Preiss and Levi[191] suggested that the lower pH may enhance the activity of certain starch hydrolases, and the increased Pi and lower ADP-Glc concentrations may stimulate starch hydrolysis by phosphorylase. However, they[191] add that definitive studies on the regulation of chloroplast starch degradation are lacking. While the four regulatory enzymes discussed above are thought to function in regulation of starch and sucrose biosynthetic pathways, reductions in the quantity of other pathway enzymes, such as aldolase, via antisense technology have been reported to cause reductions in carbon flux through the pathways.[192]

A result of the regulation of carbon partitioning between starch and sucrose is that transitory starch accumulates in the chloroplasts during the day and is degraded at night or during periods when assimilate demand exceeds current photosynthetic production.[191] Mobilization of chloroplast starch involves enzymes which cleave the (1→4)-α-D-glucosidic bonds and the (1→6)-α-D-glucosidic branches. Released sugars or sugar phosphates (G1P) may exit the chloroplast via a chloroplast membrane sugar transporter[193] or they may be converted to triose phosphates which are transported to the cytosol via the phosphate translocator.

Alpha-amylase is generally accepted as one of the most important enzymes in the hydrolysis of storage starch granules, with β-amylase and phosphorylase being less important.[191] However, phosphorylase is the primary enzyme involved in the utilization of transitory starch in pea chloroplasts.[194] Levi and Preiss[194] suggested that the small amounts of maltose found in pea chloroplasts during starch degradation may have been produced from G1P and D-glucose by maltose phosphorylase, rather than from the action of α-amylase or β-amylase. Beck[193] suggested that the enzymes responsible for breakdown of transitory starch and the products transferred out of the chloroplasts vary with species. Ludwig et al.[195] purified a starch-debranching enzyme from spinach chloroplasts which they suggested may function in hydrolysis of the (1→6)-α-D-glucosidic bonds of transitory starch.

3. Compartmentation and Regulation of Starch Synthesis in Amyloplasts

Amyloplasts are organelles specialized for the accumulation of starch in storage cells. They develop from proplastids, as do chloroplasts, and are bounded by double membranes.[17] Amyloplasts from some starchy tissues, such as those in cereal endosperm, remain colorless when exposed to light, while others such as those in potato tubers, develop into chloroplasts when placed in the light[196] and chloroplasts in aging tobacco leaves become amyloplasts.[17] In 1976, it was assumed that the nature of the amyloplast envelope is like that of the chloroplast with similar membrane transporters[197] except that, in amyloplasts, triose phosphates would be transferred into the amyloplasts to provide substrates for starch synthesis rather than out as in chloroplasts. Inner membranes of young amyloplasts from barley[151] and maize[198] have been shown to be extensively invaginated to form tubuli, stroma lamellae or vesicles. Buttrose[151] suggested that, if the inner membrane is the one limiting uptake, the increased area resulting from the invaginations of the inner membrane would allow for more rapid uptake of metabolites into the amyloplast stroma to provide substrates for starch synthesis.

Amyloplasts containing starch granules are extremely fragile, and early attempts to isolate intact amyloplasts for uptake studies were disappointing.[199] Because of the difficulty in obtaining high yields of intact amyloplasts from starchy tissues, other methods have been used to determine compartmentation of the enzymes of starch biosynthesis between amyloplasts and the cytosol. Immunolocalization of specific enzymes is the most direct approach to determining enzyme compartmentation. Kim et al.[200] reported that, in potato tubers, AGPase appeared to be closely associated with starch granules when observed at the light microscopic level, and suggested that it was a plastid localized enzyme. This was confirmed by Kram et al.[201] who, using immunogold labeling and electron microscopy, demonstrated that proteins reacting to antibodies raised against spinach leaf AGPase were dispersed throughout the amyloplast stroma of potato tubers. They also reported that the starch-branching enzyme was localized within the amyloplasts in close association with surface of the starch granules.[200] Miller and Chourey[202] reported that AGPase in maize endosperm was also localized within amyloplasts, but proteins reacting with antibodies raised against SH2 and BT2, the large and small subunits of AGPase, respectively, were also present in endosperm cell walls. Some antibody-reacting proteins were also scattered throughout the cytosol.[202] The cell wall-reacting proteins were not detected by antibodies raised against the spinach leaf AGPase. Villand and Kleczhowski[203] reported that, in barley endosperm, antibodies specific for AGPase reacted with proteins in both the cytosol and amyloplasts. AGPase in the cytosol occurred in a cluster-like pattern. They suggested that the amyloplast-localized AGPase may be a 'leaf-type' isozyme. The same antibodies used in the barley endosperm study labeled only amyloplast stromal proteins in potato tubers.[203] This difference in compartmentation of AGPase between potato tubers and cereal endosperm led Villand and Kleczhowski[203] to suggest that cereal endosperms probably have an alternative pathway for starch biosynthesis in which ADP-Glc produced in the cytosol is transferred into amyloplasts to provide substrate for starch synthesis.

There are two major drawbacks to the use of immunolocalization studies to determine enzyme compartmentation in amyloplasts. First, starchy tissues are very difficult to prepare for electron microscopic examination. For example, it is almost impossible to embed starch granules of developing maize endosperm tissues properly and thus, when thinly sectioned, the slice of starch absorbs water and 'pops' out of the plastic matrix. As a result, microscopic observations are usually restricted to regions of the tissue containing the smallest starch granules, which may not be representative of the entire tissue. Secondly, *in situ* localization is only as good as the antibody probe. For example, the SH2 and BT2 antibodies used by Miller and Chourey[202] were raised against protein bands cut from a gel and were shown by western blots to react with proteins other than SH2 and BT2, respectively.[204] The antibody raised against the spinach leaf AGPase used in the studies by Kim et al.,[200] Kram et al.[201] and Miller and Chourey[202] was more highly purified. However, these results, using an antibody to a chloroplast-specific AGPase, cannot be used to rule out the possible presence of cytosolic isozymes of AGPase not recognized by the antibody to the chloroplast-specific AGPase.

Fractionation and enzymic characterization of amyloplast from several tissues, such as suspension cultured soybean cells,[205,206] potato tuber cells[207] and sycamore cells,[208] from cauliflower buds,[209] etiolated pea epicotyls,[210] and the endosperms of wheat[206,211] and maize[212] have been reported. Although yields of amyloplasts were low, most studies supported the conclusion that amyloplasts contained AGPase and the other enzymes required for the conversion of triose phosphates into starch. However, ap Rees et al.[206,211] were unable to detect F1,6BPase in amyloplasts from wheat endosperm, and noted that, in its absence, conversion of triose phosphates to starch would not be possible. As a result, they concluded that hexose phosphates, rather than triose phosphates, are transferred into amyloplasts to provide substrate for starch synthesis. Keeling et al.[213] reached the same conclusion from a study of the redistribution of ^{13}C-label between carbon atoms 1 and 6 of glucose recovered from starch. Based on assays of a number of enzymes associated with an amyloplast-enriched preparation from barley endosperm, Williams and Duffus[214] concluded that, in barley endosperm, the production of G1P and ADP-Glc from sucrose occurs in the cytosol. However, they presented no evidence that the isolated amyloplasts were intact and indeed contained the plastid stroma enzymes. More recently, Thorbjornsen et al.[215] re-examined compartmentation of AGPase in developing barley endosperms and found that only about 15% of the endosperm AGPase was located in the amyloplasts. In contrast to the earlier study,[214] they provided evidence that the isolated amyloplasts were intact and that recovery of cellular plastid and cytosol marker enzymes and AGPase were high.[215] Similarly, Denyer et al.[216] and Pien and Shannon,[217] based on an aqueous amyloplast enrichment study and on a non-aqueous fractionation study, respectively, concluded that at least 95% of the cellular AGPase was present in the cytosol of developing maize endosperm cells. These studies have been confirmed recently with improved procedures for the isolation of amyloplasts. In maize, both cytosolic and plastid AGPase have been detected.[218] The greatest activity is found in the cytosol. Similar observations have been made for wheat[219] and endosperm in the Gaminaceous family of plants appear to differ from other starch forming organs and species in the primary localization of the AGPase in the cytosol.[220]

Since 1988, there have been numerous studies reporting metabolite uptake into amyloplasts isolated from various tissues. General methods for amyloplast isolation were recently reviewed.[199,221] The preferred metabolite taken up by isolated amyloplasts depended on the tissue from which the amyloplasts were isolated. For example, Tyson and ap Rees[222] and Tetlow et al.[223] reported that G1P was the preferred metabolite for import into amyloplasts from wheat endosperm, while G6P was more effectively taken up by amyloplasts isolated from pea roots,[224] embryos[225] and potato tubers.[226] Hexose phosphates apparently are taken into amyloplasts in exchange for Pi export via the action of a hexose-P/Pi transporter.[226] Tetlow et al.[223] reported that intact amyloplasts from wheat endosperm incorporated glucose from ADP-Glc into starch, but since broken amyloplasts were equally capable of synthesizing starch from ADP-Glc, the authors argued that ADP-Glc uptake by wheat endosperm amyloplasts is not physiologically relevant. Intact amyloplasts from maize[227] and sycamore suspension cultured cells[208] effectively take up ADP-Glc as substrate for starch synthesis. In contrast to the results of Tetlow et al.,[223] lysis of the maize amyloplast reduced incorporation by 90%. It is thought that ADP-Glc is transported into the maize amyloplast stroma in exchange for ADP (a product of starch synthase) via an adenylate translocator.[229] Several lines of evidence support the conclusion that Brittle-1 protein, BT1, is the adenylate translocator in amyloplast membranes from maize endosperm cells. For example, Sullivan et al.[230] reported that the protein with greatest similarity to the *Bt1*-encoded protein is a yeast adenylate translocator. In addition, work from this laboratory demonstrated that the *bt1*-encoded peptides are targeted to the inner membrane of the amyloplast.[231,232] The four most abundant amyloplast membrane polypeptides (38 000–42 000 molecular weight) from normal kernels were specifically recognized by antibodies raised against BT1, but amyloplast membranes from the starch deficient mutant *brittle-1* (*bt*) were missing the BT1 peptides.[229] Expression of the *bt1* in developing endosperm correlates with starch synthesis. Accumulation of the BT1 protein peaks at 14 days after pollination.[233] Liu et al.[227] reported that amyloplasts isolated from young maize kernels effectively take up ADP-Glc for starch synthesis, but not UDP-Glc. Amyloplasts from *bt* endosperms were only 25% as active in ADP-Glc uptake and incorporation into starch as amyloplasts from *normal*,[227] and contain ten times the amount of ADP-Glc as wild-type kernels.[227] Cao et al.[229] suggested that the adenylate translocator, BT1, may function *in vivo* in the transfer of ATP into amyloplasts in exchange for ADP. However, in view of the increasing evidence that ADP-Glc may be synthesized in the cytosol of cereal endosperm cells,[215–217] Pien and Shannon[217] suggest that BT1 also functions in ADP-Glc transfer into amyloplasts. Shannon et al.[234] provided additional evidence that BT1 facilitates the transfer of extraplastidial synthesized ADP-Glc into amyloplasts of maize endosperms.

Many of the metabolite uptake studies cited above rely on combined uptake and incorporation into starch. In order to separate uptake from incorporation, Schott et al.[226] extracted amyloplast membrane proteins from potato tubers and reconstituted them into liposomes. These reconstituted liposomes transported Pi, triose phosphates and G6P in a counter-exchange mode. The liposomes were ineffective in the transfer of G1P; uptake of ADP-Glc was not tested. Mohlmann et al.[236] have used a proteoliposomic system to reconstitute plastid envelope proteins. In this system, ADP-Glc is transported in exchange for AMP. Thus the more widely studied plastid ATP/ADP transporter was not responsible for ADP-Glc uptake. More recently, Bowsher et al.[237] reported that wheat endosperm amyloplasts membrane proteins reconstituted into proteoliposomes took up ADP-Glc in exchange for AMP and ADP. In addition, they showed that under conditions of ADP-Glc dependent starch biosynthesis, the efflux of ADP from intact amyloplasts was equal to that of ADP-Glc utilization by starch synthesis. The amyloplast membrane ADP-Glc/ADP transporter was a 38 000 molecular weight integral membrane protein.[237]

VII. Mutant Effects

Maize is unique among higher plants relative to the number of mutants which have been identified and examined. Several mutants affect the quantity and quality of carbohydrates in the triploid endosperm. Furthermore, these mutants often modify kernel development,[43,44] mature kernel phenotype[235] and starch granule morphology[43,90,238] The *shrunken-1* (*sh*), *shrunken-2* (*sh2*), *brittle-1* (*bt*) and *brittle-2* (*bt2*) mutants condition an accumulation of sugars at the expense of starch. The *shrunken-4* (*sh4*) mutant, which also causes a reduction in starch accumulation, was originally thought to be a phosphorylase mutant,[239] but it was later shown to affect the quantity of pyridoxal phosphate,[240] thus reducing the activities of several endosperm enzymes such as phosphorylase, which require pyridoxal phosphate. The *sh* mutant causes a reduction in sucrose synthase activity,[241] while *sh2* and *bt2* each lack[242] or have very low levels of AGPase.[243] *Bt* encodes a maize endosperm amyloplast membrane-specific polypeptide (39 000–44 000 molecular weight).[244] In its absence (as in *bt* mutant kernels), in addition to a severe reduction in starch and an increase in sucrose, there is a 12-fold increase in ADP-Glc, a product of AGPase.[28] This ADP-Glc in *bt* endosperms accumulates in the cytosol, and Shannon et al.[228,234] have suggested that the amyloplast membrane protein, BT1, is an adenylate translocator, which *in vivo* functions in the transfer of ATP and/or ADP-Glc into the amyloplasts in exchange for ADP, a product of starch synthase activity.

Mutants affecting endosperm protein production include *opaque-2* (*o2*), *opaque-6* (*o6*), *opaque-7* (*o7*), *floury-1* (*fl 1*), *fl oury-2* (*fl 2*) and *fl oury-3* (*fl 3*). These mutants all cause a reduction in the prolamin (zein) fraction of storage proteins and a compensatory increase in albumin and globulin fractions.[245] A molecular analysis of *Opaque-2* showed that it is a regulatory gene which encodes a transcriptional activator protein (Opaque-2) containing a leucine-zipper motif[246] which recognizes a specific target site on the 22 000 molecular weight zein genes.[247] The mutant *soft starch* (*h*) causes a loose packing of starch in the endosperm cells, but has not been related to any major change in storage proteins,[248] starch composition[249] or starch granule structure.[238] The greatest affect of the *h* mutant is a starch particle volume with the *h* allele being completely recessive.[250]

Because the primary focus of this section is mutant effects on polysaccharide composition, mutants in maize which cause changes from *normal* in amylose percentage and phytoglycogen production are reviewed. The maize mutants in this group include *waxy* (*wx*), *amylose-extender* (*ae*), *sugary-1* (*su*), *sugary-2* (*su2*) and *dull-1* (*du*). We will also discuss mutations for specific enzymes that have been produced by molecular techniques and not identified, in naturally occurring mutations. These mutants alone, and in various multiple mutant combinations, have dramatic effects on kernel development, starch granule development and morphology, and polysaccharide composition.[251] As will be pointed out subsequently, certain mutants cause the production of polysaccharides differing from standard amylose and amylopectin in molecular weight and degree of branching.

From 0 to 12 days post-pollination (DPP), little or no detectable differences are observed between *normal* and these mutants (except *su*) with respect to kernel and amyloplast development. The various mutant effects thus become expressed after 12 DPP during the major period of starch accumulation. The mutant *su* differs from *normal* and the other mutants by initially producing compound granules.[44,161]

Saussy[43] made an extensive survey of mutant effects on maize endosperm and starch granule development at 16 and 27 days post-pollination. 'IA5125' versions of *normal*, *ae*, *du*, *su* and *wx* singly and in double, triple and quadruple combinations (except *su wx* and *ae su wx*) were studied. *Normal* and all mutant genotypes exhibited the major gradient of starch granule development from the kernel base (least mature) to the central crown region of cells (most mature) described in Section 3.3. Two basic types of minor gradients of starch granule development were observed. The type I minor gradient is similar to that described for *normal*, with an increase in cellular maturity inward from the peripheral cells adjacent to the aleurone layer.[41,44] The type II minor gradient is similar to type I along the peripheral endosperm and toward the interior for a few cell layers (variable with the genotype), but then an abrupt decrease in the volume of cellular inclusions occurs. These differences in minor gradients and other specific mutant effects will be noted in the discussion of the mutants.

For convenience, the effects of the various mutants and mutant combinations including information on kernel phenotype, starch granule size and physical properties, water-soluble polysaccharide (WSP) concentration, amylose percentage, and relative sizes and iodine-binding capacity of polysaccharides following separation by SEC are summarized in Tables 3.3 to 3.7. The thermal behavior, gelatinization and retrogration of starches from different genotypes show wide variation and differ in inbred backgrounds.[252,253] Current information on the specific mutants singly and in combination, and information on similar mutants in other species when such mutants are known, is presented below.

1. Waxy

Waxy (*wx*) or *glutinous* (*gl*) loci have been identified in maize, sorghum, rice (different species), barley, millets and Job's tears (*Coix lachryma-jobi*).[254–257] *Waxy* mutants

Table 3.3 Mature kernel phenotype of normal and selected single, double, triple, and quadruple recessive maize genotypes[a]

Genotype	Gene expressed[b]	Kernel phenotype[c]
Normal	None	Translucent
Wx	wx	Opaque
ae	ae	Tarnished, translucent, or opaque; sometime semi-full
su	su	Wrinkled, glassy; SC:[d] not as extreme
su 2	su 2	Slightly tarnished, often etched at base
du	du	Opaque to tarnished; SC: semi-collapsed, translucent with some opaque sectors
ae wx	C	Semi-full to collapsed, translucent or glassy, may have opaque caps; SC: slightly fuller, etched, translucent to glassy
ae su	C	Not quite as full as *ae*, translucent (tarnished in SC, may have opaque caps)
ae su 2	ae	Translucent or opaque, etched base
ae du	C	Translucent, not as full as *ae*; SC: etched, translucent, or tarnished
du su	su	Wrinkled, glassy (duller than *su*); SC: extremely wrinkled, glassy
du su 2	C	Translucent, etched
du wx	C	Semi-collapsed, opaque; SC: shrunken, opaque
su wx	su	Wrinkled, glassy to opaque
su 2 wx	wx	Opaque, often etched
su su 2	su	Wrinkled, glassy
ae du su	su	Wrinkled, translucent; SC: slightly wrinkled, translucent
ae du su 2	C	Semi-collapsed, translucent
ae du wx	C	Shrunken, opaque to tarnished; SC: semi-collapsed, tarnished
ae su su 2	C	Partially wrinkled, translucent to tarnished
ae su wx	C	Semi-collapsed, opaque to translucent; SC: etched, semi-full, translucent
ae su 2 wx	C	Etched, semi-full or wrinkled, translucent
du su su 2	su	Wrinkled, glassy
du su wx	su	Wrinkled, glassy
du su 2 wx	C	Semi-collapsed, opaque, etched
su su 2 wx	su	Wrinkled, glassy
ae du su wx	C	SC: etched, semi-full, translucent to tarnished

[a] Adapted from Garwood and Creech[235]
[b] If one gene is responsible for the phenotype, that gene is listed. 'C' signifies a complementary expression giving a new phenotype differing from the phenotypes of the stocks possessing the individual genes
[c] Kernels approach full size unless indicated as semi-collapsed, shrunken, or wrinkled
[d] SC means the phenotpye observed in sweet corn inbreds

have also been isolated in diploid wheat[258] and hexaploid wheat.[259–263] These mutants produce starch granules in the endosperm and pollen which stain red with iodine and which contain nearly 100% amylopectin; however, starch granules in other plant tissues if *wx* plants contain both amylose and amylopectin and stain blue with iodine.[238] *Waxy* mutants have also been reported for different species of the dicot genus *Amaranthus*, in which starch in the perisperm is affected.[264,265] In potato, amylose-free (*amf*) mutants have been isolated after mutagenesis of diploid lines.[266] Finally, *wx*

Table 3.4 Mean starch granule size, birefringence end-point temperature (BEPT), and x-ray diffraction pattern of 14 maize genotypes 24 day post-pollination[a]

Genotype	Granule size, μm		BEPT	X-ray pattern
	Minimum	Maximum		
Normal	7.99 fgh	8.53 jkl	70.3 bcdef	A
wx	8.61 gh	9.41 l	74.3 ef	A
ae	5.56 bcd	6.32 defg	97.7 h	B
su	3.06 a	3.52 z	69.0 abcde	A
su 2	7.68 fg	9.14 kl	63.7 a	A
du	5.19 bcd	5.98 cdef	70.7 bcdef	A
ae wx	6.67 bcd	6.03 cdef	82.3 g	B
ae su	5.34 bcd	8.20 ijk	88.0 g	B
ae su 2	5.57 bcd	6.46 defg	87.7 g	B
ae du	5.42 bcd	6.54 efgh	73.3 def	B
du su	3.11 a	3.85 ab	73.7 def	A[b]
du su 2	5.97 cde	8.79 jkl	63.3 a	A
du wx	6.18 de	6.86 fgh	76.7 f	A
su su 2	2.56 a	2.98 a	67.3 abcd	–

[a]Adapted from Brown et al.[238] Means followed by the same letter are not significantly different at the 1% level using Duncan's Multiple Range Test
[b]Weak crystalline A pattern

mutants of the algae *Chlamydamonas reinhardtii* have been isolated.[32] Red-staining starches have been reported in other plant species, but these have not been characterized.[257] Floridian starch found in red algae also resembles amylopectin and lacks amylose.[19,28]

Phenotypically, *wx* kernels are full and often appear opaque (Table 3.3).[235,257,267] Starch and dry weight production in *wx* kernels are equal to that in *normal* kernels and increase at similar rates.[59,62,117,127,268,269] Sugar and WSP (Table 3.5) levels are also similar to those in *normal* in immature[270,271] and mature kernels.[272]

The *wx* mutant is epistatic to all other known mutants relative to the lack of accumulation of amylose.[90,251] Multiple mutants containing *wx* and *ae* have been reported to produce amylose (Table 3.6); but as will be pointed out in discussing the *ae wx* genotype, this apparent amylose, as measured by iodine binding, is due to loosely-branched polysaccharide molecules having long external chains.[88,89,273] Owing to the lack of amylose, *wx* granules stain reddish-purple with iodine, although some *wx* granules have blue-staining cores.[17,257,274]

Starch granules from maize, sorghum, and rice kernels homozygous for *wx* have been reported to have from 0% to 6% amylose.[112,116,119,121,124,275] This apparent amylose content may be due to the measurement technique used, to the effect of non-*waxy* starch granules from maternal tissue, to differences in the degree of branching or to the presence of some linear material as suggested by blue-staining cores. If present, this linear material is minimal, for no amylose peak is observed in chromatographic profiles of *wx* starch[88–90,276] or by other methods that fractionate starch.[277] These differences in apparent amylose content involve both cultivar and environmental effects as previously described for *normal* starch.[112,116,117,121,124,275] Alleles at the

Table 3.5 Water-soluble polysaccharide (WSP) concentration in immature and mature kernels of 26 maize genotypes[a]

Genotype	Immature		Mature
	10% ETOH[b]	HgCl$_2$[c]	10% ETOH[d]
Normal	3	0	4
wx	3	0	7
ae	4	0	6
su	28	25	19
su 2	2	–[e]	4
du	2	0	5
ae wx	6	0	5
ae su	4	0	5
ae su 2	4	–	6
ae du	7	0	5
du su	29	18	18
du su 2	3	–	4
du wx	11	2	8
su wx	29	19	17
su 2 wx	3	–	5
su su 2	31	–	19
ae du su	16	7	9
ae du su 2	10	–	8
ae du wx	4	Trace	6
ae su su 2	11	–	9
ae su wx	12	7	10
ae su 2 wx	6	–	7
du su su 2	35	–	32
du su wx	38	28	30
du su 2 wx	14	–	8
su su 2 wx	39	–	8

[a]All mutants were in a genetic background related to the single cross W23×L317. Data expressed as percentage of kernel dry weight
[b]Data adapted from Creech.[270] WSP extracted with 10% ethanol
[c]Data adapted from Black et al.[23] WSP extracted with aqueous HgCl$_2$. An aliquot was hydrolyzed with H$_2$SO$_4$ and the increase in reducing sugar determined
[d]Data adapted from Creech and McArdle.[272] WSP extracted with 10% ethanol
[e]Genotype not included in study

wx locus can also vary in amylose percentage with *wx-a* having 2–5% amylose, compared to 0% in *wx-Ref*.[278,279] *Waxy* amylopectins vary in β-amylolysis limit, average chain length and molecular size,[249,280,281] as previously described for *normal* starch. The *normal (Wx)* allele is not completely dominant to the *wx* allele, and amylose percentage is reduced by several percent in the *Wx wx wx* endosperm genotype.[127,282–287]

The increase in average granule size during kernel development of *wx* maize and the final granule morphology of *wx* granules are similar to that of *normal*.[59,238] Also, as reported for *normal*, the average size of *wx* granules varies with the cultivar[116,124,275] and environmental conditions.[275] During development, the average size of the amylopectin molecules has been shown to decrease, while the average chain length of these molecules increases.[275,287] Birefringence of *normal* and *wx* granules

Table 3.6 Apparent amylose percentages of various maize genotypes determined using iodine binding procedures[a]

Genotype	Reference		
	Kramer et al.[359]	Seckinger and Wolf[274]	Holder et al.[271]
Normal	27	27	29
wx	0	–[b]	<1
ae	61	57	60
su	29	–	33
su 2	40	28	38
du	38	35	34
ae wx	15	26	26
ae su	60	–	51
ae su 2	54	45	56
ae du	57	50	45
du su	63	13	40
du su 2	47	–	46
du wx	0	–	2
su wx	0	–	0
su 2 wx	0	–	0
su su 2	55	30	41
ae du su	41	–	28
ae du su 2	48	23	37
ae du wx	–	–	2
ae su su 2	54	–	31
ae su wx	13	4	14
ae su 2 wx	–	28	28
du su su 2	73	–	44
du su wx	0	–	0
du su 2 wx	0	–	0
su su 2 wx	0	–	0

[a]Colorimetric measurement of starch-iodine complex used to estimate apparent percentages. Genotypes were not incorporated into an isogenic background
[b]Genotype not included in study

is similar; however, iodine staining is reported to reduce the intensity of birefringence of *normal* granules, but to have little effect on that of *wx* granules.[89] However, the inability to observe birefringence in amylose-containing granules stained with iodine may simply be due to the intense absorbance of light by these granules. The BEPT of *wx* granules is similar to *normal*, and both have A-type x-ray diffraction patterns (Table 3.4).

Kernels of *wx* have the major and minor (type I) developmental gradients characteristic of *normal* kernels.[43,44] Saussy[43] observed the presence of occasional starch granules surrounded with phytoglycogen; however, this was due to the sweet corn background used in her study and not to the *wx* gene itself. Simple, spherical starch granules are initially produced in *wx* kernels, and these increase in size and, in many cells, become irregular in shape due to extensive cell packing.[43,44] Boyer et al.[44] reported that all maize starch granules are initiated at essentially the same time and that there was no evidence of additional granules (secondary granule initiation) being initiated later in development. Saussy[43] reported secondary granule initiation in *wx*,

VII. Mutant Effects

Table 3.7 Amylose percentage of starch from 16 maize genotypes determined following sepharose 2B-CL column chromatography and the peak fraction's absorbance maxima and absorptivity (a) of the polysaccharide-iodine complex[a]

Genotype	Amylose, %	Peak fraction (tube no.)	Maximum absorbance, nm	a at 615 nm
Normal	29	14	510–540	22
		31	640	121
wx	0	14	470–480	24
ae	33	13	540–550	43
		33	600	92
su	65	13	480–530	28
		28	640	97
du	55	14	480–500	35
		28	640	104
ae wx	0	13	530–540	49
		21	530–540	39
ae su	28	13	540–550	48
		21	540–560	51
		29	640	95
ae du	47	14	530–540	40
		31	640	90
du su	70	14	540–570	47
		29	640	92
du wx	0	14	470–480	20
su wx	0	14	495–505	32
ae du su	31	14	530–550	49
		23	540–560	40
		31	640	83
ae du wx	0	14	460–500	33
		24	460–500	22
ae su wx	0	13	540–550	43
		21	530–540	34
du su wx	0	14	450–480	17
		25	450–470	14
ae du su wx	0	13	<400	25
		24	450–470	19
su phytoglycogen	0	13	≤400	16
		22	≤400	6
Amylose amylopectin 1:1 mixture	51	14	470–530	31
		31	640	103

[a] Maize genotypes converted to the IA5125 sweet corn inbred background. Data adapted from Yeh et al.[90]

as well as in *normal* and most other maize mutant genotypes. Boyer et al.[44] studied *wx* in a dent background, while Saussy[43] studied it in a sweet corn background.

As noted in Section 3.6, *wx* mutants have been shown to lack the major starch granule-bound starch synthase activity.[288,289] With improved biochemical techniques and molecular biology tools, a large number of investigators have confirmed the relationship between the *waxy* gene and granule-bound starch synthases in many species.[290–295] However, multiple forms of granule-bound and soluble starch synthases have frequently been reported. For example, *wx* maize granules do contain a minor granule-bound ADPG starch synthase[296] and two soluble ADPG starch synthases.[297] These multiple enzymes have led to an occasional suggestion that the waxy protein is not a starch synthase,[298] but further studies using different methods have been able to demonstrate that the waxy protein is indeed a starch synthase.[294]

2. Amylose-extender

Mutant genes, which cause an increase in apparent amylose percentage in starch of pea cotyledons and maize, barley and rice pollen and endosperm have been reported.[25,299–301] Not all the high-amylose mutants that have been reported can be classified as one type.[299] High-amylose maize is homozygous for the *ae* gene, and the mature kernels are sometimes reduced in size (Table 3.3). High-amylose (*ae*) rice seeds also are reduced in size. High-amylose peas are homozygous for the *rugosus* (*r*) gene and have a wrinkled, collapsed phenotype,[292] while high-amylose barley kernels appear similar to *normal* (R. F. Eslick, personal communication). Starch and dry weight production are reduced and sugars increased in these high-amylose genotypes compared to non-mutant kernels or seeds. The rate of starch increase during development is also slightly reduced.[64,127,128,133,270,271] Apparent amylose content increases with increasing maize and barley kernel age and with increasing pea seed diameter, reaching values of 45–69%.[64,127,128,137] In contrast, amylose contents increase from 13–15% to 26–32% during development of *ae japonica* rice kernels and increase from 24–25% to 39–41% in *ae indica* rice.[303,304]

The *normal* alleles are not completely dominant to the recessive *ae* alleles, since two doses of the recessive allele (i.e. *Ae ae ae*) result in a 2–8% increase in apparent amylose content compared to the *normal* genotype which lacks the recessive allele.[127,285,305,306] Extensive variation in apparent amylose concentration occurs compared to the amount observed in *normal* genotypes (see Section 3.3). For example, variation is observed for amylose concentration as a function of the maize inbred crossed with *ae*,[307–311] with a range of 36.5–64.9% reported.[308] Minor modifying genes in the various inbreds have been proposed as a possible cause of the variation.[307–311] Such modifying genes have been utilized to produce a series of hybrids which differ in apparent amylose concentration from 50% to 75%.[67] Differences have also been associated with *ae* alleles arising as independent mutations, with *ae-i1* and *ae-i2* conditioning lower amylose percentages than five other alleles when compared in two isogenic backgrounds.[312] Amylose percentage also varies 17% among wrinkled-seeded pea cultivars.[59,112]

Not only does variation occur between *ae* inbred lines and hybrids (i.e. background or modifier effects) and *ae* alleles, but an 8–14% range also existed within an inbred line homozygous for *ae* and grown at a single location in a single year.[308,311] This is likely due to a combination of error in amylose determination, segregation of modifier genes which were not yet homozygous,[313] and the microenvironment of each plant.

Significant differences in *ae* amylose percentage result from both location and year of production with the effect of location considerably greater than that of years.[144,314] Later planting dates are associated with higher amylose percentages in *ae* kernels; however, poorer agronomic performance negates the value of the increase.[315] Minor mechanical damage to plants has little affect on amylose concentration with only a 1–3% reduction caused by extreme leaf defoliation.[316]

Variation in amylose concentration is also observed between butt, center and tip zones within individual *ae* ears, with the highest percentage in kernels taken from the butt of the ear and the lowest percentage in kernels from the tip zone.[317] In addition, when the endosperm tissue is divided into tip, middle and crown portions, the middle portion is highest in amylose percentage within each ear zone.[317]

The amylose percentages presented above are all based on 'blue value' tests. Yeh et al.[90] employed SEC to fractionate the starch polysaccharides from mature kernels of *normal*, *ae* and 14 other maize endosperm mutant genotypes. Column fractions were reacted with iodine, and absorbances at 560 and 615 nm were determined. Any fraction having a higher absorbance at 560 than at 615 nm was classified as amylopectin and, conversely, fractions with a higher absorbance at 615 nm were considered to be amylose. Considerable carbohydrate molecules intermediate in size between amylopectin and amylose were found. These molecules appeared to be similar to the loosely-branched amylopectin described for *ae wx* starch[88,273] and suggested for *ae* amylopectin.[318,319] Whistler and Doane[109] isolated such a polymer from *ae* starch. Low molecular weight polymers similar to the short chain amylose described by Banks and Greenwood[67] eluted near the end of the profile; these polymers had a higher absorbance at 560 nm than at 615 nm and by definition were not included as amylose. Based on Yeh's calculation, *ae* starch contains 33% amylose (Table 3.7).[90] If the low molecular weight polymers eluting after amylose are included, the amylose percentage increases to 41%, which is still much lower than amylose percentages based on blue value measurements (Table 3.6). Similar low amylose percentages were obtained following SEC after debranching by isoamylase.[276] Because the long external chains of loosely-branched polysaccharides complex iodine,[88] they contribute to the estimate of amylose percentage as measured by the blue value procedure. Although the amylose percentage based on the procedures of Yeh[90] and Ikawa et al.[276] may not be exact, they probably represent a much closer estimate of the true amylose content of *ae* starch than do blue value estimates.

Starch granule preparations from *ae* maize kernels generally contain two distinct geometric forms, spherical and irregular.[43,54,128,301,302] Irregular granules vary in shape, but often are elongated and non-birefringent. Sometimes spherical granules also develop elongated extensions of amorphous, non-birefringent starch.[57] The proportion of irregular granules in *ae* starch has been reported to vary from 0%[43,90,183] to 100%[321] and was shown to increase during kernel development,[44,128] with increasing apparent amylose content[128,301] and with the physiological age of the cells.[54] The proportion of irregular granules depends on the completeness of starch isolation, the classification criteria used[320] and the inbred background.[43,90] Average *ae* starch granule size increases with kernel development; however, *ae* granules are smaller than *normal* at all developmental stages.[57,238] Boyer et al.[49] reported a two-phase growth pattern consisting of spherical granule initiation and growth followed by a secondary initiation of irregular granules. Sandstedt[322] also reported that irregular granules in *ae* endosperm are surrounded by spherical granules within an endosperm cell. There is considerable cell-to-cell variation in the presence and proportion of irregular granules,[54,322] but in kernels harvested 36 days post-pollination, the proportion of irregular granules is highest in the more mature endosperm cells.[54]

Inbred background apparently influences the morphology of the irregular granules produced by *ae*. The elongated amorphous granules noted above occur when the *ae* mutation is incorporated into dent backgrounds.[44,54] However, when *ae* is incorporated into the sweet corn inbred 'IA5125' and the *su* mutant deleted, no elongated amorphous granules are found at 16 or 27 days after pollination[43] or at maturity.[90] Secondary granule initiation does occur, and the irregular granules are more blocky in appearance.[43] Kernels of *ae* have the major developmental gradient and type I minor gradient characteristic of *normal*.[43,44]

Starch granules from *ae* kernels have a much higher BEPT than *normal* or the other mutants (Table 3.4). Also, based on 14 genotypes studied, the B-type x-ray diffraction pattern appears to be unique to *ae* and *ae*-containing genotypes (Table 3.4).

In high-amylose barley[306] and wrinkled-seeded pea[64] starch, average granule size is less than in *normal*, with high-amylose starch granules being smaller at all stages of development. High-amylose barley starch granules are more irregular than are *normal* granules.[306] High-amylose pea starch granules often develop a very irregular system of fissures, making them superficially resemble compound granules.[17,64,157]

Based on the accumulation of loosely-branched amylopectin in *ae*[317,318] and *ae wx*[88,273] genotypes, Boyer et al.[127] suggested that the *Ae* allele affects the degree of branching of amylopectin by controlling the quantity of an effector at the site of starch synthesis, which stabilizes a starch synthase–branching enzyme complex. Subsequent work has confirmed that starches from *ae* and *rugosus* seeds have an amylopectin structure with reduced branching.[323–325] After an initial report by Boyer and Preiss[326] on the presence of three forms of branching enzyme in extracts from normal maize endosperm, similar studies showed the presence of two forms of the starch-branching enzyme in non-mutant pea seed[327–329] and up to five forms of starch-branching enzyme in *normal* rice seeds.[330,331] When the branching enzymes from *ae* kernels of maize were similarly separated, the total activity was only 20% of *normal*, and there was a complete absence of branching enzyme fraction IIb.[332] Subsequent studies with the *rugosus* mutation in peas[327–329,333] and *ae* rice[325,330] have shown a similar loss of a single form of branching enzyme. In maize, the *ae* effect was attributed to a deficiency of branching enzyme IIb.[323] Hedman and Boyer[334] reported a near-linear relationship between increasing dosage of the dominate *Ae* allele and branching enzyme IIb activity, and suggested that *ae* is the structural gene coding for branching enzyme IIb. Molecular analysis of the *ae* allele has confirmed the independence of genes encoding branching enzymes IIa and IIb.[335] Similar conclusions that the *rugosus* and *ae* genes are structural genes for starch-branching enzymes in peas and rice, respectively, have been made.[336]

The other isozymes of starch branching enzymes appear to be products of other genes.[337] Naturally occurring mutations for the other starch branching enzymes have not been identified. Guiltinan and his coworkers[338–340] have produced mutations for branching enzymes I and IIa through mutator induced insertional mutants. Mutations in BEIIa have endosperm starch that is indistinguishable from normal starch. However, leaf starch shows a highly reduced branching. This altered leaf starch probably contributes to the observed accelerated senecense in these plants as the starch is probably more resistant to degradation during the diurnal cycle. When the BEIIa mutant was combined with *ae* and *wx* in the triple mutant, the branch density and average number of branches per amylopectin cluster where higher than in starch for the *ae wx* double mutant.[339] These results were interpreted as indicating possible functional interactions between BE isoforms. This conclusion is supported by the observation that starch branching enzymes and starch synthases are phosphorylated which may regulate protein–protein interactions.[340] The degeree of the reversible phosphorylation impacts the protein to protein interactions.

Mutator induced mutations of branching enzyme I have also been developed.[341] In these mutants neither leaf nor endosperm starch differed in structure from the normal starches.

3. Sugary

The standard sweet corn of commerce is homozygous recessive for *su*. *Sugary* mutants have also been reported for sorghum and rice.[25,304,342–344] The main effect associated with *su* mutants in maize and sorghum is synthesis and accumulation of phytoglycogen to 25% or more of the kernel dry weight (Table 3.5).[25,272,342,343,345–348] Phytoglycogen is similar to amylopectin, except that phytoglycogen is more highly branched and is extracted as the major component of the water-soluble polysaccharide (WSP) fraction in sweet corn.[23,343,346,348–351]

Mature *su* sorghum and maize (Table 3.3) kernels are wrinkled and have reduced amounts of dry matter.[24,268,271,272] Their sugar content is higher and their starch content much lower than in *normal* maize[270–272,352–354] or sorghum.[24,342,355] Starch concentration in *su* maize expressed as a percentage of dry weight increases until 15–20 days post-pollination, and then remains constant.[59,270,353,356] Total polysaccharide concentration, however, increases through 30–40 days post-pollination due to increases in phytoglycogen concentration, with total carbohydrate percentage approaching that in *normal* kernels.[59,270,353,359] At maturity, the total carbohydrate percentage is equal to[354,356] or less than[223,347] that in *normal* kernels, depending on the genetic background. However, absolute amounts are reduced, reflecting the reduced dry matter in *su* kernels. In general, maize kernels from dent lines homozygous for *su* contain more sugar and less phytoglycogen and starch than kernels of a sweet corn line.[269,272] In addition, the chain lengths of *su* amylopectin have been reported to be shorter than those of amylopectin from *normal* kernels.[346,357]

The amylose percentage of starch, as measured by iodine binding, from *su* kernels averages somewhat higher than the percentage from *normal* kernels (Table 3.6), and the amylose percentage has been reported to increase with advancing kernel age.[59,358] Although the data in Table 3.6 represent data from several studies over several years, other investigators have reported widely different amylose percentages in *su* starch.[90,271,353,359–362] These have varied from 0% amylose[353] to 65% amylose.[90] The 65% amylose reported by Yeh et al.[90] (Table 3.7) was based on calculations from SEC separation of the starch polysaccharides. Similarly, the amylose percentage of starch from *su* sorghum kernels varied from near that in *normal*[116] to somewhat higher than that in *normal*.[355] The widely differing amylose percentages probably relate to kernel age and methods of starch isolation and measurement. Possible reasons for these discrepancies are discussed in more detail after considering the morphological changes occurring in the developing *su* kernel.

The morphology and development of *su* maize plastids and kernels is well established.[17,18,40,43,44,161] Immediately prior to initiation of starch synthesis in an endosperm cell, the proplastids collect around the nucleus as they do in *normal*.[43,44,161] From one to several small starch granules then form in each amyloplast.[160] During development, granules enlarge only slightly,[59,238,358] reaching an average diameter of 3.6 μm at maturity.[59] However, within the more mature cells of the central crown region, the starch granules formed intially are degraded and replaced with phytoglycogen.[43,44,161] Thus, within developing kernels, plastid types range from amyloplasts with compound starch granules, to amyloplasts containing phytoglycogen and a few small starch granules, to amyloplasts containing phytoglycogen plus many very small starch granules and/or granule fragments, to plastids containing only phytoglycogen.[43,44,161] Cells with the different plastid types are located in specific regions of the endosperm and apparently are related to the physiological age of the cells, with phytoglycogen plastids being in the most mature cells.[40,43,44] The *su* kernels go through the major and minor developmental sequence characteristic of *normal*, except that as the cells mature, they fill with phytoglycogen rather than with starch.[43,44] Later in kernel development, phytoglycogen plastids in some cells appear to rupture.[43,44,161] The released material, thought to be phytoglycogen, was described as a dense-staining 'rosette' material[161] similar in appearance to animal glycogen.[363] Thus, phytoglycogen appears to accumulate in both plastids and the cytoplasm, with that in the cytoplasm possibly arising from ruptured plastids.

Owing to the small size of *su* starch granules (Table 3.4) and their partially degraded remnants, difficulties are encountered in isolating a starch sample which is representative of that in the total population of cells found in the endosperm. With procedures involving starch-tabling, up to 90% of the starch can be lost,[364] and similar losses of the smaller granules would be expected with isolation procedures based on low-speed centrifugation or gravity sedimentation. Particles staining both red and blue with iodine have been observed *in situ* and in isolated granules.[43,364,365] Thus, granules differ from each other, and loss of small granules and granule particles probably results in granule preparations that are not representative of the total granule population. Therefore, differences in isolation procedures may explain some of the discrepancy in amylose percentages reported for *su* starch. The percentage of amylose in the starch is also affected by the completeness of phytoglycogen removal. Polysaccharide particles smaller than starch granules have been observed in *su* kernels[43] and have also been isolated from immature kernels.[362,366] These intermediate particles, composed of phytoglycogen and amylose,[362] cause a further difficulty in accurately determining the percentage of amylose in starch and the characterization of phytoglycogen. If these particles are considered to be starch granules, amylose content will be underestimated. If they are collected with the phytoglycogen fraction, amylose will be found, a phenomenon which has been reported.[366,367] Thus, kernels homozygous for the *su* gene cannot be considered to contain only phytoglycogen and starch granules, but must also be considered to have a range of particles with intermediate composition resulting from partial conversion of starch granules into phytoglycogen.

Several investigators[23,332,368–370] have reported the presence of a branching enzyme (phytoglycogen-branching enzyme) in *su* kernels, in addition to Q-enzyme, which is capable of forming a phytoglycogen-like polysaccharide from amylose *in vitro*. Black et al.[23] observed the presence of phytoglycogen-branching enzyme in all maize genotypes containing phytoglycogen and in two mutants (*du* and *wx*) which do not accumulate phytoglycogen. Boyer et al.[371] suggest that, of the three branching enzymes present in maize kernels, branching enzyme I plays a major role in phytoglycogen formation. There is a specific interaction between branching enzyme I and starch granules from *su* kernels. For example, treatment of *su* starch granules with this enzyme effects formation and release of phytoglycogen-like glucans, but no soluble glucan was released from enzyme-treated non-mutant starch granules.[371] Black et al.[23] concluded that the gene *su* is not the controlling factor in the formation of either phytoglycogen or the phytoglycogen-branching enzyme. Pan and Nelson[170] reported that *su* maize kernels had reduced levels of three fractions of debranching enzyme. Further studies revealed that enzymes with both isoamylase- and pullulanase-type activities could be isolated from normal kernels and that the pullulanase activity was reduced in *su* kernels.[372,373] Further cloning work and characterization of the enzymic activities have shown that an isoamylase activity is associated with the *su1* gene product and that a separate gene, *zpu1*, encodes pullulanase activity.[374,375] The protein sequence predicted from a clone of the *sugary* gene was found to have some sequence motifs which matched motifs in known bacterial genes, but little overall sequence homology was seen.[171] Enzymic analysis of *su* rice also demonstrated that debranching enzyme activity is reduced.[347,376] In addition, a mutant of *Chlamydamonas* which accumulates glycogen at the expense of starch has been shown to have a deficiency of debranching enzyme.[34] These results support a largely dismissed early suggestion by Erlander[377] that amylopectin is made from phytoglycogen. However, this direct conversion of phytoglycogen to amylopectin is largely unsupported. The current suggestion is that the ratio of branching and debranching activities at the surface of the growing amylopectin molecule is critical.[169] As a cluster forms and associates, further branching or debranching becomes limited by packing of the chains. The mechanisms described above still need to be further elucidated. In addition, any mechanism will need to be explained in context with the diverse genetic variation at the *sugary* locus. For example, a complex multiple allelic series exists at the *su* locus in maize, and four phenotypic categories have been established for mature kernels based on examination of 12 independently occurring mutations.[378] For most alleles, mature kernels resemble the reference allele, *su-Ref*, discussed in preceding paragraphs (Table 3.3).[378] Kernels of three alleles, including *su-am* (*amylaceous*), are near-*normal* in appearance and are best observed as double mutants with *du* or *su2*[365–380] Kernels of *su-st* (*starchy*) vary from near-*normal* to slightly wrinkled with *su-st* recessive to *su-Ref* in some backgrounds.[378,381] The fourth class, represented by *su-Bn2* (*Brawn-2*), has a kernel phenotype intermediate between *su-Ref* and *su-am*.[378] This phenotype complexity is reflected in the carbohydrate composition conditioned by these alleles with composition ranging from that

Table 3.8 Dry weight and carbohydrate composition of kernels sampled 20 days post-pollination for alleles at the sugary locus converted to the WG4A dent inbred background[a]

Sugary allele	Ears sampled, no.	Kernel weight, mg	Glucose[b]	Fructose[b]	Sucrose[b]	WSP[b]	Starch[b]
su-Ref	3	27	45	39	245	130	77
su-Bn2	8	33	41	36	177	55	241
su-st	7	27	60	54	124	122	191
su-am	7	36	60	52	78	4	356

[a] Data adapted from Garwood and Vanderslice[418]
[b] Data presented as mg per gram of dry weight

of *normal* for *su-am* to that exhibited by *su-Ref* (Table 3.8). Further variation is seen in modifier genes. An independent recessive modifier of the *su* locus, named *sugary enhancer* (*se*), has been described in the sweet corn line 'IL677a'.[382,383] The resulting *su se* genotype accumulates high sugar levels similar to *sh2*, and also high levels of phytoglycogen similar to *su-Ref*.[356,382,383]

4. Sugary-2

Kernels of the maize endosperm mutant *su2* have a slightly tarnished phenotype (Table 3.3) and are similar to *normal* in soluble sugar, WSP (Table 3.5), and starch concentrations during development[270,271] and at maturity.[272] Kernel dry weight is often,[268,270,271] but not always,[272] reduced. Starch granule size (Table 3.4)[270,360] and rate of size increase during development[238] are similar to *normal*; however, *su2* granules have extensive internal fractures.[364] Starch from *su2* endosperms is 10–15% higher in apparent amylose content than is *normal* starch (Table 3.6). Although *su2* starch composition is altered, in earlier studies, purified *su2* amylose and amylopectin were reported to have properties similar to those of normal amylose and amylopectin.[360] However, better techniques have shown that *su2* amylopectin has longer long B-chains and shorter intermediate B-chains than does normal amylopectin.[384] The *normal* (*Su2*) allele is completely dominant to *su2* for amylose content.[284–286,385] Although thermal properties of starches from kernels with differing doses of the *su2* allele were significantly different, these differences may be due to subtle variation in amylopectin structure.[385] As with other genotypes, year of production,[286] *su2* allele examined,[285] the background into which *su2* is incorporated[308,386] and different ears within an *su2* inbred[308] affect apparent amylose percentage. An extensive study of the properties of *su2* starch in a range of genetic backgrounds demonstrated that thermal properties, as measured by differential scanning calorimetry, varied significantly.[385] Singh[355] has described a sorghum mutant similar to *su2*, which also has non-mutant levels of reducing sugars, WSP and starch, but is higher in sucrose and amylose percentage. An *su2* mutant from rice was recently reported, but not well characterized.[348]

Brown et al.[238] reported that starch granules from 18- and 24-day-old *su2* kernels are weakly birefringent and have an A-type x-ray pattern (Table 3.4), in contrast to the B-type pattern reported for starch granules from mature *su2* kernels.[360,387] The BEPT of *su2* granules is lower than that of *normal* granules (Table 3.4).[359,388] Based on these granule properties, the *su2* gene has been suggested to cause a reduction in the molecular association between the starch molecules of the granule;[238] however, no genetic lesion has been established for *su2*.

5. Dull

The *du* mature kernel phenotype varies with background, ranging from full size to semi-collapsed (Table 3.3). The presence of the 'normal appearing' form is best detected in combination with *su-am*.[365,379,389] The more extreme expression may be associated with the presence of a dominant *dull-modifier* gene.[389] Mature kernel dry weight of *du* also varies, with some weights similar to those of *normal*[269,270] and others significantly less.[268] The sugar concentration is slightly higher and the starch concentration lower than that of *normal* in both immature[270,271] and mature[272] kernels.

The amylose percentage of *du* starch in a dent background is 5–10% higher than the percentage in *normal* starch (Table 3.7).[390] Yeh et al.[90] found 55% amylose in starch from mature *du* kernels in a sweet corn background (Table 3.7). These differences in reported amylose contents may be due to the SEC separation procedure used by Yeh et al.[90] or to a background effect. The *normal* (*Du*) allele is completely dominant to *du* for amylose percentage.[283,285,286] The amylose percentage is affected by the *du* allele,[285] by the background[308,391] and by the year of production.[286] Although the amylose percentage is higher than in *normal*, the polysaccharide components essentially have similar properties (Table 3.7).[335] However, as noted for *su2* above, improved techniques for examining chain lengths in amylopectins have shown that the ratio of short B-chains to long B-chains in *du* amylopectins is higher than that in *normal* and *wx* amylopectins.[351,357,392] A *du* mutant has also been reported for rice.[304,393,394] This mutant was named based on phenotypic similarities of the mutant rice kernels and maize *du* kernels. However, unlike the increase in amylose content in maize, starch from *du* rice has a reduced amylose content, and the amylose content is intermediate between those of *normal* and *wx* starches.[393,395] Therefore, some caution needs to be taken in the comparison of these mutants in rice and maize.

Most *du* granules are similar in shape, size, birefringence and iodine staining to *normal* granules;[43,360,364] however, some irregularly shaped granules and spherical granules, which have little or no birefringence, have been reported.[43,364] Average *du* starch granule size is smaller than *normal* granule size (Table 3.4).[270] Cell-to-cell variation in granule size and morphology has been reported.[364] BEPT and x-ray diffraction patterns are similar for *du* and *normal* (Table 3.4).[359]

The *du* kernels have a major developmental gradient similar to *normal* except for the presence of slender, thin-walled cells near the developing embryo which appear partially compressed.[43] Although *du* kernels in a dent background do not accumulate phytoglycogen,[23] those in a sweet corn background do have cells in the central endosperm with plastids containing phytoglycogen and one or two small starch granules.[43] Secondary initiation of granules has been observed in some cells.[43] The *du* kernels have a type-II minor developmental gradient from the outside toward the interior[43] in which there is typical starch granule initiation and enlargement for a few cell layers, followed by an abrupt reduction in number and size of starch granules. The reduction in starch is accompanied by an increase in phytoglycogen-containing plastids. All multiple mutants homozygous for *du*, but none of the others examined, had the type-II minor gradient, and Saussy[43] suggested that this property was a specific effect of the *du* gene.

Phytoglycogen-branching enzyme has been found in *du*; however, no phytoglycogen was isolated by Black et al.[23] Preiss and Boyer[396] reported that the *du* mutation lowered starch synthase II activity and also lowered branching enzyme IIa activity. Gao et al.[397] used a molecular approach to clone the *du* gene in maize endosperms and, based on amino acid sequence similarity of the predicted protein product with the soluble starch synthase III of potato,[398] concluded that *du* most likely encodes the 180 000 molecular weight, primer-dependent soluble starch synthase described previously.[399,400,401]

Additional characterization of the two major soluble starch synthases in maize endosperm indicated distinct catalytic properties.[402] Soluble starch synthase II has been reported to be encoded by the *du* gene.[397] the initially observed reduction in BEIIa is a secondary effect and may be related to protein-to-protein interactions.

6. Amylose-extender Waxy

Mature kernels of *ae wx* corn are reduced in size compared to *normal* (Table 3.3). Similarly, immature and mature kernel dry weights and starch contents are reduced almost 50%;[127,270,272] however, sugar contents are increased.[270,272,403,404] Only small amounts of material are recovered in the WSP fraction (Table 3.5).[403,404]

Apparent amylose percentages of 15–26% have been determined for *ae wx* using the blue value procedure (Table 3.6), and it was once thought that *ae wx* is the only genotype producing a significant quantity of amylose when *wx* is homozygous.[271] However, using potentiometric titration, only 1% amylose was observed, indicating the presence of little linear material.[285] This finding was confirmed by SEC separations and fine-structure analyses which showed that *ae wx* starch consisted solely of loosely-branched amylopectin with long external chains.[88,89,273] A similar loosely-branched polysaccharide of lower molecular weight is also found in *ae wx* starch in a sweet corn background (Table 3.7). In several studies, the *ae wx* amylopectin has been shown to have an increased proportion of long B-chains and a decreased proportion of short B-chains similar to that of *ae* amylopectin.[253,254,405,406] In addition, the ratio of A- to B-chains was found to be 1.5 in *ae wx* amylopectin in a dent background and around 1.0 for the *ae wx* amylopectin in a sweet corn background and other *wx* and *du wx* amylopectins in both backgrounds.[254] Thus, in this double mutant, the *wx* gene is blocking all accumulation of linear polymer, while the *ae* gene is interfering with typical branching. Apparently, the enzymes discussed under the respective single mutants are both functioning independently in the double mutant.

Increasing doses of *ae* and *wx* effect kernel phenotype[407] and amylose content.[127,285] Two or three doses of the *wx* allele significantly decrease apparent amylose, indicating tighter branching, while two or three doses of the *ae* allele significantly increase apparent amylose content, indicating looser branching, regardless of the gene dosage at the other locus.[127,285] Apparent amylose content of *ae wx* starch decreases with increasing kernel age,[57,127] indicating tighter branching. Different *ae* alleles combined with *wx* may also affect the degree of branching, since pollen from different *ae wx* combinations differs in iodine staining.[408]

Kernels of *ae wx* have the major and minor (Type I) developmental gradients characteristic of *normal*.[43,44] Starch granules are smaller than they are in *normal* (Table 3.4) and increase somewhat in size with increasing kernel age.[59,238] Considerable differences relative to starch granule and plastid development have been observed between dent and sweet corn backgrounds.[43,44] In a dent background, no secondary granule initiation, characteristic of *ae,* is observed; rather, most granules within a cell seem to develop extensions simultaneously.[44] These granules remain highly birefringent.[44] In contrast, Brown et al.[238] reported that the spherical *ae wx* granules have a polarization cross, while the irregular granules only have birefringence in the outer periphery. No phytoglycogen-containing amyloplasts are observed in *ae wx* kernels in a dent background.[44] In a sweet corn background, secondary granule initiation is observed, and many amyloplasts contain a starch granule surrounded by a non-crystalline polysaccharide.[43] Staining properties of this polysaccharide are similar to those of phytoglycogen. 'Phytoglycogen-containing' plastids of *ae wx* persist to maturity and, unlike the phytoglycogen plastids in *su* kernels, contain starch granules, many of which when isolated and 'purified' are still surrounded with the 'phytoglycogen-like' polysaccharide.[90] The nature of this polysaccharide is unknown, but it may be similar to that observed in the triple mutant *ae du wx* described later.

7. Amylose-extender Sugary

Mature kernels of *ae su* are not as full as those of *ae,* but are fuller than those of *su* (Table 3.3), and their phenotype varies with genetic background (Table 3.3).[409] Kernel dry weight and starch concentration are reduced relative to *normal*.[270,272] Sugar concentrations are slightly higher than those of *normal* in immature,[270] but not in mature,[270] kernels. Minimal WSP levels similar to those in *normal* were reported for *ae su* (Table 3.5); however, later, significant amounts of phytoglycogen were found.[43,410] Specifically, in a dent background, *ae su* endosperm contains 11% as much phytoglycogen as does *su* endosperm at 20 days post-pollination. Increasing doses of *ae* in a homozygous *su* genotype result in reduced levels of phytoglycogen.[410] Kernels of *ae su* in a sweet corn background have a large area of cells containing plastids with starch granules surrounded by a non-crystalline 'phytoglycogen-like' polysaccharide.[43] Only a few such plastids were observed in a dent background.[44] Thus, background is important in the degree of *ae* epistasis relative to the accumulation of 'phytoglycogen-like' polymers.

Starch from *ae su* kernels in a dent background consists of 51–60% amylose, as determined by the blue value procedure (Table 3.6), with the amylose percentage increasing with increasing kernel age.[57] Yeh et al.,[90] in contrast, reported that *ae su* reduced amylose concentration from 65% for *su* to 28% for *ae su*, based on SEC separation of starch polysaccharides isolated from kernels in a sweet corn background. Three fractions were obtained (Table 3.7). The first two were loosely-branched, similar to the amylopectin fractions in *ae*. Amylose from the third peak fraction was similar in iodine staining to that from *normal*; however, some short chain length amylose as found in *ae* was present. The second fraction from the SEC column eluted in the same position as phytoglycogen and may have been the non-crystalline 'phytoglycogen-like' polysaccharide shown to be present with some of the 'purified' starch granules.[90] However, the iodine complex absorbance maximum of this lower molecular weight branched component was the same as that of the first component, and similar to that of the branched components of *ae wx* (Table 3.7). Neither branched component from *ae su*, when complexed with iodine, had an absorbance maximum even close to that of *su* phytoglycogen (Table 3.7). A similar lower molecular weight loosely-branched component comprising 7.5% of *ae su* starch has been isolated.[109]

The major and minor (type I) developmental gradients characteristic of *normal*[43,44] also occur in *ae su* kernels. Starch granules are smaller than those of *normal* (Table 3.4) and increase in size with increasing kernel age.[57,238] Secondary granule initiation occurs in *ae su* kernels similar to that which occurs in *ae*[43,44] Within some cells in a dent background, granules are transformed during development into an amorphous, non-birefringent form.[43,364] Badenhuizen[18] reported that spherical granules from young kernels have an A-type x-ray pattern, but with development, irregular granules with a B-type x-ray pattern are found. Starch granules and plastid development in *ae su* kernels in a sweet corn background vary considerably from cell to cell.[43] Some cells contain irregular granules, others contain granules surrounded by 'phytoglycogen-like' polysac-charide and others have plastids with granules in various stages of fragmentation.

The effects of both genes can be seen in the double mutant. Phytoglycogen, as found in *su*, is produced; however, amounts are reduced in *ae su*, although to a lesser degree in a sweet corn background. The *ae* gene reduces branching, which is reflected in the two loosely-branched starch fractions obtained by SEC (Table 3.7). Furthermore, *ae* probably interferes with phytoglycogen-branching, as *ae su* phytoglycogen is degraded more by β-amylase than is *su* phytoglycogen.[410] In *su*, starch granules formed initially are broken down and are thought to be utilized in the production of phytoglycogen. In *ae su*, the *su* gene may be responsible for effecting the partial breakdown of the initially formed granules, but *ae* interferes with branching and amorphous irregular granules are formed in a dent corn background, along with a small amount of phyto-glycogen.[44] In the sweet corn background, more 'phytoglycogen-like' polysaccharides are formed, apparently because of modifier genes.[43]

8. Amylose-extender Sugary-2

The mature kernel phenotype of *ae su2* is similar to that of *ae* (Table 3.3). Dry weight per kernel is similar to that of *su2* and *normal*, while starch concentration is less than that of *su2*, and similar to that of *ae*.[270,272] Sugar concentrations in both immature[270] and mature[272] kernels are higher than those in either *ae* or *su2*. Amylose percentage, based on blue value determinations, is similar to that in *ae* (Table 3.6). Amylose percentage varies between *ae su2* ears,[308] although *su2* and *ae* alleles have little effect on *ae su2* amylose percentage.[281] No dosage effects are observed.[285] Starch granule sizes and x-ray diffraction patterns are similar to those in *ae*, and the BEPT approaches that of *ae* (Table 3.4).[359]

9. Amylose-extender Dull

The phenotype of mature *ae du* kernels differs from that of both *du* and *ae* (Table 3.3). Compared to *normal*, dry weight and starch concentrations are reduced, while sugar concentrations are higher in immature[270] and mature[272] kernels. The amylose percentage of *ae du*, based on blue value measurements of starch from kernels in a dent background, is similar to that in *ae* (Table 3.6). With *ae* homozygous, the apparent amylose percentage decreases with increasing doses of *du*.[285] The 47% amylose determined by the SEC of starch from *ae du* kernels in a sweet corn background is intermediate between the amount in *ae* and *du* (Table 3.7). Maximum absorbance of the iodine–amylopectin complex of *ae du* is similar to that of *ae*, while the amylose component is closer to that of *du* and *normal* (Table 3.7). The amylopectin fraction does contain an increased proportion of long B-chains to short B-chains similar to *ae* and *ae wx* amylopectins.[392] Thus, also in *ae du*, the *ae* gene appears to be interfering with the typical branching of amylopectin resulting in the production of more loosely-branched polymers.

Although low levels of WSP have been reported in *ae du* kernels (Table 3.5), Black et al.[23] concluded that no phytoglycogen accumulates in *ae du* kernels in a dent background. In contrast, kernels of *ae du* in a sweet corn background produce numerous plastids with one or two starch granules surrounded by a thick layer of non-crystalline 'phytoglycogen-like' polysaccharide.[43]

Kernels of *ae du* in a sweet corn background are slightly delayed in development, but have the *normal* major gradient of kernel development.[43] The type II-minor gradient characteristic of *du* is observed in *ae du*.[43] Saussy[43] also reported that secondary starch granule initiation occurs and that granules assume blocky, elongated and irregular shapes later in development.

In a dent background, the greatest increase in granule size occurs between 12 and 18 days post-pollination.[238] Granule size is similar to that of *ae* and *du* granules, but less than that of *normal* granules (Table 3.4). Starch granules of *ae du* have a B-type x-ray defraction pattern similar to that of *ae* (Table 3.4). In contrast, the *ae du* BEPT is similar to that of *du* (Table 3.4).[359,388] In *ae du*, the *ae* and *du* genes appear to be functioning independently with *ae* interfering with typical branching, and *du* causing the expression of the type-II minor gradient. In *ae du*, branching enzyme fractions IIa and IIb and starch synthase fraction II are considerably reduced.[397] Thus, the double mutant expresses the enzyme reductions of both individual mutants.

10. Dull Sugary

The mature kernel phenotype of *du su* is similar to that of *su*, although *du su* kernels are often more wrinkled (Table 3.3). This genotype has been extensively studied to evaluate its potential for improving sweet corn quality.[411,412] Compared to *normal*, *du su* kernels have reduced dry weight and starch concentration and increased sugar and WSP concentrations.[270,272] Sugar[270,272,365,382,411,412] and WSP (Table 3.5)[365,411] levels are similar to those in *su*, although starch[272,365,411] concentration is lower. Thus, *su* is epistatic to *du* relative to phytoglycogen accumulation.

Widely varying amylose percentages have been reported for *du su* starch samples in a dent background when measured by the blue value procedure (Table 3.6). In four other reports, *du su* amylose content ranged from 51% to 66%.[284,285,360,365] Yeh et al.[90] (Table 3.7) reported 70% amylose using SEC. Dvonch et al.[335] stated that *du su* amylopectin is intermediate in branching between glycogen and *normal* amylopectin. The *du su* amylopectin, as compared to *normal* amylopectin, has a decreased proportion of long B-chains and an increased proportion of short B- and A-chains.[391] However, based on the absorbance maximum of the iodine complex and the absorptivity, the high molecular weight branched fraction from *du su* in a sweet corn background appears to be loosely-branched with long external chains (Table 3.7). The *du su* amylose fraction is similar to that of *du* and *su* alone. No dosage effects have been observed on amylose percentage.[286]

The overall kernel and plastid development pattern in *du su* is similar to that in *su*, except that *du* causes the type-II minor gradient.[43] Compound or semi-compound granules are initially formed, followed by slight enlargement, fragmentation and accumulation of phytoglycogen. At later stages of development in some cells, the phytoglycogen plastid membrane ruptures, as in *su*, and the phytoglycogen mixes with the cytosol. Saussy[43] also reported a lack of secondary granule initiation. No increase in the average size of *du su* starch granules is observed between 12 and 24 days post-pollination.[238] Granules of *du su* are similar in size to those of *su* (Table 3.4). This lack of size increase is probably due to granule fragmentation.[43] Granules isolated from mature kernels show weak or no birefringence,[90] and those from 24-day-old kernels have a weak A-type x-ray diffraction pattern.[238]

11. Dull Sugary-2

The mature kernel phenotype of *du su2* differs from both *du* and *su2* (Table 3.3). Sugar and WSP (Table 3.5) concentrations in both immature[270] and mature[272] *du su2* kernels are similar to those in *du* and *su2* except that, in immature *du su2* kernels, the sugar concentration is higher than that in *su2*. Starch concentration is lower than that in either *du* or *su2*[270,272] and the amylose percentage, as measured by the blue value test (Table 3.6) or by potentiometric titration,[360] is higher than that in either *su2* or *du*. No dosage effects on amylose percentage have been observed.[286] Isolated *du su2* amylose and amylopectin have properties similar to those of *normal*;[360] however, Whistler and Doane[109] isolated 8.7% of *du su2* starch as a loosely-branched amylopectin fraction. Average size of *du su2* starch granules is similar to that of the single mutants (Table 3.4). The BEPT of *du su2* starch granules is similar to that of *su2* granules (Table 3.4)[388] and *du su2* granules have an A-type x-ray diffraction pattern (Table 3.4).

12. Dull Waxy

The mature *du wx* kernel phenotype differs from that of either *du* or *wx* (Table 3.3). Dry weights of mature kernels are similar to those of *du* and *wx* and slightly less than those of *normal*.[272] Sugar concentrations are higher and starch concentrations are lower than those in either *normal*, *du* or *wx* immature[270] or mature[272] kernels.

Starch in the double mutant *du wx* is essentially 100% amylopectin; thus, the *wx* mutant is epistatic to *du*. The absorbance maximum and extinction coefficient of the *du wx* branched polysaccharide–iodine complex are the same as for *wx* (Table

3.7). However, *du wx* amylopectin has a reduced proportion of long B-chains, and an increased proportion of short B-chains, similar to *du* amylopectin.[253,254,390,392,406] When the *wx-a* allele is combined with *du*, *du wx-a* starch contains 9% amylose, reflecting the increased amylose conditioned by the *wx-a* allele alone.[283]

Neither *du* nor *wx* in a dent background accumulates phytoglycogen (Table 3.5), but they both contain phytoglycogen-branching enzyme.[23] However, when combined in the double mutant *du wx*, immature kernels contain up to 11% phytoglycogen (Table 3.5). Although not quantitatively determined, Saussy[43] reported numerous phytoglycogen-containing plastids in endosperm cells of *du wx* in a sweet corn background.

Starch granule size of *du wx* at 18 and 24 days post-pollination is intermediate between *du* and *wx* (Table 3.4). The mean BEPT and A-type x-ray diffraction pattern of *du wx* starch are the same as for *normal* and the component single mutants (Table 3.4).[359]

Kernels of *du wx* in a sweet corn background have the major developmental gradient typical of *normal* and a type-II minor gradient characteristic of *du*.[43] Secondary granule initiation is observed in many cells. Granule shapes vary from spherical to irregular-blocky. Plastids containing starch granules surrounded by phytoglycogen are generally located in the more mature cells of the central endosperm region.[43]

13. Sugary Waxy

The *su wx* mature kernel phenotype is similar to that of *su* (Table 3.3). Immature[270,352] and mature[272] kernel carbohydrate composition is similar to that in *su*, except that *su wx* starch is composed of 100% amylopectin (Table 3.6). The starch component in *su wx* has properties similar to those of both *wx* starch and the amylopectin component of *su* starch (Table 3.7).[360] However, the amylopectin of *su wx* has shorter long B-chains than *normal* or *wx* amylopectins, as found in *su* amylopectin.[253] With *su* homozygous, increasing doses of *wx* reduces amylose concentration.[284] The WSP content (Table 3.5), β-amylolysis limits and chain lengths of *su wx* phytoglycogen are the same as those of the phytoglycogen from *su*.[23] In a dent background, combinations of the diverse *su* alleles described in Section 3.7.3 with *wx* were found to contain more phytoglycogen than did the single alleles.[411] Immature kernels contain phytoglycogen-branching enzyme.[23] Starch granules isolated from *su wx* are small, aggregated and compound (similar to *su* granules), and phytoglycogen is completely removed from the starch granules during isolation.[90] The granules are strongly birefringent (similar to those from *wx*).[90] Thus, *wx* is epistatic to *su* relative to the absence of amylose in the starch, and *su* is epistatic to *wx* relative to soluble sugar and phytoglycogen concentrations, kernel phenotype and starch granule size.

14. Sugary-2 Waxy

The mature kernel phenotype for *su2 wx* is similar to that for *wx* (Table 3.3). Mature[272] and immature[270] kernel dry weights and carbohydrate compositions are similar to those of the single mutants. The *wx* mutant is epistatic to *su2*, resulting in starch with approximately 100% amylopectin (Table 3.6). The amylopectin of *su2 wx* starch was found to have A:B-chain ratios and chain lengths similar to those of *normal* and *wx* amylopectins.[406]

15. Sugary Sugary-2

The *su su2* mature kernel phenotype is similar to that of *su* (Table 3.3). This genotype has been extensively evaluated for its sweet corn improvement potential.[387,412] Immature[270,411] and mature[272] kernel carbohydrate compositions approach that of *su* kernels; however, starch accumulation is reduced.[270,411] The apparent amylose percentage is similar to that of *su2* starch (Table 3.6). Apparent amylose concentration increases with increasing doses of *su2* when *su* is homozygous and with increasing doses of *su* when *su2* is homozygous.[285] Starch granule size (Table 3.4)[360] and BEPT (Table 3.4)[359] are similar to those of *su*. Thus, *su2* is epistatic to *su* for apparent amylose concentration, while *su* is epistatic to *su2* for starch granule size, carbohydrate composition and mature kernel phenotype.

16. Amylose-extender Dull Sugary

The mature kernel phenotype for the triple mutant *ae du su* is similar to that for *su* (Table 3.3). Sugar concentrations of mature[272] and immature[270] *ae du su* kernels are higher than those of either of the single mutants or the double mutants *ae du* or *ae su*, while starch concentration is similar to that in *su* and the two double mutants. The amylose percentage is close to that of *normal* when measured by either the blue value test (Table 3.6) or SEC (Table 3.7). However, in contrast to *normal*, a major proportion of the branched polysaccharide is smaller than typical amylopectin (as is that of *ae su*), and it elutes from an SEC column at an intermediate position (Table 3.7). The absorbance maximum and absorptivity (extinction coefficient) of the branched polysaccharide–iodine complexes are similar to those for *ae* and *ae su*, and are characteristic of loosely-branched polymers. The absorbance maximum of the amylose–iodine complex is similar to that of *normal*, *du*, or *su*, but the absorptivity is lower than for either (Table 3.7). No short chain amylose has been found in *ae du su* starch.[90]

Phytoglycogen accumulates in *su* and *du su* kernels, but not in *ae* or *du* kernels (Table 3.5). In the double mutant *ae su*, *ae* is epistatic to *su*, but addition of *du* allows a somewhat larger amount of phytoglycogen to accumulate (Table 3.5). Phytoglycogen-branching enzyme has been reported in *su* and *du*, but not in *ae* or *ae su*.[23] Apparently, the branching enzyme activity resulting from addition of *du* to *ae su* is sufficient to partially overcome the inhibitory effect of *ae* on phytoglycogen accumulation.

Endosperms of *ae du su* in a sweet corn background have the *normal* major developmental gradient and a type-II minor gradient characteristic of *du*.[43] Secondary starch granule initiation has been observed. Starch granules from *ae du su* are similar to those from *du su* and are weakly birefringent.[43,90] Various starch granule shapes, from simple spherical to irregular, are observed in granules from immature[43] and mature kernels.[90] Starch granule fragmentation and disappearance concomitant with increased phytoglycogen in plastids, characteristic of *su*, also are common in *ae du su*.[43] Thus, *su* is epistatic to *ae du* relative to plastid type.

17. Amylose-extender Dull Sugary-2

The mature kernel phenotype of *ae du su2* differs from each of the component single or double mutants (Table 3.3). Sugar and starch concentrations of both mature[272] and immature[270] kernels of *ae du su2* are similar to those of *ae su2* kernels. Sugar concentrations are higher than those in the single mutants or other double mutants in this combination, while starch concentration is lower.[270,272] The quantity of WSP is higher in mature and immature kernels of *ae du su2* than it is in *normal* or any of the single and double mutants in this combination (Table 3.5). However, Black et al.[23] did not detect phytoglycogen in *ae du su2*, and the nature of the WSP has not been determined. Apparent amylose percentage of *ae du su2* starch is similar to that in *du* and *su2* starch, but is lower than that in *ae* starch (Table 3.6).

18. Amylose-extender Dull Waxy

The mature kernel phenotype of the triple mutant *ae du wx* differs from any of the single mutants (Table 3.3). Starch concentration is low compared with the component single and double mutants, while sugar concentrations are several-fold higher.[270,272] WSP concentration in *ae du wx* kernels in a dent background is lower than in *du wx*, but is similar to the quantity in the single and other double mutants (Table 3.5). Little if any of this WSP is phytoglycogen.[23] In contrast, amyloplasts from *ae du wx* in a sweet corn background frequently contain one or two starch granules surrounded by a non-crystalline polysaccharide.[43,90,414] The structure of the non-crystalline polysaccharide has not been determined, but the iodine-staining property appears to be similar to that of phytoglycogen *in situ*. However, in contrast with phytoglycogen in *su* kernels, it is not readily removed from the granules during granule isolation.[90,414] Extraction of the isolated granules with 10% ethanol removes some polysaccharide.[90] This extracted material is largely composed of branched polysaccharides of intermediate size having the same polysaccharide–iodine complex absorption maximum as the granules after 10% ethanol extraction, with this maximum higher than that of *su* phytoglycogen.[90]

These genes have been incorporated into sweet corn inbreds, and a new type of vegetable corn that is intermediate in sweetness between standard sweet corn (*su*) and sweet corns based on the *sh2* mutation has been introduced.[415] This hybrid, 'Pennfresh ADX,' has the advantage of extra sweetness at harvest and sugar retention for an extended time in storage.[404]

Starch from *ae du wx* kernels is composed entirely of branched polysaccharides that are largely of intermediate size between amylopectin and amylose (Table 3.7). The absorbance maximum and absorptivities of their iodine complexes are similar to those of amylopectin from *wx* and *du*, rather than those of the loosely-branched polysaccharides of *ae*, *ae* double mutants, and other *ae*-containing triple mutants (Table 3.7). Branching patterns were examined in four backgrounds (three sweet corn and one dent) and found to vary with background.[253] However, consistent differences were observed across lines when compared to *wx* amylopectin. A shorter length was observed for longer B-chains as seen in *du wx* amylopectins, while a distinct population of longer short B-chains was observed. Both *du* and *wx* kernels contain phytoglycogen-branching enzyme.[23] In combination, they apparently overcome the effect of *ae*, resulting in the production of polysaccharides, both granular and non-granular, which are more highly branched than those of *ae* amylopectin (Table 3.7).

Endosperm of *ae du wx* in a sweet corn background has a major developmental gradient typical of *normal* and a type-II minor gradient characteristic of *du*.[43] Saussy[43] reports that starch granule and plastid development in *ae du wx* is similar to that of *du wx*.

19. Amylose-extender Sugary Sugary-2

The mature kernel phenotype of the triple mutant *ae su su2* differs from that of any of the component mutants (Table 3.3). Mature kernel dry weight is similar to that of *ae su*, and sugar and starch concentrations are intermediate between those of *su* and *su su2* and those of *ae*, *su2*, *ae su* and *ae su2*.[272] Mature and immature kernels contain intermediate levels of WSP (Table 3.5). This WSP has not been characterized and may or may not be similar to the phytoglycogen accumulating in *su* kernels. Starch from *ae su su2* kernels has been reported to contain 31–54% apparent amylose (Table 3.6). Starches from *ae su su2* have not been separated by SEC, and thus the relative sizes of the polysaccharides and degrees of branching have not been established.

20. Amylose-extender Sugary Waxy

The mature kernel phenotype of the triple mutant *ae su wx* differs from that of any of the component mutants (Table 3.3). The dry weight per mature kernel is similar to that of *ae su* and higher than that of *su*, *ae wx* and *su wx*.[272] Quantities of sugars and WSP (Table 3.5) in mature[272] and immature[270] kernels are intermediate among the component single and double mutant combinations. Starch content is relatively low, but higher than that of *su*.[271] The WSP fraction contains phytoglycogen (Table 3.5) with characteristics similar to *su* phytoglycogen.[23] Kernels of *ae su wx* contain phytoglycogen-branching enzyme.[23]

Starches of *ae su wx* are reported to contain 13–14% apparent amylose when measured by blue value tests (Table 3.6), but Yeh et al.[90] (Table 3.7) showed that the apparent amylose is due to the loosely-branched nature of the starch polysaccharides.

Amylopectin of *ae su wx* contains the higher proportion of long B-chains seen in *ae* and *ae wx* amylopectins, but these are shorter than *wx* amylopectin as seen in *su wx* amylopectin.[253] Thus, *wx* blocks amylose accumulation, *ae* influences the degree of branching and *su influences* branch chain length. The absorbance maximum and absorptivity of the polysaccharide–iodine complex is similar to that of *ae wx* (Table 3.7). Starch granules isolated from mature *ae su wx* kernels vary from large spherical granules to small aggregated and compound granules.[90] Most granules are strongly birefringent, but occasional phytoglycogen-containing plastids and non-iodine-staining and non-birefringent granule particles are present in the starch granule preparation.[90]

21. Amylose-extender Sugary-2 Waxy

The *ae su2 wx* kernel phenotype differs from each of the component mutants (Table 3.3). Mature *ae su2 wx* kernel dry weight is intermediate between that of the lighter *ae* and *ae wx* kernels and the heavier *su2*, *wx*, *ae su2* and *su2 wx* kernels.[272] The quantities of sugar in both mature[272] and immature[270] *ae su2 wx* kernels are similar to those in *ae wx*, while WSP and starch concentrations are somewhat higher. The small amount of WSP present (Table 3.5) has not been characterized, and its similarity to phytoglycogen is unknown. The *ae su2 wx* starch has been reported to contain 28% amylose (based on blue values) (Table 3.6). Although not yet determined, this apparent amylose is most likely due to the presence of a loosely-branched amylopectin similar to that present in *ae wx*.[88,273] The A:B-chain ratio and chain lengths for *ae su2 wx* amylopectin were found to be similar to those of *wx* amylopectin.[406]

22. Dull Sugary Sugary-2

The mature kernel phenotype of the triple mutant *du su su2* is similar to that of *su* (Table 3.3). Also, sugar concentrations in both mature[272] and immature[270] *du su su2* kernels are similar to those in *su*, but WSP and starch contents are higher and lower, respectively, in *du su su2*. Various amylose percentages have been reported (Table 3.6). The 77% amylose observed in one study[286] is the highest percentage observed in genotypes lacking *ae*; however, because of the low starch content, this genotype has little or no commercial value. As observed with other genotypes, amylose percentage varies with year of production.[86] Starch granules of *du su su2* are small, similar to *su*,[360] and exhibit little or no birefringence.[359,360] Starch components from *du su su2* have not been separated by SEC, so the precise nature of the polysaccharides is unknown.

23. Dull Sugary Waxy

The mature kernel phenotype of *du su wx* is similar to that of *su* (Table 3.3). The quantity of sugars is similar to that in *su*.[270,272] Addition of *du* to *su wx* causes an increase in WSP (Table 3.5) and a decrease in starch content.[272] The phytoglycogen from *du su wx* has a β-amylolysis limit and chain length similar to that of *su*. The enhanced phytoglycogen accumulation may result from the additive effect of the branching enzymes present in each of the component single mutants.[23]

Starch from *du su wx* is approximately 100% amylopectin (Table 3.6) and consists of large- and intermediate-size polymers (Table 3.7). The absorbance maximum and absorptivity of the amylopectin–iodine complex are similar to those for *wx* and *du wx* amylopectins (Table 3.7). However, the long B-chains are shorter than those in either *wx* or *du wx* amylopectin.[253] Starch granules isolated from mature *du su wx* kernels vary from small spherical to aggregated and compound granules.[90] Although most granules are strongly birefringent, non-iodine staining and non-birefringent granular particles are also observed.[90] The granular particles probably are the same as the ultra-fine starch granule fragments reported by Saussy.[43]

Young kernels of *du su wx* have the major gradient in endosperm development characteristic of *normal* and the type-II minor gradient characteristic of *du*.[43] However, later in development, much of the central endosperm consists of a non-cellular cavity containing starch granules and 'phytoglycogen' plastids.[43] Cells near the pericarp contain amyloplasts with small, compound granules, while more interior cells are filled with large 'phytoglycogen' plastids void of starch, which appear unique, in that the plastid contents are essentially unstained by iodine.[43] Since the β-amylolysis limit and mean chain lengths of phytoglycogen from *du su wx* are similar to those for *su*,[18] the reason for the difference in staining properties of phytoglycogen plastids in *du su wx* and *su* is unknown.

24. Dull Sugary-2 Waxy

The mature kernel phenotype of the triple mutant *du su2 wx* differs from that of any of the component mutants (Table 3.3). Mature kernel dry weight of *du su2 wx* kernels is similar to that of the component mutants.[272] Sugar concentrations in both mature[272] and immature[270] kernels are similar to those in *du wx* kernels. The content of WSP is slightly higher and that of starch lower in *du su2 wx* kernels as compared to *du wx* kernels (Table 3.5). The WSP has not been characterized, and its similarity to phytoglycogen is unknown. Starch from *du su2 wx* kernels is 100% amylopectin (Table 3.6), reflecting the effect of *wx*. The BEPT of *du su2 wx* starch granules is low, reflecting the influence of *su2*.[359]

25. Sugary Sugary-2 Waxy

The mature kernel phenotype of the triple mutant *su su2 wx* is similar to that of *su* (Table 3.3). Kernel dry weight and carbohydrate concentrations in both mature[272] and immature[270] *su su2 wx* kernels are similar to those in *su su2* kernels. The WSP, the concentration of which is elevated (Table 3.5), is assumed to be phytoglycogen, although it has not been characterized. The *wx* gene is epistatic to *su su2*, resulting in the accumulation of starch composed of 100% amylopectin (Table 3.6). Starch granules show little birefringence.[359] The BEPT is low, similar to that of *su2* starch.[359]

26. Amylose-extender Dull Sugary Waxy

The mature kernel phenotype of the quadruple mutant *ae du su wx* differs from each of the component mutants (Table 3.3) and varies depending on the sweet corn inbred background (Garwood, unpublished). Mature kernel dry weight is similar to that of *su* kernels.[269] Starch from *ae du su wx* consists of 100% amylopectin (Table 3.7), with most of the polysaccharides of intermediate size.[90] The degree of branching of the major component (intermediate size) is similar to that of *wx* amylopectin (Table 3.7). Aqueously isolated granules contain starch granules with associated non-birefringent polysaccharides similar to those in *ae du wx*, and extraction of the granule preparation with 10% ethanol removes 27% of the total polysaccharide.[90] The addition of *su* to *ae du wx* increases the occurrence of small, aggregated and compound granules.[90]

Endosperm development in *ae du su wx* is similar to that in *du su wx*, with the type-II minor gradient observed and the central endosperm cavity being present by 27 days post-pollination.[43] Starch granule and phytoglycogen plastid development in *ae du su wx* is similar to that in *su*, except that the quadruple mutant has greater apparent phytoglycogen content at 16 days post-pollination than does *su* or any other mutant combination.[43] However, with development, there is increasing deterioration of the plastids and central endosperm cells.[43]

VIII. Conclusions

By using mutants of maize and other species, progress has been made in understanding the pathways and enzymes involved in starch biosynthesis and the fine structure of starch polysaccharides. However, starch biosynthesis (Chapter 4) and granule formation are still not completely understood. Thus, integration of the information on polysaccharide biosynthesis (Section 3.6) with that on mutant effects (Section 3.7), is necessary to fully understand polysaccharide biosynthesis and to delineate the limits of this knowledge.

A number of maize endosperm carbohydrate mutants have been shown to influence the *in vitro* activity of particular enzymes (Table 3.9). To date, modification of a specific enzyme activity has not been related to the *su2* mutation. Effects shown in Table 3.9 need not necessarily be the primary effect of a mutant, but are the ones known at this writing. Screening for enzyme activities by earlier workers probably would not have detected changes in isozyme activities involving the multiple forms of phosphorylase, starch synthase, branching enzyme and starch debranching enzyme that exist in plants. Thus, more careful examination will be needed to identify additional enzyme lesions. In addition, molecular approaches have provided new tools to define gene–enzyme relationships. For example, the cloning of the *su* gene in maize[171] has further supported earlier observations that this gene encoded a starch-debranching enzyme.[170] A similar molecular approach has been used by Myers, James et al.[397,401] to clone the *Du* gene, which most likely encodes the 180 000 molecular weight primer-dependent soluble starch synthase.[397,401] Other experimental approaches, including transformation to over- or underexpress a given enzyme, may be needed to gain information on the precise pathway of starch biosynthesis in intact, compartmented plant cells.

Mutations such as *sh*, *sh2* and *bt2* cause major blocks in the conversion of sucrose to the sugar nucleotides UDPG and ADPG (Table 3.9), indicating the key *in vivo* roles of sucrose synthase and ADPG pyrophosphorylase in starch synthesis. The *su* mutant allows the accumulation of phytoglycogen due to the activity of phytoglycogen-branching enzyme[368,369] and the deficiency of debranching enzyme discussed above. Phytoglycogen-branching enzyme activity was also found in *wx* and *du*,[23] but these mutant kernels did not produce phytoglycogen, except when they were incorporated into a sweet corn background.[43] Thus, the presence of the branching enzyme alone is not sufficient for phytoglycogen production, and the balance of the pathway as influenced by background can alter the final glucan products. The double mutant *du wx* contains phytoglycogen-branching enzyme and also accumulates phytoglycogen.[23]

Table 3.9 Summary of mutant effects in maize where an associated enzyme lesion has been reported

Genotype	Major biochemical changes[a]			Enzyme affected
sh		↑ sugars	↓ starch	↓ sucrose synthase
sh2		↑ sugars	↓ starch	↓ ADPG-pyrophosphorylase ↑ hexokinase
bt2		↑ sugars	↓ starch	↓ ADPG-pyrophosphorylase
sh4		↑ sugars	↓ starch	↓ pyridoxal phosphate
su	↑ sugars	↑ phytoglycogen	↓ starch	↑ phytoglycogen-branching enzyme ↓ starch-debranching enzyme
wx		≅ 100% amylopectin		↑ starch-bound starch synthase ↓ phytoglycogen-branching enzyme
ae		↑ loosely branched polysaccharide ↑ apparent amylose %		↓ branching enzyme IIb
du		↑ apparent amylose %		↓ starch synthase ↓ branching enzyme IIa ↑ phytoglycogen-branching enzyme
bt1	↑ sugars	↑ ADP-Glc	↓ starch	↑ membrane bound proteins

[a] Changes relative to normal, ↑, ↓ = increase or decrease, respectively. Sugars = the alcohol-soluble sugars

Approximately 100% of amylopectin is produced in kernels homozygous for *wx* (Table 3.6). In *wx*, the major starch granule-bound starch synthase is missing, but the two soluble starch synthase activities are unaffected.[245] The *ae* mutant interferes with typical branching causing accumulation of a loosely-branched polysaccharide (Table 3.7). The presence of this polymer causes an increase in 'apparent' amylose percentage when measured by iodine-binding methods (Table 3.6). The starch-branching enzyme IIb, which co-elutes with starch synthase I from DEAE cellulose columns, is missing in *ae*, but the presence of starch-branching enzymes I and IIa are unaffected.[332]

Interaction of these mutants further clarifies the biosynthetic pathway. For example, the *wx* mutant is epistatic to all other known maize endosperm mutants and no amylose accumulates (Table 3.6). Mutants such as *sh2*, *bt2* and *su* cause major reductions in starch accumulation, but when in combination with *wx*, the starch produced is all amylopectin.[271] In the double mutant *ae wx*, *wx* prevents the production of amylose and *ae* reduces the degree of branching, resulting in the accumulation of a loosely-branched polysaccharide.[88] The *su* mutant is epistatic to *du*, *su2* and *wx* relative to accumulation of phytoglycogen, but *ae* and *sh2* are partially epistatic to *su*, causing a marked reduction in the *su* stimulated phytoglycogen accumulation (Table 3.6). The addition of *du* or *wx* to *ae su* partially overcomes the *ae* inhibitory effect, and phytoglycogen accumulates.

Obviously, our understanding of starch biosynthesis is still incomplete, since mutants occur for which the primary metabolic effect has not been determined. Continued evaluation of isozymes and effector compounds, and studies of the *in vivo* pattern and rate of ^{14}C labeling of intermediates of starch biosynthesis in *normal*, mutants and mutant combinations should aid in clarifying the nature of the mutations and the pathways of starch biosynthesis. Other aspects of starch formation also remain to be explained. For example, how are starch granules formed as the amylopectin molecules are branched and debranched? How do reserve starch granules develop species-specific shapes? Does UPTG[165] as the amylopectin molecules are branched and debranched only function in initial primer production very early in development prior to starch granule formation or does it continue to synthesize glucosyl primers in the amyloplast stroma that are necessary for further polysaccharide synthesis during development?

In spite of these limitations to our complete knowledge of starch biosynthesis, information about the pathway of starch biosynthesis gained from studies of maize endosperm mutants can probably be generalized to other plant species because related mutants have occurred in peas, sorghum, barley, rice and *Chlamydomonas*, and because the same enzymes are found in starch-synthesizing tissues in other plant species. Variation in the number of isozymes and their developmental expression, and variations in cellular compartmentation, however, could result in a range of pathways with significant differences.

IX. References

1. Boyer CD. In: Zamski E, Shaffer AA, eds. *Photoassimilate Distribution in Plants and Crops*. New York, NY: Marcel Dekker, Inc.; 1996:341 .
2. Boyer CD, Hannah LC. In: Hallaner AR, ed. *Specialty Corn*. 2nd edn. Boca Raton, FL.: CRC Press; 1994:1 .
3. Boyer CD, Shannon JC. In: Watson SA, Ramstad PE, eds. *Corn Chemistry and Technology*. St. Paul, Minnesota: American Association of Cereal Chemists; 1987:253.
4. Hannah LC. In: Larkins BA, Vasil IK, eds. *Advances in Cellular and Molecular Biology of Plants, Seed Development*. Vol. IV. Dordrecht, The Netherlands: Kluwer Academic Publishers; 1997.
5. Hannah LC, Giroux M, Boyer CD. *Scient. Hort.* 1993;55:177.
6. Nelson OE, Pan D. *Ann. Rev. Plant Physiol. Plant Mol. Biol.* 1995;46:475.
7. Preiss J. *Oxford Surveys Plant Mol. Cell Biol.* 1991;7:59.
8. Preiss JK, Bull K, Smithwhite B, Iglesics G, Kakefuda G, Li L. *Biochem Soc. Trans.* 1991;19:347.
9. Preiss J, Sivak MN. In: Zamski E, Shaffer AA, eds. *Photoassimilate Distribution in Plants and Crops*. New York: Marcel Dekker; 1996:63 .
10. Smith AM, Martin C. In: Grierson D, ed. *Biosynthesis and Manipulation of Plant Products, Plant Biotechnology*. Vol. 3. Glasgow, UK: Blackie and Son; 1993:1.
11. Wang TL, Hedley CI. In: Casey R, Davies DR, eds. *Peas: Genetics, Molecular Biology and Biotechnology*. Wallingford, UK: CAB Intl.; 1993:83 .
12. Hannah LC. *Maydica*. 2005;50:497.
13. Smith AM, Denyer K and Martin C, in *Ann. Rev. Plant Physiol. Plant Mol. Biol.* 1997; 48: 65.
14. James MG, Denyer K, Myers AM. *Current Opinion*. 2003;6:215.
15. Thomlinson K, Denyer K. *Adv. Bot. Res.* 2003;40:1.
16. Ball SG, Morell MK. *Ann. Rev. Plant Biol.* 2003;54:207.
17. Badenhuizen NP. *The Biogenesis of Starch Granules in Higher Plants*. New York: Appleton-Century-Crofts; 1969.

18. Badenhuizen NP. In: Whistler RL, Paschall EF, eds. *Starch: Chemistry and Technology*. 1st edn. Vol. 1 New York: Academic Press; 1965:65.
19. Craigie JS. In: Stewart WDP, ed. *Algal Physiology and Biochemistry*. Oxford: Blackwell Scientific Publications; 1974:206.
20. Frederick JF. *Ann. N.Y. Acad. Sci.* 1973;210:254.
21. Stacey M, Barker SA. *Polysaccharides of Micro-organisms*. Oxford: Oxford University Press; 1960.
22. Ball S, Dirick L, Decq A, Martiat JC, Matagne R. *Plant Sci.* 1990;66:1.
23. Black RC, Loerch JD, McArdle FJ, Creech RG. *Genetics.* 1966;53:661.
24. Karper RE, Quinby JR. *Hered.* 1967;54:121.
25. Satoh H, Omura T. *Jap. J. Breed.* 1981;31:316.
26. Rickson FR. *Ann. N.Y. Acad. Sci.* 1973;210:104.
27. Barker SA, Bourne EJ. In: Hunter SH, Lwoff A, eds. *Biochemistry and Physiology of Protozoa*. Vol. 2. New York: Academic Press; 1955:45.
28. Percival E, McDowell RH. *Chemistry and Enzymology of Marine Algal Polysaccharides*. London: Academic Press; 1967:73.
29. Hehre EJ. *Adv. Enzymol.* 1951;11:297.
30. Okada G, Hehre EJ. *J. Biol. Chem.* 1974;249:126.
31. Ball S, Marianne T, Dirick L, Fresnoy M, Delrue B, Decq A. *Planta.* 1991;185:17.
32. Delrue B, Fontaine T, Routier F, Decq A, Wieruszeski JM, Van den Koornhuyse N, Maddelein M-L, Fournet B, Ball S. *J. Bacteriol.* 1992;174:3612.
33. Fontaine T, D'Hulst C, Maddelein M-L, Routier F, Pepin TM, Decq A, Wieruszeski JM, Delrue B, Van den Koornhuyse N, Bossu JP, Fournet B, Ball S. *J. Biol. Chem.* 1993;268:16223.
34. Libessart N, Maddelein M-L, Van den Koornhuyse N, Decq A, Delrue B, Mouille G, D'Hulst C, Ball S. *Plant Cell.* 1995;7:1117.
35. Dodge JD. *The Fine Structure of Algal Cells*. London: Academic Press; 1973.
36. Archibald AR, Hirst EL, Manners DJ, Ryley JF. *J. Chem. Soc.* 1960:556.
37. Dickinson DB. *Plant Physiol.* 1968;43:1.
38. Artschwager EF. *J. Agric. Res.* 1924;27:809.
39. Boyer CD. *Am. J. Bot.* 1981;68:659.
40. Lampe L. *Bot. Gaz.* 1931;91:337.
41. Kiesselbach TA. *Univ. Nebr. Coll. Agric. Exp. Sta. Res. Bull.* 1948:161.
42. Shannon JC. *Cereal Chem.* 1974;51:798.
43. Saussy LA. [M.S. Thesis]. University Park, Pennsylvania: The Pennsylvania State University; 1978.
44. Boyer CD, Daniels RR, Shannon JC. *Am. J. Bot.* 1977;64:50.
45. Juliano BO. In: Houston DF, ed. *Rice Chemistry and Technology*. St. Paul, Minnesota: American Association of Cereal Chemists; 1972:16.
46. Sanders EH. *Cereal Chem.* 1955;32:12.
47. Rooney LW. In: Pomeranz Y, ed. *Industrial Uses of Cereals*. St. Paul, Minnesota: American Association of Cereal Chemists; 1973:316.
48. Freeman JE. In: Wall JS, Ross WM, eds. *Sorghum Production and Utilization*. Connecticut, Westport: Avi; 1970:28.
49. Sandstedt RM. *Cereal Chem.* 1946;23:337.

50. D'Appolonia BL, Gilles KA, Osman EM, Pomeranz Y. In: Pomeranz Y, ed. *Wheat Chemistry and Technology*. St. Paul, Minnesota: American Association of Cereal Chemists; 1971:301.
51. Simmonds DH, Campbell WP. In: Bushuk W, ed. *Rye: Production, Chemistry, and Technology*. St. Paul, Minnesota: American Association of Cereal Chemists; 1976:63.
52. Simmonds DH. In: Tsen CC, ed. *Triticale: First Man-Made Cereal*. St. Paul, Minnesota: American Association of Cereal Chemists; 1974:105.
53. Briggs DE. *Barley*. London: Chapman and Hall; 1978.
54. Boyer CD, Daniels RR, Shannon JC. *Crop Sci.* 1976;16:298.
55. Bradbury D, MacMasters MM, Cull IM. *Cereal Chem.* 1956;33:361.
56. Shannon JC. *Plant Physiol.* 1972;49:198.
57. Boyer CD, Shannon JC, Garwood DL, Creech RG. *Cereal Chem.* 1976;53:327.
58. Geddes R, Greenwood CT, Mackenzie S. *Carbohydr. Res.* 1965;1:71.
59. Wolf MJ, MacMasters MM, Hubbard JE, Rist CE. *Cereal Chem.* 1948;25:312.
60. Earley EB. *Plant Physiol.* 1952;27:184.
61. MacGregor AW, LaBerge DE, Meredith WOS. *Cereal Chem.* 1971;48:255.
62. Singh R, Juliano BO. *Plant Physiol.* 1977;59:417.
63. Bice CW, MacMasters MM, Hilbert GE. *Cereal Chem.* 1945;22:463.
64. Greenwood CT, Thompson J. *Biochem. J.* 1962;82:156.
65. Paul AK, Mukherji S, Sircar SM. *Physiol. Plant.* 1971;24:342.
66. Greenwood CT. *Adv. Cereal Sci. Technol.* 1976;1:119.
67. Banks W, Greenwood CT. *Starch and Its Components*. Edinburgh: Edinburgh University Press; 1975.
68. Greenwood CT. In: Pigman W, Horton D, eds. *The Carbohydrates*. 2nd edn. Vol. IIB New York: Academic Press; 1970:471.
69. Greenwood CT. *Adv. Carbohydr. Chem.* 1956;11:335.
70. Muetgeert J. *Adv. Carbohydr. Chem.* 1961;16:299.
71. Schoch TJ. *Adv. Carbohydr. Chem.* 1945;1:247.
72. Young AH. In: Whistler RL, BeMiller JN, Paschall EF, eds. *Starch: Chemistry and Technology*. 2nd edn. Orlando, Florida: Academic Press; 1984 [Chapter VIII].
73. Banks W, Greenwood CT, Muir DD. *Starch/Stärke*. 1974,26.73.
74. Juliano BO. *Cereal Sci. Today*. 1971;16:334.
75. Kerr RW, Trubell OR. *Paper Trade J.* 1943;117:25.
76. McCready RM, Hassid WZ. *J. Am. Chem. Soc.* 1943;65:1154.
77. Shuman CA, Plunkett RA. *Method. Carbohydr. Chem.* 1964;4:174.
78. Sowbhagya CM, Bhattacharya KR. *Starch/Stärke*. 1979;31:159.
79. Williams PC, Kuzina FD, Hlynka I. *Cereal Chem.* 1970;47:411.
80. Wolf MJ, Melvin EH, Garcia WJ, Dimler RJ, Kwolek WF. *Cereal Chem.* 1970;47:437.
81. Anderson DMW, Greenwood CT. *J. Chem. Soc.* 1955:3016.
82. Bates FL, French D, Rundle RE. *J. Am. Chem. Soc.* 1943;65:142.
83. Lansky S, Kooi M, Schoch TJ. *J. Am. Chem. Soc.* 1949;71:4066.
84. Montgomery EM, Sexton KR, Senti FR. *Starch/Stärke*. 1961;13:215.
85. Banks W, Greenwood CT, Khan KM. *Starch/Stärke*. 1970;22:292.

86. Banks W, Greenwood CT, Khan KM. *Carbohydr. Res.* 1971;17:25.
87. Perlin AS. *Can. J. Chem.* 1958;36:810.
88. Boyer CD, Garwood DL, Shannon JC. *Starch/Stärke.* 1976;28:405.
89. Yamada T, Taki M. *Starch/Stärke.* 1976;28:374.
90. Yeh JY, Garwood DL, Shannon JC. *Starch/Stärke.* 1981;33:222.
91. Banks W, Greenwood CT. *Carbohydr. Res.* 1968;6:171.
92. Wankhede DB, Shehnaz A, Raghavendra Rao MR. *Starch/Stärke.* 1979;31:153.
93. Erlander SR, French D. *J. Polymer Sci.* 1958;32:291.
94. Erlander SR, French D. *J. Am. Chem. Soc.* 1958;80:4413.
95. Takeda Y, Tomooka S, Hizukuri S. *Carbohydr. Res.* 1993;246:267.
96. Murugesan G, Shibnuma K, Hizukuri S. *Carbohydr. Res.* 1993;242:203.
97. Takeda Y, Maruta N, Hizukuri S. *Carbohydr. Res.* 1992;226:279.
98. Shibanuma K, Atkeda Y, Hizukuri S, Shibata S. *Carbohydr. Polym.* 1994;25:111.
99. Takeda Y, Maruta N, Hizukuri S. *Carbohydr. Res.* 1992;227:113.
100. Nikuni Z. *Chori Kagaku.* 1969;1:6.
101. French D. *J. Jpn. Soc. Starch Sci.* 1972;19:8.
102. Hizukuri S. *Carbohydr. Res.* 1986;147:342.
103. Hizukuri S. *Carbohydr. Res.* 1985;141:295.
104. Hanashiro I, Abe J, Hizukuri S. *Carbohydr. Res.* 1996;283:151.
105. Erlander SR, McGuire JP, Dimler RJ. *Cereal Chem.* 1965;42:175.
106. Banks W, Greenwood CT. *J. Chem. Soc.* 1959:3486.
107. Banks W, Greenwood CT. *Staerke.* 1967;19:197.
108. Whistler RL, Doane WM. *Cereal Chem.* 1961;38:251.
109. Paton D. *Starch/Stärke.* 1979;31:184.
110. Adkins GK, Greenwood CT. *Carbohydr. Res.* 1969;11:217.
111. Deatherage WL, MacMasters MM, Rist CE. *Trans. Am. Assn. Cereal Chemists.* 1955;13:31.
112. Williams JW. In: Radley JA, ed. *Starch and Its Derivatives.* 4th edn. London: Chapman and Hall; 1968:91.
113. Schoch TJ, Maywald EC. *Cereal Chem.* 1968;45:564.
114. Whistler RL, Weatherwax P. *Cereal Chem.* 1948;25:71.
115. Simek J. *Zesz. Probl. Postepow Nauk Roln.* 1974;159:87.
116. Miller OH, Burns EE. *J. Food Sci.* 1970;35:666.
117. Horan FE, Heider MF. *Cereal Chem.* 1946;23:492.
118. Medcalf DG, Gilles KA. *Cereal Chem.* 1965;42:558.
119. Goering Jr. KJ, Eslick RF, Ryan CA. *Cereal Chem.* 1957;34:437.
120. Goering KJ, Eslick R, DeHaas B. *Cereal Chem.* 1970;47:592.
121. Juliano BO, Albano EL, Cagampang GB. *Philippine Agriculturalist.* 1964;48:234.
122. Kongseree N, Juliano BO. *J. Agr. Food Chem.* 1972;20:714.
123. Raghavendra Rao SN, Juliano BO. *J. Agr. Food Chem.* 1970;18:289.
124. Reyes AC, Albano EL, Broines VP, Juliano BO. *J. Agr. Food Chem.* 1965;13:438.
125. Lii C-Y, Lineback DR. *Cereal Chem.* 1977;54:138.
126. Campbell MR, Pollak LM, White PJ. *Cereal Chem.* 1995;72:281.
127. Boyer CD, Garwood DL, Shannon JC. *J. Hered.* 1976;67:209.
128. Mercier C, Charbonniere R, Gallant D, Guilbot A. *Starch/Stärke.* 1970;22:9.

IX. References

129. Tsai CY, Salamini F, Nelson OE. *Plant Physiol.* 1970;46:299.
130. Ketiku AO, Oyenuga VA. *J. Sci. Food Agr.* 1972;23:1451.
131. Briones VP, Magbanua LG, Juliano BO. *Cereal Chem.* 1968;45:351.
132. Abou-Guendia M, D'Appolonia BL. *Cereal Chem.* 1973;50:723.
133. Jenkins LD, Loney DP, Meredith P, Fineran BA. *Cereal Chem.* 1974;51:718.
134. Kulp K, Mattern PJ. *Cereal Chem.* 1973;50:496.
135. Matheson NK. *Phytochemistry.* 1971;10:3213.
136. Wood HL. *Aust. J. Agr. Res.* 1960;11:673.
137. Banks W, Greenwood CT, Muir DD. *Starch/Stärke.* 1973;25:153.
138. Harris G, MacWilliam IC. *Cereal Chem.* 1958;35:82.
139. Duffus CM, Jennings PH. *Starch/Stärke.* 1978;11:371.
140. Banks W, Greenwood CT. *Ann. N.Y. Acad. Sci.* 1973;210:17.
141. Goering KJ, DeHaas B. *Cereal Chem.* 1974;51:573.
142. Inatsu O, Watanabe K, Maeda I, Ito K, Osanai S. *J. Jap. Soc. Starch Sci.* 1974;21:115.
143. Juliano BO. *J. Jap. Soc. Starch Sci.* 1970;18:35.
144. Fergason VL, Zuber MS. *Crop Sci.* 1965;5:169.
145. Atwell WA, Patrick BM, Johnson LA, Glass RW. *Cereal Chem.* 1983;60:9.
146. Goering KJ. *Cereal Chem.* 1967;44:245.
147. Goering KJ, Subba Rao PV, Fritts DH, Carroll T. *Starch/Stärke.* 1970;22:217.
148. Buttrose MS. *Aust. J. Biol. Sci.* 1963;16:305.
149. Stamberg OE. *Cereal Chem.* 1939;16:769.
150. May LH, Buttrose MS. *Aust. J. Biol. Sci.* 1959;12:146.
151. Buttrose MS. *J. Ultrastructure Res.* 1960;4:231.
152. Hall DM, Sayre JG. *Textile Res. J.* 1970;40:256.
153. Reichert ET. *The Differentiation and Specificity of Starches in Relation to Genera, Species, etc.* Washington, DC: Carnegie Institution of Washington; 1913 [Parts I and II].
154. Moss GE. In: Radley JA, ed. *Examination and Analysis of Starch and Starch Products.* London: Applied Science Publishers; 1976:1.
155. Kent NL. *Technology of Cereals with Special Reference to Wheat.* 2nd edn.; 1975: Pergamon Press Oxford
156. Hall DM, Sayre JG. *Textile Res. J.* 1969;39:1044.
157. Hall DM, Sayre JG. *Textile Res. J.* 1971;41:880.
158. Klassen AJ, Hill RD. *Cereal Chem.* 1971;48:647.
159. Lempiainen T, Henriksnas H. *Starch/Stärke.* 1979;31:45.
160. Duvick DN. *Am. J. Bot.* 1955;42:717.
161. Williams BR. [Ph.D. Thesis]. University Park, Pennsylvania: The Pennsylvania State University; 1971.
162. Badenhuizen NP. *K. Ned. Akad. Wet. Proc. Ser. C.* 1962;65:123.
163. Douce R, Joyard J. *Ann. Rev. Cell Biol.* 1990;6:173.
164. Flugge UI, Heldt HW. *Ann. Rev. Plant Physiol. Plant Mol. Biol.* 1991;42:129.
165. Tandecarz JS, Ardila FJ, Bocca SN, Moreno S, Rothschild A. In: Pontis HG, Salerno GL, Echeverria EJ, eds. *Sucrose Metabolism, Biochemistry, Physiology and Molecular Biology.* Rockville, Maryland: American Society of Plant Physiologists; 1995:107.

166. Shannon JC, Creech RG, Loerch JD. *Plant Physiol.* 1970;45:163.
167. Whistler RL, Young JR. *Cereal Chem.* 1960;37:204.
168. McConnell WB, Mitra AK, Perlin AS. *Can. J. Biochem. Physiol.* 1958;36:985.
169. Ball S, Guan H-P, James M, Myers A, Mouille PG, Buleon A, Colonna P, Preiss J. *Cell.* 1996;86:349.
170. Pan D, Nelson OE. *Plant Physiol.* 1984;74:324.
171. James MG, Robertson DS, Myers AM. *Plant Cell.* 1995;7:417.
172. Wu C, Colleoni C, Myers AM, James MJ. *Arch. Biochem. Biophys.* 2002;406:21.
173. Beatty MK, Rahman A, Cao H, Woodman W, Lee M, Myers AM, James MG. *Plant Physiol.* 1999;119:255.
174. Dinges JR, Colleoni C, James MG, Myers AM. *Plant Cell.* 2003;15:666.
175. Bresolin NS, Li Z, Kosar-Hashemi B, Tetlow IJ, Chatterjee M, Rahman S, Morell MK, Howitt CA. *Planta.* 2005;224:20.
176. Heldt HW, Flugge U-I, Borchert S. *Plant Physiol.* 1991;95:341.
177. Quick WP, Scheibe R, Neuhaus HE. *Plant Physiol.* 1995;109:113.
178. Heineke D, Kruse A, Flugge U-I, Frommer WB, Riesmeier JW, Willmitzer L, Heldt HW. *Planta.* 1994;193:174.
179. Riesmeir JW, Flugge U-I, Schulz B, Heineke D, Heldt HW, Willmitzer L, Frommer WB. *Proc. Natl. Acad. Sci. USA.* 1993;90:6160.
180. Barnes SA, Knight JS, Gray JC. *Plant Physiol.* 1994;106:1123.
181. Hatch MD. In: Bonner J, Varner JE, eds. *Plant Biochemistry.* 3rd edn. New York: Academic Press; 1976:797.
182. Rhoades MM, Carvalho A. *Bull. Torrey Botan. Club.* 1944;71:335.
183. Downton WJ, Tregunna EB. *Can. J. Bot.* 1968;46:207.
184. Russin WA, Evert RF, Vanderveer PJ, Sharkey TD, Briggs SP. *Plant Cell.* 1996;8:645.
185. Taiz L, Zeiger E. *Plant Physiology.* Redwood City, CA: Benjamin/Cummings; 1991 219–248.
186. Heber U. *Ann. Rev. Plant Physiol.* 1974;25:393.
187. Stitt M. *Ann. Rev. Plant Physiol. Plant Mol. Biol.* 1990;40:153.
188. Huber SC, Huber JL, Campbell WH, Redinbaugh MG. *Plant Physiol.* 1992;100:706.
189. Huber SC, Huber JL. *Ann. Rev. Plant Physiol. Plant Mol. Biol.* 1996;47:431.
190. Neuhaus HE, Quick WP, Siegl G, Stitt M. *Planta.* 1990;181:583.
191. Preiss J, Levi C. In: Gibbs M, Latzko E, eds. *Photosynthesis II, Encyclopedia of Plant Physiology, New Series.* Vol. 6. Berlin: Springer-Verlag; 1979:282.
192. Stitt M, Sonnewald U. *Ann. Rev. Plant Physiol. Plant Mol. Biol.* 1995;46:341.
193. Beck E. In: Heath RL, Preiss J, eds. *Regulation of Carbohydrate Partitioning in Photosynthetic Tissue.* Rockville, Maryland: American Society of Plant Physiologists; 1985:27.
194. Levi C, Preiss J. *Plant Physiol.* 1978;61:218.
195. Ludwig I, Ziegler P, Beck E. *Plant Physiol.* 1984;74:856.
196. Muhlethaler K. In: Gibbs M, ed. *The Structure and Function of Chloroplasts.* Berlin: Springer Verlag; 1971:7.
197. Jenner CF. In: Wardlow IF, Passioura JB, eds. *Transport and Transfer Processes in Plants.* New York: Academic Press; 1976:73.

198. Klassen AJ, Hill RD. *Cereal Chem.* 1971;48:647.
199. Shannon JC. In: Boyer CD, Shannon JC, Hardison RC, eds. *Physiology, Biochemistry, and Genetics of Nongreen Plastids*. Rockville, Maryland: American Society of Plant Physiologists; 1989.
200. Kim TK, Franceschi VR, Okita TW, Robinson NL, Morell M, Preiss J. *Plant Physiol.* 1989;91:217.
201. Kram AM, Oostergetel GT, van Bruggen EFJ. *Plant Physiol.* 1993;101:237.
202. Miller ME, Chourey PS. *Planta*. 1995;197:522.
203. Villand P, Kleczkowski LA. *Z. Naturforsch.* 1994;49c:215.
204. Giroux MJ, Hannah LC. *Mol. Gen. Genetics*. 1994;243:400.
205. MacDonald FD, ap Rees T. *Biochim. Biophys. Acta*. 1983;755:81.
206. ap Rees T, Entwistle G. In: Boyer CD, Shannon JC, Hardison RC, eds. *Physiology, Biochemistry, and Genetics of Nongreen Plastids*. Rockville, Maryland: American Society of Plant Physiologists; 1989:49.
207. Kosegarten H, Mengel K. *Physiol. Plant*. 1994;91:111.
208. Pozueta-Romero J, Frehner M, Viale AM, Akazawa T. *Proc. Natl. Acad. Sci. USA*. 1991;88:5769.
209. Journet EP, Douce R. *Plant Physiol*. 1985;79:458.
210. Gaynor JJ, Galston AW. *Plant Cell Physiol*. 1983;24:411.
211. Entwistle G, ap Rees T. *Biochem. J.* 1988;255:391.
212. Echeverria E, Boyer CD, Thomas P, Liu K, Shannon JC. *Plant Physiol*. 1988;86:786.
213. Keeling PL, Wood JR, Tyson RH, Bridges IG. *Plant Physiol*. 1988;87:311.
214. Williams JM, Duffus CM. *Plant Physiol*. 1977;59:189.
215. Thorbjornsen T, Villand P, Denyer K, Olsen O-A, Smith LM. *Plant J.* 1996;10:243.
216. Denyer K, Dunlap F, Thorbjornsen T, Keeling P, Smith AM. *Plant Physiol*. 1996;112:779.
217. Pien F-M, Shannon JC. *Plant Physiol Suppl*. 1996;111:382.
218. Denyer K, Dunlap F, Thorbjornsen T, Keeling P, Smith AM. *Plant Physiol*. 1996;112:779.
219. Beckles DM, Smith AM, ap Rees T. *Plant Physiol*. 2001;125:818.
220. Burton RA, Johnson PE, Beckles DM, Fincher GB, Jenner HL, Naldrett MJ, Denyer K. *Plant Physiol*. 2002;130:1464.
221. Shannon JC, Echeverria E, Boyer CD. In: Packer L, Douce R, eds. *Methods in Enzymology*. Vol. 148. San Diego, California: Academic Press, Inc; 1987:226.
222. Tyson RH, ap Rees T. *Planta*. 1988;175:33.
223. Tetlow IJ, Blissett KJ, Emes MJ. *Planta*. 1994;194:454.
224. Borchert S, Grobe H, Heldt HW. *FEBS Lett*. 1989;253:183.
225. Hill LM, Smith AM. *Planta*. 1991;185:91.
226. Schott K, Borchert S, Muller-Rober B, Heldt HW. *Planta*. 1995;196:647.
227. Liu K-C, Boyer CD, Shannon JC. In: Huang A, Taiz L, eds. *Molecular Approaches to Compartmentation and Metabolic Regulation*. Rockville, Maryland: American Society of Plant Physiologists; 1991:236.
228. Shannon JC, Pien F-M, Liu K-C. *Plant Physiol*. 1996;110:835.
229. Cao H, Sullivan TD, Boyer CD, Shannon JC. *Physiol. Plant*. 1995;95:176.

230. Sullivan Jr TD, Strelow LI, Illingsworth CA, Phillips RL, Nelson OE. *Plant Cell.* 1991;3:1337.
231. Sullivan T, Kaneko Y. *Planta.* 1995;196:477.
232. Li HM, Sullivan TD, Keigstra K. *J. Biochem.* 1992;267:18999.
233. Cao H, Shannon JC. *Plant Physiol.* 1997;100:400.
234. Shannon JC, Pien F-M, Cao H, Liu K-C. *Plant Physiol.* 1998;117:1235.
235. Garwood DL, Creech RG. *Crop Sci.* 1972;12:119.
236. Mohlmann T, Tjaden J, Henrichs G, Quick WP, Hausler R, Neuhaus HE. *Biochem J.* 1997;324:503.
237. Bowser CG, Scrase-Field EFAL, Esposito S, Emes MJ and Tetlow IJ, *J. Exp. Bot.*, (in press).
238. Brown RP, Creech RG, Johnson LJ. *Crop Sci.* 1971;11:297.
239. Tsai CY, Nelson OE. *Genetics.* 1969;61:813.
240. Burr B, Nelson OE. *Ann. N. Y. Acad. Sci.* 1973;210:129.
241. Chourey PS, Nelson OE. *Biochem. Genetics.* 1976;14:1041.
242. Tsai C-Y, Nelson OE. *Science.* 1966;151:341.
243. Dickinson DB, Preiss J. *Plant Physiol.* 1969;44:1058.
244. Shannon JC, Pien F-M, Liu K-C. *Plant Physiol.* 1996;110:835.
245. Nelson OE. *Adv. Cereal Sci. Technol.* 1980;111:41–71 [Chap. 2].
246. Schmidt RJ, Burr FA, Aukerman MJ, Burr B. *Proc. Natl. Acad. Sci. USA.* 1990;87:46.
247. Schmidt RJ, Ketudat M, Aukerman MJ, Hoschek G. *Plant Cell.* 1992;4:689.
248. Ma Y, Nelson OE. *Cereal Chem.* 1975;52:412.
249. Fuwa H, Sugimoto Y, Tanaka M, Glover DV. *Starch/Stärke.* 1978;30:186.
250. Wilson JA, Glover DV, Nyquist WE. *Z. Pflanzenzuecht.* 2000;119:177.
251. Creech RG. *Adv. Agron.* 1968;20:275.
252. Saunders EB, Thompson DB, Boyer CD. *Cereal Chem.* 1990;67:594.
253. Yuan RC, Thnopson DB, Boyer CD. *Cereal Chem.* 1993;70:81.
254. Tomita Y, Sugimoto Y, Sakamoto S, Fuwa H. *J. Nutr. Sci. Vitaminol.* 1981;27:471.
255. Uematsu M, Yabuno T. *Jap. J. Breed.* 1988;38:269.
256. Eriksson G. *Hereditas.* 1969;63:180.
257. Hixon RM, Brimhall B. In: Radley JA, ed. *Starch and Its Derivatives.* 4th edn. London: Chapman and Hall; 1968:247.
258. Kanazaki K, Atoda K. *Jap. J. Breed.* 1988;38:423.
259. Hoshino T, Ito S, Hatta K, Nakamura T, Yamamori M. *Breed. Sci.* 1996;46:185.
260. Miura H, Tanii S, Nakamura T, Watanabe N. *Theor. Appl. Genet.* 1994;89:276.
261. Nakamura T, Yamamori M, Hirano H. *Mol. Gen. Genet.* 1995;248:253.
262. Nakamura T, Yamamori M, Hirano H, Hidaka S. *Plant Breed.* 1993;111:99.
263. Yamamori M, Nakamura T, Nagamine T. *Breed. Sci.* 1995;45:377.
264. Irving DW, Betschart AA, Saunders RM. *J. Food Sci.* 1981;46:1170.
265. Okuno K, Sakaguchi S. *J. Hered.* 1982;73:467.
266. Hovenkamp-Hermelink JHM, Jacobsen E, Poinstein AS, Visser RGF, Vos-Scheperkeuter GH, Bijamolt EW, de Vries JN, Wiltholt B, Feenstra W. *Theor. Appl. Genet.* 1987;75:217.
267. Gorbet DW, Weibel DE. *Crop Sci.* 1972;12:378.

268. Nass HG, Crane PL. *Crop Sci.* 1970;10:276.
269. Rowe DE, Garwood DL. *Crop Sci.* 1978;18:709.
270. Creech RG. *Genetics.* 1965;52:1175.
271. Holder DG, Glover DV, Shannon JC. *Crop Sci.* 1974;14:643.
272. Creech RG, McArdle FJ. *Crop Sci.* 1966;6:192.
273. Yamada T, Komiya T, Akaki M, Taki M. *Starch/Stärke.* 1978;30:145.
274. Seckinger HL, Wolf MJ. *Starch/Stärke.* 1966;18:1.
275. Juliano BO, Nazareno MB, Ramos NB. *J. Agr. Food Chem.* 1969;17:1364.
276. Ikawa Y, Glover DV, Sugimoto Y, Fuwa H. *Carbohydr. Res.* 1978;61:211.
277. Yun SH, Matheson NK. *Carbohydr. Res.* 1992;227:85.
278. Brimhall B, Sprague GF, Sass JE. *J. Am. Soc. Agron.* 1945;37:937.
279. Nelson OE. *Science.* 1959;130:794.
280. Palmiano EP, Juliano BO. *Agr. Biol. Chem.* 1972;36:157.
281. Vidal AJ, Juliano BO. *Cereal Chem.* 1967;44:86.
282. Umemoto T, Nakamura Y, Ishikura N. *Phytochem.* 1995;40:1613.
283. Helm JL, Fergason VL, Zuber MS. *J. Hered.* 1969;60:259.
284. Kramer HH, Whistler RL. *Agron. J.* 1949;41:409.
285. Vineyard ML, Bear RP, MacMasters MM, Deatherage WL. *Agron. J.* 1958;50:595.
286. Dunn GM, Kramer HH, Whistler RL. *Agron. J.* 1953;45:101.
287. Yun SH, Matheson NK. *Carbohydr. Res.* 1993;243:307.
288. Murata T, Sugiyama T, Akazawa T. *Biochem. Biophys. Res. Commun.* 1965;18:371.
289. Nelson OE, Rines HW. *Biochem. Biophys. Res. Commun.* 1962;9:297.
290. Hylton CM, Denyer K, Keeling PL. *Planta.* 1996;198:230.
291. van der Leij FR, Visser RGF, Ponstein AS, Jacobsen E, Feenstra WJ. *Mol. Gen. Genet.* 1991;228:240.
292. Nakamura T, Yamamori M, Hirano H, Hidaka S. *Biochem. Genet.* 1993;31:75.
293. Nakamura T, Yamamori M, Hirano H, Hidaka S. *Phytochem.* 1993;33:749.
294. Sivak MN, Wagner M, Preiss J. *Plant Physiol.* 1993;103:1355.
295. Taira T, Fujita N, Takaoka K, Uematsu M, Wadano A, Kozaki S, Okabe S. *Biochem. Genet.* 1995;33:269.
296. Nelson OE, Chourey PS, Chang MT. *Plant Physiol.* 1978;62:383.
297. Ozbun JL, Hawker JS, Preiss J. *Plant Physiol.* 1971;48:765.
298. Smith AM. *Planta.* 1990;182:599.
299. Fuwa H, Okuno K, Asashiba R, Kituzaki H, Asakura M, Inouchi N, Sugimoto Y. *Starch/Stärke.* 1992;44:203.
300. Yano M, Okuno K, Kawakami J, Omura T. *Theor. Appl. Genet.* 1985;69:253.
301. Banks W, Greenwood CT, Muir DD. *Starch/Stärke.* 1974;26:289.
302. Blixt S. In: King RC, ed. *Handbook of Genetics.* Vol. 2. New York: Plenum Press; 1974:181.
303. Kaushik RP, Khush GS. *Cereal Chem.* 1991;68:487.
304. Kaushik RP, Khush GS. *Theor. Appl. Genet.* 1991;83:146.
305. Fergason VL, Helm JD, Zuber MS. *J. Hered.* 1966;57:90.
306. Walker JT, Merritt NR. *Nature.* 1969;221:482.
307. Zuber MS, Grogan CO, Deatherage WL, Hubbard JW, Schulze W, MacMasters MM. *Agron. J.* 1958;50:9.

308. Bear RP, Vineyard ML, MacMasters MM, Deatherage WL. *Agron. J.* 1958;50:598.
309. Haunold A, Lindsey MF. *Crop Sci.* 1964;4:58.
310. Loesch Jr. PJ, Zuber MS. *Crop Sci.* 1964;4:526.
311. Thomas JP. [Ph.D. Thesis]. Columbia, Missouri: University of Missouri, 1968.
312. Garwood DL, Shannon JC, Creech RG. *Cereal Chem.* 1976;53:355.
313. Helm JL, Fergason VL, Zuber MS. *Crop Sci.* 1967;7:659.
314. Fergason VL, Zuber MS. *Crop Sci.* 1962;2:209.
315. Helm JL, Fergason VL, Zuber MS. *Agron. J.* 1968;60:530.
316. Helm JL, Fergason VL, Thomas JP, Zuber MS. *Agron. J.* 1967;59:257.
317. Fergason VL, Helm JL, Zuber MS. *Crop Sci.* 1966;6:273.
318. Mercier C. *Starch/Stärke*. 1973;25:78.
319. Wolff IA, Hofreiter BT, Watson PR, Deatherage WL, MacMasters MM. *J. Am. Chem. Soc.* 1955;77:1654.
320. Wolf MJ, Seckinger HL, Dimler RJ. *Starch/Stärke*. 1964;16:375.
321. Mussulman WC, Wagoner JA. *Cereal Chem.* 1968;45:162.
322. Sandstedt RM. *Cereal Sci. Today.* 1965;10:305.
323. Boyer CD, Damewood PA, Matters GL. *Starch/Stärke*. 1980;32:217.
324. Takeda C, Takeda Y, Hizukuri S. *Carbohydr. Res.* 1993;246:273.
325. Mizuno K, Kawasaki T, Shimada H, Satoh H, Kobayashi E, Okumura S, Arai Y, Baba T. *J. Biol. Chem.* 1993;268:19084.
326. Boyer CD, Preiss J. *Carbohydr. Res.* 1978;61:321.
327. Matters GL, Boyer CD. *Biochem. Genet.* 1982;20:833.
328. Smith AM. *Planta.* 1988;175:170.
329. Edwards J, Green JH, ap Rees T. *Phytochem.* 1988;27:1615.
330. Kouichi M, Kawasaki T, Shimada H, Satoh H, Kobayashi E, Okumura S, Arai Y, Baba T. *J. Biol. Chem.* 1993;268:19084.
331. Kouichi M, Kimura K, Arai Y, Kawasaki T, Shimada H, Baba T. *J. Biochem. (Tokyo).* 1992;112:643.
332. Boyer CD, Preiss J. *Biochem. Biophys. Res. Commun.* 1978;80:169.
333. Bhattacharyya M, Martin C, Smith A. *Plant Mol. Biol.* 1993;22:525.
334. Hedman KD, Boyer CD. *Biochem. Genet.* 1982;20:483.
335. Fisher DK, Gao M, Kim KN, Boyer CD, Guiltinan MJ. *Plant Physiol.* 1996;110:611.
336. Nishi A, Nakamura Y, Tanaka N, Satoh H. *Plant Physiol.* 2001;127:459.
337. Fisher DK, Gao M, Kim KN, Boyer CD, Guiltinan MJ. *Plant Physiol.* 1996;110:611.
338. Blauth SL, Yao Y, Kluchinec JD, Shannon JC, Thompson DB, Guiltinan MJ. *Plant Physiol.* 2001;125:1396.
339. Yao Y, Thompson DB, Guiltinan MJ. *Plant Physiol.* 2004;136:3515.
340. Tetlow IJ, Wait R, Lu Z, Akkasaeng R, Bowsher CG, Esposito S, Kosar-Hashemi B, Morell MK, Emes MJ. *Plant Cell.* 2004;16:694.
341. Blauth S, Kim K-N, Klucinec J, Shannon JC, Thompson D, Guiltinan M. *Plant Mol. Biol.* 2002;48:287.
342. Watson SA, Hirata Y. *Sorghum Newsletter.* 1960;3:6.
343. Boyer CD, Liu K-C. *Phytochem.* 1983;22:2513.

344. Yano M, Isono Y, Satoh H, Omura T. *Jap. J. Breed.* 1984;34:43.
345. Wall S, Blessin CW. In: Wall JS, Ross WM, eds. *Sorghum Production and Utilization*. Westport, Connecticut: Avi; 1970:118.
346. Asaoka M, Okuno K, Sugimoto Y, Yano M, Omura T, Fuwa H. *Starch/Stärke*. 1985;37:364.
347. Chen HR, Su JC, Sung HY. *Proc. Nat. Sci. Coun. Rep. China*. 1996;20:6.
348. Koh HJ, Heu MH. *Kor. J. Crop Sci.* 1994;39:1.
349. Culpepper CW, Magoon CA. *J. Agr. Res.* 1924;28:403.
350. Greenwood CT, Das Gupta PC. *J. Chem. Soc.* 1958;703.
351. Inouchi ND, Glover DV, Fuwa H. *Starch/Stärke*. 1987;39:259.
352. Andrew RH, Brink RA, Neal NP. *J. Agr. Res.* 1944;69:355.
353. Jennings PH, McCombs CL. *Phytochem.* 1969;8:1357.
354. Laughnan JR. *Genet.* 1953;38:485.
355. Singh R. [Ph.D. Thesis]. West Lafayette, Indiana: Purdue University; 1973.
356. Gonzales JW, Rhodes AM, Dickinson DB. *Plant Physiol.* 1976;58:28.
357. Inouchi ND, Glover DV, Takaya T, Fuwa H. *Starch/Stärke*. 1983;35:371.
358. Duffus CM, Jennings PH. *Starch/Stärke*. 1978;30:371.
359. Kramer HH, Pfahler PL, Whistler RL. *Agron. J.* 1958;50:207.
360. Dvonch W, Kramer HH, Whistler RL. *Cereal Chem.* 1951;28:270.
361. Greenwood CT, Das Gupta PC. *J. Chem. Soc.* 1958;707.
362. Matheson NK. *Phytochem.* 1975;14:2017.
363. Wanson JC, Drochmans P. *J. Cell Biol.* 1968;38:130.
364. Sandstedt RM, Hites BD, Schroeder H. *Cereal Sci. Today*. 1968;13:82.
365. Cameron JW. *Genet.* 1947;32:459.
366. Boyer CD, Damewood PA, Simpson EKG. *Starch/Stärke*. 1981;33:125.
367. Peat S, Whelan WJ, Turvey JR. *J. Chem. Soc.* 1956;2317.
368. Hodges HF, Creech RG, Loerch JD. *Biochim. Biophys. Acta*. 1969;185:70.
369. Lavintman N. *Arch. Biochem. Biophys.* 1966;116:1.
370. Manners DJ, Rowe JJM, Rowe KL. *Carbohydr. Res.* 1968;8:72.
371. Boyer CD, Simpson EKG, Damewood PA. *Starch/Stärke*. 1982;34:81.
372. Doehlert DC, Knutson CA. *J. Plant Physiol.* 1991;138:566.
373. Doehlert DC, Kuo TM, Juvik JA, Beers EP, Duke SH. *J. Am. Soc. Hort. Sci.* 1993;118:661.
374. Rahman A, Wong K-S, Jane J-L, Myers AM, James MG. *Plant Physiol.* 1998;117:425.
375. Beatty MK, Rahman A, Cao H, Woodman W, Lee M, Myers AM, James MG. *Plant Physiol.* 1999;19:255.
376. Nakamura Y, Umemoto T, Takahata Y, Komea K, Amano E, Satoh H. *Physiol. Plant.* 1996;97:491.
377. Erlander S. *Enzymologia*. 1958;19:273.
378. Garwood DL, Creech RG. *Agron. Abstr.* 1972;7.
379. Mangelsdorf PC. *Genet.* 1947;32:448.
380. Garwood DL. *Maize Genet. Coop. News Letter*. 1975;49:140.
381. Dahlstrom DE, Lonnquist JH. *J. Hered.* 1964;55:242.
382. Ferguson JE, Rhodes AM, Dickinson DB. *J. Hered.* 1978;69:377.

383. Ferguson JE, Dickinson DB, Rhodes AM. *Plant Physiol.* 1979;63:416.
384. Takeda Y, Preiss J. *Carbohydr. Res.* 1993;240:265.
385. Campbell MR, White PJ, Pollack LM. *Cereal Chem.* 1994;71:464.
386. Campbell MR, White PJ, Pollack LM. *Cereal Chem.* 1995;72:389.
387. Badenhuizen NP. *Protoplasmalogia.* 1959;2:B/26.
388. Pfahler PL, Kramer HH, Whistler RL. *Science.* 1957;125:441.
389. Davis JH, Kramer HH, Whistler RL. *Agron. J.* 1955;47:232.
390. Wang YJ, White P, Pollack L, Jane J. *Cereal Chem.* 1993;70:171.
391. Katz FR, Furcsik SL, Tenbarge FL, Hauber RJ, Friedman RB. *Carbohydr. Polym.* 1993;21:133.
392. Wang YJ, White P, Pollak L, Jane J. *Cereal Chem.* 1993;70:521.
393. Ohtsubo K, Nakagahra M, Iwasaki T. *J. Jap. Soc. Food Sci. Technol.* 1988;35:587.
394. Okuno K. *Res. J. Food Agric.* 1988;11:3.
395. Kunihiro Y, Ebe Y, Shinbashi N, Kikuchi H, Tannno H, Sugawara K. *Jap. J. Breed.* 1993;43:155.
396. Preiss J, Boyer CD. In: Marshall JJ, ed. *Mechanisms of Polysaccharide Polymerization and Depolymerization.* New York: Academic Press; 1979.
397. Gao M, Wanat J, Stinard PS, James MG, Myers AM. *Plant Cell.* 1998;10:399.
398. Marshall J, Sidebottom C, Debet M, Martin C, Smith AM, Edwards A. *Plant Cell.* 1996;8:1121.
399. Dang PL, Boyer CD. *Phytochem.* 1988;27:1255.
400. Mu C, Harn C, Ko Y-T, Singletary GW, Keeling PL, Wasserman BP. *Plant J.* 1994;6:151.
401. Cao H, Imparl-Radoseviech J, Guan H, Keeling PL, James MG, Myers AM. . *Plant Physiol* 1999;120;25.
402. Commuri PD, Keeling PL, *Plant J.* 2001;25:476.
403. Wann EV, Brown GB, Hills WA. *J. Am. Soc. Hort. Sci.* 1971;96:441.
404. Garwood DL, McArdle FJ, Vanderslice SF, Shannon JC. *J. Am. Soc. Hort. Sci.* 1976;101:400.
405. Brockett E, Thompson D, Davis T, Boyer CD. *Starch/Stärke.* 1988;40:175.
406. Fuwa H, Glover DV, Miyaura K, Inouchi N, Konishi Y, Sugimoto Y. *Starch/Stärke.* 1987;39:295.
407. Garwood DL, Boyer CD, Shannon JC. *Maize Genet. Coop. News Letter.* 1976;50:99.
408. Creech RG, Kramer HH. *Am. Naturalist.* 1961;95:326.
409. Kramer HH, Whistler RL, Anderson EG. *Agron. J.* 1956;48:170.
410. Ayers JE, Creech RG. *Crop Sci.* 1969;9:739.
411. Cameron Jr. JW, Cole DA. *Agron. J.* 1959;51:424.
412. Soberalske RM, Andrew RH. *Crop Sci.* 1978;18:743.
413. Ensminger HS. [M.S. Thesis]. The Pennsylvania State University; 1988:79.
414. Liu T-TY, Shannon JC. *Plant Physiol.* 1981;67:525.
415. Garwood DL, Creech RG. *HortScience.* 1979;14:645.
416. Marshall JJ. *Wallerstein Lab. Commun.* 1972;35:49.
417. Radley JA. *Starch and Its Derivatives.* 4th edn. London: Chapman and Hall; 1968.
418. Garwood DL, Vanderslice SF. *Crop Sci.* 1982;22:367.

Biochemistry and Molecular Biology of Starch Biosynthesis

4

Jack Preiss
Department of Biochemistry and Molecular Biology, Michigan State University, East Lansing, Michigan, 48824, USA

I. Introduction	84
II. Starch Synthesis in Plants: Localization	84
1. Leaf Starch	84
2. Starch in Storage Tissues	85
III. Enzyme-catalyzed Reactions of Starch Synthesis in Plants and Algae and Glycogen Synthesis in Cyanobacteria	85
IV. Properties of the Plant 1,4-α-Glucan-Synthesizing Enzymes	87
1. ADP-glucose Pyrophosphorylase: Kinetic Properties and Quaternary Structure	87
2. Relationship Between the Small and Large Subunits: Resurrection of ADPGlc PPase Catalysis in the Large Subunit	91
3. Phylogenetic Analysis of the Large and Small Subunits	95
4. Crystal Structure of Potato Tuber ADPGlc PPase	95
5. Supporting Data for the Physiological Importance of Regulation of ADPGlc PPase	104
6. Differences in Interaction Between 3PGA and Pi in Different ADPGlc PPases	105
7. Plant ADPGlc PPases can be Activated by Thioredoxin	107
8. Characterization of ADPGlc PPases from Different Sources	108
9. Identification of Important Amino Acid Residues Within the ADPGlc PPases	111
10. Starch Synthase	114
11. Branching Enzyme	129
12. Other Enzymes Involved in Starch Synthesis	136
V. Abbreviations	138
VI. References	139

I. Introduction

This chapter reviews enzymic reactions involved in starch synthesis in higher plants and algae. The reactions of glycogen synthesis in bacteria, including cyanobacteria, are similar to those in higher plants and thus, the bacterial enzymes of glycogen synthesis are also described and compared to the analogous plant enzymes. Regulation of starch synthesis at the enzymic level is also discussed and compared to what is observed in the regulation of bacterial glycogen synthesis. In relation to this, results showing how starch content has been increased in certain plants by overexpression of biosynthetic enzymes are described. Chapter 3 of this work discusses the various maize endosperm mutants or mutant combinations that have an effect on the quantity or the nature of the starch formed. This chapter deals with some of those mutants where the biochemical process affected by the mutation is known. Previous reviews on starch biosynthesis[1–14] cover many of the same subjects described in this chapter.*

II. Starch Synthesis in Plants: Localization

1. Leaf Starch

Starch is present in most green plants and in practically every type of tissue: leaves, fruit, pollen grains, roots, shoots and stems. Starch has a negligible osmotic pressure, which allows plants to store large reserves of carbohydrate without disturbing the cell's water relations. It was demonstrated as early as the nineteenth century that illumination of leaves in bright light causes formation of starch granules in the chloroplast organelle.[15] Disappearance of this leaf starch was shown to occur in low light or during extended periods in the dark (24–48 hours). Starch accumulates due to carbon uptake during photosynthesis, and the starch synthesized under conditions of light is degraded in the dark to products generally utilized for sucrose synthesis. Mutants of *Arabidopsis thaliana* that are incapable of synthesizing starch grow as well as the wild type if placed in continuous light, because they are able to synthesize sucrose in light.[16] However, their growth rate is dramatically decreased if grown in a day–night regime, due to the starch being needed for sucrose synthesis at night, sucrose then being transported from leaf to sink tissues. Biosynthesis and degradation of starch in the leaf is, therefore, a dynamic process having diurnal fluctuations in its stored levels. Starch also plays an important role in the operation of stomatal guard cells.[17] There, it is degraded during the day while the stomata are open, and it is resynthesized in the late afternoon or evening. Leaf starch is lower in amylose content than the starch in storage tissues,[18] with the amylose polymer being smaller than that observed in reserve tissue.

---

*Abbreviations used in this chapter are explained in a list following the references

2. Starch in Storage Tissues

Starch synthesis takes place during development and maturation of storage organs, such as tubers, fruits and seeds. The synthesized starch is then degraded during the time of sprouting or germination of the seed or tuber, or in the ripening of the fruit. The metabolites arising from starch degradation are a source of both carbon and energy. Thus, the starch degradative and biosynthetic processes in the reserve tissues are separated in time. The site of starch synthesis and storage in cereal grains is the endosperm, and the starch granules are synthesized and located in amyloplasts. Starch content in potato tubers, maize endosperm, sweet potato, and roots of cassava and yam varies between 65% and 90% of the total dry weight.

Starch granules in storage tissues vary in composition, shape and size, with shape and size depending on the tissue and plant source, there is a range of sizes and shapes in each tissue. The diameter of the starch granule changes during the maturation of the reserve tissue. The botanical source of the starch can be identified by microscopic examination, due to the fine features present in each species. For example, as seen in potato starch, the 'growth rings' are spaced 4–7 μm apart and have a specific fibrillar organization.

Two polymers are present in the starch granule: amylose, which is essentially a linear polymer; and amylopectin, which is a highly branched polymer. Amylose can be isolated as linear chains of about 840 to 22 000 α-D-glucopyranosyl units linked by (1→4) bonds (molecular size ca. 1.36×10^5 to 3.5×10^6). The number of glucosyl units can vary widely with plant species and stage of development. Amylose molecules may be branched to a small extent (one branch point per 170 to 500 glucosyl units). Amylopectin usually comprises about 70–80% of the starch granule and is much more highly branched, with about 4–5% of the glucosidic linkages being branch points.

Amylopectin molecules are large flattened disks consisting of (1→4)-linked α-glucan chains joined by frequent α-(1→6)-branch points. Many models of amylopectin structure have been proposed. The most satisfactory models are those proposed by Robin et al.,[18] Manners and Matheson,[19] and Hizukuri;[20] all are known as cluster models. Reviews by Morrison and Karkalis[21] and Hizukuri[22] discuss in detail the chemical and physical aspects of the starch granule and its components, amylose and amylopectin.

III. Enzyme-catalyzed Reactions of Starch Synthesis in Plants and Algae and Glycogen Synthesis in Cyanobacteria

The sugar nucleotide utilized for synthesis of the (1→4) α-glucosidic linkages in amylose and amylopectin is ADP-glucose (ADPGlc). ADPGlc synthesis is catalyzed by ADP-glucose (synthase) pyrophosphorylase (ADPGlc PPase) (Reaction 4.1, E.C. 2.7.7.27; ATP:α-D-glucose-1-phosphate adenylyltransferase).

$$\text{ATP} + \alpha\text{-glucose 1-P} \rightleftharpoons \text{ADP-glucose} + \text{PPi} \qquad (4.1)$$

$$\text{ADP-glucose} + (1\rightarrow 4)\text{-}\alpha\text{-glucan} \rightarrow (1\rightarrow 4)\text{-}\alpha\text{-glucosyl-}(1\rightarrow 4)\text{-}\alpha\text{-glucan} + \text{ADP} \quad (4.2)$$

$$\text{Elongated}(1\rightarrow 4)\text{-linked malto-oligosaccharide chain} \rightarrow (1\rightarrow 4, 1\rightarrow 6)\text{ branched }\alpha\text{-glucan} \quad (4.3)$$

Reaction 4.2 is catalyzed by starch synthase (E.C. 2.4.1.21; ADP-glucose: 1,4-α-D-glucan 4-α-glucosyltransferase). Similar reactions are noted for glycogen synthesis in cyanobacteria and other bacteria,[24–26] but the enzyme is referred to as glycogen synthase (also E.C. 2.4.1.21).

Reaction 4.3 is catalyzed by branching enzyme [E.C. 2.4.1.18; 1,4-α-D-glucan 6-α-(1,4-α-glucano)-transferase]. Amylopectin has longer chains (~20–24 glucosyl units) and has less branching [~5% of the α-glucosidic linkages are (1→6)] compared to animal or bacterial glycogen [10–13 glucosyl units and 10% of linkages α-(1→6)]. Thus, the plant branching enzymes may have different properties with respect to size of chain transferred, or placement of branch point, than the bacterial enzyme that produces glycogen. The differences in branching between glycogen and amylopectin may be explained by different interactions of branching enzyme with the different synthases. Possibly, the interaction of starch-branching isozymes with starch synthase isozymes may be different than the interaction of the bacterial glycogen synthase with the respective bacterial branching enzyme or the chain elongating properties of the starch synthases could be different from those observed for the bacterial glycogen synthases. The differences in the catalytic properties of the different plant starch synthases and branching enzymes may also account for the variations observed in different plant starch polymer structures from different plants.

Isozymes of various plant starch synthases[2,3,27–31] and branching enzymes[2,3,27,32–37] have been reported. They are products of different genes and are proposed to play different roles in the synthesis of amylopectin and amylose. In plants,[38–43] as well as in *Chlamydomonas reinhardtii*,[44] a granule-bound starch synthase involved in catalysis of Reaction 4.2 has been shown to be involved in the synthesis of amylose. Mutants of plants defective in this enzyme are known as *waxy* mutants, and give rise to starch granules having low amounts of or no amylose.

Another enzyme, a debranching enzyme, isoamylase, most probably is involved in synthesis of the starch granule and its polysaccharide components.[45–48] Mutant plants deficient in isoamylase activity accumulate less starch granules and accumulate a soluble α-glucan termed phytoglycogen.[5,8,49] Data suggesting the role of a debranching enzyme in synthesis of amylopectin and the starch granule are discussed later (Section 4.12).

The starch synthase reaction was first reported by Leloir et al.[49] with UDP-glucose (UDPGlc) as the glycosyl donor. It was later shown that ADP-glucose was much more efficient in terms of maximal velocity and Km value.[51] Furthermore, leaf starch synthases and the soluble starch synthases of reserve tissues were later found to be specific for ADPGlc. In contrast, the granule-bound starch synthases in reserve tissues have ~1–10% activity with UDPGlc, as compared to the activity with ADPGlc. The Km values for UDPGlc are usually in the mM range, while the Km values for ADPGlc are in the μM range.

IV. Properties of the Plant 1,4-α-Glucan-Synthesizing Enzymes

1. ADP-glucose Pyrophosphorylase: Kinetic Properties and Quaternary Structure

The ADP-glucose pyrophosphorylases (ADPGlc PPase) of higher plants, green algae and the cyanobacteria are under allosteric control. The enzymes are highly activated by 3-phosphoglycerate (3PGA) and inhibited by inorganic orthophosphate (Pi), effects that are important for regulation of starch synthesis. The structural properties of the higher plant enzymes are discussed first. The potato tuber[52-65] and spinach leaf ADPGlc PPases[66-74] have been studied in most detail with respect to kinetic properties and structure. The kinetic and regulatory properties of the ADPGlc PPases from leaf extracts of barley, butter lettuce, kidney bean, maize, peanut, rice, sorghum, sugar beet, tobacco and tomato are similar to those of the spinach leaf enzyme (Table 4.1).[69]

Table 4.1 Kinetic constants from ADPGlc PPases from higher plants, green algae, and cyanobacteria

Source (reference)	Effector	Constant (mM)	n	Activation-fold
Barley endosperm purified[75]	ATP (−3PGA)	0.31	2.1	1.0
	ATP (+3PGA)	0.19	1.0	
	G1P (−3PGA)	0.12	NR[a]	
	G1P (+3PGA)	0.12	NR	
Barley leaves purified[76]	3PGA	0.005	1.0	>20
	Pi	0.025	1.0	
	ATP (−3PGA)	1.0	1.0	
	ATP (+3PGA)	0.08	1.0	
	G1P (−3PGA)	0.33	1.0	
	G1P (+3PGA)	0.11	1.0	
Maize endosperm purified[77]	3PGA	0.12	1.0	25
	3PGA (+Pi, 1 mM)	1.2	1.5	
	Pi (+3PGA, 1 mM)	0.44	1.0	
	ATP (−3PGA)	0.84	1.3	
	ATP (+3PGA)	0.11	1.0	
	G1P (−3PGA)	0.67	0.9	
	G1P (+3PGA)	0.03	1.0	
Tomato fruit[78,b]	3PGA	0.2	NR	Negligible activity in absence of 3PGA
	ATP (+3PGA)	0.12	NR	
	G1P (+3PGA)	0.086	NR	
	Pi (3PGA, 0.5 mM)	0.7		
Wheat endosperm[79]	Pi	0.7	1.3	No effect
	3PGA		No effect	
	3PGA (+Pi, 0.7 mM)	0.81	1.0	
	3PGA (+Pi, 1.5 mM)	1.51	1.4	
	3PGA (+Pi, 5 mM)	3.33	2.5	
	G1P	0.092	1.0	
	ATP	0.12	1.0	
	F6Pc		No effect	
	F6Pc (+Pi, 0.7 mM)	2.5	NR	

(continued)

Table 4.1 (continued)

Source (reference)	Effector	Constant (mM)	n	Activation-fold
Wheat leaf[79]	Pi	0.2	1.2	11
	3PGA	0.01	1.0	
	3PGA (+Pi, 2.0 mM)	1.9	2.3	
	G1P	0.45	1.1	
	G1P (+3PGA)	0.08	1.0	
	ATP	0.73	1.2	
	ATP (+3PGA)	0.22	1.1	
Chlamydomonas[80]	3PGA	0.23	1.3	15
	Pi	0.054	1.0	
	Pi (+3PGA, 2.5 mM)	0.53	1.7	
	G1P	0.22	1.7	
	G1P (+3PGA)	0.03	1.2	
	ATP	0.48	1.2	
	ATP (+3PGA)	0.08	1.3	
Rice endosperm[81]	3PGA	0.65	NR	40
	Pi (+3PGA, 1 mM)	0.40	NR	
	G1P (+3PGA)	0.17	NR	
	ATP (+3PGA)	0.18	NR	
Arabidopsis (recombinant APS1 + APL1)[d,82]	3PGA	0.34	NR	NR
	3PGA (+Pi, 2 mM)	2.7	NR	
	G1P	0.06	NR	
	ATP	0.09	NR	
Arabidopsis (recombinant APS1 + APL3)[83,d]	3PGA	0.02	NR	NR
	Pi (+3PGA, 1mM)	1.2	NR	
	G1P	0.20	NR	
	ATP	0.30	NR	
Spinach leaf[68]	3PGA	0.051	1.0	20
	Pi	0.045	1.1	
	Pi (+3PGA, 1 mM)	0.97	3.7	
	ATP	0.38	0.9	
	ATP (+3PGA)	0.062	0.9	
	G1P	0.12	0.9	
	G1P (+3PGA)	0.035	1.0	
Potato tuber[56,58]	3PGA	0.16	1.0	30
	Pi (−3PGA)	0.04	NR	
	Pi (+3PGA, 3 mM)	0.63	NR	
	ATP (+3PGA)	0.076	1.6	
	G1P (+3PGA)	0.057	1.1	
Anabaena[84]	3PGA	0.12	1.0	17
	Pi (−3PGA)	0.044	1.0	
	Pi (+3PGA, 2.5 mM)	0.46	1.7	
	G1P (−3PGA)	0.13	1.2	
	G1P (+3PGA)	0.08	1.0	
	ATP (−3PGA)	1.55	1.2	
	ATP (+3PGA)	0.46	1.1	
Synechocystis[85]	3PGA	0.81	2.0	126
	Pi (−3PGA)	0.095	1.0	

(continued)

Table 4.1 (continued)

Source (reference)	Effector	Constant (mM)	n	Activation-fold
	Pi (+3PGA, 2.5 mM)	0.57	2.2	
	G1P (−3PGA)	0.18	1.1	
	G1P (+3PGA)	0.05	1.1	
	ATP (−3PGA)	3.2	2.2	
	ATP (+3PGA)	0.80	1.0	

[a] Not reported
[b] Negligible activity in the absence of 3PGA
[c] F6P = D-fructose 6-phosphate
[d] Abbreviations: APSI, Arabidopsis small subunit. APL1 and APL3, Arabidopsis large subunits 1 and 3

The bacterial ADPGlc PPases are homotetrameric.[14] Thus, the catalytic sites, as well as the allosteric sites, reside on the same subunit. In plants and in green algae, however, the ADPGlc PPases are heterotetramers having two different, but homologous, subunits, $\alpha_2\beta_2$, of different molecular sizes.[13,14,53,71,86] The small subunit has the catalytic activity and is ca. 50–54 kDa in size. The large subunit, which is the regulatory subunit, having no catalytic activity, is approximately 51–60 kDa in size.[14,63] The large (regulatory) subunit modulates the sensitivity of the small subunit towards allosteric effectors via large subunit/small subunit interactions.[59]

The ADPGlc PPase from potato tuber has been isolated and extensively characterized.[52,53,55,56] It is composed of two different subunits, 50 and 51 kDa in size and having an $\alpha_2\beta_2$ heterotetrameric subunit structure.[53] Ballicora et al.[55] demonstrated that the small subunit of potato tuber ADPGlc PPase can be expressed as a homotetramer and is highly active in the presence of high concentrations of the activator 3PGA. The large subunit could not be expressed in an active form,[56] suggesting that one subunit, the small subunit, can be designated as the catalytic subunit.

The small subunit of many higher plant ADPGlc PPases is highly conserved among plants, with 85–95% identity.[86] The structure of potato tuber ADPGlc PPase small subunit homotetramer is representive of that of higher plant enzymes. The homotetrameric potato enzyme, composed exclusively of small subunits, has a lower apparent affinity ($A_{0.5}$ = 2.4 mM) for the activator, 3PGA, than the heterotetramer ($A_{0.5}$ = 0.16 mM), and is also more sensitive to the inhibitor P_i ($I_{0.5}$ = 0.08 mM in the presence of 3 mM 3PGA), as compared with the heterotetramer ($I_{0.5}$ = 0.63 mM).[56] The kinetic parameters of the homotetrameric small subunit appear to be nonphysiological and thus, the physiological functional activity is due to the native enzyme in the $\alpha2\beta2$ structure containing both small (catalytic) and large (regulatory) subunits.

In the case of potato tuber ADPGlc PPase, the large subunit greatly increases the affinity of the small (catalytic) subunit for 3PGA and lowers its affinity for the inhibitor, Pi.[55,56] In a plant, there may be only one conserved small (catalytic) subunit and several large (regulatory) subunits that can be distributed in different parts of the plant.[83,87] This is of physiological importance, as expression of different large subunits in different plant tissues may confer distinct allosteric properties to the ADPGlc PPase according to the different needs for starch synthesis in different tissues.

Results of this nature have been shown in the case of *Arabidopsis* ADPGlc Ppase.[83] Co-expression of its small subunit (APS1) with the different *Arabidopsis* large subunits (ApL1, ApL2, ApL3 and ApL4) resulted in heterotetramers with different regulatory and kinetic properties (Table 4.2). The heterotetramer of APS1 with ApL1, a large subunit predominant in leaves,[82,83] had the highest sensitivity to the allosteric effectors 3PGA and Pi, as well as the highest apparent affinity for the substrates ATP and glucose 1-phosphate (G1P). The heterotetrameric pairs of APS1 with either APL3 or APL4, large subunits more prevalent in sink or storage tissues,[87] had intermediate sensitivity to the allosteric effectors and intermediate affinity for the substrates ATP and G1P.[83] APL2, which is also present mainly in sink tissues, had very low affinity for either 3PGA or Pi.[83] Thus, differences in the regulatory properties conferred by the *Arabidopsis* large subunits were found *in vitro*. Distinctions noted in the source and sink large subunit proteins strongly suggest that starch synthesis is modulated in response to 3PGA and Pi, as well as to the substrate levels, in a tissue-specific manner. APS1 and ApL1 would be finely regulated in source tissues by both effectors and substrates, while in sink tissues, the hetrotetramers of APS1 with APL2, APL3 or APL4, which have lower sensitivities to effectors and substrates, would be controlled more by the supply of substrates.[83]

The pattern of expression and sugar regulation of the six *Arabidopsis thaliana* ADPGlc PPase-encoding genes (two small subunits, ApS1 and ApS2; and four large subunits, ApL1–ApL4) have been studied.[87] Based on mRNA expression, ApS1 is the main small subunit or catalytic isoform responsible for ADPGlc PPase activity in all tissues of the plant. ApL1 is the main large subunit in source tissues whereas ApL3 and, to a lesser extent, ApL4 are the main isoforms present in sink tissues. It was also found that sugar regulation of ADPGlc PPase genes was restricted to ApL3

Table 4.2 Kinetic parameters for the 3PGA of *A. thaliana* recombinant ADPGlc PPase determined in the pyrophosphorolysis direction[a,b]

	Control		0.2 mM Pi		2 mM Pi	
	3PGA A0.5, mM	nH	3PGA A0.5, mM	nH	3PGA A0.5, mM	nH
APS1[d]	1.2±0.09	1.8	5.6±0.29	2.5	ND[c]	
APS1/APL1	0.0017±0.0005	1.0	0.019±0.0019	1.9	0.48±0.005	2.7
APS1/APL2	0.219±0.024	0.8	0.820±0.12	0.9	6.95±1.59	1.5
APS1/APL3	0.029±0.009	0.6	0.105±0.025	0.8	0.29±0.048	1.0
APS1/APL4	0.030±0.003	0.9	0.110±0.008	1.0	0.80±0.167	1.2

[a]The kinetic parameters were calculated without inhibitor (Pi) and in the presence of inhibitor at 0.2 mM or 2 mM

[b]The deviation in the 3PGA A0.5 data is the difference between duplicate experiments

[c]ND = not determined

[d]APS1 is the Arabidopsis small (catalytic) subunit and APL1, APL2, APL3 and APL4, the Arabidopsis large subunits

and ApL4 in leaves.[87] Sucrose induction of ApL3 and ApL4 transcription in leaves allowed formation of heterotetramers less sensitive to the allosteric effectors, resembling the situation in sink tissues, which are regulated by an allosteric mechanism (3PGA/Pi ratio).

2. Relationship Between the Small and Large Subunits: Resurrection of ADPGlc PPase Catalysis in the Large Subunit

The similarity between the small (catalytic) and large (regulatory) subunits (~50–60% identity) suggests a common origin.[86] Most probably, gene duplication and divergence has led to different and functional roles (catalytic and regulatory) for the subunits. Subsequently, further divergence of the regulatory genes and their differential expression in plant tissue seems to be the mechanism used for regulation of starch synthesis. It would be of interest to know the evolutionary path that led to the separation of the catalytic and regulatory functions of these subunits. The ancestor of small and large subunits may have been a bacterial subunit, having both catalytic as well as regulatory function in the same subunit. This possibility is supported by the similarity between the two plant subunits with many active bacterial ADPGlc Ppases.[13,14] However, the possibility of the large subunit ancestor being originally noncatalytic as is the current large subunit remains, because of the existence of noncatalytic homologs, such as the products of the glgD genes from gram-positive bacteria.[88]

Evidence has been obtained that the large subunit of potato (*Solanum tuberosum* L.) tuber ADPGlc PPase binds substrates.[58] Thus, the plant heterotetramer, as well as bacterial homotetramers, binds four ADP-[^{14}C]glucose molecules.[58,89] Thus, it can be hypothesized that the large subunit maintained the structure of the substrate site needed for binding, and that the catalytic ability was eliminated by mutations of essential residues. To test this hypothesis, production of a large subunit having significant catalytic activity was attempted via mutating as few residues as possible.[65] Identification of critical missing residues for catalytic activity was the first step for designing a large subunit with activity. Sequence alignments of ADPGlc PPases with reported known activity focused on a few, but proven, candidates with activity. The set of invariants contain most, if not all, of the essential residues. The subset of the ones absent in the large subunit was of particular interest. Lys44 and Thr54 in the large subunit of potato tuber were selected as the best candidates to study, because the homologous residues [Arg33 and Lys43 in the small (catalytic) subunit (Table 4.3)] were completely conserved in the active bacterial and plant catalytic subunits. In addition, Lys44 and Thr54 are in a highly conserved region of ADPGlc PPases (Table 4.3). To convert the modulatory large subunit into a catalytic one, Lys44 and Thr54 were substituted by Arg44 and Lys54, respectively. The mutant, LargeK44R/T54K, was expressed in the absence of the catalytic subunit and no activity was detected. It is possible that the large subunit cannot form a stable tetramer in the absence of the catalytic (small) subunit, as previously shown with the *A. thaliana* enzyme.[83] The activity of the large subunit mutants cannot be readily tested with a co-expressed, wild type, small (catalytic) subunit because of the intrinsic activity of the latter. The regulatory-subunit mutants were co-expressed with smallD145N, an

Table 4.3 Sequence comparison of the potato tuber large subunit with ADPGlc PPases shown to be enzymatically active. Residues that are 100% conserved are in bold and the ones conserved in the small subunit but not in the large subunit are underlined[a]

R. sphaeroides	MKAQPPLRLTAQAMAFV**LAGGRGSR**LKE**LT**DRRA**KPA**VY
R. rubrum	MDQITEFQLDINRALKETLALV**LAGGRGSR**LRD**LT**NRES**KPA**VP
A. tumefaciens	MSEKRVQPLARDAMAYV**LAGGRGSR**LKE**LT**DRRA**KPA**VY
E. coli	MVSLEKNDHLMLARQLPLKSVALI**LAGGRGTR**LKD**LT**NKRA**KPA**VH
E. coli SG14	MVSLEKNDHLMLARQLPLKSVALI**LAGGRGTR**LKD**LT**NKRA**KP**T**V**H
G. stearothermophilus	MKKKCIAML**LAGGQGSR**LRS**LT**TNIA**KPA**VP
Anabaena sp. PCC7120	MKKVLAIIL**GGGAGTR**LYPLTKLRA**KPA**VP
Arabidopsis thaliana small subunit	MAVSDSQNSQTCLDPDASSSVLGII**LGGGAGTR**LYPLTKLRA**KPA**VP
H. vulgare endosperm	LPSPSKHEQCNVYSHKSSSKHADLNPHAIDSVLGII**LGGGAGTR**LYPLTKKRA**KPA**VP
	33 43
Small unit	MAVSDSQNSQTCLDPDASRSVLGII**LGGGAGTR**LYPLTKKRA**KPA**VP
Large unit	MAYSVITTENDTQTVFVDMPRLERRRANPKDVAAVI**LGGGEGT**K**L**FP**LT**SRA**T**P**A**VP
	44 54

[a] The full color version of this table can be found at www.Elsevier.books.com/

inactive small subunit in which the catalytic Asp145 was mutated.[63] In this way, the activity of the small (catalytic) subunit was reduced by more than three orders of magnitude (Table 4.4). Co-expression of the large subunit double mutant LargeK44R/T54K with smallD145N generated an enzyme with considerable activity, 10% and 18% of the wild type enzyme (smallWTLargeWT) in the ADPGlc synthetic and pyrophosphorolytic direction, respectively (Table 4.4). A single mutation (K44R), generated an enzyme (smallD145NLargeK44R) with no significant activity over the control (smallD145NLargeWT). Mutation T54K (smallD145NLargeT54K) provided about 3% of WT activity, but the combination of both mutations in the large subunit (smallD145NLargeK44R/T54K) had the most dramatic effect (Table 4.4). Therefore, it was concluded that Arg44 and Lys54 are needed for restoring catalytic activity to the large subunit. The resurrection of catalytic activity in the large subunit by incorporation of Arg44 and Lys54 predicted that the homologous residues in the small subunit are critical. Replacement of the homologous two residues with Lys and Thr in the small subunit (by mutations R33K and K43T) decreased the activity by one and two orders of magnitude, respectively, in both directions, confirming the hypothesis (Table 4.4).[65] The mutant enzymes were still activated by 3PGA and inhibited by Pi.[65] The wild type enzyme and smallD145NLargeK44R/T54K had very similar kinetic properties, indicating that the substrate site domain has been conserved. The apparent affinities for the substrates in both directions were in the same range, and the allosteric properties of smallD145NLargeK44R/T54K also resembled those of the wild type (Table 4.5).[65] The presence of 2 mM Pi decreased the apparent affinity of 3PGA for both the wild type and smallD145NLargeK44R/T54K, indicating that this new form has a similar sensitivity to Pi inhibition, and that the activator–inhibitor interaction was the same (Table 4.5).

The fact that only two mutations in the L subunit restored enzyme activity is very strong evidence that the large subunit is derived from a catalytic ancestor. To confirm

Table 4.4 Activity of small (catalytic) and large (regulatory) subunit ADPGlc PPase mutants[a]

Subunits		Units/mg	
Small (catalytic)	Large (regulatory)	ADPGlc synthesis[b]	ATP synthesis[c]
WT	WT	32 ± 1	49 ± 2
D145N	WT	0.017 ± 0.001	0.037 ± 0.002
D145N	K44R	0.031 ± 0.001	0.033 ± 0.002
D145N	T54K	0.92 ± 0.08	0.56 ± 0.03
D145N	K44R/T54K	3.2 ± 0.2	9.0 ± 0.7
R33K	WT	3.6 ± 0.2	4.1 ± 0.1
K43T	WT	0.32 ± 0.1	0.28 ± 0.01

[a]The enzyme activities of purified co-expressed small and large subunits were measured for ADPGlc synthetic activity or ADPGlc PPase (ATP synthesis) activity
[b]For ADPGlc synthesis, 4 mM 3PGA (activator), 2 mM ATP and 0.5 mM Glc-1-P were used
[c]For pyrophosphorolysis, 4 mM 3PGA, mM ADPGlc and 1.4 mM PPi were used

Table 4.5 Comparison of the kinetic properties of the wild-type ADPGlc PPase from potato tuber and the mutant smallD145NLargeK44R/T54K

ADPGlc synthesis	SmallWT	LargeWT		Activation	SmallD145N/ LargeK44R/T54K		Activation
	S0.5 /A0.5 (μM)	(nH)		(-fold)	S0.5 (μM)	(nH)	(-fold)
ATP	97 ± 7	1.6			170 ± 13	1.8	
G1P	27 ± 2	1.1			11 ± 1	1.2	
3PGA	135 ± 11	0.9		60 ± 5	13.6 ± 0.7	1.3	20 ± 2
		ADP-glucose pyrophosphorolysis					
ADPGlc	200 ± 20	1.4			70 ± 6	1.5	
PPi	37 ± 2	1.0			98 ± 10	1.0	
3PGA	1.1 ± 0.3	1.1		3.1 ± 0.3	4.0 ± 0.5	0.9	7.6 ± 1.0
3PGA (+ 2 mM Pi)	96 ± 19	1.0			101 ± 13	1.1	

that catalysis occurs in the large subunit of smallD145NLargeK44R/T54K, the substrate site was disrupted in each subunit and their kinetic properties compared. In previous experiments, replacement of Lys198 in the small subunit of the wild type enzyme decreased the substrate (G1P) affinity, whereas disruption of the homologous residue (Lys213) in the large subunit did not have the same effect.[63] In smallD145N-LargeK44R/T54K, the mutation K213R on the large subunit severely decreased the apparent affinity for G1P, whereas mutation of K198R on the small subunit did not (Table 4.6). This indicated that the large subunit double mutant, and not smallD145N, was the catalytic subunit. In the wild type enzyme, Lys213 does not seem to play any important role, but in smallD145NLargeK44R/T54K it recovered its ancestral ability to confer to the enzyme a high apparent affinity for G1P. Previous results showed that Asp145 in the small subunit of the wild type is essential for catalysis, whereas

Table 4.6 Effect of mutations on the glucose 1-P site[a]

Subunits		Glucose 1-P, S0.5		Specific activity, U/mg	
Small	Large	μM	Increase[b] (fold)	ADPGlc synthesis	ADPGlc pyrophosphorolysis
D145N+K44R/T54K/K213R		910±110	82	1.4	4.2
D145N/K198R+K44R/T54K		13.2±0.5	1.2	2.2	6.3

[a]Reaction conditions are as in Table 4.2
[b]The increase values are the ratio of the values obtained with the mutant enzymes to the wild-type enzyme

Table 4.7 The Effects of mutations of residue Asp160 on large subunit activity[a]

Subunits		Specific activity, units/mg
Small	Large	
D145N	WT	<0.001
D145N	K44R/54K	0.35±0.05
D145N	K44R/54K/D160N	<0.002
D145N	K44R/54K/D160E	<0.002

[a]ADP-glucose pyrophosphorylase activity was measured in crude extracts. Saturating concentrations used: for 3PGA, 4 mM, for ADPGlc, 2 mM. The PPi concentration was 1.4 mM

the homologous Asp160 in the Large wild type subunit is not.[63] However, mutation to D160N or D160E in the active large subunit, LK44R/T54K, abolished the activity (Table 4.7). This shows the ancestral essential role of this residue and confirms that the catalysis of smallD145NLarge K44R/T54K does occur in the large subunit.

A comparative model of LK44R/T54K illustrates the predicted role of Arg44 and Lys54 (Figure 4.1). In the model Asp160, which is homologous to the catalytic Asp145 in the small subunit and the catalytic Asp142 in *E. coli* ADPGlc Ppase,[63,90] interacts with Lys54. This type of interaction (Lys54 with Asp160) has also been observed in crystal structures of enzymes that catalyze similar reactions, e.g. dTDP-glucose pyrophosphorylase (dTDPGlc PPase) and UDP-N-acetyl-glucosamine pyrophosphorylase (UDPGlcNAc PPase), and is postulated to be important for catalysis by correctly orienting the aspartate residue.[91–93] Also, Lys54 interacts with the oxygen atom bridging the α- and β-phosphate groups as has been observed in the crystal structure of *E. coli* dTDPGlc Ppase.[93] That contact may neutralize a negative charge density stabilizing the transition state and making the pyrophosphate a better leaving group. Arg44 interacts in the model with the β- and γ-phosphates of ATP, which correspond to the PPi product (Figure 4.1). Likewise, Arg15 in *E. coli* dTDPGlc PPase was postulated to contribute to the departure of Ppi.[91] The kinetic data agreed with the predicted interaction of PPi with Arg44. A Lys in that position, in both the catalytic large subunit mutant and the small subunit, decreased the apparent affinity for

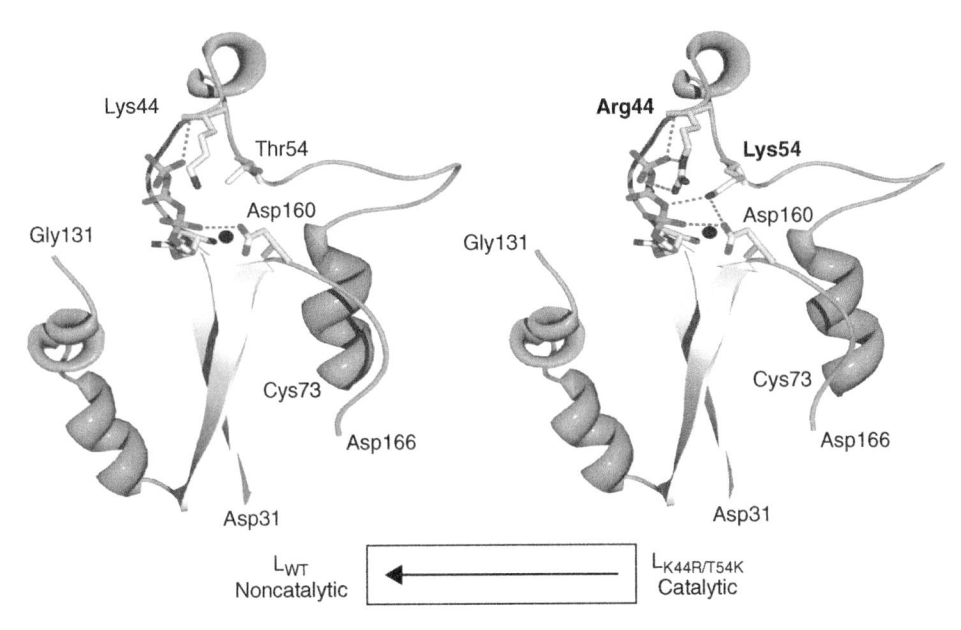

Figure 4.1 Involvement of large (regulatory) subunit of mutant K44R/T54K in enzyme catalysis.[90] The WT and double-mutant large subunits were modeled based on the dTDPGlc PPase and UDPGlcNAc PPases as indicated. Portions of residues 31–73 and 131–136 are shown. The deoxyribose triphosphate portion common to dTTP and ATP is modeled with Mg^{+2} as a blue sphere. The nitrogen atom of the adeninyl group attached to the ribosyl unit is also in blue. The dotted green lines depict hydrogen bonds. (The full color version of this figure can be found at www.Elsevier.books.com/)

PPi at least 20-fold.[65] In the model of the wild type, noncatalytic large subunit, Lys44 and Thr54 cannot interact as Arg44 and Lys54 (Figure 4.1).

3. Phylogenetic Analysis of the Large and Small Subunits

A phylogenetic tree of the ADPGlc PPases present in photosynthetic eukaryotes also sheds information about the origin of the subunits.[65] This tree showed that plant small and large subunits can be divided into two and four distinct groups, respectively. The two main groups of small subunits are from dicot and monocot plants, whereas large subunit groups correlate better with their documented tissue expression. The first large subunit group is generally expressed in photosynthetic tissues,[65] and comprises large subunits from dicots and monocots. Group II displays a broader expression pattern, whereas groups III and IV are expressed in storage organs (roots, stems, tubers and seeds). Subunits from group III are only from dicot plants, whereas group IV are seed-specific subunits from monocots. These last two groups stem from the same branch of the phylogenetic tree, and split before monocot and dicot separation.[65,94,95]

4. Crystal Structure of Potato Tuber ADPGlc PPase

The crystal structure of the potato tuber homotetrameric small (catalytic) subunit ADPGlc PPase has been determined to 2.1 Å resolution.[64] The structures of the enzyme in complex with ATP and ADPGlc have been determined to 2.6 Å and 2.2 Å resolution, respectively. Ammonium sulfate was used in the crystallization process and was found to be tightly bound to the crystalline enzyme. It was also found that the small subunit homotetrameric potato tuber ADPGlc PPase was inhibited by inorganic sulfate with an $I_{0.5}$ value of 2.8 mM in the presence of 6 mM 3PGA.[64] Sulfate is considered as an analog of phosphate, the allosteric inhibitor of plant ADPGlc PPases. Thus, the atomic resolution structure of the ADPGlc PPase probably presents a conformation of the allosteric enzyme in its inhibited state. The crystal structure of potato tuber ADPGlc PPase (Figure 4.2) allows one to determine the location of activator and substrate sites in the three-dimensional structure and their relation to the catalytic residue, Asp145. The structure also provides insights into the mechanism of allosteric regulation.

The overall fold of the potato tuber ADPGlc PPase small subunit catalytic domain is quite similar to that of two other pyrophosphorylases, viz., *N*-acetylglucosamine 1-phosphate uridylyltransferase (GlmU) from *E. coli*[92,96] and *S. pneumoniae*[97,98] and glucose 1-phosphate thymidylyltransferase (Rffh) from *P. aeruginosa*[91] and *E. coli*,[93] although their primary sequences have only very low sequence similarities. The catalytic domain is composed of a seven-stranded β sheet covered by α helices, a fold reminiscent of the dinucleotide-binding Rossmann fold.[99] At one of its ends, the central β-sheet is topped by a two-stranded β-sheet. The catalytic domain makes strong hydrophobic interactions with the C-terminal domain through an α-helix that encompasses residues 285–297 (Figure 4.2). The catalytic domain is connected to the C-terminal β-helix domain by a long loop containing residues 300–320. This loop makes numerous interactions with the equivalent region of another monomer.

The C-terminal domain comprises residues 321–451 and adopts a left-handed β-helix fold composed of six complete or partial coils with two insertions, one of which encompasses residues 368–390. The other encompasses residues 401–431.

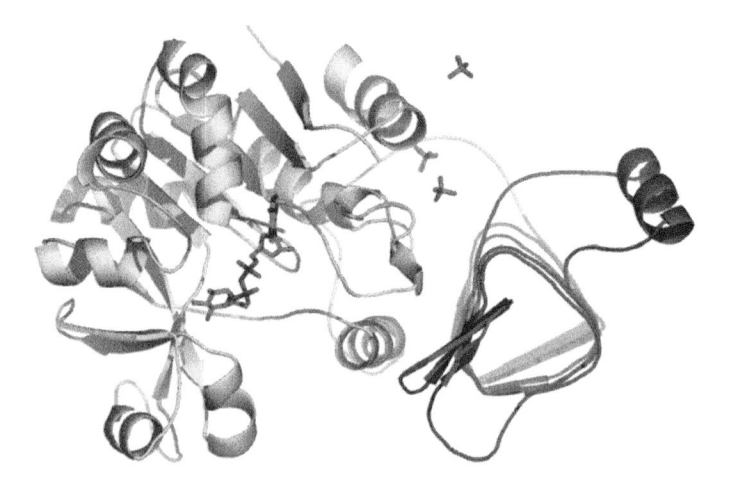

Figure 4.2 Crystal structure of potato tuber ADP-glucose small (catalytic) subunit monomer. The catalytic domain is in yellow and the beta-helix domain is in pink. ADPGlc is shown in atom type: carbon atoms are green, oxygen atoms are red, nitrogen atoms are blue, phosphorus atoms are magenta, and the sulfate group is orange. (The full color version of this figure can be found at www.Elsevier.books.com/)

This type of left-handed β-helix domain fold has been found in the structures of bacterial acetyltransferases, including *E. coli* UDP-*N*-acetylglucosamine 3-*O*-acyltransferase,[100] *Methanosarcina thermophila* carbonic anhydrase,[101] *Mycobacterium bovis* tetrahydrodipicolinate *N*-succinyltransferase[102] and GlmU,[92] and in other proteins such as T4 bacteriophage gp5.[103] However, the β-helix domain seen in the other structures is an acetyltransferase or succinyltransferase domain. In the present structure of ADPGlc PPase, the β-helix domain is involved in cooperative allosteric regulation with the N-terminal catalytic region and interactions with the N-terminal region within each monomer, and contributes to oligomerization.

Homotetramer Structure

Crystalline potato tuber ADPGlc PPase small subunit is a tetramer with approximate 222 symmetry and approximate dimensions of $80 \times 90 \times 110$ Å^3 (Figure 4.3). It can be viewed as a dimer of dimers, labeled A, A′, B and B′ (Figure 4.3). Monomers A and B interact predominantly by end-to-end stacking of their β-helix domains, although there is also a significant interface between the linker loop connecting the two domains (Figure 4.3). This interface buries 2544 Å^2 of surface area. The catalytic domains of A and B′ (and B and A′) also make an extensive interface. Several hydrogen bond and hydrophobic interactions stabilize the interface between A and B′, burying a surface area of 1400 Å^2. All residues defining dimerization interfaces are identical or similar in the large subunit. Figure 4.3 delineates all oligomerization interactions seen within

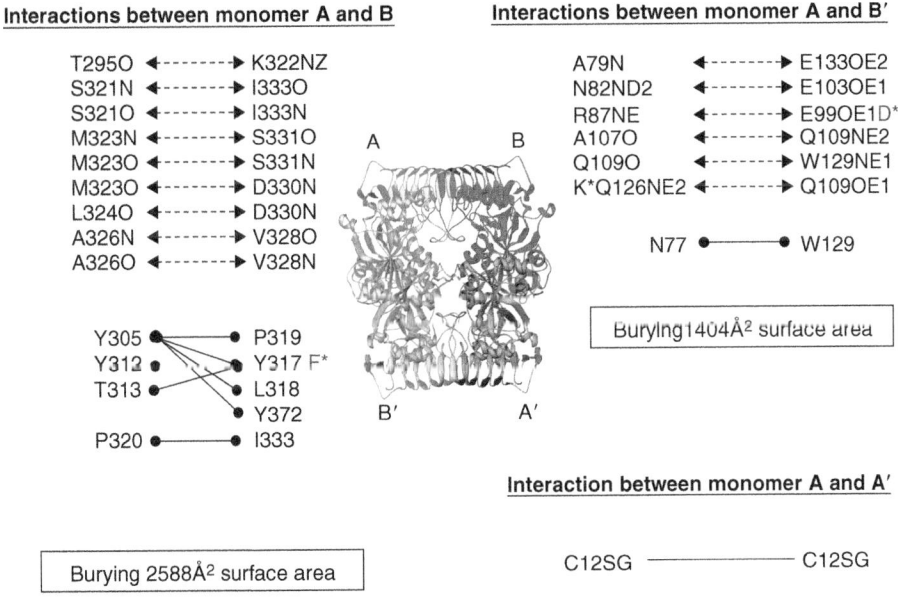

Interactions between monomer A and B

T295O ↔ K322NZ
S321N ↔ I333O
S321O ↔ I333N
M323N ↔ S331O
M323O ↔ S331N
M323O ↔ D330N
L324O ↔ D330N
A326N ↔ V328O
A326O ↔ V328N

Y305 — P319
Y312 — Y317 F*
T313 — L318
      — Y372
P320 — I333

Burying 2588Å^2 surface area

Interactions between monomer A and B′

A79N ↔ E133OE2
N82ND2 ↔ E103OE1
R87NE ↔ E99OE1D*
A107O ↔ Q109NE2
Q109O ↔ W129NE1
K*Q126NE2 ↔ Q109OE1

N77 — W129

Burying 1404Å^2 surface area

Interaction between monomer A and A′

C12SG ——— C12SG

*Residues that are not identical but similar in potato tuber ADP-glucose pyrophosphorylase large subunit.

Figure 4.3 ADPGlc PPase tetramer and interactions between the monomers in the tetramer. The figure shows amino acid interactions between the monomers and the ADPGlc PPase catalytic (small) subunit tetramer. (The full color version of this figure can be found at www.Elsevier.books.com/)

the tetramer in the asymmetric unit. Cys12 of monomer A and the equivalent cysteine residue of monomer A' make a disulfide bond, as do equivalent cysteine residues of monomers B and B'. The inter-subunit disulfide bond between the small (catalytic) subunit is preserved in the heterotetramer. However, there is no disulfide bond between the large (regulatory) subunits, as Cys12 is not conserved. This disulfide bond establishes the relative orientation of the small subunits in the heterotetramer to be like A and A' in the α_4-homotetramer structure. The disulfide bond is the only interaction made between A and A' (or B and B' subunits). Potato tuber ADPGlc PPase is redox-regulated by reduction and oxidation of the intermolecular disulfide bond between the two small subunits.[60,62] This covalent regulatory modification is discussed later.

Sulfate Binding Mimics Phosphate Inhibition

Sulfate is an inhibitor of potato tuber ADPGlc PPase small subunit homotetramer with $I_{0.5}$ = 2.8 mM in the presence of 6 mM 3-PGA.[64] The electron density map for potato tuber ADPGlc PPase small subunit suggests that there are three sulfate ions tightly bound to the enzyme. Most probably, this is due to the high sulfate concentration (150 mM) in the crystallization solution. Two sulfate ions bind within 7.5 Å of each other in a crevice located between the N- and C-terminal domains of the enzyme.[64] A third sulfate ion binds between the two subunits of the enzyme. The sulfate ions make numerous interactions with residues shown to be involved in the allosteric activator binding site, as demonstrated by chemical modification[104,105] and site-directed mutagenesis studies.[59] The structures contain 12 sulfate ions within a tetramer in the asymmetric unit (three per monomer) and are all, therefore, representative of the inhibited conformation of the enzyme.

Sulfate 1 makes hydrogen bond interactions with R41, R53, K404 and K441 (Figure 4.4).[64] The side-chain nitrogen atom of R41 makes hydrogen bond interactions with one of the sulfate ion oxygen atoms, and D403 makes a salt bridge interaction with R41 to facilitate the binding. D413 in the potato tuber enzyme large subunit (D403 in the small subunit) was identified as important for activation by 3PGA.[106] All these residues are conserved in virtually all plant ADPGlc PPases, and four of the five (all but K441) are strongly conserved in bacterial ADPGlc PPases (Figure 4.5). Site-directed mutagenesis studies have identified residues K441 and K404 in the small subunit of potato tuber as important for 3PGA activation.[59] The enzyme's affinity for 3PGA was lowered and the inhibition by Pi diminished when mutations at these residues were Ala (neutral) or Glu (negative). The kinetic parameters for the substrates, ADPGlc, PPi and the cofactor Mg^{2+} were not affected. Mutations on the homologous residues in the large subunit showed lesser or no effects on regulation of the enzyme.[63] Therefore, it was concluded that K404 and K441 in potato tuber ADPGlc PPase small subunit are important for the binding of 3PGA and Pi, and the main role of the large subunit is to interact with the small subunit and modulate its activation mechanism.[59] These studies indicate that the activator 3PGA binds at or near the inhibitor binding site defined in the structure by sulfate 1.

Sulfate 2 makes similar interactions with surrounding positively-charged residues, R53, R83, H84, Q314 and R316 (Figure 4.4). Site-directed mutagenesis studies have

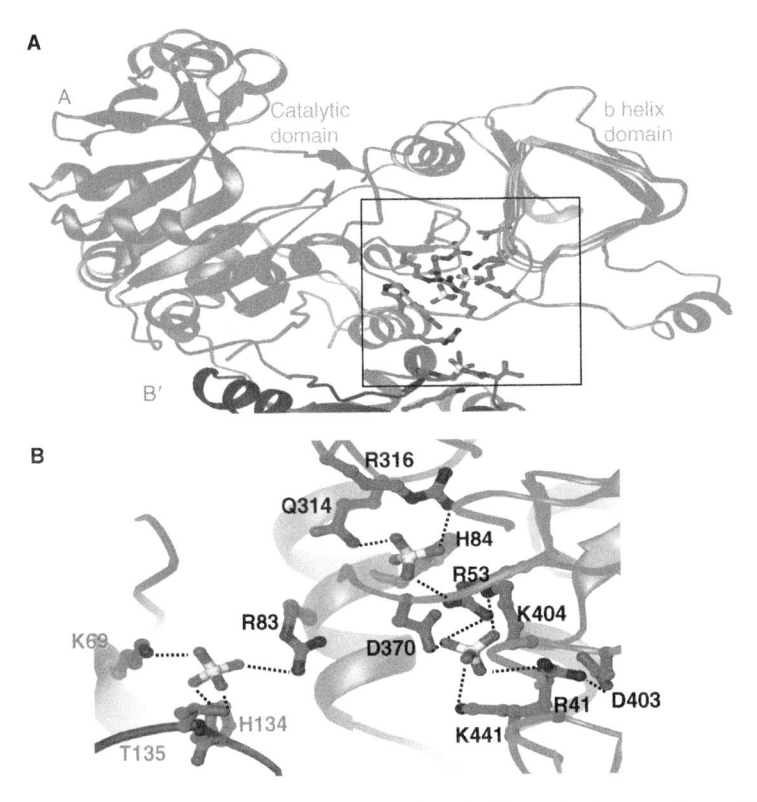

Figure 4.4 ADPGlc PPase monomer showing: (a) the sulfate binding region between the catalytic and beta-helix domains; and (b) the amino acid residues interacting with sulfate. The sulfate residues are yellow and the interacting residues are green in one subunit. The neighboring subunit and its residues are purple. (The full color version of this figure can be found at www.Elsevier.books.com/)

shown that H83 of the *E. coli* enzyme (H84 in the potato tuber enzyme small subunit) is involved in activator binding.[107] Chemical modification with phenylglyoxyl has identified R294 in the *Anabaena* sp. enzyme (R316 in the potato tuber enzyme small subunit) as an important residue for inhibition by Pi, as mutations of this residue lowered the apparent affinity for Pi more than 100-fold.[108] Mutations of this Arg residue to Ala, Glu or Lys caused a change in inhibitor selectivity such that these mutants were inhibited by NADPH or FBP.[109] Taken together, these studies confirm the importance of the sulfate ion-binding site in the allosteric regulation of the enzyme, and indicate that 3PGA may also bind near the sulfate 2 binding site.

Sulfate 3 is located between two subunits, viz., A and B'. This sulfate ion interacts with R83 of one monomer and K69, H134 and T135 of the other subunit. K69 and R83 are conserved in both the small and large subunits of all plant ADPGlc PPases. H134 is conserved in all small subunits, and T135 is conservatively replaced by Asn in other plant small subunits. The precise role of this location is not yet clear. Sulfate binding may be non-specific or it may interfere with the dimerization of the subunits, thus causing the R→T equilibrium to be more toward the T (inhibited) state. Current structural results strongly support previous data on the allosteric regulation of this enzyme, and provide some insights on how the binding of allosteric effectors could affect catalysis.

100 Biochemistry and Molecular Biology of Starch Biosynthesis

```
                                                                    H1  S1
Stu_S         ------------------------------------------MAVSDSQNSQTCLDPDASRSVLGIILGGGAGTRLYPLTKKRAKPAVPLGANYRLI  55
Ath_S_APS1    MASVSAIGVLKVPPASTSNSTGKATEAVPTRTLSFSSSVTSSDDKISLKSTVSRLCKSVVRRNPIIVS-PKAVSDSQNSQTCLDPDASSSVLGIILGGGAGTRLYPLTKKRAKPAVPLGANYRLI 124
Zma_S_endosp  -----------------------------MDMALASKASPPPWNATAAEQPIP---KRDKAAANDSTYLNPQAHDSVLGIILGGGAGTRLYPLTKKRAKPAVPLGANYRLI  79
Hvu_S_endosp  -----------------------------MDVPLASKVPLP--SPSKHEQCNV---YSHKSSS-KHADLNPHAIDSVLGIILGGGAGTRLYPLTKKRAKPAVPLGANYRLI  76
Stu_L         -----------------------------NKIKPGVAYSVITTENDT----QTVFVDMPRLERRRANPKDVAAVILGGGEGTKLFPLTSRTATPAVPVGGCYRLI  72
Eco_glgC      -----------------------------------------MVSLEKNDHLMLARQLPLKSVALILAGGRGTRLKDLTNKRAKPAVHFGGKFRII  54
Rsp_glgC      -----------------------------------MKAQP------PLRLTAQAMAFVLAGGRGSRLKELTDRRAKPAVYFGGKARII  47
Bst_glgC      -----------------------------------------MKKKCIAMLLAGGQGSRLSLTTNIAKPAVPFGGKYRII  39
Cre_S         ---------MALKMRVSQRQALGSQTFVCPHGSVVRKAVSSKARAVSRQAQVVRAQAVSTPVETKVANGVAASSAAGTGQNDPAGDISKTVLGIILGGGAGTRLYPLTKKRAKPAVPLGANYRLI 116
Ana_glgC      -----------------------------------------MKKVLAIILGGGAGTRLYPLTKLRAKPAVPVAGKYRLI  38
Atu_glgC      -----------------------------------------MSEKR------VQPLARDAMAYVLAGGRGSRLKELTDRRAKPAVYFGGKARII  47

        H2       S2       H3        S3             H4           S4            H5       S5
Stu_S         DIPVSNCLNSNISKIYVLTQFNSASLNRHLSRAYASNMGGYKNEGFVEVLAAQQSP--ENPDWFQGTADAVRQYLWLFEEHT---VLEYLILAGDHLYRMDYEKFIQAHRETDADITVAALPMDE 175
Ath_S_APS1    DIPVSNCLNSNISKIYVLTQFNSASLNRHLSRAYASNMGGYKNEGFVEVLAAQQSP--ENPNWFQGTADAVRQYLWLFEEHN---VLEYLILAGDHLYRMDYEKFIQAHRETDADITVAALPMDE 244
Zma_S_endosp  DIPVSNCLNSNISKIYVLTQFNSASLNRHLSRAYGSNIGGYKNEGFVEVLAAQQSP--DNPNWFQGTADAVRQYLWLFEEHN---VMEFLILAGDHLYRMDYEKFIQAHRETNADITVAALPMDE 199
Hvu_S_endosp  DIPVSNCLNSNISKIYVLTQFNSASLNRHLSRAYGSNIGGYKNEGFVEVLAAQQSP--DNPDWFQGTADAVRQYLWLFEEHN---VMEYLILAGDHLYRMDYEKFIQAHRETDADITVAALPMDE 196
Stu_L         DIPMSNCINSAINKIFVLTQYNSAPLNRHIARTYFGNGVSFG-DGFVEVLAANTQTPGEAGKKWFQGTADAVRKFIWVFEDAKNKNIENIVVLSGDHLYRMDYMELVQNHIDRNADITLSCAPAED 196
Eco_glgC      DFALSNCINGIRRMGVITQYQSHTLVQHIQRGWSFFNEEMN---EFVDLLPAQQRM--KGENWYRGTADAVRQTQNLDIIRRYK---AEYVVILAGDHIYKQDYSRMLIDHVEKGARCTVACMPVPI 172
Rsp_glgC      DFALSNCINMSGIRRMGVITQYQSHTLVQHIQRGWSFFNEEMN---EYLDILPASQRV--DENRWYLGTADAVRQTQNIDIVDSYD---IKYVIILAGDHVYKMDYFIMLRQHCETGADVTIGCLTVPR 165
Bst_glgC      DFTLSNCINSGIDTVGVLTQYQPLLHESYIGIGSAWDLDRRN---GGVTVLPPYSVS--SGVKWYEGTANAVKYQNINYIEQYN---PDYVLVLSGDHIYKMDYQHMLDYHIAKQADVTISVIEVPW 157
Cre_S         DIPVSNCLNSNVTKIYCLTQFNSASLNRHLSQAYNSSVGGYNSGFVEVLAASQSS---ANKSWFQGTADAVRQYMWLFEEAVREGVEDFLILSGDHLYRMDYRDFVRKHRNSGAAITIAALPCAE 239
Ana_glgC      DIPVSNCINSEIFKIYVLTQFNSASLNRHIARTYN--FSGFS-EGFVEVLAAQQTP--ENPNWFQGTADAVRQYLWMLQEWD---VDEFLILSGDHLYRMDYRLFIQRHRETNADITLSVIPIDD 155
Atu_glgC      DFALSNALNSGIRRIGVATQYKAHSLIRHLRQRGWDFFRPERN---ESFDILPASQRV--SETQWYEGTADAVRQYNIDIIEPYA---PEYMVILAGDHIYKMDYEYMLQQHVDSGADVTIGCLEVPR 165

         S6  S7     H6         S8    H7          H8    S9
Stu_S         KRATAFGLMKIDEEGRIIEFAEKPKGEQLQAMKVDTTILGLDDKRAKEMPFIASMGIVYVISKDVMNLLRDKFPGAN---DFGSEVIPGATSLGMRVQAYLYDG-----------YWEDIGTIEA 286
Ath_S_APS1    QRATAFGLMKIDEEGRIIEFAEKPKGEHLKAMKVDTTILGLDDQRAKEMPFIASMGIYVVSRDVMLDLLRNQFPGAN---DFGSEVIPGATSLGIRVQAYLYDG-----------YWEDIGTIEA 355
Zma_S_endosp  KRATAFGLMKIDEEGRIIEFAEKPKGEQLKAMMVDTTILGLDDVRAKEMPFIASMGIYVFSKDVMLQLLREQFPEAN---DFGSEVIPGATSIGKRVQAYLYDG-----------YWEDIGTIAA 310
Hvu_S_endosp  ERATAFGLMKIDEEGRIIEFAEKPKGEQLKAMMVDTTILGLDARAKEMPYIASMGIYVFSKDVMHLQLLREQFPEAN---DFGSEVIPGATSTGMRVQAYLYDG-----------YWEDIGTIEA 307
Stu_L         SRASDFGLVKIDSRGRVVQFAEKPKGFDLKAMQVDTTLVGLSPQDAKKSPYIASMGYYVFKTDVLLKLLKWSYPTSN---DFGSEIIPAAIDD-YNVQAYIFKD-----------YWEDIGTIKS 306
Eco_glgC      EEASAFGVMAVDENDKIIEFVEKP-----ANPS--M-----PNDPSKSLASMGIYVFDADYLYELLEEDDRDENSSHDFGKDLIPKITEAGL-AYAHPFPLSCVQSDPDAEPYWRDVGTLEA 282
Rsp_glgC      AEATAFGVMHVDANLRITDFLEKP------ADPPG---I-----PGDEANALASMGIYVFDWAFLRDLLIRDAEDPNSSHDFGHDLIPAIVKNGK-AMAHRFSDSCVMTGLETEPYWRDVGTIDA 275
Bst_glgC      EEASRFGIMNTNEEMEIVEFAEKP------AEP---------KS-----NLASMGIYVFNKKVLIDALTKKRKRPFAYPFEG-----------WKDVGTVKS 250
Cre_S         KEASAFGLMKIDEEGRVIEFAEKPKGEALTKMRVDTGILGVDPATAAAKPYIASMGIYVMSAKALRELLNRMPGAN---DFGNEVIPGADAGFKVQAFAFDG-----------YWEDIGTIEA 350
Ana_glgC      RRASDPGLMKIDNSGRVIDFSEKPKGEALTKMRVDTTVLGLTPEQAASQPYIASMGIYVFKKDVLIKLLKEALERT----DFGKEIIPDAAKD-HNVQAYLFDD-----------YWEDIGTIEA 264
Atu_glgC      MEATGFGVMHVNEKDEIIDFIEKP------ADPPG---I-----PGNEGFALASMGIYVFHTKMEAVRRDAADPTSSRDFGKDIIPYIVEHGK-AVAHRFADSCVRSDFEHEPYWRDVGTIDA 275

         H9                   βH1       βH2       βH3             H10         βH4
Stu_S         FYNANLGITKKPVPDFSFYDRSAPIYTQPRYLPPSKMLDADVTD------SVIGEGCVIKN-CKIHHSVVGLRSCISEGAIIEDSLLMGADYYETDADRKLLAAKGSVPIGIGKNCHIKRAIIDK 404
Ath_S_APS1    FYNANLGITKKPVPDFSFYDRSAPIYTQPRYLPPSKMLDADVTD------SVIGEGCVIKN-CKIHHSVVGLRSCISEGAIIEDSLLMGADYYETATEKSLLSAKGSVTPIGIGKNSHIKRAIIDK 473
Zma_S_endosp  FYNANLGITKKIPDFSFYDRFAPIYTQPRHLPPSKVLDADVTD------SVIGEGCVIKN-CKINHSVVGLRSCISEGAIIEDSLLMGADYYETEADKKLLAEKGGIPIGIGKNSCIRRAIIDK 428
Hvu_S_endosp  FYNANLGITKKPIPDFSFYDRSAPIYTQPRHLPPSKVLDADVTD------SVIGEGCVIKN-CKINHSVVGLRSCISEGAIIEDTLMGADYYETEADKKLLAEKGGIPIGIGKNSHIKRAIIDK 425
Stu_L         FYNASLALT-QEFPEFQFYDPKTPFYTPRFLPPTKIDNCKIKD-------AIISHGCFLRD-CSVEHSIVGERSRLDCGVELKDTFMMGADYYQTESEIASLLAEGKVPIGIENTKIRKCIIDK 423
Eco_glgC      YWKANLDLA-SVVPELDMYDRNWPIRTYNESLPPAKFVQDRSGSHGMTLNSLVSGGCVISG-SVVVQSVLFSRVRVNSFCNIDSAVLL-PE------------WVV---GRSCRLRRCVIDR 386
Rsp_glgC      FWQANIDLT-DFTPKLDLYDREWPIWPYSQIVPPAKFIHDSENRRGTAISSLVSGDCIVSG-SETRSSLLFTGCRTHSYSSSMSHVVAL-PH------------VTV---NRKADLTNCVLDR 379
Bst_glgC      LWEANMDLL-DENNELDLFDRSWRIYSVNPNQPPQYISPEAEVS------DSLVNEGCVVEG-TVERSVLFQGVRIGKGAVVKESVIM-PG-----------AAV--SEGAYVERAIVTP 348
Cre_S         FYNALALTDPEKAQFSFYDEAPIYTMSRFLPPSKVMDCDVNM------SIIGDGCVIKAGSKIHNSIIGIRSLIGSDCIIDSAMMMGSDYYETLEECEYVP---GCLPMGVGDGSIIRRAIVDK 467
Ana_glgC      FYNANLALTQQPMPPFSFYDEEAPIYTRARYLPPTKLLDCHVTE------SIIEGCILKN-CRIQHSVLGVRSRIETGCMIEESLLMGADFYQASVERQCSIDKGDIPVGIGPDTIIRRAIIDK 382
Atu_glgC      YWQANIDLT-DVVPDLDIYDKSWPIWTYAEITPPAKFVHDDEDRRGSAVSSVVSGDCIISG-SALNRSLLFTGVRANSYSRLENAVVL-PS-----------VKI--GRHAQLSNVVIDH 379

         βH5                  S10    S11      βH5     βH6
Stu_S         NARIGDNVKIINKDNVQEAARETDGYFIKSGIVTVIKDALIPSGIII
Ath_S_APS1    NARIGDNVKIINSDNVQEAARETDGYFIKSGIVTVIKDALIPTGTVI
Zma_S_endosp  NARIGDNVKILNADNVQEAAMETDGYFIKSGIVTVIKDALLPSGTVI
Hvu_S_endosp  NARIGDNVMIINVDNVQEAAARETDGYFIKSGIVTVIKDALLPSGTVI
Stu_L         NAKIGKNVSIINKDGVQEADRPEEGFYIRSGIIIILEKATIRDGTVI
Eco_glgC      ACVIPEG-MVIGENAEEDARRFYR---SEEGIVLVTREMLRKLGHKQER
Rsp_glgC      GVVVPEG-LVIGQDAEEDARWFRR---SEGGIVLVTQDMLDARARALN
Bst_glgC      DSIIPPH-SSVCPEDADDVVLV-----TAEWLKQSNEETARKDEA
Cre_S         NARIGPKCQIIINKDGVKEANREDQGFVIKDGIVVVIKDSHIPAGTII
Ana_glgC      NARIGHDVKIINKDNVQEADRESQGFYIRSGIVVVLKNAVITDGTII
Atu_glgC      GVVIPEG-LIVGEDPELDAKRFRR---TESGICLITQSMIDKLDL
```

★ ADPGlc interaction
★ ATP interaction

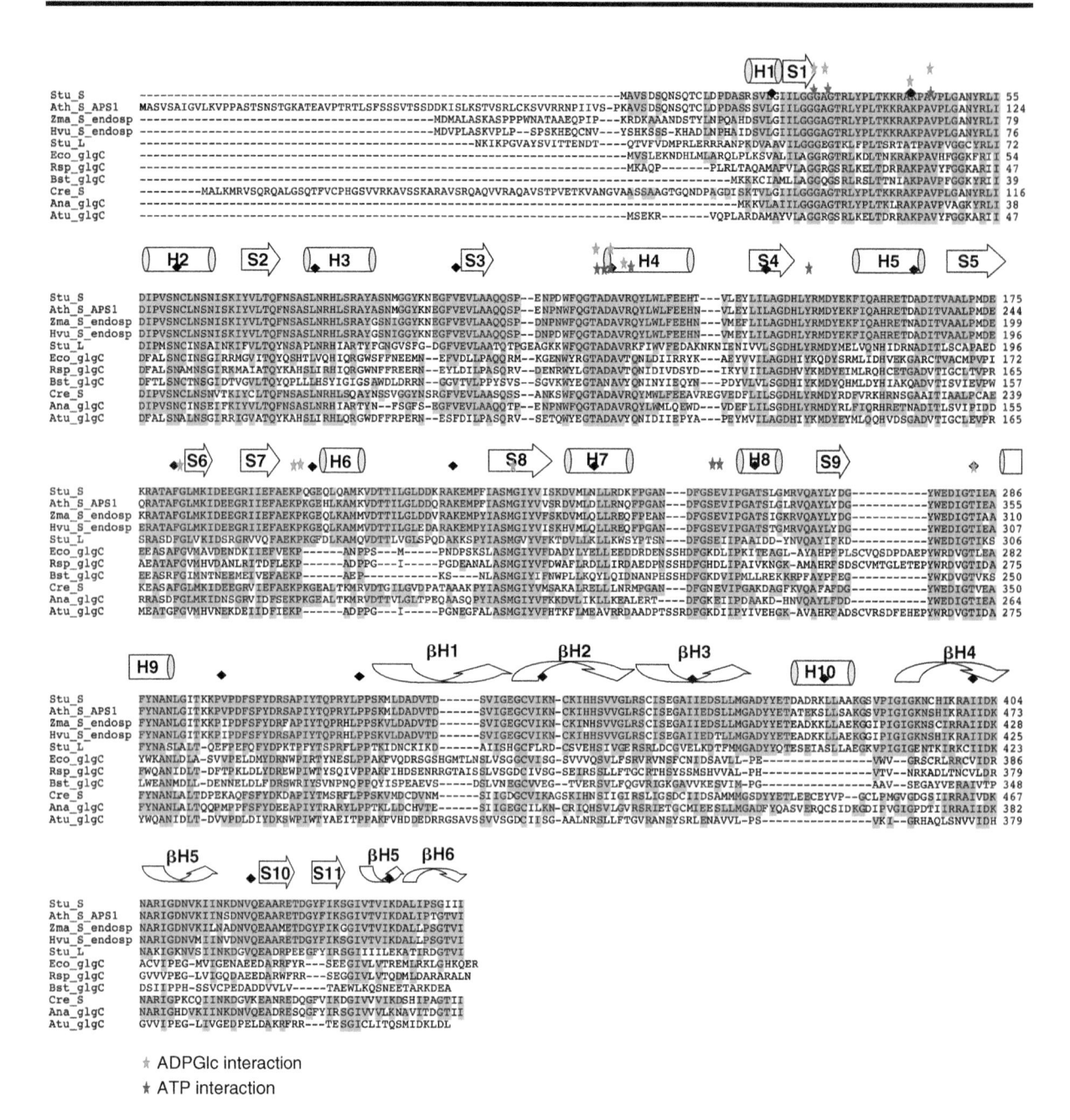

Figure 4.5 Sequence alignment of ADPGlc PPase from different species. Secondary stucture of the potato tuber enzyme is shown above the sequence. Cylinders, helices; straight block arrows, beta strands; curved block arrows, turns in the beta-helix domain. Green stars are residues interacting with ADPGlc and red stars are residues interacting with ATP. Residues that are identical are shaded in purple. Abbreviations are: Stu_s, potato tuber Small subunit; Ath_S_APS1, Arabidopsis thaliana Small subunit; Zma_S_endosp, maize Small subunit; Hvu_S_endosp, Hydra vulgaris Small subunit; Stu_L, potato tuber Large subunit; Eco_glgC, E. coli ADPGlc PPase; Rsp_glgC, R. spheroides ADP-Glc PPase; Bst_glgC, Bacillus stearothermophilus ADPGlc PPase; Cre_S, Chlamydomonas reinhardtii Small subunit; Ana_glgC, Anabaena ADP-Glc PPase; Atu_glgC, Agrobacterium tumefaciens ADP-Glc PPase. (The full color version of this figure can be found at www.Elsevier.books.com/)

ATP Binding

When ATP binds to the enzyme, both A and A' monomers undergo almost identical conformational changes. Several regions move significantly, viz., a loop region from residue 27 to 34, another loop region from residue 106 to 119, and residues K40, R41, Q75 and F76.[64] Both loop regions make direct interactions with the adenine portion of the nucleotide. Specifically, the main chain nitrogen atom of G28 makes a hydrogen bond with N3 of the adenine ring; several hydrophobic interactions are established between the adenine ring and L26 and G29; the side chain of Q118 makes a hydrogen bond with N6 of the adenine ring; and the main-chain oxygen atom of Q118 makes a hydrogen bond with N6. These residues all undergo correlated conformational change upon ATP binding. Interactions of Q75 with both G30 and W116, and interactions of the K40 side chain with P111 couple the motions of the Q75, G30 and 106–119 regions.[64]

Furthermore, ATP binding in the A and A' subunits drives conformational change in the B and B' subunits, as P111 of A and A' is packed snugly against W129 in B' and B, respectively.[64] Motion of P111 accompanying ATP binding leads directly to motion of W129 and, in fact, the entire region from 165 to 231 in the B/B' subunits.

ADP-Glucose Binding

Three of the four subunits (A, A' and B), bind ADPGlc in the ADPGlc PPase/ADP-Glc complex. The B' subunit binds neither ATP nor ADPGlc, and is conformationally more rigid than the other three subunits. A and A' bind ADPGlc identically, and ADPGlc binding produces conformational changes in A and A' identical to that which occurs when ATP binds (described above). The adenyl and ribosyl units of ADPGlc in A and A' are positioned identically to the adenyl unit of ATP, and the interactions between the enzyme and ADPGlc are also identical to those seen in the ATP complex. No electron density is seen for the glucosyl moiety of ADPGlc in the A and A' active-sites, indicating it to be disordered. This indicates that the conformational changes seen in A and A' on ATP or ADPGlc binding are due almost exclusively to the adenyl and ribosyl moieties. Both phosphate groups are also ordered. In contrast, the entire ADPGlc molecule is well-ordered in the B subunit active-site. The adenyl and ribosyl positions are very similar to those seen in the A and A' subunits through the region 112–117, which undergoes conformational change upon ATP or ADPGlc binding in A and A', and is disordered in B and B' both with and without ADPGlc in the active-site. The two phosphate groups and the glucosyl units are very well-ordered in the B active-site and adopt positions and conformations similar to that seen in other sugar nucleotide pyrophosphorylase complex structures (Figure 4.6). There are several direct interactions between the enzyme and the glucosyl unit of ADPGlc. These include hydrogen bonds between E197, S229, D280 and the glucosyl unit (Figure 4.6). In addition K198 makes a salt bridge with the phosphate group attached to the glucosyl unit.

The B subunit undergoes a very large subdomain movement in response to ADPGlc binding. Two residues, E197 and K198, are critical for binding the glucosyl and phosphate moieties of ADPGlc; K198 has been characterized as a Glc1P binding residue

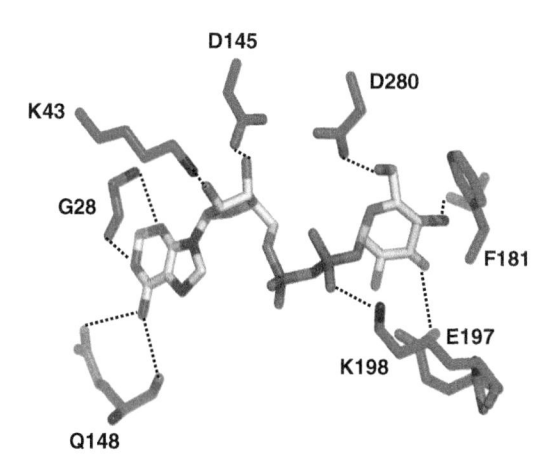

Figure 4.6 Hydrogen bond interactions between ADP-glucose and the ADPGlc PPase catalytic subunit. Protein carbon bonds are green; ADPGlc carbon bonds are yellow; oxygen atoms are red; nitrogen atoms are blue; and phosphate atoms are purple. (The full color version of this figure can be found at www.Elsevier.books.com/)

by site-directed mutagenesis[58] and both are part of a motif present in many sugar–nucleotide PPases. These two residues are shifted out of the binding pocket in the B subunit of the ATP-bound structure, while they are pulled more inward in the unbound B subunit and are pulled in significantly in the ADPGlc-bound molecule (Figure 4.6).

Implication for Catalysis

Detailed kinetic studies on ADPGlc PPase have shown that a sequential bi bi mechanism fits the data with ATP binding first.[75,89,110] Structural data from two related enzymes, GlmU and Rffh, indicate the presence of a metal ion in the active-site.[93,96] A Co^{2+} ion (in GlmU) and an Mg^{2+} ion (in GlmU and Rffh) are located in almost identical locations in the two distinct enzymes when the active-sites are aligned. In GlmU, both the Mg^{2+} and the Co^{2+} are chelated to two conserved residues (Asp105 and Asn227), and to the two phosphate groups of the product (UDP-N-acetylglucosamine). In Rffh, the Mg^{2+} is also bound to two conserved carboxylate residues (Asp223 and Asp108) and to the α-phosphate group of TTP. When the ADPGlc PPase active-site is aligned with these active-sites, two acidic residues, Asp145 and Asp280, are spatially close to the metal chelating residues in Rffh and GlmU. Mutation of Asp145 residue to Asn in the potato tuber enzyme and the equivalent Asp 142 in the *E. coli* enzyme results in a reduction in catalytic activity by four orders of magnitude.[63,90] Taken together, it is concluded that the metal-mediated catalytic mechanism proposed for RffH and GlmU is also used by ADPGlc PPase. Also concluded is that the metal ion is chelated by the residues equivalent to D145 and D280 in all ADPGlc PPases, and that the mutational sensitivity of D145 is due to the requirement for metal ion in the reaction. The absolute requirement for a metal ion has been biochemically demonstrated for ADPGlc PPase from several organisms.[13,14,111] Structural data from GlmU, RmlA and RffH have shed considerable light on the mechanism of sugar nucleotide pyrophosphorylases, and the similarity between the active-site of α_4-ADP-Glc PPase and these enzymes indicates a similar mechanism for all of these enzymes. The structure of the RffH/dTTP complex is particularly informative, because it identifies the GXGXRL loop, which is strongly conserved in all of these enzymes, to be the site of the triphosphate moiety of ATP and shows that the residue equivalent to R33 in the α subunit of ADPGlc PPase is critical for triphosphate binding. Conformational change of this loop will therefore have profound effects on the activity of the enzyme. In addition to R33, D145, D280, K43, E197 and K198 (potato tuber α_4 numbering, Figure 4.6) are also conserved in the other sugar nucleotide pyrophosphorylases of known structures, are in similar locations in the active-sites, and make similar interactions with the sugar nucleotide.

Based on results with other sugar nucleotide pyrophosphorylase structures,[91,93] the following is postulated for catalysis. The β- and γ-phosphate groups of ATP bind to the conserved loop around R33, folding back over the nucleotide and leaving the space opposite the pyrophosphate entity free for G1P binding. The location of the phosphate group in G1P and that of the α-phosphate group of ATP are close to the same positions seen for the phosphate groups in the B subunit ADPGlc complex. K198 is an absolutely conserved amino acid in ADPGlc PPases and in other sugar nucleotidyltransferases. K198 stabilizes the negative charge on the phosphate group of G1P in this model, increasing the nucleophilicity of the O1P atom. Electrostatic repulsions between the negative charges on the phosphate group of G1P and the phosphate groups of ATP are compensated for by chelation between the phosphate groups of ATP, G1P and Mg^{2++}. A number of conserved basic side chains also surround the phosphate groups at the active-site (R33, K43, K198). Additional counterbalancing charges come from the N-terminal dipole of the helical turn at R33 and the main chain amide nitrogen pocket formed by residues G27 to T32. In fact, R33 makes a close hydrogen bond with a phosphate oxygen atom of ADPGlc in the ADPGlc-bound structure. In mutagenesis studies of GlmU,[92] mutation of R15 (equivalent to R33 of ADPGlc PPase) reduced kcat almost 6000-fold, while Km was doubled, confirming that this residue has an important role in orienting and charge compensation of the pyrophosphate group. P36 within this flexible loop adopts a cis-peptide bond. The loop is located at the interface between the N-terminal catalytic domain and the C-terminal β-helix domain within the immediate vicinity of the regulator-binding site, suggesting a possible route for crosstalk between the active-site and the allosteric regulation site. While most of the residues in the active-site are conserved in the catalytically inactive large subunit, there are two changes: R33 is a lysine residue; and K43 is a threonine residue. Since K43 interacts with ADPGlc and R33 is critical for proper ATP triphosphate binding, both are likely to be important in catalytically deactivating the large subunit.

Allosteric Regulation
Crystal structure analysis identified three allosteric inhibitor-binding sites, sulfate 1, sulfate 2 and sulfate 3, two of which (sulfate 1 and sulfate 2) have been shown by numerous biochemical experiments to be involved in allosteric regulation of the enzyme. The high sequence homology of the residues in the sulfate 3 binding site suggests that it too may represent an important allosteric binding site. The major question is how these allosteric binding sites communicate with the active-site to

affect catalytic activity. Part of the answer to this question comes from careful study of the conformational changes that occur in the GXGXXG loop encompassing amino acids 27–34 on active-site binding of either ATP or ADPGlc. This loop has a similar conformation in all unbound subunits, where the loop is flipped into the active-site. A hydrogen bond between K43 and the main chain oxygen atom of either G29 or A30 contributes to the stability of this conformation. Upon binding, however, the loop is forced to move out of the active-site to accommodate ATP or ADPGlc binding and the hydrogen bond is severed. Given that K41 contacts sulfate 1 in the allosteric binding site, conformational change of K41 on replacement of the inhibitor sulfate or phosphate ion with the activator (3PGA) could lead to conformational change of K43, destabilizing the catalytically incompetent, flipped-in conformation of this loop. K43 is also pointing directly into the active-site, and makes a hydrogen bond with the adenyl unit in the ADPGlc-bound B subunit (Figure 4.6). Therefore, conformational change of K43 will also directly affect the active-site. Conformational change of this loop could also come directly from movement of the K41 region, since it is only six amino acids away from the 27–34 loop. As discussed above, the motion of the 27–34 loop is correlated with the motion of the 108–116 loop, which also undergoes conformational change on ATP binding. Conformational change in the K41 region upon activator binding could, therefore, be correlated with conformational change of the 108–116 loop, preorganizing the active-site for ATP binding. In support of this model, the P52L potato tuber large subunit mutant (equivalent to P36 in the small (catalytic) subunit) is substantially less sensitive to 3PGA activation.[57] P36 is a cis-proline residue, and mutation of it is likely to significantly alter communication between the allosteric binding site and the 27–33 GXGXGRL loop.

Several pieces of evidence indicate that inter-subunit motion may also play a role in allosteric regulation of the enzyme. Removal of the C12 disulfide bond, either by mutation or by reduction, results in an enzyme that is less activated by 3PGA,[58,60] indicating that inter-subunit interaction between the catalytic subunits is an important part of the allosteric mechanism. Release of this constraint could increase flexibility and result in more inter-subunit motion. Sulfate 3, which bridges two subunits, may inhibit inter-subunit motion, resulting in a more inhibited enzyme. The binding of ADPGlc to the B subunit, and concomitant motion of the 168–231 subdomain which is required for interaction of E197 and K198 with ADPGlc, results in differences in the subunit interface. Inter-subunit motion may, therefore, drive the motion of this subdomain upon activation, preorganizing the active-site for binding of G1P. The complete mechanism of the allosteric regulation of ADPGlc PPase cannot be deduced until the conformation of the enzyme in its activated state is elucidated. Nonetheless, current structural results strongly support previous data on allosteric regulation of this enzyme, and provide some insights into how the binding of allosteric effectors could affect catalysis.

5. Supporting Data for the Physiological Importance of Regulation of ADPGlc PPase

Ghosh and Preiss[65] were the first to show that the reaction catalyzed by ADPGlc PPase is activated by 3PGA and inhibited by Pi, and is an important step for regulation of starch synthesis in higher plants as well as in cyanobacteria.[84,85] In addition, it has been shown that, in higher plants, the enzyme activity can also be regulated by its reductive state.[60,62]

There is considerable experimental evidence now available to support the view that ADPGlc PPase is an important regulatory enzyme for the synthesis of plant starch. First, one class of a *C. reinhardtii* starch-deficient mutant has been isolated and shown to contain an ADPGlc PPase not activated by 3PGA.[112,113] Evidence for allosteric regulation by ADPGlc PPase being pertinent *in vivo* has also been obtained with *A. thaliana*.[114,115] One mutant, TL25, lacked both subunits and accumulated less than 2% of the starch seen in the normal plant,[114] which would indicate that starch synthesis is almost completely dependent on the synthesis of ADPGlc. The other mutant, TL46, was starch-deficient and lacked the regulatory 54 kDa subunit.[115] The mutant had only 7% of the wild type activity. A later report[116] showed that, in optimal photosynthesis, the starch synthetic rate of the TL46 mutant was only 9% of that of the wild type. At low light, starch synthesis in TL46 was only 26% of the wild type rate. These observations support the view that regulation of ADPGlc PPase by 3PGA and Pi is of physiological importance.

A maize endosperm mutant whose ADPGlc PPase was less sensitive to inhibition by Pi than the wild type enzyme was also isolated.[117] This mutant endosperm had 15% more dry weight and more starch than the normal endosperm. Moreover, in potato tuber[118] and wheat endosperm,[119] genetic manipulation of ADPGlc PPase activity led to an increase in starch production. This has also been shown for rice,[120] as well as for cassava root.[121] These results confirm the view that ADPGlc synthesis, the first unique reaction in plant starch synthesis, is rate-limiting. Also, data obtained with the allosteric mutant ADPGlc PPases of *C. reinhardtii*,[112] maize endosperm[117] and *Arabidopsis*,[114,115] and the resulting effects on starch synthesis provide strong evidence that the allosteric effects observed *in vitro* are operative *in vivo*.

6. Differences in Interaction Between 3PGA and Pi in Different ADPGlc PPases

Almost all plant ADPGlc PPases are activated by 3PGA and inhibited by Pi. The interaction between these allosteric effectors may vary for the different ADPGlc PPases. As previously reviewed,[13] four patterns of interactions can be distinguished between 3PGA and Pi. The four patterns are summarized in Table 4.8. Pi and 3PGA affect the enzyme separately, and increasing concentrations of 3PGA can reverse or antagonize Pi inhibition for most ADPGlc PPases in Group A.

A second pattern, exhibiting distinctive regulatory properties, is seen in Group B ADPGlc PPases, which are found in the reserve tissues of some cereals. The enzymes from pea embryos,[122] barley endosperm,[123,124] bean cotyledon[125] and wheat endosperm[79] may be considered as relatively insensitive to regulation; mainly, they exhibit no activation by 3PGA. However, full characterization of purified ADPGlc PPase of wheat endosperm shows that the enzyme is under-regulated by the coordinate action of various metabolites.[79] The wheat endosperm enzyme is allosterically inhibited by Pi, ADP and Fru 1,6-bisP (Table 4.1). In all cases, inhibition is reversed

Table 4.8 Different interaction patterns between 3PGA activation and Pi inhibition of plant ADPGlc PPases

Group	Principal effector	Secondary effector	Main effect on	Regulatory effect	ADPGlc PPase source
A	3PGA and Pi	Pi and 3-PGA	Vmax	Ultrasensitive interaction between effectors	Cyanobacteria, green algae, spinach leaf, potato tuber
B	Pi	3-PGA	Vmax	3PGA reverses inhibition caused by Pi	Wheat endosperm
C	3PGA	Pi	Vmax	Pi only inhibits the enzyme activated by 3PGA	CAM plants, maize endosperm
D	3-PGA	Pi	Km	3PGA increases affinity for the substrate, ATP, and Pi reverses the effect	Barley endosperm

by 3PGA and Fru 6-P which individually, in the absence of the inhibitors, have no effect on the enzyme's activity.[79] However, activity is affected by a specific 3PGA:Pi ratio in a unique manner (Table 4.8). Indeed, Pi inhibition is the prime signal. The importance of Pi inhibition on the wheat endosperm ADPGlc PPase and its significance on *in vivo* starch accumulation and seed yield has been shown via plant genetic transformation.[119] The reported *in vitro* properties of the wheat endosperm enzyme[79] are congruent with the fact that Pi limits starch biosynthesis in crop plants,[119] and suggest that levels of several metabolites can alter the biosynthetic pathway patterns in the endosperm tissue.

Another variation of 3PGA activation interaction with Pi inhibition is seen in Group C enzymes, represented by CAM plant leaf ADPGlc PPases from *Hoya carnosa* and *Xerosicyos danguyi*[126] and from maize endosperm (Table 4.2).[77] The enzymes can be activated about 10- to 25-fold by 2 mM 3PGA, but in the absence of 3PGA they are insensitive to Pi inhibition. In the absence of 3PGA, the maize endosperm enzyme is only inhibited ~20% by 10 mM Pi, and the CAM plant leaf enzymes ~50% by 2 mM Pi. Further addition of Pi does not increase inhibition. However, with low and subsaturating concentrations of 3PGA (~0.15–0.25 mM) these enzymes become more sensitive to Pi inhibition and become totally inhibited at 0.5–2 mM Pi. As seen with other ADPGlc PPases, 3PGA concentrations can reverse the Pi inhibition and decrease the affinity of the enzymes for Pi.

A fourth pattern of interaction (enzymes of group D) between allosteric activator and inhibitor is seen with barley endosperm. The ADPGlc PPase, which is poorly-activated by 3PGA, is inhibited by Pi (Table 4.2). However, 3PGA lowers (up to 3-fold) the $S_{0.5}$ for ATP (i.e. the apparent affinity of ATP is increased) and the Hill coefficient.[75] At 0.1 mM ATP, activation by 3PGA is about 4-fold; 2.5 mM Pi reverses the effect. Thus, in barley endosperm, the prime effect of 3PGA or Pi may be to either increase or decrease the apparent affinity of the enzyme for the substrate, ATP.

7. Plant ADPGlc PPases can be Activated by Thioredoxin

As described above, the ADPGlc PPase from potato tuber has an intermolecular disulfide bridge linking the two small subunits by the Cys12 residue. This enzyme was activated by reduction of the Cys12 disulfide linkage.[60] At low concentrations (10 μM) of 3PGA, both reduced thioredoxin *f* and *m* from spinach (*Spinacia oleracea*) leaves reduce and activate the enzyme. Fifty percent activation was obtained at 4.5 and 8.7 μM for reduced thioredoxin *f* and *m*, respectively, which are two orders of magnitude lower than obtained using dithiothreitol (DTT).[62] The activation was reversed by oxidized thioredoxin. Cys12 is conserved in the ADPGlc PPases from plant leaves and other tissues, except for monocot endosperm enzymes.[61] In photosynthetic tissues, this reduction could play a role in the fine regulation of ADPGlc PPase mediated by the ferredoxin–thioredoxin system. Until now, this is the only reported covalent modification mechanism of regulation of starch synthesis.

The potato ADPGlc PPase small subunit gene is expressed in tuber, as well as in leaf tissue.[127] Both enzymes are plastidic, with the leaf enzyme being in the chloroplast and the tuber enzyme in the amyloplast.[128] The ferredoxin–thioredoxin system is located in the chloroplast and has been fully characterized.[129] Thus, with photosynthesis, reduced thioredoxin is formed which could reduce and activate leaf ADPGlc PPase. At night, oxidized thioredoxin is formed and would oxidize the ADPGlc PPase, making it less active. This activation/inactivation during the light/dark cycle allows a fine tuning and dynamic regulation of starch synthesis in chloroplasts. Unpublished results (B. Smith-White and J. Preiss) have shown that a recombinant small subunit from the *A. thaliana* ADPGlc PPase can be activated by reducing agents.

For activation of the amyloplastic potato tuber ADPGlc PPase, regulation by thioredoxin is not clear, as a thioredoxin isoform has not been detected in the amyloplasts. However, it is extensively present in many plant tissues and in different subcellular locations, including cytosol, mitochondria, chloroplasts and nuclei.[130] The possibility is that either thioredoxin or an equivalent reductant is present in the amyloplast and is capable of reducing/activating ADPGlc PPase. Amyloplasts are capable of producing NADPH,[131] which could possibly reduce a ferredoxin-like peptide present in the plastids that in turn reduce/activate the plastid ADPGlc PPase. Thus, additional data is required in order to make any conclusion about the role of reductive activation in the regulation of potato tuber ADPGlc PPase *in vivo*. In this respect, it has been shown that removing potato tubers from growing plants results in an inhibition of starch synthesis within 24 hours after detachment.[132] This occurred despite the detached tubers having high *in vitro* ADPGlc PPase activity and high levels of the substrates ATP and G1P and with an increased 3PGA:Pi ratio. Also demonstrated in the detached tubers was that the catalytic subunit in non-reducing SDS-PAGE is solely in the dimeric form and relatively inactive, in contrast to the enzyme form in growing tubers, where it was found to be a mixture of monomers and dimers. The detached tuber ADPGlc PPase had a great decrease in affinity for the substrates, as well as for the activator. Incubation of tuber slices with either DTT or sucrose-reduced dimerization of the ADPGlc PPase catalytic subunit and stimulated starch synthesis *in vivo*. These results strongly suggest that the reductive activation, observed *in vitro*, of the tuber ADPGlc Ppase[60,62] plays a role in regulating starch synthesis.[132] There was a strong correlation between sucrose content in the tuber and the reduced/activated ADPGlc PPase. The precise mechanism of how sucrose causes a reductive activation of the ADPGlc PPase has not been shown.

8. Characterization of ADPGlc PPases from Different Sources

As previously indicated, Table 4.1 summarizes the kinetic and regulatory properties of various ADPGlc PPases, either partially purified away from interfering reactions or purified to homogeneity. Below is a summary of the properties of ADPGlc PPases from algae, leaf and plant reserve tissue.

Chlamydomonas Reinhardtii

The ADPGlc PPase from different green algae was characterized as being highly regulated by 3PGA/Pi.[80,133,134] The enzyme from *C. reinhardtii* has been purified to apparent homogeneity (81 U/mg) and characterized as a heterotetramer ($\alpha_2\beta_2$), typical of the ADPGlc PPase from eukaryotic organisms.[80] A starch deficient mutant of *C. reinhardtii* was shown to contain an ADPGlc PPase that could not be activated by 3PGA, but which exhibited sensitivity to Pi inhibition similar to that of the wild type enzyme.[112]

Spinach Leaves

The demonstration of the heterotetrameric ($\alpha_2\beta_2$) structure of plant ADPGlc PPases was first established for purified (105 U/mg) spinach leaf enzyme.[71] The rate of ADPGlc synthesis is very sensitive to activation by 3PGA and inhibition by Pi. 3PGA increases the enzyme activity 9- and 60-fold at pH 7.5 and 8.5, respectively.[66] In the absence of 3PGA, the $I_{0.5}$ for Pi is 64 and 45 µM at pH 7.3 and 8.0, respectively.[135] Inhibition by Pi is reversed by the activator 3PGA and *vice versa*. The 3PGA:Pi ratio controls the activity of the ADPGlc PPase and the synthesis of starch in chloroplasts.[12] Enzymes from several other plant leaves exhibit similar regulatory properties.[2,69]

Potato Tuber

The ADPGlc PPase from potato tuber behaves similarly to the enzyme from plant leaves. It has been purified to apparent homogeneity (56.9 U/mg), and it was the first plant heteromeric enzyme to be expressed fully active in *E. coli*.[52,53,55,56] Immunolocalization of the ADPGlc PPase in developing potato tuber cells has shown that the enzyme is localized in the amyloplast.[127]

Barley

The barley leaf ADPGlc PPase has been purified to homogeneity (69.3 U/mg), and it shows high sensitivity toward activation by 3PGA and inhibition by Pi.[75] Substrate kinetics and product inhibition studies in the synthesis direction suggested a sequential

Iso Ordered Bi Bi kinetic mechanism.[75] ATP or ADPGlc bind first to the enzyme in the synthesis or pyrophosphorolysis direction, respectively, similar to the *E. coli* enzyme.[89]

Partially purified barley endosperm ADPGlc PPase was shown to have low sensitivity to the regulators 3PGA and Pi.[123] However, 3PGA lowered up to 3-fold the $S_{0.5}$ for ATP and the Hill coefficient, as seen in Table 4.1.[76,123] At 0.1 mM ATP, activation by 3PGA was around 4-fold;[123] 2.5 mM phosphate reversed the effect. Moreover, the degree of the activation by 3PGA was found to be dependent on the conditions (mainly temperature and the presence of Mg^{2+}) under which the enzyme is maintained before assaying it for activity.[76]

A recombinant enzyme with a $(His)_6$-tag from barley endosperm was expressed using the baculovirus insect cell system.[124] It shows no sensitivity to regulation by 3PGA and Pi. However, the enzyme was assayed at saturating concentration of substrates and only in the pyrophosphorolysis direction. For ADPGlc PPases, the synthetic direction is more sensitive to activation. When the recombinant enzyme without the $(His)_6$-tag is expressed in insect cells, the heterotetrameric form still was not activated by 3PGA, nor inhibited by Pi at saturating levels of substrates.[136] Whether 3PGA had any effect on the affinity for the substrates, as shown in the enzyme purified from the endosperm, was not reported. Of interest is that the small (catalytic) subunit, when expressed alone, is very responsive to allosteric effectors.[136] This would suggest that the large subunit in barley endosperm desensitizes the small subunit to activation by 3PGA and inhibition by Pi, which is the opposite to what is seen for large subunits of potato tuber[56] and *Arabidopsis*.[82,83] The small subunit from barley seems to be encoded by a single gene that gives rise to two different transcripts. One of them was found abundantly expressed in starchy endosperm, but not in leaves, while the other was isolated from both tissues.[137]

In green algae and in leaf cells of higher plants, ADPGlc PPase has been demonstrated to reside in chloroplasts.[138] More recently, using plastids isolated from maize and barley endosperm,[139–141] the existence of two ADPGlc PPases, a plastidial form and a major cytosolic form, were found. It was proposed that, in cereals, there are two isoenzymes with different intracellular locations, and they may have different kinetic and regulatory properties.[142] Since starch synthesis occurs in plastids it was further proposed that, in cereal endosperm tissue, synthesis of ADPGlc in the cytosol requires the involvement of an ADPGlc carrier in the amyloplast envelope.[141–143]

Pea Embryos

ADPGlc PPase from developing pea embryos was purified to apparent homogeneity (56.5 U/mg) and found to be activated up to 2.4-fold by 1 mM 3PGA in the ADPGlc synthesis direction.[144] In pyrophosphorolysis, 1 mM Pi inhibited the enzyme by 50%, and 3PGA reversed this effect. The effect of 3PGA or Pi on the $S_{0.5}$ for ATP was not determined.

Three ADPGlc PPase cDNA clones have been isolated from a cotyledon cDNA library. Two of them encode small subunits and the third one a large subunit. The latter showed a greater selectivity in expression than the other two. It was undetectable in leaves, but highly expressed in sink organs.[145]

Tomato Fruit

Three clones encoding different ADPGlc PPase isoforms were isolated from a cDNA library from tomato fruit. Sequence comparison and phylogenetic analysis revealed that all of them represent different types of the large subunits of the enzyme. It was proposed that the three isoforms are organ-specific in their expressions.[146] Four clones were isolated by PCR, three corresponding to large subunits and one to a small subunit.[147] When the enzyme from tomato fruit was purified to apparent homogeneity (45 U/mg), multiple forms were detected by two-dimensional electrophoresis and immunological studies. Three polypeptides corresponded to large subunits, and two corresponded to small subunits. The purified tomato fruit enzyme was highly sensitive to 3PGA/Pi regulation, as is the enzyme from potato tuber (Table 4.3).[78,148]

Maize Endosperm

Partially purified maize endosperm ADPGlc PPase (34 U/mg) was found to be activated by 3PGA and Fru 6-P (25- and 17-fold, respectively) and inhibited by Pi.[77] The heterotetrameric endosperm enzyme has been cloned and expressed in *E. coli*, and its regulatory properties were compared to an isolated allosteric mutant less sensitive to Pi inhibition.[117] As indicated above, the increase of starch noted in the mutant maize endosperm ADPGlc PPase insensitive to Pi inhibition supports the importance of the allosteric effects of 3-PGA and Pi *in vivo*. Also as indicated above, it is believed that the major endosperm ADP-Glc PPase isoform is located in the cytosol.[141]

Wheat

Ainsworth et al.[148,149] isolated two cDNA clones that encode a large and a small subunit from wheat endosperm ADPGlc PPase. The ADPGlc PPase from wheat endosperm has been highly purified (2.44 U/mg) and characterized.[79] It showed novel regulation (Table 4.1). This form is proposed to be the major, or the only one, present in the endosperm at 28 days post-anthesis. However, cytosolic and plastidic isoforms of the enzyme in the endosperm have been separated, but no kinetic characterization of them has been performed.[151] Two isoforms of ADPGlc PPase have been reported in extracts of wheat endosperm.[152] One extract from the amyloplastidial fraction was activated two-fold by 3PGA and inhibited by Pi, while the extract from the whole endosperm exhibited sensitivity toward 3PGA only when Pi was present, as previously described, for the purified enzyme.[79,152] The ADPGlc PPase from wheat leaves was purified (59 U/mg) and characterized as a typical 3PGA/Pi regulated enzyme.[79] The kinetic data for the wheat leaf and endosperm ADPGlc PPases are given in Table 4.1.

Rice

Rice endosperm ADPGlc PPase has been purified to apparent homogeneity (43 U/mg).[153] However, electrophoretic analyses detected multiple isoforms, and no kinetic characterization was reported. However another report shows that the purified endosperm ADPGlc PPase is activated by 3PGA (40-fold) and inhibited by Pi.[81] The allosteric kinetic constants are given in Table 4.1.

Arabidopsis Thaliana

The *A. thaliana* leaf ADPGlc PPase containing the mature forms of one small subunit clone isolated from a cDNA library[82] and one large subunit (adg-2)[154] was expressed in a heterologous system and characterized. The recombinant enzyme exhibited kinetic properties similar to the enzyme purified from leaves.[82] cDNA clones of four large subunit isoforms have been isolated and co-expressed with the small subunit clone.[83] As shown in Table 4.2, the recombinant ADPGlc PPases with different large subunits had different allosteric properties (discussed in Section 4.IV.1).

Vicia Faba

Several cDNA clones encoding two different ADPGlc PPase polypeptides have been isolated from a cotyledonary library of *V. faba* L.[125] Sequences of the cDNAs were closely related to the ADPGlc PPase small subunit of other plants. One polypeptide is expressed in developing cotyledons and leaves, while the other is found solely in cotyledons. Crude extracts were assayed for ADPGlc PPase activity, and it was found that the enzyme from cotyledons is not activated by 3PGA, whereas the leaf enzyme was activated more than 5-fold. Pi inhibited both enzymes. These experiments were performed in the pyrophosphorolysis direction. Whether 3PGA could reverse Pi inhibition was not studied.[125]

9. Identification of Important Amino Acid Residues Within the ADPGlc PPases

Amino acid residues playing important roles in the binding of substrates and allosteric regulators have been identified in the ADPGlc PPases, mainly by chemical modification and site-directed mutagenesis. Thus, photoaffinity analogs of ATP and ADPGlc, viz., 8-azido-ATP and 8-azido-ADPGlc, were used to identify Tyr114 as an important residue in the enzyme from *E. coli*.[155,156] Site-directed mutagenesis of this residue rendered a mutant enzyme exhibiting a marked increase in $S_{0.5}$ for ATP, and a lower apparent affinity for G1P and the activator Fru 1,6-bisP.[136] The Tyr residue must be close to the adenine ring of ATP or ADPGlc, but probably also near the G1P and the activator regulatory sites. The homologous residue in the enzyme from plants is a Phe residue,[86] suggesting that the functionality is not given by the specific residue, but by its hydrophobicity.

Chemical modification studies on *E. coli* ADPGlc PPase that showed involvement of Lys195 in the binding of G1P[157,158] were confirmed by site-directed mutagenesis.[159] Site-directed mutagenesis was also used to determine the role of this conserved residue in the small subunit Lys198 and large subunit Lys213 of potato tuber ADPGlc PPase.[58] Mutation of Lys198 of the small subunit with Arg, Ala or Glu decreased the apparent affinity for G1P 135- to 550-fold. There is little effect on kinetic constants for ATP, Mg^{2+}, 3PGA and Pi. The results show that the Lys198 in the small subunit is directly involved in the binding of G1P. On the other hand, the homologous site residue (Lys213) in the large subunit does not seem to be involved, since similar mutations on Lys213 had little effect on the affinity for G1P.[58] This is consistent with the view that the large subunit is a modulatory subunit and does not have a catalytic role.[56]

Asp142 in *E. coli* ADPGlc PPase in modeling studies was predicted to be close to the substrate site, and this amino acid was identified as mainly involved in catalysis.[90] Site-directed mutagenesis of D142 to D142A and D142N confirmed that the main role of Asp142 is catalytic, for a decrease in specific activity of 10 000-fold was observed with no other kinetic parameters showing any significant changes.[90] This residue is highly conserved throughout ADPGlc PPases from different sources, as well as throughout the super-family of nucleotide-sugar pyrophosphorylases (NDP-sugar PPases).[63] The role of this Asp residue was also investigated by site-directed mutagenesis in the heterotetrameric potato tuber ADPGlc PPase. The homologous residues of the small subunit Asp145 and large subunit Asp160 were separately replaced by either Asn or Glu residues.[63] Mutation of the Asp145 of the small subunit rendered the enzymes almost completely inactive. D145N mutant had a four-order of magnitude decrease of activity, while the specific activity of D145E, a more conservative mutation, was decreased two orders of magnitude. The homologous mutations in the large subunit alone (D160) did not alter the specific activity, but did affect the apparent affinity for 3PGA.[63] Thus, these results agree with the view that each subunit plays a particular role, viz., catalysis for the small subunit and a regulatory role for the large subunit.

Pyridoxal 5-phosphate (PLP), which could be considered to have some structural analogy to 3PGA, was found to activate the enzymes from both spinach leaf and *Anabaena*. In spinach ADPGlc PPase, PLP bound at Lys440, very close to the C-terminus of the small subunit, as well as binding to three Lys residues in the large subunit. Binding to these sites was prevented by the allosteric effector 3PGA, which indicated that they are closely, or directly, involved in binding of this activator.[104,105]

With *Anabaena* ADPGlc PPase, PLP modified Lys419 and Lys382. That these residues were regulatory binding sites was confirmed by site-directed mutagenesis of the *Anabaena* ADPGlc PPase.[160,161] Mutation of the homologous Lys residues in the potato tuber enzyme small subunit (Lys441 and Lys404) indicated that they were also part of the 3PGA binding site in heterotetrameric ADPGlc PPases, and that they contribute additively to the binding of the activator.[59] Moreover, mutation on the small subunit yielded enzymes with lesser affinity to 3PGA (3090- and 54-fold, respectively) than the homologous mutants of the large subunit. Results indicate that Lys404 and Lys441 on the potato tuber small subunit are more important than their homologous counterparts on the large subunit (Lys417 and 455). It seems that the large subunit contributes to enzyme activation by making the activator sites already present more efficient in the small subunit, rather than providing more effective allosteric sites.[59]

Arginine residues were found in ADPGlc PPases from cyanobacteria to be functionally important (as shown by chemical modification with phenylglyoxal).[85,108] The role played by Arg294 in the inhibition of the enzyme from *Anabaena* PCC 7120 by Pi was also determined by Ala scanning mutagenesis studies.[108] More recently, it was found that replacement of this residue with Ala or Gln produces mutant enzymes with a changed pattern of inhibitor specificity, i.e. having NADPH rather than Pi as the main inhibitor.[109] All these results suggest that the positive charge of Arg294 may play a key role in determining inhibitor selectivity, rather than being specifically involved in Pi binding. However, studies on the role of Arg residues located in the N-terminal region of the enzyme from *Agrobacterium tumefaciens* demonstrated the presence of separate subsites for the activators Fru 6-P and pyruvate, and a desensitization of R33A and R45A mutants to Pi inhibition.[109]

Random mutagenesis experiments performed on potato tuber ADPGlc PPase have been useful in identifying residues that are important for the enzyme. Mutation Asp253 on the small subunit showed a specific effect of G1P on the Km, which increased 10-fold with respect to the wild type enzyme.[162] The small magnitude in the increase (only one order of magnitude) would suggest that the Asp253 residue is not directly involved in G1P binding. Remarkably, this residue is conserved in the NDP-sugar PPases whose crystal structures have been solved.[91-93,163] The alignment of Asp253 in the latter, according to the secondary structure elements, suggests that the residue is close to the substrate site without a direct interaction with G1P. This would suggest that substitution of the Asp253 causes an indirect effect on G1P binding by alteration of the structure in the G1P-binding domain. The most interesting finding in the random mutagenesis studies is related to Asp403 (described in the article as Asp413) in the small subunit and its relevance for the normal activation by 3PGA.[106] This residue, adjacent to Lys404 as indicated above, was characterized as a site for PLP binding and 3PGA activation. Also, several modifications in the C-terminal region caused modifications in the regulation of different plant enzymes.[117,164]

The identification of Lys and Arg residues involved in allosteric regulator binding as being located in the C-terminal region in ADPGlc PPases from plant and cyanobacteria constitutes a main difference from what was found for the *E. coli* and *A. tumefaciens* enzymes. In the latter, Lys39 (*E. coli*) and Arg residues in the N-terminal region (*A. tumefaciens*) were characterized as being important for interaction of the enzyme's activators and inhibitors.[111,157,158,165] These results suggest that regulatory domains are located in different sites in the bacterial and plant enzymes. However, results obtained with chimeric ADPGlc PPases from *E. coli* and *A. tumefaciens* have indicated that the C-terminal regions in bacterial ADPGlc PPases are also critical for determining effector affinity and specificity.[166] Thus, most probably regulation of ADPGlc PPases is determined by a combined arrangement and interaction between the N- and C-terminal regions in the protein.

Domain Characterization

It was observed that removal of 10 amino acids from the small subunit of ADPGlc PPase of potato tuber modified the regulatory properties by increasing the apparent affinity for the activator 3PGA and decreasing the one for the inhibitor Pi.[56] Similar results have been observed when the large (regulatory) subunit was truncated 17 amino acids at the N-terminus.[167] Truncation of 11 amino acids from the *E. coli* enzyme also affects its regulatory properties.[168,169] All data agree with the premise that the N-terminal region of the ADPGlc PPase, which is predicted to be a loop, may play a role as an 'allosteric switch' that regulates enzyme activity. This loop possibly interferes with the transition between two different conformations of the enzyme (activated and non-activated). A shorter N-terminus may favor an 'activated' conformation of the enzyme.

10. Starch Synthase

Starch synthase catalyzes the transfer of the glucosyl moiety of the sugar nucleotide ADP-glucose, either to a malto-oligosaccharide or glycogen or the growing starch polymers (amylose and amylopectin), forming a new (1→4)-α-glucosidic linkage (Reaction 4.2). Since the glucosyl unit in ADPGlc is in the form of an α-D-glucopyranosyl unit, and the newly formed glucosidic bond also has the α-D configuration, the starch synthase is considered to be a retaining GT-B glycosyltransferase, according to the nomenclature of Henrissat et al.[169] In bacteria, where the final product is glycogen, there is a similar reaction for synthesis of the (1→4)-α-glucosidic linkage, and the enzyme is referred to as glycogen synthase. There are some differences between the bacterial and plant enzymes. One, in bacteria such as *E. coli*, there is only one glycogen synthase, encoded for by one glycogen synthase gene.[171] However, since 1971, it has been known that, in every plant tissue studied, more than one form of starch synthase can be identified. This has been summarized in a number of reviews.[1–3,5,7,8,12] The plant starch synthases are, therefore, encoded for by more than one gene. Some starch synthases are retained within the starch granule, and are designated as granule-bound starch synthases (GBSS). They may be solubilized by α-amylase digestion of the granule, while others, designated as soluble starch synthases, can be found in the soluble portion of the plastid fraction.

Starch synthase or glycogen synthase activity can be measured by transfer of [^{14}C]glucose from ADPGlc into an appropriate primer, such as amylopectin or rabbit glycogen, followed by precipitation of the labeled polymer.[171,172]

Characterization of the Starch Synthases

A phylogenetic tree based on the various deduced amino acid sequences of starch synthases from crop plants and the green alga, *Chlamydomonas*, has identified five subfamilies of starch synthases.[4] These synthases are designated as granule-bound starch synthase (GBSS), starch synthase I (SSI), starch synthase II (SSII), starch synthase III (SSIII) and starch synthase IV (SSIV). There is also evidence that the SSII class may have diverged further to classes SSIIa and SSIIb.[174] Indeed, the GBSS family may have also diverged into other classes, viz., GBSSI, GBSSIb, or GBSSII.[175–177]

Soluble Starch Synthases

Multiple forms of soluble starch synthases are present in barley endosperm,[178] pea seeds,[43,179] wheat endosperm,[180–182] maize endosperm,[174,183–186] potato tuber,[187–189] *Arabidopsis*,[190] rice seed,[191,192] *Chlamydomonas reinhardtii*,[193] sorghum seeds, teosinte seeds and spinach leaf (the latter three are reviewed in references 2, 7 and 27).

Soluble Starch Synthase I

It is not clear what specific functions the starch synthase isoforms may have in synthesis of the starch granule. There has been considerable effort to isolate mutant plants specifically deficient in one of the isoforms of the starch synthase in order to determine their individual functions. These have provided some insight into the possible functions of the starch synthases, either soluble or granule-bound, in synthesis of both amylose and amylopectin. A mutant of *Arabidopsis*, produced by T-DNA insertion, was deficient in SSI activity and produced amylopectin with an altered structure.[194] Other starch metabolizing enzymes, ADPGlc PPase, branching enzyme, α- and β-amylase, maltase, α-glucanotransferase, pullulanase, phosphorylase and GBSS1, were unaffected by the mutation. The soluble starch synthase activity was reduced by 56–72% in crude extracts; the remaining activity presumably was associated with soluble starch synthases other than SSI. The starch content in the mutant was reduced to 77% that of normal, and the starch had slightly more amylose. The mutant amylopectin fraction was completely debranched with isoamylase and pullulanase, and the chain length distribution compared to the wild type *Arabidopsis* leaf starch. Whereas the wild type starch showed a typical polymodal distribution of chain length with the maximum chain length being DP 11 and DP 12, the mutant (labeled as Atss-1-1) had a unimodal chain length distribution with the most abundant chain lengths being 13–15 glucosyl units long. Further comparisons showed a marked reduction in chains of DP 8–12 in the mutant and a significant increase in chains of DP 17–20.[194] Debranching of the β-amylase limit dextrin of the mutant and wild type starches showed that the mutant starch had more chains of DP <10 and less chains that contained 11 to 24 glucosyl units. The conclusion made was that SSI is involved in incorporating glucosyl units into small chains 'filling up the cluster structure,' but is not involved in making longer interior chains.[194] These results are consistent with previous results,[195] showing that recombinant maize SSI had a reduced catalytic activity of three- to four-fold when the primer glycogen side chains had been lengthened by G1P and phosphorylase action before subjection to starch synthase action. Also, most of the label incorporated into glycogen *in vitro* by SSI was into chains <10 glucosyl units in length.[195] Thus, on the basis of these results, it appears that SSI has a role in extending the shorter A- and B-chains to a length of about 14 glucosyl units or less. Further extension or synthesis of larger chains would have to be done by the other starch synthases. Thus, starch synthase I may be mainly involved in the synthesis of the exterior A- and B-chains of amylopectin. These conclusions are also consistent with the finding that the maximal velocity of maize starch synthase I is about 1.5- to 6.5 fold greater with glycogen, which has shorter chains than amylopectin, than with amylopectin.[173,196] However, the Km for amylopectin was 13-fold lower than for glycogen, suggesting a greater affinity of maize SSI for amylopectin.[196] The Km for ADPGlc was 0.1 mM, and of interest was that citrate at 0.5 M stimulated SSI activity to form a polymer in the absence of primer.[196] This has been confirmed in later studies[183] that also found SS1 to be a monomer of 76 kDa. Another study found that maize SSI was a 77 kDa monomer.[185] The mature *Arabidopsis* SSI enzyme subunit was determined to be 67 kDa.[194] The soluble starch synthase I of rice was estimated to be 55 kDa.[191] These molecular sizes are similar to that which has been found for SSI from pea embryo SSI (77 kDa),[197] wheat endosperm (77 kDa),[181] potato tuber (79 kDa)[187] and potato leaf (70.6 kDa).[189] The expression levels of SSI in potato tubers was different from those found for other starch synthases (GBSSI, SSII, and SSIII),[189] having drastically lower expression levels during development. The expression level was appreciably higher in sink and source leaves, suggesting a minor role in starch synthesis in storage tissue. Moreover, application of the antisense technique reduced the SSI activity to non-detectable levels in tuber tissue.[189] There was no great effect on the starch content or the structure of amylopectin. Potato tuber SSI activity is minor compared to the activities of SSII and SSIII and thus, may play a minor role in tuber starch synthesis, while playing a more important role in leaves.

The gene coding for full-length maize SSI (SSI-1) and genes coding for N-terminally truncated SSI (SSI-2 and SSI-3) have been individually expressed in *E. coli*[198] and purified to homogeneity. The expressed enzymes had essentially the same properties found for other starch synthases I. In summary, with the exception of *Arabidopsis*,[194] the function of soluble SSI in plants remains to be determined.

Starch Synthase II

Starch synthases II have been characterized in several plants plants, including *Chlamydomonas*. In a maize endosperm mutant termed *dull1* (*du1*), because of the dull appearance of its mature kernels,[199] about 15% of the starch polysaccharides are in the form of a slightly altered amylopectin structure, known as 'intermediate material,' that can be distinguished from amylopectin and amylose on the basis of its starch–iodine complex.[200–202] Among many normal and mutant kernel starches studied, *du1* starch had the highest degree of branching.[201,202] If the mutant *du1* allele was combined with other starch mutants, alterations in starch structure that were more severe than those seen in the single mutants occurred[202,203] (see also Chapter 3).

The first enzymic studies done on the *du1* mutant were carried out in 1981, and both SSII and starch-branching enzyme IIa (SBEIIa) were found to have reduced activity in the endosperm compared to normal maize endosperm.[204] SSII was shown to be different from SSI.[173,205] SSII requires a primer for activity, and could not catalyze an unprimed reaction even in the presence of 0.5 M citrate, it also has less affinity for amylopectin than does SSI. However, 0.5 M citrate lowered the Km for amylopectin 17-fold. The activity with glycogen as a primer is one-half that observed with amylopectin. Therefore, glycogen is not as effective as a primer as is amylopectin, which differs from what was observed for starch synthase I. Both maize endosperm SSI and SSII had a Km for ADPGlc of 0.1 mM.[196,205]

The maize *dull1* gene has been extensively characterized.[184] A part of the *du1* locus was cloned by transposon tagging and an almost complete DU1 cDNA sequence determined. It was found that the gene coded for a predicted 1674 amino acid peptide with two regions that had 51% and 73% similarity, respectively, with corresponding regions in SSIII of potato tuber.[188,206] The deduced amino acid sequence of the DU1 cDNA codes for a putative starch synthase with a predicted molecular size of 188 kDa. It is the C-terminal portion beyond amino acid residue 1226 that is most probably where the glucosyl transferase activity resides, as the sequence of 450 residues at the C-terminus of DU1 is similar to the well-known sequences of bacterial glycogen and plant starch synthases. Gao et al.[183] conclude that the starch synthase coded by *Du1* may account for the soluble isoform identified earlier as SSII.[204,205] The only discrepancy is that the deduced molecular size of the DU1 is 188 kDa, while the SSII studied by Ozbun et al. was 95 kDa.[204] However, the deduced molecular size does match the 180 kDa molecular mass maize of SSII as reported by Mu et al.[182] The *Du1* gene product contains two repeated regions in its unique amino terminus, and one of the regions has a sequence identical to a conserved segment of starch-branching enzymes.[184] This may be related to the observation that the *du1* mutation also reduces starch-branching enzyme IIa (SBEIIa) activity. The reduction of SBEIIa activity may be a secondary effect, because of lack of interaction between SSII with SBEIIa in the mutant or because expression of SBEIIa is inhibited due to the lack of SSII.

In further experiments,[185] it was found that DU1 was one of the two major soluble starch synthases, and when the C-terminal 450 residues of DU1 were expressed in *E. coli*, it was shown to have SS activity. Of interest was that antisera prepared against DU1 detected a soluble protein in endosperm extracts of molecular size greater than 200 kDa that was absent in *du1*⁻ mutants.[185] The antisera reduced starch synthase activity by 20–30% in kernal extracts. In the same study, antisera prepared against SSI reduced starch synthase activity by 60%. In *du1*⁻ mutants, antisera prepared against SSI reduced the SS activity essentially to zero, suggesting that SSI and SSII were the only maize endosperm soluble starch synthases. Because of the high similarity in sequence of the DU1 starch synthase II to the potato SSIII, and because both are exclusively soluble, it is argued that DU1 is the evolutionary counterpart of potato SSIII.[185,188,206] It is proposed that DU1 and maize SSII should be known as maize SSIII.[184]

Mutants at the *rug5* locus of pea were isolated after chemical mutagenesis.[207] The mutation caused the peas to become wrinkled-seeded, a sign that they were starch deficient; indeed the three mutant alleles were shown to have 30–40% less starch.[179] Both mature and developing pea embryos were shown to be devoid of SSII.[179] Analysis of other enzymes involved in starch synthesis, such ADPGlc PPase, branching enzyme and starch synthase III, did not reveal any significant changes between wild type and *rug5*. However, there was a 1.7-fold increase in the specific activity of GBSS1. Other enzymes, such as phosphoglucomutase (PGM), UDPGlc pyrophosphorylase and sucrose synthase, were not affected by the mutation.

Analysis of the amylopectin of the *rug5* mutant showed that there were two forms. One co-eluted from a Sepharose column with amylopectin, while the other was of much lower molecular mass, eluting after amylose. Analysis of the mutant amylopectin fractions showed that it had more, or longer, long chains than wild type amylopectin. Debranching of the amylopectin fractions with isoamylase and size-exclusion chromatography showed that *rug5* amylopectin had a higher proportion of very long (B3) chains than did the wild type amylopectin, and that the ratio of short chains (A and B1) to long chains (B2 plus B3) was 4:1 in wild type as compared to 8:1 in the mutant. Also, the average DP of A-chains in *rug5* was lower than in wild type A-chains. Wild type amylopectin showed a maximum of chains of DP 12–15, while *rug5* amylopectin had a broader range of maximum chain size (DP 7–13).[179] Whereas very short chains of 7 to 9 glucosyl units are rare in wild type amylopectin, they are found in great numbers in the SSII mutant. Thus, lack of SSII activity in the pea embryo has a pronounced effect on the amylopectin structure that other starch synthases could not substitute for or replace. It is proposed that SSII is specifically required for synthesis of B2- and B3-chains, but another effect may be that the failure to produce the longer chains in the amylopectin affects the activities of the other starch synthase isoenzymes.

A starch synthase IIa activity has been shown to be lacking in barley *sex6* mutants.[179] The mutation is located at the *sex6* locus on chromosome 7H, and the mutant contained <20% of the amylopectin levels seen in wild type barley. Analysis showed that a 90 kDa band that was missing was attributable to SSIIa, either in the starch granule or in the cytosol. Also, there was an alteration in the distribution of other starch synthases, SSI as well as the branching enzymes BEIIa and BEIIb. In wild type barley grain, they are distributed both in the granule and in the cytosol. In the mutant, they were only present in the cytosol. However, there was no alteration in either their activity or in SSIII activity, which was also present in the cytosol.[178] The mutant had a decrease in amylose content of ~35%, but the amylopectin reduction was more drastic, being 84–91%. Thus, this mutant may be considered to be a high-amylose mutant. Analysis of the mutant amylopectin showed that it had a higher percentage of DP 6–11 chains than the wild type and a lower percentage of intermediate chain lengths (DP 12–30), resulting in a starch with a reduced gelatinization temperature. The granule morphology was also altered. It is believed that all these effects are due mainly to the complete loss of starch synthase IIa activity, but the alteration of the distribution of the other starch synthases and the branching enzymes may also play a role in the synthesis of the altered amylopectin.

This mutation also presents another approach to obtaining a high-amylose starch. As will be described later, mutations of branching enzyme can lead to production of not only more amylose, but also of amylopectin molecules with fewer and longer chains, and having properties similar to those of amylose.

Mutations of SSIIa have been observed in wheat and rice.[208] In wheat, each of the three wheat genomes were mutated to entirely eliminate expression of the SSIIa gene product, Sgp-1 protein in one line;[209] the result was reduced starch amounts and an altered starch structure. In rice, two classes of starch have been found. In Indica rices, the starch is of the long chain variety, while in Japonica, it is of the short chain variety.[208] Genetic analysis showed that the mutation in Japonica rice led to a loss of starch synthase II.[209] Thus, in higher plants, it seems that loss of SSII activity in dicots and SSIIa activity in monocots have the same results with respect to reduced starch content, due to a lowered amount of amylopectin and altered amylopectin chain size distribution. Thus, these genes may have the same function in starch biosynthesis.

A mutant of starch synthase II has also been isolated in *Chlamydomonas reinhardtii*;[193] this mutant, *st-3*, accumulated only 20–40% of the amount of starch present in the wild type. The enzyme lacking in the mutant was starch synthase II, one of the two starch synthase isoforms present in *C. reinhardtii*.[193] There was an apparent increase in the amylose content, and in a modified form of amylopectin. The changes noted in the mutant amylopectin were a decrease in the number of intermediate-size chains (DP 8–50) and an increase in short chains (DP 2–7). The conclusion made was that this starch synthase II was responsible for synthesis of intermediate-size chains, a similar conclusion reached as indicated above for the function of SSII and SSIIa in higher plants. The SSII of *Chlamydomonas* is now referred to as SSIII.[4]

Starch Synthase III
The major isoform of soluble starch synthase in potatoes, labeled as starch synthase III, has a molecular size of 139–140 kDa.[188,206] SSIII was expressed at the same level in all developmental stages, and in contrast to GBSS1 and SSII, SSIII was highly expressed in sink and source leaves.[206] In leaf disks, both GBSSI and SSIII expression was induced by sucrose addition under light.[206] Antisera to SSIII reduced the total starch synthase activity by ~75%, indicating that it was the major activity form of the soluble starch synthases.[188] A cDNA clone of the starch synthase was isolated[188,206] and analyzed; its transcript predicted a protein of 1230 amino acids with a molecular size of 139 kDa.[188] The N-terminal sequence of about 60 amino acids was suggestive of a transit peptide.[188] The N-terminal extension showed little sequence similarity to that of SSII from either pea or potato.[188] However, from amino acid residue 780 to the C-terminal end, the sequence was similar to both soluble and granule-bound starch synthases, and even bacterial glycogen synthases.[188] Indeed, the KTGG motif that is involved in ADPGlc substrate binding[210] is conserved as KVGGL at residue 794. Also conserved is the FEPCGL sequence starting at residue 1121, which has been shown to contain a Glu residue that is essential for the activity of *E. coli* glycogen synthase.[211] Using the antisense RNA approach to SSIII, it was found that SSSIII transcripts were eliminated.[188,206] There was no effect on the levels of the other starch synthases, and there was no effect on the amount of starch produced or in the amylose content in these antisense plants, as compared to the wild type.[188,206] However, what was noted in both reports was a drastic alteration in the starch granule morphology.[188,206] The conclusions made were that SSIII was a major factor for synthesis of starch in granules with normal morphology, but the exact role it plays in the synthesis of starch remains obscure.

Subsequently, the endosperm of hexaploid wheat, *Triticum aestivum* [L.], was shown to have a starch synthase III gene.[182] A cDNA was isolated and contained an open reading frame for a 1629 amino acid polypeptide. The N-terminal region started with the transit peptide region of 67 amino acids, the N-terminal region of 656 residues, the SSSIII-specific region of 470 amino acids, and then the 436 amino acid catalytic domain in the C-terminal region. These domains were compared to those seen in starch synthases III from maize (DU1 protein),[184] potato,[188,206] cowpea,[212] and *Arabidopsis*.[213]

Starch Synthase IV
A novel class of starch synthase amino acid sequences designated as SSIV has been reported in expressed sequence tag (EST) databases from several species, including *Arabidopsis*, *Chlamydomonas*, wheat and cowpea.[4] However, the SSIV gene product had not been isolated or characterized.[4] The starch synthases show high similarity to each other. Using *A. thaliana* SSIV as 100%, cowpea SSIV has 71% identity (accession number AJ006752), wheat SSIV has 58% identity (accession number

AY044844), rice SSIV-1 has 57% identity and rice SSIV-2 has 58% identity (accession numbers AY373257 and AY373258, respectively). However, BLAST analysis of the rice genome using the conserved catalytic C-terminus of starch synthase amino acid sequences then showed not only two SSIII homologous genes, but also two genes homologous to wheat SSIV.[214] cDNA clones of SSIV-1 and SSIV-2 were isolated and expressed in *E. coli*[214] and were expressed 2.7- and 2.4-fold, respectively, in activity over the baseline level of *E. coli* glycogen synthase, suggesting that the SSIV genes encoded starch synthase activity. SSIV-1 was expressed mainly in endosperm and weakly in leaves, while SSIV-2 was expressed mainly in leaves and weakly in endosperm.

The rice SSIV enzymes contained three distinct regions. A putative transit peptide region of 78 amino acids for SSIV-1 and 33 amino acids for SSIV-2, a region homologous to the Smc/myosin tail, and then a C-terminal catalytic domain region resembling bacterial glycogen synthase.[216] As seen with starch synthase III, the KTGG motif that is involved in ADPGlc substrate binding[210] is conserved as KXGGL, and the FEPCGL sequence that has been shown to contain the Glu residue important for catalysis for the *E. coli* glycogen synthase[211] is present in the C-terminal region. The SSIV gene encodes an enzyme with a predicted molecular size, including the transit peptide, of about 118 kDa.

The pattern of expression of starch synthase IV in rice was also studied and compared to the expression of the other starch synthases.[215] Ten genes were identified to be in the starch synthase gene family.[215] They were grouped on the basis of sequence analysis as SSI, SSII, SSIII, SSIV and GBSS.[215] Of interest were the different patterns of temporal expression for the various starch synthases during seed development. The early expressers that were expressed in the early stage of grain filling were SSII-2, SSIII-1 and GBSSII. Those expressed in the mid- to later-stage of grain filling, the late expressers, were SSII-3, SSIII-2 and the third group, GBSSI, SSI, SSII-1, SSIV-1 and SSIV-2 were expressed relatively constantly during the entire period of grain filling.[214]

Recently, a mutant of *A. thaliana* deficient in SSIV has been isolated.[216] As seen in rice,[214] the SSIV gene contains 16 exons separated by 15 introns.[214,216] Western blots of wild type *A. thaliana* showed a starch synthase of 112 kDa, but the two mutant alleles showed no presence of starch synthase IV in Western blots.[216] The mutant alleles showed lower growth rates under a 16-hour day/8-hour night photoregime when compared to the wild type. There was also a decrease in leaf starch of 35–40%. When the SSIV protein was restored by transformation to the mutant, both growth rate and starch levels were restored, indicating that the mutant alterations were due to SSIV deficiency. Normal growth of the mutant could also be restored by growing it in continuous light. Because of a reduced rate of starch synthesis, the levels of sucrose, fructose and glucose were higher in the mutant. The mutant alleles had no decrease in total starch synthase activity. Nor did they exhibit any significant decrease in other starch metabolizing enzymes. Upon analysis of the mutant starch, it was found that there was no change in the amylose/amylopectin ratio and there were only minor effects on the amylopectin structure (a slight decrease in the number of DP 7–10 chains). However, there were significant alterations in the morphology of the starch granules with respect to size, and in the number of granules in the chloroplast. Whereas normal chloroplasts would have 4–5 starch granules per chloroplast the SSIV mutants had only one. The mutant starch granules were considerably larger than those of the wild type. Curiously, not only was the rate of starch synthesis reduced, its rate of degradation was also reduced. This was attributed to having only one granule, with less overall surface.[216] Thus, it was proposed that the function of SSIV was to establish an initial structure for starch synthesis. The interaction of SSIV with other starch synthase isoforms would be of interest to study their specific involvement in starch granule synthesis. As will be discussed below, there is evidence that granule-bound starch synthase may also be involved in amylopectin synthesis, in addition to amylose synthesis. Thus, the mode of amylopectin synthesis by both soluble and granule-bound starch synthases is still unknown and remains an active research area.

Double Mutants of the Soluble Starch Synthases

In summary, various alterations have been observed in the amylopectin structure produced by the various starch synthase isoforms. SSI is involved in extending the shorter A- and B-chains to a maximum length of 14 glucosyl units.[194,195] SSII is involved in synthesis of B2- and B3-chains.[178,179,193,208,209] SSIII is a major factor for synthesis of starch with normal morphology. The exact role it plays, however, in the synthesis of starch remains obscure.[188,206] SSIV is proposed to form an initial structure in starch, i.e. initiate starch synthesis. However, the role or sequence or protocol in the synthesis of the starch granule is unknown for these individual starch synthases. Their specific contribution to the structure of amylopectin is unknown. Their role in starch synthesis may depend on their activity at particular times *in vivo*, on the relative activities of other starch synthases and branching enzyme isozymes, and on their particular interaction with these enzymes and the growing amylopectin structure. Some attempt to understand the starch synthase pathway of amylopectin synthesis was attempted by making transgenic potatoes where the total SSII and SSIII activities were reduced 31–80%[217] or 36–91%[218] by the antisense technique. Starch synthases II and III made up more than 90% of the total soluble starch synthase activity in the tuber;[188,206] GBSS and branching enzyme activities were not affected. In one report,[217] the starch granules of the double mutant were compared with those of wild type and single SS mutants. First, starch granule morphology of the SSII mutants were essentially similar, but the SSSIII mutant[188] and the SSII/SSSIII double mutants had abnormal morphology.[217] Granule morphology in the SSII/SSIII mutant differed from that of the single SSIII mutant in that scanning electron microscopy showed holes through the granule center that were not observed in the single SSIII mutant. An analysis of the amylopectin branch chains of the mutants showed that all had a greater concentration of shorter chains and a lower content of longer chains than did the wild type. However, the patterns of long and short chain contents were quite different. SSII and SSIII showed great enrichment of DP 9 and DP 6 chains, while SSII/SSIII had enrichments of DP 7, 8, 12 and 13 chains. There was no relationship between the chain pattern observed for SSII/SSIII and the cumulative pattern obtained for the mutant SSII and SSIII chains.[217] In addition, the gelatinization behaviors of the mutant starches were all different, signifying different alterations in granule structure.[217]

These results strongly suggest that the different isozymes of starch synthases have unique functions in the synthesis of amylopectin.[217] As indicated before, the SSII and SSIII mutants do not affect the rate of starch synthesis, but the mutants have different effects on starch structure, and the isozymes of the double mutant SSII/SSIII act in a synergistic manner in the synthesis of amylopectin and not in an independent manner. This conclusion is based on the facts that the morphology of the starch granules seen in the double mutant is unlike those of the single SSII and SSIII mutants, and that amylopectin chain lengths also differ. Moreover, this can be explained as follows: in the single mutant, SSII or SSIII, the function observed for the remaining active enzyme is different in wild type than in the double mutant.[217] This is most probably due to the activity of the starch synthase being dependent on the type of growing glucan acceptor it is presented with. Since other starch synthase isoforms are present, the glucan polymer that is synthesized in the absence of one of the starch synthases may be different, and would modify the activity of other starch synthases and even branching enzyme activity.[218]

Thus, the synthesis of amylopectin or starch is not the sum of the independent actions of the various starch synthase and branching enzyme isoforms, but is due to a complex combination of the activities of the various isoforms that vary during seed and leaf development due to varying expression of the different isozymes. Essentially, the same conclusions were reached previously.[218]

Starch Synthases Bound to the Starch Granule

Starch synthase activity was first discovered by Leloir's group, and the activity was associated with the starch granule.[50,51,219] These starch synthases are designated as granule-bound starch synthases (GBSS) to distinguish them from the starch synthases mainly found in the soluble phase of the chloroplast or amyloplast. The original characterization of GBSS[50,219] was made using UDPGlc as the glucosyl donor, but subsequently ADPGlc was found to be a superior glucosyl donor, reacting at a 10-fold faster rate than UDPGlc at 12.5 mM concentration.[51] Maize GBSS was released from the starch granule by incubating ground maize starch granules with α-amylase and glucoamylase.[172,173] The solubilized starch synthase activity was partially purified. Two peaks of activity were obtained with 80% of the activity residing with GBSSI that eluted from the DEAE-cellulose column at a lower salt concentration than did the GBSSII fraction.

The solubilized granule-bound enzyme showed a Km value for ADPGlc about 10-fold lower than that measured before solubilization (0.96 mM for the intact granule activity, unpublished results). The granule-bound enzyme is also active with UDPGlc,[172] with 1 mM UDPGlc having about 7% of the activity rate seen with 1 mM ADPGlc. If the concentration of UDPGlc is raised to 20 mM, then the rate of activity is about 73% of that of ADPGlc. Upon solubilization with α-amylase and glucoamylase, activity with UDPGlc essentially vanished. Either the activity observed with UDPGlc

Table 4.9 Kinetic parameters of the solubilized granule-bound starch synthase of maize endosperm[175]

Property	Solublized GBSS	
	SSI	SSII
K_m, ADPGlc (mM)	0.14	0.11
K_m, amylopectin (mg/mL)	1.2	0.26
Relative activity: amylopectin = 1.0		
Rabbit liver glycogen	1.6	0.8
Maltose, 1 M	2.9	0.51
Maltotriose, 0.1 M	1.6	1.4

was not solubilized, suggesting that a different enzyme catalyzed the UDPGlc activity or it became inactive during the treatment with amylase. Alternatively, the ability of the starch-bound starch synthase to utilize UDPGlc is dependent on the close association of the enzyme with the starch granule. In other words, the conformation of the GBSS is altered in the presence of starch to allow UDPGlc to be catalytically active.

The kinetic parameters of solubilized GBSS are given in Table 4.9. The solubilized, starch-free enzymes require a primer for activity. GBSSII, in contrast to SSII, has a higher apparent affinity (lower K_m) for amylopectin than the granule type-I enzyme. Soluble GBSSI has a higher activity with rabbit liver glycogen than with amylopectin, whereas soluble GBSSII has less activity with rabbit liver glycogen than with amylopectin. The solubilized granule starch synthases I and II can utilize oligosaccharide primers, as do the soluble starch synthases (Table 4.9 shows the activities with respect to maltose and maltotriose). Other malto-oligosaccharides are also used as primers. The products of reaction with maltose, maltotriose, maltotetraose, maltopentaose, maltohexaose and maltononaose are the malto-oligosaccharides with an additional glucosyl unit (for example, the primer maltotetraose when glucosylated yields maltopentaose).

The major granule starch synthase, GBSSI, has a mass of 60 kDa, while the solubilized GBSSII has a mass of ~93 kDa.[173]

Antibody prepared against SSI effectively neutralized souble SSI activity, but had no effect on either the activities of GBSSI or GBSSII, or even on SSII. Alternatively, antibody prepared against the starch granule-bound proteins effectively inhibited GBSSI activity, but had very little effect on the soluble starch synthases.[173]

Isolation of the Waxy Protein Structural Gene

Amylose content determines the degree of translucency of the endosperm (hence the name 'waxy' when amylose is absent), and it affects the cooking and eating qualities of the grains and the industrial properties of the starch extracted from those grains.

It is widely accepted that GBSS activity is a function of the protein coded by the waxy gene. The waxy locus gene product is a protein of molecular weight 58 KDa that is associated with starch granules and is similar to that found for the solubilized maize endosperm GBSSI.[173] This protein has been extracted by heating the starch with a solution of SDS or by incubation at 37°C with 9 M urea, but starch synthase activity is lost under those conditions. In mutants containing the wx allele, there is virtually no amylose, GBSS activity is very low,[38,221,222] and the waxy protein is missing.

Shure et al.[38] prepared cDNA clones homologous to Wx mRNA. Later, restriction endonuclease fragments containing part of the Wx locus were cloned from strains carrying the controlling activator (ac) wx-M9, wx-M9 and wx-M6 alleles to further characterize the ac and dissociation (ds) insertion elements.[222] Excision of the ds element from certain wx alleles produces two new alleles (S5 and S9) encoding the wx proteins having altered starch synthase activities.[223] Two of these, S9 and S5, had 53% and 32% of the starch synthase activity, respectively, of the normal endosperm. Mutant S9, with the higher starch synthase activity, had 36% of the amylose content observed in the non-mutant endosperm, while mutant S5, with an even lower starch synthase activity (32%), had only 21% of the non-mutant maize amylose content. The correlation between the amount of GBSS activity and amylose content provided additional evidence that the waxy protein is involved in amylose synthesis.

The DNA sequence of the Waxy locus of *Zea mays* was determined by analysis of both a genomic and an almost full-length cDNA clone.[224] Also, the barley Waxy locus has been cloned and its DNA sequenced.[225] The deduced amino acid sequences of the maize and barley clones can be compared with the amino acid sequence of the *E. coli* ADPGlc-specific glycogen synthase.[171] Both clones had the sequence seen in the bacterial enzyme starting at residue Lys 15, …KTGGL… As indicated before, the lysine residue in the bacterial glycogen synthase has been implicated in the binding of ADPGlc.[210,211] Moreover, the finding of similarity of sequences between bacterial glycogen synthase and the putative plant starch synthases provides more and strongly suggestive evidence that the waxy gene is indeed the structural gene for granule-bound starch synthase.

Developing pea embryo starch contains starch synthase activity that is associated with the waxy protein. The MW of the pea starch synthase is about 59 kD. The starch synthase activity was solubilized by amylase treatment and partially purified.[226] The solubilized pea GBSS preparation displayed a relatively high specific activity (>10 μmol glucose incorporated per min per mg protein). Protein staining or immunoblotting, following SDS-PAGE, with maize Wx antibody, showed only the Wx protein.[226] Thus, the biochemical examination of starch synthase present in starch granules from two species, maize and pea, strengthens the genetic evidence supporting the role of the Wx protein as a granule-bound starch synthase with a major role in the determination of the amylose content of starch.

Further Studies of Gbss and Isoforms: Their Involvement in Both Amylopectin and Amylose Synthesis

Granule-bound starch synthases have been studied in a number of plants. In many cases, two isoforms, GBSSI and GBSSII, have been identified. In pea embryos, there are two forms that are associated with starch, GBSSI and SSII. However, SSII also contributes much of the soluble starch synthase activity of the embryo;[227] its properties were discussed as a soluble starch synthase in Section 4.IV.10d. Clones of the two isoforms of GBSSI and GBSSII from pea embryo[41] and potato tubers were isolated[41,43] and characterized. Whereas the GBSSI of both potato and pea embryo were very similar in deduced amino acid sequence to the waxy proteins of maize and barley, the clones of GBSSII from potato and pea were different in sequence, but similar to each other. The difference from GBSSI was the extra N-terminal domain of 203 amino acids that is hydrophilic, having basic amino acids to give it a net positive charge and being serine-rich.[43] Both forms of GBSS had the KTGGL ADPGlc polyphosphate binding domain[210] and the important SRFEPCG-residue (E) domain for activity.[211] These clones were used to determine the temporal levels of expression during development. Pea GBSII is highly expressed earlier in development than GBSSI, and the expression is much lower in organs of pea other than the embryo. In developing potato tubers, GBSSI increased in expression during development similar to patatin, while the GBSSII expression was highest in very young tubers and declined in larger potatoes.[227] cDNAs of GBSSI and GBSSII from wheat have also been isolated and characterized.[178] GBSII was expressed in leaf, culm and pericarp tissue, but not in endosperm, where the expression of GBSSI transcripts were high. Thus, expression of the two isoforms maybe tissue- or organ-specific. Wheat GBSSI and GBSSII were 66% identical in sequence.

The pea embryo and potato tuber forms of GBSSI have been studied *in vitro* to determine the mechanism in how it may synthesize amylose.[228] Amylose was synthesized by the granules, but this was prevented by preincubating the granules with α-glucosidase, suggesting that the precursor was a soluble oligosaccharide. Amylose was then synthesized only when malto-oligosaccharides were added to reaction mixtures.[228] GBSSI had higher affinity for the malto-oligosaccharides than GBSSII (or soluble starch synthase II), and transfer of the glucosyl moiety from ADPGlc was different for the two starch synthases.[229] A series of malto-oligosaccharides were synthesized by GBSSI, suggesting that transfer was not stepwise and that more than one glucosyl unit was added before the oligosaccharide dissociated from the enzyme.[229] With GBSSII, only one glucosyl unit was added to the malto-oligosaccharide substrate.

In further studies, both potato tuber GBSSI and pea embryo SSII were expressed in *E. coli* in order to obtain them in soluble form.[230] It was immediately recognized that the GBSSI within the granules had properties that were different from those of the soluble form. The affinity for maltotriose and the ability to form amylose was less for granule-bound GBSSI compared to its soluble form. Whereas the Km value for maltotriose was 0.1 mM for GBSSII, the soluble enzyme form was not even saturated at 1 mM. The processive order of glucosyl addition was also lost. Of interest was the interaction of amylopectin with potato tuber GBSSI;[230] it acted not only as a substrate, but also as an effector. Amylopectin was an effector at lower concentrations than where it was a substrate. It increased the affinity for another substrate, maltotriose, as well as the rate of reaction with the trisaccharide, over 3-fold. The reaction mode for maltotriose reverted to processive in the presence of amylopectin. It is postulated that the amylopectin induced a conformational change in GBSSI, reverting it to a form it had when it was granule-bound.[230]

The GBSS in leaves can be differentiated from those observed in storage tissue. As has been shown, the amylose content of starch can vary from 11% to 37% in storage tissue.[203] In leaf or transitory starch, the amylose content is less than 15%. In pea, leaf starch amylose has a greater molecular weight than does storage starch amylose. Based on size-exclusion chromatography, leaf amylose had a molecular weight (MW), based on dextran standards, of 655 kDa, as compared to pea embryo amylose, which had a molecular weight of 470 kDa.[175] The cDNAs of GBSSIa from embryo and GBSSIb from leaves were isolated and compared. The predicted amino acid sequence of GBSSIb indicated a chain of 613 amino acids with a molecular weight of 67.6 kDa, and a 68.8% identity and 75.6% similarity with GBSSIa. Mature GBSSIb (after transit peptide cleavage) has a molecular weight of 58.4 kDa, while mature GBSSIIa has a molecular weight of 58.3 kDa. Both cDNAs were transformed in *E. coli*, and their properties characterized. Unfortunately, embryo GBSSIa was inactive, while embryo GBSSIb was active.[175] The GBSSIb activity was compared to potato tuber GBSSI, also expressed in *E. coli*, and was found to be similar in properties with respect to Km values for ADPGlc and amylopectin. Both GBSSIa and GBSSIb activities were highly activated when reincorporated into starch granules.[176] The two isoforms synthesized distinct isoforms of amylose with the Ia type forming a shorter type of amylose than the Ib form. These results explain why the *lam* mutation of pea embryos lacking GBSSIb still synthesizes 4–10% of the starch content, as amylose and minor amounts of Ib are present in the embryo.[207]

In *Arabidopsis* leaves, the mechanism of amylose synthesis was also studied with a view to identifying possible primers for amylose synthesis.[231] Malto-oligosaccharides, such as maltotriose, when added to *Arabidopsis* leaf starch granules and incubated with ADP[^{14}C] Glc, stimulated the synthesis of labeled amylose. These types of experiments had also been done with starch granules from pea and potato,[228] and from *C. reinhardtii*.[232] A malto-oligosaccharide accumulating *Arabidopsis* mutant, *dpe1*, also synthesized more amylose compared to the wild type.[231] This strongly suggested that malto-oligosaccharides were indeed the primers for amylose synthesis. This, however, immediately raises the question of the origin of the malto-oligosaccharides in the plant.

Finally, the contributions of studies of starch synthases made in *Chlamydomonas reinhardtii* should be noted. As indicated before SSSII, now known as SSIII, may be involved in the synthesis of the intermediate-size chains of amylopectin.[193] Also, the first evidence that GBSSI was involved in synthesis of amylopectin in addition to amylose was obtained in *C. reinhardtii*.[233,234] It was first noted that *sta2* mutants were deficient in GBSSI and had, in addition to loss of amylose, a fraction of altered amylopectin structure with longer chains.[235] It has also been shown in potatoes that GBSSI can elongate amylopectin chains.[228] Further experiments in *C. reinhardtii* showed that, in double mutants lacking both SSIII and GBSSI, an altered starch structure intermediate between glycogen and starch was formed.[234]

The cloning of *C. reinhardtii* GBSSI has been achieved and it codes for a 69 kDa protein with an extra 11.4 kDa extension at the C-terminus.[235] The *C. reinhardtii sta2* mutants transformed with the GBSSI cDNA could now synthesize amylose. An *in vitro* synthesis of amylose was studied by starch granules isolated in either WT log phase or in a growth-arrested (nitrogen-limited) ADPGlc PPase mutant.[235] The GBSSI specific activity in this system is 10–50 times higher than that seen in higher plants, and enabled the researchers to increase *in vitro* the amount of glucan in the granule ~1.6-fold. The percentage of amylose increased from 13% to 45% in the granule. X-ray diffraction studies showed an increase in the appearance of crystalline material.[235] A-type crystals remained constant and the total amount of B-type crystals increased from 7% to 33% of the total starch, suggesting that GBSSI induces *de novo* synthesis of B-type crystallites.

Amylose synthesis depends on the concentration of ADPGlc, as GBSSI has a high Km for the substrate as compared to soluble starch synthases.[236] PGM-deficient mutants do make starch, but are deficient in amylose, even though GBSS is present. The PGM deficient mutants can make amylopectin, but not amylose, as shown by detailed structure studies of the starch accumulated by the algal mutant.[237] A similar structure effect can be seen when the algae has defective ADPGlc Ppase.[236] This indicates that the relatively high Km value seen for ADPGlc for GBSSI *in vitro* is most likely also the *in vivo* Km value for GBSS. Other studies also show that a diminution of ADPGlc levels in cells causes a lowering of amylose to very low or nonexistent levels.[238,239]

Amino Acids Involved in Substrate Binding and Catalysis
Very little information is known about the structure of the catalytic site of the plant starch synthases. Because of the high expression of activity one can obtain for *E. coli* glycogen synthase, and its great similarity to the plant starch synthases, knowledge of the structure–function relationship in bacterial glycogen synthases may provide a better understanding of the nature of the catalytic- and substrate-binding residues. Up until 2004, few studies had been done to elucidate possible substrate-binding residues in either bacterial glycogen synthases or plant starch synthases. Studies showed that ADPGlc protects Lys15 in *E. coli* glycogen synthase from reacting with pyridoxal phosphate.[210] Moreover, replacement of Lys15 by Arg, Gln or Glu increases the $S_{0.5}$ 7-, 32- and 46-fold, respectively.[210] In the starch synthase IIa from maize endosperm, Arg-specific modification experiments were performed with phenylglyoxal.[240] However, the studied Arg residues are not 100% conserved, and when mutated no significant shifts in $S_{0.5}$ for ADPGlc were shown. Early experiments had shown that DTNB could inactivate *E. coli* glycogen synthase, and that this inactivation was prevented by the presence of ADPGlc.[241] In order to determine which of three Cys residues were reactive with DTNB, the three Cys residues were replaced with the amino acid, serine.[211] Substitution of Cys379 affected major loss of activity. In addition, the kinetic properties of the mutants were analyzed to ascertain the functional role of the Cys involved. Mutation of Cys379 to Ser379 caused an increase in the Km of ADPGlc 38-fold and lowered Vmax about 6-fold.[211] Since Cys379 is in putative loop with conserved residues that could be interacting with the substrate, the Cys379 region was studied by site-directed mutagenesis.[211] Table 4.10 shows that the sequence surrounding the affected Cys residue is 100% conserved in both soluble and granule-bound starch synthases. Substitution of Glu377 by Ala and Gln decreased Vmax more than 10 000-fold. The Km for ADPGlc was only affected 3.4-fold. Substitution by Asp lowered the Vmax only 57-fold. This suggested that the negative charge is essential for catalysis and that Glu377 is critical for catalytic activity. This Glu residue is also 100% conserved in the soluble SS and GBSSI enzymes.

Table 4.10 Alignment of amino acid sequences of bacterial glycogen synthases and plant starch synthases in the region of the conserved putative loop. Residues with gray background are 100% conserved. The number before the source indicates the number of amino acids of the protein, while the number before the sequence is the first aligned amino acid of each protein. The secondary structure prediction of the *E. coli* sequence is on top of the alignment: E: β-sheet; C: loop; H: helix

Accession no.		Bacterial		EEEECCCCCCCCHHHH
PO8323	(477)	*Escherichia coli*	(369)	VILVPSRFEPCGLTQL
P39670	(480)	*Agrobacterium tumefaciens*	(368)	AIIIPSRFEPCGLTQL
O08328	(485)	*Geobacillus stearothermophilus*	(370)	LFLIPSLFEPCGLSQM
Q985P2	(481)	*Mesorhizobium loti*	(368)	AIIIPSRFEPCGLTQL
Q8YVU5	(472)	*Anabaena sp.*	(378)	MIVVPSNYEPCGLTQM
Q8XPA1	(482)	*Clostridium perfringens*	(376)	IFLMPSLFEPCGLGQL
Q97QS5	(477)	*Streptococcus pneumoniae*	(370)	LFLMPSRFEPCGISQM
Q9KRB6	(484)	*Vibrio cholerae*	(376)	FFLMPSEFEACGLNQI
Q8KAY6	(488)	*Chlorobium tepidum*	(384)	ILLMTSRIEACGMMQM
		Plant (soluble)		
CAB69545	(649)	*Zea mays*	(533)	ILLMPSRFEPCGLNQL
BAA07396	(626)	*Oryza sativa*	(525)	VLLMPSRFEPCGLNQL
		Plant (granule bound)		
S07314	(605)	*Zea mays*	(473)	VLAVTSRFEPCGLIQL
AAN77103	(609)	*Oryza sativa*	(477)	VLAVPSRFEPCGLIQL
AC006424	(610)	*Arabidopsis thaliana*	(478)	FIIVPSRFEPCGLIQL
AAL77109	(603)	*Hordeum vulgare*	(471)	LLAVTSRFEPCGLIQL
CAA41359	(607)	*Solanum tuberosum*	(475)	FMLVPSRFEPCGLIQL

Further proof that Glu377 is critical for catalytic activity was done by substituting Glu377 with Cys and then modifying the Cys residue with iodoacetate. The Cys377 mutant was completely inactive, but when modified with iodoacetate, the activity was completely restored.[211] This effect was probably due to the fact that the iodoacetate-modified Cys residue had a carboxyl group that is necessary for enzyme catalysis at residue 377. Other conserved residues in the sequence, Ser374 and Gln383, were also replaced via mutagenesis with other amino acids and found to be not as essential for catalysis or substrate binding.[211]

The bacterial glycogen synthase, as well as the starch synthases, are classified as retaining glycosyl transferases since the α-linkage of the glucosyl bond of ADPGlc is retained as an α-glucosidic linkage in glycogen or starch. Modeling studies of the *E. coli* glycogen synthase shows that its active-site is similar to the active-site of retaining GT-B glycosyl transferases, such as maltodextrin phosphorylase and trehalose phosphate synthase.[211,242,243] Secondary structure prediction analyses and threading also predicted a GT-B fold for the bacterial glycogen synthase.[244,245] Recently, the crystal structure of *A. tumefaciens* glycogen synthase has been solved and shown to have a GT-B fold.[246] It has been predicted that the plant starch synthases would have this GT-B fold, and thus studies on structure–function relationships of the bacterial glycogen synthase would be important for obtaining information on the starch synthases.

11. Branching Enzyme

Purification and Characterization of Branching Enzyme Isozymes

Assay procedures: there are three accepted methods in assaying branching enzyme (BE). One measures the decrease in absorbance of the glucan-I_3^- complex that results from the branching of amylose or amylopectin. The assay mixture containing amylose or amylopectin is incubated and aliquots are taken and iodine reagent is added.[35,247] The decrease of absorbance is measured at 660 nm for amylose and at 530 nm for amylopectin.

The phosphorylase-stimulation assay[247–250] is based on the stimulation by branching enzyme of the 'unprimed reaction in absence of primer' activity seen with rabbit muscle phosphorylase *a* activity. Branching enzyme present in the reaction mixture increases the number of non-reducing ends available to phosphorylase for elongation.

In contrast to the above two assays, the branch-linkage assay[250] measures the actual number of branch chains formed by branching enzyme catalysis. BE is incubated with, a $NaBH_4$-reduced amylose, and after the reaction is stopped by heating the product containing new branches is debranched with purified *Pseudomonas* isoamylase. The reducing power of the liberated oligosaccharide chains formed by the enzyme is measured by a highly sensitive reducing sugar assay. Reduction of amylose with borohydride results in about 2% of the reducing power of the non-reduced amylose, resulting in lower blanks.

The phosphorylase-stimulation assay is the most sensitive, particularly if labeled G1P is used. The branch-linkage assay is the most quantitative assay for branching enzyme, as it measures the actual number of branched linkages formed. The iodine-binding assay (not very sensitive), does allow the assay of branching enzyme specificity with different α-1,4-dextrins. All three assays should be used when studying the properties of BE, as they provide different information with respect to the properties of the BE. Above all, before studying the properties of branching enzymes, they must be purified to the point that all hydrolytic and interfering enzymes are eliminated.

Characterization of Isozymes

Maize endosperm has three branching enzyme (BE) isoforms.[35,249,253] BE I, IIa and IIb from maize kernels[35,247,251] were purified to the point they no longer contained amylolytic activity.[35,251] Molecular weights were 82 kDa for BEI and 80 kDa for BEs IIa and IIb.[247,248] Table 4.11 summarizes the properties of the various maize endosperm BE isozymes from the studies of Takeda et al.,[249] and Guan and Preiss.[34] Of the three isoforms, BEI had the highest activity towards branching amylose as determined by the iodine assay. Its rate of branching amylopectin was ~3% that observed with amylose. The BEIIa and BEIIb isozymes branched amylopectin at twice the rate they branched amylose, and catalyzed branching of amylopectin at 2.5 to 6 times the rate observed with BEI.

Takeda et al.[249] analyzed the branched products made *in vitro* from amylose by the BE isoforms. Isoamylase was used to debranch the products of each reaction. BEIIa and BEIIb were very similar in their affinity for amylose and the size of chain transferred.

Table 4.11 Properties of maize endosperm branching isozymes[34,249]

Branching enzymes		BE I	BE IIa	BE IIb
Phosphorylase stimulation (a)		1196	795	994
Branching linkage assay (b)		2.6	0.32	0.14
Iodine stain assay (c)				
Amylose (c_1)		800	29.5	39
Amylopectin (c_2)		24	59	63
Ratio of activity	a/b	460	2484	7100
	a/c_1	1.5	27	25
	a/c_2	49.8	13.5	15.8
	c_2/c_1	0.03	2	1.6

^aPhosphorylase stimulation and branching linkage assay units are μmol/min.; the iodine stain assay, a decrease of one absorbance unit per minute

When presented with amyloses of different average chain length (CL), the branching enzymes had higher activity with the longer-chain amylose. However, while BEI could still catalyze the branching of an amylose of average CL of 197 with 89% of the activity shown with an average CL of 405, the activity of BEII dropped sharply with this change in chain length. A study of the *in vitro* reaction products indicated that the action of BEIIa and BEIIb resulted in the transfer of shorter chains than those transferred by BEI. Thus, BEI catalyzes the transfer of longer branched chains and BEIIa and IIb catalyze the transfer of shorter chains. This may suggest that BEI produces a slightly-branched polysaccharide that serves as a substrate for enzyme complexes of starch synthases and BEII isoforms to synthesize amylopectin. BEII isoforms may also play a predominat role in forming the short chains present in amylopectin. Moreover, BEI may be more involved in producing the more interior B-chains of the amylopectin, while BEIIa and BEIIb would be involved in forming the exterior (A-) chains.

Vos-Scheperkeuter et al.[251] purified a single form of branching activity of 79 kDa from potato tubers. Antibodies prepared against the native potato enzyme were found to react strongly with maize BEI and weakly with maize BEIIb. The antiserum inhibited the activities of both potato tuber BE and maize BEI in neutralization tests. Potato branching enzyme thus shows a high degree of similarity to maize BEI and a lower degree of similarity to the maize BEII isozymes.

Up to four cDNA clones have been isolated for potato BE – one for 91 to 99 kDa.[253–255] All these allelic clones have sequences similar to those of the BEI type. Also, the *sbeIc* allele codes for a mature enzyme of 830 amino acids with a MW of 95.18 kDa. The *sbeIc* BE protein product, expressed in *E. coli*, migrates as a 103 kDa protein.[253] It is of interest to note that BE isolated from other plants, bacteria and mammals have molecular masses ranging from 75 to ~85 kDa. These molecular weights are consistent with the molecular weights obtained from deduced amino acid sequences obtained from isolated genes or cDNA clones.

Potato BEII was first characterized and found to be a granule-bound protein in tuber starch.[256] In potato, BEII appears to be less abundant than BEI. Both potato tuber BEI and BEII were cloned and expressed in *E. coli*,[257,258] and the properties of the potato tuber BE isoforms were compared.[258] As seen with the maize branching enzymes, potato BEI was more active on amylose than BEII was, and BEII was more active on amylopectin than was BEI.

Mizuno et al.[31] reported four forms of branching enzyme from immature rice seeds. BEI and BE2 (composed of BE2a and BE2b) were the major forms, while BE3 and BE4 were minor forms, being less than 10% of the total branching enzyme activity. The molecular weights of the branching enzymes were BE1, 82 kDa; BE2a, 85 kDa; BE2b, 82 kDa; BE3, 87 kDa; BE4a, 93 kDa and BE4b, 83 kDa. However, BE1 and BE2a and BE2b were similar immunologically in their reaction towards maize endosperm BEI antibody. Rice seed BE1, BE2a and BE2b were very similar in their N-terminal amino acid sequences. The three BE N-terminal sequences were either TMVXVVEEVDHLPIT or VXVVEEVDHLPITDL. The latter sequence is the same as the first, but lacks the first two N-terminal amino acids. Although these activities came out in separate fractions from an anion-exchange column, most likely they are the same protein on the basis of immunology and N-terminal sequences. BE2a, however, is 3 kDa larger. It is possible that BE2a may be the less proteolyzed form. Antibody raised against BE3 reacted strongly against BE3 but not towards BE1 or BE2a,2b. Thus, rice endosperm, as noted for maize endosperm, has essentially two different isoforms of BE.

Because of the many isoforms existing for rice seed branching enzymes, Yamanouchi and Nakamura[258] studied and compared the BEs from rice endosperm, leaf blade, leaf sheath, culm and root. BE activity could be resolved into two fractions – BE1 and BE2. Both fractions were found in all tissues studied in different ratios of activity. The specific activity of the endosperm enzyme, either on the basis of fresh weight or protein, was 100- to 1000-fold greater than that of the enzyme from other tissues studied. On gel electrophoresis, rice endosperm BE2 could be resolved into BE2a and BE2b. Electrophoresis of the other tissue BE2 forms revealed only BE2b. BE2a was only detected in the endosperm tissue. In rice, it seems, there are tissue specific isoforms of BE.

Three forms of branching enzyme from developing hexaploid wheat (*Triticum aestivum*) endosperm have been partially purified and characterized.[260] Two forms are immunologically related to maize BEI and one form to maize BEII. The N-terminal sequences are consistent with these relationships. The wheat BEI_B gene is located on chromosome 7B, while the wheat BEI_{AD} peptide genes are located on chromosomes 7A and 7D. The BE classes in wheat are differentially expressed during endosperm development. BEII is constitutively expressed throughout the whole cycle, while BEI_B and BEI_{AD} are only expressed in late endosperm development.

In this respect, McCue et al.[260] showed that cDNAs encoding BE1 and BEII of wheat (*Triticum aestivum* cv Cheyenne) shared extensive identity with BE sequences reported for other plants. Using the cDNAs, they studied the steady-state RNA levels of the branching enzymes during development. For BE2, its RNA was detectable five days post-anthesis (DPA) and reached a maximum at 10 DPA. BE1 steady-state levels started rising at 10 DPA, peaking at 15 DPA. Levels of all messages declined at 20–25 DPA.

In barley, four branching enzyme isozymes were identified by separation via fast performance liquid chromatography, and three were partially purified.[262]

Subsequently, two cDNA clones encoding barley BEIIa and BEIIb were isolated.[263] The major structural difference between the two enzymes was the presence of a 94-amino acid N-terminal extension in the BEIIb precursor.

Genetic Studies on Branching Enzyme-Deficient Mutants

There are some maize endosperm mutants that appear to increase the amylose contents of starch granules. The normal maize starch granule contains about 25% of the polysaccharide as amylose, with the rest as amylopectin. In contrast, *amylose extender* (*ae*) mutants may have as much as 55–70% of the polysaccharide as amylose, and may have an amylopectin fraction with fewer branch points and with the branch chains longer in length compared to those of normal amylopectin. Results with the recessive maize endosperm mutant, amylose extender (*ae*), originally suggested that *Ae* is the structural gene for either branching enzyme BEIIa or BEIIb,[204,248,264] as the activity of BEI was not affected by the mutation. In gene dosage experiments, Hedman and Boyer[264] reported a near-linear relationship between increased dosage of the dominant *Ae* allele and BEIIb activity. Since the separation of form IIa from IIb was not very clear, it is possible that the *Ae* locus was also affecting the level of IIa.

Singh and Preiss[265] concluded that, although some homology exists between the three starch-branching enzymes, there are major differences in the structure of BEI when compared to BEIIa and BEIIb, as shown by its different reactivity with some monoclonal antibodies and differences in amino acid composition and proteolytic digest maps.

Recent studies by Fisher et al.,[267,268] from analysis of 16 isogenic lines having independent alleles of the maize *ae* locus, suggest that BEIIa and BEIIb are encoded by separate genes and that the BEIIb enzyme is encoded by the *AE* gene. They isolated a cDNA clone labeled *Sbe*2b that had a cDNA predicted amino acid sequence at residues 58 to 65 exactly the same as the N-terminal sequence of the maize BEIIb that they had purified.[268] Moreover, they did not detect any mRNA with the *Sbe*2b cDNA clone in *ae* endosperm extracts. There was some BE activity in the *ae* extracts that, on ion-exchange chromatography, migrated similarly to BEIIa.

The finding that the enzyme defect in the *ae* mutant is BEIIb is consistent with the finding that, *in vitro*, BEII is involved in the transfer of small chains. The *ae* mutant has not only an increased amount of amylose, but also the amylopectin structure is altered so that it has both longer chains and fewer total chains. In other words, there are very few short chains.

Wrinkled pea has a reduced starch level – 66–75% of that seen in the normal round seed. The amylose content is ~33% in the wild type round form, but is 60–70% in the wrinkled pea seed. Edwards et al.[268] measured several enzyme activities involved in starch metabolism at four different developmental stages of wrinkled pea and found that branching enzyme activity at its highest level was only 14% of that observed for the round seed variety. The other starch biosynthetic enzymes, as well as phosphorylase, had similar activities in wrinkled and round seeds. Smith[269] confirmed these results and found that the *r*(*rugosus*) lesion in the wrinkled pea of genotype *rr* was associated with the absence of one isoform of branching enzyme. Edwards et al.[268] postulated that the reduction in starch content observed in the BE-deficient mutant seeds was an indirect effect of the reduced BE activity through an effect on starch synthase activity. The authors suggested that, in the absence of branching enzyme activity, starch synthase forms a chain elongated with (1→4)-linked α-D-glucosyl units that was a poor glucosyl acceptor (primer) for the starch synthase substrate ADPGlc, therefore decreasing the rate of (1→4)-α-glucan synthesis. This had been previously demonstrated for rabbit muscle glycogen synthase,[271] a similar system in which it was found that continual elongation of the outer chains of glycogen caused it to become a less efficient primer with a lower Km, thus decreasing the apparent activity of the glycogen synthase. The finding that ADPGlc in the wrinkled pea accumulated to a higher concentration than in normal pea was considered evidence that the *in vivo* activity of the starch synthase was restricted. Under *in vitro* conditions, when a suitable primer like amylopectin or glycogen is added, the starch synthase activity in the wrinkled pea was equivalent to that found for the wild type.

Amylose extender mutants have been found in rice.[272] The alteration of the starch structure is very similar to that reported for maize endosperm *ae* mutants. The defect is in the BE3 isozyme, BE3 of rice being more similar in amino acid sequence to maize BEII than to BEI.[36,272] Thus, rice BE3 may catalyze the transfer of small chains, rather than long chains.

Use of gene silencing through RNA interference (RNAi) technology silenced the expression of wheat BEIIa and BEIIb and resulted in a high-amylose (>70%) phenotype.[273] The critical aspect was to suppress expression of BEIIa. Suppression of BEIIb alone had no effect. Suppression of BEIIa markedly decreased the proportion of glucose chain lengths of 4 to 12, with a corresponding increase of chain lengths greater than 12. When this high-amylose starch was fed to rats in a diet as a wholewheat, several indices of large bowel function, including short-chain fatty acid synthesis, were improved relative to a standard wholewheat. The results indicated that high-amylose wheat has a significant potential to improve human health through the high-amylose starch content resistant to digestion.[273]

Isolation of cDNA Clones Encoding the Branching Enzyme Isozyme Genes
The *r* locus of pea seed was cloned using an antibody towards one of the pea branching enzyme isoforms and screening a cDNA library.[274] However, the branching enzyme gene in the wrinkled pea contained an 800 base-pair insertion. Thus, expression of the cDNA yielded an inactive branching enzyme. The sequence of the 2.7 kB clone had more than 50% homology to the glycogen-branching enzyme of *E. coli*[275] and thus, it was proposed that the cloned cDNA corresponded to the starch-branching enzyme gene of pea seed. The *glg* B gene of a cyanobacterium has also been sequenced.[276] Its deduced amino acid sequence is 62% identical to an extensive amino acid sequence in the middle of the *E. coli* protein. This is the portion of the sequence that contains the amino acid residues critical for catalysis. Therefore, branching enzymes in nature have extensive homology irrespective of the degree of branching of their products, e.g. that which is involved in synthesis of the more highly-branched (~10% α-1,6 linkages) glycogen, the storage polysaccharide in mammals, enteric bacteria and cyanobacteria, compared to that which is involved in the synthesis of the amylopectin (~5% α-1,6 linkages) present in higher plants. It

would be of interest to determine the differences in catalysis between the BEs that are involved in synthesis of amylopectin and those involved in catalysis of bacterial or mammalian glycogen. Can it be due to different interactions of BE with the starch synthases or glycogen synthases, or are the differences inherent in the branching enzyme catalysis? The answer to this question could also resolve, in part, why starch granules in different plants are unique in their granule structure.

cDNA clones of genes representing different isoforms of branching enzyme of different plants have been isolated from potato tuber,[277–279] maize kernel,[34,35,267,268] cassava[280] and rice seeds (branching enzyme I;[32,281] branching enzyme 3[272]). The cDNA clones of maize BEI and BEII have been over-expressed in *E. coli* and purified.[34,35] The transgenic enzymes had the same properties as seen with the natural maize endosperm BEs with respect to specific activity and specificity towards amylose and amylopectin.[36]

Reserve Tissue Branching Enzyme is Localized in the Plastid

The localization of branching enzyme within the plastid has been determined in potato[282] using antibodies raised against potato BE and immunogold electron microscopy. The enzyme (which would be the equivalent of the BEI isoform of maize, as discussed above) was found in the amyloplast, concentrated at the interface between the stroma and the starch granule, rather than throughout the stroma, as is the case with the ADPGlc PPPase.[128] This would explain how amylose synthesis is possible when the enzyme responsible for its formation, i.e. the Wx protein, a granule-bound starch synthase (GBSS), is capable of elongating both linear and branched glucans. The spatial separation of branching enzyme isoforms from the granule-bound starch synthases, even if only partial, would allow the formation of amylose without it being subsequently branched by the branching enzyme. However, even if spatial separation did not exist, starch crystallization may have the same affect, i.e. it may prevent further branching. M. Morell and J. Preiss (unpublished) found that ~5% of the maize endosperm total branching enzyme activity was associated with the starch granule after amylase digestion. Whether this branching enzyme was similar to the soluble branching enzymes was not determined, but Preiss and Sivak[7] found, among the proteins present in maize and pea starch, a polypeptide of about 80 kDa that reacted with antibodies raised against maize BEl. It is worth noting, however, that small amounts of BE are expected to sediment with the starch granule, because of its affinity for the polysaccharide. The results were confirmed by Mu-Forster et al.[282]

Branching Enzyme Belongs to the α-Amylase Family

Amino acid sequence relationships between that of branching enzyme (BE) and amylolytic enzymes, such as α-amylase, pullulanase, glucosyltransferase and cyclodextrin glucanotransferase, especially at those amino acids believed to be contacts between the substrate and the amylase family enzymes, were first reported by Romeo et al.[283] Baba et al.[284] reported that there was a marked conservation in the amino acid sequence of the four catalytic regions of amylolytic enzymes in maize endosperm BEI. As shown in Table 4.12, four regions that putatively constitute the catalytic

Table 4.12 Primary structures of branching enzymes indicating the four most conserved regions of the α-amylase family
(The sequences have been derived from references referred to in the text. Two examples of enzymes from the amylase family are shown for comparison. Over 40 enzymes ranging from amylases, glucosidases, various α-1,6-debranching enzymes, and 4 examples of branching enzymes have been compared by Svensson.[286] The invariant amino acid residues believed to be involved in catalysis are in bold letters)

	Region 1	Region 2	Region 3	Region 4
Potato tuber BE	355 **D**V**V**HS**H**	424 G**FRFD**GITS	453 VTMA**E**EST	545 CVTYA**E**S**HD**
Maize endosperm BE I	277 **D**V**V**HS**H**	347 G**FRFD**GVTS	402 TVVA**E**DVS	470 CIAYA**E**S**HD**
Maize endosperm BE II	315 **D**V**V**HS**H**	382 G**FRFD**GVTS	437 VTIG**E**DVS	501 CVTYA**E**S**HD**
Rice seed BE 1	271 **D**V**V**HS**H**	341 G**FRFD**GVTS	396 TIVA**E**DVS	461 CVTYA**E**S**HD**
Rice seed BE 3	337 **D**V**V**HS**H**	404 G**FRFD**GVTS	459 ITIG**E**DVS	524 CVTYA**E**S**HD**
E. coli glycogen BE	335 **D**W**V**PG**H**	400 AL**RVD**AVAS	453 VTMA**E**EST	517 NVFLPLN**HD**
B. subtilis α-amylase	100 **D**AVIN**H**	171 G**FRFD**AAKH	204 FQYG**E**ILQ	261 LVTWV**E**S**HD**
B. sphaericus cyclodextrinase	238 **D**AVFN**H**	323 GW**RLD**VANE	350 IIVG**E**VWH	414 SFNLLGS**HD**

regions of the amylolytic enzymes are conserved in the starch-branching isoenzymes of maize endosperm, rice seed and potato tuber, and the glycogen-branching enzyme of *E. coli*. A very good and extensive analysis of this high conservation in the α-amylase family has been reported by Svensson et al.[285,286] with respect to sequence homology, and also in the prediction of (β/α)$_8$-barrel structural domains with a highly symmetrical fold of eight inner parallel β-strands surrounded by eight helices in the various groups of enzymes in the family. The (β/α)$_8$-barrel structural domain is based on crystal structures of some α-amylases and cyclodextrin glucanotransferases.

Conservation of the putative catalytic sites of the α-amylase family in the glycogen- and starch-branching enzymes may be anticipated, as BE catalyzes two continous reactions; cleavage of an (1→4)-α-glucosidic linkage in an (1→4)-α-glucan chain yielding an oligosaccharide chain which is transferred to an O-6 of the same chain or to another (1→4)-α-glucan chain, with synthesis of a new (1→6)-α-glucosidic linkage.

Amino Acid Residues that are Functional in Branching Enzyme Catalysis
As indicated in Table 4.12, four regions which constitute the catalytic regions of amylolytic enzymes are conserved in the starch-branching isoenzymes of maize endosperm, rice seed and potato tuber, and the glycogen-branching enzymes of *E. coli*.[286,287] It would be of interest to know whether the seven highly conserved amino acid residues of the α-amylase family listed in bold letters in Table 4.12 are also functional in branching enzyme catalysis. Further experiments, such as chemical modification and analysis of the three-dimensional structure of the BEs, would be needed to determine the nature of its catalytic residues and mechanism.

Indeed, the seven highly conserved amino acid residues of the α-amylase family also appear to be functional in branching enzyme catalysis. A series of experiments,[288] in which amino acids were replaced by site-directed mutagenesis, suggest that the conserved Asp residues of regions two and four and the Glu residue of region 3 (Table 4.12, in bold letters) are important for BEII catalysis. Their exact functions, however, are unknown, and additional experiments, such as chemical modification and analysis of the three-dimensional structure of the BE, would be needed to determine the precise functions and nature of its catalytic residues and mechanism. Arginine residues are also important, as suggested by chemical modification with phenylglyoxal,[289] as well as histidine residues, as suggested by chemical modification with diethyl pyrocarbonate.[290] It would also be of interest to determine the regions of the C-terminus and N-terminus which are dissimilar in sequence and in size in the various branching isoenzymes. These areas may be important with respect to BE preference for substrate (amylose-like or amylopectin-like), as well as in the size of chain transferred or the extent of branching.

12. Other Enzymes Involved in Starch Synthesis

In addition to the three starch biosynthetic enzymes, other enzymes have been shown to have some effect on starch structure.

Isoamylase

As noted in Section 4.3 (see also Chapter 3), a *su1* mutation in maize that causes a deficiency of isoamylase, an enzyme normally considered to be mainly involved in starch degradation in plants and in *C. reinhadtii*, results in accumulation of a water-soluble polysaccharide, phytoglycogen, instead of starch.[45-49,291] In *C. reinhardtii*, the mutation results in complete loss of starch, but in higher plants, the lower amount of amylopectin seen in the mutant plant may be related to the severity of the enzyme deficiency.[292,293]

It was proposed that 'maturation' or trimming of the precursor of amylopectin, 'preamylopectin,' was required in order for the amylopectin to aggregate into an insoluble granule structure.[4,49] An alternative proposal is that, during amylopectin synthesis, there is a competition between polysaccharide aggregation into granule starch and formation of water-soluble polysaccharide, and that the water-soluble polysaccharide, phytoglycogen, is consistently degraded by isoamylase.[294] If isoamylase activity is deficient, then phytoglycogen accumulates and competes with amylopectin for the starch biosynthetic enzymes. The amount of phytoglycogen accumulation is dependent on the degree of loss of isoamylase activity.[294] This competition concept between amylopectin aggregation and phytoglycogen accumulation can also explain why, in certain isoamylase mutations, the two polysaccharides, phytoglycogen and amylopectin, may be present simultaneously.[294] Although there is no question that isoamylase deficiency is the cause of phytoglycogen accumulation and lower starch accumulation in the mutant plant, the mechanism for phytoglycogen accumulation is still not completely understood.

α-1,4-Glucanotransferase

Some starch-deficient mutants of *Arabidopsis* and *Chlamydomonas* have been shown to be defective in α-1,4-glucanotransferase activity. The enzyme is also known as D-enzyme. The reaction it catalyzes is as follows:

$$\text{Maltotriose} + \text{Maltotriose} \rightarrow \text{Maltopentaose} + \text{Glucose}$$

Other oligosaccharides can also act as substrates. The transglucosylase of *Arabidopsis* leaf disproportionates maltotriose and forms higher maltodextrins of much greater size than maltooctaose.[295]

A *C. reinhardtii* mutant lacking D-enzyme activity has been characterized and has been shown to have significantly lower levels of starch.[296] Other enzymes involved in starch metabolism, such as ADPGlc PPase, granule-bound and soluble starch synthase, branching enzyme, phosphorylase, α-glucosidase, amylases or debranching enzyme activities, were not affected.[296] The starch content in the mutant was 6–13% of WT, and there was an excessive accumulation of malto-oligosaccharides up to a polymer size of 16 glucosyl units.

At present, it is not clear how D-enzyme deficiency causes a lowering of starch accumulation in *C. reinhardtii*. In fact, a previous report showed that a 98% reduction in D-enzyme activity of *Arabidopsis* had no effect on starch synthesis.[297] Also in *Arabidopsis*, it has been shown that mutants of D-enzyme overproduce starch, and that the overproduction of maltodextrins occurred only during the process of starch degradation and decreased during starch synthesis.[298] This was different from what was observed for *Chlamydomonas*, in which maltodextrin accumulation occurred during starch synthesis and decreased during starch degradation. Whether the D-enzyme plays a role both in starch synthesis and degradation is an intriguing question. It is hard to imagine that this enzyme would have different roles in oxygenic photosynthetic organisms.

In summary, much information in the past 20 years has been obtained on the enzymes involved in starch synthesis. However a number of problems remain. There is no question that the major allosteric regulation of starch synthesis occurs at the ADPGlc PPase-catalyzed reaction and is dependent on the ratio of 3PGA to Pi. The detailed process of how that ratio is regulated in storage and other plant non-photosynthetic tissue remains to be elucidated. Certainly, the detail of how the increase in sucrose concentration stimulates the reductive activation of ADPGlc PPase in potato tubers remains to be uncovered. Also, how the different isoforms of up to five or six different starch synthases coordinate their activities with the branching enzyme isoforms in synthesizing both amylose and amylopectin is still a formidable problem that requires much more detail. Of recent interest are the findings that enzymes such as isoamylase, D-enzyme and an isoform of a pastidial phosphorylase,[299] normally considered as degradative enzymes, also play some role in synthesis of starch. Their precise roles remain to be elucidated.

After submission of this chapter some new information has been reported with respect to the relationships between the large and small subunits of the ADP-glucose pyrophosphorylase.[300] As indicated in the chapter the potato tuber ADP-Glc PPase large subunit does not have catalytic activity. Here we show that in *Arabidopsis thaliana* the large subunits, APL1 and APL2, predominantly in leaf tissue, besides having a regulatory role, also have catalytic activity. Heterotetramers formed by combinations of a mutant small subunit, non-catalytic APS1, and the two large subunits showed that APL1 and APL2 exhibited ADP-Glc PPase activity with distinctive sensitivities to the allosteric activator, 3-PGA. Mutation of the large subunit glucose-1-phosphate binding site of *Arabidopsis* APLi and APL2 isoforms confirmed these observations. To determine the relevance of these activities *in vivo* a T-DNA mutant of *APS1* (*aps1*), was characterized. *aps1* is starchless, lacked ADP-Glc PPase activity, *APS1* mRNA, APS1 protein and was late flowering in long days. Transgenic lines of the *aps1* mutant, expressing an inactivate form of APS1, recovered the wild-type phenotype indicating that APL1 and APL2 have catalytic activity and may contribute to ADP-Glucose synthesis *in vivo*. Thus it is quite possible in leaf tissue that the large subunits are catalytically active as they belong in the phylogenetic tree of ADP-Glc PPases to group 1 that have the amino acids, R and K, in the large subunit residues corresponding to the potato tuber large subunit residues 44 and 54, respectively, to make the subunit catalytically active (65). As of now, no storage tissue large subunit has been shown to have catalytic activity.

V. Abbreviations*

ADP	adenosine diphosphate
ADPGlc	adenosine diphosphate glucose
ADPGlc PPase	ADP-glucose (ADPGlc) pyrophosphorylase
ATP	adenosine triphosphate
BE	branching enzyme
CAM	crassulacean acid methabolism
CL	chain length
DPA	days postanthesis
dTDPGlc PPase	deoxythymidine diphosphate glucose pyrophosphorylase
DTNB	5,5'-dithiobis(2-nitrobenzoic acid)
DTT	dithiothreitol
dTTP	deoxythimidine triphosphate
FBP	D-fructose 1,6-bisphosphate
Fru 1,6-bisP	D-fructose 1,6-bisphosphate
Fru 6-P	D-fructose 6-phosphate
GBSS	granule-bound starch synthase
Glc	D-glucose
Glc1P	α-D-glucopyranose 1-phosphate
Glucose 1-P	α-D-glucopyranose 1-phosphate
GlmU	N-acetylglucosamine 1-phosphate uridylyltransferase
mRNA	messenger RNA
NADPH	reduced nicotinamide adenine dinucleotide phosphate
NDP	nucleotide diphosphate
3PGA	3-phosphoglycerate
Pi	orthophosphate
PGM	phosphoglucomutase
PLP	pyridoxal 5-phosphate

---

*Amino acid abbreviations are the standard one- and three-letter ones

PPi	pyrophosphate
Rffh	α-D-glucopyranosyl 1-phosphate thimidylyltransferase
Rm1A	α-D-glucopyranosyl 1-phosphate thymidylyltransferase
SBE	starch-branching enzyme
SDS-PAGE	sodium dodecyl sulfate polyacrylamide-gel electrophoresis
SS	starch synthase, soluble starch synthase
TTP	thymidine triphosphate
UDPGlc	Uridine diphosphate glucose
UDPGlcNAc PPase	uridine diphosphate N-acetyl glucosamine pyrophosphorylase

VI. References

1. Martin C, Smith AM. *Plant Cell*. 1995;7:971.
2. Preiss J. Carbohydrates, Structure and Function. In: Preiss J, ed. New York, NY: Academic Press; 1988:184–249. *The Biochemistry of Plants*; Vol. 14.
3. Preiss J. In: Miflin BJ, ed. *Oxford Survey of Plant Molecular and Cellular Biology*. Vol. 7. Oxford, UK: Oxford University Press; 1991:59–114.
4. Ball SG, Morell MK. *Ann Rev Plant Biol*. 2003;54:207.
5. Sivak MN, Preiss J. Starch: Basic Science to Biotechnology. *Advan Food Nutr Res*. 1998:41.
6. Preiss J. In: Foyer C, Quick P, eds. *Engineering Improved Carbon and Nitrogen Resource Use Efficiency in Higher Plants*. London and Washington, DC: Taylor and Francis; 1997:81–104.
7. Preiss J, Sivak MN. In: Zamski E, Schaffer AA, eds. *Photoassimilate Distribution in Plants and Crops: Source-Sink Relationships*. New York, NY: Marcel Dekker; 1996:63–96.
8. Sivak MN, Preiss J. In: Kigel J, Galili G, eds. *Seed Development and Germination*. New York, NY: Marcel Dekker; 1995:139–168.
9. Smith AM, Denyer K, Martin C. *Plant Physiol*. 1995;107:673.
10. Okita TW, Nakata PA, Smith-White BJ, Ball K, Preiss J. In: Gustafson JP, ed. *Gene Conservation and Exploitation*. New York, NY: Plenum Press; 1993:161–191.
11. Preiss J, Sivak MN. In: Pinto BM, ed. *Comprehensive Natural Products Chemistry*. Oxford, UK: Pergamon Press; 1998:441–495.
12. Preiss J, Sivak MN. In: Setlow JK, ed. *Genetic Engineering, Principles and Methods*. Vol. 20. New York, NY: Plenum Press, Inc.; 1998:177–223.
13. Ballicora MA, Iglesias AA, Preiss J. *Photosynthesis Res*. 2004;79:1.
14. Ballicora MA, Iglesias AA, Preiss J. *Microb Molec Bio* Rev. 2003;67:213.
15. Sachs J. In: Ward HM, ed. *Lectures of the Physiology of Plants*. Oxford, UK: Clarendon Press; 1887:304–325.
16. Caspar C, Huber SC, Somerville C. *Plant Physiol*. 1986;79:1.
17. Ritte G, Raschke K. *New Phytologist*. 2003;159:195.
18. Matheson NK. *Carbohydr Res*. 1996;282:247.
19. Robin JP, Mercier C, Charbonniere R, Guilbot A. *Cereal Chem*. 1974;51:389.
20. Manners DJ, Matheson NK. *Carbohydr Res*. 1981;90:99.

21. Hizukuri S. *Carbohyd. Res.* 1986;147:342.
22. Morrison WR, Karkalas J. In: Dey PM, ed. *Methods in Plant Biochemistry.* London, UK: Academic Press; 1990:323–352.
23. Hizukuri S. In: Eliasson A-C, ed. *Carbohydrates in Food.* New York, NY: Marcel Dekker; 1995:347–429.
24. Preiss J, Romeo T. *Advan Bacterial Physiol.* 1989;30:184.
25. Preiss J. In: Neidhardt FC, ed. *Escherichia coli and Salmonella typhimurium: Cellular and Molecular Biology.* 2nd edn. Vol. 1. Washington, DC: American Society of Microbiology; 1996:1015–1024.
26. Preiss J, Romeo T. In: *Progress in Nucleic acid Research and Molecular Biology.* 1990;47:299–329.
27. Preiss J, Levi C. In: Preiss J, ed. *The Biochemistry of Plants, Carbohydrates, Structure and Function.* Vol. 3. New York, NY: Academic Press; 1980:371–423.
28. Denyer K, Smith AM. *Planta.* 1992;186:609.
29. Denyer K, Hylton CM, Smith AM. *Plant J.* 1993;4:191.
30. Denyer K, Hylton CM, Jenner CF, Smith AM. *Planta.* 1995;196:256.
31. Hylton CM, Denyer K, Keeling PL, Chang M-T, Smith A. *Planta.* 1996;198:230.
32. Mizuno K, Kimura K, Arai Y, Kawasaki T, Shimada H, Baba T. *J. Biochem.* 1992;112:643.
33. Bhattacharyya M, Martin C, Smith AM. *Plant Mol. Biol.* 1993;22:525.
34. Guan HP, Preiss J. *Plant Physiol.* 1993;102:1269.
35. Guan HP, Baba T, Preiss J. *Plant Physiol.* 1994;104:1449.
36. Guan HP, Baba T, Preiss J. *Cell Mol. Biol.* 1994;40:981.
37. Burton RA, Bewley JD, Smith AM, Bhattacharyya MK, Tatge H, Ring S, Bull V, Hamilton WDO, Martin C. *Plant J.* 1995;7:3.
38. Nelson OE, Chourey PS, Chang MT. *Plant Physiol.* 1978;62:383.
39. Shure M, Wessler S, Federoff N. *Cell.* 1983;35:225.
40. Sano Y. *Theor. Appl. Genetics.* 1984;68:467.
41. an der Leij FR, Visser RFG, Ponstein AS, Jacobsen E, Feenstra WJ. *Mol. Gen. Genetics.* 1991;228:240.
42. Visser RFG, Somhorst I, Kuipers GJ, Ruys NJ, Feenstra WJ, Jacobsen E. *Mol. Gen. Genetics.* 1991;225:289.
43. Dry I, Smith A, Edwards A, Bhattacharya M, Dunn P, Martin C. *Plant J.* 1992;2:193.
44. Delrue B, Fontaine T, Routier F, Decq A, Wieruszeski J-M, van den Koornhuyse N, Maddelein M-L, Fournet B, Ball S. *J. Bacteriol.* 1992;174:3612.
45. Pan D, Nelson OE. *Plant Physiol.* 1984;74:324.
46. James MG, Robertson DS, Meyers AM. *Plant Cell.* 1995;7:417.
47. Nakamura Y, Umemoto T, Takahata Y, Komae K, Amano E, Satoh H. *Physiol. Plant.* 1996;97:491.
48. Mouille G, Maddelein M-L, Libessart N, Talaga P, Decq A, Delrue B, Ball S. *Plant Cell.* 1996;8:1353.
49. Ball S, Guan H-P, James M, Myers A, Keeling P, Mouille G, Buléon A, Colonna P, Preiss J. *Cell.* 1996;86:349.
50. Leloir LF, deFekete MAR, Cardini CE. *J Biol Chem.* 1961;236:636.
51. Recondo E, Leloir LF. *Biochem Biophys Res Communm.* 1961;6:85.

52. Sowokinos JR, Preiss J. *Plant Physiol*. 69:l459 1982.
53. Okita TW, Nakata PA, Anderson JM, Sowokinos J, Morell M, Preiss J. *Plant Physiol*. 1990;93:785.
54. Nakata PA, Greene TW, Anderson JM, Smith-White BJ, Okita TW, Preiss J. *Plant Mol Biol*. 1991;17:1089.
55. Iglesias AA, Barry GF, Meyer C, Bloksberg L, Nakata PA, Greene T, Laughlin MJ, Okita TW, Kishore GM, Preiss J. *J Biol Chem*. 1993;268:1081.
56. Ballicora MA, Laughlin M, Fu Y, Okita TW, Barry GF, Preiss J. *Plant Physiol*. 1995;109:245.
57. Greene TW, Chantler SE, Kahn ML, Barry GF, Preiss J, Okita TW. *Proc Natl Acad Sci*. 1996;93:1509.
58. Fu Y, Ballicora MA, Preiss J. *Plant Physiol*. 1998;117:989.
59. Ballicora MA, Fu Y, Nesbitt NM, Preiss J. *Plant Physiol*. 1998;118:265.
60. Fu Y, Ballicora MA, Leykam JF, Preiss J. *J Biol Chem*. 1998;273:25045.
61. Ballicora MA, Fu Y, Frueauf JB, Preiss J. *Biochem Biophys Res Commun*. 1999;257:782.
62. Ballicora MA, Freuauf JB, Fu Y, Schürmann P, Preiss J. *J Biol Chem*. 2000;275:1315.
63. Freuauf JB, Ballicora MA, Preiss J. *Plant J*. 2003;33:503.
64. Jin X, Ballicora MA, Preiss J, Geiger HJ. *EMBO J*. 2005;24:694.
65. Ballicora MA, Dubay JR, Devillers CH, Preiss J. *J Biol Chem*. 2005;280:10189.
66. Ghosh HP, Preiss J. *J Biol Chem*. 1966;241:4491.
67. Preiss J, Ghosh HP, Wittkop J. In: Goodwin TW, ed. *Biochemistry of Chloroplasts*. New York, NY: Academic Press; 1967:131–153.
68. Copeland L, Preiss J. *Plant Physiol*. 1981;68:996.
69. Sanwal G, Greenberg E, Hardie J, Cameron E, Preiss J. *Plant Physiol*. 1968;43:417.
70. Ribereau-Gayon G, Preiss J. *Meth. Enzymol*. 1971;23:618.
71. Morell MK, Bloom M, Knowles V, Preiss J. *Plant Physiol*. 1987;85:185.
72. Morell M, Bloom M, Larsen R, Okita TW, Preiss J. In: Key JL, McIntosh L, eds. *Plant Gene Systems and Their Biology*. New York, NY: Alan R. Liss; 1987: 227–242.
73. Preiss J, Bloom M, Morell M, Knowles V, Plaxton WC, Okita TW, Larsen R, Harmon AC, Putnam-Evans C. *Basic Life Sciences*. 1987;41:133.
74. Preiss J, Morell M, Bloom B, Knowles VL, Lin TP. In: Biggins J, ed. *Progress in Photosynthesis Research*. Vol. III. The Hague, The Netherlands: Nijhoff; 1987:693–700.
75. Kleczkowski LA, Villand P, Preiss J, Olsen OA. *J. Biol. Chem*. 1993;268:6228.
76. Kleczkowski LA, Villand P, Olsen OA. *Z. Naturforsch*. 1993;48c:457.
77. Plaxton WC, Preiss J. *Plant Physiol*. 1987;83:105.
78. Chen BY, Janes HW. *Plant Physiol*. 1997;113:235.
79. Gomez-Casati DF, Iglesias AA. *Planta*. 2002;214:428.
80. Iglesias AA, Charng YY, Ball S, Preiss J. *Plant Physiol*. 1994;104:1287.
81. Sikka VK, Choi S-B, Kavakli IH, Sakulsingharoj C, Gupta S, Ito H, Okita TW. *Plant Science*. 2001;161:461.
82. Kavakli IH, Kato C, Choi SB, Kim KH, Salamone PR, Ito H, Okita TW. *Planta*. 2002;215:430.

83. Crevillian P, Ballicora MA, Mérida A, Preiss J, Romero JM. *J. Biol. Chem.* 2003;278:28508.
84. Iglesias AA, Kakefuda G, Preiss J. *Plant Physiol.* 1991;97:1187.
85. Iglesias AA, Kakefuda G, Preiss J. *J. Prot. Chem.* 1992;11:119.
86. Smith-White B, Preiss J. *J. Mol. Evolution.* 1992;34:449.
87. Crevillén P, Ventriglia T, Pinto F, Orea A, Mérida Á, Romero JM. *J. Biol. Chem.* 2005;280:8143.
88. Takata H, Takaha T, Okada S, Takagi M, Imanaka T. *J. Bacteriol.* 1997;179:4689.
89. Haugen TH, Preiss J. *J. Biol. Chem.* 1979;254:127.
90. Frueauf JB, Ballicora MA, Preiss J. *J. Biol. Chem.* 2001;276:46319.
91. Blankenfeldt W, Asuncion M, Lam JS, Naismith JH. *EMBO J.* 2000;19:6652.
92. Brown K, Pompeo F, Dixon S, Mengin-Lecreulx D, Cambillau C, Bourne Y. *EMBO J.* 1999;18:4096.
93. Sivaraman J, Sauve V, Matte A, Cygler M. *J. Biol. Chem.* 2002;277:44214.
94. Bremer K. *Proc. Natl. Acad. Sci. U.S.A.* 2000;97:4707.
95. Mathews S, Donoghue MJ. *Science.* 1999;286:947.
96. Olsen LR, Roderik SL. *Biochemistry.* 2001;40:1913.
97. Sulzenbacher G, Gal L, Peneff C, Fassy F, Bourne Y. *J. Biol. Chem.* 2001;276:11844.
98. Kostrewa D, D'Arcy A, Takacs B, Kamber M. *J. Mol. Biol.* 2001;305:279.
99. Rossmann MG, Liljas A, Branden CI, Banasazak LJ. In: Boyer PD, ed. *The Enzymes.* Vol. 11. New York, NY: Academic Press; 1975:61.
100. Raetz CR, Roderick SL. *Science.* 1995;270:997.
101. Kisker C, Schindelin H, Alber BE, Ferry JG, Rees DC. *EMBO J.* 1996;15:2323.
102. Beaman TW, Binder DA, Blanchard JS, Roderick SL. *Biochemistry.* 1997;36:489.
103. Kanamaru S, Leiman PG, Kostyuchenko VA, Chipman PR, Mesyanzhinov VV, Arisaka F, Rossmann MG. *Nature.* 2000;415:553.
104. Ball KL, Preiss J. *J. Biol. Chem.* 1994;269:24706.
105. Morell M, Bloom M, Preiss J. *J. Biol. Chem.* 1988;263:633.
106. Greene TW, Woodbury RL, Okita TW. *Plant Physiol.* 1996;112:1315.
107. Hill MA, Preiss J. *Biochem. Biophys. Res. Commun.* 1998;244:573.
108. Sheng J, Preiss J. *Biochem.* 1997;36:13077.
109. Frueauf JB, Ballicora MA, Preiss J. *Arch. Biochem. Biophys.* 2002;400:208.
110. Paule MR, Preiss J. *J Biol Chem.* 1971;246:4602.
111. Gomez-Casati DF, Igarashi RY, Berger CN, Brandt ME, Iglesias AA, Meyer CR. *Biochemistry.* 2001;40:10169.
112. Ball S, Marianne T, Dirick L, Fresnoy M, Delrue B, Decq A. *Planta.* 1991;185:17.
113. Iglesias AA, Charng Y-y, Ball S, Preiss J. *Plant Physiol.* 1994;104:1287.
114. Lin TP, Caspar T, Somerville C, Preiss J. *Plant Physiol.* 1988;86:1131.
115. Lin TP, Caspar T, Somerville C, Preiss J. *Plant Physiol.* 1988;88:1175.
116. Neuhaus HE, Stitt M. *Planta.* 1990;182:445.
117. Giroux MJ, Shaw J, Barry G, Cobb BJ, Greene T, Okita T, Hannah LC. *Proc. Natl. Acad. Sci. U.S.A.* 1996;93:5824.
118. Stark DM, Timmerman KP, Barry GF, Preiss J, Kishore GM. *Science.* 1992;258:287.

119. Smidansky ED, Clancy M, Meyer FD, Lanning SP, Blake NK, Talbert LE, Giroux MJ. *Proc. Natl. Acad. Sci. U.S.A.* 2002;99:1724.
120. Smidansky ED, Martin JM, Hannah LC, Fischer AM, Giroux MJ. *Planta.* 2003;216:656.
121. Ihemere U, Arias-Garzon D, Lawrence SD, Sayre R. *Plant Biotech. J.* 2006:453.
122. Hylton CM, Smith AM. *Plant Physiol.* 1992;99:1626.
123. Kleczkowski LA, Villand P, Lüthi E, Olsen OA, Preiss J. *Plant Physiol.* 1993;101:179.
124. Rudi H, Doan DNP, Olsen OA. *FEBS Lett.* 1997;419:124.
125. Weber H, Heim U, Borisjuk L, Wobus U. *Planta.* 1995;195:352.
126. Singh BK, Greenberg E, Preiss J. *Plant Physiol.* 1984;74:711.
127. Nakata PA, Anderson JM, Okita TW. *J. Biol. Chem.* 1994;269:30798.
128. Kim WT, Francheschi VR, Okita TW, Robinson NL, Morell M, Preiss J. *Plant Physiol.* 1989;91:217.
129. Wolosiuk RA, Ballicora MA, Hagelin K. *FASEB J.* 1993;7:622.
130. Jacquot JP, Lancelin JM, Meyer Y. *New Phytol.* 1997;136:543.
131. Emes MJ, Neuhaus HE. *J. Exp. Bot.* 1997;48:1995.
132. Tiessen, Hendriks JH, Stitt M, Branscheid A, Gibon Y, Farre EM, Geigenberger P. *Plant Cell.* 2002;14:2191.
133. Sanwal GG, Preiss J. 1967;119:454.
134. Nakamura Y, Imamura M. *Plant Physiol.* 1985;78:601.
135. Copeland L, Preiss J. *Plant Physiol.* 1981;68:996.
136. Doan DN, Rudi H, Olsen OA. *Plant Physiol.* 1999;121:965.
137. Thorbjørnsen T, Villand P, Kleczkowski LA, and Olsen OA. *Biochem. J.* 1996;131:149.
138. Preiss J. *Ann. Rev. Plant Physiol.* 1982;54:431.
139. Denyer K, Dunlap F, Thornbjørnsen T, Keeling P, Smith AM. *Plant Physiol.* 1996;112:779.
140. Thornbjørnsen T, Villand P, Denyer K, Olsen O-A, Smith A. *Plant J.* 1996;10:243.
141. Johnson PE, Patron NJ, Bottrill AR, Dinges JR, Fahy BF, Parker ML, Waite DN, Denyer K. *Plant Physiol.* 2003;131:684.
142. Kleczkowski LA. *Trends Plant Sci.* 1996;1:363.
143. Pozueta-Romero J, Frehner M, Viale AM, Akazawa T. *Proc. Natl. Acad. Sci. USA.* 1991;88:5769.
144. Hylton CM, Smith AM. *Plant Physiol.* 1992;99:1626.
145. Burgess D, Penton A, Dunsmuir P, Dooner H. *Plant Mol. Biol.* 1997;33:431.
146. Park SW, Chung WI. *Gene.* 1998;206:215.
147. Chen B-Y, Janes HW, Gianfagna T. *Plant Science.* 1998;136:59.
148. Chen BY, Wang Y, Janes HW. *Plant Physiol.* 1998;116:101.
149. Ainsworth C, Tarvis M, Clark J. *Plant Mol. Biol.* 1993;23:23.
150. Ainsworth C, Hosein F, Tarvis M, Weir F, Burrell M, Devos KM, Gale MD. *Planta.* 1995;197:1.
151. Burton RA, Johnson PE, Beckles DM, Fincher GB, Jenner HL, Naldrett MJ, Denyer K. *Plant Physiol.* 2002;130:1464.
152. Tetlow IJ, Davies EJ, Vardy KA, Bowsher CG, Burrell MM, Emes MJ. *J. Exp. Bot.* 2003;54:715.

153. Nakamura Y, Kawaguchi K. *Physiol. Plant.* 1992;84:336.
154. Villand P, Olsen OA, Kleczkowski LA. *Plant Mol. Biol.* 1993;23:127.
155. Lee YM, Preiss J. *J. Biol. Chem.* 1986;261:1058.
156. Lee YM, Mukherjee S, Preiss J. *Arch. Biochem. Biophys.* 1986;244:585.
157. Parsons TF, Preiss J. *J. Biol. Chem.* 1978;253:6197.
158. Parsons TF, Preiss J. *J. Biol. Chem.* 1978;253:7638.
159. Hill MA, Kaufmann K, Otero J, Preiss J. *J. Biol. Chem.* 1991;266:12455.
160. Charng Y-y, Iglesias AA, Preiss J. *J. Biol. Chem.* 1994;269:24107.
161. Sheng J, Charng Y-y, Preiss J. *Biochem.* 1996;35:3115.
162. Laughlin MJ, Payne JW, Okita TW. *Phytochem.* 1998;47:621.
163. Rost B, Sander C. *Proc. Natl. Acad. Sci. U.S.A.* 1993;90:7558.
164. Salamone PR, Kavakali IH, Slattery CJ, Okita TW. *Proc. Natl. Acad. Sci. USA.* 2002;99:1070.
165. Gardiol A, Preiss J. *Arch. Biochem. Biophys.* 1990;280:175.
166. Ballicora MA, Sesma JI, Iglesias AA, Preiss J. *Biochem.* 2002;41:9431.
167. Laughlin MJ, Chantler SE, Okita TW. *Plant J.* 1998;14:621.
168. Wu MX, Preiss J. *Arch. Biochem. Biophys.* 1998;358:182.
169. Wu MX, Preiss J. *Arch. Biochem. Biophys.* 2001;389:151.
170. Henrissat B, Coutinho PM, Davies GJ. *Plant Mol. Biol.* 2001;47:55.
171. Kumar A, Larsen CE, Preiss J. *J. Biol. Chem.* 1986;261:16256.
172. MacDonald FD, Preiss J. *Plant Physiol.* 1983;73:175.
173. MacDonald FD, Preiss J. *Plant Physiol.* 1985;78:849.
174. Harn C, Knight M, Ramakrishnan A, Guan H, Keeling PL, Wasserman BP. *Plant Mol. Biol.* 1998;37:639.
175. Edwards A, Vincken JP, Suurs LCJM, Visser RGF, Zeeman S, Smith A, Martin C. *Plant Cell.* 2002;14:1767.
176. Fujita N, Taira T. *Planta.* 1998;207:125.
177. Vrinten PL, Nakamura T. *Plant Physiol.* 2000;122:255.
178. Morell MK, Kosar-Hashemi B, Cmiel M, Samuel MS, Chandler P, Rahman S, Buleon A, Batey IL, Li Z. *Plant J.* 2003;34:173.
179. Craig J, Lloyd JR, Tomlinson K, Barber L, Edwards A, Wang TL, Martin C, Hedley CL, Smith AM. *Plant Cell.* 1998;10:413.
180. Hawker JS, Jenner CF. *Austr. J. Plant Physiol.* 1993;20:197.
181. Denyer K, Hylton CM, Jenner CF, Smith AM. *Planta.* 1995;196:256.
182. Li Z, Mouille G, Kosar-Hashemi B, Rahman S, Clarke B, Gale KR, Appels R, Morell MK. *Plant Physiol.* 2000;123:613.
183. Mu C, Harn C, Ko Y-T, Singletary GW, Keeling PL, Wassermann BP. *Plant J.* 1994;6:151.
184. Gao M, Wanat J, Stinard PS, James MG, Myers AM. *Plant Cell.* 1998;10:399.
185. Cao H, Imparl-Radosevich J, Guan H, Keeling PL, James MG, Meyers AM. *Plant Physiol.* 1999;120:205.
186. Zhang X, Colleoni C, Ratushna V, Sirghie-Colleoni M, James MG, Myers AM. *Plant Mol. Biol.* 2004;54:865.
187. Edwards A, Marshall J, Sidebottom C, Visser RGF, Smith A, Martin C. *Plant J.* 1995:283.

188. Marshall J, Sidebottom C, Debet M, Martin C, Smith AM, Edwards A. *Plant Cell.* 1996;8:1121.
189. Kossmann J, Abel GJW, Springer F, Lloyd JR, Willmitzer L. *Planta.* 1996;208:503.
190. Zhang X, Myers AM, James MG. *Plant Physiol.* 2005;138:663.
191. Baba T, Nishihara M, Mizuno K, Kawasaki T, Shimada H, Kobayashi E, Ohnishi S, Tanaka K, Arai Y. *Plant Physiol.* 1993;103:565.
192. Dian W, Jiang H, Wu P. *J. Exptl. Bot.* 2005;56:623.
193. Fontaine T, D'Hulst C, Maddelein M-l, Routier F, Pepin TM, Decq A, Wieruszeski J-M, Delrue B, Van den Koornhuyse N, Bossu J-P, Fournet B, Ball S. *J. Biol. Chem.* 1993;268:16223.
194. Delvalle D, Dumez S, Wattebled F, Roldan I, Planchot V, Berbezy P, Colonna P, Vyas D, Chatterjee M, Ball S, Merida A, D'Hulst C. *Plant J.* 2005;43:398.
195. Commuri PD, Keeling PL. *Plant J.* 2001;25:475.
196. Pollock C, Preiss J. *Arch. Biochem. Biophys.* 1980;204:578.
197. Denyer K, Smith AM. *Planta.* 1992;186:609.
198. Imparl-Radosevich JM, Li P, Zhang L, McKean AL, Keeling PL, Guan H. *Arch. Biochem. Biophys.* 1998;353:64.
199. Mangelsdorf PC. *Genetics.* 1947;32:448.
200. Wang Y-J, White P, Pollak L, Jane J-L. *Cereal Chem.* 1993;70:521.
201. Wang Y-J, White P, Pollak L, Jane J-L. *Cereal Chem.* 1993;70:171.
202. Nelson OE, Pan D. *Ann. Rev. Plant. Physiol. Plant Mol. Biol.* 1995;46:475.
203. Shannon JC, Garwood DL. In: Whistler RL, BeMiller JN, Paschall EF, eds. *Starch: Chemistry and Technology.* Orlando, FL: Academic Press; 1984:25–86.
204. Boyer CD, Preiss J. *Plant. Physiol.* 1981;67:1141.
205. Ozbun JL, Hawker JS, Preiss J. *Plant Physiol.* 1971;48:765.
206. Abel GJW, Springer F, Willmitzer L, Kossmann J. *Plant J.* 1996;10:981.
207. Wang TL, Hadavizideh A, Harwood A, Welham TJ, Harwood WJ, Faulks R, Hedley CL. *Plant Breed.* 1990;105:311.
208. Umemoto T, Yano M, Satoh H, Shomura A, Nakamura Y. *Theor. Appl. Genet.* 2002;104:1.
209. Yamamori M, Fujita S, Hayakawa K, Masuki J, Yasui T. *Theor. Appl. Genet.* 2000;101:21.
210. Furukawa K, Tagaya M, Inoye M, Preiss J, Fukui T. *J. Biol. Chem.* 1990;265:2086.
211. Yep A, Ballicora MA, Sivak MN, Preiss J. *J. Biol. Chem.* 2004;279:8359.
212. GenBank accession no. AJ225088.
213. GenBank accession no. AC007296.
214. Dian W, Jiang H, Wu P. *J. Exptl. Bot.* 2005;56:623.
215. Hirose T, Terao T. *Planta.* 2004;220:9.
216. Roldán I, Wattebled F, Lucas MM, Delvallé D, Planchot V, Jiménez S, Pérez R, Ball S, D'Hulst C, Mérida A. *Plant J.* 2007;49:492.
217. Edwards A, Fulton DC, Hylton CM, Jobling SA, Gidley M, Rössner U, Martin C, Smith AM. *Plant J.* 1999;17:251.
218. Lloyd JR, Landschütze V, Kossmann J. *Biochem. J.* 1999;338:515.
219. de Fekete MAR, Leloir LF, Cardini CE. *Nature.* 1960;187:918.
220. Nelson OE, Rines HW. *Biochem. Biophys. Res. Commun.* 1962;9:297.

221. Tsai CY. *Biochem. Genet.* 1974;11:83.
222. Federoff N, Wessler S, Shure M. *Cell.* 1983;35:235.
223. Wessler SR, Baran G, Varagona M, Dellaporta SL. *EMBO J.* 1986;5:2427.
224. Klösgen RF, Gierl A, Schwarz-Sommer Z, Saedler H. *Mol. Gen. Genet.* 1986;203:237.
225. Rohde W, Becker D, Salamini F. *Nucl. Acids Res.* 1988;16:7185.
226. Sivak MN, Wagner M, Preiss J. *Plant Physiol.* 1993;103:1355.
227. Denyer K, Sidebottom C, Hylton CM, Smith AM. *Plant J.* 1993;4:191.
228. Denyer K, Clarke B, Hylton C, Tatge H, Smith AM. *Plant J.* 1996;10:1135.
229. Denyer K, Waite D, Motawia S, Møller BL, Smith AM. *Biochem. J.* 1999;340:183.
230. Denyer K, Waite D, Edwards A, Martin C, Smith AM. *Biochem. J.* 1999;342:647.
231. Zeeman SC, Smith SS, Smith AM. *Plant Physiol.* 2002;128:1069.
232. Van de Wal M, D'Hulst C, Vincken JP, Buléon A, Visser R, Ball S. *J. Biol. Chem.* 1998;273:22232.
233. Delrue B, Fontaine T, Routier F, Decq A, Wieruszeski J-M, Van den Koornhuyse N, Maddelein M-L, Ball S. *J. Bacteriol.* 1992;174:3612.
234. Maddelein M-L, Libessart N, Bellanger F, Delrue B, D'Hulst C, Van den Koornhuyse N, Fontaine T, Wieruszeski J-M, Decq A, Ball S. *J. Biol. Chem.* 1994;269:25150.
235. Watteblad F, Buléon A, Bouchet B, Ral J-P, Liénard L, Devallé D, Binderup K, Dauvillée D, Ball S, D'Hulst C. *Eur. J. Biochem.* 2002;269:3810.
236. Van den Koornhuyse N, Libessart N, Delrue B, Zabawinski C, Decq A, Iglesias A, Carton A, Preiss J, Ball S. *J. Biol. Chem.* 1996;271:16281.
237. Libessart N, Maddelein ML, Van den Koornhuyse N, Decq A, Delrue B, Ball S. *Plant Cell.* 1995;7:1117.
238. Clarke BR, Denyer K, Jenner CF, Smith AM. *Planta.* 1999;209:324.
239. Lloyd JR, Springer F, Buléon A, Müller-Röber B, Willmitzer L, Kossmann J. *Planta.* 1999;209:230.
240. Imparl-Radosevich JM, Keeling PL, Guan H. *FEBS Lett.* 1999;457:357.
241. Holmes E, Preiss J. *Arch. Biochem. Biophys.* 1982;216:736.
242. Yep A, Ballicora MA, Preiss J. *Biochem. Biophys. Res. Commun.* 2004;316:960.
243. Yep A, Ballicora MA, Preiss J. *Arch. Biochem. Biophys.* 2006;453:188.
244. MacGregor EA. *J. Protein Chem.* 2002;21:297.
245. Bourne Y, Henrissat B. *Curr. Opin. Struct. Biol.* 2001;11:593.
246. Buschiazzo A, Ugalde JE, Guerin ME, Shepard W, Ugalde RA, Alzari PM. *EMBO J.* 2004;23:3196.
247. Boyer CD, Preiss J. *Carbohydr. Res.* 1978;61:321.
248. Boyer CD, Preiss J. *Biochem. Biophys. Res. Commun.* 1978;80:169.
249. Hawker JS, Ozbun JL, Ozaki H, Greenberg E, Preiss J. *Arch. Biochem. Biophys.* 1974;160:530.
250. Takeda Y, Guan HP, Preiss J. *Carbohydr. Res.* 1993;240:253.
251. Singh BK, Preiss J. *Plant Physiol.* 1985;78:849.
252. Vos-Scheperkeuter GH, De Wit JG, Ponstein AS, Feenstra WJ, Witholt B. *Plant Physiol.* 1989;90:75.
253. Khoshnoodi J, Blennow A, Ek B, Rask L, Larsson H. *Eur. J. Biochem.* 1996;242:132.

254. Kossmann J, Visser RGF, Müller-Röber BT, Willmitzer L, Sonnewald U. *Mol. Gen. Genet.* 1991;230:39.
255. Poulsen P, Kreiberg JD. *Plant Physiol.* 1993;102:1053.
256. Larsson C-T, Hofvander P, Khoshnoodi J, Ek B, Rask L, Larsson H. *Plant Sci.* 2001;117:9.
257. Jobling SA, Schwall GP, Westcott RJ, Sidebottom CM, Debet M, Gidley MJ, Jeffcoat R, Safford R. *Plant J.* 1999;18:163.
258. Rydberg U, Andersson L, Andersson R, Åman P, Larsson H. *Eur. J. Biochem.* 2001;268:6140.
259. Yamanouchi H, Nakamura Y. *Plant Cell Physiol.* 1992;33:985.
260. Morell MK, Blennow A, Kosar-Hashemi B, Samuel MS. *Plant Physiol.* 1997;113:201.
261. McCue KF, Hurkman WJ, Tanka CK, Anderson OD. *Plant Molec. Biol. Report.* 2002;20:191.
262. Sun CX, Sathish P, Ahlandsberg S, Dieber A, Jansson C. *New Phytol.* 1997;137:215.
263. Sun CX, Sathish P, Ahlandsberg S, Dieber A, Jansson C. *Plant Physiol.* 1998;118:37.
264. Preiss J, Boyer CD. In: Marshall JJ, ed. *Mechanisms of Saccharide Polymerization and Depolymerization.* New York, NY: Academic Press; 1980:161–174.
265. Hedman KD, Boyer CD. *Biochem. Genet.* 1982;20:483.
266. Singh BK, Preiss J. *Plant Physiol.* 1985;78:849.
267. Fisher DK, Boyer CD, Hannah LC. *Plant Physiol.* 1993;102:1045.
268. Fisher DK, Gao M, Kim K-n, Boyer CD, Guiltinan MJ. *Plant Physiol.* 1996;110:611.
269. Edwards J, Green JH, aP Rees T. *Phytochem.* 1988;27:1615.
270. Smith AM. *Planta.* 1988;175:270.
271. Carter J, Smith EE. *Carbohydr. Res.* 1978;61:395.
272. Mizuno K, Kawasaki T, Shimada H, Satoh H, Kobayashi E, Okamura S, Arai Y, Baba T. *J. Biol. Chem.* 1993;268:19084.
273. Regina A, Bird A, Topping D, Bowden S, Freeman J, Barsby T, Kosar-Hashemi B, Li Z, Rahman S, Morell M. *Proc. Natl. Acad. Sci. U.S.A.* 2006;103:3546.
274. Bhattacharyya MK, Smith AM, Noel-ellis TH, Hedley C, Martin C. *Cell.* 1990;60:115.
275. Baecker PA, Greenberg U, Preiss J. *J. Biol. Chem.* 1986;261:8738.
276. Kiel JAKW, Boels JM, Beldman G, Venema G. *Gene.* 1990;89:77.
277. Khoshnoodi J, Blennow A, Ek B, Rask L, Larsson H. *Eur. J. Biochem.* 1996;242:132.
278. Kossmann J, Visser RGF, Müller-Röber BT, Willmitzer L, Sonnewald U. *Mol. Gen. Genet.* 1991;230:39.
279. Poulsen P, Kreiberg JD. *Plant Physiol.* 1993;102:1053.
280. Salehuzzaman SNIM, Jacobsen E, Visser RGF. *Plant Mol. Biol.* 1992;20:809.
281. Nakamura Y, Yamanouchi H. *Plant Physiol.* 1992;99:1265.
282. Kram AM, Oostergetel GT, Van Bruggen EFJ. *Plant Physiol.* 1993;101:237.
283. Mu-Forster C, Huang R, Powers JR, Harriman RW, Knight M, Singletary GW, Keeling P, Wasserman BP. *Plant Physiol.* 1996;111:821.
284. Romeo T, Kumar A, Preiss J. *Gene.* 1988;70:363.

285. Baba T, Kimura K, Mizuno K, Etoh H, Ishida Y, Shida O, Arai Y. *Biochem. Biophys. Res. Commun.* 1991;181:87.
286. Svensson B. *Plant Mol. Biol.* 1994;25:141.
287. Jesperson HM, Macgregor EA, Henrissat B, Sierks MR, Svensson B. *J. Protein Chem.* 1993;12:791.
288. Kuriki T, Guan H, Sivak M, Preiss J. *J. Prot. Chem.* 1996;15:305.
289. Cao H, Preiss J. *J. Prot. Chem.* 1996;15:291.
290. Funane K, Libessart N, Stewart D, Michishita T, Preiss J. *J. Prot. Chem.* 1998;17:579.
291. Rahman S, Wong K-S, Jane JL, Myers AM, James MG. *Plant Physiol.* 1998;117:425.
292. Dinges JR, Colleoni C, Myers AM, James MG. *Plant Physiol.* 2001;125:1406.
293. Nakamura Y, Kubo A, Shimamune T, Matsuda T, Harada K, Satoh H. *Plant J.* 1997;12:143.
294. Zeeman SC, Umemoto T, Lue WL, Auyeung P, Martin C, Smith AM, Chen J. *Plant Cell.* 1998;10:1699.
295. Lin T-P, Preiss J. *Plant Physiol.* 1988;86:260.
296. Colleoni C, Dauvillée D, Mouille G, Buléon A, Gallant D, Bouchet B, Morell M, Samuel M, Delrue B, d'Hulst C, Bliard C, Nuzillard J-M, Ball S. *Plant Physiol.* 1999;120:993.
297. Takaha T, Critchley J, Okada S, Smith SM. *Planta.* 1998;205:445.
298. Critchley J, Zeeman SC, Takaha T, Smith AM, Smith SM. *Plant J.* 2001;26:89.
299. Dauvillée D, Chochois V, Steup M, Haebel S, Eckermann N, Ritte G, Ral J-P, Colleoni C, Hicks G, Wattebled F, Deschamps P, D'Hulst C, Liénard L, Cournac L, Putaux J-L, Dupeyre D, Ball SG. *Plant J.* 2006;48:274.
300. Ventriglia T, Kuhn ML, Ruiz MT, Ribeiro-Pedro M, Valverde F, Ballicora MA, Preiss J, Romero JM. *Plant Physiol.* 2008;148:65.

Structural Features of Starch Granules I

5

Serge Pérez[1], Paul M. Baldwin[2] and Daniel J. Gallant[3]

[1]Centre de Recherches sur les Macromolécules Végétales (affiliated with the Université Joseph Fourier, Grenoble), CNRS, Grenoble, France
[2]Centre de Recherches Agro-Alimentaires, INRA, Nantes, France
[3]Centre de Recherches Agro-Alimentaires, INRA, Nantes, France

I. Introduction	149
II. Granule Architecture	153
1. An Overview of Granule Structure	153
2. Molecular Organization of Crystalline Structures	153
3. Crystalline Ultrastructural Features of Starch	158
4. The Supramolecular Organization of Starch Granules	160
III. The Granule Surface	167
1. Starch Granule Surface and Chemistry and Composition	168
2. Surface-Specific Chemical Analysis	169
IV. Granule Surface Imaging	170
1. Granule Imaging by SEM Methods	170
2. Principles of AFM	171
3. Sample Preparation for AFM Imaging of Granular Starch	172
4. Surface Detail and Inner Granule Structure Revealed by AFM	173
5. Interpretation of AFM Images with Respect to Granule Structure	175
6. Discussion of Granule Surface Imaging by Scanning Probe Microscopy (SPM)	177
7. Future Prospects of SPM of Starch	179
V. A Hypothesis of Starch Granule Structure: The Blocklets Concept	180
VI. Location and State of Amylose Within Granules	184
VII. Surface Pores and Interior Channels of Starch Granules	186
VIII. Conclusions	187
IX. References	188

I. Introduction

As this chapter and Chapter 6 are read it should be remembered that, as far as is known, the amylose and amylopectin molecules, the granule structure, and the natures and amounts of the lipid and protein molecules present in granules vary with the botanical source of the starch, i.e. are unique to each type of starch. Therefore, what has been discovered about the structural features of one type of starch does not necessarily apply to other types of starch. At this time the generalities that can be attributed to all starch granules, i.e. features that apply to all granules of all starches, are unknown.

The starch granule is nature's chief way of storing energy over long periods in green plants. The granule is well-suited to this role, being insoluble in water and densely packed, but still accessible to the plant's catabolic enzymes. Starch granules are mainly found in seeds, roots and tubers, but are also found in stems, leaves, fruits and even pollen. Starch granules occur in all shapes and sizes (spheres, ellipsoids, polygons, platelets, irregular tubules); their long dimensions range from 0.1 to at least 200 μm, depending on the botanical source.[1] Differences in external granule morphology are generally sufficient to provide unambiguous characterization of the botanical source, via optical microscopy.

Native starch granules have a crystallinity varying from 15% to 45%,[2] thus, most native starch granules exhibit a Maltese cross when observed under polarized light (Figure 5.1). Theoretically, the positive birefringence indicates a radial orientation of the principle axis of the crystallites. However, birefringence remains unchanged on both polar and equatorial sections of elongated starch granules,[1] indicating that crystallites are extremely small and exhibit multiple orientations, which interfere during observations. From the level of starch crystallinity, it is clear that most starch polymers in the granule (on average ~70%) are in an amorphous state.[3] Native granules do, nevertheless, yield x-ray diffraction patterns, which although they are generally of

Figure 5.1 Raw starch granules observed by scanning electron microscopy: (a) potato; (b) cassava; and (g) rice starches. The corresponding granules under polarized light are shown in insets. Lowest figure shows SEM of *in situ* granules in potato parenchyma cell. (Reproduced with permission from reference 1)

I. Introduction

low quality, can be used to identify the several allomorphs.[4] Classification based on diffractometric spectra does not follow the morphological classification, but is able to group most starches conveniently according to their physical properties. Generally, most cereal starches give the so-called A-type pattern; some tuber starches (e.g. potato) and cereal starches rich in amylose yield the B-type pattern, while legume starches generally give a C-type pattern.

The two major macromolecular components of starch are amylose and amylopectin (Figure 5.2). They can be identified only after separation following solubilization

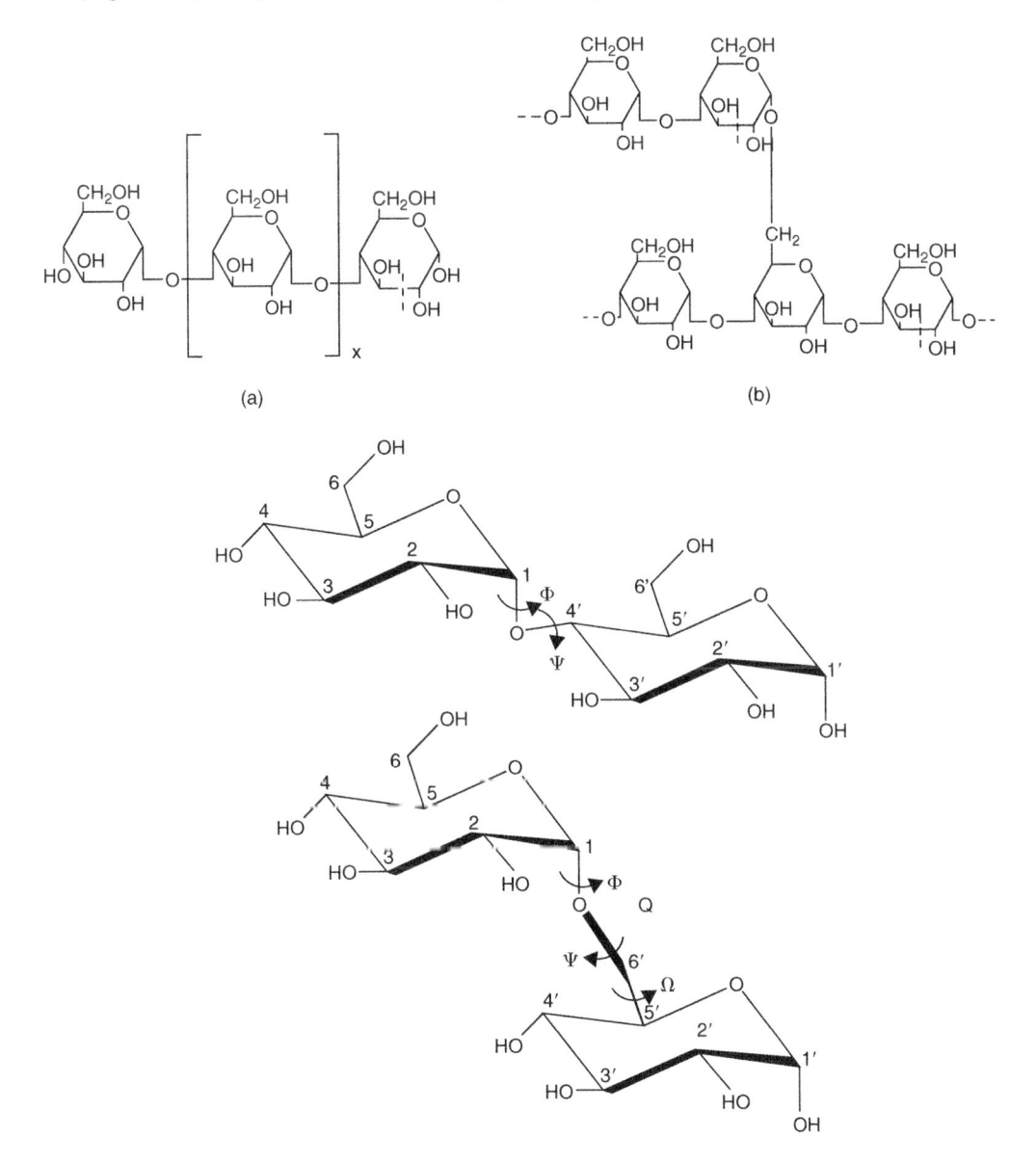

Figure 5.2 Schematic diagram of (a) amylose; and (b) amylopectin with a branch point at the O6 position. (c) Schematic representation of the disaccharide components of starch: maltose [αGlc-(1→4)Glc] and isomaltose [αGlc(1→6)Glc] along with the torsion angles that define the conformations at the glycosidic linkage between two contiguous residues: Φ = O-5-C-1-O-1-C-4', Ψ = C-1-O-1-C-4'-C-5', Ω = O-1-C-6'-C-5'-O-5'.

of the granule. Amylose is the predominantly linear (1→4)-linked α-glucan and can have a degree of polymerization (DP) as high as 600. Amylopectin, α(1→4)-linked α-glucan with α-(1→6) branch points, is the major component of the granule (30– > 99%). With a molecular weight ranging from $50 - 500 \times 10^6$, it is one of the largest natural polymers known. Amylopectin contains about 5% of branch points, which, as compared to amylose, impart profound differences in physical and biological properties. Following sequential analysis of the macromolecule, it becomes apparent that the molecule possesses several populations of polymer chains which can be classified as short chains (12 < DP < 20), long chains (30 <DP <45), and very long chains having an average DP > 60. The chains are further classified into A-, B- and C-chains, where A-chains do not carry any other chains, B-chains carry one or more chains, and the C-chain is the original chain carrying the sole reducing end (Figure 5.3). The current and widely accepted model is the cluster model.[5–11] In this model, the branch points in the amylopectin molecules are not randomly distributed, but are 'clustered.' Clusters of many short linear chains (with DP between 12 and 70)[12–14] are thought to be more crystalline than the branching regions, and to form thin crystalline lamellae (5–7 nm thick) which alternate with less crystalline (3–4 nm thick) regions composed of the branch points. These two domains appear to exist in two main directions, with a relative angle between them of about 25°.[3] Hizukuri[15] demonstrated that B-chains of amylopectin can participate in more than one crystalline side chain cluster. They, therefore, proposed a revised model of amylopectin structure and classified the B-chains according to the number of side chain clusters in which they participate. B1-chains participate in one cluster; B2- and B3-chains extend into two and three clusters respectively, and B4-chains link four or more clusters.

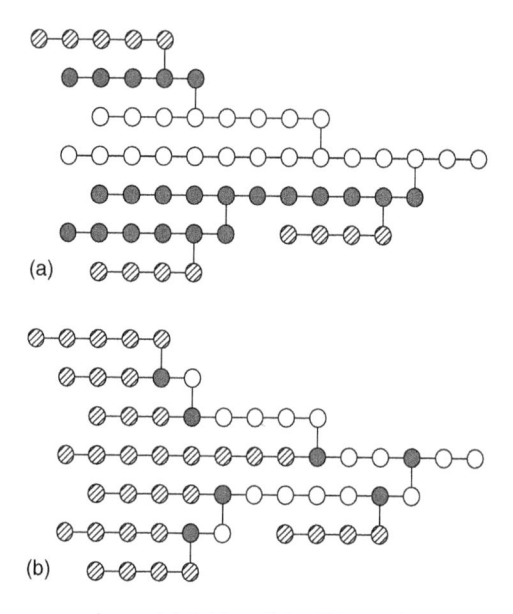

Figure 5.3 Schematic representation and definition of the different chains constituting the amylopectin macromolecule: (a) gray circles represent A-chains, dotted circles Ba-chains, and white circles Bb-chains; (b) gray circles represent external chains, black circles branch points, and white circles internal chains.

It has been estimated by Manners[16] that 80–90% of the total number of chains in an amylopectin molecule are part of side chain clusters, while the remaining 10–20% of chains form inter-cluster connections. Thus, substantial progress in understanding the basic structure of amylopectin has been made. Although the three-dimensional structure of amylopectin in the granule is not yet known, there is evidence that it is a two-dimensional ellipsoid.[17–19]

II. Granule Architecture

1. An Overview of Granule Structure

At the lowest level of structure, most starch granules are made up of alternating amorphous and crystalline shells which are between 100 and 400 nm thick.[11,12,20] These structures are termed 'growth rings.' Radial organization of amylopectin molecules within such structures is thought to cause optical polarization, since the visible optical polarization is in the order of the wavelength of visible light (100 to 1000 nm).[21] At a higher level of molecular order, x-ray diffraction investigations[22–24] in association with electron microscopy[20,25] indicate a periodicity of 9–10 nm within the granule. The periodicity is interpreted as being due to the crystalline and amorphous lamellae formed by clusters of side chains branching off from the radially arranged amylopectin molecules, and appears to be a universal feature of starch granules, independent of botanical source. Furthermore, it suggests a common mechanism for starch deposition.[26]

2. Molecular Organization of Crystalline Structures

Good quality powder diffraction patterns can be obtained from starch granules subjected to mild acid-catalyzed hydrolysis to remove amorphous portions. Good powder diffraction patterns have also been obtained from crystallized short amylose chains (DP <50), either in the form of spherulites[27] or lamella single crystals.[28] These powder diffraction patterns are difficult to interpret, because of the complexity of the polymer structures. Fiber diffraction may complement the paucity of the data from powder diffraction. One fiber diffraction study was performed using the radial axis of a giant granule,[29] but in most cases the samples originate from a film cast from solutions of high DP amylose. Stretching the films aligns the crystallites' axes. The filaments give the same characteristic A- and B-diffraction patterns[30,31] as the short branch segments of amylopectin in granules. Similar observations can be made for single crystals grown *in vitro*[28] from monodisperse fractions of amylose having DP 15 and DP 30, which give rise to powder diffraction patterns typical of the A-type and B-type patterns, respectively (Figure 5.4).

The structure of A-type starch crystals was derived through the joint use of electron diffraction of single crystals, x-ray powder patterns decomposed into individual peaks, x-ray fiber diffraction data and extensive molecular modeling[32] (Figure 5.5). The density calculated for the crystalline region (d = 1.48) is reasonably close to the observed density, and indicates that there are 12 glucosyl units and 4 water molecules in the unit cell. Intra- and inter-molecular energy calculations showed that

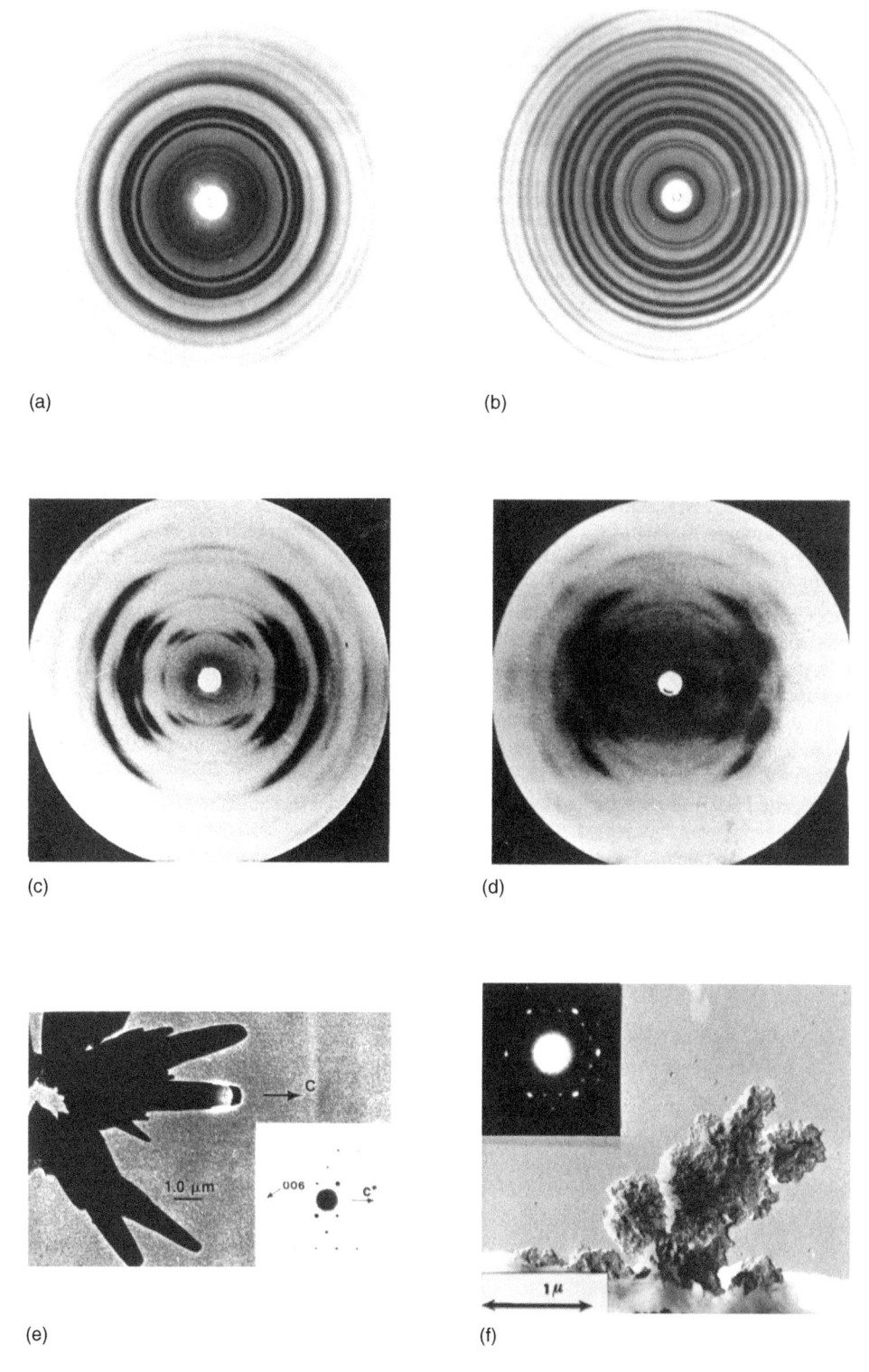

Figure 5.4 X-ray powder diffractogram recorded for: (a) A-type amylodextrins; and (b) B-type amylodextrins grown as spherulites. X-ray fiber diffraction patterns (fiber axis vertical) for: (c) A-amylose (fiber spacing 1.04 nm); and (d) B-amylose (fiber spacing 1.05 nm). (Reproduced with permission from references 30 and 31). Microcrystal of: (e) A-starch; and (f) B-starch observed by low dose electron microscopy. Inset: the electron diffraction diagrams recorded under frozen wet conditions (e). (Reproduced with permission from references 32 and 34)

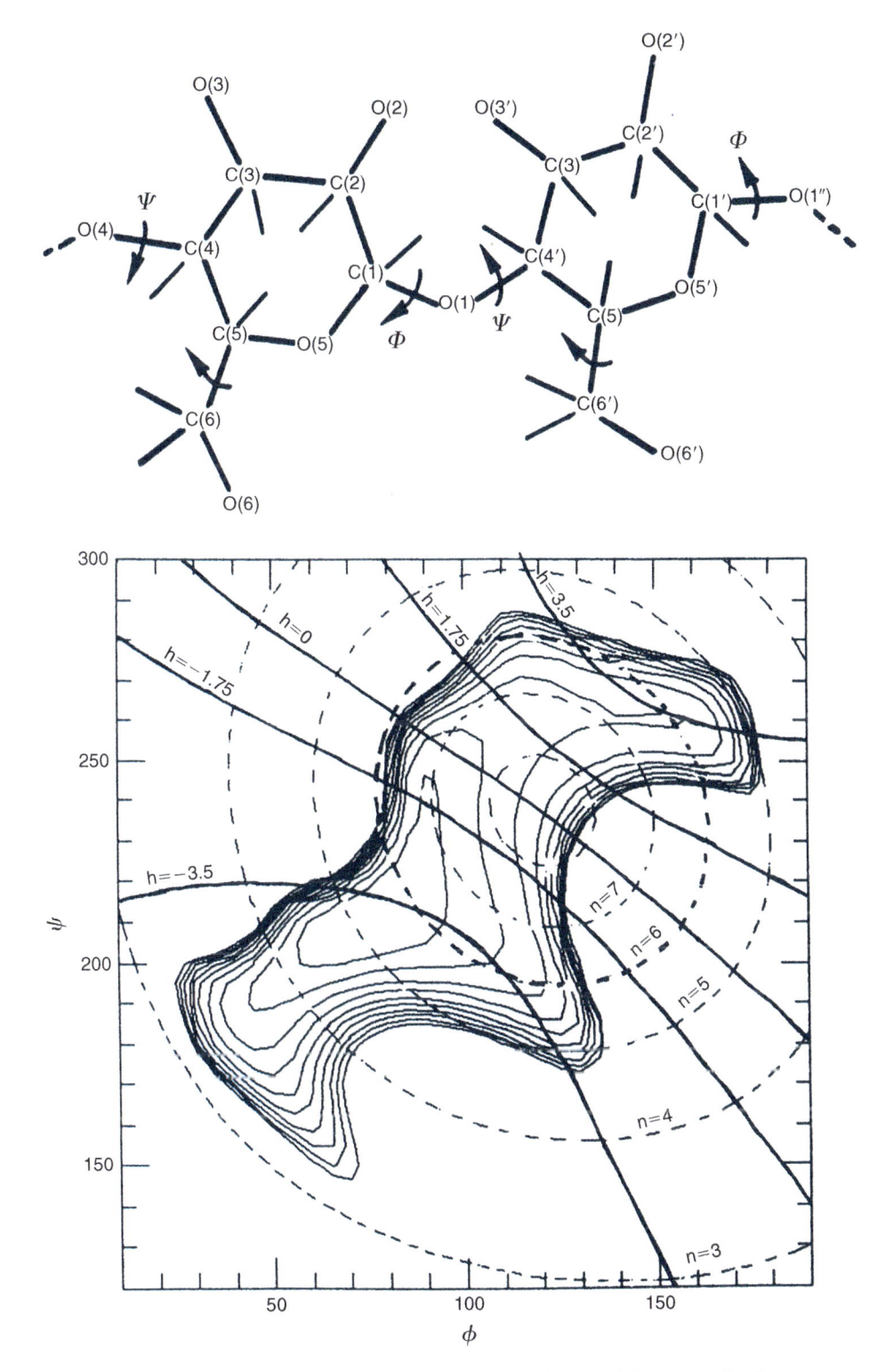

Figure 5.5 Selected iso-n and iso-h contours superimposed on the potential energy surface for maltose computed as a function of Φ and Ψ glycosidic torsion angles. Iso-energy contours are drawn by interpolation of 1 kcal/mol with respect to the energy minimum (*). The iso-$h = 0$ contour divides the map into two regions corresponding to right-handed ($h > 0$) and left-handed ($h < 0$) chirality.

the only suitable models for the chain structure were left-handed, parallel-stranded, double helices. Each strand repeats in 2.138 nm, but is related to the other strand by a two-fold axis of rotation, yielding the apparent fiber repeat distance of 1.069 nm. There are no intra-chain hydrogen bonds, but there is an O-2 ... O-6 hydrogen bond between the two strands. The double helix is very compact, and there is no space for water or any other molecule in the center (Figure 5.6a).

The monoclinic space group B_2 ($a = 2.124$ nm, $b = 1.172$ nm, $c = 1.069$ nm, $\gamma = 123.5°$) requires that the asymmetric unit contains a maltotriosyl unit, and that the packing contains one double helix at the corner and another at the center of the unit cell. Synchrotron radiation microdiffraction data has confirmed these crystallographic assignments.[33] The space group also demands that all helices be parallel (Figure 5.6b). There are hydrogen bonds between these helices, either direct or through the four water molecules in the unit cell. These water molecules are buried deep in the crystal structure, and it is impossible to remove them without complete destruction of the crystalline structure (Figure 5.6).

The structure of B-type starch crystals was established by combining the set of experimental data derived from x-ray fiber and electron diffraction crystallography via an appropriate molecular modeling technique.[34] The chains in B-type starch are also organized in double helices, but the structure differs from A-type starch in crystal packing and water content, the latter ranging from 10% to 50%. The crystalline unit cell is hexagonal ($a = b = 1.85$ nm, $c = 1.04$ nm), space group $P6_1$. Double helices are connected through a network of hydrogen bonds that form a channel inside the hexagonal arrangement of six double helices (Figure 5.6c). This channel is filled with water molecules, half of which are bound to amylose by hydrogen bonds and the other half to other water molecules. Thus, with a hydration of 27%, 36 water molecules are located in the unit cell between the six double helices, creating a column of water surrounded by the hexagonal network. There is no indication of disorder of these water molecules, agreeing with an NMR investigation that indicates that 'freezable water' can be observed only when the hydration is above 33%.[35]

The structural features of A-type and B-type starch crystallites can be compared at the molecular level. The double helices in both A- and B-type starches are left-handed, almost perfectly six-fold structures, with a crystallographic repeat distance of about 1.05 nm. The geometry of the single strands is similar to the geometry of KOH amylose and amylose triacetate. However, in the KOH and triacetylated structures, the amylose chain exists as a single strand. In the A and B allomorphs, the observed space group imposes parallel arrangement of all double helices. Double helices of both forms are packed in hexagonal or pseudohexagonal arrays. The void in the lattice of B-type starches, which accommodates numerous water molecules, is not present in A-type starches. In both arrangements there is a pairing of double helices that corresponds to 1.1 times the distance between the axes of the two double helices. A relative translational shift of 0.5 nm along the orientation of the chains allows very close nesting of the crests and troughs of the paired double helices. Such a dense association, which is strengthened by O-2 ... O-6 and O-4 ... O-3 hydrogen bonding, corresponds to the most energetically favored interactions between two double helices, as shown by theoretical calculations.[36]

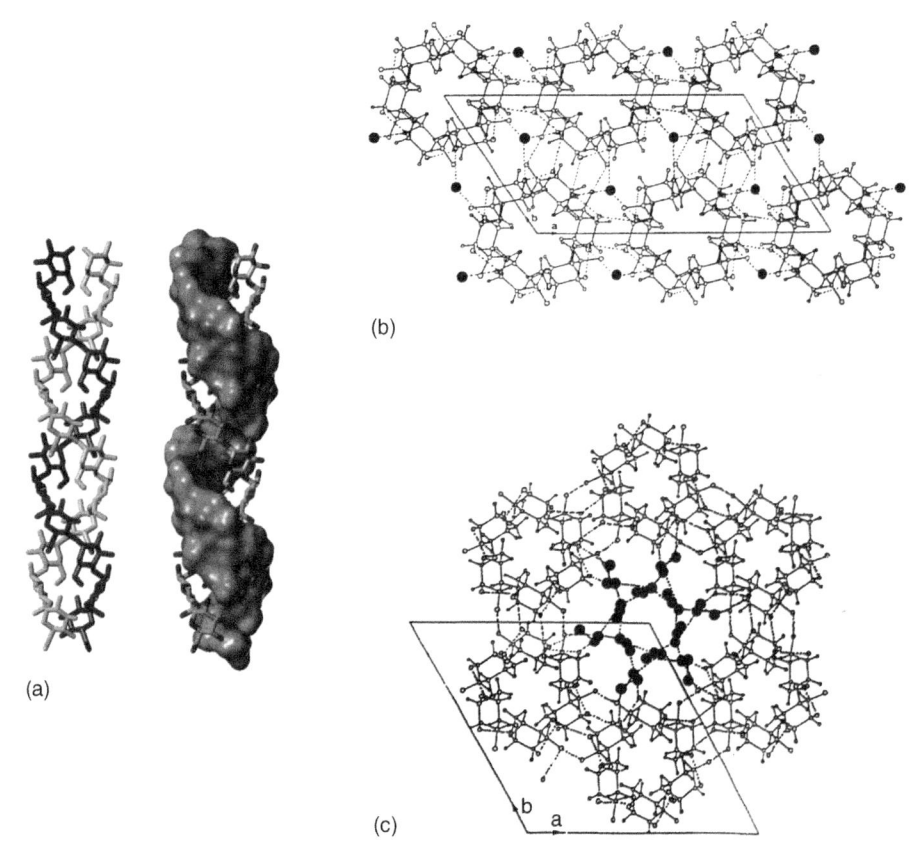

Figure 5.6 (a) Molecular drawing for the double helix found in A and B starches. Each single strand of an amylosic chain is in the left-handed conformation having six-fold symmetry, repeating in 2.1 nm; the double helix is generated by association of two single strands through two-fold symmetry. (b) Structure of A starch. Chains are crystallized in a monoclinic lattice. In such a unit cell, 12 glucopyranosyl units are located in two left-handed, parallel-stranded double helices, packed in a parallel fashion. For each unit cell, four water molecules (closed circles) are located between the helices. Projection of the structure onto the (a, b) plane. Hydrogen bonds are indicated as broken lines. (Reproduced with permission from reference 32) (c) Structure of B starch. Chains are crystallized in a hexagonal lattice, where they pack as an array of left-handed parallel-stranded double helices in a parallel fashion. 36 water molecules represent 27% hydration. Half the water molecules are tightly bound to double helices; the remainder form a complex network centered around the six-fold screw axis of the unit cell. Projection of the structure onto the (a, b) plane. Hydrogen bonds are indicated as broken lines. (Reproduced with permission from reference 33)

Conditions required to generate A- and B-type crystal conformations are reasonably well-understood. Under cool, wet conditions (such as in a potato tuber), B-type starch crystals form, while in warmer, drier conditions (e.g. in a cereal grain), the A allomorph is preferred. Chain length also affects the selection of crystal form, chains of DP <10 do not crystallize, chains with a DP from 10 to 12 tend to form A-type crystals, and chains with a DP >12 tend to yield the B form.[37,38] This is likely the result of differences in loss of entropy on crystallization experienced by chains of different lengths.[39] An irreversible transition from B-type starch to A-type starch can be accomplished under conditions of low humidity and high temperature. This so-called

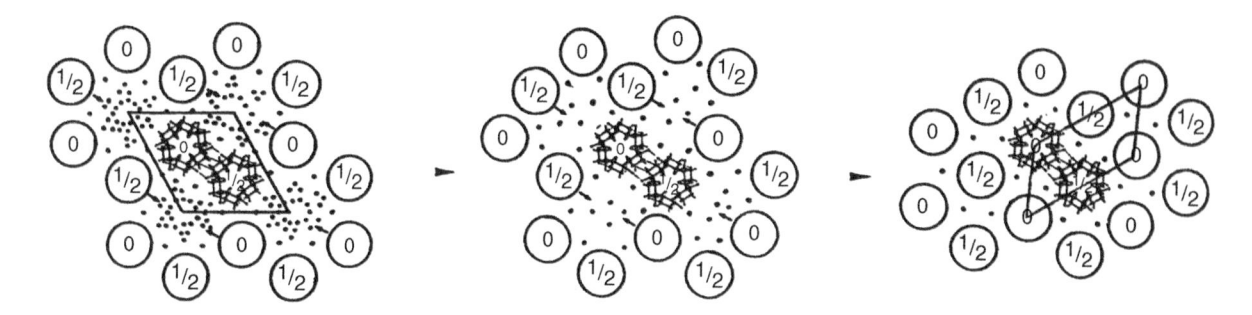

Figure 5.7 Model of the polymorphic transition from B-type to A-type starch in the solid state. The parallel double helices which form the duplex are labeled 0 and ½, indicating their relative translation along the c axis. Water molecules are shown as dots. (Reproduced with permission from reference 159)

heat–moisture treatment involves rearrangement of the pairs of double helices[14,40] (Figure 5.7).

Crystal form is also considered to be an important contributing factor in determining overall granule properties. Thus, in starches which display the A-type crystal form, the gelatinization temperature generally tends to increase with increasing overall crystallinity of the granule, while the reverse is generally true for starches displaying the B-type crystal form.

3. Crystalline Ultrastructural Features of Starch

Most common starch granules are much too small to be studied individually by diffraction analysis or solid-state polymer techniques such as spectroscopy. Until recently, there was only a single case where an oriented x-ray diagram had been obtained from an isolated starch granule. This diffraction diagram was obtained in 1951[29] with an x-ray micro-camera from a gigantic granule extracted from the pseudobulb of the orchid *Phajus grandifolius*. The results showed that the molecular orientation of the diffracting material was perpendicular to the growth rings of this unusually large granule. Characterization of starch granules that are elliptical in shape with their long axis never greater than 100 µm has now become possible through use of microfocus x-ray diffraction from synchrotron radiation. An x-ray beam of 2 µm full width, having flux of about 10^{10} photons/second/µm^2 (available at the European Synchrotron Radiation Facility in Grenoble, France), has been used to map the occurrence of crystalline regions within a single granule without subjecting the granule to any sample preparation. Oriented two-dimensional diffraction patterns can be obtained, yielding key information about the nature of the crystalline structure, along with its location and orientation with respect to the granule. Diffraction patterns can be collected at 10-µm steps across a single granule, thereby providing a complete mapping of the crystalline components (schematically represented in Figure 5.8). Individual wheat and potato starch granules have been subjected to such an investigation.[41,42] Individual diffraction patterns having unit cell dimensions in very good agreement with those of A- and B-type crystal structures[32,34] have been recorded on individual granules. The results establish, without any ambiguity, the

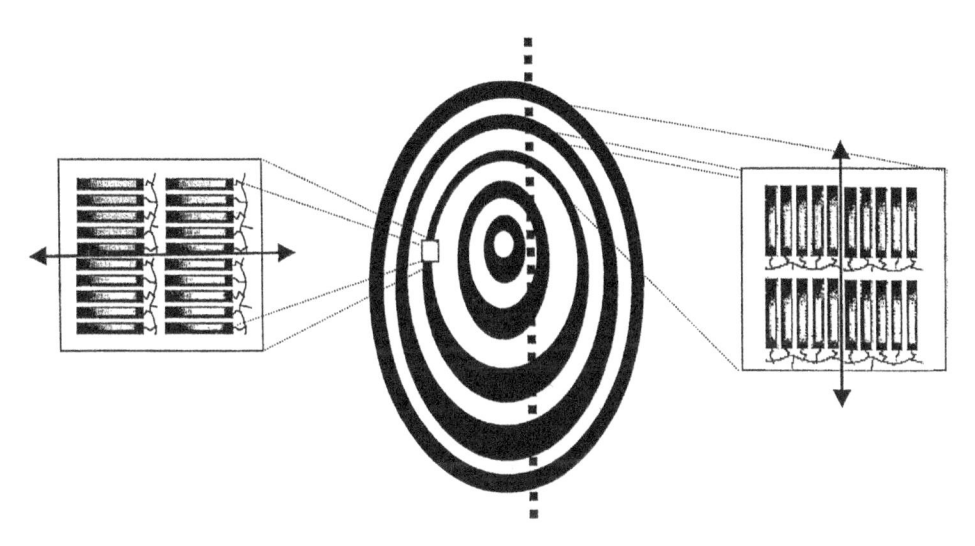

Figure 5.8 Granule in cross-section showing the orientation of amylopectin double helices in the crystalline lamellae. The dashed line indicates the path followed to obtain the diffraction diagram using a microfocus x-ray diffraction beam having a diameter of 2 μm. (Adapted with permission from reference 42)

orientation of amylopectin double helices in the crystalline lamellae (Figure 5.8). The most accurate mapping has been recorded for potato starch. Interestingly, these double helices do not point toward a single focus, but instead toward the surface of an inner ellipsoid. The helices are radially oriented, i.e. they are found to be perpendicular to the surface of the granule. This is in agreement with information derived from birefringence studies. At 10 μm resolution, there was no discontinuity of orientation (disclination or granule boundaries). This indicates a gradual change in the direction of the helices between 10 μm steps, consistent throughout with radial orientation.[42]

Microfocus synchrotron wide-angle diffraction mapping was also used to decipher the crystalline microstructure of granules from smooth pea, exhibiting so-called C-type polymorphism.[43] The specimen contained 60% A-type structure and 40% B-type structure, and these two crystalline phases co-existed within the same granule (Figure 5.9). The A allomorph was essentially located in the outer part of the granules, whereas the B-type was found mostly near their center. The diffraction diagrams for the A component were always poorly-oriented fiber patterns with the fiber axis systematically oriented toward the center of the granule. For the B allomorph in the center of the granule (where B-type cystallites are located), only powder diagrams could be observed. In in-between areas, the B component was much better oriented than the A component. These observations confirm the results of Bogracheva et al.,[44] who followed the loss of birefringence of granules during gelatinization in aqueous KCl solution, and found that the central part of smooth pea starch granules contained the B allomorph and the outer part the A allomorph. Together, these results provide a definite answer to the nature of the so-called C allomorph, i.e. that it is a combination of A- and B-crystalline constituents. This corroborates previous conclusions derived from molecular modeling which showed that the A and B arrangements were the only two possible crystalline arrangements of low energy.[36]

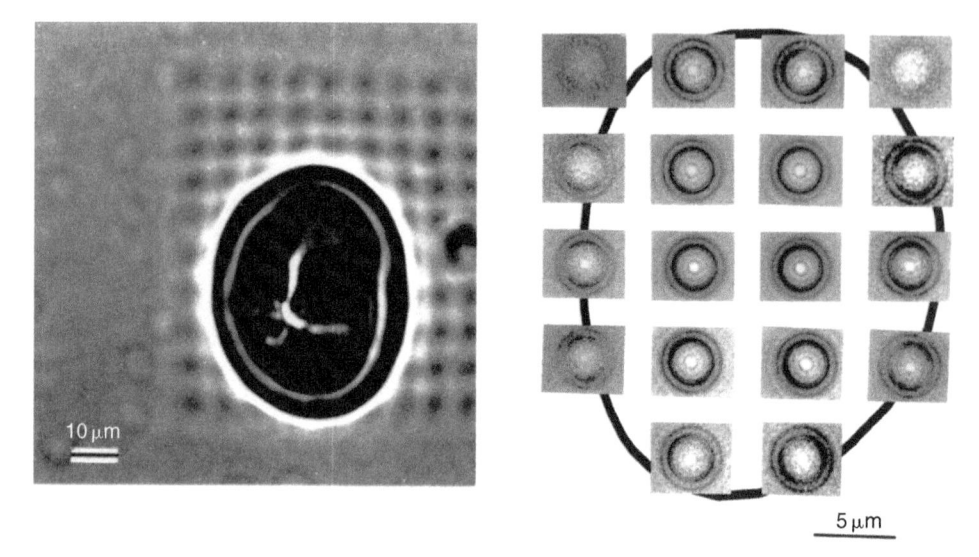

Figure 5.9 Distribution of the crystalline domains in pea starch: (a) optical micrograph of a typical sample of smooth pea starch after a 7μm step irradiation with an x-ray beam of 2μm diameter, each step consisting of a 16s exposure; (b) set of microfocus x-ray diffraction patterns recorded on a smooth pea starch granule. Each diagram corresponds to a diffraction area of ~3μm²; steps of 7μm separate the diagrams. (Reproduced with permission from reference 41)

At the present time, the synchrotron microbeam has a diameter of around 2μm. Such a size is small enough to assess gross ultrastructural features of the diffracting materials of individual starch granules. However, this beam size is still too large to assess the details of the crystalline micro-morphology of individual concentric layers (~500nm in thickness) that are located within the granule. During scanning of a granule, several of these layers are diffracting at the same time. In the case of starch from smooth pea, it is impossible to assess whether the A- and B-phases are present within the same single layer of the granule or along a single cluster of amylopectin.

4. The Supramolecular Organization of Starch Granules

Even though some detailed information regarding starch polymer structures has been simulated at the atomic level by computer modeling, the structure of granules at the level of crystalline and amorphous domains (both on the lamella scale and that of the 'growth' rings) is less well-understood. Such knowledge is important for an understanding of the physical properties of starch. As stated above, small angle x-ray diffraction and electron microscopy have revealed a periodicity in the granule of about 10nm, which has been explained by stacks of alternating thin crystalline and amorphous lamellae, as proposed in the cluster model of amylopectin.[9,16]

The structural characteristics of the branching areas of amylopectin are difficult to assess, since they are thought to be located in the amorphous regions between crystallites and since they constitute only a small fraction of the total molecule. Some basic structural features have, however, been established through computer modeling.[45] In particular it was found that, among the low-energy arrangements, one had the side chains folded back onto the carrying chain, thereby producing dense, three-dimensional structures in which a parallel arrangement is achieved. Furthermore, the branching between the two strands of double helices was investigated. It was found that one particular set of conformations about the glycosidic linkages in the two strands could result in an arrangement such that double helical strands could be connected through an α−(1→6) linkage with a minimum of distortion (Figure 5.10). This indicates that the branch points do not induce extensive defects in the double helical structure, but instead may serve to initiate the crystalline arrangement.

The concept of amylopectin forming double helices easily integrates into the currently-accepted cluster model, with the short linear chains of the branches being intertwined into double helices, while the branch points are located in the more amorphous regions between the clusters of double helices. Understanding that parts of amylopectin molecules are capable of forming double helices explains the apparent anomaly that a branched polymer is the source of structural order within granules.

According to Hizukuri,[15] amylopectin molecules of A-type starches have shorter constitutive chains and a larger short-chain fraction than amylopectin molecules of the B-type starches. Jane et al.[46] determined that A-type starches had branch points scattered in both amorphous and crystalline regions, while B-type starch had the most branch points clustering in amorphous areas. It was concluded that the branching pattern of amylopectin played a key role in determination of the type of crystallinity.[44] This point of view was confirmed through an investigation of the relationship of the distance between branching points in a cluster to the crystal pattern.[47] The study was performed on two allomorphs, maize double mutants of A- and B-types (*wxdu* and *aewx*, respectively). It was shown that the branching zone of clusters in A-type clusters was larger, but with a shorter distance between branching points, than in the B-type branching zone of clusters. These results indicate that both chain length and (1→6) linkage distribution are determinants of cluster features.

A further interesting observation regarding starch granule molecular order was gained by solid-state ^{13}C-NMR investigations,[48] which allow an assessment of the ratio of single to double chains in a given starch sample. The investigation suggested that the level of helical order is often significantly greater than the extent of crystalline order, which means that starch granules contain many double helices that are not part of extended crystalline arrays. This observation is assumed to be relevant to the amylopectin component of the granule (since amylose is not thought to be in the double helical state), and suggests that much of the amylopectin in the more semicrystalline shells of the granule is in a double helical form. As mentioned above, computer modeling[45] supports such a view, i.e. that double helices in amylopectin are sustainable in terms of molecular energetics, even at branch points that are located in the amorphous regions of lamellae (Figure 5.11).

Recent advances in molecular modeling offer new possibilities to investigate further the nature and importance of both branch points and amorphous/crystalline interfaces in starch granules. In approaching these questions, it is wise to adopt the attitude that one is not searching for single answers, but for all indications that would lead to a clearer image of how starch macromolecules are arranged and interact in nature. To this end, subunits of single helices, double helices and branch points were

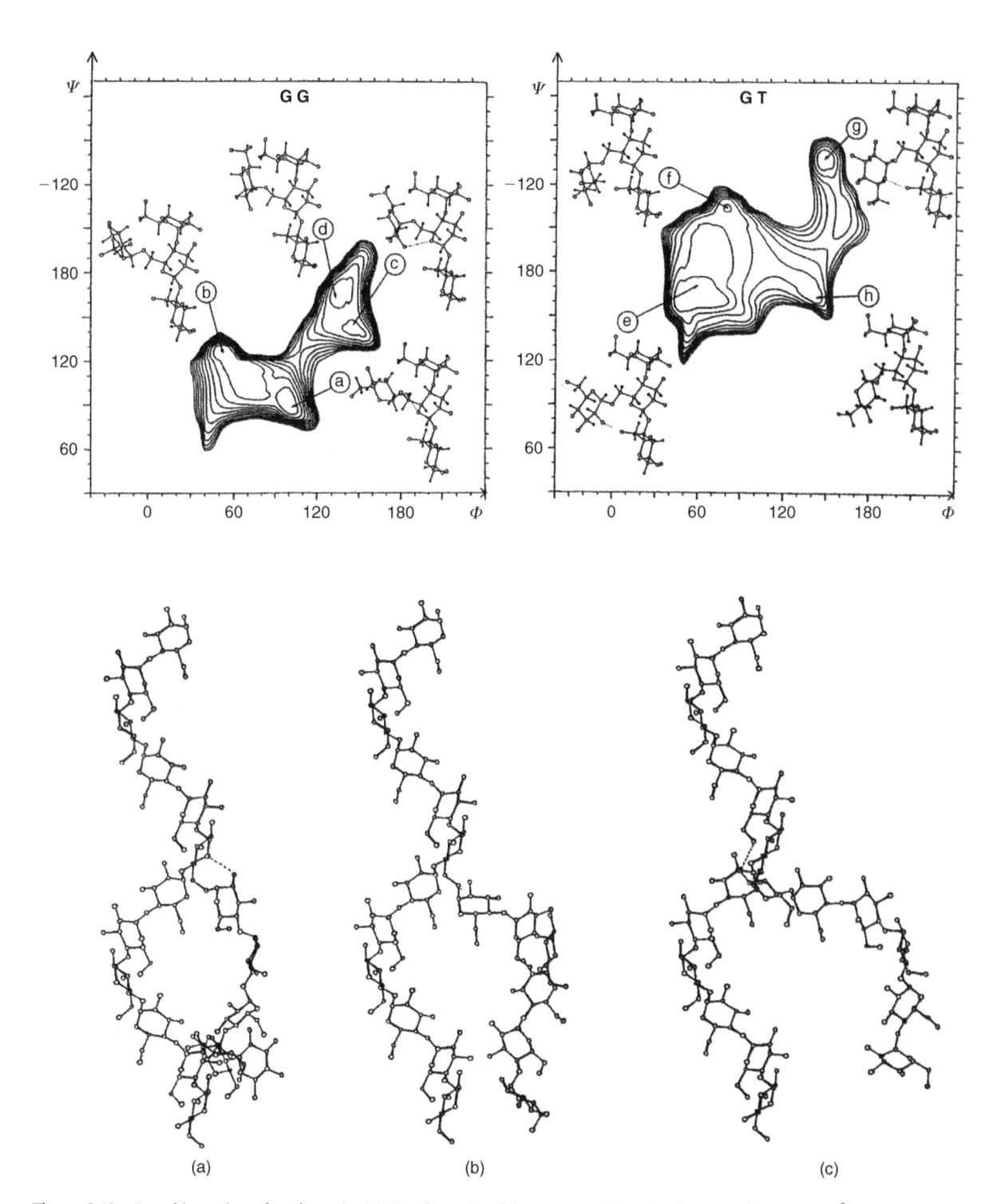

Figure 5.10 Branching points of amylopectin. (a) Two-dimensional, iso-energy surfaces for the (1→6) linkage of a 6^2-α-D-glucosylmaltotriose molecule. The maltotriose moiety was kept in the conformation observed for crystalline starch. Two orientations for the Ω angles were taken into account (GG; *gauche-gauche* orientation: $\Omega = -60°$) and (GT, *gauche-trans* orientation: $\Omega = 60$). Iso-energy contours were drawn by extrapolation of 1 kcal/mol with respect to the absolute minimum of each map. Four molecules, corresponding to conformations of lowest energy are shown on each map. (b) Three models of singly-branched amylopectin obtained by propagating amylosic chains from three low-energy conformations of the tetrasaccharide model of the branch point. The main chains have 12 glucosyl units and the side chain 6.

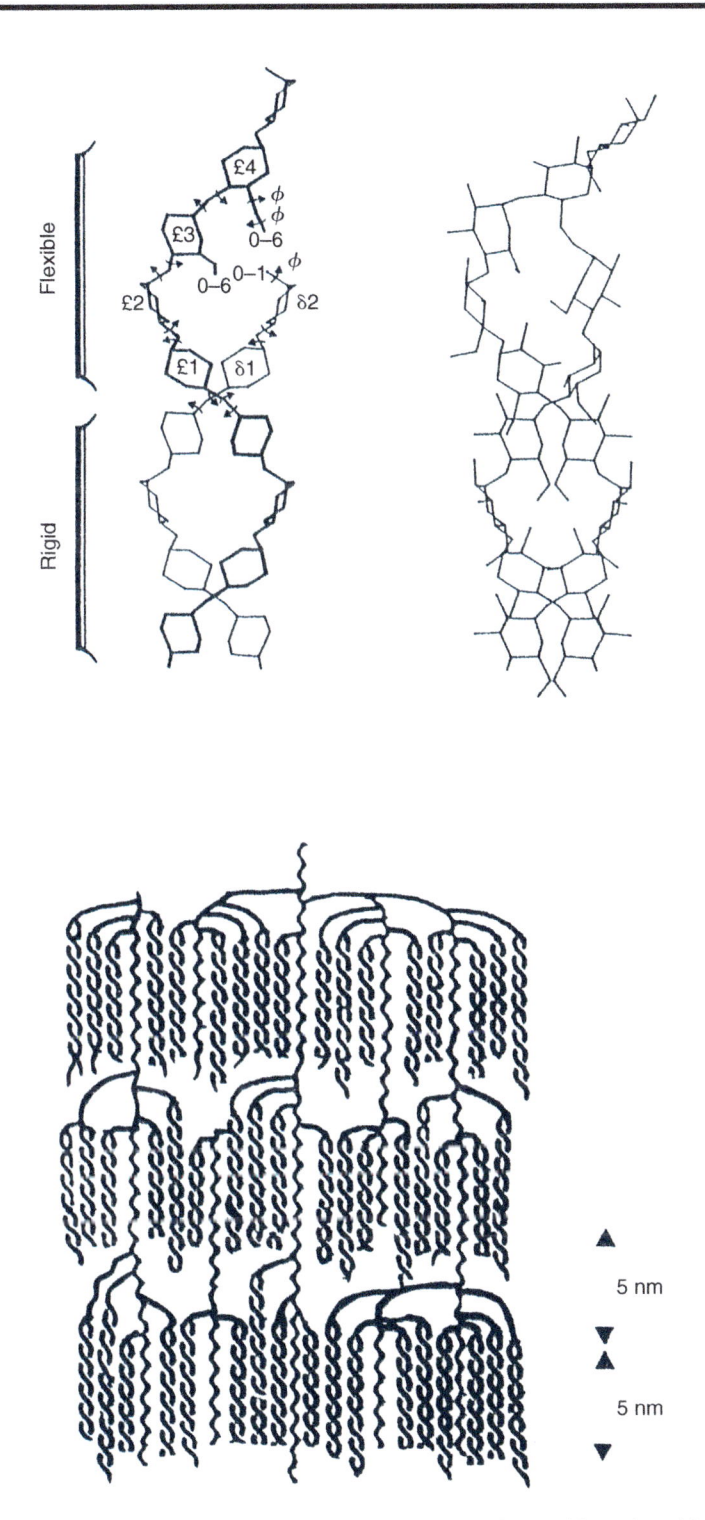

Figure 5.11 Representation of the double helix of crystalline starch after modeling a branching point between two strands. Schematic cluster model of amylopectin molecule incorporating the double helical fragments. (Reproduced with permission from reference 45)

used as building blocks of larger systems.[49] The possible make-up of amylopectin unit clusters was investigated via a series of models, including single–single, double–single, and double–double helix systems. The lengths of the single helix section linking two branch points (internal chains) was systematically varied between values of 0 and 10 glucosyl units. It was found that certain internal chain lengths led to parallel double helices. It can, therefore, be postulated that the length of internal chains may determine the degree of local crystallinity. Furthermore, it was noted that some low-energy arrangements of double helices could be superimposed on either of the two adjacent or non-adjacent double helices of crystalline A and B starch polymorphs (Figure 5.12). In other instances, the distance between the double helices is so large that it may in fact be a model for the branching separating two amylopectin crystals or unit clusters. Results of such a modeling exercise strongly suggest that the branching in amylopectin is not random. This, in a sense, is not a surprising feature, as the branching process is controlled by the specific enzymes involved. The location and the length of branching segments necessarily reflect the three-dimensional structure and specificity of the branching enzymes, along with the spatial and temporal availability of the oligosaccharide complexes involved in synthesis (Chapter 4). It might, therefore, be emphasized that the characteristics of branching pattern are equally as important as chain length distributions, and that they play a determinant role in some physicochemical properties of starch.

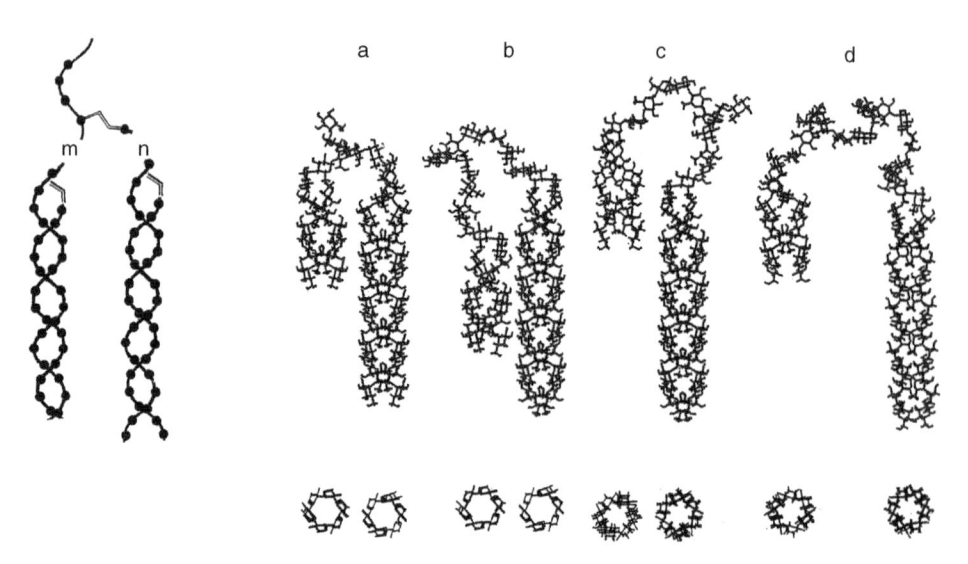

Figure 5.12 Schematic representation showing some parameters underlying branching between two double helices. The number of α-D-Glc$p$ units on the reducing side and the non-reducing side of the (1→6) linkage are designated as n and m, respectively. Molecular drawings of some low-energy arrangements as a function of m and n: (a) $m = 1$, $n = 3$; (b) $m = 4$, $n = 6$; (c) $m = 6$, $n = 4$; (d) $m = 7$, $n = 7$. The $m = 1$ and $n = 3$ model (a) is easily superimposed on two adjacent double helices as found in the A allomorph. The $m = 4$ and $n = 6$ model b superimposes equally well on the A- and B-crystalline starch structures. The $m = 6$, $n = 4$ model c corresponds to two non-adjacent double helices of crystalline B-starch. (Reproduced with permission from reference 49)

The domain structure in starch has been investigated by the use of enzyme-catalyzed hydrolysis, followed by either scanning or transmission electron microscopy, with the aim of characterizing the mosaic composition of starch resulting from the presence of 'hard' and 'soft' material.[50] During α-amylolysis, most crystalline (hard) regions are less digested than are semicrystalline (soft) regions (Figure 5.13). Examination of the three-dimensional structure of pig pancreatic α-amylase reveals that the binding site cannot accommodate such large and stiff fragments as the double helices found in the crystalline A- and B-allomorphs[51] (Figure 5.14).

The susceptibility of starch granules to hydrolysis catalyzed by α-amylase can be quantified by following the degree and the manner in which erosion and corrosion take place. Most starch granules are first hydrolyzed superficially. Granules of some

Figure 5.13 Interaction of an amylose chain in the vicinity and in the hydrolytic site of pig pancreatic α-amylase. The impossibility of fitting a double helix in the hydrolytic site has been clearly established, along with determination of the direction of binding of the amylosic substrate in the cleft of the enzyme. (Reproduced with permission from reference 51)

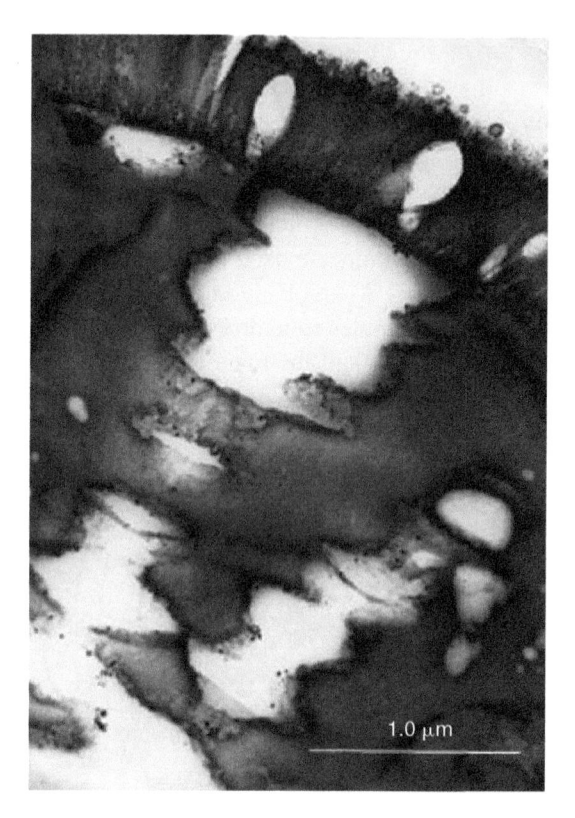

Figure 5.14 Transmission electron micrograph of waxy maize starch granule after α-hydrolysis showing internal canal of corrosion. In the outer shell diameter, the size of the thin canalicles is about 25 nm. (Gallant, unpublished)

starches have specific openings and/or susceptible surface zones, which become enlarged and/or form pits due to endocorrosion. As the pores or pits become larger, canals of endoerrosion sink into the granules.[1,11] In maize starch at least, and probably also in other starches with endogeneous pores and channels, channel openings appear to be enlarged, not only from the outside in, but also from the inside out. Scanning electron microscopy observations at high magnification of starch granules that have been treated by amylases indicate that granules of at least some starches are composed of small (50–500 nm diameter), more or less spherical 'blocklets.'[1,11,52] These blocklet structures are shown in Figure 5.15, and are further discussed with respect to starch granule structure in Sections 5.4 and 5.5.

The super helical structure in starch has been characterized by transmission electron microscopy of small, negatively-stained granule fragments. With proper preparation conditions, the lamella structure is not disrupted.[24] Using results of electron tomography on negatively-stained potato starch granule fragments and cryo-electron diffraction of frozen-hydrated granule fragments, a model for the structural organization of amylopectin in potato starch granules has been proposed.[3] In this model, helices form a continuous, regular crystalline network, which appears to be a framework skeleton around which the rest of the granule is built. The crystalline domains

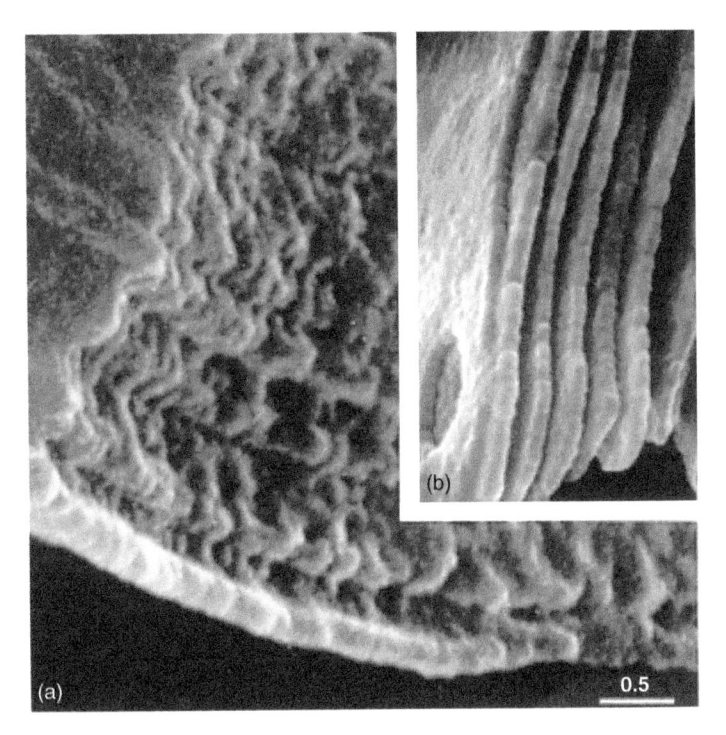

Figure 5.15 Scanning electron micrographs of starch granules after mild α-amylolysis showing the occurrence of spherical blocklet-like structures. (a) potato; and (b) wheat starch granules. (Adapted with permission from reference 1)

containing the double helical linear segments in the amylopectin molecules form a continuous network consisting of left-handed helices packed in a tetragonal array. Since neighboring helices interpenetrate each other, the crystalline lamellae form a more or less continuous superhelical structure. Such a semicrystalline structure is built up from more or less continuous left-handed fragments, and has a diameter of approximately 180 nm and a pitch of 10 nm (Figure 5.16). A central cavity within the super helices would have a diameter of about 8 nm. It is proposed that the pitch of the helix in starch originates from the clustering of branch points and may be characteristic for the botanical source. Helical pitch would therefore be determined directly by the specificity of the branching enzymes involved in the synthesis of amylopectin (Chapter 4).

III. The Granule Surface

Characterization of the state, nature and structure of the starch granule surface has been somewhat limited. Over the past few years, however, substantial progress in starch surface chemical analysis and surface imaging has been made. The following two sections detail recent developments in these two new areas of starch research.

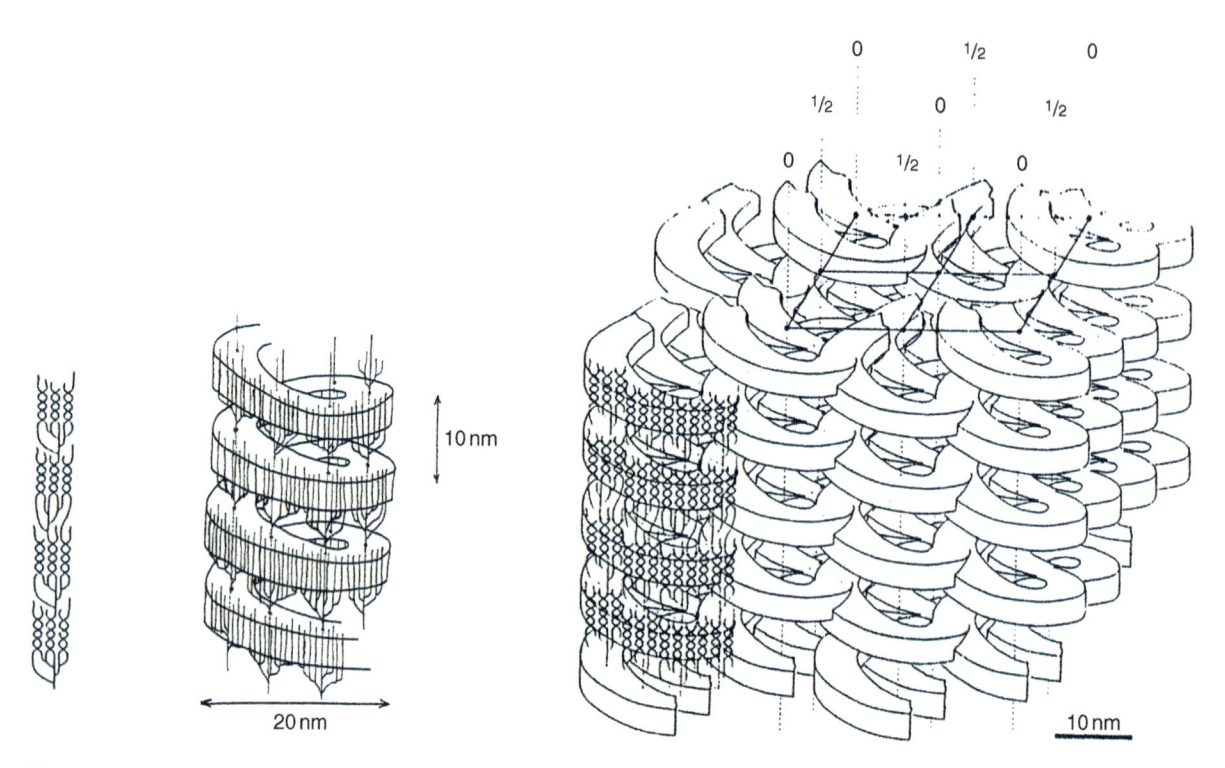

Figure 5.16 Schematic model for the arrangement of amylopectin in potato starch. Crystalline layers containing double helical linear segments in amylopectin molecules form a continuous network consisting of left-handed helices packed in tetragonal arrays. Neighboring molecules are shifted relative to each other by half the helical pitch. (Adapted with permission from reference 3)

1. Starch Granule Surface and Chemistry and Composition

For most starches, the external surface of starch granules is the first barrier to processes such as granule hydration, enzyme attack and chemical reaction with modifying agents. Consequently, it is becoming recognized that the nature of the granule surface, and particularly the presence of surface proteins and lipids, may have significant effects on the properties of the starch.[53–58] For example, the presence of lipids at the granule surface appears to influence the rheological properties of starch–water pastes,[57] and changes to both the protein and lipid at the granule surface are implicated in the improved baking performance of chlorine-treated flour.[54,56,58,59]

Starch granule-associated protein and lipid are by far the most abundant of the minor components of starch.[60,61] These components are thought to be incorporated into the granule during its synthesis, although in a cereal starch sample which has not been well washed, some protein from the endosperm may adhere to the granule. The true granule-associated proteins are distinct from the endosperm (gluten) proteins and many are believed to be enzymes which were involved in granule synthesis (hence, such proteins are found even in tuber starches).[60,62] The exact origin of the true starch lipids is not fully known, although it has been hypothesized that free fatty acids and lysophospholipids are normal membrane degradation products that are rendered metabolically inactive by incorporation in the granule (i.e. complexation with amylose – see below).[61] Furthermore, it has been hypothesized that some lipids may have a direct role in the function of granule-bound enzymes, and that lipo-protein complexes may exist.

The exact quantity of protein and lipid present depends on the species and variety of starch; however, typical well washed cereal starch samples contain ~0.3% protein and up to 1.0% lipid, while a typical root or tuber starch may contain ~0.05% protein and 0.05–0.1% lipid.[63] These values result from bulk analysis of starch; in reality, the actual distribution of the components is not uniform, with both protein and lipid enrichment believed to be present towards and at the granule surface.[58,64,65]

2. Surface-Specific Chemical Analysis

To date, surface-specific analysis of the starch granule has been limited. Until recently, the only substantial work in this field was the XPS (x-ray photoelectron spectroscopy) study of native and chlorine-treated starch granules performed in 1987 by Russell et al.[66] Recently, however, substantial surface specific analysis of a range of native starch granules has been performed using the complementary techniques of XPS[58] and time-of-flight secondary-ion mass spectrometry (TOF-SIMS).[67] The semi-quantitative TOF-SIMS study allowed extensive characterization of the various carbohydrates and lipid species found in the outer 1–2 nm of native (and chlorine-treated) wheat, rice and potato starch granules.[67] Peaks assigned to carbohydrate structures included a series of peaks (at m/z 221-, 383-, 545-, 707-, 869- and 1031-) corresponding to dimers, trimers, etc. of the basic glucosyl monomer unit and a peak at m/z 405- which corresponds to a dimer with one ionic phosphate group attached (at either the O-1, O-3 or O-6 position) (O-1 should, however, be excluded). Characterization of the lipid peaks identified the free fatty acids, glycerides and phosphoglycerides at the surface of each sample, including precise determination of their hydrocarbon chains. The presence of nitrogen-containing species was also evidenced, and attributed to peptide fragments, although the precise identification of peptide species was not possible with the instrument used. The complementary quantitative XPS study of native (and chlorine-treated) starch granule surfaces allowed the elemental composition of the outer 5–10 nm of the granule surface to be reproducibly and accurately determined. The results, which have yet to be fully published, build on and confirm the study by Russell et al.,[66] and lend considerable weight to the view that substantially higher levels of protein and lipid are found at or near the surface of the granule.[58] Mathematical modeling from the elemental percentages provided by XPS experiments suggests that the bulk (~90% in cereal starches and ~95% in potato starch) of the granule surface is carbohydrate in nature.[58] Granule surface protein was found to represent around 5% of the cereal starch surface and 0.05% of the potato starch surface. Both starch types may have up to 5% lipid at granule surfaces.[58]

Initial progress in characterizing and understanding the surface chemistry of starch granules has thus been made. In association with such chemical characterization, significant advances in the investigation of the structure of the starch granule surface have also been made, due to recent advances in surface imaging techniques.

IV. Granule Surface Imaging

Since the early 1990s, surface imaging has rapidly and extensively advanced, due to the development of a wide range of scanning probe microscopy (SPM) techniques.[68] The development of the atomic force microscope (AFM) in 1986,[69] which stemmed from the invention of the scanning tunneling microscope (STM), revolutionized the field by permitting molecular and even atomic resolution imaging of non-electrically conducting samples. The technique allows samples to be imaged under ambient conditions, either in air or under liquids, and usually requires minimal sample preparation compared to other forms of microscopy. Furthermore, a range of AFM modes and derivative techniques exist,[68,70] allowing mapping of the mechanical and chemical properties of the sample, as well as producing images. Consequently, prospects of imaging under virtually 'native' conditions at superior resolution to that obtained using current scanning electron microscopy (SEM) techniques has sparked substantial interest in AFM imaging of non-electrically conducting biological samples, such as plant cell walls, polysaccharide gel networks and living cells.[70–76] Applications of AFM imaging to starch research have been limited in number,[58,59,75,76] but they have successfully demonstrated the potential of the AFM technique in starch research by revealing important new insights into starch granule structure, molecular organization and degradation kinetics. This section aims to demonstrate the significant advantages that AFM imaging of starch offers over SEM imaging, and to highlight the important future role of the AFM technique in furthering understanding of both starch granule architecture and its component polysaccharide polymers.

1. Granule Imaging by SEM Methods

Until the 1990s, research aimed at imaging starch granule surfaces was largely limited to conventional scanning electron microscopy (SEM). Starch granules are not, however, ideal specimens for SEM imaging, due to their particulate, biological, non-electrically conducting nature. Thus, while SEM has proved to be (and continues to be) invaluable in increasing our knowledge of the starch granule surface, internal structure, patterns of enzyme attack and transformations,[11,78–85] it can rarely be used to its full potential with biological samples (in terms of image resolution), due to charging and subsequent damage of the electrically insulating starch granules. Furthermore, samples for conventional SEM can never be imaged in their original state, since the imaging is performed in a high vacuum and techniques of preparation associated with it may change the structure of the products and introduce undesirable artifacts.

Environmental scanning electron microscopes (ESEM) are new variable pressure SEMs working at moderate vacuum in the sample chamber (in the range of 1 to 50 Torr) while the electron column is maintained under high vacuum.[86,87] This SEM provides imaging capabilities with uncoated dry, moist or oily samples, thus reducing the risks of damaging samples as well as creating artifacts. Létang et al.[87] studied bread dough at different water contents (50–56% moisture). McDonough and Rooney[88] observed that the starch granules in a fully-hydrated wheat starch–water dough appeared rounder and plumper than when viewed with conventional SEM under high vacuum. Fannon et al.[84] used ESEM in order to investigate specific surface openings of some starch granules, with samples being examined wet or dry in a water vapor environment of 2–10 Torr. However, ESEM experiments have not been extensively developed for, or applied to, starchy products; even structural studies of the behavior of starch granules during hydrothermic processing and enzyme treatments could be exploited using this technique.

Using conventional SEM methods, drying or freezing (cryo-SEM) is employed to either dehydrate the samples or fix the water phase temporarily as a solid. In order to achieve high-quality images of starch, the specimen has to be metal-coated (e.g. with gold or platinum) to avoid sample charging and to improve the secondary electron signal strength (although it is still notoriously difficult to obtain quality high-resolution images of starch granule surfaces, even after coating).

Used in order to follow the behavior of starchy components during thermal processing (water uptake, granule swelling, leaching of amylose, retrogradation) cryo-SEM has revealed structural features not seen by classic cyto-techniques.[89] Some of the highest resolution images of starch granules were acquired in the 1970s using an SEM instrument capable of working with extremely low beam currents (1×10^{13} A).[1,50,90,91] High-resolution images of gold-coated (dried), α-amylase-attacked starch granules were achieved, revealing information about granule internal structure (see below).[79,81,92,93] Low-voltage scanning electron microscopes (LVSEM) also offer considerable advantages over conventional SEMs for imaging of biological samples, because the primary electron beam in a LVSEM is generated via field emission rather than by thermo-ionic emission. Therefore, such instruments are capable of achieving high-quality images (resolutions of a few nanometers are quoted for modern instruments) at very low accelerating voltages (e.g. 500 V).[94] The technique is still performed in a vacuum, but sample charging and damage are considerably reduced, permitting most electrically insulating samples to image without coating. This is of advantage in high-resolution studies, since the thin metallic surface coats required by conventional SEM are particulate in nature (particle size can be up to 10 nm) and hence, there is a potential problem of misinterpretation of results at high resolution.[74] LVSEM has been extensively applied for imaging of a range of insulating biological specimens,[94–96] but its application to starch granule imaging has been limited. However, the usefulness of this technique for providing superior images of native uncoated starch granule surfaces, as compared to other forms of SEM, which complement and validate higher resolution AFM studies, has been demonstrated.[59,60,97]

2. Principles of AFM

Detailed technical descriptions of the AFM technique, also known as scanning force microscopy (SFM), can be found elsewhere.[69,98,99] In brief, AFM is a lens-less microscopy technique in which a sample surface is brought into such close proximity with a fine crystal tip mounted on a flexible cantilever that atomic interaction forces between the tip and the sample cause the cantilever to bend. Bending of the cantilever in the z (height) plane is monitored and, by scanning the tip across the sample, a three-dimensional image of the sample surface topology is acquired (Figure 5.17). The AFM

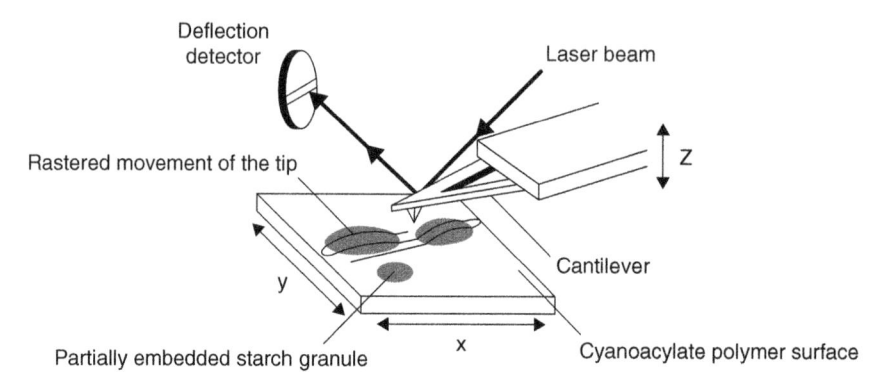

Figure 5.17 Schematic diagram representing AFM imaging of partially-embedded starch granules.

instrument is capable of molecular or even atomic resolution on appropriate samples, with vertical resolutions in the region of 1–2 Å (0.1–0.2 nm) and lateral resolution of 10 Å (1 nm) or less.[75] It is widely recognized, however, that 'soft' biological materials, such as organic molecules, generally are problematic with respect to higher resolutions.[72,100] In support of this, Weihs et al.[101] have speculated that the limit of imaging resolution of 'soft' (e.g. biological) samples may be 2–3 nm (20–30 Å) at best with small forces applied in air. Nevertheless, it is clear that such resolution coupled with the minimal need for uncoated sample preparation, the ability to image in a range of environments (ambient, air, liquid, cryo, heated), and the extensive range of imaging AFM modes offers opportunities for further investigation of starch granules.

3. Sample Preparation for AFM Imaging of Granular Starch

Sample preparation for AFM imaging is critical for achieving reproducible, high-quality and meaningful results. Although routine methods have been reported for preparation of isolated polysaccharide chains on flat supports,[70] AFM imaging of starch granules (particles that have diameters in the range of ~0.1 to 200 μm) presents a number of problems directly connected with the size and morphology of the samples. In particular, these problems merge with curved or dotted surfaces, or when special methods for material preparation are required in order to expose the bare inner surface of the starch granule.

Ideally, AFM should be performed on flat surfaces, due to the need to raster the cantilever across the sample at a micrometric scale, and because large sample height differences (greater than a few micrometers) may either prevent tip scanning, due to the limits of the vertical motion of the cantilever or cause 'self-imaging' of the tip, due to the sides of the tip becoming involved in image generation,[100,102,103] Non-fixed particles may also suffer problems of probe-induced particle movement or 'sweeping' as the probe moves across the sample.[103] It is for these reasons that almost all published AFM studies of whole starch granules[59,67,77,105] have involved embedding of the granules in order to immobilize them, and to reduce the overall height variation of the sample surface. It was reported[106] that physical destruction using a glass homogenizer was more suitable for observation of the bare inner surfaces of the starch granules than was using digestion or cutting, but the new tendency is to study thin sections of granules encased in resins using the contact mode.[107–110]

4. Surface Detail and Inner Granule Structure Revealed by AFM

Using suitable preparation techniques, reproducible AFM imaging of starch granules has been performed. While studies performed to date had different aims and resulted in images of the starch granules at different resolutions, they are complementary in nature and demonstrate both the versatility and applicability of the AFM technique to starch research. It is, nevertheless, important to recognize the primary sources of artifacts in AFM images.[104,111]

Real-Time AFM Studies of Degradation by Enzymes

The AFM study of wheat starch granules from the cultivar Timmo by Thomson et al.[77] is the earliest recorded usage of AFM to image the starch granule surface. The aim of the study was to demonstrate the potential of the AFM technique to image a biological process in real-time, using starch granule digestion by α-amylase as the example system. From images of the granules taken in air (before introduction of an aqueous solution), Thomson et al.[77] reported that the surface of the starch granule appeared to be 'dotted with features between 50 and 450 nm across,' and that a highly elongated structure was seen on the surface of a granule in one image. No explanation of the 50 to 450 nm structures was given, but they postulated that the elongated structure may have been a contaminant, possibly a protein molecule. Some granules were reported to have 'cracked' surfaces, and it was postulated that this may have arisen from milling of the wheat kernels during extraction of the flour. On introduction of the enzyme solution (after any swelling had finished), the surface of the granules appeared to become covered by the enzyme rapidly, although the binding appeared to be weak, since the enzyme was easily displaced by the AFM tip. They then followed the attack of the starch granules by enzyme molecules by recording images at 105-second intervals. The granule appeared to be attack by α-amylase preferentially (and rapidly) at sites of damage (i.e. weak spots) via the classic 'pin-hole' attack pattern.[1] A pile-up of material seen at the edge of the hole was attributed to possible interference of the tip in the hole formation. As the hole became deeper, the shape of the hole appeared to become square, due to the inability of the pyramidal AFM tip to accurately image inside the hole, i.e. the square shape of the hole reflects that of the AFM tip.

High-Resolution AFM (and LVSEM) Studies of Starch Granule Surface Structure

In parallel with the surface-specific chemical analysis of the native starch granule surface (detailed above), high-resolution imaging of native starch granule surfaces was performed using both LVSEM and AFM.[58,59,97] The aim was to visualize at high-resolution the surface structure of native starch granules with known differences in granule size, structure and minor components.

Observation of potato starch granule surfaces: in general, in LVSEM images (Figure 5.18a), the potato starch granule surface had a rough, undulating appearance with occasional indentations and numerous small, raised nodules that were approximately 100 to 300 nm in diameter (evident as small bright areas). The roughness of the surface can be clearly seen in the images (Figure 5.19a), where the presence of the small raised nodules (approximately 50 to 300 nm in diameter) is highly apparent. Higher-resolution images[97] indicate that the raised nodules appear to protrude above a smoother surface which consists of smaller semi-spherical structures (~20–50 nm in diameter). Large surface 'blebs' (gross surface 'lumps' occurring mainly at the 'poles' of a granule,[112] occasionally evident on some potato starch granules), have been imaged by both LVSEM and AFM.[104]

Observation of wheat starch granule surfaces: in comparison with the potato starch surface, the surface of both a and b wheat starch granules appeared to be relatively smooth (Figure 5.18b), possessing only a few surface blemishes and an occasional raised nodule. From the corresponding AFM images (Figure 5.19b), it is evident that the wheat starch granule surface appears to consist of a mass of similarly-sized structures (20–50 nm in diameter). Some larger structures (50–300 nm in diameter) corresponding to the raised nodules of the potato starch are present, although their

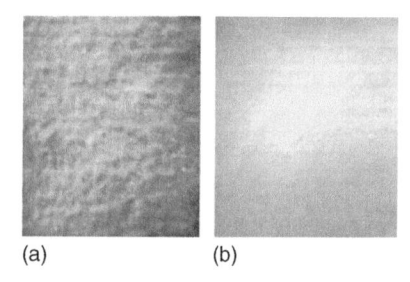

(a)   (b)

Figure 5.18 (a) LVSEM micrograph of a typical potato starch granule surface region (magnification 10000×, scale bar represents 3 μm); (b) LVSEM micrograph of a typical region of a Riband wheat starch granule surface. (Magnification 10000×, scale bar represents 3 μm)

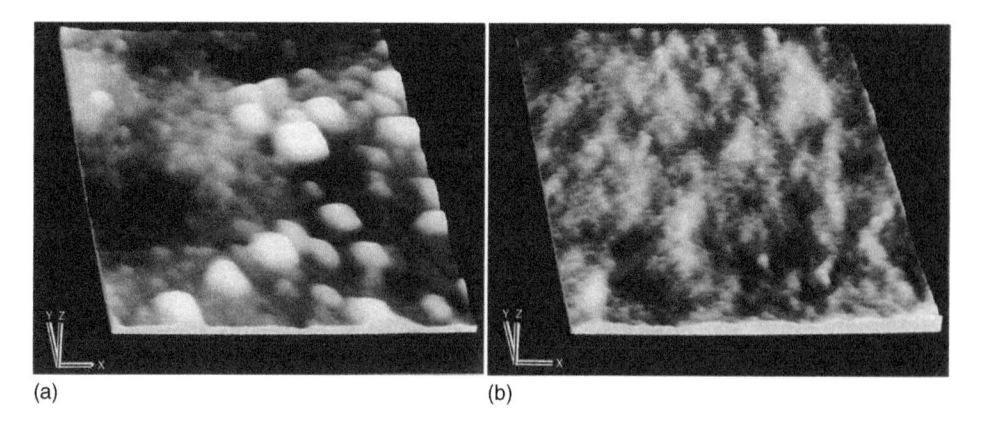

(a)   (b)

Figure 5.19 (a) AFM image of a typical surface region of a potato starch granule (scan size 1000 nm², z height difference 73.8 nm); (b) AFM image of a typical surface region of a Riband wheat starch granule (scan size 1000 nm², z height difference 52.6 nm)

occurrence is far less frequent than on potato starch granules. The same observations were reported for barley, oat, maize, waxy maize and triticale granules.[113,114]

Another method used for starch is the non-contact mode, where the starch granules are simply spread onto an adhesive tape fixed on the AFM sample holder. Using this technique, both nodules (20–50 nm in diameter) and smooth areas without any visible features were observed at the surface of potato starch granules.[114] Multiple freeze–thaw cycles have revealed an internal, lamellar structure of chain clusters bundled into 'blocklets' 50–300 nm in diameter.[115]

Observation of inner structures of starch granules: as noticed by Ohtani et al.[106] AFM, which gives greater resolution as compared with electron microscopy, needs very precise tools. In particular, the amplitude mode image may be used to enhance edge contrast of samples, but resolution of the probe remains limited by the radius of the crystal tip and its finite thickness. This very important point and the choice of method of sample preparation are the main limiting factors for imaging accuracy. AFM images in the tapping mode obtained from starch granules from different botanical sources[106] showed no large size aggregates but, in all cases, similar small particles of approximately 30 nm in diameter, often arranged in straight, chain-like structures. Particularly visible on the AFM images of crushed sweet potato starch granules, such chains appeared bundled into rods or larger columns. Similarly, ring-like protrusions were seen (non-contact mode) in potato starch granules frozen in excess water, then dried.[116] These structures were interpreted as individual 'blocklets' or single clusters, because their size was comparable to that of 'blocklets.'

5. Interpretation of AFM Images with Respect to Granule Structure

Observation of two starch types by AFM and LVSEM demonstrates that, on a macromolecular scale, the surfaces of starch granules differ, i.e. there are significant differences in granule surfaces between starches from different botanical sources. Observation of the raised nodules (50–300 nm in diameter) on granule surfaces of granules corresponds well with the observation of Thompson et al.[77] of 'features between 50 and 450 nm across,' although the structures are thought to be carbohydrate in nature.[97] This deduction arises from the facts that: (a) the features are far more common on potato starch granule surfaces than on the surfaces of wheat starch granules (the surface protein content of potato starch granules is much less than that of wheat starch granules); and (b) protein molecules are far smaller than the 50 to 300 nm structures, e.g. one hemoglobin molecule which has a molecular weight of ~66 kD (similar to starch granule-bound starch synthase SGBSSI) has a diameter of only 5.5 nm. Thus, extremely large numbers of protein molecules (e.g. multi-molecular complexes) would have to be present to explain the quantity and size of the structures seen on the surface of potato starch granules by AFM. Such a quantity of protein does not, however, fit with the observation that ~95% of the granule surface is carbohydrate in nature. Moreover, potato starch granules (which have far more of the raised nodules than do wheat starch granules) have only ~0.05% surface protein (i.e. about tenfold less protein than does wheat starch).[58]

Observations regarding the surface structure of potato and wheat starch granules in the AFM study of Baldwin et al.[58,59,97] support the SEM observations of Gallant et al.[1,11,50] who described the structure of starch as being composed of 'more or less spherical blocklets,' which were observed both within and at the surface of starch granules degraded by α-amylase. In general, the 'blocklets' were found to be larger (400–500 nm in diameter) in starches of the B- (e.g. potato starch) and C-crystalline types than in starches with the A-crystalline type (e.g. wheat starch, in which the blocklets were 25–100 nm in diameter). Full discussion of the findings with respect to current (and historic) views on the structure of starch granules has been made;[11] they are further presented in Section 5.5. However, in brief, it was shown that, in wheat starches, larger blocklets (100 nm) are found in the hard crystalline 'shells' of the granule than in the softer amorphous shells (blocklet size is ~25 nm in the amorphous shells). In potato starch, very large (400–500 nm) blocklets are found in the outer 10 μm of the granule, with smaller blocklets towards the granule center. Such observations (in association with other isolated evidence in the literature; see Section 5.5) have demonstrated the real presence of a blocklet structure in starch, and have led to the hypothesis that large blocklet size contributes to starch resistance due to locally increased levels of crystalline structure.[11] Observation of this crystalline blocklet structure in starch granules leads to obvious questions as to whether the blocklets represent single molecules of amylopectin, and if so, whether the different-sized blocklets represent molecules of amylopectin with different molecular weights and/or different degrees of branching. Currently, such questions cannot be definitely answered from data and images available. However, Martin and Smith[117] stated that the average diameter of an amylopectin molecule is 200 to 400 nm, and that such a molecule would contain 20 to 40 side chain clusters. It is, therefore, clear that these average dimensions for the amylopectin molecule are congruent with the average size of the raised nodule (blocklets) seen at the surface of, and within, starch granules as investigated by AFM[106] and SEM.[1,11,50] Furthermore, from published values relating to the size of a glucosyl unit and to the size of amylopectin side chain clusters, it is possible to estimate the number of glucosyl units and amylopectin side chain clusters present in the raised nodule (blocklet) structures. Firstly, since the diameter of a glucosyl unit has been estimated to be approximately 0.4 nm,[2] it is evident that a raised nodule structure with a diameter of 20 nm could contain an absolute maximum of 50 glucosyl units across its diameter (if they were placed directly adjacent to one another). The glucosyl units are not, however, side-by-side, but arranged into A- and B-type crystalline lattices, as demonstrated by computer modeling simulations of the starch polymers.[118] The computer models were based on amylose chains; however, the crystalline arrangements are thought to be relevant to the side chains (A- and B-chains) of amylopectin molecules which form crystalline clusters. As previously stated, the clusters are alternatively grouped with the branching regions of the amylopectin to form layers of crystalline and amorphous lamellae.[16] The average diameter of an amylopectin side chain cluster has been estimated to be 10 nm.[16] It therefore follows, that the smallest structures seen at the surface of starch granules by AFM, which are ~10 nm in diameter, can be attributed to a single amylopectin side chain cluster, and that the slightly larger structures (or blocklets, 20–50 nm in diameter) are on average composed of between 2 and

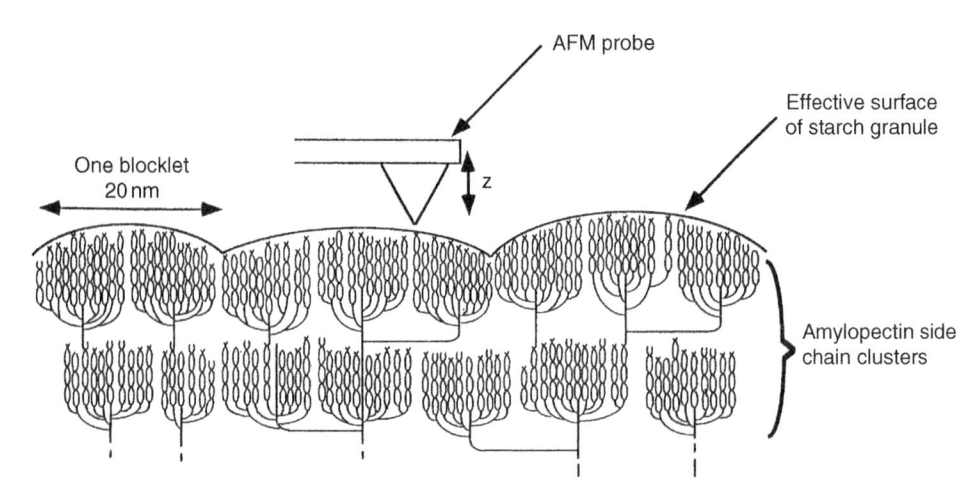

Figure 5.20 Schematic diagram of amylopectin side chain clusters forming the 10–50 nm raised nodule ('blocklet') structures seen at the starch granule surface.

5 amylopectin side chain clusters. The substantially larger blocklets (50 to 500 nm in diameter) would therefore contain between 5 and 50 amylopectin side chain clusters. On this basis, it has been postulated[97] that the structures revealed in the AFM images are amylopectin side chain clusters grouped into blocklets seen from above (Figure 5.20), and that the near-molecular resolution of potato and wheat starch granule surfaces has, therefore, been achieved by AFM imaging.

6. Discussion of Granule Surface Imaging by Scanning Probe Microscopy (SPM)

Representative Sampling and Imaging Artifacts

Due to the very nature of the high-resolution imaging provided by AFM, the total sampling area in AFM studies is inherently very low. Consequently, it is highly important to ensure that sample representivity is achieved in the resulting images. As detailed above, low-voltage (field-emission) SEM can be viewed as a complementary technique to AFM, due to its applicability to biological samples after minimal sample preparation and its ability to provide high-resolution images of uncoated samples. The technique facilitates imaging of comparatively large areas on numerous samples, and provides images that normally have sufficiently high resolution to allow confirmation of the AFM images. Furthermore, the technique provides a prior knowledge of the sample at high resolution, thus giving extra security in ensuring correct interpretation of subsequent AFM images. This is evident in the study of Baldwin et al.,[97] where there is clearly a high correlation between the AFM and LVSEM images with respect to granule surface topology of the two starch types, allowing the conclusion that subsequent AFM images are representative.

While the use of a separate LVSEM instrument to ensure sampling representivity provides high-resolution imaging of uncoated samples, various other approaches also exist. For example, there are now scanning probe microscopes located within scanning electron microscopes[119] and AFM instruments which attach to inverted optical microscopes.[68] These systems allow SEM imaging of large surface areas (e.g. 1 cm^2), followed by 'zooming-in' to analyze the area by SPM. Thus, not only is exactly the same sample imaged by both techniques, thereby providing a full range of resolutions, but the state of the SPM tip can be regularly checked by SEM. Clearly, however, SPM imaging in such instruments is performed in a vacuum, and the use of conventional SEM requires biological samples to be metal-coated. The use of a combined LVSEM/AFM instrument would, however, be of interest for biological samples.

Instruments in which an AFM head is attached to an inverted optical microscope offer optical and SPM imaging of the same biological samples in a range of environmental conditions, and are particularly powerful when coupled to a high-resolution optical microscope such as a confocal laser scanning microscope (CLSM). Complementary images of samples can thus be obtained at a range of resolutions to ensure representivity (and correct AFM positioning). The images may also contain some chemical information, due to the feasibility of fluorescence optical imaging and some internal structural detail of the sample, due to the optical sectioning ability of the CLSM.

AFM Imaging Mode, Imaging Environment and Improving Resolution
In general, non-contact or tapping-mode AFM is recommended for biological specimens,[74,120] since these modes reduce the potentially damaging forces (both lateral and vertical) experienced by the sample due to the close proximity of the AFM probe. This advice holds for small, easily-displaced biological samples (e.g. individual molecules) and for soft, easily-deformed biological samples (e.g. viscous liquids and living animal cells), but experience has shown that non-contact or tapping-mode AFM is not necessarily required for high-resolution imaging of whole, native starch granules. At ambient conditions (~22°C, ~10% water content) starch granules are below their glass transition point[121] and are, therefore, comparatively rigid and robust compared to other biological specimens (e.g. live cells). The AFM examinations detailed above demonstrate that the constant-force, contact-mode of AFM in air (and in an aqueous environment) can be successfully employed for imaging of partially-embedded starch granules, without the generation of gross imaging artifacts.

The AFM imaging environment chosen is often governed by the type of experiment required. As stated above, AFM is unique in its ability to offer high-resolution imaging under a variety of imaging environments, e.g. gas, vacuum, liquid (aqueous or non-aqueous). Thomson et al.[77] demonstrated imaging of starch granules in air and in an aqueous environment. Baldwin et al.[97] acquired high-resolution images of starch granule surfaces in air. This choice was made because a starch granule surface under ambient conditions in air was deemed to be in as 'native' a state as possible, but it is possible that even higher-resolution imaging of starch granules is possible if imaging is performed under other conditions. This is because any moisture in the air has a tendency to condense onto both the sample surface and the AFM tip. Capillary action between the tip and the surface may then cause large attractive forces between the two in the order of 10^{-6} to 10^{-8} N, which tends to increase the chances of sample damage and decrease image resolution.[70,122] Consequently, imaging of biological samples, particularly 'soft' samples or isolated molecules, is often performed under an alcohol (e.g. 1-butanol), where problems of water absorption are not experienced, hence permitting imaging with smaller, less damaging contact forces.[70,123] Whole starch granules appear to be relatively robust, thus allowing comparatively high-resolution imaging in air without significant sample damage, but imaging under alcohol might be considered in future work as a means of achieving higher resolution. Similarly, there is increasing interest in the imaging of biological samples in vacuum, as suggested in the original AFM paper of Binnig et al.[69] Although imaging in vacuum may cause potentially damaging sample dehydration, image resolution is often improved, due to the reduction in contact forces obtainable under such conditions.

So far, AFM imaging of starch granules has only been performed with pyramidal Si_3N_4 tips. While these tips are standard, high-quality tips, other types of tips exist which, due to their higher aspect ratios, can generally be used to acquire higher-resolution images than those obtained with the standard Si_3N_4 tips. The properties of most commonly available AFM tips (as well as most common AFM imaging artifacts) have been extensively reviewed.[102,104] In general, it is held that by reducing the diameter at the tip of the probe, image resolution is improved, due to reduction in interaction forces between the sample and the probe tip.[101,102] However, very thin tips are more easily damaged, and the risk of damaging the sample increases as tips are sharpened.[100] Clearly, therefore, in achieving higher resolution imaging of starch, the choices of tips, imaging mode and imaging environment need to be optimized. However, the use of higher aspect tips, in non-contact/tapping mode, in air or under alcohol, would seem to represent logical steps in the right direction.

7. Future Prospects of SPM of Starch

The preceding sections demonstrate that scanning-probe microscopy of starch is in its infancy. The few SPM studies performed to date, however, demonstrate the potential of AFM for observing the starch granules at high resolution under a range of conditions, and it is evident that it is becoming possible to observe structures which previously could only be investigated by x ray scattering techniques. To date, however, except for a few cases, AFM imaging of starch has been limited to contact-mode imaging in air. Under such conditions, AFM imaging of 'soft' biological samples has been predicted to have an imaging resolution limit of 2–3 nm at best, in spite of the improved method of image acquisition known as 'AFM by error signal image,' which considerably enhances the high spatial frequencies and sharpens the contrast of features. Starch granules are not typical soft biological samples, and hence, it seems probable that by employing the best imaging conditions (as outlined above) this limit may be surpassed. Furthermore, the origin of image contrast in AFM was recently identified.[109,124] A comparative study of sectioned pea starch granules, either embedded in a non-penetrating matrix of rapid set Araldite or in a hydrophilic melamine resin before sectioning, with sections submitted to hydration during sample preparation, showed that, in contact mode images, contrast is introduced and enhanced by swelling of softer and easier to compress hydrated material. Consequently, structural features are visible in higher resolution AFM images, because the procedure highlights amorphous versus crystalline materials. Nevertheless, it is clear that an image resolution of 2–3 nm, coupled with the minimal need for sample preparation and the ability to image in a range of environments (ambient, air, liquid, cryo, vacuum) offers considerable scope for further AFM investigation of both starch granule surface topology and internal structure.

V. A Hypothesis of Starch Granule Structure: The Blocklets Concept

As stated above, AFM microscopy of starch is in its infancy; however, in either the contact or the non-contact mode, results on the whole provide substantial support of earlier SEM observations and suggest that a level of crystalline structure between that of the large 'growth' rings and the amylopectin lamellae exists in starch granules. There is however, substantial other evidence (both from microscopy and non-microscopy techniques) in favor of such a level of granule structure, which has been termed the 'blocklet level of structure.'[11] This section details this evidence and presents the current re-emergence of the 'blocklet concept' of starch granule structure.

The idea of crystalline units in starch is not new, and can be originally traced back to the prescience of Nägeli,[125] although it was Badenhuizen[126] who first demonstrated the presence of natural resistant units of material in chemically degraded starch. Badenhuizen described these resistant blocks as 'blökchen Strüktur,' from which the term 'blocklet concept' is derived. At the same time, a fibrillar concept of starch granule structure was developed[127,128] and continued, even though it was proven that features supposed to be fibrillar structures at the TEM scale were artifacts.[129] It now appears, however, that both concepts of granule structure (the blocklet and the fibrillar) were in fact founded in truth, with the fibrillar concept being somewhat related to the radial organization of the starch polymers, while the blocklet concept relates to a higher order of crystalline organization within granules.

While SEM[1,11,50] and AFM[58,97,106–108,113–116,124,130,131] images provide strong visual evidence for a blocklet structure of starch, a considerable amount of other supporting evidence from electron microscopy and enzyme treatments exists. Helbert and Chanzy,[132] using hydrophilic melamine resin for preparation of ultra-thin starch granule sections for transmission electron microscopy (TEM), also obtained images in which the outline of individual blocklets with dimensions of a few hundred nanometers were seen. Such structures have been imaged at higher resolution in sectioned corn starch granules contrasted with periodic acid (thiosemicarbazide) silver (PATAg). At a very low degree of oxidation,[50,133] the PATAg marker, which penetrates and thus highlights only the amorphous regions of the granule, revealed the presence of roughly ellipsoidal regions of 20 to 500 nm in diameter, which were less easily penetrated by the marker (silver ions) that corresponded to roughly spherical crystalline blocklets within the granule.[11] Furthermore, the high resolution of the TEM revealed evidence of alternating crystalline and semi-amorphous amylopectin lamellae within the blocklets.[11]

V. A Hypothesis of Starch Granule Structure: The Blocklets Concept

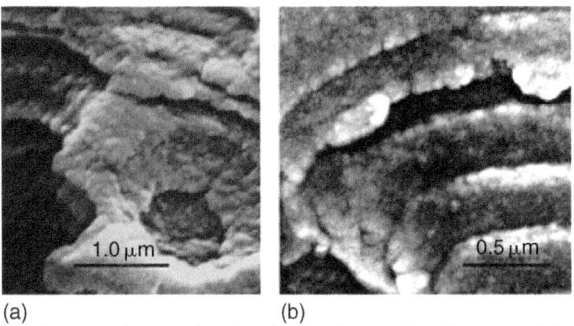

Figure 5.21 Scanning electron micrographs of maize starch granules after α-amylolysis showing resistant shells composed of blocklet-like structures: (a) at internal canal of corrosion level; and (b) at shells level. (Gallant, unpublished)

Further evidence of an intermediate level of granule structure is evident from a number of enzymic degradation studies of starch (Figure 5.21). Bertoft[134] found that the molecules of amylopectin from large and small barley starch granules and from waxy maize starch granules are composed of 'super-clusters' with molecular weights of approximately 10^5. These he interpreted as arising from 'highly ordered regions of amylopectin.' Furthermore, specific (1→4)-α-D-glucosidic linkages between the super-clusters appeared to undergo preferential degradation during initial stages of α-amylolysis, thus releasing the super-clusters. Such observations fit with the idea that an extra level of amylopectin crystallization exists in starch granules, and it has been hypothesized by Gallant et al.[11] that the super-clusters relate to the blocklet structure of starch. The preferentially degraded (1→4)-α-D-glycosidic linkages between the super-clusters must, therefore, be located in the more amorphous 'channel' regions between the blocklets and, thus, are more readily accessible to degradation.

Other approaches to structural determination of starch have been performed by Yamaguchi et al.,[20] by Kassenbeck,[25,135] and by Oostergetel and van Bruggen.[3,24] Yamaguchi et al.[20] used negative staining of crushed, lintnerized starch granules and described 'worm-like ripple structures' in maize starch granules, which they interpreted as 5-nm thick, crystalline lamellae created by association of double helices perpendicular to the plane of the lamellae. Oostergetel and van Bruggen, using a more sophisticated procedure, studied lintnerized wheat[24] and potato starch[3] granules. Three-dimensional reconstructions of the residual crystallites in potato starch were done using negatives taken from a tilt series in the TEM treated by a low-pass Fourier filter. The helical structure observed three dimensions in stereo mounts and revealed that the organization of lamellae was much more complex than was previously thought to be the case. As a result, they proposed the concept of a 'super helical' structure, which clearly represents a level of structure between that of stacks of lamellae and the granule 'growth' ring.

Debranching studies of starch lends further support to the idea that a blocklet level of structure exists. Hizukuri[15] demonstrated that B-chains of amylopectin can participate in more than one crystalline amylopectin side chain cluster, and proposed a revised model of amylopectin structure, classifying the B-chains according to the number of side chain clusters in which they participate. From this work, it is evident that B-chains may link the amylopectin side chain clusters to form larger crystalline units. Such evidence corresponds well with the blocklet concept of starch structure. Furthermore, Hizukuri[15] demonstrated that the connecting B-chains were more abundant in potato starch, and proposed that they are probably characteristic of starches with the B-crystal pattern, since such starches have higher amounts of B2–B4 chain fractions. This observation fits with the observation of Gallant et al.[11] that, in general, blocklet size is larger in starches with the B-crystal pattern.

The evidence to date in favor of the blocklet concept of starch granule structure, therefore, indicates that the amylopectin lamellae are organized into effectively spherical blocklets which range in diameter from 20 to 500 nm, depending on starch botanical type and their location in the granule.[1,11,50,136] This organization fits with the current knowledge of starch granule structure and is represented schematically in Figure 5.22. In general, the blocklets are larger (400–500 nm in diameter) in starches of the B- (e.g. potato starch) and C-crystalline types than in starches with the A-crystalline type (e.g. wheat starch, in which the 'blocklets' were 25 to 100 nm in diameter). The large blocklets in potato starch appear to predominate near the granule surface. Consequently, it has been hypothesized that, while granule resistance appears to be linked to several interacting factors, the size of the blocklets (i.e. the degree of local crystallinity) may play a role in starch granule resistance to acid- and enzyme-catalyzed hydrolysis.[11] Blocklet size may, therefore, play a role in the relative resistance of the outer shell of potato starch, and also in the relative resistances of crystalline and semicrystalline growth shells.

Results from every AFM study support the blocklet model. However the blocklet model was newly refined[110] after it was found that contrast, in AFM, was mainly due to localized absorption of water, leading to selective swelling of the matrix material within the starch granule.[124,137] This discovery was the result of a series of studies[107,109,110,124,131] with a goal of comparing near-isogenic pea starches differing in amylose:amylopectin ratio. Structural differences were characterized with AFM, allowing a totally new interpretation of the internal structure of the starch granule. General morphology of the blocklets was imaged using shaded topography and error signal mode images, the matrix being characterized by the force modulation image. The new information leads to the conclusion that the starch granule is composed of continuous hard blocklets dispersed in a softer matrix material. Growth rings are often visible in pea starch granules (wild type, *lam*, and *rug3* low-amylose mutants), but sometimes are missing (high-amylose *r* mutant). Growth rings possibly originate from localized defects in the blocklet production around the surface of spheroidal shells during biosynthesis. In some cases, such as in the *lam* and *rug3* mutants, blocklets are not easily distinguished, and the spheroidal features in the matrix are smaller than the blocklets in parental starch granules. The blocklets may be embedded in a harder interconnecting matrix, possibly amylose in partial crystalline form (high-amylose *r* mutant), or may be enclosed in a hard fine network which encloses a group of blocklets (*rrb* high-amylose mutant).

Thus, each type of starch has its own structural features that now are well-characterized by AFM, using the new, high-resolution technique, which reveals structures much more accurately than in the past. The difficulty remains to determine the

V. A Hypothesis of Starch Granule Structure: The Blocklets Concept 183

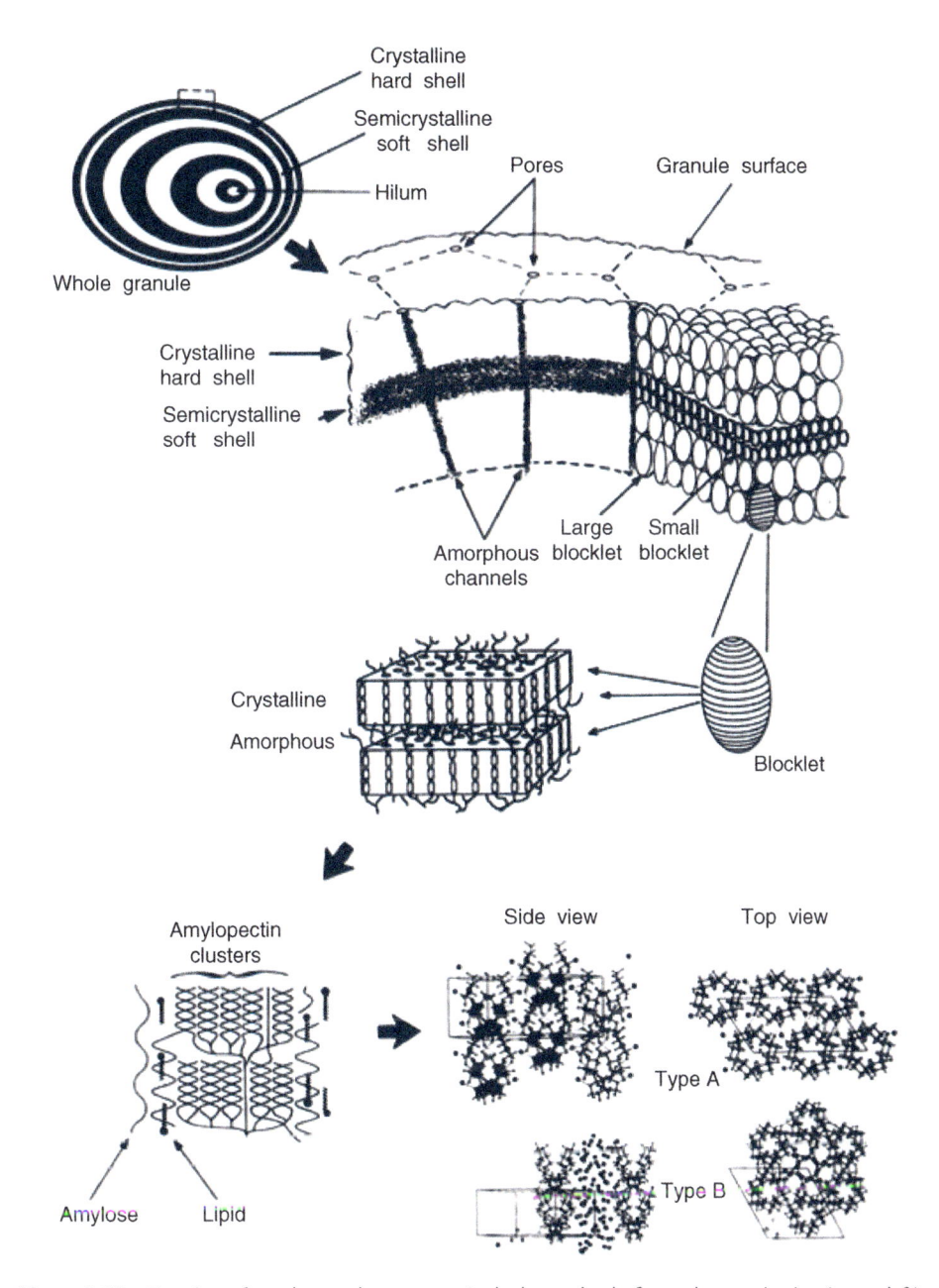

Figure 5.22 Overview of starch granule structure. At the lowest level of granule organization (upper left), alternating crystalline (hard) and semicrystalline (soft) shells are shown (dark and light colors, respectively). Shells are thinner towards the granule exterior (due to increasing surface area to be added to by constant growth rate) and the hilum is shown off-center. At a higher level of structure, the blocklet structure is shown in association with amorphous radial channels. Blocklet size is smaller in the semicrystalline shells than in the crystalline shells. At the next highest level of structure, one blocklet is shown containing several amorphous crystalline lamellae. In the next diagram, amylopectin is shown in the lamellae. The next image is a reminder of the importance of amylose–lipid (and protein) components in the organization of amylopectin chains. At the highest level of order, crystal structures of the starch polymers are shown. (Adapted with permission from reference 1; redrawn in reference 1 from references 32 and 34)

relationship between biochemical data (which gives mean values for the bulk) and data which comes from restricted areas selected under the microscope. Biochemical analysis of smooth pea starch (35% amylose) showed that these starch granules were quite resistant to hydrolysis catalyzed by α-amylase, but a substantial part (amorphous or weakly crystalline) that was more easily solubilized was restricted to the inner parts between the growth rings.[138] Analysis of the β-limit dextrins indicated a significant amount of intermediate branched material (intermediate between amylose and amylopectin) with units of clusters regularly interconnected.[139] The smooth pea granule, a C-type, is a mixture of A-type and B-type crystalline orders. Recording of juxtaposed micro-x-ray diffraction patterns has been performed (thanks to the accuracy of the synchrotron x-ray beam) which demonstrated that the two crystalline phases co-exist within the same granule (Figure 5.9). In the same manner, we may imagine that the accuracy of AFM will soon permit imaging very close areas within such starch granules, thus enabling different types of crystallites to be visualized. Such correspondence between AFM and biochemical data is easier with wrinkled pea starch (64–76% amylose, B-type). Internal areas of the granules are in an amorphous state and parts of the amylose molecules are anchored in the crystalline area.[140] A greater amount of long chain lengths, lower cluster sizes and the presence of intermediate material[141] are elements which could be linked to understand AFM imaging of wrinkled pea starch.

The exact location and organization of amylose in relation to the blocklet organization is not yet known; however, substantial information regarding the location and state of amylose within the granule is beginning to be gathered and is summarized in Section 5.6. The presence of locally more amorphous regions between the blocklets has, however, been visualized by TEM.[11] These regions appear to form amorphous channels between the blocklets. It can be hypothesized that a locally higher concentration of amylose (particularly short chains) may be present in such regions, but consideration of the ratio of amylopectin to amylose (e.g. 75% to 25%) in starch granules and the overall crystallinity (15% to 45%) of starch granules leads to the conclusion that amorphous amylopectin is certainly also present in the amorphous regions of the starch granule.

It is evident from the above discussion that considerable evidence exists for a blocklet level of granule organization between that of the lamellae and the growth rings. Clearly, this level of granule structure has significant implications for the internal architecture (and consequently properties) of starch granules.

VI. Location and State of Amylose Within Granules

Until recently, the location and state of amylose within granules was one of the most important questions remaining to be answered. Three main hypotheses for the location of amylose within starch granules have been put forward. The first hypothesis is that amylose is laid down tangentially to the radial orientation of amylopectin in order to minimize the amylose–amylopectin helical interactions.[142] There is, however, no experimental basis for such a model, which merely considers the need for amylose and amylopectin to have minimal helical interactions. The other two hypotheses advocate radial deposition of the amylose, either in bundles[43,143] or as individual chains, which are randomly interspersed amongst amylopectin clusters in both crystalline and semicrystalline regions.[144,145] Of the three hypotheses, the third hypothesis appears to be the most sustainable, since it has been demonstrated, via a crosslinking experiment using corn starch, that amylose molecules do not crosslink to one another, but do crosslink to amylopectin chains.[144,145] In order for this crosslinking of amylose with amylopectin to have occurred, hydroxyl groups from the two chains must have been within 7.5 Å of each other.[145] No crosslinking of amylose chains with other amylose chains was observed, which rules out the second hypothesis that amylose chains occur in bundles. The currently accepted model of amylose location in starch granules is, therefore, as individual, radially-orientated chains randomly distributed among the radial amylopectin chains.

There is now substantial evidence indicating that there is an enrichment of amylose towards the periphery of the granule,[64,146–149] and that the amylose molecules found near the surface of the granule have shorter chain lengths than those located nearer the center of the granule.[143] Ring et al.[150] demonstrated that the majority of the amylose in granules can be leached out of granules at temperatures just below the gelatinization temperature. They further demonstrated that the majority of the leached amylose chains were in the single helical state, rather than in a double helical state. The single helical state is, thus, thought to be the predominant state of amylose chains within native starch granules. It is clear, however, from the evidence that amylose can only be completely extracted from granules at temperatures above 90°C,[53] that some large amylose molecules are present within the starch granule, and it is further hypothesized that these larger amylose chains that are not easily leached may participate in double helices with amylopectin.[145]

In association with the enrichment of amylose towards the granule surface, an enrichment of lipid towards (and at) the granule surface is also believed to exist. Evidence in support of this arises from x-ray microanalysis (EDX) and x-ray photoelectron spectroscopy (XPS) which have demonstrated an enrichment of phosphorus (and hence phospholipids at least) towards and at the granule surface.[64,65,151,152] Furthermore, it is now well-established that there is a good correlation between amylose and lipid contents of all normal (i.e. non-mutant) cereal starches.[61,153] Waxy cereal starches (as well as legume and tuber starches) have small amounts of surface lipids and little or no internal lipids, while high-amylose cereal starches tend to have more lipid (most of which is internal) than do the corresponding normal starches.[154] Thus, the quantity of lipid in starch granules appears to be linked to the amylose content, and the lipid appears to be distributed similarly to the amylose. Furthermore, evidence exists that lipid, like the amylose (and amylopectin) polymers, is aligned radially within the granule.[155,156] Both the distribution of the lipid molecules in the granule (which is similar to that of amylose) and its probable radial orientation fit well with evidence that a small proportion of the amylose chains are involved in amylose–lipid complexes. Until recently, evidence for the existence of amylose–lipid complexes in native starch was entirely indirect, arising principally from observations of the conditions required to extract lipids from starches,[146,157] and from the V-type x-ray diffraction pattern which can be observed with native starches under the right conditions. Direct evidence for the presence of amylose–lipid complexes in native starches has recently been provided by use of (^{13}C-cross-polarization/magic-angle spinning nuclear magnetic resonance (^{13}C-CP/MAS-NMR).[158] Using this technique, it has been demonstrated that amylose–lipid complexes occur in many native cereal starches, including wheat, barley, maize, oat and rice. Comparison of the amylose content of most normal cereal starches (i.e. 25–30%) with the total lipid content of waxy starches (maximum ~1%) suggests that only a small proportion of the amylose in such starches is complexed with lipid. Similarly, it is hypothesized that not all lipid in starch is involved in amylose–lipid complexes and that a free-lipid fraction exists.[156]

In summary, therefore, substantial information regarding the location and state of amylose has been obtained. Individual amylose chains are believed to be randomly located in a radial fashion among the amylopectin molecules. The concentration of amylose (and lipid) increases towards the surface of the granule, with smaller (leachable) amylose chains predominating near the surface. Amylose chains are believed to be in a single helical state, although a small proportion may be involved in lipid complexes.[147] Some of the larger (non-leachable) amylose chains may be involved in double helical interactions with amylopectin.[159]

VII. Surface Pores and Interior Channels of Starch Granules

Small pores, also called 'pin holes', have been observed under SEM on the surface of starch granules. Fannon et al.[84] showed that pores could be found on the granules of corn, sorghum and millet starches, as well as along the equatorial groove of the large granules of wheat, rye and barley starches. Pores were not found on compound starches (rice and oat), nor on tuber and root starches (tapioca, arrowroot, canna and potato). Using SEM and ESEM, they proved in a logical fashion that pores were not artifacts but real features (about 1 μm in diameter) that might be elaborated during granule growth and might be under genetic control.

On the other hand, the surface of the starch granules has been studied by AFM[113–115,130] at high resolution in the non-contact mode, focusing on 'protrusions,' 'depressions' and 'pores.' Depressions 1 μm diameter or less in potato and tapioca starches and pores <100 nm in rice and <40 nm in oat starch granules and in granules of all the cereal starches examined were seen. However, it is impossible to know exactly the depth of the depressions because the 'z' measurements were done from the gray values in the images, unfortunately without any true reference point. But one can wonder if the resolution of this new tool (AFM) is opening a new world of discoveries. Important differences may be observed between high-resolution surface images of starch granules, either performed in the contact mode or in the non-contact mode. In the contact mode, blocklets appear to be of regular shape; in the non-contact mode, the surfaces appear to be very irregular in shape, with plenty of free space or voids.

Fannon et al.[160] suggested that surface pores are openings to channels (5 to 400 nm in diameter) connecting an internal cavity at the granule hilum to the external surface. Studies by Huber and BeMiller[161,162] proved this concept unequivocally. Using waxy maize and sorghum starches and different techniques, e.g. detection of fluorescent dyes and colloidal gold particles by fluorescence microscopy, confocal scanning laser microscopy and by SEM backscattered electron imaging, they proved that these channels and cavities were not filled with amorphous material, but were voids. As mentioned above, only certain starch granules show such openings, suggesting that channels are a predetermined system. One can wonder 'What is the benefit of pores and channels to the plant?' It has been reported that developing and expanding pores were openings to internal corrosion channels with saw-tooth patterns,[90] and that 'these pores may be the site of initial enzyme attack, openings that allow enzyme molecules direct access to the granule interior (hilum) or both.'[84] During the later stages of wrinkled pea starch synthesis, exceptional increase in long chain length occurs. During this stage, the starch granules acquire central fissures. Bertoft et al.[140] suggested that 'soluble starch synthase becomes entrapped at the cracks where it then maintains the synthesis of amylose.' It is possible that pores and channels are related to the control of starch conversion during seed germination[84] or other metabolic activities.

VIII. Conclusions

The intent of this chapter was to start with the polymer composition of starch and continue through the various levels of increasing structural complexity, in order to yield some understanding of the levels of starch granular structure. It has been shown that, with the aid of ever-improving analytical techniques, significant progress has been made in understanding the higher degrees of granule order. The entire picture of starch granule architecture appears to be becoming better understood; however, it should be emphasized that much of the progress made consists of small pieces of information regarding specific features of certain granule structures, and many important questions remain unanswered. For example, it should be remembered that while detailed atomic level models of starch crystal structure have been developed, these only represent a small portion of the granule structure (i.e. the ordered crystalline lamellae containing the double helical amylopectin side chains), since much of the amylopectin in starch is in an amorphous state. From the work of Gidley and Bociek,[48] it is evident that some of this 'amorphous' amylopectin may be in a double helical state; however, the structure of such 'amorphous double helical' amylopectin remains to be described. Similarly, while substantial information regarding the state and location of amylose chains within the granule is now known, numerous further questions regarding the state and location of starch polymers remain. Among these are:

1. Do double helical interactions between amylose and amylopectin chains occur within the granule?
2. Do the semi-amorphous 'growth' shells have a local increase in amylose content (while respecting the general increase in amylose towards the granule surface)?

3. Does amylose disrupt the ordered packing of the amylopectin clusters (i.e. does it play a role in blocklet size)?
4. Are the amylopectin side chains capable of complexing lipids as it was shown in maize starches examined by DSC.[163]

In the course of this chapter, a number of studies of starch have been reviewed. These studies, by revealing substantial new information regarding both the surface chemistry and the structure of the starch granule, highlight the usefulness of applying novel techniques (such as AFM and TOF-SIMS) to examinations of starch. Consequently, it is clear that both surface-specific analysis and scanning-probe microscopy of starch are techniques that can be applied to add to our knowledge of granule structures. Other novel domains may also be considered. In particular, one can envisage using the microdiffraction facilities that synchrotron radiation provides to map the occurrence and the relative orientations of crystalline domains within the same granule. Parallel to these developments, use of chromatographic techniques, in conjunction with enzyme-catalyzed digestion, has provided a detailed description of some essential repeating motifs occurring in starch granules of a given botanical origin. These are the distributions of the lengths of the primary and secondary branches, as well as their attachments onto a main amylosic branch. Unfortunately, existing molecular modeling techniques are not capable of simulating the three-dimensional features of these structures, since they encompass several tens of thousands of atoms. Computing technology is rapidly developing, however, and one can hope that ever-more complex simulations will become possible, thus aiding in the elucidation of the three-dimensional structures of the starch polymers. Via incorporation of biological considerations derived from the structural and functional biology of the different steps of starch biosynthesis (Chapter 4), those computational methods will greatly help in establishing a consistent description and understanding of the several levels of starch structural organization.

As early as 1858, Carl Nägeli stated; 'The starch grain … opens the door to the establishment of a new discipline … the molecular mechanics of organized bodies.'[125] Thus, despite the progress made in understanding starch granule structure, one can appreciate 'Pandora's box' has been opened only slightly, and we are only beginning to explore its boundaries.

IX. References

1. Gallant DJ, Bouchet B, Buléon A, Pérez S. *Eur. J. Clini. Nutr.* 1992;46:S3.
2. Zobel HF. *Starch/Stärke.* 1988;40:44.
3. Oostergetel GT, van Bruggen EFJ. *Carbohydr Polym.* 1993;21:7.
4. Buléon A, Colonna P, Planchot V, Ball S. *Int J Biol Macromol.* 1998;23:85.
5. Nikuni Z. *Chori Kagaku.* 1969;2.
6. French D. *J. Jpn. Soc. Starch Sci.* 1972;19:8.
7. Nikuni Z. *Starch/Stärke.* 1978;30:105.
8. Robin JP, Mercier C, Charbonnière R, Guilbot A. *Cereal. Chem.* 1974;51:389.
9. Manners DJ, Matherson NK. *Carbohydr. Res.* 1981;90:99.
10. Lineback DR. *J. Jpn. Soc. Starch Sci.* 1986;30:80.
11. Gallant DJ, Bouchet B, Baldwin PM. *Carbohydr. Polym.* 1997;32:177.

12. French D. In: Whistler RL, BeMiller JN, Paschall EF, eds. *Starch, Chemistry and Technology*. 2nd edn. Orlando, FL: Academic Press; 1984:184–247.
13. Hizukuri S, Takeda Y, Maruta N, Juliano BO. *Carbohydr. Res.* 1989;189:227.
14. Bertoft E. *Carbohydr. Res.* 1991;212:229.
15. Hizukuri S. *Carbohydr. Res.* 1986;147:342.
16. Manners DJ. *Carbohydr. Polym.* 1989;11:87.
17. Callaghan PT, Lelievre J. *Biopolym.* 1985;24:441.
18. Lelievre J, Lewis JA, Marsden K. *Carbohydr. Res.* 1986;153:195.
19. Callaghan PT, Lelievre J, Lewis JA. *Carbohydr. Res.* 1987;162:33.
20. Yamaguchi M, Kainuma K, French D. *J. Ultrastruc. Res.* 1979;69:249.
21. Gidley MJ, Cooke D. *Biol. Soc. Trans.* 1991;19:551.
22. Sterling CJ. *Polym. Sci.* 1962;56:10.
23. Blanshard JMV, Bates DR, Muhr AH, Worcester DL, Higgins JS. *Carbohydr. Polym.* 1984;4:427.
24. Oostergetel GT, van Bruggen EFJ. *Starch/Stärke*. 1989;41:331.
25. Kassenbeck P. *Starch/Stärke*. 1978;30:40.
26. McDonald AML, Stark JR, Morrison WR, Ellis RP. *J. Cereal Sci.* 1991;13:93.
27. Ring SG, Miles MJ, Morris VJ, Turner R, Colonna P. *Int. J. Biol. Macromol.* 1987;9:158.
28. Buléon A, Duprat F, Booy FP, Chanzy H. *Carbohydr. Polym.* 1984;4:161.
29. Kregger DR. *Biochem. Biophys. Acta.* 1951;6:406.
30. Wu HCH, Sarko A. *Carbohydr. Res.* 1978;61:27.
31. Wu HCH, Sarko A. *Carbohydr. Res.* 1978;61:7.
32. Imberty A, Chanzy H, Pérez S, Buléon A, Tran V. *J. Mol. Biol.* 1988;201:365.
33. Popov D, Burghammer M, Buléon A, Montesanti N, Putaux JL, Riekel C. *Macromol.* 2006;39:3704.
34. Imberty A, Pérez S. *Biopolym.* 1988;27:1205.
35. Lechert HT. In: Rockland LB, Stewards GF, eds. *Water Activity: Influences on Food Quality*. London, UK: Academic Press; 1981:223–245.
36. Pérez S, Imberty A, Scaringe RP. In: French AD, Brady JW, eds. *Computer Modeling of Carbohydrate Molecules*. Washington, DC: American Chemical Society; 1990:281–289.
37. Pfannemüller B. *Int. J. Biol. Macromol.* 1987;9:105.
38. Gidley MJ, Bulpin PV. *Carbohydr. Res.* 1987;161:291.
39. Gidley MJ. *Carbohydr. Res.* 1987;161:301.
40. Sair L. *Heat-moisture Treatment of Starches*. New York, NY: Academic Press; 1981.
41. Buléon A, Pontoire B, Riekel C, Chanzy H, Helbert W, Vuong R. *Macromol.* 1997;30:3952.
42. Waigh TA, Hopkinson I, Donald AM, Butler MF, Heidelbach F, Riekel C. *Macromol.* 1997;30:3813.
43. Buléon A, Gérard C, Riekel C, Vuong R, Chanzy H. *Macromol.* 1998;31:6605.
44. Bogracheva TY, Morris VJ, Ring SG, Hedley CL. *Biopolym.* 1998;45:323.
45. Imberty A, Pérez S. *Int. J. Biol. Macromol.* 1989;11:177.
46. Jane JL, Wong KS, McPherson AE. *Carbohydr. Res.* 1997;300:219.
47. Gérard C, Planchot V, Colonna P, Bertoft E. *Carbohydr. Res.* 2000;326:130.
48. Gidley MJ, Bociek SM. *J. Am. Chem. Soc.* 1985;107:7040.
49. O'Sullivan AC, Pérez S. *Biopolym.* 1999;50:381.

50. Gallant DJ. [*Doctorat d'Etat*]. Université de Paris VI; 1974.
51. Casset F, Imberty A, Haser R, Payan F, Pérez S. *Eur. J. Biochem.* 1995;232:284.
52. Duprat F, Gallant DJ, Guilbot A, Mercier C, Robin JP. In: Monties B, ed. *Les polymères Végétaux – Polymères pariétaux et alimentaires non azotés*. Paris, France: Gauthier Villars; 1980:176–231.
53. Banks W, Greenwod CT. *Starch and Its Components*. Edinburgh, UK: Edinburgh University Press; 1975.
54. Cauvain SP, Gough BM, Whitehouse ME. *Starch/Stärke*. 1977;29:91.
55. Bowler P, Towersey PJ, Waight SG, Galliard T. In: Hill RD, Münck L, eds. *New Approaches to Research on Cereal Carbohydrates*. Amsterdam, The Netherlands: Elsevier Science; 1985:71–79.
56. Greenwell P, Evers AD, Gough BM, Russell PL. *J. Cereal. Sci.* 1985;3:279.
57. Nierle W, Bayâ AWE, Kersting HJ, Meyer D. *Starch/Stärke*. 1990;42:471.
58. Baldwin PM. [Ph.D. thesis]. University of Nottingham, UK; 1995.
59. Baldwin PM, Davies MC, Melia CD. *Int. J. Biol. Macromol.* 1997;21:103.
60. Schofield JD, Greenwell P. In: Morton ID, ed. *Cereals in a European Context*. Chichester, UK: Ellis Horwood; 1987:407–420.
61. Morrison WR. *J. Cereal Sci.* 1988;8:1.
62. Goldner WR, Boyer CD. *Starch/Stärke*. 1989;41:250.
63. Swinkels JJM. *Starch/Stärke*. 1985;37:1.
64. Morrison WR, Gadan HJ. *J. Cereal Sci.* 1987;5:263.
65. Malouf RB, Lin WDA, Hoseney RC. *Cereal Chem.* 1992;69:169.
66. Russell PL, Gough BM, Greenwell P, Fowler A, Munro HS. *J. Cereal Sci.* 1987;5:83.
67. Baldwin PM, Melia CD, Davies MC. *J. Cereal Sci.* 1997;26:329.
68. Newman A. *Anal. Chem.* 1996;68:267A.
69. Binnig G, Quaate CF, Gerber C. *Phys. Rev. Lett.* 1986;56:930.
70. Kirby AR, Gunning AP, Morris VJ. *Trends Food Sci. Technol.* 1995;6:359.
71. Hoh JH, Lal R, John SA, Revel J-P, Arnsdorf MF. *Science*. 1991;253:1405.
72. Yang J, Tamm LK, Somlyo AP, Shao Z. *J. Microscopy*. 1993;171:183.
73. Putman CAJ, van der Werf KO, Grooth BGd, van Hulst NF, Greve J. *Biophys. J.* 1994;67:1749.
74. Gunning AP, Kirby AR, Morris VJ, Wells B, Brooker BE. *Polymer Bull.* 1995;615.
75. Yang J, Shao Z. *Micron*. 1995;26:35.
76. Kirby AR, Gunning AP, Morris VJ. *Biopolym.* 1996;38:355.
77. Thomson NH, Miles MJ, Ring SG, Shewry PR, Tatham AS. *J. Vac. Sci. Technol. B, Microelectron. Nanometer Struc.* 1994;12:1565.
78. Hall DM, Sayre JG. *Textile Res. J.* 1969;39:1044.
79. Evers AD. *Starch/Stärke*. 1969;21:96.
80. Hall DM, Sayre JG. *Textile Res. J.* 1970;40:256.
81. Evers AD. *Starch/Stärke*. 1971;23:157.
82. Gallant DJ, Mercier C, Guilbot A. *Cereal Chem.* 1972;49:354.
83. Bowler P, Williams MR, Angold RE. *Starch/Stärke*. 1980;32:186.
84. Fannon JE, Hauber RJ, BeMiller JN. *Cereal Chem.* 1992;69:284.
85. Baldwin PM, Adler J, Davies MC, Melia CD. *Starch/Stärke*. 1994;46:341.
86. Li MJ, Rogers K, Rust CA. *Adv. Mater. Process*. 1995;148:24.
87. Létang C, Piau M, Verdie C. *J. Food Eng.* 1999;41:121.

88. McDonough CM, Rooney LW. *Cereal Food World.* 1999;44:342.
89. Verrez-Bagnis V, Bouchet B, Gallant DJ. *Food Structure.* 1993;12:309.
90. Gallant DJ, Derrien A, Aumaitre A, Guilbot A. *Starch/Stärke.* 1973;25:56.
91. Gallant DJ, Guilbot A. *Starch/Stärke.* 1973;25:335.
92. Fuwa H, Glover D, Sugimoto Y, Tanaka M. *Nutr. Sci. Vitaminol.* 1978;24:437.
93. Fuwa H, Sugimoto Y, Tanaka M, Nikuni Z. *Carbohydr. Res.* 1979;70:233.
94. Pawley J. *J. Microscopy.* 1984;136:45.
95. Hefter J. *Scanning Microscopy.* 1987;1:13.
96. Müllerova I, Lenc M. *Ultramicroscopy.* 1992;41:399.
97. Baldwin PM, Adler J, Davies MC, Melia CD. *J. Cereal Sci.* 1997;21:255.
98. Jahanmir J, Haggar BG, Hayes JB. *Scanning Microscopy.* 1992;6:625.
99. Quate CF. *Surface Sci.* 1994;299–300:980.
100. Blackford BL, Jericho MH, Mulher PJ. *Scanning Microscopy.* 1991;5:907.
101. Weihs TP, Nawaz Z, Jarvis SP, Pethica JB. *Appl. Phys. Lett.* 1991;59:3536.
102. Schwartz UD, Haefke H, Reimann P, Güntherodt HJ. *J. Microscopy.* 1994;173:183.
103. Shakesheff KM, Davies MC, Jackson DE, Roberts CJ, Tendler SJB, Brown VA, Watson RC, Barrett DA, Shaw PN. *Surf. Sci. Lett.* 1994;304:L393.
104. West P, Starostina N. *Advan. Mat., Nanoparticle Technol.* Santa Clara, CA: Pacific Nanotechnology; 2000:1–12.
105. Baldwin PM, Frazier RA, Adler J, Glasbey TO, Keane MP, Roberts CJ, Tendler SJB, Davies MC, Melia CD. *J. Microscopy.* 1996;184:75.
106. Ohtani T, Yoshino T, Hagiwara S, Maekawa T. *Starch/Stärke.* 2000;52:150.
107. Ridout MJ, Gunning AO, Wilson RH, Parker ML, Morris VJ. *Carbohydr. Res.* 2000;50:123.
108. Baker AA, Mervyn JM, Helbert W. *Carbohydr. Res.* 2001;330:249.
109. Morris VJ, Ridout MJ, Parker ML. *Prog. Food Biopolym. Res.* 2005;1:28.
110. Ridout MJ, Parker ML, Hedley CL, Bogracheva TY, Morris VJ. *Carbohydr. Polym.* 2006;65:64.
111. West P, Starostina N. *Advan. Mat., Nanoparticles Technol.* Santa Clara, CA: Pacific Nanotechnology; 2006:1–10.
112. Baldwin PM, Adler J, Davies MC, Melia CD. *Starch/Stärke.* 1995;47:247.
113. Juszczak L. *Electronic J. Polish Agric. Univ.* 2003;6:1.
114. Juszczak L, Fortuna T, Krok F. *Starch/Stärke.* 2003;55:1.
115. Szymonska J, Krok F. *Int. J. Biol. Macromol.* 2003;33:1.
116. Krok F, Szymonska J, Tomasik P, Szymonski M. *Appl. Surf. Sci.* 2000;157:382.
117. Martin C, Smith AM. *Plant Cell.* 1995;7:971.
118. Imberty A, Chanzy H, Pérez S, Buléon A, Tran V. *Macromol.* 1987;20:2634.
119. Ermakov AV, Garfunkel EL. *Rev. Sci. Instrum.* 1994;65:2853.
120. You HX, Lowe CR. *Curr. Opin. Biotechnol.* 1996;7:78.
121. Ollet AL, Kirby AR, Clarke SA, Parker R, Smith AC. *Starch/Stärke.* 1993;45:51.
122. Roberts CJ, Davies MC, Jackson DE, Tendler SJB. *Nanobiol.* 1992;2:73.
123. Goodman FO, Garcia N. *Phy. Rev. B.* 1991;43:4728.
124. Ridout MJ, Parker ML, Hedley CL, Bogracheva TY, Morris VJ. *Biomacromol.* 2004;5:1519.
125. Nägeli CW. In: Schülthess F, ed. *Pflanzenphysiologische Untersuchungen.* Zurich, Switzerland; 1858.

126. Badenhuizen NP. *Protoplasma.* 1937;28:293.
127. Sterling CJ. In: Radley JA, ed. *Starch and Its Derivatives.* 4th edn. London, UK: Chapman and Hall; 1968;139–171.
128. Wetzstein HY, Sterling C. *Stärke.* 1977;29:365.
129. Gallant DJ, Guilbot A. *Stärke.* 1971;23:244.
130. Juszczak L, Fortuna T, Krok F. *Starch/Stärke.* 2003;55:8.
131. Ridout MJ, Parker ML, Hedley CL, Bogracheva TY, Morris VJ. *Carbohydr. Res.* 2003;338:2135.
132. Helbert W, Chanzy H. *Starch/Stärke.* 1996;48:185.
133. Gallant DJ, Guilbot A. *Starch/Stärke.* 1969;21:156.
134. Bertoft E. *Carbohydr. Res.* 1986;149:397.
135. Kassenbeck P. *Starch/Stärke.* 1975;27:217.
136. Gallant DJ, Bouchet BB. *J. Food Microstructure.* 1986;5:141.
137. Morris VJ, Ridout MJ, Parker ML. *Prog. Food Biopolym. Res.* 2005;1:28.
138. Bertoft E, Manelius R, Qin Z. *Starch/Stärke.* 1993;45:215.
139. Bertoft E, Qin Z, Manelius R. *Starch/Stärke.* 1993;45:377.
140. Bertoft E, Manelius R, Qin Z. *Starch/Stärke.* 1993;45:258.
141. Bertoft E, Qin Z, Manelius R. *Starch/Stärke.* 1993;45:420.
142. Gidley MJ. In: Phillips GO, Williams PA, Wedlock DJ, eds. *Gums and Stabilizers for the Food Industry 6.* Oxford, UK: Oxford University Press; 1992:87–99.
143. Zobel HF. In: Alexander RJ, Zobel HF, eds. *Developments in Carbohydrate Chemistry.* St. Paul, MN: American Association of Cereal Chemists; 1992:1–36.
144. Jane JL, Shen JJ. *J. Carbohydr. Chem.* 1993;247:279.
145. Kasemsuwan T, Jane JL. *Cereal Chem.* 1994;71:282.
146. Boyer CD, Shannon JC, Garwood DL, Creech RG. *Cereal Chem.* 1976;53:327.
147. Morrison WR. In: Pomeranz Y, ed. *Advances in Cereal Science and Technology.* St Paul, MN: American Association of Cereal Chemists; 1978;224–348.
148. Schoch TJ. *Meth. Carbohydr. Chem.* 1964;4:25.
149. Shannon JC, Garwood DL. In: Whistler RL, BeMiller JN, Paschall EF, eds. *Starch, Chemistry and Technology.* 2nd edn. Orlando, FL: Academic Press; 1984:26–86 .
150. Ring SG, l'Anson KJ, Morris VJ. *Macromol.* 1985;18:182.
151. Morrison WR. *Starch/Stärke.* 1981;33:408.
152. Morrison WR, Milligan TP. In: Inglett GE, ed. *Maize: Recent Progress in Chemistry and Technology.* New York, NY: Academic Press; 1982:1–18.
153. South JB, Morrison WR, Nelson OE. *J. Cereal Sci.* 1991;14:267.
154. Morrison WR, Milligan TP, Azudin MN. *J. Cereal Sci.* 1984;2:257.
155. Hizukuri S, Nikuni Z. *Nature.* 1957;180:436.
156. Blanshard JMV. *Starch: Properties and Potential, Critical Reports on Applied Chemistry.* Vol. 13. London, UK: Academic Press; 1985:16–54.
157. Morrison WR, Coventry AM. *Starch/Stärke.* 1985;37:83.
158. Morrison WR, Tester RF, Snape CE, Law R, Gidley MJ. *Cereal Chem.* 1993;70:385.
159. Imberty A, Buléon A, Tran V, Pérez S. *Starch/Stärke.* 1991;43:375.
160. Fannon JE, Shull JM, BeMiller JN. *Cereal Chem.* 1993;70:537.
161. Huber KC, BeMiller JN. *Cereal Chem.* 1997;74:537.
162. Huber KC, BeMiller JN. *Carbohydr. Polym.* 2000;41:269.
163. Villwock VK, Eliasson A-C, Silverio J, BeMiller JN. *Cereal Chem.* 1999;76:292.

Structural Features of Starch Granules II

Jay-lin Jane

Department of Food Science and Human Nutrition and the Center for Crops Utilization Research, Iowa State University, Ames, Iowa, USA

I. Introduction	193
II. General Characteristics of Starch Granules	194
1. Granule Shapes, Sizes and Distributions	194
2. Porous Structures of Starch Granules	195
3. Shapes of Gelatinized Starch Granules	200
III. Molecular Compositions of Starch Granules	201
1. Amylopectin and Amylose	201
2. Intermediate Material and Phytoglycogen	202
3. Lipids and Phospholipids	204
4. Phosphate Monoesters	205
IV. Structures of Amylose and Amylopectin	205
1. Chemical Structure of Amylose	205
2. Single-Helical Structures (V-Complexes) of Amylose	208
3. Double Helical Structures of Amylose	211
4. Chemical Structure of Amylopectin	212
5. Cluster Models of Amylopectin	218
6. Effects of Growing Temperature and Kernel Maturity on Starch Structures	224
V. Locations of Molecular Components in the Granule	225
VI. References	227

I. Introduction

Starch is produced by green plants for energy storage and is synthesized in a granular form. Biosynthesis of starch granules takes place primarily in the amyloplast. The biosynthesis of the granule is initiated at the hilum and the starch granule grows by apposition.[1] Starch granules are densely packed with semicrystalline structures and have a density of about $1.5\,g/cm^3$.[2] Because of this stable semicrystalline structure, starch granules are not soluble in water at room temperature. Without gelatinization, potato starch can absorb up to 0.48–0.53 g of water per gram of dry starch.[3,4] The water sorbtion is reversible.

Starch granules are commonly found in seeds, roots, tubers, stems and leaves. Grain seeds, such as maize kernels, contain up to 75% starch on a dry starch basis (dsb). The stored starch granules can be converted by enzymes (amylases) to glucose, and the glucose is utilized to generate energy during germination or whenever energy is needed. In the granular form, starch can be isolated easily by gravity sedimentation, centrifugation and filtration, and can be subjected to various chemical, physical and enzymatic modifications with subsequent washing and processing. Consequently, starch is produced as one of the most economical commodity products.

There are two major starch polymers: amylopectin and amylose. Amylopectin is a highly branched molecule, with (1→4)-linked α-D-glucosyl units in chains joined by (1→6) linkages. Amylose is primarily linear with α-1-4 linked glucosyl units. Some amylose molecules, particularly those of large molecular weight, may have up to 10 or more branches.[5] Amylose and amylopectin have different properties. For example, amylose has a high tendency to retrograde and produce tough gels and strong films, whereas amylopectin, in an aqueous dispersion, is more stable and produces soft gels and weak films. Entanglements between amylose and amylopectin, particularly with the presence of lipids or phospholipids, have been demonstrated to significantly affect the pasting temperature, paste viscosity, stability and clarity, as well as the retrogradation rate of pastes.

Minor components found in starch granules include polymers of sizes and with properties intermediate between those of amylose and amylopectin, starch lipids (including phospholipids), monostarch phosphate ester groups and proteins, particularly granule-bound starch synthase. Phytoglycogen is found in certain cereal grains, such as the sugary-1 mutant of maize and rice. Some minor components, especially phospholipids,[6,7] free fatty acids and phosphate ester groups, although at low concentrations, can drastically affect the properties of starch pastes and gels. Phosphate monoester derivatives, carrying negative charges at neutral pH, are found in many starches. Potato starch, which has up to 0.09% phosphate monoester content, produces high paste viscosity and clarity, and has a relatively low gelatinization temperature. The unique properties of potato starch are attributed to the charge repelling between the covalently attached phosphate monoester groups.

An understanding of the internal organization of starch granules is crucial for scientists and engineers to optimize reaction conditions for chemical, physical and enzymatic modifications. The knowledge of the internal organization can help us understand the functionalities and the transformation behavior of starch, and improve the properties and stability of starch products. This knowledge will also help biochemists reveal the mechanism by which starch granules are developed during biosynthesis. In this chapter, some recent advances in the understanding of the structure of starch granules are reviewed. Readers are encouraged to refer to the review chapters of French[8] for earlier work in starch organization and Hizukuri[9] for analytical aspects of amylose and amylopectin.

II. General Characteristics of Starch Granules

1. Granule Shapes, Sizes and Distributions

Starches isolated from different botanical sources display characteristic granule morphology.[10–16] Starch granules vary in shape, including spherical, oval, polygonal, disk (lenticular), elongated and kidney shapes, and in size from <1 μm to 100 μm in diameter. Normal and waxy maize starches are spherical and polygonal in shape (Figures 6.1a and 6.1b). Potato starch has both oval and spherical shapes (Figure 6.1c). Wheat (Figure 6.1d), triticale, barley and rye starches have bimodal size distributions. In these starches, the large (A) granules have a disk shape, whereas the small (B) granules have a spherical shape. A recent study on molecular structures of the large (A) and the small (B) granules of wheat, barley and triticale starches shows that the amylopectin of the large, disk-shaped A granules consists of substantially more B2-chains and lesser A- and B1-chains than that of the small, spherical B granules.[11] The authors suggest that amylopectin molecules consisting of more B2- and fewer A- and B1-chains have cylindrical shapes, are easily aligned in parallel and lead to disk-shaped granules (see Section 6.2.3). In contrast, amylopectin molecules consisting of more A- and B1-chains display a cone shape, which tends to develop into spherical granules.[11] Sorghum starch (Figure 6.1e) also has a bimodal size distribution, but the shapes are different; large granules of sorghum starch are polygonal and spherical, instead of disk shaped. Sweet corn maize starch granules have different granule size (Figure 6.1f).[12] Granules of amaranth starch are small (1–2 μm) (Figure 6.1g).[10,13] High-amylose maize starches have elongated, filamentous granules, in addition to polygonal and spherical granules; some granules also have granule appendages (Figure 6.1h).[10,14] The greater the amylose content of the starch, the greater the number of filamentous granules found in the high-amylose maize starch.[17] Development of normal and high-amylose maize starch granules during maturation from the fifteenth to the 72nd day after anthesis was investigated using polarized light microscopy.[14] Morphological evolution of the starch granules in relation to amylose content is shown in Figure 6.2. Diffenbachia starch has an elongated shape with protuberances.[10] Shoti starch granules are in the shape of a disk with a sharp edge.[10] Almost all legume starches have a characteristic indentation of bean-like shapes on their granules.[10]

Most starch granules are produced individually in separate amyloplasts; however, some starches, such as rice, waxy rice, oats and wrinkled pea, contain compound granules (Figure 6.3). In those cases, more than one granule is produced simultaneously in a single amyloplast. The compound starches have granules that are tightly packed together and difficult to separate. The shapes of the compound starch granules are mostly polyhedral, possibly as a result of space constraints during the development of starch granules.

Diameters of starch granules vary from submicron, such as amaranth and small pigweed, to more than 100 μm (canna starch).[10] Small wheat starch granules have diameters of 2–3 μm, large wheat granules are 22–36 μm, potato granules are 15–75 μm, maize granules are 5–20 μm, rice granules range from 3–8 μm, and legume granules are 10–45 μm.[10] Small granule starches, such as amaranth,[13,18–21] cow cockle,[22,23] pigweed,[24] taro[25,26] and quinoa[27] have diameters between sub-micrometers to 2 μm.

2. Porous Structures of Starch Granules

Granules of sorghum, millet and corn starches have openings (pores) on the surface of the granules.[28] Pores are also found in the equatorial groove of the large granules

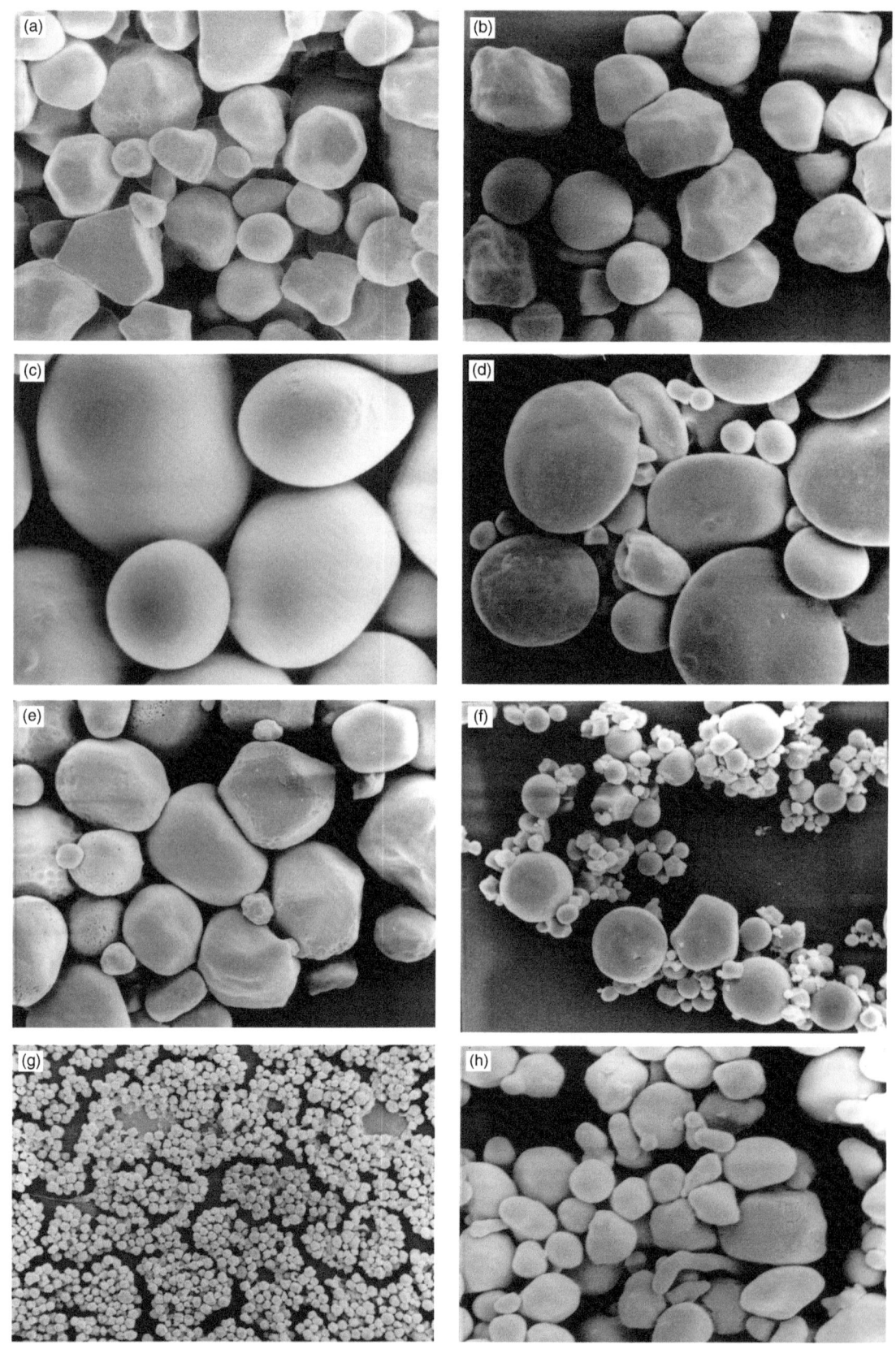

Figure 6.1 Scanning electron micrographs of starches: (a) normal maize; (b) waxy maize; (c) potato; (d) wheat; (e) sorghum; (f) sweet corn; (g) amaranth; (h) high-amylose maize 7.[10]

II. General Characteristics of Starch Granules 197

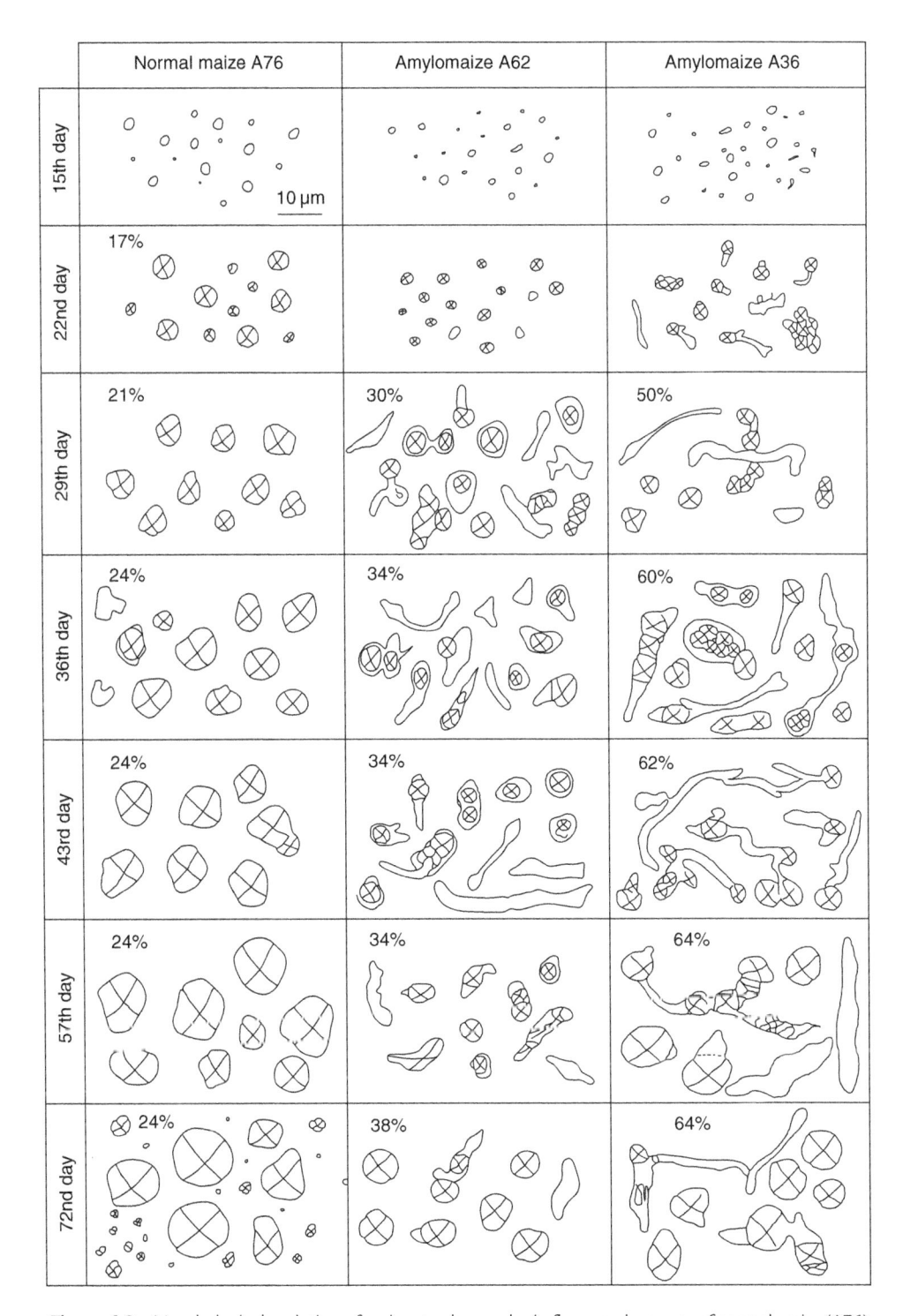

Figure 6.2 Morphological evolution of maize starch granules in floury endosperms of normal maize (A76) and high-amylose maize (A62 and A36) during maturation from the fifteenth to the 72nd day after anthesis, as seen under polarized light. Note that the filamentous starch granules of the high-amylose maize show Maltese crosses only in some nucleations. Percentages are amylose contents during maturation.[14]

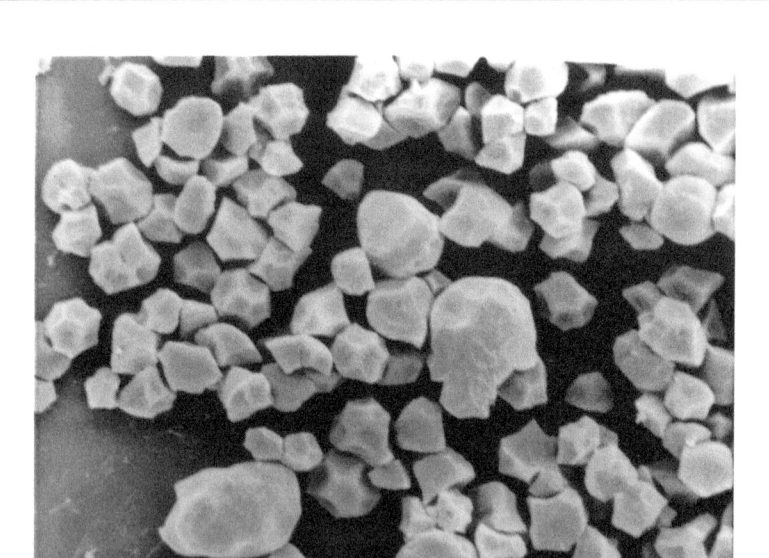

Figure 6.3 Waxy rice starch, a compound starch.[10]

of wheat, rye and barley starches,[28] but not in the small granules.[29] Pores are also found on tapioca,[15] rice and waxy rice starch granules.[10] In contrast, potato, arrowroot and canna starch granules have smooth, nonporous surfaces. Fannon et al.[30] first found channels penetrating into the granule, a finding confirmed by Huber and BeMiller.[31] The diameters of the external openings of the pores found on the surface of sorghum starch were estimated to be about 0.1–0.3 µm, and that of the internal channels about 0.07–0.1 µm). These pores have been proven to be normal structures and not to be caused by drying.[28] The pores and channels are congruent with the report by MacGregor and Ballance[29] that enzymes 'attack at discrete points on the granule surface, form tunnels into the granule interior, and then hydrolyze the granule from the inside out.' Enzyme-catalyzed hydrolysis of starch granules from the inside out has also been observed by others.[32–38] Therefore, the frequency and distribution of the pores are likely related to raw granule digestibility.

Recent studies have shown that pores on maize starch granules develop during a late maturation stage, i.e. 30 days after pollination (DAP).[39] Starch granules isolated from maize kernels harvested on 30 DAP show few or no pores, whereas those from kernels harvested on 45 DAP (fully matured and dried in the field) show a large number of pores.[39] Heavily pitted starch granules are observed close to the germ of maize kernels in a dormant state.[40]

After starch molecules have been removed from the periphery of granules by surface gelatinization using a concentrated neutral salt solution,[41,42] scanning electron micrographs (SEM) show that the starch structure at the hilum of the granule is loosely packed (Figure 6.4a). Light microscopy of neutral salt solution-treated wheat starch revealed that its granules gelatinize and swell at the equatorial groove,[43] whereas granules of other starches gelatinize evenly on the surface. It is known that drying starch generates cavities in starch granules.[44,45] Baldwin et al.[46] found that the

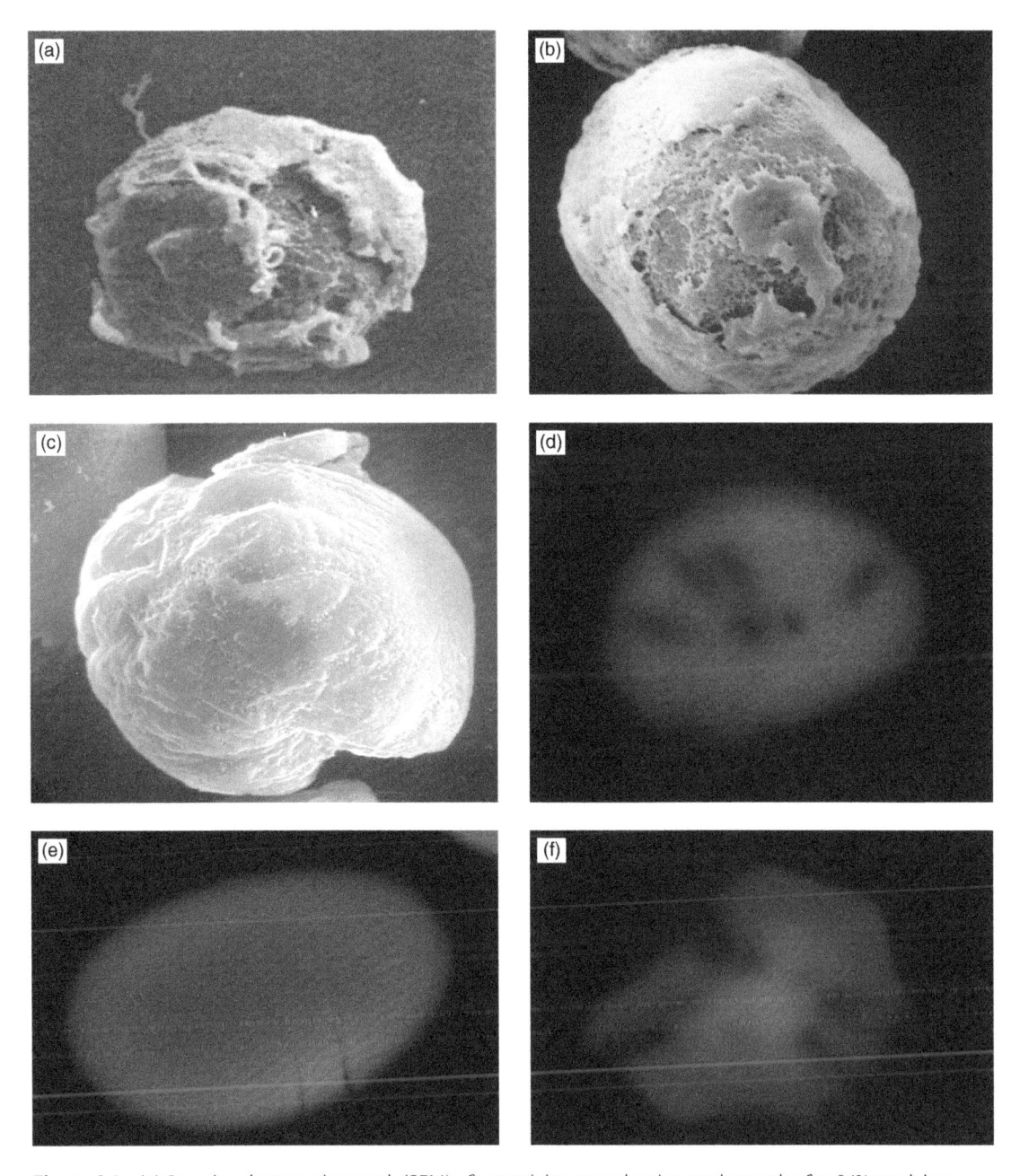

Figure 6.4 (a) Scanning electron micrograph (SEM) of a remaining normal maize starch granule after 84% starch has been removed by surface gelatinization. It shows a loose structure at the hilum of the granule.[42] (b) SEM of a remaining normal maize starch after 65% starch has been removed by surface gelatinization; the remaining starch granules show porous structures.[42] (c) SEM of a remaining potato starch granule after 52% starch has been removed.[41] (d) Confocal laser-light scattering micrograph (CLSM) of a normal maize starch granule; the image shows a porous internal structure of the starch granule.[55,56] (e) CLSM of a potato starch granule; the image shows a uniform internal structure.[55,56] (f) CLSM of a waxy maize starch granule; the image shows voids at the periphery of the starch granule.[55,56]

cavities at the hilum of potato, rice and wheat starch granules were spherical. They also found an increase in the number of cavities as a function of drying, the temperature of drying, and the variety of starch. Cavities at the hilum of maize starch granules are not spherical, but are irregular in shape, and are often star shaped.[31] There is no obvious relationship between the size of the cavity and size of the granule. After heat–moisture treatment (10–30% moisture, 100°C for 16 hours), starch granules also develop a hollow structure at the hilum of the granule, and display an increase in gelatinization temperature[47–52] and a decrease in enthalpy change, an indication of an annealing effect on the crystalline structure.[47] Heat–moisture treatments also affect B-type starch by changing the B-type x-ray pattern toward A-type, but have little effect on A-type starch granules.[48–50] This indicates that heat and moisture induce a rearrangement of starch molecule packing from the B- to the A-crystalline structure (a monoclinic unit cell). The starch molecules loosely packed at the hilum of the native starch granule are pushed aside by the pressure generated during heating. Ultra-high pressure treatments of starch granules (up to 690 MPa) in the presence of water transforms A-type starch into B-type-like starch.[53,54] The transformation is attributed to water being introduced into the crystalline packing unit under pressure, converting the monoclinic packing unit into hexagonal packing. There is little effect on the crystallinity of B-type starches by the same treatment.

SEM micrographs of surface-treated remaining maize starch granules show porous structures (Figure 6.4b), whereas those of the treated potato starch granules exhibit smooth surface structures (Figure 6.4c). These images are in agreement with the confocal laser light scanning micrographs (CLSM) of normal maize (Figure 6.4d) and potato (Figure 6.4e) starch granules. The CLSM images of waxy maize starch granules (Figure 6.4f) show more porous structures at the periphery of the granules than the normal maize starch granules.[55] All the results indicate that A-type starch granules have porous internal structures. In contrast, the B- and C-type starches exhibit uniform internal structures.[55,56] (See Section 6.4.5 in this chapter for the development of porous structures.)

3. Shapes of Gelatinized Starch Granules

After gelatinization in the presence of sufficient moisture,[57] starch granules swell up to 50 times of their original volume, depending on the temperature and the starch variety.[58] Different starches display different thermal transitions and swelling patterns.[59–66] For example, disk-shaped granules of wheat and barley starches were transformed into saddle shapes and then into highly puckered swollen disk shapes.[59] The mechanism of the structural change is that granules swell radially in the flat (xy) plane of the disk granule to about three times the original diameter at about 50°C. Very little swelling occurs in the thin (z) direction throughout the whole process. The granule then swells tangentially in the xy plane to pucker out of the plane.[67] The starch molecular chains in the granule are perpendicular to the surface (parallel to the z-axis).[67] Because the covalent bonds of the molecular chains have definite lengths and are not extendable, there is little swelling on the thin (z) direction. Neutral salt-solution treated wheat starch, however, shows that starch gelatinizes and swells at the equatorial groove, the edge of the xy plane.[43] There is little change shown on the surface of the xy plane. This phenomenon is attributed to the known fact that starch molecules perpendicular to the xy plane are aligned in parallel and are in better order, whereas molecules at the groove are radially arranged and are more susceptible to interaction with neutral salt solutions and to enzyme-catalyzed hydrolysis. Other starch granules, such as those of maize, oats and potato starches, swell radially and are gelatinized evenly on their surface by neutral salt solutions.[41–43] Various mutants of maize starch develop into different shapes, sizes and iodine-stained colors, which have been used to identify the cultivars.[61] SEM of crosslinked starch granules have also been investigated, and show hard shells on the surface (granule remnants).[68–70]

III. Molecular Compositions of Starch Granules

Substances commonly found in starch granules are amylopectin, amylose, molecules intermediate between amylose and amylopectin, lipid (including phospholipids and free fatty acids), phosphate monoester and proteins/enzymes. The contents and the structures of amylopectin and amylose play major roles in the functional properties of starch. However, lipids, phospholipids and phosphate monoester groups have significant effects on starch functional properties, even though they are minor constituents.

1. Amylopectin and Amylose

Normal starches, such as normal maize, rice, wheat and potato, contain 70–80% amylopectin and 20–30% amylose. The iodine affinity of solution of a defatted starch[71] and the blue color derived from the amylose complex of iodine[72–76] are the two methods commonly used to determine the apparent amylose content of starch. Results obtained from these methods, however, are affected by the presence of lipids[77] and the structure of amylopectin,[78–80] for the long branch chains of amylopectin also can complex with iodine and give additional iodine affinity and blue color.[78,79] Complexes of amylose with lipids reduce its ability to complex iodine. Most amylose content data reported in the literature are based on apparent amylose content. Apparent amylose and absolute amylose content (the iodine affinity of amylopectin have been subtracted from the iodine affinity of the starch) of selected starches are shown in Table 6.1. The so-called waxy starches contain little or no amylose: waxy maize starch contains <1% amylose; waxy barley contains up to 8% amylose. High-amylose starch contains >50% apparent amylose determined by iodine binding.[81–84] Absolute amylose contents of high-amylose maize and many other starches are lower than the values obtained by the iodine affinity or blue color methods. The discrepancy is caused by the long branch-chains of amylopectin, which also bind iodine and produce blue color.

Mutants of maize starch varieties have been extensively studied.[12,85–100] Amylose, amylopectin, and intermediate material contents of various single and double mutants have been analyzed; the results are shown in Table 6.2. Relationships between structures of the starch polymers and functional properties of these mutant starches have also been reported.[12,85–100]

Table 6.1 Iodine affinities and amylose contents of starches[79]

Source	Iodine affinity[a]		Amylose content (%)[b]		A − B
	Starch	Amylopectin	Apparent (A)[c]	Absolute (B)[d]	
A-type starch[e]					
Normal maize	5.88 ± 0.14[sd]	1.78 ± 0.03	29.4	22.5	6.9
Rice	5.00 ± 0.02	1.11 ± 0.00	25.0	20.5	4.5
Wheat	5.75 ± 0.15	0.8 ± 0.20	28.8	25.8	3.0
Barley	5.1 ± 0.30	0.50 ± 0.04	25.5	23.6	1.9
Cat-tail millet	3.97 ± 0.06	1.08 ± 0.03	19.8	15.3	4.5
Mung bean	7.58 ± 0.06	2.07 ± 0.03	37.9	30.7	7.2
Chinese taro	2.75 ± 0.06	0.00 ± 0.00	13.8	13.8	0.0
Tapioca	4.7 ± 0.20	1.39 ± 0.07	23.5	17.8	5.7
B-type starch					
Amylomaize V	10.4 ± 0.10	6.79 ± 0.04	52.0	27.3	24.7
Amylomaize VII	13.6 ± 0.20	9.3 ± 0.10	68.0	40.2	27.8
Potato	7.20 ± 0.01	4.6 ± 0.10	36.0	16.9	19.1
Green leaf canna	8.64 ± 0.00	5.31 ± 0.03	43.2	22.7	20.5
C-type starch					
Water chestnut	5.79 ± 0.19	3.08 ± 0.02	29.0	16.0	13.0

[a] Iodine affinities were averaged of 3 replications of defatted sample

[b] Iodine affinity for pure amylose was 20%

[c] Apparent amylose contents were calculated from the equation: $C = 100 \times IA_S/0.20$. IA_S is the iodine affinity of the whole defatted starch

[d] Absolute amylose contents were calculated from the equation: $C = (IA_S - IA_{AP+IC})/\{0.20 - (IA_{AP+IC}/100)\}$. C is the percentage of real amylose content. IA_S is the iodine affinity of whole defatted starch. IA_{AP+IC} is the iodine affinity of the amylopectin and the intermediate component mixture

[e] Determined by x-ray diffraction

[sd] Standard deviation

2. Intermediate Material and Phytoglycogen

Intermediate material is found in starch granules; this material has structures and properties in between those of the essentially linear amylose molecules and those of the larger and more highly branched amylopectin molecules. The intermediate material has a heterogeneous nature.[101] Lansky et al.[102] fractionated amylose and reported that some amylose molecules were branched and had a lower affinity for iodine.

A fraction of a branched molecule with molecular weight smaller than amylopectin and similar to amylose has been isolated from starch granules. These molecules are mainly found in high-amylose maize starch (starch from maize mutants and double mutants containing the *ae* gene)[86,87,98,103–110] and starch from corn carrying the *sugary–2* gene.[12,85,111] The branch chains (DP 52 and 21 for long and short branches, respectively) of this type of intermediate material isolated from high-amylose maize starch are longer than those of the amylopectin isolated from the same starch source (DP 45 and 19).[104,109,110] A group of small (DP ~95) but linear molecules are also found in the high-amylose maize starches.[98,108,109] The small and linear molecules (~27%, by weight, of amylopectin) remain in the supernatant with amylopectin after

Table 6.2 Percentage compositions of polysaccharides[a,b] and chain length distribution[b] of starches from 17 mutant genotypes of Oh43 inbred[96]

Genotype[c]	GPC[d], native			GPC, debranched				CL[k] at peak of		Intermediate materials[l] (%)
	Fraction I[e] (%)	Fraction II[f] (%)	IA[g] (%)	Fraction I[h] (%)	Fraction II[i] (%)	Fraction III[j] (%)	Fraction III: fraction II	Fraction II[i]	Fraction III[j]	
Normal	70.1	29.9	27.8	26.7	19.2	54.1	2.8	43	15	3.2
	(0.6)	(0.6)	(1.1)	(0.3)	(0.1)	(0.1)	(0)	(2)	(1)	
ae	38.7	61.3	56.4	46.0	27.2	26.8	1.0	48	20	15.3
	(0.1)	(0.1)	(1.1)	(0.6)	(0.2)	(0.4)	(0)	(1)	(1)	
bt1	73.3	26.7	23.2	24.9	18.9	56.2	3.0	42	14	1.8
	(1.3)	(1.3)	(0.8)	(0.7)	(0.6)	(0.1)	(0.1)	(3)	(0)	
bt2	73.2	26.8	26.9	24.7	22.1	53.2	2.4	41	15	2.1
	(0.6)	(0.6)	(0)	(0.6)	(0.3)	(0.3)	(0)	(1)	(0)	
du1	54.3	45.7	31.4	30.5	10.6	58.9	5.6	51	14	15.2
	(0.1)	(0.1)	(0.4)	(0.2)	(0.4)	(0.3)	(0.2)	(1)	(0)	
h	71.3	28.7	26.4	28.1	19.2	52.7	2.8	41	15	0.6
	(2.5)	(2.5)	(0.8)	(0.1)	(1.0)	(1.1)	(0.2)	(1)	(1)	
sh2	72.6	27.4	28.8	30.1	14.7	55.2	3.8	40	13	−2.7
	(0.6)	(0.6)	(0)	(0.6)	(0)	(0.6)	(0.1)	(3)	(1)	
su1	59.5	40.5	37.4	31.2	12.3	56.5	4.6	41	14	9.3
	(0.4)	(0.4)	(1.5)	(0.6)	(0.1)	(0.7)	(0.1)	(4)	(0)	
Wx	100.0	0	0	0	27.7	72.3	2.6	39	15	0
	(0)	(0)	(0)	(0)	(0.4)	(0.4)	(0)	(1)	(1)	
ae bt1	45.1	54.9	30.5	32.4	24.5	43.1	1.8	49	16	22.5
	(0.1)	(0.1)	(0.1)	(0)	(0.4)	(0.4)	(0)	(2)	(1)	
ae du1	23.8	76.2	57.8	57.3	19.8	22.9	1.2	49	19	18.9
	(0.1)	(0.1)	(1.4)	(0.1)	(0.1)	(0.1)	(0)	(0)	(1)	
du1 su1	53.8	46.2	39.9	34.5	13.0	52.5	4.0	47	13	11.7
	(0.4)	(0.4)	(0.7)	(1.3)	(0.4)	(0.9)	(0.1)	(1)	(0)	
h sh2	70.5	29.5	25.4	26.4	20.0	53.6	2.7	44	14	3.1
	(0.4)	(0.4)	(0.1)	(0.2)	(0.2)	(0.4)	(0)	(2)	(0)	
h wx	100.0	0	0	0	29.2	70.8	2.4	43	16	0
	(0)	(0)	(0)	(0)	(0.1)	(0.1)	(0)	(2)	(1)	
sh2 bt1	71.7	28.3	27.6	27.7	15.4	56.9	3.7	49	13	0.6
	(0.2)	(0.2)	(0.4)	(0.2)	(0)	(0.1)	(0)	(1)	(0)	
sh2 wx	100.0	0	0	0	30.6	69.4	2.3	42	13	0
	(0)	(0)	(0)	(0)	(0)	(0)	(0)	(2)	(1)	
wx du1	100.0	0	0	0	26.3	73.7	2.8	39	15	0
	(0)	(0)	(0)	(0)	(0.4)	(0.4)	(0)	(0)	(0)	

[a] The division of each fraction is described in reference 96
[b] Values are the average of two separate determinations, with SD in parentheses
[c] ae = Amylose extender; bt = brittle; du = dull; h = horny; sh = shrunken; su = sugary; and wx = waxy
[d] Gel-permeation chromatography
[e] Amylopectin
[f] Amylose and intermediate materials
[g] Amylose percentage calculated from iodine affinity
[h] Amylose
[i] Long B-chains of amylopectin
[j] A- and short B-chains of amylopectin
[k] Average chain lengths of isoamylase-debranched starch measured at the apex of the peak from each fraction. Expressed as number of glucose units
[l] Calculated as the difference between fraction II of the native starch and fraction I of the debranched starch GPC elution profiles

fractionation by 1-butanol treatment.[108–110] These molecules, found in the intermediate fraction of high-amylose maize starch, mutants result from the lack of branching enzyme IIb that preferentially transfers short chains.[109–112]

Phytoglycogen, a water-soluble polysaccharide, is found in the *sugary–1* mutants of maize and rice endosperm.[98,113–120] Phytoglycogen has an average branch chain length of DP 10.3, which is shorter than that of waxy maize amylopectin (DP 18.5).[89] A size-exclusion chromatogram of isoamylase-debranched phytoglycogen shows a single peak, rather than a bimodal distribution as is found in debranched amylopectin. Phytoglycogen has an interior chain length (DP 7.4) similar to that of waxy maize amylopectin (DP 8.5), but a much shorter exterior chain length (DP 2.9) than waxy maize amylopectin (DP 10).[88] A similar structure, but of slightly longer chain length (average chain length of DP 12 and exterior chain length of DP 6), was reported by Yun and Matheson.[115] The lack of a *sugary–1* gene-coded debranching enzyme in *sugary–1* maize[114] and rice[120] mutants results in the highly branched structure and water solubility of phytoglycogen.

3. Lipids and Phospholipids

Lipids are common minor constituents found in cereal starches.[121] Maize starch contains small amounts of free fatty acids and little phospholipid;[121] normal rice starch contains substantial amounts of phospholipids and some free fatty acids. Wheat, barley, rye and triticale starches contain exclusively phospholipids and an insignificant amount of starch phosphate monoesters.[121] Thus, the phospholipid contents in these starches can be quantified by multiplying the starch phosphorus content by 16.3.[122] ^{31}P-nmr spectroscopy has been used for qualitative and quantitative analyses of phospholipids and phosphate monoesters in starch.[123–127] Phospholipids are not found in most waxy starches except *dull–waxy* maize.[124] The presence of phospholipids in *dull–waxy* maize starch may result in a very high retrogradation of the starch.[79] Most cereal starches contain amounts of lipids proportional to the apparent amylose content of the starch.[80,128,129] In waxy barley starch, the phospholipid content is linearly proportional to the apparent amylose content, but in non-waxy barley starch the linear plot intercepts at 19% amylose content, indicating that 19% free amylose is present in the starch.[129] Lipids and phospholipids are known to form stable complexes with long chains of starch, with both amylose and long branch chains of amylopectin, which results in the restricted swelling of granules.[130–133] Waxy cereal starches contain small amounts of lipids, whereas high-amylose starches contain the most lipids. Tuber and root starches contain very little lipid. (The structure of the amylose–lipid complex is discussed in Section 6.4.1.)

Normal wheat starch has a higher pasting temperature (90.6°C) and produces a lower peak viscosity (96 RVU) than does normal maize starch (81.5°C and 159 RVU, respectively). In comparison, waxy wheat starch has a lower pasting temperature (62.5°C), but produces a higher peak viscosity (230 RVU) than does waxy maize starch (69.8°C and 200 RVU, respectively). The extraordinarily large differences between normal and waxy wheat starches (28.1°C and 134 RVU) are attributed to an amylose–phospholipid complex present in normal wheat starch.[134]

4. Phosphate Monoesters

Almost all the starches investigated contain some phosphorus.[124–126] In addition to phospholipids, phosphorus is also commonly found in starch as monostarch phosphate esters.[6,7] Inorganic phosphate is present in some starches.[124–126] Monostarch phosphate esters and phospholipids have different effects on starch paste properties.[133,135] Monostarch phosphate esters, found in potato, shoti and other starches,[6,7,136–140] increase paste clarity and paste viscosity.[79] Little phosphate monoester is found in cereal starch.[121,124,141,142] Phospholipids, found in normal cereal starches (e.g. wheat, rice and maize) decrease paste clarity and viscosity.[133] ^{31}P-nmr spectroscopy has been useful in determining the structural types and the contents of phosphorus in starch.[123–126,139,140,143,144]

Monostarch phosphate esters in native starches, such as potato and rice, are primarily found on amylopectin molecules,[6,7,142] only trace amounts are found on amylose molecules. One report states that ~61% of the phosphate monoester groups in potato starch are on the O-6, ~38% on O-3, and possibly 1% on O-2 of the glucosyl units.[7] In a waxy rice starch, 80–90% of the phosphate monoester groups were found on O-6.[7] ^{31}P-Nmr studies of potato starch have revealed that ~80% of the phosphate monoester content is on O-6 and about 20% on O-3.[143] Among common starches, potato starch has the largest phosphate monoester content (up to 0.09% P),[124,125,139,140] which is second only to shoti starch (0.18% P).[139] Potato amylopectin contains one phosphate monoester per 317 glucosyl units,[145] equivalent to one phosphate group per 13 branch chains.[146] Surface gelatinization studies have shown that phosphate monoesters are more concentrated at the core (hilum) than in the periphery of potato starch,[41] and that small granules have greater phosphate content than large granules.[41,147]

The phosphate monoester content of potato starch is inversely proportional to the crystallinity[148] and to the gelatinization enthalpy of the starch,[149] indicating that the phosphate monoesters are present within the crystalline region of the starch granule. This result is in agreement with the phosphate monoesters being located more than nine glucosyl units away from the branch points of amylopectin.[145]

IV. Structures of Amylose and Amylopectin

1. Chemical Structure of Amylose

Although all amylose molecules were once considered to be linear, many amylose molecules cannot be completely hydrolyzed by β-amylase. With a concurrent or mixed action of pullulanase and beta-amylase, however, amylose can be completely hydrolyzed to maltose.[150,151] These results rule out the theory that the incomplete hydrolysis of amylose by β-amylase is a result of retrogradation, i.e. junction zone formation. It is now clear that the incomplete hydrolysis of an amylose preparation by β-amylase is due to branching of some molecules. The β-amylolysis limit of amylose varies from 72% to 95%[152,153] compared with 55–61% for amylopectin. Amylose of most cereal starches, such as maize,[154] rice,[155,156] wheat[157] and barley,[158] give >80% β-amylolysis

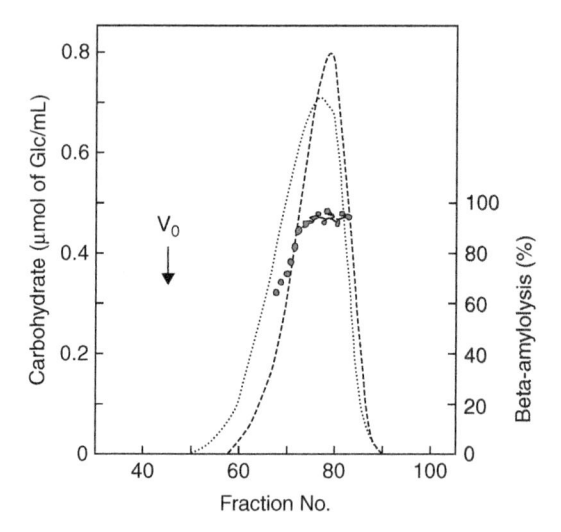

Figure 6.5 Gel permeation chromatogram of isoamylase-treated potato amylose on Toyopearl HW-75F. Potato amylose (50 mg in 5 mL) was incubated at 45°C with 5.5 U of *Pseudomonas* isoamylase in 20 mM acetate buffer (pH 3.5) for 2.5 h. An aliquot (1 mL) was applied to the column: --- and •, carbohydrate concentration and beta-amylolysis limit, respectively, of the isoamylase-treated amylose; ....., carbohydrate concentration of the native amylose.[159]

limits. The amylose of high-amylose maize starches has an ~75% β-amylolysis limit.[110] The percentage of the β-amylolysis limits of tuber and root starch amyloses varies from 72% for sweet potato amylose[152,159] to 95% for water chestnut amylose.[153] The extremely large degree of β-amylolysis of water chestnut amylose is attributed to its small and unbranched molecules. Both the molecular weight and branch number of amylose extracted by hot water increase with the extraction temperature.[160]

Many studies have demonstrated the multi-branched nature of amylose.[5,151,161–168] Hizukuri et al. investigated and reported that potato amylose molecules with DPs of 4850 and 6340 have 9 and 12 branches, respectively; tapioca amylose (DP 2660–3390) has 7.8 to 20 branches depending on the variety, kuzu amylose (DP 1590) has 9 branches,[5] lily amylose (DP 2300) has an average of 4.9 chains[5] and rice amyloses have 3.4 to 7.6 chains depending on the variety.[142] The same authors reported that potato amylose can be 30% debranched with isoamylase and 43% with pullulanase.[142] The results suggest the presence of maltosyl branches on amylose molecules, because isoamylase cannot remove maltosyl stubs. Isoamylase hydrolysates of potato amylose consist of short chains, such as maltotetraose and maltopentaose, and long chains (DP > 100).[159] Because of these short branch stubs, the average chain lengths of amylose molecules, calculated by dividing the molecular weight by the number of chains, can be substantially smaller than the real chain length of the molecule, as indicated by gel permeation chromatograms (Figure 6.5).[159]

Amyloses of different origins and molecular weights have been extensively studied.[151] Results indicate that cereal amyloses are smaller than other amyloses. Amyloses isolated from high-amylose maize starches display substantially smaller molecular weights (average DP 690–740 and chain lengths 215–255).[161] There is about an equal

Table 6.3 Properties of selected amylose[151]									
	Corn	Rice	Wheat	Chestnut	Kuzu	Nagaimo	Lily	Tapioca	Sweet potato
Iodine affinity (g/100 g)	21.1	20.6	19.9	19.9	20.0	19.9	20.2	20.0	19.7
Blue value	1.39	1.40	1.40	1.41	1.46	1.55	1.49	1.47	1.50
λ_{max} (nm)	644	658	664	655	658	658	648	662	660
DPa	960	1110	1290	1690	1460	2000	2300	2660	3280
Chain length	335	320	270	375	310	525	475	340	335
Number of chains per molecule	2.9	3.5	4.8	4.5	4.7	3.8	4.9	7.8	9.8
Beta-amylolysis (%)	82	81	82	91	76	86	89	75	76
Beta-amylolysis with pullulanase (%)	99	103	101	102	100	102	100	99	99

aAverage degree of polymerization

molar ratio of branched molecules (with 5–6 branch chains) and small, unbranched molecules. Selected features of these amyloses are given in Table 6.3. The molar fraction of branched molecules can be calculated by the equation (NCa − 1)/(NCb − 1). In this equation, NCa is the average number of branch chains in amylose, and NCb is the average number of branch chains in the β-limit dextrins of the amylose. The value of NCb is larger than NCa because the linear molecules of amylose have been completely hydrolyzed by β-amylase.[151] Determined by this approach cereal amyloses, in general, have a smaller molar fraction of branched molecules (0.27–0.44) than tuber and root amyloses (0.34–0.70).[151] Some amyloses, such as those of maize, wheat and chestnut, have mainly small branch chains (DP ~18), whereas other amyloses, such as potato and sweet potato, contain both long and short branch chains. After isoamylase treatments, maize and wheat amyloses show no significant change in the molecular weight; potato and sweet potato amyloses show slight changes. There are no distinguishable differences between japonica and indica rice amyloses.[155,156]

Structures of subfractions of maize amylose obtained by crystallization from an aqueous solution of 10% 1-butanol at 40°C have been studied.[167] Yields of the amylose subfraction remaining in the supernatant, the subfraction which precipitated, and the amylopectin were 2.7, 16.7 and 73.3%, respectively. The subfraction that remained in the supernatant displayed a lower iodine affinity (i.a. 16.0 g/100 g) and a larger molecular weight than the precipitated subfraction (i.a. 20.5 g/100 g). The branched molecules remaining in the supernatant had an average of 20 chains per molecule, while their counterparts in the precipitate contained six chains per molecule. Isoamylase hydrolyzates of the branched molecules had short branches (DP 16–18) with the shortest chain being DP 6, as well as long and extremely long chains. On the basis of these structural data, the authors proposed a structure of the branched amylose molecule carrying immature clusters (Figure 6.6).[167] The short clustered branches could be transferred by branching enzyme. Maize and rice amyloses fractionated by molecular weight are composed of large molecules (DP ~2500) that are mainly branched molecules (maize 66%, rice 61%) and a small amylose molecule (DP ~400) fraction that contains lesser amounts of branched molecules (maize

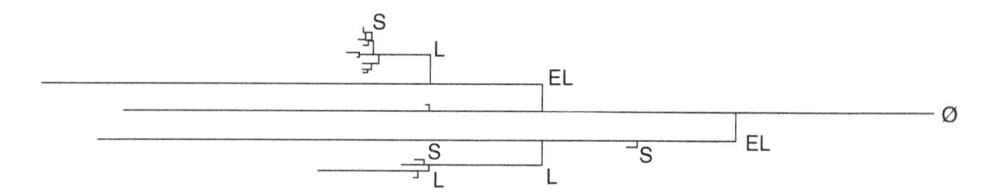

Figure 6.6 Proposed structure of the branched amylose molecule comprising immature clusters. EL, extremely long; L, long; and S, short chains; Ø, reducing end.[167]

29%, rice 25%).[156,166] Maize and rice amyloses of the same molecular weight subfraction had similar structures.

Using results of these kinds of studies, the characteristic structure of amylose can be differentiated from that of amylopectin. Amylose has a small number of branches and crystallizes and precipitates when complexed with 1-butanol. The iodine affinity of amylose is much greater (i.a. 18.5 to 21.1) than that of amylopectin (i.a. 0.0 to 6.6),[79,152–158,163,169–174] and the iodine affinity of amylose β-limit dextrin is similar to that of the parent amylose. The average chain length of amylose β-limit dextrins is much larger than that of the amylopectin β-limit dextrin.[160]

2. Single-Helical Structures (V-Complexes) of Amylose

In a freshly prepared aqueous solution, amylose is present as a random coil.[175] The random coil conformation, however, is not stable. Amylose tends to form either single-helical (inclusion) complexes with suitable complexing agents[176] or to form double helices among themselves when no suitable complexing agent is available.[177–179] The transition from coil to a single or double helix is attributed to the chemical structure of the (1→4) linked α-D-glucopyranosyl chains. In a single-helical complex, the linear portion of the starch molecule has its hydrophobic side of the molecule facing the cavity of the helix and interacting with the non-polar moiety of the complexing agent, such as the hydrocarbon chains of 1-butanol and fatty acids. This complex structure resembles that of a cyclodextrin–guest molecule complex. With a complexing agent readily available in an aqueous solution, the single-helical complex forms instantaneously.[176,180,181] Differential scanning calorimetry has revealed that amylose–lipid complex formation is instantaneous and reversible. The amylose–lipid complex is melted with an endothermic peak during heating to the melting temperature, but the complex is reformed during cooling, reflected by an exothermic peak at a similar temperature. Without any complexing agent present, the linear portion of starch molecules will pair up to have their hydrophobic sides folded inside the double helix. Formation of a double helix requires an alignment of two molecules and thus, is a time-consuming process. Both single and double helices result in lower energy states and are thermodynamically favorable.

The single-helical (inclusion) complex was first reported by Katz with cooked dough.[178] After cooking, dough develops a diffraction pattern which differs from the original A, B or C pattern.[178] Katz named it the V-pattern from a German word 'Vaklinestorone' (gelatinization). Structures of the amylose single-helical complexes have been studied by x-ray crystallography and show varied sizes of helices. Molecules with small cross-sections, such as 1-butanol,[182,183] fatty acids[184,185] and glycerol monostearate,[186] complex with amylose to form helices of six glucose units per turn. Molecules with larger cross-sections, such as iso- and tertiary-butyl alcohols, benzoic acid, dimethyl sulfoxide and tetrachloroethane, complex with amylose and form helices of seven glucose units per turn.[187] Molecules of even larger cross-sections, such as 1-naphthol, form helices of eight glucose units per turn.[188] Electron microscopic studies have shown lamellar structures of these amylose helices with a lamellar thickness of about 10 nm.[182,183,187,188] Electron micrographs show that the lamella of an amylose helix can go through a transformation when the complexing agent is replaced.[182] Enzymic studies of the structures of amylose helical complexes of 1-butanol, tert-butyl alcohol and 1,1,2,2-tetrachloroethane[189] using porcine pancreatic and human salivary α-amylase reveal resistant folding chain lengths of DP 75, 92 and 90, respectively, corresponding to a resistant helical chain lengths of about 10 nm and 6, 7 and 7 glucosyl units per turn, respectively (see Figure 7.15). These results further confirm the lamellar structures of amylose helices with different numbers of glucosyl units per turn. The enzymic method is less successful with the amylose-1-naphthol complex[189] and the crystalline amylose-glycerol monostearate complex,[190] which may be attributed to their rigid structures. The amorphous folding sites are possibly hindered and thus, are less susceptible to enzyme-catalyzed hydrolysis.

A quantitative study of amylose/fatty acid ratio in the helical complex shows that there are 87 glucosyl units involved in each helix of the amylose–fatty acid complexes.[191] The authors postulated that about 12 glucosyl units are present at the amorphous folding end (the surface of the lamella) on the basis of a six-member helix, with a pitch of 0.8 nm and 75 glucosyl units per folding length of 10 nm.[191] Both porcine pancreatic α-amylase[189,190,192] and human salivary α-amylase[189,192] have a binding subsite of five glucosyl units; *Bacillus subtilis* α-amylase has a binding subsite of nine glucosyl units.[189,190,193] Thus, there is a need for a sufficient chain length of amylose to fill the binding subsites to facilitate the enzyme-catalyzed reaction. The amorphous loop of 12 glucosyl units provides abundant chain length for the binding subsites of porcine pancreatic and human salivary α-amylases (five glucosyl units),[191] but may be a little constrained for *B. subtilis* α-amylase (nine glucosyl units). This may also explain the fact that porcine pancreatic and human salivary α-amylases hydrolyze the helical lamella and produce results that fit well with the theoretical values, but *B. subtilis* α-amylase fails to do so.[189] Amylose helical complex formation has also been studied by using high-resolution liquid ^{13}C-nmr spectroscopy.[194] The conformational changes of coil–helix transformation are monitored by the substantial chemical shifts of the C1 and C4 signals, which are congruent with the chemical shifts of cyclodextrins.[194] On chemical reduction of iodine molecules to iodide ions to destroy the amylose–iodine helix, the downfield chemical shift changes and the suppression of the C-1 signal disappears.[194] A similar pattern of downfield chemical shift change is also observed with methyl α-maltoside, indicating the hydrophobic interaction between maltose and the complexing agent. The same pattern of chemical shift changes for C1 and C4 of the amylose single helical complex is also observed by solid ^{13}C-cross polarization-magic angle spinning nuclear magnetic resonance ^{13}C-CP-MAS NMR.[195]

Amylose–iodine complexes have a deep blue color, which is a result of an electron relay on the polyiodide ions.[196] The helix of amylose provides a tunnel for iodine molecules to align. Stability of the amylose–iodine complex has been studied.[197] Iodine has been widely used for quantification of amylose contents despite the fact that the blue color development is affected by many factors, including temperature, pH and mechanical mixing. Several improved methods have been reported (see Section 6.III.1).

Differential scanning calorimetry (DSC) thermograms of normal cereal starches (e.g. maize and wheat) show an additional thermal transition peak at a higher temperature than the starch gelatinization peak.[134,198,199] This peak is not found in all-amylopectin (i.e. waxy) starches, disappears after the lipid of the starch is removed, and reappears on addition of a complexing lipid[181] and thus, is attributed to the melting of the amylose–lipid complex.[181,200–203] The starch–lipid complex in the native starch granule is amorphous (type 1) and can be annealed into a more ordered semicrystalline (presumably lamellar) form which displays birefringence and a V-type x-ray pattern (type 2).[190,204–207] The type-1 amorphous complex is found in most native cereal starches, which display a melting temperature at about 94–100°C. The melting temperature of the amylose–fatty acid complex increases with the chain length of the fatty acid.[190,203,204,207,208] The type-2 lamellar crystalline form melts at 100–125°C.[207,209,210] Therefore, the V-type x-ray pattern is rarely seen in native cereal starch, except for high-amylose maize starches.[211] Monoglycerides of short chain fatty acids, such as capric and lauric acids, complex with amylose directly to the type-2 semicrystalline structure,[190,204] whereas monoglycerides of longer-chain fatty acids (C-14 and longer), which complex with amylose rapidly, give the type-1 amorphous form. At a low moisture content (<70%) the type-1 form can easily be annealed by heating at 95°C or rescanned in a DSC and converted to the type-2 complex.[190,204,205]

^{13}C-CP-MAS NMR produces a broad resonance with a chemical shift of 31.2 ppm,[129] a characteristic of mid-chain methylene carbons of fatty acids in the V-amylose complex. The results showed that up to 43% of amylose in non-waxy rice starch, 33% in oat starch, and 22% in normal maize and wheat starch granules are complexed with lipids at a single helical conformation, and the remaining amylose is free of lipids and is in a random coil conformation.[212] Up to 60% of apparent amylose in waxy barley starch is complexed with lipids.[212]

Phosphoglycerides, monoglycerides and fatty acids are good starch complexing agents. Saturated fatty acids and trans-unsaturated fatty acids are more effective in forming complexes than are cis-unsaturated fatty acids.[191] This is attributed to the fact that molecules of saturated and trans-unsaturated fatty acids are straight, whereas molecules of cis-unsaturated fatty acids are bent. Amylose–fatty acid complex formation at pH 12 is a cooperative and stoichiometric reaction. Quantitative analyses have shown that the crystalline precipitate of amylose–fatty acid complex prepared from amylose solutions with different molar ratios of fatty acids is saturated and displays a constant amylose:fatty acid ratio, but with different yields. The amylose:fatty acid ratio decreases with the increase of fatty acid chain length.[191]

Amylose single helical complexes are metastable.[182,189,190] When the complexing agents are lost, the amylose single helices are prone to convert to random coils[189] or to double helices.[177,178]

3. Double Helical Structures of Amylose

Without the presence of complexing agents, amylose molecules gradually associate and form double helices.[213–219] Kinetic studies have shown that the rate of double helix formation depends on the molecular size of the amylose, its concentration, and the temperature. A minimum chain length of DP 10 is required for double helix formation in a pure oligosaccharide solution.[219,220,221] However, oligomers as short as maltohexaose (a single turn of the double helix) can co-crystallize with longer chains.[220] This fact is interesting when placed next to the fact that DP 6 is the shortest branch chain length found in amylopectins.[79,222–224] The rate of retrogradation and the temperature of turbidity onset increase with the increase of amylose molecular size up to DP 90–110[225–228] and steadily decrease when the DP is larger than 250.[226] Amyloses of DP <110 precipitate from aqueous solutions at all temperatures. Amyloses of DP 250–660 precipitate or gel, depending on the concentration and temperature. Amyloses of DP >1100 predominantly form gels rather than precipitate.[226]

Structures of double helical amyloses obtained by retrogradation have been investigated by x-ray crystallography[2,177,178,229] and found to display A-, B- and C-type x-ray patterns. ^{13}C-CP-MAS NMR has been used to study the conformations of starch molecules.[66,230,231] Enzyme- and acid-catalyzed hydrolysis of retrograded amylose produces resistant amylodextrins with narrowly distributed molecular sizes (DP ~35).[189,232,233] The peak chain lengths of the resistant amylodextrins after extensive hydrolysis by acid, human salivary α-amylase, porcine pancreatic α-amylase and *Bacillus subtilis* α-amylase are DP 31, 42, 44 and 50, respectively. These differences are attributed to the binding subsites of various amylases, which leave stubs at both ends of the resistant double helix, i.e. five glucosyl units for human salivary and porcine pancreatic α-amylases and 9 glucosyl units for *Bacillus subtilis* α-amylase[189] (see Figure 7.15).

Morphologies of retrograded amylose produced in a low concentration (3.5 mg/mL) solution differ with incubation temperature. Retrograded amylose developed at 5°C displays interconnected nodules, whereas that developed at 15°C and higher temperatures displays a gel-like structure.[232] At concentrations greater than 15 mg/mL, amylose retrogrades to form gels, even at 1°C.[217] Electron micrographs show a structure of interconnected filaments of about 10–20 nm wide (Figure 6.7).[217] The filaments are composed of double helices of amylose with segments of chain length of DP 26–31 which are orientated obliquely to the filament axis.[217] The double helix content of the retrograded amylose decreased from 58.8% to 7.1%, and the lamellar chain length of the double helix increased from DP 34 to DP 40 when the incubation temperature was increased from 5°C to 45°C.[232] The effect of incubation temperature on double helix formation coincides with the effect of growth temperature on the yield of maize starch and the chain length of amylopectin. This result seems to favor the hypothesis that double helix formation is a requisite for branching reactions and the biosynthesis of starch.[234]

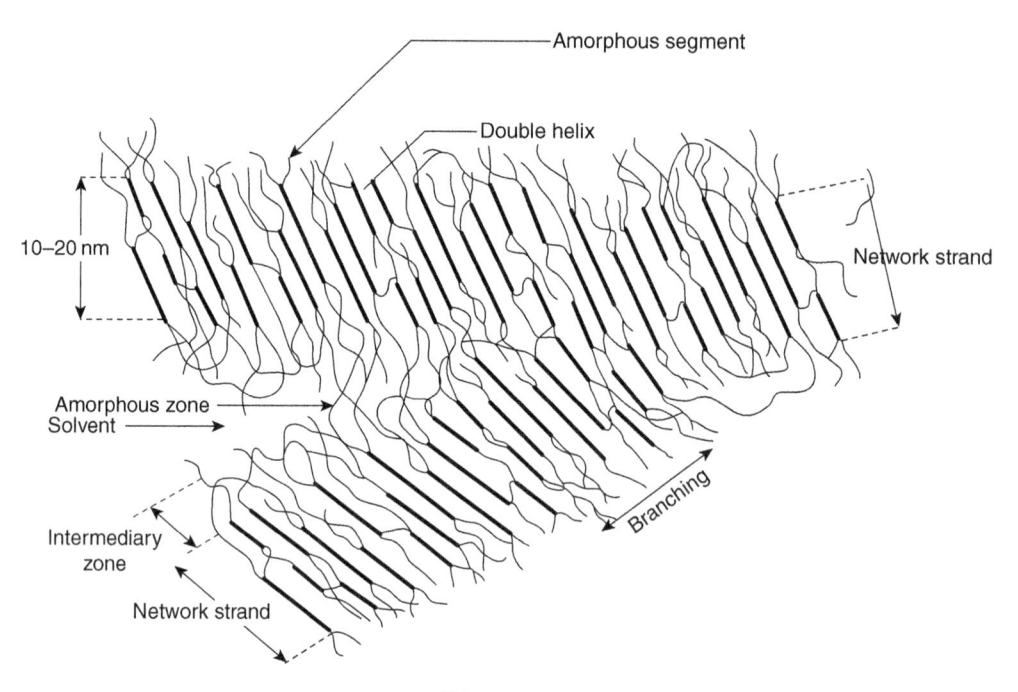

Figure 6.7 Continuous model for amylose gels.[217]

Retrograded amylose double helices are known to be resistant to the action of amylolytic enzymes[189,232,233,235] and thermally resistant with a melting temperature between 130°C and 170°C.[236–240] These properties are desirable for low-calorie bulking agents having functions of dietary fiber, known as type 3 resistant starch.[73,197,235–256] The type 3 resistant starch is defined as the starch that resists degradation by amylolytic enzymes *in vitro* and *in vivo*. The resistant starch can be produced by repeated cycles of autoclaving and cooling.[237–239] Amylose content, processing temperature and water content are factors known to influence the yields of resistant starch.[237,238] Resistant starches have been obtained in high yields from high-amylose maize starch (23–48%) and from debranched potato and waxy maize amylopectins (47% and 34%, respectively).[237] Without debranching, waxy maize and potato amylopectins produce very little resistant starch (0.2% and 4.2%, respectively).[237]

4. Chemical Structure of Amylopectin

Amylopectin is a highly-branched molecule, consisting of three types of branch chains. A-chains are those linked to other chains (B- or C-) by their reducing ends through α-D-(1→6) linkages, but they are not branched themselves. B-chains are those linked to another B-chain or a C-chain, but B-chains are branched by A-chains or other B-chains at O-6 of a glucosyl unit. Each amylopectin molecule has only one C-chain, which carries the sole reducing end of the molecule (Figure 6.8).

The average branch chain length of amylopectin has a bimodal distribution that differs from the single modal distribution of that of glycogen.[257] The average branch chain

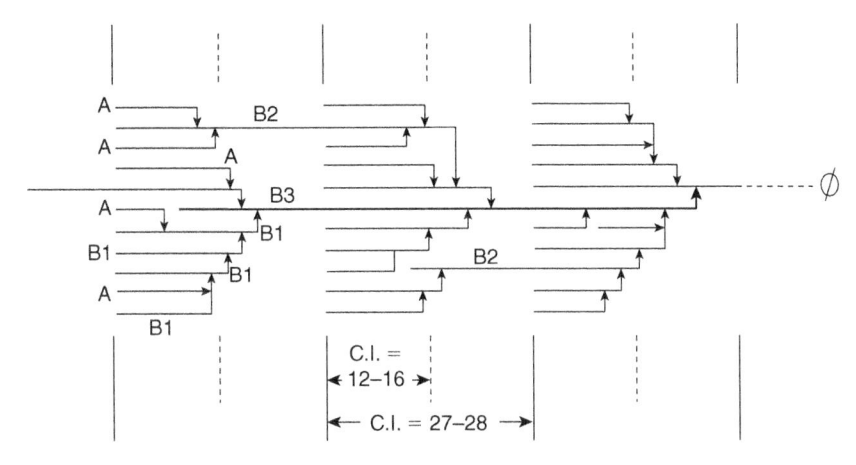

Figure 6.8 A cluster model of amylopectin proposed by Hizukuri with A, and B1–B3 chains. The chain carrying the reducing end (Ø) is the C chain, (1→4)-α-D-glucan chain; α-(1→6) linkage.[263]

length of amylopectin varies with the origin and maturity of the starch and the location of molecules in the granule. Hizukuri et al.[258,259] determined average chain lengths of the starches and their relationship to the crystalline starch's x-ray diffraction pattern for a variety of starches. The weight-average chain lengths of the amylopectins of the A-, B- and C-type starches were in DP ranges of 19–28, 29–31, and 25–27, respectively (Table 6.4).[79] In general, amylopectin molecules of A-type starches have both long and short chains that are smaller than those of the B-type starch, and have a larger proportion of the short-chain fractions than that of B-type starch. The molar ratios of short and long branch-chains were 8–12 for A-type starches, 7–9 for C-type starches and 3–7 for B-type starches.[259] C-type starches, such as banana,[81] smooth pea,[260,261] sweet potato[140] and ginkgo[262] starch consist of mixtures of A- and B-type x-ray patterns and A- and B-type branch chain patterns, and have both very long and very short chains.

With improved chromatographic separation, Hizukuri[263] further separated amylopectin branch chains and obtained polymodal distributions: A, B1, B2, B3 and B4 (Figures 6.9a and 6.9b). Hizukuri stated that A and B1 chains are in a single cluster while B2, B3 and B4 chains extend into 2, 3 and 4 or more clusters, respectively. The sum of A and B1 chains is 89–91% of total chains for waxy rice, tapioca and kuzu amylopectins, and 82% for potato amylopectin. The length of a complete cluster is about 27–28 glucosyl units,[263,264] equivalent to 9.7 nm. The span lengths between the branch linkages are about 10 glucosyl units.[145] These results agree with those obtained by scattering.[265] Hanashiro et al.[222] further defined A chains as having DP 6–12, B1 chains as having DP 13–24, B2 chains as having DP 25–36 and B3 chains as having DP > 37. The authors also reported that the differences in the amount of each chain between arrowhead and other amylopectins display periodic waves in the distribution which divided the abscissa at intervals of DP 12 (15 for edible canna and yam). Branch chain length of amylopectin, after isoamylase hydrolysis, can be analyzed by HPAEC-PAD quantitatively by using a post-column enzyme reactor of amyloglucosidase.[224,266–268] Examples of branch chain length distribution are shown in

Table 6.4 Branch chain length distributions of amylopectins[79]								
Source	Peak DP		Average	% distribution				Highest detectable DP
	I	II	CL	DP 6–9	DP 6–12	DP 13–24	DP 25–36	DP ≥ 37	
A-type starch[a]									
Normal maize	13	48	24.4	3.85	17.9	47.9	14.9	19.3	80
Waxy maize	14	48	23.5	6.94	17.0	49.4	17.1	16.5	73
Du waxy maize	14	51	23.1	1.21	16.7	51.9	17.4	14.0	80
Normal rice	12	46	22.7	4.10	19.0	52.2	12.3	16.5	80
Waxy rice	12	41	18.1	8.57	27.4	53.4	12.6	6.6	66
Sweet rice	12	45	21.6	7.59	23.5	48.7	13.7	14.0	78
Wheat	12	41	22.7	5.18	19.0	41.7	16.2	13.0	77
Barley	12	43	22.1	4.92	20.8	48.9	17.7	12.6	75
Waxy amaranth	12	43	21.8	6.31	25.1	47.7	12.5	14.7	80
Cat-tail millet	13	45	21.5	4.21	20.2	53.8	12.7	13.3	74
Mung bean	13	48	24.8	2.45	15.6	47.6	18.3	18.5	74
Chinese taro	13	45	23.4	5.99	18.8	48.7	14.8	17.7	71
Tapioca	12	49	27.6	4.68	17.3	40.4	15.6	26.7	79
B-type starch									
Ae waxy maize	16	53	29.5	2.29	10.4	43.5	18.1	28.0	84
Amylomaize V	16	48	28.9	1.90	9.7	43.9	20.3	26.1	86
Amylomaize VII	16	48	30.7	1.81	8.5	40.7	21.3	29.5	86
Potato	14	52	29.4	3.53	12.3	43.3	15.5	28.9	85
Green leaf canna	15	52	28.9	3.41	11.7	45.3	16.2	26.8	85
C-type starch									
Lotus root	13	52	25.4	4.57	16.4	47.2	15.4	21.0	83
Water chestnut	13	50	26.7	5.86	17.8	43.7	15.3	23.2	80
Green banana	13	48	26.4	5.25	16.8	46.3	12.9	24.0	79

[a] Determined by x-ray diffraction

Figures 6.10a and 6.10b for the A- and B-type starches, respectively. Average branch chain lengths of selected A-, B- and C-type starches are shown in Figure 6.10c.[56]

Amylopectins isolated from indica rice,[156] normal maize,[111,154] normal barley,[31,269] wheat[134,270] and high-amylose maize[110] starches are known to have some branch chains, which are much longer than any of those found in waxy maize starch amylopectin.[62,134,156] Amylopectins isolated from indica rice (IR48 and IR64) consist of >7% extra-long branch chains of DP 140–150.[156] The iodine affinity and the limiting viscosity numbers of the starches are dependent on the longest chain component.[62,156] Characterization of maize and wheat amylopectins shows extremely long branch chains with DP 700–1000, which are not found in either potato, sweet potato or waxy wheat amylopectins.[134,168] In contrast, amylopectin of sugar cane starch consists of predominantly A- and short B-chains; long branch chains appear as a shoulder on the size-exclusion chromatography profile.[271]

Amylopectins of high-amylose maize, maize mutants containing ae[79,86–91,96–98,108,110,258] and dominant mutant Ae genes,[109] and high-amylose mutants of rice[272,273] and others are all known to have longer branch chains than their normal starch counterparts. Amylopectins isolated from various high-amylose maize starches

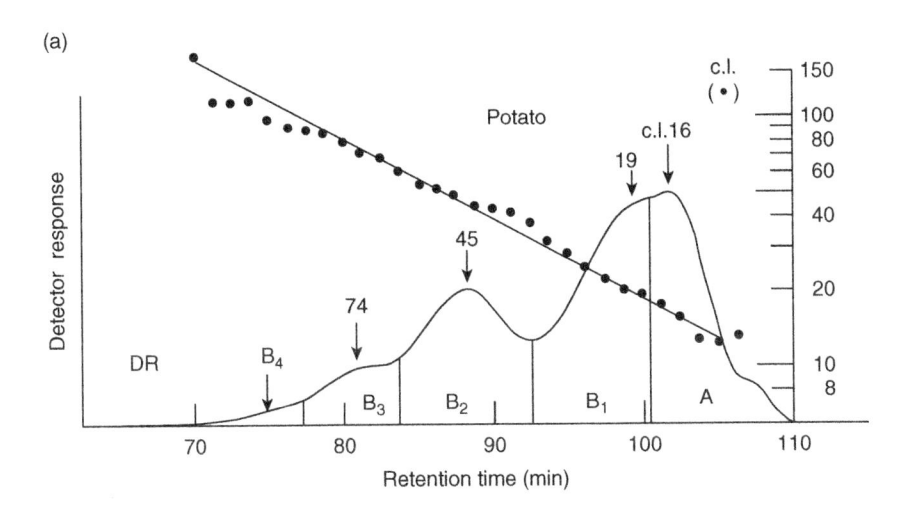

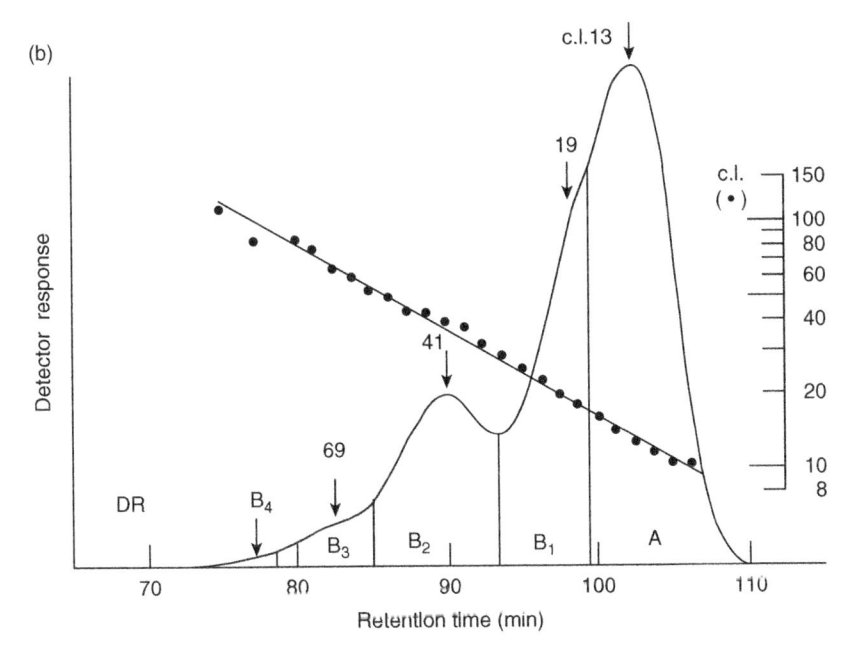

Figure 6.9 Gel-permeation hplc of isoamylase-debranched (a) potato; and (b) waxy rice amylopectin.[263]

(*ae* mutants),[110] and dominant mutant *Ae* maize[109] have iodine affinities between 3.6 and 6.6 g/100 g, and λ_{max} between 573 and 575 nm.[110] Amylopectin of Hylon 5 gives the greatest blue value.[274] The long chain branches of high-amylose maize amylopectin are attributed to the absence of the starch-branching enzyme IIb that transfers short chains.[112,275,276] When biosynthesis of starch-branching enzymes isoform II[81–83] or isoforms I and II[84] is interrupted, starches that consist of up to 90% apparent amylose are produced. These low-amylopectin starches consist of mainly amylose and intermediate components, and have little or no large molecular weight amylopectin.

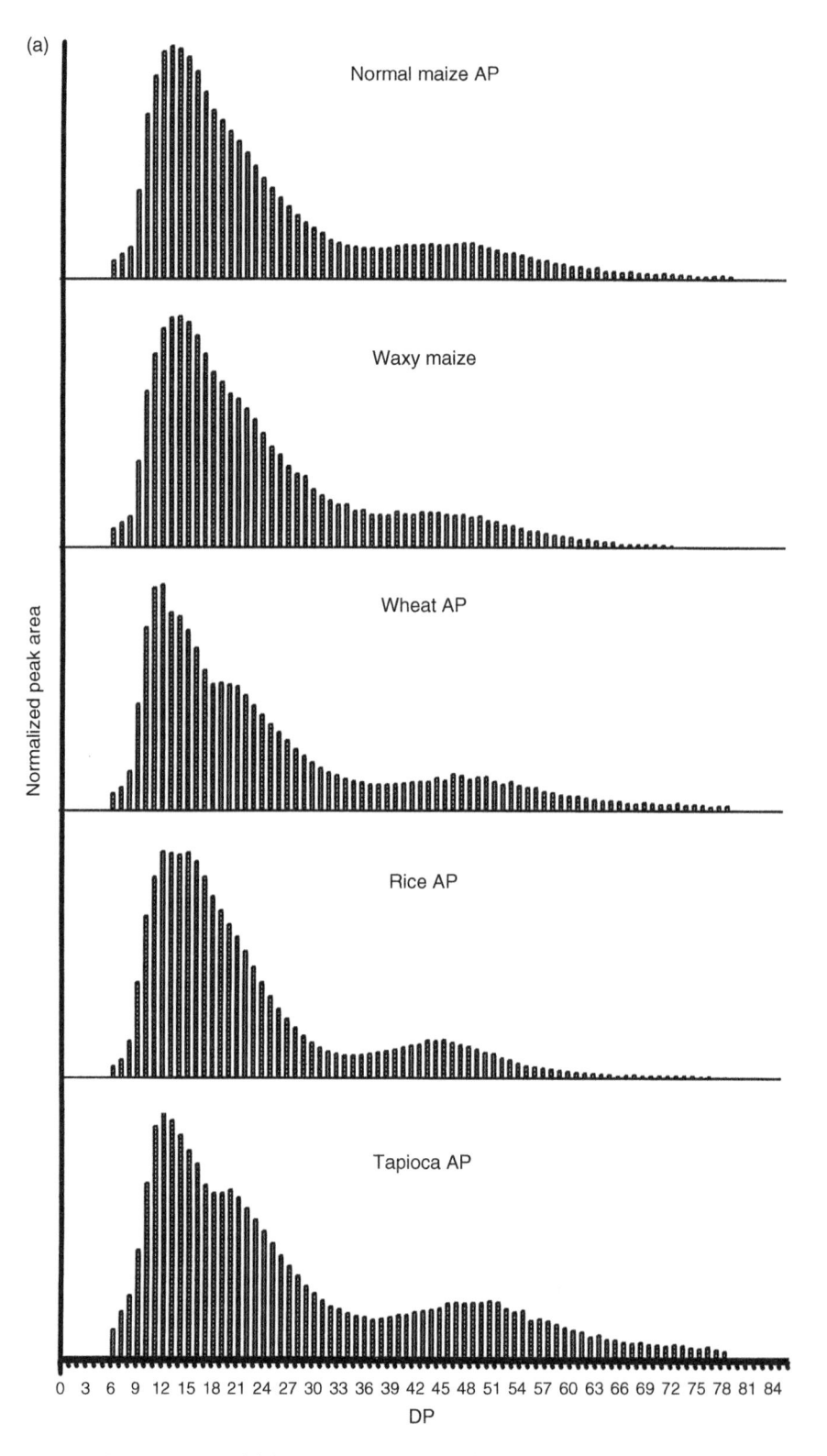

Figure 6.10 Chromatograms of debranched amylopectins of (a) A-type starches; (b) B-type starches analyzed using anion-exchange chromatography a pulsed amperometric detector and an amyloglucosidase enzyme reactor (HPAEC-ENZ-PAD); and (Continued overleaf)

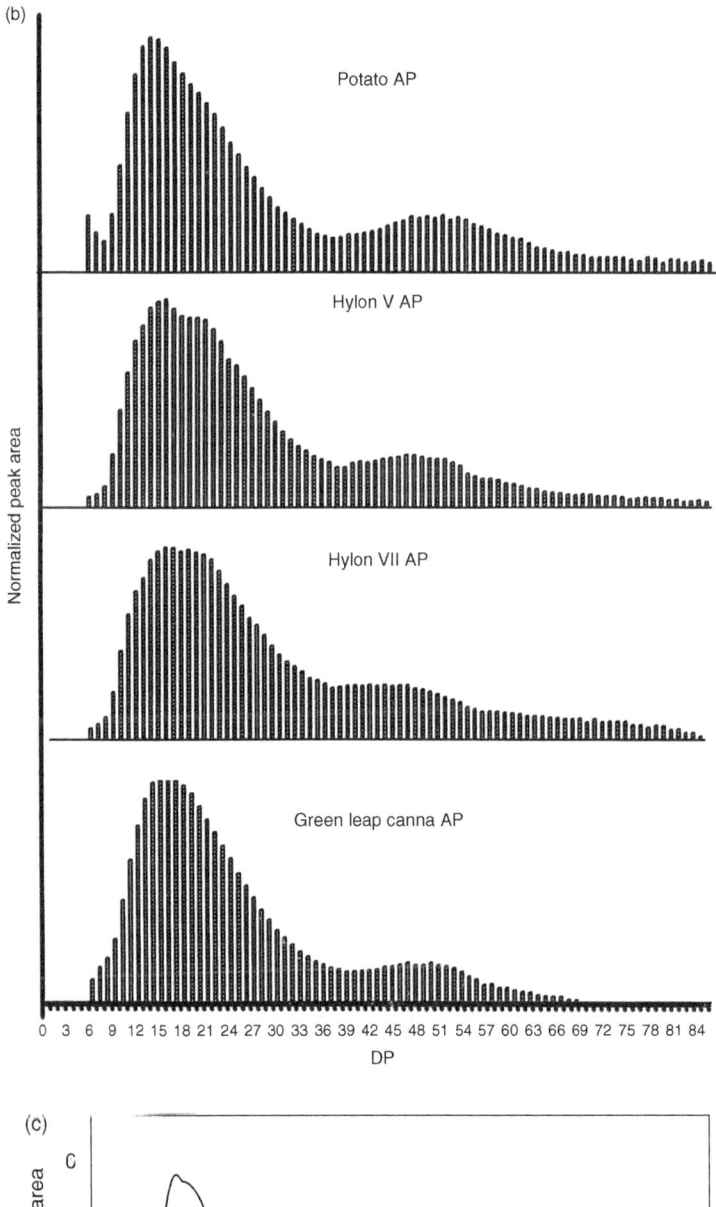

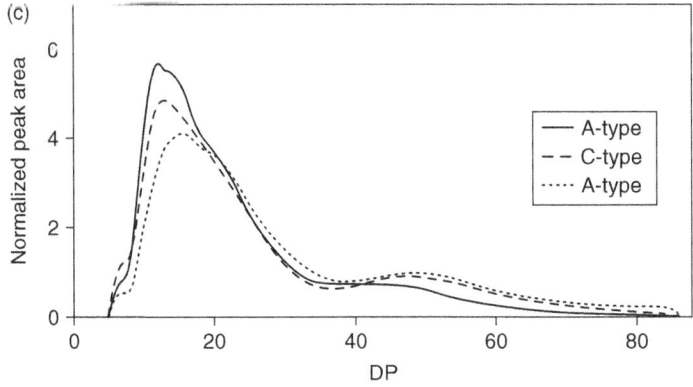

Figure 6.10 (Continued) (c) average chain-length distributions of selected A-, B-, and C-type starch.[79]

Because of its very large molecular weight, large polydispersity and susceptibility to shear degradation, amylopectin molecular weight determination is difficult, and results vary with the analytical technique and method used to disperse the sample.[274–279] Weight-average molecular weights of amylopectins determined by using size-exclusion chromatography and multi-angle laser-light scattering and refractive index detectors vary between 7×10^7 and 5.7×10^9 g/mole.[280] Molecular weight of waxy starch is larger than that of the normal starch counterpart.[280] The lower molecular weight of normal starch amylopectin is attributed to carbon flux partitioning between amylopectin and amylose molecules. Some normal starch amylopectins have number-average molecular weights that are substantially smaller than their weight-average molecular weights, which indicate the presence of small molecules. For example, wheat amylopectin has DP_n between 5000 and 9400.[157] Indica rice has a relatively small amylopectin (DP_n 4700–5800) compared with japonica rice (DP_n 8200–11 000).[281] Barley starch amylopectin has DP_n 7800–8700.[158] After the removal of intermediate material, high-amylose maize has large amylopectin molecules of DP_n up to 40 000.[110] Sago starch has large amylopectin molecules of DP_n 11 800 and 40 000 for low- and high-viscosity starch varieties, respectively.[163] Beta-amylolysis limits of amylopectin vary between 55 and 61%.

5. Cluster Models of Amylopectin

The cluster model of amylopectin, proposed independently by Nikuni[282] and French (Figure 6.11a),[67] has the branch points located in clusters and the external branch chains present in a double helical crystalline structure. The structure of the amylopectin molecule consists of alternating crystalline and amorphous lamellae. Kassenbeck[283] studied enzyme-treated starch granules and reported a repeating distance of 7.0 nm. Yamaguchi et al.[284] examined wet-meshed and acid-hydrolyzed waxy maize starch granules and reported a repeating distance of 7.0 ± 1.0 nm, and an acid-resistant lamellae thickness of ~5.0 nm. They proposed that alternating crystalline and amorphous regions of 5.0 nm and 2.0 nm, respectively, are arranged in the amylopectin structure of waxy maize starch.

Robin et al.[285] derived a model of amylopectin (Figure 6.11b) on the basis of structural analysis of Lintnerized potato starch. The crystallinity of granular starch is enhanced after acid hydrolysis.[140,285,286] Amylose is quickly depolymerized during the hydrolysis. This result indicates that amylose is present in an amorphous structure in the granule and that amylopectin chains are primarily responsible for the crystallinity of starch.[285] After extensive hydrolysis with acid, the resistant chains display two chain lengths: DP 25 and 15. The DP 25 chains can be debranched to give two chains each of DP 15, indicating the branch point in the amylopectin molecule after Linterization is close to the reducing end.[67,285,287] On the basis of these results, the crystalline region of potato starch is proposed to consist of 15 glucosyl units,[285] close to 5.3 nm in length. Calculated angular dependence of light scattering data show the best fit model is one in which interconnecting long B-chains carry an average of 1.4 clusters.[288]

IV. Structures of Amylose and Amylopectin 219

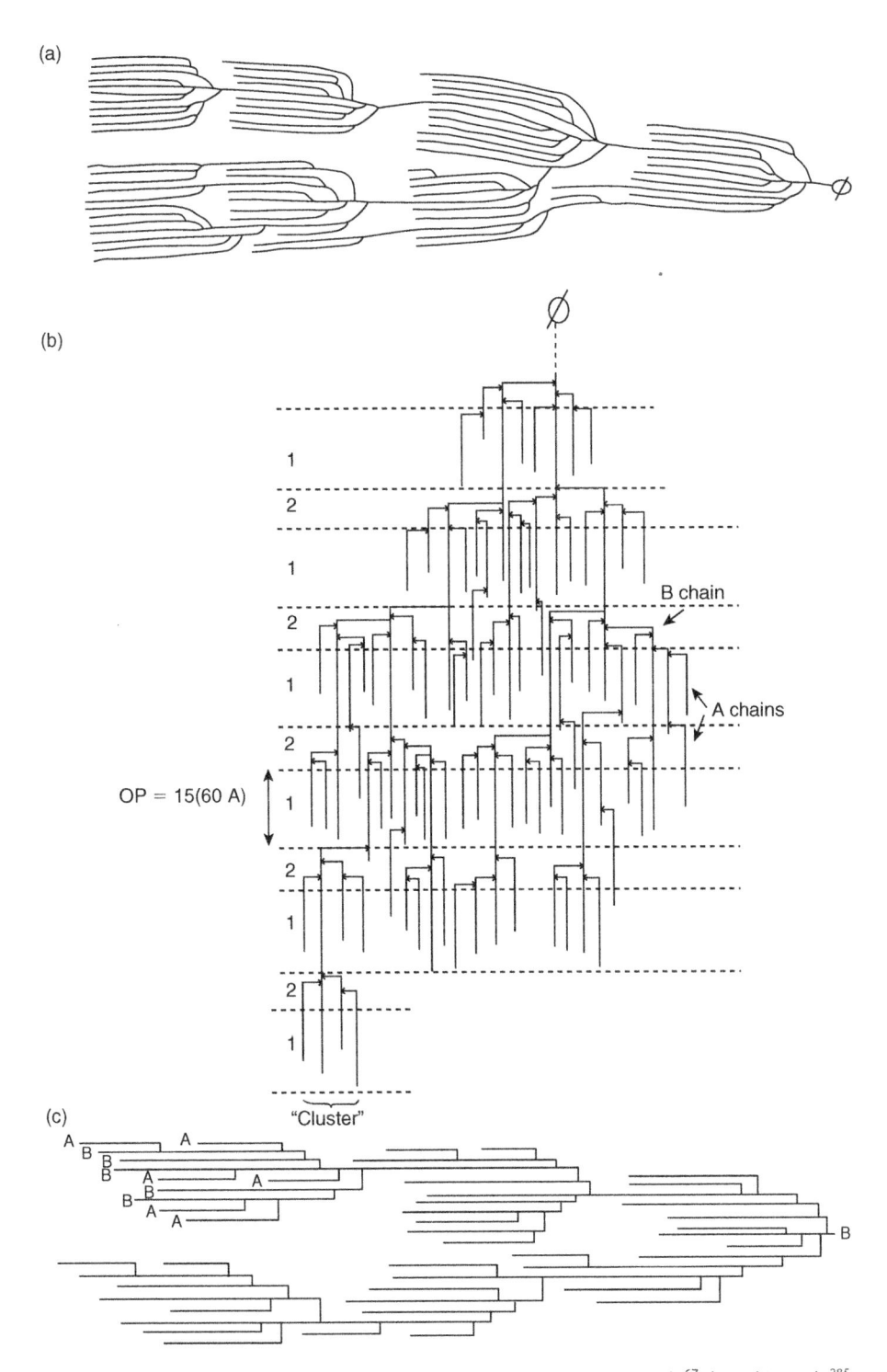

Figure 6.11 Structural cluster models for amylopectin as proposed by (a) French;[67] (b) Robin et al.;[285] and (c) Manners and Matheson.[290] 1 and 2 in the model represent crystalline and amorphous regions, respectively.

Fine-structures of waxy maize starch Naegeli dextrins analyzed by degradation with α-amylase and pullulanase confirm that the branching points are located close to the reducing ends of other branch chains, with a hairpin-like structure.[287] This result supports a double helical structure.[287] Lintnerized rice starches also produced two major distributions of chain length DP 24 and DP 13.[289] Debranching of chains of DP 24 are hydrolyzed to two chains with a peak DP of 15. Some rice varieties, which have higher amylose contents, produce resistant chains of DP 30, which coincides with the chain length of resistant fragments of retrograded amylose.[189,217,289] Thus, these chains can be attributed to retrograded amylose derived during the 15 days of acid treatment.[289] The x-ray diffraction pattern of the Lintnerized rice starch and Naegeli dextrins of *sugary–2* maize starch also give a significant V-pattern, indicating the presence of a resistant amylose–lipid complex.

On the basis of A:B chain ratio, Manners and Matheson[290] further revised the cluster models of French[67] and Robin et al.[285] This revised model (Figure 6.11c) emphasizes that a proportion of the B-chains must carry more than one A-chain to match the ratio of A:B chain of 1.0 (± 0.1):1.[291,292] Using a sequential treatment with β-amylase, isoamylase and β-amylase, Hizukuri and Maehara[270] determined that wheat amylopectin had an A:B chain ratio of 1.26:1. The $B_a:B_b$ (number of B-chains carrying one or more A-chains:number of B-chains carrying no A-chains) ratio was 1.5:1, indicating that ~40% of the B-chains carry no A-chains.[269] With the knowledge of multiple chain length distribution, Hizukuri further refined the structure of amylopectin (Figure 6.8).[263]

Amylopectin molecules hydrolyzed by various enzymes such as *B. amyloliquefaciens* α-amylase,[261,293–303] *P. stutzeri* amylase,[304] and cyclodextrin glycosyltransferase (CGT)[305] produce different limit dextrins. The consistency of the structures of the limit dextrins produced by each enzyme indicates the nature of the enzyme-catalyzed hydrolysis is not random. The orderly production of limit dextrins (Figure 6.12) further confirms the cluster structure of amylopectin. A unit cluster, however, is difficult to define because different enzymes and different hydrolytical methods produce limit dextrins with different structures. Using *B. amyloliquefaciens* α-amylase-catalyzed hydrolysis, Bertoft et al.[261,293–301] found that waxy barley,[300] smooth pea,[261] waxy maize[298] and potato starch[301] have clusters in the DP ranges 65–85, 32–55, 150–200 and 30–70, respectively. Potato starch has smaller clusters, which is in agreement with the larger proportion of long B-chains.[261] (The long B-chains are more susceptible to hydrolysis by *B. amyloliquefacians* α-amylase.) The molar ratio of short chains to long chains in potato starch is 6, whereas it is 11 in waxy maize starch.[259] The branching patterns of these amylopectins may also affect the susceptibility of the amylopectin to the α-amylase. Potato limit dextrins have substantially more long B-chains resistant to CGT reaction remaining. This resistance was attributed to branch chains carried by the long B-chains. The authors, however, did not comment on whether there might be effects of phosphate derivatives on the long B-chains.[305] The non-random nature of amylopectin branching has been discussed by Thompson.[306]

After extensive acid-catalyzed hydrolysis, the crystallites of starch granules are detached. With mild attrition, the starch granules break into small particles.[307] The size of the small-particle starch varies with the conditions of the acid hydrolysis;

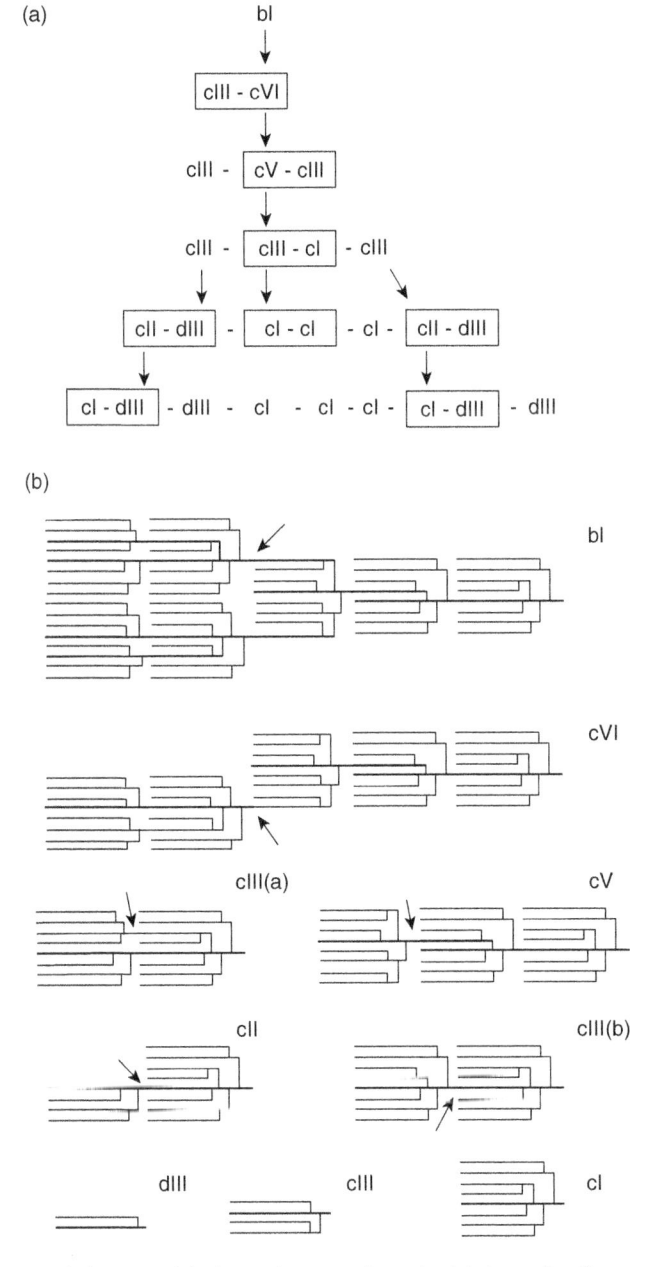

Figure 6.12 Proposed cluster model of smooth pea amylopectin: (a) the mode of interconnection of small structural units to build up dextrin b1. Arrows trace the sub-pieces (included in boxes) of the larger dextrins; (b) the fine structure of the dextrins showing how the clusters of short chains (—) are interconnected by long B-chains.[261]

the more extensive the acid hydrolysis, the smaller are the particle sizes produced. The small-particle starch produced from normal maize starch also displays strong birefringence and an enhanced A-type crystallinity. The production of small-particle starch further demonstrated the cluster structure of starch molecules in the granule.[286,308]

Blanshard et al.[309] observed a Bragg peak at approximately 10.0 nm by using small-angle neutron scattering studies of starch granules. The Bragg peak disappeared on gelatinization. X-ray diffraction studies of normal and waxy maize, wheat, potato and tapioca starches gave similar results (9.7 to 10.3 nm).[179] Recent studies of amylopectin structural periodicity by using small-angle x-ray scattering have also shown a constant repeating distance of between 8.7 and 9.2 nm for starch varieties of wheat, potato, tapioca, rice, corn and barley.[310] Results obtained from these two scattering studies show larger repeating distances than those determined by electron microscopy.[283,284] Blanshard et al.[309] attributed the differences to possible shrinkage of starch during the drying process required for the electron microscopy.

The consistent repeating distance of about 9 to 10 nm agrees with a complete cluster composed of 27–28 glucose units (equivalent to about 9.7 nm).[260] The distance is also congruent with the distance of the amylose double helical crystallite (about 31 glucose units, 10.8 nm)[189,232,233] and the lamellar distance of an amylose single helix complexed with various organic substances (~10 nm).[182,183,187–189]

Whether the length of the amylopectin cluster is governed by the size of double helix formed during biosynthesis is of great interest. Retrogradation of amylose at different temperatures has demonstrated that amylose crystallites produced at a higher incubation temperature consist of longer chain lengths.[232] This concurs with the growth temperature effect on the branch chain length of amylopectin; amylopectin of the starch developed at higher temperature consists of more long B-chains[311–314] (see Section 6.IV.6). Further studies are needed to reveal the mechanism of starch biosynthesis and granule development.

Nägeli and Lintner dextrins prepared from A-, B- and C-type starches consist of linear chains with peak chain lengths of DP 12–17, and singly-branched chains with peak DPs of 24–27.[12,139,140,287,315–318] The proportion of the singly-branched chains of Naegeli dextrins, prepared with extensive (60–70%) hydrolysis, decreases for A > C > B-type starches.[140,316,317] Examples of Nägeli dextrins of A-, C- and B-type starches are shown in Figure 6.13.[316] These results show that the branch linkages of B-type starch amylopectins are more susceptible to acid-catalyzed hydrolysis than are those of C- and A-types. The differences are attributed to the amylopectin structures of the B-type starches, which consist of fewer B1- and A-chains and have most of the branch linkages located in or close to the amorphous region. Thus, the branch linkages of B-type amylopectins are more susceptible to acid-catalyzed hydrolysis. Amylopectins of A-type starches, consisting of larger proportions of short B1- and A-chains, have branch linkages scattered in both amorphous and crystalline regions. Those branch linkages located in crystalline regions are protected from acid-catalyzed hydrolysis and remain as singly-branched chains. These results are in agreement with the report that most hydroxypropyl groups are found in the region of (1→6) branch linkages of chemically modified tapioca starch, with a few located at the non-reducing ends of short branch chains; half of the short chains of DP 15 contain no modifying groups because they are located in the crystalline region.[319]

The large proportion of short chains (B1- and A-chains) present in A-type starch amylopectin is located within one cluster, with their non-reducing ends free and unrestricted. These short chains have more freedom to move around and are likely

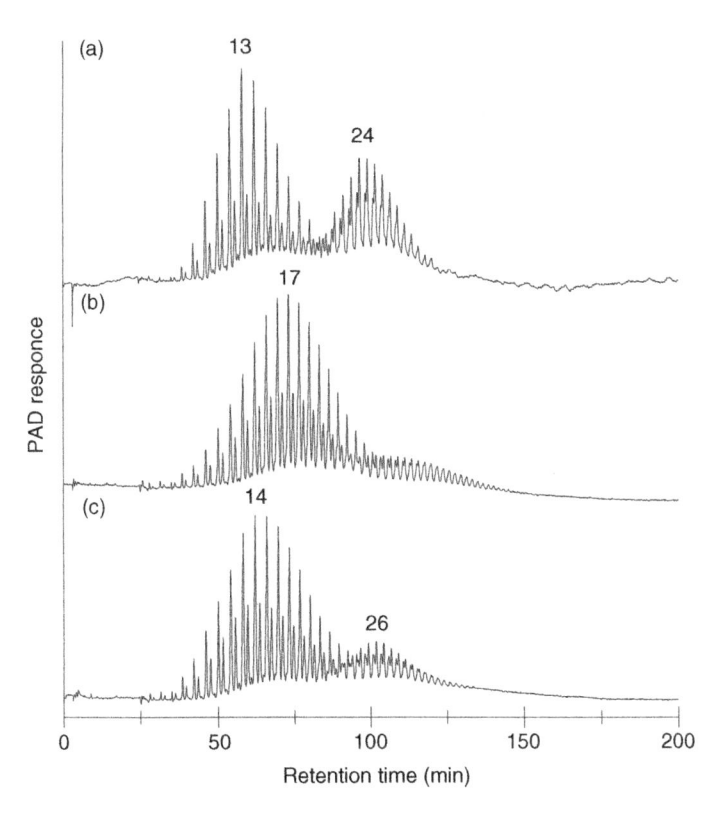

Figure 6.13 HPAEC-EN-PAD chromatograms of Naegeli dextrins of: (a) waxy maize (A-type); (b) *ae waxy* maize (B-type); and (c) banana (C-type) starches.[316]

to rearrange to thermodynamically favorable monoclinic unit cells from kinetically favored hexagonal unit packing.[320] Consequently, A-type starch granules develop porous internal structures, as revealed by confocal laser-light scattering micrographs (CLSM) (Figures 6.4d and 6.4f) and surface gelatinization (Figures 6.4b and 6.4c). The porous internal structure is possibly the result of rearrangement of the crystalline structure through annealing during starch granule development.[321] In contrast, B-type starch granules, consisting of more long B-chains (B2-, B3- and B4-chains) that extend through multiple clusters and are restricted from moving and rearranging, maintain the hexagonal crystalline structure. Thus, B-type starch granules do not display porous internal structures (Figures 6.4c and 6.4e). The porous structures of A-type starch granules agree with the results that maize starch granules consist of more total surface area than the external surface area of the granules determined by using a gas absorption method.[322] The total surface area of potato starch granules, however, is in agreement with the area of external granule surface. Internal channels have been reported in maize and sorghum starch granules using CLSM.[323,324]

The porous internal structures of A-type starch granules agree with known features; for example, A-type starches exhibit faster chemical penetration and derivatization reactions,[325] have weak points, are more susceptible to enzyme-catalyzed hydrolysis,[38] and display pinholes on the granule surface.[37,38,70] Details of the enzyme digestibility of starch are discussed in Chapter 7. The rearrangements of branch chains and crystalline structures in A-type starch granules could also result in the smooth surface of starch granules as observed by atomic force microscopy, which is different from the large blocks observed on the surface of potato starch granules[14,326,327] (see Figures 5.10 and 5.11).

6. Effects of Growing Temperature and Kernel Maturity on Starch Structures

The environmental temperature during anthesis and the grain-filling period has profound effects on the structure[313,314] and content[311–314,328,329] of amylose and on the fine structure of amylopectin.[311–314,328,329] When the temperature increases from 25° to 35°C, the amylose molecular weight and content decrease,[311–314] the proportion of long B-chains increases, and the proportion of short (A- and B1-chains) chains decreases. These changes are more prominent during the period 5–15 days after flowering (DAF). For high-amylose mutants of rice, the amylose content decreases with no change in the branch chain length of amylopectin when the growth temperature increases during the first 20 days after anthesis.[272] The amylose content of wheat starch increases slightly, the lipid content increases and swelling power decreases, with an increase in growth temperature.[330] The effects of developmental temperatures on the yields and structures of maize starches of different genetic backgrounds vary.[314] Some plants suffer yield loss at a high developmental temperature, whereas other plants can minimize the yield loss by modifying the chemical structures of the starch polymers.[314]

The iodine affinities, blue values, λ_{max}, beta-amylolysis limits and branch chain length distributions are similar for the starch samples harvested at different stages (7 to 30 days DAF) of development of waxy rice grains,[331] japonica rice[328] and indica rice.[332] The gelatinization temperature of maize starch increases from 12 to 24 DAF, but is reduced at 36 DAF.[333]

Structures of starch granules isolated from developing maize kernels harvested on different days after pollination (DAP) have been studied.[39] Results show that starch granules are first detected in the endosperm on 5 DAP. The number of starch granules 1–4μm in diameter increases significantly, but the starch content by weight is only about 2% (dsb) on 12 DAP. The starch content of the endosperm increases quickly from 10.7% (dsb) on 14 DAP to 88.9% on 30 DAP. The amylose content of the maize endosperm starch increases from 9.2% on 12 DAP to 24.2% on 30 DAP.[39] The branch chain-length distributions of amylopectin vary with kernel development; the average branch chain length increases from DP 23.6 on 10 DAP to a maximum of DP 26.9 on 14 DAP, then decreases to DP 25.4 on 30 DAP.[39] Amylose and amylopectin contents and branch chain length distributions of amylopectin agree with the expressions of granular-bound starch synthase, isoforms of soluble starch synthases, and branching enzymes during the development of the kernels.[334] The onset gelatinization temperature of the endosperm starches also increases from 61.3°C on 8 DAP to 69.0°C on 14 DAP, and then decreases to 67.4°C on 30 DAP and 62.8°C on 45 DAP (mature and dried).[39] Starch isolated from the pericarp of the maize kernel remains similar in granule size (diameter 1–4 μm), structure and properties. The starch content of the pericarp maintains at about 10% (dsb) during maize kernel development.[39]

V. Locations of Molecular Components in the Granule

Amylose leaches easily from swollen granules at a temperature slightly above the gelatinization temperature.[160] The molecular weight of the extracted amylose increases when the extraction temperature increases. Amylose does not contribute to the total crystallinity of starch granules.[285] Thus, Kassenbeck[283] proposed that amylose is in an ordered radial arrangement present in an amorphous structure in granules. In contrast to the dichroism observed in iodine-stained wheat starch granules, a weak or absent dichroism was found in iodine-stained potato starch granules. On this basis, Blanshard[306] and Zobel[179] suggest that amylose may partly co-crystallize with amylopectin chains in potato starch, but they are separated in maize starch granules.

Normal maize starch, which contains amylose, maintains granule integrity during cooking much better than does waxy maize starch. This phenomenon suggests that linear amylose molecules are intertwined with amylopectin to maintain the integrity of the granule during heating in water.[335,336] The same phenomenon is found in barley starch.[131] Jane et al.[144,325] conducted crosslinking reactions of normal maize and potato starch granules using epiclorohydrin, adipic acetic mixed anhydride and phosphoryl chloride (chain lengths of 4–7.5 Å) at low concentrations, to study the proximity of amylose molecules to each other and to amylopectin molecules in the granule. Results show that amylose is crosslinked onto amylopectin, but not crosslinked to itself. This suggests that amylose is located adjacent to or intertwined with amylopectin, but not in close proximity (4–7.5 Å) to other amylose molecules. Small-angle x-ray scattering studies of the effect of varying amylose content on the internal structure of maize, barley and pea starch have shown that amylose disrupts the structural order within the amylopectin crystallites.[337,338] This further suggests that amylose is intertwined with amylopectin.

Amylose is synthesized by granular-bound starch synthase, whereas amylopectin is synthesized by soluble starch synthase (Chapter 4).[334,339] Because amylose is synthesized by the granular-bound starch synthase in a progressive manner,[340] the amylose molecule is likely confined in the granule and has little opportunity to interact and form double helices with other starch molecules to facilitate branch formation. Branching reactions do occur on some amylose molecules, but at a much lower frequency than with amylopectin, and result in slightly branched amylose molecules.

Amylose contents of starch granules increase with kernel maturity and with an increase in granule size (Table 6.5).[39,41,42,269,334,341] Surface gelatinization of starch granules using concentrated neutral salt ($CaCl_2$ or $LiCl$) solutions has revealed that amylose is more concentrated at the periphery.[41] These results are consistent with the amylose content increasing with increases in granule size and kernel maturity.[39]

Table 6.5 Amylose contents of potato starch with different granular sizes and at different radial locations[41]

Sample	Amylose content (%)[a]
Native potato starch	20.2 ± 0.1
Potato starch (<20 μm[b])	16.9 ± 0.2
Potato starch (<30 μm[b])	17.5 ± 0.1
Potato starch (30–52 μm[b])	20.3 ± 0.1
Potato starch (>52 μm[b])	20.6 ± 0.1
Remaining granular starch after 80% chemical gelatinization	18.8 ± 0.1
Remaining granular starch after 52% chemical gelatinization	19.6 ± 0.1
Chemically gelatinized starch (52% chemical gelatinization)	21.1 ± 0.4
Chemically gelatinized starch (10% chemical gelatinization)	22.0 ± 0.1

[a] The amylose content was calculated by dividing the iodine affinity of the sample by 19.9
[b] Diameter

Table 6.6 Branch chain length of amylopectin debranched with isoamylase[a][41]

	Branch chain length, dp[b]	
	Long chain	Short chain
Native potato starch	41.2 ± 1.3	13.2 ± 0.3
Potato starch (<20 μm[c])	44.7 ± 1.3	14.7 ± 0.7
Potato starch (30–52 μm[c])	41.2 ± 1.8	13.2 ± 0.4
Potato starch (>52 μm[c])	34.0 ± 1.2	13.4 ± 0.2
Remaining granular starch after 80% chemical gelatinization	42.5 ± 1.8	13.1 ± 0.1
Chemically gelatinized surface starch (20% chemical gelatinization)	32.0 ± 0.8	13.1 ± 0.7

[a] Data reported are averages of duplicate sample and chemical analyses, except for the long chain of large granules (>52 μm) (one sample and duplicate chemical analysis)
[b] Determined with the three peak fractions; dp, degree of polymerization
[c] Diameter

The surface gelatinization method has also been employed to investigate the structures of amylopectin molecules at different radial locations (e.g. the core and the periphery) within the granule.[41,42] Results show that amylopectin isolated from the inner part of the granule has substantially longer long B-branches compared with the amylopectin molecules isolated from the outer regions of the granule. The amylopectin isolated from the periphery of the granule has the shortest long B-chains (Table 6.6). These results agree with those obtained during kernel development.[39] Using the same surface gelatinization technique, phosphate ester groups were found to be more concentrated in the inner portion of potato starch granules.[41]

Bogracheva et al.[342] reported that the B polymorphs are located at the center of granules and are surrounded by the A polymorphs in C-type starches. Gelatinization of pea starch (C-type) gives a double transition with the gelatinization of the B polymorphs located at the center of the granule melting at a lower temperature and the A polymorphs on the periphery at a higher temperature.[342]

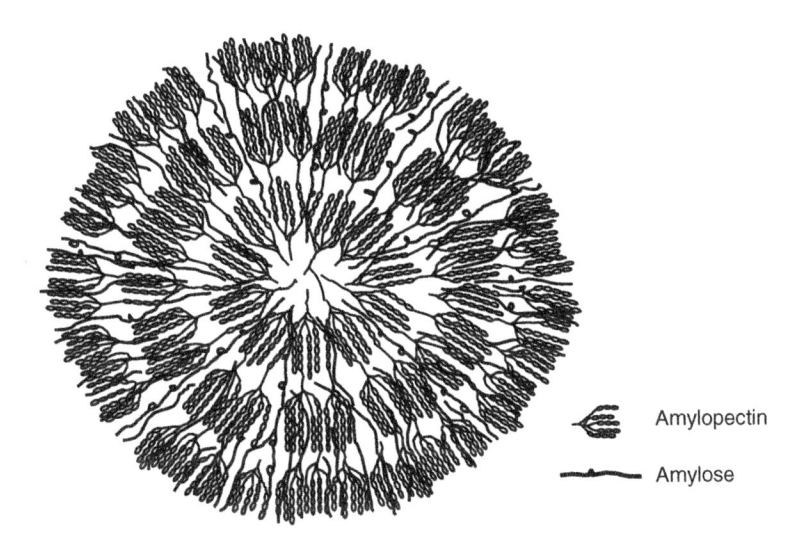

Figure 6.14 Schematic of the organization of a starch granule.

With knowledge available at the time, Lineback[343] proposed a schematic model of a cross-section of the starch granule. In the diagram, amylose is present in the amorphous structure and part of the amylose is present as a helical complex with the lipids naturally present in cereal starches. Part of the outer chains of amylopectin are shown as double helices. The surface is a boundary with the appearance of a 'hairy billiard ball,' having chains of different lengths protruding through the surface. Since this diagram was proposed, several studies have confirmed the proposed structures. It has been reported that the non-reducing ends of starch chains on the surface of the starch granule can be extended by transferring glucose units from ADP-glucose or UDP-glucose, catalyzed by a starch synthase.[344] Solid-nmr studies have shown that amylose is partly complexed with lipids.[129,212] Using Lineback's model,[343] and adding recent advances in the understanding of granule structures, including: (a) amylose present in close proximity to and, possibly, co-crystallized with amylopectin;[144,167,325] (b) amylose more concentrated at the periphery;[39,41,42] and (c) amylopectin in the inner region of granules having longer branch chains,[39,41,42] an updated diagram of the cross-section of a starch granule is proposed in Figure 6.14.

VI. References

1. Badenhuizen NP, Dutton RW. *Protoplasma*. 1956;XLVII:156.
2. Imberty A, Buleon A, Tran V, Perez S. *Starch/Stärke*. 1991;43:375.
3. Brown SA, French D. *Carbohydr. Res.* 1977;59:203.
4. BeMiller JN, Pratt GW. *Cereal Chem.* 1981;58:517. **60**, 254 (1983).
5. Hizukuri S, Takeda Y, Yasuda M, Suzuki A. *Carbohydr. Res.* 1981;94:205.
6. Schoch TJ. *J. Am. Chem. Soc.* 1942;64:2954.
7. Hizukuri S, Tabata S, Nikuni Z. *Stärke*. 1970;22:338.

8. French D. In: Whistler RL, BeMiller JN, Paschall EF, eds. *Starch: Chemistry and Technology*. 2nd edn. Orlando, FL: Academic Press; 1984:183.
9. Hizukuri S. In: Eliasson A-C, ed. *Carbohydrates in Food*. New York, NY: Marcel Dekker; 1996:347.
10. Jane J, Kasemsuwan T, Leas S, Zobel HF, Robyt JF. *Starch/Stärke*. 1994;46:121.
11. Ao Z, Jane J. *Carbohydr. Polym.* 2007;67:46.
12. Perera C, Lu Z, Sell J, Jane J. *Cereal Chem.* 2001;78:249.
13. Raosavljevic M, Jane J, Johnson LA. *Cereal Chem.* 1998;75:212.
14. Gallant DJ, Bouchet B. *Food Microstructure*. 1986;5:141.
15. Hall DM, Sayre JG. *Text. Res. J.* 1969;39:1044. **40**, 256 (1970).
16. Hood LF, Liboff M. In: Bechtel DB, ed. *New Frontiers in Food Microstructure*. St. Paul, MN: American Association of Cereal Chemists; 1983:341.
17. Mercier CC, Charbonniere R, Gallant DJ, Guilbot A. *Stärke*. 1970;22:9.
18. Zhao J, Whistler RL. *Cereal Chem.* 1994;71:392.
19. Yanez GA, Messinger JK, Walker CE, Rupnow JH. *Cereal Chem.* 1986;63:273.
20. Irving DW, Betschart AA, Saunders RM. *J. Food Sci.* 1981;46:1170.
21. Sugimoto Y, Yamada K, Sakamoto S, Fuwa H. *Starch/Stärke*. 1981;33:112.
22. Biliaderis CG, Mazza G, Przybylski R. *Starch/Stärke*. 1993;45:121.
23. Mazza G, Biliaderis CG, Przybylski R, Oomah BD. *J. Agr. Food Chem.* 1992;40:1520.
24. Goering KJ. *Starch/Stärke*. 1978;30:181.
25. Jane J, Shen L, Chen J, Lim S, Kasemsuwan T, Nip WK. *Cereal Chem.* 1992;69:528.
26. Sugimoto Y, Nishihara K, Fuwa H. *J. Jpn. Soc. Starch Sci.* 1986;33:169.
27. Atwell WA, Patrick BM, Johnson LA, Glass RW. *Cereal Chem.* 1982;60:9.
28. Fannon JE, Hauber RJ, BeMiller JN. *Cereal Chem.* 1992;69:284.
29. MacGregor AW, Ballance DL. *Cereal Chem.* 1980;57:397.
30. Fannon JE, Shull JM, BeMiller JN. *Cereal Chem.* 1993;70:611.
31. Huber KC, BeMiller JN. *Cereal Chem.* 1997;74:537.
32. Schwimmer S. *J. Biol. Chem.* 1945;161:219.
33. Nikuni Z. *J. Agr. Chem. Soc. Jpn.* 1956;30:A131.
34. Nikuni Z, Whistler RL. *J. Biochem (Tokyo)*. 1957;44:227.
35. Leach HW, Schoch TJ. *Cereal Chem.* 1961;38:34.
36. Fuwa H, Glover DV, Sugimoto Y, Tanaka M. *J. Nutr. Sci. Vitaminol.* 1978;24:437.
37. Hood LF, Liboff M. In: Bechtel DB, ed. *New Frontiers in Food Microstructure*. St. Paul, MN: American Association of Cereal Chemists; 1983:341–370.
38. Helbert W, Schuelein M, Henrissat B. *Int. J. Biol. Macromol.* 1996;19:165.
39. Li L, Blanco M, Jane J. *Carbohydr. Polym.* 2007;67:630.
40. Jane J, Ao Z, Duvick SA, Wiklund M, Yoo S-H, Wong K-S, Gardner C. *J. Appl. Glycosci.* 2003;50:167.
41. Jane J, Shen JJ. *Carbohydr. Res.* 1993;247:279.
42. Pan DD, Jane J. *Biomacromol.* 2000;1:126.
43. Koch K, Jane J. *Cereal Chem.* 2000;77:115.
44. Whistler RL, Spencer WW, Goatley JL, Nikuni Z. *Cereal Chem.* 1958;35:331.
45. Whistler RL, Goatley JL, Spencer WW. *Cereal Chem.* 1959;36:84.
46. Baldwin PM, Adler J, Davies MC, Melia CD. *Starch/Stärke*. 1994;46:341.
47. Donovan JW, Lorenz K, Kulp K. *Cereal Chem.* 1983;60:381.

48. Hoover R, Vasanthan T. *Carbohydr. Res.* 1994;252:33.
49. Lorenz K, Kulp K. *Starch/Stärke*. 1983;35:123.
50. Stute R. *Starch/Stärke*. 1992;44:205.
51. Lorenz K, Kulp K. *Starch/Starke*. 1982;34:50.
52. Hagiwara S, Esaki K, Kitamura S, Kuge T. *Denpun Kagaku*. 1991;3:241.
53. Katopo H, Song Y, Jane J. *Carbohydr. Polym.* 2002;47:233.
54. Hibi Y, Matsumoto T, Hagiwara S. *Cereal Chem.* 1993;70:671.
55. Jane J, Atichokudomchai N, Suh D-S. In: Tomasik P, Yuryev VP, Bertoft E, eds. *Starch: Progress in Structural Studies, Modifications and Applications*. Cracow Poland: Polish Society of Food Technology; 2004:147.
56. Jane J. *J. Appl. Glycosci.* 2006;53:205.
57. Atwell WA, Hood LF, Lineback DR, Varriano-Marston E, Zobel HF. *Cereal Foods World*. 1988;33:306.
58. Gudmundsson M, Eliasson A-C. *Starch/Stärke*. 1992;44:379.
59. Bowler P, Williams MR, Angold RE. *Starch/Stärke*. 1980;32:186.
60. Williams MR, Bowler P. *Starch/Stärke*. 1982;34:221.
61. Obanni M, BeMiller JN. *Cereal Chem.* 1995;72:436.
62. Ziegler GR, Thompson DB, Casasnovas J. *Cereal Chem.* 1993;70:247.
63. Yeh A-I, Li J-Y. *J. Cereal Sci.* 1996;23:277.
64. Tester RF, Morrison WR. *Cereal Chem.* 1990;67:551.
65. Liu J, Zhao S. *Starch/Stärke*. 1990;42:96.
66. Cooke D, Gidley MJ. *Carbohydr. Res.* 1992;227:103.
67. French D. *J. Jpn. Soc. Starch Sci.* 1972;19:8.
68. Hood LF, Serifried AS, Meyer R. *J. Food Sci.* 1974;39:117.
69. Chabot JF, Hood LF, Allen JE. *Cereal Chem.* 1976;53:85.
70. Fannon JE, BeMiller JN. *Cereal Chem.* 1992;69:456.
71. Schoch TJ. *Meth. Carbohydr. Chem.* 1964;4:157.
72. Morrison WR, Laignelet B. *J. Cereal Sci.* 1983;1:9.
73. Chrastil J. *Carbohydr. Res.* 1987;159:154.
74. Knutson CA, Cluskey JE, Dintzis FR. *Carbohydr. Res.* 1982;101:117.
75. Knutson CA. *Cereal Chem.* 1986;63:89.
76. Knutson CA, Grove MJ. *Cereal Chem.* 1994;71:469.
77. Bolling H, El Baya AW. *Chem. Mikrobiol. Technol. Lebensm.* 1975;3:161.
78. Morrison WR, Karkalas J. In: Dey PM, ed. *Methods in Plant Biochemistry*. Vol. 2. London, UK: Carbohydrates, Academic Press; 1990.
79. Jane J, Chen YY, Lee LF, McPherson AE, Wong KS, Radosavljevic M, Kasemsuwan T. *Cereal Chem.* 1999;78:629.
80. Tester RF, Morrison WR. *Cereal Chem.* 1992;69:654.
81. Sidebottom C, Kirkland M, Strongitharm B, Jeffcoat R. *J. Cereal. Sci.* 1998;27:279.
82. Shi Y-C, Capitani T, Tizasko P, Jeffcoat R. *J. Cereal Sci.* 1998;27:289.
83. Case SE, Capitani T, Whaley JK, Shi YC, Trasko P, Jeffcoat R, Goldfarb HB. *J. Cereal Sci.* 1998;27:301.
84. Schwall GP, Stafford R, Westcott RJ, Jeffcoat R, Tayal A, Shi YC, Gidley MJ, Jobling SA. *Nature Biotechnol.* 2000;18:551.
85. Campbell MR, White PJ, Pollak LM. *Cereal Chem.* 1994;71:464.

86. Ikawa Y, Glover DV, Sugimoto Y, Fuwa H. *Starch/Stärke*. 1981;33:9.
87. Inouchi N, Glover DV, Takaya T, Fuwa H. *Starch/Stärke*. 1983;35:371.
88. Inouchi N, Glover DV, Sugimoto Y, Fuwa H. *Starch/Stärke*. 1984;36:8.
89. Inouchi N, Glover DV, Fuwa H. *Starch/Stärke*. 1987;39:259.
90. Inouchi N, Glover DV, Fuwa H. *Starch/Stärke*. 1987;39:284.
91. Fuwa H, Glover DV, Miyaura K, Inouchi N, Konishi Y, Sugimoto Y. *Starch/Stärke*. 1987;39:295.
92. Inouchi N, Glover DV, Sugimoto Y, Fuwa H. *Starch/Stärke*. 1991;43:468.
93. Inouchi N, Glover DV, Sugimoto Y, Fuwa H. *Starch/Stärke*. 1991;43:473.
94. Li J, Berke TG, Glover DV. *Cereal Chem*. 1994;71:87.
95. Wang Y-J, White P, Pollak L. *Cereal Chem*. 1992;69:328.
96. Wang Y-J, White P, Pollak L, Jane J. *Cereal Chem*. 1993;70:171.
97. Wang Y-J, White P, Pollak L. *Cereal Chem*. 1993;70:199.
98. Wang Y-J, White P, Pollak L, Jane J. *Cereal Chem*. 1993;70:521.
99. Katz FR, Furesik SL, Tenbarge FL, Hauber RJ, Friedman RB. *Carbohydr. Polym*. 1995;21:133.
100. Shi YC, Seib PA. *Carbohydr. Polym*. 1995;26:141.
101. Banks W, Greenwood CT. *Starch and Its Components*. Edinburgh, UK: Edinburgh University Press; 1975:51.
102. Lansky S, Kooi M, Schoch TJ. *J. Am. Chem. Soc*. 1949;71:4066.
103. Whistler RL, Doane WM. *Cereal Chem*. 1961;38:251.
104. Baba T, Arai Y. *Agric. Biol. Chem*. 1984;48:1763.
105. Ikawa Y, Glover DV, Sugimoto Y, Fuwa H. *Carbohydr. Res*. 1978;61:211.
106. Yuan RC, Thompson DB, Boyer CD. *Cereal Chem*. 1993;70:81.
107. Yeh JY, Garwood DL, Shannon JC. *Starch/Stärke*. 1981;33:222.
108. Jane J, Chen J. *Cereal Chem*. 1992;69:60.
109. Kasemsuwan T, Jane J, Schnable P, Stinard P, Robertson D. *Cereal Chem*. 1995;71:457.
110. Takeda C, Takeda Y, Hizukuri S. *Carbohydr. Res*. 1993;246:273.
111. Takeda Y, Preiss J. *Carbohydr. Res*. 1993;240:265.
112. Takeda Y, Guan H-P, Preiss J. *Carbohydr. Res*. 1993;240:253.
113. Pan D, Nelson OE. *Plant Physiol*. 1984;74:324.
114. James MG, Robertson DS, Myers AM. *Plant Cell*. 1995;7:417.
115. Yun S, Matheson NK. *Carbohydr. Res*. 1993;243:307.
116. Doehlert DC, Kuo TM, Juvik JA, Beers EP, Duke SH. *J. Am. Soc. Hort. Sci*. 1993;118:661.
117. Headrick JM, Pataky JK, Juvik JA. *Phytopathol*. 1990;80:487.
118. Boyer CD, Simpson EKG, Demewood PA. *Starch/Stärke*. 1982;34:81.
119. Dickinson DB, Boyer CD, Velu JG. *Phytochem*. 1983;22:1371.
120. Nakamura Y, Kubo A, Shimamune T, Matsuda T, Harada K, Karusatoh H. *Plant J*. 1997;12:143.
121. Morrison WR. *J. Sci. Food Agr*. 1978;29:365.
122. Morrison WR. *J. Cereal Sci*. 1988;8:1.
123. Mührbeck P, Tellier C. *Starch/Stärke*. 1991;43:25.
124. Lim S-T, Kasemsuwan T, Jane J. *Cereal Chem*. 1994;71:488.

125. Kasemsuwan T, Jane J. *Cereal Chem.* 1996;73:702.
126. Jane J, Kasemsuwan T, Chen JF, Juliano BO. *Cereal Foods World.* 1996;41:827.
127. Song Y, Jane J. *Carbohydr. Polym.* 2000;41:365.
128. South JB, Morrison WR, Nelson OE. *J. Cereal Sci.* 1991;14:267.
129. Morrison WR, Tester RF, Snape CE, Law R, Gidley MJ. *Cereal Chem.* 1993;70:385.
130. Morrison WR, Law RV, Snape CE. *J. Cereal Sci.* 1993;18:107.
131. Morrison WR, Tester RF, Gidley MJ, Karkalas J. *Carbohydr. Res.* 1993;245:289.
132. Batres LV, White PJ. *J. Am. Oil Chem. Soc.* 1986;63:1537.
133. Medcalf DG, Young VL, Gilles KA. *Cereal Chem.* 1968;45:88.
134. Yoo SH, Jane J. *Carbohydr. Polym.* 2002;49:297.
135. Lim S, Seib PA. *Cereal Chem.* 1993;70:137.
136. Tabata S, Hizukuri S. *Stärke.* 1971;23:267.
137. Abe J, Takeda Y, Hizukuri S. *Biochim. Biophys. Acta.* 1982;703:26.
138. Abe J, Nagano H, Hizukuri S. *J. Appl. Biochem.* 1985;7:235.
139. Blennow A, Bay-Smidt AM, Wischmanin B, Olsen CE, Moeller BL. *Carbohydr. Res.* 1998;307:45.
140. McPherson AE, Jane J. *Carbohydr. Polym.* 1999;40:57.
141. Tabata S, Nagata K, Hizukuri S. *Stärke.* 1975;27:333.
142. Hizukuri S, Shirasaka K, Juliano BO. *Starch/Stärke.* 1983;35:348.
143. Lim S-T, Seib PA. *Cereal Chem.* 1993;70:145.
144. Kasemsuwan T, Jane J. *Cereal Chem.* 1994;71:282.
145. Takeda Y, Hizukuri S. *Carbohydr. Res.* 1982;102:321.
146. Hizukuri S, Abe J. In: Meuser F, Manners DJ, Seibel W, eds. *Plant Polymeric Carbohydrates.* London, UK: The Royal Society of Chemistry; 1993:16.
147. Nielsen TH, Wischmann B, Enevoldsen K, Moeller BL. *Plant Physiol.* 1994;105:111.
148. Mührbeck P, Svensson E, Eliasson A-C. *Starch/Stärke.* 1991;43:466.
149. Mührbeck P, Eliasson A-C. *J. Sci. Food Agr.* 1991;55:13.
150. Banks W, Greenwood CT. *Arch. Biochem. Biophys.* 1966;117:674.
151. Takeda Y, Hizukuri S, Takeda C, Suzuki A. *Carbohydr. Res.* 1987;165:139.
152. Takeda Y, Tokunaga N, Takeda C, Hizukuri S. *Starch/Stärke.* 1986;38:345.
153. Hizukuri S, Takeda Y, Shitaozono T, Abe J, Otakara A, Takeda C, Suzuki A. *Starch/Stärke.* 1988;40:165.
154. Takeda Y, Shitaozono T, Hizukuri S. *Starch/Stärke.* 1988;40:51.
155. Takeda Y, Hizukuri S, Juliano BO. *Carbohydr. Res.* 1986;148:299.
156. Takeda Y, Maruta N, Hizukuri S, Juliano BO. *Carbohydr. Res.* 1989;189:227.
157. Shibanuma K, Takeda Y, Hizukuri S. *Carbohydr. Polym.* 1994;25:111.
158. Schulman AH, Tomooka S, Suzuki A, Myllarinen P, Hizukuri S. *Carbohydr. Res.* 1995;275:361.
159. Takeda Y, Shirasaka K, Hizukuri S. *Carbohydr. Res.* 1984;132:83.
160. Hizukuri S. *Carbohydr. Res.* 1991;217:251.
161. Takeda C, Takeda Y, Hizukuri S. *Cereal Chem.* 1989;66:22.
162. Takeda Y, Hizukuri S, Juliano BO. *Carbohydr. Res.* 1989;186:163.
163. Takeda Y, Takeda C, Suzuki A, Hizukuri S. *J. Food Sci.* 1989;54:177.
164. Hizukuri S. *Denpun Kagaku.* 1988;35:185.

165. Takeda Y, Tomooka S, Hizukuri S. *Carbohydr. Res.* 1993;246:267.
166. Takeda Y, Maruta N, Hizukuri S. *Carbohydr. Res.* 1992;226:279.
167. Takeda Y, Shitaozono T, Hizukuri S. *Carbohydr. Res.* 1990;199:207.
168. Murugesan G, Shibanuma K, Hizukuri S. *Carbohydr. Res.* 1993;242:203.
169. Takeda C, Takeda Y, Hizukuri S. *Denpun Kagaku.* 1987;34:31.
170. Suzuki A, Hizukuri S, Takeda Y. *Cereal Chem.* 1981;58:286.
171. Suzuki A, Takeda Y, Hizukuri S. *Denpun Kagaku.* 1985;32:205.
172. Suzuki A, Kaneyama M, Shibanuma K, Takeda Y, Abe J, Hizukuri S. *Cereal Chem.* 1992;69:309.
173. Suzuki A, Kaneyama M, Shibanuma K, Takeda Y, Abe J, Hizukuri S. *Denpun Kagaku.* 1994;40:41.
174. Takeda C, Takeda Y, Hizukuri S. *Cereal Chem.* 1983;60:212.
175. Hayashi A, Kinoshita K, Miyake Y. *Polym. J.* 1981;13:537.
176. Takeo K, Tokumura A, Kuge T. *Stärke.* 1973;25:357.
177. French AD, Murphy VG. *Cereal Foods World.* 1977;22:61.
178. Zobel HF. *Starch/Stärke.* 1988;40:1.
179. Zobel HF. *Starch/Stärke.* 1988;40:44.
180. French D, Pulley AO, Whelan WJ. *Stärke.* 1963;15:349.
181. Bulpin PV, Welsh EJ, Morris ER. *Starch/Stärke.* 1982;34:335.
182. Manley RSJ. *J. Polym. Sci.* 1964;A2:4503.
183. Yamashita Y. *J. Polym. Sci.* 1965;A3:3251.
184. Codet MC, Buleon A, Tran V, Colonna P. *Carbohydr. Polym.* 1993;21:91.
185. Codet MC, Bouchet B, Colonna P, Gallant DJ, Buleon A. *J. Food Sci.* 1996;61:1196.
186. Hulleman SHD, Helbert W, Chanzy H. *Int. J. Biol. Macromol.* 1996;18:115.
187. Yamashita Y, Hirai N. *J. Polym. Sci.* 1966;A4:161.
188. Yamashita Y, Monobe K. *J. Polym. Sci.* 1971;A9:1471.
189. Jane J, Robyt JF. *Carbohydr. Res.* 1984;132:105.
190. Biliaderis CG, Galloway G. *Carbohydr. Res.* 1989;189:31.
191. Karkalas J, Raphaelides S. *Carbohydr. Res.* 1986;157:215.
192. Robyt JF, French D. *Arch. Biochem. Biophys.* 1967;122:8.
193. Robyt JF, French D. *Arch. Biochem. Biophys.* 1963;100:451.
194. Jane J, Robyt JF, Huang D-H. *Carbohydr. Res.* 1985;140:21.
195. Gidley MJ, Bociek SM. *J. Am. Chem. Soc.* 1988;110:3820.
196. Yu S, Houtman C, Atalla RH. *Carbohydr. Res.* 1996;292:129.
197. Moulik SP, Gupta S. *Carbohydr. Res.* 1984;125:340.
198. Donovan JW, Mapes CJ. *Starch/Stärke.* 1980;32:190.
199. Kugimiya M, Donovan JW, Wong TY. *Starch/Stärke.* 1980;32:265.
200. Biliaderis CG, Page CM, Slade L, Sirett RR. *Carbohydr. Polym.* 1985;5:367.
201. Hoover R, Hadziyev D. *Starch/Stärke.* 1981;33:390.
202. Kugimiya M, Donovan JW. *J. Food Sci.* 1981;46:765.
203. Eliasson A-C, Krog N. *J. Cereal Sci.* 1985;3:239.
204. Biliaderis CG, Page CM, Maurice TJ. *Carbohydr. Polym.* 1986;6:269.
205. Biliaderis CG, Seneviratne HD. *Carbohydr. Polym.* 1990;13:185.
206. Biliaderis CG. *Can J. Physiol. Pharmacol.* 1991;69:60.

207. Raphaelides SR, Karkalas J. *Carbohydr. Res.* 1988;172:65.
208. Stute R, Konieczny-Janda G. *Starch/Stärke*. 1983;35:340.
209. Kowblansky M. *Macromol.* 1985;18:1776.
210. Karkalas J, Ma S, Morrison WR, Pethrick RA. *Carbohydr. Res.* 1995;268:233.
211. Zobel HF. In: Alexander RJ, Zobel HF, eds. *Developments in Carbohydrate Chemistry*. St. Paul, MN: Amerian Association of Cereal Chemists; 1992:1.
212. Morrison WR. In: Shewry PR, Stobart AK, eds. *Seed Storage Compounds: Biosynthesis, Interactions and Manipulation*. Oxford, UK: Oxford University Press; 1993.
213. Miles MJ, Morris VJ, Ring SG. *Carbohydr. Polym.* 1984;4:73.
214. Miles MJ, Morris VJ, Ring SG. *Carbohydr. Res.* 1985;135:257.
215. Miles MJ, Morris VJ, Orford PD, Ring SG. *Carbohydr. Res.* 1985;135:271.
216. Leloup VM, Colonna P, Buleon A. *J. Cereal Sci.* 1991;13:1.
217. Leloup VM, Colonna P, Ring SG, Roberts K, Wells B. *Carbohydr. Polym.* 1992;18:189.
218. Ring SG, Miles MJ, Morris VJ, Turner R, Colonna P. *Int. J. Biol. Macromol.* 1987;9:158.
219. Wild DI, Blanshard JMV. *Carbohydr. Polym.* 1986;6:121.
220. Pfannemüller B. *Int. J. Biol. Macromol.* 1987;9:105.
221. Gidley MJ, Bulpin PV. *Carbohydr. Res.* 1987;161:291.
222. Hanashiro I, Abe J, Hizukuri S. *Carbohydr. Res.* 1996;283:151.
223. Koizumi K, Fukuda M, Hizukuri S. *J. Chromatogr.* 1991;585:233.
224. Wong KS, Jane J. *J. Liq. Chromatogr.* 1997;20:297.
225. Pfannemüller B, Mayerhofer H, Schulz RC. *Biopolym.* 1971;10:243.
226. Gidley MJ, Bulpin PV. *Macromol.* 1989;22:341.
227. Gidley MJ. *Macromol.* 1989;22:351.
228. Gidley MJ, Bulpin PV, Kay S. *Gums and Stabilisers for the Food Industry*. Vol. 3. London, UK: Elsevier; 1986:167.
229. Gidley MJ. *Macromol.* 1989;22:351.
230. Gidley MJ, Bociek SM. *J. Am. Chem. Soc.* 1985;107:7040.
231. Marchessault RH, Taylor MG, Fyfe CA, Veregin RP. *Carbohydr. Res.* 1985;144:C1.
232. Lu T-J, Jane J, Keeling P. *Carbohydr. Polym.* 1997;33:19.
233. Ring SG, Colonna P, I'Anson KJ, Kalichevsky MT, Miles MJ, Morris VJ, Orfore PD. *Carbohydr. Res.* 1987;162:277.
234. Borovsky D, Smith EE, Whelan WJ, French D, Kikumoto S. *Arch. Biochem. Biophys.* 1979;198:627.
235. Earlingen RC, Deceuninck M, Delcour JA. *Cereal Chem.* 1993;70:345.
236. Eberstein K, Hopcke R, Konieczny-Janda G, Stute R. *Starch/Stärke*. 1980;32:397.
237. Berry CS. *J. Cereal Sci.* 1986;4:301.
238. Sievert D, Pomeranz Y. *Cereal Chem.* 1989;66:342.
239. Sievert D, Pomeranz Y. *Cereal Chem.* 1990;67:217.
240. Gruchala L, Pomeranz Y. *Cereal Chem.* 1993;70:163.
241. Sievert D, Würsch P. *Cereal Chem.* 1993;70:333.
242. Berry CS, I'Anson K, Miles MJ, Morris VJ, Russell PL. *J. Cereal Sci.* 1988;8:203.
243. Earlingen RC, Crombez M, Delcour JA. *Cereal Chem.* 1993;70:339.

244. Brumosvky JO, Thompson DB. *Cereal Chem.* 2001;78:680.
245. Leeman AM, Karlsson ME, Eliasson A-C, Bjorck IME. *Carbohydr. Polym.* 2006;65:306.
246. Cairns P, Leloup V, Miles MJ, Ring SG, Morris VJ. *J. Cereal Sci.* 1990;12:203.
247. Leloup V, Colonna P, Ring SG. *J. Cereal Sci.* 1992;16:253.
248. Russell PL, Berry CS, Greenwell P. *J. Cereal Sci.* 1989;9:1.
249. Ring SG, Gee JM, Whittam M, Orford PD, Johnson IT. *Food Chem.* 1988;28:97.
250. Ring SG, Colonna P, I'Anson KJ, Kalichevsky MT, Miles MJ, Morris VJ, Orford PD. *Carbohydr. Res.* 1987;162:277.
251. Gidley MJ, Cooke D, Darke AH, Hoffmann RA, Russell PL, Greenwell P. *Carbohydr. Polym.* 1995;28:23.
252. Ranhotra GS, Gelroth JA, Glaser BK. *J. Food Sci.* 1996;61:453.
253. Annison G, Topping DL. *Ann. Rev. Nutr.* 1994;14:297.
254. Cairns P, Morris VJ, Botham RL, Ring SG. *J. Cereal Sci.* 1996;23:265.
255. Faisant N, Champ M, Colonna P, Buleon A. *Carbohydr. Polym.* 1993;21:205.
256. Cairns P, Sun L, Morris VJ, Ring SG. *J. Cereal Sci.* 1995;21:37.
257. Calder PC. *Int. J. Biochem.* 1991;23:1335.
258. Hizukuri S, Kaneko T, Takeda Y. *Biochem. Biophys. Acta.* 1983;760:188.
259. Hizukuri S. *Carbohydr. Res.* 1985;141:295.
260. Gernat S, Radosta S, Damaschun G, Schierbaum F. *Starch/Stärke.* 1990;42:175.
261. Bertoft E, Qin Z, Manelius R. *Starch/Stärke.* 1993;45:377.
262. Spence KE, Jane J. *Carbohydr. Polym.* 1999;40:261.
263. Hizukuri S. *Carbohydr. Res.* 1986;147:342.
264. Ong MH, Jumel K, Tokarczuk PF, Blanshard JMV, Harding SE. *Carbohydr. Res.* 1994;260:99.
265. Cameron RE, Donald AM. *Polymer.* 1992;33:2628.
266. Larew LA, Johnson DC. *Anal. Chem.* 1988;60:1867.
267. Wong KS, Jane J. *J. Liq. Chromatogr.* 1995;18:63.
268. Wong KS, Jane J. In: Townsend RR, Hutchkiss A, eds. *Techniques in Glycobiology.* New York, NY: Marcel Dekker; 1997:553–566.
269. MacGregor AW, Morgan JE. *Cereal Chem.* 1984;61:222.
270. Hizukuri S, Maehara Y. *Carbohydr. Res.* 1990;206:145.
271. Blake J, Clarke ML. *Carbohydr. Res.* 1985;138:161.
272. Asaoka M, Okuno K, Sugimoto Y, Yano M, Omura T, Fuwa H. *Starch/Stärke.* 1986;38:114.
273. Yano M, Onuno K, Kawakami J, Satoh H, Omura T. *Theor. Appl. Genet.* 1985;69:253.
274. Chen J, Jane J. *Cereal Chem.* 1994;71:623.
275. Boyer CD, Preiss J. *Biochem. Biophys. Res. Commun.* 1978;80:169.
276. Boyer CD, Preiss J. *Carbohydr. Res.* 1978;61:321.
277. Millard MM, Dintzis FR, Willet JL, Klavons JA. *Cereal Chem.* 1997;74:687.
278. Han J-A, Lim S-T. *Carbohydr. Polym.* 2004;55:193.
279. Han J-A, Lim S-T. *Carbohydr. Polym.* 2004;55:265.
280. Yoo SH, Jane J. *Carbohydr. Polym.* 2002;49:307.
281. Takeda Y, Hizukuri S, Juliano BO. *Carbohydr. Res.* 1987;168:79.

VI. References

282. Nikuni Z. *Starch/Stärke*. 1978;30:105.
283. Kassenbeck VP. *Starch/Stärke*. 1978;30:40.
284. Yamaguchi M, Kainuma K, French D. *J. Ultrastruc. Res.* 1979;69:249.
285. Robin JP, Mercier C, Charbonniere R, Guilbot A. *Cereal Chem.* 1974;51:389.
286. Kainuma K, French D. *Biopolym.* 1971;10:1673.
287. Umeki K, Kainuma K. *Carbohydr. Res.* 1981;96:143.
288. Thurn A, Burchard W. *Carbohydr. Polym.* 1985;5:441.
289. Maningat CC, Juliano BO. *Starch/Stärke*. 1979;31:5.
290. Manners DJ, Matheson NK. *Carbohydr. Res.* 1981;90:99.
291. Manners DJ. *Carbohydr. Polym.* 1989;11:87.
292. Manners DJ. *Cereal Foods World*. 1985;30:461.
293. Bertoft E, Henriksnas H. *J. Inst. Brew* 1982;88:261.
294. Bertoft E. *Carbohydr. Res.* 1986;149:379.
295. Bertoft E, Spoof L. *Carbohydr. Res.* 1989;189:169.
296. Bertoft E. *Carbohydr. Res.* 1989;189:181.
297. Bertoft E. *Carbohydr. Res.* 1989;189:195.
298. Bertoft E. *Carbohydr. Res.* 1991;212:229.
299. Bertoft E. *Carbohydr. Res.* 1991;212:245.
300. Bertoft E, Avall A-K. *J. Inst. Brew*. 1992;98:433.
301. Zhu Q, Bertoft E. *Carbohydr. Res.* 1996;288:155.
302. Bertoft E, Manelius R. *Carbohydr. Res.* 1992;227:269.
303. Bertoft E, Koch K. *Carbohydr. Polym.* 2000;41:121.
304. Finch P, Sebesta DW. *Carbohydr. Res.* 1992;227:C1.
305. Bender H, Siebert R, Stadler-Szoke A. *Carbohydr. Res.* 1982;110:245.
306. Thompson DB. *Carbohydr. Polym.* 2000;43:223.
307. Jane J, Shen L, Wang L, Maningat CC. *Cereal Chem.* 1992;69:280.
308. Kainuma K, French D. *Biopolym.* 1972;11:2241.
309. Blanshard JMV. In: Blanshard JMV, Frazier PJ, Galliard T, eds. *Chemistry and Physics of Baking*. London, UK: The Royal Society of Chemistry; 1986:1.
310. Jenkins PJ, Cameron RE, Donald AM. *Starch/Stärke*. 1993;45:417.
311. Asaoka M, Okuno K, Fuwa H. *Agr. Biol. Chem.* 1985;49:373.
312. Asaoka M, Okuno K, Sugimoto Y, Kawakami K, Fuwa H. *Starch/Stärke*. 1984;36:189.
313. Asaoka M, Okuno K, Konishi Y, Fuwa H. *Agr. Biol. Chem.* 1987;51:3451.
314. Lu T, Jane J, Keeling PL, Singletary GW. *Carbohydr. Res.* 1996;282:157.
315. Shi YC, Seib PA. *Carbohydr. Res.* 1992;227:131.
316. Jane J, Wong K, McPherson AE. *Carbohydr. Res.* 1997;300:219.
317. Watanabe T, Akiyama T, Takahashi H, Akachi T, Matsumato A, Matsuda K. *Carbohydr. Res.* 1982;109:221.
318. Watanabe T, Akiyama T, Matsumato A, Matsuda K. *Carbohydr. Res.* 1983;112:171.
319. Hood LF, Mercier C. *Carbohydr. Res.* 1978;61:53.
320. Gidley MJ. *Carbohydr. Res.* 1987;161:301.
321. Nakazawa Y, Wang Y-J. *Carbohydr. Res.* 2003;338:2871.
322. Hellman NN, Melvin EH. *J. Am. Chem. Soc.* 1950;72:5186.
323. Huber KC, BeMiller JN. *Carbohydr. Polym.* 2000;41:269.

324. Gray JA, BeMiller JN. *Cereal Chem.* 2004;81:278.
325. Jane J, Xu A, Rodosovljivic M, Seib PA. *Cereal Chem.* 1992;69:405.
326. Baldwin PM, Frazier RA, Adler J, Glasbey TO, Keane MP, Roberts CJ, Tendler SJB, Davies MC, Melia CD. *J. Microscopy.* 1996;184:75.
327. Hatta T, Nemoto S, Kainuma K. *J. Appl. Glycosci.* 2003;50:159.
328. Asaoka M, Okuno K, Sugimoto Y, Fuwa H. *Agr. Biol. Chem.* 1985;49:1973.
329. Umemoto T, Nakamura Y, Saloh H, Terashima K. *Starch/Stärke.* 1999;51:58.
330. Shi Y-C, Seib PA, Bernardin JE. *Cereal Chem.* 1994;71:369.
331. Murugesan G, Hizukuri S, Fukuda M, Juliano BO. *Carbohydr. Res.* 1992;223:235.
332. Enevoldsen BS, Juliano BO. *Cereal Chem.* 1988;65:424.
333. Ng KY, Duvick SA, White PJ. *Cereal Chem.* 1997;74:288.
334. Fujita N, Yoshida M, Ohdam N, Yiyao A, Firochika H, Nakamura Y. *Plant Physiol.* 2006;140:1070.
335. Jane J, Craig S, Seib PA, Hoseney C. *Starch/Stärke.* 1986;38:258.
336. Chen J, Jane J. *Cereal Chem.* 1994;71:618.
337. Jenkins PJ, Donald AM. *Int. J. Biol. Macromol.* 1995;17:315.
338. Cameron RE, Donald AM. In: Dickinson E, ed. *Food Polymer, Gels, and Colloids.* London, UK: The Royal Society of Chemistry; 1991:301.
339. Preiss J, Sivak M. In: Zamski A, Schaffer AA, eds. *Photoassimilate Distribution in Plants and Crops.* New York, NY: Marcel Dekker; 1995:63 [Section A].
340. Denyer K, Waite D, Motawia S, Moeller BL, Smith AM. *Biochem. J.* 1999;340:183.
341. Yun SH, Matheson NK. *Carbohydr. Res.* 1992;227:85.
342. Bogracheva TYa, Morris VJ, Ring SG, Hedley CL. *Bioplymers.* 1998;45:323.
343. Lineback DR. *Bakers' Digest.* 1984;3:16.
344. Baba T, Yoshii M, Kainuma K. *Starch/Stärke.* 1987;39:52.

Enzymes and Their Action on Starch

7

John F. Robyt
Laboratory of Carbohydrate Chemistry and Enzymology, Department of Biochemistry, Biophysics, and Molecular Biology, Iowa State University, Ames, Iowa, 50011, USA

I.	Introduction	238
II.	Amylases	238
	1. Action of Endo-Acting α-Amylases	238
	2. Action of Exo-Acting β-Amylases	244
	3. Amylases Producing Specific Maltodextrin Products	246
	4. Action of Isoamylases	247
	5. Archaebacterial Amylases	248
	6. Action of Cyclomaltodextrin Glucanosyltransferase	250
III.	Relation of Structure with Action of the Enzymes	253
	1. Relation of Structure with Action of Endo-Acting α-Amylases	253
	2. Structure and Action of Soybean β-Amylase	257
	3. Structure and Action of Glucoamylases	257
	4. Specific Amino Acids at the Active-Site Involved in Catalysis and Substrate Binding	261
	5. Structure and Function of Domains in Amylolytic Enzymes	263
IV.	Mechanisms for the Enzymatic Hydrolysis of the Glycosidic Bond	264
V.	Action of Amylases on Insoluble Starch Substrates	267
	1. Action of α-Amylases on Amylose-V Complexes and Retrograded Amylose	267
	2. Action of Amylases with Native Starch Granules	269
VI.	Inhibitors of Amylase Action	272
VII.	Action of Phosphorylase and Starch Lyase	276
	1. Plant Phosphorylase	276
	2. Starch Lyase	277
VIII.	Enzymic Characterization of Starch Molecules	278
	1. Determination of the Nature of the Branch Linkage in Starch	279
	2. Identification and Structure Determination of Slightly Branched Amyloses	280
	3. Formation of β-Amylase Limit Dextrins of Amylopectin and Determination of their Fine Structure	282
IX.	References	284

I. Introduction

Enzymes involved in the breakdown of starch chains are primarily of four types:

1. those that hydrolyze (1→4) α-D-glucosidic bonds (amylases);
2. those that hydrolyze (1→6) α-D-glucosidic bonds (isoamylases);
3. those that transfer (1→4) α-D-glucosidic bonds (glucanosyltransferases); and
4. branching enzymes [α-(1→4)α-(1→6) transferases].

The amylases can be divided into three classes:

1. the endo-acting α-amylases;
2. the exo-acting β-amylases; and
3. isoamylases.

II. Amylases

1. Action of Endo-Acting α-Amylases

The α-amylases comprise a broad collection of different kinds of enzymes from all biological classes – bacteria, fungi, plants and animals. In the latter case, α-amylases may be found in different locations or organs in the animal, such as in mammalian salivary glands and mammalian pancreas. These α-amylases are not identical, having different product specificities and producing specific kinds of malto-oligosaccharide products. One of the earliest recognized amylases was called ptyalin and was found in human saliva. This is now known as human salivary α-amylase and has been highly purified and crystallized.[1] Another mammalian α-amylase was one of the enzymes in porcine pancreatin that is now known as porcine pancreatic α-amylase, which also has been highly purified and crystallized.[2,3] Other pancreatic α-amylases from human pancreas[4] and from rat pancreas[5] have also been purified and crystallized. Microbial α-amylases are known and have been highly purified and crystallized from *Bacillus amyloliquefaciens*,[i] [6,7] *B. subtilis* var. *saccharitikus*,[9] *B. coagulans*,[10] *Pseudomonas saccharophila*[11] and *Aspergillus oryzae*.[12]

The α-amylases are distinguished by two features, the formation of products that have the α-configuration at the reducing end anomeric carbon of the newly formed products and an endo-mechanism of attacking the starch chain at glucose residues in the interior parts of the starch chain. In the exo-mechanism (discussed in the next section), the enzyme binds glucose residues at the non-reducing ends of the starch chains, and frequently produces products that have the β-configuration at the anomeric carbon of the newly formed reducing end of the products. The endo-mechanism of α-amylases rapidly fragments starch chains into smaller chains and thereby rapidly decreases the starch–iodine–iodide blue color and the viscosity, and endo-acting

---

[i]The organism that elaborates this α-amylase was originally called *Bacillus subtilis*. It was reclassified as *Bacillus amyloliquefaciens* in 1967.[8]

amylases are, therefore, sometimes called liquefying amylases. These particular features of α-amylase attack have produced misunderstandings that their action is random. A comparison of the percentage drop of the blue starch–iodine color versus the percentage increase in reducing value for several different kinds of α-amylases showed that each of the α-amylases had different curves, and that these curves were also different from those produced by acid-catalyzed hydrolysis, which should be a random process or close to one.[13] Another misunderstanding is that the final products of α-amylase action are glucose and maltose. Alpha-amylases from different sources produce different products in different amounts, with glucose being invariably a very minor final product and almost never an early product.

The first reported detailed action pattern of an α-amylase was that of *Bacillus amyloliquefaciens*[i] α-amylase.[14] Maltodextrin products that were formed in the early stages of the reaction for this enzyme were maltotriose (G3), maltohexaose (G6) and maltoheptaose (G7). Later products included glucose (G1), maltose (G2), maltotriose (G3), maltotetraose (G4), maltopentaose (G5) and maltohexaose (G6), with G3 and G6 predominating and with G1, G2 and G4 formed in much lesser amounts. The latter products plus G5 result from the secondary hydrolysis of G6, G7 and G8. G6 is hydrolyzed extremely slowly at the first glycosidic bond from the reducing end to give G1+G5; G7 is hydrolyzed faster at the first and second glycosidic bonds to give G1+G6 and G2+G5, respectively, and G8 is hydrolyzed at the second and third glycosidic bonds to give G2+G6 and G3+G5, respectively.

When the substrate was amylopectin β-amylase limit dextrin, a different pattern of products was formed, namely G1, G2 and G3, with no G6, G7 or higher sized dextrins.[14] Reaction with the β-limit dextrin indicated that G1, G2 and G3 are formed from the chains between the α-(1→6) branch linkages as the outer chains were removed by the action of β-amylase (see Section 7.2 for a discussion of the action of β-amylase). It further indicated that G6 and G7 from amylopectin were formed exclusively from longer unbranched outer chains. It also indicated that the number of glucosyl units between the branch linkages of amylopectin were sufficiently few that they could not yield the larger G6 and G7 products, but could give the smaller G1, G2 and G3 products.

When shellfish glycogen was the substrate, only G6 and G7 were chromatographically observed products. When the β-limit dextrin of glycogen was the substrate, no chromatographically observable products were formed, although some hydrolysis did occur. The G6 and G7 produced from glycogen were formed from the outer non-branched chains; the chain lengths between the branch points of glycogen and its

---

[ii] Complex oligosaccharides are named using the IUPAC system, in which individual glucose units in the longest saccharide of the molecule are assigned Roman numerals, starting at the reducing end and used as superscripts; substitutions at specific carbon atoms are indicated by the number of the carbon atom in the monosaccharide unit. For example, panose is 6^2-α-D-glucopyranosylmaltose, in which an α-D-glucopyranosyl unit is linked to O-6 of the second glucosyl unit from the reducing end of maltose. Isopanose is 6^1-α-D-glucopyranosylmaltose, in which an α-D-glucopyranosyl unit is linked to O-6 of the reducing end glucose unit of maltose. Replacement of various hydroxyl groups of oligosaccharides uses the same type of numbering.[100] For example, substitution of an amino group for the hydroxyl group at C-6 on the second glucosyl unit of maltotriose gives 6^2-amino-6^2-deoxymaltotriose, and substitution at C-4 on the third glucosyl unit of maltotriose gives 4^3-amino-4^3-deoxymaltotriose.

β-limit dextrin are too short to give any chromatographically observable products. Therefore, only limited hydrolysis of single (1→4)-D-glucosidic linkages occurred between the α-(1→6) branch linkages of the glycogen.

To explain the relatively high initial yields of G3 and G6 and the reactions of the maltodextrins, amylopectin β-limit dextrin, shellfish glycogen and glycogen β-limit dextrin, Robyt and French[14] postulated a nine unit binding subsite that produced products by a 'dual product specificity.' The site of hydrolysis by the catalytic groups was proposed to be at the glycosidic linkage between the glucosyl units bound at subsites III and IV (Figure 7.1a). Their hypothesis stated that either one of the two dextrin chains resulting from the first hydrolytic cleavage can diffuse from the active-site and fill the open glucosyl unit-binding subsites. In the one instance, the chain to the right of the catalytic groups diffuses away and the remaining chain is repositioned to fill the three open subsites. Maltotriose is then formed in the next hydrolytic cleavage. In the other instance, the chain to the left of the catalytic groups diffuses away and the chain to the right is repositioned to fill the six open subsites and maltohexaose is formed. Figure 7.2 gives a summary of the proposed dual product action to form G3 and G6.

The nine glucosyl unit-binding subsites with the catalytic groups oriented to attack the glycosidic bond between the third and fourth glucosyl unit subsite was confirmed by Thoma et al.[15] by determining the Michaelis parameters of individual maltodextrins, G3 to G12, and applying the concept of the unitary free energy of subsite binding. The subsite energy profile that was obtained was consistent with a nine glucosyl unit-binding subsite (Figure 7.1a).

Later, MacGregor and MacGregor[16] reported that barley malt α-amylase had an action pattern that was similar to that of *B. amyloliquefaciens* α-amylase, forming

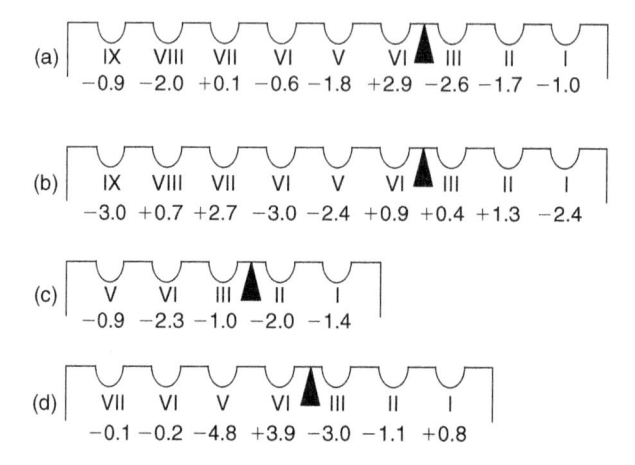

Figure 7.1 Glucose-binding subsites at the active-site of four endo-acting α-amylases. Subsites for endo-acting amylases are numbered with Roman numerals, beginning with the subsite that would bind the reducing-end glucosyl unit. Arabic numbers indicate binding energies in kcal/mole, with increasing negative values indicating the strongest binding; ▲ represents the location of the catalytic groups in relation to the glucose-binding sites. (a) Nine subsites for *Bacillus amyloliquefaciens* α-amylase after Robyt and French;[14] binding-site energies from Thoma et al.;[15] (b) nine subsites for barley malt α-amylase after MacGregor and MacGregor;[16] (c) five subsites for porcine pancreatic α-amylase after Robyt and French;[19] binding energies from Seigner et al.[23] (d) seven subsites for *Aspergillus oryzae* α-amylase after Suganuma et al.[24]

relatively high amounts of G3 and G6. They also postulated a nine glucosyl unit-binding subsite and determined the glucosyl binding energies of the individual subsites (Figure 7.1b).

The action of human salivary and porcine pancreatic α-amylases on starch also gives several products: G1, G2, G3, G4 and α-limit dextrins that have one and two α-(1→6) linkages.[13,17,18] Figure 7.3 summarizes the structures of these products. Maltotriose and maltotetraose are slowly hydrolyzed by these enzymes in a secondary reaction to yield glucose and maltose. Three of the α-limit dextrins ($B6_1$, $B6_2$ and $BB8_2$) are also hydrolyzed slowly at specific bonds, producing the smallest α-limit dextrin (B4) that is completely resistant to further hydrolysis; $BB8^1$ is also slowly hydrolyzed to give the B4 structure at the reducing end, but the product BB7 (essentially a double B4 structure, Figure 7.3) is resistant to further hydrolysis, and along with B4 and B5, constitute the porcine pancreatic and human salivary α-amylase completely resistant, α-limit dextrins.

In these α-amylase reactions, glucose is not the primary product. Glucose is formed as a secondary product in the hydrolysis of maltotriose, maltotetraose and branched dextrins. Using ^{14}C-reducing end labeled maltodextrins, G3 to G8, Robyt and French[19] showed that porcine pancreatic α-amylase has different specificities for hydrolyzing the different linkages of the maltodextrins. The different frequencies for

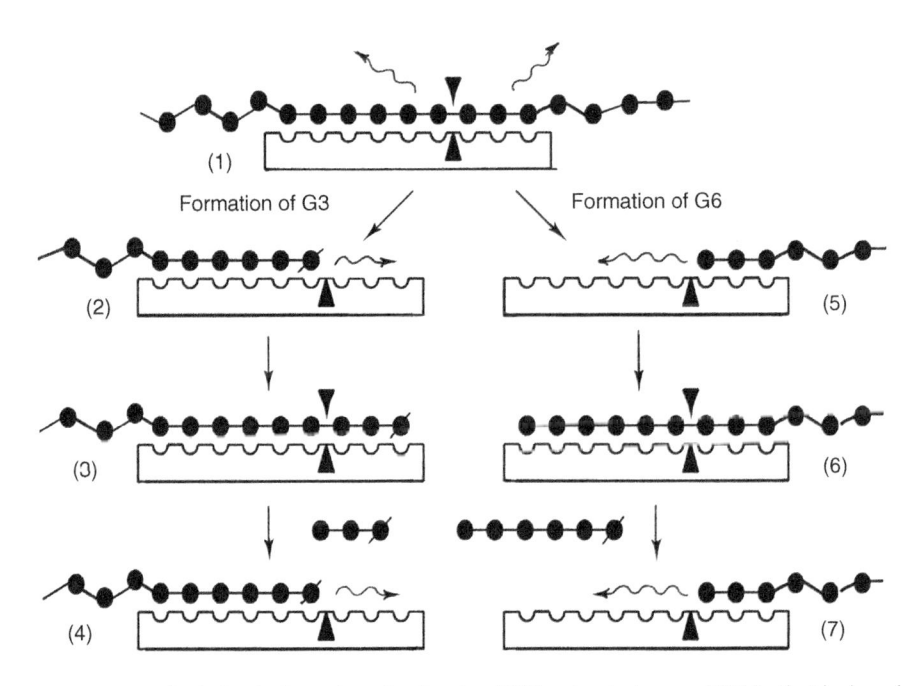

Figure 7.2 Hypothesis for the formation of maltotriose [G3] and maltohexaose [G6] by the 'dual-product specificity' mechanism proposed by Robyt and French[14] for *Bacillus amyloliquefaciens* α-amylase. In step (1), the chain is cleaved and the right-hand chain dissociates from the active-site, giving (2). The chain that remains at the active-site then repositions to fill the subsites, giving (3), which dissociates to give (4), and the process repeats or the chain dissociates from the active-site. Similarly, when cleavage occurs in step (1), the left-hand chain can dissociate to give (5). The chain that remains at the active-site repositions to fill the subsites, to give (6). Cleavage occurs to give G6 and (7), from which the process can repeat or the chain dissociates from the active-site.

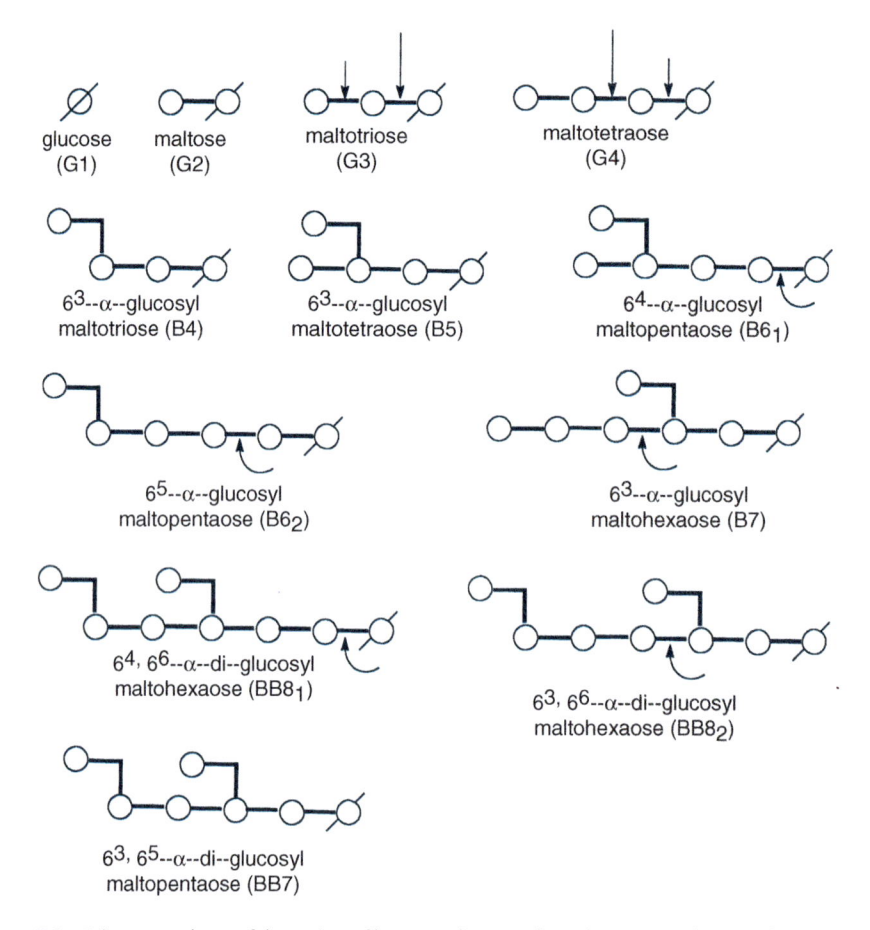

Figure 7.3 Primary products of the action of human salivary and porcine pancreatic α-amylases acting on starch. Seven of the primary products are hydrolyzed slowly at specific bonds (indicated by arrows) to give secondary products that are limit dextrins. (From Robyt and French[13,19] and Kainuma and French[17,18])

hydrolysis of the bonds of these maltodextrins are given in Figure 7.4. The overall rates of hydrolysis of G3 and G4 were quite slow (0.001 to 0.004, respectively) compared with the rates of hydrolysis of G5 and larger maltodextrins. With maltotriose, bond 1 was hydrolyzed more than 5 times more frequently than bond 2, and with maltotetraose, bond 2 was hydrolyzed 2.3 times more frequently than bond 1, while bond 3 was not hydrolyzed at all. With maltopentaose, only the glycosidic bond second from the reducing end was hydrolyzed to give G3+G2. With maltohexaose, maltoheptaose and malto-octaose, the first and last bonds were not hydrolyzed and the frequency shifted from bond 2 to bond 3 as the size of the maltodextrins increased.

At high concentrations of G3 and G4, condensation reactions were observed with the formation of G6 and G8, respectively, which were then hydrolyzed at their specific bonds to change the product distribution observed for G3 and G4 under dilute conditions. Condensation reactions occurred via formation of nonproductive complexes with G3 and G4 that occur with the subsites to the left of the catalytic groups. Thus, porcine pancreatic α-amylase does not hydrolyze the (1→4)-α-D-glucosidic bonds randomly.

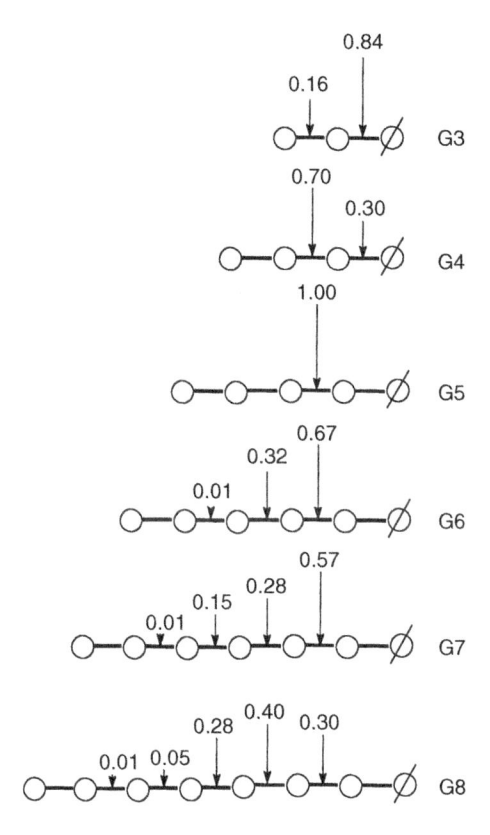

```
                    0.84
             0.16  ↓
           O—O—⌀  G3
              0.70
                  0.30
           O—O—O—⌀  G4
                1.00
                 ↓
          O—O—O—O—⌀  G5
                 0.67
            0.32 ↓
       O—O—O—O—O—⌀  G6
       0.01
             0.28 0.57
         0.15  ↓   ↓
    O—O—O—O—O—O—⌀  G7
    0.01
              0.40
         0.28  ↓  0.30
    0.01 0.05  ↓      ↓
  O—O—O—O—O—O—O—⌀  G8
```

Figure 7.4 Bond frequency for hydrolysis of maltodextrins [G3 to G8] by porcine pancreatic α-amylase. (From Robyt and French[19])

Robyt and French[19] postulated that porcine pancreatic α-amylase has five glucosyl-unit-binding subsites with the catalytic groups oriented for attack at the glycosidic bond between the second and third subsites (Figure 7.1c). This hypothesis has been corroborated by x-ray crystallography[20] and by individual subsite binding-energy determinations.[21,22]

Porcine pancreatic α-amylase has a high degree of what is called multiple attack, with an average of seven hydrolytic cleavages occurring per productive encounter between the starch chain and the active-site of the enzyme.[13] Human salivary α-amylase produces an average of three hydrolytic cleavages per chain encounter. To account for multiple attack, it was postulated that, after the first hydrolytic cleavage, one of the starch chains diffuses away from the active-site, leaving the other chain still associated.[13] This chain is then repositioned, filling the empty subsites, and a second hydrolytic cleavage occurs to give specific low molecular weight products, such as maltose or maltotriose, before the chain completely diffuses away from the active-site. For porcine pancreatic α-amylase, this process occurs, on average, six times per enzyme–substrate encounter and twice for human salivary α-amylase.

Suganuma et al.[24] determined that *Aspergillus oryzae* α-amylase (Taka-amylase A or AoA) has seven glucosyl-unit-binding subsites, with the catalytic groups located between the third and fourth subsites from the reducing end (Figure 7.1d).

A. oryzae α-amylase has about the same degree of multiple attack as human salivary α-amylase, i.e. an average of three hydrolytic cleavages per chain encounter. *A. oryzae* α-amylase is the only common α-amylase known to hydrolyze the cyclic α-(1→4)-linked maltodextrins to any extent.[25,26] A kinetic analysis revealed that the catalytic constant, k_2, increased significantly from 3.3 to 270 to 3270 min^{-1} for cyclomaltohexaose, cyclomaltoheptaose and cyclomalto-octaose, respectively, while the K_m was 4.7 M, 10.2 M and 2.4 M, respectively.[26]

Active-sites of the endo-acting α-amylases are located in a groove or cleft that accommodates the binding of the inner (endo) glucosyl units of the starch chain. Products produced by α-amylases are determined by the number of glucosyl unit-binding subsites, their relative affinities for binding glucosyl units, and their location in relation to the position of the catalytic groups that produce the actual hydrolytic cleavage of the α-(1→4) glucosidic bond. Products are also determined by whether the substrate is amylose, amylopectin, glycogen, β-amylase limit dextrin of amylopectin or glycogen, or specific individual maltodextrins, linear or branched. Thus, different α-amylases produce different products, depending on the natures of the enzymes' active-sites and the substrates.

2. Action of Exo-Acting β-Amylases

The β-amylases catalyze starch hydrolysis by a mechanism that gives inversion of configuration at the anomeric center. All known β-amylases have an exo-mechanism and act on the non-reducing ends of starch polymer chains or starch polymer-derived chains. There are two general classes of β-amylases, those that are classically known as β-amylases and produce β-maltose, and those that are known as glucoamylases and produce β-D-glucose.

Maltose β-amylases are primarily found in plants and have been isolated from sweet potatoes,[27] soybeans,[28] barley[29] and wheat.[30] Maltose β-amylases are also elaborated by bacteria, e.g. by *Bacillus polymyxa*,[31] *B. megaterium*,[32] *B. cereus*[33] and *Pseudomonas* sp. BQ6.[34] These β-amylases all produce β-maltose and a high molecular weight β-limit dextrin. The limit dextrins result when the enzyme reaches an α-(1→6) branch linkage, which it cannot pass. Approximately half of an amylopectin molecule is converted to β-maltose; the remaining half is the β-limit dextrin.

All known glucoamylases are elaborated by fungi, such as *Aspergillus niger*,[35] *A. awamori*,[36] *A. awamori* var. *kawachi*,[37] *Rhizopus delemar*[38] and *R. niveus*.[39] In contrast to other amylases, glucoamylases can catalyze the hydrolysis of both (1→4) and (1→6) α-D-glucosidic bonds in starch, although the α-(1→4) linkage is hydrolyzed approximately 600 times faster than the α-(1→6) linkage.[40] Glucoamylase, thus, can completely convert starch to D-glucose, and has a prominent role in the industrial conversion of starch to glucose syrups (see Chapter 21).

Exo-acting β-amylases act by binding the non-reducing ends of starch chains into a pocket in their structure rather than into a cleft, as do endo-acting α-amylases. Like α-amylases, however, β-amylases also have a number of glucosyl unit-binding subsites. Soybean β-amylase has at least five subsites[41] with the catalytic groups located between the second and third glucosyl-unit-binding subsites from the non-reducing end. Individual subsite energies for binding glucosyl units by maltose soybean β-amylase have been determined (Figure 7.5a).

Glucoamylases have been found to have five to seven glucosyl unit-binding subsites with the catalytic groups located between the first and second subsites from the

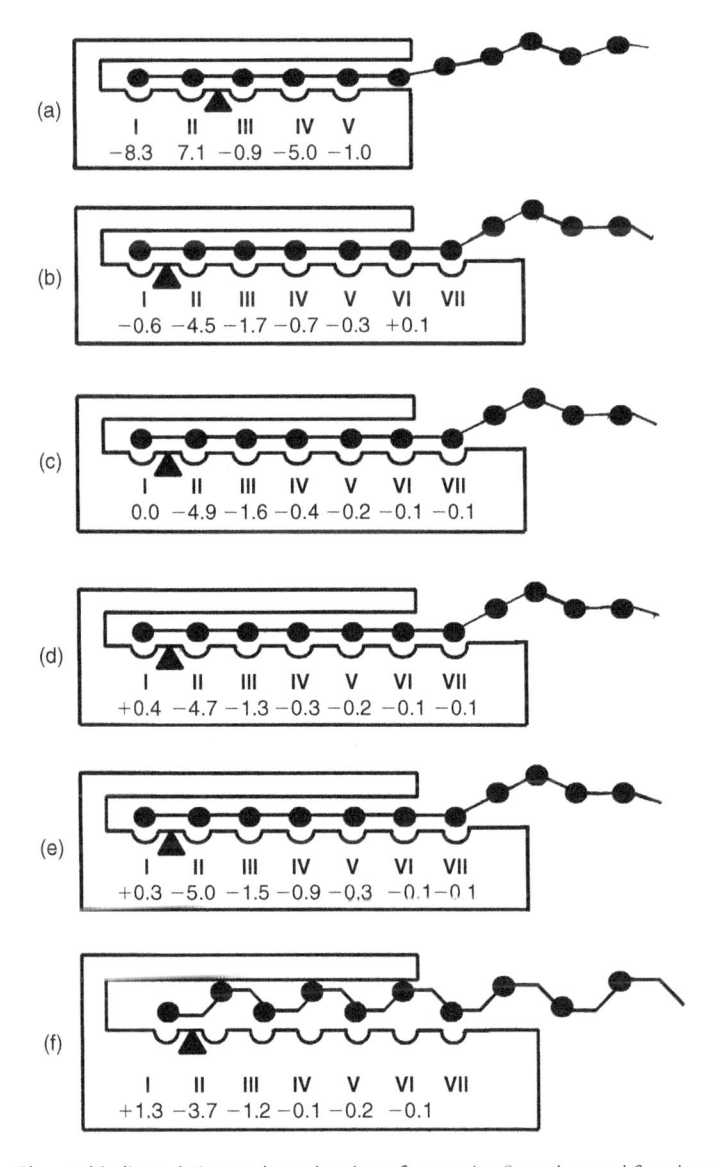

Figure 7.5 Glucose-binding subsites at the active-sites of exo-acting β-amylase and five glucoamylases with subsite binding energies. The subsites for exo-acting amylases are numbered with Roman numerals, beginning with the subsite binding the non-reducing end glucosyl unit. Arabic numbers indicate the binding energies in kcal/mol, with increasing negative values indicating the strongest binding; ▲ indicates the location of the catalytic groups in relation to binding-subsites. (a) Soybean maltose β-amylase, from Kato et al.;[41] (b) *Aspergillus awamori* glucoamylase, from Savel'ev et al.;[37] (c) *Rhizopus delemar* glucoamylase, from Hiromi;[38] (d) *R. nievus* glucoamylase, from Tanaka et al.;[42] (e) *A. niger* glucoamylase binding maltodextrins, from Meagher et al.;[40] (f) *A. niger* glucoamylase binding isomaltodextrins, from Meagher et al.[40]

non-reducing end. Subsite energies for binding maltodextrins have been determined for glucoamylases from several fungal sources.[37–40,42] Binding energies for these enzymes are similar, in that the first subsite, the product subsite adjacent to the non-reducing side of the catalytic groups, has low affinity for glucose. The second subsite, which is toward the reducing side, has the highest affinity for glucose. Subsite III has considerably lower affinity, although much higher than subsite I. For three of the enzymes (from *A. awamori*, *A. niger* and *R. delemar*), subsite IV has lower affinity than subsite III, but still has significant binding energy. For two of the enzymes (from *R. niveus*, binding maltodextrins and *A. niger*, binding isomaltodextrins), subsite IV has relatively low binding affinity, as do subsites V, VI and VII for all five glucoamylases. Meagher et al.[40] determined the subsite energies for isomaltodextrins binding to *A. niger* glucoamylase and concluded that the isomaltodextrins are bound and hydrolyzed in a different fashion than are the maltodextrins. This is not surprising, as contiguous α-(1→6) linkages found in isomaltodextrins do not occur in starch chains, have different chain conformations than do the α-(1→4)-linked maltodextrins, and hence should bind with lower affinity and be hydrolyzed more slowly than the maltodextrins. Figures 7.5b–f summarize the binding energies for the glucose subsites determined for glucoamylases from different fungi.

In addition to differences in the location of the catalytic groups and the strict specificity for the exclusive hydrolysis of (1→4) α-D-glucosidic linkages by maltose β-amylases and the broad specificity for the hydrolysis of both (1→4) and (1→6) α-D-glucosidic linkages by glucoamylase, the two types of β-amylases have widely different turnover numbers; maltose β-amylase has a very high turnover number of 2550 sec^{-1}, while glucoamylase has a very low turnover number of 34 sec^{-1}. Further, sweet potato β-amylase produces four hydrolytic cleavages per enzyme-chain encounter for low molecular weight amyloses,[44] while the glucoamylases do not exhibit multiple attack.[45] It can be speculated, based on their structures, why the two types of β-amylases differ so widely with respect to their turnover numbers and multiple attack (see Section 7.6.3).

3. Amylases Producing Specific Maltodextrin Products

In 1971, Robyt and Ackerman[46] reported a novel amylase from *Pseudomonas stutzeri* (NRRL B-3389) that formed maltotetraose as a specific product. This was the first observed amylase that produced a single, specific product, other than glucose or maltose. Like the β-amylases, it is an exo-acting amylase, but produces maltotetraose with an α-anomeric configuration. It specifically hydrolyzes the fourth α-(1→4) glucosidic linkage from the non-reducing end of a starch chain and gives 42% α-maltotetraose and 58% limit dextrin.[46] The enzyme does not bypass branch linkages of amylopectin. Later, a similar enzyme was reported from *Pseudomonas saccharophila*,[47] and Fogarty et al.[48] reported a *Pseudomonas* sp. (IMD 353), maltotetraose-producing amylase that was constitutive for the amylase, producing the enzyme on a glucose medium. Relatively high yields of the enzyme (29 U/mL) were obtained. The enzyme had optimum activity at pH 7 and 50°C.

In 1978, Wako et al.[49] described an enzyme from *Streptomyces griseus* that produced primarily maltotriose from the non-reducing ends of starch chains. Amylose was converted to maltotriose, with only small amounts of glucose and maltose being formed. Waxy maize starch, oyster glycogen and phytoglycogen gave 51%, 40% and 20% maltotriose, respectively, and corresponding amounts of limit dextrins, indicating that this enzyme was also exo-acting. Takasaki[51] reported an α-amylase from a *B. subtilis* strain that produced maltotriose, and another from a *B. circulans* strain[51] that produced maltotetraose and maltopentaose in relatively high yields.

An α-amylase elaborated by *B. licheniformis* gives relatively high yields (33%) of maltopentaose and a range of other maltodextrins from G1 to G12.[52] The organism is a thermophile, producing an enzyme with an optimum temperature in the range of 70°C to 90°C.[52] Kainuma et al.[53,54] discovered an exo-acting amylase. The enzyme elaborated by *Aerobacter aerogenes* produced a specific product, maltohexaose. Later, Taniguchi et al.[55] discovered a maltohexaose-forming amylase elaborated by a strain of *B. circulans* that degraded potato starch granules from the surface of the granules (Section 7.9.2). These maltohexaose-producing α-amylases are similar to *Pseudomonas stutzeri* maltotetraose α-amylase in that their action is exo and they produce a single, specific product, maltohexaose, which has an α-anomeric configuration.

Fogarty et al.[56] described an α-amylase elaborated by *B. caldovelox* that preferentially produces maltohexaose in yields of 40–44% and glucose is not a product. Maltoheptaose is the smallest malto-oligosaccharide hydrolyzed by this enzyme, with specific hydrolysis of only one glycosidic linkage, the second from the reducing end, to give G2+G5. With malto-octaose, bonds two and three are cleaved with about equal frequency.

Halobacterium sp.,[57] *Halobacterium halobium*[58] and *Micrococcus halobius*[59] elaborate amylases that require the presence of Na^+, K^+ and Ca^{+2} for activity and stability. Relatively high salt concentrations (2 M NaCl or 1 M KCl) are required in the culture medium for activity. Purified *M. halobius* amylase requires 0.25 M NaCl or 0.75 M KCl for maximum activity.

An α-amylase from a hyperthermophilic archaebacterium, *Pyrococcus furiosus*, was purified and found to have optimal activity, with substantial thermal stability, at 100°C.[60] Unlike other α-amylases, it is not dependent on calcium ions for activity or stability. The enzyme produces G4, G5 and G6 as the primary products and hydrolyzes these products, as well as G3, but at a much reduced rate, characteristic of α-amylase secondary reactions.

Using molecular biology techniques, Conrad et al.[61] produced hybrids of the α-amylases from *B. amyloliquefaciens* and *B. licheniformis*. Thirty-three hybrids were formed. They consisted of the entire α-amylase sequence with variable proportions from *B. amyloliquefaciens* α-amylase and *B. licheniformis* α-amylase. The hybrid enzymes fell into six groups that retained the extra-thermostability of the *licheniformis* enzyme. A specific hybrid sequence (residues 34–76) was correlated with the enzymes' product specificity for forming and accumulating maltohexaose. Two of the hybrids were less thermostable than either of the parent types, while two others were enzymatically inactive.

4. Action of Isoamylases

Isoamylase is the name given to enzymes that hydrolyze the branch linkages of starch. These enzymes act by an endo-mechanism, hydrolyzing the branch linkages located in the interior parts of the amylopectin molecule to give linear maltodextrins with retention of the α-configuration. The first recognized enzyme of this class was the so-called 'R-enzyme' isolated from broad beans.[62] The enzyme hydrolyzed some of the α-(1→6) linkages of amylopectin, amylopectin β-amylase limit dextrin and various branched α-amylase limit dextrins. It did not act on glycogen or its β-amylase limit dextrin.[62]

Pullulanase, a (1→6) α-D-glucanohydrolase, is elaborated by *Aerobacter aerogenes*, growing on a medium that uses pullulan as a carbon source.[63] In addition to catalyzing hydrolysis of the α-D-glucosidic linkages of pullulan, a linear molecule, it also hydrolyzes the α-(1→6) branch linkages of amylopectin and glycogen.[64]

A bacterial isoamylase elaborated by *Pseudomonas amylodermosa*[65–67] has been crystallized and its specificity determined.[68,69] Similar to pullulanase, *Pseudomonas* isoamylase is unable to remove single branch glucosyl units from branched maltodextrins, and acts only very slowly on maltosyl units attached by α-(1→6) linkages. It differs from pullulanase by being able to hydrolyze the α-(1→6) linkages of amylopectin and glycogen completely and at rates of 7 times and 124 times, respectively, the rates of hydrolysis of these substrates by pullulanase.[68,69] Hydrolysis of the α-(1→6) branch linkages of short chains occurs at a rate of 10% or less of the rate of hydrolysis of the branch linkages of the larger chains. *Pseudomonas* isoamylase hydrolyzes the α-(1→6) linkages of maltotriosyl chains much faster than those of the α-(1→6)-linked maltose chains. Further, *Pseudomonas* isoamylase hydrolyzes the α-(1→6) linkages of pullulan [a linear polysaccharide of α-(1→6) linked maltotriose units] only very slowly. *Pseudomonas* isoamylase catalyzes hydrolysis of both the inner and the outer branch linkages of amylopectin and glycogen, whereas pullulanase can only hydrolyze the outer branch linkages. With glycogen, *Pseudomonas* isoamylase hydrolyzed all of the branch linkages, while pullulanase could hydrolyze only a few. *Pseudomonas* isoamylase, thus, is a true isoamylase or debranching enzyme for starch and glycogen, although it does not release single glucosyl units and releases maltosyl units very slowly. Its specificity for longer chain lengths and its ability to hydrolyze the interior α-(1→6) branch linkages suggests that the active-site of *Pseudomonas* isoamylase binds two chains (the A- and B- or the A- and C-chains).

A starch-debranching isoamylase, obtained from sugary 1 (*sul*) maize, has been cloned by James et al.[70] and found to have 32% sequence identity with *Pseudomonas* isoamylase. Other isoamylases or starch debranching enzymes have been isolated from *Cytophaga* sp.,[71] *Streptomyces* sp.,[72] *Flavobacterium* sp.,[73] a yeast, *Lipomyces kononenkoae*,[74] potato tubers,[75] *B. circulans*[76] and an alkaline isoamylase with a pH optimum of 9 from an alkalophilic *Bacillus* sp.[77]

5. Archaebacterial Amylases

Unusual forms of bacteria were discovered growing in unusual environments of high temperatures and/or high salt in the 1970s (see previous section and references 57–59). Many of these bacteria elaborate enzymes that have high temperature optima, and therefore have special industrial interest in the conversion of starch to D-glucose or specific maltodextrins.[78]

A large number of hyperthermophilic *Archaebacteria*, especially the deep sea *Thermococcale* and *Sulfolobus* species elaborate α-amylases.[79–82] Many have been cloned and sequenced.[78] *Pyrococcus furiosus*,[83,84] *Thermococcus profundus*,[85] *Thermococcus hydrothermalis*,[78] *Sulfolobus solfataricus* and *Sulfolobus acidocaldarius*[86] secrete thermophilic α-amylases. The α-amylases of all of these organisms have optimal enzyme activity at 90°C or higher and often only begin to show activity at 40°C or 50°C. *Pyrococcus furiosus* secretes an α-amylase with an optimum temperature of 100°C and a maximum temperature of 140°C.[87] The optimum pH values vary between 5 and 9. Table 7.1 summarizes the names of the organisms, the optimum temperature, and optimum pH values for several of these enzymes.

Thermophilic archaeal α-amylases do not differ superficially from other α-amylases in their molecular size and amino acid composition. They are primarily monomeric enzymes with molecular weights of 40–70 kDa.[84,88–90] Some of these amylases produce specific maltodextrin products, which are also summarized in Table 7.1.

Several *Archaebacteria* from *Pyrococcus furiosus*, *Pyrococcus woesei*, *Thermococcus aggregans*, *Thermococcus celer*, *Thermococcus guaymagensis*, *Thermococcus hydrothermalis* and *Thermococcus litoralis* also produce thermophilic isoamylases (amylopullulanases).[91–97] The genes for these thermophilic isoamylases have been cloned and sequenced.[95,98,99] These thermophilic isoamylases have optimum activity between 80°C and 100°C, with pH optima between 5.5 and 6.6.[91–95] The isoamylase elaborated by *Thermococcus litoralis* is an exception, with an unusually high temperature optimum of 125°C. The thermophilic isoamylases have molecular

Table 7.1 Thermophilic archaebacterial amylolytic enzymes[78]

Bacterium	Enzymes	Optimum temperature of activity, °C	Optimal pH of activity	Main products	Location of enzyme[a]
Desulfurococcus mucosus	α-amylase	100	5.5	–	–
	pullulanase	100	5.0	–	–
Pyrococcus furiosus	α-amylase	100	5.0	G4, G5, G6	I
	α-glucosidase	105–115	5.0–6.0	G1	E
Pyrococcus woesei	α-amylase	98	5.5	G2, G5	E
	α-glucosidase	100	5.0–5.5	G1	E
Pyrococcus sp. KOD1	α-amylase	100	5.5	G2, G3	E
Thermococcus hydrothermalis	α-amylase	75–85	5.0–5.5	G3, G4	E
	α-glucosidase	110	6.0	G1	I
	pullulanase	100		α-(1→6) debranching	I
Thermococcus litoralis	isoamylase	125	5.0–5.5	α-(1→6) debranching	E
Thermococcus profundus	α-amylase	80	5.5–6.0	G2, G3	E
Thermococcus zilligii	α-glucosidase	75	7.0	G1	E
	pullulanase	80–95	5.0–5.5	α-1→6 debranching	E

[a] E = extracellular; I = intracellular

weights of 90–100 kDa. *Pyrococcus woesei* isoamylase (90 kDa) forms an extremely compact and rigid structural form at temperatures below 80°C with a molecular weight of 45 kDa, indicating dissociation of a dimer at the lower temperatures.[95]

6. Action of Cyclomaltodextrin Glucanosyltransferase

Cyclomaltodextrin glucanosyltransferase (CGTase) catalyzes the breakdown of starch into cyclic, non-reducing, α-(1→4)-linked cyclomaltodextrins (commonly called cyclodextrins; see Chapter 22). Several different CGTases are elaborated by different species of bacteria. The first to be studied was that elaborated by *Bacillus macerans*, which initially was called *macerans* amylase,[100] because it reduced the blue iodine–iodide color of starch. Its action, however, is not that of an amylase, in that it does not catalyze hydrolysis of α-(1→4) α-D-glucosidic linkages, but it catalyzes an intramolecular transfer of α-(1→4) linkages to form cyclic α-(1→4)-linked maltodextrins, containing primarily six, seven and eight α-D-glucopyranosyl units, called cyclomaltohexaose (α-cyclodextrin), cyclomaltoheptaose (β-cyclodextrin) and cyclomaltooctaose (γ-cyclodextrin), respectively. To do this, the action of the enzyme must be exo and capable of binding six, seven or eight glucosyl units at the non-reducing end of a starch polymer chain. After binding the starch chain, the enzyme catalyzes cleavage of the α-(1→4) linkage and transfers the chain to O-4 of the non-reducing end glucosyl unit to form an (1→4) α-D-glucosidic linkage and, thus, a cyclic dextrin (Figure 7.6). The mechanism now has been completely elucidated. It has been shown to operate by a ping-pong mechanism[101a] that is consistent with forming a covalent glucanosyl–enzyme complex. This mechanism is further suggested by the fact that the enzyme readily catalyzes acceptor reactions (reactions involving the transfer of the dextrinyl chain to exogenous carbohydrate acceptors) and disproportionation reactions between two maltodextrin chains, as well as the intramolecular cyclization reactions. The

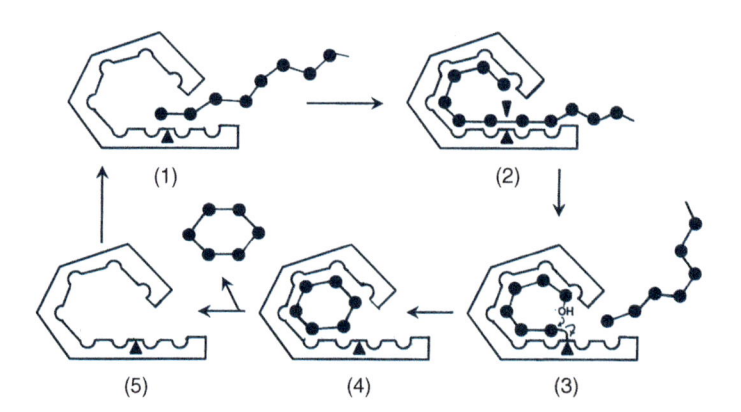

Figure 7.6 Proposed mechanism for the formation of cyclomaltohexaose from starch by cyclomaltodextrin glucanosyltransferase. (1) Non-reducing end of starch chain enters the active-site pocket; (2) six glucopyranosyl units bind with the subsites and cleavage of the (1→4) linkage takes place; (3) the C-4 hydroxyl group of the non-reducing end glucopyranosyl unit of maltohexaose makes a nucleophilic attack onto C-1 of the maltohexaosyl–enzyme complex, forming an α-(1→4) linkage and releasing the covalently linked glucopyranosyl unit from the enzyme; (4) the cyclomaltodextrin that was formed leaves the active site to give free enzyme (5), which is ready to repeat step (1).

formation of a covalent intermediate was subsequently confirmed by trapping a maltohexaosyl unit attached to the enzyme during reaction with cyclomaltohexaose.[101b]

The formation of higher cyclomaltodextrins, containing nine, ten, eleven and twelve (1→4)-linked α-D-glucopyranosyl units has been reported.[102,103] Cyclomaltodextrins having 9–13 D-glucosyl units (δ-, ε-, ζ-, η- and θ-cyclomaltodextrins) have been isolated.[104,105] Some of these cyclic dextrins were shown to be branched. The amount of unbranched cyclomaltodextrins decreased as the size increased; δ-cyclomaltodextrin was essentially unbranched, ε-cyclomaltodextrin was 50% branched, ζ-cyclomaltodextrin was 75% branched, and η- and θ-cyclomaltodextrins were almost 100% branched.

A study of the time course of the action of CGTase with a completely unbranched amylose gave very large cyclomaltodextrins that were preferentially formed in the initial stages of the cyclization reaction and then subsequently converted into the smaller cyclomaltodextrins (α- and β-cyclomaltodextrin). When the reaction of CGTase with the completely unbranched amylose was terminated in the early stage and the reaction digest was treated with glucoamylase, glucoamylase-resistant dextrins with DP values of 9 to 60 were found in addition to the traditional α-, β- and γ-cyclomaltodextrins.[106a] B. macerans CGTase transglycosylation reactions with D-glucose and cyclomaltohexaose forms, in addition to a series of maltodextrins of DP 2 to 75, also forms a series of cyclomaltodextrins of DP 7–25.[106b]

CGTases from B. megaterium,[107] B. circulans,[108] B. stearothermophilus,[109] Klebsiella pneumoniae[110] and Brevibacterium sp. have also been reported.[111] Some of these enzymes differ in pH and temperature optima and in the relative amounts of each cyclomaltodextrin formed. Most of the CGTases have pH optima of 5.0–5.5. The exception is the enzyme from Brevibactenum sp. that has an alkaline pH optimum of 10.0. Most of the CGTases have a temperature optimum of 50–55°C, with the exception of the enzyme from B. stearothermophilus, which has a higher temperature optimum of 70–75°C. The enzymes elaborated by B. macerans, K. pneumoniae and B. stearothermophilus primarily form α-cyclomaltodextrin as the main product, with lesser amounts of β- and γ-cyclomaltodextrins; the enzymes elaborated by B. megaterium and B. circulans primarily form β-cyclomaltodextrin as the main product, and the enzyme elaborated by Brevibacterium sp. primarily forms γ-cyclomaltodextrin as the main product. The CGTase from B. macerans has been highly purified and crystallized.[112,113]

As indicated above, the CGTases also catalyze acceptor or transfer reactions between cyclic maltodextrins and a carbohydrate acceptor.[114] In this reaction, the enzyme opens the cyclic ring and most probably forms a covalent enzyme–maltodextrin complex that undergoes reaction with the acceptor to give transfer of the maltodextrin to the acceptor. The acceptor product can then undergo further disproportion reactions. A linear acceptor product binds at the active-site and undergoes cleavage at one of its interior linkages. Then a second linear acceptor product comes into the active-site and reacts with the covalent dextrinyl–enzyme complex. The acceptor displaces the dextrin from the active-site and forms an α-(1→4) glycosidic linkage with the dextrin, giving a disproportionation reaction (Figure 7.7). Several different reactions can occur to give a wide range of maltodextrins with different chain lengths. The different amounts and kinds of maltodextrins are dependent on the molar ratios of the acceptor and the cyclomaltodextrin.[106b] The many varied disproportionation reactions are illustrated by the reactions of G7 shown in Example 7.1.

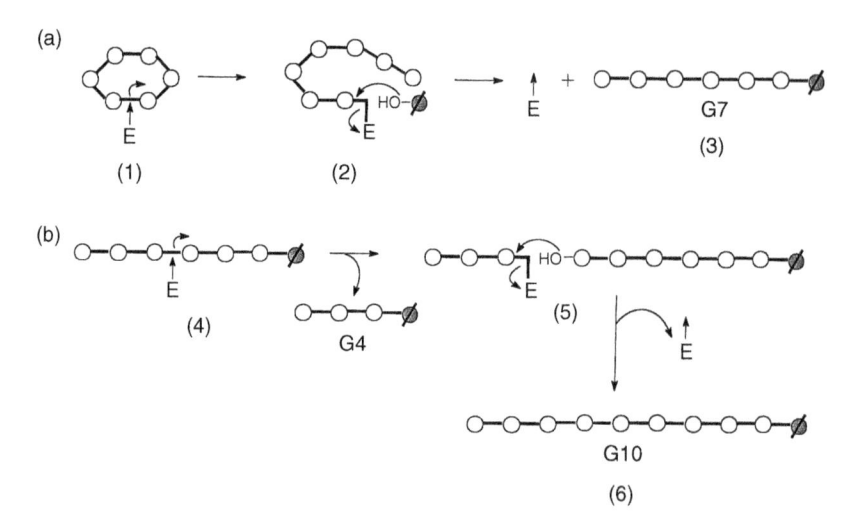

Figure 7.7 Proposed mechanism for the acceptor and the disproportionation reactions catalyzed by cyclomaltodextrin glucanosyltransferase. Acceptor reaction: (1) enzyme reacts with cyclomaltodextrin to open the ring and forms a covalent enzyme-maltohexaosyl product; (2) an acceptor (labeled glucose) reacts with the glucopyranosyl covalent product, displacing the enzyme and forming an α-1→4 glucosidic linkage, giving an acceptor terminated maltodextrin (3). Disproportionation reaction: (4) maltodextrin is cleaved by the enzyme to give a smaller dextrin [G4] and a covalent maltodextrinyl-enzyme product is formed; (5) another maltodextrin partially enters the active-site pocket so that its C-4 hydroxyl group at the non-reducing end makes a nucleophilic attack onto C-1 of the enzyme-product, displacing the enzyme and forming an α-(1→4) glucosidic linkage, giving a new maltodextrin [G10] (6). The open circles represent non-labeled glucosyl units and the shaded circles represent labeled glucosyl units. The mechanism shows how the label always appears at the reducing-end glucosyl unit, when glucose is the labeled acceptor.

$$G_7 + G_7 \xrightarrow{CGTase} \begin{Bmatrix} G_1 + G_{13} \\ G_2 + G_{12} \\ G_3 + G_{11} \\ G_4 + G_{10} \\ G_5 + G_9 \\ G_6 + G_8 \end{Bmatrix} \text{various disproporionation products from reaction of two maltoheptaose molecules}$$

Each of the products of the above disproportionation reactions can themselves undergo reactions, either with itself or with another product, to give a complex mixture of homologous maltodextrins. An important consequence of this disproportionation reaction is that, if the starting reactants are cyclomaltodextrin and ^{14}C-labeled glucose, the labeled glucose will always appear exclusively at the reducing end of all resulting maltodextrins. This reaction has permitted the specific synthesis of reducing end glucose-^{14}C-labeled maltodextrins[115] (Figure 7.7).

Acceptors other than D-glucose may also react if the non-reducing terminal unit of the acceptor is D-glucose.[116a] For example, isomaltose, sucrose or panose can be the acceptors, giving products with the acceptors attached at the reducing end of maltodextrin chains (Figure 7.8). The acceptor product can then undergo disproportionation reactions to give a homologous series of products with the acceptor located exclusively at the reducing end of maltodextrin chains. More recently, it also has

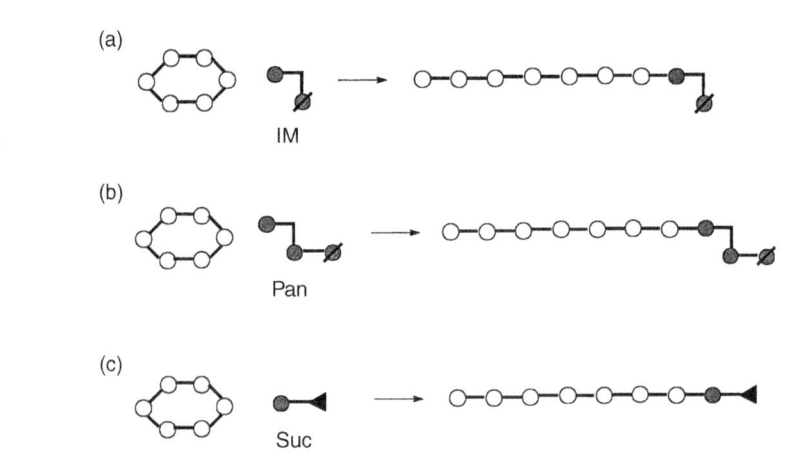

Figure 7.8 Reaction of cyclomaltodextrin with different acceptors and cyclomaltodextrin glucanosyltransferase: (a) reaction of cyclomaltohexaose with isomaltose to give an octasaccharide with maltohexaose attached by an α-(1→4) linkage to the non-reducing end of isomaltose, giving isomaltose at the reducing end of the product; (b) reaction of cyclomaltohexaose with panose to give maltohexaose attached to the non-reducing end of panose by an α-(1→4) linkage, giving panose at the reducing end of the product; (c) reaction of cyclomaltohexaose with sucrose to give maltohexaose attached to the glucosyl unit of sucrose by an α-(1→4) linkage.

been found that *B. macerans* CGTase can glycosylate non-carbohydrate, phenolic hydroxyl groups by reaction with cyclomaltohexaose.[116b]

III. Relation of Structure with Action of the Enzymes

One of the major advances in the last part of the twentieth century in the understanding of starch-degrading enzymes resulted from the application of modern biochemical and biophysical techniques of protein sequence analysis, chemical modification of specific amino acid groups, x-ray crystallography, recombinant DNA cloning and sequencing, and site-directed mutagenesis. These techniques, combined with the enzymological studies described in the previous sections, have given us a much more complete understanding of how the enzyme protein structure provides specificity and performs catalysis.

1. Relation of Structure with Action of Endo-Acting α-Amylases

The α-amylases from *A. oryzae*, porcine pancreas and barley malt have had their structures determined by sequence analysis and x-ray crystallography. Porcine pancreatic α-amylase is a single, N-acetylated polypeptide of 496 amino acids of known sequence, containing five disulfide linkages and two sulfhydryl groups, with a molecular weight of $55\,000 \pm 1000$.[117] The enzyme contains one chloride ion that acts as an allosteric activator, increasing the V_m 30-fold, but having no effect on the K_m.[118]

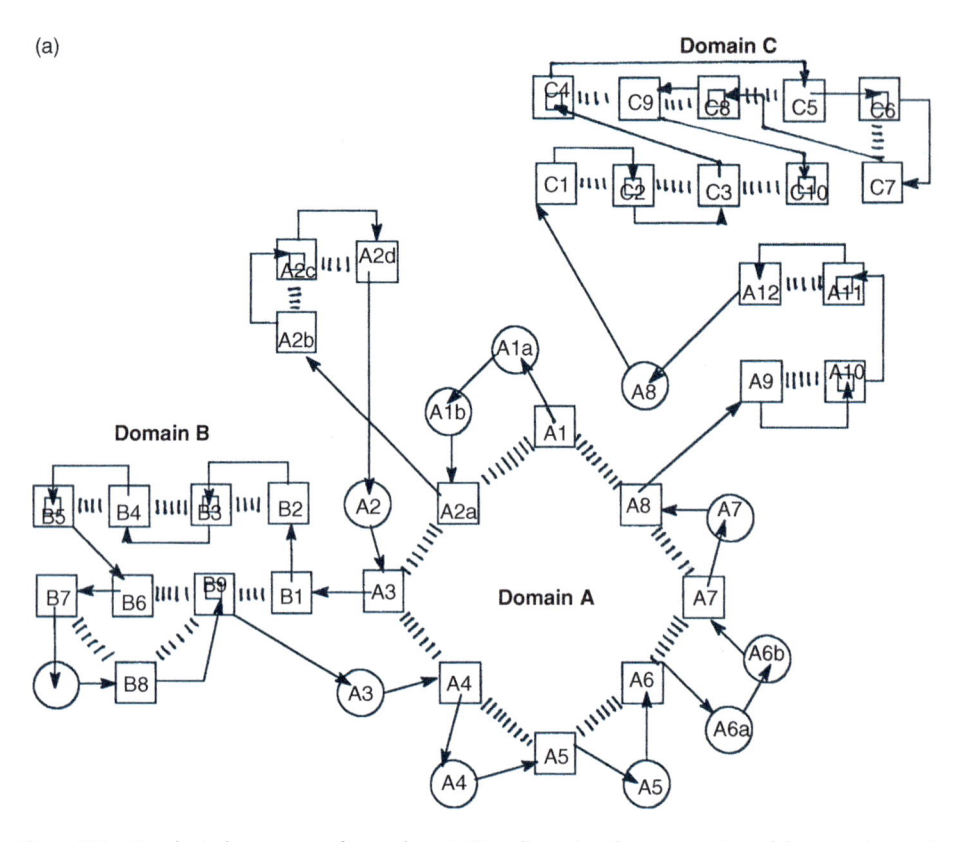

Figure 7.9 Topological structures of α-amylase: A. Two-dimensional representation of the secondary and domain structures of porcine pancreatic α-amylase. Alpha helices are represented as circles and β-strands in the up-direction as squares, and in the down direction as double squares. The $(\alpha/\beta)_8$–TIM barrel comprises domain A. Hydrogen bonds between β-strands are shown by dashed lines. The α-helices and β-strands are identified in the various domains by A, B and C. (Reprinted by permission of the authors M. Qian et al.[120]) Two-dimensional representation of the secondary and domain structures of barley malt α-amylase (AMY2-2). Alpha helices are represented as cylinders and β-strands as arrows. The $(\alpha/\beta)_8$–TIM barrel comprises domain A, with eight β-strands and an equivalent of eight α-helices. The active-site is composed of the loops that connect the C-termini of the β-strands to the N-termini of the peripheral α-helices. (Adapted from A. Kadziola et al.[121])

It contains one tightly bound calcium ion that is essential for enzymic activity by maintaining the tertiary structure of the protein.[119]

X-ray crystallographic analysis shows that porcine pancreatic α-amylase has three domains. The first 403 amino acyl units form a large domain (domain A) that comprises eight parallel β-strands with eight α-helices joining the β-strands together. The whole structure is rolled into a circular array known as a TIM-barrel or $(\beta/\alpha)_8$-barrel[120] (Figure 7.9). The β-strands are hydrogen bonded together to give a circular β-sheet. Domain B is inserted between the third β-strand and the third α-helix of domain A. It begins with residue 100 and ends with residue 169. This domain has four short β-strands, three anti-parallel and one parallel. Domain C forms a distinct globular unit that extends from residues 405 to 496 and consists of ten anti-parallel

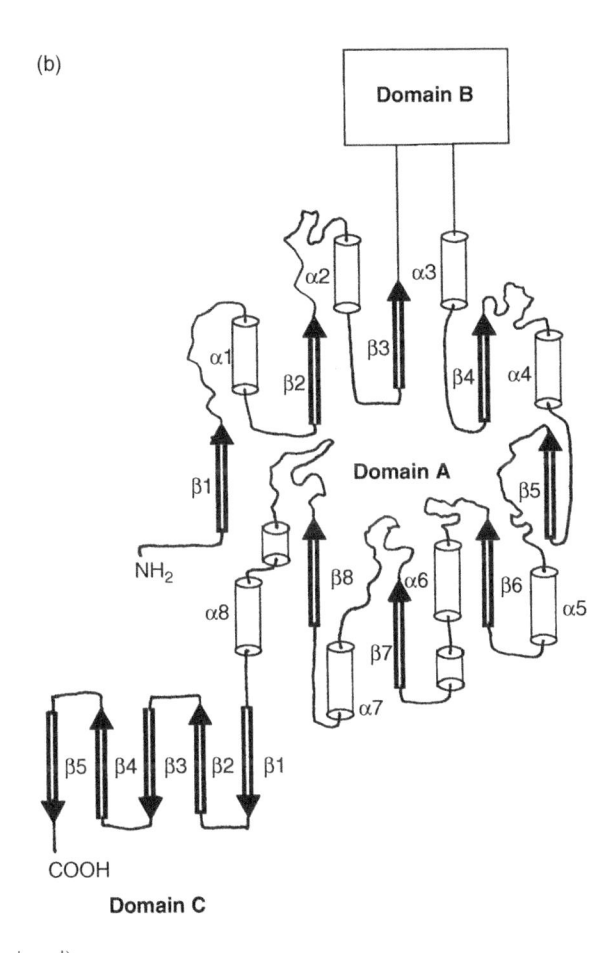

III. Relation of Structure with Action of the Enzymes 255

(b)

Figure 7.9 (Continued)

β-strands. The single, tightly bound calcium ion is bound, in part, by ligands in both domains A and B, tying the two together.[12]

X-ray crystallographic analysis of the enzyme binding a maltodextrin analog inhibitor, acarbose (Figure 7.17) revealed that acarbose occupied five subsites in a cleft made at the C-terminal end of the TIM barrel in domain A.[21] The active-site region is a V-shaped depression in the structure, whose walls are mainly composed of the loops between the third β-strand and the third α-helix, and the seventh β-strand and the seventh α-helix of the barrel. Three acidic amino acid residues (Asp197, Glu233 and Asp300) are close together in the center of the active-site cleft, located between subsites II and III. In the enzyme–acarbose complex, the third subsite is occupied by the cyclohexene ring, which can assume a half-chair conformation similar to the postulated transition-state of the glucopyranosyl unit undergoing glycosidic bond cleavage. There are four histidine residues (imidazolium groups) present in the active-site. Two of these (His101 and His299) are involved in hydrogen bonding with glucosyl units of the carbohydrate substrate. His101 is hydrogen-bonded to the C-6 hydroxyl group of the cyclohexene ring of acarbose; His299 forms hydrogen bonds with the C-2 and C-3 hydroxyl groups of the same ring. The cleft also contains aromatic groups on Trp58, Trp59 and Tyr62 that bind glucosyl units of the substrate at the other subsites. There is a flexible loop (residues 304 to 309) that has a glycine-rich sequence of –Gly-His-Gly-Ala-Gly-Gly-. On binding of acarbose in the enzyme cleft, this loop moves, covering the active-site cleft, and produces a narrowing of the cleft giving partial protection of the substrate from the solvent and a more tightly bound carbohydrate chain in the active-site cleft. Qian et al.[120] have speculated that, after cleavage of the glycosidic linkage, the flexible loop moves away and exposes the product(s) to the aqueous solvent, giving an opportunity for one of the cleaved chains to be 'pulled' from the active-site by hydration. The loop then closes again, and the remaining hydrolyzed chain in the active-site moves into the open subsites, thereby promoting multiple attack. Multiple attack is primarily observed for mammalian pancreatic α-amylases. A sequence comparison of several mammalian pancreatic α-amylases from porcine, human, mouse and rat has shown a high degree of amino acid homology,[117] indicating that the flexible loop is most probably conserved and present near the active-site cleft for all these mammalian α-amylases. The flexible loop is not present in fungal, plant or bacterial α-amylases that are known to have low or no multiple attack action.

X-ray crystallographic study of barley malt α-amylase has shown that it also has three domains in which the largest domain, domain A, also has a $(\beta/\alpha)_8$-barrel structure, composed of amino acid residues 1 to 88 and 153 to 350.[122] Domain B is composed of 64 residues (88 to 152) that are an excursion from the third β-strand and the third α-helix, similar to the structure of porcine pancreatic α-amylase. Domain C occurs at the C-terminus and is composed of 53 residues (351 to 403) that are arranged into a five-strand β-sheet.

The $(\beta/\alpha)_8$-barrel motif appears in all α-amylases studied to date[121–123] and makes up the active-sites of these enzymes. The β-strands and α-helices alternate along the peptide chain and are joined together by the so-called irregular loops that link the C-terminus end of the adjacent α-helix (Figure 7.9). The β/α structures are rolled into a circular barrel structure. It is the loops that link the β-strands and α-helices that make up the active-site clefts. The $(\beta/\alpha)_8$-barrel motif has been observed in other starch-degrading enzymes, such as the cyclomaltodextrin glucanosyltransferases,[124,125] and has been predicted from amino acid sequences for isoamylases and pullulanase,[126] and starch-branching enzymes.[127,128]

Differences in product specificities of these enzymes are due to differences in the number and relative affinities for the glucosyl units of the starch chain. These differences occur in the amino acid compositions and in the folding of the different loops joining the β/α structures. For example, differences in the folding of loop 2 in the *A. oryzae* and porcine pancreatic α-amylases form an extra subsite in the former and prevent its occurrence in the latter.[127,128] Thus, differences in the length, folding and amino acid sequences of loops 1–8 account for the differences in the number of subsites, and in the types of bonds that are either cleaved or formed. MacGregor[129] speculated that a peptide loop could fold and completely block one end of an active-site cleft to form a pocket rather than a groove, giving an exo-acting α-amylase rather than an endo-acting enzyme. If this blocked groove had loops that folded to create only two subsites from the end of the pocket to the position of the catalytic groups, an enzyme specific for forming maltose would result, and if the loops folded to give four subsites from the end of the pocket to the catalytic groups, an enzyme specific for forming maltotetraose would result, and so forth. As indicated previously, the presence of a flexible loop that can move back and forth over the active-site cleft as the substrate is bound would give an enzyme displaying multiple attack.

2. Structure and Action of Soybean β-Amylase

Soybean β-amylase consists of a single polypeptide chain with 495 amino acid residues and a molecular weight of 56 000. Similar to the α-amylases and other starch degrading enzymes, it has a large domain that contains a $(\beta/\alpha)_8$-barrel structure.[130] A small domain extends from the third and fourth β- and α-strands of domain A. Also, similar to the α-amylases, the active-site is located on the C-terminal side of the TIM-barrel at the interface between the two domains. X-ray crystallographic analysis of its maltose complexes revealed two maltose molecules binding in tandem in the active-site, similar to what would be expected for the binding of maltotetraose during catalysis. The non-reducing ends of the maltose molecules were oriented toward the base of the enzyme's active-site pocket. The catalytic center, located between two bound maltose molecules, has two oppositely positioned carboxyl groups from two glutamic acid residues (Glu186 and Glu380). One of these carboxyl groups (Glc380) is ionized and acts as a nucleophilic base that initiates cleavage of the glycosidic linkage, and the other carboxyl group (Glc186) is protonated and donates its proton to O-4 of the leaving (reducing end) maltosyl unit. Binding of the two maltose molecules in the active-site also produces a significant (10 Å) conformational shift in a flexible peptide segment that borders the active-site pocket.[131] It was speculated that this flexible loop in this enzyme exists in equilibrium with several different conformations.[131] For the free enzyme, the predominant conformation has the peptide loop extending out into the solvent. With the introduction of a substrate molecule, the loop closes over the active-site and shields the catalytic groups, the reaction center and the bound substrate from the solvent, giving a tightly bound substrate and an ordered water molecule located near the catalytic groups. The catalytic groups then catalyze hydrolysis of the glycosidic linkage of the substrate. After hydrolysis occurs, the loop then opens at the end of the pocket, allowing the product (maltose) to leave. The loop then closes again, allowing repositioning of the starch chain into the open subsites at the end of the active-site pocket. This is similar to the functioning of the flexible loop in the active-site of porcine pancreatic α-amylase. The 'hinged' loops in both porcine pancreatic α-amylase and in soybean β-amylase may also facilitate multiple attack. In β-amylase, the flexible loop also contributes to its relatively high turnover number by providing a 'trap door' for the release of the product (maltose) from the active-site without prior dissociation of the starch chain from the active-site pocket.

3. Structure and Action of Glucoamylases

Two forms of glucoamylase (often incorrectly called amyloglucosidase) have been observed in the culture supernatants of *Aspergillus niger*, *A. awamori* and

A. awamori var. *kawachi*.[132,133] The two glucoamylases differ significantly in their molecular weights and in their ability to bind and hydrolyze whole starch granules. Gluoamylase I (GA-I) has 616 amino acid residues, a molecular weight of 74 000, and is capable of binding to and hydrolyzing native starch granules. Glucoamylase II (GA-II) consists of a mixture of peptides having 512–514 amino acid residues and a molecular weight of 62 000 Da and is not capable of binding to or hydrolyzing native starch granules. GA-II is formed by post-translational proteolysis of GA-I.[134]

The proteinase subtilisin cleaves *A. niger* GA-I into two fragments: a fragment consisting of the first 470 amino acid residues that can act on solubilized starch but not on native starch granules,[135] and a fragment consisting of 146 amino acid residues that can bind starch granules but cannot catalyze their hydrolysis.[136,137] Belshaw and Williamson[138] purified the latter fragment and found that it had a molecular weight of 25 100, contained 38% (w/w) carbohydrate, and corresponded to residues 471–616 from the C-terminus of GA-I. When the peptide was bound to starch granules, it inhibited the hydrolysis of granules by GA-I, but it had no effect on the hydrolysis of solubilized starch. This starch-binding fragment could bind two molecules of β-cyclomaltodextrin with a $K_d = 1.68\,\mu M$.[139] The catalytically-active fragment (GA-II) had no interaction with β-cyclomaltodextrin, which competitively inhibited the adsorption of GA-I with native starch granules. It was also found that the starch-binding fragment could bind maltodextrins (G2-G11) with a stoichiometry of 1 mol/mol of protein, and that affinity for these maltodextrins increased as the size of the maltodextrins increased up to G9, after which no further increase in affinity was observed,[114] suggesting that the starch-binding fragment has nine glucosyl-unit binding subsites.

Three different peptides were isolated from the hydrolysis of *A. niger* GA-I by *B. licheniformis* proteinase.[139] The peptides corresponded to amino acids 471 to 616 (146 residues), amino acids 499 to 616 (118 residues) and amino acids 509 to 616 (108 residues) from the C-terminus of GA-I. The three peptides contained different proportions of O-linked oligosaccharides, and all three could bind native starch granules. The smallest peptide had almost no glycosylated carbohydrate, but could bind tightly to native starch granules. Previously, Hayashida et al.[140,141] reported that the O-glycosylated region played an important role in breaking hydrogen bonds between two adjacent chains in the starch granules, and that this was an important role that the starch-binding domain played in the hydrolysis of the starch granules by glucoamylase. Williamson et al.,[139] however, found that the differences in the binding of the three peptides was very small (about 1 kJ/mole). It was concluded that one of the major functions of the O-glycosylated region in glucoamylase was to provide an extended or linker peptide backbone between the catalytic domain and the starch-binding domain, and that it was not involved in the binding or granule hydrolysis. *A. niger* GA-I, thus, consists of three domains: a large catalytic domain of about 440 amino acids;[142] an O-glycosylated domain of 72 amino acids (residues 441 to 512); and a starch-binding domain of 104 amino acids (residues 513–616).[143] GA-II arises by the proteolytic removal of the O-glycosylated linker from the starch-binding domain from GA-I (residues 441 to 616). Schematic structures of GA-I and GA-II are given in Figure 7.10. Amino acid sequences of GA-I from the various *Aspergillus* species show a very high degree of homology, differing only in 6% of their residues.[144]

III. Relation of Structure with Action of the Enzymes 259

(a)

Aspergillus niger GA-I

NH₂ — Catalyic domain 471 AA, MW = 56 500 — Linker domain 43 AA, Mw = 5200 — Starch-binding domain 102 AA, Mw = 12 240 — COOH

(b)

Aspergillus niger GA-II

NH₂ — Catalyic domain 471 AA, MW = 56 500 — COOH

(c)

Rhizopus nievus GA-I

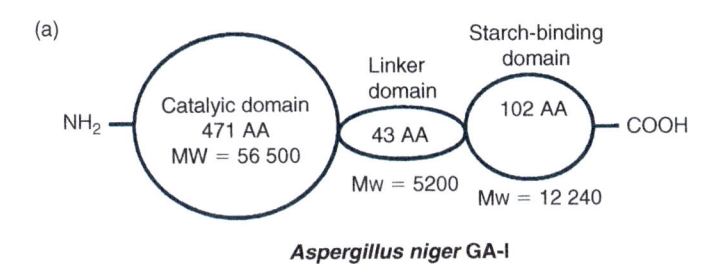

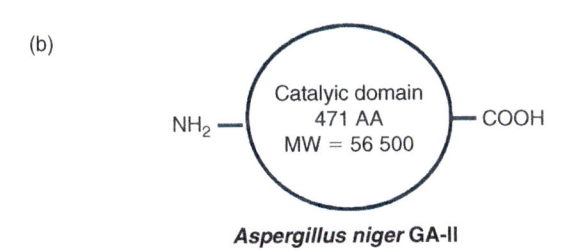

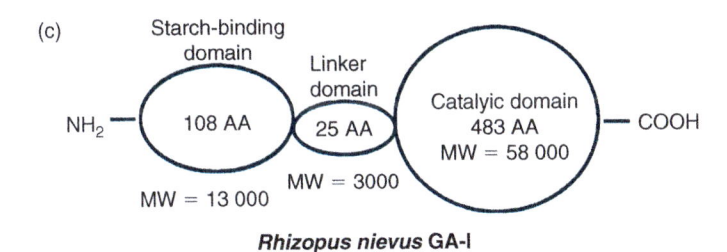

NH₂ — Starch-binding domain 108 AA, MW = 13 000 — Linker domain 25 AA, MW = 3000 — Catalyic domain 483 AA, MW = 58 000 — COOH

Figure 7.10 Schematic representation of the domain structure of glucoamylases: *Aspergillus niger* glucoamylase-1, with starch-binding, linker and catalytic domains; *Aspergillus niger* glucoamylase-2, with only the catalytic domain; *Rhizopus nievus* glucoamylase-I, with starch binding, linker and catalytic domains

GA-II binds a single molecule of acarbose, which occupies the first four glucosyl unit-binding subsites in an active-site pocket.[145] The four units of acarbose are hydrogen bonded to Arg54, Asp55, Arg305, a carbonyl group of residue 177, a main chain amide of residue 121, Glu179, Glu180 and a carbonyl group of residue 179.[146] The carboxylate group of Glu179 forms a salt linkage to the positively charged imino linkage between the last two acarbose residues, and most probably contributes to the very tight binding of acarbose to the active-site.[147] The enzyme's domain has the general shape of a doughnut in which the 'hole' of the doughnut consists of hydrophobic residues at the center that separate two water-filled cavities, one of which serves as the active-site.[143–145] The catalytic domain of *A. niger* glucoamylase is very different from the $(\beta/\alpha)_8$-barrel observed for α-amylases, β-amylase, isoamylase and cyclomaltodextrin glucanosyltransferases. It consists of an α/α-barrel composed of 12 α-helices (51% of the residues in the entire catalytic domain) with limited regions of antiparallel β-strands in the vicinity of the active-site.[48] A similar structure has been observed in cellulaseD from *Clostridium thermocellum*.[149] The structure of glucoamylase is a nontrivial variation of the α/β-barrel structure. It has a core of six, mutually-parallel α-helices. The inner cores of the α-helices are connected to each other by a set of six peripheral α-helices. The peripheral helices are mutually parallel to each other, but approximately antiparallel to the inner core helices. The (β/α)$_8$-barrel, in contrast, has an inner core of eight parallel β-strands connected by eight peripheral, parallel α-helices. (Figure 7.11 presents a topological comparison of the α/α-barrel structure and the α/β-barrel structure.) Another difference between the structures of glucoamylase and the α-amylases and related enzymes is that the active-sites of the α/β-barrel enzymes are formed by loops bridging the C-termini of the β-strands to the N-termini of the peripheral α-helices, while the packing void between the inner core of the six α-helices of glucoamylase makes up the active-site, of which one end is blocked off and the other end is open, forming the entrance to the active-site pocket.

Three forms of glucoamylase occur in the culture supernatants of *Rhizopus* sp. They have molecular weights of 74 000, 58 000 and 61 000, and have been designated GA-I, GA-II and GA-III.[150] GA-I bound and hydrolyzed native starch granules, but GA-II and GA-III neither bound nor hydrolyzed native starch granules.[151] The three glucoamylases had the same C-terminal end, but differed in their N-terminal ends.[152] Two

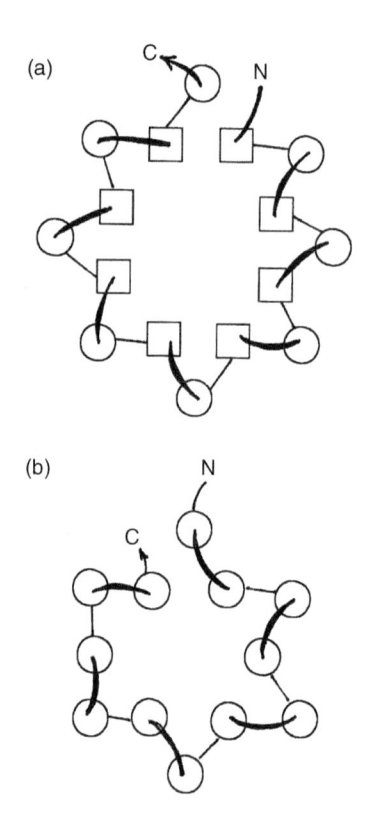

Figure 7.11 Topological comparison of the (α/β)$_8$–TIM barrel structure of the α-amylases (A) and the α/α-barrel structure of glucoamylase (B). α-Helices are represented as circles and β-strands as squares. (From Aleshin et al.;[147] reprinted by permission)

glycopeptide fragments were obtained by proteolytic hydrolysis from the N-terminus of GA-I to give GA-II and GA-III. Papain converted *Rhizopus* sp. GA-I into GA-II of molecular weight 57 000, and chymotrypsin converted GA-I into GA-III of molecular weight 64 000 by hydrolysis of a glycopeptide from the N-terminus of GA-I.

These transformations indicated that the starch binding domain of *Rhizopus* sp. glucoamylase was located at the N-terminus, and attached to the catalytic domain by a small linker domain (Figure 7.10). This is in contrast with the *A. niger* glucoamylases that have an opposite structure, with the catalytic domain at the N-terminus and the starch binding domain at the C-terminus.

Although starch binding domains are not found in most endo-acting α-amylases, they are present in other starch degrading enzymes such as *B. macerans*,[151] *B. circulans*[125,152] and *B. stearothermoplilus*[153] cyclomaltodextrin glucanosyltransferases, two exo-acting α-amylases,[154,155] a β-amylase,[156] *B. circulans* maltohexaose starch granule-degrading amylase[157] and *Ps. amyloderamosa* isoamylase.[158]

The active-sites of the glucoamylases are pockets that do not have a flexible, mobile loop that can act as a trap door to let the product (glucose) out. After the glycosidic bond is cleaved by glucoamylase, the remaining starch chain must dissociate and leave the active-site before glucose can leave. This explains why glucoamylase has such a low turnover number, and why it does not have a multiple attack mechanism.

4. Specific Amino Acids at the Active-Site Involved in Catalysis and Substrate Binding

Three carboxyl groups (Asp206, Glu230 and Asp297 using *A. oryzae* α-amylase numbering) of α-amylases[159–163] and cyclomaltodextrin glucanosyltransferase[164] are essential for catalysis and are postulated to be involved in hydrolysis and/or the breaking and forming of glucosidic bonds. Similar carboxyl groups (Asp179 and Glu180) were likewise identified by mutagenesis of *A. awamori* glucoamylase.[165]

Mutation of Trp83 to Leu in *Saccharomycopsis* α-amylase resulted in the loss of hydrolytic activity on maltoheptaose, but an increase in the transglycosylation reactions to give higher maltodextrins.[166] A similar tryptophan residue has been recognized in other amylolytic enzymes.[152] It has been suggested that the Trp indole ring is involved in binding of the glycon part of the cleaved substrate, and that it controls its release from the active-site.[167] Mutations of several histidines to asparagines in barley α-amylase, *Bacillus* α-amylase and human pancreatic α-amylase produced drastic drops in activity,[168] because histidine residues are critical in the binding of glucosyl units in specific glucosyl unit-binding subsites.

Several *Aspergillus niger* glucoamylase mutants have been produced that differ from the parent type in only one amino acid, for example, Glu179 to Gln and Glu180 to Gln,[163] Trp120 to Phe, Leu, His and Tyr,[169] Asp55 to Gly[170] and Tyr48 to Trp.[171] All these mutations gave significant loss of activity. X-ray diffraction studies[146–148] indicated that the mutated residues were important in forming hydrogen bonds or effecting hydrophobic interactions with the substrate at active-site subsites. Trp120 was postulated to be important for stabilization of the transition state during catalysis.[168,171,172] Glu180 was found to be important for substrate binding at subsite II,[165] the subsite with the highest affinity for binding glucose in glucoamylase.

In addition to site-directed mutations, deletion mutants that eliminated some, or all, of the linker domain have been produced for *A. niger* glucoamylase.[173] Deletions of up to 30 amino acids from the C-terminus of the linker domain had no effect on the activity of either glucoamylase-I or glucoamylase-II.[173]

X-ray diffraction studies have shown that the glucosyl units of starch substrates are bound by two different modes of interaction. The most common mode is hydrogen bonding between hydroxyl groups of glucosyl units and polar groups very specifically positioned at the active-site. The second mode involves stacking or sandwiching of glucose CH groups with aromatic side chain groups, such as the indole ring of tryptophan and the phenol ring of tyrosine. The hydrogen bonding of the OH groups of glucosyl units at the active-site is shown in Figure 7.12 for the complex of *A. oryzae* α-amylase with maltoheptaose that is bound in a ribbon-like conformation.[98] Twelve charged amino acid residues are involved in hydrogen bonding with uncharged glucosyl unit OH groups in the *A. oryzae* α-amylase complex.

Trp120 and Tyr116 are involved in the binding of glucosyl units at the active-site of *A. niger* glucoamylase, along with tryptophan residues at the starch-binding domain.[175] Using dipeptides and maltotriose, Otsuki et al.[175] have found that glucosyl units most probably interact with l-Trp-l-Trp and l-Trp-l-Tyr by π-CH interactions, in a stacking structure in which a glucosyl unit is sandwiched between the two aromatic rings and held there by a combination of hydrophobic and van der Waals forces (Figure 7.13).

5. Structure and Function of Domains in Amylolytic Enzymes

Primary, secondary and tertiary structures of amylolytic enzymes from a wide variety of sources and functions (the α-amylases, bacterial cyclomaltodextrin glucanosyltransferases, isoamylases and starch-branching enzymes) have been found to be closely related, and have been placed into the so-called structural α-amylase family.[176,177] These enzymes have been studied with regard to the number, structure, organization and function of domains.[178]

As discussed above, glucoamylases consist of three domains: a catalytic domain; a linker domain; and a starch-binding domain. These domains can be in different orders in which the catalytic domain is located at the N-terminus or the C-terminus and conversely the starch-binding domain is at the C-terminus or the N-terminus, with the linker domain between the two.

A. oryzae α-amylase consists of three domains (A, B and C).[179,180] Domain A has an amino-terminal $(\beta/\alpha)_8$-barrel structure, followed by domain C, consisting of β-strands that are folded into a Greek motif. Domain B is inserted between the third β-strand and the third α-helix of the $(\beta/\alpha)_8$-barrel. This is a highly variable domain in both its length and amino acid sequence, depending on the source and type of the enzyme.[181,182] Cyclomaltodextrin glucanosyltransferases generally consist of five domains (A, B, C, D and E). Domains A, B and C consist of the same catalytic domains found in

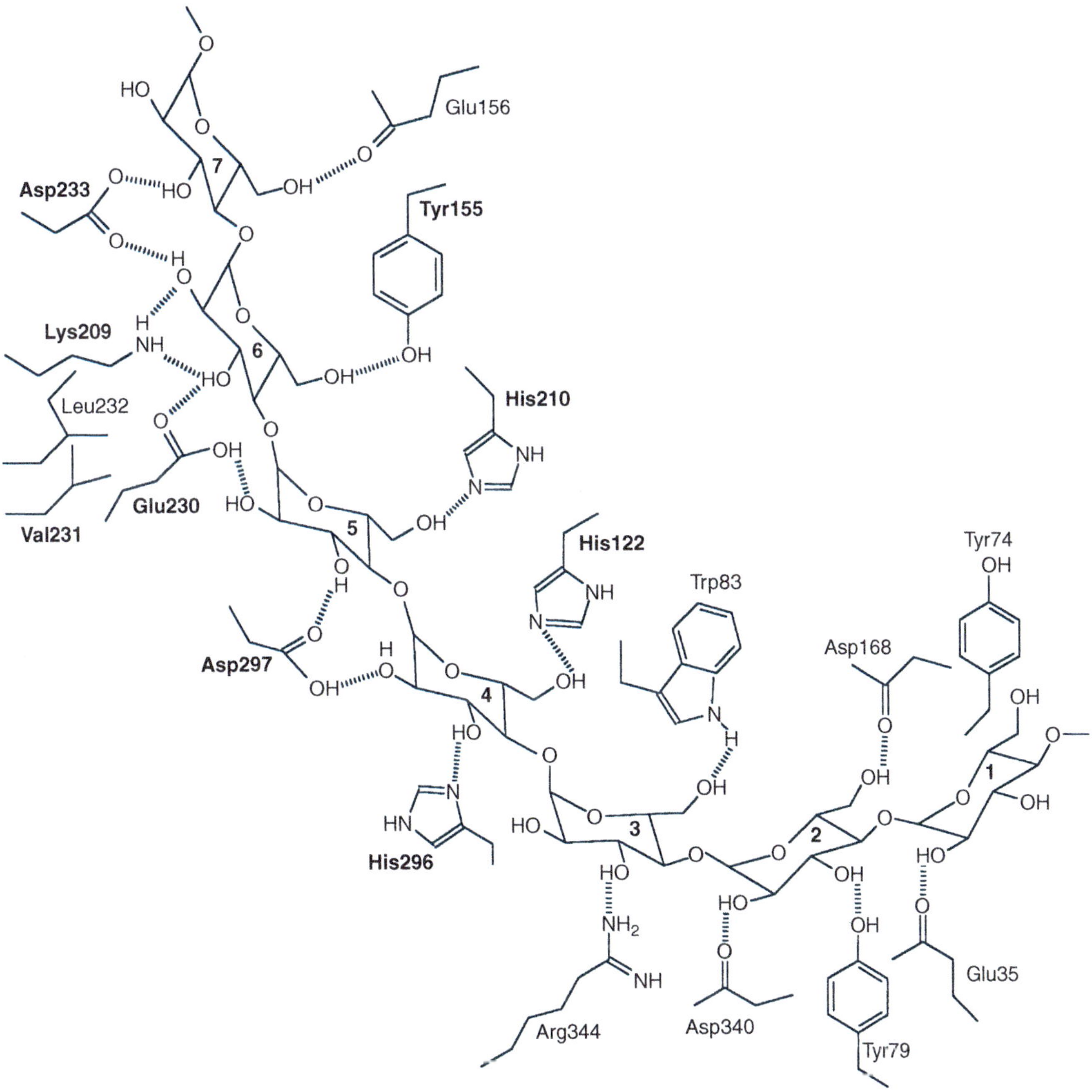

Figure 7.12 Identification of the amino acid residues involved in the seven subsite binding of the substrate in the active-site of *Aspergillus oryzae* α-amylase. The bold faced amino acid residues are those conserved in the active-site of porcine pancreatic α-amylase. (From Matsuura et al.;[123] reprinted by permission)

A. oryzae α-amylase.[178,182–185] The structure of domain E is typical of enzymes that bind and react with raw starch granules.[186] Domain D contains antiparallel β-barrels,[183–185] but nothing is known about its function.

A few chimeric enzymes have been constructed by adding one or more domains from one amylolytic enzyme to another. Some of these chimeric enzymes have been studied in regard to the secretion of the enzyme,[187] substrate specificity[188] and product specificity.[189] A starch binding domain from a *Bacillus* sp. cyclomaltodextrin glucanosyltransferase was fused with *B. subtilis* α-amylase, and an α-amylase was

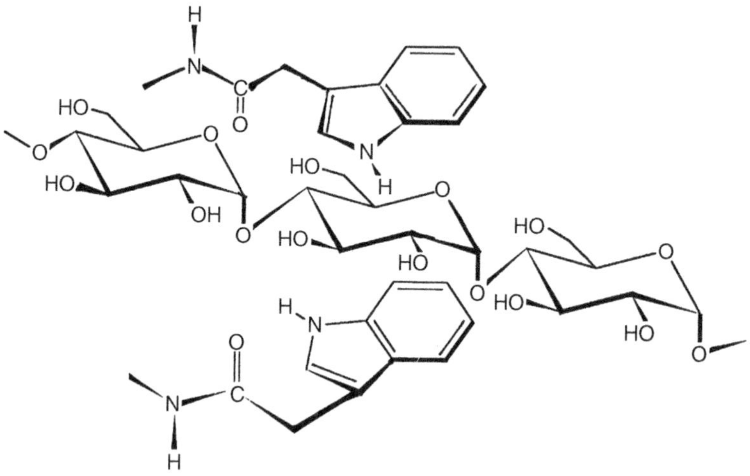

Figure 7.13 Hydrophobic van der Waals interaction of the CH-π bonds of a maltodextrin molecule sandwiched between two tryptophan indole rings. (Adapted from Otsuki et al.[175])

produced that had high binding and hydrolytic activity with raw starch granules.[190] Product specificity, pH and temperature optima, and stability were not altered, but the ability to hydrolyze raw starch granules was obtained for the modified α-amylase.

IV. Mechanisms for the Enzymatic Hydrolysis of the Glycosidic Bond

Mechanisms involved in catalyzing hydrolysis of glycosidic bonds must take into account the two stereochemical pathways: retention of configuration and inversion of catalytic configuration. Two types of catalytic groups, an acid and a base, at the active-site effect hydrolysis of the glycosidic linkage by acid–base catalysis in both mechanisms. These groups, for the most part, have been identified as a carboxylate ion (a base or nucleophile) and a carboxylic acid group (an acid or proton donor).

Formation of a covalent glucanyl–enzyme complex is indicated for these enzymes that catalyze transfer and condensation reactions.[19,100,114] Evidence for the covalent glucanyl–enzyme complex was provided for porcine pancreatic α-amylase, using cryogenic ^{13}C-NMR.[191a] [1-^{13}C]-enriched maltotetraose was used as a substrate for porcine pancreatic α-amylase at subzero temperatures (−20°C) in the presence of 40% (v/v) dimethylsulfoxide. The ^{13}C-NMR spectrum gave a broad resonance peak, whose chemical shift, relative signal intensity and time-course appearance corresponded to formation of a β-carboxyl-acetal ester (the postulated covalent enzyme complex).[191a] The formation of a covalent carbohydrate–enzyme intermediate has now been directly demonstrated by trapping it during catalysis by rapid denaturation of the α-amylase reacting with ^{3}H-labeled methyl α-maltodextrin glycosides.[191b] The covalent intermediate was then analyzed by ^{1}H NMR saturation-transfer analysis, which showed that the carbohydrate was β-linked to the denatured protein. This is what would be expected for an α-amylase that catalyzes the hydrolysis of α-(1→4) glucosidic linkages by a two-step S_N2 double-displacement reaction to give an α-anomeric carbon at the reducing end of the products.[191b]

Evidence for the formation of covalent intermediates for *B. circulans*[251] cyclomaltodextrin glucanyltransferase (CGTase), which primarily forms cyclomaltoheptaose, was obtained by trapping the intermediate using α-4^3-deoxymaltotriosyl fluoride.[192a] This substrate has a good leaving group, the fluoride ion, which facilitates formation of the intermediate. Once formed, it cannot undergo the disproportionation transglycocylation reactions, as the requisite hydroxyl groups at the C-4 position are missing. More direct evidence for the formation of a covalent intermediate has been obtained for *B. macerans* CGTase by trapping it, using rapid denaturation during its reaction with ^{3}H-labeled cyclomaltohexaose.[192b]

For these enzymes that retain the configuration of the α-glycosidic bond (the α-amylases, isoamylase and cyclomaltodextrin glucanosyltransferases), a classical S_N2, double-displacement mechanism has been demonstrated. In this mechanism (Figure 7.14a), the carboxylate base initiates cleavage of the bond via nucleophilic attack at C-1 of the glucopyranosyl unit undergoing hydrolysis. Simultaneously, the glycosidic oxygen is displaced and protonated by the carboxylic acid group. The attacking carboxylate group forms a covalent β-linked acetal-ester, giving a glucopyranosyl–enzyme intermediate. This high-energy linkage is subsequently hydrolyzed by water. Reaction with water is facilitated by the second aspartate carboxylate group (see Section 7.3.4. above) at the active-site of these enzymes. The carboxylate group abstracts a proton from water, making the water more nucleophilic for its attack on the β-linked acetal-ester. The anomeric carbon atom of the resulting glucopyranosyl unit that is released from the enzyme complex, thus, has the α-configuration, giving retention of configuration of the product at its reducing end.

An identical mechanism can be postulated for hydrolysis of the α-(1→6) branch linkage by isoamylases and for cyclomaltodextrin glucanosyltransferase. For the latter enzyme, the water molecule is replaced by the C-4 hydroxyl group on the nonreducing end glucosyl unit of the starch chain (Figure 7.6).

Formation of the covalent glucanosyl–enzyme complex is suggested for these enzymes by the transfer and condensation reactions they catalyze.[19,100,114] Direct evidence for the covalent glucanosyl–enzyme complex was provided for porcine pancreatic α-amylase using cryogenic ^{13}C-NMR.[191a] [1-^{13}C]-enriched maltotetraose was used as a substrate at subzero temperatures in the presence of dimethylsulfoxide. The ^{13}C-NMR spectrum gave a broad resonance peak, whose chemical shift, relative signal intensity and time-course appearance corresponded to formation of a β-carboxyl acetal-ester (the postulated covalent enzyme complex).[191a] Direct evidence for the formation of covalent intermediates for cyclomaltodextrin glucanosyltransferase was obtained by trapping the intermediate, using α-4^3-deoxymaltotriosyl fluoride.[192a] This substrate has a good leaving group, a fluoride ion, which facilitates formation of the intermediate. Once formed, it cannot undergo the transglycosylation reaction as the requisite hydroxyl group at the C-4 position is missing.

Understanding the mechanism of glycosidic bond hydrolysis by the inverting β-amylases has been more difficult. With these enzymes, carboxylate-base and

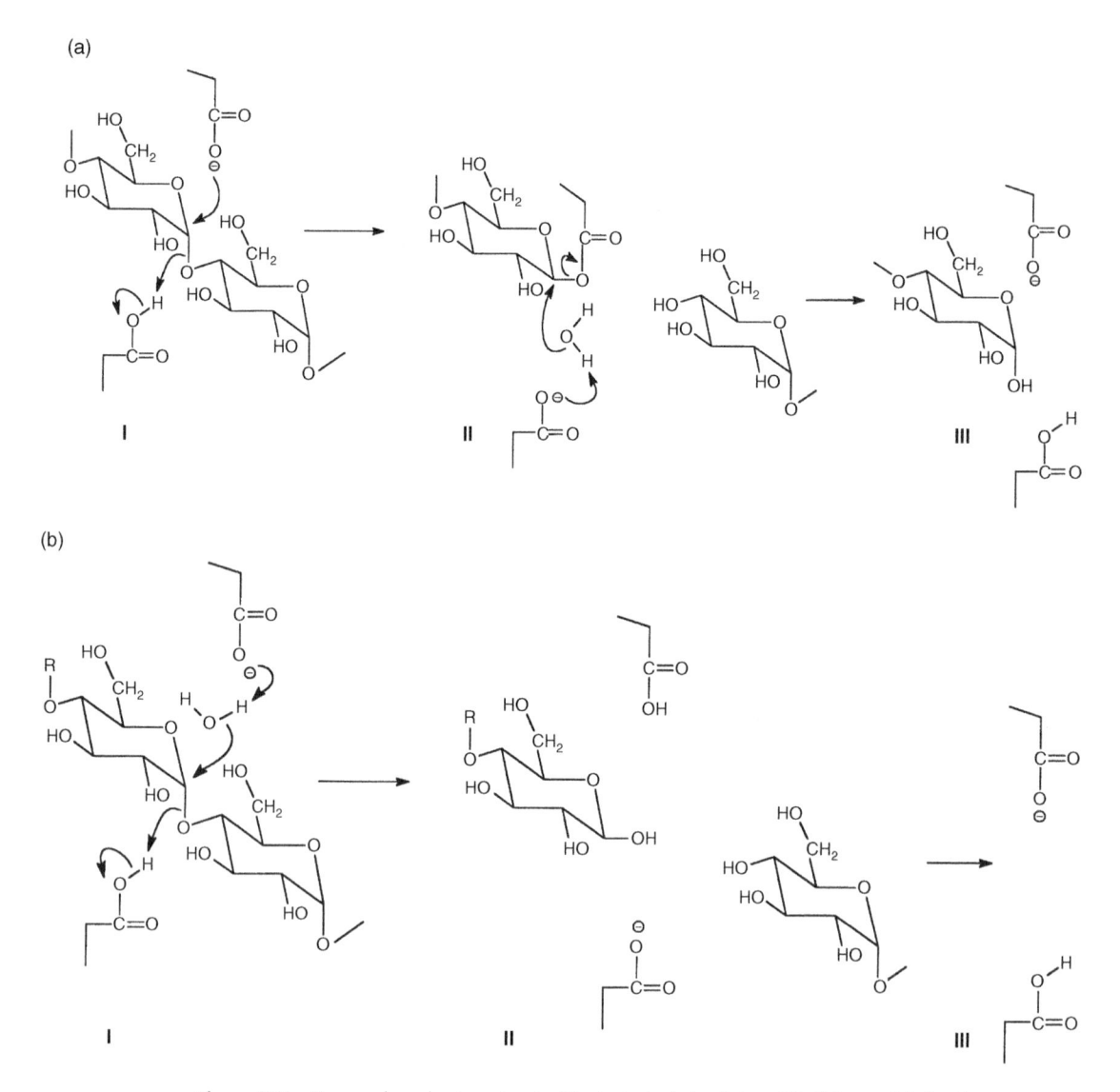

Figure 7.14 Proposed mechanisms involved in the hydrolysis of glycosidic linkages. (a) double displacement, S_N2, mechanism, giving retention of configuration (α-amylases). I. Attack on C-1 by a carboxylate group and donation of a proton to the leaving oxygen atom by a carboxyl group. II. Attack on water on the covalent β-acetal-carboxyl-ester to give product III with retained configuration at the anomeric end; (b) direct displacement, S_N1, mechanism, giving inversion of configuration (β-amylases). I. Direct attack on C-1 by water, assisted by a carboxylate group and donation of a proton to the leaving oxygen atom by a carboxyl group. II. Product with inverted configuration. III. Regeneration of the catalytic groups at the active-site by proton exchange between the two carboxyl groups.

carboxylic acid groups also have been identified at the active-site by x-ray crystallography and site-directed mutagenesis.[132,164,165] *A. awamori* glucoamylase-II has been found to have a carboxylate group (Glu400) hydrogen-bonded with a water molecule that can make a direct nucleophilic attack on C-1 of the glucosyl unit bearing the glycosidic linkage.[170] The reaction (Figure 7.14b) displaces the glycosidic oxygen atom that is simultaneously protonated by the carboxylic acid group of Glu179 via a single displacement, SN1-type mechanism, inverting the configuration of the anomeric carbon atom of the newly formed reducing end. While oxycarbonium (carboxonium) ion intermediates can be postulated for both mechanisms, their actual existence is highly unlikely, except as very fleeting transition-state intermediates, and no evidence has ever been presented for such a mechanism.

V. Action of Amylases on Insoluble Starch Substrates

1. Action of α-Amylases on Amylose-V Complexes and Retrograded Amylose

Jane and Robyt[193] studied the action of α-amylases from human saliva, porcine pancreas, and *Bacillus amyloliquefaciens* on retrograded amylose and amylose-V complexes formed with 1-butanol, *tert*-butyl alcohol (1,1-dimethylethanol), 1,1,2,2-tetrachloroethane and 1-naphthol. It was found that amylase-resistant, amylodextrin fragments were formed from each of the complexes. The fragments were of relatively uniform size, and dependent on the type of complex. The degree of polymerization (DP) of the peak fractions from the action of human saliva and porcine pancreatic α-amylases was 75 ± 4 for the 1-butanol complex, 90 ± 3 for the *tert*-butyl alcohol complex and 123 ± 2 for the 1-naphthol complex. *B. amyloliquefaciens* α-amylase gave somewhat higher values (104 ± 5, 110 ± 5 and 14 ± 8, respectively) for the three amylose-V complexes. These DP values correspond to helices of 6 glucosyl units per turn with a repeating length of 10 nm, 7 glucosyl units per turn with a length of 10 nm, and 9 glucosyl units per turn with a length of 10 nm, respectively, for the three types of amylose complexes. The DP of the resistant amylodextrin fragments, thus, depended on the number of glucosyl units per turn of the helix and the folding length of the particular amylose complex. The data suggests that α-amylases hydrolyzed the amorphous areas on the turns of the lamella of packed helical chains that have a length of 10 nm (Figure 7.15a), and that the α-amylases did not hydrolyze the relatively regular chains in the helical, crystalline regions of the structure. The differences in the DP values that resulted from the action of human salivary and porcine pancreatic α-amylases and from *B. amyloliquefaciens* α-amylase can be attributed to differences in the number of glucosyl unit-binding subsites at the active-sites of the two types of α-amylases. *B. amyloliquefaciens* α-amylase has nearly twice as many binding-subsites as human salivary and porcine pancreatic α-amylases (see Section 7.2 and Figures 7.1a and 7.1c).

Action of the α-amylases on retrograded amylose also gave resistant amylodextrin fragments whose size was dependent on the type of α-amylase used. Human salivary α-amylase gave a fragment with an average DP of 42, porcine pancreatic α-amylase gave a fragment with an average DP of 44 and *B. amyloliquefaciens* α-amylase gave a fragment with an average DP of 50. Hydrolysis of retrograded amylose with 16% (v/v) sulfuric acid at 25°C for 20 days gave an amylodextrin fragment with an average DP of 33, hydrolysis for 40 days gave a fragment with an average DP of 31.

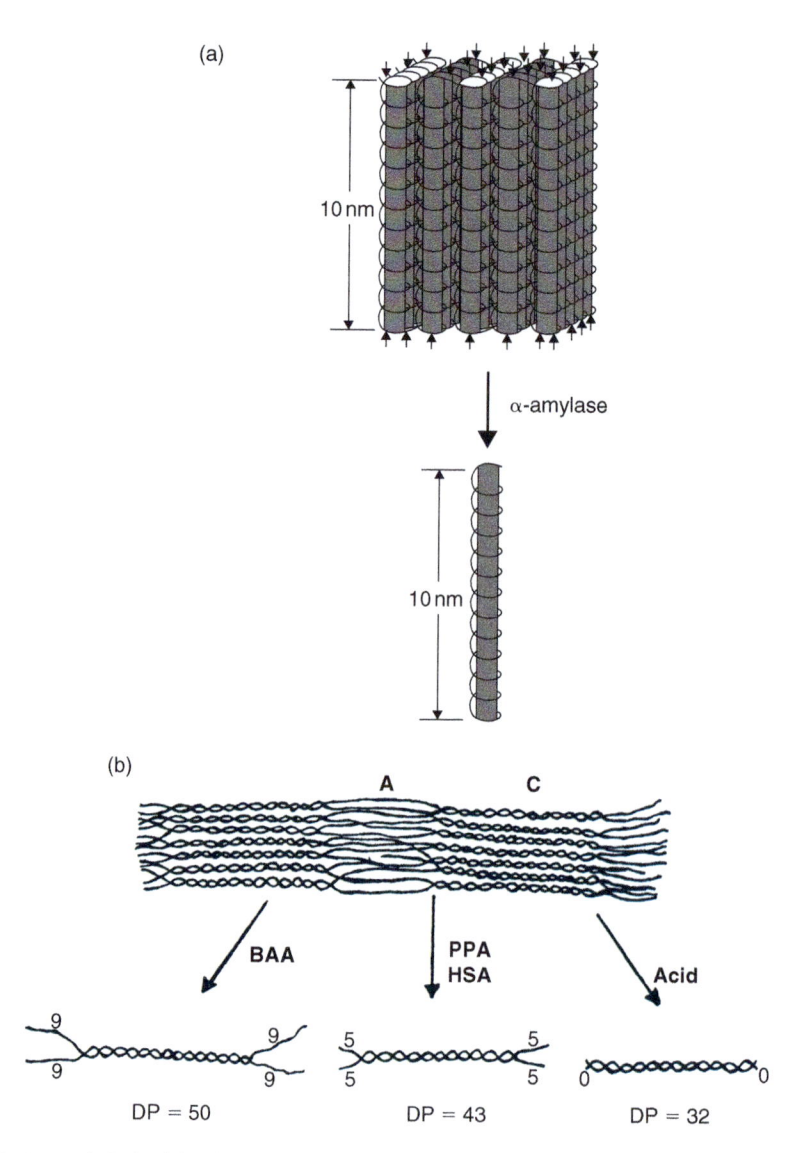

Figure 7.15 Hydrolysis of the glycosidic linkage of water-insoluble starch complexes. Alpha-amylase hydrolysis of the lamella-structured, crystalline sheets of amylose–alcohol complex. The arrows on the top and the bottom of the sheets are the proposed amorphous sites that α-amylase hydrolyzes to give resistant amylodextrin fragments of relatively uniform size. Mechanism of hydrolysis of the proposed double helical structure of retrograded amylose by acid and α-amylases; 'A' is the amorphous area, and 'C' is the crystalline area. BaA is *Bacillus amyloliquefaciens* α-amylase; PPA is porcine pancreatic α-amylase; and HSA is human salivary α-amylase. Differences in the DP values resulting from the action of the different catalysts reflect differences in their specificities: acid-catalyzed hydrolysis cleaves the amorphous regions right up to the juncture of the double helix; porcine pancreatic and human salivary α-amylases cleave amorphous regions up to five glucosyl units from the juncture, giving a DP of 10 glucosyl units longer than what acid gives; *B. amyloliquefaciens* α-amylase cleaves the amorphous regions up to nine glucosy units from the juncture, giving a DP of 18 units longer than what acid gives. (From Jane and Robyt;[193] reprinted by permission)

A double helical structure for retrograded amylose with double helical regions of 10 nm length interspersed with amorphous regions was postulated.[193] Results of α-amylase- and acid-catalyzed hydrolysis suggest that both catalysts were hydrolyzing glycosidic bonds located in the amorphous regions of the retrograded amylose, leaving the crystalline, double helical regions intact (Figure 7.15b). Differences in the sizes of the resistant amylodextrin fragments were due to differences in the number of glucosyl unit-binding subsites of the hydrolyzing agents. Acid, having no binding-subsites, catalyzed hydrolysis right up to the ends of the crystalline regions, whereas α-amylases hydrolyzed the amorphous regions up to some point that was several glucosyl units away from the crystalline regions, leaving 'stubs' on the ends of the amylodextrin chains. Sizes of the stubs were dependent on the number of binding-subsites of the α-amylases, with *B. amyloliquefaciens* α-amylase having twice as many subsites as mammalian α-amylases, thus giving stubs that were about twice as long as those produced by mammalian enzymes.

2. Action of Amylases with Native Starch Granules

Native starch granules are generally thought to be resistant to amylase action, although as early as 1879[194] starch granules were reported to be digested by amylases. In 1939, Stamberg and Bailey[195] reported a 10% conversion of wheat starch granules in 24 hours with α-amylase. In 1954, Sandstedt and Gates[196] reported the digestion of starch granules by four sources of α-amylase (from malt, bacteria, fungi and pancreas) and found that pancreatic α-amylase was the most effective, followed in order by malt, bacterial and fungal α-amylases. In 1960, Leach and Schoch[197] reported the action of bacterial (*B. subtilis*, later reclassified as *B. amyloliquefaciens*) α-amylase on starch granules from various botanical sources. They found that the different types of starches had widely different susceptibilities, with waxy maize starch granules being the most susceptible and high-amylose corn starch the least susceptible. In 1971, however, Manners[198] reported that *A. niger* glucoamylase only hydrolyzed starch granules to a very limited extent. Rasper et al.[199] then showed that degradation of starch granules by glucoamylase does take place, but how much hydrolysis occurs is dependent on the type of starch. Smith and Lineback[200] followed the reaction of *Rhizopus niveus* glucoamylase on wheat and normal maize starch granules using the scanning electron microscope. They found that, indeed, glucoamylase was degrading these granules. Manners most likely had been using a glucoamylase preparation that was mostly glucoamylase-II in which its starch binding domain had been removed and, hence, it had no or very little action on the starch granules (see Section 7.4). It is now generally agreed that ungelatinized starch granules do undergo some hydrolysis by amylases, although there are differences between the types of starches and between the different types of amylases. Differences in the hydrolysis of gelatinized starch and starch granules by amylases are of the order of 100- to 1000-times greater.

Kimura and Robyt[201] made a kinetic study of the hydrolysis of seven types of starch granules from normal maize, waxy maize, barley, tapioca, amylomaize-7, shoti and potato by *R. nievus* glucoamylase-I. The different types of starch granules had a relatively broad representation of starch types: A-, B- and C- x-ray pattern types; starches of commercial and of non-commercial importance; starches with widely varying amounts of amylopectin and amylose; and starches from cereal and tuber sources. Four potato starches that had been modified by acid-catalyzed hydrolysis in four different kinds of alcohols were also studied. Three different concentrations (2-, 20- and 200-units/mL) of enzyme were used. The starch granules exhibited a wide variation in their susceptibility to glucoamylase hydrolysis. They divided into three groups: (a) very susceptible, which was waxy maize starch granules being converted into 50%, 95% and 98% glucose in 32 hours for the three concentrations of enzyme, respectively; (b) intermediate susceptibility, with barley, maize and tapioca starch granules being converted into 12–18%, 58% and 75–80% glucose in 32 hours for the three concentrations of enzyme, respectively; and (c) the least susceptible, with amylomaize-7, shoti and potato starch granules being converted into 2–8%, 5–10% and 12% glucose in 32 hours for the three concentrations of enzyme, respectively. The percentage conversion into glucose for the modified potato starches was 13%, 17%, 21% and 27% for modification in the four types of alcohols (methanol, ethanol, 2-propanol and 1-butanol, respectively) in 32 hours of reaction with 200 units/mL of enzyme. For the most part, the number and size of the granules did not change drastically. Granules converted 50% or more to glucose in the first two groups had the classical 'Swiss cheese' appearance in which there were many deep holes, but deeper into the granules retained a shell-like structure (Figure 7.16a–c).

It was apparent from the study that starch granules were hydrolyzed by glucoamylase, but that granules from different sources had widely different susceptibilities to glucoamylase hydrolysis. Further, the 10- and 100-fold increases in the amount of enzyme did not give 10- and 100-times the amount of hydrolysis. While waxy maize starch was nearly completely converted to glucose by the 10-fold amount of enzyme, the other starches were converted to a lesser extent in 32 hours of reaction.[201] There are about three orders of magnitude difference in the reaction with gelatinized starch and with starch granules. Variation in the susceptibility of the various types of starch granules was not related to the relative ease of gelatinization. For glucoamylase to hydrolyze the granules at all, it must have the starch binding domain.

Kim and Robyt[202] modified waxy maize starch granules by reaction with glucoamylase *in situ* to give 100% retention of the glucose inside the granules. Starch granules, containing 5% to 50% (w/w) glucose inside the granules were prepared by stopping the enzyme reaction at various periods of time. A similar reaction of waxy maize starch with cyclomaltodextrin glucanosyltransferase and isoamylase gave 3.4% (w/w) cyclomaltodextrins completely retained inside the granules.[203]

Gallant et al.[204] and Valentudie et al.[205] studied the action of *B. amyloliquefaciens* α-amylase and porcine pancreatic α-amylase on starch granules. They found that the most crystalline parts of the granules were only slightly digested, and that the majority of the digestion occurred in the amorphous regions. They also found that α-amylases produced pits in several types of granules and that these pits became enlarged and were channels of corrosion into the granules. These channels were deep and enlarged throughout the granule, giving a granule shell with various layers. The kinetics showed that *B. amyloliquefaciens* α-amylase (168 IU/g of starch) gave

Figure 7.16 Scanning electron micrographs of the action of amylases on native starch granules. Extensive action of *Rhizopus nievus* glucoamylase-1 on: (a) waxy maize starch; (b) maize starch; and (c) barley starch. (From Kimura and Robyt[201] reprinted by permission) Action of *Bacillus circulans* α-amylase on potato starch granules: (d) early stage of the reaction in which the enzyme has eroded the external surface of the granule, showing early concentric rings proceeding down the granule from the hilum; (e) intermediate state of the reaction, showing the concentric furrows proceeding down the length of the granule; (f) late stage of the reaction, showing an elongated granule with changed shape. (From Taniguchi et al.;[55] reprinted by permission) Action of *Bacillus amyloliquefaciens* α-amylase on potato starch granules: (g) early stage of the reaction that has started at the hilum, showing lamallear, concentric rings (reprinted by permission of Kimura and Robyt); (h) more extensive reaction showing the lamellar structure produced by the enzyme; (i) high magnification of the lamellar structure of h. (From Hollinger and Marchessault;[208] reprinted by permission)

a limit of about 50% hydrolysis for cassava (tapioca) starch, 30% for sweet potato starch, 11% for potato starch and 5% for yam starch.[206]

Taniguchi et al.[55] reported an unusual α-amylase that is elaborated by *B. circulans* F2. The enzyme is specifically induced in the culture by either solubilized potato starch or by potato starch granules. It completely digested potato starch granules, which are usually quite resistant to amylase digestion. Scanning electron microscopy of partially digested granules showed that the enzyme started from a single point at the end of the granule surface and proceeded around and down the length of the granule, producing ridges and furrows over the granule surface (Figures 7.16d and 7.16e). The enzyme eventually changed the shape of the potato granule, giving an elongated granule (Figure 7.16f).

Following this discovery, several other bacteria were found to elaborate α-amylases that could digest potato starch granules: an anaerobic *Clostridium butyricum*;[206] a non-sulfur purple photosynthetic bacterium, *Rhodopseudomonas gelatinosa*;[205] and two other *Bacillus* species.[206,207]

Hollinger and Marchessault[208] used *B. amyloliquefaciens* α-amylase to study the internal structure of potato starch granules. Scanning electron micrographs revealed that, after amylase digestion, the interior of the granule contained highly ordered, enzyme resistant, lamella (Figures 7.16g and 7.16h).

VI. Inhibitors of Amylase Action

Amylase inhibitors fall into one of two categories: (a) naturally-occurring proteins, that are thought to act in a defense mechanism against predators such as insects; and (b) mono- or oligosaccharides that have structural features favorable for binding relatively strongly to the active-site of the enzymes. It has been hypothysized that the oligosaccharide inhibitors mimic the transition state of the enzyme-catalyzed reaction sufficiently to bind and inhibit the enzyme. Plant protein inhibitors primarily inhibit α-amylases. Wheat produces an α-amylase inhibitor that has been known for over 50 years as Kneen's Inhibitor. Rye produces an α-amylase inhibitor similar to Kneen's Inhibitor, kidney beans and other legumes produce an α-amylase inhibitor known as phaseolamin, and oats produce a protein inhibitor that inhibits both α- and β-amylases. These plant inhibitors were reviewed in the second edition of this work.[209]

The small molecule inhibitors are usually site-directed, substrate analogs. Of these, acarbose is probably the best recognized. It is produced by *Actinomyces* sp. and is a transition-state analog inhibitor for several carbohydrases. Acarbose has a non-reducing terminal unit that is an unsaturated cyclohexitol. The cyclohexitol is linked to the nitrogen atom of 4-amino-4,6-dideoxy-D-glucosyl (4-amino-4-deoxy-D-quinovose).[2] This pseudodisaccharide is known as acarviosine and is linked α-(1→4) to maltose to give acarbose (Figure 7.17). Acarbose is a potent inhibitor of glucoamylase[146] with a K_I value of 1×10^{-12} M, and a good inhibitor of cyclomaltodextrin glucanosyltransferase[210] and porcine pancreatic α-amylase[211] with K_I values of 1×10^{-6} M.

Acarbose itself is hydrolyzed by *B. stearothermophilus* maltogenic amylase at the first glycosidic linkage from the reducing end to give D-glucose and acarviosine-glucose

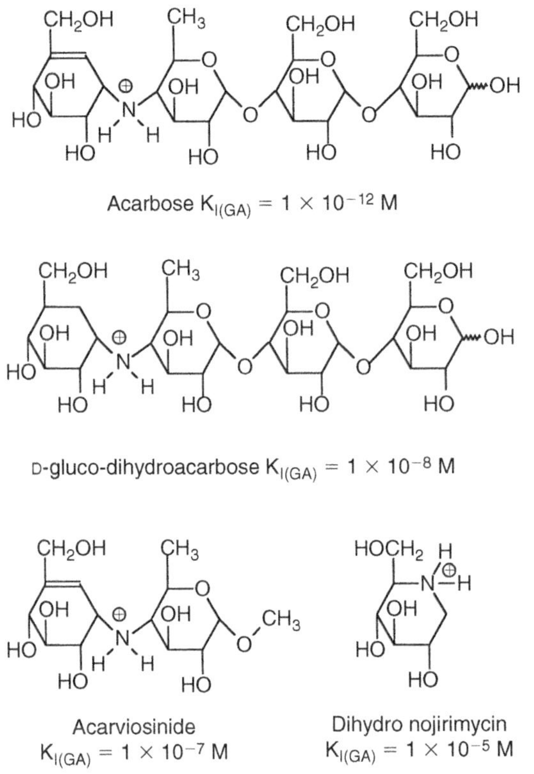

Figure 7.17 Structures of active-site directed amylase inhibitors. The K_I values are for glucoamylase (GA), porcine pancreatic α-amylase (PPA) and cyclomaltodextrin glucanosyltransferase (CGTase).

(Figure 7.17).[212] The enzyme also catalyzes transglycosylation reactions in which acarviosinyl-glucose was transferred to the O-6 of D-glucose to give α-acarviosinyl-α-D-glucopyranosyl-(1→6)-D-glucose (isoacarbose).[213] Isoacarbose is a more potent inhibitor of α-amylase and cyclomaltodextrin glucanosyltransferase than is acarbose, inhibiting 15.2- and 2.0-times more, respectively.[214]

Dihydronojirimycin [1,5-anhydro-5-amino-5-deoxy-D-glucitol] (Figure 7.17) is a relatively good inhibitor for glucoamylase with a K_I of 1×10^{-5} M.[215] Lehmann et al.[216] synthesized a number of 6-amino-6-deoxy derivatives of maltose and maltotriose and studied their inhibition of porcine pancreatic α-amylase. The most effective inhibitors were 6^1-amino-6^1-deoxy- and 6^2-amino-6^2-deoxy-maltotriose, with K_I values of 2×10^{-6} M. These maltotriose derivatives were bound 900–950 times better than maltotriose itself and 1500–1600 times better than methyl α-maltotrioside glycoside. Lehmann et al.[216] also synthesized three maltodextrin analogs containing photoaffinity probes that were active-site-directed inhibitors for porcine pancreatic α-amylase. These probes joined two maltodextrin segments together with a linker (Figure 7.18). Lehmann and Schmidt-Schuchardt[217] synthesized di-, tri- and tetrasaccharides that had a spacer or linker containing azido or amino groups joining

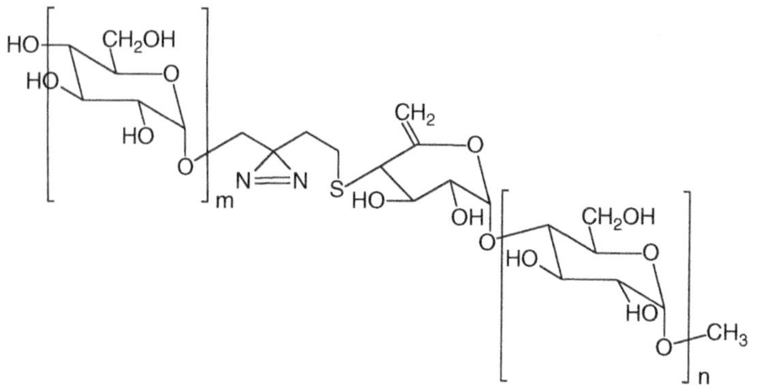

Figure 7.18 Photoactive active-site directed inhibitors of porcine pancreatic α-amylase. Three inhibitors were synthesized by Lehmann et al.,[215] with (m = 1 and n = 2), (m = 2 and n = 1) and (m = 3 and n = 0).

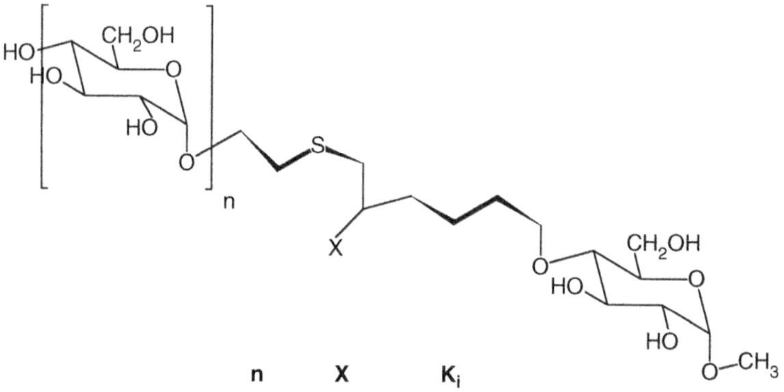

	n	X	K_i
I	1	N_3	36.6 mM
II	2	N_3	8.1 mM
III	3	N_3	2.4 mM
IV	4	NH_2	11.0 mM
V	5	NH_2	0.53 mM
VI	6	NH_2	0.42 mM

Figure 7.19 Oligosaccharide active-site directed inhibitors for porcine pancreatic α-amylase, synthesized by Lehmann and Schmidt-Schuchardt.[218]

glucosyl, maltosyl and maltotriosyl units to methyl α-D-glucopyranoside at O-4. The structures and K_I values of these compounds are given in Figure 7.19. Although these compounds were relatively good inhibitors for porcine pancreatic α-amylase when compared with maltotriose or methyl α-maltotrioside, they were not as good as 6-amino-6-deoxymaltose or 6-amino-6-deoxymaltotriose derivatives. The K_I values of the various inhibitors of porcine pancreatic α-amylase are summarized in Table 7.2.

Table 7.2 Specificity of active-site-directed inhibitors and their KI values for porcine pancreatic α-amylase

Inhibitor	K_I (PPA)[a]
Isoacarbose	0.04 μM[b]
Acarviosine glucose	0.54 μM[b]
Acarbose	0.64 μM[b]
6^1-Amino-6^1-deoxy maltotriose	1.9 μM[c]
6^2-Amino-6^2-deoxy maltotriose	2.0 μM[c]
Acarbose	9.7 μM[c]
6^1-Amino-6^1-deoxy maltose	88.0 μM[c]
6^3-Amino-6^3-deoxy maltotriose	175.0 μM[c]
Methyl α-6^2-amino-6^2-deoxymaltotrioside	360.0 μM[c]
Maltotriose	1800.0 μM[c]
Methyl α-maltotrioside	3000.0 μM[c]
α-6^2-Amino-6^2-deoxymaltosyl-(1→4)-1,5-anhydro-D-glucitol	7600.0 μM[c]
Methyl β-maltotrioside	9000.0 μM[c]

[a]PPA = porcine pancreatic α-amylase
[b]K_I value from reference 215
[c]K_I values taken from reference 217

About 25 years ago, there was an obscure report that suggested ascorbic acid was an inhibitor for α-amylases.[218] Abell et al.[219] recently studied a series of ascorbic acid derivatives and ascorbic acid analogs as inhibitors of several α-amylases from different sources. They found that both ascorbic acid (the l-isomer) and isoascorbic acid (the d-isomer) (Figure 7.17) were potent inhibitors of malt α-amylase. At 5 mM isoascorbic acid inhibited bacterial and fungal α-amylases to the extent of 98–99%, and inhibited pancreatic and salivary α-amylases 100%. Various substitutions of fatty acids and acetyl groups at C-6 of ascorbic acid did not appreciably affect inhibition by either ascorbic acid or isoascorbic acid. The introduction of methyl groups at O-2 or O-3 and reduction of the C-2—C-3 double bond decreased the inhibitory potency to 29% and 4%, respectively. Substitution at O-6 or the formation of an acetone ketal derivative at O-5 and O-6 did not greatly change the potency. A kinetic study of the ascorbic acid compounds showed that they were competitive inhibitors and that the K_I for barley malt α-amylase was 43 μM. The enediol structure was critical for inhibition; dihydroxyfumaric acid was also an inhibitor.[220a]

Uchida et al.[220b] reported the synthesis of 6^{III}-deoxy maltodextrins by *B. macerans* CGTase transglycosylation reactions between mono-6-O-*p*-toluene sulfonyl cyclomaltohexaose and maltose. The best inhibitors for human salivary α-amylase and human pancreatic α-amylase were 6^{III}-deoxy maltotetraose and 6^{III}-deoxy maltopentaose.

The most potent inhibitors of α-amylases to date are acarbose analogs in which maltohexaose (G6) and maltododecaose (G12) have been added to the C4-hydroxyl group at the non-reducing end cyclohexitol of acarbose, by the *B. macerans* CGTase catalyzed reaction of cyclomaltohexaose with acarbose.[220c] The maltohexaose analog was 1.6- to 2.2-times more potent than the maltododecaose analog. All of the

K_I values, however, were in the nanomolar range, with the K_I for the maltohexaosyl acarbose, 33 nM for *A. oryzae* α-amylase, 37 nM for *B. amyloliquefaciens* α-amylase, 14 nM for human salivary α-amylase and 7.0 nM for porcine pancreatic α-amylase. These inhibition constants represented relative inhibitory potency over acarbose of 8200-, 351-, 90- and 114-times, respectively for the four α-amylases.[220d]

VII. Action of Phosphorylase and Starch Lyase

1. Plant Phosphorylase

Phosphorylase is a starch degrading enzyme produced by many plants. It is an exo-acting enzyme that removes single glucosyl units from the non-reducing ends of starch chains by reaction with inorganic phosphate (P_i) to give α-d-glucopyranose 1-phosphate (α-Glc 1-P) according to the following reaction:

$$P_i + \underset{\text{starch chain}}{\text{G-G-G-G-G}} \cdots \underset{\text{synthetic}}{\overset{\text{degradative}}{\rightleftarrows}} \underset{\alpha\text{-Glc-1-p}}{\text{G-P}} + \underset{\text{so-called primer}}{\text{G-G-G-G-}} \cdots$$

When phosphorylase reacts with amylopectin, the action of the enzyme stops when the chain is degraded down to the vicinity of the α-(1→6) branch linkage. The result is formation of a limit dextrin plus Glc 1-P. The limit dextrin has four glucosyl units in the A-chains that have the α-(1→6) branch linkage and four glucosyl units in the B-chains to which the A-chains are attached.[221] The limit dextrin has the structure shown in Example 7.2, in which the circles represent glucosyl units, the horizontal lines α-(1→4) linkages and the bent vertical lines α-(1→6) branch linkages.

potato phosphorylase limit dextrin

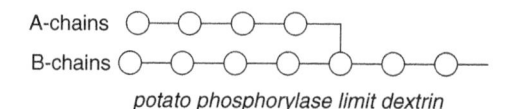

When Hanes[222] first studied potato phosphorylase, it was thought to be the enzyme involved in the biosynthesis of starch chains. Hanes observed that starch chains could be elongated *in vitro* by the transfer of the glucosyl group from Glc 1-P to the non-reducing ends of starch molecules or a maltodextrin, which were absolutely required for synthesis. It was from similar observations that Cori and Cori,[223] and Swanson and Cori,[224] studying glycogen phosphorylase, developed the hypothesis of the requirement of a primer molecule for the biosynthesis of polysaccharides. It was later shown that phosphorylase was a degradative enzyme, rather than a synthetic enzyme, and catalyzed the reaction of P_i with C-1 of the non-reducing end glucosyl unit of a polysaccharide chain to form α-Glc 1-P.[225] The synthetic reaction was the reverse of the degradative reaction, as indicated in the reaction above. When the phosphorylase reaction was conducted in the synthetic direction by starting with α-Glc 1-P and a starch primer chain, the reaction added glucosyl residues to the non-reducing ends of the chains. The reaction, however, would slow down relatively rapidly and stop after the addition of only a few glucosyl residues. It was shown that the equilibrium ratio of P_i to α-Glc 1-P at pH 6.8 was 3.6,[226] and that the concentration of P_i in plant tissue was 7.5-times higher than that of α-Glc 1-P.[227] So, the *in vivo* conditions for the synthesis of starch by phosphorylase are quite unfavorable, and it is now recognized that phosphorylase catalyzes the degradation reaction rather than the synthetic reaction.

Because the reaction is degradative, it requires a starch chain, and the product of a single event is α-Glc 1-P and a starch chain that has had one or more glucosyl residues removed. It is, therefore, not surprising that the reaction in the reverse synthetic direction also requires a starch chain or so-called primer. The primer hypothesis for the synthesis of polysaccharide was, thus, developed from an enzyme that was actually a degradative enzyme and not a synthetic enzyme. Nevertheless, the primer-dependent mechanism has been incorrectly assumed for the biosynthesis of starch for over 60 years.

The mechanism of cleavage of the α-(1→4) linkage by phosphorylase to give α-Glc 1-P might be expected to go by a double displacement, i.e. a covalent glucosyl–enzyme intermediate, to give a product with a retained α-configuration. Although this intermediate has never been demonstrated for the plant starch phosphorylases, it has been found for bacterial sucrose phosphorylase.[228] Thus, it would be expected that phosphorylase cleaves the α-(1→4) linkage of non-reducing end glucosyl units to give a β-glucosyl-enzyme intermediate. The P_i then attacks C-1 of the β-glucosyl-enzyme intermediate, releasing α-Glc 1-P.

Potato phosphorylase is a relatively large enzyme with a molecular weight of 207 000.[229] It contains two moles of pyridoxal 5-phosphate per mole of enzyme that are essential for activity. Potato phosphorylase consists of four subunits, two with molecular weights of 40 000 and two with molecular weights of 60 000.[230] Sweet potato phosphorylase was shown to have a similar size of 220 000 with two subunits of 110 000 each containing one pyridoxal 5-phosphate.[231] Potato phosphorylase is inhibited by arsenate (AsO_4^{-3}), which is chemically similar to phosphate (PO_4^{-3}) in size, geometry and charge. Arsenate can replace phosphate and form glucose 1-arsenate, which has a very transient existence and is quickly hydrolyzed. Bisulfite is a competitive inhibitor of Glc 1-P, P_i and AsO_4^{-3}.[232] The K_I is 1.7 mM for potato phosphorylase compared with a Km of 2.5 mM for the formation of α-Glc 1-P. Glucose and several glucose derivatives, (e.g. α-D-glucopyranosyl fluroride, α-D-glucopyranosylamine, 1-thio-D-glucopyranose, 5-thio-D-glucopyranose and nojirimycin) are competitive inhibitors for the binding of α-Glc 1-P to the active-site of potato phosphorylase.[233]

2. Starch Lyase

Starch lyase, discovered in 1993 in red seaweeds[234] and located in the chloroplasts,[235] has been postulated to be involved in starch metabolism in red seaweeds. It is a single polypeptide with a molecular weight of 111 000, has a wide optimum pH range of 3.5 to 7.5 and an optimum temperature of 50°C. It is an exo-acting enzyme that degrades maltose, maltodextrins, amylose, amylopectin and glycogen from the non-reducing end to form 1,5-anhydro-D-fructose. The enzyme is highly specific for cleaving α-(1→4) linkages. When a linear α-(1→4) glucan was the substrate, 1,5-anhydro-D-fructose

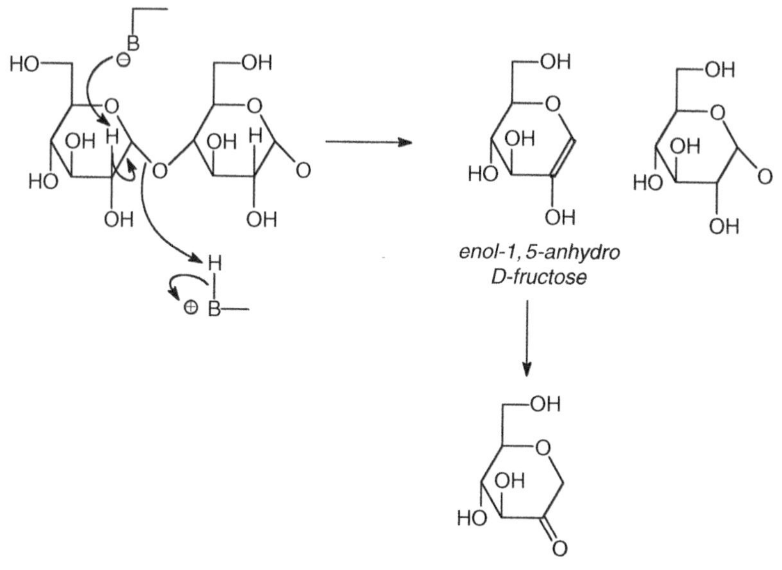

Figure 7.20 Mechanism for the formation of 1,5-anhydro-D-fructose from starch by the exo-acting starch lyase.

was successively formed until only one glucosyl unit remained. When the branched substrate 6^2-α-maltosyl maltotriose was the substrate, one glucosyl unit was removed from the A-chain to give 1,5-anhydro-D-fructose and 6^2-α-glucopyranosyl maltotriose. The glucosyl unit attached to the 4-position of the 4,6-branched unit was not removed.[236]

The mechanism of cleavage of the α-(1→4) linkage by starch lyase involves removal of the proton on C-2, with subsequent formation of a double bond between C-2 and C-1, and simultaneous displacement of the protonated glycosidic oxygen atom. This results in the release of the starch chain with one less glucosyl unit and the formation of the enol of 1,5-anhydro-D-fructose (Figure 7.20).

VIII. Enzymic Characterization of Starch Molecules

Enzymes are used in the determination of polysaccharide structures. Because enzymes have specificity, i.e. they produce specific products of low molecular weight and can cleave specific kinds of bonds; they can be used to determine the fine structure of starch. A first premise, however, is that the action patterns and specificity must need to be thoroughly investigated and elucidated before reliable information can be obtained about the structure of the polysaccharide or oligosaccharide being studied.

In its simplest application, the use of an enzyme to characterize a polysaccharide involves formation, separation and identification of the products formed. The products can be both of low molecular weight (monosaccharides and oligosaccharides) and high-molecular weight fragments formed from the polysaccharide (limit dextrins and resistant polysaccharide fragments). From determination of the structures of the products, the more complex structures from which they were derived can be inferred or deduced. Frequently, enzymes are used to determine the 'fine' structure after the 'gross' structural features have been determined by other procedures. Sometimes two enzymes can be used in a sequential manner, analyzing the products after each reaction. At other times, two enzymes can be deliberately used together to determine some specific structural features of the polysaccharide.

Use of enzymes can lead to both qualitative and quantitative data, depending on the types of analyses made. Quantitative data has been obtained by the use of reducing value determinations and the determination of the amount of a specific product, such as D-glucose, by using the quantitative glucose oxidase method of analysis.[237] Other specific products can be determined by using quantitative thin-layer chromatography to separate and quantitate linear and branched maltodextrins of DP 2–12,[238] and high pressure liquid chromatography (HPLC) can separate 20 to 40 or more maltodextrins.[239] Although a large number of oligosaccharides can be separated, their identity, structure and quantitative amounts are not readily obtained. The DP of an individual saccharide can usually be inferred by counting, starting from glucose or maltose or some other known compound whose retention time is known; however, the presence or absence of branch linkages in the dextrins affect their retention time and can throw the counting off by giving two or more peaks for a single DP value. The usual method of detection of carbohydrates in a HPLC analysis is refractive index. This method is neither quantitative nor very sensitive. Another method that has gained popularity in recent years is the pulse-amperometric detection system (PAD). Neither of these detection systems, however, are quantitative; equal weights of oligosaccharides give a response that is inversely related to the DP of the dextrin, i.e. as the size of the dextrin is increased, the size of the response is decreased.[240–242] These quantitative HPLC problems for determining maltodextrins have been overcome by the use of immobilized glucoamylase in a post-column reaction.[242] The maltodextrins are converted to glucose that can then be quantitatively determined by PAD or by a glucose electrode.

The major enzymes that have been used for studying structures of starch polymers and fragments from them are the endo-acting α-amylases, the exo-acting glucoamylase and β-amylases, and the debranching enzymes, isoamylases and pullulanases. These enzymes have varied and diverse specificities that have been extensively studied (see previous sections).

1. Determination of the Nature of the Branch Linkage in Starch

Although partial acid-catalyzed hydrolysis had ascertained that the major linkage in starch was α-(1→4), with lesser amounts of α-(1→6) branch linkages, controversies developed over the possible presence of other types of linkages.[243–245] α-(1→3), β-(1→4), and β-(1→6) linkages were inferred from the formation of specific disaccharides that contained the linkages, viz., nigerose, cellobiose and gentiobiose, that were produced in small amounts during acid-catalyzed hydrolysis. It is now recognized that these disaccharides were produced as artifacts from acid-catalyzed reversion reactions.[246] Definitive proof of the nature of the branch linkage in starch was obtained by French,[246] who used highly purified human salivary and porcine pancreatic α-amylases to produce a series of oligosaccharides containing both α-(1→4) and α-(1→6) linkages. Kainuma and French[17,18] determined the structures of these oligosaccharides and showed that they were maltodextrins that had one or more α-(1→6) branch linkages. Enzymes usually do not introduce anomalous linkages in their products, especially if the concentrations of the products are kept low, and no anomalous linkages were found. These studies by Kainuma and French closed the door on the possibility of the presence of anomalous linkages in starch.

Structures of the α-limit dextrins that are obtained from the action of porcine pancreatic α-amylase on amylopectin are shown in Figure 7.3. The smallest α-limit dextrin is 6^3-α-D-glucopyranosylmaltotetraose. The action of B. amyloliquefaciens α-amylase on waxy maize starch (amylopectin) produced a branched pentasaccharide, 6^2-α-maltosyl maltotriose,[248,249] isomeric with the porcine pancreatic α-amylase branched pentasaccharide, 6^3-α-maltosyl maltotetraose. The structures of both of these α-limit dextrins, produced by different α-amylases, are very important in that they both display the structural features of starch-branching that occurs in amylopectin. Treatment of the B. amyloliquefaciens α-amylase limit dextrin with glucoamylase gave two products, 6^2-α-D-glucopyranosylmaltotriose and 6^2-α-D-maltosylmaltose. Both products were subsequently converted to panose (6^2-D-glucopyranosylmaltose) by glucoamylase. Action of a debranching enzyme on 6^2-α-maltosyl maltotriose confirmed the structure by producing maltose and maltotriose (Figure 7.21).

The doubly branched dextrins, $6^4,6^6$-di-α-D-glucopyranosylmaltohexaose and $6^3,6^5$-di-α-D-glucopyranosylmaltopentaose, isolated by Kainuma and French[17] after action of porcine pancreatic α-amylase on waxy maize starch, established that the α-(1→6) branch linkages could be as close as having one glucosyl unit between them. No saccharides were found that indicated that the branch linkages could be adjacent to each other.

In a similar use of α-amylase, Parrish and Whelan[249] treated potato starch with crystalline human salivary α-amylase and obtained a phosphorylated maltotetraose that had previously been reported by Posternak[250] and that was the smallest phosphodextrin formed. They determined its structure to be 6^3-phosphomaltotetraose, similar in structure to the smallest α-limit dextrin, 6^3-α-D-glucopyranosylmaltotriose, formed by this enzyme and porcine pancreatic α-amylase.

2. Identification and Structure Determination of Slightly Branched Amyloses

Initially, the action of maltose β-amylase on amylose gave complete conversion into maltose,[235] and amylose was considered to be a completely linear α-(1→4) glucan. However, when highly purified crystalline sweet potato β-amylase was used, the

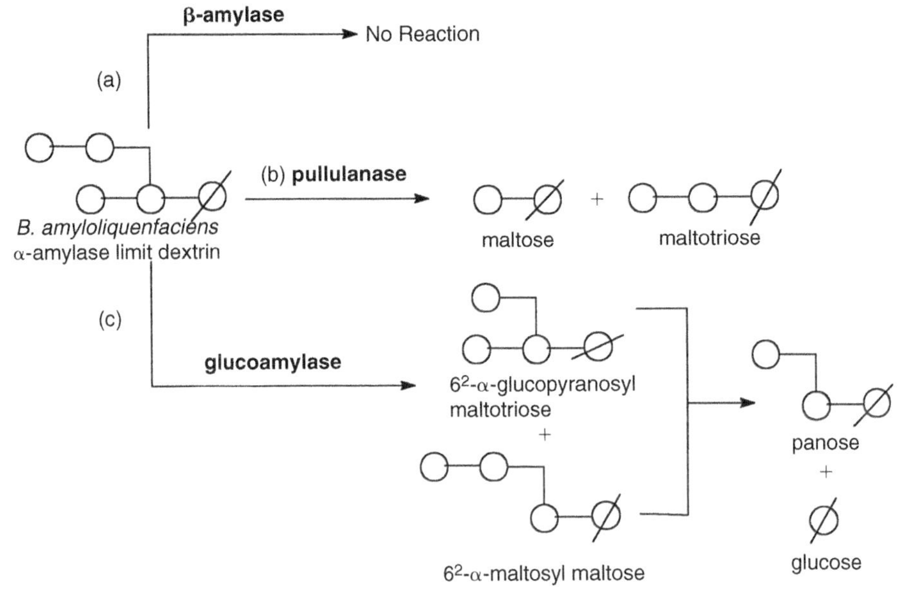

Figure 7.21 Determination of the structure of Bacillus amyloliquefaciens α-amylase limit dextrin, using enzymes: no reaction with β-amylase; (b) reaction with pullulanase to give maltose + maltotriose; (c) reaction of glucoamylase to give two tetrasaccharides, both of which are eventually converted into panose + glucose. Analysis of the reactions can be made by thin layer chromatography[239].

conversion into maltose was incomplete, at only 70%.[251–253] The resistant material gave an iodine color that was dark blue, similar to amylose. The β-amylase resistance of the native amylose fraction was due to the presence of a small amount of α-(1→6) branch linkages, which were demonstrated by the action of a debranching enzyme. After treatment of the amylose with a yeast isoamylase, the percentage conversion into maltose by β-amylase was increased from 70% to 90%,[254] and treatment of the β-amylase limit dextrin with isoamylase increased the yield of maltose to 95%.[255] Treatment of the amylose with pullulanase, followed by β-amylase gave quantitative conversion to maltose.[256]

Hizukuri et al.[257] studied the structures of the branched amyloses. They treated potato amylose with pullulanase and with Pseudomonas isoamylase, and found that about 30% of the branch linkages were hydrolyzed and that the average DP dropped from 6340 to 1470. The amylose was completely converted into maltose when a mixture of β-amylase and pullulanase was used. Hizukuri et al.[257] interpreted the incomplete debranching with pullulanase as due to retrogradation of the amylose during the reaction. The incomplete hydrolysis of the branch linkages by isoamylase was interpreted as due to the presence of branch chains that contained only two glucosyl units. Takeda et al.[258] determined that Pseudomonas isoamylase liberated short (maltotetraose) to long amylodextrin chains of DP >100 from potato amylose. In another study, Takeda et al.[259] showed that the branch chains of slightly branched amyloses obtained from starches of various botanical origins had distinct characteristics of molecular size, inner-chain lengths and number of chains per molecule, and a different percentage

Table 7.3 Properties of slightly branched amyloses from different botanical sources[259]

	Maize	Rice	Wheat	Water chestnut	Lily bulb	Tapioca	Sweet potato
Number average d.p.	960	1100	1290	1690	2300	2660	3280
% branched amylose	44	31	27	34	39	42	70
% unbranched amylose	56	69	73	66	61	58	32
Average chain length	335	320	270	375	475	340	335
Number of chains/molecule	2.9	3.5	4.8	4.5	4.9	7.8	9.8

of branched and unbranched molecules. These characteristics for branched amyloses from seven botanical sources are summarized in Table 7.3.

A wheat starch sample had an amylose that was the least branched, with only 27% of the molecules having a branch linkage. These branched amyloses had approximately five branch chains per molecule with an average DP of 270. The most highly-branched amylose was that obtained from sweet potato starch in which 70% of the molecules were branched, with 10 chains per molecule and an average DP of 335.

Hizukuri[260] distinguished the slightly branched amyloses from amylopectin as having: (a) a lower percentage of α-(1→6) linkages per molecule; (b) average chain lengths of their β-amylase limit dextrins of 60–200 glucosyl units; (c) capability to form precipitable complexes with 1-butanol; (d) iodine affinities of their β-limit dextrins similar to, or only slightly less than, those of unbranched amyloses; and (e) molecular weights of their β-limit dextrins that are only slightly lower than those of the starting amylose molecules.

3. Formation of β-Amylase Limit Dextrins of Amylopectin and Determination of their Fine Structure

The action of maltose β-amylase stops when the enzyme nears the α-(1→6) branch linkages, which cannot be bypassed. With most amylopectins, β-amylase converts 50–60% of the molecule into maltose, with the remaining material being a high molecular weight β-limit dextrin. There are four structures that result around the outermost branch linkages, depending on whether the two types of chains (the A-chains and the B-chains) have an odd or an even number of glucosyl units. These structures are represented in Example 7.3.

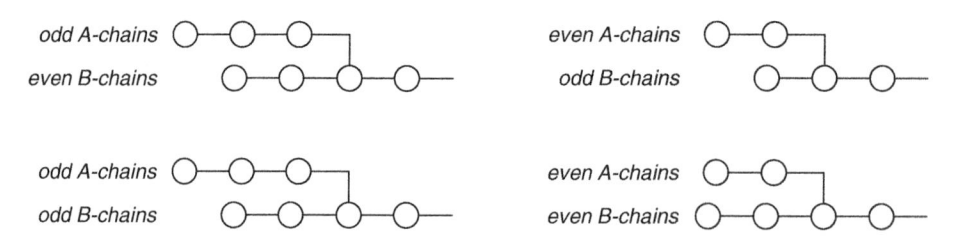

odd A-chains
even B-chains

even A-chains
odd B-chains

odd A-chains
odd B-chains

even A-chains
even B-chains

The β-amylase removes all glucosyl units that are part of the outer chains of amylopectin, leaving the inner part intact. The outer peripheral chains, thus, represent approximately 50% of the amylopectin molecule. The relatively large amount of maltose that is formed tends to inhibit the enzyme, slowing the reaction down and giving incomplete formation of the limit dextrin. To obtain complete β-amylolysis and a real β-limit dextrin, this inhibition must be overcome. This is most easily accomplished by conducting the reaction in a dialysis bag, and continuously changing the dialysis buffer to remove the maltose from the reaction site inside the bag. After reaction the β-limit dextrin, which is retained inside the dialysis bag, can be precipitated with 2 volumes of ethanol.

The fine structure of amylopectin has been studied by determining the structure of the β-limit dextrin. Theoretical considerations predict three kinds of chains: A-chains that have no chains attached to them; B-chains that have one or more chains attached by α-(1→6) linkages; and a C-chain that is essentially a B-chain terminated by a reducing end glucose unit. Marshall and Whelan[261] developed a method for determining the relative numbers of A- and B-chains in amylopectin that uses the action of isoamylase and pullulanase on the β-limit dextrin. For the β-amylase limit dextrins, half of the A-chains are maltosyl units and the other half are maltotriosyl units (see the diagram above). Pullulanase hydrolyzes the α-(1→6) linkages of both maltosyl and maltotriosyl chains, giving release of all A- and B-chains. Reaction with isoamylase hydrolyzes the α-(1→6) linkage of maltotriosyl and longer branch chains, thus releasing all B-chains and half of the A-chains. Action of pullulanase gives a reducing value X; action of isoamylase gives a reducing value Y. The difference between the two reducing values $(X - Y)$ gives the reducing value for half of the A-chains, or Z. So, the reducing value for the B-chains is $Y - 2Z$ and the reducing value for the hydrolysis of the A-chains is $2Z$. Therefore, the relative ratio of A- to B-chains is $(2Z) \div (Y - 2Z)$.

Reducing values following action of isoamylase and pullulanase on the β-limit dextrins of several amylopectins from different botanical sources were measured and the A:B chain ratios determined as described above. The ratio of A:B chains for wheat and rice amylopectins was 1.5, for maize amylopectin 2.0, and for waxy maize and waxy sorghum amylopectins 2.6. Thus, every B-chain for the amylopectins examined carried a range of 1.5 to 2.6 A-chains attached by an α-(1→6) linkage. The procedure does not take into account the possibility of steric effects of buried A-chains that cannot be completely reduced to maltosyl or maltotriosyl units by β-amylase and are, therefore, counted as B-chains. They, however, have a very small affect on the relative ratio. The method takes advantage of the specificities of β-amylase, isoamylase and pullulanase and is relatively simple to perform. The reducing-value analysis can be performed on a micro scale.[237]

Further characterization of the structure of amylopectin molecules involves determination of the average unit chain length. This can be determined enzymically on milligram quantities, using the simultaneous action of β-amylase and pullulanase.[262] All branch linkages are hydrolyzed by pullulanase, which in turn renders all unit chains susceptible to the action of β-amylase. Beta-amylase hydrolyzes the chains containing an even number of glucosyl units entirely to maltose, and it hydrolyzes the chains containing an odd number of glucosyl units to maltose and one molecule of glucose. If it is assumed that the number of even and odd chains are the same (i.e. they have been synthesized on a random basis), there will be one molecule of glucose formed for every two chains. Specific measurement of the amount of glucose formed then permits the determination of the average chain length. The analysis is performed by dissolving the amylopectin and determining the total carbohydrate on an aliquot,[237] and then reacting the amylopectin with a mixture of pullulanase and β-amylase. The amount of glucose formed in this reaction is then quantitatively determined by using glucose oxidase,[237] and the average chain length can be computed by the following:

$$\text{Average chain length} = \frac{\text{total carbohydrate as glucose (from phenol-sulfuric acid)}}{2 \times \text{the amount of glucose (from glucose oxidase)}}$$

Using this procedure, the average chain length for potato amylopectin was found to be 23 and for waxy maize and waxy sorghum amylopectin 20.

Bertoft[263] has made extensive use of alpha- and beta-amylases and phosphorylase, along with differential precipitation of the resulting amylodextrins, in determining the fine structures of amylopectins.[264] *B. amyloliquefaciens* α-amylase limit dextrins were prepared to determine the fine structures of amylopectins from waxy maize starch,[265] potato starch,[266] and the demonstration of the structural heterogeneity of waxy rice starch.[267,268] Additional studies on the relationship between the density of the α-(1→6) branch linkages and the crystalline structures of A- and B-types of x-ray diffraction patterns of starch granules for different maize starch mutants were also obtained by using α-, β-amylase and phosphorylase degradations, followed by differential separations and debranching with isoamylase hydrolysis.[268,269]

Bertoft also determined the nature of the categories of amylopectin chains of waxy maize and waxy potato starches by isoamylase debranching of the *B. amyloliquefaciens* α-amylase limit dextrins.[270] The external chain lengths and the chain distributions of phosphorylase- and β-amylase limit dextrins were determined and classified into characteristic categories. For waxy potato starch, the shortest A-chains had DP values of 6–8 and the long A-chains had DP values ≥33, and were found to be located in the amorphous regions of the granules. A two-dimensional backbone model was hypothesized from these studies in which the clusters of α-(1→6) branch linkages are connected to a backbone that extends in a nearly perpendicular direction and are found in the amorphous regions of the granules. The hypothesized model was further shown to involve a possible super-helical structure that is made up of single amylopectin molecules.[270]

IX. References

1. Fischer EH, Stein EA. *Biochem Preps*. 1961;8:27.
2. Meyer KH, Fischer EH, Bernfeld P. *Helv Chim Acta*. 1974;30:64.
3. Caldwell ML, Adams M, Keng JT, Toralballa GC. *J Am Chem Soc*. 1952;74:4033.
4. Fischer EH, Duckert F, Bernfeld P. *Helv Chim Acta*. 1950;33:1060.
5. Olavarris J, Torres M. *J. Biol. Chem*. 1962;237:1746.
6. Stein EA, Fischer EH. *Biochem. Preps*. 1961;8:34.
7. Welker NE, Campbell LL. *Biochemistry*. 1967;6:3681.

8. Welker NE, Campbell LL. *J. Bacteriol.* 1967;94:1124.
9. Fukumoto J, Yamamoto T, Ichikawa K. *Proc. Japan Acad.* 1951;27:352.
10. Campbell LL. *J. Am. Chem. Soc.* 1954;76:5256.
11. Markovitz A, Klein HP, Fischer EH. *Biochim. Biophys. Acta.* 1956;19:267.
12. Fischer EH, De Montmollin R. *Helv. Chim. Acta.* 1951;34:1987.
13. Robyt JF, French D. *Arch. Biochem. Biophys.* 1967;122:8.
14. Robyt JF, French D. *Arch. Biochem. Biophys.* 1963;100:451.
15. Thoma JA, Rao GVK, Brothers C, Spradlin J. *J. Biol. Chem.* 1971;246:5621.
16. MacGregor EA, MacGregor AW. *Carbohydr. Res.* 1985;142:223.
17. Kainuma K, French D. *FEBS Lett.* 1968;5:257.
18. Kainuma K, French D. *FEBS Lett.* 1969;6:182.
19. Robyt JF, French D. *J. Biol. Chem.* 1970;245:3917.
20. Buisson G, Duee E, Haser R, Payan F. *EMBO J.* 1987;6:3909.
21. Qian M, Haser R, Buisson G, Duée E, Payan F. *Biochemistry.* 1994;33:6284.
22. Prodanov E, Seigner C, Machis-Mouren G. *Biochem. Biophys. Res. Commun.* 1984;122:75.
23. Seigner C, Prodanov E, Machis-Mouren G. *Biochim. Biophys. Acta.* 1987;913:200.
24. Suganuma T, Matsuno R, Ohinishi M, Hiromi K. *J. Biochem. (Tokyo).* 1978;84:293.
25. Abdullah M, French D, Robyt JF. *Arch. Biochem. Biophys.* 1969;144:595.
26. Suetsugu N, Koyama S, Takeo K, Kuge T. *J. Biochem. (Tokyo).* 1974;76:57.
27. Balls AK, Walden MK, Thompson RR. *J. Biol. Chem.* 1948;173:9.
28. Mikami B, Shibata T, Hirose M, Aibara S, Sato M, Katsube Y, Morita Y. *J. Biochem. (Tokyo).* 1992;112:541.
29. Piguet A, Fischer EH. *Helv. Chim. Acta.* 1952;36:257.
30. Meyer KH, Spahr PF, Fischer EH. *Helv. Chim. Acta.* 1953;36:1924.
31. Robyt JF, French D. *Arch. Biochem. Biophys.* 1964;104:338.
32. Higashihara M, Okada S. *Agric. Biol. Chem.* 1974;38:1023.
33. Ikkasaki Y. *Agric. Biol. Chem.* 1976;40:1515.
34. Shinke R, Kunimi Y, Mshira H. *J. Ferment. Technol.* 1975;53:698.
35. Pazur JH, Ando T. *J. Biol. Chem.* 1959;234:1966.
36. Ueda S. *Bull. Agr. Chem. Soc. Jap.* 1956;20:148.
37. Savel'ev AN, Sergeev VR, Firsov IM. *Biochem. (USSR).* 1982;47:330.
38. Hiromi K. *Biochem. Biophys. Res. Commun.* 1970;40:1.
39. Tsujisaka Y, Fukumoto J, Yamamoto T. *Nature.* 1958;181:94.
40. Meagher MM, Nikolov ZL, Reilly PJ. *Biotechnol. Bioeng.* 1989;34:681.
41. Kato M, Hiromi K, Morita Y. *J Biochem. (Tokyo).* 1974;75:563.
42. Tanaka A, Yamashita T, Ohinishi M, Hiromi K. *J. Biochem. (Tokyo).* 1983;93:1037.
43. Bailey JM, French D. *J. Biol. Chem.* 1956;266:1.
44. French D, Youngquist RW. *Stärke.* 1963;12:425.
45. Savel'ev AN, Tsendina ML, Firsov IM. *Biokhimiya.* 1985;50:1743.
46. Robyt JF, Ackerman RJ. *Arch. Biochem. Biophys.* 1971;145:105.
47. Zhou J, Baba T, Ikkano T, Kobayashi S, Arai Y. *FEBS Lett.* 1989;255:37.
48. Fogarty WM, Bourke AC, Kelly CT, Doyle EM. *Appl. Microbiol. Biotechnol.* 1994;42:198.
49. Wako K, Hashimoto S, Kubomura S, Yolota K, Aikawa K, Kanaeda J. *Denpun. Kagaku.* 1979;26:175.

50. Ikkasaki Y. *Agric. Biol. Chem.* 1985;49:1091.
51. Takasaki Y. *Agric. Biol. Chem.* 1983;47:2193.
52. Morgan FJ, Priest FG. *J. Appl. Bacteriol.* 1981;50:107.
53. Kainuma K, Kobayashi S, Ito T, Suzuki S. *FEBS Lett.* 1972;26:281.
54. Kainuma K, Wako K, Kobayashi S, Nogami A, Suzuki S. *Biochim. Biophys. Acta.* 1975;410:333.
55. Taniguchi H, Odashima F, Igaashi M, Maruyama Y, Nakamura M. *Agric. Biol. Chem.* 1982;46:2107.
56. Fogarty WM, Bealin-Kelly F, Kelly CI, Doyle EM. *Appl. Microbiol. Biotechnol.* 1991;36:184.
57. Nachum R, Bartholomew JW. *Bacteriol. Proc.* 1969;137:137.
58. Good WA, Hartman PA. *J. Bacteriol.* 1970;104:601.
59. Onishi H. *J. Bacteriol.* 1972;109:570.
60. Laderman KA, Davis BR, Krutzsch HC, Lewis MS, Griko YV, Privalow PL, Anfinsen CB. *J. Biol. Chem.* 1993;286:24394.
61. Conrad B, Hoang V, Polley A, Hofemeister J. *Eur. J Biochem.* 1995;230:481.
62. Hobson PN, Whelan WJ, Peat S. *J. Chem. Soc.* 1951:1451.
63. Bender H, Wallenfels K. *Biochem. Z.* 1961;334:79.
64. Abdullah M, Cately BJ, Lee EYC, Robyt JF, Wallenfels K, Whelan WJ. *Cereal Chem.* 1966;43:111.
65. Harda T, Yokobayashi K, Misaki A. *Appl. Microbiol.* 1968;16:1493.
66. Yokobayashi K, Misaki A, Harada T. *Agric. Biol. Chem.* 1969;33:625.
67. Yokobayashi K, Misaki A, Harada T. *Biochim. Biophys. Acta.* 1970;212:458.
68. Harada T, Misaki A, Akai H, Yokobayashi K, Sugimoto K. *Biochim. Biophys. Acta.* 1972;268:497.
69. Kainuma K, Kobayashi S, Harada T. *Carbohydr. Res.* 1978;61:345.
70. James MG, Robertson DS, Myers AM. *The Plant Cell.* 1995;7:417.
71. Gunja-Smith Z, Marshall JJ, Smith EE, Whelan WJ. *FEBS Lett.* 1970;12:96.
72. Ueda S, Yagisawa M, Sato Y. *J. Ferment. Technol.* 1971;49:552.
73. Sato HH, Park YK. *Starch/Stärke.* 1980;32:132.
74. Spenser-Martin I. *Appl. Environ. Microbiol.* 1982;44:1253.
75. Ishizaki Y, Taniguchi H, Maruyama Y, Nakamura M. *Agric. Biol. Chem.* 1983;47:771.
76. Castro GR, Garcia GF, Fiñeriz J. *Appl. Bacteriol.* 1992;73:520.
77. Katsutoshi A, Saeki K, Ito S. *J. Gen. Microbiol.* 1993;139:781.
78. Lévêque E, Janecek S, Haye B, Belarbi A. *Enzyme Microb. Technol.* 2000;26:3.
79. Leuschner C, Antranikian G. *World J. Microb. Biotech.* 1995;11:95.
80. Sunna A, Moracci M, Rossi M, Antranikian G. *Extremeophiles.* 1997;1:2.
81. Bragger JM, Daniel RM, Coolbear T, Morgan HW. *Appl. Microbiol. Biotechnol.* 1989;31:556.
82. Legin E, Ladrat C, Godfroy A, Barbier G, Duchiron F. *C. R. Acad. Sci. Paris.* 1997;320:893.
83. Dong G, Vieille C, Savchenko A, Zeikus JG. *Appl. Environ. Microbiol.* 1997;63:3569.
84. Jorgensen S, Vorgias CE, Antranikian G. *J. Biol. Chem.* 1997;272:16335.

85. Lee JT, Kanai H, Kobayashi T, Akiba T, Kudo T. *J. Ferment. Bioeng.* 1996;82:432.
86. Kobayashi K, Kato M, Miura Y, Kettoku M, Komeda T, Iwamatsu A. *Biosci. Biotech. Biochem.* 1996;60:1720.
87. Koch R, Zablowski P, Spreinat A, Antranikian G. *FEMS Microbiol. Lett.* 1990;71:21.
88. Tachibana Y, Leclere LM, Fujuvara S, Takagi M, Imanaka T. *J. Ferment. Bioeng.* 1996;82:224.
89. Koch R, Spreinat A, Lemke K, Antranikian G. *Arch. Microbiol.* 1991;155:572.
90. Chung YC, Kobayashi T, Kanai H, Akiba T, Kudo T. *Appl. Environ. Microbiol.* 1995;61:1502.
91. Brown SH, Costantino HR, Kelly RM. *Appl. Environ. Microbiol.* 1990;56:1985.
92. Canganella F, Andrade CM, Antranikian G. *Appl. Microbiol. Biotechnol.* 1994;42:239.
93. Brown SH, Kelly RM. *Appl. Environ. Microbiol.* 1993;59:2614.
94. Schuliger JW, Brown SH, Baross JA, Kelly RM. *Mol. Marine Biol. Biotech.* 1993;2:76.
95. Rüdiger A, Jorgensen PL, Antranikian G. *Appl. Environ. Microbiol.* 1995;61:567.
96. Gantelet H, Duchiron F. *Appl. Microbiol. Biotechnol.* 1998;49:770.
97. Gantelet H, Ladrat C, Godfroy A, Barbier G, Durchiron F. *Biotechnol. Lett.* 1998;20:819.
98. Dong C, Vieille C, Zeikus JG. *Appl. Environ. Microbiol.* 1997;63:5377.
99. Erra-Pujada M, Debeire P, Durchiron F, O'Donohue MJ. *J. Bacteriol.* 1999;181:3284.
100. French D. *Adv. Carbohydr. Chem.* 1975;12:189.
101a. Nakamura A, Haga K, Yamane K. *FEBS Lett.* 1994;337:66.
101b. Lee S-B, Robyt JF. *Carbohydr. Res.* 2001;336:47.
102. Pulley AO, French D. *Biochem. Biophys. Res. Commun.* 1961;5:11.
103. French D, Pulley AO, Effenberger JA, Rougvie MA, Abdullah M. *Arch. Biochem. Biophys.* 1965;111:153.
104. Endo T, Ueda H, Kobayashi S, Nagai T. *Carbohydr. Res.* 1995;269:369. Endo T, Nagase H, Ueda H, Kobayashi S, Nagai T. *Chem. Pharm. Bull.* 1997;45:532.
105. Terada Y, Yanase M, Takata H, Takaha T, Okada S. *J. Biol. Chem.* 1997;272:15729.
106a. Robyt JF. In: Friedman RB, ed. *Biotechnology of Amylopectin Oligosaccharides*. Washington, DC: American Chemical Society 1991:98–110 *ACS Symp. Ser.* 458.
106b. Yoon S-H, Robyt JR. *Carbohydr. Res.* 2002;337:2245.
107. Kitahata S, Tsuyama N, Okada S. *Agric. Biol. Chem.* 1974;38:387.
108. Kitahata S, Okada S. *Denpun Kagaku.* 1982;29:13.
109. Kitahata S, Okada S. *Denpun Kagaku.* 1982;29:7.
110. Bender H. *Arch. Microbiol.* 1977;111:271.
111. Mori S, Hirose S, Takaichi O, Kitahata S. *Biosci. Biotech. Biochem.* 1994;58:1968.
112. Kobayashi S, Kainuma K, Suzuki S. *Carbohydr. Res.* 1978;61:229.
113. Stravn A, Granum PE. *Carbohydr. Res.* 1979;75:243.
114. French D, Levine ML, Norberg E, Nordin P, Pazur JH, Wild GM. *J. Am. Chem. Soc.* 1954;76:2387.
115. Norberg E, French D. *J. Am. Chem. Soc.* 1950;72:1202.

116a. Pasero L, Mazzev-Pierron Y, Abadie B, Chicheportiche Y, Machis-Mouren G. *Biochim. Biophys. Acta.* 1986;869:147.
116b. Yoon S-H, Fulton DB, Robyt JF. *Carbohydr. Res.* 2004;339:1517.
117. Levitzki A, Steer ML. *Eur. J. Biochem.* 1974;41:171.
118. Vallee BL, Stein EA, Sumerwell WN, Fischer EH. *J. Biol. Chem.* 1959;234:2901.
119. Farber GK, Petsko GA. *Trends Biochem. Sci.* 1990;15:228.
120. Qian M, Haser R, Payan F. *J. Mol. Biol.* 1993;231:785.
121. Kadziola A, Abe J, Svensson B, Haser R. *J. Mol. Biol.* 1994;239:104.
122. Matsuura Y, Kunusoki M, Harada W, Kakudo M. *J. Biochem. (Tokyo).* 1984;95:697.
123. Kubota M, Matsuura Y, Saki S, Katsube Y. *Protein Eng.* 1990;3:328.
124. Klein C, Schulz GE. *J. Mol. Biol.* 1991;217:737.
125. Jespersen HM, MacGregor EA, Sierks MR, Svensson B. *Biochem. J.* 1991;280:51.
126. Kossmann J, Visser RGF, Müller-Röber B, Willmitzer L, Sonnewald U. *Mol. Gen. Genet.* 1991;230:39.
127. Mizuno K, Kimura K, Arai Y, Kawasaki T, Shimada H, Baba T. *J. Biochem.* 1992;112:643.
128. Buisson G, Duee E, Haser R, Payan F. *EMBO J.* 1987;6:3909.
129. MacGregor EA. *Starch/Stärke.* 1993;45:232.
130. Mikami B, Sato M, Shibata T, Hirose M, Aibara S, Katsube Y, Morita Y. *J. Biochem.* 1992;112:541.
131. Mikami B, Degano M, Hehre EJ, Sacchettini JC. *Biochemistry.* 1994;33:7779.
132. Lineback DR, Russell IJ, Rasmussen C. *Arch. Biochem. Biophys.* 1969;134:539.
133. Saha BC, Zeikus JG. *Starch/ Stärke.* 1989;41:57.
134. Svensson B, Larsen K, Gunnarsson A. *Eur. J. Biochem.* 1986;154:497.
135. Stoffer B, Frandsen TP, Busk PK, Schneider P, Svendsen I, Svensson B. *Biochem. J.* 1993;292:197.
136. Belshaw NJ, Williamson G. *FEBS Lett.* 1990;269:350.
137. Belshaw NJ, Williamson G. *Biochim. Biophys. Acta.* 1991;1078:117.
138. Belshaw NJ, Williamson G. *Europ. J. Biochem.* 1993;211:717.
139. Williamson G, Belshaw NJ, Williamson MP. *Biochem. J.* 1992;282:423.
140. Hayashida S, Nakahara K, Kanlayakrit W, Hara T, Teramoto Y. *Agric. Biol. Chem.* 1989;53:143.
141. Hayashida S, Kuroda K, Ohta K, Kuhara S, Fukada K, Sakaki Y. *Agric. Biol. Chem.* 1989;53:923.
142. Evans R, Ford C, Sierks M, Nikolov Z, Svensson B. *Gene.* 1990;91:131.
143. Svensson B, Larsen K, Svensson I, Boel E. *Carlsberg Res. Commun.* 1983;48:529.
144. Neustroyev KN, Firsov LM. *Biokhimiya.* 1990;55:776.
145. Aleshin A, Golubev A, Firsov LM, Honzatko RB. *J. Biol. Chem.* 1992;267:19291.
146. Aleshin AE, Firsov LM, Honzatko RJ. *J. Biol. Chem.* 1994;269:15631.
147. Aleshin AE, Hoffman C, Firsov LM, Honzatko RB. *J. Mol. Biol.* 1994;238:575.
148. Juy M, Adolfo GA, Alzari PM, Poljak RJ, Claeyssens M, Béguin P, Aubert J-P. *Nature.* 1992;357:89.
149. Takahashi T, Tsuchida Y, Irie M. *J. Biochem.* 1978;84:1183.
150. Saha BC, Ueda S. *J. Ferment. Technol.* 1983;61:67.

151. Svensson B. *FEBS Lett.* 1988;230:72.
152. Villette JR, Krzewinski FS, Looten PJ, Sicard PJ, Bouquelet SJ-L. *Biotechnol. Appl. Biochem.* 1992;16:57.
153. Kimura K, Katoka S, Ishii Y, Takano T, Yamane Y. *J. Bacteriol.* 1987;169:4399.
154. Diderichsen B, Christiansen L. *FEMS Microbiol. Lett.* 1988;56:53.
155. Fujita M, Torigoe K, Nakada K, Tsusaki K, Kubota M, Sakai S, Tsujisaka Y. *J. Bacteriol.* 1989;171:1333.
156. Kitamoto N, Yamagata H, Kato T, Tsukagoshi N, Udaka S. *J. Bacteriol.* 1988;170:5848.
157. Kim C-H, Kwon S-T, Taniguchi H, Lee D-S. *Biochim. Biophys. Acta.* 1992;1122:243.
158. Fang T-Y, Lin L-L, Hsu W-H. *Enzyme Microbiol. Technol.* 1994;16:247.
159. Holm L, Lehtovaara AK, Hemaninki PM, Knowles JKC. *Protein Eng.* 1990;3:181.
160. Vihinen M, Helin S, Mäntsälä P. *Mol. Eng.* 1991;1:267.
161. Vihinen M, Ollikka P, Niskaner J, Meyer P, Suominen I, Karp M, Holm L, Knowles JKC, Mäntsälä P. *J. Biochem.* 1990;107:267.
162. Nagashima T, Toda S, Kitamoto K, Gomi K, Kumagai C, Toda H. *Biosci. Biotech. Biochem.* 1992;56:207.
163. Takase K, Matsumoto T, Mizuno H, Yamane K. *Biochim. Biophys. Acta.* 1992;1120:281.
164. Nakamura A, Haga K, Ogawa S, Kuwano K, Kimura K, Yamane K. *FEBS Lett.* 1992;296:37.
165. Sierks MR, Ford C, Reilly PJ, Svensson B. *Protein Eng.* 1990;3:193.
166. Matsui I, Ishikawa K, Miyairi S, Fukui S, Honda K. *Biochim. Biophys. Acta.* 1991;1077:416.
167. Svensson B, Sogaard M. *J. Biotechnol.* 1993;29:1.
168. Sierks MR, Ford C, Reilly PJ, Svensson, B. *Protein Eng.* 1989;2:621.
169. Sierks MR, Svensson B. *Biochemistry.* 1993;32:1113.
170. Frandsen TP, Dupont C, Lembek J, Stoffer B, Sierks MR, Honzatko RB, Svensson B. *Biochemistry.* 1994;33:13808.
171. Clarke AJ, Svensson B. *Carlsberg Res. Commun.* 1984;49:559.
172. Sierks MR, Svensson B. *Protein Eng.* 1992;5:185.
173. Evans R, Ford C, Sierks MR, Nikolov Z, Svensson B. *Gene.* 1990;91.131.
174. Svensson B, Sierks MR. *Carbohydr. Res.* 1992;227:29.
175. Otsuki J, Kobayashi K, Toi H, Aoyama Y. *Tet. Lett.* 1993;34:1945.
176. Kuriki T, Okada S. *Enzyme Chemistry and Molecular Biology of Amylases and Related Enzymes.* Boca Raton, FL: CRC Press; 1995 87:92.
177. Kuriki T, Imanaka T. *J. Biosci. Bioeng.* 1999;87:557.
178. Jespersen HM, MacGregor EA, Sierks MR, Svensson B. *Biochem. J.* 1991;280:51.
179. Boel E, Brady L, Brzozowski A, Derewenda Z, Dodson G, Jensen V, Petersen S, Swift H, Thim L, Woldike H. *Biochemistry.* 1990;29:6244.
180. Matsuura Y, Kusunoki M, Harada W, Kakudo M. *J. Biochem. (Tokyo).* 1984;95:697.
181. Janecek S, Svensson B, Henrissat B. *J. Mol. Evol.* 1997;45:322.
182. Jespersen HM, MacGregor EA, Henrissat B, Sierks MR, Svensson B. *J. Prot. Chem.* 1993;12:791.
183. Klein C, Schulz GE. *J. Mol. Biol.* 1991;217:737.

184. Kubota M, Matsuura Y, Sakai S, Katsube Y. *Denpun Kagaku*. 1991;38:141.
185. Larson CL, Vanmontfort R, Strokopytov B, Rozeboom HJ, Kalk KH, Devries GE, Penninga D, Dijkhuizen L, Dijkstra BW. *J. Mol. Biol.* 1994;236:590.
186. Svensson S, Jespersen H, Sierks MR, MacGregor EA. *Biochem. J.* 1989;264:309.
187. Juge N, Søgaard M, Chaix JC, Martin-Eauchlaire MF, Svensson B, Machis-Mouren G, Guo XJ. *Gene*. 1993;130:159.
188. Terashima M, Hosono M, Katoh S. *Appl. Microbiol. Biotechnol.* 1997;47:364.
189. Wind RD, Buitelaar RM, Dijkhuizen L. *Eur. J. Biochem.* 1998;253:598.
190. Ohdan K, Kuriki T, Takata H, Kaneko H, Okada S. *Appl. Environ. Microbiol.* 2000;66:3058.
191a. Tao BY, Reilly PJ, Robyt JF. *Biochim. Biophys. Acta*. 1989;995:214.
191b. Yoon S-H, Robyt JF. *Carbohydr. Res.* 2007;342:55.
192a. Mosi R, He S, Uitdehaag J, Dijkstra BW, Whithers SG. *Biochemistry*. 1997;36:9927.
192b. Lee S-B, Robyt Jf. *Carbohydr. Res.* 2001;336:47.
193. Jane J-I, Robyt JF. *Carbohydr. Res.* 1984;132:105.
194. Brown HT, Heron J. *J. Chem. Soc.* 1879;34:596.
195. Stamberg OE, Bailey CH. *Cereal Chem.* 1939;16:319.
196. Sandstedt RM, Gates RL. *Food Res.* 1954;19:190.
197. Leach HW, Schoch TJ. *Cereal Chem.* 1961;38:34.
198. Manners DJ. *Biochem. J.* 1971;123:1.
199. Rasper V, Perry G, Duitschaever CL. *Can. Inst. Food Sci. Technol.* 1974;7:166.
200. Smith JS, Lineback DR. *Stärke*. 1976;28:243.
201. Kimura A, Robyt JF. *Carbohydr. Res.* 1995;277:87.
202. Kim Y-K, Robyt JF. *Carbohydr. Res.* 1999;318:129.
203. Kim Y-K, Robyt JF. *Carbohydr. Res.* 2000;328:509.
204. Gallant DJ, Bouehet B, Buléon A, Perez S. *Eur. J. Clin. Nutr.* 1992;46(Suppl. 2):S3.
205. Valetudie J-C, Colonna P, Bouchet B, Gallant DJ. *Starch/Stärke*. 1993;45:270.
206. Tanaka T, Ishimoto E, Shimoaura Y, Taniguchi M, Oi S. *Agric. Biol. Chem.* 1987;51:399.
207. Buranakarl L, Ito K, Izaki K, Takahashi, H. *Enzyme Microbiol. Technol.* 1988;10:173.
208. Hollinger G, Marchessault RH. *Biopolymers*. 1975;14:265.
209. Robyt JF. In: Whistler RL, BeMiller JN, Paschall EF, eds. *Starch: Chemistry and Technology*. 2nd edn. New York, NY: Academic Press; 1984:108–110.
210. Strokopytov B, Penninga D, Roseboom HT, Kalk KH, Dijkhuizen L, Dijstra BM. *Biochemistry*. 1995;34:2234.
211. Brzozowski AM, Davis MJ. *Biochemistry*. 1997;36:10837.
212. Kang GJ, Kim MJ, Kim JW, Park KH. *J. Agric. Food Chem.* 1997;45:4168.
213. Park KH, Kim MJ, Lee HS, Han NS, Kim D, Robyt JF. *Carbohydr. Res.* 1998;313:235.
214. Kim MJ, Lee SB, Lee H-S, Lee S-Y, Baek J-S, Kim D, Moon T-W, Robyt JF, Park KH. *Arch. Biochem. Biophys.* 1999;371:277.
215. Lehmann J, Schmidt-Schuchardt M, Steck J. *Carbohydr. Res.* 1992;237:177.
216. Lehmann J, Marchis-Mouren G, Schlitz E, Schmidt-Schuchardt M. *Carbohydr. Res.* 1994;265:19.

217. Lehmann J, Schmidt-Schuchardt M. *Carbohydr. Res.* 1995;276:43.
218. Pella J-C, Verrier JV. *J. Ann. Technol. Agric.* 1974;23:151.
219. Abell AD, Ratcliffe MJ, Gerrard J. *Bioorg. Med. Chem. Lett.* 1998;8:1703.
220a. Gerrard JA, Prince MJ, Abell AD. *Bioorg. Med. Chem. Lett.* 2000;10:1575.
220b. Uchida R, Nasu A, Tokutake S, Kasai K, Tobe K, Yamaji N. *Carbohydr. Res.* 1998;307:69.
220c. Yoon S-H, Robyt JF. *Carbohydr. Res.* 2002;337:509.
220d. Yoon S-H, Robyt JF. *Carbohydr. Res.* 2003;338:1969.
221. Whelan WJ, Bailey JM. *Biochem. J.* 1954;58:560.
222. Hanes CS. *Proc. Roy. Soc.* 1940;B129:174.
223. Cori GT, Cori CF. *J. Biol. Chem.* 1939;131:397.
224. Swanson MA, Cori CF. *J. Biol. Chem.* 1948;172:815.
225. Stetten Jr D, Stetten MR. *Physiol. Revs.* 1960;40:513.
226. Trevelyan WE, Mann PFE, Harrison JS. *Arch. Biochem. Biophys.* 1952;39:419.
227. Liu T-T, Shannon JC. *Plant Physiol.* 1981;67:525.
228. Voet JG, Abeles RH. *J. Biol. Chem.* 1970;245:1020.
229. Lee Y-P. *Biochim. Biophys. Acta.* 1960;43:18.
230. Iwata S, Fukui T. *FEBS Lett.* 1973;36:222.
231. Ariki M, Fukui T. *Biochim. Biophys. Acta.* 1975;386:301.
232. Kamogawa A, Fukui T. *Biochim. Biophys. Acta.* 1973;302:158.
233. Ariki M, Fukui T. *J. Biochem.* 1977;81:1017.
234. Yu S, Kenne L, Pedersén M. *Biochim. Biophys. Acta.* 1993;1156:313.
235. Yu S, Pedersen M. *Planta.* 1993;191:137.
236. Yu S, Ahmad T, Kenne L, Pedersén M. *Biochim. Biophys. Acta.* 1995;1244:1.
237. Fox JD, Robyt JF. *Anal. Biochem.* 1991;195:93.
238. Robyt JF, Mukerjea R. *Carbohydr. Res.* 1994;251:187.
239. Ammeraal RN, Delgado GA, Tenbarge FL, Friedman RB. *Carbohydr. Res.* 1991;215:179.
240. Koizumi K, Kubota Y, Tanimoto T, Okada Y. *J. Chromatog.* 1989;464:365.
241. Koizumi K, Fukuda M, Hizukuri S. *J. Chromatog.* 1991;585:223.
242. Wong KS, Jane J-L. *J. Liq. Chrom. Rel. Technol.* 1997;20:297.
243. Wolfrom ML, Thompson A. *J. Am. Chem. Soc.* 1957;79:4212.
244. Thompson A, Anno K, Wolfrom ML, Inatome M. *J. Am. Chem. Soc.* 1954;76:1309.
245. Whelan WJ. In: Ruhland W, ed. *Encyclopedia of Plant Physiology*. Berlin, Germany: Springer Verlag; 1958;6:154.
246. French D. *Biochem. J.* 1966;100:2.
247. Hughes RC, Smith EE, Whelan WJ. *Biochem. J.* 1963;88:63.
248. French D, Smith EE, Whelan WJ. *Carbohydr. Res.* 1972;22:123.
249. Parrish FW, Whelan WJ. *Stärke.* 1961;13:231.
250. Posternak IA. *J. Biol. Chem.* 1951;188:317.
251. Hassid WZ, McCready RM. *J. Am. Chem. Soc.* 1943;65:1157.
252. Peat S, Whelan WJ, Pirt SJ. *Nature.* 1949;164:499.
253. Thomas GJ, Whelan WJ, Peat S. *Biochem. J.* 1950;47:1x.
254. Peat S, Pirt SJ, Whelan WJ. *J. Chem. Soc.* 1952;705.

255. Kjolberg O, Manners DJ. *Biochem. J.* 1963;86:258.
256. Banks W, Greenwood CT. *Arch. Biochem. Biophys.* 1966;117:674.
257. Hizukuri S, Takeda Y, Yasuda M, Suzuki A. *Carbohydr. Res.* 1981;94:205.
258. Takeda Y, Shiraska K, Hizukuri S. *Carbohydr. Res.* 1984;132:83.
259. Takeda Y, Hizukuri S, Takeda C, Suzuki A. *Carbohydr. Res.* 1987;165:139.
260. Hizukuri S. *Carbohydr. Res.* 1991;217:251.
261. Marshall JJ, Whelan WJ. *Arch. Biochem. Biophys.* 1974;161:234.
262. Lee EYC, Whelan WJ. *Arch. Biochem. Biophys.* 1966;116:162.
263. Bertoft E. *Carbohydr. Res.* 1989;189:181–195.
264. Bertoft E, Spoof L. *Carbohydr. Res.* 1989;189:169.
265. Bertoft E. *Carbohydr. Res.* 1991;212:229.
266. Zhu Q, Bertoft E. *Carbohydr. Res.* 1996;288:155.
267. Bertoft E, Zhu Q, Andtfolk H, Junger M. *Carbohydr. Polym.* 1999;38:349.
268. Bertoft E, Koch K. *Carbohydr. Polym.* 2000;41:121.
269. Gérard C, Planchot V, Colonna P, Bertoft E. *Carbohydr. Res.* 2000;326:130.
270. Bertoft E. *Carbohydr. Polym.* 2004;57:211.

Structural Transitions and Related Physical Properties of Starch

Costas G. Biliaderis
Department of Food Science and Technology, Aristotle University, Thessaloniki, Greece

I. Introduction	293
II. Starch Structure, Properties and Physical Methods of Analysis	295
1. Ordered and Amorphous Structural Domains (See Also Chapters 5 and 6)	296
2. Physical Properties of Starch in Water	301
III. State and Phase Transitions	310
3. Glass Transitions of Amorphous Structural Domains	311
4. Annealing and Structural Modifications by Heat–Moisture Treatments	320
5. Melting Transitions of Crystallites in Granular Starch	323
6. Gelation and Retrogradation of Starch and its Polymeric Components	332
7. Phase Transitions and Other Properties of V-Structures	354
IV. References	359

I. Introduction

Starch is a component of foods such as rice, bread, other bakery items, pasta and potatoes. Besides being an essential energy source in the human diet, starch constitutes an excellent raw material for modifying the texture of many processed foodstuffs. Starch has also found numerous non-food uses, e.g. in fabric stiffeners, as adhesives and binders, as a sizing agent, as a fermentation substrate, etc. New applications of starch are steadily emerging, including low-calorie fat mimetics, 'biodegradable' packaging materials, thin films, carrier matrices for controlled release of agrochemicals, and thermoplastic materials with improved thermal and mechanical properties. Many food and industrial uses of starch rely on the colloidal properties of its two structurally distinct α-D-glucan components, amylose and amylopectin. These components are not single molecular entities, but rather each constitutes a family of related molecular species with diversity in size, fine structure and function.

Most starches contain about 70–75% amylopectin. However, mutants having starch with essentially no amylose (waxy) or much higher amylose content have been developed by conventional breeding, particularly for corn, barley and rice (Chapters 3, 9, 13 and 16).[1-3] Developments in plant biotechnology offer new molecular approaches to manipulate the composition and fine structure of starch polymers (Chapter 4).[4-6] In native starches, amylose and amylopectin molecules, along with small amounts of water, are densely packed into partially crystalline, water-insoluble, granular particles (0.5–175 µm) whose shape, size, chemical composition and structural organization are characteristic of the plant source (Chapters 5 and 6). It is thus possible to identify most starches from their morphological features when viewed under a light[7] or scanning electron microscope.[8] Granule composition, morphology and supermolecular organization are, to a large extent, under genetic control,[1,9] although environmental conditions during starch synthesis could also affect the packing of starch polymers within granules.

It has long been recognized that the functional properties of starch depend on a number of integrated factors which include polymer composition, molecular structure, interchain organization, and minor constituents such as lipids, phosphate ester groups (typical of potato amylopectin) and proteins.[1,3,10-12] As a result, starches from various botanical origins differ in their physical and functional properties. Furthermore, chemical, enzymic and physical modification of starch, with either preservation or destruction of the native granule, broadens the functionality-imparting properties of different starches.

Although little attention has been paid to the structural features of starch in processed food products, in such products there is a diversity of structures depending on starch source, amylose–amylopectin ratios, lipid, moisture and plasticizer contents, and thermo-mechanical histories. In addition to difficulties in describing and quantifying the structural morphology of starch materials, the ultrastructural level of starch also presents a great challenge. Recent information derived with the use of sensitive analytical probes of polymer structure (e.g. calorimetry, small amplitude dynamic mechanical testing, electron and neutron scattering and x-ray diffraction, electron microscopy, solid-state ^{13}C-NMR, Fourier transform infrared spectroscopy) provide evidence for various levels of chain organization, not only within the granules, but also in heated aqueous starch dispersions and processed foods (Chapters 5 and 6). Such measurements also point to the non-equilibrium (metastable) nature of starch structures and the relevance of the phase transition behavior of starch materials to the glass and melting transitions of synthetic polymers.[9,14-18] For concentrated starch dispersions, equilibrium states and processes are hardly reached, and thus physical properties are often time-dependent and sensitive to the thermal and processing history of the system.

Heat–moisture mediated disruption of the ordered structures in granular starch (gelatinization) is generally a prerequisite for its utilization. Gelatinization of granules in excess water causes large changes in the rheological properties of the system and has a major influence on the behavior and functionality of starch-containing systems. Formation of new structures upon cooling and storage of starch dispersions, such as retrogradation and complexation of starch molecules with lipids, may be either detrimental or beneficial to end-product quality. Even more important is a growing recognition of the relationships between internal structure (molecular and supermolecular) and the macroscopic properties of starch.[19,20] The greatest challenge is to relate physicochemical properties and functions of starch with information on various levels of structure.

This chapter focuses on some aspects of phase transition behavior and other material properties of starch, particularly as they pertain to the structural order and interactions of the starch polysaccharides with water, lipids and other solutes. Understanding the thermally induced structural transitions of starch is helpful in controlling its physical properties and processing behaviors (e.g. plasticization, viscosity), as well as in designing products with improved properties (e.g. texture, stability).

II. Starch Structure, Properties and Physical Methods of Analysis

Stability, transformations and physical properties of starch materials are largely dependent on the nature of the amorphous and crystalline domains present in native granules or processed starch materials. Many physical techniques have been applied to probe the structural order of starch polymers. Among them, light and electron microscopy, wide-angle x-ray scattering and diffraction (WAXD), small angle x-ray scattering (SAXS), solid-state ^{13}C-NMR, various viscometric techniques and differential scanning calorimetry (DSC) are the most widely used (see Chapters 5 and 6). Each of these methods is sensitive to a different level of structure and is applicable over a specific range of distances present in a starch system. For example, with optical microscopy structural features spanning fairly large distances (1–4 μm, e.g. growth rings) can be viewed. WAXD is sensitive to crystalline order over distances as short as 0.3–2 nm, although the minimum dimensions for significant diffraction effects to be observed are probably of the order of 5–10 nm. SAXS gives information on longer range periodicity, ca. 9–10 nm. Cross-polarization magic-angle spinning ^{13}C-NMR (^{13}C-CP/MAS-NMR) gives characteristic detailed spectra of A- and B-type crystal polymorphs of starch, as well as of V-amylose complexes. Whereas structural information given by x-ray diffraction is rather long-range (crystallinity), NMR is considered a short distance range probe, measuring order at the level of individual helices, i.e. it distinguishes helices either in or out of crystalline registry. Information about starch structures with short-range order can also be obtained by calorimetry. Electron microscopy is important for studying structures ranging from single macromolecules up to larger macromolecular assemblies in the range of 0.1–1.0 μm. Spectroscopic methods are generally non-destructive in nature, and thus permit comparison with other analytical tests using the same sample. However, thermal analysis and large deformation mechanical tests can be used only once with a particular specimen. Some of these methods require dilute solutions or dispersions, while others are applicable to more concentrated systems or even dry materials. In this context, it is important to be aware of the concentration-dependent effects of hydrated starch matrices. Since high-, intermediate- and low-moisture systems have distinct behavioral characteristics results obtained at one moisture level may not be applicable to another moisture regime.

1. Ordered and Amorphous Structural Domains (See Also Chapters 5 and 6)

Native starch granules are birefringent when viewed under polarized light, suggesting molecular orientation of some sort in the granules. Birefringence patterns (interference cross, known as a 'Maltese cross') for granular starches are consistent with a radial direction of the macromolecules, i.e. normal to the growth rings and the surface of the granule.[21] Further evidence for radial ordering of polymer chains, which is typical of spherulitic organization, is obtained from small-angle light scattering (SALS) measurements.[11,22,23] Optical birefringence is also exhibited by acid-treated starches[9] and spherulitic particles of short-chain amyloses crystallized from solution.[24] Loss of birefringence on heating, indicative of disordering processes, is used for determining gelatinization temperatures.

Evidence for molecular order in granular starch is also provided by calorimetry; ordered chains give rise to endothermic transitions upon heating. Application of differential scanning calorimetry (DSC) to starch has been useful in inferring differences in starch structures, changes in physical states of starch, and interactions of starch polymers with other constituents in model systems and composite food matrices.[25] DSC measures the direction and extent of heat energy flow while a small sample is exposed to constant change in temperature. It can detect both first-order (melting, crystallization) and second-order (glass) transitions of starch. As a structural probe of starch, DSC is sensitive to the molecular order of chains (helical conformation), irrespective of their involvement in crystalline arrays. Other thermal analysis techniques used to characterize starch materials are thermal mechanical analysis (TMA) in the thermodilatometry mode, and dynamic mechanical thermal analysis (DMTA). Because of the limiting sensitivity of DSC to detect minor changes in heat capacity at the glass transition temperature (T_g), attempts have been made to determine T_g by following other material properties that are more sensitive to changes at the glass transition temperature, e.g. changes in thermal expansion by TMA[26] and viscoelasticity by DMTA.[27] For DMTA, samples are submitted to sinusoidal deformation at a certain frequency over a range of temperatures. Analysis of the signal into the in- and out-of-phase stress components to the strain function permits computation of the dynamic storage or elastic (E') modulus, and the loss or viscous (E'') modulus. The dimensionless ratio of energy lost (mainly as heat) to the energy recovered per deformation cycle, expressed as $\tan\delta = E''/E'$, is also used to describe a second-order transition ($\tan\delta$ peak).

Long-range ordering (crystallinity) in starch granules is demonstrated by x-ray diffraction (Chapter 6). WAXD patterns of starches correspond to one of the two limiting polymorphs (A or B) or the combination C form, the C-pattern being a superposition of A- and B-patterns.[28,29] The A form is typical of cereal starches, while the B-pattern is given by tuber starches, as well as by high-amylose and some waxy starches with the *ae wx* genotype (Chapter 3) and retrograded starch.[30] The most detailed analysis of the A and B polymorphs has been based on x-ray and electron diffraction data from crystalline amylose fibers and modeling studies.[31–35] Both structures are composed of ordered arrays of double helices, and the polymorphic variation is based on packing differences of the hexagonally arranged chain duplexes; the B structure is more spacious, accommodating larger amounts of water. Water is an integral part of the crystal lattices of starch polymorphs (ca. 0.1 and 0.25 g of water per g of dry starch for the A and B forms, respectively). Relative crystallinity of starches can be obtained from the intensities of characteristic diffraction lines.[36] Values of 15–45% crystallinity based on wide-angle x-ray diffractometry have been reported for various granular starches.[23] Since regular maize (~75% amylopectin) and waxy maize (~98% amylopectin) starches (both giving A patterns) have similar crystallinities,[23] and granule crystallinity is preserved after leaching out of amylose,[37] amylopectin seems to be the dominating crystalline component in granular starch. Crystalline regions in amylopectin molecules are considered as being bundles of double helices formed by adjacent short DP chains (14–20 glucose units or ~50Å); the chain duplexes are stabilized by H-bonds and numerous van der Waals interactions. A cluster-type molecular structure for amylopectin has been suggested[21,38] which features alternating crystalline arrays of double helices, and amorphous zones of dense branching and cluster-interconnecting, long-chain segments. A superhelical, lamellar structure for amylopectin which was based on electron microscopy and electron diffraction data of non-disrupted granule fragments of potato starch has been proposed.[39] According to this model crystalline lamellae are arranged in a superhelical fashion, leaving voids of ~8 nm diameter.

Another interesting aspect of starch granule structure is the observation that starches with amylopectins of short chain length (<20 glucose residues) exhibit A-type crystallinity, whereas those with amylopectins of longer chain length give the B-pattern.[40,41] The postulate that chain length is a major determinant of crystalline polymorphism of starch was further supported by crystallization studies of malto-oligosaccharides used as model compounds.[42,43] According to Gidley et al.[40,41] the A polymorph (the most thermodynamically stable form) is favored over the B structure (kinetically preferred form) under conditions of: (a) shorter α-D-glucan chain length; (b) higher crystallization temperature; (c) higher polymer concentration; (d) slower crystallization conditions; and (e) the presence of alcohol. The reverse is true for the B polymorph, which crystallizes readily from pure water. When hydrated granular starches are viewed with small angle x-ray[45–47] and neutron scattering,[48] a Bragg peak is seen at a value of 9–10 nm, which is believed to arise from the alternating crystalline and amorphous lamellae of amylopectin. This peak,[48–51] as well as the WAXD pattern,[52] disappears during starch gelatinization.

Other WAXD starch patterns, known as V-types, are associated with the amylose fraction. In V-amylose, the chain conformation is a left-handed single helix with six residues per turn (V_6) for complexes with aliphatic alcohols and monoacyl lipids; with ligands bulkier than a hydrocarbon chain, helices of seven or eight glucose residues per turn are feasible.[31] Aliphatic chains within amylose helices are rather locked in a solid-like state (i.e. with a fixed conformation that is prevented from rotation due to steric conflicts), as suggested by Raman spectroscopy,[53] ^{13}C-CP/MAS-NMR[54,55] and molecular modeling.[54] If water is located between helices, the crystalline lattice is expanded, a structure designated as V_h, the anhydrous form being denoted as V_a. The V-pattern is not usually observed with common native starches. It has been detected, however, in high-amylose (>40%) starches (e.g. wrinkled pea starch, high-amylose corn, and other maize genotypes with the recessive genes *su* and *du*). Lack of V-type diffraction peaks does not necessarily imply the absence of amylose–lipid complexes in granular starch, but rather indicates the absence of organized helices into well-developed, three-dimensional structures (long-range order). The question whether amylose inclusion complexes are present in normal cereal starches has been resolved after a long debate. Besides all circumstantial evidence for some form of association of amylose with granular monoacyl lipids (lysophospholipids, free fatty acids) *in situ*,[12] ^{13}C-CP/MAS-NMR analysis provided proof of the presence of the V-conformation in granules of maize, oat, barley, rice and wheat starches.[54,55,57,58] The two spectral features of non-waxy cereal starches indicative of single V-amylose helices were: (a) the appearance of a broad resonance peaking at 31 ppm (corresponding to mid-chain methylene carbon atoms of monoacyl lipids) which reflects an almost solid-state structure of lipids due to steric constraints in the helical cavity; and (b) a signal of C-1 at 103–104 ppm attributed to the V-conformation. These resonances are enhanced in acid-treated (Lintnerized) starches.[55] Although the hydrocarbon chain in the helical cavity of amylose is in the solid state,[53,55] it has been suggested that the amylose–lipid complexes are present mainly as amorphous structures.[54] X-ray scattering data of enzymically-degraded wheat starch (to increase the proportion of ordered domains in the enzyme-resistant residues) also revealed the V-structure, confirming the presence of amylose–lipid complexes in native granules.[59] Recently, WAXS studies suggested the coexistence of A-, B- and V-type polymorphs in some starches,[59] e.g. the crystalline regions of wheat and rye starch (A-type starches) granules were found to contain between 7–10% B-type structures, while the crystalline phase of high-amylose starches (amylomaize, wrinkled pea starch) consisted of 74.6–84.6% B-type and 15.4–22.6% V-type structures.

Development of the V-polymorph can be readily induced by heat–moisture treatment of starch (18–45% moisture, 90–130°C for 1–16 hours),[30] extrusion cooking[60–62] or simply by gelatinization and cooling of starch dispersions. Under such hydrothermal conditions, there is increased mobility of amylose chains complexed with naturally occurring monoacyl lipids or externally added ligands (monoglycerides, fatty acids), leading to formation of larger molecular assemblies detectable by x-rays. Extrusion of cereal starches[60] and blends of cassava starch with 2–4% monoacyl lipids[61] at 22% initial water content led to formation of two types of crystal structure depending on extrusion temperature. Below 170°C, the x-ray diffraction patterns of extrudates were typical of the V_h structure. However, when starch was extruded at >185°C and the final moisture content was <13%, a new structure called 'extruded' or E_h-type was observed. The latter was characterized by a slight displacement of all three diffraction peaks of V_h to lower angles; e.g. the main diffraction peak of E_h occurred at è ≈ 9°03′, compared to 9°54′ for the V_h structure. Differences between the two structures were attributed to different interaxial distances of the helices;[60] distances between adjacent helices were 13.8 Å and 15.0 Å for V_h and E_h, respectively. It was also shown that the E_h structure is fairly labile and irreversibly transforms to the V_h structure by reconditioning the extrudates at 30% moisture. This suggests that the E_h form, with the less well-packed chains, is a metastable structure. Non-equilibrium states, in the form of E_h crystals, were also detected in foams prepared by extrusion of maize grits at high temperatures and low moisture contents,[60] and their formation was related to the development of a macrostructure with spherical rather than polyhedral pores. In starch-based foods, V-amylose can exist in various states of aggregation depending on processing conditions. Calorimetry and x-ray diffraction have undoubtedly advanced our knowledge of the supermolecular structure and properties of amylose–lipid complexes.[63–71] With monoglycerides,[65–68] internal rice starch lipids,[70] fatty acids[71] and linear alcohols[72] as complexing ligands, at least two structurally distinct forms were identified, and a morphological model for the solid-state organization of chains in these polymorphs (I, II) has been postulated.[66] Form I (of low melting temperature, T_m), obtained under conditions favoring rapid nucleation, was described as an aggregated state where ordered (helical) chain segments have insufficient packing for crystalline order to be shown by x-ray diffraction (i.e. it gives an amorphous pattern). In contrast, form II (of high T_m) appeared as a polycrystalline aggregate with well-developed long-range order, showing the typical reflections of the V-pattern. Understanding the supermolecular structures, stabilities and transformations between the various forms of amylose–lipid complexes is of great fundamental and technological importance, considering the multifunctional role of lipids in starch-based products.

An important advance in starch structure analysis has been the application of high-resolution, solid-state ^{13}C-NMR to studying molecular conformations of polymer chains in granules or gels.[73–77] Being a non-invasive spectroscopic method, this technique is considered a short distance range probe of molecular order;[78] it can distinguish between double helices (short-range order) and single helices (amorphous chains), and give estimates of their relative amounts. Amorphous starch shows broader ^{13}C-CP/MAS-NMR spectral features than ordered structures.[73,76,78] The most distinct differences in spectra are observed in the C-1 and C-4 regions; signal intensities at 81–83 ppm (C-4), 94–98 ppm and 102–105 ppm (C-1) can be assigned to amorphous sites. By combining spectra of crystalline and amorphous model materials at various proportions, Gidley and Bociek[73] simulated the spectra of several starches and quantified their double helix to amorphous (single-chain) ratios. For granular starches from various botanical origins, the double helical content was found to be between 38% (high-amylose corn) and 53% (waxy maize), values substantially higher than the respective crystallinity range, 25–35%.[73,79] Results of ^{13}C-CP/MAS-NMR agree with the postulation that granular order resides mainly in the amylopectin fraction. However, C-1 and C-4 of amorphous starch had chemical shifts similar to those of the corresponding signals of the V-structure,[73,75] thus making the distinction between mobile amorphous regions (lipid-free amylose and branching regions of amylopectin) and solid-like V-amylose chains difficult. The ^{13}C-CP/MAS-NMR spectra of V_a and V_h amyloses also indicate that these structures differ from those of the A and B polymorphs.[75] Improved spectral resolution made possible identification of peak multiplicities at C-1 for the A- (triplet at approximately 99.3, 100.4 and 101.5 ppm) and B- (doublet at approximately 100.0 and 100.9 ppm) structures,[73,75,76] corroborating differences in packing symmetry of the respective crystals.[34,35]

A fast and direct method, based on IR spectroscopy, for the quantitative measurement of short-range order has been developed.[80] The absorbance bands at 1047 and 1022 cm^{-1} were shown to be sensitive to the amount of ordered starch and amorphous material, respectively. Changes in the short-range structure of potato starch with 18% and 26% (w/w) water during heating were congruent with the structural changes measured by DSC and WAXD. It was also observed that the IR spectrum was very sensitive to changes in water content. With further development of this method and that of solid-state ^{13}C-NMR, structure–function relationships and molecular processes involved in many applications of starch materials could be explored.

The exact nature of the amorphous regions in native starch is largely unknown, although they constitute the major portion of the granular material and, as will be discussed later in this chapter, the amorphous material exerts a great influence on the properties of starch. Amorphous areas are less dense and more susceptible to chemical and enzymic modification than are crystalline regions.[9,81,82] Diffusion of small, water-soluble molecules (molecular weight of <1000) in starch granules can only occur through the amorphous phase. In terms of composition it appears that, besides the regions of dense branching in amylopectin, the majority of amylose is included in this domain. Several arguments to support the view that amylose is separated from amylopectin in the granules of corn and wheat starch, and mixed (partly co-crystallized) with amylopectin in potato starch have been presented.[22,23] Light crosslinking of granular starches (corn, potato) and characterization of the products by size-exclusion chromatography indicated that amylose became crosslinked to amylopectin, and that there was no crosslinking between amylose molecules.[83,84] These observations suggest that, in granular starch, amylose molecules are not present as bundles in the amorphous regions, but rather are interspersed among amylopectin molecules. It is also plausible that some amylose molecules participate in double helices with amylopectin, and thereby become less prone to aqueous leaching or complexation with iodine, particularly in the case of potato starch.

When starch granules are immersed in water, they undergo limited swelling (reversible), presumably due to hydration and swelling of the amorphous regions. Potato starch granules have been reported to contain 0.48–0.54 g of absorbed water per gram of dry starch in the fully hydrated state.[85–87] At these water levels, the amorphous phase of granular starch is fully plasticized by the solvent. In a critical review on starch–water relationships, van den Berg[85] was first to stress the importance of the plasticizing action of water on starch. Zobel[22] reported that sorption of water in the amorphous phase decreases their x-ray scattering relative to that of the crystalline areas. Moreover, the plasticizing water relieves intercrystalline strains. These effects result in sharper x-ray diffraction peaks and a lower baseline with increasing water content, particularly for B-type starches.[89–91]

Amorphous regions alternating with crystalline material in a form of concentric shells have been revealed by transmission and scanning electron microscopy (TEM and SEM).[93] By modeling SAXS data from wheat starch slurries, Cameron and Donald[48] identified three types of structure in granular starch, viz., stacks of alternating crystalline and amorphous lamellae embedded in 'background' amorphous material. In terms of electron distribution functions, the embedding medium had an electron density slightly higher than that of the alternating amorphous lamellae.[49,50] During gelatinization the background material, because of its less physical constraints, was found to absorb water at a faster rate and to a greater extent than did amorphous intercluster lamellae.[49] In view of these findings, the cluster-like structure of amylopectin, and the fact that most of the amylose molecules appear to exist in a disordered state, it is not unreasonable to accept that there are no sharp boundaries between crystalline and amorphous domains in granular starch, i.e. that some chains will run continuously from one region to the next. Thus, a range of structures would be expected between well-developed crystallites and fully disordered regions. In this type of supermolecular organization, amorphous and crystalline phases are interdependent.

2. Physical Properties of Starch in Water

Granule swelling, gelatinization, pasting and retrogradation are important aspects of starch functionality. When starch granules are heated in excess water to progressively higher temperatures, a point is reached where the polarization cross starts to disappear and the granules begin to swell irreversibly. These phenomena, associated with the disruption of granular structure, are called 'gelatinization.' In this context, gelatinization can be defined as the collapse (disruption) of molecular orders (breaking of H-bonds) within the granule, along with all concomitant and irreversible changes in properties such as water uptake, granular swelling, crystallite melting, birefringence loss, starch solubilization and viscosity development. Based on these property changes, a diversity of chemical, enzymic and physical methods have been applied in investigations of starch gelatinization, as discussed in several review articles.[9,11,94,95] These include light microscopy,[79,96,114] electron microscopy,[97–100] light transmission,[94] viscometry,[101] swelling and solubility determinations,[102] particle size analysis with laser light scattering,[103,104] ^{1}H-NMR,[105–109] high-resolution ^{13}C-, ^{17}O- and ^{23}Na-NMR,[110] ^{13}C-CP/MAS-NMR (79), SALS,[11,111,112] FTIR,[80] ESR,[113] WAXD,[52,79,92,114] SAXS,[49–51] thermal analysis such as DSC[25,115–117] and TMA,[25,117] and enzymic tests.[118,119]

Measurement of birefringence loss by the use of the Kofler hot stage microscope with polarized light is a sensitive and relatively simple test.[96] The birefringence end point temperature (BEPT, i.e. the temperature at which 98% of the granules have lost their birefringence) varies among starches of different origin depending on their composition and the structural organization of their granules. Even within a given starch granule population, the gelatinization temperatures of individual granules might span a 10–15°C range. Loss of birefringence occurs over a broader temperature interval and at higher temperatures as water becomes limiting; e.g. for wheat starch, birefringence was still detectable after heating in a calorimeter to 232°C at a water content of 1.5%.[120] In general, there is a good agreement between DSC transition temperatures (T_m) and BEPT temperatures for granular starches of different botanical origins with sufficient water for plasticization.[25,116,117,121–123]

At temperatures above the gelatinization temperature, there is considerable increase in swelling of the granules. The swelling usually starts at a temperature corresponding to the onset temperature (To) of the DSC endothermic transition (Figure 8.1a) and continues well above the concluding temperature (Tc). Different starches exhibit different swelling behaviors, both in terms of the final gel volumes obtained and in the temperature response. Potato and tapioca starches give very large gel volumes, compared to their cereal counterparts.[102] Swelling seems to be primarily a property of intact amylopectin molecules;[124,125] waxy (all-amylopectin) varieties give the highest gel volumes when starches differing in amylose content are compared. This is also evident from thermo-mechanical analysis (TMA) data of rice starches with varying amylose contents heated at 50% water content (Figure 8.1b).[117] Increased shearing was found to increase swelling,[126] whereas granular lipids in cereals starches seemed to inhibit swelling.[54,124,127] For some starches (e.g. potato and corn), swelling occurs to a similar degree in all directions of the granule, whereas for wheat, rye and barley starches, there is preferential expansion in one direction, resulting in multiple folding of the granules.[128,129]

Swelling of starch granules is also accompanied by leaching of starch molecules from the granules. This material is largely amylose, although amylopectin may also solubilize depending on the nature of the starch and the severity of heat-shear conditions employed.[126,30,131] The molecular weight of the solubilized amylose increases with increasing temperature.[1] As will be discussed in detail later in this chapter, leaching of amylose is a requirement for true gel formation.

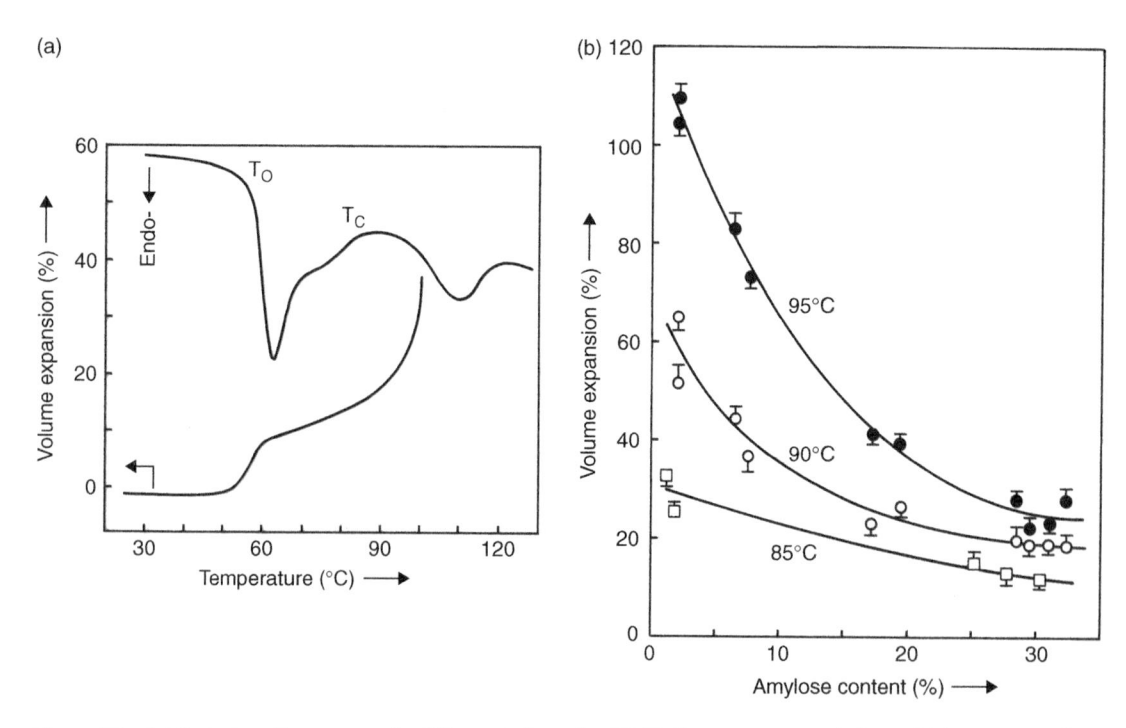

Figure 8.1 (a) Typical swelling pattern of a 50% starch dispersion (TMA dilatometry pattern) along with its corresponding DSC trace; (b) relationship between swelling and amylose content of granular rice starches.

Changes in rheological behavior of a heated starch suspension as a result of granule swelling and solubilization (leaching) of macromolecules upon gelatinization can be monitored by viscosity measurements. The Brabender ViscoAmylograph was (up to 1995–1999) traditionally used worldwide in the characterization of the pasting behavior of starches.[101] The Brabender ViscoAmylograph was largely replaced by the Rapid ViscoAnalyzer (RVA) for routine analyses.[132] A Brabender equivalent of the RVA is also available. With either instrument, changes in viscosity (consistency) of starch–water dispersions are monitored continuously under constant stirring and following a programmed heating and cooling cycle; i.e. for the Brabender ViscoAmylograph heating from 30° to 95°C at 1.5°C/minute, maintaining the sample at 95°C for 20, 30 or 60 minutes, then cooling to 50° or 25°C at a rate of 1.5°C/minute, and finally holding the starch paste at this temperature for 1 hour. Brabender pasting curves are characteristic of each type of starch (Figure 8.2). Six points on these curves are generally recognized and used to compare samples:[94] (a) pasting temperature (initiation of paste formation); (b) peak viscosity, the highest apparent viscosity obtained, irrespective of the temperature at which it was attained; (c) viscosity at 95°C (ease of cooking indicated by the apparent viscosity at 95°C in relation to peak viscosity); (d) viscosity after one hour at 95°C (paste stability during cooking; when compared to (b), the degree of breakdown); (e) viscosity at 50°C (when

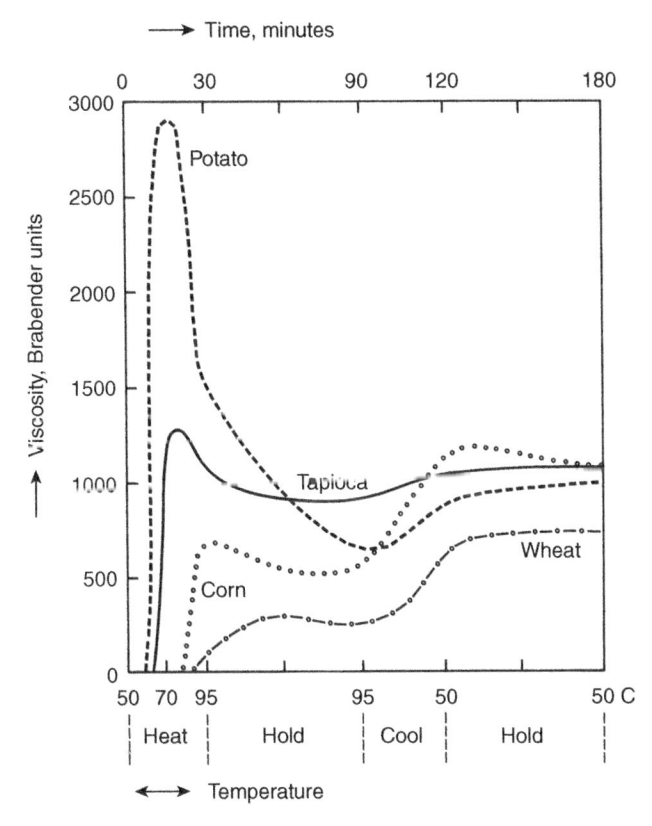

Figure 8.2 Typical Brabender viscosity curves of 8% granular suspensions of common starches.[130]

compared to (d), the degree of setback); and (f) viscosity after one hour at 50°C (stability of the cooked paste). The pasting temperature is usually higher than the gelatinization temperature measured by the loss in birefringence. Tuber and root starches show a sharper rise in viscosity during cooking and have a higher peak viscosity than do common cereal starches; potato starch gives the highest, whereas wheat starch gives the lowest peak viscosity among common starches.[133] The peak viscosity is a measure of the thickening power of a starch. Potato, tapioca and waxy starches swell to a greater extent than regular corn starch. However, pastes from tuber and waxy starches break down more readily than those from cereal starches. The increase in viscosity during cooling of a paste is a measure of retrogradation (setback) due to reassociation of the starch molecules. Amylographs of most starch dispersions at pH 5–7 show essentially the same consistency responses; substantial breakdown occurs at pH 3.5–4.0 for normal amylose content cereal starch pastes.[94] Hot pastes of waxy starches, and particularly of potato starch, show rapid losses in viscosity at pH <5.0; this functional drawback is minimized or eliminated by light crosslinking.[134] Although cooking curves have been widely used to characterize the pasting properties of native and modified starches and to examine the effects of additives (sugars, lipids, etc.), interpretation of the physical events responsible for the observed changes in consistency has been difficult and often challenged. As the paste is continuously torn apart by the weak shear imparted by the viscosity-measuring device, only limited information is given about its structure, and the results are not comparable with those from other rheological tests. Starch pastes, being regarded as composite materials in which a continuous phase is interspersed with particles (swollen granules), have properties that are dependent on a number of factors, including the rheological properties of the continuous phase, the volume fraction and deformability (rigidity) of the particles, and the interactions between the dispersed and continuous phases.[135,136] These in turn are not only dependent on intrinsic properties related to the starch itself, but also on extrinsic parameters related to the conditions used for paste preparation (upper heating temperature, stirring and heating rates, cooking time, pH, water quality and the presence of solutes). For example, Doublier[127] has presented a study of the viscographic and swelling solubility patterns of wheat, corn, faba bean and smooth pea starches, and concluded that the effects of the pasting conditions may dominate those of the botanical origin of the starch. The impact of various parameters affecting starch paste rheology has been discussed in two comprehensive reviews by Launay et al.[133] and Doublier.[134] Another operationally similar instrument is the Ottawa starch viscometer.[139]

Paste properties of native starches from different botanical origins have been reviewed.[92,133] Relevant to practical usage of starch, the most important paste properties are viscosity, texture, paste transparency, resistance to shear and tendency to retrograde. In terms of texture, the translucent potato starch pastes can be described as cohesive, long-bodied, stringy and rubbery. Other root, tuber and waxy starches give pastes of similar texture to that of potato starch, but are generally less cohesive. On the other hand, pastes from common cereal starches are opaque and can be described as noncohesive and short- and heavy-bodied. These and other properties of several native starches are summarized in Table 8.1.

Table 8.1 Gelatinization and pasting properties of native starches[130]

Property	Potato	Corn	Wheat	Tapioca	Waxy maize	Rice	Sorghum
Gelatinization temperature range(°C)[e]	58–63–68	62–67–72	58–61–64	59–64–69	63–68–72	68–74–78	68–74–78
Brabender peak viscosity (BU)[b]	2900	700	250	1200	1100	500	700
Swelling power at 95°C[c]	1150	24	21	71	64	19	22
Critical concentration at 95°C[d]	0.1	4.4	5.0	1.4	1.6	5.6	4.8
Paste viscosity	Very high	Medium	Medium-low	High	Medium-high	Medium-low	Medium
Cold paste texture	Long, stringy	Short, congealing	Short	Long	Long	Short	Short
Paste clarity	Translucent	Opaque	Opaque	Translucent	Translucent	Opaque	Opaque
Resistance to shear	Medium-low	Medium	Medium	Low	Low	Medium	Medium
Retrogradation rate	Medium	High	High	Low	Very low	High	High

[a]Determined by Kofler hot stage microscopy (onset-midpoint-end)
[b]BU at 8% starch concentration
[c]Weight of sedimented swollen granules per gram of dry starch
[d]Critical concentration value: grams of dry starch per 100 ml of water required to produce a paste at 95°C

Because of the empirical nature of the viscographic tests and their inherent limitations with regard to interpretation of the rheological data on a structural basis, there has been an increasing tendency to rely more on fundamental methods in rheological studies of aqueous starch dispersions. The forces and deformations from fundamental measurements have the benefit of being well-defined without relation to any specific equipment, and the values, given in SI units, are independent of the size and shape of the sample. Fundamental tests are carried out with either a constant stress or a constant strain or with either varying as a function of time. This leads to the following methods: (a) creep tests (constant stress); (b) stress relaxation tests (constant deformation); (c) viscometry (increasing deformation rate with time); and (d) dynamic or oscillatory tests (stress and/or strain change with time). A comprehensive description of the various rheological techniques and the mathematical theories behind them can be found in classical rheology texts.[140]

The best way of measuring the non-Newtonian flow properties of dilute starch dispersions is to employ a rotational viscometer with coaxial cylinder or cone-and-plate geometries.[137] Flow curves (shear stress or viscosity versus shear rate) can thus be generated, allowing for a complete description of the rheological behavior of the sample over a range of shear rates. For more concentrated gels or solid starch specimens, large deformation tests are relevant, yielding information about stresses and strains that must be applied for fracture to take place.[141–145] Data from large deformation tests are correlated with sensory textural evaluations. Uniaxial compression or tension tests are usually conducted at different rates of deformation. From the measured data and the actual size of the sample, stress–strain (σ–ε) curves are obtained and these are used to determine the fracture stress (maximum stress the sample could sustain before failure) and Young's modulus, $E = (\partial\sigma/\partial\varepsilon)_{\varepsilon \to 0}$. Large deformation measurements are more relevant to ultimate (failure) properties of the material. A comparison among the various methods of studying large deformation properties of food materials, e.g. 10% starch gels, has been presented.[146]

Although the use of small-strain, dynamic rheometry (also known as mechanical spectroscopy) in the study of viscoelastic properties of starch is not new,[147,148] the development of commercial oscillatory rheometers in recent years has prompted wider use of this technique, and provided fundamental information on structural aspects of aqueous starch systems at a molecular level.[149] Small deformation measurements are invaluable for monitoring the 'curing' events of gelling systems. In such an experiment, the term dynamic implies the application of a continuously changing stress or strain (usually in a sinusoidal form) and measurement of the viscoelastic responses of the material. If a small shear strain is applied, the two controlled variables are the frequency (ω) and the maximum amplitude (γ_0) of the strain, whereas the measured responses are the maximum amplitude of the shear stress (σ_0) and the phase difference (δ) between the applied strain and the stress wave function. For purely viscous and elastic materials, representing the two extreme limits of viscoelastic response, the phase shift between shear strain and shear stress are 90° and 0°, respectively. Instead, a viscoelastic material responds with an intermediate phase shift between 0° and 90°. Thus, with a shear strain function of $\gamma(t) = \gamma_0 \sin(\omega t)$, the corresponding sinusoidal stress function for a viscoelastic material is $\sigma(t) = \sigma_0 \sin(\omega t + \delta)$ or $\sigma(t) = \sigma_0 [\sin(\omega t) \cos\delta + \sin\delta \cos(\omega t)]$, or $\sigma(t) = \gamma_0 [\sigma_0/\gamma_0 \sin(\omega t) \cos\delta + \sigma_0/\gamma_0 \sin\delta \cos(\omega t)]$. With the latter equation, the stress function is separated into two components, a portion of stress in phase with the strain (elastic component), and a portion of stress out of phase with the strain (viscous component). This relationship may be then written as $\sigma(t) = \gamma_0 [G' \sin(\omega t) + G'' \cos(\omega t)]$.

G', called the dynamic shear storage modulus, is a measure of the energy recovered per cycle of deformation and can be taken as an indicator of the solid or elastic character of the material. The loss modulus (G'') is an estimate of the energy dissipated as heat per cycle of deformation and is an indicator of the viscous properties of the material. For a gel network structure with non-permanent crosslinks, as most of the polysaccharide gels are, strong interchain associations (i.e. those having long relaxation times) contribute to G', whereas weak and rapidly relaxing bonds contribute only to G''. Interactions with relaxation times in the timescale of the measurements contribute to both G' and G''. The results from dynamic tests may be also presented in the form of the complex modulus (G^*) defined as $G^* = G' + iG'' = (G'^2 + G''^2)^{1/2}$. Another parameter which is useful in characterizing the physical state of a viscoelastic material is the loss tangent or $\tan\delta = G''/G'$. This ratio is a more sensitive indicator than G' and G'' to changes in the viscoelastic character of a polymer network structure; a low loss tangent means that the material behaves more like a solid than a liquid.

Small deformation dynamic tests should be always made under conditions that do not alter the structure of the sample. It is, therefore, essential to verify experimentally that the test conditions satisfy the requirements of the linear viscoelasticity theory, i.e. that the stress is linearly proportional to the strain. Linear viscoelastic dynamic moduli are functions of frequency and, as such, are sensitive probes of the structure of biopolymer solutions and gels.[150,151] Modern rheometers, capable of dynamic rheological testing, allow for a description of the frequency and strain dependence of the viscoelastic parameters over a frequency range of two or more decades. The mechanical properties of starch systems using dynamic rheometry have been examined both during gel formation[135,152–155] (Figure 8.3) and storage.[156–165] The findings of these and other studies are discussed in relation to structure development of starch gels in following sections of this chapter.

On cooling sufficiently concentrated starch dispersions, amylose and amylopectin molecules begin to reassociate, eventually forming crystallites of the B-type. This

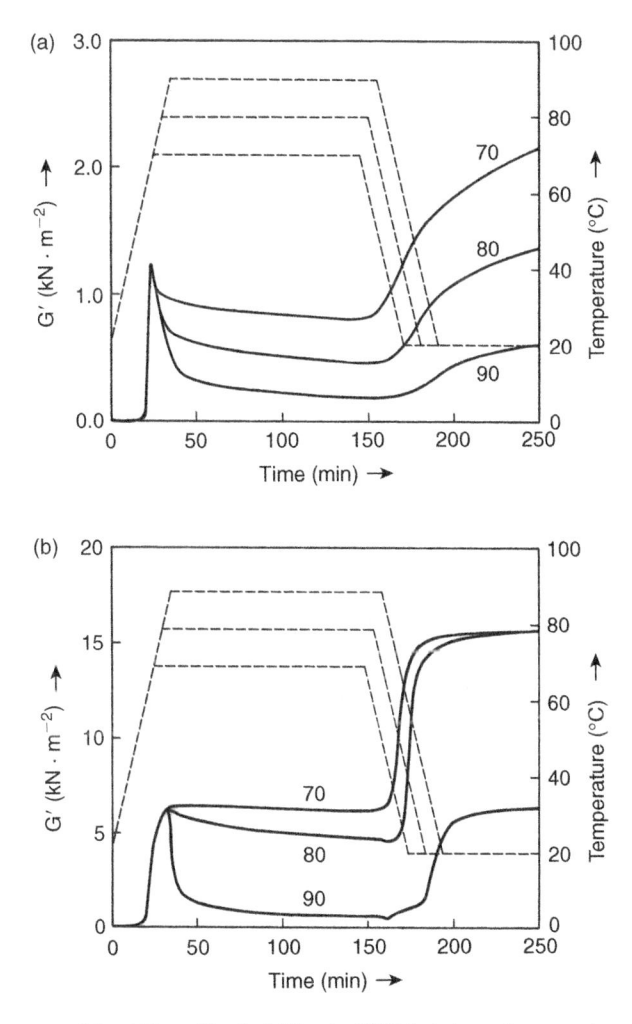

Figure 8.3 Storage modulus (G′) profiles (solid lines) of 15% (w/w) starch suspensions as a function of time during a heating and a cooling cycle (dotted lines): (a) potato starch; (b) wheat starch. The maximum heating temperature is indicated in each curve.[153]

is accompanied by gradual increases in rigidity and a phase separation between solvent and polymer chains. The latter process, causing a decrease in water-holding capacity (development of syneresis), is fairly common for aging gels of most native starches. In contrast to granular starch, where amylopectin is the main crystalline component, both starch polymers are involved in structure development of aging starch pastes and gels. Post-gelatinization changes accompanying the restoration of molecular order in starch, which are collectively described by the term 'retrogradation,' exert a major influence on the texture of foods rich in starch. A considerable amount of experimental data is available to suggest that retrogradation is a major factor in the staling of bread and other bakery products,[166–171] although other factors and alternative mechanisms have been also proposed.[171–174] Retrogradation is a complex phenomenon and depends on many factors, such as the source of starch, starch concentration, cooking and cooling temperature regimes, pH, and the presence of solutes such as lipids, electrolytes and sugars;[170] e.g. freeze–thaw cycles drastically accelerate retrogradation and syneresis.[175] As retrogradation involves the formation of chain entanglements, short-range molecular order and crystallization of double helical aggregates, a variety of physical techniques have been employed to monitor these phenomena and other associated changes. The most common methods are DTA and DSC,[15,156–160,163,176–181] x-ray diffraction,[156–158,161,182,183] small-[156–158,160,163,184,185] and large-deformation mechanical tests,[141,143,144,158,186–190] and several spectroscopic methods, including NMR,[191,192] FTIR[193,194] and Raman spectroscopy.[195] These methods do not monitor the same starch gel properties and are not sensitive to the same fractions or structural elements developing in an aging gel network. For example, thermal analysis is sensitive to chain ordering of the amylopectin fraction, a process manifested by the development of the retrogradation endotherm (T_m = 45–60°C, Figure 8.4c). X-ray diffraction measures crystalline order of both amylose and amylopectin combined, i.e. evolution of the characteristic d-spacing at 15.8 Å (100 reflection) of the B-pattern. However, it is important to remember that this reflection is highly dependent on the water content of the sample.[35,89–91] FTIR and Raman spectroscopy probe conformation- and crystallinity-dependent vibrational frequencies of chemical linkages, whereas NMR monitors chain segmental motions, conformational-dependent chemical shifts and short-range order. Finally, mechanical tests may be sensitive to chain entanglements with or without formation of crystallites. As a result, the time course of retrogradation events in starch gels varies with the measuring technique.

Structural studies have been carried out to establish the length and nature of polymer chains involved in interchain associations or crystallites of retrograded amylopectin and amylose.[15,196,197] Gels were submitted to heterogeneous acid-catalyzed hydrolysis to degrade the amorphous regions, and the acid-resistant amylodextrins were characterized by chromatography and enzymic analysis. Crystallites in retrograded amylopectin gels were found to consist of singly- or multiply-branched amylodextrins originating from the outer, short-DP chain clusters. Because of the relatively low DP segments (~15) involved in the crystallites of amylopectin, their melting transition starts at about 37–40°C, and this explains the susceptibility of retrograded amylopectin to α-amylolysis *in vitro* and *in vivo*.[198–200] In contrast, the chain length over which associations between amylose chains occur in retrograded

II. Starch Structure, Properties and Physical Methods of Analysis 309

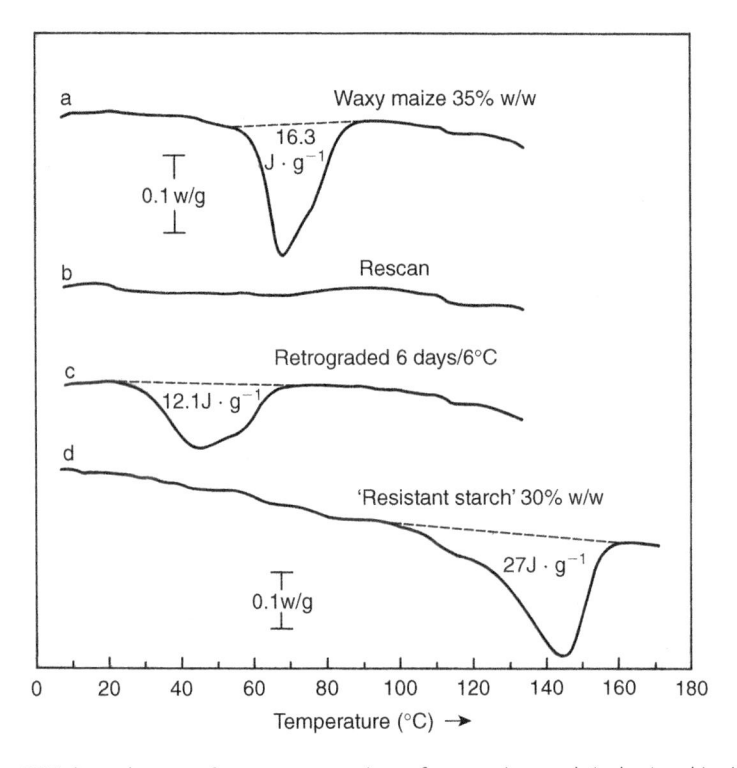

Figure 8.4 DSC thermal curves of aqueous suspensions of waxy maize starch (a, b, c) and 'resistant starch' (retrograded amylose) (d).

amylose is relatively longer, with a DP of ~30–50.[196,201,202] High-resolution, solid-state ^{13}C-NMR confirmed the presence of aggregated double helices of the B-type in amylose gels,[203,204] while calorimetry showed that these structures were thermally resistant (T_m ~130–170°C).[63]

One consequence of conformational ordering and aggregation of double helical chain segments in retrograded amylose is their resistance to acid- and enzyme-catalyzed hydrolysis.[199,202] The amylase-resistant linear amylodextrin (originating from retrograded amylose) constitutes what is known as 'resistant starch' (RS); i.e. starch resisting digestion in the small intestine and becoming available for fermentation in the large intestine of humans.[205,206] The main techniques used to characterize the molecular and physical structure of RS are size-exclusion chromatography,[199,202,207–209] DSC,[20,210–214] x-ray diffraction[208,209,215,216–219] and recently solid-state ^{13}C-NMR.[204,219] Resistant starch readily forms during thermal processing and cooling of starch-containing foods as a result of amylose retrogradation (e.g. in bread, breakfast cereals, cooked potatoes, canned beans, etc.). Depending on starch source, sample preparation and enzyme-catalyzed hydrolysis conditions, the reported average degree of polymerization (DP) of linear dextrins of RS ranges between 20 and 65.[199,201,202,207–209] With most chromatographic systems employed, there seems to be a rather broad distribution of chain lengths in enzyme-treated retrograded starches. Gidley et al.,[216] using high performance anion exchange chromatography, have reported chain distribution profiles with DP between 10 and ~100. The wide range of chain lengths is consistent with the broad melting endotherm (spanning from about 100 to 170°C, and having a T_{max} of 150–160°C, Figure 8.4d) observed for most preparations of RS heated in excess water.[20,61,166,208,209,212,217] The amount of crystallinity for RS, estimated by x-ray diffraction, is dependent on the conditions employed for sample preparation (starch type and concentration, heating-cooling regime, acid or enzyme hydrolysis conditions, etc.). Estimates of crystallinity for enzyme-resistant residues range between 25% and 60%,[215,217,218] while quantitative analysis of the molecular order (double helix content), carried out by solid-state ^{13}C-NMR, gives much higher values (60–95%).[202,217] The high level of double helical order in RS is consistent with the resistance of this material to enzymic hydrolysis, as was first reported by Jane and Robyt.[199] Several models depicting chain packing in the partially crystalline structure of retrograded amylose and RS have been proposed.[199,200,207] Based on x-ray diffraction, ^{13}C-CP/MAS-NMR and chromatographic data, Gidley and coworkers[216] have presented a structural model for enzyme-resistant starch derived from retrograded wheat and amylomaize starches. According to their description, there is a substantial amount of double helical order (~60–70%) in RS, with the chains (DP: 10–100) arranged into aggregates of limited crystalline packing (~25–30% crystallinity). The disordered non-double helical material, interspersed between aggregates, may consist of folding regions, chain ends and single chains trapped within the aggregates. There is also a small fraction (5–8%) of amylose–lipid inclusion complexes in the solid-state matrix. It was further suggested that the broad distribution of linear chains observed in RS is a consequence of the highly aggregated structure of gelled amylose (high double helix content, low crystallinity and a broad range in DP of chain segments at junction zones), and the reduced accessibility of α-amylase to the domains of aggregated double helices due to steric reasons. There is presently considerable interest in the resistant forms of starch in foods, as they seem to have an influence on reduced plasma glucose and insulin levels,[221] increased fecal bulking,[222] and some beneficial effects on the health of colonic epithelial cells.[223] As a result, several resistant starch-based commercial products have entered the marketplace as functional ingredients, offering an equivalent of about 30% total dietary fiber and being free of the undesirable flavor and color of most cereal fiber.[224]

III. State and Phase Transitions

Structural transformations, related to changes in molecular mobility and the order ↔ disorder processes starch polymers undergo upon heating and cooling of aqueous starch dispersions, have been receiving increased attention during the past decade because of their significance to the processing, quality and stability of starch-based products.[11,15–20,22,25,225,226] It is now widely accepted that gelatinization and retrogradation of starch can be viewed as melting and chain ordering/recrystallization processes facilitated by the solvent. The structural changes accompanying these events have been identified using a wide range of analytical probes, among which the most common are thermal analysis, x-ray diffraction and various rheological methods. Thus, much has been learned about the non-equilibrium nature of starch structures

III. State and Phase Transitions

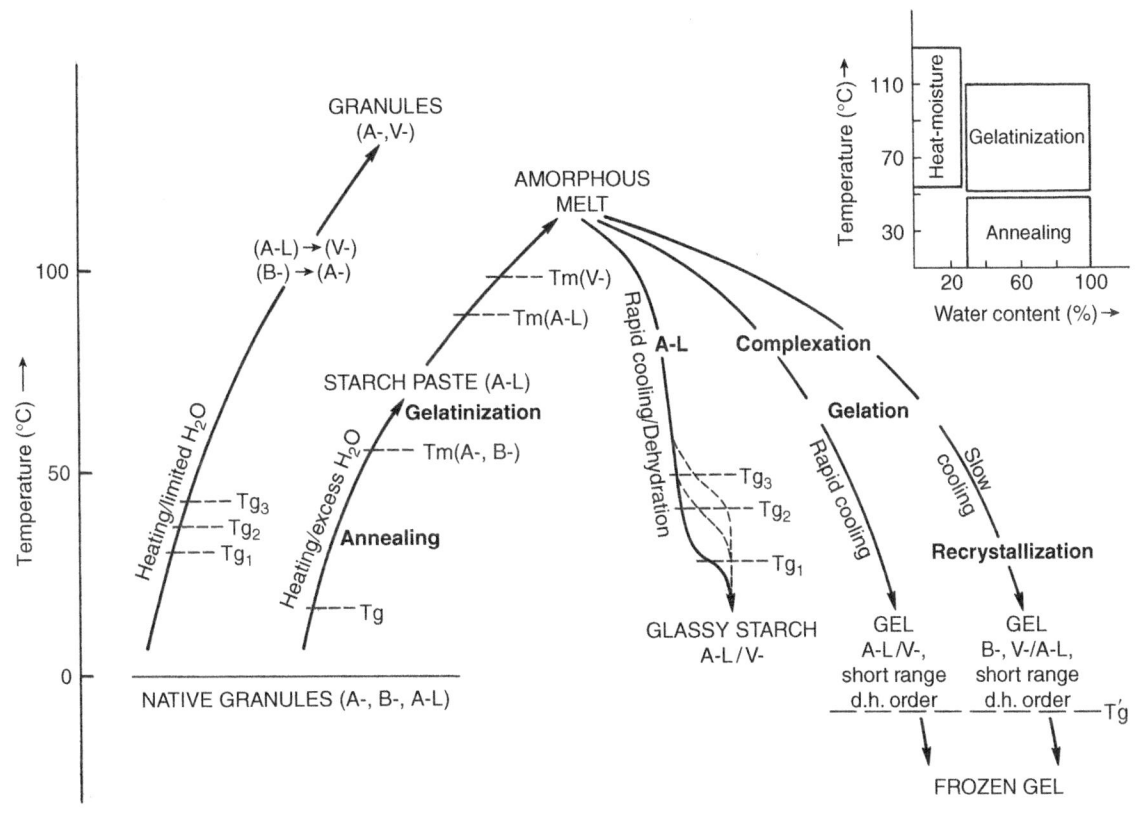

Figure 8.5 Schematic diagram of state and phase transitions of starch; note the effect of moisture content and time on the various states. Tg_1, Tg_2 and Tg_3 represent the glass transition at different moisture content levels. A-L and V-structures denote short- and long-range order of amylose–lipid complexes, whereas d.h. order corresponds to short-range B-type structures.

and the critical role of water in controlling the stability and state or phase transitions of starch polymers. In this respect, fundamental similarities between starch–water and synthetic polymer–plasticizer systems were realized. The broader implications and practical significance of such findings on the macroscopic properties of starch materials (product performance during processing, texture, aging effects, shelf life, etc.) have been also recognized.

The following sections focus on the description of the state and phase transition behavior of starch systems, as schematically illustrated in Figure 8.5, with an emphasis on their molecular organization and their response to various environments (temperature, solvent, other co-solutes, etc.). Selected material properties are also discussed in an effort to demonstrate structure–function relationships of this biopolymer mixture in pure systems and in real food products.

3. Glass Transitions of Amorphous Structural Domains

According to Eisenberg,[225] the glass transition is perhaps the most important single parameter related to applications of amorphous polymers, as it greatly affects their mechanical properties, processing behavior and stability. Amorphous and partially crystalline polymers at sufficiently low temperatures exist in a solid-like or glassy state, where large-scale molecular motions stop, i.e. the chains are locked into a non-periodic and non-symmetric network structure of very high viscosity ($\eta > 10^{12}$–10^{14} Pa.s). By applying thermal energy in a glassy polymer, large-scale segmental motions ('entanglement slippage') commence at a characteristic temperature, called the glass transition temperature, T_g. Although translational motions become constrained at sub-T_g temperatures, side group motions (small-scale rotational motions) still persist, giving rise to subsidiary transitions (known as β and γ relaxations). There are large changes in material properties (thermal, mechanical, dielectric) as the T_g is exceeded and the physical state is shifted from a brittle, glassy-like solid into a more mobile, highly viscous rubber. In the vicinity of the glass transition, there is a gradual change in slope of the plots of temperature versus some thermodynamic function, such as thermal expansivity and heat capacity. Moreover, there are large changes in the viscosity and mechanical properties of the material accompanying this transition (e.g. a typical drop in modulus of $\sim 10^3$). The functional behavior is in turn affected as a result of the dramatic acceleration in molecular mobility and all diffusion-controlled processes when the material is going through its T_g, e.g. crystallization of amorphous solids, collapse and shrinkage of amorphous matrices during drying, and enhanced reactivity due to increased molecular encounter between reactants or enzymes and substrates. In this respect, the physical properties and stability of low-moisture foods and frozen products have been successfully related to their T_g. The pertinence of the glass transition to various food process operations (agglomeration, baking, extrusion, dehydration and freezing) and quality preservation during storage has been discussed in several reviews.[17,18,226,229–234]

Several theoretical approaches have been used to describe different aspects of glass transition behavior, including free volume theories, thermodynamic theories and kinetic theories.[228,235] However, although the phenomenology is fairly well-known, a complete theoretical understanding of the glass transition is not yet available. A variety of thermal analysis techniques (DSC, DMTA, TMA) are used to determine the glass transition temperature of amorphous polymers. An important observation made in studying volumetric or thermal properties of glasses is that the precise location of T_g is dependent on the rate of cooling. A slower rate of cooling results in a lower T_g, as a more dense glass is produced. This implies that glasses usually exist in a non-equilibrium state, and that the equilibrium T_g lies below the measured values. It further signifies that relaxation of glasses is possible by aging (or annealing) at temperatures approaching the T_g. For pure amorphous polymers, the glass transition entails a property change over a temperature range of about 5–20°C. However, in semicrystalline materials, the presence of microcrystals and strained chain segments between crystallites tends to increase the temperature and range over which this transition occurs.[236] Each amorphous or partially crystalline material will have its own T_g (often specified as the temperature at which the polymer is midway between the immobile glass and the rubbery liquid), depending on several factors such as molecular size, side-branching, degree of crosslinking (including crystallinity) and diluent content.[228,236,237] For the latter, it is known that addition of a low molecular

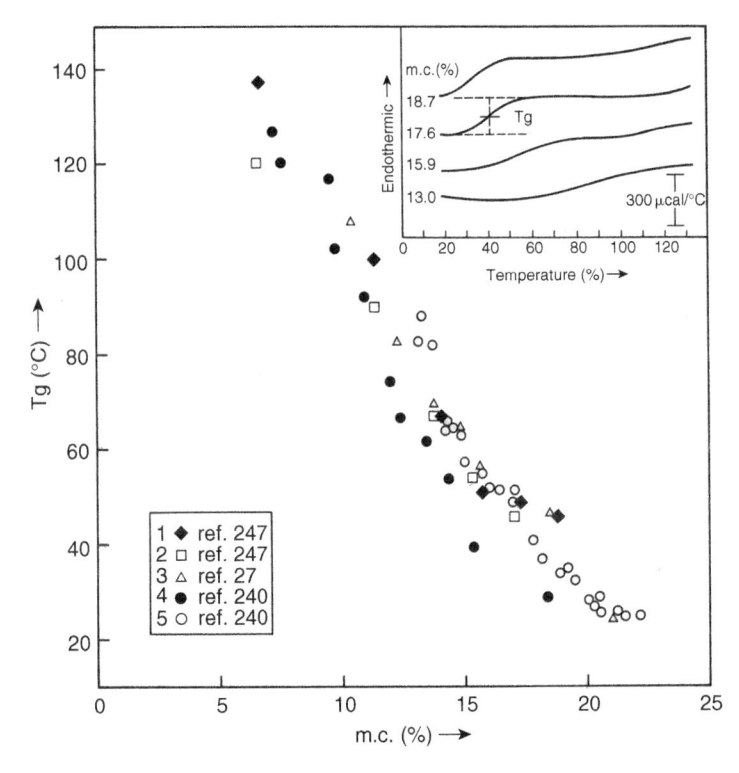

Figure 8.6 Variation in glass transition temperature of starch (DSC data) as a function of water content: 1, 2 and 4 correspond to gelatinized potato, rice and wheat starch; 3 and 5 correspond to amorphous amylopectin and granular wheat starch, respectively. Inset shows the DSC traces of native wheat starch at various water contents.[240]

weight compound, or plasticizer, to a glassy polymer lowers its T_g. Water, for example, is a very effective plasticizer of many synthetic and natural polymers, including starch (Figure 8.6), causing a large depression in T_g.[17,18,226,238] The addition of small amounts of water under isothermal conditions has the same effect on molecular mobility as increasing the temperature, i.e. there is an increase in chain segmental mobility of the polymer, compared with the non-plasticized state, at any given temperature above T_g. This is attributed to an increase in free volume and/or the breaking or weakening of some intermolecular bonds by the plasticizer.[238] There are numerous examples in the literature of the strong T_g-depressing effect of water on food macromolecules and amorphous sugars.[17,225,226] The ability of plasticizers to depress T_g of a polymer decreases as their molar mass increases.

Some of the first evidence suggesting the presence of a glass transition for heated starch specimens came from changes in heat capacity seen on DSC traces. An incremental change in C_p (~0.11 J g^{-1}.K^{-1}) prior to the melting endotherm of starch crystallites[16,117] and a minor baseline shift upon superposition of initial scan (granular starch) and rescan of the same sample[239] were attributed to the glass transition of the amorphous domains of starch granules. However, it appears that the ΔC_p accompanying the major endothermic peak[239] is not due to a glass transition, as argued by Lelievre.[237] Noel and Ring[238] also reached a similar conclusion by finding that the heat capacity increment observed on gelatinization of granular starches is comparable to that observed on melting and dissolution of highly crystalline A and B polymorphs of short chain amylose; this implies that there is no apparent contribution to ΔC_p arising from a glass transition involving the amorphous regions of the granules. For native wheat starch, Zeleznak and Hoseney[239] have shown DSC traces with a ΔC_p, typical of a glass transition, at lower temperatures than just prior to the melting endotherm of starch crystallites (Figure 8.6, inset). For a protonated potato starch sample (containing 20% w/w water), Willenbücher et al.[240] detected a T_g at ~40°C only if the sample had first been heated to at least 110°C. It would appear that the discontinuity in C_p at T_g is small and not easily detectable calorimetrically for most granular starches. Despite the relatively low crystallinity (15–45%) of these materials,[23] the difficulty in identifying and assigning a definite T_g most likely originates in the nature of the amorphous granular phase. For partially crystalline polymers, the reduction in ΔC_p due to increasing crystallinity does not obey a simple additive rule of a two-phase system ('fringed-micelle' model) where crystallites and amorphous matrix behave independently of each other.[244,245] The crystalline regions, acting as physical crosslinks, cause changes in the thermal properties of the amorphous material and decrease ΔC_p more than would be anticipated, assuming an ideal two-phase model structure, e.g. for polycarbonate with a degree of crystallinity as low as 23%, there is no detectable ΔC_p at the T_g.[245] To account for such thermal behavior of granular starch, a simplified 'three-microphase' model was proposed.[117] In addition to starch crystallites, this model incorporates two types of non-crystalline material: (a) bulk, mobile-amorphous regions giving rise to a heat capacity change (e.g. amylose); and (b) intercrystalline, rigid-amorphous material that does not contribute to the ΔC_p observed at T_g as it is motionally restricted by its proximity to crystallites (e.g. regions of dense branching in amylopectin). The sequence of thermal transitions in such a system would be $T_{g\,(mobile\,amorphous)} < T_{g\,(rigid\,amorphous)} < T_{m\,(rigid\,crystalline)}$. Although this polymeric view of starch granule structure is convenient to explain the lack of a discernible glass transition on many DSC thermograms, it is likely that, instead of two distinct amorphous microphases, a range of non-ordered structural domains exist in granular starch. Each of these regions could manifest its own T_g and exhibit a different plasticization behavior by water. This would spread the glass transition over a broad range of temperatures, and account for the difficulty in detecting it by calorimetry. Even for amorphous amylopectin, the glass-to-rubber transition extends over a broad range of temperatures, particularly at low moisture contents.[27] This suggests that, as water becomes limiting, starch becomes even more heterogeneous from one region to the next, due to heterogeneous hydration, giving rise to a broad distribution of relaxation times for the various structural domains (the width of the glass transition reflects the distribution of relaxation times). For gelatinized starch, Vodovotz and Chinachoti,[244] using DMA, reported at least two separate and broadly-distributed domains at moistures <30%, which seemed to undergo a range of transitions over a temperature range of 50–100°C. The glass–rubber transition is more readily detected for gelatinized materials[242,248–250] or isolated starch polymers[27,241,251–255] by thermal analysis techniques. An estimate of the heat capacity increment at T_g for dry amorphous amylopectin is ~0.47 J.g^{-1}.K^{-1}.[241,254] It has also been noted that the T_g increases with increasing crystallinity,[27,242] which is in agreement with observations made on synthetic polymers,[245] denoting the stiffening action of crystallites on the molecular mobility of chains in amorphous domains. Moreover, the temperature location of the glass transition of starch is dependent on the thermal history of the sample, as well as the timescale of the experiment, i.e. the heating rate (DSC) and/or the frequency used in a relaxation type of experiment [dynamic mechanical thermal analysis (DMTA) or dielectric measurements]. Thus, for amorphous starch and amylopectin, the T_g values determined from the midpoint of the heat capacity change at a rate of 10°C/minute and usually fall between the drop in modulus and the tanδ peak obtained with DMTA at a heating rate of 2°C/minute and a frequency of 1 Hz.[27,249] For all these reasons and in view of the strong dependence of T_g on moisture content, it is not surprising that there is considerable variation among the reported values of T_g for starch materials in the literature.

The most important feature of amorphous starch, with many technological implications and of scientific interest, is its plasticization by water and other low molecular weight co-solutes. Molecular interactions of water with starch have indeed been a longstanding matter of investigation. These interactions are thought to be very strong and localized during the early stages of sorption,[88] whereas at high water contents, some of the water has typical liquid-like properties.[241,256] With increasing water content in a starch matrix, there is a redistribution of hydrogen bonds. The number of starch-to-water hydrogen bonds is increased, while intra- and interchain hydrogen bonds are decreased, resulting in a decrease in the interaction energy between starch chains.[257] Moreover, because of the incorporation of water molecules, there is an increase of the distance between starch chains. Both these effects are manifestations of the plasticizing action of water on starch at a molecular level. Figure 8.6 presents the moisture content dependence of T_g using DSC data of several starches from different literature reports. Note that the strong depressing effect of water on T_g occurs within 0–20% water content. For granular starch, it has been shown that, above 30% moisture content, T_g remains constant as water forms a separate phase outside the granules.[9,86,117] For a fully-plasticized amorphous starch matrix, in excess water, the T_g has been reported[15,239] to fall below 0°C (to $T_g' \approx -5°C$, the T_g of the maximally freeze concentrated system). In the dry state, starch is strongly networked by hydrogen bonds, giving rise to a high T_g. The T_g of dry amorphous starch is experimentally inaccessible because of thermal degradation of starch polymers at elevated temperatures. Two indirect procedures have been employed to estimate the T_g of dry starch: (a) study of the T_g of starch with varying amounts of water as a plasticizer and extrapolation to zero moisture content;[117] and (b) study of T_g for dry malto-oligosaccharides and extrapolation to a high molecular weight limit.[254] Applying these methods, the estimated T_g of dry starch is in the range of 225–235°C, much higher than the value of 132°C reported earlier by van den Berg.[85] In studies on the glass transition of waxy maize starch films, plasticized by water[27,259] or by water–sugar,[260,261] an approximately hundred-fold drop in modulus (E′) was observed on heating through the T_g; the magnitude of the drop in E′ and the tanδ peak height were also found to decrease with decreasing moisture content.

The composition dependence of T_g of binary (starch–water) or ternary (starch–sugar–water) blends, assuming component miscibility and the absence of crystallinity effects, has been modeled with the classical thermodynamic treatment of Couchman–Karasz[259] as modified for polymer/diluent mixtures:[263] $T_g = (\sum_i^n w_i \cdot \Delta Cp_i \cdot Tg_i)/(\sum_i^n w_i \cdot \Delta Cp_i)$, where w_i, ΔCp and Tg_i are the weight fraction, change in heat capacity and glass transition temperature of the i component, respectively. Application of this equation to water/starch-containing blends requires the use of published values for T_g (−139°C) and ΔCp (1.94 J g^{-1}.K^{-1}) of water.[263,264] Predictions of T_g made with this expression had a reasonable[27,254,260] or limited[62,244,260,261,265,266] degree of success, depending on the ΔCp and T_g values used for the pure components (starch, sugar) and the level of plasticizer present in the blends. This theoretical approach was also employed to test the glass transition behavior of water-plasticized amylopectin blends with gluten or casein.[267] Another empirical expression, known as the Gordon–Taylor equation,[265] has been also employed to fit the T_g data of amorphous starch or sugars at various water contents:[229,230,260] $T_g = (w_1 \cdot Tg_1 + K \cdot w_2 \cdot Tg_2)/(w_1 + K \cdot w_2)$, where Tg_1 and Tg_2 are the glass transition temperatures of the polymer and diluent, respectively, w_1 and w_2 are the corresponding weight fractions of the polymer and diluent, and K is a constant related to the strength of polymer–diluent interaction (the larger the value of K, the lower is the plasticization effect).

DSC and DMTA scans for some amylopectin–gluten blends[267] provided clear evidence for component immiscibility (phase separation) as two distinct T_g values were obtained, particularly between 10% and 15% water content. Also, for amylopectin–sugar–water blends, at high sugar contents, there seems to be a multiphase system containing amylopectin-rich and sugar-rich domains.[259–261] As the sugar content is increased in a composite amylopectin–polyol blend, the mechanical behavior of the system also seems to be dominated by the lower glass transition of the sugar.[259–261] Sugars were found to exert a strong plasticizing effect on amorphous amylopectin under conditions of reduced water contents (7–10%).[259–261] The degree of plasticization of amylopectin by different sugars, as assessed by both DSC and DMTA, was of the order: fructose > xylose > glucose > sucrose;[260] i.e. it followed the order of increasing T_g for the respective sugars ($T_{g\ fruct.}$ = 7°C, $T_{g\ xyl.}$ = 13°C, $T_{g\ gluc.}$ = 38°C, $T_{g\ sucr.}$ = 70°C).

Bizot et al.[252] explored, using DSC, the impact of variations in structure on T_g of several α-D-glucans differing in molecular size, degree of branching and type of glycosidic linkage. The T_g-moisture content data, modeled by the Couchman–Karasz equation, seemed to follow parallel trends in the range of 5–25% water content. At any given moisture content, the T_g followed the order: amylose (MW ~5.9 × 10^5) > potato starch > amylopectin > phytoglycogen > lintnerized potato starch (DP of ~15). Extrapolated T_g values for dry amylose and amylopectin from their fitted data were 332°C and 285°C, respectively, values much higher than those (225–237°C) reported earlier.[117,254] The effect of branching as a depressor of T_g was confirmed from their data, as in the presence of α-(1→6) linkages, there would be increased total free volume and chain mobility.

The relationship between T_g and molecular weight, M, is often described by the Fox and Flory equation:[266] $T_g = T_g^{(\infty)} - K/M$, where $T_g^{(\infty)}$ is the glass transition temperature of the polymer with infinite molecular weight and K is a constant characteristic of the polymer. For the malto-oligomer series, a double reciprocal plot of

T_g versus degree of polymerization was linear.[254] Similarly, for a series of commercial starch hydrolysis products (polydisperse solutes), there was a linear relationship between T_g' (T_g of the maximally freeze-concentrated solute glass) and $1/MW$.[270]

Calorimetric measurements on amorphous starch and other polysaccharides at low moisture contents (8–20%) have often revealed small endothermic transitions near or below the T_g.[27,248,271–273] The magnitude of such endothermic events is dependent on the moisture content and increases with aging time at sub-T_g temperatures.[248,271,272] This behavior is typical of many synthetic amorphous polymers and reflects an enthalpy relaxation of the glassy state which occurs on storage at temperatures below T_g.[274,275] Materials in the glassy state are generally regarded as solidified supercooled liquids whose free volumes, enthalpies and entropies are greater than they would be at equilibrium at any given temperature. As a result, glasses continue to undergo relaxation toward equilibrium, even during storage at sub-T_g temperatures,[228] which implies that molecular mobility does exist below the apparent glass transition temperature of the material. This process is frequently referred to as glassy state relaxation or simply as physical aging.[276] Physical aging affects several properties of a glassy material such as density, stress-relaxation, creep-compliance, dielectric constant and dielectric loss.[276,277] Physical aging of glassy polymers can be explained qualitatively by a gradual shift and a narrowing of the relaxation time distribution to longer times; the short time portion of structural relaxations can reach equilibrium fairly rapidly, whereas slower relaxations continue over longer times. The rate of relaxation during annealing is dependent on the state of the material when annealing commences, i.e. cooling rate and thermal history of the sample prior to annealing. The endothermic effect can be manifested either as a distinct sub-T_g peak, when aging occurs at temperatures far below the T_g, or as an enthalpy overshoot at T_g, when aging takes place at temperatures approaching the T_g. For amorphous starch–water systems, the sub-T_g endothermic events are intensified with increasing moisture content and annealing time or temperature.[248,271,272,273] Further to the plasticizing action of water (entropic effect), specific interactions (energetic associations) between water and the hydroxyl groups of the amorphous carbohydrate matrix (i.e. establishment of a long-range hydrogen bonding network upon aging) have been suggested as contributing to the sub-T_g endothermic events.[272,273] Clearly, more work is required to further explore the nature, extent and temperature- or moisture content-dependence of the DSC endothermic peak of low-moisture, amorphous carbohydrate systems, including starch. It is also important to examine the kinetics of changes in other material properties associated with aging effects. As for synthetic polymers, physical aging of amorphous polysaccharides systems would certainly have an imprint on their end use quality and stability.

Besides water, plasticization of starch is also effected by polyols, such as sugars and alditols, including glycerol and ethylene glycol. A number of investigations have had as their goal producing thermoplastic materials using conventional techniques employed for commodity plastics, such as extrusion, compression molding, injection molding, and blow molding. Several new thermoplastic materials have been patented (Chapter 19).[278] The mechanical properties,[62,248,266] fracture behavior[279,280–282] and other physical properties (density, compressibility and other textural attributes)[62,283–285] of starch extrudates are a function of starch type, as well as processing parameters (amount of plasticizer, temperature, shear), cooling regime (i.e. how fast the polymer matrix reaches its T_g as it cools down) and the state of the end product. The latter is controlled by the extent of starch plasticization with water or other polyhydroxy compounds. Thus, with increasing moisture content, the fracture mechanism of starch extrudates changes from a rapid, brittle fracture to slow, plastic fracture (tearing);[227,280–282] fracture stress also decreases with increasing water content (8–20%).[227,280,281] Co-plasticizing solutes (sugars, polyols) also decrease the flexural modulus (Young's modulus determined from three-point bend tests) and fracture stress of thermoplastic starch at any given water content.[227,250,266,279,286] The fall in modulus with the addition of polyols at the glass transition range becomes somewhat smoother and shifts to lower water contents[266,279,280,286] (Figure 8.7). At high temperatures and low water contents during extrusion, conditions leading to a high T_g for the starch extrudate, the material (foam) solidifies more quickly as it emerges from the die, yielding a low-density product.[62] In this respect, it has been shown that a rapid setting of the foam also favors the formation of non-equilibrium structures; e.g. for V-amylose complexes, the E_h is preferred compared to the V_h form.[60–62] The amylose component at a level of 5–20% provides elasticity and crispness to extruded starch products,[287] whereas maximum expansion of starch extrudates is observed at 50% amylose content.[288] Moreover, the strength of water- or water/glycerol-plasticized films increases with amylose content,[289] while the tensile strength of starch films decreases and elongation at break increases with the amount of plasticizer.[289,290] Excellent films can be made from casting starch-poly(vinyl alcohol) (PVA) mixtures in the presence of

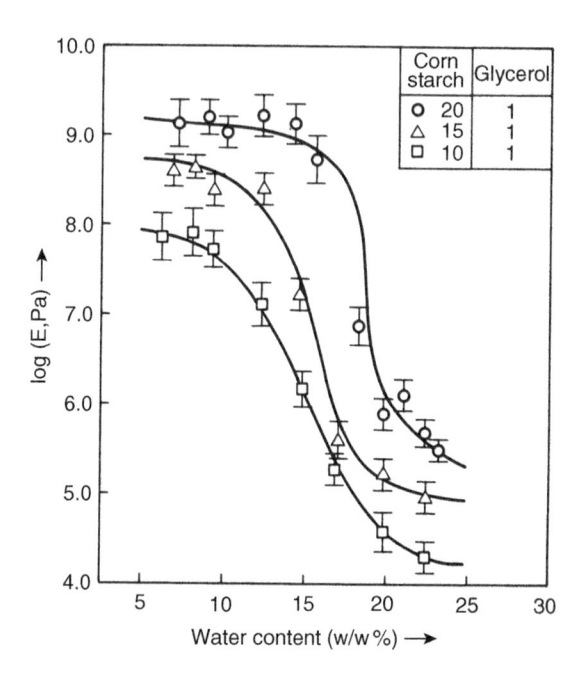

Figure 8.7 Flexural modulus of gelatinized corn starch plasticized by glycerol as a function of water content.[284]

glycerol and poly(ethylene-co-acrylic acid)[291] (see also Chapter 20). Compared to waxy starches or starches containing normal levels of amylose, films made with high-amylose corn starch and PVA had good elongations, tensile strength and tear resistance, and showed stability in their properties on storage at different relative humidity environments.[289] Gas permeability data for starch films indicate that incorporation of water and polyols results in extensive plasticization of the polymer matrix (widening of the structure) and enhances the diffusion rate of gases through the film.[249,286] Gas permeability and diffusivity plots versus 1/T (K) show an inflection point in the region of glass transition, allowing for an indirect estimate of T_g of the plasticized films.

Thermo-mechanical analysis (TMA)[26] and DMTA[292] have been used to study the viscoelastic properties of bread, a composite two-phase system. The moisture dependence of T_g was similar to that of pure amorphous starch and gluten.[242,282,293–295] At the moisture content of bread, water exerts its full plasticization effect on the composite polymer matrix, reducing the effective T_g to about $-10°$ to $-12°C$,[26,292] well below normal shelf life storage temperatures of this product.

Intensive research activity on the glassy behavior of starch and starch-based products over the past decade has provided invaluable information on the glass transition and material properties of starch. The recognition that water and other small molecular weight compounds can effectively plasticize starch, and lower its glass transition temperature, has many practical implications on processing behavior, physical properties and stability of the end product. For example, baked wheat products, immediately after baking and during cooling–dehydration, undergo structural transformations via a broad distribution of non-equilibrium rubbery states of their polymeric components (starch and gluten) to leathery and possibly to glassy states, depending on product composition (sugar and/or moisture content) and processing conditions.[22] Such state transitions are accompanied by drastic changes in the mechanical properties (texture) and stability of the composite network. Also, many processed cereal foods (e.g. breakfast cereals, wafers, biscuits) have a moisture content in the range where the T_g is close to storage temperatures and, therefore, small deviations in moisture content of the product or its temperature are accompanied by changes in the rheological properties of several decades in magnitude. The result of going through the glass–rubber transition for such products may be an undesirable texture, perceived as loss in crispness or crunchy character. The importance of recognizing the T_g depression by diluents and the need to identify an operative (effective) glass transition temperature zone of starch-based products lies in the control by T_g over processes such as annealing, gelatinization and (re)crystallization of starch.[11,15,16,20,22,117,225,226] All these phenomena can proceed only within the temperature range defined by the T_g and T_m; e.g. below T_g, crystal growth (starch recrystallization) rates practically cease, while above T_m, the melting rate is greater than the crystallization rate. Manipulation of T_g (via control of moisture or product formulation) in relation to storage conditions (temperature, relative humidity environment) is a means of maintaining quality and extending the shelf life of food products.[17,226,231,232] For example, pasta, and other extruded and thermally-processed cereal products of low moisture content, exhibit an elevated T_g, and frozen foods are kept at freezer temperatures below the effective T_g of the product.

4. Annealing and Structural Modifications by Heat–Moisture Treatments

Starch granules, being partially crystalline entities, are prone to modest molecular reorganization (annealing) when held in excess of water at temperatures above the T_g and below the equilibrium T_m of the crystallites, i.e. softening of the amorphous granular regions causes crystallite growth and/or perfection. This is manifested by an increase in gelatinization temperature and a narrowing of the temperature range over which the heat-treated granules melt. The effects of annealing on wheat starch gelatinization were first demonstrated by Gough and Pybus[293] using Kofler hot stage microscopy. Their findings were confirmed later by DSC measurements.[14,125,297–302] For corn starches, a more efficient multi-step process, which allows for higher annealing temperatures than a single-step treatment, was proposed.[299] Depending on the annealing temperature, time and starch type, there might be an increase in the gelatinization enthalpy of starch granules,[125,297–299,302–306] implying a net increase in the level of short-range molecular order. For corn starches from normal and mutant genotypes, the ΔH increase is inversely proportional to the degree of crystallinity of the native starch.[297,299] For potato starches, the shift in gelatinization temperature was higher with samples of low degree of phosphorylation, whereas the ΔH increase was largest with samples of high degree of phosphorylation.[306] However, if the temperature of annealing exceeds a certain limit, there is a reduction in ΔH, presumably due to partial melting of the most unstable crystallites or granules.[14,299,301] Annealing is also accompanied by a progressive loss in the inter-cluster distance of ~9 nm detected by SAXS,[49,301] a substantial loss in long-range radial order (birefringence)[301] and no apparent change[14,301] or a slight increase[307] in crystallinity as assessed by WAXD. A change in the polymorphic form, as a result of annealing, has not been reported, i.e. the degree of order increases without altering the type of crystal lattice. The d-spacings of the B-pattern remained unchanged upon annealing of potato starch.[306]

Swelling and solubility are both affected by annealing, and usually decrease.[308,309] Viscoamylography of annealed starches frequently shows increased consistencies of the cooled pastes, a feature attributed to granular stiffening.[302] However, observed changes in the pasting properties of annealed starches and the leaching behavior of amylose during heating do vary with the source of starch and the hydrothermal treatment used for their preparation.[302,310]

An analogous hydrothermal process, known to induce major structural alterations in granular starches, is that described as 'heat–moisture treatment.' Typical conditions for such treatments are temperature 90–110°C, moisture content 18–27% and conditioning period 12–16 hours. Early studies[311–316] have shown that the physicochemical properties of heat–moisture (H/M)-treated starches change dramatically without any gelatinization occurring. H/M treatment of cereal, legume and tuber starches increases the gelatinization temperature and its range, the water binding capacity, the stability of starch pastes to heat and shear, and in most cases granule susceptibility to enzyme-catalyzed hydrolysis.[310–320] The same process also decreases the swelling power and, for most starches, their hot paste consistency and solubility (amylose leaching).[312,318,319] The gelatinization enthalpy of wheat, oat and lentil starches remained unchanged, but those of potato and yam starches decreased on H/M treatment.[318] The magnitude of these changes are dependent on the moisture content and starch type; the effects are particularly marked with potato and other tuber starches, which become more like wheat starch in terms of rheological behavior.[312] The changes in physical properties of H/M-treated potato starch have an improving effect on its baking performance,[313] whereas the functional properties of cereal starches usually decline.[315] Electron microscopy revealed a large hollow region at the center of the H/M-treated granules of potato and corn starches,[320] implying major structural rearrangements of the granular material. In a DSC study, Donovan et al.[318] studied the structural changes induced in wheat and potato starches by H/M treatment (100°C, 16 hours, moisture content between 18% and 27%). Thermal profiles indicated a progressive shift from a single melting endotherm (native granules) to broader and biphasic transitions (H/M-treated starches) as the moisture level increased during the heat treatment stage. Defatting of the starches prior to the hydrothermal treatment resulted in sharper peaks and a narrower range for the biphasic transitions. These observations clearly pointed to an extensive chain reorientation in the crystallites of the granular starches.

At a molecular level, x-ray diffraction patterns of cereal starches (A-type) indicate that their chain packing remains unchanged after H/M treatment, except for the development of V-type structures. In contrast, the B-pattern of tuber starches shifts toward an A-pattern, as readily recognized by the loss of the characteristic d-spacing at 15.8 Å.[1,30,52,311,317,318,320] For cassava (tapioca) starch, a change from a C-type to an A-type pattern has been observed.[314] These structural alterations caused by H/M treatments indicate a closer packing of the double helices, i.e. there is a shift from the less stable polymorphs, B and C, to the most stable one, A. The crystallinity for cereal and tuber starches is usually reported to be reduced,[314,321] although Hoover and Vasanthan[315] have shown improved x-ray patterns for H/M-treated cereal starches. The polymorphic transformation of B- to B+A- or A-type has recently been examined by calorimetry and x-ray diffraction using short DP amylose of the B polymorph (potato lintners with a DP ~15).[322] The endothermic heat flow regimes of the lintners, heated at moisture contents above 20% (db), showed the presence of a minor transition ('T,' Figure 8.8) preceding the main melting peak ('M') of the crystallites. This endotherm (T_m ~90–100°C, moisture content 20–42%) was not observed with A-type lintners. The structural origin of the 'T' endotherm was related to the polymorphic phase transition of B- to A-type, which involves a change in the packing arrangements of chain duplexes in the crystal lattice. The energy level involved in this type of lattice disruption was small (ΔH ~1–4 J/g) compared to the melting enthalpy of the crystallites (~20 J/g). The interpretation for the structural origin of the 'T' endotherm was further supported by x-ray diffraction measurements performed on the potato lintners after their heating stepwise at increasing temperatures (e.g. between 60°C and 152°C, at 35% H_2O content, db, Figure 8.8). Below 20% water content, the 'T' endotherm was no longer observed in the DSC curves, and there was no evolution of the B-type d-spacings in the x-ray diffraction patterns of the thermally treated lintners up to temperatures where thermal degradation occurred. A minimum water content of ~20% is thus required to facilitate the structural transition from the hexagonal metastable B structure into the more compact, monoclinic A structure. In this context, a molecular model for the solid-state transition of B- to A-starch, involving the progressive removal of interhelical water and a unidirectional slipping of the chain duplexes (Figure 8.8), has been proposed by Perez et al.[320]

Changes in the physical properties of heat-treated starches under limited water conditions have not been exploited commercially to any degree. Production of H/M-treated starches has been largely based on an empirical design of processing conditions (temperature, moisture content, time). Processing protocols designed from well-defined state diagrams of granular starches, in which T_g and T_m versus moisture content relationships are described, would appear to be more appropriate for generation of materials with consistency in their functional properties. An increase in the gelatinization temperature, obtained by hydrothermal treatments, could be an advantage in thermal processing of some foods. Since starch swelling and solubilization are restricted by H/M treatments, there might be a change in the competition

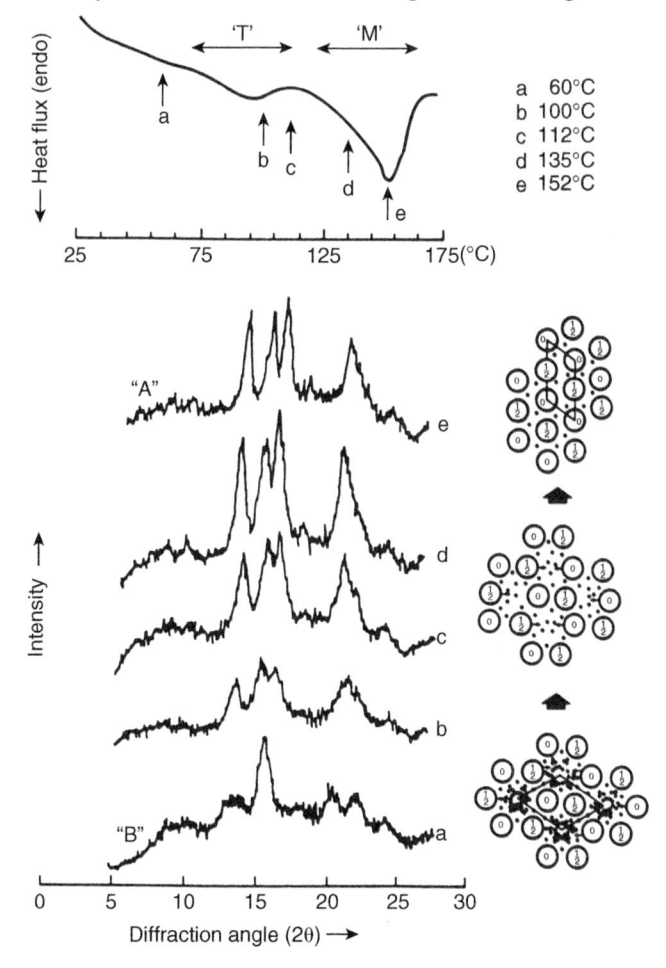

Figure 8.8 The B- to A-type transition of B-type potato starch lintners (DP 15) at 35% water content (d.b.). The arrows on the DSC thermal curve specify the temperatures at which the lintners were heated prior to x-ray analysis.[320] The model of Perez et al.[320] for the solid state polymorphic transition B to A is also shown, illustrating the progressive removal of water molecules (small dots) and the change in packing of chain duplexes; the chain duplexes marked with 0 and ½ indicate their relative translation along the c axis.

for available water between starch gelatinization and protein denaturation processes during heating of a composite food matrix; this in turn could affect the structure development and the textural attributes of the end product. For bread and other bakery items, a delayed gelatinization may also prolong the period of loaf expansion. A greater softness in bread baked with annealed starches has been reported.[324]

5. Melting Transitions of Crystallites in Granular Starch

Application of DSC to starch has prompted new interest in, and provided new insights into, the nature of gelatinization and granule structure. There are several endothermic events detectable by heat flow calorimetry on heating aqueous suspensions of granular starches. For example, the DSC trace of Figure 8.9 for a rice starch sample (22% amylose) at 50% (w/w) moisture content shows four distinct endothermic transitions. Despite the existing controversy as to the exact structural origin of the events responsible for the low temperature (<100°C) M_1 and M_2 endotherms, there is no doubt that these transitions (thermally irreversible on immediate reheating) reflect melting of ordered chain domains of amylopectin.[16,19,52,87,115–117,240] Total losses of granule birefringence[120] and crystalline properties[79,92] are associated with passage through the M_2 endotherm. On the other hand, the two high-temperature (100–130°C) transitions, M_3 and M_4 (reversible on immediate reheating), have been attributed to order→disorder processes of amylose–lipid complexes differing in the degree of helical chain organization in the solid state;[65–69,321,325] waxy and defatted

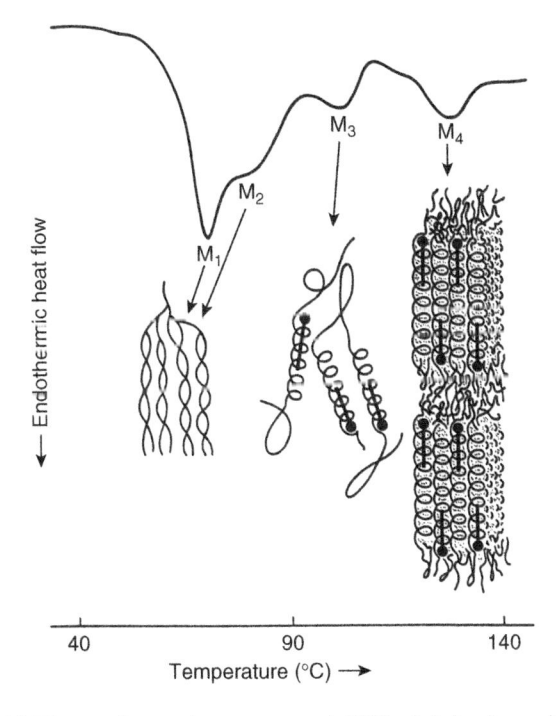

Figure 8.9 A typical DSC trace of a cereal non-waxy starch (50% w/w) showing multiple melting transitions and the respective structural domains undergoing a phase change.

or low lipid content starches do not show the M_3 and M_4 transitions. The structural changes underlying the latter endotherms are discussed in a subsequent section.

By analogy with synthetic polymers, the melting behavior of starch crystallites, manifested by the M_1 and M_2 transitions, has been examined as a function of the diluent (water) content for many granular starches of different botanical origin.[14,87,115–117,120,321,326–333] Typical DSC thermal curves for a waxy rice starch of varying water content are presented in Figure 8.10. At relatively high water levels, where the volume fraction of starch is less than about 0.45, a single melting endotherm (often denoted as the gelatinization endotherm, G) is observed at a temperature characteristic of the starch specimen being investigated.[25] When the volume fraction of starch is increased, a multiple melting profile is observed, implying that the granular structures respond differently at low moisture environments. The first transition (M_1) occurs at the same temperature as before. The development of the M_2 transition at intermediate water levels, and its progressive shift to higher temperatures as the moisture content decreases, have led to several hypotheses for the structural origin of the observed peak multiplicity in the thermal profiles.[11,16,240] Donovan[112] was the first to demonstrate the complex interplay between starch gelatinization temperature, moisture content and overall melting profile of the structural domains in granular starch. According to his view, when starch is heated in the presence of excess moisture, crystallites are destabilized and melt cooperatively as a result of tensions exerted from the adjacent amorphous regions which are fully hydrated and swollen; this gives rise to the M_1 endotherm. As water becomes a limiting factor, only part of the crystallites destabilize with this mechanism, whereas the remaining give rise to the second endotherm, M_2, in accordance with the Flory–Huggins theory of polymer–diluent interactions.[334] This theory, which is based on equilibrium thermodynamics, describes the depression of the equilibrium melting point, T_m^0, of a polymer to a new value, T_m, as a function of the volume fraction of the diluent (solvent), v_1: $(1/T_m) - (1/T_m^0) = (R/\Delta H_u) \cdot (V_u/V_1) \cdot (v_1 - \chi v_1^2)$, where V_u and V_1 are the molar volumes of the polymer repeating unit and diluent, respectively, ΔH_u is the enthalpy of fusion per repeating unit, and χ is the Flory–Huggins interaction parameter whose value is determined by the solubility of polymer and diluent in one another. From this equation, it is predicted that the smaller the molar volume of the diluent and the stronger the interaction with the carbohydrate, the greater the depression in T_m^0 with increasing volume fraction of diluent. Since this theory applies only to equilibrium crystals, its application to phase transitions of starch–water systems has achieved only limited success. As with other partially crystalline polymers, most starch structures exist as metastable states which are amenable to perfection (annealing) on plasticization with water and thermal treatments.[14,16,25,117,298–300,303,304] Lelievre,[332,333] using hot stage microscopy, demonstrated that equilibrium conditions in heated aqueous starch systems can be approached only under very slow heating rates (~1°C/3 hours), far below those commonly used in DSC experiments. Slade and Levine[14,236,223] have questioned the Flory–Huggins treatment of DSC melting data of starch on the grounds that true equilibrium conditions are not met during gelatinization, and that melting of starch crystallites is controlled by the glass–rubber transition of the amorphous regions, i.e. the amorphous granular material has a depressing effect on crystallite perfection and melting temperature. Whittam et al.[262] suggested that estimates of T_m^0 for starch crystallites (of about 257°C), using the

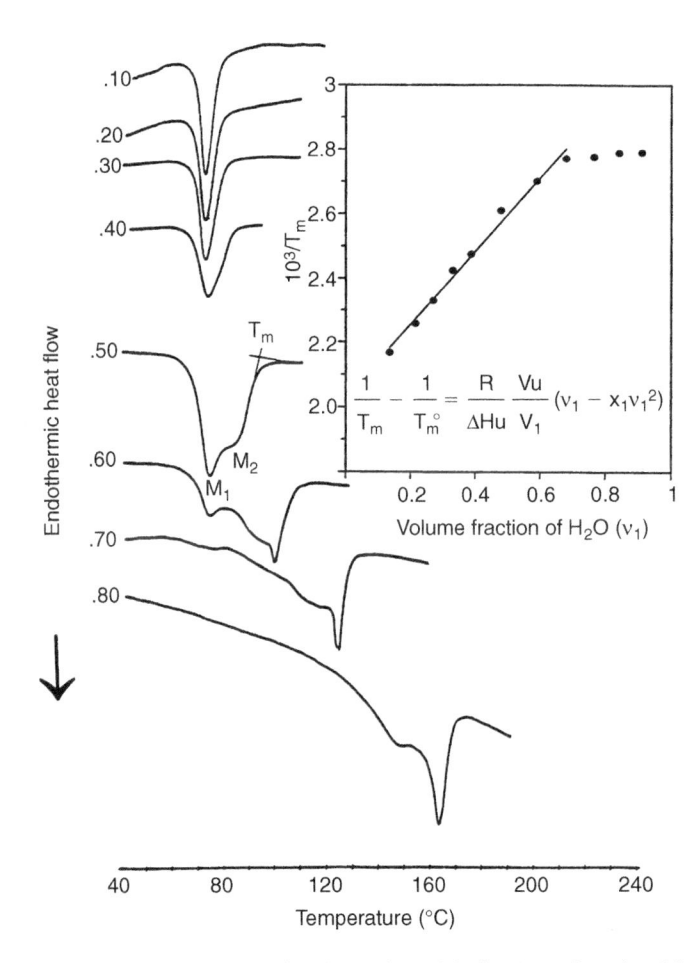

Figure 8.10 DSC traces of a waxy rice starch at increasing weight fractions of starch solids; the inset shows a plot of the melting data according to the Flory–Huggins theory.[20]

Flory–Huggins approach, may be somewhat low even with highly crystalline preparations of A- or B-type starch lintners (DP ~15). Despite its theoretical limitations, the Flory analysis (as illustrated in Figure 8.10 inset) can be used to simulate the melting behavior of starch in practical applications (e.g. extrusion cooking, baking) and to compare the thermal stability of different starch materials under dynamic heating at various moisture conditions.[20,25,240,337]

An alternative explanation for the biphasic endothermic events at intermediate water contents was given by Evans and Haisman,[84] who suggested that crystalline zones in different granules vary in their thermal stabilities. Granules with the least stable crystallites melt cooperatively first (M_1), and this results in absorption of more water by the disordered chains. This means that the remaining crystallites and granules have less water available (effective volume fraction of diluent is reduced) and, therefore, will melt at a higher temperature (M_2). This view has been subsequently supported by Liu et al.,[111,335] who claimed that the positions of melting endotherms in DSC thermal curves are governed by the melting of granules with different thermal stabilities, heated at continuously changing volume fractions of available diluent.

Other interpretations for starch structure loss in limited water systems invoked crystallite disruption followed, at a higher temperature, by double helix melting[124,328] or disordering of double helices associated with short-range order (M_1) followed by melting of crystallites (M_2).[330] These latter models, however, are not consistent with the findings that both double helical order (assessed by solid-state ^{13}C-NMR or DSC) and crystalline structure (x-ray diffraction) are lost concurrently during heating for most granular starch systems examined.[50,79,92,114,240] Recently, Svensson and Eliasson[89] attributed the biphasic endotherm to a slow plasticization of the amorphous granular regions under restricted water conditions, which shifts the melting of crystallites to higher temperatures.

Tomka et al.[240,280] examined the structural changes occurring during heating of potato starch in the range of 10–20% water content using several complementary analytical probes (DSC, x-ray diffraction and solid-state ^{13}C-NMR). At 18% water content, the heat flow-temperature diagrams were characterized by at least four irreversible endothermic transitions, which were denoted by G (in the range of 60–70°C), S_1 (at 110°C), S_2 (at 155°C) and H (in the range of 165–200°C, depending on the ionic environment). ^{13}C-CP/MAS-NMR and WAXD data gave no indication of any significant changes in the fraction of double helical structures or the relative crystallinity by heating through the G endotherm; this transition merely reflects the loss of birefringence and hydration/swelling of the amorphous granular phase. The melting process of the crystalline domains were attributed to both S_1 and S_2 transitions; a progressive loss of crystallinity and helical order occurred by going through these two endotherms. Thus, a two-stage melting process was proposed: first, there is disintegration of large crystallites into smaller units along with a concurrent loss of ~50% of the native helical conformations (S_1 endotherm). At this stage there is considerable molecular mobility, and it encompasses the conditions under which structural rearrangements of the chain duplexes might occur, i.e. the well-known 'B-' to 'A-' transformation. In the second stage (S_2 endotherm), all crystalline and double helical structural domains eventually disappear. More precise investigations are required to test this melting mechanism under varying moisture content regimes involving granular starches of both A- and B-type. The changes in short-range structure (FTIR) and crystallinity (WAXD) during melting of potato starch at 18% or 26% water contents were also studied by van Soest et al.[77] Although the losses in both levels of structure seemed to follow parallel paths, their data indicated that long-range order was lost slightly before short-range order in the heated granular material.

In many studies of starch melting behavior, the non-equilibrium character of native starch structures and the kinetic constraints related to the state transformations of starch in the presence of water have often been ignored. Blanshard[10] has thoroughly reviewed the non-equilibrium aspects of starch gelatinization. SALS measurements of heated aqueous dispersions of granules revealed that, within partially gelatinized granules, rearrangements of the polymer chains in the amorphous phase and the remaining crystallites might occur.[339] In view of the inherent molecular heterogeneity and the non-equilibrium nature of crystallites in native starch,[14–16,88,117] the interdependence between amorphous and crystalline domains,[16,20,117] and the thermal behavior of V-amylose complexes,[63–66,68,69] it was proposed that the multiple melting profiles observed during heating of starch at intermediate to low water contents may reflect melting and reorganization events occurring concurrently in a dynamic DSC experiment.[16,25,117] The multiple melting transitions (Figure 8.10) may thus be a direct manifestation of composite thermal effects, due to partial disruption of ordered structures (endothermic) and chain reorganization (exothermic) events. The structural reordering is of rather short-range scale, non-detectable by x-ray diffraction. In this respect, the melting behavior is governed by non-equilibrium factors (crystallite size, perfection, time-scale of the process, etc.). Certainly, this interpretation does not preclude any possibility for development of water gradients in the sample during heating, which would affect the melting process of the more resistant granules or crystallites due to changes in the effective diluent concentration, and thereby result in a broad and multiple endothermic heat flow–temperature profile. Much remains to be learned about the mechanism of starch structure melting behavior, particularly in low moisture granular systems. In future studies, it is important that a combination of techniques is applied to probe changes in the structural order of starch in order to elucidate underlying molecular events.

The temperature interval over which the loss in structure occurs during heating of dilute starch granule suspensions has been the subject of several investigations. Quantitative measurements can be made using various probes of structural order (of short- and long-range), which usually show different temperature dependence patterns for the disappearance of granular order (Table 8.2). For waxy maize starch, residual birefringence was consistently at a lower level than the extent of crystallinity (x-ray diffraction) or the short-range order (DSC, solid-state ^{13}C-NMR).[79,340]

Table 8.2 Relative losses[a] in structural order upon heating starch granules in excess water, as probed by different techniques;[77,338] values across each line give levels of residual order after heating at a constant temperature

Starch source	Birefringence loss,%	X-ray crystallinity loss,%	Molecular order loss (^{13}C-CP/MAS NMR),%	Molecular order loss (DSC),%
Waxy maize	10	3	14	14
	50	24	24	24
	70	35	38	41
	80	51	41	50
	90	71	62	77
Regular maize	n.d.[b]	4	9	6
	n.d.	15	19	29
	n.d.	30	35	36
	n.d.	51	42	45
	n.d.	67	53	62
Potato	n.d.	19	5	6
	n.d.	54	37	30
	n.d.	74	50	55
	n.d.	79	52	60
	n.d.	90	58	75

[a] Expressed relative to the values determined for the native granular starches
[b] n.d. = not determined

The observed reductions in crystallinity and molecular order, as assessed by both solid-state NMR and DSC, indicated more or less similar levels of change over the temperature range studied. Thus, for waxy maize starch at least, Gidley and Cooke[76] were unable to conclude whether crystallite disruption or loss in double helical order occurs first during gelatinization. For corn starch, the decrease in crystallinity commenced even before birefringence began to disappear, and continued well after all birefringence was lost.[114] With the possible exception of potato starch, other granular starches exhibit a parallel loss of crystalline and molecular order;[79] for potato starch granules, complete loss in crystallinity occurs at a lower temperature than that for disappearance of short-range order.[79,170]

Attempts to compile DSC gelatinization data of native starches[94,116,117,125,170,328,330,341–345] from the same or different botanical origin are of questionable significance, if one considers that, besides the granule structure organization effects, the melting transition parameters (T_m, ΔH_{gel}) are influenced by the heating rate, moisture content, mechanical damage, thermal history of the sample (annealing effects), extraction conditions of starch, etc. According to the calorimetric studies of Shiotsubo and Takahashi[343] on potato starch, the gelatinization peak temperature increases with increasing heating rate (superheating effect); a rise in heating rate to 20°C/minute resulted in an increase of the transition peak by about 10°C. It was also shown that, only at heating rates below 0.5°C/minute (well below those found in conventional DSC studies), the melting peak temperature remains constant, thus approaching equilibrium conditions. As to the structural origin of the DSC endothermic enthalpy values, the general consensus is that they primarily reflect loss of double helical order, rather than loss of crystalline registry.[79,265,347] This infers that the forces responsible for the structural stability of starch granules are largely at the double helical level and that chain packing energy contributions are very small. For highly crystalline preparations of A- and B-type spherulites, obtained by crystallization of aqueous solutions of potato starch lintners (DP ~15) or debranched glycogen under different conditions, the DSC melting data indicated a surprisingly similar enthalpy value for the two polymorphs (~35J/g),[79,265] and a difference in melting temperatures of 15–20°C at a fixed water content.[265] The more closely packed A-crystals, possessing a more stable structure,[44] had a higher T_m than the less densely organized B polymorph; in excess water, the B-spherulites of potato starch lintners melted at 75°C.[265] Despite the observed difference in T_m between the model crystalline materials, melting temperatures of granular starches do not necessarily follow this rule, e.g. wheat starch granules (A-type) usually exhibit lower melting temperatures than those of potato starch (B-type).[30,79,94,341,342,348] Obviously, crystallite size and perfection of ordered chains, as well as the depressing influence of the amorphous regions, have a major impact on the thermal stability of the ordered chain domains in granular starch.

Reported gelatinization enthalpies of native starches are generally in the range 5–20J/g. The gelatinization enthalpy (ΔH_{gel}) of granular starch represents net thermodynamic quantities of different events: granule swelling and crystallite melting (endothermic); hydration and chain reordering (exothermic).[17,321] There is generally a good agreement between DSC transition temperatures and birefringence end point temperatures of native starches.[116,117,162,345] Large variations in DSC melting parameters (ΔH_{gel}, transition temperatures) among different genotypes of corn,[297,343–345,349–353] rice,[125,163] barley,[54] amaranth,[354] sorghum,[355] legume[116,328,332] and potato[356] starches have been reported. For starches of similar genetic origin, handled under identical processing protocols to minimize contributions from annealing effects, amylose content does not seem to correlate with T_m or melting enthalpies of starch crystallites.[163] Most studies involving thermal properties of corn starches have concentrated on the effect of mutant endosperm genes, such as waxy (*wx*), amylose-extender (*ae*), dull (*du*) and sugary-2 (*su2*) (see Chapter 3). Genetic variability in physical properties of starch, as quickly detected by DSC in a microscale test, may be useful in breeding programs aimed at screening germ plasm for desired properties of starch. For example, among three *wx*-containing genotypes (*wx*, *ae wx* and *du wx*), *ae wx* was found to have the highest ΔH and gelatinization temperature.[352] Moreover, starches from corn mutants with the *ae* gene frequently show broad endothermic peaks or even a double melting transition for starch crystallites in excess water, indicative of the heterogeneity of ordered structures inside their granules.[352,353] In contrast, the *su2* allele in corn results in starches with lower gelatinization temperature and melting enthalpy than normal starch.[350,351] Environmental factors (e.g. temperature during grain filling, soil fertility, planting date and location) may also influence the structure and the thermal properties of starch from corn,[357] wheat[358] and potato[356] genotypes.

The partial loss of granular integrity when starch[359] or flours[341,359,360] are subjected to shear or compressive forces (e.g. milling) is reflected by a reduction in transition temperature and melting enthalpy of the mechanically damaged granules. The overall loss in order, as measured by x-ray diffraction, DSC and solid-state ^{13}C-NMR, is dependent on the severity of milling conditions and milling time.[359,360] Stevens and Elton[338] have attributed differences in ΔH_{gel} of wheat flours to varying levels of mechanical damage. The results of Nishita and Bean[358] on rice flours, as well as those from studies on hard red spring farina,[360] concur with this view. For composite flours and fines, a negative correlation was shown between gelatinization ΔH_{gel} and starch damage, as assessed by an enzymic method.[360]

In most applications of starch, other components (sugars, salts, other hydrocolloids) are also present and, as a result, affect the gelatinization and the pasting behavior of starch. Understanding these interactions is important for better process control and for improving the texture and other quality attributes of starch-based products. Sugars and polyhydric alcohols, in general, have long been known to impede granule swelling, raise the temperature of gelatinization and elevate the temperature of viscosity rise (pasting).[87,189,362–370] There might even be restrictive effects on the extent of conformational disordering of starch chains on gelatinization of wheat starch, as inferred by the reduced enthalpy values observed in the presence of sugars.[363,367,368] However, in other studies, the ΔH_{gel} has been reported to be unaffected[87,370] or even increased.[110,369] Seow and Vasanti-Nair[301] have shown that the effect of sucrose on the ΔH_{gel} is dependent on the physical condition of the glassy regions in the granules. Sucrose added to unannealed rice starch resulted in increased ΔH_{gel}. In contrast, addition of sugar to physically aged granules (possessing glassy regions of lower energy) did not seem to affect the observed ΔH_{gel}. At the concentration used in cakes (55–60%), sucrose delays gelatinization of wheat starch from 57°C to 92°C.[362,368] Sugars also narrow the gelatinization temperature range.[368,369] The double endothermic transition observed when starch is heated at intermediate water contents is transformed into a single melting peak with the addition of sucrose.[369] Furthermore, granules isolated from high sucrose content baked products exhibit decreased granule deformation, compared to those from products with low sugar levels.[371,372] The magnitude of these effects increases with the sugar or polyol concentration, whereas at an equivalent solute concentration polyols with higher numbers of hydroxyl groups are the most effective.

Several hypotheses have been put forward to explain the influence of sugars in stabilizing the supermolecular structure of the starch granule and thereby delaying gelatinization. These include competition for water between starch and sugar (depression of a_w),[373] sugar–starch interactions[278,336,364] and an 'anti-plasticization' mechanism by the sugar–water co-solvent.[15,239,374] The antiplasticizing effect of sugars has been explained with a decrease in the free volume, relative to the plasticizing action of water alone. The co-solvent, having a greater molecular weight than water itself, is less effective in increasing the free volume and in depressing the T_g of the amorphous material in starch granules. The reduced molecular mobility of the amorphous domains would then result in a higher gelatinization temperature. For ternary starch–sugar–water systems, Slade and Levine[14,371] presented linear relationships for the dependence of gelatinization temperature on the amount of sucrose and the inverse molecular weight of the co-solvent (water + sugar) using a series of sugars as co-solvents. With a series of 1:1:1 sugar–water–starch mixtures, the gelatinization temperature increased in the following order: water alone < galactose < xylose < fructose < mannose < glucose < maltose < lactose < maltotriose < DE10 maltodextrin < sucrose.[15] The NMR data of Chinachoti et al.[107] also showed that solvent mobility, as measured by ^{17}O relaxation times (T_2) in a starch–D_2O–sucrose mixture, decreases drastically upon addition of the sugar. It was then suggested that such reduced mobility of the solvent in the presence of sugar may play a role in lowering the effectiveness of plasticization of starch by the co-solvent. However, it was subsequently shown by the same research group that the reduced mobility of D_2O is limited to only some concentrations of the sugar.[375] On the basis of these findings, it was suggested that both solvent mobility and a decrease in free volume (related to the molecular weight of the co-solvent) affected by the added sugars may contribute to the inhibitory effect these solutes have on starch gelatinization. van Soest[275] has recently proposed that sugars act as 'crosslinking' agents stabilizing the polymer chains in the granular structure via intermolecular hydrogen bonding. Attempts have also been made to explain and predict the thermal responses of granular starch in the three component system of water–starch–sugar on purely thermodynamic grounds by using an extension of the Flory equation.[11,87,336] Although this approach appeared to provide a convenient framework for analyzing starch melting data, its development is based on equilibrium melting and on making simplifying assumptions regarding the magnitude of interaction parameters between solvent, solute and polymer. However, in view of the non-equilibrium nature of starch gelatinization, as discussed above, as well as the uncertainty concerning the role of solute on processes like annealing, such treatment of gelatinization data is inappropriate.

Starch–gum interactions during heating of their aqueous dispersions are also important for processing and end product quality attributes for products such as bakery items and low-calorie foods. Polysaccharide hydrocolloids (e.g. CMC, xanthan, gellan, guar and locust bean gums) at low concentrations (0.5–2.0%), despite a large increase in the viscosity of the composite starch paste matrix,[376,377] cause very little change in the DSC gelatinization parameters (ΔH_{gel}, T_m) of starch.[377–379] However, at high concentrations, the effects of hydrocolloids on phase transition behavior of starch are similar to those of sugars at equivalent concentration.[380,381]

Effects of salts on gelatinization are more complex than those of nonionic solutes, as they are solute-specific and concentration-dependent. For most electrolytes, an increase in gelatinization temperature, T_m, is found at low salt concentrations. At high salt concentrations, T_m decreases and can even drop below the initial value. This behavior was shown for potato starch with increasing concentration of sodium chloride and sodium acetate,[87] and for several cereal starches in the presence of sodium chloride.[370] Granules of A-type starches are more resistant to the decline in their T_m as the concentration of NaCl is raised above 4 M than are tuber starches, presumably because of their more compact structure;[382] however, the multiple melting profiles of starches at limited water levels were not affected by this salt. A similar trend was also shown for calcium chloride which, at intermediate concentrations (~3 M), brings about swelling and gelatinization of starch granules at room temperature (cold gelatinization). Starch granule suspensions in >4 M $CaCl_2$ exhibit exothermic transitions on heating, which suggests that the starch is dehydrated at these salt concentrations and that, on melting, the exothermic process of hydration obscures the endothermic effect of crystallite structure disruption.[87] In general, the effect of neutral salts on starch gelatinization follows the order of the classical Hofmeister (lyotropic) series, particularly in the case of anions.[87,383] Thus, sodium sulfate strongly inhibits swelling and gelatinization of starch, and this makes it suitable as a stabilizing solute in chemical derivatization reactions involving granular starch.[87,84] Sodium bromide, on the other hand, decreases the gelatinization temperature of potato starch.[384] Cations of different chloride salts exhibit a more complicated behavior as evidenced with polarizing microscopy.[385] According to Oosten,[383,384] the effects of electrolytes are related to the weak acid character of starch. The latter stems from the dissociation of one or more hydroxyl groups according to the equation starch–OH $\leftrightarrow$ starch–O$^-$ + H$^+$. At the equilibrium, the starch carries a negative charge and the water phase a positive charge. The resulting potential between the starch granules and the water phase, known as the Donnan potential, tends to prevent the penetration of negative anions in the granules. Oosten[384] claimed that anions are the actual gelatinizing agents, whereas cations exert a protective effect on granule structure. Jane[385] has attributed the electrolyte effects on starch gelatinization to: (a) indirect effects of the ions on water structure (enhancement or disturbance of the hydrogen-bonded network of water molecules depending on the charge density and size of the ion); and (b) electrostatic interactions between the ions and the polymer chains. Undoubtedly, additional information should be acquired to fully elucidate specific interactions between ions and starch, and thus predict the direction and magnitude of salt effects on starch gelatinization.

Another structural modification of starch, characterized by a change in supermolecular order while maintaining its granular form, is that found in the production of granular cold-water-swelling starches (GCWS). The unique characteristic of these materials is that, although in a granular form, they provide an 'instant' paste in cold or warm water. The GCWS can be prepared by several methods: (a) heating starch in aqueous monohydric alcohol solutions (e.g. 75% ethanol, w/w) at high temperatures (150–170°C) under pressure;[389,390] (b) heating starch in aqueous polyhydric alcohol (e.g. propan-1,2-diol) solutions at elevated temperatures (>115°C, atmospheric pressure) and exchanging the solution with ethanol;[391] and (c) treating starch in a slurry with alcoholic alkali solutions at low temperatures (25–35°C).[392] The aqueous alcohol medium used during the thermal treatment restricts swelling and molecular dispersion of the granules, while facilitating melting of the crystallites. Cold water swelling of these materials has been attributed to the conversion of the double helical crystalline domains in native granules into single helices; single helical amylose and amorphous starch have been reported to be water-soluble at 25°C.[31] Indeed, the x-ray diffraction diagrams of non-waxy GCWS starches show the characteristic d-spacings of the V-polymorph.[393–395] The structural transformation of granular starch (native or chemically modified) for its conversion to a GCWS product thus involves melting of crystallites and formation of helical V-type complexes.[390,391]

6. Gelation and Retrogradation of Starch and its Polymeric Components

Gelation can be characterized as the process by which dispersed starch molecules in an aqueous medium begin to reassociate and form three-dimensional network structures. Interchain associations may involve both short- and long-range order depending on the extent of lateral associations between ordered chain segments. These structures constitute what are known as 'crosslinks' or 'junction zones' in a gel network. For physically crosslinked systems (non-covalent crosslinking), as those of polysaccharide gels, the density and stability of the intermolecular junction zones determine the mechanical properties of the network structure. For starch gels, non-covalent chain interactions are generally believed to involve intermolecular double helix formation stabilized by hydrogen bonding. The ordered chain microdomains that are produced are far from equilibrium, and will always be influenced by the thermal and processing history, as well as molecular composition (amylose, amylopectin) and polymer fine structure. Annealing processes at the crosslinking sites also occur on storage of starch gels, as long as the temperature is kept between T_g and the T_m of the ordered chain domains. Thus, any examination of starch gel structures and physical properties must be considered in the context of temperature- and time-dependence of gel curing events.

Amylose Gels

Aqueous solutions of amylose are quite unstable, and on cooling to room temperature show turbidity development and eventually form a precipitate or a gel depending on concentration and molecular size of the polymer chain. Gidley and Bulpin[393] have thoroughly examined the phase behavior and aggregation kinetics of nearly monodispersed amyloses of varying DP in aqueous solutions (0.2–5.0%). Short chain amyloses (DP <110) were found to precipitate at all concentrations up to 5%. On the other hand, gelation has always occurred with amyloses of DP >1100 in solutions above 1%. For chains of intermediate length (250 < DP < 660), either gelation or precipitation was observed depending on concentration and cooling rate. In general, precipitation (occurring for DP <1100) was favored by shorter chain lengths, lower concentrations and slower cooling rates, whereas gelation (occurring for DP >250) was favored by longer chain lengths, higher concentrations and faster cooling rates. The rate of aggregation and the physical state of the aggregated material were also found to be dependent on amylose chain length. The observation (from turbidity measurements on 0.5–2.0% amylose solutions) that maximum aggregation rates occurred at a DP ~100 was in reasonable agreement with earlier findings of Pfannemüller et al.[394] on aggregation kinetics of amylose in dilute solution (<0.1%); these authors, using light scattering, have found a maximum aggregation rate at a DP ~80. The variation of phase behavior and aggregation with chain length has been rationalized on the basis of interchain associations occurring over chain lengths of 80–100 glucosyl units.[396]

The gelation process of aqueous solutions of amylose has been the subject of many investigations,[77,156,160,182,193,197,398–402] as the linear component of starch profoundly influences its gelling properties. Kinetic experiments with amylose concentrations between 0.3–2.0% are usually carried out to resolve the sequential steps leading to the development of an infinite network structure from the homogeneous sol. In this respect, various molecular probes of structural order (e.g. turbidity, dilatometry, light scattering, x-ray scattering or diffraction, optical rotation, NMR, FTIR, viscometry and small deformation rheometry), responding to different levels of macromolecular organization, have been applied. The evolution of shear modulus (G') for amylose solutions reveals two stages[156,197,160,374,401] (Figure 8.11a): an initial rapid rise in G'; followed by a step of slower increases in modulus (pseudoplateau region). The initial rise in G' and attainment of the plateau region occur over shorter times with increasing amylose concentration.[400,401] The kinetics of G' are also dependent on amylose chain length. For monodisperse amyloses in the DP range of 250–1100 (at 2% solutions, 25°C), Clark et al.[398] observed that a plateau G' value was attained faster with amyloses of shorter chain length. Rheological data also indicated that amyloses of very high DP (2550, 2800) exhibit a very slow development in G'' under identical gelation conditions (Figure 8.11a). For such long amylose chains, it is possible that initial formation of relatively few crosslinks significantly retards chain mobility and thereby slows gelation kinetics. In other studies on amylose gelation,[399] it was also shown that a gel formed by rapid solution (3.4%, DP 3080) quenching to 25°C and a cure time of 24 hours had a lower limiting modulus of $\sim 3.5 \times 10^3$ Pa than a slowly cooled (4 hours to approach 25°C) system which reached a plateau value of $\sim 5.3 \times 10^3$ Pa. All these observations clearly point to the dynamic (non-equilibrium) nature of amylose gel structures, i.e. diffusion-controlled chain aggregation processes are important determinants of the network structure and thereby control its mechanical properties.

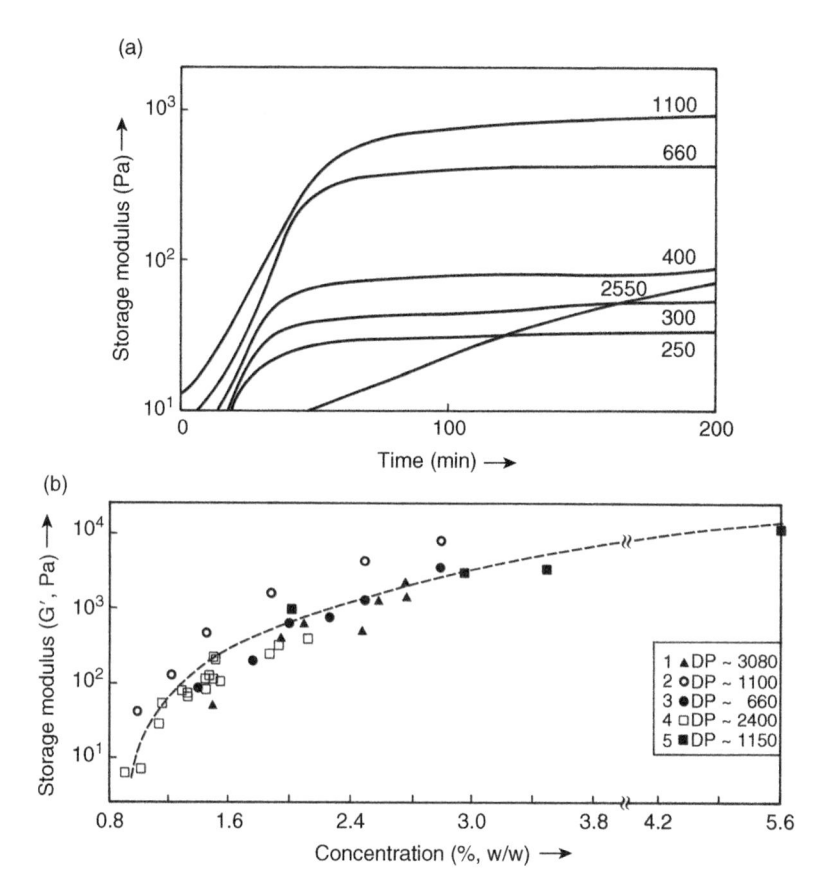

Figure 8.11 (a) Storage modulus (G′) versus time for 2.0% solution of nearly monodisperse amyloses of varying DP;[399] (b) master plot of limiting storage modulus data (G′) as a function of amylose concentration: 1;[397] 2 and 3;[399] 4;[398] 5.[158]

Gelation of amylose has been noted even at concentrations below the coil overlap concentration (C* ~1.45%), a parameter determined by viscosity measurements.[400,401] A minimum amylose concentration for gelation, C_o of ~0.8–1.1%, has been identified to be independent of the molecular size of the amylose.[401] It was, therefore, suggested that molecular entanglements do not necessarily determine the critical gelling concentration of this polysaccharide, as initially proposed by Miles et al.[194] This behavior is not consistent with what is expected for gelation of synthetic polymers for which significant chain overlap is a prerequisite for gel formation.[403] According to Clark and Ross-Murphy,[148] gelation at sub-coil overlap concentrations is possible when the crosslinks have finite energy, and this condition is probably met for amylose gels which are quite thermoresistant, melting at temperatures of ~140–160°C.[63]

Besides the formation of an elastic network, amylose gelation is also characterized by the development of opacity, which is generally attributed to chain aggregation.[197,400,401] For a polydisperse amylose preparation (DP 3080, 2.4% solution, quenched to 32°C), the increase in turbidity slightly preceded the onset of G′ development.[197] Crystallization, as detected by x-ray diffraction (intensity of the 100 diffraction peak), was shown to be the slowest process. In similar experiments with amylose solutions (1.78%, DP 2400) in 0.5 M KCl at 25°C, a fairly good correspondence between cloud time (onset of turbidity) and gelation time (crossover point of the G' and G" curves) was observed.[400] These findings supported the notion that network formation in amylose solutions is initiated by phase separation (demixing) in the homogeneous sol, which yields polymer-rich regions interspersed with polymer-deficient regions. Depending on polymer concentration, molecular size, solvent quality and quenching temperature, such liquid–liquid demixing may be a relatively slow process. Phase separation brings about enhancement of the local concentration of amylose in the polymer-rich regions and leads to interchain associations. Crystallization in the polymer-rich microphase occurs at a much slower rate than that of demixing. Using both SAXS and WAXD measurements, with synchrotron and conventional x-ray radiation, I'Anson et al.[179] have probed gelation of a 6% amylose solution after quenching from 75°C to 22°C over short and long timescales. Their findings suggested that, if gelation is initiated by a phase separation, this is closely followed by crystallization in the polymer-rich phase. The SAXS data, using a high-density synchrotron beam, indicated that nucleation and limited growth of rod-shape microcrystals occur during network formation, at short times after quenching. On the other hand, turbidimetric and rheological measurements on solutions (2%) of monodisperse amyloses[401] indicated that turbidity development may either precede or lag behind increases in G', depending on the chain length; e.g. for amylose with DP 250, turbidity developed before that of storage modulus, whereas the reverse was true for amyloses with DP >1100. As gelation was noted even at concentrations significantly lower than those leading to molecular entanglements, Gidley[74] suggested that gelation is due to interchain associations in the form of double helices and aggregation of helices which act as junction zones. It is these cooperative noncovalent interchain interactions that bring about a phase separation in the developing network. Moreover, ^{13}C-CP/MAS-NMR and ^{1}H-NMR T_2 measurements of a 10% amylose gel provided evidence for the existence of two conformational states having different mobilities,[77] rigid double helical chain segments (of B-type with ^{1}H T_2 ~10 μs) and more mobile amorphous interjunction single chains (^{1}H T_2 of 1–10 ms). Estimates of double helix content in 10% amylose gels gave 67%, 67% and 83% for potato (polydisperse), DP 2800 (monodisperse) and DP 300 (monodisperse) amyloses, respectively.[77,203] An x-ray scattering study[402] of the structural features of a 2% wheat amylose solution during gelation has favored the gelation mechanism proposed by Gidley,[74] i.e. crystallization is not preceded by a phase separation stage. The scattering features of the hot (70°C) amylose solution revealed the presence of only V-helices, and it was only after cooling the solution to 40°C that a mixture of B- and V-form nanocrystallites was detected. Scattering data thus supported the view that B-form crystals grow by dimerization of single V-type helices, which undergo an accompanying conformational shift. The V- and B-nanocrystals at the early stages of the gelation process did not seem to be included in a phase-separated network, as they scatter like independent particles with relatively sharp electron density boundaries. However, after ripening the gel (storage at 21°C for 5 weeks) the unstable V-nanocrystals disappeared, with a concomitant increase in the number and size of B-type crystals. The aged amylose networks would thus consist of double helices, small aggregates of double helices (junction zones), B-nanocrystals and their aggregates.[402] The notion that conformational transitions occur before phase separation is also supported by optical rotation, light scattering and fluorescence depolarization data.[404] In an FTIR study of amylose gelation,[193] spectral changes of a 10% polymer concentration system indicated that a fast short-range ordering takes place over a similar timescale to development of the gel network. Thus, double helices form at the start or early on in the gelation process, whereas subsequent aggregation of helices yields crystalline domains. Morris[402] has also suggested that amylose gel networks can develop in different ways, depending on molecular size, polymer concentration and gelation conditions. For high molecular weight amyloses, poor chain matching and associations occurring over relatively short chain segment lengths favor the formation of a fine network that coarsens thereafter. In contrast, shorter chains may initially foster lateral interchain associations, due to better matching on double helix formation, followed by interlinking of the coarse aggregates to form a network. The microstructural aspects of amylose gels, such as porosity,[201,406] molecular order and crystallinity,[77,217] and lengths of chain segments in the ordered regions of the network structure,[201,202] have been examined. A structural model for the organization of the polysaccharide chains in amylose gels, with crystallite lengths of 8–18 nm, was also presented.[201] Amylose gels exhibit a macroporous structure (mesh size 100–1000 nm) containing filaments of 20 ± 10 nm wide; these filaments are assemblies of many double helical chain segments partially organized in B-type crystalline arrays. According to Cairns et al.,[214] the extent of crystallinity for a 10% amylose gel, as assessed by x-ray diffraction, is below 25%. Double helices are linked to others by single non-helical loops of chain segments which, along with chain ends, constitute the disordered fraction of the gel network structure.

Amylose gels behave as elastic solids under small deformation, the G' value being at least one order of magnitude greater than G'' and both moduli being essentially independent of frequency. The shear modulus of amylose gels also shows a very strong power law dependence on concentration. Depending on origin, molecular size, degree of polydispersity, concentration range and gel curing conditions, G' was found to vary with amylose concentration as C^7 (DP 3080, 1.5–7 wt%),[399] $C^{4.4-5.9}$ (DP 300–1100, 1–3 wt%)[401] and $C^{3.1}$ (DP 1150, 1.9–8.8 wt%).[160] These different exponent values suggest that a simple power law expression is not sufficient to describe the concentration dependence of gel modulus. A masterplot of G'-concentration was presented by Biliaderis,[146] summarizing published data for amyloses of different origin and molecular size, which by extrapolation to the concentration axis yielded a critical concentration for gelation of ~0.9 wt% (Figure 8.11b). It is also notable that the strongest dependence of modulus on concentration occurs within the polymer concentration range of 1–2%. As pointed out by Clark and Ross-Murphy,[148] high exponent values represent a clear deviation from the theoretical relationship of $G' \propto C^2$ predicted for biopolymer gels. This deviation, however, is not uncommon at low concentrations, and particularly where C/C_o is less than 10. Modulus-concentration data for monodisperse amyloses have also been analyzed by the method of Clark and Ross-Murphy,[404] which can give an accurate prediction of the critical gelling concentration, C_o, within experimental error, estimates of C_o in the range of 0.8–1.1% (w/w) for amyloses of DP between 300 and 2550 were obtained.[401] Thus, C_o was found to be independent of the molecular size of amylose chains, as evidenced also by the data of Figure 8.11b.

Amylopectin Gels

Although amylopectin solutions are generally regarded as stable, under proper conditions (usually low temperature and high polymer concentration) they can form gels. Amylopectin gelation is a much slower process than that observed for amylose (Figure 8.12), and occurs at substantially higher polysaccharide concentrations (usually >10%) than the coil overlap concentration ($C^* \sim 0.9\%$) of this biopolymer.[160,196,408] Moreover, the kinetics of amylopectin gelation are more sensitive to storage temperature (1–25°C) than those of amylose. A range of physical, spectroscopic and chemical techniques have been employed to monitor structure development in amylopectin dispersions.[160,191,193,195,196,408–413] Ring et al.[193] observed gelation for amylopectin solutions at concentrations as low as 10% (at 1°C), the gel modulus taking, however, several weeks to reach a limiting value. With increasing polymer concentration gelation proceeded at a faster rate. At the early stages of the gelation process, G' and G'' are very close and exhibit a strong dependence on frequency,[149] indicative of viscous-like behavior. Upon attainment of a limiting modulus value, however, G' is raised by at least one order of magnitude above G'', and both storage and loss moduli become independent of frequency (typical mechanical spectrum of a gel). Amylopectin gels behaved as Hookean-like solids at strains below 0.1.[196]

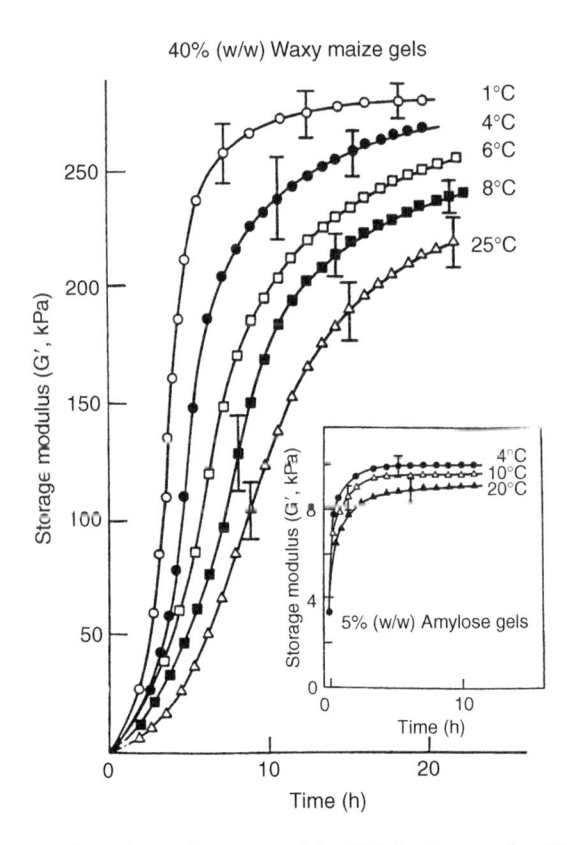

Figure 8.12 Temperature dependence of storage modulus (G') development for 40% waxy maize starch and 5% amylose dispersions.[158]

In a comprehensive study of amylopectin gelation,[196] it was noted that turbidity data for 20% amylopectin dispersions (at 1°C) reached limiting values after 4–5 days, whereas other structural changes, probed by dilatometry, shear modulus measurements, calorimetry and WAXD, seemed to follow parallel time paths over much longer periods (30–40 days). These findings were interpreted as reflecting chain aggregation (increases in turbidity) prior to gelation, and the changes monitored by the other techniques as being related to crystallization of the short DP (~15–20) chains of the molecule. The concurrent growth of G' and ΔH has also been noted during the early stages of gelation of a 40% amylopectin dispersion;[160,412] however, over a longer timescale (3–48 hours), the development of G' progresses faster than does the DSC endotherm. Treatment of amylopectin with α-amylase (production of α-limit dextrin) diminishes its gelling ability, implying that the outer short DP chains are necessary for gel formation.[72] Amylopectin gelation is thermoreversible by heating the gel to temperatures below 100°C;[156,412] with DSC, a broad endotherm (45–60°C) is observed corresponding to melting of the gel structure (Figure 8.4c). The range of temperatures over which the crystallinity index decreases and the modulus values dramatically fall coincides with that of the DSC melting transition.[412] The strong dependence of rigidity development of 40% amylopectin gels on storage temperature, as shown in Figure 8.12, further supports the notion that gelation of amylopectin follows typical nucleation kinetics, and that a network-based structure of amylopectin is largely related to chain reordering and crystallization of the short DP chains of this polymer. Raman[195] and FTIR spectroscopic studies[193,413] revealed a multi-stage process for gelation of amylopectin. First, at the very early stage, chain conformational ordering (coil $\rightarrow$ double helix transitions) was suggested to occur. This is followed by a slower aggregation stage (giving rise to turbidity) and growth of crystallites. Formation of interchain associations (short-range order), without turbidity development, was also realized by dynamic stress rheometry of dilute amylopectin solutions (3–10% w/w).[411] There were no visible inhomogeneities in such weak ($G' < 10\,\mathrm{N\,m^{-2}}$) gel networks over the timescale of the experiments (i.e. the gels remained clear), implying a gelation mechanism other than long-range ordering (crystallization) of the amylopectin chains. Wu and Eads,[188] using ^{1}H cross-polarization and high-resolution NMR, have identified three different motional states of the polymer chains in aging waxy maize starch gels: (a) a highly mobile fraction, with correlation times of nearly isotropic segmental motion ($<10^{-9}$ s), producing fairly narrow resonances in high-resolution ^{1}H-NMR; (b) the most rigid regions, corresponding to crystalline domains; and (c) an intermediate in mobility fraction, giving rise to a narrow component of cross-relaxation spectra, and attributed to unassociated or partially associated material in the vicinity of crystallites. Their data indicated that part of the starch material converts from a highly mobile liquid-like state into the immobile solid-like state, whereas the intermediate mobility component seemed to be unaffected by storage. The rate of this conversion during aging increased with polymer concentration of the gel.

In contrast to amylose, amylopectin gel stiffness does not show a strong dependence on concentration; for waxy maize amylopectin gels (10–25% w/w), stored for six weeks at 1°C, the modulus varied linearly with concentration.[196] Durrani and Donald[409] reported a $C^{1.6}$ relationship between modulus and concentration. The Clark and Ross-Murphy model was also used to fit their G'-concentration data in the range of 22–50% (w/w), giving reasonably good fits and a predicted critical gel concentration value, C_o, of ~18% (w/w). The limiting moduli values are generally larger for gels made from high molecular weight undegraded amylopectin.[408,412] The fine structure of amylopectin also seems to influence its gelation kinetics. Cereal amylopectins (wheat, barley, maize) of shorter average chain length exhibited slower rates in modulus development than those from tuber starches (potato, canna).[408] The acceleration in retrogradation kinetics for amylopectins with higher proportions of longer chains has been also reported by others.[352,409,410]

The kinetics of amylopectin gelation and retrogradation are also influenced by the presence of low molecular weight solutes (e.g. sugars, salts, lipids). Understanding these effects is important for improving the texture and shelf life of formulated bakery products, e.g. high-sugar items such as cakes and cookies. Effects of polyhydroxy compounds (PHCs) on structure formation in aging amylopectin gel networks have been investigated by DSC[164,165,410,414] and dynamic rheometry.[165] Observed changes in retrogradation kinetics by the addition of PHCs at a weight ratio of waxy maize starch: PHC:water of 1:0.5:1.5 were solute specific[187] (Figure 8.13). In the homologous series of glucose oligomers, maltotriose exhibited the greatest inhibitory effect, while malto-octaose promoted amylopectin retrogradation. Among various pentose and hexose sugars, ribose and talose, respectively, were the most effective in retarding structure development in aging gels. Fructose led to increased rates of chain ordering, particularly at high concentrations. Reduced rates of firmness (large deformation measurements) and x-ray crystallinity (attributed mainly to the amylopectin component) of aging wheat starch gels containing ribose, glucose, or sucrose at the nominal ratio of starch:sugar:water of 1:1:1 have been also reported by I'Anson et al.[412] The effectiveness of sugars in reducing firmness and crystallinity followed the order ribose > sucrose > glucose. Ribose, at the level used in the composite gel, completely suppressed crystallinity and firmness development over one month of storage at 20°C. In a subsequent study,[416] the effect of sugar concentration on the crystallization rate of amylopectin in wheat starch gels was investigated by WAXD. The pentoses ribose and xylose were found to suppress amylopectin retrogradation with increasing level of sugar, whereas fructose promoted this phenomenon. The unusual behavior of fructose was first noted by Slade and Levine[14] using DSC to follow retrogradation of wheat starch gels (starch:sugar:water at 1:1:1) at 25°C. They reported that the extent of amylopectin recrystallization decreased in the order of fructose > mannose > water alone > galactose > glucose > maltose > sucrose > maltotriose > xylose. They have also suggested that sugars, in general, behave as antiplasticizing solutes (relative to the plasticizing action of water alone), thus reducing chain mobility (by elevating the T_g) and crystallization of amylopectin. However, this theory cannot explain the profound accelerating effect certain solutes (e.g. fructose) have on retrogradation kinetics if one assumes a homogeneous solvent–polymer–co-solute matrix. Another factor that seems to play a dominant role in the kinetics of retrogradation is related to solute compatibility with the water structure. Some strong relationships between retrogradation rate and hydration characteristics of PHCs were observed.[164] Solutes which seem to cause very little perturbation in the hydrogen-bonded network of water have a stabilizing effect on

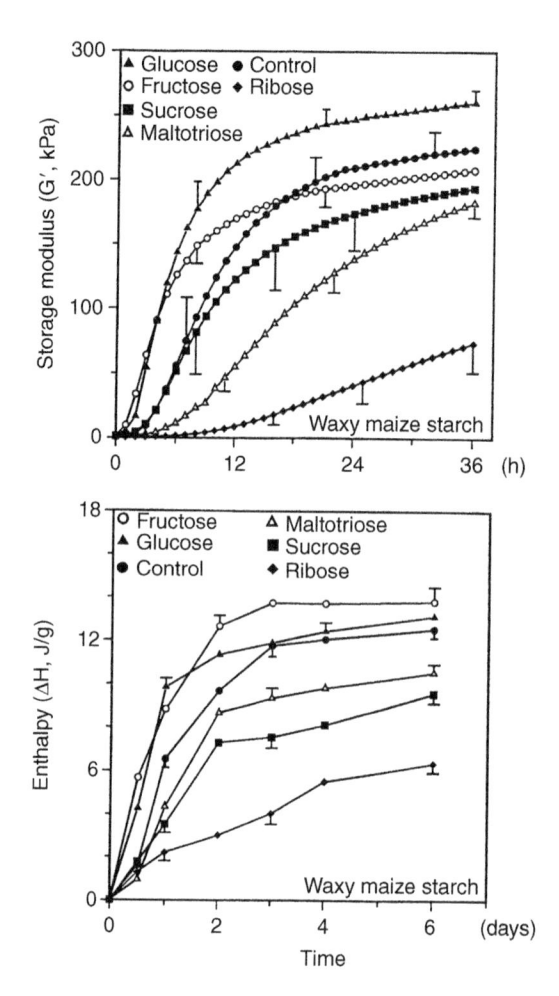

Figure 8.13 Effect of added sugars on retrogradation kinetics (G′, ΔH) of waxy maize starch gels at a weight ratio of starch:sugar:water of 1.0 : 0.5 : 1.5 w/w.[162]

the polymer chains, thus inhibiting retrogradation. In contrast, PHCs which greatly disturb the 'normal' water structure (e.g. fructose), forming strong hydration layers around them, promote chain ordering and crystallization of amylopectin. The kinetics of short-range conformational ordering during the early stages of retrogradation in a waxy maize starch gel, assessed by FTIR spectral changes, were found to be greatly retarded in the presence of small amounts of glycerol (1–5% w/w).[413] This effect was interpreted on the basis of stabilization of water and a decreasing mobility of the starch chains in the water–starch–glycerol system. Further studies on the relationships between the solution behavior and the physicochemical properties of PHCs in a composite starch–solute–water matrix are required to explain and predict the retrogradation phenomena of starch.

Interactions of Amylose and Amylopectin

Of importance to the microstructure and viscoelastic properties of starch gels is the identified thermodynamic incompatibility of starch polysaccharides.[417] Amylose and amylopectin in aqueous solution have been shown to be immiscible at moderate concentrations (e.g. 6%) leading to phase separation of the two polymers; one phase being enriched in amylose, and the other containing mostly amylopectin. When phase separation occurs, the effective concentration of both polymers in their respective microphases is raised, which may have a profound influence on the mechanical properties of the composite system; e.g. it is possible to observe viscosity enhancement or gelation at concentrations where neither one of the two polysaccharides is capable separately of such a phenomenon. Such observations were made by German et al.,[415] who found that introduction of amylopectin into an amylose–water system facilitates amylose aggregation and crystallization. A decrease in the elastic modulus of the ternary [amylose (5.8%)–amylopectin (of varying concentration between 0 and 35%)–water] system was observed when the concentration of the amylopectin was in the range 10–22%. This complicated dependence of modulus on amylopectin concentration was attributed to amylose aggregation and development of crystallites in the composite network structure.

Polymer compositions of the continuous and dispersed phases are dependent on the ratio of amylose:amylopectin ratio (r). At a particular value of this ratio (called the phase inversion point), the continuous phase becomes the discontinuous dispersed filler and *vice versa*. Structural characteristics of gels formed with amylose and amylopectin at different ratios (r) have been studied at a fixed total polysaccharide concentration of 8%.[198] Mixed gels were found to behave like pure amylopectin for r <0.43, and similar to pure amylose for higher r values. In both cases, it was suggested that the structural morphology of the gel consists of a continuous phase containing inclusions of the other phase and that the particular value of r ~ 0.43 reflects the inversion point between the two types of gel. Small amplitude oscillatory shear measurements of mixed amylose–amylopectin systems (at a total polymer concentration of 4%) also confirmed that their rheological properties strongly depend on the weight ratio of the two polymers.[419] The viscoelastic behavior of the gels seemed to be governed by a phase separation process, yielding a composite system. Drastic changes in network properties, identified with the point of phase inversion, occurred at r ≈ 0.17. It is questionable, however, whether one can draw a direct relationship between the rheological properties of native starches and their amylose/amylopectin ratios by taking into account the findings on mixed amylose–amylopectin gels. Different properties would be expected for the starch gel, depending on the extent of macromolecular demixing and the type of solubilized material during pasting and storage. High soluble amylose levels and swelling powers have been found to increase elasticity, whereas high levels of soluble amylopectin seem to be detrimental to gelation.[420] Amylose does tend to separate from amylopectin during heating of aqueous dispersions of starch granules, as evidenced by light microscopy of starch pastes and gels stained with iodine solution.[152,421,422] However, the extent to which this demixing process occurs would depend on starch concentration, temperature–time, shear, molecular size and degree of branching of amylose, as well as on the structure of the amylopectin crystalline network in starch granules and its interactions (chain entanglements) with amylose chains. Even for mixtures of molecularly dispersed biopolymers in an aqueous solution, complete phase separation (demixing) does not occur in either phase after equilibrium.[423] In cereal starch pastes produced by cooking under low shear conditions, amylose seemed to diffuse outside the granules, forming a continuous network,[152,421] whereas amylose was retained more in the peripheral layers of the swollen granules in potato starch pastes.[422] Even when gels were prepared under high shear, wheat starch amylopectin did not appear to solubilize (still localized in the granular fragments), compared to potato starch pastes where the continuous phase was a mixture of co-solubilized amylose and amylopectin.[421,422] This led to the suggestion that cereal starches form stronger gels than potato starch, because of the more efficient separation of starch molecules during heating, which promotes the gelation of the linear starch fraction.[422] It is also likely that gelatinized cereal starch granules (because of their lower swelling capacity and deformability) are stiffer than those of potato starch, thus increasing the firmness of the gel. According to Steeneken,[421] for concentrated suspensions of gelatinized starch granules, where individual granular particles are confined within discrete physical boundaries by neighboring particles, the thickening power is mainly governed by the rigidity of closely packed granules. This points to the highly complex rheological behavior of starch gels and to the need for more fundamental work to relate the structural aspects and the physical properties of starch gel networks. On the practical side, gels with high amylose content ($>30\%$) where amylose forms the continuous phase exhibit thermal stability, high elasticity and resistance to α-amylolysis.

In a retrogradation study of composite (nongranular material) gels (at 50% total solids) made with varying amylose:amylopectin ratios, synergistic interactions between the two polymers were inferred from DSC data.[425] For mixtures with less than 50% amylopectin, melting enthalpies of the aged gels higher than those anticipated by simply considering the weight fraction of the branched polysaccharide were observed. The possibility for co-crystallization between the two polymers has been proposed,[181] and this could be promoted when amylose is included in large amounts in a composite gel. Similar calorimetric findings were reported by Sievert and Würsch.[423] Chain reordering of amylose on cooling amylose–amylopectin mixtures (22% total solids in water) with varying weight ratios of the two polymers was restricted in the presence of amylopectin. These responses were attributed to either dilution or steric hindrance effects caused by the interspersed amylopectin molecules in the composite dispersion.

Starch Pastes and Gels

Several events take place during heating of starch granules in water that bring about large changes in rheological behavior and are responsible for the establishment of a gel network structure. These include swelling of the granules, crystallite melting, (partial) demixing of amylose and amylopectin, changes in the extent of molecular entanglements and rearrangements in the granules and the exuded material, and (partial) fragmentation of the swollen granules. Starch pastes are thus composites, in which particulate inclusions (swollen-deformable gelatinized granules and occasionally granular fragments) are embedded in a continuous matrix of entangled amylose molecules leached during gelatinization of starch.[184,427] Thus, any quantitative model for the description of the rheological properties of such a complex system has to take into account contributions from both the particulate and the solubilized material. For starch dispersions, in particular, there are many factors which render the development of a quantitative rheological model extremely difficult: (a) the solubilized material varies in nature and amount with the type of starch and the pasting conditions (temperature, shear, starch concentration); (b) gelatinized swollen granules are geometrically very complicated, highly folded and deformable particles; (c) interaction effects between solubilized material and swollen granular remnants are difficult to quantify; (d) retrogradation effects in both the amylose- and the amylopectin-rich regions continuously alter the rheology of the system. Nevertheless, several attempts for a theoretical and/or semiempirical treatment of viscosity and elasticity data of aqueous starch suspensions have been made by Bagley et al.,[425–429] Evans et al.,[430,431] Wong and Lelievre,[432–434] Steeneken,[421] Doublier et al.[127,132] and others.[348,438,439] Starch granules influence the rheological properties of pastes and gels by their phase volume and deformability, as well as their adhesion to the continuous matrix. These parameters exert their effects in a rather complicated manner and can account for the existence of concentration-dependent 'crossover' phenomena.[424] The latter is related to the observation that low-swelling starches (e.g. cereal starches) exhibit lower viscosity than their high-swelling counterparts (e.g. potato starch) at low starch concentrations, whereas at high concentrations this situation is reversed. Apparently, high-swelling starches would be able to form a more closely packed network of granules in a dilute dispersion, and therefore yield a higher viscosity. On the other hand, for concentrated systems, the inherent deformability of the granules becomes the dominant factor and low-swelling starches, having low deformability, produce a thicker dispersion. Factors such as granule size and uniformity, amylose content, granular structure, minor starch constituents (lipids, phosphate ester groups), presence of other solutes (salts, sugars, lipids), pH, starch concentration and shear–temperature–time regimes employed during gel preparation are all important determinants of starch viscoelasticity and can account for the differences in rheological properties among various starches, as well as the variation often reported for gels of starches from the same botanical origin.[136,137,149] This chapter does not give a detailed analysis of rheological data concerning starch pastes and gels in terms of theoretical or empirical models, as this has been a subject of two comprehensive reviews.[136,137] Instead, the following discussion focuses on the effects of various parameters on structure development and mechanical properties of starch gels.

Mechanical properties of potato and wheat starch suspensions during heating and cooling in the concentration range of 10–30% (w/w) were recently examined by dynamic rheometry.[155] The higher stiffness (G′) of wheat starch suspensions at concentrations above 15% (tightly packed gel systems), compared to potato starch, was attributed to the lower swelling capacity and the firmer gelatinized granules of wheat starch. These differences may be due to the presence of lipids and a larger number of entanglements in the granules of wheat starch.

Sensitivity to shear would also be a function of the extent of molecular entanglements in the swollen granules. Svegmark and Hermansson[150,437] have found that warm cereal starch pastes are less sensitive to shearing than similarly treated potato starch pastes (Figure 8.14); the complex shear modulus of the sheared (HS) 10%

pastes decreased to ~50% for wheat and maize starches and to ~2.5% for potato starch of the respective modulus values of the low-shear (LS) control pastes. In terms of gel network properties, the potato starch pastes (4–10%) changed from a weak gel (LS, δ:6–13°) to a more viscous state (HS, δ:25–27°). Moreover, the mechanical properties of potato starch gels are sensitive to the pH and ionic strength of the medium, because of the polyelectrolyte nature of potato starch.[441] Phosphate ester groups on some glucosyl residues of amylose, and especially on amylopectin, greatly affect the swelling behavior of potato starch; i.e. repulsion between the charged groups promotes swelling and thereby affects gel stiffness.[155] Added electrolytes compress the double layer around the charged groups, lowering starch granule swelling.

Preparation temperature also affects the mechanical properties of concentrated starch gels; for 30% (w/w) systems, larger reductions in fracture stress were found for potato than for wheat starch gels when the heating temperature was raised from 70° to 120°C.[190] Microscopic examination of heated barley and oat starch dispersions at 95°C indicated substantial granule fragmentation compared to heating at 90°C.[152] This was reflected in the viscoelastic properties of the gels for both starches; gelation

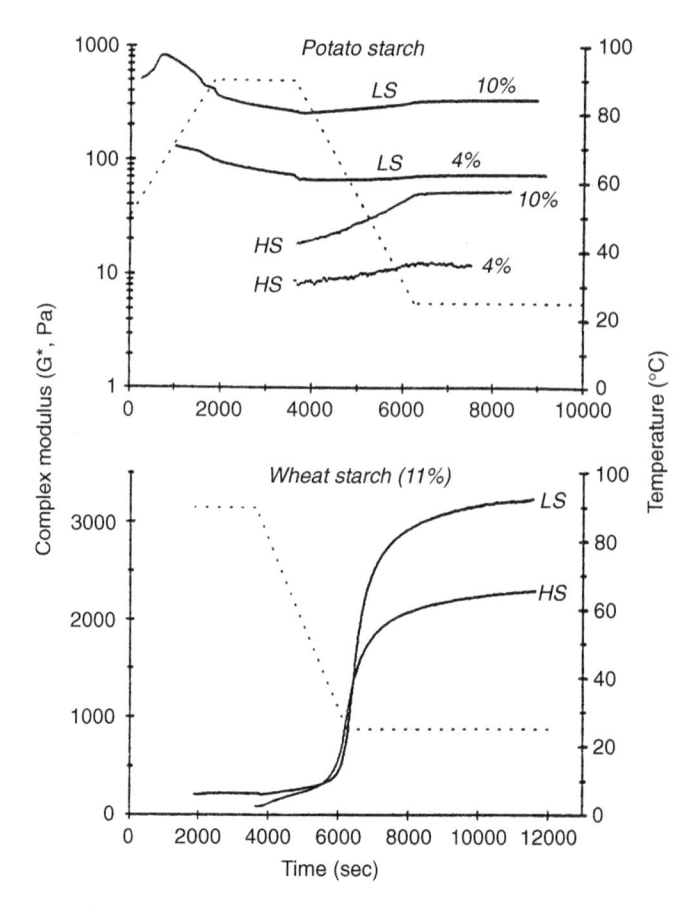

Figure 8.14 Changes in the complex modulus (G*) (solid lines) of low shear (LS) and high shear (HS) starch pastes as a function of time during heating and cooling (dotted lines indicate the temperature profiles).[151]

at 95°C yielded networks of a lower storage modulus and a higher tanδ than heating at 90°C. For dilute gelatinized suspensions of corn starch heated at a fixed temperature between 70°C and 90°C, there was a change in the flow behavior depending on the heating temperature; dilatancy was observed between 70°C and 80°C and shear thinning above 85°C.[439]

The concentration dependence of starch gel rigidity has been examined. Linear relationships between the rigidity modulus (as determined with a Ward and Saunder's U-tube) and concentration (6–30%) were reported for potato, wheat and corn starch gels.[184] A linear or almost linear concentration dependence of G′ was observed for wheat flour, potato and corn starch gels (4–10%) using small amplitude oscillatory shear testing.[430] A slightly stronger concentration dependence ($C^{1.4}$) was shown for wheat starch pastes (2–4%) using an oscillatory viscometer.[435] Over a broader concentration range (8–40%), much stronger power law relationships were found for wheat, legume and rice starches using dynamic rheometry,[160,163,442,443] $G′ \propto C^{2.1-3.2}$. Exponential dependence of gel moduli (G′ and Young's modulus) on concentration of wheat starch (15–40%) and potato starch (10–40%) were also reported for gels tested at 90°C and after aging.[442]

The amylose–amylopectin ratio has a great influence on the rheological properties of starch pastes and gels. Starches from different rice cultivars, because of their uniform granules (shape, size and size distributions), offer a unique model system to explore the relationship between amylose content and mechanical properties of starch gels. Figure 8.15 shows the dependence of G′ and tanδ on amylose content for freshly prepared 25% (w/w) gels of rice starches from 43 different varieties.[163] Isolation of the granular starches, gel casting conditions[442] and testing protocols were identical, to minimize other contributions to the rheological properties of the gels. An inverse linear relationship between tanδ and amylose content was observed (r = 0.75,

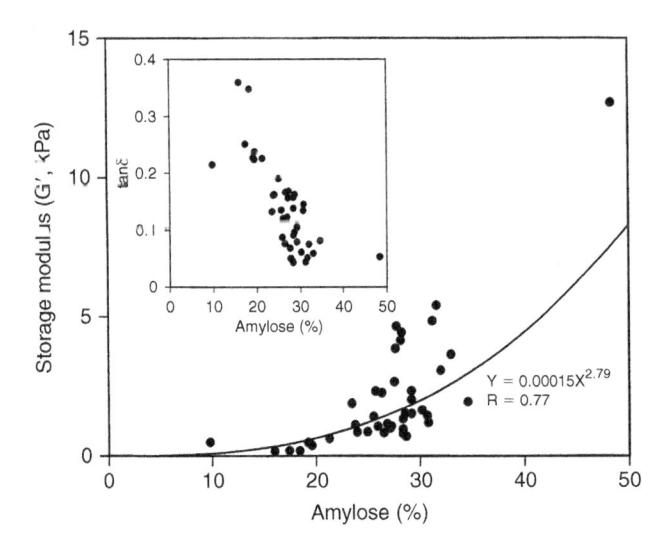

$Y = 0.00015X^{2.79}$
$R = 0.77$

Figure 8.15 Storage modulus and tanδ values of rice starch gels (25% w/w) with varying amylose content.[161]

p < 0.01). Also, there was an exponential trend for the dependence of G' on amylose content; G' ∝ (A%)$^{2.72}$, r = 0.76, p < 0.01. However, the data scatter in these plots implies that, in addition to amylose, other factors contribute to the mechanical properties of such composite networks.

Mechanical properties of concentrated starch gels are greatly affected by storage, as the starch molecules are prone to retrograde, forming aggregates and crystallites which stiffen the network structure. Similar changes also occur in the structure of most products with high starch content, and these are responsible, at least in part, for the decrease in eating quality of bakery items; e.g. staling of bread.[168–170] The time course of retrogradation for starch gels has been shown to consist of at least two separate phases. Short time (hours) increases in rigidity (modulus), short-range molecular order and crystallinity have been ascribed to gelation of the solubilized amylose in the intergranular spaces.[156] In contrast, changes that occur over longer periods (days) have been attributed to chain reordering and crystallization of amylopectin.[156,160,196] It is only this latter portion of the changes in rigidity, molecular order, and crystallinity that can be reversed by heating above the T_m of retrograded amylopectin.[156] In this respect, toasting of staled bakery items (refreshening) brings about a substantial reduction in the amount of retrograded amylopectin and lowers the firmness of the crumb.[169] On the basis of time-dependent FTIR spectral changes of 10% potato and waxy maize starch gels, as well as other spectroscopic data reports,[193,195] van Soest et al.[191] postulated a multi-stage process for starch retrogradation. The first stage was ascribed to conformational ordering (double helices) of amylose and may also involve short-range ordering of amylopectin side chains. Amylose aggregation and crystallization mostly occurs in this period. The second stage was described as an induction time for the onset of amylopectin helix aggregation. In stage three, amylopectin aggregation and crystallization proceeds, while stage four involves phase separation of water (syneresis) due to excessive retrogradation of starch chains. The observation, with various techniques, that amylose retrogradation is almost completed within the very first few hours of storage[156,160,193,194,197] was confirmed by isothermal microcalorimetry on aqueous dispersions of amylose and starch.[444]

A comparison of the initial rate of modulus development for 30% gelatinized starch dispersions (at 20°C) from various botanical sources revealed the order of pea > corn > wheat > potato, and was related to the amount of amylose solubilized during gelatinization.[157] However, the rate of long-term modulus increases followed the order pea > potato > corn > wheat. Higher firming rates for pea and potato may reflect the lack of endogenous lipids from these starches, which are known to retard amylopectin recrystallization. In a comparative study on retrogradation kinetics of eight starches, using DSC and rheological tests, Roulet et al.[156] found that potato and pea starches were the most prone and that waxy starches were the least sensitive to retrogradation. Structural differences in the amylopectin component (chain length) may also account for the variation in retrogradation rates of starches; among different *wx*-containing corn genotypes, *ae wx*, having the largest proportion of longer chains in its amylopectin, also exhibited the highest retrogradation rate.[352] Aging effects of starch gels are reflected in their mechanical behavior at large deformations. Stress–strain curves of 30% (w/w) wheat and potato starch gels showed an increase of stress at fracture and Young's modulus during storage, whereas the strain at fracture decreased.[144,154,190] The increase in Young's modulus is mainly due to increasing stiffness of the swollen granules caused by recrystallization of the amylopectin. An increase in B-type crystallinity (from 5% to 30%) of thermoplastic potato starch during storage, as a result of amylopectin crystallization, led to an increased elastic modulus and tensile strength of the material.[445]

Retrogradation is sensitive to the water content of the starch gel or the baked product. Water is an important parameter for firming and recrystallization processes, because of its action as a plasticizer of amorphous regions. DSC studies have indicated that maximum crystallinity occurs in gels[177,180] or bread crumb[180] at 35–45% water content (Figure 8.16). This is the moisture range of most bakery items, indicating that kinetics of deteriorative physicochemical processes in these products is likely to be favored. At these intermediate water contents, formation of B-type crystallites might also reduce the water content of the surrounding amorphous phase and thereby increase its rigidity.

Bread and other baked products are complex heterogeneous systems consisting of starch and gluten polymers in structurally diverse domains with their own distinct sensitivity to water plasticization.[446,447] DMTA analysis of bread indicates a rather broad transition temperature zone at the T_g, particularly at moisture contents below 30%. Crystallization in gels[177,180] and bread[180] did not occur below 10% and above 80% solids level; the overall response of enthalpy (retrogradation endotherm) to moisture content followed a bell-shaped curve (Figure 8.16). Acceleration of retrogradation with increasing water content from 10% to 50% merely reflects the plasticizing effect of

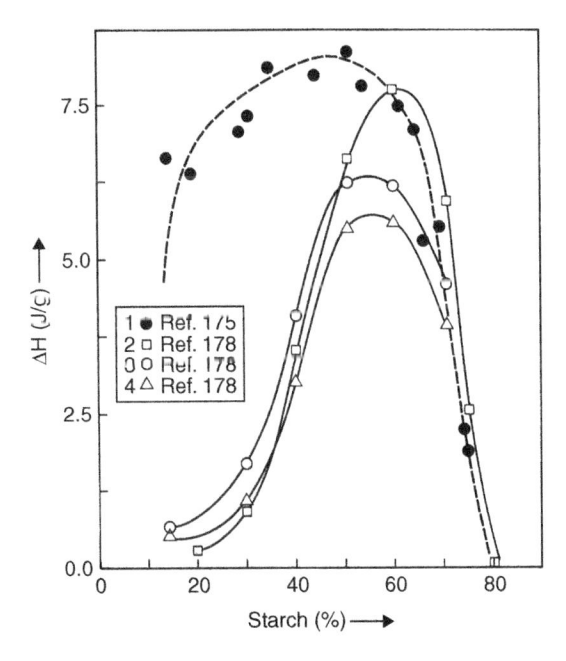

Figure 8.16 Effect of moisture content on retrogradation kinetics of starch in starch gels or bread: 1 wheat starch gels stored at 4°C for 14 days;[175] 2, 3 and 4 correspond to wheat starch gels, bread and bread with 0.5% monoglyceride, respectively, stored at 25°C for seven days.[178]

water in depressing T_g of the gelatinized starch; i.e. it brings the effective $T_g - T_m$ into a temperature region, relative to storage temperature, where nucleation and crystallite growth proceed at high rates. On the other hand, the rate of starch crystallization is reduced with increasing moisture content above 50%, presumably due to dilution of the crystallizable component in the water-plasticized starch matrix. The melting enthalpy increase for retrograded starch shows that crystallization occurs at a faster rate and to a greater extent when a lower single storage temperature is employed in the range of $-1°$ to 43°C.[176–178] Furthermore, in retrograded potato starch[179] and cooked wheat grains,[186] storage at refrigeration temperatures resulted in formation of less perfect crystallites (lower T_m and broader melting transitions) than those formed on storage at room temperature. These results can be understood easily with classical polymer crystallization theory,[15,448] which predicts growth of less perfect crystals with increasing degree of supercooling ($T_m - T$). Crystallization follows the classical three-step mechanism of nucleation → growth (propagation) → crystal ripening (perfection).[15,448] Within the range of $T_g - T_m$, both nucleation and propagation exhibit an exponential dependence on temperature such that the nucleation rate increases with decreasing temperature down to the T_g, while the propagation rate increases with increasing temperature up to the T_m.[15,405,448] Single[15,177,178,186] or multi-step[449] temperature crystallization studies showed that the rate-limiting step is nucleation rather than propagation. The DSC kinetic data of Slade and Levine[14] further indicated that the effective temperature range where starch crystallization (in 50% w/w gels) can take place is confined between $-5°$C (T_g of fully water-plasticized amorphous starch) and $\sim 60°$C (T_m of B-type crystallites of amylopectin), as the theory predicts. In order to accelerate the retrogradation process, it is advantageous to store the product first at a low temperature to foster nucleation, and then bring it to a higher temperature to promote crystal growth. The dynamic rheological properties also vary with the storage temperature.[160,185] The G' of wheat starch pastes increased depending on the storage temperature in the order of 30°C < 20°C < 15°C < 10°C.[185] Similar responses were also seen for concentrated (20% w/w) wheat starch gels in the temperature range of 1°C to 20°C.[160]

Kinetic data of starch retrogradation have been often analyzed using the Avrami equation,[158,159,162,176,181,183,185–189,191,194,195,413,446,450–454] which relates the fraction of the total change in the measured property (e.g. crystallinity, modulus, spectral feature, enthalpy of retrogradation endotherm) still to occur (φ) to time t: $\varphi = (A_\infty - A_t)/(A_\infty - A_o) = \exp(-kt^n)$ or $\log[-\ln \varphi] = n \log t + \log k$, where k is the rate constant and n is the Avrami exponent. A_o, A_t and A_∞ are experimental values of the measured property at time zero, time t, and infinity, respectively. The kinetic data of these studies were shown to fit reasonably well the Avrami model, with calculated exponent values close to one. However, application of Avrami analysis to concentrated polymer–solvent systems, such as starch gels, has been criticized, due to the non-equilibrium nature of the recrystallization process and the inherent limitations in interpretation of the exponent value in terms of nucleation type and crystal morphology.[15,455] Crystallization in aqueous starch systems is known to proceed via a three-step mechanism (nucleation → growth → maturation) controlled by the amount of water present, and involves two separate polymers, i.e. microcrystallites formed in gels upon aging are not uniform and certainly far from equilibrium crystals. Despite these theoretical limitations, the Avrami equation provides a convenient means to quantify retrogradation kinetics and make comparisons among similar starch systems submitted to different processing treatments. For example, Russell reported that addition of glyceryl monostearate during dough mixing[453] or spraying bread loaves with ethanol after baking[454] led to a reduction in the rate constant for the development of the 'staling endotherm,' while they had little effect on the limiting enthalpy value. In a similar study,[452] it was shown that the use of different flour grists had no effect on staling kinetics of bread and that the rate constant was independent of the specific loaf volume. Acceleration in storage modulus (G') evolution of aging potato starch pastes (12.5 wt%) with increasing NaCl concentration (above 0.01M),[162] retardation of short-range structural order development in waxy maize starch gels (10% w/w) by the addition of glycerol,[413] and changes in retrogradation kinetics (compression modulus) of concentrated (50% w/v) wheat starch gels caused by various ions[188] have also been evaluated by the Avrami analysis.

Retrogradation kinetics can be modified by the presence of other solutes. The effects of low molecular weight compounds or structure forming hydrocolloids on gelation and retrogradation of starch is of practical significance, in view of the complex nature of most starch-based foods and the need to control texture and extend shelf life. For instance, monoacyl lipids have long been known to affect gelatinization, retrogradation and texture of baked products, due to their interaction with the starch chains;[456,457] formation of helical inclusion complexes (V-type) is most likely the underlying mechanism for retarding interchain associations and aggregation of starch molecules. The small amount of lipids present in cereal starches affects their properties similar to added lipids. There appears to be a reduction in the pasting temperature (viscoamylography)[458–460] and an increase in swelling[124] of cereal starches following solvent extraction of their endogenous lipids; interestingly, this is not accompanied by changes in the characteristic melting temperatures of the granules as assessed by DSC.[124,460,461] When present as adjuncts, monoacyl lipids reduce the gel volume (granule swelling) and amylose leaching.[462–465] The extent and mode of interactions between granular starch and added lipids also depend on the chemical nature of the lipid. Thus, while swelling and gelatinization are generally restricted with fatty acids and most emulsifiers,[462–464,466] addition of some ionic surfactants (e.g. sodium dodecyl sulfate, SDS) promotes granule swelling.[442,464,467–469] These effects and the resultant rheological behavior are modulated by the type of starch, concentration of the various components, pH, presence of salts, and heat-shear regime used during pasting and cooling. This explains the large variability in published data concerning the influence of lipids on starch rheology. Moreover, most of the earlier studies were done using empirical rheological tests where, due to the temperature–shear cycles employed, it is difficult to separate the impact of mechanical damage of the gel network structure from contributions of the lipid additive itself. Consequently, a whole gamut of effects have been observed, ranging from large increases in gel consistency, to no effect, to large viscosity reductions.[457–459,461,462,470,471] Therefore, it is important to study the rheological properties of such systems under controlled conditions of thermal and mechanical treatments.

Viscoelastic behavior during gelatinization of aqueous starch dispersions (10% w/w, wheat, corn, potato and waxy barley) in the presence of emulsifiers was investigated by dynamic rheometry.[472] Glycerol monostearate (GMS) and sodium stearoyl lactylate (SSL) at 1% (dry starch basis) delayed the onset of G' and G" increases on gelatinization. This effect was less pronounced with waxy barley starch. G" of the hot starch pastes (except for wheat starch) was increased in the presence of GMS, SSL and SDS; this resulted in a higher tanδ value, compared to the lipid-free gels. In a subsequent study,[472] a cationic surfactant (cetyltrimethylammonium bromide, CTAB) was found to increase the G' of native and modified corn starch pastes (7.5% w/w, lipid at 2% w/w of starch, testing temperature 25°C); instead, saturated monoglycerides increased the G' for pastes of native corn starch, decreased the G' of crosslinked waxy maize starch, and had no effect on an acetylated high-amylose corn starch derivative. Interactions between monoacyl lipids (SDS, GMS, CTAB and lysophosphatidylcholine or LPC) at 2–4% (starch basis) were also investigated in thermoset gel networks at high starch concentrations (20–35% w/w) by dynamic rheometry and DSC.[442] All lipids reduced the apparent gelatinization enthalpies of the granular starches, suggesting complexation with the starch molecules on heating. Although rice and wheat starch gels (21–24% amylose) exhibited higher G' values when lipids were included, smaller changes in the viscoelastic properties were observed for legume starch gels (31–33% amylose), i.e. the higher amylose content of pea and garbanzo bean starches seemed to dominate the gel network properties. Among the lipids tested, LPC (one of the main endogenous lipids of cereal starches) exerted the greatest effect in increasing the G' and lowering the tanδ of cereal starch gels. Kinetic experiments on the development of G' and starch crystallization (ΔH of the retrogradation endotherm) of aging rice and pea starch gels (35% w/w) indicated that lipids retard both processes.[442] The antistaling role of native granular lipids or added surfactants in starch gels or bread has been repeatedly confirmed by DSC[169,180,329,425,451,474–476] and mechanical measurements.[142,169,453,476] The antifirming action of lipids may be caused by amylose complexation, which in turn weakens the cohesion of the amylopectin-enriched granules in the composite gel network.[142] For bread, a combination (1:1) of monoglycerides and diacetyl tartaric acid esters of monoglycerides was shown to give optimum reduction of firmness over long storage periods.[169] However, the reduction in the rate of amylopectin staling (ΔH) during aging of waxy maize starch gels[74,72] or gels made with blends of varying amylopectin–amylose ratios,[425,474] as well as other indirect evidence,[113,457,469,477] support the view that amylopectin may also interact with lipids. Reduced gelatinization enthalpies were observed on heating waxy maize starch in the presence of SDS and CTAB, implying amylopectin-lipid complexation.[457,473] X-ray diffraction analysis of non-granular amylopectin–surfactant (SDS and CTAB) mixtures, following heat treatment (105°C), revealed diffraction lines typical of both B- and V-patterns.[425] Isothermal microcalorimetric studies have recently demonstrated that added monoglycerides or SDS inhibit the crystallization processes of both starch polymers during the first 24 hours of aging starch gels.[444]

Sugars and other PHCs are also known to affect starch gel rheology and the rates of starch retrogradation. However, there is considerable confusion in the literature as to the exact role these solutes play in starch chain ordering and aggregation events in gels or products rich in starch. Such inconsistencies may arise from differences in the proportion of starch to sugar, the conditions for preparation and storage of the sample, and the different techniques used to monitor the rheological properties and the underlying structural changes. For sucrose, a maximum of viscosity, yield stress and dynamic rigidity of 3.5–4.0% w/w wheat starch pastes was observed at a sucrose concentration of 20% w/w.[478] Maxwell and Zobel,[138] using large deformation mechanical tests, showed a marked increase in the rate of firming of starch gels by fructose, a slight increase by glucose, and essentially no affect by sucrose (wheat starch:sugar:water 1:1:1 w/w). In similar rheological studies, Germani et al.[186] observed that the disaccharides maltose and sucrose were generally more effective in increasing the firming rate of corn starch gels than glucose. In both these investigations, however, the elastic moduli of sugar-containing gels were lower than those of pure starch gels (starch:water at 1:1 w/w), presumably due to the greater overall plasticization (H_2O+sugar) of the polymer matrix. The influence of three different sugars on crystallization of amylopectin in aging wheat starch gels (starch:sugar:water 1:1:1 w/w) was investigated using WAXD and firmness measurements.[415] The effectiveness of sugars in reducing firmness and crystallinity followed the order ribose > sucrose > glucose. Whereas ribose and xylose were found to inhibit crystallization of the amylopectin component in wheat starch gels, fructose promoted both thermally-reversible (amylopectin) and thermally-irreversible (amylose) crystallization.[416] The magnitude of these effects was highly dependent on sugar concentration. The structure promoting behavior of fructose in aging starch gels, compared to other PHCs, has also been confirmed by DSC at starch:sugar:water ratios of 1:1:1 w/w,[15] 1.0:0.5:1.5 w/w[164,165] and 0.2:0.09:0.71 w/w.[479] However, in other calorimetric studies on retrogradation of sweet potato starch gels (33% w/w)[480] fructose was found to prevent retrogradation, although it was not as effective as glucose or sucrose. On the basis of viscoelastic responses (creep compliance) of starch gels (~33% w/w) containing sugars, at ~6.0% (w/w on a starch basis), Katsuta et al.[478,479] and Miura et al.[473] argued that the ability of sugars to impede retrogradation is related to the mean number of equatorial hydroxyl groups, i.e. sugars with large numbers of equatorial —OHs were more effective in stabilizing the amorphous and entangled matrix of starch chains in the gel. The relative ranking of sugars in decreasing the rate of retrogradation was maltotriose > maltose > sucrose > glucose > fructose > xylose, ribose > control gel. However, among several monosaccharides tested, hexoses were found more effective than pentoses in inhibiting retrogradation; this contradicts other findings.[164,165,415,416] A comparative study of the effect of sugars on the thermal and mechanical properties of aging waxy maize, wheat, potato and pea starch gels (starch:sugar:water 1:0.5:1.5 w/w) revealed that sugars inhibited chain organization in the order of ribose > sucrose > maltotriose > water alone, glucose > fructose.[165] Kinetic experiments on the development of the staling endotherm (ΔH) and the evolution of gel rigidity (G') indicated that modification of starch gel structure caused by the sugars was most distinct with starches of low amylose content, i.e. amylose in high amounts dominates the gel network properties, masking the effects of sugars. These findings, and the calorimetric data of the effect of a large number of PHCs on the stability of waxy maize starch gels,[164] have led to the postulation that starch chain ordering in a composite starch–sugar–water matrix is governed by the stereochemistry of the sugar. Solutes which fit well into the hydrogen-bonded network structure of water have a stabilizing effect on the polymer chains, thus retarding retrogradation.[164,165] The reverse is true for incompatible solutes that greatly disturb the water structure. In designing a small carbohydrate solute as an effective antistaling agent, emphasis must be placed on the stereochemistry of its main conformer(s) in solution, particularly in relation to water structure.

In search of effective retrogradation inhibitors, the influence of *n*-alkyl glucosides on structure development in aging waxy maize and wheat starch gels was recently examined by DSC and dynamic rheometry.[414] Rheological and thermal properties of concentrated gels (40% w/w) indicated that *n*-alkyl glucosides are as effective as other commonly used emulsifiers in retarding starch retrogradation; glucosides with intermediate alkyl chain lengths (6–8 carbon atoms) were the most effective in this respect. There were minor differences in the relative ranking of the alkyl α- and β-D-glucopyranosides as retrogradation inhibitors. In other studies, amylopectin retrogradation (DSC endotherm) seemed to be accelerated in the presence of low DE starch hydrolyzates in corn[479] and wheat[160] starch gels. In contrast, storage modulus data of these composite gels showed much weaker network structures,[160] presumably due to interruption of the interchain associations by short DP dextrins. These findings clearly point to the different structural elements monitored by the various techniques in following aging phenomena in starch gels and related products. Frequently, development of ordered microdomains (DSC and x-ray diffraction) does not follow trends in firmness indices.[168,171] Rogers et al.[480] found that bread crumb firming rate was inversely proportional to its moisture content, within a moisture content range of 22–37%. In breads of low moisture content (e.g. 22%), starch did not retrograde noticeably, yet those breads firmed rapidly. Obviously, starch retrogradation and crumb firming do not correlate with each other across varying moisture contents. Any variation in moisture content of the end product, as a result of changes in formulation, may therefore lead to different trends in crystallization kinetics and sensory attributes. This has been also shown for breads fortified with wheat endosperm, water-soluble arabinoxylans.[484] Breads supplemented with these polysaccharides (at a level of 0.3–0.9% w/w of flour) retained higher moisture content during storage and exhibited greater rates in starch retrogradation (ΔH) than control breads; yet the fortified bread crumbs, because of their higher water content, had an overall softer texture. Water-soluble polysaccharides (added at 1% on a flour weight basis) decreased the rate of firming of bread.[446] It was suggested that these polymers may alter the plasticity of the amorphous regions, either through better water retention or by inhibiting gluten–starch interactions in the composite network. Apparently, the final firmness of an aging composite starch matrix would not only depend on the level of crystallinity of the starch fraction, but also on the extent of network plasticization with water.

It is well-known that comparatively low concentrations of other food hydrocolloids can dramatically alter the pasting and gelling properties of starch dispersions.[485–487] Observed viscosity enhancement effects in starch-hydrocolloid mixtures are generally thought to arise from modification in granule swelling and leaching of amylose during gelatinization (affecting granule deformability), as well as starch–hydrocolloid interactions in the continuous phase. The gelation process of corn starch dispersions (4%) with added galactomannans (0.1–0.5%) was examined by constant stress rheometry.[488]

Galactomannans accelerated gelation kinetics and resulted in slightly higher G' values for the final networks, compared to starch alone. Mixed gels were also characterized by greater tanδ values, implying less elastic networks, whereas their G' values showed a slightly greater dependence on frequency. These findings were rationalized in terms of thermodynamic incompatibility between amylose and galactomannan in the continuous phase of the gel, i.e. there is an increase in concentration of the hydrocolloid in the continuous phase as a result of phase separation processes. More pronounced effects were found for starch–xanthan systems. In similar studies, the kinetics of gel formation in 7% mixed starch–hydrocolloid dispersions (6.65% corn starch+0.35% gum) were shortened considerably in the case of guar gum, locust bean gum and κ-carrageenan;[489] instead, ι-carrageenan seemed to interfere with the network development. However, the final gel strength (G') of pure starch gels was never reached by the composite starch–gum gels, implying that the hydrocolloid prevented formation of the same number of crosslinks that would occur in a pure starch paste. Addition of these hydrocolloids also led to an increase in the viscous component (G'') of the gel, thus raising the tanδ values for the mixed gel, with the exception of ι-carrageenan which showed the opposite effect. The acceleration of gelation was explained by increasing concentrations of starch and gum in their respective microdomains due to phase separation, whereas the exceptional behavior of ι-carrageenan was attributed to intermolecular interactions between starch and this hydrocolloid. In a related study, the influence of purified wheat and rye arabinoxylans (at 2% on an α-D-glucan basis) on retrogradation kinetics of concentrated amylose (20% w/w), waxy maize (40% w/w) and wheat starch (40% w/w) gels was examined by dynamic rheometry, DSC and WAXD.[161] Amylose- and wheat starch-arabinoxylan mixtures exhibited higher G' and a slightly greater rate of structural order development than gels of amylose and starch alone. Similar kinetic responses were also obtained for the arabinoxylan–waxy maize gels in terms of molecular order development (ΔH and crystallinity), but these systems showed a markedly reduced gelation rate (G'). These observations suggested that arabinoxylans interfere with the intermolecular processes responsible for the establishment of a network structure of amylopectin. Obviously, there is need for more systematic and fundamental work on structure formation and the rheological properties in mixed starch–hydrocolloid gels. Such studies would provide the basis for predicting and controlling the structure and texture of processed formulated products containing starch.

The above discussion on gelation and retrogradation of starch polymers points to several possibilities one may use to retard starch retrogradation and firming kinetics:[19,25] (a) storage of the product at temperatures below its effective T_g (e.g. baked products promptly packaged and stored at freezer temperatures); (b) modification of the starch matrix (T_g, plasticization behavior) by water content and distribution, as well as ingredient formulation to make the aggregation and crystallization of the starch chains less favorable (e.g. extruded products of low moisture content exhibiting low molecular mobility at ambient conditions); (c) modification of the starch structure to reduce intermolecular associations and interrupt the continuity of the gel network (e.g. controlled hydrolysis with α-amylase supplements to partially cleave interconnecting chain segments between crystallites or lipids to complex with individual starch chains in the V-polymorph); and (d) incorporation of water-soluble hydrocolloids to modify water retention and facilitate a softer network structure. Various aspects of aging of bread and other bakery items, as well as the critical factors influencing the stability and the shelf life of these products have been reviewed.[170,171,490]

7. Phase Transitions and Other Properties of V-Structures

Although the extent to which amylose and lipids are complexed in granular starch is a matter of debate it is beyond doubt that, in thermally processed products, helical V-amylose exists in various states of aggregation. From a practical viewpoint, it is also important to recognize that functionality and quality attributes of starch containing foods are greatly dependent on the physical state of amylose. Modification of the solid-state structure of this polymer by complexation with monoacyl lipids is commonly practiced to control development of firmness in baked products, solubility and cooking losses of cereals, surface stickiness of pasta and processed rice, and texture of a wide range of starch-based products.

During the past fifteen years, the use of WAXD and DSC in studies of amylose–lipid (A–L) complexes, either in isolated form or in actual products, has contributed new knowledge about the structure and phase transition behavior of these materials.[16,63–72,117,325,329,457,491–501] A–L complexes constitute a distinct structural domain, as their thermal dissociation occurs in a temperature region ($T_m \sim 100–120°C$) different from those of the endothermic transitions of native A- or B-crystallites of amylopectin and retrograded amylopectin ($T_m \sim 40–55°C$) or amylose ($T_m \sim 140–160°C$); e.g. their melting endotherms in DSC traces of non-waxy cereal starches are those identified as M_3 and M_4 transitions in Figure 8.9. Complex formation is thermoreversible, as demonstrated by exothermic transitions in the cooling curves of the melts of these materials.[63,492,494–496] Formation and thermal stability of the complexes are highly dependent on the nature of the ligand,[63–65,71,456,493–497,500] the chain length of amylose[64,500] and the conditions (temperature–solvent–time) employed during their preparation.[63,66,68–71,493,495–498,500] With fatty acids as ligands, the T_m of A–L complexes increases with the length of the saturated aliphatic chain[463,495,497] and it decreases with increasing amounts of cis-unsaturation.[71,493,496,497] A minimum of 4–6 and an optimum chain length between 12 and 18 carbon atoms is reported in most cases. Saturated monoglycerides are effective complexing agents, since they are usually added in a physical state (lamellar liquid–crystalline phase) that easily reacts with the amylose.[170] Monoglycerides with the *trans*-configuration give more stable complexes than do the *cis*-forms, presumably because the former have a straight chain conformation.[496,502] The T_m of highly crystalline preparations of amylose–fatty acid complexes, crystallized in dilute solutions of H_2O-DMSO mixtures, increased with increasing chain length of the amylose from a DP 30 to 900.[500] The observed trend in T_m was corroborated with x-ray diffraction data of the complexes. Improved crystalline features, and especially crystallite size, can account for the high T_m complexes of the long DP chains. Complexes formed by cooling from a melt at high polymer concentrations (10–40% w/w) show broader melting transitions and lower enthalpies than their solution grown counterparts.[63] In general, the reported melting enthalpies of isolated A–L complexes vary within 8 and 32J/g, depending on ligand type and complexation conditions.

The melting behavior of complexes is also influenced by water content.[63,64,321,329] At high moisture contents, single cooperative transitions are observed (Figure 8.17a), implying zero-entropy production melting (i.e. melting without an apparent change in the metastability of the system on heating). As the water content is decreased (<70% w/w), however, the peak temperature is elevated and a second endotherm at higher temperatures progressively develops. In most of the DSC curves at low moisture levels, an exothermic effect between the two endotherms was observed, indicative of secondary reordering/crystallization processes during heating. The magnitude of this effect was found to depend strongly on the heating rate.[63–65] These DSC thermal profiles are typical of non-equilibrium melting of metastable macromolecular structures. Overall, the influence of water on phase transitions of V-complexes can be explained by considering several parameters operating simultaneously:[63–66] solvent–polymer interactions (thermodynamic and free volume effects manifested by solvent-mediated depression

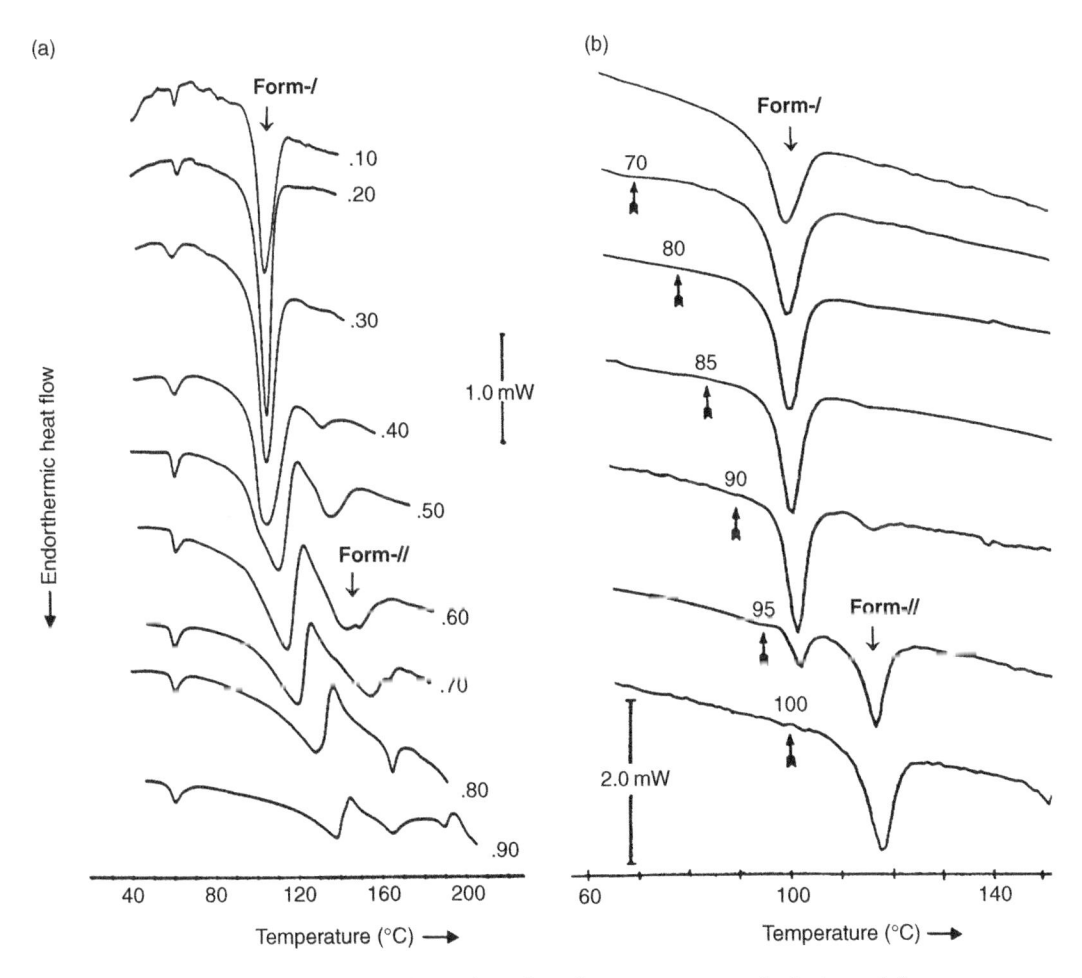

Figure 8.17 (a) Melting profiles of the amylose–glycerol monostearate complex (polymorph *I*) as a function of weight fraction of the complex;[62] (b) effect of annealing treatments (2 hours) at several temperatures specified by the arrows on state transformation of the polymorph *I*.[66]

of T_m and T_g); morphological effects (related to structural organization of the complexes); and kinetic effects (rate of reorganization/annealing versus melting kinetics).

Multiple melting transitions for V-structures have been often reported,[63,65,72,321,457,492,493,495,497,498] implying a relationship between thermal properties and supermolecular structures of the complexes. The structural origin of the multiple transitions has been resolved using amylose-monoglyceride complexes prepared under various crystallization conditions.[66,68,503] Several supermolecular structures were thus obtained (see Figure 8.18, *I*, *IIa* and *IIb*) which were characterized by calorimetry, WAXD and by enzymic structural analysis methods. For any given monoglyceride used (1-C14, 1-C16, 1-C18) as ligand, forms *I* and *IIa* were characterized by constant melting temperatures, regardless of the degree of supercooling ($\Delta T = T_m - T_c$) employed for their formation from dilute amylose solutions (0.25% w/v, amylose:ligand 5:1 w/w).[66] Form *I*, obtained under conditions favoring rapid nucleation (at low T_c), had a low T_m and in the hydrated state gave an amorphous-like x-ray diffraction pattern, i.e. it had only short-range order. In contrast, form *IIa* (preferred polymorph under high T_c) appeared as a polycrystalline specimen, giving the typical reflections of the V_h complex. At intermediate crystallization temperatures, the solid-state structures of the complexes were characterized by various ratios of forms *I* and *IIa*; no intermediate form was detected. These findings were later confirmed by similar studies on amylose complexes with linear alcohols (4–6 carbon atoms)[72] and fatty acids.[71] Crystallization and annealing studies on amylose–monoglyceride complexes further suggested that the structural forms *I* and *II* belong to two distinct free-energy domains which are separated by high-energy barriers.[66,68] Transformation of the kinetically preferred form *I* into the thermodynamically favored form *II* (state of lower free energy) occurs only after partial melting of the former structure, as evidenced by isothermal annealing at increasing temperatures (Figure 8.17b).[68] It is likely that interhelical chain segments of form *I* are under considerable strain, and thus exhibit an elevated T_g very close to the T_m of the helices. This situation leads to a kinetically stable system. However, once some helical chain segments of form *I* melt, sufficient molecular mobility is acquired and reorganization can proceed rapidly around the remaining helices which act as nucleation sites. Annealing of the type *IIa* complex at high temperatures gives rise to more perfected V-crystals (*IIb* of higher T_m), typical of metastable partially crystalline polymers.[66,71] The thermal stabilities and transitions between forms *I* and *II* were also examined in the presence of various solutes.[66,67] The observed conformational responses in stabilizing (Na_2SO_4, sucrose, CsCl) and destabilizing (urea, guanidine hydrochloride) solvent environments further suggested that interconversion between the various structural forms of complexes can be explained by considering changes at two levels of ordered structures:[68] association ↔ dissociation of aggregated helices; and V-helix ↔ coil transitions. Among various electrolytes tested, CsCl at high concentrations (>3.0 M) was found to disrupt the crystallites of the *II* form without affecting conformations of individual helices. A lamellar morphology has been suggested for V-amylose based on electron and SAXD data on single crystals and polycrystalline aggregates.[504–508] With long-chain amyloses, both chain folding and a 'fringed micelle' type of chain organization in crystallite formation have been proposed;[509]

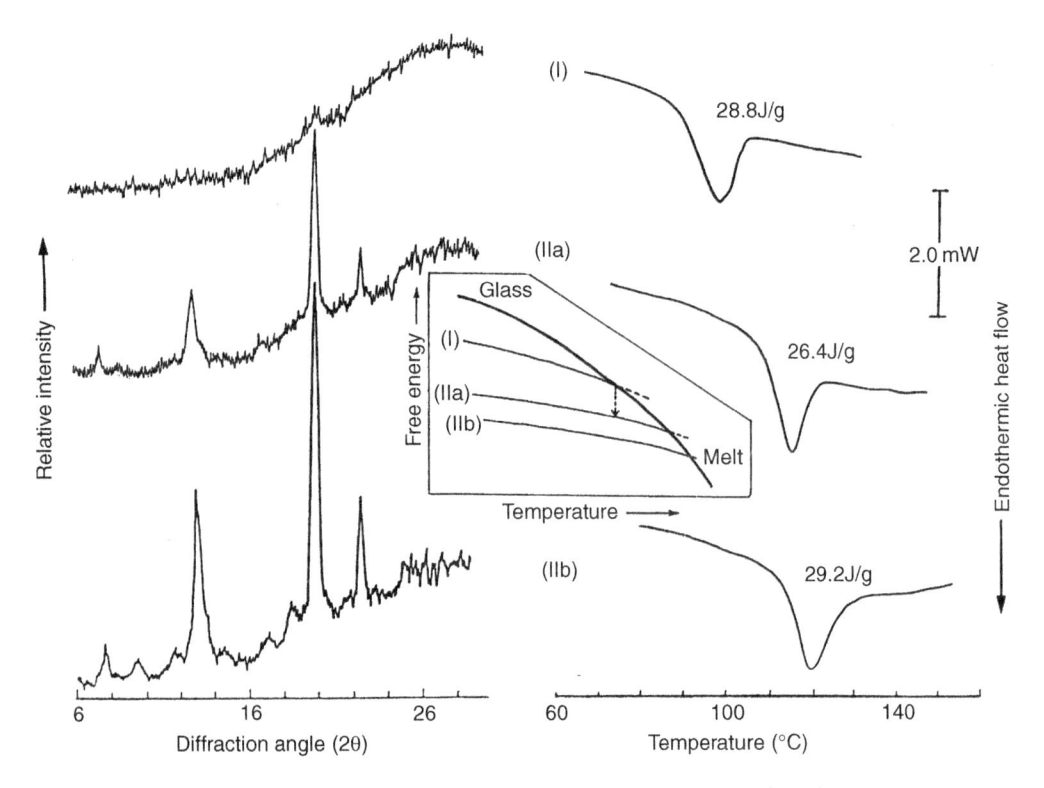

Figure 8.18 DSC (at 30% solids) and x-ray diffraction diagrams of three amylose–glycerol monostearate complexes differing in the degree of packing of their helices in the solid-state. The inset depicts the relationships between Gibbs' free energy and temperature for the three structural forms, *I*, *IIa* and *IIb*; the arrow indicates the melt-mediated transformation of metastable form *I* to *IIa*.[20]

according to the latter model, chains can participate in several successive lamellae. Enzyme-catalyzed digestion of forms *I* and *II* of the amylose-1-C18 monoglyceride complex, in a heterogeneous reaction mixture, revealed some differences in the chain length of the resistant amylodextrin fragments;[66] the *II* polymorph had ordered chain segments of larger length. However, the gross structural difference between the polymorphs *I* and *II* is mainly the extent of chain packing in a three-dimensional crystalline lattice; form *I* is an aggregated state with random distribution of helical chain segments, whereas form *II* has a compact structure with well-developed crystalline packing arrangements of the helices. The rate of enzyme-catalyzed degradation of V-complexes is also dependent on the degree of their long-range ordering; e.g. with *Bacillus subtilis* and porcine pancreatic α-amylases, the rates of α-amylolysis in the solid state of amylose–monostearin complexes followed the order of $I > IIa > IIb$.[503]

Melting enthalpies of amylose–lipid complexes show very little dependence on the extent of long-range ordering of chains in the supermolecular structure;[66,68,71] e.g. for the amylose-1-C18 complexes, the ΔH of forms *I* and *II* were 22 and 23 J/g, respectively. This implies that measured enthalpic changes mainly represent contributions from breaking of intrahelical H-bonds on conformational disordering of the V-helices, i.e. as for granular starch, DSC is a molecular probe of short-range order of the V-complexes. A small difference in melting enthalpies between the *IIa* (ΔH 27.5 J/g) and *IIb* (ΔH 32.2 J/g) forms of the amylose–stearic acid complex has been attributed to dissociation energy of the numerous van der Waals contacts between helices in the crystallites and limited interhelical H-bonding.[71] The melting enthalpy of A–L complexes, when determined at increasing amounts of lipid, reaches a limiting plateau value at a weight ratio of lipid–amylose∼0.06–0.15, depending on the ligand type.[457,510] Viscosimetric data for complexes with saturated fatty acids suggested that saturation of amylose chains occurs at a molar ratio of∼0.05–0.08 fatty acid (C22–C12)/glucosyl unit.[513] Rapid calorimetric methods for determination of the amylose content of starches have been developed on the basis of the melting[510,512] or crystallization[511] enthalpies measured for complexes formed with saturating amounts of L-α-lysophosphatidylcholine (LPC); the transition enthalpy is proportional to the amount of amylose. There is a generally good agreement between the DSC and colorimetric tests for amylose determination.[512]

In view of V-structure polymorphism, attempts have been made to relate product quality attributes to the solid-state structure of A–L complexes.[20] Earlier extrusion work by Mercier et al.[57,58] indicated that formation of crystalline amylose–monoglyceride complexes reduces the water solubility, stickiness and extent of retrogradation after repeated freeze–thaw cycles of starch extrudates. Studies of the influence of glyceryl monostearate addition on extrusion of soft wheat flour revealed inverse relationships between the amount of *II*-polymorph and extrudate properties such as solubility, water absorption, digestibility by enzymes and degree of expansive puffing.[67] Eating quality characteristics of parboiled rice were also related to the type and amount of A–L complexes formed during parboiling.[70] Cooking losses in the form of soluble amylose and total solids were minimized when the parboiling conditions favored extensive development of the *II*-polymorph. In fact, several structural domains of starch have been identified in parboiled rice that depend on rice variety and severity of parboiling conditions.[70,513] These include ungelatinized-annealed A-crystallites, B-type retrograded amylopectin and V-type structures with varying degrees of helix packing. The firmer texture and restricted swelling of pressure parboiled rice could be due to some of these structural domains; e.g. the helical regions of the *II*-polymorph, being thermally stable at 100°C, may act as crosslinks to stabilize the rice kernels during cooking in boiling water. In high-amylose corn starch films, prepared by drying of autoclaved starch dispersions (2%), different types of structures (identified by WAXD) were found, depending on the drying temperature.[514,515] B-type crystals were mainly detected in films dried at temperatures below 60°C, whereas a mixture of B-, A- and V-crystallites was obtained when drying was done between 60°C and 100°C. Films dried at 100°C contained only A-type crystallites. Attempts have also been made to explain the macroscopic properties of the films on the basis of molecular organization of amylose; formation of crystalline amylose–lipid complexes was found to reduce the maximum elongation of the films in stress–strain experiments.[515] The texture (firmness, surface characteristics and stickiness) of cooked pasta was recently shown to be strongly dependent on amylose–lipid interactions. Addition of glycerol monostearate (1.0 g/100 g semolina) to a pasta formulation led to a significant decrease in product stickiness on cooking, due to complexation of the available amylose in the form of the *II* polymorph.[501]

All the above examples clearly demonstrate that the physical properties of starch-based materials are related to the molecular structure and the state of starch. Undoubtedly, there is a need for more work in the area of structure–property relationships of starches in model systems and in composite matrices of real products. Such studies would be useful in optimizing product formulation and in refinement of processing conditions to improve end-product characteristics and increase shelf life.

IV. References

1. Banks W, Greenwood CT. *Starch and Its Components*. Edinburgh, UK: Edinburgh University Press; 1975.
2. Lineback DR. *Baker's Dig.* 1984;58:16.
3. Preiss J. In: Miflin BJ, ed. *Oxford Surveys of Plant Molecular and Cell Biology*. Vol. 7. Oxford, UK: Oxford University Press; 1991:59.
4. Willmitzer L, Koâmann J, Muller-Riber B, Sonnewald U. In: Meuser F, Manners DJ, Seibel W, eds. *Plant Polymeric Carbohydrates*. Cambridge, UK: Royal Society of Chemistry; 1993:33.
5. Mouille G, Maddelein M-L, Libessart N, Talaga P, Decq A, Delrue B, Ball S. *Plant Cell.* 1996;8:1353.
6. Fitt LE, Snyder EM. In: Whistler RL, BeMiller JN, Paschall EF, eds. *Starch: Chemistry and Technology*. 2nd ed. New York, NY: Academic Press Inc.; 1984:675.
7. Jane J-J, Kasemsuwan T, Leas S, Zobel H, Robyt JF. *Starch/Stärke*. 1994;46:121.
8. French D. In: Whistler RL, BeMiller JN, Paschall EF, eds. *Starch: Chemistry and Technology*. 2nd edn. New York, NY: Academic Press Inc.; 1984:184.
9. Swinkels JJM. *Starch/Stärke*. 1985;37:1.
10. Blanshard JMV. In: Galliard T, ed. *Starch: Properties and Potential*. New York, NY: John Wiley; 1987:16.
11. Galliard T, Bowler P. In: Galliard T, ed. *Starch: Properties and Potential*. New York, NY: John Wiley; 1987:55.
12. Morrison WR. *Cereal Foods World*. 1995;40:437.
13. Maurice TJ, Slade L, Page C, Sirett R. In: Simatos D, Multon JL, eds. *Properties of Water in Foods*. Dordrecht, The Netherlands: Martinus Nijhoff; 1985:211.
14. Slade L, Levine H. In: Stilva SS, Crescenzi V, Dea ICM, eds. *Industrial Polysaccharides*. New York, NY: Gordon and Breach Science; 1987:387.
15. Biliaderis CG. In: Levine H, Slade L, eds. *Water Relationships in Foods*. New York, NY: Plenum; 1991:251.
16. Slade L, Levine H. In: Blanshard JMV, Lillford PJ, eds. *The Glassy State in Foods*. Loughborough, UK: Nottingham University Press; 1993:35.
17. Slade L, Levine H. *J. Food Eng.* 1994;22:143.
18. Biliaderis CG. *Can J. Physiol. Pharmacol.* 1991;69:60.
19. Biliaderis CG. *Food Technol.* 1992;46(6):98.
20. French D. *J. Jpn. Soc. Starch Sci.* 1972;19:8.
21. Blanshard JMV. In: Blanshard JMV, Frazier PJ, Galliard T, eds. *Chemistry and Physics of Baking*. London, UK: Royal Society of Chemistry; 1986:1.

22. Zobel HF. *Starch/Stärke*. 1988;50:44.
23. Ring SG, Collona P, Miles MJ, Morris VJ, Turner R. *Int. J. Biol. Macromol.* 1987;9:158.
24. Biliaderis CG. In: Harwalkar VR, Ma C-Y, eds. *Thermal Analysis of Foods*. London, UK: Elsevier Applied Science; 1990:168.
25. LeMeste M, Huang VT, Panama J, Anderson G, Lentz R. *Cereal Foods World*. 1992;37:264.
26. Kalichevsky MT, Jaroszkiewicz EM, Ablett S, Blanshard JMV, Lillford PJ. *Carbohydr. Polym.* 1992;18:77.
27. Zobel HF. *Starch/Stärke*. 1988;50:1.
28. French AD, Murphy VG. *Cereal Foods World*. 1977;22:61.
29. Wu HCH, Sarko A. *Carbohydr. Res.* 1978;61:7.
30. Wu HCH, Sarko A. *Carbohydr. Res.* 1978;61:27.
31. Imberty A, Chanzy H, Perez S, Buleon A, Tran V. *J. Mol. Biol.* 1988;201:365.
32. Imberty A, Perez S. *Biopolymers*. 1988;27:1205.
33. Nara S, Mori A, Komiya T. *Stärke*. 1978;30:111.
34. Montgomery EM, Senti FR. *J. Polym. Sci.* 1958;28:1.
35. Robin JP, Mercier C, Charbonniere R, Guilbot A. *Cereal Chem.* 1974;51:389.
36. Oostergetel GT, van Bruggen EFJ. *Carbohydr. Polym.* 1993;21:7.
37. Hizukuri S. *Carbohydr. Res.* 1986;147:342.
38. Hizukuri S, Kaneko T, Takeda Y. *Biochim. Biophys. Acta*. 1983;760:188.
39. Pfannemüller B. *Int. J. Biol. Macromol.* 1987;9:105.
40. Gidley MJ, Bulpin PV. *Carbohydr. Res.* 1987;161:291.
41. Gidley MJ. *Carbohydr. Res.* 1987;161:301.
42. Sterling C. *J. Polym. Sci.* 1962;56:S10.
43. Oostergetel GT, van Bruggen EFJ. *Starch/Stärke*. 1989;41:331.
44. Jenkins PJ, Cameron RE, Donald AM. *Starch/Stärke*. 1993;45:417.
45. Blanshard JMV, Bates DR, Muhr AH, Worcester DL, Higgins JS. *Carbohydr. Polym.* 1984;4:427.
46. Cameron RE, Donald AM. *Polymer*. 1992;33:2628.
47. Cameron RE, Donald AM. *J. Polym. Sci., Part B: Polym. Phys.* 1993;31:1197.
48. Cameron RE, Donald AM. *Carbohydr. Res.* 1993;244:225.
49. Zobel HF, Young SN, Rocca LA. *Cereal Chem.* 1988;65:443.
50. Carlson TL-G, Larsson K, Dinh-Nguyen N, Krog N. *Stärke*. 1979;31:222.
51. Morrison WR, Tester RF, Snape CE, Law RV, Gidley MJ. *Cereal Chem.* 1993;70:385.
52. Morrison WR, Law RV, Snape CE. *J. Cereal Sci.* 1993;18:107.
53. Godet MC, Tran V, Delage MM, Buleon A. *Int. J. Biol. Macromol.* 1993;15:11.
54. Morrison WR, Tester RF, Gidley MJ, Karkalas J. *Carbohydr. Res.* 1993;245:289.
55. Morgan KR, Furneaux RH, Larsen NG. *Carbohydr. Res.* 1995;276:387.
56. Garnet C, Radosta S, Anger H, Damaschun G. *Starch/Stärke*. 1993;45:309.
57. Mercier C, Charbonniere R, Gallant D, Guilbot A. In: Blanshard JMV, Mitchell JR, eds. *Polysaccharides in Food*. London, U.K.: Butterworths; 1979:152.
58. Mercier C, Charbonniere R, Grebaut J, de la Gueriviere JF. *Cereal Chem.* 1980;57:4.
59. Donald AM, Warburton SC, Smith AC. In: Blanshard JMV, Lillford PJ, eds. *The Glassy State in Foods*. Loughborough, UK: Nottingham Univ. Press; 1993:375.
60. Biliaderis CG, Page CM, Slade L, Sirett RR. *Carbohydr. Polym.* 1985;5:367.

61. Biliaderis CG, Page CM, Maurice TJ. *Carbohydr. Polym.* 1986;6:269.
62. Biliaderis CG, Page CM, Maurice TJ. *Food Chem.* 1986;22:279.
63. Biliaderis CG, Galloway G. *Carbohydr. Res.* 1989;189:31.
64. Galloway G, Biliaderis CG, Stanley DW. *J. Food Sci.* 1989;54:950.
65. Biliaderis CG, Seneviratne HD. *Carbohydr. Polym.* 1990;13:185.
66. Biliaderis CG, Seneviratne HD. *Carbohydr. Res.* 1990;208:199.
67. Biliaderis CG, Tonogai JR, Perez CM, Juliano BO. *Cereal Chem.* 1993;70:512.
68. Karkalas J, Ma S, Morrison WR, Pethrick RA. *Carbohydr. Res.* 1995;268:233.
69. Whittam MA, Orford PD, Ring SG, Clark SA, Parker ML, Cairns P, Miles MJ. *Int. J. Biol. Macromol.* 1989;11:339.
70. Gidley MJ, Bociek SM. *J. Am. Chem. Soc.* 1985;107:7040.
71. Veregin RP, Fyfe CA, Marchessault RH, Taylor MG. *Macromolecules.* 1986;19:1030.
72. Horii F, Yamamoto H, Hiray A, Kitamaru R. *Carbohydr. Res.* 1987;160:29.
73. Gidley MJ, Bociek SM. *J. Am. Chem. Soc.* 1988;110:3820.
74. Gidley MJ. *Macromolecules.* 1989;22:351.
75. Gidley MJ. In: Alexander RJ, Zobel HF, eds. *Developments in Carbohydrate Chemistry*. St. Paul, MN: American Association of Cereal Chemists; 1992:163.
76. Cooke D, Gidley MJ. *Carbohydr. Res.* 1992;227:103.
77. van Soest JJG, Tournois H, de Wit D, Vliegenthart JFG. *Carbohydr. Res.* 1995;279:201.
78. Hood LF, Mercier C. *Carbohydr. Res.* 1978;61:53.
79. Biliaderis CG. *J. Agric. Food Chem.* 1982;30:925.
80. Jane J, Xu A, Radosavljevic M, Seib PA. *Cereal Chem.* 1992;69:405.
81. Kasemsuwan T, Jane J. *Cereal Chem.* 1994;71:282.
82. Brown SA, French D. *Carbohydr. Res.* 1977;59:203.
83. BeMiller JN, Pratt GW. *Cereal Chem.* 1981;58:512.
84. Evans ID, Haisman DR. *Starch/Stärke.* 1982;34:224.
85. van den Berg C. *Vapour Sorption Equilibria and Other Water-starch Interactions; A Physico-chemical Approach* [Ph.D. Thesis]. The Netherlands: Agricultural University of Wageningen; 1981.
86. Cleven R, van den Berg C, van der Plas L. *Stärke.* 1978;30:223.
87. Nara S, Mori A, Komiya T. *Starch.* 1978;30:111.
88. Buleon A, Bizot H, Delage MM, Multon JL. *Starch/Stärke.* 1982;34:361.
89. Svensson E, Eliasson A-C. *Carbohydr. Polym.* 1995;26:171.
90. Guilbot A, Mercier C. In: Aspinall GO, ed. *The Polysaccharides*. Vol. 3. New York, NY: Academic Press; 1985:209.
91. Zobel HF. In: Whistler RL, BeMiller JN, Paschall EF, eds. *Starch: Chemistry and Technology*. 2nd edn. New York, NY: Academic Press Inc.; 1984:285.
92. Zobel HF, Stephen AM. In: Stephen AM, ed. *Food Polysaccharides and Their Applications*. New York, NY: Marcel Dekker; 1995:19.
93. Watson SA. *Meth. Carbohydr. Chem.* 1964;4:240.
94. Hill RD, Dronzek BL. *Starch.* 1973;25:367.
95. Hoseney RC, Atwell WA, Lineback DR. *Cereal Foods World.* 1977;22:56.
96. Liu J, Zhoa S. *Starch/Stärke.* 1990;42:96.

97. Valetudie J-C, Colonna P, Bouchet B, Gallant DJ. *Starch/Stärke*. 1995;47:298.
98. Shuey WC, Tipples KH. *The Amylograph Handbook*. St. Paul, MN: American Association of Cereal Chemists; 1980.
99. Leach HW, McCowen LD, Schoch TJ. *Cereal Chem*. 1959;36:534.
100. Ziegler GR, Thompson DB, Casasnovas J. *Cereal Chem*. 1993;70:247.
101. Okechukwu PE, Rao MA. *Starch/Stärke*. 1996;48:43.
102. Jaska E. *Cereal Chem*. 1971;48:437.
103. Hennig HJ, Lechert H, Goemann W. *Stärke*. 1976;28:10.
104. Hennig HJ, Lechert H. *J. Colloid Interf. Sci*. 1977;62:199.
105. Lelievre J, Mitchell J. *Stärke*. 1975;27:437.
106. Mendes da Silva CE, Ciacco CF, Barberis GE, Solano WMR, Rettori C. *Cereal Chem*. 1996;73:297.
107. Chinachoti P, White VA, Lo L, Stengle TR. *Cereal Chem*. 1991;68:238.
108. Marchant JL, Blanshard JMV. *Stärke*. 1978;30:257.
109. Blanshard JMV. In: Blanshard JMV, Mitchell J, eds. *Polysaccharides in Foods*. Boston, MA: Butterworths; 1979:139.
110. Biliaderis CG, Vaughan DJ. *Carbohydr. Polym*. 1987;7:51.
111. Liu H, Lelievre J, Ayoung-Chee W. *Carbohydr. Res*. 1991;210:79.
112. Donovan JW. *Biopolymers*. 1979;18:263.
113. Biliaderis CG, Maurice TJ, Vose JR. *J. Food Sci*. 1980;45:1669.
114. Biliaderis CG, Page CM, Maurice TJ, Juliano BO. *J. Agric. Food Chem*. 1986;34:6.
115. Chiang BY, Johnson JA. *Cereal Chem*. 1977;54:429.
116. Varriano-Marston E, Ke V, Huang G, Ponte J. *Cereal Chem*. 1980;57:242.
117. Burt DJ, Russell PL. *Starch/Stärke*. 1983;35:354.
118. Lund D. *CRC Crit. Rev. Food Sci. Nutr*. 1984;20:249.
119. Russell PL, Juliano BO. *Starch*. 1983;35:382.
120. Wada K, Takahashi K, Shirai K, Kawamura A. *J. Food Sci*. 1979;44:1366.
121. Tester RF, Morrison WR. *Cereal Chem*. 1990;67:551.
122. Tester RF, Morrison WR. *Cereal Chem*. 1990;67:558.
123. Doublier JL. *J. Cereal Sci*. 1987;5:247.
124. Tester RF, Morrison WR. *Cereal Chem*. 1992;69:654.
125. Bowler P, Williams MR, Angold RE. *Starch/Stärke*. 1980;32:186.
126. Williams MR, Bowler P. *Starch/Stärke*. 1982;34:211.
127. Doublier JL. *Starch*. 1981;33:415.
128. Prentice RDM, Stark JR, Gidley MJ. *Carbohydr. Res*. 1992;227:121.
129. Swinkels JJM. In: van Beynum GMA, Roels JA, eds. *Starch Conversion Technology*. New York, NY: Marcel Dekker; 1985:15.
130. Rutenberg MW, Solarek D. In: Whistler RL, BeMiller JN, Paschall EF, eds. *Starch: Chemistry and Technology*. 2nd edn. New York, NY: Academic Press; 1984:311.
131. Eliasson A-C. *J. Text. Stud*. 1986;17:253.
132. Doublier JL, Llamas G, LeMeur M. *Carbohydr. Polym*. 1987;7:251.
133. Launay B, Doublier J-L, Cuvelier G. In: Mitchell JR, Ledward DA, eds. *Functional Properties of Food Macromolecules*. London, UK: Elsevier Applied Science; 1986:1.
134. Doublier J-L. In: Faridi H, Faubion JM, eds. *Dough Rheology and Baked Product Texture*. New York, NY: Van Nostrand Reinhold; 1990:111.

135. Voisey PW, Paton D, Timbers GE. *Cereal Chem.* 1977;54:534.
136. Ross AS, Walker CE, Booth RI, Orth RA, Wrigley CWS. *Cereal Foods World.* 1987;32:827.
137. Whorlow RW. *Rheological Techniques.* Chichester, UK: Ellis Harwood; 1980.
138. Maxwell JL, Zobel HF. *Cereal Foods World.* 1978;23:124.
139. Conde-Petit B, Escher F. *Starch/Stärke.* 1994;46:172.
140. Luyten H, van Vliet T. *J. Text. Stud.* 1995;26:281.
141. Keetels CJAM, van Vliet T, Walstra P. *Food Hydrocolloids.* 1996;10:355.
142. van Soest JJG, Benes K, de Wit D. *Starch/Stärke.* 1995;47:429.
143. Luyten H, van Vliet T, Walstra P. *J. Text. Stud.* 1992;23:245.
144. Myers RR, Knauss CJ, Hoffman RD. *J. Appl. Polym. Sci.* 1962;6:659.
145. Myers RR, Knauss CJ. In: Whistler RL, Paschall EF, eds. *Starch: Chemistry and Technology.* Vol. 1. New York, NY: Academic Press; 1965:393.
146. Biliaderis CG. In: Alexander RJ, Zobel HF, eds. *Developments in Carbohydrate Chemistry.* St. Paul, MN: American Association of Cereal Chemists; 1992:87.
147. Morris ER. In: Phillips G, Wedlock DJ, Williams PA, eds. *Gums and Stabilisers for the Food Industry 2.* Vol. 2. New York, NY: Pergamon Press; 1984:57.
148. Clark AH, Ross-Murphy SB. *Adv. Polym. Sci.* 1987;83:60.
149. Autio K. *Food Structure.* 1990;9:297.
150. Svegmark K, Hermansson A-M. *Carbohydr. Polym.* 1991;15:151.
151. Keetels CJAM, van Vliet T. In: Dickinson E, Walstra P, eds. *Food Colloids and Polymers: Stability and Mechanical Properties.* Cambridge, UK: Royal Society of Chemistry; 1993:266.
152. Keetels CJAM, van Vliet T, Walstra P. *Food Hydrocoll.* 1996;10:343.
153. Miles MJ, Morris VJ, Orford PD, Ring SG. *Carbohydr. Res.* 1985;135:271.
154. Orford PD, Ring SG, Carroll V, Miles MJ, Morris VJ. *J. Sci. Food Agric.* 1987;39:169.
155. Roulet Ph, MacInnes WM, Würsch P, Sanchez RM, Raemy A. *Food Hydrocoll.* 1988;2:381.
156. Roulet P, MacInnes WM, Gumy D, Wursch P. *Starch/Stärke.* 1990;42:99.
157. Biliaderis CG, Zawistowski J. *Cereal Chem.* 1990;67:240.
158. Biliaderis CG, Izydorczyk MS. In: Phillips GO, Wedlock DJ, Williams PA, eds. *Gums and Stabilisers for the Food Industry 6.* Oxford, UK: IRL Press; 1992:227.
159. Mita T. *Carbohydr. Polym.* 1992;17:269.
160. Biliaderis CG, Juliano BO. *Food Chem.* 1993;48:243.
161. Biliaderis CG, Prokopowich DJ. *Carbohydr. Polym.* 1994;23:193.
162. Prokopowich DJ, Biliaderis CG. *Food Chem.* 1995;52:255.
163. Schoch TJ, French D. *Cereal Chem.* 1947;24:231.
164. Hellman NN, Fairchild B, Senti FR. *Cereal Chem.* 1954;31:495.
165. Kulp K, Ponte JG. *Crit. Rev. Food Sci. Nutr.* 1981;15:1.
166. Krog N, Olesen SK, Toernaes H, Joensson T. *Cereal Foods World.* 1989;34:281.
167. Eliasson A-C, Larsson K. *Cereals in Breadmaking: A Molecular Colloidal Approach.* New York, NY: Marcel Dekker; 1993.
168. Ponte JG, Payne JD, Ingelin ME. In: Charalambous G, ed. *Shelf Life Studies of Foods and Beverages.* Amsterdam, The Netherlands: Elsevier Science Publishers; 1993:1143.

169. Erlander SR, Erlander LG. *Starch/Stärke*. 1969;21:305.
170. Martin ML, Zeleznak KJ, Hoseney RC. *Cereal Chem*. 1991;68:498.
171. Ovadia DZ, Walker CE. *Starch/Stärke*. 1996;48:137.
172. Radley JA. *Industrial Uses of Starch and its Derivatives*. London, UK: Applied Sciences Publishers; 1976.
173. Colwell KH, Axford DW, Chamberlain N, Elton GAH. *J. Sci. Food Agric*. 1969;20:550.
174. Longton J, LeGrys GA. *Starch/Stärke*. 1981;33:410.
175. Eliasson A-C. In: Hill RD, Munck L, eds. *New Approaches to Research on Cereal Carbohydrates*. Amsterdam, The Netherlands: Elsevier Science Publishers B.V.; 1985:93.
176. Nakazawa F, Noguchi S, Takahashi J, Takada M. *Agicr. Biol. Chem*. 1985;49:953.
177. Zeleznak KJ, Hoseney RC. *Cereal Chem*. 1986;63:407.
178. Russell PL. *J. Cereal Sci*. 1987;6:147.
179. I'Anson KJ, Miles MJ, Morris VJ, Ring SG. *Carbohydr. Polym*. 1988;8:45.
180. Marsh RDL, Blanshard JMV. *Carbohydr. Polym*. 1988;9:301.
181. Ring SG. *Starch/Stärke*. 1985;37:80.
182. Wong RBK, Lelievre J. *Starch/Stärke*. 1982;34:231.
183. Jankowski T, Rha CK. *Starch/Stärke*. 1986;38:6.
184. Kim SK, Ciacco CF, D'Appolonia BL. *J. Food Sci*. 1976;41:1249.
185. Ciacco CF, Fernades JLA. *Starch/Stärke*. 1979;31:51.
186. Germani R, Ciacco CF, Rodriguez-Amaya DB. *Starch/Stärke*. 1983;35:377.
187. Keetels CJAM, van Vliet T, Walstra P. *Food Hydrocoll*. 1996;10:363.
188. Wu JY, Eads TM. *Carbohydr. Polym*. 1993;20:51.
189. Teo CT, Seow CC. *Starch/Stärke*. 1992;44:288.
190. Goodfellow BJ, Wilson RH. *Biopolymers*. 1990;30:1183.
191. van Soest JJG, de Wit D, Tournois H, Vliegenthart JFG. *Starch/Stärke*. 1994;46:453.
192. Bulkin BJ, Kwak I, Dea ICM. *Carbohydr. Res*. 1987;160:95.
193. Ring SG, Colonna P, I'Anson KJ, Kalichevsky MT, Miles MJ, Morris VJ, Orford PD. *Carbohydr. Res*. 1987;162:277.
194. Miles MJ, Morris VJ, Ring SG. *Carbohydr. Res*. 1985;135:257.
195. Leloup VM, Colonna P, Buleon A. *J. Cereal Sci*. 1991;13:1.
196. Leloup VM, Colonna P, Ring SG. *J. Cereal Sci*. 1992;16:253.
197. Siljestrom M, Bjorck I, Eliasson A-C, Lonner C, Nyman M, Asp NG. *Cereal Chem*. 1988;65:1.
198. Leloup VM, Colonna P, Ring SG, Roberts K, Wells B. *Carbohydr. Polym*. 1992;18:189.
199. Jane J, Robyt JF. *Carbohydr. Res*. 1984;132:105.
200. Gidley MJ. In: Phillips GO, Welock DJ, Williams PA, eds. *Gums and Stabilisers for the Food Industry 4*. Oxford, UK: IRL Press; 1988:71.
201. Colquhoun IJ, Parker R, Ring SG, Sun L, Tang HR. *Carbohydr. Polym*. 1995;27:255.
202. Englyst HN, Cummings JH. *Am. J. Clin. Nutr*. 1987;45:423.
203. Englyst HN, Kingman SM, Cummings JH. *Eur. J. Clin. Nutr*. 1992;46:S33.
204. Russell PL, Berry CS, Greenwell P. *J. Cereal Sci*. 1989;9:1.

205. Siljestrom M, Eliasson A-C, Bjorck I. *Starch/Stärke*. 1989;41:147.
206. Eerlingen RC, Deceuninck M, Delcour JA. *Cereal Chem*. 1993;70:345.
207. Sievert D, Pomeranz Y. *Cereal Chem*. 1989;66:342.
208. Sievert D, Pomeranz Y. *Cereal Chem*. 1990;67:217.
209. Szczodrak J, Pomeranz Y. *Cereal Chem*. 1992;69:217.
210. Sievert D, Würsch P. *Cereal Chem*. 1993;70:333.
211. Gruchala L, Pomeranz Y. *Cereal Chem*. 1993;70:163.
212. Eerlingen RC, Crombez M, Delcour JA. *Cereal Chem*. 1993;70:339.
213. Eerlingen RC, Cillen G, Delcour JA. *Cereal Chem*. 1994;71:170.
214. Cairns P, Leloup V, Miles MJ, Ring SG, Morris VJ. *J. Cereal Sci*. 1990;12:203.
215. Sievert D, Czuchajowska Z, Pomeranz Y. *Cereal Chem*. 1991;68:86.
216. Gidley MJ, Cooke D, Drake AH, Hoffman RA, Russell AL, Greenwell P. *Carbohydr. Polym*. 1995;28:23.
217. Cairns P, Sun L, Morris VJ, Ring SG. *J. Cereal Sci*. 1995;21:37.
218. Ranganathan S, Champ M, Pechard C, Blanchard P, N'Guyen M, Colonna P, Krempf M. *Am. J. Clin. Nutr*. 1994;59:879.
219. Gee JM, Johnson IT, Lund EK. *Eur. J. Clin. Nutr*. 1992;46:S125.
220. Cummings JH, Edwards Ch, Gee JM, Nagengast F, Mathers J. Final Report of European Flair Concerted Action on Resistant Starch; 1995.
221. Alexander RJ. *Cereal Foods World*. 1995;40:455.
222. Slade L, Levine H. *Carbohydr. Polym*. 1993;21:105.
223. Slade L, Levine H. *Adv. Food Nutr. Res*. 1995;38:103.
224. Attenburrow GE, Davies AP. In: Blanshard JMV, Lillford PJ, eds. *The Glassy State in Foods*. Loughborough, UK: Nottingham University Press; 1993:317.
225. Eisenberg A. In: Mark JE, Eisenberg A, Graessley WW, Mandelkern L, Koenig JL, eds. *Physical Properties of Polymers*. Washington, DC: American Chemical Society; 1984:55.
226. Roos Y, Karel M. *Food Technol*. 1991;45(12):66.
227. Roos Y, Karel M. In: Blanshard JMV, Lillford PJ, eds. *The Glassy State in Foods*. Loughborough, UK: Nottingham University Press; 1993:207.
228. Karel M, Buera MP, Roos Y. In: Blanshard JMV, Lillford PJ, eds. *The Glassy State in Foods*. Loughborough, UK: Nottingham University Press; 1993.13.
229. Noel TR, Ring SG, Whittam MA. *Trends Food Sci. Technol*. 1990;1(9):62.
230. Roos Y. *Food Technol*. 1995;49(10):97.
231. Genin N, Rene F. *J. Food Eng*. 1995;26:391.
232. Mansfield ML. In: Blanshard JMV, Lillford PJ, eds. *The Glassy State in Foods*. Loughborough, UK: Nottingham University Press; 1993:103.
233. Wunderlich B. *Thermal Analysis*. Boston, MA: Academic Press; 1990.
234. Billmeyer FW. *Textbook of Polymer Science*. 3rd edn. New York, NY: Wiley Interscience; 1984.
235. Sears JK, Darby JR. *The Technology of Plasticizers*. New York, NY: Wiley Interscience; 1982.
236. Slade L, Levine H. *Carbohydr. Polym*. 1988;8:183.
237. Lelievre J. In: Alexander RJ, Zobel HF, eds. *Developments in Carbohydrate Chemistry*. St. Paul, MN: American Association of Cereal Chemists; 1992:137.

238. Noel TR, Ring SG. *Carbohydr. Res.* 1992;227:203.
239. Zeleznak KJ, Hoseney RC. *Cereal Chem.* 1987;64:121.
240. Willenbücher RW, Tomka I, Müller R. Carbohydrates in industrial synthesis. In: Clarke MA, ed. *Proceedings of the Symposium Div. Carb. Chem., Am. Chem. Soc.* Berlin, Germany: Dr. Albert Bartens; 1992:93.
241. Ellis TS, Jin X, Karasz FE. *Polym. Prep.* 1984;25:197.
242. Jin X, Ellis TS, Karasz FE. *J. Polym. Sci.: Polym. Phys. Ed.* 1984;22:1701.
243. Wissler GE, Crist B. *J. Polym. Sci.: Polym. Phys. Ed.* 1980;18:1257.
244. Vodovotz Y, Chinachoti P. *J. Food Sci.* 1996;61:932.
245. Shogren RL. *Carbohydr. Polym.* 1992;19:83.
246. Arvanitoyannis I, Kalichevsky M, Blanshard JMV, Psomiadou E. *Carbohydr. Polym.* 1994;24:1.
247. Georget DMR, Smith AC. *Carbohydr. Polym.* 1995;28:305.
248. Nakamura S, Tobolsky AV. *J. Appl. Polym. Sci.* 1967;11:1371.
249. Cowie JMG, Toporowski PM, Costaschuk F. *Makromol. Chem.* 1969;121:51.
250. Cowie JMG, Henshall SAE. *Eur. Polym. J.* 1976;12:215.
251. Orford PD, Parker R, Ring SG, Smith AC. *Int. J. Biol. Macromol.* 1989;11:91.
252. Bizot H, LeBail P, Leroux B, Davy J, Roger P, Buleon A. *Carbohydr. Polym.* 1997;32:33.
253. Tanner SF, Hills BP, Parker RJ. *J. Chem. Soc., Faraday Trans.* 1991;87:2613.
254. Trommsdorff U, Tomka I. *Macromol.* 1995;28:6138.
255. Trommsdorff U, Tomka I. *Macromol.* 1995;28:6128.
256. Kalichevsky MT, Blanshard JMV, Marsh RDL. In: Blanshard JMV, Lillford PJ, eds. *The Glassy State in Foods*. Loughborough, UK: Nottingham University Press; 1993:133.
257. Kalichevsky MT, Jaroszkiewicz EM, Blanshard JMV. *Polymer*. 1993;34:346.
258. Kalichevsky MT, Blanshard JMV. *Carbohydr. Polym.* 1993;20:107.
259. Couchman PR, Karasz FE. *Macromol.* 1978;11:117.
260. ten Brinke G, Karasz FE, Ellis TS. *Macromol.* 1983;16:244.
261. Sugisaki M, Suga H, Seki S. *Bull. Chem. Soc. Jpn.* 1968;41:2591.
262. Whittam MA, Noel TR, Ring SG. In: Dickinson E, ed. *Food Polymers, Gels and Colloids*. Cambridge, UK: Royal Society of Chemistry; 1991:277.
263. Ollett A-L, Parker R, Smith AC. In: Dickinson E, ed. *Food Polymers, Gels and Colloids*. Cambridge, UK: Royal Society of Chemistry; 1991:537.
264. Kalichevsky MT, Blanshard JMV. *Carbohydr. Polym.* 1992;19:271.
265. Gordon M, Taylor JS. *J. Appl. Chem.* 1952;2:493.
266. Fox TG, Flory PJ. *J. Appl. Phys.* 1950;21:581.
267. Levine H, Slade L. In: Mitchell JR, Blanshard JMV, eds. *Food Structure – Its Creation and Evaluation*. London, UK: Butterworths; 1988:149.
268. Appelqvist IAM, Cooke D, Gidley MJ, Lane SJ. *Carbohydr. Polym.* 1993;20:291.
269. Gidley MJ, Cooke D, Ward-Smith S. In: Blanshard JMV, Lillford PJ, eds. *The Glassy State in Foods*. Loughborough, UK: Nottingham Univiversity Press; 1993:303.
270. Yuan RC, Thompson DB. *Carbohydr. Polym.* 1994;25:1.
271. Berens AR, Hodge IA. *Macromol.* 1982;15:756.
272. Berens AR, Hodge IA. *Macromol.* 1982;15:762.
273. Hodge IA. *Science*. 1995;267:1945.

274. Struik LCE. *Physical Aging in Amorphous Polymers and Other Materials*. New York, NY: Elsevier Scientific; 1978.
275. van Soest JJG. *Starch Plastics: Structure–Property Relationships* [Ph.D. Thesis]. University of Utrecht, The Netherlands 1996.
276. Kirby AR, Clark SA, Parker R, Smith AC. *J. Mater. Sci.* 1993;28:5937.
277. Ollett A-L, Parker R, Smith AC. *J. Mater. Sci.* 1991;26:1351.
278. Attenburrow GE, Davies AP, Goodband RM, Ingman SJ. *J. Cereal Sci.* 1992;16:1.
279. Nicholls RJ, Appelqvist IAM, Davies AP, Ingman SJ, Lillford PJ. *J. Cereal Sci.* 1995;21:25.
280. Tomka I. In: Levine H, Slade L, eds. *Water Relationships in Foods*. New York, NY: Plenum; 1991:627.
281. Sala R, Tomka I. In: Blanshard JMV, Lillford PJ, eds. *The Glassy State in Foods*. Loughborough, UK: Nottingham University Press; 1993:475.
282. Kaletunc G, Breslauer KJ. *Cereal Chem.* 1993;70:548.
283. Arvanitoyannis I, Psomiadou E, Nakayama A, *Carbohydr Polym.* 1997. In press.
284. Matz SA. *Snack Food Technology*. Westport, CT: AVI; 1976.
285. Chinnaswamy R. *Carbohydr. Polym.* 1993;21:157.
286. Lourdin D, della Valle G, Colonna P. *Carbohydr. Polym.* 1995;27:261.
287. Potente H, Schnoppner V, Rucker A. *Starch/Stärke*. 1991;43:231.
288. Lawton JW. *Carbohydr. Polym.* 1996;29:203.
289. Hallberg LM, Chinachoti P. *J. Food Sci.* 1992;57:1201.
290. Hoseney RC, Zeleznak K, Lai CS. *Cereal Chem.* 1986;63:285.
291. Noel TR, Parker R, Ring SG, Tatham AS. *Int. J. Biol. Macromol.* 1995;17:81.
292. Gontard N, Ring S. *J. Agric. Food Chem.* 1996;44:3474.
293. Gough BM, Pybus JN. *Starch/Stärke*. 1971;23:210.
294. Krueger BR, Walker CE, Knutson CA, Inglett GE. *Cereal Chem.* 1987;64:187.
295. Krueger BR, Knutson CA, Inglett GE, Walker CE. *J. Food Sci.* 1987;52:715.
296. Knutson CA. *Cereal Chem.* 1990;67:376.
297. Larsson I, Eliasson A-C. *Starch/Stärke*. 1991;43:227.
298. Derek R, Prentice M, Stark JR, Gidley MJ. *Carbohydr. Res.* 1992;227:121.
299. Jacobs H, Eerlingen RC, Clauwaert W, Delcour JA. *Cereal Chem.* 1995;72:480.
300. Seow CC, Teo CH. *Starch/Stärke*. 1993;45:345.
301. Seow CC, Vasanti Nair CK. *Carbohydr. Res.* 1994;261:307.
302. Yost DA, Hoseney RC, *Starch/Stärke*. 1986;38:289.
303. Muhrbeck P, Svensson E, *Carbohydr Res.* 1997. In press.
304. Ahmed M, Lelievre J. *Stärke*. 1978;30:78.
305. Lorenz K, Kulp K. *Starch/Stärke*. 1978;30:333.
306. Lorenz K, Kulp K. *Starch/Stärke*. 1984;36:116.
307. Stute R. *Starch/Stärke*. 1984;44:205.
308. Sair L. *Cereal Chem.* 1967;44:8.
309. Kulp K, Lorenz K. *Cereal Chem.* 1981;58:46.
310. Lorenz K, Kulp K. *Cereal Chem.* 1981;58:49.
311. Lorenz K, Kulp K. *Starch/Stärke*. 1982;34:50.
312. Lorenz K, Kulp K. *Starch/Stärke*. 1982;34:76.
313. Lorenz K, Kulp K. *Starch/Stärke*. 1983;35:123.

314. Collona P, Buleon A, Mercier C. In: Galliard T, ed. *Starch: Properties and Potential.* New York, NY: John Wiley; 1987:55.
315. Hoover R, Vasanthan T. *Carbohydr. Res.* 1994;252:33.
316. Hoover R, Vasanthan T, Senanayake N, Martin AM. *Carbohydr. Res.* 1994;261:13.
317. Kawabata A, Takase N, Miyoshi E, Sawayama S, Kimura T, Kudo K. *Starch/Stärke.* 1994;46:463.
318. Donovan JW, Lorenz K, Kulp K. *Cereal Chem.* 1983;60:381.
319. LeBail P, Bizot H, Buleon A. *Carbohydr. Polym.* 1993;21:99.
320. Perez S, Imberty A, Scaringe RP. Computer Modelling for Carbohydrate Molecules. *ACS Symp. Ser.* 1990; 430:281.
321. Lorenz K, Kulp K. *Stärke.* 1978;32:181.
322. Kugimiya M, Donovan JW, Wong RY. *Starch/Stärke.* 1980;32:265.
323. Donovan JW, Mapes CJ. *Starch/Stärke.* 1980;32:190.
324. Eliasson A-C. *Starch/Stärke.* 1980;32:270.
325. Colonna P, Mercier C. *Phytochem.* 1985;24:1667.
326. Paton D. *Cereal Chem.* 1987;64:394.
327. Russell P. *J. Cereal Sci.* 1987;6:133.
328. Ghiasi K, Hoseney RC, Varriano-Marston E. *Cereal Chem.* 1982;59:258.
329. Hoover R, Sosulski F. *Starch/Stärke.* 1985;37:181.
330. Biliaderis CG, Mazza G, Przybylski R. *Starch/Stärke.* 1993;45:121.
331. Flory PJ. *Principles of Polymer Chemistry.* Ithaca, New York, NY: Cornell University Press; 1953.
332. Lelievre J. *J. Appl. Polym. Sci.* 1973;18:293.
333. Lelievre J. *Polymer.* 1976;17:854.
334. Munzing K. *Thermochim. Acta.* 1991;193:441.
335. Liu HL, Lelievre J. *Carbohydr. Polym.* 1992;17:145.
336. Marchant JL, Blanshard JMV. *Stärke.* 1978;30:257.
337. Gidley MJ, Cooke D. *Biochem. Soc. Trans.* 1991;19:551.
338. Stevens DJ, Elton GAH. *Stärke.* 1971;23:8.
339. Wootton M, Bamunuarachchi A. *Starch/Stärke.* 1979;31:201.
340. Katz FR, Furcsik SL, Tenbarge FL, Hauber RJ, Friedman RB. *Carbohydr. Polym.* 1993;21:133.
341. Li J, Berke TG, Glover DV. *Cereal Chem.* 1994;71:87.
342. Campbell MR, Pollak LM, White PJ. *Cereal Chem.* 1995;72:281.
343. Shiotsubo T, Takahashi K. *Agric. Biol. Chem.* 1984;48:9.
344. Whittam MA, Noel TR, Ring SG. *Int. J. Biol. Macromol.* 1990;12:359.
345. Noel TR, Ring SG, Whittam MA. In: Dickinson E, Walstra P, eds. *Food Colloids and Polymers: Stability and Mechanical Properties.* Cambridge, UK: Royal Society of Chemistry; 1993:126.
346. Inouchi N, Glover DV, Sugimoto Y, Fuwa H. *Starch/Stärke.* 1984;36:8.
347. Campbell MR, White PJ, Pollack LM. *Cereal Chem.* 1994;71:464.
348. Campbell MR, White PJ, Pollack LM. *Cereal Chem.* 1995;72:389.
349. Yuan RC, Thompson DB, Boyer CD. *Cereal Chem.* 1993;70:81.
350. Kasemsuwan T, Jane J, Schnable P, Stinard P, Robertson D. *Cereal Chem.* 1995;72:457.
351. Uriyapongson J, Rayas-Duarte P. *Cereal Chem.* 1994;71:571.

352. Akingbala JO, Gomez MH, Rooney LW, Sweat VE. *Starch/Stärke*. 1988;40:375.
353. Kim YS, Wiesenborn DP, Orr PH, Grant LA. *J. Food Sci.* 1995;60:1060.
354. Campbell MR, Pollack LM, White PJ. *Cereal Chem.* 1994;71:556.
355. Tester RF, Morrison WR, Ellis RH, Piggott JR, Batts GR, Wheeler TR, Morison JIL, Hadley P, Ledward DA. *J. Cereal Sci.* 1995;22:63.
356. Morrison WR, Tester RF, Gidley MJ. *J. Cereal Sci.* 1994;19:209.
357. Scanlon MG, Dexter JE, Biliaderis CG. *Cereal Chem.* 1988;65:486.
358. Nishita KD, Bean MM. *Cereal Chem.* 1982;59:46.
359. Bean MM, Yamazaki WT. *Cereal Chem.* 1978;55:936.
360. Wootton M, Bamunuarachchi A. *Starch/Stärke*. 1979;31:126.
361. Spies RD, Hoseney RC. *Cereal Chem.* 1982;59:128.
362. Abboud AM, Hoseney RC. *Cereal Chem.* 1984;61:34.
363. Ghiasi K, Hoseney RC, Varriano-Marston E. *Cereal Chem.* 1983;60:58.
364. Buck JS, Walker CE. *Starch/Stärke*. 1988;40:353.
365. Kim CS, Walker CE. *Cereal Chem.* 1992;69:212.
366. Eliasson A-C. *Carbohydr. Polym.* 1992;18:131.
367. Bello-Perez LA, Paredes-Lopez O. *Food Chem.* 1995;53:243.
368. Hoseney RC, Atwell WA, Lineback DR. *Cereal Foods World*. 1977;22:56.
369. Hoseney RC, Lineback DR, Seib PA. *Baker's Dig.* 1978;52:11.
370. D'Appolonia BL. *Cereal Chem.* 1972;49:532.
371. Slade L, Levine H. In: Millane RP, BeMiller JN, eds. *Frontiers in Carbohydrate Research – 1: Food Applications*. London, UK: Elsevier Applied Science; 1989:215.
372. Chinachoti P, Kim-Shin M-S, Mari F, Lo L. *Cereal Chem.* 1991;68:245.
373. Christianson DD, Hodge JE, Osborne D, Detroy RW. *Cereal Chem.* 1981;58:513.
374. Liu H, Lelievre J. *Cereal Chem.* 1992;69:597.
375. Kohyama K, Nishinary K. *J. Food Sci.* 1992;57:128.
376. Gudmundsson M, Eliasson A-C. *Starch/Stärke*. 1991;43:5.
377. Kim K-O, Hansen L, Setser C. *J. Food Sci.* 1986;51:1095.
378. Kim SS, Setser CS. *Cereal Chem.* 1992;69:447.
379. Lii CY, Lee BL. *Cereal Chem.* 1993;70:188.
380. Metcalf DG, Gilles KA. *Stärke*. 1966;18:101.
381. Takahashi K, Wada K. *J. Food Sci.* 1992;57:1140.
382. Gough BM, Pybus JN. *Stärke*. 1973;25:123.
383. Oosten BJ. *Starch/Stärke*. 1982;34:233.
384. Oosten BJ. *Starch/Stärke*. 1990;42:327.
385. Jane J. *Starch/Stärke*. 1993;45:161.
386. Eastman JE, Moore CO. US Patent 4 465 702; 1984.
387. Jane J, Craig SAS, Seib PA, Hoseney RC. *Starch/Stärke*. 1986;38:258.
388. Rajagopalan S, Seib PA. *J. Cereal Sci.* 1992;16:13.
389. Chen J, Jane J. *Cereal Chem.* 1994;71:618.
390. Jane J, Craig SAS, Seib PA, Hoseney RC. *Carbohydr. Res.* 1986;150:C5.
391. Rajagopalan S, Seib PA. *J. Cereal Sci.* 1992;16:29.
392. Chen J, Jane J. *Cereal Chem.* 1994;71:623.
393. Gidley MJ, Bulpin PV. *Macromol.* 1989;22:341.
394. Pfannemüller B, Mayerhöfer H, Schulz RC. *Biopolym.* 1971;10:243.

395. Miles MJ, Morris VJ, Ring SG. *Carbohydr. Polym.* 1984;4:73.
396. Ellis HS, Ring SG. *Carbohydr. Polym.* 1985;5:201.
397. Doublier J-L, Choplin L. *Carbohydr. Res.* 1989;193:215.
398. Clark AH, Gidley MJ, Richardson R, Ross-Murphy SB. *Macromol.* 1989;22:346.
399. Müller JJ, Gernat C, Schulz W, Müller E-C, Vorwerg W, Damaschun G. *Biopolym.* 1995;35:271.
400. Boyer RF, Baer E, Hiltner A. *Macromol.* 1985;18:427.
401. Hayashi A, Kinoshita K, Miyake Y, Cho C-H. *Agric. Biol. Chem.* 1983;47:1699.
402. Morris VJ. *Trends Food Sci. Technol.* 1990;1:2.
403. Leloup VM, Colonna P, Ring SG. *Macromol.* 1990;23:862.
404. Clark AH, Ross-Murphy SB. *Brit. Polym. J.* 1985;17:164.
405. Kalichevsky MT, Orford PD, Ring SG. *Carbohydr. Res.* 1990;198:49.
406. Shi Y-C, Seib PA. *Carbohydr. Res.* 1992;227:131.
407. Bello-Perez LA, Paredes-Lopez O. *Starch/Stärke.* 1995;47:83.
408. Cameron RE, Durrani CM, Donald AM. *Starch/Stärke.* 1994;46:285.
409. Durrani CM, Donald AM. *Polym. Gels Networks.* 1995;3:1.
410. van Soest JJG, de Witt D, Tournois H, Vliegenthart JFG. *Polymer.* 1994;35:4722.
411. Biliaderis CG, Prokopowich DJ, Jacobson MR, BeMiller JN. *Carbohydr. Res.* 1996;280:157.
412. I'Anson KJ, Miles MJ, Morris VJ, Besford LS, Jarvis DA, Marsh RA. *J. Cereal Sci.* 1990;11:243.
413. Cairns P, Miles MJ, Morris VJ. *Carbohydr. Polym.* 1991;16:355.
414. Kalichevsky MT, Ring SG. *Carbohydr. Res.* 1987;162:323.
415. German ML, Blumenfeld AL, Guenin YV, Yuryev VP, Tolstoguzov VB. *Carbohydr. Polym.* 1992;18:27.
416. Doublier J-L, Llamas G. In: Dickinson E, Walstra P, eds. *Food Colloids and Polymers: Stability and Mechanical Properties.* Cambridge, UK: Royal Society of Chemistry; 1993:138.
417. Hansen LM, Hoseney RC, Faubion JM. *Cereal Chem.* 1991;68:347.
418. Langton M, Hermansson A-M. *Food Structure.* 1989;8:29.
419. Svegmark K, Hermansson A-M. *Food Structure.* 1989;10:117.
420. Durrani CM, Prystupa DA, Donald AM, Clark AH. *Macromol.* 1993;26:981.
421. Steeneken PAM. *Carbohydr. Polym.* 1989;11:23.
422. Gudmundsson M, Eliasson A-C. *Carbohydr. Polym.* 1990;13:295.
423. Sievert D, Würsch P. *J. Food Sci.* 1993;58:1332.
424. Ring SG, Stainsby G. *Prog. Food Nutr. Sci.* 1982;6:323.
425. Bagley EB, Christianson DD. *J. Text. Stud.* 1982;13:115.
426. Bagley EB, Christianson DD. *Starch/Stärke.* 1983;35:81.
427. Bagley EB, Christianson DD, Beckwith AC. *J. Rheol.* 1983;27:503.
428. Taylor NW, Bagley EB. *J. Appl. Polym. Sci.* 1974;18:2747.
429. Taylor NW, Bagley EB. *J. Appl. Polym. Sci.* 1974;21:113.
430. Evans ID, Haisman DR. *J. Text. Stud.* 1979;10:347.
431. Evans ID, Lips A. *J. Text. Stud.* 1992;23:69.
432. Wong RBK, Lelievre J. *Rheol. Acta.* 1981;20:299.
433. Wong RBK, Lelievre J. *Starch/Stärke.* 1982;34:231.

434. Wong RBK, Lelievre J. *J. Appl. Polym. Sci.* 1982;27:1433.
435. Ellis HS, Ring SG, Whittam MA. *J. Cereal Sci.* 1989;10:33.
436. Okechukwu PE, Rao MA. *J. Text. Stud.* 1995;26:501.
437. Svegmark K, Hermansson A-M. *Carbohydr. Polym.* 1990;13:29.
438. Muhrbeck P, Eliasson A-C. *Carbohydr. Polym.* 1987;7:291.
439. Biliaderis CG, Tonogai JR. *J. Agric. Food Chem.* 1991;39:833.
440. Lii C-Y, Shao Y-Y, Tseng K-H. *Cereal Chem.* 1995;72:393.
441. Silverio J, Svensson E, Eliasson A-C, Olofsson G. *J. Therm. Anal.* 1996;47:1179.
442. van Soest JJG, Hulleman SHD, de Wit D, Vliegenthart JFG. *Carbohydr. Polym.* 1996;29:225.
443. Davidou S, Le Meste M, Debever E, Bekaert D. *Food Hydrocoll.* 1996;10:375.
444. Vodovotz Y, Hallberg L, Chinachoti P. *Cereal Chem.* 1996;73:264.
445. Wunderlich B. *Macromolecular Physics: Crystal Nucleation, Growth, Annealing.* New York, NY: Academic Press; 1976.
446. Fearn T, Russell PL. *J. Sci. Food Agric.* 1982;33:537.
447. McIver R, Axford AWE, Colwell KH, Elton GAH. *J. Sci. Food Agric.* 1968;19:560.
448. Eliasson A-C. *J. Cereal Sci.* 1983;1:207.
449. Russell P. *J. Cereal Sci.* 1983;1:285.
450. Russell P. *J. Cereal Sci.* 1983;1:297.
451. Russell P. *Starch/Stärke.* 1983;35:277.
452. Mandelkern L. *Crystallization of Polymers.* New York, NY: McGraw Hill; 1964.
453. Krog N. *Stärke.* 1971;23:206.
454. Evans ID. *Starch/Stärke.* 1986;38:227.
455. Goering KJ, Jackson LL, Dehaas BW. *Cereal Chem.* 1975;52:493.
456. Melvin MA. *J. Sci. Food Agric.* 1979;30:738.
457. Eliasson A-C, Carlsson TL, Larsson K, Miezis Y. *Starch/Stärke.* 1981;33:130.
458. Takahashi S, Seib PA. *Cereal Chem.* 1988;65:474.
459. Gray VM, Schoch TJ. *Stärke.* 1962;14:239.
460. Eliasson A-C, Larsson K, Miezis Y. *Starch/Stärke.* 1981;33:231.
461. Eliasson A-C. *Starch/Stärke.* 1985;37:411.
462. Hoover R, Hadziyev D. *Starch/Stärke.* 1981;33:346.
463. Osman EM, Dix MR. *Cereal Chem.* 1960;37:464.
464. Harbitz O. *Starch/Stärke.* 1983;35:198.
465. Gough BM, Greenwell P, Russell PI. In: Hill RD, Munck L, eds. *New Approaches to Research on Cereal Carbohydrates.* Amsterdam, The Netherlands: Elsevier Science Publishers B.V.; 99.
466. Svensson E. *Crystalline Properties of Starch.* Sweden: Lund University; 1996. [Ph.D. Dissertation].
467. Krog N. *Starch.* 1973;25:22.
468. Nierle W, Baya AW. *Starch/Stärke.* 1990;42:268.
469. Eliasson A-C. *J. Text. Stud.* 1986;17:357.
470. Eliasson A-C, Finstad H, Ljunger G. *Starch/Stärke.* 1988;40:95.
471. Eliasson A-C, Junger G. *J. Sci. Food Agric.* 1988;44:353.
472. Rao P, Nussinovitch A, Chinachoti P. *Cereal Chem.* 1992;69:613.
473. Miura M, Nishimura A, Katsuta A. *Food Structure.* 1992;11:225.

474. Batres LV, White PJ. *J. Am. Oil Chem. Soc.* 1986;63:1537.
475. Cheer RL, Lelievre J. *J. Appl. Polym. Sci.* 1983;28:1829.
476. Wang Y-J, Jane J. *Cereal Chem.* 1994;71:527.
477. Kohyama K, Nishinary K. *J. Agric. Food Chem.* 1991;39:1406.
478. Katsuta K, Nishimura A, Miura M. *Food Hydrocoll.* 1992;6:387.
479. Katsuta K, Nishimura A, Miura M. *Food Hydrocoll.* 1992;6:399.
480. Rogers DE, Zeleznak KJ, Hoseney RC. *Cereal Chem.* 1988;65:398.
481. Biliaderis CG, Izydorczyk MS, Rattan O. *Food Chem.* 1995;53:165.
482. Christianson DD, Hodge JE, Osborne D, Detroy RW. *Cereal Chem.*, 1981; 58, 513.
483. Sajjan SU, Rao MRR. *Carbohydr. Polym.* 1987;7:395.
484. Bahnassey YA, Breene WM. *Starch/Stärke.* 1994;46:134.
485. Alloncle M, Doublier J-L. *Food Hydrocoll.* 1991;5:455.
486. Eidam D, Kulicke W-M, Kuhn K, Stute R. *Starch/Stärke.* 1995;47:378.
487. Ovadia DZ, Walker CE. *Starch/Stärke.* 1996;48:137.
488. Hoover R, Hadziyev D. *Starch/Stärke.* 1981;33:290.
489. Bulpin PV, Welsh EJ, Morris ER. *Starch/Stärke.* 1982;34:335.
490. Stute R, Konieczny-Janda G. *Starch/Stärke.* 1983;35:340.
491. Eliasson A-C. *Thermochim. Acta.* 1985;95:369.
492. Kowblansky M. *Macromol.* 1985;18:1776.
493. Eliasson A-C, Krog N. *J. Cereal Sci.* 1985;3:239.
494. Raphaelides S, Karkalas J. *Carbohydr. Res.* 1988;172:65.
495. Eliasson A-C. *Carbohydr. Res.* 1988;172:83.
496. Le Bail P, Bizot H, Pontoire B, Buleon A. *Starch/Stärke.* 1995;47:229.
497. Godet MC, Bizot H, Buleon A. *Carbohydr. Polym.* 1995;27:47.
498. Dalvand CC. *Investigations on Starch and Starch–Emulsifier Interactions in Durum Wheat Pasta.* Zurich, Switzerland: Swiss Federal Institute of Technology; 1995. [Ph.D. Dissertation].
499. Riisom T, Krog N, Eriksen J. *J. Cereal Sci.* 1984;2:105.
500. Seneviratne HD, Biliaderis CG. *J. Cereal Sci.* 1984;13:129.
501. St.J. Manley R. *J. Polym. Sci.: Part A.* 1964;2:4503.
502. Yamashita Y. *J. Polym. Sci.: Part A.* 1965;3:3251.
503. Zobel HF, French AD, Hinkle ME. *Biopolym.* 1967;5:837.
504. Buleon A, Duprat F, Booy FP, Chanzy H. *Carbohydr. Polym.* 1984;4:161.
505. Helbert W, Chanzy H. *Int. J. Biol. Macromol.* 1994;16:207.
506. Godet MC. *Mechanismes de Formation des Complexes Amylose-Acide Gras.* Nantes, France: Universite de Nantes; 1994. [These de Doctorat].
507. Kugimiya M, Donovan JW. *J. Food Sci.* 1981;46:765.
508. Karkalas J, Raphaelides S. *Carbohydr. Res.* 1986;157:215.
509. Sievert D, Holm J. *Starch/Stärke.* 1993;45:136.
510. Ong MH, Blanshard JMV. *Starch/Stärke.* 1995;47:7.
511. Bader HG, Göritz D. *Starch/Stärke.* 1994;46:229.
512. Bader HG, Göritz D. *Starch/Stärke.* 1994;46:435.

Corn and Sorghum Starches: Production

9

Steven R. Eckhoff[1] and Stanley A. Watson[2]
[1]Department of Agricultural Engineering University of Illinois,
Urbana, Illinois, USA
[2]Ohio Agricultural Research and Development Center
The Ohio State University, Wooster, Ohio, USA (Retired)

I. Introduction	374
II. Structure, Composition and Quality of Grain	375
1. Structure	376
2. Composition	381
3. Grain Quality	385
III. Wet-milling	391
1. Grain Cleaning	392
2. Steeping	394
3. Milling and Fraction Separation	408
4. Starch Processing	421
5. Product Drying, Energy Use and Pollution Control	421
6. Automation	423
IV. The Products	423
1. Starch	423
2. Sweeteners	423
3. Ethanol	424
4. Corn Oil	425
5. Feed Products	426
V. Alternative Fractionation Procedures	427
VI. Future Directions in Starch Manufacturing	429
1. Continued Expansion into Fermentation Products	429
2. Biosolids as Animal Food	429
3. Processing of Specific Hybrids	430
4. New Corn Genotypes and Phenotypes via Biotechnology and Genetic Engineering	430
5. Segregation of the Corn Starch Industry	430
VII. References	431

Starch: Chemistry and Technology, Third Edition
ISBN: 978-0-12-746275-2

Copyright © 2009, Elsevier Inc.
All rights reserved

I. Introduction

The word 'corn' is used in the United States as the common name for the cultivated member of the grass family (Gramineae) known to botanists as *Zea mays* L. More specifically, 'corn' here means the seed produced by this plant. Outside of the United States, this crop is commonly known as maize. Corn is believed to be a product of domestication in central Mexico beginning 5000–7000 years ago. The evidence is now quite strong that domestic corn was derived by human selection from mutants of the grass, teosinte (*Zea mays,* sp. *Mexicana*) which grows wild in the Mexican central highlands.[1] Recent discovery of a distinct new species, *Zea diploperennis* Iltis,[2] a wild perennial teosinte with the same number of chromosomes as corn, has given new validity to the origin of teosinte. Corn has reached its present state of development through continual mutations, hybridizations, segregations and selections by random, natural processes, and by conscious selection. A number of types have developed by this process, which differ primarily in the structure of the grain. Examples are popcorn, sweet corn, dent corn, flint corn and flour corn.

Corn has been the staple food for countless generations of natives of North and South America. The diverse seed types were probably selected and cultivated by primitive farmers in response to food preferences. 'Mahys' specimens were taken to Spain by Columbus on an early voyage, who introduced corn to Europe. The first recorded planting of corn was near Seville in 1494,[3] from where it subsequently spread to all of Europe, Asia Minor and eventually to the Far East. It is interesting that a mutant of considerable industrial importance, waxy maize, was first discovered in China in 1909.[4] Flint types of corn gradually became an important crop across southern Europe and Turkey. However, it was the development of American dent types and their eventual hybridization that propelled corn to agricultural dominance as the most cost-effective feed grain, so important in the development of today's highly specialized animal agriculture.

The ready availability of corn at relatively low and steady prices, its storability from season to season, its ease of transportation and handling, and its high starch content led naturally to the development of commercial processes for recovery of corn starch. From the early nineteenth century, when inventors first discovered that corn starch was fairly easily recovered by grinding the soaked grain, the process has gradually evolved into today's highly sophisticated automated process which produces a multitude of useful food and industrial products. Mechanical innovations developed by trial and error during full-scale operations stimulated much of the early growth of the corn starch industry.[5] Today process and product improvements more commonly follow research and engineering studies and pilot plant evaluations.

Process innovations have resulted from pressures to improve worker efficiency and the work place environment, to reduce energy consumption, to reduce initial capital expenditure, to reduce air and water pollution, to improve end product quality, and to introduce new products. In the last 40 years, older facilities have been completely redesigned and expanded, and since 1970, 23 completely new wet-milling facilities have been built in the United States. Such installations now require less space than the units they replaced; working conditions and plant sanitation have vastly improved and continuous operation and product monitoring have improved product quality and uniformity.

Grain sorghum (*Sorghum bicolor Moench*) is a cereal grain also known in some localities as milo, milo maize, or kaffir corn. Sorghum culture probably began in eastern Africa 5000–7000 years ago and spread to all of Africa, Europe, Asia and eventually the United States in the mid-nineteenth century.[6] The sorghum plant resembles the corn plant, but the grains, about the size and shape of No. 5 shot, are borne on a terminal, bisexual rachis (head).[7] Some sorghum varieties have juicy stalks of high sugar content and are grown for syrup production on very limited acreage. They bear small, inedible seeds and should not be confused with the grain sorghums.

Although grain sorghums are a major world crop, they are generally considered to be inferior to corn for food, feed or industrial uses. They require less water for growth than corn and, therefore, are grown in more arid regions. In this chapter, some space will be devoted to grain sorghum starch production, but it is of marginal interest because the only known manufacturer of sorghum starch in the United States stopped production in 1975. The wet-milling of grain sorghum is similar to that of corn and has been thoroughly described elsewhere,[8] although novel methods of extracting starch from sorghum have recently been investigated.[9–11]

II. Structure, Composition and Quality of Grain

Wet-milling, as a process to recover starch, is essentially a method of disrupting the corn or sorghum kernel in such a way that the component parts can be separated in an aqueous medium into relatively pure fractions. Dry-milling using screening and air classification for component separation of corn and sorghum achieves less efficient fractionation.[12] Satisfactory starch recovery cannot be attained practically by dry-milling methods, even when fine grinding and air classification processing are used.[13] Only wet-milling can achieve a commercially satisfactory yield and quality of starch from corn or grain sorghum. Attempts to combine the wet and dry-milling processes have been reported,[14–16] and it is possible to recover starch from corn flour using alkaline solutions,[17,18] and from corn grits using solutions of sulfur dioxide.[19] Alkaline debranning of corn enhanced the dry-milling recovery of germ,[20] but still did not achieve the recovery level of wet-milling. The energy saving obtained by dry separation of germ and fiber and wet processing of only the endosperm would be useful. However, no practical application has been achieved because germ (oil) yield by dry-milling is significantly lower than the yield from the wet-milling process. Use of a standard dry-milling process to separate the germ and pericarp from the endosperm showed that good starch yield was achieved in an eight-hour slurry steep. However, low yield of germ, and hence oil, completely offset the large energy savings that would be achieved by not needing to dry one-third of the germ fiber fraction.[21] Alternative methods of corn fractionation are discussed later in this chapter.

1. Structure

Mature corn or sorghum kernels are unique, well-organized entities that exist for the purpose of reproducing the species. Descriptions of their structures and compositions are helpful in understanding the process of disruption that is achieved during the wet-milling process.

Corn

The corn kernel is a one-seeded fruit classified as a grain,[22] borne on a female inflorescence commonly known as the ear. Each ear is comprised of a central stem, the cob, on which up to 1000 seeds develop. The seeds (kernels) mature about 60 days after pollination and are harvested in late summer or early fall (in the northern hemisphere) when kernel moisture content has dropped below 30% (wet weight basis).

The mature corn kernel is composed of four principle parts: tip cap, 0.8%; pericarp, 5.3%; germ, 11.9%; and endosperm, 82% (Figure 9.1; Table 9.1).[23] The tip cap is the remnant of the pedicel which attaches the seed to the cob. It is composed of insoluble fibrous elements arranged in a spongy structure well adapted for rapid water adsorption.[24] When the tip cap is removed, a black tissue known as the hilar layer, which apparently functions as a sealing layer across the base of the germ upon kernel maturation, is revealed.[24,25] The tip cap acts as the major pathway for the entrance of water or other liquids or gases into the kernel.[24,26,27]

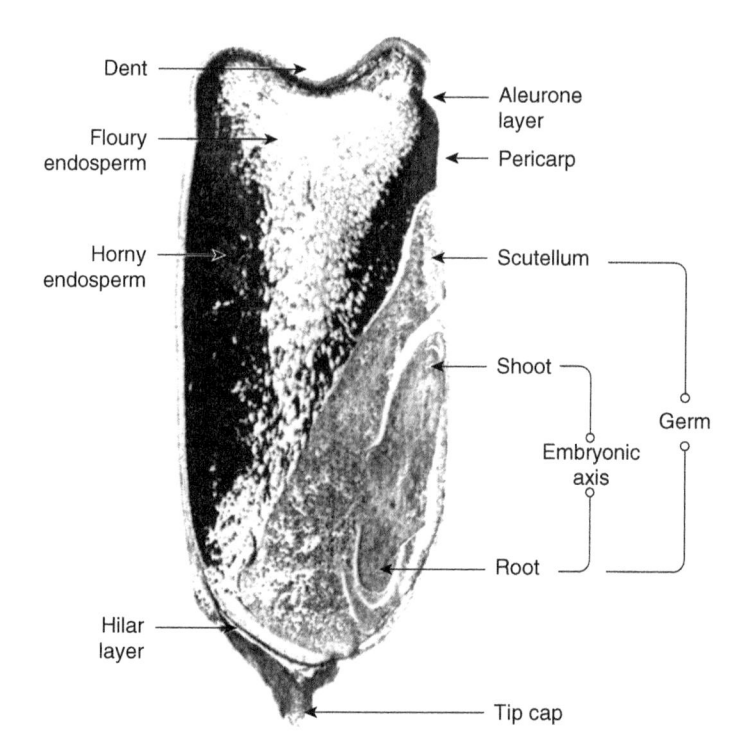

Figure 9.1 Longitudinal 10-μm bisection of a steeped dent corn kernel (magnified ×6, iodine stained). Note how starch has been lost from floury endosperm cells as a result of sulfur dioxide action.

The pericarp is the smooth, dense, outer covering of the kernel.[24] The pericarp is often referred to in the industry as the 'hull,' which is a misnomer since a hull is a distinctly different morphological entity. The pericarp's outer layer has a thin coating of wax-like cutin, beneath which are several layers of dead, hollow, elongate cells compressed into a dense tissue. Beneath this layer is a spongy tissue known as the tube cell and a cross-cell layer which acts as a natural channel for absorption of water. Directly beneath the spongy layer is a very thin suberized membrane known as the seed coat or testa. This layer is thought to act as a semipermeable membrane, limiting passage of large molecules into or out of the endosperm and germ. Liquids or gases which enter the tip cap region diffuse rapidly through the cross and tube cells, so that the region between the outer pericarp and the seed coat quickly becomes saturated with the liquid or gas. Diffusion into the endosperm then occurs across the seed coat from around the total periphery, rather than from just the tip cap end. This pathway gives the appearance that the diffusivity of the pericarp is much higher than it actually is. A second major pathway of passage of gases and liquids into the endosperm via naturally occurring connections between the germ and the center of the soft endosperm has also been identified.[26,27] Diffusion through the germ is 18 times higher than through the pericarp;[28] this pathway effectively provides for endosperm penetration from the inside out. Results from recent research do not support this mode of entry by water.[29]

Immediately beneath the testa is the first layer of the endosperm, the aleurone, a single layer of highly proteinaceous cells having thick, tough cell walls. It is possible that the aleurone's thick dense cell walls may be responsible for the observed semipermeability attributed to the seed coat simply by a size limiting diffusional process. Aleurone comprises about 3% of the weight of the kernel.[30,31] Although structurally part of the endosperm, the aleurone, along with the pericarp and tip cap, is included in the 'fiber' fraction from wet-milling and in the bran fraction from dry-milling. In alkali debranning, the aleurone layer is not removed with the pericarp,[32] because the alkali dissolves the weak bonds between the loose cells of the cross and tube cell layer.

Table 9.1 Weight and composition of component parts of dent corn kernels from eleven Midwest hybrids[49]

		Percent dry weight of whole kernel	Composition of kernel parts (% d.b.)				
			Starch	Fat	Protein	Ash	Sugar
Germ	Range	10.5–13.1	5.1–10.0	31.1–38.9	17.3–20.0	9.38–11.3	10.0–12.5
	Mean	11.5	8.3	34.4	18.5	10.3	11.0
Endosperm	Range	80.3–83.5	83.9–88.9	0.7–1.1	6.7–11.1	0.22–0.46	0.47–0.82
	Mean	82.3	86.6	0.86	8.6	0.31	0.61
Tip cap	Range	0.8–1.1	–	3.7–3.9	9.1–10.7	1.4–2.0	–
	Mean	0.8	5.3[a]	3.8	9.7	1.7	1.5
Pericarp	Range	4.4–6.2	3.5–10.4	0.7–1.2	2.9–3.9	0.29–1.0	0.19–0.52
	Mean	5.3	7.3	0.98	3.5	0.67	0.34
Whole kernels	Range	–	67.8–74.0	3.98–5.8	8.1–11.5	1.27–1.52	1.61–2.22
	Mean	100	72.4	4.7	9.6	1.43	1.94

[a] Composite

Mature endosperm is comprised of a large number of cells, each packed with starch granules embedded in a continuous matrix of protein; all the cell contents are surrounded by a cellulosic cell wall (Figure 9.2).[30,33,34] Mature endosperm of dent corns contains a central core of soft or floury endosperm extending to the crown, which shrinks on drying, causing a 'dent.' The floury endosperm is surrounded by a glassy-appearing region known as horny or hard endosperm. The average ratio of floury to horny regions is about 1:2, but it varies considerably depending on the protein content of the kernel.[23,35] Flour corns contain virtually no horny endosperm, while flint corn and popcorn endosperms are comprised of a small core of floury endosperm entirely surrounded by horny endosperm which resists denting when the kernel dries. Although the dividing line between horny and floury regions of the

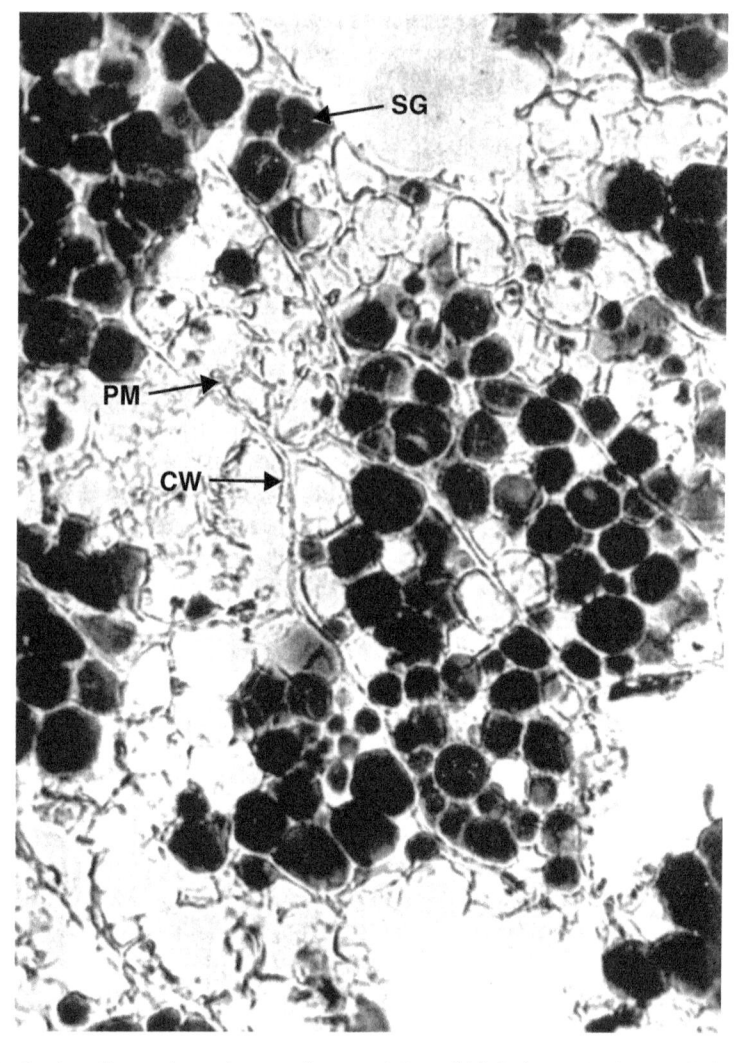

Figure 9.2 Section of steeped corn horny endosperm (10 μm thick, iodine stained, magnified ×612; SG, starch granules; PM, protein matrix; CW, cell wall).

endosperm is morphologically indistinct, the floury region is characterized by larger cells, large spherical starch granules and a relatively thin protein matrix.[30] The thin strands of protein matrix rupture during kernel drying, causing air pockets, which give the floury region a white opaque appearance[33] and a porous texture which results in easy starch recovery.[36]

In the horny endosperm, the thicker protein matrix shrinks during drying, but does not rupture. The resulting pressure produces a dense glassy structure in which starch granules are forced into an angular, close-packed conformation. The dense nature of horny endosperm requires adequate steeping to ensure recovery of starch. The horny endosperm contains only ~1.5–2.0% higher protein content[31,35,37] and a greater concentration of the yellow carotenoid pigments than the floury region.[37] Just under the aleurone layer is a dense row of cells known as subaleurone or dense peripheral endosperm[38] containing as much as 28% protein.[31] These small cells comprise probably less than 5% of the endosperm. They contain very small starch granules and a thick protein matrix. Starch from this area can be difficult to isolate and purify.[38]

The germ is comprised of two major parts: the scutellum and the embryonic axis. The embryonic axis is the structure that grows into the seedling on germination and makes up only ~10% of the weight of the germ.[39] The scutellum functions as a storage organ from which nutrients can be quickly mobilized during initial seedling growth. The surface of the scutellum adjacent to the endosperm is covered by secretory epithelium, a layer deeply furrowed by canals or glands lined with elongated secretory cells. The function of these cells is to secrete enzymes[40] which diffuse into the endosperm where they digest starch and other constituents to provide nourishment for the embryo.

The scutellar epithelium adheres to the endosperm by an insoluble cementing substance, apparently consisting of degradation products of crushed cells and composed largely of pentoglycans and protein.[41] This layer provides a strong bond that resists many chemical and physical means of separating the germ and endosperm; hence, prolonged steeping is required for an effective separation of whole intact germ from the endosperm. The major portion of the scutellum is composed of thick-walled isodiametric cells densely packed with cytoplasm. Oil droplets may be seen in sections of such cells and have been shown to be nearly identical to the spherosome bodies[42] observed in the repository of oil in cottonseed and peanut cotyledons.[43]

A simple mental picture of the corn kernel can be created by visualizing the pericarp as a plastic bag similar to those used to package groceries. The bag, like the pericarp, protects the contents inside by limiting exposure to the external elements. In the bag is the germ which, to a miller, is like a large sponge soaked with oil. The majority of the bag is filled with marbles, which are starch granules. Visualize glue poured into this marble and oil-laden, sponge-filled bag, so that all void areas are filled. When the glue hardens, it totally encapsulates the marbles (starch), and adheres the bag (pericarp) to the marbles (starch), the bag to the oily sponge (germ), and the oily sponge to the marbles. The purpose of wet-milling is to chemically loosen the glue and mechanically (primarily by density) separate the kernel into nearly homogeneous piles of oily sponges (germs), marbles (starch), glue (gluten meal) and bags (gluten feed).

Grain Sorghum

The structure and composition of the sorghum kernel is quite similar to that of corn. The kernels are flattened spheres measuring about 4.0 mm long, 3.5 mm wide and 2.5 mm thick. The weight of individual kernels ranges from 8 to 50 mg, with an average of 28 mg. Colors of the prehybrid varieties range from white through pale orange, tan and red, to dark red-brown.[6] Most grain now in commercial channels is a brownish-red color, because one or both parents of the best hybrids have that color.

A longitudinal section of a typical sorghum grain (Figure 9.3) is nearly identical to a corn kernel in gross morphology. The structure of the germ is identical to that of corn germ. The endosperm differs from that of corn only in the relative proportions of floury and horny endosperms; all sorghums are like flint corn in that the horny endosperm entirely surrounds the floury region. The dense peripheral endosperm layer comprises a larger proportion of the kernel than it does in corn, which results in much greater problems during starch purification.[38] Common sorghums have white endosperm. However, yellow (xanthophyll pigmented) endosperm types were introduced from Africa for breeding purposes. Several hybrids have been developed that

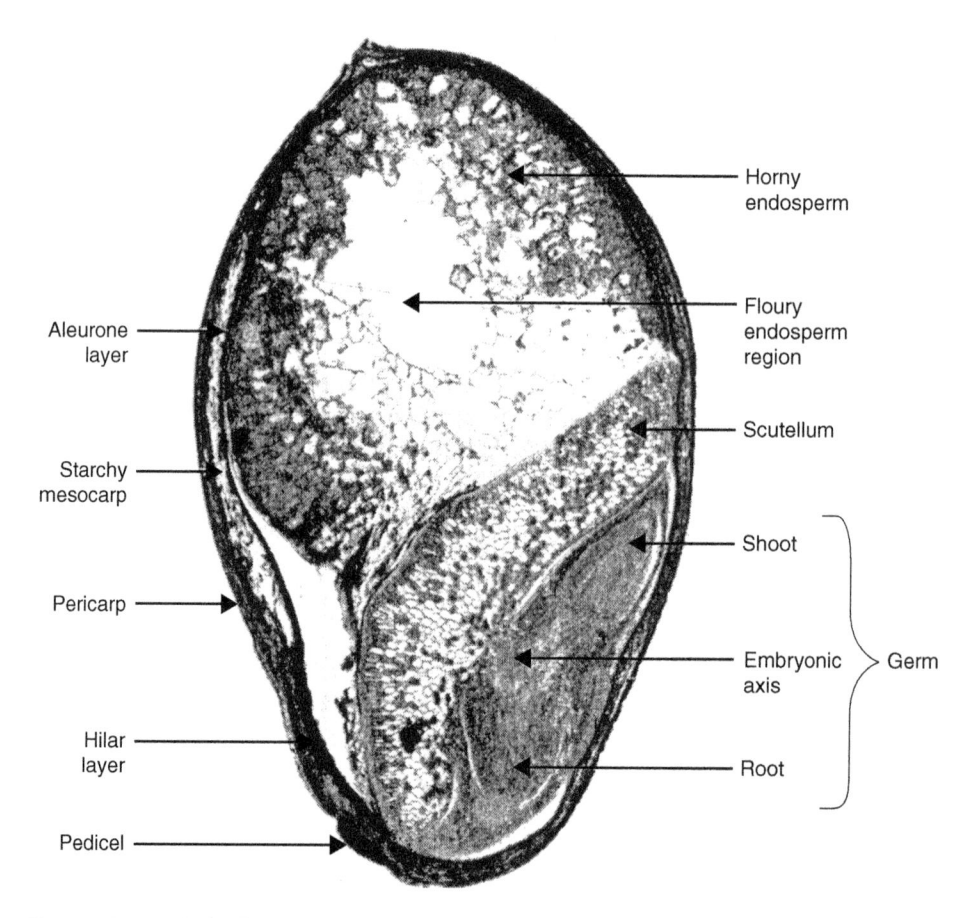

Figure 9.3 Longitudinal 10-μm bisection of a steeped grain sorghum kernel (magnified ×22, iodine stained).

have a high percentage of horny endosperm where the pigment is concentrated, but they have not been successful because the carotenoid pigments bleach out rapidly in the field before harvest.

The pericarp of grain sorghums shows the most differences from that of corn. It is covered with a thick layer of wax;[44] most varieties have a thick mesocarp layer in the pericarp containing very small, unrecoverable starch granules. Some varieties have a thick, orange-pigmented testa layer[45] which shatters during wet-milling and adds colored particles to the starch. The epidermis layer of pericarp in some varieties contains water-soluble, flavone-type pigments ranging in color from red to orange. In some varieties, a purple flavonoid contained in the glume which surrounds each seed leaches into the seed and gives a gray cast to isolated starch. The wax extracted from sorghum has been found to be similar to commercial Carnauba wax in some physical and chemical properties, and may be a substitute for some applications.[46]

2. Composition

Corn

The chemical composition of commercial corn based on proximate analysis is shown in Table 9.1. The average values for main components are averages for corn purchased by several large wet-milling plants in 1970–1974. The range of values covers analyses from numerous sources.[47] Commercial corn, which is a thorough mixture from many sources, has a much narrower range. Two fiber values are given in Table 9.2. The crude fiber value is a crude measure of cellulose, but is in disrepute because of non-specificity. A preferred method measures neutral detergent residue (NDR) which includes all insoluble, cell wall polysaccharides.[48] Acid detergent residue (ADR) measures the sum of cellulose and lignin. About 90% of the kernel NDR fiber content is in the pericarp and tip cap. Because lignin content is negligible, the ADR value is

Table 9.2 Proximate analysis of corn grain

Fraction	Range[a]	Average[b]
Moisture (%, wet basis)	7–23	16.7
Starch (%, dry basis)	64–78	71.3
Protein[c] (%, dry basis)	8–14	9.91
Fat (%, dry basis)	3.1–5.7	4.45
Ash (oxide) (%, dry basis)	1.1–3.9	1.42
Pentoglycans (as xylose) (%, dry basis)	5.8–6.6	6.2
Crude fiber (%, dry basis)	1.8–3.5	2.66
Fiber (neutral detergent residue) (%, dry basis)	8.3–11.9	9.5
Cellulose + lignin (acid detergent residue) (%, dry basis[d])	3.3–4.3	3.3
Sugars, total (as glucose) (%, dry basis)	1.0–3.0	2.58
Total carotenoids (mg/kg)	5–40	30

[a] Numerous sources, including reference 47 and personal data
[b] Moisture, starch, protein and fat values average of corn purchased on open market 1970–1974
[c] N × 6.25
[d] Lignin content less than 0.1%[248]

essentially a measure of the cellulose content. The difference between the two values (NDR − ADR) is approximately equal to the pentoglycan (hemicellulose) content.

Data for chemical compositions of the kernel parts given in Table 9.1 were obtained by hand dissecting kernels of eleven typical corn belt hybrids.[49] These data show that endosperm contains >98% of the total starch. The endosperm also contains 74% of the kernel protein and 16% of the fat. Endosperm lipids, mostly bound to cell contents, are only 18% triglycerides (triacylglycerols), whereas germ lipids are nearly all triglycerides.[50] Thus, the 84% of the kernel fat that is in the germ is all that can be economically recovered. Germ also contains 22% of the kernel protein, 82% of the total ash and 65% of the sugar. Germ contains all the phytin phosphorus in corn,[51] probably as a magnesium salt, and a small, unrecoverable quantity of starch.

There are three classes of corn recognized in the US Grain Standards: yellow, white and mixed. White corn and mixed white and yellow corn classes are generally not purchased by the wet-milling industry because yellow corn is what the vast majority of farmers produce, as white corn hybrids generally give lower yields. Furthermore, the yellow pigment which is concentrated in the gluten meal fraction is highly valued as poultry feed, and commands a price approximately 1.5 times the value of starch. Even at comparable production yields, wet-millers would prefer to purchase yellow hybrids. Currently, some specialty starch wet-milling companies process white corn to produce starches with enhanced whiteness. Processing corn of mixed classes is not desirable.

The average oil content of commercial corn hybrids planted in the corn belt has declined over the last 50 years from 4.8–5.0% to 4.4%.[52] The apparent cause of this decline has been widespread use of one or two high yielding, low oil content corn inbreds. At the same time, the iodine value of corn oil has increased from 122–124 to 128–130, equivalent to a linoleic acid content increase from 55% to 60%. This change is apparently associated with the decline in oil content.[53]

The composition of corn can be altered by plant breeding methods. Corn selection studies at the University of Illinois[54,55] have shown that the oil or protein content can be increased or decreased over a wide range. Breeding corn for higher oil content has been seriously pursued for many years and has resulted in development of good performing hybrids containing 6–9% of oil (dry basis).[56] The high oil hybrids that were available commercially until recently gave significantly lower grain yields, which limited their acceptability. Recently, a method of producing higher oil corn having acceptable grain yields has been developed; this method employs a top-cross system using a male-sterile counterpart of a standard hybrid blended with 8–10% of a high oil pollinator. The resulting grain, produced as Optimum™ High Oil Corn, gives an increase of 2.4% oil over standard hybrids under average agronomic practices.[57] Published data[57] show that oil contents of five different High Oil TC Blends™ (five different hybrid seed parents) ranged from 6.8–7.2% (6.9% average) compared to 3.9–5.6% oil (4.1% average) for the same five hybrids open pollinated. Grain yields averaged only 6% lower for the Optimum™ High Oil Corn. These high oil corn products found immediate value in livestock rations, because they deliver about 4% higher metabolizable energy, 6% more available lysine and 4% more protein than a standard hybrid grown under comparable conditions. Other corn genetics companies are enhancing the yield potential of conventional high oil hybrids. However, the abundant availability and the low cost of white grease and contracting problems have reduced the amount of high oil corn grown to the point where it is now difficult to find.

Higher-oil corn has been strongly desired by the wet-milling industry[52,53] because oil is a product of high value, but as high oil corn become available commercially in large quantities, processors are not readily accepting it. Wet-millers are concerned about the premium price set for high oil corn above the commodity corn price, the residual oil content of the starch, and the potential that high oil corn has in disrupting the mill balance. High oil corn production grew rapidly during the start of the twenty-first century for use in export and domestic livestock production, but was not used in wet-milling. However, it has been reported that seed companies have discovered how to increase the oil content of the germ without increasing its size. This could be a boost to wet-millers who would get more oil revenue without increasing the capacity of their germ recovery system.

Increasing the total protein content of corn by breeding has not resulted in development of high protein, commercial hybrids, partly because the use of low or high levels of nitrogen fertilizers can also cause protein levels to vary from 8% to 12%. Furthermore, the increased protein is predominantly zein[51,58] which is nearly devoid of two essential amino acids, lysine and tryptophan.

Search for corn of better nutritional protein value led to the discovery[59] that the floury type of endosperm mutants, termed opaque 2 and floury 1, are significantly higher in lysine and tryptophan than normal dent corn (see Chapter 3). A great deal of breeding work has since been conducted, but commercially acceptable hybrids have not yet been developed. One major disadvantage of the initial opaque 2 variety was that it contained high percentages of floury or soft endosperm, which fragmented on harvesting and handling, resulting in excessive levels of broken kernels. Varieties having denser endosperm and normal agronomic properties, while retaining high lysine content, have been developed by a Mexican research team.[60] However, if high lysine hybrids are eventually developed, they probably will not be of much significance to the wet-milling industry because of poor economics and little product advantage.[61] Laboratory studies indicate that higher steepwater yields would be expected to lower yields of gluten meal and starch, when conventional steep times were used.[62] Use of an 8 hour steep time for opaque 2 corn was found to give steepwater solids and starch yields comparable to those of regular dent corn hybrids.[62] Even with comparable yields, opaque 2 corn will be more difficult to wet-mill satisfactorily, because of increased breakage during handling. The higher lysine content of the gluten meal would be of little economic significance.

The most important genetic change in corn for starch production was development of waxy maize (corn) hybrids.[63] Starch from waxy maize, which contains 100% of the branched starch fraction, amylopectin, has become a major food starch (see Chapters 3 and 20). This corn must be grown in fields isolated from dent corn to prevent cross-pollination, and generally will be grown under identity-preserved contracts to ensure delivery of uncontaminated shipments of maize to the wet-milling establishment and to provide adequate supply. The milling properties of waxy maize are similar to those of regular corn, except that the yield of solubles is a little higher, starch yield is lower and starch and gluten slurries filter and dry a little more slowly than those from normal dent corn. This characteristic has been shown to be caused by the presence of

Table 9.3 Proximate analysis of grain sorghum

Fraction	Range[a] % dry basis	Average[b] % dry basis
Water (% wet basis)	8–20	15.5
Starch	60–77	74.1
Protein (N × 6.25)	6.6–16	11.1
Fat (CCl$_4$ extract)[c]	1.4–6.1	3.7
Ash	1.2–7.1	1.5
Crude fiber	0.4–13.4	2.6
Pentoglycans (as xylose)	1.8–4.9	2.5
Sugars, total (as glucose)	0.5–2.5	1.8
Tannin	0.003–0.17	0.1
Wax	0.2–0.5	0.3

[a] Numerous sources including reference 47 and personal data
[b] Average of grain purchased in south and west Texas, 1959–1970, for use in wet-milling
[c] Includes wax

a small amount of a phytoglycogen. The addition of a small amount of an amylolytic enzyme to the slurries results in normal filtering and drying characteristics.[64]

High-amylose corn having starch which contains 55–80% of the linear starch fraction, amylose, has also been developed[65] (see Chapter 3) and must also be grown under contract. Production of amylomaize is even smaller than it is for waxy hybrids because of limited uses for the starch. Starch yields of high-amylose corn are lower than those of dent corn.[66]

Grain Sorghum

The proximate chemical composition of commercial grain sorghum is shown in Table 9.3. The average values for the major components were obtained on grain purchases from south and west Texas for wet-milling at Corpus Christi, Texas, 1959–1970. Grain sorghum differs from corn in minor ways. Moisture content is generally lower because the grain dries more in the field before harvest. Compared with corn, starch and protein contents are 1–2% higher, while total fat is always lower because the germ constitutes a smaller proportion of the kernel than it does in corn (Table 9.4). Furthermore, the pericarp wax amounts to ~8% of the total fat[67] and must be removed in refining the crude oil. Analysis of the wax indicates that it is comprised of saturated fatty acids, aldehydes and alcohols of 27–30 carbon atoms.[68] All sorghum varieties contain tannins, but the dark-brown, bird-resistant varieties contain tannin levels so high as to reduce digestibility when fed.[69] It is also known that, in sorghums devoid of tannins which have been cooked, the proteins are poorly digested by humans, apparently caused by disulfide cross-bonding.[70,71]

As with corn, the only recoverable crude oil is in the germ; it amounts to 76% of the total fat (Table 9.4). Total fat in sorghum is 80–85%, whereas that of corn and includes 8% wax (Table 9.3). Germ contains 62% of the total ash in the kernel. About 92% of the total kernel starch is in the endosperm. Unlike corn, another 3.7% of the total

II. Structure, Composition and Quality of Grain

Table 9.4 Weight and composition of component parts of kernels from five grain sorghum varieties[67]

		Percent dry weight of whole kernel	Composition of kernel parts (% dry basis)			
			Starch	Protein	Fat	Ash
Germ	Range	7.8–12.1	–	18.0–19.2	26.9–30.6	–
	Mean	9.8	13.4[a]	18.9	28.1	10.36[a]
Endosperm	Range	80.0–84.6	81.3–83.0	11.2–13.0	0.4–0.8	0–0.44
	Mean	82.3	82.5	12.3	0.6	0.37
Bran	Range	7.3–9.3	–	5.2–7.6	3.7–6.0[b]	–
	Mean	7.9	34.6[a]	6.7	4.9	2.02[a]
Whole kernel	Range	–	72.3–75.1	11.5–13.2	3.2–3.9[b]	1.57–1.68
	Mean	–	73.8	12.3	3.6	1.65

[a] Composite
[b] Includes wax

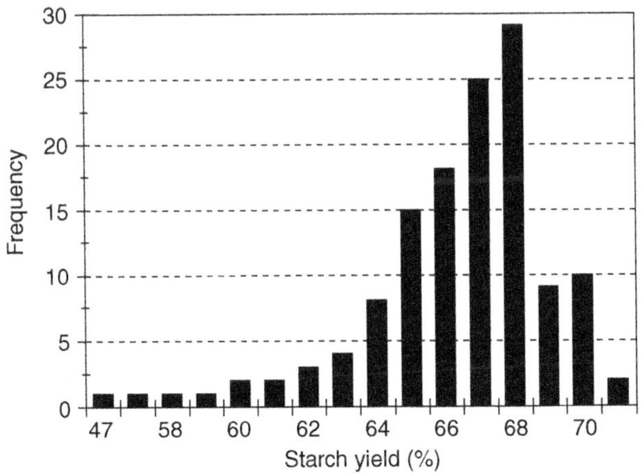

Figure 9.4 Starch yield frequency distribution for 131 corn samples representing the range of commercially available hybrids in the central Illinois area.

starch is in the mesocarp layer (Figure 9.3) and is not recoverable because of the small size of the granules.

Plant breeding methods have also been used to modify grain sorghum kernel properties. Waxy varieties and hybrids, which contain starch composed of 100% amylopectin, have been developed.[6,72] White pericarp, waxy hybrids were processed for production of waxy sorghum starch in the United States between 1955 and 1970. High-amylose sorghum has not been found, but a high lysine mutant has.[73]

3. Grain Quality

Corn

Optimal quality corn for wet-milling is affected by three main factors: genetics; environmental or growing conditions; and postharvest handling and storage. Freeman[74] has described quality parameters that have a bearing on wet-milling performance. Wet-milling yields and wet-millability (the ease of separation) have been shown to be heritable.[75,76] Figure 9.4 shows the range of starch yields resulting from laboratory wet-milling of 133 commercial yellow dent hybrid samples in 1994. The samples were acquired from central midwestern seed companies, and no attempt was made to prevent duplication of genetic material (hybrids identified by company number only) or to control production or harvesting practices. The average of all hybrids tested was approximately 65.5%, with the peak frequency at 68%. Starch yields as high as 71.5% occurred. Using an estimated starch price of $0.10 per lb and a gluten feed price of $0.04 per lb, the value of increased starch yield can be estimated at approximately $0.03 per percentage starch yield increase. The variability between hybrids indicates value differences of more than $0.30 per bushel ($0.012/kg). The increased value between the top hybrid and the average of all hybrids was $0.18 per bushel ($0.0071/kg).

Other research has shown that starch yield can be affected by as much as 4% due to growing region alone.[77] However, the starch yield ranking of the hybrids studied was not affected by the growing region. A high starch yielding hybrid was found to be the highest or near the highest at all locations, even if the starch yield for all hybrids was affected by the environment.

One method of measuring corn quality is provided by the US Grade Standards.[78] The official standards (Table 9.5) are based on four major criteria plus miscellaneous minor criteria which define the distinctly undesirable sixth grade designated 'Sample Grade.' These criteria are designated 'The Official Methods of Determination.' Grade is determined at the marketing point by licensed inspectors who issue official inspection

Table 9.5 United States, grading standards for corn[a,76]

Grade	Minimum test weight per bushel (lb)	Broken corn and foreign material (%)	Maximum limits damaged kernels	
			Total (%)	Heat-damaged kernels (%)
1	56	2.0	3.0	0.1
2	54	3.0	5.0	0.2
3	52	4.0	7.0	0.5
4	49	5.0	10.0	1.0
5	46	7.0	15.0	3.0
Sample grade[b]	–	–	–	–

[a]Grades and grade requirements for the classed yellow corn, white corn and mixed corn

[b]Sample grade is corn that:
(a) does not meet the requirements for the grades US No. 1 to No. 5; or
(b) contains 8 or more stones which have an aggregate weight in excess of 0.20% of the sample weight, 2 or more pieces of glass, 3 or more crotalaria seeds (*Crotalaria* spp.), 2 or more castor beans (*Ricinus communis* L.), 4 or more particles of an unknown foreign substance(s) or a commonly recognized harmful or toxic substance(s), 8 or more cockleburs (*Xanthium* spp.) Or similar seeds singly or in combination, or animal filth in excess of 0.20% in 1000 grams; or
(c) has a musty, sour or commercially objectionable foreign odor; or
(d) is heating or otherwise of distinctly low quality

certificates. The grade factor, which is the lowest of the four criteria, determines the grade of a particular lot of corn. Additionally, there are special grade standards for flint and waxy corn, viz., that each must contain no less than 95% of these special specified classes. The same applies to yellow and white corn classes.

As corn moves from farm to end user, the lots become more uniform owing to extensive blending, both conscious and circumstantial, by each handler. As a result, corn moving toward the end user tends to average about US No. 2 grade. Nevertheless, differences in grade do exist, and buyers apply discounts for each grade factor lower than US No. 2. The discount is related to the negative effect which that factor may have on the wet-milling yields and quality of products.[74]

It is important to realize that the four grade-determining factors for corn are related to quality interests for all end uses of corn and do not directly relate to potential wet-milling yields. Commercial wet-millers have long known that the purchase of US No. 1 corn does not increase either the millability of the corn or the yield of starch compared with lower grade corn. The selection of US No. 2 corn as the primary purchase grade by wet-millers is based on relative price and availability and by concern for public perception if they purchase large quantities of what is perceived as off-quality corn (US No. 3 or lower). In the absence of any rapid test which would estimate true millability or starch yield, the grade standards provide wet-millers with a basis for purchasing corn (for a more detailed discussion of the importance of quality factors, references 79–82 should be consulted).

Moisture content was formerly a major grade criterion but, in 1984,[83] moisture content was removed from being grade-determining to a 'standard' which must be listed on the certificate of inspection. Moisture discount schedules used for adjusting the price are a matter of dilution of the dry weight of the grain substance; however, the presence of moisture above 14.5% may result in the development of mold if grain temperature exceeds 22–24°C over a long period of time. Corn containing over 18% moisture rapidly becomes moldy at favorable temperatures. Molds, usually species of *Aspergillus* or *Penicillium*, grow on moist kernels if temperature and moisture conditions are right for germination of their ubiquitous spores.[84] Mycelium growth is especially vigorous in the nutrient-rich germ. The resulting decomposition causes reduced germ recovery and increased oil refining losses, due to fragmentation of damaged germs during milling and hydrolysis of glycerides to free fatty acids.[85] The damage factor in US Grade is a visual estimation of the percentage of kernels having mold invaded germs.

Another undesirable result of mold growth on corn is the production of mycotoxins, especially the carcinogenic aflatoxins caused by the growth of *Aspergillus flavus*.[86] Frequency of occurrence in the US midwest is low,[87,88] but corn grown in southern climates is especially vulnerable,[89] mainly because of damage and inoculation of kernels in the fields[90,91] and climatic conditions which favor aflatoxin production by the invading microorganisms. Laboratory wet-milling tests have shown that fractionation of corn containing aflatoxin results in starch that is essentially devoid of the mycotoxin, with the aflatoxin being concentrated in the co-product streams of gluten meal and gluten feed.[92]

More recent laboratory studies[93] showed that, for dry grind ethanol, adding aflatoxin B1 toxin (at levels of 0, 100, 200, 350 and 775 ppb) to toxin-free corn did not affect fermentation rates or final ethanol concentration. Corn infected with *A. flavus* shows a blue–green–yellow (BGY) fluorescence under ultraviolet light.[94] Wet-milling companies monitor incoming corn for the presence of aflatoxin using an ultraviolet fluorescence test[95] and by use of amino assay kits for verification and quantification of aflatoxin level.

Other mycotoxins of concern in corn in some years are zearalenone, deoxynivalenol (vomitoxin) and fumonusin. These compounds are produced by ear-rotting molds of the Fusarium (Gibberella) genus,[86] which frequently invade corn in the field but may become epidemic during years of cold, wet fall weather.[96] Laboratory wet-milling tests have shown that zearalenone in corn is concentrated in corn gluten feed and is not present in starch.[97,98] Fumonusin (FB1 and FB2) was not found in starch recovered by wet-milling corn naturally contaminated with 13.9 ppm total fumonusin.[99] Fumonusin was present in all other wet-milled fractions.

Wet-millers avoid purchasing obviously moldy corn, but mycotoxins can be present in corn with low levels of inoculum. Blending moldy corn with uncontaminated corn, as occurs in the market channel, results in contamination of larger quantities of corn than originally existed from the field.

The grade factor of 'Test Weight' is an indirect measure of grain maturity in most years. An early frost stops maturation of corn, if the corn has not reached the black layer (hilar layer) stage, resulting in incomplete filling of the kernel. Corn maturity by itself has only a slight effect on starch yield.[100,101] Samples harvested near 50% w.b. moisture content, well before physiological maturity, showed only approximately a 3% decrease in starch yield compared to samples harvested after black layer formation at approximately 25% w.b. moisture content. Both sets of samples were ear dried using laboratory ambient air.

Similarly, test weight by itself does not affect starch yield.[102] Low test weight resulting from an early frost necessitates the use of high temperature drying from higher than normal moisture contents. Frost kills the corn plant, which stops filling of the kernels and, more importantly, eliminates use of corn plant as a conduit for moisture removal from the kernel. The corn kernels must then dry in the field by diffusing the moisture through the kernel pericarp and through the husk. This dry down pathway is considerably slower than back down through the stalk, resulting in the need to harvest the corn wetter than desired. The low starch yields associated with low test weight corn is primarily due to the effect of high temperature drying.

To a wet-miller, the most damaging effect of low test weight corn on their operation can be decreased steep capacity. Steeping is a volume-limited process and low test weight decreases the weight of corn which can be put into specific tanks. Researchers[102] calculated that wet-millers should discount low test weight corn by $0.03 per bushel per lb/bu if the test weight is below 56 lb/bu. This discount is over three times the normal market discount for test weight, and is based on decreased capital utilization, not on starch yield loss. High test weight corn (above 60 lb/bu) is not desired for wet-milling because it indicates the presence of a higher percentage of horny endosperm in the kernels. Horny endosperm is more difficult to steep, necessitating the use of longer steep times.[62]

The grade category of 'broken corn, foreign material and other grains' (BCFM) is predominantly broken corn which is removed at the wet-milling plant before steeping, because it interferes with water flow in the steeps. It is usually added to the gluten feed fraction. If the foreign material is removed, the broken corn can be steeped in a separate slurry tank of sulfur dioxide solution for 8–12 hours, then added to the millhouse stream for starch recovery.[103,104]

High levels of broken corn in commercial channels result from the way it is harvested and dried. Practically all marketed corn is now harvested with field shellers or combines at a moisture content of 22–28%.[105] Corn at this high moisture level is quickly attacked by spoilage organisms and, therefore, must be dried quickly. Many kinds of corn dryers are in use on farms and in elevators.[106] The preferred dryers are continuous-flow, heated-air dryers. Corn that has been heated to temperatures greater than 60°C during drying produces starch of lowered yield and paste viscosity on wet-milling.[107–109] Corn dried from highest moisture (28–30%) at temperatures above 82°C also exhibits reduced oil yield and lower protein in steepwater. The latter observation is an indication that the adverse wet-milling separations obtained with high temperature dried corn result from protein denaturation.[110,111]

Unfortunately for the wet-miller, dryer damage is difficult to detect and does not adversely affect the grade designation. One US wet-miller during the fall of 1993 received a load of corn, graded US No. 1, which had been harvested at over 35% moisture and rapidly dried. This corn did not hydrate above 30% moisture, even after 48 hours steeping. It was not possible to mill the corn to produce starch of acceptable quality. Because drying damage cannot be readily determined by any rapid test, such corn cannot be penalized by applying price discounts. However, extensive publicity by the wet-milling industry about the adverse effects of excessive temperatures in drying corn and higher energy costs resulting from poor thermal efficiency have reduced the severity of the problem.

The heat damage category in the US Grade Standard (Table 9.5) is a measure of visual kernel and/or germ darkening caused by severe microbial heating and should not be confused with the heat damage caused by rapid artificial drying. Damage caused by rapid artificial drying is usually not visually detectable. Only when using severe drier temperatures will the kernel darken sufficiently to result in a change of grade.

The most adverse result of heated air drying is kernel breakage that occurs during subsequent handling in commercial channels.[112–114] Kernel brittleness can result from drying to low moisture content, but it is primarily caused by formation of internal stress cracks in the horny endosperm[115] due to a high rate of drying and rapid cooling. Two methods of measuring corn breakage susceptibility are: (a) visual estimation of the percentage of kernels having internal stress cracks in the horny endosperm; and (b) quantification of fines produced when a small sample is processed in a breakage testing instrument.[115] Recently, the FGIS has added visual stress crack analysis as additional data available during grading. High levels of stress cracks can indicate rapid drying and potential starch yield loss (unpublished data). Stress cracks by themselves do not cause starch yield loss, but they are indicative of high temperature dried corn, which does cause starch yield loss.

Brown et al.[116] developed a steeping index which predicts factory performance of dryer-damaged corn based on the rate and extent of hydration. Their test apparently was developed on corn dried at high temperatures. The ability of the test to delineate moderate dryer damage is apparently not good, and the test has not been adopted by the wet-milling industry. Rausch[117] evaluated a procedure which utilized an electric resistance and electric capacitance moisture tester in combination to identify dryer damage. He found that it was too sensitive to growing season and to hybrid differences to be of any commercial value.

Laboratory wet-milling is the most accurate way to estimate milling characteristics of corn samples.[118–123] The major disadvantages inhibiting widespread use of laboratory milling are the slow turnaround for results (2–3 days), the cost and the low number of samples that can be milled per day. Wehling et al.[124] showed that determination of starch yield may be possible using NIR or NIT technology. Such a procedure would be fast and low cost, but would require extensive calibration, and as an indirect method of measurement it may not be able to account for all the variables affecting wet-milling starch yield.

Grain Sorghum

The quality parameters of grain sorghum for wet-milling are similar to those of corn. The US Grades for sorghum (Table 9.6) are similar to those of corn, except that test weight levels are a little higher, moisture levels a little lower, total damage a little lower, heat damage a little higher, and 'broken kernels and foreign material' (BKFM) twice as high. Buffo[125] examined 46 commercial sorghum hybrids for laboratory wet-milling properties and found significant differences with respect to various physical and chemical properties.

Table 9.6 US grading standards for grain sorghum[a78]

Grade	Minimum test weight per bushel (lb)	Moisture (%)	Maximum limits of		Broken kernels, foreign material, and other grains (%)
			Damaged kernels		
			Total (%)	Heat-damaged kernels (%)	
1	57	13.0	2.0	0.2	4.0
2	55	14.0	5.0	0.5	8.0
3[b]	53	15.0	10.0	1.0	12.0
4	51	18.0	15.0	3.0	15.0
Sample grade[c]	–	–	–	–	–

[a]Grades and grade requirements for the classes yellow grain sorghum, white grain sorghum, brown grain sorghum, and mixed grain sorghum

[b]Grain sorghum which is distinctly discolored shall not be graded higher than No. 3

[c]Sample grade shall be grain sorghum which does not meet the requirements of any of the grades from No. 1 to No. 4, inclusive; or which contains stones; or which is musty, sour or heating; or which is badly weathered; or which has any commercially objectionable foreign odor except of smut; or which is otherwise of distinctly low quality

The lower level of total damage in commercial grain sorghum is related to the fact that most grain sorghum is seldom harvested at moisture contents greater than 18%. This low moisture content results because the kernels are individually exposed to wind and sun; thus, in the more arid climates, where major production is located, kernels dry rapidly. Since little artificial drying of sorghum is practiced, there is little dryer damage and less stress cracking. However, some stress cracking probably takes place in the field due to wetting and drying, as was shown to be the case with wheat.[126]

Harvesting of sorghum grain at low moisture levels also makes it less vulnerable to mold damage. However, grain that is to be stored must be dried to 13–14% moisture to prevent mold growth. The same mold species that attack corn will invade grain sorghum and produce mycotoxins. Germ damage and mycotoxins in grain sorghum are just as detrimental in starch production as they are in corn, but frequency of occurrence is low.[127] A much greater problem with commercial sorghum in warmer climates where much sorghum is grown is infestation by weevils and other storage insect pests.

The major quality problem with grain sorghum for wet-milling relates to pericarp pigmentation. The US Grade divides grain sorghum into four classes: yellow grain sorghum; white grain sorghum; brown grain sorghum; and mixed grain sorghum. Yellow grain sorghum is defined as those varieties which have yellow, salmon-pink or red seedcoats, and which contain <10% of grain sorghum of other colors. 'Yellow' here does not refer to yellow (carotenoid-containing) endosperm, as is the case with corn. Yellow endosperm types have never achieved a sufficiently wide use in commercial channels to be recognized as a grade.

Colors of finished sorghum starch are related to the intensity of pericarp colors.[128] Steepwater color, and probably its tannin content, varies with the intensity of the pericarp pigments. Some varieties contain pigments that are colorless at acid pH, but are bright yellow at pH 9–10. These pigments cause starch to have a gray cast which can be improved by bleaching.[129]

As mentioned previously, some sorghum varieties have a highly-pigmented testa or subcoat. Some varieties have a white pericarp masking a brown subcoat. Such varieties are not classified as white grain sorghum, but must be classed as brown grain sorghum. Any variety with a brown subcoat or dark brown pericarp, or both, is undesirable for starch production.

III. Wet-milling

The basic principles of steeping and milling corn for separation of starch are universal, because they are dictated by the nature of the corn kernel and the properties of its components, but the machinery assembled to accomplish starch recovery has changed greatly over the years. The current commercial corn wet-milling method is an elegant and practical process which has been gradually developed over a period of about 150 years. The process evolved at first by trial and error methods by entrepreneurs seeking an advantage over competitors in processing costs, simplicity, and/or product quality and innovation. In the last 75 years, process improvements have been sought through more scientific and engineering studies for the same competitive reasons. Technologies developed for other industries or developed specifically for the corn wet-milling process have been adapted. Current processes are marvels of technological control and simplicity compared with those in use as little as 40–50 years ago. Individual facilities around the world differ somewhat in the kinds of machinery employed and the configurations of product flow, because of differing product outputs, machinery preferences, degree of automation and age. Most US plants have been modernized to optimize worker output, and the process described is that which is now in most general use. The generalized flow diagram of the starch recovery process is given in Figure 9.5. References 5 and 130–132 are good technical references about the technology of starch production from corn.

1. Grain Cleaning

Shelled grain is received in bulk at wet-milling plants and can contain large and small pieces of cob, chaff, sand, weed seeds, other cereal grains and other undesirable foreign material. Cleaning the corn is an important first step in the wet-milling process, because the presence of small pieces of broken kernels can alter the normal flow of steepwater through the corn mass, resulting in non-uniform steeping. Furthermore, starch granules, eroded from the exposed endosperm surfaces, are washed into the steepwater and are gelatinized during evaporation, resulting in viscous steepwater.

The degree to which the broken corn and foreign material are removed from the corn mass depends on the relative value of the corn to the value of the removed material as livestock feed. Wet-milling facilities in countries which import corn tend to remove less broken corn from the corn mass, due to economics and because of the high levels of broken corn resulting from the handling. In the United States, the standard size screen is 4.8 mm (12/64 in), while in many importing countries a smaller screen size (3.2 mm; 8/64 in) is often used. Reciprocating screening equipment removes about 50% of the smaller pieces of material when the equipment is operated at capacity. Aspiration is used to remove dust and light chaff. A magnetic separator is often used to recover metal trash (bolts, tools, scrap iron, etc.) from the corn in order to protect mills from damage.

Wang[133] studied the effect of broken and damaged kernels on the water absorption rate and steepwater composition in recirculated batch steeping. He observed six differences in the steeping of damaged kernels (damage created by use of a Stein breakage tester) and whole intact kernels: (a) rate of water sorption was much faster for the damaged kernels, averaging 8% per hour compared to 5% per hour for intact kernels during the first 4 hours of steeping; (b) final moisture equilibrium was not affected by damage and both samples achieved equilibrium in about 12 hours; (c) rate of solids release into the steepwater was much faster for damaged kernels with 3% total solids in the steepwater after only 5 hours, compared to intact kernels taking 36 hours to achieve the same level; (d) higher solids levels (3.8% versus 3%) resulted from steeping damaged kernels even after 48 hours of steeping; (e) steepwater pH increased more rapidly and maintained a higher level throughout steeping; and (f) protein content of the steepwater was almost double after 12 hours of steeping with damaged kernels and was approximately 60% higher after 48 hours of steeping. Similar observations were made for broken corn.

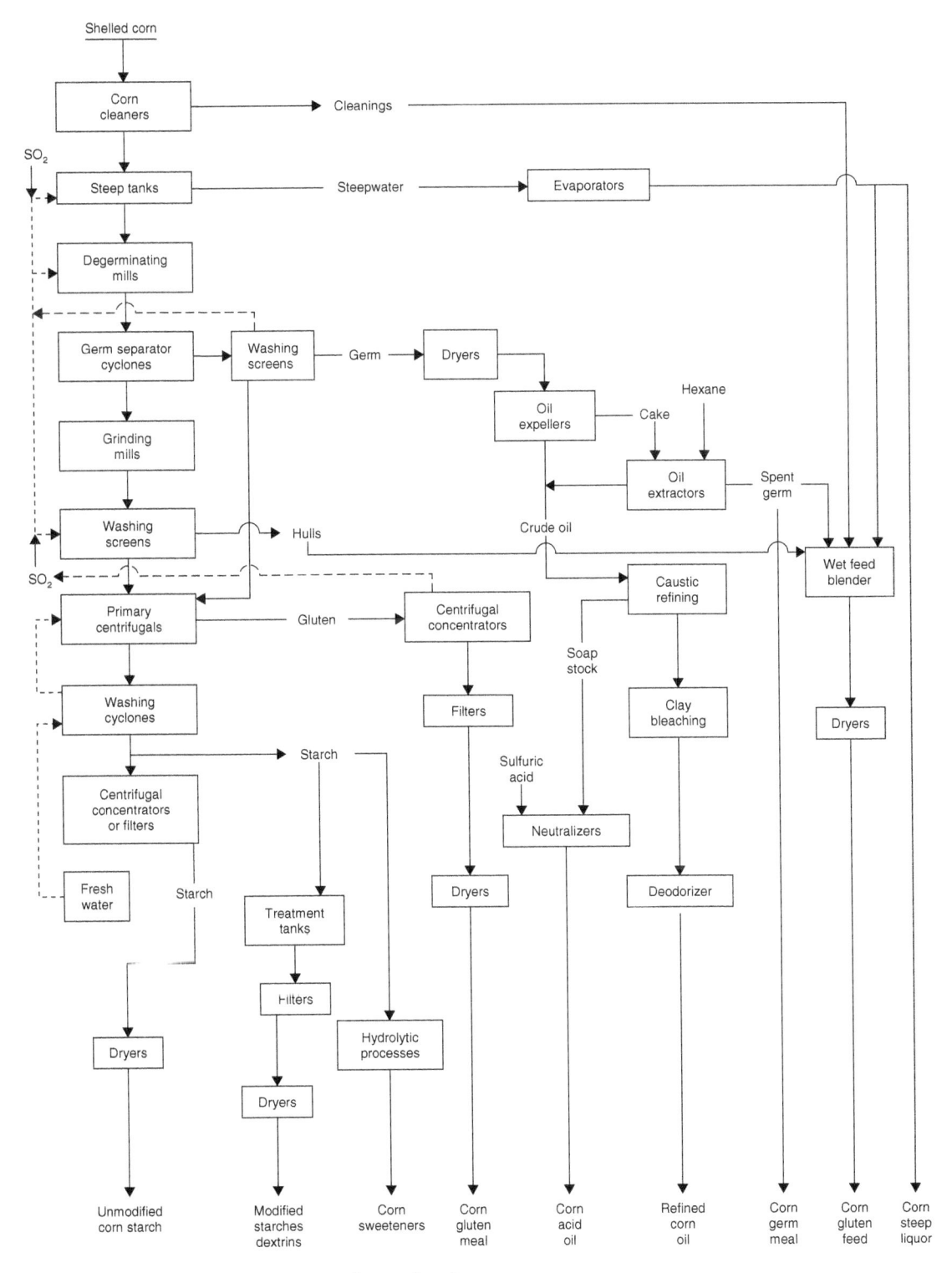

Figure 9.5 Flow diagram of the corn starch manufacturing process.

2. Steeping

Principles

Prior to wet-milling, the corn must be softened by a steeping process developed specifically to produce optimum milling and separation of corn components. Steeping is more than simple water soaking of corn. It involves maintaining the correct balance of water flow, temperature, sulfur dioxide concentration and pH. Corn is normally steeped for 24–40 hours at a temperature of 48°C to 52°C. By the end of the steeping period, the kernels should have: (a) absorbed water to about 45% (wet basis); (b) released about 6.0–6.5% of their dry substance as solubles into the steepwater; (c) absorbed about 0.2–0.4 g of sulfur dioxide per kg; and (d) become sufficiently soft to yield when squeezed between the fingers. At this point, the kernel can be pulled apart easily with the fingers. The germ is easily liberated intact and free of adhering endosperm or pericarp. When the endosperm is macerated with water, starch easily separates as a white floc and gluten is obtained as a yellow floc.

Conventional countercurrent steeping can be divided into three relatively equal stages. The first stage of steeping (8–12 hours) is the fermentation phase where there is an active *Lactobacillus* sp. fermentation producing lactic acid concentration of up to 2% in the steepwater. The fermentation phase is not only characterized by active fermentation, but also by the near-complete hydration of the kernels. The length of the fermentation phase is controlled by the sulfur dioxide level in the steepwater. When the sulfur dioxide level exceeds a critical level (100–300 ppm) the fermentation stops. The end of the first stage and the beginning of the second stage is determined by when lactic acid fermentation ceases.

The second stage of steeping (8–12 hours) is the sulfur dioxide diffusion stage. The critical item characterizing the second stage of steeping is the sulfur dioxide which diffuses into the corn kernel. It is the sulfur dioxide which diffuses into the kernel during the second stage of steeping which actually performs the reactions leading to starch release in the endosperm. Because it takes 8–12 hours for the sulfur dioxide to diffuse through the stagnant interstitial water into the hard-to-reach interior cells of the horny endosperm, the sulfur dioxide the corn is exposed to in the third stage is of little value with regard to endosperm reactions. Little sulfur dioxide diffuses into the kernel in the first stage because of the low concentration in the steepwater, and the sulfur dioxide that does diffuse into the kernel reacts quickly with germ and pericarp components before it has a chance to diffuse into the endosperm. It is this second stage which provides a sufficient sulfur dioxide flux to achieve the necessary reactions. The end of the second stage is determined by the start of the third stage. The third stage starts when the next molecule of sulfur dioxide is absorbed by the kernel and does not have sufficient time to diffuse into it. At this point, none of the remaining sulfur dioxide absorbed is helping the steeping, except that it is necessary to keep the concentration of sulfur dioxide high in this stage in order to provide enough sulfite to stop the fermentation stage (discussed below).

The third stage of steeping (the last 8–12 hours) is known as the sulfur dioxide dominated stage because it is characterized by high concentrations of sulfur dioxide (up to 2500 ppm) in the process water being added to the oldest corn. The third stage needs this high level of sulfur dioxide to ensure that there is adequate sulfur dioxide in the second stage. Because the process is countercurrent, a majority of the sulfur dioxide is absorbed by the corn in the third stage and carried with the corn into subsequent milling operations. Proper steep battery operation maintains a proper balance between these three stages.

Mechanics of Commercial Steeping

Steeps are large tanks that may be constructed of any material resistant to the corrosive action of solutions of sulfur dioxide and lactic acid at pH 3–4, but stainless steel construction is now the most common. Steep tanks will vary in size and number depending on the capacity of the facility. It is common in newer US facilities for tanks to hold 250–600 metric tons (10 000–25 000 bushels), but tanks as small as 60 metric tons (2500 bushels) of corn are still in use in some plants, with even smaller tanks being used in some other countries. The number of tanks in each separate steep battery varies, ranging from 6 to over 50.

Steep tanks are filled with raw grain from an overhead conveyor and are emptied through an orifice at the apex of the conical bottom. The inside surface of the cone bottom is covered with a strainer (slatted wood or stainless steel) for drainage. Each steep is equipped with piping and a pump to move steepwater from one steep to another, to recirculate steepwater back to the same tank, or to withdraw it from the system. Each tank, or every 2–3 tanks (depending on the facility design), are equipped with a heat exchanger which helps maintain the desired temperature in the steep by adding heat to the steepwater as it is being recirculated or pumped forward.

To prevent the loss of valuable organic matter and to avoid sewage disposal problems, careful management of the reuse of water is essential. For this reason, the entire milling and steeping installation is operated as a countercurrent washing unit. Fresh water, usually a mixture of steam condensate and demineralized tap water, is introduced at the final starch washing step, just prior to drying or conversion, at a rate of 1200–1480 L/t (8–10 gal/bu). Water of low hardness must be used to prevent the formation of calcium sulfate haze in starch hydrolyzates. The filtrate from the final starch washing is used to slurry the starch for the next to last washing operation. Water works its way from starch washing to steeps (Figure 9.6), gradually increasing in solubles content during passage through a series of dilution and reconcentration steps, and finally attaining a solubles level of 0.5–1.5% (more at low total water rates). It is then ready to be used in steeping corn.

Steeps are normally operated as countercurrent batteries of 6–12 tanks although, as mentioned earlier, much longer batteries are sometimes used. Process wash water containing 0.1–0.2% sulfur dioxide is placed on corn that has been in the steeps longest and, therefore, has the lowest residual solubles content. This process water is sequentially moved from tank to tank (from oldest corn to newest) while it decreases in sulfur dioxide content and increases in solubles.

Figure 9.7 shows the sequencing of steepwater from steep to steep, which allows for continuous operation. Each steep tank is equipped with a pump which continuously pulls the steepwater down through the tank. The steepwater is either sent

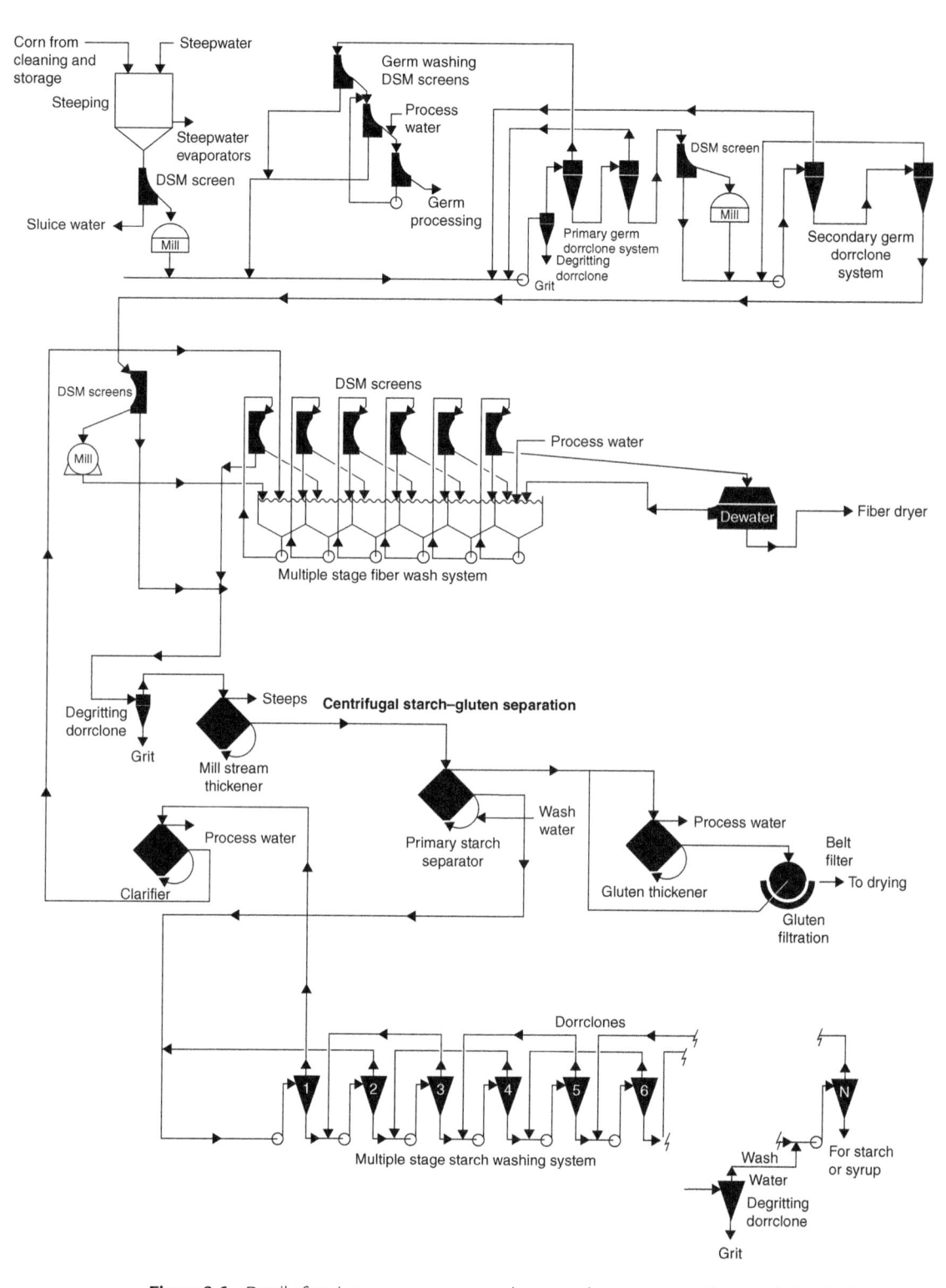

Figure 9.6 Detail of equipment arrangement and water and process stream flow employing the high-density, four-stage separation centrifugation system. (Courtesy of Dorr-Oliver, Inc., Stamford, Connecticut)

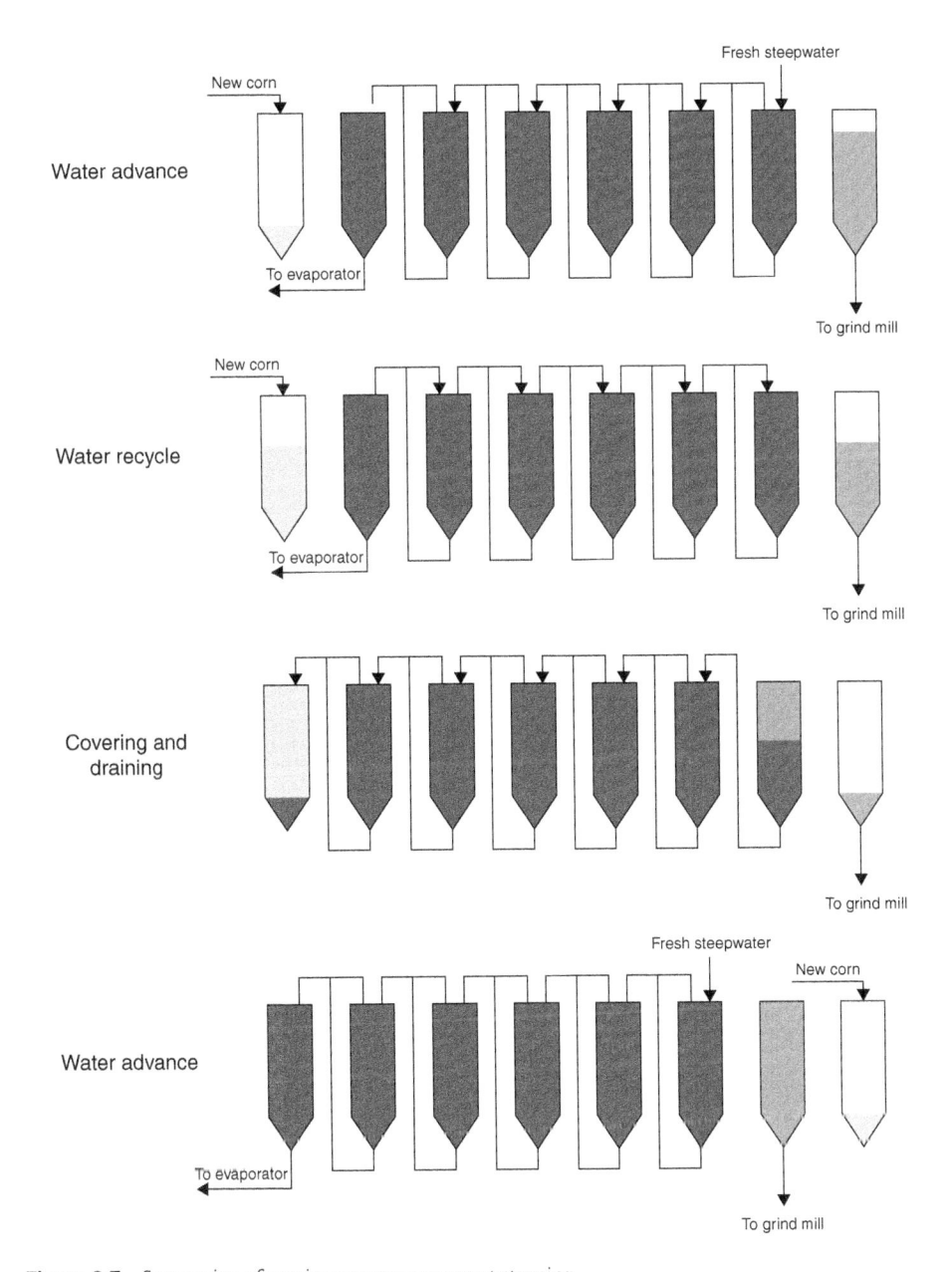

Figure 9.7 Sequencing of continuous countercurrent steeping.

forward to the next tank in the battery or is recycled back into the same tank. The steepwater is heated as needed, to compensate for heat loss by the steep tanks, by pumping it through a heat exchanger every tank or every 2–3 tanks, depending on the facility design.

Most steep systems are operated in a 'pull' fashion, where the rate of water movement through the steeps is controlled by the rate of light steepwater evaporator draw (steepwater leaving the newest corn) or the rate of water movement needed to cover new corn. As the water level in the steep with the newest corn decreases, a lower level switch is activated, opening a valve and causing steepwater from the tank with the second newest corn to be pumped forward. The lower level switch on the tank with the second newest corn is activated calling for water from the tank with the third newest corn. This process continues until the lower level switch on the tank with the oldest corn is activated. This activates a valve which pumps in the fresh sulfur dioxide-treated process water. 'Push' systems can be used where the rate of fresh sulfur dioxide-treated process water is constant and upper level switches activate the transfer of steepwater between tanks; the risk of overflowing steep tanks is greater with the push system, although the push system ensures that the corn is always covered, even when one or more of the tank pumps have reduced capacity. The other disadvantage of the push system is that the evaporator use rate is not held constant.

The pump on each tank runs continuously, while the tank is active, either pumping water forward to the next tank or recycling water back to the same tank. The percentage of time the steep tank is in the recycle mode depends on pump capacity, water use rate in starch washing and evaporator draw rate. Generally no two wet-milling facilities have the same recycle rate. The effect of the recycle rate, and concurrently the interstitial velocity in the tank on process yields and processing characteristics is not known. A laboratory-size steep battery which offers the potential for elucidating such engineering design effects has been developed.[134]

As mentioned earlier, fine material can cause considerable problems in steeping by plugging the screens at the bottom of the steep tank. This decreases flow through the tank and, if the plugging is severe enough, can result in uncovered corn in some tanks and a build-up of process water throughout the plant. Excess fine material can also cause channeling in the steep tanks, resulting in non-uniform steeping.

Some older facilities operate with no recycle, and transfer of water occurs as either batch advance or continuous advance. Most commonly, the systems are continuous advance where the water is moved by a small pump attached to each steep tank (Figure 9.8). In older 'batch advance' systems, one large pump serves an entire battery.

Several US facilities have adopted a continuous steeping system which allows the complete steeping operation to occur in each tank separately.[135] Corn is continuously added to the top of each steep, and steeped corn is pulled off at the bottom of each tank (Figure 9.9). Fresh steepwater (containing the highest level of sulfur dioxide) is pumped into the bottom of the tank at a flow rate low enough to give a residence time of 20–40 hours. The steepwater flows upward through the tank, and light steepwater is drawn off the top of each steep tank and sent to the steepwater evaporator. At steady-state, plug flow conditions, each layer within the tank will remain constant, i.e. the level of solids, sulfur dioxide and lactic acid will not change in the layer. The corn will move down through each layer during its residence time and will experience the complete steeping profile.

Continuous steeping offers several major advantages over conventional continuous-batch countercurrent steeping. Since each tank operates independent of all other steep tanks, pump failures or other mechanical problems do not significantly affect plant operation as occurs within a conventional system. Steep capacity can be expanded easily and discretely, and does not need to be tied with the old steep tanks.

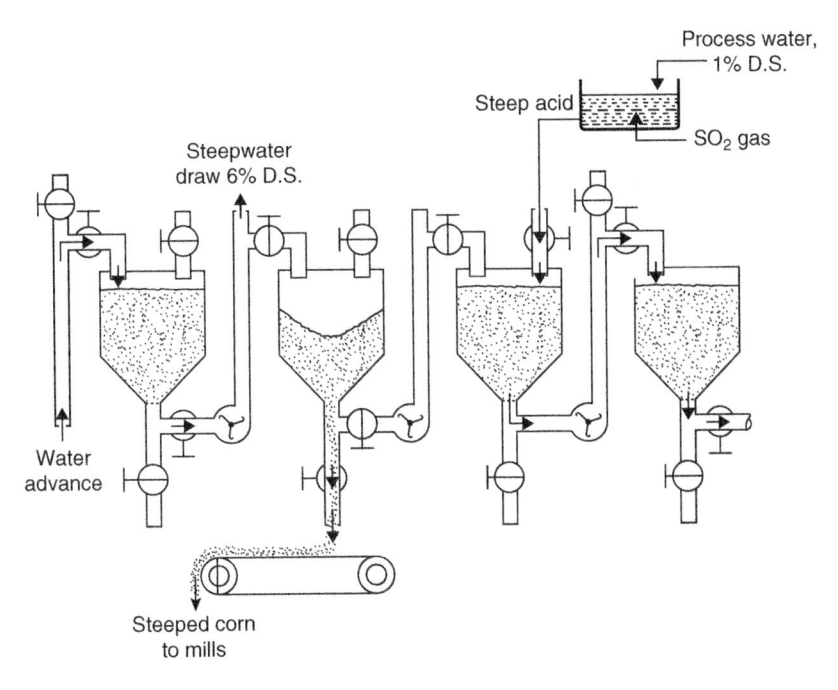

Figure 9.8 Details of countercurrent continuous advance steep battery operation.

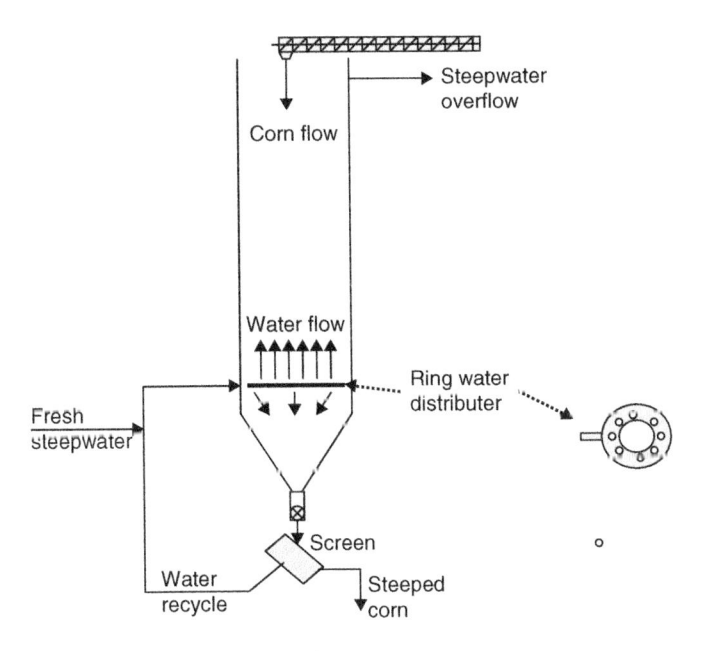

Figure 9.9 Design of tank for continuous steeping.

Little public information is known about the operating performance of the continuous steeping system, although it is generally recognized that at least one company has used the process successfully for more than 15 years. The major concern with using the continuous steeping process is starch quality. Problems could be encountered with uniformity, since channeling would be difficult to prevent and plug flow of the corn would be difficult to achieve. Care would also need to be taken to ensure that fine material is uniformly distributed across the steep tank. It would be difficult to remove corn continuously from the tank in a uniform manner. Unless measures are taken to equalize drag forces caused by the sidewalls, corn flowing down the center of the steep tank will have a shorter steep time than corn near the wall.

In the conventional countercurrent steeping process, the oldest light steepwater, now containing 5–7% solubles, is continuously pumped out of the steep containing the newest corn into the steepwater 'draw' tank and then to the evaporator at a rate of about 596–894 L/t (4–6 gal/bu). Instead of being withdrawn from the system, it may be bypassed through a heater and passed into the adjacent steep when necessary, to cover and heat new corn just added to the system. The system should be maintained at a temperature of 48–52°C (118–125°F). Addition of steep acid to the oldest corn is stopped in time to be able to pump out (advance) all the water and uncover the oldest corn for grinding. About 2–3 times the volume of water passes through the battery for each volume of corn. The corn is not moved except for loading and unloading the steeps. Although it has been claimed that the steeping process is faster if both corn and water are moved,[136,137] the evidence is not convincing and the cost in energy and equipment is prohibitive.

Figure 9.10 is a plot of steepwater composition taken from successive steeps across a ten steep, continuous-flow battery in 1950 when process water used for steep acid provided a rich bacterial inoculum. Today's continuous countercurrent steep batteries have approximately the same types of changes in every property except relative

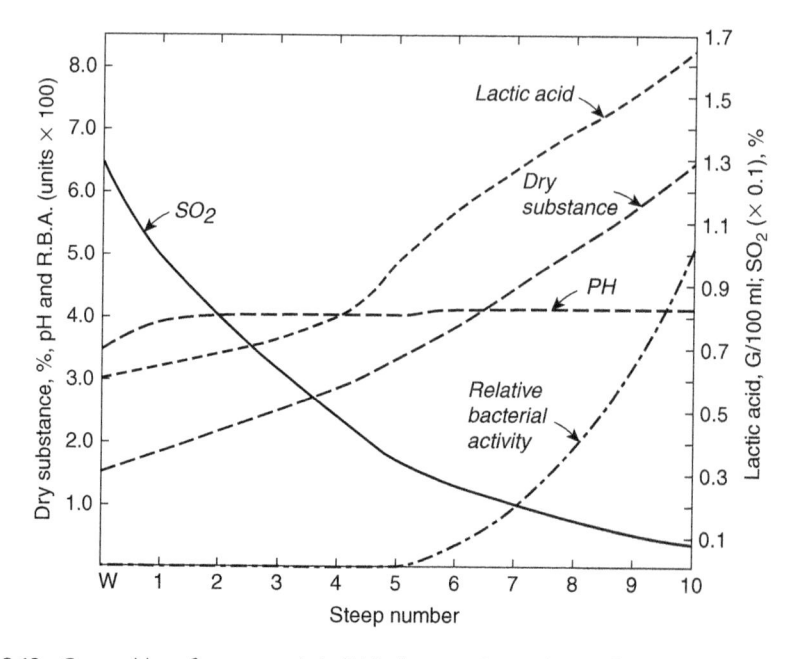

Figure 9.10 Composition of steepwater in individual steeps of a continuous flow countercurrent steep battery. Ten steeps sampled simultaneously. 'W' under steep number is composition of steep acid (input water). Water flow is from left to right. Steep number 10 contains newest corn.

bacterial activity, the number of actively reproducing bacteria, which today will be at a much lower level. Therefore, a lower level of lactic acid is observed in steeps 6 to 10, but the steepwater has a fairly constant pH of 4.0. Steady-state pH will vary from facility to facility, from 3.5–4.3, with very little apparent affect on yield or operational performance. Differences are probably due to differences in the incoming water used for starch washing and on the level of sulfite added downstream from the steeps.

In some factories, grind increase is accomplished by adding additional steeps to a continuous countercurrent system, eventually achieving from 15 to 50 steeps in a single battery. To operate a long battery, several steeps are simultaneously ground. The sulfur dioxide is added in liquid form by injection into the water entering 2–4 oldest corn steeps, but may be preceded by a 'washing phase,' in which process water with low sulfur dioxide content is passed over the oldest corn in preparation for grinding. The sulfur dioxide concentration in the steepwater declines as the water moves over newer corn, dropping to a level that is tolerated by lactic acid bacteria in the three or four steeps at the new corn end of the battery, termed the first or 'fermentation phase.'

A steep battery having only one large pump is termed a 'batch advance system.' In this system, steeping is quiescent except during periods of water movement which may occur once, commonly twice, or possibly three times during the period when a steep is being ground. To prepare a steep for grinding, a 'draw' of the volume scheduled (4–6 gal/bu, 0.60–0.90 L/kg) is made by pumping all the water off the newest corn and enough from the second oldest to make up the required draw volume. Water is then advanced on to newer corn in 'leapfrog' fashion to cover the corn uncovered by the draw pumping. Water is advanced in this manner until older corn steeps are emptied in succession. Empty steeps, except for the one next in line to be ground, are then covered with steep acid containing fresh sulfur dioxide. If a second or third advance is made during grinding of each steep, then the draw volume is proportioned accordingly, and all old steeps are covered with steep acid. The advance pump is usually attached to a heat exchanger that maintains the desired temperature each time water is moved. This pump must deliver a high volume so that water movement is rapid, and considerable aeration occurs when water is advanced into the next steep. No marked differences in type of lactic acid bacteria or milling efficiency have been observed when compared with a continuous-advance or pull system.

Water Absorption and Solubles Removal
Raw corn entering a steeping battery will start at 14–16% moisture. Higher moisture levels are available on the market for several months after the start of the harvest season when good quality, high-moisture corn is available at a discount.

Water enters the kernel mostly at the tip cap and moves quickly through voids in the pericarp by capillary action or through the germ into the endosperm, as discussed earlier in this chapter.[24,26,27,137] Fan et al.[138] showed that the effect of sulfur dioxide dissolved in the water is to increase the rate of water diffusion into dent corn kernels. Increasing the temperature up to 60°C increases the rate of water diffusion. After 17 hours steep at 49°C, the moisture increase was 25% higher in the sulfur dioxide solution than in water alone. However, in conventional steeping, the level of sulfur dioxide is low during the period of greatest water absorption.

Water moves quickly up the sides of the kernel and enters through the dent region into the porous floury endosperm region, as it does in wheat.[139] Water penetrates quickly into the germ because the aleurone layer is thin and the testa is absent. The germ becomes wet in about 4 hours at 49°C and the endosperm in about 8 hours, but the kernel is not appreciably softened.[24,137] Work with wheat[139] indicates that a dense, proteinaceous sub-aleurone layer appreciably slows penetration of water into the endosperm, which may explain why grain sorghum requires longer steeping than corn.

Although mass movement of water into kernel parts is relatively rapid, the hydration of cellular components required for thorough softening is slower and takes longer, possibly 12–18 hours. For this reason, the addition of wetting agents or scratching the pericarp[140] in an effort to increase the rate of water penetration does not decrease steeping time appreciably. Corn artificially dried at elevated temperatures attains lower final water content during steeping than unheated corn.[109] This probably results from the reduction of water-binding sites by protein denaturation. Another factor which might slow water adsorption is the formation of gas bubbles which can be observed evolving from the tip cap for many hours after corn has been immersed in water. Initially a small volume of gas evolved may result from replacement of air as water enters the voids, but formation of carbon dioxide by glutamic acid decarboxylase is probably the major source.[141]

Light steepwater, into which the raw corn is initially immersed, is a complex solution of organic and inorganic molecules (see Section 9.4.5). Major components are peptides and amino acids, lactic acid and cations, mainly K^+, Mg^{++} and H^+. Lactate salts generated by bacteria growing in the steepwater buffer the steepwater to a pH of 3.8–4.1. Lactic acid is absorbed into the kernel with the water and reduces internal pH in about 12 hours, which in concert with the high temperature soon results in death of all living cells in the germ. As a result, cell membranes become porous, and soluble sugars, amino acids, proteins, minerals and large assortment of organic molecules required for growth of living cells leach into the steepwater. Total dry substance extraction is most rapid in the first 12–20 hours,[142,143] but may continue at a slower rate as the solubles, formed by reaction of kernel proteins with sulfur dioxide, diffuse. Calculations show that about one-half of the steepwater solubles come from the germ, which represents only 11–12% of the kernel weight (Table 9.1). Verification of this is seen in the increase in fat content of the raw germ, from 30–38% to 55–60% in steeped germ. The other half of the solubles comes from the endosperm. Calculations indicate that naturally soluble substances in the germ account for 95% of its contribution to solubles. About 60% of the germ protein is soluble in salt solutions, but only ~9% of the endosperm protein is soluble.[51,58] Since sugars, ash and soluble proteins can account for less than half of the solubles contribution of the endosperm, the remainder must be made during steeping, as will be discussed later.

Water absorption by both corn and sorghum kernels in water or sulfur dioxide solutions causes a volume expansion of 55–65%.[144,145] However, the volume expansion of the two grains in the large bulk is different because of differences in the shape of the kernels. Corn in a large steep does not appear to increase in volume, because the swelling kernels reorient and fill the irregular voids between kernels. Spherical sorghum kernels, being initially closely packed, exhibit a volume increase during the water-absorption phase of steeping and create a force great enough to break steeps, so grain sorghum must be introduced into a steep in several increments to prevent rupture of the tank. Intact corn kernels steeped in an excess volume of sulfur dioxide solution, such as in the laboratory semisynthetic steeping method,[38] are quite turgid, but shrink when placed in salt solution. This indicates that the kernel has a semipermeable membrane which retains soluble molecules. Unpublished experiments indicate that these molecules are large peptides or soluble proteins.

Effect of Sulfur Dioxide

Sulfur dioxide was first used in corn steeping, probably to prevent growth of putrefactive organisms. Eventually it was found indispensable for obtaining maximum starch yield. Cox et al.[146] were the first to examine the effects of steeping agents by microscopic observations of thin sections of corn kernels. They demonstrated that, during steeping with sulfur dioxide over a 24 hour period at 50°C, the protein matrix gradually swells, becomes globular, and finally disperses. The degree of protein globularity was shown to be directly related to the ease of starch recovery on grinding or on agitation of the thin sections of the endosperm from the steeped kernels. Protein dispersion increased as the sulfur dioxide concentration increased to 0.4%. Lactic acid produced some apparent softening, but other acids tested had little effect. Other reducing agents can replace sulfur dioxide,[139,147] but have no practical value.

Wagoner[148] found that the rate of protein swelling is faster and dispersion more complete with freshly harvested corn than with old dry corn. Industry finds that starch yields decrease throughout the year and that the greatest operational problems are often encountered during the summer. This is probably due to degradative postharvest storage factors and adulteration of sound corn with damaged corn, as allowed and promoted by the US Grade Standards, rather than any natural intrinsic biological change in the kernel, although microscopic physical changes have been observed[149,150] for storage at elevated temperatures (60°C). Tests (unpublished data) in which three different hybrids of corn were either stored at 5°C or stored outside in ambient conditions showed no loss in starch yield or processing characteristics over a one year period. The outside stored samples were monitored closely for insect and microbial infestation.

Sulfur dioxide dissolved in water forms an equilibrium mixture:

$$H_2SO_3 \rightleftharpoons H^+ + HSO_3^- \quad K_a = 1.54 \times 10^{-2}$$
$$HSO_3^- \rightleftharpoons H^+ + SO_3^{2-} \quad K_a = 1.02 \times 10^{-7}$$

The reaction of the bisulfite ion with endosperm protein is completed in whole kernels in 6–10 hours.[142,143] Both SO_2^- and HSO_3^- ions are capable of reducing disulfide bonds; pH conditions inside the kernel probably determine which reaction predominates. The kernel pH of 3.6–4.0 reported by Wahl[151] would favor HSO_3^- as the reactant, but pH values can be closer to 4.0–4.5 in commercial steeps with a solubles level of 1–2% in the steep acid (unpublished data).

Steeping sections of horny endosperm (10-μm thick) bathed in a solution of constant bisulfite ion concentration provides a means of studying the rate of the sulfur dioxide reaction under conditions where diffusion is not limiting.[152] These data showed that loss of starch granules from the protein matrix on gentle agitation was 50–70% complete in 1 hour and complete in 4 hours at 52°C in 0.2% sulfur dioxide solution. Under these conditions, the rate and extent of starch release from dent corn endosperm sections increased with increasing bisulfite concentration (0.05% to 0.2%) and with increasing temperature (52°C to 60°C). Further experiments showed that at pH <9, adjusted by aqueous addition of hydrochloric acid or sodium hydroxide, the pH of the medium did not have a significant effect on starch release. Other acids, including lactic acid, did not affect starch release from the thin endosperm sections, even on agitation for 24 hours at 52–60°C. This research emphasized the unique role of sulfur dioxide in starch release. Figure 9.2 shows horny endosperm cells from which most starch granules have been released by the action of sulfur dioxide, and clearly shows the globular nature of the protein strands which surround the granules.

Similar effects have been observed on steeping isolated corn endosperm (degerminated corn meal) in bisulfite solutions buffered at pH 4 with lactic acid–potassium lactate (Y. Hirata and S. Watson, unpublished data). During the first 2 hours of contact with the bisulfite solution at 50°C, the endosperm loses more nitrogenous material than control endosperm steeped in lactate buffer alone. The bulk of the dissolved nitrogen is precipitated when the steepwater is adjusted to pH 7 and heated to boiling, indicating a relatively high molecular weight protein is dissolved from the endosperm by the action of bisulfite. The appearance of this protein in solution coincides with the release of starch from the endosperm cells.

Swan[153] demonstrated that sulfite or bisulfite ion reacts with sulfhydryl (SH) groups of cystine in wool protein, reducing the disulfide linkage and giving a protein fraction (P′) having one cysteine SH and a second protein (P′) on which a S-sulfo derivative of the cysteine moiety has been formed:

$$P'S - SP' + HSO_3^- \rightarrow P'SH + P'SSO_3$$

This reaction has been confirmed for corn[154,155] and grain sorghum.[156] The S-sulfo reaction product permanently increases protein solubility by preventing reformation of the disulfide bond and because of the ionic nature of the 'P' protein. The reaction rate is greatest at pH 5.0,[157] but the pH inside the corn kernel in a countercurrent battery is 4.0–4.5 in the old corn end of the battery where the sulfur dioxide concentration is highest. Since the reactions occur throughout the second and third stages of steeping as the sulfur dioxide diffuses deeper into the endosperm, it is possible that the pH is closer to 5.0 at the reactions site. Nierle[158] estimates that only 5.7% of the sulfur dioxide used in steeping is absorbed by the corn, with 45% in the endosperm protein, 2% in the starch, and 40% in the germ. Only 12% of the absorbed sulfur dioxide reacts with protein.

Knowledge of the protein components of corn has been clarified considerably in recent years and has been thoroughly reviewed by Wall[159] and Paulis.[160] Normal corn

Table 9.7 Corn endosperm proteins extracted with different solvents

Solvent	Protein class	Percent of total nitrogen
Water	Albumin plus NPN[a]	4–5
0.5 M N_2SO_4	Globulin	3–4
71% Ethanol	Zein	40–50
0.2 M NaOH	Glutelin	30–40
(Remains insoluble)	Glutelin	10–20

[a]Non-protein nitrogen

endosperm contains a mixture of proteins which can be extracted with a succession of solvents[56,161,162] with the representative results shown in Table 9.7.

Albumins and globulins are cellular enzymes and are quickly dissolved in the steeping process.[162] Zein occurs in spherical bodies[33,34,163] embedded in a glutelin matrix. Treatment of corn endosperm with 71% ethanol to remove zein does not improve starch separation over untreated endosperm, but steeping in an alkaline solution does liberate starch. This indicates that the matrix surrounding starch granules is composed of glutelin. Wall[159] has characterized glutelin as a large and complex protein molecule composed of about 20 different protein subunits ranging in molecular weight from 11 000 to 12 700. These subunits are united through disulfide bridges in such a complex way that native glutelin molecular weight may not be definable. Hydrogen bonds also take part in the supermolecular structure, because only a combination of reducing agents and hydrogen bond breaking chemicals (NaOH, urea, detergent) will completely dissolve glutelin. The glutelin of grain sorghum is similar to that of corn.[159]

The action of sulfur dioxide in steeping is to weaken the glutelin matrix by breaking inter- and intramolecular disulfide bonds (Figure 9.2). Some zein is solubilized during steeping owing to disulfide crossbond disruption by sulfur dioxide, but this solubilization probably doesn't contribute much to matrix weakening. The solubilized polypeptides resulting from these reactions are nondialyzable[154] and too large to pass through cellular and seed coat membranes. They are retained inside intact corn kernels until liberated in first break milling. Corn that has been dried at excessively high temperatures yields less soluble protein on steeping[110,111] than undried corn, presumably because of the increasing the degree of intramolecular hydrogen and non-polar bonding. However, temperatures ordinarily used in commercial corn drying have little effect on protein solubility.

Role of Lactic Acid Bacteria

Although corn can be adequately steeped in sterile, aqueous solutions of sulfur dioxide or sodium, magnesium or potassium bisulfite (hydrogen sulfite) in the laboratory,[164–166] commercial steeping involves microorganisms. Raw corn carries natural populations of bacteria, yeasts and molds which are capable of rapid multiplication in aqueous systems. Wet-millers learned early that corn steeped at temperatures of 45–55°C was 'sweet,' but that putrefaction and butyric acid or alcohol production occurred at lower temperatures. Steeping at 45–55°C is now known to favor development of lactic acid bacteria. The produced lactic acid lowers the pH of the medium, thereby restricting growth of most other organisms.

Steeping at temperatures >55°C can result in a favorable environment for the propagation of *Clostridium* spp., bacteria which produce acetic acid. Some strains of *Clostridium* are highly exothermic and can result in a self-regulated steep temperature of ~59°C, even with all heat exchangers turned off. Total acidity does not seem to be much different than in a *Lactobacillus* spp. fermentation, except that the primary acid produced is acetic rather than lactic acid. There does not seem to be any discernible effect on starch yield or quality.

Substrate for the microorganisms is the sugar that quickly leaches from the corn. Sucrose is quickly hydrolyzed to D-glucose and D-fructose, each of which is converted almost completely to two moles of lactic acid with the release of energy needed for bacterial development. The amino acids required by the bacteria are also supplied by corn, partly as free amino acids, but largely as solubilized protein. Although lactic acid bacteria generally do not show much proteolytic activity, the organisms endemic to the steeps do affect significant protein degradation.[143] Only about one-tenth of the nitrogen in corn is in the form of non-protein nitrogen (NPN) which, therefore, accounts for only about one-fourth of the soluble nitrogen obtained during steeping. Only enzymic degradation of dissolved protein can account for the fact that 85% of the nitrogen in incubated steepwater is dialyzable NPN. Kerpisci[167] showed proteolytic activity to occur naturally in corn kernels, even though the level is much less than in other cereal grains.

Microbial species existing in commercial steeps have been neither positively identified nor seriously studied. Under steeping conditions that produce optimum conversion of sugar to lactic acid, elongate rods that have been identified as homofermentative *Lactobacilli* are observed as the dominant organism. They appear to closely resemble *Lactobacillus bulgaricus*. Also, steepwater does appear to contain nutrients which stimulate the growth of lactic *Streptococci*.[168] The grain is obviously the primary source of the initial microbial inoculum. However, because specific steeping conditions are maintained on a continuous basis in a commercial steephouse, natural selection or adaptation is probably quickly established for each particular environment, resulting in biota that may differ significantly from traditional species descriptions.

Examination of fermentation characteristics in commercial steeping in 1952–1955 showed that raw corn incubated at 50°C (122°F) with fresh milling process water develops a vigorous culture of lactic acid bacteria in about 32 hours.[143] *Lactobacilli* normally do not function well at pH 4 which exists in a steep battery, so it was not surprising when Wahl[151] found 70% of the bacterial activity to be inside the corn kernels in the new grain end of a battery, where the pH is closer to 5 and sugar concentration is highest (probably in the spaces inside the tip cap).

In today's automated process, the water is a poor source of inoculum and light steepwater has a low population of bacteria. Conversion of sugar to lactic acid is not complete at the time steepwater is drawn off for evaporation, but is more complete than expected from observed populations of organisms. A higher population would probably be found inside the kernel. The cause of the low steep population can be found in the millhouse. In 1952, starch was still separated from gluten on large surface, open, wood tables, and gluten was recovered from the dilute slurries by allowing it to settle by gravity in large tanks for five days at ambient temperature. Sulfur dioxide was added to prevent development of putrefactive organisms. Today, starch is separated on a continuous basis and gluten is concentrated in centrifugal machines. The entire process requires a transit time of 2–3 hours in stainless steel equipment maintained at 43–45°C (110–113°F) in the presence of sulfur dioxide. All tanks are small and act as surge reservoirs for process streams. Thus, there is little time for an inoculum to develop in the millhouse process water used in steeping.

Although every wet-milling plant has different operational parameters, the principles of bacterial development and their influence on steeping, as expounded above, are still important for optimum performance. A process water pH of 4.0–5.0 generally is most favorable for optimum separation of components, especially the starch–gluten separation; the effectiveness of sulfur dioxide in preventing the growth of undesirable microorganisms is also most operative in that pH range. Maintaining a pH of 3.8–4.2 in light steepwater for evaporation is important because, at higher pH levels, mineral scales may be deposited on the heat exchange surfaces, reducing the rate of water evaporation. Generation of acidity from sugars in the corn is a low-cost source of acidity. Steepwater is so highly buffered that a large volume of mineral acid would be required to achieve adequate acidity. In addition to producing acids by fermentation, lactic acid bacteria also affect hydrolysis of high molecular weight, soluble proteins. If these proteins are not degraded, they produce stable foams that can interfere with steepwater evaporation and may be deposited as gelatinous precipitates on the heat exchange surfaces of the evaporator, resulting in a reduced evaporation rate and more frequent shutdowns for evaporator clean out.

Several remedies for the problem caused by low natural bacterial inoculum have been tried. The simplest solution is to ensure that the sulfur dioxide concentration in the fermentation zone of the battery is not so high as to inhibit bacterial growth and to maintain the steeping temperature between 48°C and 52°C (118°F to 125°F) for optimum growth of the desired bacterial species. Additional incubation after light steepwater is withdrawn from the steeps may also be necessary. A temperature of 46–48°C and a pH maintained at 4.5 by addition of ammonia gives most rapid fermentation. The time required to achieve the desired degree of lactic acid development depends on local conditions, but probably is not less than four hours. The problem of steepwater foaming and gel formation may be solved by adding a proteolytic enzyme to the steeps. A low dosage of a crude enzyme preparation is quite adequate and apparently has no effect on the corn.[169] A more complete solution has been demonstrated by Balana and Caixes,[170] who prepared active cultures of one of several *Lactobacillus* species, such as *L. delbrukii* or *L. leichmanii*. Inocula prepared with synthetic culture media were added to steeps and allowed to pass through the battery with the advancing water. As expected, the concentration of reducing sugars was lower, the pH was lower, the concentration of lactic acid was higher and the concentrated steepwater gave no evaporation problems; in addition, the concentrated steepwater gave improved penicillin yield when used in commercial media.[171] Claims that inoculation permitted reduction of steeping time from 40–60 to 25 hours, and that the sulfur dioxide concentration in steep acid could be reduced from 0.15–0.075%, were not confirmed on subsequent evaluation.[172]

Lactic acid has been shown to increase starch yields,[173–177] averaging ~4%, although the addition of other organic and inorganic acids as adjuncts to sulfur dioxide can achieve similar results.[175] A study of 19 hybrids showed that lactic acid increased starch yield by 2–12%, depending on the hybrid genetics.[178] The compilation of data indicates that the effect of the production of lactic acid on increasing starch yield in commercial systems appears to be due to the lactic acid alone and not due to the fermentation, although the fermentation performs several helpful tasks, as already discussed. This is an important distinction which may allow for the development of more rapid and cost-effective steeping procedures which would not have a fermentation phase.

The concentration of lactic acid in the steep has been shown to affect both the yield of starch[178] and the quality of the resulting starch.[179] Starch yields increased with increasing lactic acid concentration up to ~0.5%, and decreased when more than ~2% lactic acid was used. Shandera and Jackson[179] tested starch quality when steeped at 57°C with either 0.2% or 1.5% lactic acid and 0.05% or 0.3% sulfur dioxide. The level of sulfur dioxide had a small effect on starch quality, but the level of lactic acid had a pronounced reduction effect on peak viscosity, shear thinning viscosity, set-back viscosity, pasting viscosity and starch water solubility.

3. Milling and Fraction Separation

Component Yields

The objective of the milling process is to provide as complete a separation of parts as is possible and practical. While commercial yield figures are not available, laboratory steeping and milling procedures[119–121,164,179–184] give results close to those obtained in commercial wet-milling. Data are given in Table 9.8 for regular dent hybrid corn, a high oil corn variety and regular red grain sorghum. All three samples were steeped 48 hours at 52°C using a 0.1% sulfur dioxide solution at pH 4 and processed as described in reference 164. Starch separation was conducted by tabling. In tabling, the denser starch settles on the table and the lighter gluten containing many of the small starch granules flows off the end of the table. The surface of the starch is then gently hosed (washed) or 'squeegeed' to remove traces of gluten. The material washed off the table is termed the 'squeegee' fraction; it has a high starch content. In some of the samples, the squeegee starch is added to the gluten fraction.

Data in Table 9.8 show that product yields are closely related to initial grain composition. For example, high oil corn gives a greater germ and oil yield than dent corn does, while red milo gives a lower germ and oil yield. Likewise, high oil corn gives lower starch and gluten yields because the proportion of endosperm is less than in dent corn. For the same reason, red milo gives higher gluten yield, but lower starch yield is obtained because starch is more difficult to separate from other components.

Gluten obtained by tabling seldom contains more than 50% protein. To approximate the results obtained by continuous commercial centrifugation, gluten and squeegee data were recalculated to 70% protein, which increases starch yield. Similar

Table 9.8 Yields and composition of laboratory wet-milled fractions of regular and high oil dent corn and red grain sorghum (Milo)[a]

	Fraction	Regular dent corn (%)	High oil dent corn (%)	Red milo (%)
Whole grain analysis				
	Moisture	14.3	13.6	14.9
	Starch	71.5	67	73.1
	Protein	10.5	10.4	13
	Fat	5.1	8.0	3.6
	Wax	Trace	Trace	0.3
Solubles				
	Yield[b]	7.6	10.8	7.2
	Protein[c]	46.1	46.9	41.5
Starch				
	Yield[b]	3.7	59.7	60.2
	Protein	0.30	0.26	0.32
	Fat	0.02	0.03	0.03
Germ				
	Yield[b]	7.3	10.9	6.2
	Starch	7.6	7.2	19.1
	Protein	10.7	7.2	11.9
	Fat	58.9	65.5	39.6
	Oil yield[b]	4.3	7.14	2.4
Fiber				
	Yield[b]	9.5	9.8	9.3
	Starch	11.4	12.3	36.7
	Protein	11.3	11	19.7
	Fat	1.8	2.7	3.8
Gluten				
	Yield[b]	7.4	6.3	9.6
	Starch	25.8	32	39.9
	Protein	50.7	42.3	47.2
	Fat	3.7	4.4	5.4
Squeegee				
	Yield[b]	3.9	3.6	5.6
	Starch	91.7	93.8	74.8
	Protein	6.1	3.6	20.7
	Fat	0.3	0.4	1.6
	Total dry substance[b]	99.4	101.0	98.0
Recalculation to expected centrifugal results				
Starch				
	Yield	68.5	65.7	67.2
Gluten				
	Yield	5.8	4.0	8.1
	Protein	70.0	70.0	70.0

[a]All percentages other than moisture are expressed on a dry basis
[b]Percentage of original grain
[c]Analytical values expressed as percentage of the fraction

experiments with Argentine flint corn have shown that component separation is about equal to that of dent corn, even though steeped flint corn kernels appear to be much less soft.[185]

Many simplified laboratory procedures have been developed[38,74,148,152] to determine the effectiveness of steeping or to predict the performance of a corn sample in a commercial process. In the end, the analysts agree with Kempf[186] that the complete laboratory fractionation method is slow and cumbersome and there is no good substitute for adequate evaluation of wet-milling properties of grain. A review of the various laboratory wet-milling procedures[178] showed that, although there are significant differences between some of the laboratory procedures, most provide an acceptable means of comparing different genotypes and phenotypes of corn and studying process modifications. The major criteria for getting reproducible estimates of component yield seem to be the care taken by the miller to minimize variability in the test procedure.

Table 9.9 compares starch yields of duplicate samples of four different types of commercial corn phenotypes. The samples were steeped with 0.2% sulfur dioxide and 0.55% lactic acid for 24 hours at 52°C using the procedure outlined in reference 120. The samples represent one commercial hybrid from each phenotype and are not shown to be indicative of all hybrids in that phenotype. However, some of the observations which can be made from the data in Table 9.9 are generally true. High oil, waxy and high-amylose hybrids almost always have lower starch yields than dent corn, with the high-amylose hybrids yielding the least amount of starch and being the most difficult to process. High oil corn starch yields vary greatly, as can be seen by comparing the data in Table 9.8 with that in Table 9.9. Steepwater solubles are usually greater for high oil corn because of the larger germ. Other phenotypes, such as various genetic mutants[187] and those with genetically engineered starches,[188] will probably also be more difficult to mill than ordinary dent corn, because alteration to the ratio of amylose and amylopectin seems to affect the shape and surface characteristics of the corn starch granule.

Germ Separation

After steeping in the commercial process, the grain is coarsely ground or 'pulped' with water in an attrition mill in preparation for degermination. The most commonly used

Table 9.9 Wet-milling yields[a] of different corn phenotypes

	Normal dent corn	High-oil corn	High-amylose corn	Waxy corn
Steepwater	3.53 ± 0.03	4.50 ± 0.04	3.84 ± 0.10	3.86 ± 0.00
Germ	6.78 ± 0.13	8.55 ± 0.12	6.18 ± 0.10	6.08 ± 0.42
Fiber	13.37 ± 0.42	15.02 ± 0.08	18.92 ± 0.05	15.90 ± 0.09
Gluten	10.91 ± 0.21	22.39 ± 0.39	27.54 ± 0.09	12.63 ± 0.41
Starch	65.29 ± 0.16	49.65 ± 0.49	43.41 ± 0.69	60.42 ± 0.62
Total solids	99.88 ± 0.04	100.11 ± 0.00	99.89 ± 0.35	98.91 ± 0.12

[a]Average of duplicate determinations

mill has one stationary and one rotating milling surface, similar to the mill shown in Figure 9.11. When used for degermination, the plates are covered with pyramidal knobs and the impact ring is absent. The bulk of the germ is freed in the first pass, but a second pass is usually provided after free germ has been removed. The mill gap is adjusted to give the maximum free germ with minimum germ breakage. Any germ cells that are cut or disrupted in any way during any step in the process will lose oil, which is mostly absorbed by gluten and cannot be recovered. Over half of the starch and gluten is also freed in this first milling step. The high density difference between the oil-rich germ and the heavier kernel components provides a basis for easy separation. This is accomplished by continuous flow through liquid cyclones or hydroclones (hydrocyclones). The type used for germ separation is a conical tube 15.2 centimeters (6 inches) in diameter at the top of its approximately 1 meter (3 feet) length (Figure 9.12). Recently equipment manufacturers have increased the size of hydrocyclones used for germ recovery to 20.3–22.5 centimeters (8–9 inches) in diameter, with a proportional increase in length. These larger cyclones increase capacity and reduce the likelihood of plugging. Pulped corn adjusted to 7–8° Baume (12–14% dry solids) with suspended starch is forced into the tube under pressure. The orifice angle, aperture and pump pressure are chosen to produce a rotational velocity sufficient to cause a separation of the particles of differing density.[189–191] The heavier endosperm and fiber particles pass out the bottom of the tube at a dry solids concentration of 20–24%, while the lighter germ is drawn off the top of the vortex. A bank of germ hydroclones is shown in Figure 9.13.

The cyclone-type of germ recovery equipment occupies less floor space, is easier to maintain and clean, and allows greater response to changes in operating conditions than do the earlier germ flotation methods.[5] As shown in Figure 9.6, recovered germ

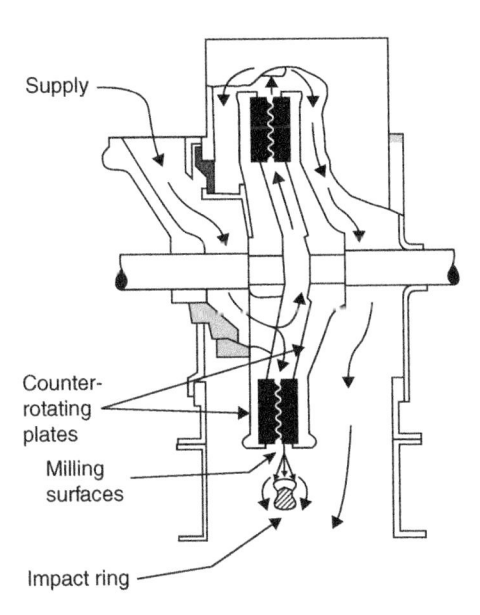

Figure 9.11 Diagram of a Bauer attrition mill. Configuration shown is used for fine milling degerminated residue. With wider spacing of interlocking teeth and no impact ring, it is used for degermination (first-break) milling.

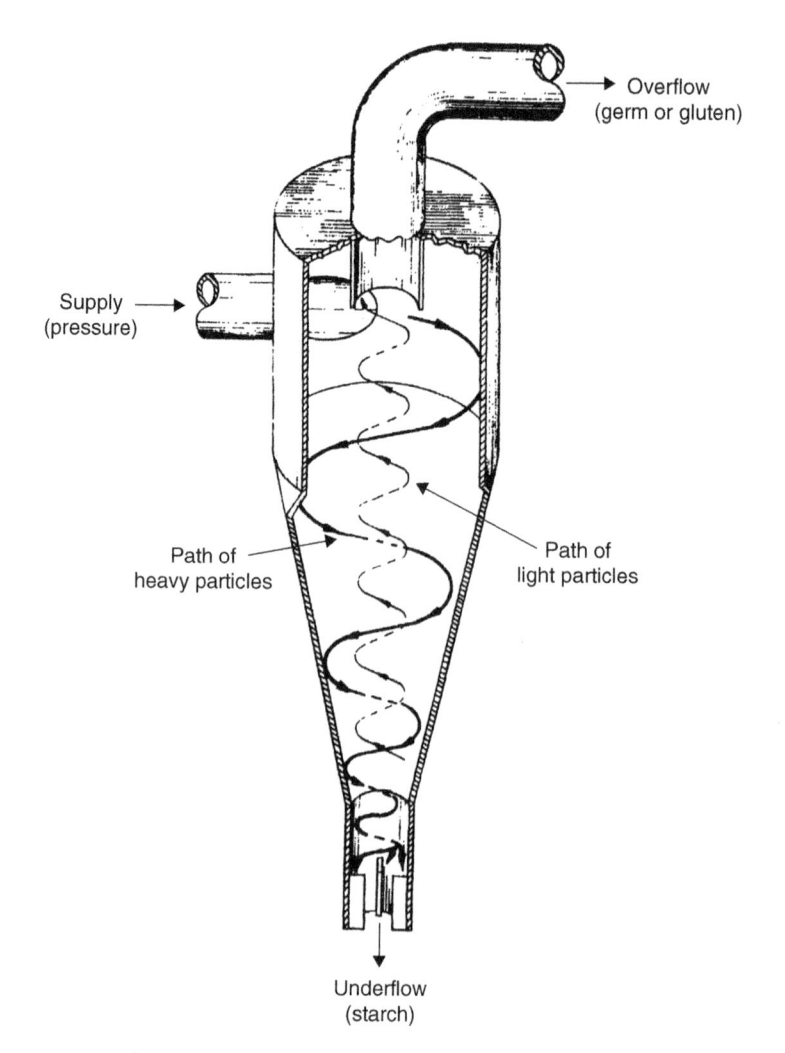

Figure 9.12 Cutaway diagram of a hydrocyclone separator.

is counter-currently washed 2–3 times with process water on screen bend devices to remove occluded starch (see Section 9.3.3c). The washed germ is dewatered to 50–55% water content (wet weight basis) by passing it through mechanical squeezers prior to drying in preparation for oil recovery.

The free starch and protein, which comprise about half of the dry substance in the germ cyclone underflow, are separated from the unmilled endosperm and fiber by screening over a screen bend device to reduce the load of solids and water to subsequent milling operations. The slurry of starch and gluten obtained by screening at this point is called 'prime mill starch,' because it derives mainly from the floury endosperm. Prime mill starch is lower in protein than whole starch. Laboratory work has shown that it is easier to separate starch from gluten using this product as opposed to the final mill starch combined from both milling steps (unpublished data).

One company in the United States has utilized the prime starch phenomena in a unique way. In this patented process,[192] prime starch is recovered and marketed while

Figure 9.13 Battery of hydroclones used for separation of germ from milled, steeped corn. (Courtesy of Dorr-Oliver, Inc., Stamford, Connecticut)

the starch retained in endosperm pieces and fiber is converted to ethanol. The process is identical to the process being described here, except that the endosperm milling step is omitted and 'starch yield' is maximized by enzymatically digesting the horny endosperm starch, which is the most difficult to recover by milling. The resulting glucose is fermented to ethanol.

Second Milling and Fiber Separation
In processes where all the starch is to be recovered the underflow from the germ cyclones, containing fiber and pieces of horny endosperm, is more thoroughly milled ('grinding mills' in Figure 9.5) to recover the maximum yield of starch. One type,

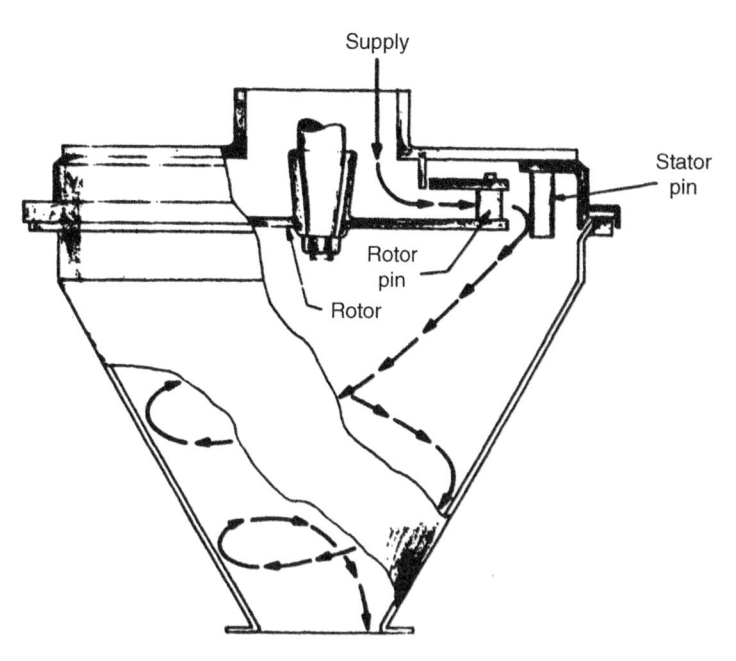

Figure 9.14 Cutaway diagram of an Entoleter mill used for fine milling degerminated residue.

the Bauer mill (Figure 9.11), employs a combination of attrition and impact milling.[193] The corn material must pass between counter-rotating grooved plates made of a hardened steel alloy and on discharge strike an outer impact ring.[194]

An impact-type mill known as the Entoleter mill is preferred by many operators[195] (Figure 9.14). Endosperm slurry dropped onto a rotating horizontal disk is flung with great force against both rotating and stationary pins. Rapid and complete starch release is obtained with minimum fiber attrition. The larger pieces of fiber permit occluded starch to be recovered by washing more efficiently. These two types of mills produce roughly equivalent results. They have advantages of high throughput rates, low maintenance cost, uniform operation, and improved quality and yield of starch over the formerly-used Buhr stone mills.[5] The Entoleter mill has an added advantage that there is no mill gap adjustment and, therefore, less operator attention is required.

Starch and gluten released by milling must be separated from fiber. This is best achieved by taking advantage of the difference between the fine particle sizes of starch granules and gluten particles, and the larger endosperm and pericarp particles. A few dense pieces of horny endosperm that are not disrupted in milling always remain. For many years, this separation was accomplished over fine nylon screens attached to a horizontally agitated frame. These types of devices (shakers, Rotex, etc.) required large areas for installation and continual attention to maintenance and sanitation. Now all modern wet-milling establishments use a fixed concave screen arranged in a vertical position, over which the slurry to be screened is pumped with considerable force (Figure 9.15). The degree of concavity of the screen is varied to suit each screening problem and is designated in angular degrees as 50°, 120°, etc.

Figure 9.15 DSM fiber washing unit (120° model) employing wedge-bar screening surface. (Courtesy Dorr-Oliver, Inc., Stamford, Connecticut)

This device has been made possible by the development of wedge-bar screening material of uniform slit width[196,197] (Figure 9.16), which has a unique slicing action that affords high capacity and eliminates clogging. These devices are known as screen bends, pressure fed screens or proprietorially as DSM (Dutch State Mines) screens, and have the advantage of affecting very sharp separations. They also have the advantages of low maintenance, minimal operator attention and high capacity, thus requiring much less floor space and fewer employees than shaker stations. They are manufactured with screens having slit widths of 50, 75, 100 and 150 µm. A recent innovation is a screen assembly which allows for easy replacement or adjustment of a fine wire screen.[198] A 50° pressure-fed screen with large bar spacing is used to wash germ discharging from the germ recovery cyclones as described above.[199]

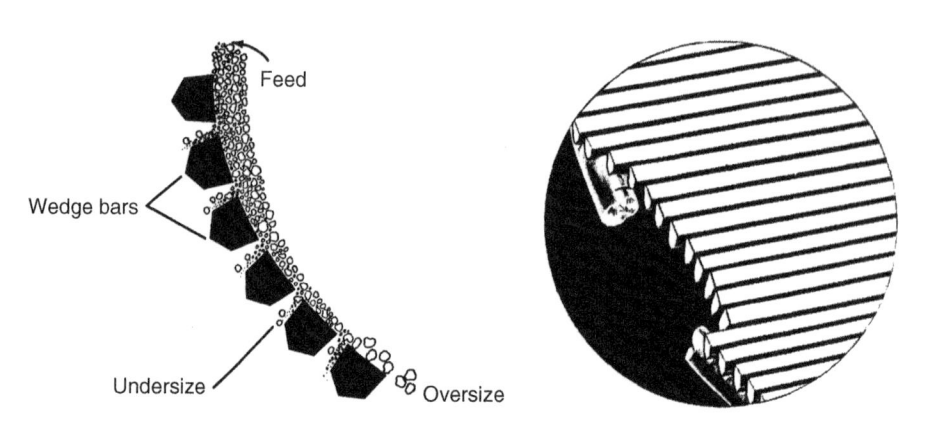

Figure 9.16 Detail of wedge-bar screening surface (right) and simulation of the unique slicing action of this screen.

A 120° pressure-fed screen with a bar screen slit width of 50 μm is used for washing fiber free of occluded starch and gluten particles. The fiber mat discharging from the screen bend surface is reslurried in process water and passed over a second screen bend. As shown in Figure 9.16, the process of slurrying and screening is repeated 5–6 times. Most often the first pressure fed screen is equipped with a 50 μm screen, while the last 4–5 units are equipped with 75 μm screens in order to reduce the number of units, since larger screens have higher capacity. The 50 μm screen is considered a fiber block which prevents fine fiber from entering the centrifuges and starch hydrocyclones. However, this system presents a problem for fiber particles in the 50–75 μm range. The particles must either adhere to large particles and be carried out with the fiber or be reduced in size to pass through the 50 μm screen. It is the authors' contention that such systems result in higher levels of fine fiber in the starch than would occur if all pressure fed screens were equipped with 50 μm screens.

The pressure fed screens are connected in a manner that gives a countercurrent passage of water with respect to fiber in order to achieve high water use efficiency. Several different operating configurations for using these devices in a starch factory have been described.[199–201] The number of successive units may vary with the manufacturer's goal of complete starch recovery. Typically, finished fiber contains 15–20% starch, about half free and half bound.

Following the final screening, the fiber is further dewatered in a centrifugal screen, mechanical squeezers, and/or a horizontal solid bowl, continuous discharge centrifuge.[199] Typically, final moisture content is 65–75% (wet weight basis). The fiber is blended with concentrated steep liquor, either before or after partial drying, and then is further blended with corn cleanings, spent (defatted) germ flakes and starch hydrolyzate residue, when available, and dried to make corn gluten feed (Figure 9.5), typically containing 18–22% protein.

Starch–Gluten Separation
The defibered mixture of starch and protein, known as 'mill starch,' carries 5–8% insoluble protein content, depending on the protein content in the original corn or

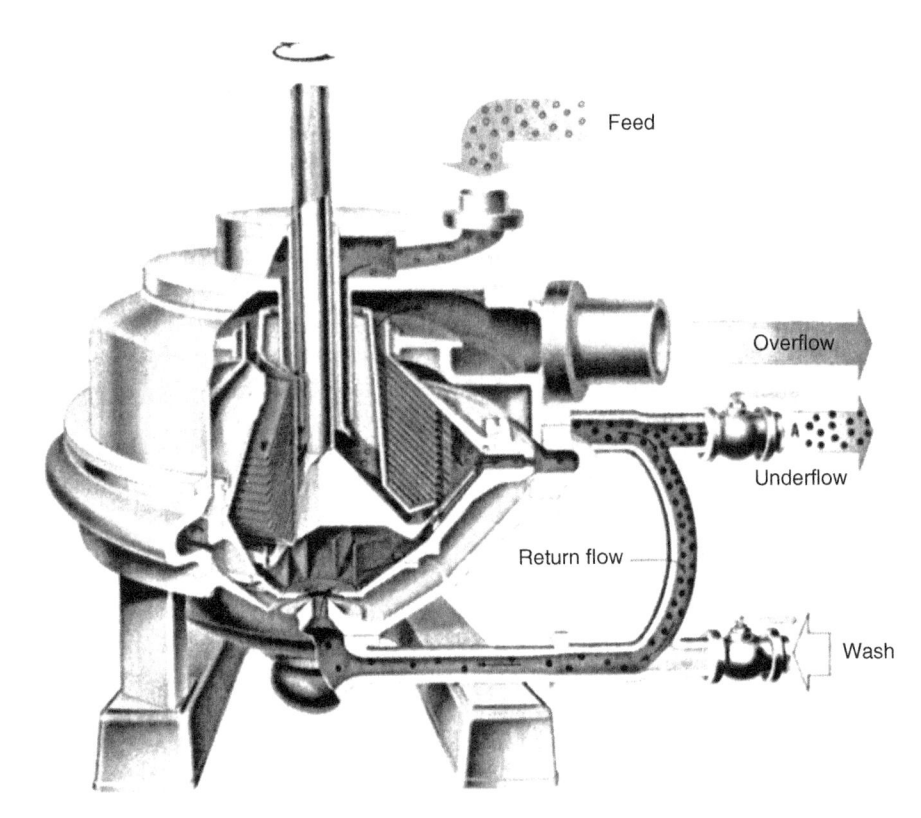

Figure 9.17 Cutaway diagram of a Merco starch separation centrifuge. (Courtesy of Dorr-Oliver, Inc., Stamford, Connecticut)

sorghum endosperm. The mill starch streams from both the degermination and fiber washing steps are combined and centrifugally concentrated to reduce solubles and to adjust the concentration of solids in preparation for the final step of starch separation. Any one of several bowl type, nozzle discharge centrifuges are used for this purpose, especially Westfalia, deLaval and Merco centrifuges (Figure 9.17). For this use, the centrifuge is equipped with a clarifier assembly as a decanter to permit discharge of water in the overflow and all solids through the nozzles.[202,203]

The low density of hydrated gluten particles ($1.1\,g/cm^3$) as compared with starch ($1.5\,g/cm^3$) permits their ready separation by settling or centrifugation. A container of mill starch obtained from adequately steeped corn will deposit a white layer of starch with a layer of yellow gluten on top in about 10 minutes. In about 1 hour, the gluten layer will have further settled to leave a layer of supernatant liquid. This separation is accomplished commercially with the same type of centrifuges used for clarification, but equipped for particle classification. These machines contain a stack of conical disks (Figure 9.17), each separated by a narrow space to accentuate separation of discrete particles with distinctly different specific gravities.[202,204] The heavier starch granules are thrown to the periphery of the centrifuge bowl and ejected through nozzles. The lighter gluten particles are carried up between the disks by a stream of water and are ejected at a low solids concentration and a protein content of 68–75%

(dry basis). The starch discharge contains 1–2% protein depending on the mode of operation. The mill starch is preferably supplied to the centrifuge at 10–12° Baume (18–21% dry solids), while the starch slurry discharges at a density of 20–24° Baume (35–42% dry solids).

The light gluten discharges from the centrifuge at 1–2% dry solids concentration and must be concentrated in centrifuges to 12–15% for efficient solids recovery. One means of gluten recovery is by filtration. Formerly, manually discharged filter presses were used. Now recovery may be achieved with rotary vacuum filters arranged for continuous discharge of cake from the filter cloth belt and equipped with a mechanism to continuously wash the filter cloth to prevent blinding by the very fine gluten particles. An alternative gluten recovery system dewaters the concentrated gluten in a horizontal decanter solids discharge centrifuge.[205] The rate of gluten sedimentation has a sharp peak between pH 4.5 and 5.5, requiring addition of sodium hydroxide solution to the gluten slurry to maximize operation of the decanter. In order to prevent microbial growth, the process temperature must be maintained above 45°C, which also coincides with minimum viscosity.

Fox[206] studied factors affecting gluten filtration by laboratory tests using a hand-held filtration block and gluten from a wet-milling plant. Gluten slurry was found to form a highly compressible cake and too high a vacuum level decreased cake formation. Vacuum from 356–559 mm Hg did not affect the amount of gluten deposited in the filter cake, the filtrate volume, or the level of filtrate solids, but did decrease cake moisture content. Dry-to-form time ratios greater than 2 were required to get filter cake sufficiently dry for good cake discharge. The operational parameters which most significantly affected filtering rate were temperature, increasing solids content of the gluten slurry and the age of the gluten. Decreasing gluten temperature from 40°C to 20°C significantly increased the amount of gluten filtered and the ease of cake discharge. Increasing gluten solids by 70% resulted in a 126% increase in the amount of gluten deposited in the cake. Gluten stored at room temperature for up to 24 hours was much easier to filter, but resulted in a significant loss of solids into the filtrate, primarily due to proteolytic activity. Starch addition to the gluten slurry increased the filtration rate by increasing porosity.

The next step in the process is removal of remaining soluble and insoluble protein. Starch slurry discharging from the primary centrifugal separator must be diluted with process water to a slurry density of 10–12° Baume (18–21% dry solids). The starch may then be further purified in a second centrifugal separator to a final insoluble protein level of <0.38% dry basis (preferably 0.27–0.32% dry basis). However, since solubles content of the slurry must next be reduced by filtration of centrifugal decantation, the second step currently preferred is to utilize 8 to 14 stages of liquid cyclones which simultaneously remove residual gluten and wash the starch.[207–209]

Individual hydroclones are molded plastic devices with a tangential port for entry of the slurry into the top of a cone-shaped separation chamber which has two outlets. One outlet is an overflow vortex finder port for discharging water and the low-density fraction from the top of the chamber. The other opening at the apex of the cone is a discharge port for the high-density fraction. The inside diameter is 10 mm at the top of the cone, and the cone length is usually 16 mm.

Figure 9.18 Removable metal (RC model) hydroclone starch washing separation modules. Individual 10 cm hydroclone tubes appear as 'spokes' in the circular modules. (Courtesy of Dorr-Oliver, Inc., Stamford, Connecticut)

The hydroclones are assembled into manifolds of several different designs holding numerous tubes. The standard in the past has been the 'clamshell' unit, in which the manifold containing several hundred tubes is covered top and bottom, forming a three partitioned unit. In the newer 'RC' design (Figure 9.18) rapidly being adopted as the standard by the US industry, individual tubes are radially oriented into metal modules which may be 'stacked' into compact units having any desired number of stages.[210] The metal modules have replaced earlier plastic modules which were more susceptible to cracking and warping. In each type of unit, the starch–gluten mixture enters the central chamber and is forced by means of pump pressure (normally 5.4–6.8 atm, 80–100 psi) simultaneously into each of the individual tubes. The pressure energy creates rotational motion to the liquid entering the conical chamber, producing a centrifugal force throwing the heavier starch granules out the underflow port into a common collection chamber; the gluten particles discharge through the vortex finder into an overflow side collection chamber.[199,207] In Figure 9.18, the region between the hub and the outer rim is pressurized with the starch–protein slurry; the starch-rich fraction is collected in the hub region, and the protein-rich fraction is collected around the outside of the module. The concentrated starch stream is rediluted with water overflow from a succeeding hydroclone and passes into a second set of hydroclones. Six to ten sets of hydroclones arranged in countercurrent sequence provide a final opportunity to remove insoluble protein and give a complete washing operation.

As has been previously described, the only fresh water entering the milling operation contacts the starch in the last hydroclone stage. Overflow from each cyclone stage is used to dilute underflow entering the next to last stage. Water progresses back through the hydroclone station from stage to stage, and when it emerges from the first stage it carries away solubles, gluten, very fine particles of fiber, small and damaged starch granules, and some larger occluded starch granules. This stream is termed 'middlings' and analyzes ~2% insoluble and ~0.5% soluble dry substance. The dry substance is composed of ~85% starch and ~10% protein. Starch is typically recovered by recycling it into the dilute mill starch stream for another pass through the separation system. Bonnyay[203] recommends that the middlings stream be concentrated in a clarifier centrifuge, thus employing the high-density, four-stage centrifugation system depicted in Figure 9.6. This system gives the purest form of a countercurrent system possible. Gluten concentrator overflow water is used for germ washing and steeping, and middlings concentrator overflow is used for injection into the primary separation centrifuge and for fiber washing.

The high-density four centrifuge arrangement shown in Figure 9.6 is common to most US wet-milling plants, although a few plants operate with a low-density three centrifuge system, where no mill stream thickener is used. Other two and three centrifuge arrangements are also possible,[211] and selection of an arrangement depends on plant capacity, the type of products being produced, and the product value relative to raw material costs. Recently, a new high speed washing centrifuge has been introduced which offers the opportunity to cascade primary centrifuges to efficiently decrease starch soluble content.[212]

Centrifuges do have disadvantages of high capital cost, significant cleaning time and frequent operational adjustment. A method has been disclosed by which the entire starch–protein separation system can be completed in a hydroclone station, thus eliminating the primary separator centrifuge.[213] Furthermore, the hydroclones operate at normal mill starch concentrations of 7.5–8.5° Baume (13–15% dry solids), thus eliminating the need for centrifugal concentrators ahead of the primary separation step. The system described utilizes two or three primary hydroclone stages for gluten and solubles recovery, followed by 9–10 starch washing stages. Higher pump pressure differentials of 8.1–12.2 atm (120–180 psi) must be used for primary stages, and pressure differentials of 6.8–10.1 atm (100–150 psi) for washing stages. Operation of this system will normally produce gluten of 70% protein (dry weight basis) and starch of 0.33% protein (dry weight basis) at lower capital and operating costs than the presently used centrifuge hydroclone system.

Grain Sorghum Processing Innovations

In wet-milling of grain sorghum, the starch and gluten in mill starch usually do not separate sharply into two components (Table 9.8). Extensive laboratory studies[214] have indicated that the sorghum pericarp is the cause of this difficulty, because it is much more fragile than that of corn. Small particles of fiber apparently interfere with starch and gluten separation, and may also give the starch a pink color. Mold infection of the pericarp in field 'weathering' accentuates this tendency. A solution to this problem is to dehull the kernels before steeping, preferably using an aqueous slurry technique.[214,215] Dry dehulling techniques can be used, but starch losses are much higher.[216] When dehulled grain is steeped in the conventional manner, the starch gluten separation is much sharper, and gives a starch of lower color and lower protein content; gluten has a higher protein content. Germ separation is also improved, indicating a possible 20% increase in oil recovery.[214]

Alternatively, dehulled kernels can be degerminated prior to steeping by subjecting them to a pressure just sufficient to rupture the endosperm, e.g. by passing the kernels through a roller mill gapped at 0.114–0.165 mm.[215] Gentle milling of the cracked kernels allows the germs to be readily separated by flotation. The isolated germs may be used for oil recovery or as a human food item. Steeping of the endosperm pieces for easy recovery of high-quality starch can then be accomplished in sulfur dioxide solution for 8–10 hours.

4. Starch Processing

The finished starch slurry from the final hydroclone stage may be further processed in alternative ways: (a) dried directly and sold as unmodified corn starch; (b) modified by chemical or physical treatment in a way that preserves the granule structure, then rewashed to remove residual reactants, and dried; (c) gelatinized and dried; (d) chemically or physically modified and then dried; (e) hydrolyzed either completely to D-glucose or partially hydrolyzed to give mixtures of soluble oligosaccharides and D-glucose (see Chapter 21), which can be fermented to yield ethanol and other products.

5. Product Drying, Energy Use and Pollution Control

Proper operation of a wet-milling system requires the use of 1420–1830 L of fresh water per metric ton of corn (15–22 gal per 100 lb), all added at the final hydroclone stage. This water must eventually be removed from the products by evaporation, since most wet-milled products are marketed at 10–12% moisture content.

Flash drying is now most universally used for drying unmodified starch, gluten and gluten feed. In the case of starch or gluten, finished slurry is dewatered in a centrifuge or filter, and the filter cake is injected into the bottom of a stream of rapidly moving air heated to temperatures of 93–127°C (200–260°F) with natural gas or steam. The particles are dried instantly and are collected in cyclones. Baunack[217] has described several flash dryer arrangements. He reports that drying conditions in flash dryers can be varied to control particle size and bulk density of starch, gluten and fiber products. Rotary steam tube dryers are used for drying gluten and gluten feed. Fluidized-bed dryers, such as the device described by Idaszak,[218] are coming into use for drying starch, especially chemically modified types. Steepwater for use as a feed ingredient is generally concentrated to a viscous fluid containing 45–50% moisture in triple-effect evaporators with forced feed to reduce the problem of fouling of the tubes and to improve energy efficiency. Most new and modernized corn plants have installed vapor recompression evaporators to reduce energy use in the initial evaporation of dilute solutions. As reported by Wu,[219] a combination of microfiltration and reverse osmosis is the most recent alternative being considered for concentration of dilute streams. Further, micro-, ultra- and nano-filtration of streams, such as light steepwaters, enables evaporation to higher concentrations prior to drying.[220] Based on discussions with membrane manufacturers, there has been a recent flurry of activity in applying membranes to dewatering and purification of various wet-milling streams.

Evaporation and drying require large inputs of electrical and heat energy. The corn wet-milling industry has been characterized as the second most energy intensive of all the food industries in the United States.[221] In 1972, the average energy consumption was 1×10^6 kilojoules per metric ton (433 214 Btu per 100 lb of corn), of which ~11% was used for feed drying, ~4% for starch drying, ~21% for evaporation, ~33% for process steam and heating, ~15% for mechanical power and lighting, ~17% for boiler losses and ~9% miscellaneous.[222] These figures explain why one of the major activities in recent years in the industry has been the development of methods for energy reduction.

For many years the control of air and water pollution from biodegradable streams and streams which contain hazardous chemical residues, as well as recovery of unaccounted dry substance, has been a major concern for the wet-milling industry. These problems have been reduced gradually by the development of new by-products and by treatment of wastewater streams, so that they may be returned to the process, put into community sewers or sent to dedicated waste treatment facilities. In the 1930s, the practice of disposing of steepwater and unwanted process water in waterways was stopped, resulting in the development of the 'bottled up' process.[5,223]

Although the 'bottled up' process increased plant dry substance yields to 99%, many dilute streams containing organic and inorganic solutes that cannot be returned to the process remained. Stavenger[224] estimated that the average wet-mill discharges about 7.0 g BOD_5 per metric ton of corn processed. This figure has fallen considerably in the intervening years. Standard municipal sewer treatment methods have not always proved adequate, because of large surges in pH, volume of effluents and variations in the type of organic matter contained. One corn wet-milling plant using an aerobic digestion system[225] found it necessary to receive plant effluents in a large basin equipped for agitation, pH and temperature control. Biomass develops in surface aerated tanks and is recovered by gravity and flotation collectors, to reduce BOD and suspended solids in the water to the limits established by the US Environmental Protection Agency. Current practice is to treat effluents by this standard primary treatment followed by a secondary treatment. Sludge is dewatered by mechanical decanters, and the overflow is discharged to a sewer or waterway.

New methods of processing product streams, waste streams or dilute process water streams incorporate membrane systems for separation and/or concentration at all levels, including microfiltration, ultrafiltration, nanofiltration and reverse osmosis.[220,226,227] A list of possibilities includes separation of starch and water, and starch and gluten, regular and modified starch washing, continuous glucose production, recovery of sludge and stillbottoms, yeast recovery and syrup refining. Micro-, ultra- and nano-filtration membranes are manufactured with different materials, including polymers, ceramics, carbon and most recently, stainless steel. Different porosities provide high selectivity in separation of molecules of different sizes, which have been found to be useful in processing streams containing soluble and insoluble materials.[220,228] Stainless steel membranes have excellent potential, because of uniform pore size, resistance to heat, ease of back flushing and longer life than some of the other materials.[229] For example, many wet-milling plants use microfiltration to separate sugars in starch hydrolyzates from residual fat and protein, gaining substantial reductions in refining costs and problems associated with disposal of spent diatomacious earth.[229] Membrane systems are also being used for final filtration of corn syrups to remove any residual particulate or biological contamination.[229]

Exhaust gases from feed co-product dryers contain particulates, aerosols and gases. The gases contain low molecular weight compounds that are volatilized from the steepwater. These compounds have characteristic odors that may be offensive to some people. Air pollution problems can be ameliorated by reducing drying temperatures, by passing exhaust gas through scrubber columns containing potassium permanganate solution[230] or by incineration[231] with recovery of heat.

6. Automation

Changes in the wet-milling process during the past 20 years have been mainly in the direction of drastically reducing the number of operating personnel per unit volume ground. This change has required the development of automated computer process control stations, as reviewed by Simms.[232] Sweetener processes have been more easily computerized than milling processes, because of the availability of liquid analysis systems such as polarimetry, refractive index measurements and pump automated analysis systems. However, the technology of analysis of solid material by infrared reflectance analysis[200] now makes computerized automation of the entire milling and product drying process possible. All plants are operated with continuous flow on a 24-hour-per-day basis and steep time has been reduced to 24–35 hours.

IV. The Products

1. Starch

Dry, unmodified corn starch is a white powder with a pale yellow cast; grain sorghum starch has a faint pink tint. Absolute whiteness of either starch, such as is required for aspirin tablets, must be achieved by bleaching. Table 9.10 gives general properties of regular corn and sorghum starch, and of waxy corn starch.[233–237] The starch produced by different manufacturers is quite uniform with respect to most of these properties. The most important properties with respect to utilization of starch are largely related to properties of pastes produced by cooking.

2. Sweeteners

Many different types of starch hydrolyzate products are produced for a multitude of uses, mostly in food products (see Chapter 21). Table 9.11 provides definitive data on

Table 9.10 Analysis and properties of powdered corn and sorghum starches[a]

| | Corn | | Sorghum |
	Waxy	Normal	Normal
Starch (%)	88	88	88
Moisture (%)	11	11	11
Protein (N × 6.25) (%)	0.28	0.35	0.37
Ash (%)	0.1	0.1	0.1
Fat (by ether extraction) (%)	0.04	0.04	0.06
Lipids, total (%)[b]	0.23	0.87	0.72
SO2 (mg/kg)[c]	–	49	–
Crude fiber (%)	0.1	0.1	0.2
pH	5	5	5
Linear starch fraction (amylose) (%)	0	28	28
Branched starch fraction (amulopectin) (%)[d]	100	72	72
Granule size (microns)[e]	–	5–30	4–25
Average granule size (microns)[e]	–	9.2	15
Granule gelatinization temperature range (°C)[f]	63–72	62–72	68–75
Swelling power at 95°C[g]	64	24	22
Solubility at 95°C[g]	23	25	22
Specific gravity	1.5	1.5	1.5
Weight per cubic foot (pounds)	44–45	44–45	44–45

[a]Reference 233, except as noted. Values for waxy corn and sorghum from unpublished data except as noted

[b]Reference 235

[c]Reference 237

[d]Percentage of carbohydrate

[e]Reference 236

[f]Initial and end temperatures for loss of microscopic birefringence[234,236]

[g]Reference 234

properties of the most popular of the liquid sweetener products.[238] Three other products are normally sold in dry form: D-glucose (dextrose) in the monohydrate and anhydrous crystal forms; very low DE (22–30) corn syrup; and maltodextrins (5–20 DE). The latter two products are sold as amorphous powders.

3. Ethanol

Ethanol produced by fermentation of starch hydrolyzates is regarded legally as equivalent to grain alcohol and may be used in beverages. It also qualifies for tax-exempt status when blended with gasoline at a level of 10% for use as a motor fuel. Ethanol is a renewable commodity when produced from a biological material, has a current net energy ratio (energy from ethanol:energy to produce corn and ethanol) of 2.51:1,[239] and offers societal benefits when compared to petroleum-based products. Ethanol production in the US increased dramatically in a three-year period around 2005, to the point that use of corn for ethanol production became almost twice that used for starch production.

Table 9.11 Properties of commercial corn syrups[234]

	Acid	Conversion, DE level		Acid-enzyme			Enzyme–enzyme		
Commercial Baume	43°	43°	43°	43°	43°	43°	–	–	
Solids (%)	80	80.3	81	80.3	82	82.2	71	71	
Moisture (%)	20	19.7	19	19.7	18	17.8	29	29	
Dry basis									
Dextrose equivalent	37	42	52	42	62	69	96	(95)	
Ash (sulfated) (%)	0.4	0.4	0.4	0.4	0.4	0.4	0.03	0.03	
Carbohydrate composition									
Monosaccharides (%)									
D-Glucose (%)	15	19	28	6	39	50	93	52	46
Fructose (%)	0	0	0	0	0	0	0	42	55
Disaccharides (%)	12	14	17	45	28	27	4	3	2
Trisaccharides (%)	11	12	13	15	14	8			
Tetrasaccharides (%)	10	10	10	2	4	5			
Pentasaccharides	8	8	8	1	5	3	3	3	2
Hexasaccharides (%)	6	6	6	1	2	2			
Higher saccharides (%)	38	31	18	30	8	5			
Viscosity, centipoises at:									
24E	15000	56000	31500	56000	22000	–	–	–	–
37.7E	30000	14500	8500	14500	6000	–	–	–	–
44E	8000	4900	2900	4900	2050	–	–	–	–

There are two basic processes for ethanol production. One is traditional wet-milling; the other is the dry grind process, sometimes referred to as dry-milling. Until the early part of the twenty-first century, wet-milling was the preferred means of producing ethanol because the co-products of wet-milling are of greater value than those from the dry grind process. Because ethanol had a low price, the value of the co-products was the difference between profitability and negative revenue. However, as the price of ethanol increased, the dry grind process became more profitable than wet-milling, because of its lower capital requirements and higher yield of ethanol per unit weight of corn. Modified dry grind processes have been proposed and offer to increase the co-product value.[240–243]

4. Corn Oil

About 70 kg of crude corn oil is recovered from the germ isolated from a metric ton of corn (1260 bushels). The crude oil is refined by standard methods to reduce the content of free fatty acids, waxes, phospholipids, color and miscellaneous unsaponifiable substances. Its low solidifying point, low smoke point and slightly 'corny' flavor make it a preferred oil for household use, where 50–60% of the production is utilized. Nearly all the remainder is used in the manufacture of oleomargarine. The high level of linoleic acid is claimed to be a dietary advantage. The low level of linolenic acid and an adequate level of tocopherols contribute to corn oil's good oxidative stability.[53] Grain sorghum oil is similar in fatty acid composition to corn oil; the crude oil has a higher wax content and is more difficult to refine.

5. Feed Products

Total co-product volume amounts to about one-third of the total mill output. Most is sold as animal feed ingredients, except for the corn oil and a small amount of steep liquor used in antibiotic fermentation media.[244] The composition, nutritional values, and animal feeding uses have been thoroughly described by Schroder and Heiman.[245] Listed in the approximate order of volume of sales, the major feed by-products are briefly described as follows.

Corn gluten feed, which is priced at 21% protein, but often sells in the 18–22% range, is composed of fiber (pericarp and cellular fiber), steep liquor and screenings, plus corn germ residue at locations where germ is processed. The product is dried to 11% moisture. Use of the term 'gluten' in the product descriptions, viz., corn gluten feed and corn gluten meal, are misnomers since the products contain no gluten. Gluten is a specific wheat protein not found in corn. Corn gluten feed is often used as a dairy ration ingredient, but is a suitable feed product for most ruminants. The average corn gluten feed contains 25–30% steep liquor solids, but feed products containing up to 50% steep liquor solids are marketed.

Corn gluten meal is the 60% (w.b.) protein overflow stream from the first starch separation step dried to 11% moisture content. It is bright yellow as result of the 200–400 mg/kg of xanthophyll pigments it contains. A small amount of 41% corn gluten meal is made by blending with finely ground gluten feed. The primary use of gluten meal is in broiler and laying hen rations as a protein concentrate and to supply yellow pigmentation. It is also a pet food ingredient.

Corn germ meal (21% protein) is the dried residue left from oil recovery from germ. It has high absorbency for liquids, such as molasses and tallow, but the primary use is as an ingredient in corn gluten feed.

Corn starch molasses is the concentrated mother liquor remaining from dextrose crystallization. It is used with cane molasses in cattle feeding regimes.

Concentrated steep liquor is officially identified for feeding purposes as 'Condensed corn fermentation extractives.' This product carries 25% protein content at 48% moisture. Its only feeding use is as an ingredient in liquid feeds for cattle, but it is also an excellent nitrogen source in nutrient media for antibiotic production.[244]

Corn bran is the dried fiber fraction. The protein content is about 10%. Its only use is in beef or dairy cattle feeds. A purified form of corn fiber has been offered to the food industry for use in high fiber food products.[246]

Hydrolyzed vegetable oil (HVO) or acid oil is a by-product of alkali refining of crude corn oil, and is obtained by acidulating alkaline soapstock. HVO must contain at least 92% total fatty acids. It is used to control dust and as an energy source in beef and poultry rations.

Although no longer manufactured, feed products produced during grain sorghum starch production have names and protein contents similar to those of corn starch feed by-products.[8,247] Milo (sorghum) steep liquor (condensed milo fermentation extractives) has been used as an ingredient in liquid feed supplements for beef cattle feeding. Because grain sorghums are brown in color and contain condensed tannins, all these feed products are dark brown in color. Therefore, they were used almost exclusively in ruminant feeds.[247,248]

V. Alternative Fractionation Procedures

The high capital and energy costs associated with conventional corn wet-milling, and the continued growth of the industry, has renewed interest in developing alternative starch production methods. These new processes attempt to lower capital and energy requirements and, in some cases, may result in new value-added products. The alternative processes can be divided into processes which decrease diffusion time, which use mechanical shear to enhance separation of starch and protein, and which use a different chemistry to create separation of starch and protein.

Because steeping is a diffusion-limited process which consumes such a large proportion of the total process time, researchers have explored dry-milling or dry grinding of the corn prior to steeping[15,103,249] to reduce particle size and decrease diffusion time. Grinding of the corn kernel dry results in an extensive amount of broken germ, which is difficult to remove from endosperm pieces without also removing a considerable amount of endosperm. Generally, dry-milling will recover no more than 19.7 kg oil/t (1.0 lb oil/bu)[245,250] compared to wet-milling, which will recover 29.5–31.4 kg oil/t (1.5–1.6 lb oil/bu). Broken germ not only represents an economic loss, it presents problems for downstream processing to recover starch. Oil which is left in the endosperm fraction will affect starch quality, coat process equipment, decrease operating efficiency, increase maintenance costs and affect evaporator fouling. The advantage of combined wet and dry-milling is that steep times can be reduced to 1–8 hours compared to the conventional 24–36 hours.

Watson and Stewart[251] developed a degermination procedure where corn was soaked in water for a short period of time, then cracked using a roller mill. They found that average kernel moisture of 37% was required for the germ to be pliable enough to withstand shearing without breaking. Germ comparable in quality to wet-milled germ could be recovered by soaking corn in water for 3–12 hours, then processed using conventional wet-milling germ recovery equipment.[241,252–255] This 'Quick Germ' process has the greatest potential application in dry grind ethanol production, but may be useful as a front end for other wet-milling procedures.

The intermittent milling and dynamic steeping process (IMDS) was developed and analyzed by Lopes-Filho[256] as a means of recovering starch quickly by reducing the diffusional distance into the endosperm. The difference between the IMDS process and other particle size reduction processes is that the germ is not recovered until steeping is completed. Lopes-Filho found that corn soaked in water (no sulfur dioxide or lactic acid) for 1–2 hours, lightly milled in a Bauer-type mill, and then steeped for 3 hours, with additional milling after each hour of steeping (total 5 hours process time), yielded ~1% more starch than corn steeped conventionally for 36 hours. By gradually reducing the particle size of the endosperm, using the intermittent milling steps, the germ is not damaged and diffusional flux is maximized.

Another way to get around the diffusional limitations of whole kernel steeping is to hydrate the corn using pressure.[257,258] While pressure hydration accelerates the rate of water uptake, steeping involves more than water hydration (in conventional wet-milling hydration occurs primarily during the first 12 hours). Pressure hydration must be used in conjunction with some method to disrupt the protein matrix. The major disadvantages of pressure steeping to a wet-miller are the liability and maintenance problems associated with pressure vessels.

A third method of overcoming the diffusional limitation encountered in whole kernel steeping is to apply gaseous sulfur dioxide to dry corn prior to immersion in water.[259] Gaseous sulfur dioxide diffuses into the corn kernel approximately 100 times faster than diffusion as sulfurous acid (sulfur dioxide in water) and can penetrate the kernel in <5 minutes.[26,260] The concept of gaseous addition of sulfur dioxide is to get adequate sulfur dioxide throughout the kernel, absorbed by the cellular moisture (13.5–15%) found in market channel corn, so that it can initiate the starch releasing reactions. The diffusing steepwater will achieve near complete penetration in 8–12 hours, at which time the reaction is complete. Laboratory tests[261] showed that steep times of >8 hours with gaseous addition of sulfur dioxide had comparable starch yields to conventional 24 hour steeping. It has been reported both that the starch peak amylograph viscosity was considerably higher for gaseous-treated samples[26] and that there was no difference.[259] As with all processes which do not have a fermentation period to produce lactic acid, lactic acid or another organic or inorganic acid need to be added to maximize starch yields.

A recent development in the production of starch from corn is the use of high mechanical shear rates to effect the separation of starch and protein instead of using a chemical to facilitate separation. The Westfalia 'HD Process' is a combination of high-pressure hydration with high-pressure disintegration (high shear) to release the starch from the endosperm protein matrix.[262–265] In the HD Process, the third grind mill is replaced by two high-pressure disintegration valves used in series. No sulfur dioxide is used. Starch–protein separation occurs as a result of the high mechanical shear which occurs as endosperm pieces pass through the valve. Total processing time can be reduced to 3–5 hours by using pressure hydration. Analysis of the HD Process[266] suggested that ambient pressure hydration could substitute for high-pressure hydration with total process time increasing to only 8–12 hours. Use of the pressure disintegration valve in conjunction with the 'Quick Germ' process would be a compatible combination. It was also suggested that application of low levels of sulfur dioxide in conjunction with high-pressure disintegration might be more suitable than using water hydration alone.[266] Recent studies on the use of pourer ultrasound have shown that shear alone can disrupt the protein matrix of the endosperm enough to achieve starch release.[267–270]

It is possible to recover various corn components using solvent extraction.[271–273] The sequential extraction (SE) process first grinds the corn using a hammer mill, extracts oil using 95% ethanol (while at the same time dehydrating the ethanol), and then uses a mixture of ethanol and sodium hydroxide to extract the zein and glutelin.[272–274] The remaining mash can be fermented into ethanol or used for other fermentation purposes. It may also be possible to process the residual mash to produce a saleable starch product. The SE process maximizes protein and oil yields, recovering 12.5% more oil and 3% more protein, recovers zein and a food grade protein separately, and increases oxidative stability of the oil because it has not been exposed to sulfur dioxide. An economic evaluation of the process[5,272] showed that it was economically viable, but that profitability was contingent on the value of the improved protein fraction.

Alkali was the first chemical used in corn wet-milling,[275] but was quickly abandoned in favor of sulfur dioxide. Researchers over the years have tried to refine the use of alkali for starch processing,[17,18,276] because of the ease with which alkali solubilizes endosperm protein. An alkali wet-milling process which yields ~2% more starch and has ~40% lower capital cost has been developed.[277]

An enzyme milling procedure has been developed.[278–281] In it, corn is soaked only 2–4 hours prior to grinding. After light grinding to open up the kernels and reduce diffusional limitations, the corn mash is treated with a proprietary enzyme preparation[282] that works directly on the endosperm protein matrix, similarly to the approach described in reference 174. The process has been successfully piloted in a small commercial facility and is being implemented in several larger commercial facilities. The pilot test resulted in a starch yield of >70% and good germ, fiber and gluten recovery.

VI. Future Directions in Starch Manufacturing

Starch manufacturing seems poised to make major changes. Over the past 35 years, the industry has gone through an almost 500% increase in capacity, added two major new commodities (fructose and ethanol), and developed a host of smaller but expanding markets in products such as lysine and citric acid (see Chapter 2).[283] Major changes in the next two decades are not likely to be explosive expansion and major new commodities, but will likely revolve around increasing processing efficiency, utilizing lower-cost fractionation methods that minimize sulfur dioxide use, and dividing the industry to service different market segments. Specific directions of potential change are discussed in the following sections.

1. Continued Expansion into Fermentation Products

Continued growth is expected in fermentation of glucose into saleable products, with the production of additional amino acids, vitamin feed additives, lactic acid or biopesticides.[283] Not all companies will expand into these products. Ethanol production will continue to expand, but not at the rate that dry grind facilities did during the period from 2004–2007. For wet-mills, the addition of ethanol capacity will be as part of a waste treatment plan to minimize recycle streams. If polylactic acid[284–287] becomes economically competitive, the demand for lactic acid could be explosive.

2. Biosolids as Animal Food

Increased utilization of fermentation will mean a host of new biosolids (cell mass solids and fermentation solubles) which will need to be utilized. The logical use of these biosolids is as an animal feed,[288] but animal feed use will require greater control of plant operations to ensure consistent quality products acceptable to the livestock industry. Because each facility will have a different mixture of fermentation products, it will not be possible for all facilities to produce the same set of livestock feed products, as is currently done with corn gluten meal and corn gluten feed. Branded animal feed products will provide market delineation and increased revenue for some wet-millers.

3. Processing of Specific Hybrids

Processing a limited number of hybrids reduces process variability, improves capital utilization and reduces maintenance costs.[289] The major deterrent to hybrid-specific processing has been the ability to segregate the desired corn through the market channel. Recent opportunities to export hybrid-specific corn or value-enhanced corn[290,291] have spurred the market channel into exploring options for low cost delivery of such corn to US companies. High extractable starch and high fermentable hybrids are being marketed for the wet-milling and ethanol industries, respectively. With the current high price of corn and the potential for wide price swings due to the growth of the ethanol industry, accessibility to corn may become a problem for some wet-millers. Contract production or similar options may become attractive.

4. New Corn Genotypes and Phenotypes via Biotechnology and Genetic Engineering

Biotechnology and genetic engineering techniques have the potential to accelerate modification of elite line hybrids into phenotypes with unique starch characteristics or dent hybrids with enhanced production and/or processing characteristics.[292,293] The desired results of ongoing research and development in this area are to create new products and to enhance the profitability of existing products. Such developments will also increase genetic diversity in the corn produced and increase the need for low cost methods to move the special corn genotypes and phenotypes through the market channel.

5. Segregation of the Corn Starch Industry

In the past 35 years, the corn wet-milling industry has changed from an industry where there was considerable commonality in processes and products between companies and plants to an industry in which there are companies and plants that concentrate in producing only specialty starches for the food and/or paper industries and companies who only produce commodities such as ethanol or sweeteners. Although the companies and plants produce different product lines, they use essentially the same fractionation (milling) process; it is just operated differently to achieve the desired benefits and savings. Because of the growth of the ethanol industry and the advent of newer fractionation methods, there is the potential that dry grind ethanol processors may try to divert part of their production into starch. This is particularly true where a wet modified dry grind process, such as the quick germ/quick fiber process coupled with enzyme milling, is used. The modified dry grind processes are essentially low-cost wet-mills. All that is needed is equipment for the separation of starch and protein (gluten) sufficient to isolate prime mill starch (as discussed earlier).[192]

VII. References

1. Beadle GW. In: Walden DB, ed. *Maize Breeding and Genetics*. New York, NY: Wiley; 1978:93–112.
2. Iltis HH, Doebley JF, Guzman MR, Pazy B. *Science*. 1979;203:186.
3. Trifunovic V. In: Walden DB, ed. *Maize Breeding and Genetics*. New York, NY: Wiley; 1978:41–58.
4. Collins GN. *US Bur. Plant Inds.* 1909;161:31.
5. Kerr RW. *Chemistry and Industry of Starch*. New York, NY: Academic Press; 1942 [2nd edn. 1950].
6. Martin JH. In: Wall JS, Ross WM, eds. *Sorghum Production and Utilization*. Westport, CT: AVI Publishing Co.; 1970:1–27.
7. Freeman JE. In: Wall JS, Ross WM, eds. *Sorghum Production and Utilization*. Westport, CT: AVI Publishing Co.; 1970:28–72.
8. Watson SA. In: Wall JS, Ross WM, eds. *Sorghum Production and Utilization*. Westport, CT: AVI Publishing Co.; 1970:602–626.
9. Yang P, Sieb PA. *Cereal Chem.* 1996;73:751.
10. Xie XJ, Seib PA. *Cereal Chem.* 2000;77:392.
11. Wang FC, Chung DS, Seib PA, Kim YS. *Cereal Chem.* 2000;77:478.
12. Brekke OL. In: Inglett GE, ed. *Corn: Culture, Processing, Products*. Westport, CT: AVI Publishing Co.; 1970:262–291.
13. Headley VE, Spanheimer J, Freeman JE, Heady RE. *Cereal Chem.* 1972;49:142.
14. Powell EF. US Patent 3 909 288; 1975; *Chem. Abstr.* 1975;83:195629.
15. Chwalek VP, Olson RM. US Patent 4 181 748; 1980; *Chem. Abstr.* 1980;92:109–438.
16. Mistry AH, Eckhoff SR. *Wet-milling Notes*. No. 6. Urbana, IL: University of Illinois; 1992.
17. Mistry AH, Schmidt SJ, Eckhoff SR, Sutherland JW. *Starch/Stärke*. 1992;44:284.
18. Mistry AH, Eckhoff SR. *Cereal Chem.* 1992;69:296.
19. Eckhoff SR, Jayasena WV, Spillman CK. *Cereal Chem.* 1993;70:257.
20. Mistry AH, Eckhoff SR. *Cereal Chem.* 1992;69:82.
21. Olson RM. CPC International, Inc., Argo, IL; 1972, personal communication.
22. Hill JD, Overholt LO, Popp HW. *Botany. A Textbook for Colleges*. York, PA: McGraw Hill; 1936:249.
23. Wolf MJ, Buzan CL, MacMasters MM, Rist CE. *Cereal Chem.* 1952;29:321.
24. Wolf MJ, Buzan CL, MacMasters MM, Rist CE. *Cereal Chem.* 1952;29:334.
25. Kieselbach TA, Walker ER. *Am. J. Bot.* 1952;39:561.
26. Eckhoff SR, Okos MR. *Cereal Chem.* 1989;66:30.
27. Ruan R, Litchfield JB, Eckhoff SR. *Cereal Chem.* 1992;69:600.
28. Syarief AM, Gustafson RJ, Morey RV. *Trans. ASAE*. 1987;30:522.
29. Ramos G, Pezet-Valdez M, O'Connor-Sánchez A, Placenia C, Pless RC. *Cereal Chem.* 2004;81:308.
30. Wolf MJ, Buzan CL, MacMasters MM, Rist CE. *Cereal Chem.* 1952;29:349.
31. Hinton JJC. *Cereal Chem.* 1953;30:441.

32. Mistry AH, Eckhoff SR. *Cereal Chem.* 1992;69:202.
33. Duvick D. *Cereal Chem.* 1961;38:374.
34. Wolf MJ, Khoo V. *Cereal Chem.* 1975;52:771.
35. Hopkins CB, Smith LH, East EM. *Univ. of Illinois Agr. Exp. Sta. Bull.* 1903;87.
36. Fox FJ, Eckhoff SR. *Cereal Chem.* 1993;70:402.
37. Blessin CW, Beicher JD, Dimler RJ. *Cereal Chem.* 1963;40:582.
38. Watson SA, Sanders EH, Wakely RD, Williams CB. *Cereal Chem.* 1955;32:165.
39. Wolf MJ, Buzan CL, MacMasters MM, Rist CE. *Cereal Chem.* 1952;29:362.
40. Dure LS. *Plant Physiol.* 1963;35:925.
41. Seckinger HL, Wolf MJ, MacMasters MM. *Cereal Chem.* 1960;37:121.
42. Jacks TJ, Yatsu LY, Altschul AM. *Plant Physiol.* 1967;42:585.
43. Yatsu L. Southern Regional Research Center, USDA-ARS, New Orleans, LA; 2007, personal communication.
44. Kummerow FA. *Oil Soap.* 1946;23:167–273.
45. Sanders EH. *Cereal Chem.* 1955;32:12.
46. Saraiva RA. *Sorghum Wax and Selected Applications* [M.S. Thesis]. Lincoln, NE: University of Nebraska; 1995.
47. Miller DF. *Composition of Cereal Grains and Forages.* Vol. 585. Washington, DC: National Academy of Science, National Research Council; 1958.
48. Van Soest PJ, Wind RH. *J. Off. Anal. Chem.* 1968;51:780.
49. Earle FR, Curtis JJ, Hubbard JE. *Cereal Chem.* 1946;23:504.
50. Tan SL, Morrison WR. *J. Am. Oil Chem. Soc.* 1979;56:759.
51. Hamilton TH, Hamilton BC, Johnson BC, Mitchell HH. *Cereal Chem.* 1951;28:163.
52. Reiners RA. In: Inglett GE, ed. *Corn: Culture, Processing, Products.* Westport, CT: AVI Publishing Co.; 1970:243.
53. Weber EJ, Alexander DE. *J. Am. Oil Chem. Soc.* 1975;52:370.
54. Watson SA, Freeman JE. In: *Proceedings of the Thirtieth Annual Corn and Sorghum Research Conference.* Washington, DC: American Seed Trade Association; 1975:251–275.
55. Dudley JW, Lambert RJ, Alexander DE. In: Dudley JW, ed. *Seventy Generations of Selection for Oil and Protein Concentration in the Maize Kernel*: Crop Science Society of America, Special Publication; 1974:181–212.
56. Alexander DE, Creech JR. In: Sprague GF, ed. *Corn and Corn Improvement.* 2nd edn. Madison, WI: American Society of Agronomy; 1977:336–390.
57. Thomison PR. *Agronomy Facts-AGF133.* Columbus, OH: Ohio State Extension; 1996.
58. Schneider EO, Early EB, DeTurk EE. *Agron. J.* 1949;41:30.
59. Mertz ET, Bates LS, Nelson OE. *Science.* 1964;145:279.
60. Anon. In: *ACIMMT Research Highlights 1984.* International Maize and Wheat Improvement Center; 1985:29–35.
61. Watson SA, Yahl KR. *Cereal Chem.* 1967;44:488.
62. Fox EJ, Eckhoff SR. *Cereal Chem.* 1993;70:402.
63. Hixon RM, Sprague GF. *Ind. Eng. Chem.* 1944;34:959.
64. Freeman JE, Abdullah M, Bocan BJ. US Patent 3 928 631; 1975; *Chem. Abstr.* 1976;84:88301.

65. Bear RP, Vineyard MM, MacMasters MM, Deatherage WL. *Agron. J.* 1958;50:598.
66. Anderson RA, Vojonovich C, Griffin Jr., EL. *Cereal Chem.* 1961;38:84.
67. Hubbard JE, Hall HH, Earle FR. *Cereal Chem.* 1950;27:415.
68. Bianchi G, Avato P, Mariani G. *Cereal Chem.* 1979;56:491.
69. Maxon ED, Rooney LW. *Cereal Chem.* 1972;49:719.
70. Hosney RC, Varriano-Marston E, Dendy DA. *Advan. Cereal Sci. Technol.* 1981;4:94.
71. Oria MP, Hamaker BR, Shull J. *J. Agric. Food Chem.* 1995;43:2148.
72. Karper RD. *J. Heredity.* 1933;24:257.
73. Singh R, Axtell JD. *Crop Sci.* 1973;13:535.
74. Freeman JE. *Trans. Am. Soc. Agr. Engr.* 1973;16:671.
75. Zehr BE, Eckhoff SR, Nyqyist WE, Keeling PL. *Crop Sci.* 1996;36:1159.
76. Zehr BE, Eckhoff SR, Singh SK, Keeling PL. *Cereal Chem.* 1995;72:491.
77. Singh N. *Hydrocyclone Recovery of Waxy Corn Hybrids Starch as Affected by Planting Location* [M.S. Thesis]. Urbana, IL: University of Illinois; 1994.
78. *Official Grain Standards of the United States.* Washington, DC: USDA, Grain Inspection, Packers and Stockyards Administration, Federal Grain Inspection Service; March, 1995.
79. Watson SA. In: Watson SA, Ramstadt PE, eds. *Corn: Chemistry and Technology.* St. Paul, MN: American Association of Cereal Chemists; 1987:1125–1183.
80. Hill LD. *Grain Grades and Standards.* Urbana, IL: University of Illinois Press; 1990.
81. Eckhoff SR, Paulsen MR. In: Henry J, Kettlewell P, eds. *Cereal Grain Quality.* London, UK: Chapman and Hall; 1996.
82. Eckhoff SR, Litchfield JB, Paulsen MR. In: Macrae R, Robinson R, Sadler M, eds. *Encyclopedia of Food Science, Food Technology, and Nutrition.* Vol. 4. London, UK: Academic Press; 1993:2825–2831.
83. Federal Grain Inspection Service. *Fed. Register.* 1984;49:35743–35745.
84. Christenson CN, Kaufmann HH. *Grain Storage: The Role of Fungi in Quality Loss.* Minneapolis, MN: The University of Minnesota Press; 1969:153.
85. Freeman JE, Heatherwick HJ, Watson SA. In: *Proceedings of a Conference on Research in Corn Quality.* University of Illinois, IL; April 28–29, 1970 [AE-4251].
86. Interactions of Mycotoxins in Animal Production. Proceedings of the Symposium. Washington, DC: National Academy of Sciences; July 13, 1978 (Publ. 1979).
87. Shotwell OL, Goulden ML, Hesseltine CM. *Cereal Chem.* 1973;51:492.
88. Watson SA, Yahl KR. *Cereal Sci. Today.* 1971;16:153.
89. Lillehoj EB, Wolek WFK, Horner ES, Widstrom NW, Josephson LM, Franz AO, Catalano EA. *Cereal Chem.* 1980;57:255.
90. Zuber MS, Lillehoj EB. *J. Environ, Qual.* 1979;8:1.
91. Anderson HW, Nehring EW, Wichser WR. *J. Agr. Food Chem.* 1975;23:775.
92. Yahl KR, Watson SA, Smith RJ, Barabolak R. *Cereal Chem.* 1971;48:385.
93. Shotwell OL, Hesseltine CW. *Cereal Chem.* 1981;58:124.
94. Murthy GS, Townsend DE, Meerdink GL, Bargren GL, Tumbleson ME, Singh V. *Cereal Chem.* 2005;82:302.
95. Barabolak R, Colburn CR, Just DE, Schleichert EA. *Cereal Chem.* 1978;55:1065.

96. Tuite J, Shaner G, Rambo G, Foster J, Caldwell RW. *Cereal Sci. Today.* 1974;19:238.
97. Bennett GA, Vandergraft EE, Shotwell OL, Watson SA, Bocan BJ. *Cereal Chem.* 1978;55:455.
98. Bennett GA, Anderson RA. *J. Agr. Food Chem.* 1978;26:1055.
99. Bennett GA, Richard JL, Eckhoff SR. In: Jackson LS, DeVries JW, Bullerman LB, eds. *Fumonisins in Food*: American Chemical Society; 1995:317–322.
100. Jennings. SD, Meyers DJ, Johnson LA. *Cereal Foods World.* 1995;40(Abstract 151):664.
101. Jennings SD. *The Effect of Corn Maturity on the Quantity and Quality of Starch Produced by Wet-milling* [M.S. Thesis]. Ames, IA: Iowa State University; 1996.
102. Eckhoff SR, Denhart R. *Wet-milling Notes.* Vol. 10. Urbana, IL: University of Illinois; 1994.
103. Eckhoff SR, Jayasena WV, Spillman CK. *Cereal Chem.* 1993;70:257.
104. Powell LP, McGeorge GG. US Patent 3 909 288; 1975.
105. Shove GC. *Cereal Chem.* 2000;77(392):60–72.
106. Bakker-Arkema FW, Brook RC, Lerew LE. *Adv. Cereal Sci. Technol.* 1977;2:1.
107. Watson SA, Hirata Y. *Cereal Chem.* 1962;39:35.
108. Foster GH. In: *Twentieth Hybrid Corn Industry – Research Conference.* Washington, DC: American Seed Trade Association; 1965:75–85.
109. Vojonovich C, Anderson RA, Griffin Jr., EL. *Cereal Foods World.* 1975;20:333.
110. Wall JS, James C, Donaldson GL. *Cereal Chem.* 1975;52:779.
111. Wight AW. *Starch/Stärke.* 1981;33:165.
112. Foster GH, Holman LE. Grain Breakage Caused By Commercial Handling Methods. *USDA. Marketing Report.* 1976;968:13.
113. Stroshine RL. *Fine Material in Grain, OARDC Special Circular 141.* Wooster, OH: Ohio Agricultural Research and Development Center; 1992.
114. Thompson RA, Foster GH. Stress Cracks and Breakage in Artificially Dried Corn. *USDA Marketing Research Report.* 1963;631 [Agr. Marketing Service].
115. Watson SA. In: Watson SA, Ramstadt PE, eds. *Corn: Chemistry and Technology.* St. Paul, MN: American Association of Cereal Chemists; 1987:153–156.
116. Brown RB, Fulford GN, Daynard TB, Meiering AG, Otten L. *Cereal Chem.* 1979;57:529.
117. Rausch KD. *Evaluation of the Resistance–Capacitance Method for Detection of Reduced Corn Wet-milling* [Ph.D. Thesis]. Quality Urbana, IL: University of Illinois; 1993.
118. Anderson RA. *Cereal Sci. Today.* 1963;8:190.
119. Eckhoff SR, Rausch KD, Fox EJ, Tso CC, Wu X, Pan Z, Buriak P. *Cereal Chem.* 1993;70:723.
120. Eckhoff SR, Singh SK, Zehr BE, Rausch KD, Fox EJ, Mistry AK, Haken AE, Niu YX, Zou SH, Buriak P, Tumbleson ME, Keeling PL. *Cereal Chem.* 1996;73:54.
121. Fox SR, Johnson LA, Hurburgh Jr., CR, Dorsey-Redding C, Bailey TB. *Cereal Chem.* 1992;69:191.
122. Pelshenke PF, Lindemann E. *Stärke.* 1954;6:177.
123. Shandera DL, Parkhurst AM, Jackson DS. *Cereal Chem.* 1995;72:371.
124. Wehling RL, Jackson DS, Hooper DG, Ghaedian AR. *Cereal Chem.* 1993;70:720.

125. Buffo RA. *Optimization of Sorghum Kernel Component Separation and Kafirin Utilization* [M.S. Thesis]. Lincoln, NE: University of Nebraska; 1955.
126. Grosh GM, Milner M. *Cereal Chem.* 1959;36:260.
127. Shotwell OL, Hesseltine CW, Burmeister HR, Kwolek FW, Shannon GM, Hall HH. *Cereal Chem.* 1969;46:446.
128. Watson SA, Hirata Y. *Agron. J.* 1955;47:11.
129. Freeman JE, Watson SA. *Cereal Sci. Today.* 1971;16:378.
130. Radley JA. ed. *Starch Production Technology*. London, UK: Applied Science Publishers, Ltd; 1976.
131. Knight JW. *The Starch Industry*. London, UK: Pergamon Press; 1969.
132. Blanchard PH. *Technology of Corn Wet-milling and Associated Processes*. Amsterdam, The Netherlands: Elsevier; 1992.
133. Wang D. *Effect of Broken and Pericarp Damaged Corn on Water Absorption and Steepwater Characteristics* [M.S. Thesis]. Urbana, IL: University of Illinois; 1994.
134. Yaptenco KF, Fox EJ, Eckhoff SR. *Cereal Chem.* 1996;73:249.
135. Randall JR, Langhurst AK, Schopmeyer HH, Seaton RL. US Patent 4 106 487; 1978.
136. Kempf W. *Stärke*. 1971;23:89.
137. Veger HJ. British Patent 1 238 725; 1971.
138. Fan LT, Chen HC, Shellenberger JA, Chung DS. *Cereal Chem.* 1965;42:385.
139. Stenvert NL, Kingswood K. *Cereal Chem.* 1976;53:141. 1977;54:627.
140. Roushdi M, Ghali Y, Hassanean A. *Stärke*. 1979;31:78.
141. Bautista GM, Linko P. *Cereal Chem.* 1962;39:455.
142. Franzke C, Wahl G. *Stärke*. 1970;22:64.
143. Watson SA, Hirata Y, Williams CB. *Cereal Chem.* 1955;32:382.
144. Fan LT, Chu PS, Shellenberger JA. *Biotechnol. Bioeng.* 1962;4:311.
145. Anderson RA. *Cereal Chem.* 1962;39:406.
146. Cox MJ, MacMasters MM, Hilbert GE. *Cereal Chem.* 1964;21:447.
147. Montgomery EM, Sexson KR, Dimler RJ. *Stärke*. 1964;16:314.
148. Wagoner JA. *Cereal Chem.* 1948;25:354.
149. McDonough CM, Floyd CD, Rooney LW. *Cereal Foods World*. 1995;40(Abstract 134):660.
150. Sullins BD, Almeida-Dominguez HD, Rooncy LW. *Cereal Foods World*. 1995;40(Abstract 152):664.
151. Wahl G. *Stärke*. 1969;21:77.
152. Watson SA, Sanders EH. *Cereal Chem.* 1961;38:22.
153. Swan JN. *Nature*. 1957;180:643.
154. Boundy JA, Turner JW, Wall JS, Dimler RJ. *Cereal Chem.* 1967;44:281.
155. Wall JS. *J. Agr. Food Chem.* 1971;19:619.
156. Beckwith AC. *J. Agr. Food Chem.* 1972;20:761.
157. Kolthoff JH, Anastasi A, Tan BH. *J. Am. Chem. Soc.* 1960;82:4147.
158. Nierle W. *Stärke*. 1972;24:345.
159. Wall JS. *Adv. Cereal Sci. Technol.* 1978;2:135.
160. Paulis JW. *J. Agr. Food Chem.* 1982;30:14.
161. Landry J, Moureaux T. *Bull. Soc. Chim. Biol.* 1970;52:1021.
162. Paulis JW, Wall JS. *Cereal Chem.* 1969;46:263.

163. Burr B, Burr FA. *Proc. Natl, Acad. Sci. USA*. 1976;73:515.
164. Watson SA. *Meth. Carbohydr. Chem.* 1964;4:3.
165. Suzuki T, Sugimoto M. *Denpun Kogyo Gakkaishi*. 1973;20:161.
166. Rausch KD, Singh SK, Eckhoff SR. *Cereal Chem.* 1993;70:489.
167. Kerpisci MR. *Effect of Variety and Drying Temperature on Proteolytic Enzyme Activity of Yellow Dent Corn* [M.S. Thesis]. Manhattan, KS: Kansas State University; 1988.
168. Johnson IC, Gilliland SE, Speck ML. *Appl. Microbiol.* 1971;21:316.
169. Heady RE. Corn Products International, unpublished data.
170. Balana RC, Caixes AM. US Patent 4 086 135; 1978; *Chem. Abstr.* 1976;84:57–393.
171. Balana R. Corn Products International, personal communication.
172. Gillece CD. Corn Products International, personal communication.
173. Du L, Li B, Lopes-Filho JF, Daniels CR, Eckhoff SR. *Cereal Chem.* 1996;73:96.
174. Eckhoff SR, Tso CC. *Cereal Chem.* 1991;68:248.
175. Ling D, Jackson DS. *Cereal Chem.* 1991;68:205.
176. Singh N, Eckhoff SR. *Cereal Chem.* 1995;72:344.
177. Steinke JD, Johnson LA, Wang C. *Cereal Chem.* 1991;68:12.
178. Singh V, Haken AE, Paulsen MR, Eckhoff SR. *Starch/Stärke*. 1997;50:81.
179. Shandera DL, Jackson DS. *Cereal Chem.* 1996;73:632.
180. Watson SA, Williams CB, Wakely RD. *Cereal Chem.* 1951;28:105.
181. Lindemann E. *Stärke*. 1954;6:274.
182. Saint-Lebe L, Joussoud M, Andre C. *Staerke*. 1965;17:341.
183. Singh N, Eckhoff SR. *Cereal Chem.* In press.
184. Singh SK, Johnson LA, Pollak LM, Fox SR. *Cereal Chem.* 1996;73:659.
185. Watson SA, Hirata Y. Corn Products International, unpublished data.
186. Kempf W. *Stärke*. 1973;25:376.
187. Wang YJ, White P, Pollak L. *Cereal Chem.* 1992;69:328.
188. Mauro DJ. *Cereal Foods World*. 1996;41:776.
189. Stamicarbon BV, Heerlen, Netherlands, British Patent 701 613; 1953.
190. Stavenger PL, Wuth DE, US Patent 2 913 112; 1959; *Chem. Abstr.* 1960;54:2844.
191. Bradley D. *The Hydroclone*. Oxford, UK: Pergamon Press; 1965.
192. Smith NB, McFate HA, Eubanks EE. US Patent 3 236 740; 1966; *Chem. Abstr.* 1966;64:14401.
193. Ginaven ME. US Patent 3 040 996; 1962.
194. Dill RR, Ginaven ME. US Patent 3 118 624; 1964.
195. Dowie DW, Martin D. US Patent 3 029 169; 1962; *Chem. Abstr.* 1962;57:3684.
196. Fontein FJ. US Patent 2 916 142; 1959.
197. Fontein FJ. US Patent 2 975 068 (1961); *Chem. Abstr.* 55,12898 (1961).
198. Probstmeyer H. US Patent 5 246 579; 1993.
199. Bier H, Eisken JC, Honeychurch RW. *Stärke*. 1974;26:23.
200. Bell WL. *Cereal Foods World*. 1983;28:249.
201. Chwalek VP. US Patent 3 813 298; 1974.
202. Huster H. *The Use of Separators and Decanters in the Starch Industries*. Oelde, West Germany: Westfalia Separator AG; undated.
203. Bonnyay L. *Stärke*. 1978;30:61.

204. Peltzer Sr., A. US Patent 2 973 896; 1961.
205. Hiepke CH, Huster H. *Stärke*. 1976;28:14.
206. Fox EJ. *Factors Affecting Corn Gluten Filtration* [Ph.D. Thesis]. Urbana, IL: University of Illinois; 1993.
207. Beger HJ. US Patent 2 689 810; 1954; *Chem. Abstr.* 1954;48:14624.
208. Beger HJ. US Patent 2 778 752; 1957; *Chem. Abstr.* 1957;51;4745.
209. Singh N, Eckhoff SR. *Cereal Foods World*. 1996;41:676.
210. Lewis KD, Charlton AP, Nyrops P, Merediz TO. US Patent 4 260 480; 1981.
211. Chiang WC. *Sugar y Azucar*. 1993;88:23.
212. Chiang WC, Lee A. *New Developments for Improving Starch–Gluten Separation*. Milford, CT: Dorr-Oliver; 1995.
213. Chwalek BP, Schwartz CW. US Patent 4 144 087; 1979; *Chem. Abstr.* 1978;89:7960.
214. Freeman JE, Watson SA. *Cereal Sci. Today*. 1969;14:10.
215. Freeman JE. US Patent 3 477 855; 1969.
216. Zipf RL, Anderson RA, Slotter RL. *Cereal Chem.* 1950;27:463.
217. Baunack F. *Stärke*. 1963;15:299.
218. Idaszak LR. US Patent 4 021 927; 1977; *Chem. Abstr.* 1977;86:92222.
219. Wu IYV. *Cereal Chem.* 1988;65:105.
220. Cheryan M. In: Singh RP, Wiraktusuman MA, eds. *Advances in Food Engineering*. Boca Raton, FL: CRC Press; 1992.
221. Industrial Energy Study of Selected Food Industries for the Federal Energy Office. US Department of Commerce, Final Report; 1974.
222. Casper ME. *Energy Saving Techniques for the Food Industry*. Park Ridge, NJ: Noyes Data Corp.; 1977:83–94.
223. Pulfrey AL, Kerr RW, Reintjes HR. *Ind. Eng. Chem.* 1929;21:205.
224. Stavenger PL. *Stärke*. 1979;31:81.
225. Bensing HO, Brown DR, Watson SA. *Cereal Sci. Today*. 1972;17:304.
226. Sourirajan S. *Reverse Osmosis*. New York, NY: Academic Press; 1970.
227. Cheryan M. In: Heldman DR, Lund DB, eds. *Handbook of Food Engineering*. Basel, Switzerland: Marcel Decker; 1992.
228. Brock TD. *Membrane Filtration: A User's Guide and Reference Manual*. Madison, WI: Science Tech, Inc.; 1983.
229. Müller H. German Patent 2 618 131; 1976; *Chem. Abstr.* 1977;86:31230.
230. Healey VE. *Trans. Am. Soc. Agr. Engr.* 1982;25:788.
231. Grant A, Tailor TR, Powers J. *Chem. Processing*. 1975;38(2):86.
232. Simms RL. *Sugar y Asucar*. March 1978;50.
233. Corn starch. 3rd edn. Washington, DC: Corn Industries Research Foundation; 1964.
234. Leach HW. In: Whistler RL, Paschall EF, eds. *Starch: Chemistry and Technology*. Vol. I. New York, NY: Academic Press; 1965:289–307.
235. Morrison WR. *Adv. Cereal Sci. Technol.* 1976;1:221.
236. Schoch TJ, Maywald EC. *Anal. Chem.* 1956;28:382.
237. Bergthaller W, Tegge G. *Stärke*. 1972;24:348.
238. Newton JM. *Proceedings of the 16th Annual Symposium*. St. Louis, MO: Central States Section. Am. Assoc. Cereal Chemists; 1975.

239. Lorenz D, Morris D. *How Much Energy Does It Take to Make a Gallon of Ethanol?* Washington, DC: Institute for Local Self-Reliance; 1995.
240. Lewis S. *Proceedings of the Corn Utilization Technology Conference.* Dallas, TX; 2006:10–14.
241. Singh V, Eckhoff SR. *Cereal Chem.* 1996;73:716.
242. Singh V, Johnston DB. *Adv. Food Nutr. Res.* 2004;48:151.
243. Wahjudi J, Xu L, Wang P, Singh V, Buriak P, Rausch KD, McAloon AJ, Tumbleson ME, Eckhoff SR. *Cereal Chem.* 2000;77:640.
244. Bowden JP, Peterson WH. *Arch. Biochem.* 1946;9:387.
245. Schroder JD, Heiman V. In: Inglett GE, ed. *Corn: Culture, Processing, Products.* Westport, CT: AVI Publishing Co. 1970 [Chapter 12].
246. Kickle HL, Ball WJ, Schanefelt RV. US Patent 4 181 747; 1980; *Chem. Abstr.* 1980;92:109437.
247. Grain Sorghum By-Product Feeds for Farm Animals. *Texas Agr. Expt. Stn., Bull.* 1951;743:1–32 [College Station].
248. VanSoest PJ, Fadel J. Sniffen CJ. *Proc. Cornell Nutr. Conf.* Ithaca, NY: Cornell University; 1979: 63–75.
249. Gillenwater DL, Pfundstein GB, Harvey AR. US Patent 3 597 274; 1971; *Chem. Abstr.* 1971;75:119379.
250. Brekke OL. In: Inglett GE, ed. *Corn: Culture, Processing, Products.* Westport, CT: AVI Publishing Co.; 1970:277.
251. Watson SA, Stewart CW. US Patent 3 474 722; 1969.
252. Singh V. *A Germ Recovery Process for Dry Grind Ethanol Facilities* [M.S. Thesis]. Urbana, IL: University of Illinois; 1994.
253. Singh V, Eckhoff SR. *Cereal Chem.* 1996;73:716.
254. Singh V, Eckhoff SR. *Cereal Chem.* 1997;74:462.
255. Singh V, Eckhoff SR. *Wet-milling Notes. No. 13.* Urbana, IL: University of Illinois; 1995.
256. Lopes-Filho JF. *Intermittent Milling and Dynamic Steeping Process for Corn Starch Recovery* [Ph.D. Thesis]. Urbana, IL: University of Illinois; 1995;13.
257. Muthukumarappan K, Gunasekaran S. *Trans. ASAE.* 1992;35:1885.
258. Gunasekaran S, Farkas DF. *Trans. ASAE.* 1988;31:1589.
259. McKinney JC. *Gaseous Sulfur Dioxide Steeping for Corn Wet-milling* [M.S. Thesis]. Urbana, IL: University of Illinois; 1996.
260. Eckhoff SR. *Sorption and Reaction of Sulfur Dioxide on Yellow Dent Corn* [Ph.D. Thesis]. West Lafayette, IN: Purdue University; 1983.
261. Eckhoff SR, Tso CC. *Cereal Chem.* 1991;68:319.
262. Meuser F, Wittig J, Huster H. *Starch/Stärke.* 1989;41:225.
263. Meuser F, Wittig J, Huster H, Holley W. In: Morton ID, ed. *Cereals in a European Context.* Weinheim, Germany: VCH; 1987.
264. Meuser F, German H, Huster H. In: Hill RD, Munck L, eds. *New Approaches to Research on Cereal Carbohydrates.* Amsterdam, The Netherlands: Elsevier Science Publishers; 1985.
265. Huster H, Meuser F, Hoepke GH. US Patent 4 416 701; 1983.
266. Eckhoff SR, Wu X. *Wet-milling Notes.* Vol. 2. Urbana, IL: University of Illinois; 1989.

267. Zhang Z, Feng H, Niu Y, Eckhoff SR. *Cereal Chem.* 2005;82:447.
268. Zhang Z, Feng H, Niu YX, Eckhoff SR. *Starch/Stärke.* 2005;57:240.
269. Wang L, Wang YJ. *Cereal Chem.* 2004;81:140.
270. Cameron DK, Wang YJ. *Cereal Chem.* 2006;83:505.
271. Chen L, Hoff JE, US Patent 4 716 218; 1987.
272. Johnson LA, Hojilla-Evangelista MP, Myers DJ, Chang DI. *Proceedings of the Corn Utilization Conference.* St. Louis, MO: V. National Corn Growers Association; 1994.
273. Hojilla-Evangelista MP, Johnson LA, Myers DJ. *Cereal Chem.* 1992;69:643.
274. Hojilla-Evangelista MP. *Sequential Extraction Processing: Alternative Technology for Corn Wet-milling* [Ph.D. Thesis]. Ames, IA: Iowa State University; 1990.
275. Jones O. US Patent 2 000; 1841.
276. Hansen DW. US Patent 2 472 971; 1949.
277. Eckhoff SR, Du L, Yang P, Rausch KD, Wang DL, Li BH, Tumbleson ME. *Cereal Chem.* 1999;76:96.
278. Johnston DB, Singh V. *Cereal Chem.* 2001;78:405.
279. Johnston DB, Singh V. *Cereal Chem.* 2001;81:626.
280. Wang P, Singh V, Xu L, Johnston DB, Rausch KD, Tumbleson ME. *Cereal Chem.* 2006;83:455.
281. Murthy GS, Singh V, Johnston DB, Rausch KD, Tumbleson ME. *Cereal Chem.* 2006;83:455.
282. Genencor, Prosteep product literature, No. 2629, Revision 02/07; 2007.
283. Nelson D; 1994. *1994 Corn Annual.* Washington, DC: Corn Refiners Association:7–11.
284. Datta R, Tsai SP. *Proceedings of the Corn Utilization Conference VI.* St. Louis, MO: National Corn Growers Association; 1996.
285. Chang YN, Yang PC, Jin Z, Chellapa H, Xing C, Mueller R, Anderson M, Clarke A, Iannotti E. In: Hill RD, Munck L, eds. *New Approaches to Research on Cereal Carbohydrates.* Amsterdam, The Netherlands: Elsevier Science Publishers; 1985.
286. Lunt J. In: Hill RD, Munck L, eds. *New Approaches to Research on Cereal Carbohydrates.* Amsterdam, The Netherlands: Elsevier Science Publishers; 1985.
287. Verser DW, Schilling KH, Chen XM. In: Hill RD, Munck L, eds. *New Approaches to Research on Cereal Carbohydrates.* Amsterdam, The Netherlands: Elsevier Science Publishers; 1985.
288. Belyea RL, Clevenger TE, Van Dyne DL, Eckhoff SR, Wallig MA, Tumbleson ME. *Proceedings of the First Biomass Conference.* Burlington, VT: Americas Energy, Environment, Agriculture, and Industry; 1993.
289. Eckhoff SR. *Wet-milling Notes.* Vol. 11. Urbana, IL: University of Illinois; 1995.
290. Agricultural Education and Consulting. *Merchandising IP, Specialty, and Other Value-Enhanced Corn in Japan*: US Feed Grains Council; 1994.
291. Agricultural Education and Consulting. *1995–1996 Value-Enhanced Corn Quality Report*: US Feed Grains Council; 1996.
292. Nilles D. *Ethanol Producer Mag.* Aug. 2007;90:94.
293. Anon. *Pioneer Growing Points.* 2003; 1(Sept./Oct.):5.

Wheat Starch: Production, Properties, Modification and Uses

10

C.C. Maningat,[1] P.A. Seib,[2] S.D. Bassi,[3] K.S. Woo[4]
and G.D. Lasater[5]

[1] MGP Ingredients Inc., Atchison, Kansas, USA
[2] Department of Grain Science and Industry, Kansas State University, Manhattan, Kansas, USA
[3] MGP Ingredients Inc., Atchison, Kansas, USA
[4] MGP Ingredients Inc., Atchison, Kansas, USA
[5] MGP Ingredients Inc., Atchison, Kansas, USA

I. Introduction	442
II. Production	442
III. Industrial Processes for Wheat Starch Production	444
1. Conventional Processes	446
2. Hydrocyclone Process (Dough–Batter)	448
3. High-pressure Disintegration Process	450
IV. Properties of Wheat Starch and Wheat Starch Amylose and Amylopectin	451
1. Large Versus Small Granules	452
2. Fine Structures of Amylose and Amylopectin	457
3. Partial Waxy and Waxy Wheat Starches	465
4. High-amylose Wheat Starch	470
5. A Unique Combination of Properties	471
V. Modification of Wheat Starch	475
1. Crosslinking	475
2. Substitution	478
3. Dual Derivatization	479
4. Bleaching, Oxidation and Acid-thinning	480
VI. Uses of Unmodified and Modified Wheat Starches	481
1. Role in Baked Products	481
2. Functionality in Noodles and Pasta	485
3. Other Food Uses	488
4. Industrial Uses	489
VII. References	491

I. Introduction

Cultivation of wheat began in prehistoric times and bread, the primary food from wheat, has been a staple in human diets throughout recorded history.[1,2] Although it is difficult to determine with certainty when starch was first recognized as a distinct substance, the first isolated starch was undoubtedly wheat starch.[3] Manufacture of starch from wheat is believed to have started in ancient Egypt and Greece. It is theorized that starch was liberated by fermentation of non-carbohydrate components.[3–9] Natives of Chios, a Greek island, were considered the pioneers in the manufacture of wheat starch. In 130 BCE, their unique efforts in the separation of starch and protein from wheat flour were reported in the writings of Pliny, who stated that the highest grade of wheat starch came from Chios, with inferior products originating from Crete and Egypt.[3–10] Commercial manufacture of wheat starch presumably started in England in the sixteenth century, when wheat starch was introduced into that country during the reign of Elizabeth I.[11]

In 1745, Professor Jacobo Beccari of the Academy of Bologna separated wheat flour into an 'amylaceum' fraction, which possessed characteristics similar to sugars, and a glue-like 'glutinosum' fraction, which was similar to substances of animal origin.[12,13] These two fractions correspond to the present day wheat starch and wheat gluten, respectively, and the aqueous process used to separate them forms the basis for modern day industrial isolation of wheat starch and wheat gluten.[14]

II. Production

Wheat provides more than 20% of the calories for the world population of 6.6 billion people, and will continue to be center stage in universal efforts to feed a growing population. The Food and Agriculture Organization of the United Nations estimate that the world population will reach 9 billion in the year 2040.[15] Wheat is widely grown in five continents and in some 108 countries, and is the leading cereal grain produced, consumed and traded in the world today;[1] areas of cultivation range as far north as Finland and as far south as Argentina. The largest growing area is in the temperate zone of the northern hemisphere, between 30° and 60° latitude, with smaller amounts being grown between 27° and 40° latitudes in the south. It is probably the only crop being harvested somewhere in the world in any given month during the year.[16,17]

From 1986–1990 the average annual world wheat production was 533 million tons. During the same period, yields of corn (maize), rice, barley and other grains averaged 451, 331, 178 and 168 million tons, respectively.[1] World wheat production grew steadily from 1960 to 1990 averaging 273, 378 and 494 million tons per year per decade during the 1960s, 1970s and 1980s, respectively. The 150% increase in production was made possible by the introduction of high-yielding, fertilizer-responsive wheat varieties and better cultural and management practices, despite an increase of only 14% in the world's area of wheat harvested.

During the period 1992–1995 world wheat production averaged 544 million tons/year, while the corn production figure was 514 million tons/year.[18] However, in

2004–2007 world corn production increased to 721 million tons (see Chapters 2 and 9) compared to 611 million tons of wheat[19] (Table 10.1). World production of other cereals in 2004–2007 was rice 415 million tons, barley 140 million tons, sorghum 59 million tons, oat 25 million tons and rye 15 million tons.[20] The sharp increase in corn production is attributable to its recent use for fuel ethanol. Currently three-fourths of world wheat production occurs in Europe, Far East Asia, and North and Central America.

Utilization of wheat can be divided into four categories: food, feed, seed and industrial.[1] Food is the major use for wheat, accounting for 67% of total consumption, while feed and seed utilization represent 20% and 7%, respectively. Industrial uses of wheat, which include wet-processing into starch and gluten, consume about 6% of the production.

Wheat starch and wheat gluten are important and valuable co-products of the wet-processing of wheat flour.[21–26] These two products are entering a new development phase, mainly due to the preponderance of value-added ('green') products[27–39] that are currently being commercialized because of their sustainability. In the wheat starch market,[40] the products available for food and industrial uses are large-granule starch, low-moisture starch, resistant starch, and cook-up and pregelatinized versions of unmodified starch and those starches chemically modified according to the provisions of the US Code of Federal Regulations (Title 21 CFR Part 172.892) or similar regulations in other countries and regions. Value-added products from gluten consist of wheat protein isolates and texturized, fractionated, deamidated, reduced, complexed, quaternized and hydrolyzed wheat gluten with applications in food, feed, paper and cosmetic industries.[26,29,30,40]

In 1993, the European Union (EU) began reforming its Common Agricultural Policy, which resulted in a decline in their floor price for wheat. The EU also set the floor price of corn at a level higher than that of wheat, and put a production quota on potato starch, thereby making wheat the most profitable raw material for manufacturing starch. As a result, EU manufacturers embarked on an expansion of wheat starch

Table 10.1 Estimated world wheat production in million tons[19]

Country or region	2004	2005	2006	2007
European Union	146.9	132.4	124.8	120.9
China	92.0	97.5	104.5	106.0
India	72.2	68.6	69.4	74.9
United States	58.7	57.3	49.3	56.3
Russian Federation	45.4	47.7	44.9	48.0
Pakistan	19.5	21.6	21.7	23.0
Canada	24.8	25.8	25.3	20.6
Australia	21.9	25.4	9.9	13.0
Turkey	18.5	18.5	17.5	15.5
Argentina	16.0	14.5	15.2	15.5
Iran	14.5	14.5	14.8	15.0
Others	96.5	97.9	96.3	94.6
World	626.9	621.7	593.6	603.3

production to where, in 2005, 36% of the 9.6 million tons of starch produced in the EU was from wheat. Corn starch production was 46% and potato starch 18%.[41]

Estimated annual world wheat starch production was 2.11 million tons in 1993 and increased to 2.36 million tons in 1996.[28] Because of significant expansion in manufacturing capacity in several countries, wheat starch production rose to 4.67 million tons in 2007 (Table 10.2).[42] However, this production figure for wheat starch represents only a small fraction of the total world starch production of 60 million tons.[41] By raw materials, wheat represents only 8% of world starch production, while potato, cassava and other crops contribute 4%, 14% and 1%, respectively; corn dominates at 73%. Approximately 52% of world starch is produced in the USA, 17% in the EU and 31% in other places.[43] Wheat starch (<0.5% protein) is commercially produced in about 59 manufacturing plants in some 30 countries. The top 10 producers are France, the United States, Germany, The Netherlands, Australia, the United Kingdom, Belgium, Canada, Japan and China.

III. Industrial Processes for Wheat Starch Production

Before the 1980s, mostly bread wheats were cited as suitable for commercial production of wheat starch and wheat gluten, but because of new separation technologies and economic factors, hard, soft, spring and winter wheats are now also being processed.[3–9,21,44–61] The usual practice is to dry-mill the wheat to separate bran and germ from the endosperm, which is ground into flour. The flour can be a 72% extraction or higher if a short-milling process is employed,[52,53] and no malt, enrichment or bleaching agent is added to the flour. Straight-grade or second-clears flour may be blended in small proportions with no apparent adverse effects on the separation, yield or quality of the starch and gluten.[24] A good quality wheat flour suitable for commercial wet processing has high protein (>11%), minimal starch damage, low ash (low bran) and no alpha-amylase (high Falling Number), and is low priced. Additionally, the gluten proteins in the flour should agglomerate readily in excess water, thereby forming protein-rich particles.[21,45–47] Potable water of desired hardness is the other

Table 10.2 Estimated world wheat starch production

Year	Production,[a] millions of tons
1993	2.12
1996	2.36
2003	2.52
2004	3.12
2005	3.94
2006	4.24
2007	4.67
Total	22.96

[a]Production calculated by multiplying wheat gluten production by 5.25.[42] Gluten data from International Wheat Gluten Association

major raw material for industrial isolation of gluten and starch. No chemicals such as sulfurous acid (sulfur dioxide), sodium sulfite, alkali or organic acids are used in the process. Optionally, enzymes with hemicellulase/pentosanase activities may be added to improve starch and gluten separation, reduce processing time, increase product purity, and increase gluten and starch yield.[21,62–65] The amount of xylanase required to improve the processing of wheat flour is impacted by the level of xylanase and xylanase inhibitors present in the flour.[66–68]

In designing and managing a wheat starch process, Barr[52] addressed four major issues to consider: raw materials; products; cost; and operability. Production conditions frequently require high-protein wheat flour, and the process must be flexible to changes in raw materials from various vendors without affecting starch and gluten. The qualities and uses of the products, namely, A-starch,[i] wheat gluten, B-starch[i] and by-products, require serious consideration. Finally, yield, water consumption and effluent disposal demand careful operation of the plant. In contrast to the corn wet-milling process, currently existing wheat wet-milling processes do not involve steeping in warm water over an extended period of time. Hence, unlike corn starch, commercially-produced wheat starch is not an annealed starch.[69]

In comparison to isolation of starches from other botanical sources, the industrial isolation of wheat starch is quite unique because different processing techniques may be employed.[21,45,70] Kempf and Röhrmann[54] described 15 different processes for industrial production of wheat starch and wheat gluten using wheat kernels or wheat flour as the raw material (Table 10.3). Only three of the processes in Table 10.3 are actually used.

Table 10.3 Processes for industrial production of wheat starch and wheat gluten[a]

Process	Raw material	Separation	Gluten quality
Alsatin	Wheat kernels	Mechanical	Vital
Halle	Wheat kernels	Fermentative	Non-vital
Alsace	Wheat kernels	Mechanical	Non-vital
Martin	Wheat flour	Mechanical	Vital
Fesca	Wheat flour	Mechanical	Vital
Batter	Wheat flour	Mechanical	Vital
Dimler	Wheat flour	Chemical	Non-vital
Longford-Slotter	Wheat kernels	Chemical	Non-vital
Phillips-Salans	Wheat flour	Chemical	Non-vital
Alfa-Laval/Raisio	Wheat flour	Mechanical	Vital
Verberne-Zwitzerloot	Wheat flour	Mechanical	Vital
Weipro	Wheat flour	Mechanical	Vital
Far-Mar-Co	Wheat kernels	Chemical	Vital
Pillsbury	Wheat kernels	Chemical	Non-vital
Tenstar	Wheat flour	Mechanical	Vital

[a]Adapted from reference 54.

---

[i]A-starch, the prime starch, contains the large, lenticular granules and some of the small, more spherical, granules (see Section 10.4.1). The B-starch fraction, also called tailings starch or squeegee starch, is composed of small granules, damaged granules, nonstarch polysaccharides and low levels of protein. Up to 20% of the total starch can be in the lower-value B-starch fraction, and processes are being developed to purify the B-starch for use in cosmetics and paper coating, among others.

According to Schofield and Booth,[71] there are six general techniques of wheat starch and gluten separation: dough; batter; aqueous dispersion; chemical dispersion; wet-milling of kernels; and non-aqueous separation. The last three techniques[48,54,72–74] are not used because of poor product quality, high operating cost, effluent problems and poor efficiency. The processes used commercially all begin with wheat flour and water, and are distinguished largely by the degree of gluten agglomeration in the initial mixing step.[21,45] They are named the dough, dough–batter and batter processes. Several other processes begin with an aqueous dispersion of flour with initial mixing done at different solids content, shear and temperature. Factors that complicate the isolation of wheat starch compared to corn starch include the presence of viscous arabinoxylans, small granules (<10 μm), damaged granules and some water-soluble protein. Wheat flour contains 2–7% cell walls that are comprised mostly of arabinoxylans, along with low levels of mixed-linkage β-glucan and traces of arabinogalactan and cellulose.[75] Approximately 25% of the arabinoxylans are water-extractable and they account for ~95% of the viscosity of water extracts of flour. Laboratory tests that mimic commercial processes have been reviewed,[45–47] and a method to isolate starch from half a wheat kernel is known.[76] In this chapter, the various commercial processes are discussed chronologically.

1. Conventional Processes

Prior to the mid 1970s, two processes dominated the production of wheat starch and vital wheat gluten: the Martin process, based on separating the starch from gluten by washing a stiff dough, and the batter process, based on separation of gluten and starch from a thin flour batter.[53]

Fundamentally, the more than 170-year-old 'dough ball' or Martin process involves mixing flour (10 parts) and water (6–7 parts) to make a 'developed' dough that contains a continuous matrix of intermeshed long-gluten strands. The dough is then washed in a variety of mixing troughs with very large amounts of water to get purified gluten and diluted starch milk, which is then concentrated and separated into A- and B-starch. Typically, the process uses at least 10–15 parts of fresh water per part of flour, which complicates starch recovery and results in high levels of dilute effluent.[52,53] Several options are available to handle the effluents. Among them are anaerobic/aerobic digestion, spray irrigation, concentration by evaporation and drying for animal feed, fermentation for ethanol production, and discharge to a municipal sewer system.[56,58] Modifications and/or improvements to the Martin process progressed somewhat slowly until the twentieth century, when other starch and gluten separation technologies were developed, patented and/or commercialized.[24,48–61] A modified Martin process is described by Maningat and Bassi[24] in which the process flow-chart incorporates the principles of hazard analysis and critical control points. Typical yields of starch (<1% protein) and gluten (>75% protein) are 45–60% and 10–15% of flour weight, respectively.

The batter process was devised independently in 1944 in Canada by Shewfelt and Adams[77] and in the US by Hilbert et al.[78] Wheat flour (1 part) and water (1.8 parts) are mixed at 43°C into a smooth batter that contains millimeter-long gluten strands.[45] Using a relatively warm temperature and stirring the batter with more water causes rapid agglomeration of the gluten strands into curds (7–13% of flour) that can be removed from the starch slurry (milk) by screening. The starch slurry is further refined through a series of screens, sieves and centrifuges into prime starch (68–77%), low-grade starch and water solubles. Improvements in the Martin and batter processes by recycling effluent water reduced fresh water consumption to 5–7 parts of water per part of flour.[53]

In the early 1970s, Fellers et al.[79–81] modified and scaled-up the old but poorly defined Fesca process developed in 1928 by Eynon and Lane.[82] The Fesca process involves centrifugal separation of starch from a pumpable batter, leaving a fluid protein concentrate that contains ~22% protein and ~67% starch after drying. In forming a smooth batter from hard wheat flour (1 part) and water (1.6–1.7 parts) or soft wheat flour (1 part) and water (1.3–1.5 parts), Fellers and his colleagues[79,80] found no rapid gluten agglomeration at pH 7–8 if the temperature was maintained near 25–30°C during blending and high shearing. Adjustment of the pH of the flour slurry from 5.9 to 7.5–8.0 using sodium or ammonium hydroxide reduced its consistency at 25–30°C so that centrifugation at an optimized speed in laboratory tests gave essentially two layers. The bottom layer was almost pure starch, which accounted for 70% of the starch in the flour. The supernatant layer was a concentrated protein phase, and there were little or no tailings fraction on top of the starch. Furthermore, the protein concentrate possessed gluten functionality that matched hand-washed, freeze dried gluten. The gluten could be easily coagulated and isolated when water was mixed with the protein concentrate.[80]

The Alfa-Laval/Raisio process appears to have been developed from the modified Fesca process, principally because the process was recognized to require relatively low amounts of water, i.e. 3 parts of fresh water per part of flour. A flour–water dispersion with ~30 weight percent of solids is mixed at high shear to produce a batter at ~30°C.[21,45,83] The batter is separated by a decanter centrifuge into a bottom starch fraction with about 1% protein, and a light gluten fraction with about 40% protein. The starch is purified using rotating conical screens and decanters, and the gluten is matured, agglomerated in a disk-type disintegrator, screened, dewatered and dried. Filtrates coming from gluten screens are rich in B-starch plus solubles and small amounts of A-starch. The B-starch and solubles are recovered using a decanter, and the A-starch fraction is recycled for further refining. The B-starch is concentrated in a nozzle or solids-ejecting centrifuge and further dewatered in a decanter centrifuge to 35–40% solids prior to drying.

Processes in Table 10.3, starting with wheat kernels as raw material, avoid the cost of dry-milling, but require steeping of the grain.[51,53–55] Wet-milling of whole wheat has the inherent advantages of starting with the entire endosperm region and without any damaged starch. However, there are often drawbacks. The isolated starch usually has low whiteness and excessive bran contamination, and the process yields a high volume of effluent with a low concentration of solids and possibly added acid. Another disadvantage is that some agglomeration of gluten occurs before all the bran can be removed from the endosperm, so the gluten becomes contaminated with excessive bran and is of poor quality, as indicated by color, fiber, ash and fat content.[53]

Although some initial successes were reported, the whole-wheat processes were short-lived, principally because the separation of fiber from gluten is difficult, and because drying of wet bran is energy-intensive. Batey[84] opined that it seems unlikely that any wet-milling of wheat kernels will be economical, since only 6–7% of the kernels' starch is lost in dry-milling.

While the major aim in industrial wheat starch production is to produce a refined grade of A-starch, the production of a purified B-starch may also have commercial significance because of its unique uses, as described later in this chapter. In Europe, a new process was developed to separate B-starch into two fractions: a high-purity, small granular starch and a feed fraction.[53] The process involves enzyme treatment followed by high-pressure treatment and purification on fine screens, separators and decanters. Large and small wheat starch granules are marketed in Japan.[85]

The majority of isolated wheat starch is dried by conventional spray drying or flash drying technology to yield a 'cook-up' starch. The most rapid drying can be achieved by flash drying where starch is dried in a stream of hot air and collected in dust cyclones. Instant or pregelatinized wheat starch is produced by the traditional drum drying method.[86] Jane[87] described several technologies to produce cold-water-swelling starch, including gelatinization with steam in two fluid nozzles and then spray drying,[88] alcoholic alkali treatment,[89] thermal treatment in aqueous alcohol solution under pressure[90] and treatment with polyhydric alcohols at atmospheric pressure.[91]

2. Hydrocyclone Process (Dough–Batter)

Two basic technological advances characterize the latest developments in the production of wheat gluten and wheat starch from wet-processing of flour: initiation of agglomeration of gluten with high-shear homogenizers and centrifugal separation of a gluten-rich fraction from starch using hydrocyclones or decanters.[55] In the early 1980s, the hydrocyclone process was the principal process installed throughout the wheat industry worldwide, probably because the equipment needed is compact and relatively inexpensive.[52] Other benefits include reduced water usage, absence of moving parts and a wide range of operating parameters.

A hydrocyclone is a cone-shaped separation chamber with an entry port at the top (wide) end and two outlets. One outlet is an overflow port at the top for discharging the light stream, and the other at the apex (bottom) of the cone is a discharge port for the denser stream[92,93] (see Chapter 9). Hydrocyclones are widely used in corn wet-milling (Chapter 9) to fractionate particles of different compositions when those particles suspended in water have different settling rates. The hydrocyclone process presented in Figure 10.1 starts with a high-extraction wheat flour typically produced by a short flow milling process.[52] The wheat flour is conveyed into a dough kneader where it is mixed with warm water to form a slack dough. After maturing for 10–12 minutes, the dough is transferred into a tank where it is vigorously agitated with water to form gluten threads contaminated with a few lumps. The whole slurry is then passed through a strainer and pumped to a set of hydrocyclones, where the gluten spontaneously agglomerates in the light stream leaving the cone. Agglomeration of the gluten strands occurs in the exit port because of increased protein concentration and intense mixing. The agglomerated gluten in the overflow fraction is isolated by screening, washed to remove B-starch, bran and cell-wall materials, and then dewatered prior to drying.[52,60] A set of hydrocyclones is used to separate, wash and concentrate the A-starch in the underflow fraction in a countercurrent fashion. Further purification of A-starch from fiber involves static and rotary screening. The remaining starch slurry separated in the overflow fraction of the set of hydrocyclones undergoes screening to remove fine fibers, and is then separated into B-starch and a second A-starch fraction. Wet coarse and fine fibers are removed as a separate stream. The remaining solubles constitute the effluent stream. Pentosans and fiber adversely affect the operation of hydrocyclones by increasing the viscosity of the slurry being subjected to the shear forces in the liquid vortex.[21]

The hydrocyclone process is applicable for processing hard and soft wheat flours, and perhaps even wholemeal flour.[92] In addition to its versatility in processing different wheat classes, other advantages of the hydrocyclone process compared to the Martin process include an increased yield of gluten, lower water and energy requirements, and manageable levels of effluent. To ensure the food safety of wheat starch, quality programs in many companies dictate the implementation of hazard analysis and critical control points in the hydrocyclone process.[24]

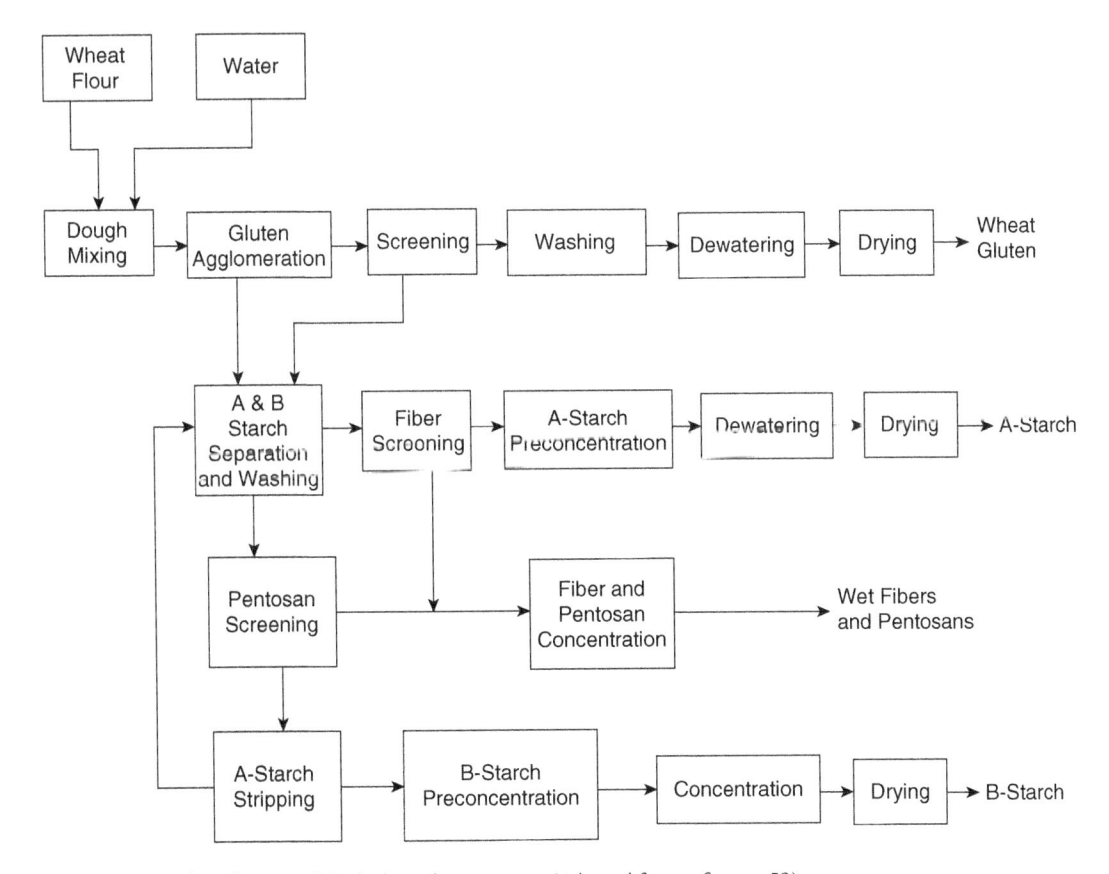

Figure 10.1 Flow diagram of the hydrocyclone process. (Adapted from reference 52)

3. High-pressure Disintegration Process

The latest technological developments in wheat starch production are the high-pressure disintegration (HD)[53,94–97] and the Tricanter®[97] processes. The HD process, which is based on a highly sheared batter and on centrifugal forces for separation, was jointly developed by the Technical University of Berlin and Westfalia Separator.

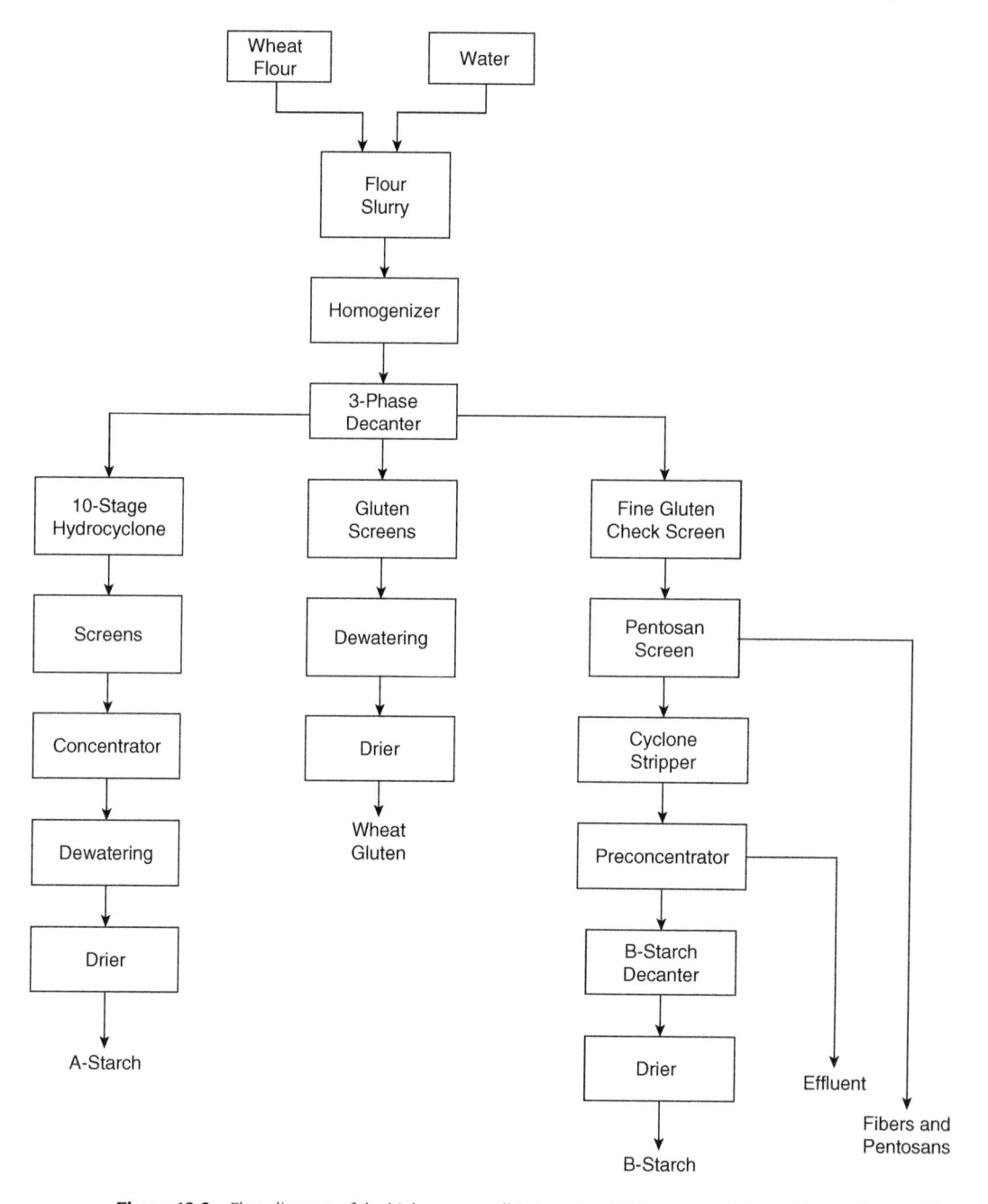

Figure 10.2 Flow diagram of the high-pressure disintegration (HD) process. (Adapted from reference 53)

Its use was first reported in 1985 by Meuser et al.[98] for extracting starch from corn. The HD process for commercial production of wheat starch and wheat gluten is used in many countries in Europe and in Australia, Canada, Mexico, India and Venezuela.

In the HD process (Figure 10.2), and in the Decanter® process, wheat flour (1 part) is mixed with water (0.85–0.95 parts) at 35°C in a continuous dough mixer to achieve a smooth batter.[53,94–97] The batter is conveyed into a high-pressure pump operating at pressures of up to 100 bar (10^5 kPa), which forces the batter through a homogenizing valve. The high shear forces release starch granules from the hydrated endosperm particles and cause the gluten threads to aggregate into micrometer-sized particles. A high shearing force is important to achieve high yields of high-quality starch and gluten with low freshwater usage in the process.

The sheared batter from the high-pressure pump is diluted with water to ~30% solids and fed into a three-phase decanter centrifuge designed to rapidly: (a) separate and concentrate the A-starch from the slurry and discharge the A-starch from the decanter; (b) separate the B-starch and the gluten in the medium density stream; and (c) separate the viscous pentosans and other solubles in the least dense, top-flow fraction. Removing the soluble pentosans, mainly arabinoxylans, enhances subsequent agglomeration of gluten in the medium-dense stream. That stream is pumped into a sieve drum where B-starch is washed from the gluten. Fine gluten particles still remaining with the B-starch fraction and the pentosan fraction are recovered using vibrating screens.

The A-starch slurry from the centrifugal decanter is screened and then refined either with hydrocyclones or with separators and decanters. Purification and concentration of A-starch is accomplished in multistage hydrocyclones, or the A-starch slurry is separated into A and B fractions in a nozzle-type centrifuge and the A-starch fraction is finally refined in a centrifugal decanter. The B-starch stream is passed through vibrating screens and concentrated in a decanter. Pentosans and other solubles are concentrated and either dried or co-fermented with the B-starch for ethanol production.

Two major advantages of the HD process are reduction in water consumption to as low as three parts per part of wheat flour, and a 10% increase in A-starch yield. In addition, this process is not as sensitive to flour quality as the other processes, and is also applicable for potato, bean, pea and corn starch production.

IV. Properties of Wheat Starch and Wheat Starch Amylose and Amylopectin

Starch comprises 54–72% of the dry weight of wheat kernels.[99,100] Starch content is positively associated with grain yield, but inversely related to protein content.[101–103] Growing environment[103] caused less change in starch concentration than in protein concentration among Canadian spring wheat cultivars. Among 21 Australian wheat varieties covering wide ranges of wheat hardness and flour protein content, the yield of extracted starch had a significant negative correlation with flour protein, whereas the yield of extracted gluten was highly correlated with flour protein.[102] Starch isolated from different US wheat classes showed that granule size, swelling power and solubility decreased in the order soft red winter > hard red winter > hard red spring > durum.[104] Conversely, amylose and phosphorus contents and the intensity of the DSC endotherm for the amylose–lipid complex increased in the order soft red winter < hard red winter < hard red spring < durum. Amylose content, both total and lipid-free, and gelatinization temperature increased with the growing temperature of the wheat.[105,106] Starch properties of club and soft winter wheats are strongly influenced by varietal differences and growing location.[107] Among soft wheats, higher flour yields during milling were produced from cultivars that had higher starch contents and larger starch granules.[108] Large-granule wheat starches were isolated from nine soft, eleven medium–hard and ten hard wheats grown in east China.[109] The starches differed in average granule size, apparent amylose content, gelatinization temperature, and enthalpy and pasting parameters. The properties and functionality of starches, in general, are discussed in a number of reviews and monographs.[110–125]

1. Large Versus Small Granules

Endosperm starch is synthesized in a specialized plastid called an amyloplast[126–130] (see Chapter 5). In mature wheat, two distinct classes of granules occur that differ in size and shape, i.e. large, lenticular (A-type) and small, spherical (B-type) granules. A size–distribution curve of wheat starch which has been isolated quantitatively is bimodal when plotted in differential form of volume (mass) versus granule diameter. The curve shows a minimum at $10\,\mu m$ by the electrozone method (Coulter Counter)[131–133] and laser diffraction technology,[134–135] but that minimum is observed at $15–16\,\mu m$ by light microscopy[136] and by microscopic image analysis.[135] In the electrozone and laser technology methods, the volumes of both the large and small granules of wheat starch are modeled as spheres,[134,135,137] but in image analysis, the small ($<5\,\mu m$) granules are modeled as spheres and the medium ($5–15\,\mu m$) and large ($>15\,\mu m$) granules as oblate spheroids with a thickness of $5.0\,\mu m$ at maturity.[126,135] Granule size distributions of four wheat starches by laser diffraction gave peak diameters for the large ($17.0–20.2\,\mu m$) and small ($3.9–4.2\,\mu m$) granules that were 40–50% smaller than the values determined by image analysis ($28.5–34.0\,\mu m$ and $8.2–10.0\,\mu m$).[135] Quantitative separations of wheat starch granules followed by weighing indicated that wheat starch contains approximately 30% by weight of B granules.[136,138–141] A range of 23–50 weight percent of B granules was found in 130 lines of Australian wheat, with the lowest levels in soft wheat;[134] seven European wheats contained 20–22% of B granules.[142] Preparative separation of A and B granules has been accomplished by microsieving,[133,143] density-gradient centrifugation,[143,144] gravity-sedimentation in a column of water[136,140,145,146] and centrifugal sedimentation in water.[147] The sharpest separation is by gradient centrifugation[144] or microsieving.[143] The contrasting biosynthetic, physical and chemical properties of A and B granules[126–129,131–133,137,140,141,142,144,148–159] are summarized in Table 10.4. Two starch granule-bound proteins (SGP), SGP-140 and SGP-145, appear to be preferentially associated with the development of A granules in endosperm tissue.[159]

Table 10.4 Composition and properties of large and small wheat starch granules[126-129,131-133,137,140,141,144,148-159]

Property	Large (A) granules	Small (B) granules
Period of synthesis	First week after anthesis	2-3 weeks after anthesis
Granule development	Asymmetric	Symmetric
Granule-bound proteins	60, 80, 92, 100, 108, 115, 140, 145 kD	60, 80, 92, 100, 108 kD
Shape	Lenticular	Spherical
Surface pores and channels	Present	Inner voids
Amylase susceptibility	Less susceptible	More susceptible
Mode of digestion by alpha-amylase	Pitting and erosion of core; localized in certain granules	Surface erosion; localized in certain granules
Swelling pattern	Preferentially in the equatorial plane	All directions
Density at 12% and 43% moisture	1.57 and 1.35 g/cm^3	1.57 and 1.35 g/cm^3
Swelling power (90°C) (g/g)	10	12
Swelling factor (70°C) (v/v)	7.1 ± 0.2	5.7 ± 0.1
X-ray pattern, dominant and traces	A-type, V- and B-type	A-type, V- and B-type
X-ray crystallinity (%)	32-36	32-36
Crystallite shape and size (nm)	Spherical and 10	Spherical and 10
Mean volume (μm3)	1824 ± 221	56.6 ± 11.6
Mean diameter (sphere) (μm)	14.1 ± 0.6	4.12 ± 0.28
Modal diameter (μm)	15-19	5-7
Percentage by number	3-10	90-97
Percentage by volume	65-80	20-35
Specific surface area (m^2/g)	0.265 ± 0.011	0.788 ± 0.058
Amylose content (%)	Apparent 34.0 Absolute 30.9	Apparent 27.0 Absolute 25.5
Lysophospholipid content (%)	0.845 ± 0.065	1.062 ± 0.090
Free fatty acid content (%)	0.084 ± 0.025	0.162 ± 0.050
Enthalpy of gelatinization (J/g)	10-14	9-13
Onset temperature of gelatinization (°C)	52-62	51-58
Peak temperature of gelatinization (°C)	56-64	58-65
Conclusion temperature of gelatinization (°C)	61-70	64-72
Enthalpy of gelatinization of amylose-lipid complex (J/g)	0.9	1.3-4.2

During growth of endosperm tissue in a hard wheat, a single population of granules, destined to become the large A-type granules (>16 μm in diameter by light microscopy) is initiated four days after anthesis with one granule per amyloplast.[126,128] A second burst of starch synthesis occurs about 10 days after anthesis and results in B-type granules (5–16 μm in diameter). At 21 days after anthesis, a third burst of synthesis forms C-type granules (diameter < 5 μm at endosperm maturity). The C-type granules in wheat starch increase starch surface area (8%), but have less impact on mass (<3%). Soft, as well as hard, wheats show trimodal granule size distributions at maturity.[137] Baruch et al.[158] reported a bimodal deposition of wheat starch granules with the A-type initiated approximately 4–12 days after anthesis, and the B-type 16–22 days after anthesis. Biosynthesis of small granules occurs either in narrow protrusions that project from A-type amyloplasts or in the stroma of the parent A-type amyloplasts in association with the peripheral groove.[129,151] Confocal laser-scanning micrographs of living cells 13 days after anthesis showed protrusions (l = 2–30 μm and d = 0.5–1.5 μm) on all amyloplasts. The protrusions, which entrained several to many small granules, also interconnected the amyloplasts, suggesting coordinated plastid activity.[160]

Size distributions of starch granules from mature wheat have been measured by electrozone,[131,132,140,143,148,156] laser diffraction,[137,142,161] centrifugal sedimentation,[144] polarization microscopy[162] and field-flow fractionation;[163,164] most data suggest a bimodal size distribution. This bimodal distribution, which is also characteristic of related starches from barley, rye and triticale, is commonly accepted among starch scientists. However, based on the timing of biosynthesis, the distribution is trimodal as shown by transmission electron microscopy and image analysis techniques[126,135] and by mathematical modeling of granule size distribution curves.[165]

Among 59 wheat varieties, variation in granule size distribution appears to be largely under genetic control, although edaphic factors, such as drought and cultural practices, may decrease the proportion of mature A granules with diameters > 20 μm.[132] However, environments with a high level of accumulated temperature above 30°C during anthesis produced grain with a high proportion of A granules.[166] This temperature effect is consistent with the earlier findings of Blumenthal et al.,[167,168] in which a heat shock of 35°C during grain development reduced B granule content. Using electrozone separation, other researchers have also found varietal differences and soil and climatic effects on wheat starch granule dimensions.[156] Raeker et al.[137] reported that both cultivar and environmental factors affect the granule size distribution of soft wheats.

A granules, which develop asymmetrically, increase in number and size during the early stages of development when wheat endosperm cells are dividing. After cell division, their number remains constant, but their size continues to increase throughout the grain-filling (cell-growth) period.[129,148] A granules are built from an initial spherical nucleus by forming a thin plate in the equatorial plane. Then, the plate is enlarged mostly above and below the plane until the granule assumes the familiar lenticular or disk shape seen in scanning electron micrographs of A-type granules.[100,169,170] The disk shape of the large wheat starch granules is shared by mature starch granules from barley, rye, triticale, ginger, shoti and dieffenbachia.[171] Small (B) wheat starch granules appear to develop symmetrically, and they increase in number until they account for over 90% by number of total starch granules at maturity, but only 20–35% by volume (Table 10.4). A granules typically have an equatorial groove. Deposition of new starch molecules by apposition occurs at the face of the granule[148] and appears to proceed in a diurnal manner,[157] producing the characteristic alternating layers ('growth rings'). In contrast to cereal starches, growth rings in potato starch are attributed to a complex interplay of circadian rhythms, physical mechanisms and perhaps, diurnal rhythms.[172]

The surface of mature, undamaged wheat starch granules appears to be smooth when viewed with a scanning electron microscope.[169,170] Pores and channels are apparent, especially along the equatorial groove of the A granules.[173–175] B granules do not develop an equatorial groove,[157] but they contain inner voids that penetrate to the surface.[175] Practically all starch granules show a small void at their hilum. In corn and sorghum starch granules, surface pores were discovered to be openings to channels that penetrate radially into the hilum.[174,176,177] Pores and channels in starch granules appear to be related to the accessibility of reagents to starch granules, and may affect reactivity when starch is subjected to enzymic and chemical modification.[141] Raw wheat and corn starches are slowly but completely digested by α-amylase, whereas potato starch, which lacks granule pores, and which also gives a B- instead of an A-crystal pattern, is initially much more resistant to α-amylase.[178,179]

Indentations on wheat starch granules are evidence of the tight packing of A and B granules in zones of the wheat endosperm.[157] Using a combination of low-voltage scanning electron microscopy and atomic force microscopy, Baldwin et al.[180] showed that the generally smooth surface of wheat starch granules consists of a mass of regularly spaced structures of 20–50nm in diameter. Maize and potato granules have similar sized structures of 20–50nm height;[181] those on rice starch granules are about 100nm in height.[182] These nodules on a granule's surface are believed to be the ends of amylopectin clusters and the source of soluble carbohydrate released on digestion of the granule surface by isoamylase[183,184] and by mechanical damage.[185,186] The existence of nodules and regularly-spaced structures on the wheat granule surface lends support to the 'hairy billiard ball' model of the starch granule proposed by Lineback[187,188] (see Chapter 6) and the 'blocklet' concept proposed by Gallant et al.[189,190] (see Chapter 5).

Digestion of undamaged A granules by alpha-amylase leads to pitting, because the enzyme penetrates the granules, where it preferentially digests the internal architecture of the amorphous region.[149,178] Pitting preferentially occurs in the equatorial groove where surface pores of about 1000Å in size are found to be prevalent.[173,174] Surface pores and channels in B granules were visualized by scanning electron and confocal laser scanning microscopy.[175] B granules are digested more rapidly by alpha-amylase than are A granules,[141,191] although not because of their small size but possibly because the B granules contain less amylose (Table 10.4).

Large (A) and small (B) wheat starch granules differ in composition, properties, reactivity and uses. A granules contain somewhat more amylose[137,141,144,145,147] and more internal lipid.[137,140,143,192] Starch deposited on the granules as they grow and increase in size contains more amylose and lysophospholipid.[148] The model of a mature A granule is depicted to have a low-amylose, low-lipid central region.[148] This model could explain why A granules, when heated in excess water, swell preferentially at their equatorial groove and assume a puckered or convoluted shape.[193–194]

The amylose of native wheat starch granules consists of lipid-complexed amylose[ii] and lipid-free amylose.[195] Lipid-free amylose and lipid-complexed amylose represent

---

[ii]Iodometric assay of amylose on cereal starches is most often done on lipid-extracted starch, and the amylose content from that assay is termed 'apparent amylose'[145] or 'total amylose.'[123] When the iodometric assay is performed on a lipid-extracted starch and the amylose content is corrected for interference from extra-long chains on amylopectin, the assay gives 'absolute amylose' or 'real amylose.'[145] Iodometric assay of starch that has not been lipid-extracted gives 'lipid-free' (uncomplexed) amylose. Subtraction of 'total amylose' from 'lipid-free' amylose gives 'lipid-complexed' amylose.[123] The use of the term 'apparent amylose' to designate 'lipid-free' amylose is avoided in this chapter. All amylose percentages are calculated based on starch.

21.5–25.9% and 5.0–7.1% of the granule, respectively, with the latter amounting to 18–22% of total amylose. Lipids inside granules amount to 0.7–1.4%;[149] 89–94% of these lipids are lysophospholipids comprised of 70% lysophosphatidylcholine, 20% lysophosphatidylethanolamine and 10% lysophosphatidylglycerol. Fatty acid analysis indicates 35–44% 16:0, <2% 18:0, 1–14% 18:1, 44–52% 18:2 and 1–4% 18:3, or predominantly palmitic and linoleic acids. A and B granules from 23 bread wheats contained 29.2% and 27.4% total (apparent) amylose, respectively, of which 5.5% and 7.5%, respectively, was lipid-complexed amylose.[140] Total lipids in the large and small granules were 0.7–0.8% and 0.9–1.4%, respectively.[192]

Amylopectin was isolated and purified from A and B granules, and was debranched with isoamylase.[145] The chain-length distribution analysis (DP 6–74) showed that B granules contained a higher weight percent of A + B_1 chains (DP 6–12) and a lesser amount of long B chains (DP 37–74), as compared to A granules. In agreement with those findings, Liu et al.[147] found that B granules had a higher weight percent of chains with DP 6–12 and a lower level with DP 13–54. Sahlstrom et al.[142] reported that B granules have a higher weight percent of chains with DP 3–7 and a lower weight percent of chains with DP 24–30, and Vermeylen et al.[146] reported that B granules have a higher mol percent of chains with DP 8–24 and lower mol percent of DP >24 chains. Crystallinity of B granules determined by wide-angle x-ray diffraction was somewhat higher (36% versus 32%) than in A granules,[145,152] although the opposite was found by Vermeylen et al.,[146] probably because of a different method of calculating crystallinity. In excess water, the gelatinization endotherm for B granules was at a somewhat higher temperature or occurred over a broader temperature range[140,142,144,146,147,152,154] and the gelatinization enthalpy was often 2–3 J/g lower for B granules,[140,144,154] but sometimes equal to that (10–14 J/g) of the A granules.[145,146] The amylose–lipid melting endotherm of B versus A granules had a higher enthalpy by 1–2 J/g,[142,145,146,147,154] which is explained by B granules containing a higher level of lysophospholipids.[192]

Small angle x-ray scattering at 42% moisture indicated that B granules possess a denser crystalline lamella,[146] which is in agreement with the B granules having an increased level of A + B_1 chains.[145] Granule densities (1.55–1.59 g/cm^3) of six wheat starches at 10% moisture failed to correlate with their proportions (79–93%) by number of small granules (<12 μm).[152] A density of 1.50 g/cm^3 has been used[153] to calculate the mass of A and B granules from volume data at ~10–13% moisture;[133,192] their hydrated (43% moisture) density was taken at 1.35 g/cm^3.[148]

Pasting curves of A and B granules were different when recorded on a Rapid ViscoAnalyzer at 8–15% starch solids.[142,143,145,147] For B granules, the pasting temperature was 3–7°C higher compared to that of A granules, whereas the peak, breakdown, setback and final viscosities were lower. The reduced viscosity generated by B granules is attributed to their elevated level of lipid-complexed amylose, which suppresses granule swelling. The higher setback and final viscosities of A granule pastes are explained by the elevated amylose in A granules, and to the increased long chains in their amylopectin molecules. The size difference between A and B granules was discounted as a factor in their viscosity differences during pasting at 8% starch solids.[145] Above ~90°C, small granular wheat starch shows higher swelling power (14 versus 12 g/g) than the large granules.[196] The amylose–lipid complex is elevated in the small granules, and when that complex begins to dissociate above ~85°C, lysophosphatidyl choline with its quaternary amine group is released into the medium.[197] Gray and Schoch[198] demonstrated that addition of a cationic surfactant, dodecyltrimethyl ammonium chloride, to a slurry of corn or sorghum starch increased its swelling and 'thickening' power, especially at temperatures above 70°C.

The proportion of small wheat starch granules is of particular importance to starch yield in industrial manufacturing, because those granules are difficult to purify and recover on a large scale.[53] The proportion and thermal properties of A and B wheat starch granules have been reported to influence dough properties, bread volume, and crumb grain and softness, although the optimum proportions of A and B granules has varied between laboratories.[142,199,200] Hydroxypropylation of A and B granules with 10% propylene oxide (starch basis) at alkaline pH gave modified A and B granules with hydroxypropyl levels of 4.7 and 3.8%, respectively.[201] Acetylation with 8% acetic anhydride at pH 8 also gave ~10% more acetyl groups on A granules, but crosslinking with 0.025% phosphoryl chloride followed by acetylation with 8% acetic anhydride gave more acetyl groups (1.73%) on B compared with that (1.61%) on A granules.[85] Crosslinking of wheat starch with phosphoryl chloride (0.4–1.2% starch basis), followed by fractionation of the modified granules, showed that the B granules reacted with ~10% more reagent than the A granules,[141] but others found no difference.[85] Reacting wheat starch with 2–4% propylene oxide, and then separating the A from the B granules, showed no difference in the extent of hydroxypropylation with granule size for maize, potato and wheat starches.[202] Economically, if the small granules of wheat starch can be purified, they may be used as fat mimetics and flavor carriers and in cosmetic powders, biodegradable films and paper coatings.[144,161,203] Purified A granules of wheat starch have been used for years as 'stilt particles' in carbonless copy paper.[204] Those 'stilts' prevent inadvertent crushing of the microencapsulated ink spheres prior to use of the copy paper.

2. Fine Structures of Amylose and Amylopectin

Normal wheat starch is comprised of 25% amylose and 75% amylopectin. Amylose is a mixture of linear and lightly branched (0.2–0.8% of linkages) molecules having a number average DP of 1000–5000, whereas amylopectin is monodisperse and highly branched (4–6% of linkages) with a DP of ~10000 or more. The detailed structural features of amylose and amylopectin are dependent on the source of starch, even among wheats. Moreover, the proportions of amylose and amylopectin in starch may vary within a plant source, due to genetic mutation. Amylose and amylopectin are readily isolated and purified, but branched and linear amylose molecules from wheat starch have not been separated and isolated.

Amyloses of five commercially important starches (wheat, corn, rice, tapioca and potato) were purified by fractional crystallization of their amylose–alcohol complexes and shown to be free of amylopectin.[111,205] All five amyloses showed comparable iodine affinity, blue value and λ_{max} of their polyiodide complexes (Table 10.5), which is consistent with their being (1→4)-linked α-glucans with an average DP_n of ~500 or above.[206] The average size of amylose molecules in cereal starches is smaller than those in tuber or root starches (Table 10.5). The branched molecules in amylose constitute approximately 25% to 50% by number of the total amylose. In wheat amylose, the branched molecules are ~60% by weight compared to 80% for corn.[205] Wheat amylose contains somewhat more branches than corn or rice amylose, but approximately the same as tapioca amylose. Hizukuri et al.[207] reported that branched amylose molecules, and even their β-limit dextrins, behave much like linear amylose, presumably because the branched molecules are 1.5–3.0 times larger than the linear ones and their branches are mostly short chains that may occur in small clusters.

On digestion with β-amylase, the amyloses in Table 10.5 released 73–85% maltose, and all gave quantitative yields of maltose when digested simultaneously with β-amylase and pullulanase.[111,205] Debranching wheat, corn and potato amylose with isoamylase removed 50%, 65% and 94%, respectively, of their branched chains, which increased β-amylolysis to 94%, 90% and 86%.[208] The molecular weight distributions of the debranched amyloses showed that the debranched amylose molecules of wheat and corn starches underwent almost no change in size, that the branch chains released by enzymolysis were high in number-percent, but low in weight-percent, and that they were of very low molecular weight. However, the debranched amylose from potato (and also from sweet potato) decreased in size, indicating the presence of long branches in those tuber amyloses. Other data indicates that many of the branches in amylose are found near its reducing end.[111,205,208]

Table 10.5 Molecular properties of amylose from wheat and other starches of commerce[111,205]

Property	Corn	Rice (Indica)	Wheat	Tapioca	Potato
Iodine affinity (g/100 g)	20.0–20.1	20.0–21.1	19.0–19.5	20.0	20.0–20.9
Blue value	1.35–1.39	1.40–1.45	1.13–1.31	1.47	1.41–1.48
λ_{max} (nm)	644	653–57	640–648	662	660
Limiting viscosity (mL/g)	183	180–249	118–237	384	368–384
Average size (DP_n)	990	920–1040	980–1570	2660	2110–4920
Apparent DP_n distribution	400–14700	320–11700[a]	460–18200[a]	580–22400	840–21800
Number of chains/molecule	2.4	1.3–3.3	4.8–5.5	6.8	3.9–6.3
Number of chains/branched molecule	4.4	4.7–8.7	11.9–19.2	16.1	–
Beta-amylolysis limit (%)[b]	82	73–84	79–85	75	80–87
Beta-amylolysis + pullulanase (%)	99	103	101	99	100
Branched molecules (mole%)	48	25–49	29–44	42	–

[a]Mean of 5 cultivars
[b]Beta-amylolysis limit of branched amylose fraction estimated to be 40%

The classic method of determining the mass distributions of α-(1→4)-linked unit chains in starch is achieved by size-exclusion column chromatography with refractive index detection. Alternatively, for chains below DP ~40–80, separation and quantitation is done by high-performance anion-exchange chromatography with pulsed amperometric detection.[111] In these classic methods, the chain lengths of individual fractions and reference standards are determined by colorimetry. More recently, the fine structure of amylose has been probed by fluorescent labeling of the reducing end using reductive amination with 2-aminopyridine.[209,210] High-performance size-exclusion chromatography with fluorescence and refractive index detectors in tandem enables the direct determination of molar- and weight-based distributions. Besides 2-aminopyridine, negatively charged fluorophores have also been attached to the reducing end of unit chains of DP 3 to 135, and the labeled molecules separated and counted by capillary electrophoresis and fluorimetry.[211,212] Fluorophore-assisted carbohydrate electrophoresis (FACE) has not been established for amylose chains above DP_n 135.[209,211]

Table 10.6 compares the average size (DP_n) and size distributions of six laboratory-purified amyloses and one commercial sample of potato amylose, which were determined by classic colorimetric and fluorescent-labeling techniques using 2-aminopyridine. The data by the two techniques are consistent and show that wheat and other cereal amyloses are smaller in size than those from root and tuber starches. The molar distribution technique indicated that wheat amylose contained two molecular species, compared with one for rice and corn amyloses.[209,210] Moreover, the molar size distributions for the cereal amyloses are much narrower than those of the tuber amyloses, and the cereal amyloses contain a preponderance of molecules of $DP_n \leq 1000$ whereas the tuber amyloses contain 78–95% of molecules with $DP_n \geq 1000$, and even 3–5% above DP_n 10 000. None of the amylose samples in Table 10.6 showed molecules with less than DP_n 200, possibly because they had been purified as alcohol-inclusion complexes.[209]

Table 10.6 Number-average degrees of polymerization and DP_n distributions of amyloses determined by colorimetry and by end-group labeling with 2-aminopyridine[209]

Amylose	$\overline{DP_n}$		Apparent DP_n distribution
	Labeling	Colorimetric	
Wheat	1220	1230	190–3130
Rice (indica)	1410	1190	290–3100
Rice (japonica)	1150	963	230–2750
Corn	1030	891	230–2370
Sweet Potato	3230	3320	440–3880
Potato	4370	4400	970–9770
Potato (commercial)[a]	807	796	220–1730

[a]Commercial sample, Type II, Sigma Chemical Co., St. Louis, Missouri

Amylopectin accounts for 70–80% of most starches; it is responsible for the crystallinity, birefringence and growth rings observed in starch granules, and for the swelling and gelatinization of granules (see Chapters 5 and 6). Amylopectin is also exclusively the molecule in starch that is esterified with low levels of phosphate groups. Recent light-scattering data[213–215] indicates a molecular weight of 2.6–7.0 × 10^8 g/mol for wheat amylopectin compared to 2.8 × 10^8 and 3.4 × 10^8 g/mol for corn and rice amylopectin, respectively.

Amylopectin molecules are radially oriented in a granule and are constructed of unit chains containing an average of 18–25 (1→4)-linked α-D-glucopyranosyl units per chain. Hundreds of these unit chains are linked together by α-(1→6)-linkages to form amylopectin. Animal glycogen and algal floridian starch contain the same two glycosidic linkages as amylopectin, but glycogen has approximately twice as many α-(1→6)-linkages as amylopectin or floridian starch. Table 10.7 compares the properties of amylopectin from wheat with those from other common sources.

The α-(1→6) branch points in amylopectin occur in clusters, and the high flexibility of those branch points in the presence of water allows parallel clustering of the unit chains. The branched regions of an amylopectin molecule are amorphous regions, but the unit chains in the linear portions intertwine into left-handed, parallel-stranded, double helices, and those helices organize into a crystalline array (see Chapters 5 and 6). The double helical content, or so-called short-range molecular order, as well as the amorphous content of starch, can be determined by solid-state ^{13}C-NMR spectroscopy, whereas crystallinity, or long-range order, is measured by wide-angle x-ray diffraction.[216] The anomeric ^{13}C-signals indicate 18–47% double helical content in starches (9–10% moisture level) from corn, waxy maize, high-amylose (Hylon V) corn and rice,[216] and 39% in wheat starch.[104] The amorphous contents of those starches were determined to be 53–75%.[216] At the same moisture

Table 10.7 Molecular properties of amylopectins from wheat and other starches of commerce[104,111,210]

Property	Corn	Wheat	Potato	Rice (Indica)	Rice (Japonica)
Iodine affinity (g/100g)	1.10	0.93–1.10	0.06–0.08	0.63–2.57	0.39–0.87
λ_{max} (nm)	554	551–560	560	542–575	531–542
Limiting viscosity (mL/g)	171	116–148	125	152–165	134–137
Number-average degree of polymerization	8200	4800	9800	4700–15000	8200–13000
Beta-amylolysis limit (%)	59	56–59	56	56–59	58–59
P_o (ppm)	14	9–20	650–900	11–29	8–13
Number-average chain length	22	20–21	23	20–22	19–20
Number of chains per molecule	373	228–240	426	213–750	410–684
Average exterior chain length	15	13–14	15	13–15	12–13
Average interior chain length	6	5–6	8	5–6	5

IV. Properties of Wheat Starch and Wheat Starch Amylose and Amylopectin

level, crystallinity was reported to be 13–30% for corn and rice starches and 12.0–14.5%,[213] 20%[104] and up to 36% (Table 10.4) for normal wheat starch.

The crystalline phase in starch is usually the A- or B-type. The A polymorph contains double helices that are densely packed in a monoclinic lattice with ~4% water of crystallization, whereas the B polymorph has double helices packed in a hexagonal lattice with ~25% water. The A-type pattern, with main reflections at $2\Theta \sim 15°$ and $23°$ and a doublet at $17°$ and $18°$, is observed with wheat and other cereal starches, where the amylopectin molecules have a short average chain length ($\overline{CL}_w\ 23-29$) and where the starch is deposited in a warm environment. On the other hand, the B-type pattern, with reflections at $2\Theta \sim 5°$ and $17°$ and a doublet at $22°$ and $24°$, is observed in many tuber and root starches, which have a long average chain length ($\overline{CL}_w\ 30-44$) and where the starch is deposited underground at a cool temperature.[217,218] Legume starches, and other starches with intermediate average chain lengths or with a variable temperature of deposition, quite often show a mixed A- and B-type x-ray pattern, which is referred to as the C-type pattern. Malto-oligosaccharides crystallized from a 20% aqueous solution at 25°C display polymorphic crystallization, depending on their chain length. Those chains with DP 6–8 fail to crystallize in months, whereas those with DP 10–12 give A-type crystals and those with DP 15 and above give B-type crystals. Crystals of chains with DP 9 and DP 13–14 are transitory.[219,220] The A-type pattern of a non-waxy cereal starch sometimes contains a low background of a V-pattern ($2\Theta \sim 7°$, $13°$ and $20°$), indicating the presence of a low level of the single helical complex of amylose with monoacyl lipid.[190,207,214] A wheat starch at 9% moisture content and 20% crystallinity contained 2% V-type crystallinity.[221]

{AQ2}

The crystalline regions of double helical chains and the amorphous branch-point regions in amylopectin are organized into alternating lamellae with a total repeat distance of approximately 9 nm. The 9 nm lamellae are stacked generally in the radial direction of a granule. The branch-amorphous region (~3 nm) in one lamella is about one-half the thickness of the alternating crystalline region (~6 nm). At the next structural level of a starch granule, the lamellae appear to be organized into blocklets of diameter 20–500 nm depending on the botanical source or the location in a granule.[180–182,190] The blocklets may be the same as the super helices that are observed in potato starch.[222] The partially crystalline blocklets are embedded in growth rings of thickness 140–400 nm. The growth rings appear to be alternating crystalline hard layers and semicrystalline soft layers, where a hard layer contains a much lower concentration of amorphous material.[223]

Structurally, the unit chains in amylopectin are classified as A, B or C chains, depending on whether one or more α-(1→6)-linked branch points is attached to the backbone of a chain. The A-type chain is linear without any branch point along its backbone and is linked through its potential reducing end to another chain by an α-(1→6)-glucosidic linkage. The B chain carries one or more A and/or B chains α-(1→6)-linked along its backbone. The C chain carries the single reducing end in an amylopectin molecule; it is usually counted with the B chains. The ratio of A to B chains in wheat amylopectin is ~1.2:1.0, which is similar to the ratio in corn, rice or potato amylopectin.[111]

The chain length (DP_n 6 to 70–80) distributions of amylopectins from wheat and other sources are polymodal. Size-exclusion chromatograms of debranched amylopectin with refractive index (mass) detection show 2–5 peaks depending on column resolution, whereas high-performance anion-exchange chromatograms that display individual chains indicate 4 periodicities in different lengths of the unit chains.[111,224,225] Extra long chains of DP ~1000 also are detected in debranched amylopectins of wheat, corn and rice, but not in those of waxy starches, taro or potato.[111,208,213] The extra long chains in amylopectins correlate with their iodine affinities, and their absence in potato amylopectin is consistent with its low iodine affinity (Table 10.7). A high level of extra long chains in wheat starch appears to decrease paste viscosity.[226]

The polymodal distribution of unit chains and their proportions in amylopectin support the cluster model. The cluster model of amylopectin suggests a repeat structure made up of one population of A chains, another of relatively short B chains (B_1), and three populations or more of long B chains (B_2, B_3, B_4, B_5). The various populations of chains appear to be separated with a periodicity of 12, so that A chains have been assigned to the group of chains with DP_n 6–12, B_1 chains = DP 13–24, B_2 chains = DP 25–36 and $(B_n)_{n \geq 3}$ to chains > 37.[225] The weight-percent

Table 10.8 Chain-length distributions of amylopectins determined after debranching[a]

Source	Crystal pattern	Average chain length	Weight% of chains[b]				Mole% of chains[c]			(A + B_1 chains)/(B_2 + B_3 chains)[c]
			DP 6–12 A	DP 13–24 B_1	DP 25–36 B_2	DP > 37 $(B_n)_{n \geq 3}$	A	B_1	B_2 + B_3	
Wheat	A	22.7	20.0	42.9	15.9	18.0	64.9	27.8	7.3	12.3–12.9
Rice (japonica)	A	22.7	19.0	52.2	12.3	16.5	68.0	23.3	8.7	10.1–10.8
Normal maize	A	24.4	17.9	47.9	14.9	19.3	65.1	25.8	9.1	10.0
Tapioca	A	27.6	17.3	40.4	15.6	26.7	–	–	–	–
Sweet potato[d]	A	22.1	21	47	17	15	59.7	30.5	9.8	8.9–9.5
Potato	B	29.4	12.3	43.3	15.5	28.9	51.8	33.7	14.5	5.4–6.4
Potato[d]	B	23.6	18	48	15	18	51.8	33.7	14.5	5.4–6.4
Yam[d]	B	23.2	18	56	15	11	86.4	13.6	6.4	

[a]Except otherwise referenced, size-distribution data determined by high-performance anion-exchange chromatography with post-column enzymolysis to D-glucose, and then pulsed-amperometric detection.[224] Wheat data is the mean of 13 samples.[213,214,224]

[b]The unit chains with DP_n 6–12, 13–24, 25–36, and >37 designated, respectively, to be A, B_1, B_2 and $(B_n)_{n \geq 3}$ chains.[225]

[c]Ratio of number of short chains (A + B_1) in one cluster to number of long chains (B_2 + B_3) that traverse two or more clusters determined by fluorescence labeling followed by high-performance size-exclusion chromatography with fluorescence and refractive index detection.[210,226] The experimentally measured average chain-lengths ($\overline{CL}_n$) for the different starches ranged as follows; A chains 12–13, B_1 22–27, and B_2 + B_3, 53–65. B_4 chains not detected.

[d]Determined by same method as (a) except quantitation was done by relative peak areas.[225] Phosphorylated (13% and 3%) amylodextrins were removed from debranched potato and sweet potato amylopectins before chromatography.

and number-percent of the various groups of chains in wheat and six other amylopectins are given in Table 10.8.

The A plus B_1 chains dominate the chain length distributions in amylopectins both by weight (58–74%) and by number (85–93%), compared with the sum of the long B chains (see Table 10.8). The A and B_1 chains are believed to occur within one cluster, while the B_2 and B_3 chains transverse two and three clusters, respectively. The B_4 chains and beyond are small in number and were not detected by the end group labeling protocol. However, the weight-percent of B_4 chains with DP_n above ~50 (DP_w ~80) is measurable.[217,224,225]

The last column in Table 10.8 shows the ratio of the number of (A + B_1 chains)/(B_2 + B_3 chains). That ratio is an indication of the relative proportion of chains in one cluster to those in two or more clusters. The amylopectin from the starches with an A-crystal pattern in Table 10.8 (cereal starches and sweet potato starch) had almost twice the number of (A + B_1) chains attached to (B_2 + B_3) chains as the amylopectins from the B-type starches. Wheat amylopectin showed the highest proportion of A + B_1 chains in its amylopectin, which may explain its slow rate of retrogradation at 25% solids and 1°C[224,227] or at 15% solids and 4°C,[215] and its ~7° lower gelatinization temperature compared to normal corn and rice starches.[224] The C chains in wheat amylopectin showed a molar distribution of chains between DP_n 10 and 130, with two-thirds spanning two clusters and one-third within one cluster. Moreover, 65% of the C chains in wheat amylopectin were linear. The C chains from five other botanical sources of amylopectin showed almost the same chain length distributions as wheat with peaks at DP_n 40–49 and 21–27, except high-amylose (70%) maize with a single peak at DP_n 80.[228]

The crystallites in wheat starch granules have been estimated by x-ray diffraction and transmission electron microscopy to be 7–10 nm in size,[229,230] which seem to be the size of one cluster of chains in the amylopectin model. The double helix of two chains in an A-crystal of starch has a repeat distance of 2.1 Å for every six α-D-glucopyranosyl units per chain.[231] Within one crystallite, three turns of a double helix contain 18 α-D-glucopyranosyl units (chain length 18) and occupy a length of 6.3 nm. That length equals ~75% of the maximum chain length of 24 assigned to the B_1 chains within one cluster. The remaining 25% of length of the chains would be in the amorphous repeat zone of ~3 nm within a lamella. The diameter of 7–10 nm for wheat starch crystallites indicates a bundle of 7–10 double helices making up one cluster, where a double helix has a diameter of ~1 nm.[231]

Shibanuma et al.[232] examined the molecular structures of starch isolated from three Japanese wheat varieties, one Australian standard white wheat and one US western white wheat. The data presented in Tables 10.9 and 10.10 again indicate that the properties and structural features of amylose and amylopectin are dependent on the starch source. The molecular sizes of amylose and amylopectin were larger in the US wheat compared to the corresponding starch fractions from the Australian and Japanese wheat starches. Among the five wheats, the two preferred for salt noodles in Japan, the Japanese variety Chihoku and the Australian standard white, contained a higher proportion of branched amylose and a lower number of chains per amylose molecule. In addition, the starches from the two noodle wheats contained 2% less amylose than the other three wheats. With the exception of size, little difference was reported between the amylopectins, all of which contained 4–6% by weight of extra long chains.

Table 10.9 Molecular properties of wheat amylose and its beta-limit dextrins as affected by starch source[232]

Property	Amylose			Beta-limit dextrin		
	Australia	Japan	US	Australia	Japan	US
Iodine affinity (g/100 g)	19.5	19.0–19.5	19.0	16.9	15.3–18.1	17.2
Blue value	1.24	1.13–1.25	1.31	1.25	1.14–1.25	1.31
λ_{max} (nm)	648	636–648	647	640	630–645	645
Limiting viscosity (mL/g)	183	118–185	237	–	–	–
Number-average degree of polymerization	1180	830–1080	1570	930	700–1050	1430
Number-average chain length	200	135–195	255	71	50–67	69
Number of chains per molecule	5.8	5.5–6.5	6.2	13.0	12.9–20.2	20.7
Beta-amylolysis limit (%)	81	79–83	85	–	–	–
Weight-average degree of polymerization	3480	2360–4880	5450	3080	1670–4120	5880
Linear molecules (mole%)	60	56–71	74	–	–	–
Branched molecules (mole%)	40	29–44	26	–	–	–

Table 10.10 Molecular properties of wheat amylopectins as affected by starch source[232]

Property	Source of wheat starch		
	Australia	Japan	US
Iodine affinity (g/1000 g)	0.89	0.93–1.12	0.66
Blue value	0.098	0.099–0.115	0.078
λ_{max} (nm)	552	551–560	547
Limiting viscosity number (mL/g)	145	116–148	145
Number-average degree of polymerization	6200	5000–6700	9400
Number-average chain length	20	19–21	20.5
Beta-amylolysis limit (%)	57	58–59	56
Weight-average chain length after isoamylolysis			
Whole	85	93–112	72
A chain	11	10–13	10
B_1 chain	21	20–23	22
B_2 chain	41	41–43	43
B_3 chain	68	70–91	91
Long chain	1600	1600–1800	1200

3. Partial Waxy and Waxy Wheat Starches

Granule-bound starch synthase (GBSS) has been known since the 1960s to be associated with the biosynthesis of amylose in cereals, which is discussed in reviews of starch biosynthesis[233–240] (see Chapter 4). GBSS is also termed the waxy protein and is encoded in the *wx* gene. The term waxy is a descriptor for all non-amylose starches, such as those from rice, barley, sorghum, wheat, potato, pea and amaranth, even though the term was derived from a single waxy cereal, that of non-amylose corn whose endosperm has a waxy appearance. Waxy grain is described as glutinous when cooked and eaten because it has a cohesive (sticky) texture. In wheat and other monocots, GBSSI is active in endosperm, whereas its isoform GBSSII is found in granules of transitory starch that is deposited in non-storage tissues, such as pericarp, leaf and stem tissues. GBSSII is under separate genetic control, and transitory starch granules have a different size, shape and amylose content compared to reserve endosperm starch.[241] For example, the pericarp of waxy wheat kernels when immature contains starch that has 19% amylose, compared to 0.8% in the endosperm starch.[242]

Common or hexaploid wheats have three sets of chromosomes, the A, B and D genomes, so hexaploid wheat may be thought of as three plants hybridized into one. Each genome produces a waxy protein differing slightly in structure. The three isozymes, viz., the Wx-A1, Wx-B1 and Wx-D1 proteins, were separated in 1993.[243,244] At the same time, the term partial waxy wheat was coined to indicate hexaploid wheats lacking one or two of the waxy proteins in the endosperm starch. In the 1990s, thousands of wheats were assayed for the three waxy proteins, and wheats in Australia and India showed a 40–50% frequency of single nulls lacking the Wx-B1 protein, while those in Japan, Turkey and Korea showed a 10–50% frequency lacking the Wx-A1 protein. Only one wheat (three are now known) from China was found lacking the Wx-D1 protein. No phenotype devoid of all three waxy proteins was found in the international sampling of wheat, since spontaneous mutation had not overcome the redundancy in the multiple genomes of wheats.

Studies on amylose levels and waxy proteins in wheat began in the 1970s in Japan, Australia and New Zealand after it was discovered that the preferred wheat flour for the Japanese white-salted (sodium chloride) noodle contained a high-swelling starch that had levels of amylose 3–5% below those of traditional wheat starches. Japanese wheat breeders at that time were charged with improving domestic wheats for noodles. A chronology of events[245] leading to wheats with low to very low amylose levels[243–268] is given in Table 10.11.

Eight combinations (types 1–8, Table 10.12) of Wx-A1, Wx-B1 and Wx-D1 proteins are possible in common wheats.[257] Near-isogenic lines of the eight types of wheats showed apparent amylose levels of 3% to 25%, as determined by the blue color of the amylose–iodine complex.[266] The amylose level in wheat kernels increases positively with the level of waxy proteins.[257,269–271] In the double nulls, the Wx-A1 protein in type 5 wheat produces ~3% less amylose than the Wx-B1 and Wx-D1 proteins (types 6 and 7, respectively). In the single nulls, compared to the wild type (type 1), absence of Wx-B1 (type 3) reduces the amylose content by ~2% compared to ~1% for the absence of either of the other two waxy proteins (types 2

Table 10.11 Events leading to development of waxy and partial waxy wheats[245]

Year	Location	Event	References
1975	Christchurch, New Zealand	Starches from selected Australian wheats with 2–3% less amylose than New Zealand wheats had a high pasting peak	246
1977	Tokyo, Japan	Flour of acceptable quality for salt noodles milled from Western Australian wheat produced an amylogram with low pasting temperature and high pasting peak	247
1980	North Ryde, Australia and Atsugi, Japan	High amylograph pasting peak or rapid time to reach pasting peak for starch from salt noodle flour was associated positively with eating quality of salt noodles and negatively with apparent amylose levels between 19% and 32%	248, 249
1984	North Ryde, Australia	Wheat starch with high amylose level had a low pasting peak and no breakdown	250
1987	Chorleywood, UK	Granule-bound starch synthase (waxy protein) was confirmed in wheat starch	251
1989	Morioka, Japan	Japanese wheats Kanto 79 and Kanto 107 had 22% amylose in flour compared with $\geq$27% in 116 other Japanese wheats	252
1989	South Perth, Australia	Flour swelling power and swelling volume tests were developed for breeders to identify noodle wheats	253, 254
1990s	Australia, Japan, Korea and Hong Kong	Investigators verified that the quality of salt noodle flour was associated with enhanced swelling of wheat starch and reduced amylose level	250, 255, 256
1993	Morioka, Okinawa and Tsukuba, Japan	Waxy proteins in wheat starch produced by Wx-A1, Wx-B1, and Wx-D1 loci in hexaploid wheats were separated; the term 'partial waxy' wheat was coined; and waxy wheat was predicted	243, 244
1994	Morioka, Okinawa and Nara, Japan	Wheat cultivars ($\approx$2000) were assayed for waxy proteins; 17% of wheats with a single null allele at Wx locus were identified (only one wheat had a null for Wx-D1 locus); 0.5% double nulls at Wx-A1 and Wx-B1 were found, but no double nulls involving Wx-D1 locus; no triple null was found; and distributions of waxy proteins in a total of eight types of wheat starches were assigned	257
1994	Okinawa and Morioka, Japan	A durum (tetraploid) wheat Aldura was crossed with Kanto 107 (hexaploid with double null for Wx-A1 and Wx-B1 proteins) to produce the first waxy tetraploid wheat; Chinese Bai Huo wheat (null for Wx-D1 protein) was crossed with Japanese Kanto 107 and Saikai 173 (null for Wx-A1 and Wx-B1 proteins) to obtain waxy hexaploid wheat	258–260
1996	Camden, Australia	An improved one-dimensional gel electrophoresis method was used to separate three waxy proteins in wheat starch	261
1998	Western Australia and New South Wales, Australia and Nebraska, USA	Quality of wheat for salt noodle predicted by 1) separation of waxy proteins isolated from starch; 2) polymerase chain reaction-based assay for null Wx-B1 locus in leaf DNA; 3) immunoassay (enzyme-linked immunosorbent assay) for total waxy proteins dissolved from starch	262–264
1999	Saskatoon, Canada	Wheat was produced with a double null for Wx-B1 and Wx-D1 proteins; starch contained 12.4% amylose	265
2002	Obihiro, Tsukuba and Nagoya, Japan	Near-isogenic lines of hexaploid wheats were produced with all combinations of waxy proteins; double nulls contained up to 6–9% less amylose compared with the wild-type	266
2002	Morioka, Japan	Rapid classification of partial waxy and waxy wheats was developed using PCR-based markers	267
2005	Seattle and Davis, USA	Near-waxy hexaploid wheat produced through TILLING (targeting induced local lesions in genomes).	268

Table 10.12 Granule-bound starch synthase I (Wx-A1, Wx-B1, and Wx-D1 proteins) and amylose levels in near-isogenic lines[266] and isolines[273] of wheat

Type	Wx-A1	Wx-B1	Wx-D1	Apparent amylose (%)[a] Near-isogenic	Apparent amylose (%)[a] Isolines[b]	Unit chains in near-isogenic lines (mol%) DP 6-12	DP 13-24	DP 25-36
Wild type								
1	+	+	+	25	28	47.6	43.6	8.8
Single null								
2	−	+	+	23	27	47.7	43.9	8.4
3	+	−	+	22	26	46.6	45.0	8.5
4	+	+	−	23	27	46.5	44.9	8.7
Double null								
5	+	−	−	16	20	47.9	43.5	8.6
6	−	+	−	19	24	47.3	44.5	8.3
7	−	−	+	19	22	46.9	44.2	8.9
Waxy								
8	−	−	−	3	<1	49.2	42.6	8.2

[a]Wheats grown for three years, and mean amylose levels determined by colorimetry. Data rounded to two digits.

[b]Isolines of Type 1, 2, 4 and 6 carry a variant Wx-B1 protein from an allele Wx-B1e instead of the standard Wx-B1a allele.

and 4). Two near-isogenic lines of a partial waxy and a waxy wheat were produced from the Japanese cultivar Kanto 107.[272] The partial waxy wheat was a double null with a functional Wx-D1 protein, and its starch contained 23–25% apparent amylose compared to 3% for the waxy line.

Other workers[271,273] have produced eight lines of wheat having all the possible combinations of waxy proteins. The amylose levels of the isolines shown in Table 10.12 fall in the same order as those found in the near-isogenic lines. Another set of eight wheats that had a soft wheat background[271] contained 0.9% amylose for the triple null (waxy), 19.6–21.1% for the double nulls, 23.7–24.5% for the single nulls and 25.8% for the wild type. A double-null partial waxy wheat containing only the Wx-A1 protein produced 12–13% amylose in its starch, as determined by the concanavalin-A method and by high-performance size-exclusion chromatography of debranched starch,[265] whereas a double null (cv. Ike) containing only the Wx-D1 protein contained 21–23% amylose by the concanavalin-A method.[274]

In 1994, Makoto Yamamori and Toshiki Nakamura in Japan produced the first waxy tetraploid and hexaploid wheats from crosses of partial waxy wheats[258–260] (Table 10.11). Shortly thereafter, waxy wheat was produced by chemical[275] and somaclonal[276] mutation to silence the *Wx-D1* gene in the double-null cultivar Kanto 107. The amylose level in wheat is independent of grain hardness, so soft and hard waxy wheats have been produced.[277,278] Waxy wheats produced by hybridization, mutagenesis and somaclonal mutation are not genetically modified organisms (GMO). Recently, a reverse-genetic method, termed TILLING (targeted induced local lesions in genomes) has also been employed to produce a non-GMO waxy phenotype from a bread wheat.[268] In TILLING, a library of thousands of DNA samples can be screened at a high rate to identify mutagen-induced single nucleotide polymorphisms (SNPs) within a gene. TILLING permits genetic variation at the *wx* or other loci in an existing commercial cultivar, which eliminates the introduction of genetic heterogeneity into elite cultivars when they are crossed with less exotic wheat. High-speed identification of a desired mutation in an exotic wheat can result in rapid development of a desired genotype.

The apparent amylose level of waxy wheat starch has been reported to be between 0–5%. In general, amylose determined by the concanavalin-A procedure gave higher levels (1–5%) of amylose[265,275,279,280] than that determined by colorimetry (0–1.0%),[271,281–283] iodine affinity (0.2–2.0%)[275,283,284] and high-performance size-exclusion chromatography (none detected).[265] Soaking waxy wheat starch in aqueous 1.4 M (25%) potassium iodide/0.4 M (10%) iodine caused the large starch granules to take on a ghost-like appearance with a black-brown central zone and a surrounding red-brown zone.[286–288] Other waxy starches from corn, rice, millet and barley gave only a single zone on swelling.[287] The two different colored zones in swollen waxy wheat starch were separated by sonication, and the black-brown zone was found to contain ~7% amylose, whereas the red-brown zone contained ~1% amylose.[288] The two different-colored zones are congruent with the two-stage growth of large wheat starch granules.[170,171] The size, shape and size distribution of granules from waxy, partial waxy and wild type wheats all appear to be the same.[271,289,290]

All wheat starches give the same A-type crystal pattern, but their degree of crystallinity is inversely proportional to amylose content. In addition, the intensity of one reflection of an amylose–lipid complex at 2Θ 20–23° (d 4.4Å) increases proportionally to the amylose level.[265,271,281,284,289] The lipid content of waxy wheat starch is much lower (0.02–0.17%) than that in non-waxy wheat starch (0.6–0.7%).[286,291] However, the crude fat content of waxy wheat flour is elevated by 20–40%; β-glucan and pentosan are increased by ~30%.[291,292] The degree of crystallinity of waxy wheat starch is reported to be 18–21% versus 13–16% for the wild type.[284,289] Increases of 35% for waxy wheat starch, 23–28% for double nulls, 20–23% for single nulls and 22% for the wild type have also been reported.[271]

Despite its elevated crystallinity, the water-retention capacity of waxy wheat starch at room temperature was much higher (80–81%) than the 55–62% found for normal wheat starch.[293] These results agree with the elevated water absorption (30–35%) of waxy corn starch compared to that (27–30%) of normal corn starch.[294] Waxy starches lack amylose, lipid and extra long chains in their amylopectin, all of which are thought to restrict granule swelling in water. In waxy wheat flour, the increased absorption of the waxy granules, accompanied by increased starch damage, partly explains the flour's high dough absorption[289,295] and its high alkaline water retention value.[271]

The gelatinization temperature of waxy wheat starch is some 3–4°C higher and its gelatinization enthalpy 2–4 J/g higher than for normal wheat starch,[265,280,281,284,285,289,290] but is some 5–6°C and 1 J/g lower than for waxy corn starch.[281,289] The gelatinization temperature range (T_o–T_p–T_c) for waxy wheat starch in excess water averages approximately 57°–62°–68°C with an enthalpy of gelatinization of 13 J/g. The chain length distribution in waxy wheat has been measured by high-performance anion-exchange chromatography with pulsed amperometric detection. Yasui et al.[285] reported no difference in the structure of amylopectin from waxy wheat and their parent lines. On the other hand, Yoo and Jane[284] found that waxy wheat starch contained ~2 weight percent more chains of DP >37, ~1% less of DP 6–12 and ~1% less of DP 13–24, compared to commercial wheat starch. Also, the waxy wheat starch contained no extra long chains with average DP ~770. Sasaki et al.[279] also reported that waxy wheat starch had more chains with DP >37 and fewer of DP 6–12. Other workers[296] reported a lower level (18%) of B_2 plus B_3 chains in waxy wheat amylopectin compared to 20–21% in both waxy corn and waxy rice.

The pasting profile of waxy wheat starch displays a relatively low pasting and low peak temperature, a high peak viscosity, high breakdown, low setback, high paste clarity[270] and a high loss of viscosity in the presence of alpha-amylase.[289] The swelling power of waxy wheat starch at 40–90°C was almost double that of normal wheat starch,[283] in agreement with the negative correlation of amylose content with swelling power (at ~100°C) of wheat starches.[279] The low temperatures of pasting and peak viscosity of waxy wheat starch are explained by its low gelatinization and swelling temperatures, whereas the other pasting characteristics are common to all waxy starches.

Wheat is used in a broad range of foods, feeds and industrial products. The waxy trait in wheat must be introduced into wheat along with desirable agronomic and end use properties. It is not clear whether there is a yield penalty in growing waxy versus non-waxy wheats. Some[266,278] have reported no difference in yield, but others[275,276] have noted a lower 1000-kernel weight. Waxy kernels from mutated Japanese wheat have been reported to have less starch (~61 versus 64%, db) than their parent lines,[291] but a waxy wheat made by hybridization with a US wheat contained about the same level of starch (77% versus 76%, db).[295] The starch-related properties of waxy wheats are much more stable to genetic and environmental variation than their protein and grain traits.[278]

Flour yields from 20 lines of waxy bread wheats were lower (~57% versus 68%) and flour ash levels higher (0.51 versus 0.37%) compared to four control wheats, perhaps because of the 20–40% increase in crude fat found in waxy wheat flour,[291,292] which makes flour sieving less efficient. The protein contents of the two sets of flours were comparable, but protein quality tests for bread flour showed that the waxy flours were inferior. Also, the pasting (cooking) qualities of the two sets of flours differed.[278] Flour yield from a soft waxy versus a soft wild type wheat was also lower (70% versus 73%).[271] Waxy wheat flours have ~5% higher starch damage,[270] because of the fragility of the waxy granules during dry- and wet-milling.[297]

Uses of partial-waxy and waxy wheat in Asian noodles, bread and tortillas have been reviewed.[269,270,298] French bread made from double-null partial waxy wheat flour retained 1–2% more moisture in its crumb in 24 hours compared to bread from a wild type flour. French bread made from the mutant wheat had a softer crumb immediately and up to 48 hours after baking.[299] Wheat noodles are made from flour and low levels of salts, so starch plays a major role in noodle quality. Flours from partial waxy wheats with ~10% protein and 21–24% apparent amylose (starch basis) are favored for white salted (sodium chloride) noodles because partial waxy starch swells during cooking somewhat more than non-waxy wheat starch. For good appearing noodles, the flour should be from a white wheat low in polyphenol oxidase, and the flour also should be of low extraction rate. The added swelling and release of amylose of the partial waxy wheat starch increases surface smoothness in the cooked noodle, and causes the eating texture to resemble a soft gel. In contrast, yellow alkaline noodles are made from non-waxy wheat flour with ~12% protein and 25–28% apparent amylose (starch basis) and soluble carbonate salts. The starch in the yellow alkaline noodle undergoes limited swelling during cooking because of its elevated amylose content, and because of the additional protein matrix and carbonate salt. The cooked yellow alkaline noodle has a hard and noncohesive eating texture.

Soups, gravies and sauces may be thickened with waxy wheat flour that has undergone heat-moisture treatment to inhibit formation of a 'long stringy' texture.[300] Bakery foods with a soft crumb, such as bread and cakes, may benefit from the addition of waxy wheat flour as an ingredient. The starch in waxy wheat flour possesses nearly the same thermal properties and granule size and shape as starch from wild type flour, but it has increased swelling and moisture-holding capacity that may spare fat, increase softness and extend shelf life. In dry cereal foods, such as instant breakfast cereals, crackers and salty snacks, addition of waxy wheat flour as an ingredient would be expected to soften a product's bite and increase its friability.

Wet-processing of waxy wheat flour to produce waxy wheat starch and vital wheat gluten appears to be feasible by one of the new processes starting from a batter[292] (see Section 10.3 of this chapter). However, gluten agglomeration in the Martin dough-washing process was impaired for many waxy lines developed in the US.[292] The impairment of gluten agglomeration may result from the increased levels of fat, beta-glucan and hemicellulose[291,292] in waxy wheat flour. Waxy and partial waxy wheat starches can be modified[301,302] to produce thickeners with a low pasting temperature and a low rate of retrogradation.[279,281,302] Additionally, the low lipid content of waxy wheat starch implies that waxy wheat is suitable for conversion to maltodextrins[303] and to D-glucose (dextrose).

4. High-amylose Wheat Starch

Wheat starch with 31–38% apparent amylose, or 4–8% above wild types, was produced by crossbreeding three hexaploid wheats, each of which was deficient in one isoform of starch synthase IIa.[304] The triple mutant also had an altered amylopectin structure, apparently because starch synthase IIa elongates short chains on amylopectin[305] and because the mutation decreases the levels of starch synthase I and starch-branching enzyme IIb. The starch content of the kernels of this wheat was 50% compared to 61% for the wild type, and its large granules appeared deformed and cracked at the hilum. High-performance anion-exchange chromatography revealed that the amylopectin of the mutant wheat contained more short chains of DP 6–10, fewer of DP 11–25, a slight increase of DP 29–36 and a slight decrease of DP 42–54.[304] More recent work[306] on the mutant's amylopectin showed 9.5 weight percent of chains having DP >100, chains that were absent in the wild type. The molar ratio of $(A + B_1)/(B_2 + B_3)$ was 13.8 versus 13.1 for the mutant and normal starches, respectively; the DP_n of amylose was 800 versus 1000. The mutant's high proportion of short chains explains its low crystallinity (9% versus 28%) and low gelatinization temperature with T_o–T_p–T_c of 47–52–56°C and enthalpy of 1.7 J/g compared to the wild type with T_o–T_p–T_c of 55–61–67°C and ΔH 6.8 J/g. The mutant did display an enhanced (ΔH 2.5 versus 1.1 J/g) amylose–lipid peak at T_o–T_p–T_c 89–96–101°C.[283]

A doubled haploid wheat population was produced by crossing a hexaploid wheat null in starch synthase IIa (SSIIa) with an Australian wheat having functional *ssIIa* alleles on all three genomes.[307] In the eight possible genotypes for SSIIa, an increase of 9% amylose was found for the triple null (44%) compared to the wild type lines (35%). The triple nulls also showed lower (10%) starch and consequently higher (3%) protein content, fewer (4%) B-type granules, fewer (23%) double helices, less (3%) crystallinity and more (28%) V-type polymorph and deformed granules. Other properties of the eight genotypes, affected mostly by amylopectin structure, changed gradually from wild type through single, double and triple nulls. As the amylopectin molecule lost chains of DP 11–25 and gained chains of DP 6–10, the starch gelatinization temperature, swelling power and pasting parameters gradually declined.

A more effective method of increasing amylose content during biosynthesis of starch is to reduce the level of branching enzyme. Cereals have three isoforms of starch-branching enzyme (SBE), SBE I, SBE IIa and SBE IIb.[233,308] SBE IIb is the major isoform in the endosperm of corn and rice, whereas SBE IIa is the major one in wheat. The amylose-extender mutation in corn, which is caused by the loss of SBE IIb, gives high-amylose corn with 50–90% amylose, whereas that mutation in rice increases amylose only up to ~35%. Recently, the levels of SBE IIa and SBE IIb in wheat endosperm were reduced by more than 90% using the RNA interference (RNAi) method, which produced a high-amylose wheat with 70–80% amylose.[233,308,309] Eliminating either SBE IIb or SBE I in wheat did not yield mutants with elevated amylose; instead the starch matched the wild type. These results show that eliminating the various isoforms of SBE affects starch synthesis differently, depending on the plant species.[308] High-amylose wheats are being sought to increase the level of resistant starch in foods and to modify eating texture. Interestingly, a 'sweet wheat' phenotype with 10-fold more maltose plus sucrose per unit of fresh weight results from a double mutation that elminates both GBSS I and starch synthase IIa.[310]

5. A Unique Combination of Properties

Wheat starch possesses a unique combination of properties that is important to its utilization in various food and industrial products. Those properties are related to its color, purity, flavor, paste viscosity, paste clarity, paste texture and gel strength.[27,28,311–315]

Wheat starch is bright white in color and is probably devoid of pigments. Low levels of a yellow xanthophyll pigment (lutein) and colorless flavonoids and other phenolic compounds, as well as sterols, occur in wheat flour,[316–322] but presumably they become bound to wheat gluten or are lost as water solubles during wet-milling. (Some of the yellow pigments in corn kernels are retained by corn starch and some gray color, possibly browning reaction products, is present in potato starch. Bleaching starch with suitable reagents can improve the whiteness of starch, but adds cost.) The whiteness of commercial wheat starch is about 97% of the whiteness of magnesium oxide powder.[28]

Wheat starch is low in protein, ash and fiber, and contains no residual sulfites. Sulfur dioxide is detrimental to the viscoelastic character of wheat gluten and is not used in the commercial production of wheat starch.[28,323] The protein content of ten samples of unmodified and modified wheat starches ranged from 0.06–0.22%.[28] A 0.23% protein (0.0404% nitrogen) level in wheat starch essentially indicates a gluten-free starch, as confirmed by an enzyme immunoassay.[324] That purity of wheat starch is important in diets for celiac individuals. Wheat starch-based, gluten-free flour products were not harmful in the treatment of celiac sprue and dermatitis herpetiformis,[325] although traces of an immunoreactive gliadin can be found in wheat starch.[326]

The protein in/on wheat starch is contributed mainly by granule-bound starch synthase I (GBSSI)[241,327] and the polypeptide friabilin (15 kDa molecular weight) located on the surface of the granule and associated with the mechanical properties of wheat endosperm.[328–335] Friabilin is constructed of a mixture of proteins, predominantly puroindoline-a and puroindoline-b (MW~13 kDa), which are 60% similar in sequence and both of which have a tryptophan-rich domain that binds membrane lipids. Mutation in either gene that encodes puroindoline-a or -b, which eliminates their expression or modifies their structure, produces hard-textured kernels,[332,335] while the wild type wheat containing unaltered puroindoline-a and -b proteins is soft.

In addition to the prominent GBSSI (60 kDa), Rahman et al.[336] found four other proteins embedded in the starch granule; namely, soluble starch synthase (75 kDa); branching enzyme (85 kDa); and two polypeptides (100 kDa and 105 kDa) that appear to have starch synthase activity based on the work of Denyer et al.[337] By comparison, Japanese researchers[327,338] discovered one major protein (61 kDa) corresponding to GBSSI and five minor proteins, four of which have soluble starch synthase activity (115 kDa, 108 kDa, 100 kDa and 80 kDa) and one with starch-branching enzyme activity (92 kDa). Gao and Chibbar[339] categorized the major polypeptides embedded in wheat starch granules into four groups, the 100–115 kDA starch synthases;[327,338,340,341] the 87 kDa starch-branching enzymes;[336,342–344] the 75–77 kDa starch synthase I;[340,345] and the 57–63 kDa granule-bound starch synthase I. A fifth group may be represented by the two starch granule-bound proteins (140 kDa and 145 kDa) thought to be variants of starch-branching enzyme Ic; these proteins may impact granule size distribution.[159]

Lysophospholipids that are trapped in wheat starch granules contain nitrogen from the choline and ethanolamine groups, so they contribute to the apparent protein content of wheat starch. Well-purified starches from soft wheats contain 0.014–0.019% lysophospholipid nitrogen, 0.0034–0.0086% friabilin nitrogen and 0.0130–0.0140% integral nitrogen from GBSSI.[329] Since the composition of lysophospholipids in wheat starch is practically constant, the phosphorus level of wheat starch can be multiplied by 16.4 to measure the lysophospholipid content.[346] The surface protein on wheat starch becomes more hydrophobic with age, heat or chlorine treatment, which enhances the oil-emulsification of wheat starch and baking performance.[347,348]

Wheat starch often has a blander flavor compared to other cereal starches, presumably because of differences in lipids. It contains around 1% lipids, which occur mainly as lysophospholipids.[346,349–351] The amount of lipids in wheat starch, as determined by hydrolysis, is affected by the type of modification.[28] The raw cereal flavor in some cereal starches may be due to the oxidation of free fatty acids that are not complexed with amylose due to their limited solubility. Many volatile compounds in wheat and other starches were confirmed to be degradation products of lipid peroxidation.[352] Soaking wheat starch in aqueous sodium hydroxide at pH 11.5–12.3 diminishes the cereal odor.[353] An electronic nose was employed to measure the characteristic odors of starches from wheat, corn, potato and tapioca starches.[28]

Wheat starch lysophospholipid forms tiny liposomes in water that could readily be transported into the interior of a starch granule during its development. Solid-state nuclear magnetic resonance spectroscopy suggests that the phospholipids in wheat starch are predominantly complexed with amylose in an amorphous form in the granules.[354–356]

As indicated by amylography, the paste viscosity of wheat starch is lower than that of corn starch.[312,314] This property offers advantages in soups, sauces and gravies, and in wallpaper paste, where high solids content without excessive thickening is desired. The viscosity difference is explainable by the higher swelling power, larger molecular size of amylopectin and lower phospholipid content of corn starch relative to wheat starch.[205,311,357] The viscosity of an aqueous suspension of gelatinized starch granules is dependent on the volume fraction occupied by the swollen granules, their shape and their deformability; the dissolved starch polysaccharides in the continuous phase making a relatively minor contribution.[358] Both the swelling and deformability of gelatinized wheat starch granules exhibit a strong dependence on the heating rate during pasting. Wheat starch generates higher viscosity than corn starch when cooked at 4.5°C/minute (a high heating rate) in a Rapid ViscoAnalyzer.[359] The viscosity-enhancing effect of a faster heating rate was corroborated by Jacobs et al.,[360] Maningat and Seib,[28] and Yun and Quail.[361] Others previously reported that the swelling and solubilization of wheat starch increases with increased heating rate, as determined in an Ottawa Starch Viscometer and a coaxial cylinder viscometer.[358,362] It was postulated that viscosity is enhanced either because wheat starch granules experienced shorter shearing time at faster heating rates,[360] or because of concomitant increases in both granular swelling and leached solubles with an accelerated heating rate.[358]

Several factors affect the paste viscosity of wheat starch.[28,104,311] Notable among them are wheat class, amylose content, cultivar or genetics, pH, granule size, viscosity-measuring instrument, starch concentration, heating rate, chemical modification and interactions with other ingredients. There is good evidence that paste viscosity is negatively associated with total amylose content in wheat starch between 23% and 43%,[363,364] which can be attributed to the negative effect of amylose on starch swelling.[365] However, the final paste viscosity of waxy wheat starch is less than that of those containing amylose, because of the high breakdown of the waxy starch paste. Nonionic surfactants increase the pasting temperature and decrease the hot paste viscosity of non-waxy wheat starch, presumably because they inhibit granule swelling, but those surfactants have little effect on pastes of waxy wheat starch.[366] Wheat starch also exhibits the phenomenon of crossover or convergence of viscosity with that of a different starch in the Rapid ViscoAnalyzer (RVA) as the concentrations of two wheat flours or wheat starches are increased.[28,226,367,368]

Shibanuma et al.[226] reported a strong positive correlation between RVA maximum viscosity and number-average and weight-average degree of polymerization of wheat amylose and number-average degree of polymerization of amylopectin. They suggested that the large molecular size of amylose and amylopectin in conjunction with low amounts of extra long chains of amylopectin contribute to the increased viscosity of pasted wheat starch suspensions. The greater proportion of long chains of amylopectin of DP >35 contributed to increased starch swelling.[365] Reconstituted starches made with mixtures of amylose and amylopectin from different sources and at different proportions showed that long branch-chain amylopectin and intermediate molecular size amylose produce the greatest synergistic effect on viscosity.[369]

Unlike essentially spherical granules, swollen A-type wheat starch granules assume different shapes depending on the degree of cooking or degree of swelling. The folded, to saddle, to convoluted, to complex puckered shapes are readily visible using a scanning electron microscope.[194,195] Rheological and microscopic examination of wheat starch pastes illuminates heat-triggered changes in starch granular and molecular properties.[370,371] Wheat starch paste consistently shows higher clarity than corn starch paste, as indicated by percent transmittance at a wavelength of 650 nm.[315]

Wheat starch pastes in foods have traditionally been known to provide a 'melt-in-your-mouth' sensation. A fruit filling made from hydroxypropylated phosphate crosslinked wheat starch had a short, smooth mouthfeel with quick release from the palate.[313] The mouthfeel of fillings made from similarly modified waxy corn starch were more gummy and sticky, with a slower release. Because of fast release, fillings made from modified wheat starch also seem to have more intense flavor. More solubles are leached from hydroxypropylated wheat distarch phosphate than from a correspondingly modified waxy corn starch. It is speculated that these solubles, which are modified amylose molecules, act to disperse swollen granules. Water from the mouth quickly dilutes the soluble phase, leading to a thinning of the paste. On the contrary, the swollen granules of modified waxy corn starch give a cohesive network that is difficult to disperse. Another possible mechanism of the sticky versus clean mouthfeel of a starch paste may be related to the stiffness of the swollen granules. It seems plausible that swollen modified wheat starch granules may be less pliable than comparable swollen modified waxy corn starch granules, and that highly swollen granules can contort, conform and adhere to structures in the mouth, such as a tooth's surface.

Starch gels from wheat are stronger than those from corn starch if the starch concentration is above 6%.[314] Gel strength develops quickly (within an hour) upon cooling a starch paste; below concentrations of 6% starch solids, corn starch gives a firmer gel than wheat starch. Wheat amylose contains more linear molecules[205] than corn amylose which helps explain the higher gel strength of wheat starch relative to corn starch at 6–25% solids. Starch gels are depicted as composite structures which gelatinized starch granules are embedded in, and reinforce, a continuous phase amylose gel matrix.[372] The stiffness of the gel is dependent on the stiffness of the amylose matrix gel, the deformability of the gelatinized granules, and interactions between the continuous and discontinuous phases.

V. Modification of Wheat Starch

Native starch is abundant and economical, but there are limitations to its functional properties that render it unsuitable for the many demanding conditions of modern food and industrial processes. When starch is used as a thickener or gelling agent, for example, shear forces and/or changes in temperature, pH or salinity and the presence of other substances may cause instability of the hydrated granules and molecules, resulting in altered viscosity, texture or degree of syneresis, for example. These problems can be ameliorated, and in most cases overcome, by modification of the starch.[373–389] BeMiller[390,391] offered several promising possibilities for making novel chemically-modified starch products, viz., control of reaction sites within granules and on amylose and/or amylopectin molecules, use of new commercial sources of starch, and biological modification of existing base starches.

The five major starches of commerce (waxy maize, tapioca, potato, corn and wheat) are commonly modified to improve rheological, processing and storage performance, in food and non-food systems. In countries where wheat is the major cereal crop, such as in Europe and Australia, chemical modification of wheat starch has been practiced for years.[6] In the US, this technology has been implemented on a substantial scale only since the 1990s.[27,28] Knight and Olson[6] reported that, in Australia, wheat starch has been modified to produce granular or pregelatinized forms of oxidized, acid-treated, etherified and dextrinized products. Other types of modified wheat starch and their uses are discussed by Cornell and Hoveling.[49] In a series of studies, laboratory-prepared and commercially-modified wheat starches from Australia were characterized for their water-binding capacities, gelatinization properties and digestibilities.[392–398] Crosslinked, substituted, doubly-modified, bleached and oxidized wheat starches have been analyzed for their paste and gel properties, and assayed for their substituent groups using ^{1}H- and ^{31}P-NMR spectroscopy,[27,399–404] NMR,[404–406] infrared spectroscopy,[407,408] x-ray microanalysis,[409] gas chromatography-mass spectrometry,[406,410,411] Raman spectroscopy;[412] enzymic and titration[413–416] methods have been used to determine the types and amounts of modifying groups. The changes in the properties of modified wheat starches resemble those of other botanical starches with similar modifications and levels of amylose.

1. Crosslinking

Large granule wheat starch can be crosslinked in aqueous slurry at pH 11–12 by reaction with phosphoryl chloride (phosphorus oxychloride, $POCl_3$) at 0.005–0.2% reagent level (starch basis, sb) in the presence of sodium chloride or sulfate at 1–10% (sb).[27,417] The gelatinization temperature (starch:water = 1:4 w/w) of wheat distarch phosphate prepared with 0.03% $POCl_3$ is almost the same as that of native or control starch with T_o, T_p and T_c at 62°, 67° and 73°C (ΔH 6.8 J/g), whereas the product prepared with 0.2% phosphoryl chloride gives a somewhat increased gelatinization temperature, viz., 63°, 68° and 76° (ΔH 7.4 J/g).[417] The distarch phosphate prepared with 0.005% $POCl_3$ gives reduced hot water solubles, and its amylograph pasting curve at 10% solids is characterized by high peak, setback and cold-paste viscosities and a low degree of breakdown.[27] Increasing the crosslinking level, from 0.008% to 0.03% POCl$_3$, results in an increase in pasting temperature, an increase in thickening power and a decrease in hot water solubility (see Table 10.13). Wheat starch reacted with 0.01% POCl$_3$ has the highest peak and cold paste viscosities at 10% solids. A further increase in the amount of crosslinker above ~0.04% (sb) results in a decrease in the final viscosity of the paste at 50°C. Products made with 0.02% and 0.03% POCl$_3$ have low breakdown viscosities and, therefore, high resistance to shear thinning.[27]

When plotted as a function of the level of crosslinker, both setback and final viscosity at 50°C exhibit a curvilinear relationship.[27] An increase in setback viscosity and final viscosity at 50°C is observed with an increasing level of POCl$_3$ only up to ~0.01% POCl$_3$, beyond which any increase in crosslinking agent results in their decrease. A negative linear dependence of hot water solubility with the level of crosslinking agent is observed,[27] although the soluble starch exerts a relatively minor effect on paste viscosity.[358] The weight percent of macromolecules leached into 98.5 parts of water at 95–97°C is reduced from 34% for unmodified wheat starch to 23% for distarch phosphate made with 0.03% POCl$_3$. The extent of crosslinking of wheat starch with 0.01–0.1% phosphoryl chloride can be followed by the increase in fluidity (2% solids) and by the decrease in clarity (0.2% solids) of the distarch phosphate swollen in 0.25 M sodium hydroxide at 25°C.[418]

A starch paste is a composite structure of swollen starch granules suspended in a dilute solution of mostly amylose.[419] It is hypothesized that crosslinking of starch influences paste viscosity by two main mechanisms that operate in opposite directions. Crosslinking inhibits the loss of solubles from the granules and mechanically stiffens them, which builds paste viscosity, yet crosslinking reduces swelling of the granules and reduces their volume fraction in an aqueous medium, which diminishes viscosity. Wheat starch crosslinked with ~0.01–0.03% POCl$_3$ gives a high peak viscosity because of the prominence of the stiffening mechanism. However, no peak viscosity is observed in the heating cycle of amylograms for 0.04, 0.05, 0.06, 0.075 and 0.10% POCl$_3$-treated wheat starch. Instead, the viscosity is low when the paste temperature reaches 95°C, but it continues to rise as the paste is held at 95°C, apparently because the anti-swelling mechanism predominates. Prolonged heating of the paste at 95°C allows slightly increased swelling of the granules crosslinked with 0.04–0.10%

Table 10.13 Effect of degree of crosslinking on pasting properties of large granule wheat starch at 10% solids[27]

Phosphoryl chloride added	Pasting temperature °C	Amylograph viscosity, BU				
		Peak	Heated to 95°C	Final at 95°C	Cooled to 50°C	Final at 50°C
0	78	1130	1110	890	1780	1915
0.005%	78	1180	1170	1040	2020	2120
0.008%	81	1210	1205	1070	2040	2140
0.010%	82	1230	1220	1100	2100	2240
0.020%	83	1190	1170	1100	1955	2065
0.030%	83	1190	1150	1120	1870	2010

POCl$_3$. Under the microscope, the crosslinked granules heated to 80°C appear folded, but not convoluted, which corroborates restricted granule swelling.[27]

The size distribution of granules of unmodified and crosslinked wheat starches were measured by laser diffraction particle-size analysis, and changes were followed with time as each starch (0.2%) was heated in water at 80°C.[417] The unmodified A granules increased in mean diameter from 20 μm to 45 μm and on to 54 μm after heating for 1 minute and 30 minutes, respectively, then the granules shrank back to a mean of 47 μm after heating for 360 minutes. In contrast, the crosslinked (0.1% POCl$_3$) wheat granules increased in mean diameter from 20 μm to 37 μm after heating at 80°C for 1 minute, then their size remained constant on heating for 360 minutes. Those results show that crosslinking of wheat starch strengthens its granules against excess swelling in hot water, thereby preventing the thermal/mechanical breakdown of its hot paste viscosity.

Phosphoryl chloride is a fuming, pungent and toxic liquid, whereas sodium trimeta- and tripolyphosphates are non-toxic solids. Those three reagents can be used to prepare distarch phosphate, although their order of reactivity is phosphoryl chloride >>> sodium trimetaphosphate >> sodium tripolyphosphate. Crosslinking of wheat starch (~33% in water) with 2% POCl$_3$ at pH 11.5 for 1 hour at 25°C in the presence of 15% (sb) sodium sulfate gave distarch phosphate with 0.28% phosphorus, while crosslinking with 10% sodium trimetaphosphate at pH 11.5 for 3 hours at 45°C in the presence of 10% sodium sulfate gave distarch phosphate with 0.32% phosphorus.[420] The phosphorus added onto the starch, beyond the 0.06% in the control, amounted to 54% for phosphoryl chloride and 9% for sodium trimetaphosphate. The two crosslinked wheat starches gave 76–86% dietary fiber by the Prosky method, and 60–65% resistant starch by the Englyst method. Crosslinked resistant wheat starch, which is in the Type 4 class of resistant starch,[421] has a low swelling power (~3 g/g) in water (0.17% solids) at 95°C compared to control wheat starch (~15 g/g), and gives no pasting curve. The RS4 wheat starch retains the A-type crystal pattern and gelatinizes at temperatures 4–11°C above the control; it is insoluble in 1 M potassium hydroxide or 95% dimethylsulfoxide. The ratio of distarch monophosphate (crosslinks) to monostarch monophosphate (grafts) determined by ^{31}P-NMR is 2:1 in the RS4 wheat starch containing 0.38% P,[401] which agrees with results on a model crosslinking reaction between methyl α-D-glucopyranoside and sodium trimetaphosphate at pH 10–13 at 27°C for 4–16 hours.[422] The ratio of distarch to monostarch monophosphates was 4:1 in a crosslinked wheat starch (0.35% phosphorus) prepared with POCl$_3$.[402]

Sodium trimetaphosphate and phosphoryl chloride were compared as crosslinking agents for waxy maize starch.[423] Sodium trimetaphosphate, and another slowly reacting crosslinking agent, epichlorohydrin, when reacted with starch at equimolar concentrations (less than 2×10^{-6} mol/g starch), produced a crosslinked starch with higher swelling volume than starch crosslinked with the fast-reacting phosphoryl chloride. In addition, the starch crosslinked with phosphoryl chloride produced the highest viscosity pastes, even though the starch crosslinked with phosphoryl chloride had the lowest swelling volume. It was suggested that the fast-reacting phosphoryl chloride created a surface crust and extra rigidity to granules or caused flocculation of granules because of a roughened surface and reduced surface hydration.

2. Substitution

Acetyl and hydroxypropyl substituents on starch chains generally decrease gelatinization temperature and enthalpy, increase hydration and swelling of granules, reduce retrogradation and syneresis of starch pastes, and lower the strength of starch gels. Hydroxypropyl ether groups are much more stable to hydrolysis than acetyl ester groups, and food regulations allow a higher level of hydroxypropylation. However, hydroxypropylated starches are more expensive to produce. Hydroxypropylated (MS 0.18, 6.1% hydroxypropyl) wheat starch absorbs more moisture at 93% relative humidity and 25°C compared to unmodified wheat starch (26% versus 21%), and hydroxypropylated starch molecules show higher mobility as determined by pulsed ^{1}H-NMR.[424] The same hydroxypropylated (MS 0.18) wheat starch gelatinized at T_o, T_p and T_c of 52°, 56° and 66°C, which is 8–10°C below that of the control. The enthalpy of gelatinization was also lower (1.5 J/g) than that of the control (6.8 J/g).[417] These gelatinization properties confirm that granule swelling is promoted by hydroxypropyl groups,[417] and the added swelling accelerates cooperative melting of starch crystals in the presence of water.

Monosubstitution of wheat starch by acetyl or hydroxypropyl groups[27] without crosslinking yields products with a reduced pasting temperature, and yields pastes with enhanced clarity and viscosity but with increased stringiness (Table 10.14). The usefulness of this product is somewhat limited and requires special applications. Amylose has been found to be hydroxypropylated ~20% more than amylopectin in corn and potato starches at a molar substitution level of 0.1 (3.5% hydroxypropyl).[425,426]

When an anionic group such as the succinate half-ester is attached to wheat starch by reaction with 4% succinic anhydride, the modified product exhibits gelatinization properties similar to those of native waxy cereal and tuber starches.[27] Sodium starch succinate swelled extensively with a high peak viscosity, but the paste viscosity was reduced considerably during cooking and yielded a stringy and cohesive paste. Addition of sodium chloride (2.3%) during amylography of sodium starch succinate produced a paste characterized by a reduced peak viscosity, reduced breakdown viscosity and slightly reduced cold-paste viscosity. By tripling the amount of salt added (7%), a further decline in hot and cold-paste viscosities was observed, but increasing the salt ten-fold more did not significantly alter the amylograph profile. Starch succinates are polyelectrolytes and, therefore, their macromolecules are highly expanded in solution. However, the viscosities of their aqueous pastes are very sensitive to soluble salts, which drastically reduce viscosity. Wheat starch phosphates give viscous pastes that are also sensitive to salts and low pH and possess little cold temperature stability.[427] Cationic wheat starches can be produced for use in the paper, textile and cosmetic industries.[428]

Compared to native wheat starch, the sodium starch l-octenylsuccinate derivative gives much higher peak and cold-paste viscosities, but with considerably reduced paste viscosity during the cooking cycle.[27] This product also exhibits emulsifying properties. When the l-octenylsuccinate derivative is complexed with aluminum ion, a modified wheat starch that flows freely when dry and resists wetting by water

Table 10.14 Comparison of amylograph viscosities and pasting temperatures of modified wheat starches (7.5% solids)[a]

Modified wheat starch	Treatment or substituent	Amylograph viscosity, BU					
		Pasting temperature °C	Peak	Heated to 95°C	Final at 95°C	Cooled to 50°C	Final at 50°C
Native	None	85	290	245	260	520	495
Acetylated	2.1% Acetyl	61	770	495	310	830	705
Acetylated distarch phosphate	0.03% $POCl_3$ 1.73% Acetyl	77	695	595	550	1400	1435
Acetylated distarch adipate	0.10% Adipic anhyd. 2.2% Acetyl	76	770	690	600	910	955
Hydroxypropyl distarch phosphate	0.01% $POCl_3$ 3.2% Hydroxypropyl	62	825	630	565	1110	1135
Sodium starch succinate	4% Succinic anhydride	57	2055	900	800	1725	1600
Sodium starch (1-octenyl)succinate	3% 1-Octenyl-succinic anhydride	62	1090	1055	625	845	780
Cationic	5% N-(3-chloro-2-hydroxypropyl)-trimethylammonium chloride	59	1240	715	425	935	765
Acid-thinned	2% HCl	61	70	50	25	60	70
Bleached	0.1% $Ca(OCl)_2$	85	155	140	130	300	290
Bleached/Acid-thinned	0.82% Chlorine/3% HCl	61	65	25	15	30	35
Oxidized/Acid-thinned	2% Chlorine/3% HCl	56	280	65	10	15	20
Oxidized	1.2% Chlorine	60	110	75	15	45	50
Oxidized	5.5% Chlorine	44	50	5	5	5	5

[a]Adapted from reference 27

is produced. In dry form, individual granules remain discrete and do not cake or agglomerate into clumps. It pours easily when dropped from a funnel and forms a flat cone with a low angle of repose. When placed in water and stirred, it resists forming a homogeneous suspension; rather the starch rises to the surface as a dry mass.

3. Dual Derivatization

Crosslinking of wheat starch with 0.01% $POCl_3$ reduces hot water solubles, increases pasting temperature, and increases amylograph viscosity (10% solids) throughout the heating and cooling cycles compared to unmodified wheat starch.[27] When that crosslinking reaction is followed by treatment with acetic anhydride, the resulting acetylated distarch phosphate (DS=0.078) exhibits enhanced clarity with a 14°C lower pasting temperature, and a more viscous paste than the native and crosslinked alone starches.

Acetylated wheat distarch phosphate pastes had improved clarity and better storage stability than the pastes of unmodified and crosslinked alone starches when 1% aqueous dispersions were heated in a boiling water bath for 30 minutes and stored at two different temperatures (4°C and 25°C) for 9 days.[27] Storage at 4°C of 5% pastes of native and crosslinked wheat starch revealed that both were unstable after one day, the soft texture of the initial pastes being transformed into an opaque and rigid gel that exhibited syneresis. In contrast, the acetylated distarch phosphate was stable for at least one month at 4°C. It did not develop undesirable properties under severe conditions of prolonged exposure to subnormal temperature. In freeze–thaw stability tests, syneresis appeared in a 5% paste of acetylated distarch phosphate after 4 freeze–thaw cycles. Improvement in viscosity, paste clarity and stability to low temperature and freeze–thaw storage of acetylated distarch phosphate was attributed mainly to the acetyl groups which promote swelling of the granules and retard association of starch chains.

Large granule wheat starch modified with 0.01% or 0.03% POCl$_3$ followed by an acetic anhydride treatment to DS values of 0.061 and 0.065, respectively, showed high solubility and swelling power and the two-stage swelling characteristic of acetylated distarch phosphate derivatives of cereal starches.[27] Microscopic observation at 80°C revealed highly swollen granules in folded and puckered shapes, a phenomenon described in detail during heating of large, lenticular (A) wheat starch granules.[194,195]

Acetylated distarch adipate prepared from large granule wheat starch exhibited varying properties depending on the level of crosslinking.[27] Treatment with 0.02 or 0.04% adipic acetic mixed anhydride, followed by acetylation to a DS of 0.087, resulted in high amylograph viscosity and 16–18°C lower pasting temperature, but the paste was stringy. By increasing the level of crosslinking reagent to 0.10–0.14% and with an acetyl DS of 0.083–0.089, the paste had comparable viscosity to that made with 0.02–0.04% adipic acetic mixed anhydride, but the more highly crosslinked product produced a smoother and shorter texture and had an 8–9°C decrease in pasting temperature.

Hydroxypropylated and crosslinked large granule wheat starch with a hydroxypropyl MS of 0.09–0.10 also has properties sensitive to the amount of crosslinking agent.[27] For example, addition of 0.01% POCl$_3$ to hydroxypropylated wheat starch resulted in a product that produced a viscous paste. Increasing the POCl$_3$ amount to 0.02–0.03% drastically reduced paste viscosities in all stages of amylography by as much as 50%. Viscosity was not enhanced by increasing the hydroxypropyl substitution to MS 0.13, if the amount of crosslinker was 0.03% (POCl$_3$). Compared to unmodified starch, a substantial reduction in pasting temperature (19–23°C) was attained during dual modification. Ono et al.[429] reported a lower pasting temperature, increased swelling and solubility, and reduced storage modulus for crosslinked and hydroxypropylated wheat starch.

Reddy and Seib[301,302] measured the paste and gel properties of crosslinked and substituted partial-waxy and waxy wheat starches. Those modified starches exhibited advantages over similarly modified regular wheat or waxy corn starch. Pastes of a partial waxy wheat starch with a low level of hydroxypropylation (1.5–2.5%) and crosslinked with 0.025% POCl$_3$ had higher viscosity than that of a wild type wheat starch modified in the same manner.[301] An hydroxypropylated (~2%) waxy wheat distarch phosphate (0.008–0.013% POCl$_3$) produced a pasting curve with higher viscosity at 6.25% solids and better freeze–thaw stability compared with a similarly-modified waxy corn starch.[302]

4. Bleaching, Oxidation and Acid-thinning

Bleaching of large granule wheat starch in powder form with solid calcium hypochlorite (0.1%) for seven days at room temperature yielded a product with amylograph properties[27] that differed from those of wheat starch oxidized with sodium hypochlorite (5.5 g chlorine per 100 g starch) in aqueous slurry at pH 8. Bleaching with calcium hypochlorite decreases the viscosity of pastes, but oxidation with sodium hypochlorite effects extensive granule fragmentation and solubilization, and reduces the viscosity and the pasting temperature to a greater extent (Table 10.14).

Low levels of oxidation (1.2% chlorine) yield intermediate paste viscosities compared to the above modifications. Acid-thinning with 2% HCl produces low viscosities, comparable to those given by oxidation with 1.2% chlorine. Sequential bleaching (0.82% chlorine) and acid-thinning (3% HCl), or oxidation (2% chlorine) and acid-thinning (3% HCl) produces low hot and cold-paste viscosities, although the latter treatment produces modified wheat starch that gives a high peak viscosity (Table 10.14).

Oxidation of wheat starch using a concentration range of 0.4–2.4% available chlorine in sodium hypochlorite solution at pH 8 causes the amylograph peak viscosity to decrease at first, then to increase with increasing chlorine concentration used for oxidation. In addition, the temperature at peak viscosity and the final viscosity at 95°C decrease as the chlorine concentration is increased.

VI. Uses of Unmodified and Modified Wheat Starches

The uses of starch permeate the entire economy[430] and a preponderance of uses of wheat starch are apparent from materials and products found in households, bakeries, restaurants, schools, offices and supermarkets.

1. Role in Baked Products

Bakery products remain the predominant application for wheat starch. The properties of wheat starch closely match those of endogenous starch in wheat flour, and wheat flour is the major ingredient around which bakery formulas and processes have been developed. In yeast-leavened bread,[431–435] wheat starch dilutes the wheat gluten to an appropriate consistency, provides maltose for fermentation through the action of amylase, provides a surface for strong bonding with wheat gluten, provides flexibility for loaf expansion during partial gelatinization while baking, sets the loaf structure by providing a rigid network to prevent the loaf collapsing on cooling, gives structural and textural properties to the baked product, holds or retains water by acting as a temperature-triggered water sink, and contributes to staling.

The loaf volumes of breads baked from wheat starch coming from different classes of wheat, be it hard red winter, hard red spring, soft red winter or soft white, were similar.[436] By contrast, club wheat starch produced a larger loaf volume and durum wheat starch a smaller loaf volume. Other investigators found a range in the loaf volumes of breads baked from wheat starches isolated from different classes of wheat.[437] Fractionation and reconstitution studies revealed that rye and barley starches can substitute for wheat starch in producing bread of satisfactory volume. Starches from corn, sorghum, oats, rice and potato produce inferior bread. Barley starch behaved functionally similar to wheat starch in Arabic bread.[438] In a separate study, it was found that breads containing wheat, potato and waxy maize starches showed anisodiametric swelling of the granules, whereas those baked with rice, tapioca, corn and amylomaize starches exhibited isodiametric swelling.[439] Microscopic examination of bread crumb showed that the swollen granules of wheat starch appeared to be more responsive to the tensile forces in gas cell walls when the dough is baking, since they align in a more closely parallel array than those of other starches. The elongated, swollen granules are aligned parallel to a cell wall surface, and some are oriented and partly fused with neighboring granules.[440] Amylose accumulates near the hilum of large starch granules. Outside the starch granules, amylose-rich zones are also observed along the starch–protein interface. In hearth bread, loaf volume was not significantly affected by waxy, partially waxy or wild type wheat flour, but waxy wheat flour produced a more open crumb grain structure and an unacceptable appearance.[441]

The unique role of wheat starch granules in controlling the expansion of dough during baking is related to its gelatinization temperature and to the integrity of swollen granules in the gluten matrix.[442] Setting of the crumb structure of bread is influenced by the amylose fraction.[443] The limited expansion of the large lenticular granules into narrow, platelike disks, and their orientation at the surface of gas cells could account for the firm, spongy texture of breads containing wheat starch. The ability of rye and barley starches to produce bread of satisfactory loaf volume can, therefore, be attributed to the presence of large lenticular granules with physical properties similar to those found in wheat starch. Limited, intermediate and advanced swelling of large granules are depicted in scanning electron micrographs of pie crust, cinnamon rolls and angel-food cake.[444] Greenwood described the stages of progressive disordering of wheat starch granules in a number of baked products, varying from a maximal effect in wafers to a minimal change in Scottish shortbread.[445]

Others suggested that the functionality of wheat starch in breadmaking is related to the adhering matter on the granule surface that serves as a linkage between starch and other components.[446] This adhering matter facilitates interaction of gluten and starch in the formation of wheat flour dough for making a satisfactory bread.

The influence of granule size distribution on bread characteristics has been the subject of several investigations. A significant effect of granule size distribution on amylograph pasting and paste properties, dough behavior during mixing in the farinograph, and viscoelastic properties of dough in the extensograph has been reported.[142,143,148,199,447,448] Small granules increase the extensibility of the dough, whereas large granules increase the resistance to extension.[449] It was reported that small granule starch from hard red winter wheat possesses bread baking characteristics nearly equal to the starch in control wheat flour.[436] Others, however, reported that small granule wheat starch is inferior in breadmaking quality and that loaf volume is inversely related to the relative number of small wheat starch granules.[140,450,451] Soulaka and Morrison[140] found an optimum level (25–35% by weight, sb) of small granules required to produce satisfactory loaf volume beyond which bread volume decreases. An optimum range of B granules in wheat cultivars to produce the best crumb grain score was shown to be 20–23% in a pup-loaf bread study. Crumb grain and texture were highest when the flour was reconstituted with 30% small granules and 70% large granules in straight dough bread.[452,453] Lelievre et al.[454] found that the optimum starch granule size varied with protein content. Wheat starch greatly affects the toughness of the crumb in a lean-formula product, but affects soft-textured bakery foods in a complex manner dependent on the protein level. The effect of granule size on bread characteristics is altered by process conditions, such as work input and mixing speed.[455] Hayman et al.[456] demonstrated that the presence of a greater proportion of large starch granules in dough results in gas coalescence and an open crumb grain in bread. Small granule starch from European wheats imparted interesting properties to bakery products, batters and coatings.[457] It served as a good binder in coatings and batters, because of the large contact surface between the starch granules and the coated product.

Hahn[458] reviewed the use of starch in other bakery products (pies, cakes, biscuits, crackers and meringues). The unique functional properties of wheat starch are demonstrated in other baked goods in which the product quality is balanced or improved by the addition of wheat starch. It makes possible the use of hard wheat flour for cake production, and the improvement in the quality of cakes relative to those baked with soft wheat flour alone. Replacement of 30% of the cake flour with wheat starch in angel-food and other foam-type cakes results in significant improvements in volume, grain, texture, eating quality and freshness retention,[459,460] and it contributes to the structure of angel-food cake when using a reduced egg white formula.[461] Wheat starch out-performs potato, corn and tapioca starches in the production of high-quality angel-food cake. Wheat and potato starches have a comparable performance in white layer cakes that is superior to that of corn and tapioca starches. Addition of up to 30% wheat starch to low-grade cake flour improves the volume and quality of both angel-food and white layer cakes. Complete elimination of chlorinated wheat flour in high ratio cake formulations has been achieved by substituting a blend of untreated wheat flour (hard or soft) and wheat starch.[462] A blend of 40% wheat starch and 60% non-chlorinated wheat flour has been used successfully to make high ratio layer, angel-food, pound, Texas sheet and cup cakes.[463] Cauvain and Gough[464] were successful in substituting a wheat starch–carboxymethylcellulose blend for flour in a high ratio yellow cake. Wheat and potato starches, but not corn starch, produced acceptable high ratio cakes.[465]

Reconstituted flours containing unbleached rye starch gave surprisingly good quality cakes, whereas barley and potato starches produced cakes with better volume and a higher quality score than those made with wheat starch.[466] Rice and corn starches produced poor quality cakes. In Japan, wheat starch improves the quality of sponge cakes,[467–469] and pregelatinized hydroxypropylated wheat starch, in particular, imparts a soft texture with good moisture retention and retards firming of the crumb.[470] Contrasting results were obtained in two studies when the functionality of wheat, potato and corn starches was evaluated in German sand cakes.[471,472]

Substitution of wheat starch for 30% of the flour in high fat products, such as pastries, increases product tenderness such that a 17–20% reduction in shortening is possible.[473] To prevent cracking of European biscuits and warping of wafers, wheat starch at a level of 5–30% based on flour is added.[457] Ungelatinized wheat starch increases cookie spread when used to replace 30% of the flour; cookie spread is decreased with pregelatinized wheat starch.[474] Rye, barley and potato starch in reconstituted flours produced cookies almost equal in diameter to wheat starch cookies.[466] Marginal products were obtained using corn or rice starch in reconstituted flours for cookie baking. Wheat starch does not form a continuous structure in cookies, as the granules remain in their ungelatinized condition after baking.[475] Thus, staling by retrogradation of wheat starch cannot occur during cookie storage. Moisture interchange among cookie components contributes to observed textural changes.

In the US, unmodified wheat starch is used in doughnut sugar, pie crusts, cake and doughnut mixes, snack cakes, cup cakes, doughnuts, cookies, cheese analogs, angel-food cakes, icings, French crullers and glazes.[40] It functions as a tenderizer in doughnuts and other bakery mixes, and serves as a carrier for enrichment premixes used in flour fortification. It also serves as a bulking agent or carrier for cheese and meat flavors. Modified wheat starch has special applications in dusting, waffle mixes, biscuit dry mixes, cereals, surimi, dairy products, biscuits, muffins, brownies, cakes, fruit fillings, gravies, sauce mixes, pizza crust, cake mixes, snack cakes and doughnut mixes. In Japan, modified wheat starches have particular applications in fish cakes (kamaboko), sausages, cake mixes, frozen foods and Chinese dumplings (gyoza). Takahashi et al.[313] reported the benefits of using wheat starch and its hydroxypropyl distarch phosphate derivative in pie fillings, steamed bread, udon noodles and flexible starch sheet for shrimp dumplings.

RS4-type resistant wheat starch is conveniently incorporated into bakery and non-bakery products to increase dietary fiber, to reduce calories and/or to reduce fat. In particular, high protein, high fiber white and whole wheat breads formulated with RS4-type resistant wheat starch displayed greater volume and softer crumb than the control bread.[476] Furthermore, crumb firming of high protein, high fiber breads decreased over time, as shown by their softer texture compared to control breads during a 10-day storage period. Commercially, RS4-type resistant wheat starch is used in formulations for flour tortillas, breads, pizza crusts, bagels, English muffins, cookies, pastries, muffins, pretzels and pasta. RS-3 type resistant starch has been prepared from wheat starch containing 39.5% amylose.[477]

It has been stated earlier that wheat starch in baked products contributes to the staling process. Bread staling, as indicated by crumb firming, has been linked for years to amylopectin recrystallization within swollen wheat starch granules.[478–482] Results from fractionation and reconstitution studies[483] using crosslinked waxy barley starch in place of wheat starch suggest that amylose dilutes the effect of amylopectin, the recrystallization of which is at least partially responsible for the firming of bread crumb. Blanshard[484] acknowledged that while staling may involve changes in the crystallinity of amylopectin inside gelatinized granules, amorphous molecules in the intergranular space may also exert a profound influence on the overall rheological behavior of the baked product. On the other hand, it has been hypothesized that the reorganization of the intragranular amylose fraction enhances the rigidity of starch granules during bread staling.[440]

A model of bread firming proposes that the continuous gluten protein matrix is denatured and crosslinked by entanglements and/or hydrogen bonds to the discontinuous remnants of the swollen starch granules.[485–487] Morgan et al.[488] provided a contrasting view and concluded that bread firming can be explained by changes in starch–starch interactions, because gluten-free starch breads firmed at the same rate as regular wheat breads. According to Every et al.,[489] increasing bread firmness results from molecular chains of partially-leached amylose and amylopectin attached to swollen starch granules forming hydrogen bonds with other swollen starch granules and, to a lesser extent, with gluten fibrils. The mechanical properties of starch breads are determined by cell geometry and by the mechanical properties of lamellae and beams formed by irregularly-shaped, partly-swollen starch granules.[490,491]

Another laboratory invoked the role of water in bread staling involving a zipper mechanism that results in the formation of interchain crosslinks.[492] Moisture redistribution from crumb to crust played a significant role in crumb firming and amylopectin recrystallization for breads stored for more than seven days.[493] According to Piazza and Masi,[494] to inhibit staling it is more important to slow the dehydration phenomena than it is to increase the initial moisture content in the bread.

When modified wheat starch was used to replace 20% of the flour, the rate of crumb firming was delayed by hydroxypropylated wheat starch, but enhanced by oxidized, aluminum starch 1-octenylsuccinate and oxidized-hydroxypropylated wheat starches.[495] Other additives such as alpha-amylase from microbial or cereal origins and emulsifiers have been shown to retard the staling process, as measured by crumb firming.[486,496,497] It was concluded that the anti-staling effect of amylases is due to depolymerization of starch molecules or to modification of the starch interfering with starch–gluten interactions, and not to the production of dextrins.[498,499] Hydrolytic activity results in weaker mechanical properties of the three-dimensional network, which then decreases firmness.[500] Several researchers, however, postulated that the anti-firming effect of amylases was related to the presence of dextrins or maltodextrins,[485,486,501–504] and the magnitude of the effect was related to the size and ratio of maltodextrins.[505] The anti-staling effect of emulsifiers may be due to interaction with amylopectin chains,[506–508] thereby interfering with amylopectin recrystallization.

2. Functionality in Noodles and Pasta

Wheat starch is the major component of pasta and represents about 70% of its weight. However, wheat protein is the major determinant of spaghetti cooking quality and starch plays a secondary role.[509–511] Starch functionality in pasta is somewhat ignored, as evidenced from the scant information available in the literature. Delcour et al.[512] provided insight into the role of durum gluten and starch interactions in spaghetti quality using fractionation and reconstitution experiments.

The effect of spaghetti processing and cooking on the starch component of semolina or spaghetti has been investigated. Loss of semolina shape during extrusion was shown by microscopic examination.[513] Matsuo et al.[514] found an alignment of starch granules in the direction of flow for the material taken at the first kneading plate of the pasta extruder. In the extruded spaghetti, granules are embedded within a protein matrix and those granules on the surface of spaghetti strands are coated with a smooth protein film. Examination by amylography of starch isolated from semolina or raw spaghetti revealed that semolina starch has a higher peak viscosity than spaghetti starch.[515] The different profile may be attributed to an annealing process or to the presence of damaged starch in spaghetti.

Wheat starch undergoes changes during the drying of spaghetti, especially when high temperature or ultra high temperature drying cycles are employed. High temperature drying caused modifications in the starch that significantly improved spaghetti cooking quality.[516] It was also observed that the amount of resistant starch significantly increased with high temperature treatment.[517] Furthermore, the starch isolated from the spaghetti exhibited a higher gelatinization temperature than the starch isolated from semolina.[517,518] Results of the studies on the effect of high temperature drying of spaghetti on starch gelatinization range, enthalpy of gelatinization and paste viscosity have been inconsistent.[517–520]

Water penetration during the cooking of spaghetti is mainly a function of protein content. Starch gelatinization takes place in an inward direction and occurs at a rapid rate at a low protein concentration.[521,522] When viewed by scanning electron microscopy, cooked spaghetti exhibits a filamentous network near the outer surface that is a starch-coated protein network interconnected by starch fibrils.[523] Resmini and Pagani[524] reported a physical competition between protein coagulation into a continuous network and starch swelling and gelatinization during cooking of pasta. They postulated that, if protein coagulation prevails, starch particles will be trapped in the protein network, promoting firmness in pasta. If starch swelling and gelatinization prevail, the protein will coagulate in discrete masses rather than a continuous framework, resulting in soft and sticky pasta.

Starch changes during cooking of pasta are reported to vary from a hydration-driven gelatinization process in the outer layer to a heat-induced crystallite melting in the center.[525] It is speculated that both the state of the starch and the surface structure contribute to the development of the elastic texture and stickiness of pasta. Interactions between starch and the surrounding protein matrix are evident in the outer and intermediate layer. In the center of cooked pasta, wheat starch granules retain their shape due to limited water diffusion, and the protein network remains dense.

Interchange of starch isolated from durum wheat and a hard red spring wheat had little effect on macaroni cooked weight, cooking water residue or firmness of cooked macaroni.[526] The importance of amylose content in determining spaghetti cooking quality has been emphasized.[527] Strands of cooked spaghetti made with low-amylose starch lacked resilience. Blends of gluten and non-wheat starches, such as waxy barley or waxy maize starches, resulted in spaghetti of poor cooking quality. By contrast, normal barley starch appeared to impart better cooking quality compared to other starches. It was speculated that other starch properties may impart better cooking quality to spaghetti once a certain level of amylose is present. In a reconstitution study, the optimum level of amylose was determined to be 32–44%.[528] This amylose range yielded an extensible dough with increased spaghetti firmness and decreased water absorption. Total absence of amylose is detrimental, as confirmed in a reconstitution study where waxy wheat starch yielded poor quality spaghetti, judging from the decreased firmness and increased stickiness of the cooked spaghetti.[529] In a related study, waxy durum wheat produced softer pasta with higher cooking loss than pasta made from traditional durum wheat cultivars.[530]

Nelson[531] investigated the properties of spaghetti made from blends of modified starch and semolina. Addition of up to 10% of commercially-modified starches from normal corn, waxy maize and tapioca starches, and laboratory-modified durum starch did not adversely affect spaghetti color and strand diameter. Firmness scores decreased as the percentage of starch in the blend increased. The level of starch added did not significantly affect spaghetti cooked weight and cooking loss. The type of commercially modified starch was found to have a significant effect on spaghetti cooked weight, but the type of modified durum starch did not significantly affect cooked weight or cooking loss. Using a reconstitution approach, small granule wheat starch imparted the highest cooked spaghetti firmness when added at 10–15% above the level normally found in durum wheat starch.[528]

Wheat noodles are consumed in large amounts in China, Japan, Korea, Malaysia, Taiwan, Singapore, Hong Kong and the Philippines.[509] Oriental wheat noodles may be divided broadly into two classes: the Japanese white salted noodle; and the Chinese (Cantonese) yellow alkaline noodle. The first class is made from wheat flour (100 parts), water (35 parts) and table salt (1–2 parts); whereas the second class is made by replacing table salt with a mixture of sodium and potassium carbonates, often referred to as 'kansui.' Within the two classes, the five popular types of noodles are: raw; wet (boiled); dried; instant fried; and steamed and dried.[509,532,533]

Because wheat flour, which comprises 70–75% starch, accounts for >95% of the dry solids in Oriental noodles, it is not surprising that the quality of noodles varies with starch properties.[534] Quality differences of Chinese yellow noodles produced from Australian and US wheats are attributable to their starch properties.[535] The water-holding capacity of starch was strongly correlated with the viscoelastic properties of Japanese noodles.[536] Numerous researchers have provided evidence that moderately high swelling wheat starch in flour is important to the quality of Japanese salt noodles.[313,534,537–549] By contrast, wheat flours with low-swelling starch are preferred for alkaline noodles in Japan.[550]

Nagao et al.[537] were the first to report that good Japanese salt noodle flour from Australian standard white wheat gave a relatively low amylograph pasting temperature and a high pasting peak, which was later confirmed by Moss.[538] Oda et al.[539] reported that the eating quality of Japanese salt (udon) noodles correlated positively with rapid swelling of the starch in the amylograph. Others[541] showed that the starch from Australian standard white wheat flour gave a low pasting temperature and a relatively high paste consistency, and that the Australian standard white wheat starch had a higher swelling power compared to the starch from a representative Japanese wheat. High swelling of starch in flour also affects the texture of alkaline noodles and instant fried noodles.[532–535,551–553]

Crosbie[543] measured the swelling power of wheat flours and starches from 13 cultivars grown in Western Australia and found that the desired texture of cooked salt noodles was correlated positively with swelling power, which agrees with the results of other researchers.[541,545] He also found a significant positive correlation between wholegrain flour swelling volume from 16 cultivars and total texture score of cooked salt noodles.[534] Addition of 10% doubly-modified wheat starch with good swelling properties to hard wheat flour increased the cutting stress and surface firmness of cooked Oriental noodles.[540] The modified granules swelled to fill many voids between gluten fibrils and yet avoided disintegration because of their resistance to the shearing action of boiling water. Wang and Seib[554] found that the swelling power at 92.5°C of three top quality Australian flours was greater (20–21 g/g) than that of 12 US wheats (15–19 g/g) representing six classes.

Peak paste viscosities of flours determined with the Rapid ViscoAnalyzer were reported to correlate significantly with salt noodle eating quality.[546] Konik and Moss[555] found that Rapid ViscoAnalyzer peak viscosities for starches and flours from 49 wheat varieties correlated positively with salt noodle eating quality, and that setback and final viscosity at 50°C correlated negatively with eating quality. Significant negative correlation was discovered to exist between the DP 5 oligosaccharides, most likely resulting from sprout damage, and noodle eating quality.[556] Furthermore, an optimum amylose content of ~22% is required for good quality salt noodles. Using blends of natural wild type wheat flour and waxy wheat flour, Guo et al.[557] confirmed that the optimal flour amylose content range for Asian salt noodles was 21–24%. Park and Baik[558] described the effect of amylose content on properties of instant noodles prepared using waxy, partial waxy and regular wheat flours and reconstituted flours with starches of various amylose content. Hardness of cooked instant noodles was positively correlated with the amylose content of the starch.

3. Other Food Uses

In the United States, modified and unmodified wheat starches are used in batters and breadings, breakfast cereals, sugar grinding, ice cream toppings, retorted soups, gravies, vitamins and flavors.[28,40] Wheat starch functions in these products to provide adhesion, structure, moisture control and/or thickening. Otherwise, wheat starch functions simply as a flow aid or a carrier. One of the major uses of wheat starch in Europe and Australia is in the production of crystalline D-glucose (dextrose) and D-glucose syrups[49,53,559,560] (see also Chapter 21). Much of the wheat starch converted to syrups is used in foods, beverages and confectionery.[566]

Minor components of wheat starch[561–564] have deleterious effects on both processing (filterability) and on final glucose syrup quality (color). Due to precipitations that occur during liquefaction and saccharification of wheat starch, hydrolyzates are difficult to filter using standard equipment, and the resulting filtrate is unacceptably cloudy.[565] The problem is caused primarily by the presence of phospholipids, arabinoxylans and nitrogenous materials. Arabinoxylans possess a high viscosity in aqueous solution, and they become insoluble by oxidative crosslinking of ferulic acid esterified to the hemicellulose. Although phospholipids are eventually removed at the final purification stage, their presence together with arabinoxylans and nitrogenous materials during early stages of syrup processing may contribute to the observed color and flavor of wheat starch syrups. Treatment with mixed phospholipase, xylanase and beta-glucanase enzymes improves the filterability of the solution and the clarity of the filtrate.[565] The treatment also reduces foaming of the hydrolyzate and makes the filtrate more amenable to ion-exchange and carbon treatment. Enzymes derived from *Aspergillus niger* rich in lysophospholipase and pentosanase activities assist in wheat starch processing for glucose syrup production.[566] Nebesny et al.[567] obtained optimal hydrolysis using a glucoamylase/lysophospholipase preparation in combination with a xylanase and a protease.

Water washed, salt solution washed, cold solvent washed and sieved wheat starches did not improve filterability and final color of glucose syrups compared to those of well washed large granule wheat and corn starches.[561] When using a well washed, large granule wheat starch fraction, the ease of processing and final syrup quality are similar in most respects to that of corn syrups.

Maltodextrins with dextrose equivalents (DE) of 1 and 10 were produced from waxy wheat starch.[568] The spray-dried wheat maltodextrin with a DE value of 10 did not develop a rancid odor on storage in a closed container, as did normal wheat maltodextrins with dextrose equivalents of 2–3.[569]

Wheat starch has been formulated in a variety of snack foods, whether they are extruded, expanded, microwaved or in the form of a stick or half products.[570–574] In the case of yogurt or custard-type milk desserts, modified wheat starches were used for their thickening or gelling properties.[575,576]

In meat products, starch is added primarily to bind water.[577] Increased levels of starch decrease the firmness of sausages and bologna, probably because of increased water retention.[577–581] The gelatinization temperature of starch is also important in determining the final product properties of comminuted sausages containing wheat, corn, tapioca or modified potato starches.[578]

Addition of 3% potato starch or a combination of potato starch and potato flour produced frankfurters that were comparable to those being successfully marketed that contained wheat starch.[582] Native wheat starch and its acetylated distarch phosphate and hydroxypropyl distarch phosphate derivatives, when incorporated at a level of 3.5%, reduced cooking loss and firmness of low fat, high added water sausages cooked to 70°C or 80°C.[583]

In heat-induced surimi gel, incorporation of 5% wheat starch increased cohesiveness and chewiness and decreased syneresis.[584] The effects of heating temperature and the water-absorbing ability of various starches on properties of kamaboko gels were described by Yamazawa[585,586] and others.[587,588] Gel strength increased when kamaboko was heated up to 90°C, although the temperature for maximum gel strength is dependent on the type of starch added. Starch that maintained its granular integrity even after heat processing, and that had high water absorption, imparted high tensile strength to kamaboko. Modified wheat starch also finds applications in other Japanese food products, such as imitation crab legs and rice confectioneries, and in special, clear-type Chinese dumplings.[28]

4. Industrial Uses

Starch, in general, is used in a number of non-food applications.[589–596] Worldwide, the majority of wheat starch is sold in unmodified form to manufacturers of paper, textile and food products.[4,6,28,49] Because of worldwide interest in ethanol and other biofuels as alternative sources of energy, a significant amount of wheat starch is utilized as a fermentation substrate in the production of alcohol[28,597] (see Chapter 9). Paper applications (see Chapter 18) include its use as a surface coating agent, wet end additive, wallpaper paste and an adhesive for corrugated board. Cationic wheat starch gives dry paper strength development comparable to that of a commercial cationic corn starch.[598] Because of its lower gelatinization temperature, wheat starch has a cost advantage over corn starch for use in corrugating adhesives, because less caustic soda is needed to reduce the gelatinization temperature of wheat starch.[6] Adhesive applications of acid-thinned/hydroxypropylated and oxidized/hydroxypropylated wheat starches are described by Chung and Seib.[599]

In Europe, wheat starch has become indispensable in the paper, glue and corrugated board industries where it has a favorable influence on the production costs and quality of fine graphic and packaging paper, cardboard, corrugated board, acoustic fiber board and single or multi-walled paper bags.[600–602] It is also used in the production of molds for the metal industry, as a binding agent in charcoal briquettes, as an additive in various glues, for flotation, for reinforcement of textile yarn, in fermentation, and in pharmaceutical tablets and cosmetic products. Furthermore, wheat starch is used on a large scale as a source of carbohydrates in calf milk replacers and additionally as a binding agent and carbohydrate source in pet and fish foods.

In laundry sizing and cotton finishing, wheat starch is considered to produce a superior stiff finish that is attributed to the bimodal granule size distribution. The smaller granules penetrate the fibers of the fabric, whereas the larger ones coat the exterior surface. Wilhelm[161] described the use of unmodified and crosslinked small starch granules from wheat, oat and amaranth as a graphic paper coating. Pregelatinized B granules are used as a milk replacer in calf feed and as a core binder.[53]

Large wheat starch granules are particularly useful as a replacement for the relatively scarce and more expensive arrowroot starch as protective (stilt) material in coated carbonless duplicating paper and as an anti-offset lithographic powder.[603–607] The original carbonless copy paper[204] transferred images according to the following scheme: in a three-sheet carbonless paper set, the back side of the top sheet was coated with a mix consisting of a micro-encapsulated leuco dye, stilt material (large wheat starch granules) and binder. The front and back sides of the middle sheet had a receptor coat and a microencapsulated coat, respectively. An acidified mineral clay was used as the receptor coat. The bottom sheet of the set had a receptor coat only on the front side. When pressure was applied to the top sheet of the set by means of typing or writing, an image was formed by the reaction between the chemicals liberated from the collapsed capsules and the contacting receptor coat.

It was determined that the ideal stilt material should have a size 1.5–2.5 times the diameter of the microcapsules or microcapsule clusters. Since the microcapsules are often agglomerated into clusters of a maximum size of 10 µm, the optimal stilt diameter should vary from 15 to 25 µm, with an average of 20 µm.[204] In addition, a round stilt material is more efficient than a polygonal material which can fracture the microcapsules by contact. It also should possess a certain rigidity to provide protection to the microcapsules during normal manipulation of the carbonless paper, but not be so rigid as to resist writing pressure. Large granule wheat starch (sometimes called calibrated wheat starch) provided the best results for optimal functionality as a stilt material, because of the appropriate granule size and smooth surface. Thermoresistant large granule wheat starch (presumably highly crosslinked) is required for modern high-speed coaters running at 700–800 meters/minute and equipped with infrared driers to provide intensive drying due to a short residence time in the drying section of the coaters.[204]

One of the primary industrial applications of pregelatinized wheat starch is in the oil field services industry to thicken drilling fluids.[4-6] In the building industry, both modified and unmodified wheat starches in pregelatinized form are used to texturize wall and ceiling coatings, as bonding agents in joint compounds for embedding joint tape, and in finishing gypsum panel joints, nail heads and metal corner beads.[28] Starch in a glassy state was prepared from wheat starch by twin-screw extrusion for use as an environmentally-friendly abrasive grit to remove paints from aircraft surfaces.[608-611] Another application of wheat starch is in cosmetics, where it was found to be non-toxic, non-irritating, non-sensitizing, and functional.[612,613] Starch, in general, is useful in cosmetic powders because of its small size, enormous surface area, mobility, porosity, slip property and absorptive capacity.[613] Wheat starch can enhance the softness and smoothness of face and body powders.[614] Several modified wheat starches have been successfully formulated in creams, lotions, depilatories, hair relaxers, liquid make-up and cosmetic powders.

Interest has been increasing in recent years to develop degradable plastics from starch, especially for disposable applications.[615-621] The early efforts in the 1970s focused on using starch granules as fillers.[589,593] More recent developments include thermoplastics that are intimate blends of starch molecules with hydrophilic vinyl polymers, such as poly(ethylene-co-acrylic acid) and poly(ethylene-co-vinyl alcohol), and with poly(ethylene glycol), polylactic acid and polycaprolactone[622-628] (see Chapter 19). The function of the compatibilizer and the relationship of composition and morphology to mechanical properties of starch–polyolefin blends are the subject of several studies.[629-633] Rigid and flexible foams, films and cushioning materials containing starch have also been developed.[634-649] Native and modified wheat starches have been included in these investigations.

VII. References

1. Olsen BT. In: Bushuk W, Rasper VF, eds. *Wheat: Production, Properties and Quality*. London, UK: Chapman and Hall; 1994 [Chapter 1].
2. Olewnik MC. In: Chung OK, Lookhart GL, eds. *Third International Wheat Quality Conference*. Manhattan, KS: Grain Industry Alliance; 2005 [Session I].
3. Radley JA. *Starch and Its Derivatives*. London, UK: Chapman and Hall; 1968.
4. Knight JW. *Wheat Starch and Gluten*. London, UK: Leonard Hill; 1965.
5. Knight JW. *The Starch Industry*. New York, NY: Pergamon Press; 1969.
6. Knight JW, Olson RM. In: Whistler RL, BeMiller JN, Paschall EF, eds. *Starch: Chemistry and Technology*. New York, NY: Academic Press; 1984 [Chapter 15].

7. Anderson RA. In: Whistler RL, Paschall EF, eds. *Starch: Chemistry and Technology*. Vol. 2. New York, NY: Academic Press; 1967 [Chapter 2].
8. Anderson RA. In: Inglett GE, ed. *Wheat: Production and Utilization*. Westport, CT: AVI Publishing Co.; 1974 [Chapter 13].
9. Cornell H. In: Eliasson A-C, ed. *Starch in Food: Structure, Function and Applications*. Boca Raton, FL: CRC Press LLC; 2004 [Chapter 7].
10. Plinus Secundus, Gaius, Naturalis Historia.
11. Overton M. *Agricultural Revolution in England*. London, UK: Cambridge University Press; 1996.
12. Bailey CH. *Cereal Chem*. 1941;18:555.
13. De frumento. In: *De Bononiensi Scientarum et Artium Instituto Atque Academia*. Vol. 2, Part 1; 1745:122–127.
14. Shewry PR, Tatham AS, Barro F, Barcelo P, Lazzeri P. *Biotechnol*. 1995;13:1185.
15. US Census Bureau. *International Data Base*; 2007.
16. Orth RA, Shellenberger JA. In: Pomeranz Y, ed. *Wheat: Chemistry and Technology*. Vol. I. St. Paul, MN: American Association of Cereal Chemists; 1988 [Chapter 1].
17. Percival J. *The Wheat Plant*. New York, NY: Dutton; 1921.
18. International Grains Council, London, UK; 1996.
19. US Department of Agriculture, Foreign Agricultural Service, World Wheat Production Consumption and Stocks. Available at: www.fas.usda/psdonline.
20. US Department of Agriculture, Foreign Agricultural Service, World Corn, Rice, Barley, Rye or Oats Production, Consumption, and Stores. Available at: www.fas.usda.gov/psdonline.
21. van der Borght A, Goesaert H, Veraverbeke WS, Delcour JA. *J. Cereal Sci.* 2005;41:221.
22. De Baere H. *Starch/Stärke*. 1999;51:189.
23. Gordon I. *Starch/Stärke*. 1999;51:193.
24. Maningat CC, Bassi S. In: *Proceedings of the International Starch Technology Conference*. Urbana, IL, June 7–9, 1999.
25. Maningat CC, Bassi S, Hesser JM. *AIB Technical Bull*. 1994;16:1.
26. Maningat CC, Bassi S. In: *Proceedings of Expanding Agriculture Co-Product Uses in Aquaculture Feeds*. Des Moines, IA, December 5–7, 1994.
27. Maningat CC. [Ph.D. Dissertation]. Manhattan, KS: Kansas State University; 1986.
28. Maningat CC, Seib PA. In: *Proceedings of the International Wheat Quality Conference*. Manhattan, KS, May 18–22, 1997:261–284.
29. Maningat CC, DeMeritt Jr. GK, Chinnaswamy R, Bassi SD. *Cereal Foods World* 1999;44:650.
30. Maningat CC, Bassi S. *PBI Bulletin*. 1997;September 6–7.
31. Maningat CC, Bassi S, Lasater GD, Seib PA. *Cereal Foods World (Abstract)*. 1994;39:622.
32. Maningat CC, Chinnaswamy R, Nie L, Bassi S. *Cereal Foods World (Abstract)*. 1995;40:641.
33. Bassi S, Maningat CC, Chinnaswamy R, Gray DR, Nie L. US Patent 5 610 277. 1997.
34. Bassi S, Maningat CC, Chinnaswamy R, Nie L. US Patent 5 665 152. 1997.

35. Bassi S, Maningat CC, Chinnaswamy R, Nie L. US Patent 5 747 648. 1998.
36. Bassi S, Murphy L, Maningat CC, Nie L. US Patent 5 780 013. 1998.
37. Bassi S, Murphy L, Maningat CC, Nie L. US Patent 5 945 086. 1999.
38. Bassi S, Maningat CC, Chinnaswamy R, Nie L, Weibel MK, Watson JJ. US Patent 5 965 708. 1999.
39. Bassi S, Maningat CC, Chinnaswamy R, Nie L. US Patent 5 977 312. 1999.
40. MGP Ingredients, Inc. Wheat Starch and Wheat Gluten Product Brochures. Atchison, KS; 2008.
41. Daniel JR, Whistler RL, Röper H, Elvers B. *Ullmann's Encyclopedia of Industrial Chemistry*. 7th edn. Weinheim, Germany: Wiley-VCH Verlag Gmbh; 2007 [Chapter 'Starch'].
42. International Wheat Gluten Association, Prairie Village, KS; 2008.
43. Authorn WJ. In: Holik H, ed. *Handbook of Paper and Board*. Weinheim, Germany: Wiley-VCH Verlag Gmbh KGaA; 2006 [Chapter 3].
44. Feiz L, Martin JM, Giroux MJ. *Cereal Chem.* 2008;85:44.
45. Sayaslan A. *Lebensm.-Wiss.-Technol.* 2004;37:499.
46. Bergthaller WJ, Witt W, Seiler M. In: Chung OK, Steele J, eds. *Proceedings of International Wheat Quality Conference II*. Manhattan, KS: Grain Industry Alliance; 2003:381–401.
47. Lindhauer M, Bergthaller W. In: Yuryev VP, Cesaro A, Bergthaller W, eds. *Starch and Starch Containing Origins: Structure, Properties and New Technologies*. Hauppauge, New York, NY: Nova Science Publishers; 2002 [Chapter 27].
48. Robertson GH, Cao TK, Ong I. *Cereal Chem.* 1999;76:843.
49. Cornell HJ, Hoveling AW. *Wheat Chemistry and Utilization*. Lancaster, PA: Technomic Publishing Co.; 1998.
50. Kema P, Helmens HJ, Steeneken PAM. *Starch/Stärke*. 1996;48:279.
51. Meuser F. In: Bushuk W, Rasper VF, eds. *Wheat: Production, Properties, and Quality*. London, UK: Chapman & Hall; 1994 [Chapter 13].
52. Barr DJ. In: Pomeranz Y, ed. *Wheat is Unique*. St. Paul, MN: American Association of Cereal Chemists; 1989 [Chapter 29].
53. Zwitserloot WRM. In: Pomeranz Y, ed. *Wheat is Unique*. St. Paul, MN: American Association of Cereal Chemists; 1989 [Chapter 30].
54. Kempf W, Röhrmann C. In: Pomeranz Y, ed. *Wheat is Unique*. St. Paul, MN: American Association of Cereal Chemists; 1989 [Chapter 31].
55. Meuser F, Althoff F, Huster H. In: Pomeranz Y, ed. *Wheat is Unique*. St. Paul, MN: American Association of Cereal Chemists; 1989 [Chapter 28].
56. Grace G. In: Applewhite TH, ed. *Proceedings of the World Congress on Vegetable Protein Utilization in Human Foods and Animal Foodstuffs*. Champaign, IL: American Oil Chemists Society; 1989.
57. Wadhawan CK [Ph.D. Dissertation]. Winnipeg, Canada: University of Manitoba; 1988.
58. Witt W, Kröner H. *Starch/Stärke*. 1988;40:139.
59. Dahlberg BI. *Starch/Stärke*. 1978;30:8.
60. Verberne P, Zwitserloot W. *Starch/Stärke*. 1978;30:337.
61. Jackson A. In: Radley JA, ed. *Starch Production Technology*. London, UK: Applied Science Publishers; 1976 [Chapter 9].

62. Christophersen C, Andersen E, Jakobsen TS, Wagner P. *Starch/Stärke*. 1997;49:5.
63. Hamer RJ, Weegels PL, Marseille JP, Kelfkens M. In: Pomeranz Y, ed. *Wheat is Unique*. St. Paul, MN: American Association of Cereal Chemists; 1989 [Chapter 27].
64. Weegels PL, Marseille JP, Hamer RJ. *Starch/Stärke*. 1992;44:44.
65. Roels SP, Courtin CM, Delcour JA. *J. Cereal Sci*. 2001;33:53.
66. Frederix SA, Courtin CM, Delcour JA. *J. Cereal Sci*. 2004;40:41.
67. Gys W, Gebruers K, Sorensen JF, Courtin CM, Delcour JA. *J. Cereal Sci*. 2006;39:363.
68. Beaugrand J, Gebruers K, Ververken C, Fierens E, Dornez E, Goddeeris BM, Delcour JA, Courtin CM. *J. Agric. Food Chem*. 2007;55:7682.
69. Krueger BR, Knutson CA, Inglett GE, Walker CE. *J. Food Sci*. 1987;52:715.
70. Radley JA. *Starch Production Technology*. London, UK: Applied Science Publishers; 1976.
71. Schofield JD, Booth MR. In: Hudson BJF, ed. *Developments in Food Proteins – 2*. New York, NY: Applied Science Publishers; 1983:1–65.
72. Phillips KL, Sallans HR. *Cereal Sci. Today*. 1966;11:61.
73. Finley JW. *J. Food Sci*. 1976;41:882.
74. Robertson GH, Cao TK. US Patent 5 851 301. 1998.
75. Saulnier L, Sado P-E, Branlard G, Charmet G, Guillon F. *J. Cereal Sci*. 2007;46:261.
76. Mohammadkhani A, Stoddard FL, Marshall DR, Uddin MN, Zhao X. *Starch/Stärke*. 1999;51:62.
77. Shewfelt L, Adams GA. *Can. Chem. Process Ind*. 1944;27:502.
78. Hilbert GE, Dimler RJ, Rist CE. *Am. Miller Process*. 1944;72:32.
79. Fellers DA, Johnston PH, Smith S, Mossman AP, Shepherd AD. *Food Technol*. 1969;23:162.
80. Johnston PH, Fellers DA. *J. Food Sci*. 1971;36:649.
81. Johnston PH, Fellers DA. US Patent 3 574 180. 1971.
82. Eynon L, Lane JH. *Starch: Its Chemistry, Technology, and Uses*. Cambridge, UK: W. Heffner and Sons; 1928.
83. Maijala M. *Food Eng. Intl*. 1976;48:41.
84. Batey I. *Starch/Stärke*. 1985;37:118.
85. Hung PV, Morita N. *Starch/Stärke*. 2005;57:413.
86. Powell EL. In: Whistler RL, Paschall EF, eds. *Starch: Chemistry and Technology*. Vol. II. New York, NY: Academic Press; 1967 [Chapter 22].
87. Jane J. *Trends Food Sci. Technol*. 1992;3:145.
88. Pitchon E, O'Rourke JD, Joseph TH. US Patent 4 280 851. 1981.
89. Jane J, Seib PA. US Patent 5 057 157. 1991.
90. Eastman JE, Moore CO. US Patent 4 465 702. 1984.
91. Rajagopalan S, Seib PA. *J. Cereal Sci*. 1992;16:13.
92. Esch FV. *Starch/Stärke*. 1991;43:427.
93. Jansma W, Mars J, Stoutjesdijk PG, Vegter HJ. US Patent 4 494 530. 1985.
94. Centrico, Inc., *The Processing of Wheat Starch*. Elgin, IL; 1987.
95. Witt W. In: *Proceedings of the International Wheat Quality Conference*. Manhattan, KS; May 18–22, 1997:231–248.

96. Svonja G. In: Campbell GM, Webb C, McKee SL, eds. *Cereals: Novel Uses and Processes*. New York, NY: Plenum Press; 1997: 177–183.
97. Flottweg GmbH, Flottweg Tricanter® Process for Wheat Starch, Vilsbiburg, Germany; 1992.
98. Meuser F, German H, Huster H. In: Hill RD, Munck L, eds. *New Approaches to Research on Cereal Carbohydrates*. Amsterdam, The Netherlands: Elsevier Science Publishers B.V.; 1985:161–180.
99. Pomeranz Y. In: Pomeranz Y, ed. *Wheat Chemistry and Technology*. 3rd edn. Vol. II. St. Paul, MN: American Association of Cereal Chemists; 1988 [Chapter 4].
100. Lineback DR, Rasper VF. In: Pomeranz Y, ed. *Wheat: Chemistry and Technology*. Vol. I. St. Paul, MN: American Association of Cereal Chemists; 1988 [Chapter 6].
101. Hopkins CY, Graham RP. *Can. J. Res.* 1935;12:820.
102. Wooton M, Mahdar D. *Starch/Stärke*. 1993;45:255.
103. Hucl P, Chibbar RN. *Cereal Chem*. 1996;73:756.
104. Seib PA. *Oyo Toshitsu Kagaku (J. Appl. Glycosci.)*. 1994;41:49.
105. Shi Y-C, Seib PA, Bernardin J. *Cereal Chem*. 1994;71:369.
106. Tester RF, Morrison WR, Ellis RH, Piggott JR, Batts GR, Wheeler TR, Morison JIL, Hadley P, Ledward DA. *J. Cereal Sci*. 1995;22:63.
107. Lin PY, Czuchajowska Z. *Cereal Chem*. 1997;74:639.
108. Gaines CS, Raeker MO, Tilley M, Finney PL, Wilson JD, Bechtel DB, Martin RJ, Seib PA, Lookhart GL, Donelson T. *Cereal Chem*. 2000;77:163.
109. Hou HX, Dong HZ, Zhang H, Song XQ. *Cereal Chem*. 2008;85:252.
110. BeMiller JN. *Carbohydrate Chemistry for Food Scientists*. 2nd edn. St. Paul, MN: AACC International; 2007.
111. Hizukuri S, Abe J, Hanashiro I. In: Eliasson A-C, ed. *Carbohydrates in Foods*. 2nd edn. Boca Raton, FL: CRC Press, Taylor and Francis Group; 2006 [Chapter 9].
112. Eliasson A-C, Gudmundsson M. In: Eliasson A-C, ed. *Carbohydrates in Foods*. 2nd edn. Boca Raton, FL: CRC Press, Taylor and Francis Group; 2006 [Chapter 10].
113. Björck I. In: Eliasson A-C, ed. *Carbohydrates in Foods*. 2nd edn. Boca Raton, FL: CRC Press, Taylor and Francis Group; 2006 [Chapter 11].
114. Zobel HF, Stephen AM. In: Stephen AM, Phillips GO, Williams PA, eds. *Food Polysaccharides and Their Applications*. 2nd edn. Boca Raton, FL: CRC Press, Taylor and Francis Group; 2006 [Chapter 2].
115. Liu Q. In: Cui SW, ed. *Food Carbohydrates: Chemistry, Physical Properties and Applications*. Boca Raton, FL: CRC Press, Taylor and Francis Group; 2005 [Chapter 7].
116. Eliasson A-C, ed. *Starch in Food: Structure, Function and Application*. Boca Raton, FL: CRC Press; 2004.
117. Yuryev VP, Cesaro A, Bergthaller W, eds. *Starch and Starch Containing Origins: Structure, Properties and New Technologies*. Hauppauge, New York, NY: Nova Science Publishers; 2002.
118. Parker R, Ring SG. *J. Cereal Sci*. 2001;34:1.
119. Thomas DJ, Atwell WA. *Starches*. St. Paul, MN: Eagan Press; 1999.
120. Buleon A, Colonna P, Planchot V, Ball S. *Intern. J. Biol. Macromol*. 1998;23:85.
121. Frazier PJ, Richmond P, Donald AM. *Starch Structure and Functionality*. Cambridge, UK: The Royal Society of Chemistry; 1997.

122. Meuser F, Manners DJ, Seibel W, eds. *Plant Polymeric Carbohydrates*. Cambridge, England: Royal Society of Chemistry; 1993. [and Progress in Plant Polymeric Research. Hamburg, Germany: B. Behr's Verlag Gmb; 1995]
123. Morrison WR. In: Shewry PR, Stobert K, eds. *Seed Storage Compounds: Biosynthesis, Interactions, and Manipulations*. Oxford, UK: Clarendon Press; 1993.
124. Alexander RJ, Zobel HF. *Developments in Carbohydrate Chemistry*. St. Paul, MN: American Association of Cereal Chemists; 1992.
125. Morrison WR, Karkalas J J. Methods in Plant Biochemistry. In: Dey PM, ed. New York, NY: Academic Press; 1990. *Carbohydrates*; vol. 2.
126. Bechtel DB, Zayas I, Kaleikau L, Pomeranz Y. *Cereal Chem.* 1990;67:59.
127. Bechtel DB, Wilson J. *Cereal Chem.* 2000;77:401.
128. Briarty LG, Hughes CE, Evers AD. *Ann. Bot.* 1979;44:641.
129. Parker ML. *J. Cereal Sci.* 1985;3:271.
130. Balmer Y, Vensel WH, Dupont FW, Buchanan BB, Hurkman WJ. *J. Exp. Bot.* 2006;57:1591.
131. Karlsson R, Olered R, Eliasson A-C. *Starch/Stärke.* 1983;35:335.
132. Dengate H, Meredith P. *J. Cereal Sci.* 1984;2:83.
133. Evans AD, Lindley J. *J. Sci. Food Agric.* 1977;28:98.
134. Stoddard FL. *Cereal Chem.* 1999;76:145.
135. Wilson JD, Bechtel DB, Todd TC, Seib PA. *Cereal Chem.* 2006;83:259.
136. Meredith P. *Starch/Stärke.* 1981;33:40.
137. Raeker MO, Gaines CS, Finney PL, Donnelson T. *Cereal Chem.* 1998;75:721.
138. Evers AD. *Stärke.* 1973;25:303.
139. Dengate H, Meredith P. *J. Cereal Sci.* 1985;2:83.
140. Soulaka AB, Morrison WR. *J. Sci. Food Agric.* 1985;36:709.
141. Bertolini AC, Souza E, Nelson JE, Huber KC. *Cereal Chem.* 2003;80:544.
142. Sahlstrom S, Baevre AB, Brathen E. *J. Cereal Sci.* 2003;37:285.
143. Shinde SV, Nelson JE, Huber KC. *Cereal Chem.* 2003;80:91.
144. Peng M, Gao M, Abdel-Aal E-SM, Hucl P, Chibbar R. *Cereal Chem.* 1999;76:375.
145. Ao Z, Jane JL. *Carbohydr. Polym.* 2007;67:46.
146. Vermeylen R, Goderis B, Reynaers H, Delcour JA. *Carbohydr. Polym.* 2005;62:170.
147. Liu Q, Gu Z, Donner E, Tetlow I, Emes E. *Cereal Chem.* 2007;84:15.
148. Morrison WR, Gadan H. *J. Cereal Sci.* 1987;5:263.
149. Morrison WR. In: Pomeranz Y, ed. *Wheat is Unique*. St. Paul, MN: American Association of Cereal Chemists; 1989 [Chapter 12].
150. Eliasson A-C. In: Pomeranz Y, ed. *Wheat is Unique*. St. Paul, MN: American Association of Cereal Chemists; 1989 [Chapter 11].
151. Eliasson A-C, Larsson K. *Cereals in Breadmaking*. New York, NY: Marcel Dekker; 1993.
152. Wong RBK, Lelievre J. *Starch/Stärke.* 1982;34:159.
153. Lelievre J. *Starch/Stärke.* 1975;27:2.
154. Eliasson A-C, Karlsson R. *Starch/Stärke.* 1983;35:130.
155. Bechtel DB, Zayas I, Dempster R, Wilson JD. *Cereal Chem.* 1993;70:238.
156. Morrison WR, Scott DC. *J. Cereal Sci.* 1986;4:13.

157. Rahman S, Li Z, Batey I, Cochrane MP, Appels R, Morell M. *J. Cereal Sci.* 1999;31:91.
158. Baruch DW, Meredith P, Jenkins LD, Simmonds LD. *Cereal Chem.* 1979;56:554.
159. Peng M, Gao M, Båga M, Hucl P, Chibbar RN. *Plant Physiol.* 2000;124:265.
160. Langeveld SMJ, van Wijk R, Stuurman N, Kijne JW, dePeter S. *J. Exptl. Bot.* 2000;51:1357.
161. Wilhelm EC. In: Meuser F, Manners DJ, Seibel W, eds. *Plant Polymeric Carbohydrates*. Cambridge, UK: The Royal Society of Chemistry; 1993 [Chapter 14].
162. Sebecic B, Sebecic B. *Nahrung.* 1995;39:106.
163. Moon MH, Giddings JC. *J. Food Sci.* 1993;58:1166.
164. Contado C, Reschigilian P, Faccini S, Zattoni A, Dondi F. *J. Chromatogr. A.* 2000;871:449.
165. Baruch DW, Jenkins LD, Dengate HN, Meredith P. *Cereal Chem.* 1982;60:32.
166. Panozzo JF, Eagles HS. *Austr. J. Agric. Res.* 1998;49:757.
167. Blumenthal C, Wrigley CW, Batey IL, Barlow EWR. *Austr. J. Plant Physiol.* 1994;21:901.
168. Blumenthal C, Bekes F, Gras PW, Barlow EWR, Wrigley CW. *Cereal Chem.* 1995;72:539.
169. Evers AD. *Starch/Stärke.* 1969;21:96.
170. Evers AD. *Starch/Stärke.* 1971;23:157.
171. Jane J-L, Leas S, Zobel H, Robyt JF. *Starch/Stärke.* 1994;46:121.
172. Pilling E, Smith AM. *Plant Physiol.* 2003;132:365.
173. Fannon JE, Hauber RJ, BeMiller JN. *Cereal Chem.* 1992;69:284.
174. Fannon JE, Shull JM, BeMiller JN. *Cereal Chem.* 1993;70:611.
175. Kim HS, Huber KC. *J. Cereal Sci.* 2008;48:159.
176. BeMiller JN, Huber KC. *Cereal Chem.* 1997;74:537.
177. BeMiller JN, Huber KC. *Carbohydr. Polym.* 2000;41:269.
178. Planchot V, Colonna P, Gallant DJ, Bouchet B. *J. Cereal Sci.* 1995;21:163.
179. Jane J-L, Ao Z, Duvick SA, Klund MW, Yoo S-H, Wong K-S, Garder C. *J. Appl. Glycosci.* 2003;50:167.
180. Baldwin PM, Davies MC, Melia CD. *Int. J. Biol. Macromol.* 1997;21:103.
181. Hatta T, Nemoto S, Kainuma K. *J. Appl. Glycosci.* 2003;50:159.
182. Dang JMC, Copeland L. *J. Cereal Sci.* 2003;37:165.
183. Lynn A, Stark JR. *Carbohydr. Res.* 1992;227:379.
184. Kimura A, Robyt JF. *Carbohydr. Res.* 1996;287:255.
185. Craig SAS, Stark JR. *Carbohydr. Res.* 1984;117:125.
186. Tester RF, Patel T, Harding SE. *Carbohydr. Res.* 2006;341:130.
187. Lineback DR. *Baker's Dig.* 1984;58:16.
188. Lineback DR. *J. Jap. Soc. Starch Sci. (Denpun Kagaku).* 1986;33:80.
189. Gallant DJ, Bouchet B, Buleon A, Perez S. *Eur. J. Clin. Nutri.* 1992;46:S3.
190. Gallant DJ, Bouchet B, Baldwin PM. *Carbohydr. Polym.* 1997;32:177.
191. Manelius R, Qin Z, Avall A-K, Andtfolk H, Bertoft E. *Starch/Stärke.* 1997;49:142.
192. Meredith P, Dengate HN, Morrison WR. *Starch/Stärke.* 1978;30:119.
193. Bowler P, Williams MR, Angold RE. *Starch/Stärke.* 1980;32:186.
194. Williams MR, Bowler P. *Starch/Stärke.* 1982;34:221.

195. Morrison WR. *Cereal Foods World.* 1995;40:437.
196. Kulp K. *Cereal Chem.* 1973;50:666.
197. Ghiasi K, Variano-Marston E, Hoseney RC. *Cereal Chem.* 1982;59:86.
198. Gray VM, Schoch T-J. *Stärke.* 1962;14:239.
199. Sebecic B, Sebecic B. *Nahrung.* 1996;40:256.
200. Park S-H, Chung OK, Seib PA. *Cereal Chem.* 2005;82:166.
201. Hung PV, Morita N. *Carbohydr. Polym.* 2005;59:239.
202. Stapley JA, BeMiller JN. *Cereal Chem.* 2003;80:550.
203. Lindebrom N, Chang PR, Tyler RT. *Starch/Stärke.* 2004;56:89.
204. Nachtergaele W, van Nuffel J. *Starch/Stärke.* 1989;41:386.
205. Takeda Y, Hizukuri S, Takeda C, Suzuki A. *Carbohydr. Res.* 1987;165:139.
206. Banks W, Greenwood CT. *Starch and Its Components.* New York, NY: Halsted Press, Wiley J. & Sons; 1975:77–82.
207. Hizukuri S, Takeda Y, Abe J, Hanashiro I, Matsunobu G, Kiyota H. In: Frazier PJ, Richmond PR, Donald AM, eds. *Starch Structure and Functionalities.* Cambridge, UK: The Royal Society of Chemistry; 1997:121–128.
208. Murugasan G, Shibanuma K, Hizukuri S. *Carbohydr. Res.* 1993;242:203.
209. Hanashiro I, Takeda Y. *Carbohydr. Res.* 1998;306:421.
210. Takeda Y, Hanashiro I. *J. Appl. Glycosci.* 2003;50:163.
211. Morell MK, Samuel MS, O'Shea MG. *Electrophoresis.* 1998;19:2603.
212. Wattebled F, Dong Y, Dumez S, Devalle D, Planchot V, Berbeay P, Vyas D, Colonna P, Chatterjee M, Ball S, D'Hulst C. *Plant Physiol.* 2005;138:184.
213. Yoo SH, Jane JL. *Carbohydr. Polym.* 2002;49:297.
214. Franco CML, Wong KS, Yoo SH, Jane JL. *Cereal Chem.* 2002;79:243.
215. Chung JH, Han JA, Yoo B, Seib PA, Lim ST. *Carbohydr. Polym.* 2008;71:365.
216. Tan I, Flanagan BM, Halley PJ, Whittaker AK, Gidley MJ. *Biomacromol.* 2007;8:885.
217. Hizukuri S. *Carbohydr. Res.* 1985;141:295.
218. Hizukuri S. *Carbohydr. Res.* 1986;147:342.
219. Gidley MJ, Bulpin PV. *Carbohydr. Res.* 1987;161:291.
220. Pfannemüller B. *Starch/Stärke.* 1986;38:401.
221. Chanvrier H, Uthayakumaran S, Applelqvist IAM, Gidley MJ, Gilbert EP, Lopez-Rubio A. *J. Agric. Food Chem.* 2007;55:9883.
222. Waigh TA, Donald AM, Keidelbach F, Gidley MJ. *Biopolym.* 1999;49:91.
223. Ridout MJ, Parker ML, Hedley CL, Bogracheva TY, Morris VJ. *Biomacromol.* 2004;5:1519.
224. Jane J, Chen YY, Lee LF, McPherson AE, Wong KS, Radosavljevic M, Kasemsuwan T. *Cereal Chem.* 1999;76:629.
225. Hanashiro I, Abe J, Hizukuri S. *Carbohydr. Res.* 1996;283:151.
226. Shibanuma Y, Takeda Y, Hizukuri S. *Carbohydr. Polym.* 1996;29:253.
227. Kalichevsky MT, Orford PD, Ring SG. *Carbohydr. Res.* 1990;198:49.
228. Hanashiro I, Tagawa M, Shibahara S, Iwata K, Takeda Y. *Carbohydr. Res.* 2002;337:1211.
229. Mührbeck P. *Starch/Stärke.* 1991;43:347.
230. Cameron RE, Donald AM. *Polymer.* 1992;33:2628.
231. Imberty A, Chanzy H, Perez S, Buleon A, Tran V. *J. Mol. Biol.* 1988;201:365.

232. Shibanuma K, Takeda Y, Hizukuri S, Shibata S. *Carbohydr. Polym.* 1994;25:111.
233. Rahman S, Bird A, Regina A, Li Z, Ral JP, McMaugh S, Topping D, Morell M. *J. Cereal Sci.* 2007;46:251.
234. Blennow A. In: Eliasson A-C, ed. *Starch in Food*. 2nd edn. Boca Raton, FL: CRC Press; 2004 [Chapter 3].
235. Jobling S. *Curr. Opin. Plant Biol.* 2004;7:210.
236. Smith AM. *Biomacromol.* 2001;2:335.
237. Morell MK, Rahman S, Regina A, Appels R, Li Z. *Euphytica.* 2001;119:55.
238. Kossmann J, Loyd J. *Crit. Rev. Biochem. Mol. Biol.* 2000;35:141.
239. Morell MK, Li Z, Regina A, Rahman S, d'Hulst C, Ball SG. Control of Primary Metabolism in Plants. In: Plaxton WC, McManus MT, eds. Oxford, UK: Blackwell; 2006:258–289. *Annual Plant Reviews*; vol. 22.
240. Tetlow IJ, Morell MK, Emes MJ. *J. Exptl. Bot.* 2004;55:2131.
241. Vrinten PL, Nakamura T. *Plant Physiol.* 2000;122:255.
242. Nakamura T, Vrinten P, Hayakawa K, Ikeda J. *Plant Physiol.* 1998;118:451.
243. Nakamura T, Yamamori M, Hirano H, Hidaka S. *Plant Breed.* 1993;111:99.
244. Nakamura T, Yamamori M, Hirano H, Hidaka S. *Biochem. Gen.* 1993;31:75.
245. Seib PA. *Cereal Foods World.* 2000;45:504.
246. Loney PP, Jenkins LD, Meredith P. *Starch/Stärke.* 1975;27:145.
247. Nagao S, Ishibashi S, Imai S, Sato T, Kambe T, Kaneko Y, Otsubo H. *Cereal Chem.* 1977;54:198.
248. Moss HJ. *Cereal Res. Commun.* 1980;8:297.
249. Oda M, Yasuda Y, Okazaki S, Yamaguchi Y, Yokayama Y. *Cereal Chem.* 1980;57:253.
250. Moss HJ, Miskelly DM. *Food Technol. Aust.* 1984;36:90.
251. Schofield JD, Greenwell P. In: Morton ID, ed. *Cereals in a European Context.* Chichester, UK: VCH, Weinheim and Ellis Horwood; 1987.
252. Kuroda A, Oda S, Miyagawa S, Seko H. *Jpn. J. Breed.* 1989;39:142.
253. Crosbie GB. *Proceedings of the 39th Annual Conference of the Royal Australian Chemical Institute.* Parkville, Victoria, Australia: Cereal Chemistry Division, RACI; 1989.
254. Crosbie GB. *J. Cereal Sci.* 1991;13:145.
255. Morris CF, King GE, Rubenthaler GL. *Cereal Chem.* 1997;74:147.
256. Morris CF, Shackley BJ, King GE, Kidwell KK. *Cereal Chem.* 1997;74:16.
257. Yamamori M, Nakamura T, Endo TR, Nagamine T. *Theor. Appl. Genet.* 1994;89:179.
258. Yamamori M, Nakamura T, Nagamine T. *Jpn. J. Breed.* 1994;44:242.
259. Yamamori M, Nakamura T. *Gamma Field Symp.* 1994;33:63.
260. Nakamura T, Yamamori M, Hirano H, Hidaka S, Nagamine T. *Mol. Gen. Genet.* 1995;248:253.
261. Zhao XC, Sharp PJ. *J. Cereal Sci.* 1996;23:191.
262. Zhao XC, Batey IL, Sharp PJ, Crosbie G, Barclay I, Wilson R, Morell MK, Appels R. *J. Cereal Sci.* 1998;28:7.
263. Briney A, Wilson R, Potler RH, Barclay I, Crosbie G, Appels R, Jones MGK. *Mol. Breed.* 1998;4:427.

264. Graybosch RA, Shemmerhorn KJ, Skerritt JH. *J. Cereal Sci.* 1999;30:159.
265. Demeke T, Hucl P, Abdel-Aal E-SM, Baga M, Chibbar RN. *Cereal Chem.* 1999;76:694.
266. Miura H, Wickramasinghe MHA, Araki E, Komae K. *Euphytica.* 2002;123:353.
267. Nakamura T, Vrinten P, Saito M, Konda M. *Genome.* 2002;45:1150.
268. Slade AJ, Fuerstenberg SI, Loeffler D, Steine MN, Facciotti D. *Nature Biotech.* 2005;23:75.
269. Graybosch R. *Trends Food Sci. Technol.* 1998;9:135.
270. Chibbar RN, Chakraborty M. In: Abdel-Aal E, Wood P, eds. *Specialty Grains for Food and Feed.* St. Paul, MN: American Association Cereal Chemists; 2005 [Chapter 6].
271. Kim W, Johnson JW, Graybosch RA, Gaines CS. *J. Cereal Sci.* 2003;37:195.
272. Hung PV, Yasui T, Maeda T, Morita N. *Starch/Stärke.* 2008;60:34.
273. Yamamori M, Quynh NT. *Theor. Appl. Genet.* 2000;100:32.
274. Seib PA, Liang X, Guan F, Liang YT, Yang HC. *Cereal Chem.* 2000;77:816.
275. Yasui T, Sasaki T, Matsuki J, Yamamori M. *Breeding Sci.* 1997;47:161.
276. Kiribuchi-Otobe C, Nagamine T, Yanagisawa T, Ohnishi M, Yamaguchi I. *Cereal Chem.* 1997;74:72.
277. Morris CF, Konzak CF. *Crop Sci.* 2001;41:934.
278. Graybosch RA, Souza E, Berzonsky W, Baenziger PS, Chung O. *J. Cereal Sci.* 2003;38:69.
279. Sasaki T, Yasui T, Matsuki J, Satake T. *Cereal Chem.* 2002;79:861.
280. Sasaki T, Yasui T, Matsuki J. *Cereal Chem.* 2000;77:58.
281. Hayakawa K, Tanaka K, Nakamura T, Endo S, Hoshino T. *Cereal Chem.* 1997;74:576.
282. Yamamori M, Nakamura T, Nagamine T. *Breeding Sci.* 1995;45:377.
283. Hung T, Maede PV, Morita N. *Starch/Stärke.* 2007;59:125.
284. Yoo SH, Jane JL. *Carbohydr. Polym.* 2002;49:297.
285. Yasui T, Matsuki J, Sasaki T, Yamamori M. *J. Cereal Sci.* 1996;24:131.
286. Seguchi M, Yasui T, Hosomi K, Imai T. *Cereal Chem.* 2000;77:339.
287. Seguchi M, Hosomi K, Yamauchi H, Yasui T, Imai T. *Starch/Stärke.* 2001;53:140.
288. Hayashi M, Yasui T, Kiribuchi-Otobe C, Seguchi M. *Cereal Chem.* 2004;81:589.
289. Abdel-Aal ESM, Hucl P, Chibbar RN, Han HL, Demeke T. *Cereal Chem.* 2002;79:458.
290. Fujita S, Yamamoto H, Sugimoto Y, Morita N, Yamamori M. *J. Cereal Sci.* 1998;27:1.
291. Yasui T, Sasaki T, Matsuki J. *J. Sci. Food Agric.* 1999;79:687.
292. Sayaslan A, Seib PA, Chung OK. *J. Food Eng.* 2006;72:167.
293. Baik BK, Lee MR. *Cereal Chem.* 2003;80:304.
294. BeMiller JN, Pratt GW. *Cereal Chem.* 1981;58:517.
295. Guo G, Jackson DS, Graybosch RA, Parkhurst AM. *Cereal Chem.* 2003;80:437.
296. Chung J-H, Han J-A, Yoo B, Seib PA, Lim S-T. *Carbohydr. Polym.* 2008;71:365.
297. Bettge AD, Giroux MJ, Morris CF. *Cereal Chem.* 2000;77:750.
298. Hung PV, Maeda T, Morita N. *Trends Food Sci. Technol.* 2006;17:448.
299. Park CS, Baik BK. *Cereal Chem.* 2007;84:437.

300. Messager A, Despre D. US Patent Appl. Publ., US 2002/0037352 A1, March 28. 2002.
301. Reddy I, Seib PA. *Cereal Chem.* 1999;76:341.
302. Reddy I, Seib PA. *J. Cereal Sci.* 2000;31:25.
303. Lumbudwong N, Seib PA. *Starch/Stärke.* 2001;53:605.
304. Yamamori Y, Fujita S, Hayakawa K, Matsuki J, Yasui T. *Theor. Appl. Genet.* 2000;101:21.
305. James MG, Denyer K, Meyer AM. *Curr. Opin. Plant Biol.* 2003;6:215.
306. Hanashiro I, Ikeda I, Yamamori M, Takeda Y. *J. Appl. Glycosci.* 2004;51:217.
307. Konik-Rose C, Thistleton J, Chanvrier H, Tan I, Halley P, Gidley M, Kosar-Hashemi B, Wang H, Larroque O, Ikea J, McMaugh S, Regina A, Rahman S, Morell M, Li Z. *Theoret. Appl. Gen.* 2007;115:1053.
308. Regina A, Bird AR, Li Z, Rahman S, Mann G, Chanliand F, Berbezy P, Topping D, Morell MK. *Cereal Foods World.* 2007;52:182.
309. Regina A, Bird A, Topping D, Bowden S, Freeman J, Barsby T, Kosar-Hashemi B, Rahman S, Morell MK. *Proc. Natl. Acad. Sci. USA.* 2006;103:3546.
310. Nakamura T, Shimbata T, Vrinten P, Saito M, Yonemaru J, Seto Y, Yasuda H, Takahama M. *Genes Genet. Syst.* 2006;81:361.
311. Shi Y-C, Seib PA. In: Pomeranz Y, ed. *Wheat is Unique.* St. Paul, MN: American Association of Cereal Chemists; 1989 [Chapter 13].
312. Takahashi S, Maningat CC, Seib PA. *Cereal Chem.* 1989;66:499.
313. Takahashi S, Maningat CC, Seib PA. *ASEAN Food J.* 1993;8:69.
314. Takahashi S, Seib PA. *Cereal Chem.* 1988;65:474.
315. Craig SAS, Maningat CC, Seib PA, Hoseney RC. *Cereal Chem.* 1989;66:173.
316. Anderson JA, Perkin AG. *J. Chem. Soc.* 1931;4:2624.
317. Lepage M, Sims RPA. *Cereal Chem.* 1968;45:600.
318. Maga JA, Lorenz K. *Lebens.-Wiss. -Technol.* 1974;7:273.
319. Sosulski F, Krygler K, Hogge L. *J. Agric. Food Chem.* 1982;30:337.
320. Barron C, Serget A, Rouau X. *J. Cereal Sci.* 2007;45:88.
321. Asenstorfer RE, Wang Y, Mares DJ. *J. Cereal Sci.* 2006;43:108.
322. Nystrom L, Paasonen A, Lampi AM, Piironen V. *J. Cereal Sci.* 2007;45:106.
323. Galliard T. *Starch: Properties and Potential.* New York, NY: John Wiley & Sons; 1987.
324. Skerritt JH, Hill AS. *Cereal Chem.* 1992;69:110.
325. Kaukinen K, Collin P, Holm K, Rautala I, Vuolteenaho N, Reunala T, Maki M. *Scand. J. Gastroent.* 1999;34:163.
326. Chartrand LJ, Russo PA, Duhaime AG, Seidman EG. *J. Amer. Diet. Asso.* 1997;97:612.
327. Takaoka M, Watanabe S, Sassa H, Yamamori M, Nakamura T, Sasakuma T, Hirano H. *J. Agric. Food Chem.* 1997;45:2929.
328. Greenwell P, Schofield JD. *Cereal Chem.* 1986;63:379.
329. Sulaiman BD, Morrison WR. *J. Cereal Sci.* 1990;12:53.
330. Bettge AD, Morris CF, Greenblatt GA. *Euphytica.* 1995;86:65.
331. Greenblatt GA, Bettge AD, Morris CF. *Cereal Chem.* 1995;72:172.
332. Oda S, Schofield DJ. *J. Cereal Sci.* 1997;26:29.

333. Bloch HA, Darlington HF, Shewry PR. *Cereal Chem.* 2001;78:74.
334. Wanjugi HW, Martin JM, Giroux MJ. *Cereal Chem.* 2007;84:540.
335. Xia L, Geng H, Chen X, He Z, Lillemo M, Morris C. *J. Cereal Sci.* 2008;47:33.
336. Rahman S, Kosar-Hashemi B, Samuel MS, Hill A, Abbott DC, Skerritt JH, Preiss J, Appels R, Morell MK. *Austr. J. Plant Physiol.* 1995;22:793.
337. Denyer K, Sidebottom C, Hylton CM, Smith AM. *Plant J.* 1993;4:191.
338. Yamamori M, Endo TR. *Theor. Appl. Genet.* 1996;93:275.
339. Gao M, Chibbar RN. *Genome.* 2000;43:768.
340. Denyer K, Hylton CM, Jenner CF, Smith AM. *Planta.* 1995;196:256.
341. Li Z, Chu X, Mouille G, Yan L, Kosar-Hashemi B, Hey S, Napier J, Shewry P, Clarke B, Appels R, Morell M, Rahman S. *Plant Physiol.* 1999;120:1147.
342. Morell MK, Blennow A, Kosar-Hashemi B, Samuel MS. *Plant Physiol.* 1997;113:201.
343. Baga M, Glaze S, Mallard CS, Chibbar RN. *Plant Mol. Biol.* 1999;40:1019.
344. Baga M, Repellin A, Demeke T, Caswell K, Leung N, Abdel-Aal ES, Hucl P, Chibbar RN. *Starch/Stärke.* 1999;51:111.
345. Li Z, Rahman S, Kosar-Hashemi B, Mouille G, Appels R, Morell M. *Theor. Appl. Genet.* 1999;98:1208.
346. Morrison WR. *J. Cereal Sci.* 1988;8:1.
347. Hayashi M, Seguchi M. *Cereal Chem.* 2004;81:621.
348. Seguchi M. *Cereal Chem.* 1993;70:362.
349. Morrison WR. In: Pomeranz Y, ed. *Wheat is Unique*. St. Paul, MN: American Association of Cereal Chemists; 1989 [Chapter 19].
350. Morrison WR. In: Bushuk W, Rasper VF, eds. *Wheat: Production, Properties, and Quality*. London, UK: Chapman & Hall; 1994 [Chapter 9].
351. Kasemsuwan T, Jane J-L. *Cereal Chem.* 1996;73:702.
352. Chung OK, Sayaslan A, Seib PA, Seitz LM. *Cereal Chem.* 2000;77:248.
353. Matsunaga N, Seib PA. *Cereal Chem.* 1997;74:851.
354. Morgan KR, Furneaux RH, Larsen NG. *Carbohydr. Res.* 1995;276:387.
355. Morrison WR, Law RV, Snape CE. *J. Cereal Sci.* 1993;18:103.
356. Morrison WR, Tester RF, Snape CE, Law R, Gidley MJ. *Cereal Chem.* 1993;70:385.
357. Melvin MA. *J. Sci. Food Agric.* 1979;30:731.
358. Ellis HS, Ring SG, Whittam MA. *J. Cereal Sci.* 1989;10:33.
359. Deffenbaugh LB, Walker CE. *Cereal Chem.* 1989;66:493.
360. Jacobs H, Eerlingen RC, Delcour JA. *Starch/Stärke.* 1996;48:266.
361. Yun S-K, Quail K. *Starch/Stärke.* 1999;51:274.
362. Doublier JL. *J. Cereal Sci.* 1987;5:247.
363. Zeng M, Morris CF, Batey IL, Wrigley CW. *Cereal Chem.* 1997;74:63.
364. Blazek J, Copeland L. *Carbohydr. Polym.* 2008;71:380.
365. Sasaki T, Matsuki J. *Cereal Chem.* 1998;75:525.
366. Mira I, Persson K, Villwock VK. *Carbohydr. Polym.* 2007;68:665.
367. Allen HM, Blakeney AB, Oliver JR. In: Martin DJ, Wrigley CW, eds. *Cereals International*. Victoria, Australia: Cereal Chemistry Division, Royal Australian Chemical Institute, Parkville; 1991:159–160.

368. Morris CF, King GE, Rubenthaler GL. *Cereal Chem.* 1997;74:147.
369. Jane J-L, Chen J-F. *Cereal Chem.* 1992;69:60.
370. Dengate HN. In: Pomeranz Y, ed. *Advances in Cereal Science and Technology*. Vol. 6. St. Paul, MN: American Association of Cereal Chemists; 1984 [Chapter 2].
371. Langton M, Hermansson A-M. *Food Microstructure.* 1989;8:29.
372. Noel TR, Ring SG, Whittam MA. In: Dickinson E, Walstra P, eds. *Food Colloids and Polymers: Stability and Mechanical Properties*. Cambridge, UK: The Royal Society of Chemistry; 1993:126.
373. Singh J, Kaur L, McCarthy OJ. *Food Hydrocoll.* 2007;21:1.
374. Wurzburg OB. In: Stephen AM, Phillips GO, Williams PA, eds. *Food Polysaccharides and Their Applications*. 2nd edn. Boca Raton, FL: CRC Press; 2006:87–118.
375. Xie SX, Liu Q, Cui SW. In: Cui SW, ed. *Food Carbohydrates: Chemistry, Physical Properties and Applications*. Boca Raton, FL: CRC Press; 2005:357–405.
376. Tharanathan RN. *Crit. Rev. Food Sci. Nutr.* 2005;45:371.
377. Taggart P. In: Eliasson AC, ed. *Starch in Food: Structure, Function and Applications*. Boca Raton, FL: CRC Press; 2004:368–392.
378. Tomasik P, Schilling CH. *Adv. Carbohydr. Chem. Biochem.* 2004;57:175.
379. Mauro DJ, Abbas IR, Orthoeffer FT. In: White PJ, Johnson LA, eds. *Corn; Chemistry and Technology*. 2nd edn. St. Paul, MN: American Association of Cereal Chemists; 2003:605–634.
380. Light JM. *Cereal Foods World.* 1990;35:1081.
381. Wurzburg OB. *Modified Starches: Properties and Uses*. Boca Raton, FL: CRC Press; 1987.
382. Rogols S. *Cereal Foods World.* 1986;31:869.
383. Fleche G. In: Van Beynum GMA, Roels JA, eds. *Starch Conversion Technology*. New York, NY: Marcel Dekker; 1985 [Chapter 4].
384. Rutenberg MW, Solarek D. In: Whistler RL, BeMiller JN, Paschall EF, eds. *Starch Chemistry and Technology*. New York, NY: Academic Press; 1984 [Chapter 10].
385. Smith PA. *Food Carbohydrates*. Westport, CT: AVI Publishing Co.;1982:237.
386. O'Dell JD. *Food Manuf.* 1976;51:17.
387. Nappen B. *South Afr. Food Rev.* 1975;2:15.
388. Howling D. *Food Technol. Austr.* 1974;26:464.
389. O'Dell J. *Austr. Food Manuf. Dist.* 1974;43:28.
390. BeMiller JN. *Starch/Stärke.* 1997;49:127.
391. BeMiller JN. *J. Appl. Glycosci.* 1997;44:43.
392. Wooton M, Bamunuarachi A. *Starch/Stärke.* 1978;30:306.
393. Wooton M, Chaudhry MA. *Starch/Stärke.* 1979;31:224.
394. Wooton M, Bamunuarachi A. *Starch/Stärke.* 1979;31:201.
395. Wooton M. and Haryadi. *J. Cereal Sci.* 1991;14:179.
396. Wooton M. and Haryadi. *J. Cereal Sci.* 1992;15:181.
397. Wooton M, Mahdar D. *Starch/Stärke.* 1993;45:337.
398. Stahl H, McNaught RP. *Cereal Chem.* 1970;47:345.
399. Ostergard K, Bjorck I, Gunnarsson A. *Starch/Stärke.* 1988;40:58.
400. Xu A, Seib PA. *J. Cereal Sci.* 1997;25:17.

401. Monzon BE. *Ion (Madrid)*. 1972;32:90.
402. Sang Y, Prakash O, Seib PA. *Carbohydr. Polym.* 2007;67:201.
403. Duanmu J, Gamstedt EK, Rosling A. *Starch/Stärke*. 2007;59:523.
404. Heinze T, Rensing S, Koschella A. *Starch/Stärke*. 2007;59:199.
405. Kasemsuwan T, Jane J. *Cereal Chem.* 1994;71:282.
406. Cui SW. In: Cui S, ed. *Food Carbohydrates*. Boca Raton, FL: CRC Press, Taylor and Francis Group; 2005 [Chapter 3].
407. Klaushofer H, Berghofer E, Diesner L. *Starch/Stärke*. 1976;28:298.
408. Berghofer E, Klaushofer H. *Starch/Stärke*. 1977;29:296.
409. Sanders P, Brunt K. *Starch/Stärke*. 1996;48:448.
410. Park PW, Goins RE. *J. Agric. Food Chem.* 1995;43:2580.
411. Phillips DL, Liu H, Pan D, Corke H. *Cereal Chem.* 1999;76:439.
412. Mitchell G, Wijnberg AC. *Starch/Stärke*. 1995;47:46.
413. Radley JA. *Examination and Analysis of Starch and Starch Products*. London, UK: Applied Science Publishers; 1976.
414. Whistler RL. *Meth. Carbohydr. Chem.* 1964;4:28.
415. National Academy of Sciences. *Food Chemicals Codex*. 5th edn. Washington, DC: National Academy Press; 2004.
416. Joint FAO/WHO Expert Committee on Food Additives. *Specifications for Identity and Purity of Thickening Agents, Anticaking Agents, Antimicrobials, Antioxidants, and Emulsifiers*. Rome, Italy: FAO, UN; 1978:59–71.
417. Choi SG, Kerr WL. *Starch/Stärke*. 2004;56:181.
418. Woo K, Seib PA. *Carbohydr. Polym.* 1997;33:263.
419. Steeneken PAM. *Carbohydr. Polym.* 1989;11:23.
420. Woo KS, Seib PA. *Cereal Chem.* 2002;79:819.
421. Thompson DB. In: Biliaderis CG, Izydorczyk MS, eds. *Functional Food Carbohydrates*. Boca Raton, FL: CRC Press, Taylor and Francis Group; 2007 [Chapter 2].
422. Lack S, Dulong V, Picton L, Le Cerf D, Condamine E. *Carbohydr. Res.* 2007;342:943.
423. Hirsch JB, Kokini J. *Cereal Chem.* 2002;79:102.
424. Choi SG, Kerr WL. *Lebensm. Wiss. Technol.* 2003;36:105.
425. Kavitha R, BeMiller JN. *Carbohydr. Polym.* 1998;37:115.
426. BeMiller JN, Shi X. *Carbohydr. Polym.* 2000;43:333.
427. Lim S, Seib PA. *Cereal Chem.* 1993;70:137.
428. Heinze T, Haack V, Rensing S. *Starch/Stärke*. 2004;56:288.
429. Ono K, Seib PA, Takahashi S. *Nippon Kasei Gakkaishi*. 1998;49:985.
430. Lawton JW. In: Wrigley C, Corke H, Walker CE, eds. *Encyclopedia of Grain Science*. 1st edn. Vol. 3. Oxford, UK: Elsevier; 2004.
431. Sandstedt RM. *Bakers' Dig.* 1961;35:36.
432. Medcalf DG. *Bakers' Dig.* 1968;42:48.
433. Hoseney RC, Lineback DR, Seib PA. *Bakers' Dig.* 1978;52:11.
434. Dennett K, Sterling C. *Starch/Stärke*. 1979;31:209.
435. Batey IL, Miskelly DM, Konik CM. *Chem. Austr.* 1991;Sept.:362.
436. Hoseney RC, Finney KF, Pomeranz Y, Shogren MD. *Cereal Chem.* 1971;48:191.

437. D'Appolonia BL, Gilles KA. *Cereal Chem.* 1971;48:625.
438. Toufeili I, Habbal Y, Shadarevian S, Olabi A. *J. Sci. Food Agric.* 1999;79:1855.
439. Dennett K, Sterling C. *Starch/Stärke.* 1979;31:209.
440. Hug-Iten S, Handschin S, Conde-Petit B, Escher F. *Lebens. -Wiss. -Technol.* 1999;32:255.
441. Sahlstrom S, Baevre AB, Graybosch R. *Cereal Chem.* 2006;83:647.
442. Kusunose C, Fujii T, Matsumoto H. *Cereal Chem.* 1999;76:920.
443. Sahlstrom S, Mosleth E, Baevre AB, Gloria H, Fayard G. *Carbohydr. Polym.* 1993;21:169.
444. Hoseney RC, Atwell WA, Lineback DR. *Cereal Foods World.* 1977;22:56.
445. Greenwood CT. In: Pomeranz Y, ed. *Advances in Cereal Science and Technology.* Vol. 1. St. Paul, MN: American Association of Cereal Chemists; 1976 [Chapter 3].
446. Kulp K, Lorenz K. *Bakers' Dig.* 1981;55:24.
447. Sebecic B, Sebecic B. *Nahrung.* 1996;40:209.
448. Sebecic B, Sebecic B. *Nahrung.* 1995;39:117.
449. Larsson H, Eliasson A-C. *J. Text. Stud.* 1997;28:487.
450. Ponte JG, Titcomb ST, Cerning J, Cotton RH. *Cereal Chem.* 1963;40:601.
451. Kulp K. *Cereal Chem.* 1973;50:666.
452. Park SH, Wilson JD, Chung OK, Seib PA. *Cereal Chem.* 2004;81:699.
453. Park SH, Chung OK, Seib PA. *Cereal Chem.* 2005;82:166.
454. Lelievre J, Lorenz K, Meredith P, Baruch DW. *Starch/Stärke.* 1987;39:347.
455. Sahlstrom S, Brathen E, Lea P, Autio K. *J. Cereal Sci.* 1998;28:157.
456. Hayman A, Sipes K, Hoseney RC, Faubion JM. *Cereal Chem.* 1998;75:585.
457. Tammenga W. *Food Technol. Europe.* 1995;2:75.
458. Hahn RR. *Bakers' Dig.* 1969;43:48.
459. Dubois DK. *Bakers' Dig.* 1959;33:38.
460. Dubois DK. *Amer. Soc. Bakery Eng.* 1961;274.
461. Miller LL, Setser C. *Cereal Chem.* 1983;60:62.
462. Kulp K, Hoover WJ. US Patent 4 294 864. 1981.
463. Donelson JR, Gaines CS, Finney PL. *Cereal Foods World (Abstract).* 1995;40:642.
464. Cauvain SP, Gough BM. *J. Sci. Food Agric.* 1975;26:1861.
465. Kim CS, Walker CE. *Cereal Chem.* 1992;69:206.
466. Sollars WF, Rubenthaler GL. *Cereal Chem.* 1971;48:397.
467. Fujii T, Danno G-I. *J. Jpn. Soc. Food Sci. Technol.* 1988;35:684.
468. Fujii T, Kuyama S, Danno G-I. *J. Jpn. Soc. Food Sci. Technol.* 1990;37:619.
469. Fujii T, Kuyama S. *J. Jpn. Soc. Food Sci. Technol.* 1992;39:524.
470. Takahashi S, Suzuki M, Yoshida E, Ono K, Seib PA. *Nippon Kasei Gakkaishi.* 1998;49:1099.
471. Ludewig H-G, Tegge G, Starke H. *Getreide Mehl Brot.* 1977;31:107.
472. Witt W. *Brot Backwaren.* 1981;29:166.
473. Huron Bakery Series, No. 100, Wilmington, DE: Hercules Powder Co.
474. Huron Bakery Series, No. 104, Wilmington, DE: Hercules Powder Co.
475. Kulp K, Olewnik M, Lorenz K. *Starch/Stärke.* 1991;43:53.
476. Maningat C, Bassi S, Woo K, Dohl C, Gaul J, Stempien G, Moore T. *AIB Tech. Bull.* 2005;27:1.

477. Wasserman LA, Krivandin AV, Kiseleva VI, Schiraldi A, Blaszczak W, Fornal J, Sharafetdinov KK, Gapparov MG, Yuryev VP. In: Yurgev V, Tomasik P, Bertoft E, eds. *Starch: Achievements in Understanding of Structure and Functionality*. Hauppauge, New York, NY: Nova Science Publishers; 2007 [Chapter 11].
478. Schoch TJ, French D. *Cereal Chem*. 1947;24:231.
479. Morgan KR, Furneaux RH, Stanley RA. *Carbohydr. Res*. 1992;235:15.
480. Hebeda RE, Zobel HF. *Baked Goods Freshness: Technology, Evaluation, and Inhibition of Staling*. New York, NY: Marcel Dekker; 1996.
481. Chinachoti P, Vodovotz Y. *Bread Staling*. Boca Raton, FL: CRC Press; 2000.
482. Gray JA, BeMiller JN. *Comp. Rev. Food Sci. Food Safety*. 2003;2:1.
483. Inagaki T, Seib PA. *Cereal Chem*. 1992;69:321.
484. Blanshard JMV. In: Blanshard JMV, Frazier PJ, Galliard T, eds. *Chemistry and Physics of Baking*. London, UK: The Royal Society of Chemistry; 1988 [Chapter 1].
485. Martin ML, Zeleznak KJ, Hoseney RC. *Cereal Chem*. 1991;68:498.
486. Martin ML, Hoseney RC. *Cereal Chem*. 1991;68:503.
487. Hoseney RC, Miller R. *AIB Tech. Bull*. 1998;20:1.
488. Morgan KR, Gerrard J, Every D, Ross M, Gilpin M. *Starch/Stärke*. 1997;49:54.
489. Every D, Gerrard JA, Gilpin MJ, Ross M, Newberry MP. *Starch/Stärke*. 1998;50:443.
490. Keetels CJAM, Visser KA, Van Vliet T, Jurgens A, Walstra P. *J. Cereal Sci*. 1996;24:15.
491. Keetels CJAM, Van Vliet T, Walstra P. *J. Cereal Sci*. 1996;24:27.
492. Schiraldi A, Piazza L, Riva M. *Cereal Chem*. 1996;73:32.
493. Baik M-Y, Chinachoti P. *Cereal Chem*. 2000;77:484.
494. Piazza L, Masi P. *Cereal Chem*. 1995;72:320.
495. Midwest Grain Products, Inc., *The Effect of Modified Starches on Crumb Firming Rate of No-Time Dough White Pan Bread*. Manhattan, KS: American Institute of Baking; 1993.
496. Dragsdorf RD, Varriano-Marston E. *Cereal Chem*. 1980;57:310.
497. Kamel BS. *AIB Tech. Bull*. 1993;15:1.
498. Colonna P, Buleon A, Champenois Y, Della Valle G, Planchot V. *Sci. Aliments*. 1999;19:471.
499. Gerrard JA, Every D, Sutton KH, Gilpin MJ. *J. Cereal Sci*. 1997;26:201.
500. Champenois Y, Valle GD, Planchot V, Buleon A, Colonna P. *Sci. Aliments*. 1999;19:471.
501. Every D, Mann JD, Ross M. In: Humphrey-Taylor VJ, ed. *Proceedings of the 42nd Australian Cereal Chemistry Conference*. Sydney, New South Wales, Australia: Cereal Chemistry Division, Royal Australian Chemical Institute; 1992.
502. Min B-C, Yoon S-H, Kim J-W, Lee Y-W, Kim Y-B, Park HA. *J. Agric. Food Chem*. 1998;46:779.
503. Ziobro R, Gambus H, Nowotna A, Bala-Piasek A, Sabot R. *Zywnosc*. 1998;5:251.
504. Defloor I, Delcour JA. *J. Agric. Food Chem*. 1999;47:737.
505. Akers AA, Hoseney RC. *Cereal Chem*. 1994;71:223.
506. Huang J, White P. *Cereal Chem*. 1993;70:42.
507. Villwock VK, Eliasson A-C, Silverio J, BeMiller JN. *Cereal Chem*. 1999;76:292.

508. Jang J-K, Lee Y-H, Lee S-H, Pyun Y-R. *Kor. J. Food Sci. Technol.* 2000;32:500.
509. Kruger JE, Matsuo RR, Dick JW. *Pasta and Noodle Technology*. St. Paul, MN: American Association of Cereal Chemists; 1996.
510. Dick JW, Matsuo RR. In: Pomeranz Y, ed. *Wheat Chemistry and Technology*. Vol. II. St. Paul, MN: American Association of Cereal Chemists; 1988 [Chapter 9].
511. Matsuo RR. In: Bushuk W, Rasper VF, eds. *Wheat: Production, Properties, and Quality*. London, UK: Chapman & Hall; 1994 [Chapter 12].
512. Delcour JA, Vansteelandt J, Hythier MC, Abecassis J, Sindic M, Deroanne C. *J. Agric. Food Chem.* 2000;48:3767.
513. Banasik OJ, Hader TA, Seyam AA. *Macaroni J.* 1976;58:18.
514. Matsuo RR, Dexter JE, Dronzek BL. *Cereal Chem.* 1978;55:744.
515. Lintas C, D'Appolonia BL. *Cereal Chem.* 1973;50:563.
516. Dalbon G, Pagani M, Resmini R, Lucisano M. *Getreide Mehl Brot.* 1985;39:183.
517. Yue P, Rayas-Duarte P, Elias E. *Cereal Chem.* 1999;76:541.
518. Vansteelandt J, Delcour JA. *J. Agric. Food Chem.* 1998;46:2499.
519. Cubadda R, Pasqui LA, Acquistucci R. *Riv. Soc. Ital. Sci. Aliment.* 1988;17:235.
520. Zweifel C, Conde-Petit B, Escher F. *Cereal Chem.* 2000;77:645.
521. Marshall S, Wasik R. *Cereal Chem.* 1974;51:146.
522. Grzybowski RA, Donnelly BJ. *J. Food Sci.* 1977;42:1304.
523. Dexter JE, Dronzek BL, Matsuo RR. *Cereal Chem.* 1978;55:23.
524. Resmini P, Pagani MA. *Food Microstruct.* 1983;2:1.
525. Cunin S, Handschin S, Escher F, Walther P. *Lebensm-Wiss. -Technol.* 1995;28:323.
526. Sheu R-Y, Medcalf DG, Gilles KA, Sibbitt LD. *J. Sci. Food Agric.* 1967;18:237.
527. Dexter JE, Matsuo RR. *Cereal Chem.* 1979;56:190.
528. Soh HN, Sissons MJ, Turner MA. *Cereal Chem.* 2006;83:513.
529. Gianibelli MC, Sissons MJ, Batey IL. *Cereal Chem.* 2005;82:321.
530. Vignaux N, Doehlert DC, Elias EM, McMullen MS, Grant LA, Kianian SF. *Cereal Chem.* 2005;82:93.
531. Nelson JM. [Ph.D. dissertation]. Fargo, ND: North Dakota State University; 1982.
532. Miskelly DM, Moss HJ. *J. Cereal Sci.* 1985;3:379.
533. Konik CM, Mikkelsen LM, Moss R, Gore PJ. *Starch/Stärke.* 1994;46:292.
534. Seib PA. In: *Proceedings of the International Wheat Quality Conference*. Manhattan, KS; May 18–22, 1997:61–82.
535. Akashi H, Takahashi M, Endo S. *Cereal Chem.* 1999;76:50.
536. Toyokawa H, Rubenthaler GL, Powers JR, Schanus EG. *Cereal Chem.* 1989;66:387.
537. Nagao S, Ishibashi S, Imai S, Sato T, Kanbe Y, Kaneko Y, Otsugo H. *Cereal Chem.* 1977;54:198.
538. Moss HJ. *Cereal Res. Commun.* 1980;8:297.
539. Oda M, Yasuda Y, Okazaki S, Yamauchi Y, Yokoyama Y. *Cereal Chem.* 1980;57:253.
540. Rho KL, Chung OK, Seib PA. *Cereal Chem.* 1989;66:276.
541. Endo S, Okada K, Nagao S. *J. Cereal Sci.* 1989;10:33.
542. Crosbie GB. *Cereal Foods World.* 1989;34:678.
543. Crosbie GB. *J. Cereal Sci.* 1991;13:145.
544. Crosbie GB, Lambe WJ, Tsutsui H, Gilmour RF. *J. Cereal Sci.* 1992;15:271.
545. Toyokawa H, Rubenthaler GL, Powers JR, Schanus EG. *Cereal Chem.* 1989;66:387.

546. McCormick KM, Panozzo JF, Hong SH. *Aust. J. Agric. Res.* 1991;42:317.
547. Konik CM, Miskelly DM, Gras PW. *J. Sci. Food Agric.* 1992;58:403.
548. Konik CM, Miskelly DM, Gras PW. *Starch/Stärke.* 1993;45:139.
549. Konik CM, Mikkelsen LM, Moss R, Gore PJ. *Starch/Stärke.* 1994;46:292.
550. Jun WJ, Seib PA, Chung OK. *Cereal Chem.* 1998;75:820.
551. Moss HJ, Gore PJ, Murray IC. *Food Microstruct.* 1987;6:63.
552. Kim WS, Seib PA. *Cereal Chem.* 1993;70:367.
553. Crosbie GB, Ross AS, Moro T, Chin PC. *Cereal Chem.* 1999;76:328.
554. Wang L, Seib PA. *Cereal Chem.* 1996;73:167.
555. Konik CM, Moss R. *Proceedings of the 42nd RACI Cereal Chemistry Conference.* Christchurch, New Zealand: Royal Australian Chemical Institute, Cereal Chemistry Division; 1992 [209–212].
556. Batey IL, Gras PW, Curtin BM. *J. Sci. Food Agric.* 1997;74:503.
557. Guo G, Jackson DS, Graybosch RA, Parkhurst AM. *Cereal Chem.* 2003;80:437.
558. Park CH, Baik BK. *Cereal Chem.* 2004;81:521.
559. Van Beynum GMA, Roels JA. *Starch Conversion Technology.* New York, NY: Marcel Dekker; 1985 [Chapter 1].
560. Cornell H. In: Eliasson A-C, ed. *Starch in Food.* Boca Raton, FL: CRC Press; 2004 [Chapter 7].
561. Bowler P, Towersey PJ, Waight SG, Galliard T. In: Hill RD, Munck L, eds. *New Approaches to Research on Cereal Carbohydrates.* Amsterdam, The Netherlands: Elsevier Science Publishers; 1985 [Chapter 8].
562. Galliard T, Bowler P, Towersey PJ. In: Pomeranz Y, ed. *Wheat is Unique.* St. Paul, MN: American Association of Cereal Chemists; 1989 [Chapter 15].
563. Matser AM, Steeneken PAM. *Cereal Chem.* 1998;75:241.
564. Matser AM, Steeneken PAM. *Cereal Chem.* 1998;75:289.
565. Derez FGH, de Sadeleer JWGC, Reeve AL. US Patent 4 916 064. 1990.
566. Konieczny-Janda G, Richter G. *Starch/Stärke.* 1991;43:308.
567. Nebesny E, Rosicka J, Sucharzewska D. *Zywnosc.* 1998;5:181.
568. Lumdubwong N. [Ph. D. dissertation]. Manhattan, KS: Kansas State University; 2000.
569. McPherson AE, Seib PA. *Cereal Chem.* 1997;74:424.
570. Breen MD, Seyam AA, Banasik OJ. *Cereal Chem.* 1977;54:728.
571. Dove G, Giddeu C, Menzi R, Tzanos D. British Patent 1 516 628. 1978.
572. Anon., *New Prod. Prom. Trends.* 1986;16:4.
573. Anon., *Snack Food.* 1987;76:21.
574. Whalen P. US Patent 5 102 679. 1990.
575. Rudin RE. US Patent 4 624 853. 1986.
576. Mottar J. In: *Proceedings of Food Ingredients Europe.* Paris, France; September 27–29, 1989:108–111.
577. Mittal GS, Usborne WR. *Food Technol.* 1985;44:121.
578. Skrede G. *Meat Sci.* 1989;25:21.
579. Comer FW. *Can. Inst. Food Sci. Technol. J.* 1979;12:157.
580. Claus JR, Hunt MC. *J. Food Sci.* 1991;56:643.
581. Bonnefin G, Baumgartner PA. *Proc. Intern. Cong. Meat Sci. Technol.* 1988;34:333.

582. Bushway AA, Belyea PR, True RH, Work TM, Russell DO, McGann DF. *J. Food Sci.* 1982;47:402.
583. Payne CA. [Ph.D. dissertation]. Manhattan, KS: Kansas State University; 1993.
584. Kim JM, Lee CM. *J. Food Sci.* 1987;52:722.
585. Yamazawa M. *Nippon Suisan Gakkaishi.* 1990;56:505.
586. Yamazawa M. *Nippon Suisan Gakkaishi.* 1991;57:965.
587. Yamashita T, Yoneda T. *Nippon Shokuhin Kogyo Gakkaishi.* 1989;36:214.
588. Ojima T, Ozawa T, Yamawa I. *J. Jpn. Soc. Starch Sci.* 1985;32:267.
589. Doane WM. In: Pomeranz Y, ed. *Wheat is Unique.* St. Paul, MN: American Association of Cereal Chemists; 1989 [Chapter 36].
590. Kearney RL, Maurer HW. *Starch and Starch Products in Paper Coating.* Atlanta, GA: TAPPI Press; 1990.
591. Radley JA. *Industrial Uses of Starch and Its Derivatives.* London, UK: Applied Science Publishers; 1976.
592. Kirby KW. In: Alexander RJ, Zobel HF, eds. *Developments in Carbohydrate Chemistry.* St. Paul, MN: American Association of Cereal Chemists; 1992:371.
593. Koch H, Roper H, Hopcke R. In: Meuser F, Manners DJ, Seibel W, eds. *Plant Polymeric Carbohydrates.* Cambridge, UK: The Royal Society of Chemistry; 1993 [Chapter 13].
594. Koch H, Roper H. *Starch/Stärke.* 1988;40:121.
595. Munck L, Rexen F, Haastrup L. *Starch/Stärke.* 1988;40:81.
596. Maurer HW, Kearney RL. *Starch/Stärke.* 1998;50:396.
597. Maningat CC, Bassi SD. In: Wrigley C, Corke H, Walker CE, eds. *Encyclopedia of Grain Science.* 1st edn. Vol. 1. Oxford, UK: Elsevier; 2004.
598. Craig SAS, Seib PA, Jane J. *Starch/Stärke.* 1987;39:167.
599. Chung KM, Seib PA. *Starch/Stärke.* 1991;43:441.
600. Latenstein Zetmeel BV. *Latenstein, the Specialist in Wheat Starch and Wheat Protein Products.* The Netherlands: Nijmegen; 1994.
601. Neisser W, Thomann R. *Feldwirtschaft.* 1991;32:176.
602. Batchelor SE, Booth EJ, Entwistle G, Walker KC, Morrison I, Mackay G, Rees T, Hacking A. *Outlook Agric.* 1996;25:43.
603. Bond JL, Rogols S, Salter JW. US Patent 3 951 948. 1976.
604. Rogols S, Salter JW. US Patent 4 139 505. 1979.
605. Best RW. US Patent 4 141 747. 1979.
606. Bond JL, Rogols S, Salter JW. US Patent 3 901 725. 1975.
607. Johnson DL, Bond JL, Rogols S, Salter JW. US Patent 4 280 718. 1981.
608. Lane CC, Lenz RP, Athanassoulias C. US Patent 5 066 335. 1991.
609. Lane CC, Lenz RP, Athanassoulias C. US Patent 5 360 903. 1994.
610. Lane CC, Lenz RP, Athanassoulias C. US Patent 5 367 068. 1994.
611. Ward M. *New Scientist.* 1997;153:21.
612. CIR Experts. *J. Environ. Pathol. Toxicol.* 1980;4:19.
613. Lower E. *Soap Perfum. Cosmet.* 1996;69:27.
614. Anon., *Happi Mag.* May 1992:14.
615. Roper H, Koch H. *Starch/Stärke.* 1990;42:123.
616. Wiedmann W, Strobel E. *Starch/Stärke.* 1991;43:138.

617. Doane WM. *Cereal Foods World.* 1994;39:556.
618. Mayer JM, Kaplan DL. *Trends Polym. Sci.* 1994;2:227.
619. Nawrath C, Poirier Y, Somerville C. *Molecular Breed.* 1995;1:105.
620. Ching C, Kaplan DL, Thomas EL. *Biodegradable Polymers and Packaging.* Lancaster, PA: Technomic Publishing Co.; 1993.
621. Chum HL. *Polymers from Biobased Materials.* Park Ridge, NJ: Noyes Data Corp.; 1991.
622. Lim S-T, Jane J-L, Rajagopalan S, Seib PA. *Biotechnol. Prog.* 1992;8:51.
623. Shi B, Seib PA. *J. Macromol. Sci. – Pure Appl. Chem.* 1996;A33:655.
624. Shogren RL. *Carbohydr. Polym.* 1993;22:93.
625. Averous L, Moro L, Dole P, Fringant C. *Polymer.* 2000;41:4157.
626. Abbes B, Ayad R, Prudhomme J-C, Onteniente J-P. *Polym. Eng. Sci.* 1998;38:2029.
627. Arvanitoyannis I, Psomiadou E, Biliaderis CG, Ogawa H, Kawasaki N, Nakayama A. *Starch/Stärke.* 1997;49:306.
628. Ke T, Sun X. *Cereal Chem.* 2000;77:761.
629. Bikiaris D, Aburto J, Alric I, Borredon E, Botev B, Panayiotou C. *J. Appl. Polym. Sci.* 1999;71:1089.
630. Bikiaris D, Panayiotou C. *J. Appl. Polym. Sci.* 1998;70:1503.
631. Prinos J, Bikiaris D, Theologidis S, Panayiotou C. *Polym. Eng. Sci.* 1998;38:954.
632. Thakur IM, Iyer S, Desai A, Lele A, Devi S. *J. Appl. Polym. Sci.* 1998;74:2791.
633. Sailaja RN. *Polymer Intern.* 2004;54:286.
634. Shi B, Cha JY, Seib PA. *J. Environ. Poly. Deg.* 1998;6:133.
635. Cha JY, Chung DS, Seib PA. *Trans. ASAE.* 1999;42:1765.
636. Cha JY, Chung DS, Seib PA. *Trans. ASAE.* 1999;42:1801.
637. Onteniente JP, Abbes B, Safa LH. *Starch/Stärke.* 2000;52:112.
638. Wang W, Flores RA, Huang CT. *Cereal Chem.* 1995;72:38.
639. Wang W, Flores RA. *Appl. Eng. Agric.* 1996;12:79.
640. Buttery RG, Glenn GM, Stern DJ. *J. Agric. Food Chem.* 1999;47:5206.
641. Yu PC, Wood DF, Orts WJ, Glenn GM. *Scanning Microscopy.* 1999;21:115.
642. Psomiadou E, Arvanitoyannis I, Biliaderis CG, Ogawa H, Kawasaki N. *Carbohydr. Polym.* 1997;33:227.
643. Neumann PE, Seib PA. US Patent 5 208 267. 1993.
644. Gaulin S, Lourdin D, Forssell PM, Colonna P. *Carbohydr. Polym.* 2000;43:33.
645. Glenn GM, Irving DW. *Cereal Chem.* 1995;72:155.
646. Bhatnagar S, Hanna MA. *Cereal Chem.* 1996;73:601.
647. Lawton JW. *Carbohydr. Polym.* 1996;29:203.
648. Lawton JW, Shogren RL, Tiefenbacher KF. *Cereal Chem.* 1999;76:682.
649. Riaz MN. *Cereal Foods World.* 1999;44:705.

Potato Starch: Production, Modifications and Uses

11

Hielko E. Grommers[1] and Do A. van der Krogt[2]
[1]AVEBE U.A., P.O. Box 15, 9640 AA, Veendam, The Netherlands
[2]AVEBE U.A., P.O. Box 15, 9640 AA, Veendam, The Netherlands

I.	History of Potato Processing in The Netherlands.	512
II.	Starch Production	514
	1. World Starch Production	514
	2. Potato Starch Production in Europe	514
III.	Structure and Chemical Composition of the Potato	515
	1. Formation and Morphology of the Tuber	515
	2. Anatomy of the Tuber	516
	3. Chemical Composition	518
	4. Differences Between Commercial Starches	519
	5. New Development: The All-amylopectin Potato	521
IV.	Potato Starch Processing.	522
	1. Grinding	525
	2. Potato Juice Extraction	525
	3. Fiber Extraction.	526
	4. Starch Classification	527
	5. Starch Refinery	529
	6. Sideline Extraction.	530
	7. Removal of Water from the Starch	532
	8. Starch Drying and Storage	533
V.	Potato Protein	534
	1. Environmental Aspects	534
	2. Protein Recovery	535
	3. Properties and Uses.	535
VI.	Utilization	535
	1. Substitution (See Also Chapters 17 and 20)	535
	2. Converted Starches (See Also Chapters 17 and 20)	536
	3. Crosslinked Starches (See Also Chapters 17 and 20)	536
	4. The Preference for Potato Starch in Applications	537
VII.	Future Aspects of Potato Starch Processing.	538
VIII.	References.	538

Starch: Chemistry and Technology, Third Edition
ISBN: 978-0-12-746275-2

Copyright © 2009, Elsevier Inc.
All rights reserved

I. History of Potato Processing in The Netherlands

The potato originated in South America[1] where the Incas cultivated and consumed it about 13 000 years ago,[2] therefore the history of the potato in Europe and elsewhere in the world began after the discovery of the Americas. When the Spaniards, commanded by Pizarro, conquered Peru and Chile (1525–1543) they found both huge amounts of gold and the potato. They brought this new kind of food to Europe. From 1570 onwards, the potato spread throughout Europe.[3]

When Clusius,[1] a botanist born in Arras (France), was appointed to the position of professor at Leiden University (The Netherlands) in 1593, the potato was introduced into The Netherlands. From there, it slowly spread over the rest of The Netherlands, being mainly used as food. About 1700 farmers began, on a small scale, to produce starch from potatoes for the purpose of starching linen. Beside this application, potato starch was mainly used by the monarchies in Europe to powder faces and wigs, a very fashionable custom at that time.

About the middle of the eighteenth century, the potato was introduced into the north of The Netherlands, where the soil was quite suitable for this new type of vegetable. There was also a good infrastructure for transporting the potatoes, i.e. the canals that had been dug to transport blocks of peat.

In the eighteenth century, potato processing in factories made its appearance. In the first factories, malt spirit was produced from potatoes instead of potato starch. Some jenever (Dutch gin) distilleries switched to using potatoes instead of cereal grain as a raw material. In this process, the potatoes were boiled in wooden barrels, crushed between rollers and mixed with malt (pregerminated barley) and water in closed barrels. Saccharification then took place. Subsequently, yeast was added to generate alcohol from the sugars. After a few days, alcohol could be obtained by distillation.

In The Netherlands, the production of potato starch in factories began in Gouda for the purpose of converting it to syrup. Jacobus Hendrikus van de Wetering was able to build on the invention of Kirchhoff (Chapter 21) and received permission to construct a syrup factory in Gouda in 1819.

In 1839, W.A. Scholten started the production of potato starch in the northern part of The Netherlands. At the end of the nineteenth century, it was discovered that the growth frequency of potatoes on a field can be increased from once in seven years to once in two years by the application of fertilizer. In 1897, W.A. Scholten, along with 13 other starch producers, formed a cartel to reduce the price to be paid to the farmers for potatoes.

In the same year, two agricultural trade unions were founded by farmers in response. A year later, the first potato starch factory owned by farmers, a cooperative, was opened. The farmers became competitors to the starch producers. In 1904, an agricultural cartel was founded, which in 1916 consisted of 22 farm cooperatives. In 1919, this agricultural cartel founded a potato starch sales office (in Dutch: Aardappelzetmeel Verkoop Bureau), which came to be known as AVEBE. Due to

Figure 11.1 Potato starch factories in nineteenth-century Netherlands (Foxhol above, Veendam below).[1]

AVEBE, the starch producers' cartel declined rapidly. Their companies were either closed or taken over by the farmers. In 1948, it was decided that AVEBE should increase its market base and derivatization of potato starch was started. Today, AVEBE modifies about 70% of the potato starch it produces.

From 1839 onwards, over 50 potato starch factories were established in the north of The Netherlands; most are now closed. The smallest factories produced one ton of starch per day. Today, the largest factory produces ~44 tons of starch from 250 tons of potatoes per hour. Because of takeovers, environmental problems and scale enlargement, there are two AVEBE factories currently producing potato starch in the north of The Netherlands. The potato harvest and processing starts in August and ends in March or April. AVEBE has also bought potato starch factories in Germany and France. The total AVEBE annual potato starch production is about 648 000 tons (about 25% of the total potato starch market in the world) from a supply of about 3 200 000 tons of potatoes.

II. Starch Production

1. World Starch Production

The total production of potato starch is small compared to the total amount of starch produced in the world. An overview of the total production of starch worldwide is given in Table 11.1.

2. Potato Starch Production in Europe

For effective production of potato starch, potatoes should contain as much starch as possible. Therefore, in modern potato starch factories in Europe, only special species of industrial potatoes are used. These potatoes are not very tasty due to the high amount of starch and, therefore, are not consumed as food. The potatoes are harvested and processed in Europe between August and April; this period is referred to as the starch campaign.

(Culled) food potatoes are not normally used because of their low dry matter content (glassy potatoes) and relatively small starch granules,[5] which are more difficult to process. When the price of food potatoes is low due to overproduction (culled) food potatoes are sometimes processed between the starch campaigns in Europe. Reclaimed potato starch, which is starch recovered from the process waters from other potato processing industries (French fry, chip/crisp and potato puree manufacturers) is processed on a small scale in Europe and North America.

Potatoes are cultivated in areas where they can be stored throughout the winter in temperature-conditioned sheds. For economical potato processing, the length of the campaign should be an optimum between the capacity of the factory, and thus the costs of investment and processing, and starch losses during potato storage, particularly

Table 11.1 Production figures of commercial starches in the world in 2003 (12% moisture)[4]

Source	Potato	Maize	Wheat	Tapioca
Synonyms		Corn		Cassava Manioc
World, 10^6 t/year	2.49	45.8	4.9	7.5
EUIS, 10^6 t/year	1.6	3.9	3.4	0.0
Main producers	Netherlands	USA	France	Thailand
	Germany	Japan	Germany	Indonesia
	France	China	USA	Brazil
	China	South Korea	China	China

due to freezing temperatures in winter and higher temperatures in the spring. Therefore, the average temperature during the winter should not differ too much from the storage temperature of the potatoes (to prevent high costs), which is 4–7°C. In Europe, potatoes are processed in the region between the south of Sweden and the north of France, a region too cold for maize cultivation. Due to overproduction in the past, the central European Union regulates the amount of starch which can be produced in each country and by each company (Table 11.2). Quotas are normally fully used.

III. Structure and Chemical Composition of the Potato

1. Formation and Morphology of the Tuber

During growth of the potato plant, a number of lateral shoots are formed below soil level at the base of the main stem.[7] These stems usually remain below soil level (stolons) and, unlike the stems above ground, they mainly grow horizontally. During growth, a horizontal stolon begins to swell and a potato is formed. Each variety of potato has a characteristic tuber shape. The uniformity of tuber shape of many potato species is determined not only by varietal characteristics, but also by plant spacing. Closer plant spacing gives more uniform tubers. Due to 'stress' during tuber growth, the developing tuber may become malformed. A stress situation may occur, for example, as a result of a period of drought followed by heavy rain, causing rapid new (secondary) growth. Apart from the reduction in market value due to tuber malformation, the quality of a tuber in which secondary growth occurs is adversely affected. The secondary parts of the tubers often grow at the expense of the first, so that the first part may have a very low dry matter content. It is possible for the latter to be almost completely stripped of starch. These 'glassy' tubers or tuber parts cannot be detected easily.

Table 11.2 Production figures of commercial starches in Europe[6]

Potato starch production quota in EU in 2003, tons/year (18% moisture)	
Germany	656 300
Netherlands	507 400
France	265 400
Denmark	168 200
Others	164 900
Potato starch production in EU in 2003, tons/year (18% moisture)	
AVEBE	648 000
Emsland Stärke	340 000
Roquette	200 000
KMC	132 000
Sudstärke	110 000
Others	180 000

2. Anatomy of the Tuber

The anatomy of the tuber is shown diagrammatically in Figure 11.2. Parts that can be distinguished are the skin (periderm) with the lenticels, the eyes, the bud and stem ends, the cortex, the ring of vascular bundles, the perimedullary zone, and the pith with medullary rays which are homologous with the medulla of the stolon.[7]

During development of a stolon tip into a tuber, a suberized layer of cells constituting the skin or periderm is formed on the outside of the tuber. As soon as the tips of the stolon swell during tuber growth, the formation of the periderm (Figure 11.3) starts. Periderm formation usually begins at the stem end of the young tuber. The outer cell layer (epidermis) divides from the outside inwards, while the layer below the epidermis (the hypodermis) divides from the inside outwards. This dividing cell layer is called phellogen. The newly-formed cells (phellem) constitute the periderm and become suberized. The cells of the hypodermis (inside the vascular ring) may deposit parenchyma cells (phelloderm) towards the inside of the tuber. The major part of the tuber tissue consists of parenchyma cells, those of the cortex and of the perimedullary zone. The parenchyma cells contain starch granules as reserve material (storage parenchyma). The potato tuber parenchyma cells are generally of a uniform shape in several (or all) directions. Tuber growth is caused primarily by the increase of the perimedullary zone. The phellogen is active throughout growth of the tuber. The skin of the potato usually consists of six to ten suberized cell layers.

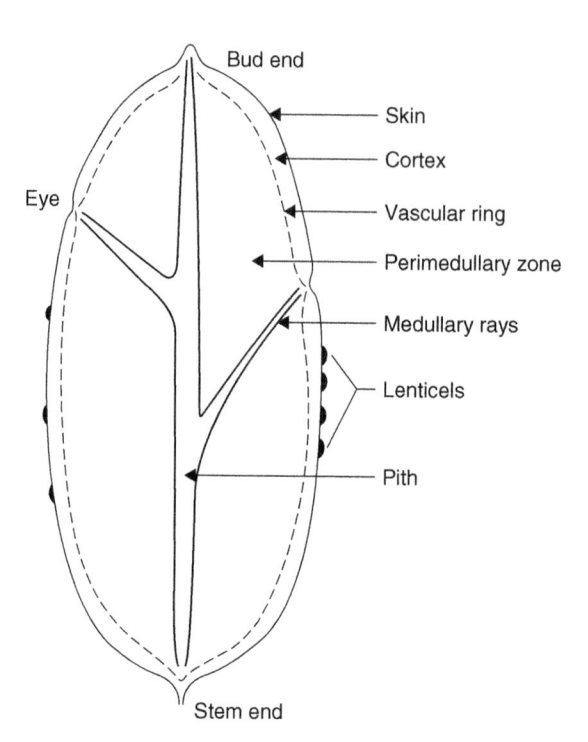

Figure 11.2 Longitudinal diagram of a potato tuber.[7]

III. Structure and Chemical Composition of the Potato

Lenticels are formed in the periderm on the sides of the young tuber where a stoma had originally been formed.[8] These stomata are formed on the stolons, which are, in fact, underground stems. The lenticels are essential for respiration of the tuber, since almost no carbon dioxide or oxygen can penetrate the skin itself. An average-sized potato has about 250 lenticels.

Cell walls consist mainly of microfibrillae of cellulose and pectic substances.[7] The thickness of a cell wall is about 1 μm. The cell walls are linked together by means of a middle lamella, which consists mainly of pectin. The middle lamella is responsible for cohesion between the cell walls. Middle lamellae are often interrupted, thus forming intercellular spaces. These spaces are interconnected and allow gas exchange with the environment for respiration via the lenticels.

Most (~77%) of the dry matter of a potato tuber consists of starch granules. Starch forms the reserve material in the tuber for use in respiration and sprouting. Starch granules are formed from the very early stages of tuberization, as soon as the stolon

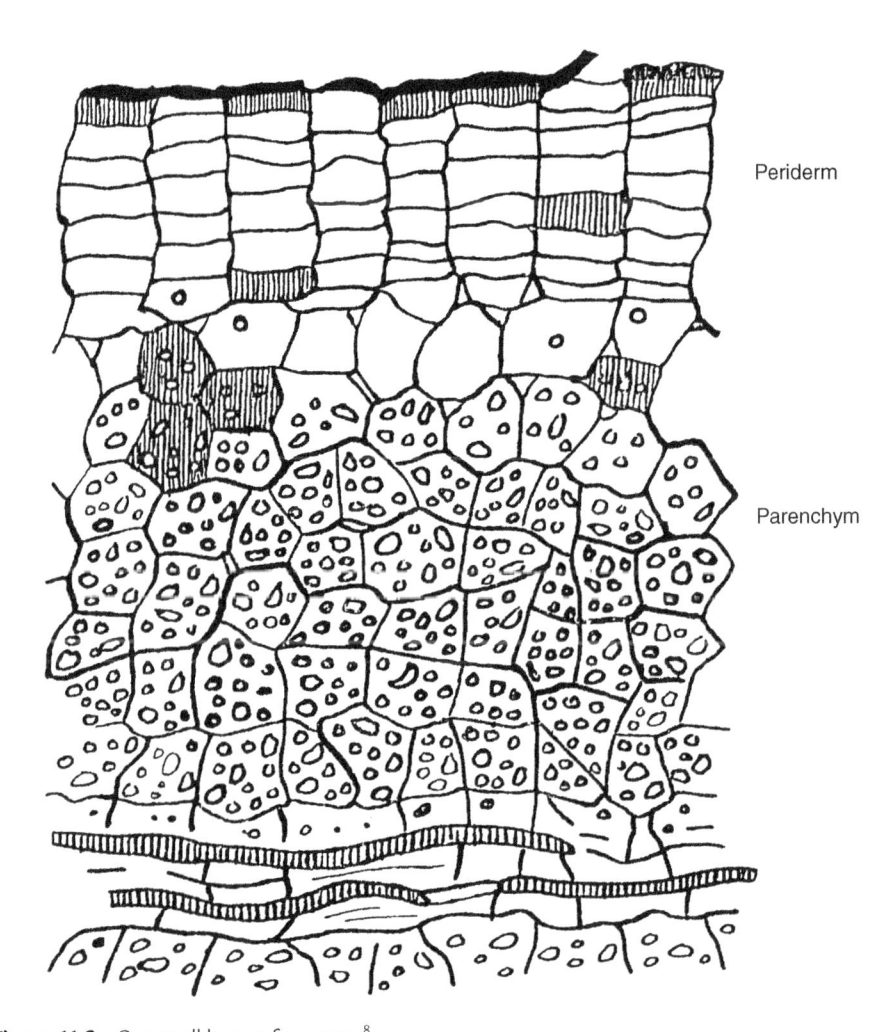

Figure 11.3 Outer cell layers of a potato.[8]

tip begins to swell. The starch content increases during tuber growth. This increase is caused both by an increase in the number of granules and by enlargement of granule size. The size (long axis) of the starch granules in a mature tuber may range from 1 µm to about 120 µm. The largest granules are often present in the large cells of the perimedullary zone. The small granules largely occur in the tissue around the vascular ring. The cells of the cortex contain the largest number of granules per cell, and those of the pith and medullary rays the smallest.

3. Chemical Composition

The chemical composition of the potato is very important to the potato starch industry. Potato processors must take account of factors such as dry matter, starch and protein content. The dry matter content in potatoes is a very important factor. This is mainly determined genetically and thus depends on the variety. Since most of the dry matter consists of starch, there is a major correlation between the dry matter content and the starch content of the tuber. Climate, soil and addition of fertilizer all affect the growth and dry matter content of the tuber. In The Netherlands, potatoes with the highest dry matter content are obtained from silty soil, while sandy soil gives the lowest content.[7] This is related to the availability of water to the plant. There is an obvious influence of the weather on the dry matter content. Warm, dry weather is beneficial to high dry matter content, while cold, wet weather tends to reduce it. The availability of potassium may also be relevant. The amount of starch in the tubers is determined when the potatoes are delivered to the factory. Within limits, this can be done very simply by weighing tubers under water. The under water weight (UWW) is the weight under water (in grams) of 5000 g of potatoes (use of 5050 g is

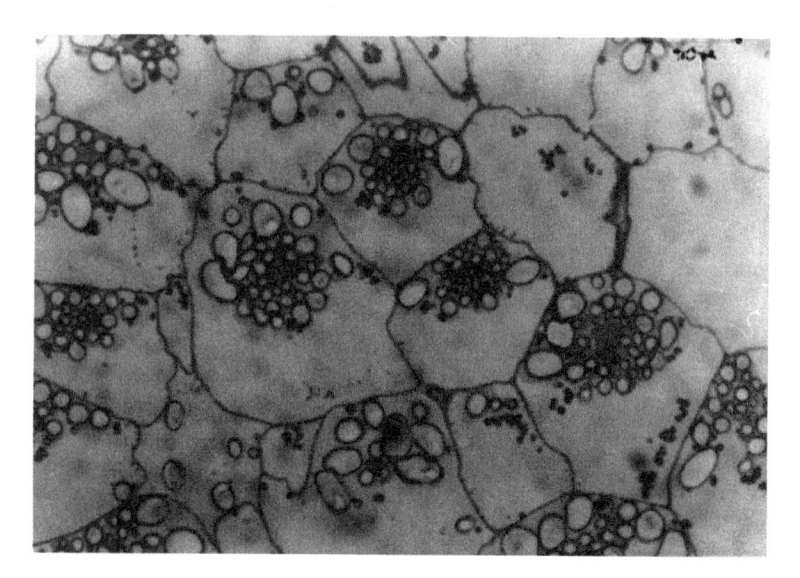

Figure 11.4 Cells with starch.[9]

an EU standard). Starch content for mature and non-stored potatoes can be obtained directly by applying the formula:[10]

$$\% \text{ starch content} = 0.0477(\text{UWW} - 70.3)$$

Tubers from a single species can have UWW, starch, sugar and protein content values that vary substantially from year to year (Figures 11.5 and 11.6). The chemical composition of the dry matter in potatoes (average 22.3%) can vary substantially according to variety, conditions during growth and the degree of maturity. Changes also occur during storage. The average composition (major components) of dry matter of a potato is given in Table 11.3.

The starch consists of amylose (21–25%) and amylopectin (75–79%). The principal sugars are the reducing sugars D-glucose and D-fructose and the non-reducing disaccharide sucrose. Considerable variations in sugar content may occur at the expense of the starch content during storage. Cell walls consist mainly of pectins, cellulose, hemicellulose and lignin. The pectins consist mainly of partially esterified straight polygalacturonic acid chains.

4. Differences Between Commercial Starches

While starches from all sources have many similar properties, they also differ in many aspects.[12,39] Table 11.4 shows the major differences between the most important commercial native starches.

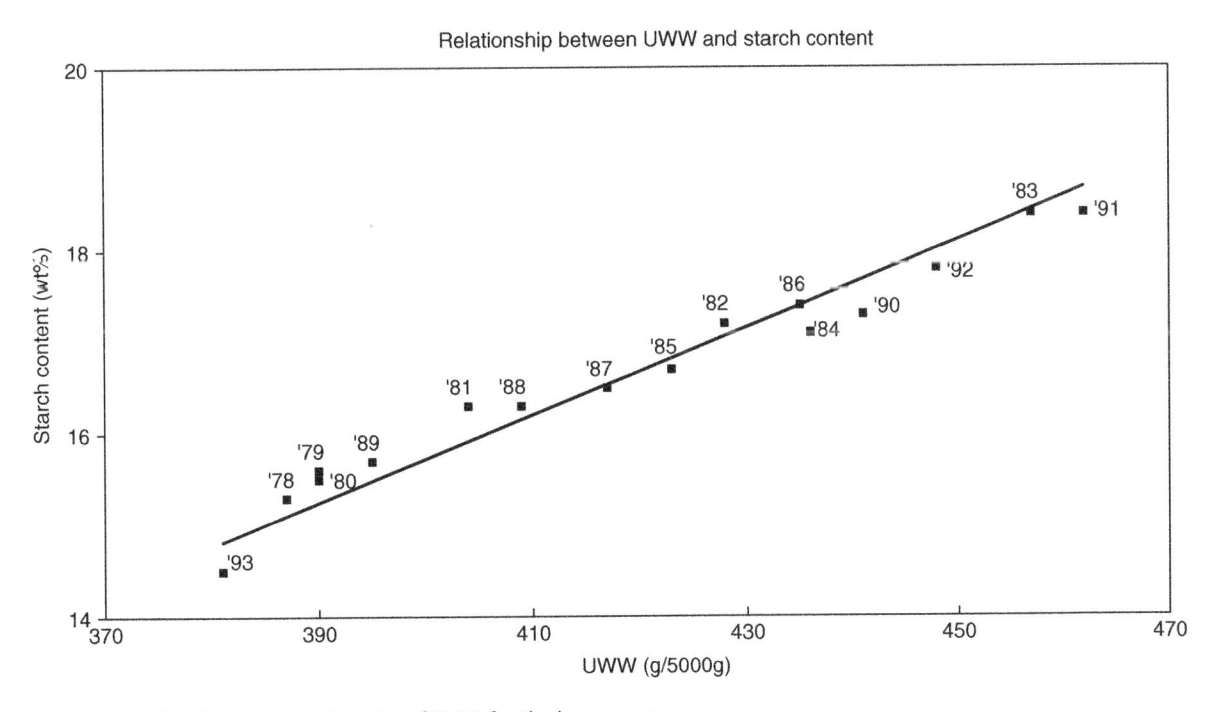

Figure 11.5 Starch content as a function of UWW for Ehud-type potatoes.

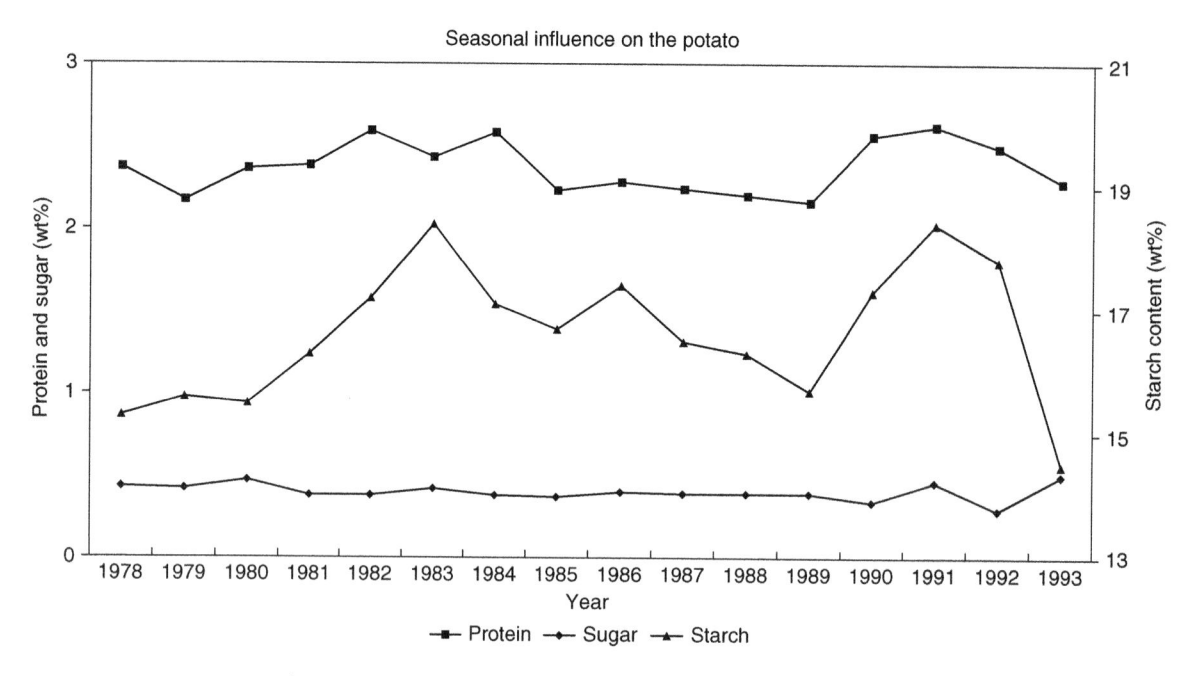

520 Potato Starch: Production, Modifications and Uses

Figure 11.6 Variation in composition of the Ehud potato in The Netherlands.[11]

Table 11.3 Composition of the potato		
Component		Percentage of dry matter
Starch		76.6
Coagulable protein	5.0	
Free amino acids	1.3	
Amides	1.8	
Peptides	1.3	
Total N compounds		10.7
Cell walls		4.0
Fructose	0.5	
Glucose	1.1	
Sucrose	1.2	
Total sugars		2.7
Citric acid	2.0	
Malic acid	0.5	
Oxalic acid	0.15	
Total organic acids		2.8
Potassium	1.8	
Magnesium	0.1	
Total ash content		4.0
Phosphorus		0.2
Lipids		0.4
Solanidine	0.03	
Total glycoalkaloids		0.05
Ascorbic acid (Vitamin C)		0.1
Phenolic compounds		0.13
Chlorine		0.14
Sulfur		0.15
Nitrate		0.04

III. Structure and Chemical Composition of the Potato

Table 11.4 Differences between commercial starches

Composition and characteristics	Potato	Maize	Wheat	Tapioca	Waxy maize	Waxy potato
Shape of granules	oval spherical	round polygonal	round lenticular	truncated round	round polygonal	oval spherical
Diameter, range (μm)	5–100	2–30	0.5–45	4–35	2–30	5–100
Diameter, number average (μm)	23	10	8	15	10	23
Diameter, weight average (μm)	45	15	25	25	15	45
Number of granules per gram starch $\times 10^6$	100	1300	2600	500	1300	100
Amount of taste and odor substances (gelatinized starch)	low	high	high	very low	medium	low
Number average DP of amylose	4900	930	1300	2600	–	–
Pasting temperature (°C)	60–65	75–80	80–85	60–65	65–70	63–67
Brabender peak viscosity[a]	3000	600	300	1000	800	2000
Swelling power at 95°C	1153	24	21	71	64	1500
Solubility at 95°C (%)	82	25	41	48	23	90
Starch paste viscosity	very high	medium	low	high	high	very high
Paste texture	long	Short	short	long	long	long
Paste clarity	almost clear	opaque	cloudy	quite clear	reasonably clear	clear
Paste, resistance to shear	low	medium	medium	low	low	low
Paste, rate of retrogradation	medium	high	high	low	very low	very low
Film clarity and gloss	high	low	low	high	high	high
Film strength	high	low	low	high	high	high
Film flexibility	high	low	low	high	high	high
Film solubility	high	low	low	high	high	high

[a] Average Brabender peak viscosity (5% starch concentration) in Brabender units

5. New Development: The All-amylopectin Potato

In some potato starch applications, as with applications of other native starches, the presence of amylose is undesirable.[13] After gelatinization, it forms crystals which reduce paste clarity. Amylose retrogradation can be prevented by chemical modification; however, the costs of the chemical process step used and waste water treatment are high. A new type of potato has been bred to solve this problem. This potato contains only amylopectin, it can be developed in two ways. One way consists of treating a mutation-sensitive potato with radiation or chemicals and selecting the desired variety. After succesful selection the normal plant cross-fertilization method to produce starch potatoes is used. The second way involves genetic modification of the potato with an antisense technique to eliminate the formation of amylose.[14] By this method, no materials are introduced into the potato which are not normally present, and the amylopectin potato starch contains no genetically manipulated substances. This method avoids the time consuming (10 years) cross-fertilization program. Amylopectin starches, like waxy maize starch, are mostly used by the food industry. However, amylopectin potato starch has specific properties which are not available in current commercial starches.

Starch properties which may determine applications are the size and swelling power of the granules, the glass transition temperature, taste and the presence of amylose. Potato starch is characterized by large granules with considerable swelling power, a low glass transition temperature, paste clarity and a reasonably neutral taste.

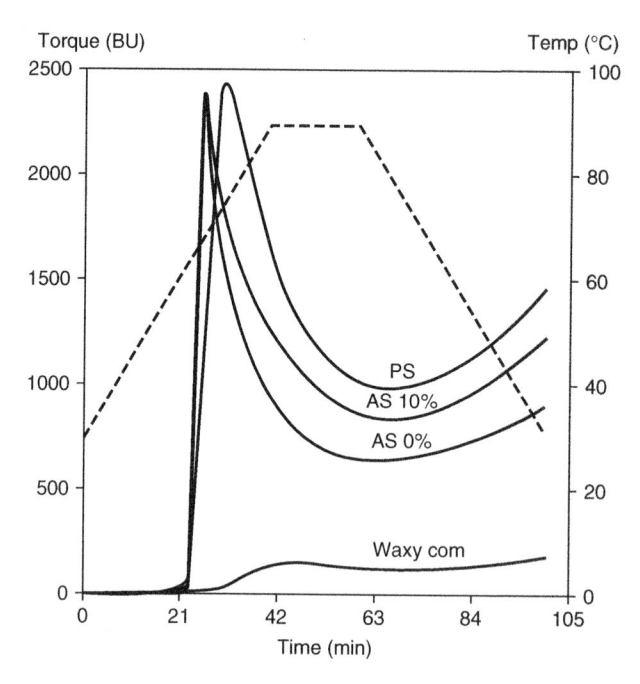

Figure 11.7 Brabender ViscoAmylograph curves for potato starch (with 20 [PS] and 10 [AS 10%] percent amylose), amylopectin potato starch (AS 0%) and waxy maize/corn starch.[14]

The viscosity of a starch paste is mainly a function of the size of swollen granules. Waxy maize starch has smaller granules, so its swelling power is less than that of amylopectin potato starch. The peak viscosity of amylopectin potato starch, in comparison to potato starch and waxy maize starch, is shown in Figure 11.7. The gelatinization properties of amylopectin starch resemble those of potato starch. With reduced amylose content, the peak viscosity occurs as a lower temperature as swollen granules are weak and can be ruptured more easily with shear.

IV. Potato Starch Processing

To obtain starch from potatoes, one of several processing routes can be chosen. The first steps, starting with the delivery of potatoes and moving on through sampling, storage, washing and grinding, are in most cases more or less equivalent, and need not be discussed. With the grinding of the potatoes, the cells (as shown in Figure 11.3) are broken open and a mixture of starch granules, broken cell walls, and the remainder of the cell content, which is mostly water containing soluble proteins, amino acids, sugars and salts (the so-called 'potato juice') is formed. Table 11.5 shows an average composition (in weight percent) of ground potatoes.

For further processing, physical separation of the two solid components (starch and fiber) from each other and from the potato juice is essential. Separating these three components can be done in three ways (Figure 11.8). In process I, the potato juice is first separated from the ground potatoes, and in the second step the starch and fiber are separated. In process II, the starch is first removed from the ground

IV. Potato Starch Processing

Table 11.5 Composition of potatoes

	Average composition, %	Standard deviation
Starch	19	5.5
Fiber	1.6	
Protein (including amino acids)	2	1.0
Sugars	1.1	
Salts (including sand and dirt)	1.2	0.45
Lipid	0.15	
Water	75	6

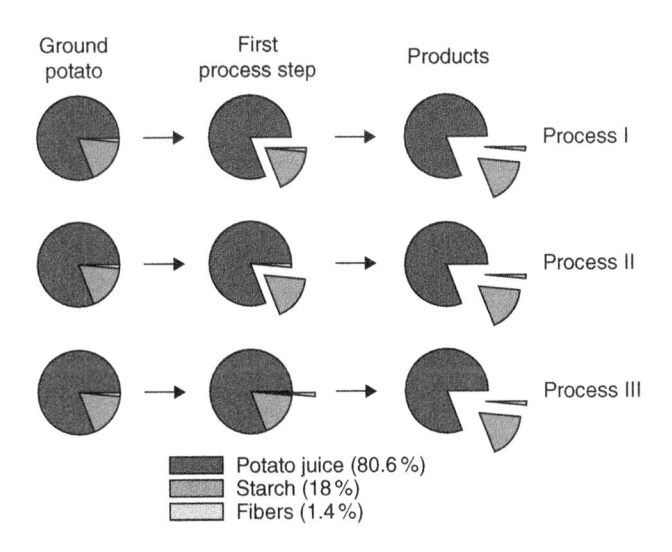

Figure 11.8 Three methods of potato processing.

Table 11.6 Evaluation of the three starch processes

	Process I	Process II	Process III
Investment	average	low	high
Starch yield	average	poor	high
Protein yield	average	high	average
Protein quality	good	average	poor
Process costs	average	low	average
Water consumption	average	low	low
Process control	average	easy	sophisticated

potatoes, and in the second step the potato juice is separated from the fiber. In process III, fiber is first removed from the ground potatoes, and in the second step the potato juice is separated from the starch. All three ways of processing ground potatoes are currently in use. The choice of process depends on a number of factors. The pros and cons of the different processes are presented in Table 11.6. To build the optimal potato starch factory, one has to decide which factors have the greatest influence on the overall economic picture over a long period.

524 Potato Starch: Production, Modifications and Uses

A possible potato starch process which attempts to combine the best elements of the described processes (Figure 11.8) and to find an optimum which can be implemented is shown in Figure 11.9. This process is described in more detail in the rest of the chapter.

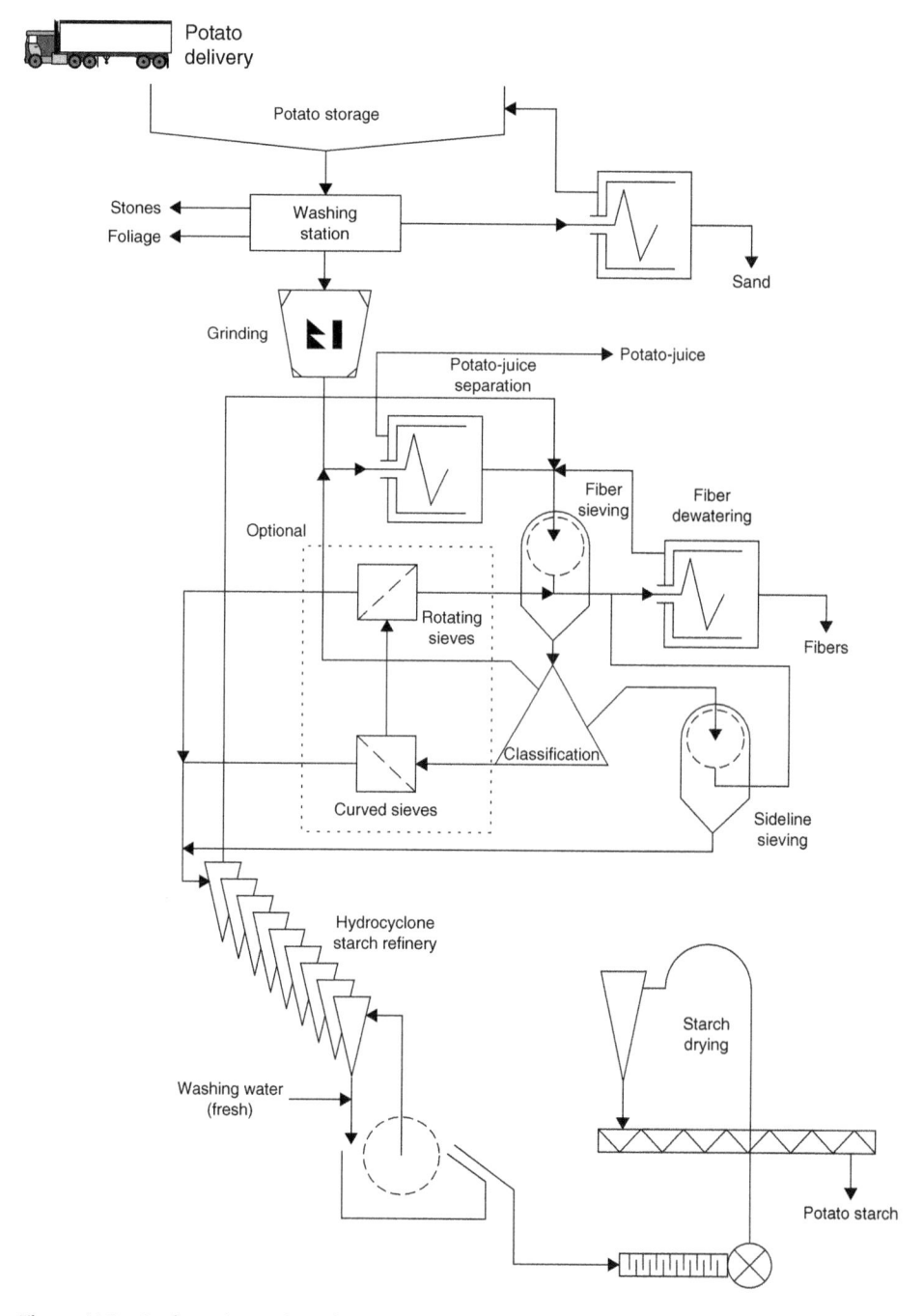

Figure 11.9 A schematic overview of a potato starch process.

In the potato starch industry, the starch itself represents >90% of the sales value, while the co-products (fiber, protein and concentrated deproteinized potato juice) represent <10% of the sales value. Nevertheless, protein quality is becoming increasingly important to the economics of the total process.

1. Grinding

The goal of this step is to break the cells of the potato and liberate the starch granules. Thus, the first step is the actual shredding of the potatoes. It is done in ultra-rasps (Figure 11.10). In these drum rasps, the potatoes are rasped by sawblades between the drum and a perforated plate. The ground potato slurry flows through the sieve plate and is collected in a tank. To prevent undesirable coloring of the potato juice, an antioxidant is added. This antioxidant represses the oxidation of tyrosine, dihydroxyphenylalanine and/or chlorogenic acid as catalyzed by polyphenoloxidase (a potato enzyme) and prevents formation of melanin, which has a red-brown color.[15,16] Another way to prevent coloring of the potato juice is by grinding in a vacuum system, i.e. removing oxygen. However, one of the problems with vacuum grinding is that the rest of the process also has to be closed to prevent the reaction from occurring later in the process. During grinding, 98% of the starch granules are freed from cells, yielding an average of 177 kg/ton of potatoes with an UWW of 450 g per 5050 g potatoes.

2. Potato Juice Extraction

The goal of this step is to remove the potato juice (containing the proteins) from the ground potatoes. This is done in a decanter centrifuge, a continuous centrifuge consisting of a cylindrical drum with a screw inside (Figure 11.11). The drum and screw produce about 3000 g so that the starch granules (density ~1600 kg/m^3) and fiber (density ~1100 kg/m^3) are separated from the potato juice (density ~1000 kg/m^3). The solid starch and fiber are pressed to the drum wall, and the potato juice flows over the front of the machine. A stationary centrifugal pump is mounted in the overflow, and the torque of the decanter is used to pump out the potato juice. The screw

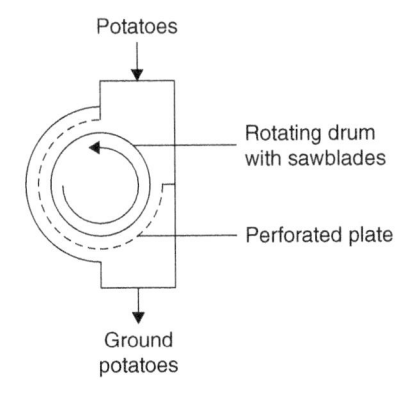

Figure 11.10 Rasp used for shredding potatoes.

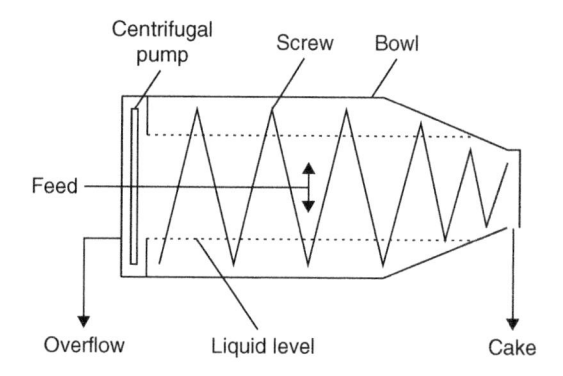

Figure 11.11 Continuous, decanter-type centrifuge.

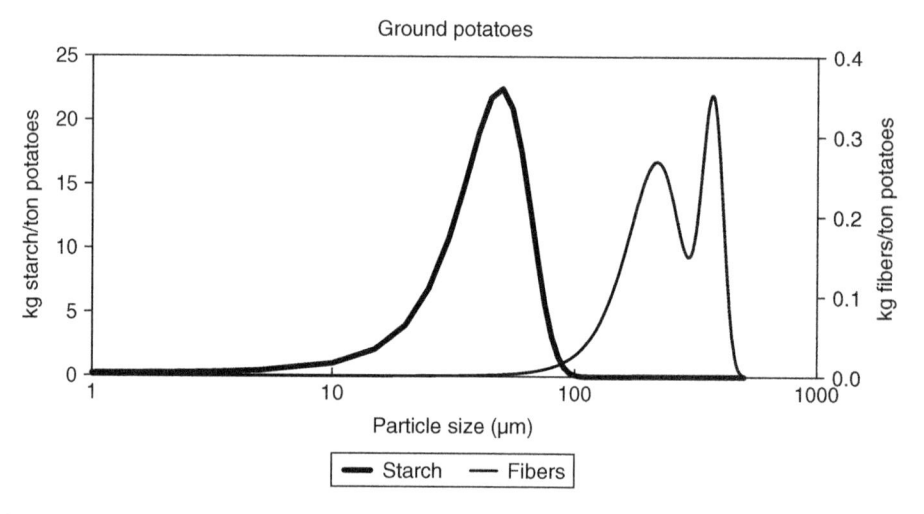

Figure 11.12 Particle size (diameter) of ground potatoes.

has a slightly different speed than the bowl of the decanter so the solids are screwed out at the back part of the machine. The products are a starch-fiber cake of about 40% dry matter and a potato juice overflow, free of solids, that is treated in the protein factory. The effectiveness of potato juice removal depends on the amount of dilution water and the number of water removal stages used.

3. Fiber Extraction

This process involves a physical separation by sieving, based on particle size distribution. Figure 11.12 displays a particle size distribution (PSD) of ground potatoes. Starch granules range between 1 and 120 μm in diameter, while fiber particles have diameters between 80 and 500 μm. Fiber particles have a distribution corresponding to two Gaussian curves. The smaller particles are the ground cell walls; the larger particles are the more woody fibers and whole plant cells. The area under the curves gives the amount of starch (177 kg) and fiber (14 kg) per ton of potatoes with an underwater weight of 450 g per 5050 g potatoes.

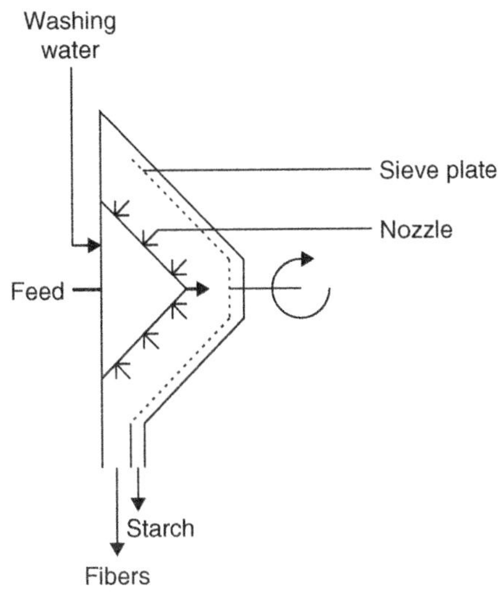

Figure 11.13 Conical centrifugal sieves.

The goal of this step is to separate fiber from the starch-fiber cake with an acceptable starch loss. Some small fiber particles may remain in the rough starch. For fiber removal, diluted cake from the potato juice extraction is sieved in so-called centrifugal sieves. This type of sieve consists of a conical rotating sieve with perforations of 125 μm (Figure 11.13).

Figure 11.15 shows that, with these sieves, a part of the fiber passes through the sieve together with starch granules. The starch in the fiber comes from potato cells which remained unbroken during the grinding process (~2% of the total starch). The effectiveness of the sieving depends on the sieve area installed, the amount of washing water used and the number of sieve stages. Generally, the amount of starch passing through the sieve is ~97%. This means a total starch loss of 2–3%. In the starch milk, there is about 1.2% fiber. The fiber is dehydrated to 17% dry matter and sold as feed. To describe the separation steps in a potato starch factory, the theory of reduced efficiency curves is used.[17] This theory gives a description of solids classification based on particle size distribution. Figure 11.14 displays a typical set of reduced efficiency curves for a potato starch factory. The rough starch milk has a particle size distribution as shown in Figure 11.15 and is moved on to the classification step.

4. Starch Classification

After fiber extraction, a slurry containing starch, small fiber particles that were not removed during extraction and some remaining protein is obtained. The goal of the classification step is to free the starch slurry of fiber. This classification is done in a separator centrifuge. Figure 11.16 shows a schematic diagram of this type of separator.

A separator centrifuge is a disk-type, continuous centrifuge. The starch is concentrated in the separator due to centrifugal force. Starch granules with the highest density

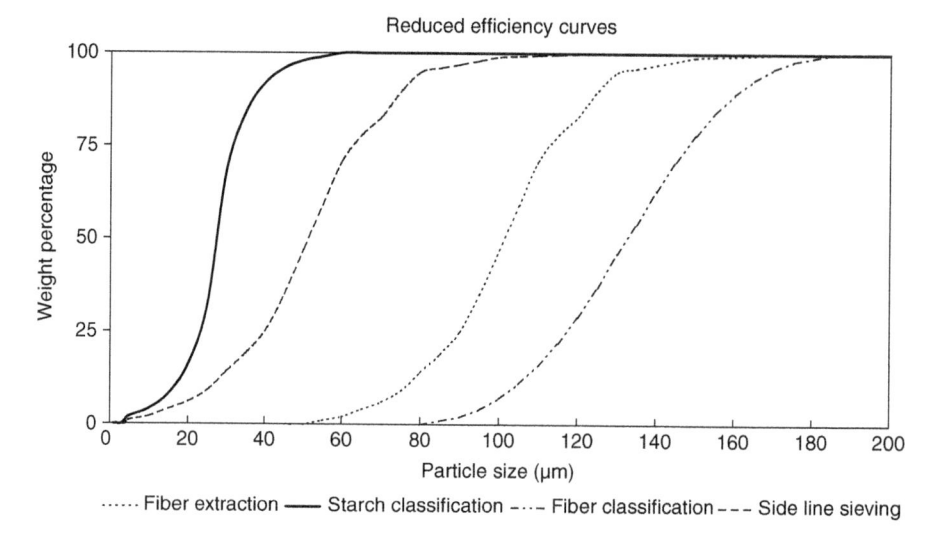

Figure 11.14 Reduced efficiency curves [G'(X)] for a potato starch process.

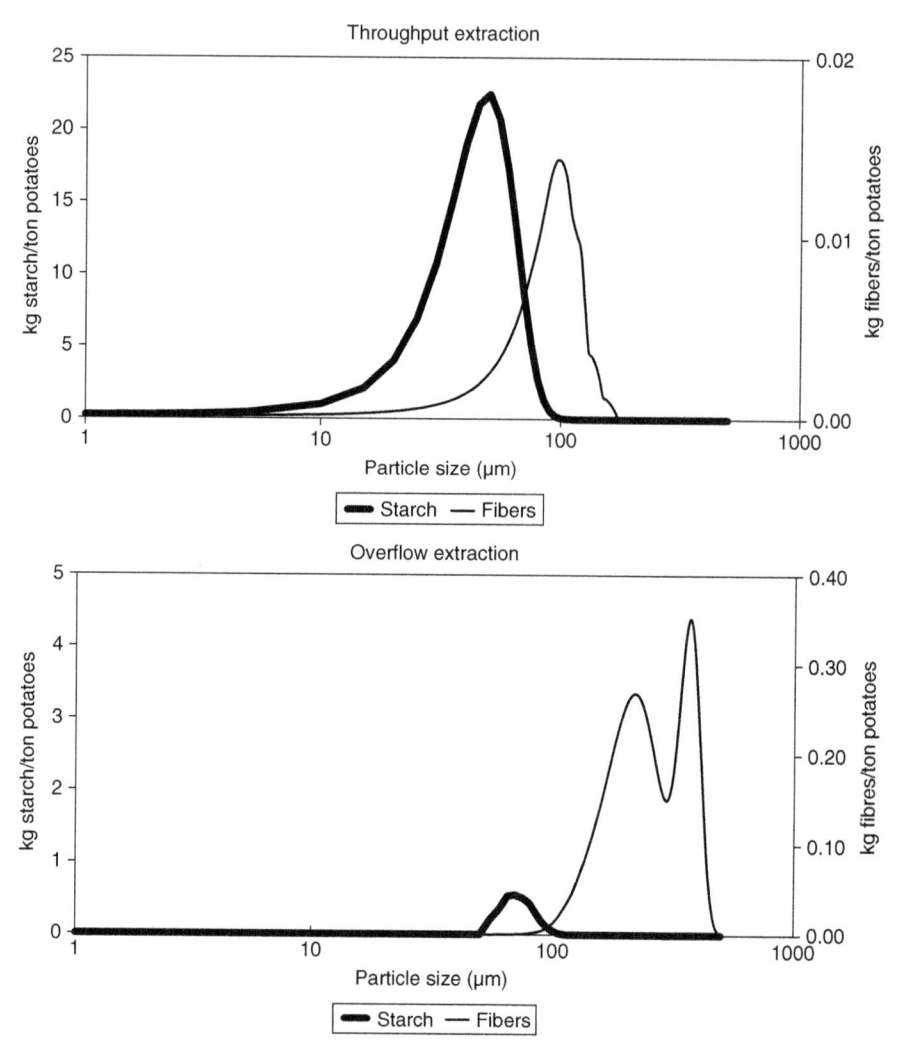

Figure 11.15 Particle size distributions of the starch and fiber fractions after fiber extraction.

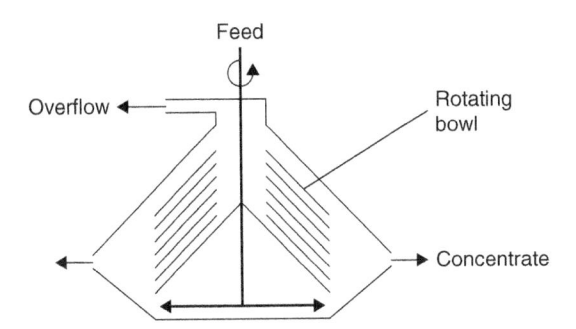

Figure 11.16 Disk-type, continuous centrifuge.

are concentrated in front of the nozzles. This concentrate is removed from the separator in a fixed amount (determined by the nozzle diameter) with a starch concentration of 40 weight percent. Smaller starch granules and the fiber (which is less dense) go to the overflow, along with the process water and the soluble components. Figure 11.14 gives efficiency curves for starch and fiber in the classification. Figure 11.17 shows particle size diameters in the concentrate and the overflow. The areas under the curves show that the fraction of starch in the concentrate is ~85% and in the overflow the fraction of starch is ~12%. About 1.0% of the total fiber (1.2%) is concentrated in the overflow. Thus, the only impurities left in the starch are the soluble components, such as protein, sugars, salts and amino acids. Further purification takes place in the starch refinery.

5. Starch Refinery

After classification, the slurry contains starch and some solubles that were not removed during potato juice extraction. The function of the starch refinery is to remove the soluble protein. The protein is removed by diluting with water, then concentrating the starch in a multi-stage, countercurrent flow system (hydrocyclones). Figure 11.18 displays a single cyclone and the network of a 9-stage countercurrent starch refinery. The cyclone creates tangential flow.[17,18] Thus, heavier particles (starch granules) flow towards the hydrocyclone wall and the flow direction is from top to bottom. Concentrated starch slurry escapes at the bottom of the cyclone. In the middle of the cyclone, the flow direction is reversed, going from bottom to top. Starch-free process water escapes at the top of the cyclone. Because most of the hydrocyclone types used industrially do not have geometries suitable for the potato starch industry, AVEBE has developed its own hydrocyclones.[19–22] The advantage of these cyclones compared to other starch cyclones lies in the higher concentration of potato starch granules at the outlet, so that less washing stages and/or less washing water is required.

An optional sieving block is displayed in the schematic diagram of the starch process (Figure 11.9). Depending on the origin of the potatoes, the starch slurry can be contaminated (with wood, sand, peat, etc.) with particles with sizes of 90 μm and larger, and with densities the same as or higher than that of starch. There has been no separation step for this kind of material (for the fraction with diameters <125 μm)

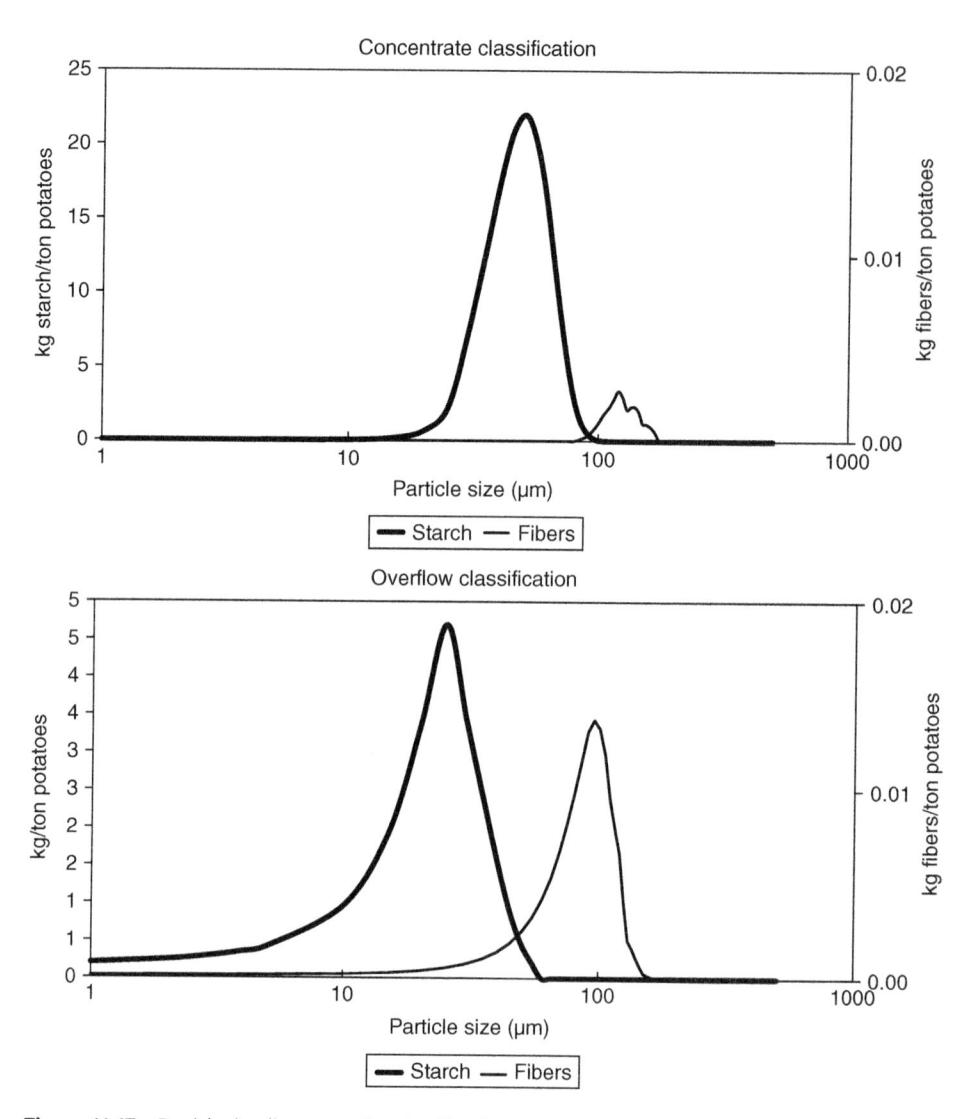

Figure 11.17 Particle size diameters after classification.

in the process up until now. To separate this dirt element from the starch, an optional sieving can be used. The first step is to remove all the dirt (along with some starch). The second step is to recover as much starch as possible to keep the yield as high as possible.

6. Sideline Extraction

The goal of sideline extraction is to regain the starch (fine granules) that has been lost in the overflow of the classification. Figure 11.19 indicates that the particle sizes of starch and fiber are sufficiently different to enable separation by sieving. This is done by the same conical rotating sieves that were used in the fiber extraction, but with a

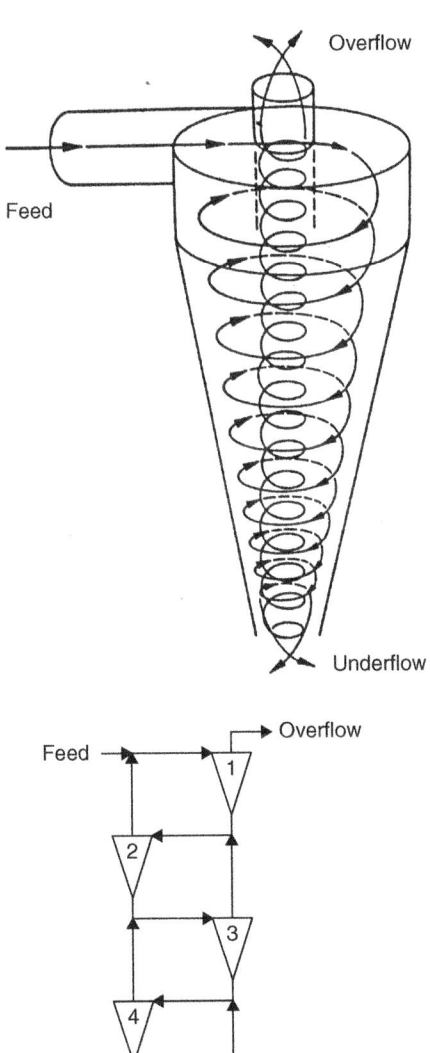

Figure 11.18 A single hydrocyclone[23] and a countercurrent, 9-stage starch refinery.

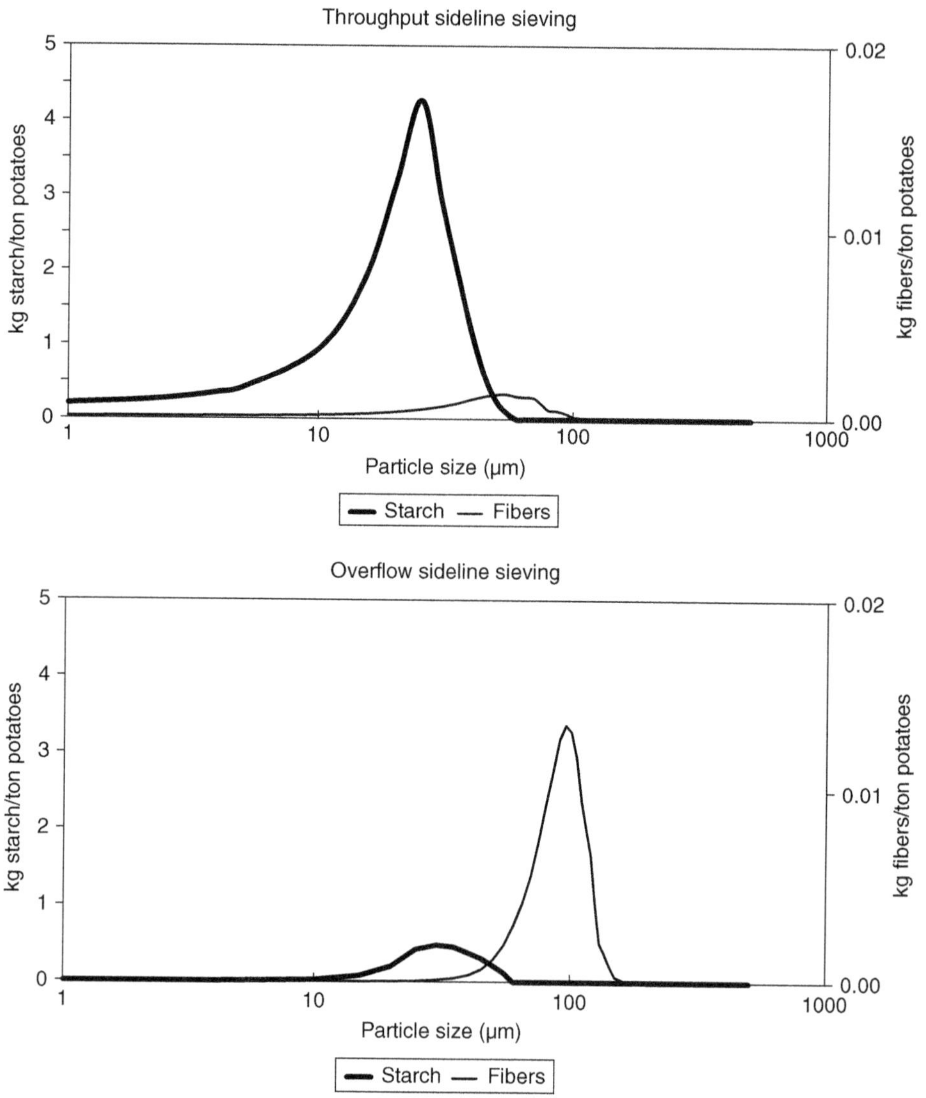

Figure 11.19 Particle size distribution after sideline sieving.

smaller mesh size (~70 μm). The effect of this sieving can be seen from the separation curves in Figure 11.14. The fiber part is then mixed with the fiber from the fiber extraction. The starch part, which is 11% of the total starch and which still has an excessively high protein concentration, is refined in the same way as the rest of the starch. The total starch recovery is ~96%, 2% being lost during the grinding process and 2% in the rest of the process.

7. Removal of Water from the Starch

The goal of this process unit is to remove as much water from the starch as possible. The more water mechanically removed in this step, the less that will have to be

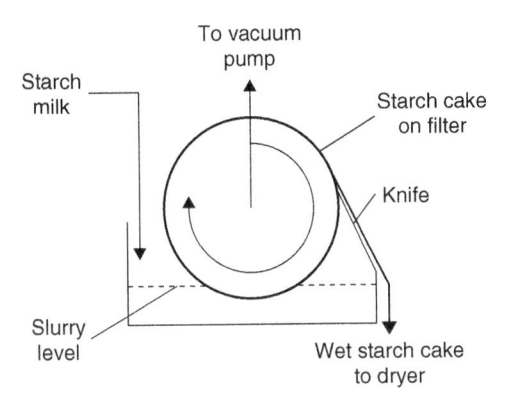

Figure 11.20 Vacuum drum filter.

removed by evaporation in a dryer. The water from the concentrate from the starch refinery is removed on rotating vacuum drum filters. A starch cake with 40% moisture is produced. In Figure 11.20, a schematic diagram of the filter is shown. The starch slurry is fed into a tank. In this tank, a drum with a filter cloth rotates. This is due to the pressure difference across the cloth. The starch forms a cake on the cloth and the process water passes through the cloth. The starch cake on the drum is scraped off with a knife while the filter drum is rotating. The filter is continuously cleaned with water through spray nozzles. The filter has a running time of about four hours, after which time the cake is removed and the filter surface is cleaned. The wet starch cake is transported (by a screw or a conveyer belt) to the dryer.

8. Starch Drying and Storage

The goal of this process is to dry the starch to its equilibrium moisture content of ~20% moisture. The dryer is a pneumatic ring dryer (Figure 11.21). The dryer has four distinct parts. The first part involves heating the air to 150°C. Generally, this is done in a heat-exchanger using steam. The second part, the heart of the dryer, is a tube in which the starch is dried by means of hot air. The residence time (and thus the drying time) is ~2 seconds. The third part is the separation section, where the dried starch is separated from the air. This is done in dust cyclones. The fourth part of the dryer is the transportation element. Here, kinetic energy for the transportation of the air and starch is brought into the unit by a ventilator. The dried starch is mostly cooled with air (to prevent condensation problems) and transported pneumatically to a storage silo. The starch is stored in bulk silos with an average volume of $30\,000\,m^3$ (Figure 11.22). The silos are filled in the center. The starch is subsequently spread over the surface by means of a rotating screw. After the silos have been filled during the campaign, they are emptied for sale and modification during the 'inter-campaign' period. This emptying of the silos is done in the same way as the filling, only in reverse. Thus, the storage of starch is done on the principle of 'last in, first out.'

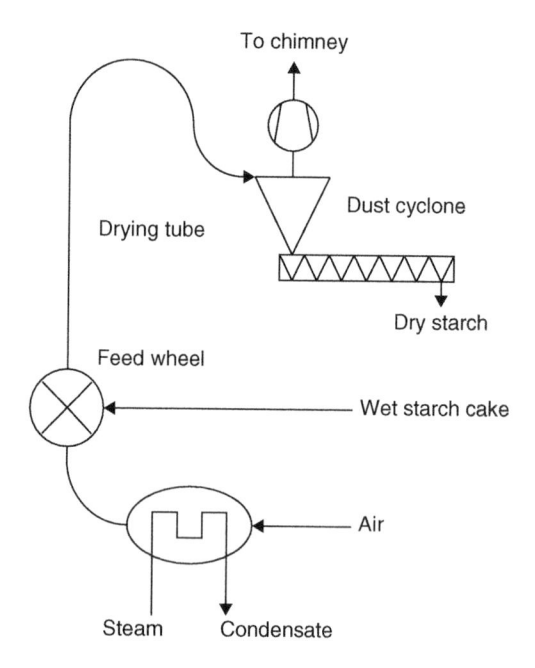

Figure 11.21 Schematic diagram of a pneumatic starch dryer.

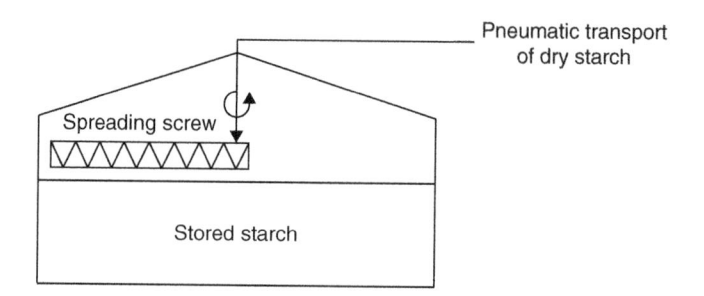

Figure 11.22 Bulk storage of potato starch.

V. Potato Protein

1. Environmental Aspects

At the beginning of industrial potato processing,[24,25] the potato juice with the proteins was drained into canals. The environmental pollution, in the form of odor and huge amounts of foam in the canals, along with the technological possibilities to recover protein from the potato juice, led to protein processing beginning in 1977. A potato starch factory which processes 250 tons of potatoes per hour produces 200 tons per hour of potato juice with 3% protein. After recovery of the protein by heat coagulation, half of the protein remains in the deproteinized potato juice. In Sweden, France and Germany, this liquid is sprayed on fields, a practice banned by law in The Netherlands.

Deproteinized potato juice can be treated in two ways. One consists of an anaerobic water treatment in which 6.8 MNm^3/year of biogas is generated, but 50% of the original protein content remains. This liquid must be treated in an aerobic waste water treatment facility. The second way to treat the deproteinized potato juice is to reduce the liquid amount by evaporation. This produces concentrated potato juice with a dry substance of 60 weight percent which can be sold as animal feed. The remaining condensate and small amounts of waste liquid still contain about 5% of the original protein and are treated in an aerobic waste water treatment plant. Thus, by protein recovery and evaporation, the protein content is reduced by 95% and the resulting feed products can be sold.

2. Protein Recovery

During processing in the potato starch factory, much air is introduced into the potato juice. Subsequently, the potato juice is stored in defoaming tanks to raise the density from 500 to 1000 kg/m^3. To reduce the energy consumption for protein processing, the protein concentration in the potato juice can be increased by hyperfiltration after defoaming. After preheating, the pH is set between 4.8 and 5.6, depending on the quality of the potato juice. Then, the juice is jet-cooked (102–115°C) to coagulate about half of the protein. Solid protein flakes are separated from the slurry by means of a decanter (Figure 11.11) to produce a cake with 60% moisture. The protein is dried in a pneumatic flash dryer to give a dry matter content of 90%.

3. Properties and Uses

Potato protein dry matter content should be ~90% in order to produce a balance between microbiological deterioration, energy consumption and fine powder content. The protein content of the dry matter is ~85% and the bulk density is ~550 kg/m^3. Potato protein contains a high amount of protein, as compared to fishmeal, milk powder and soy protein. Also, the amount of important amino acids, such as lysine, methionine and cystine, is relatively high. Potato protein is mainly sold as cattle feed.

VI. Utilization

Native potato starch, which is used in the food, paper and textile industries, is often not optimal for a particular application. Modifications are done to obtain the properties needed for specific uses.[26–28] More than 500 modifications of potato starch are currently known. This chapter presents an overview of the major commercial modifications rather than a complete bibliography of all potato starch reactions. With a distinction based on the chemical character of the product, three groups can be distinguished. Some derivatives are made from starch by a combination of reactions.

1. Substitution (See Also Chapters 17 and 20)

Substituted products of starch are the esters and ethers of starch which prevent the formation of ordered structures in a starch paste and retard retrogradation.

Monostarch phosphate and starch acetate are examples of commercial esters, while hydroxyalkyl ethers, cationic starches and carboxymethylstarch are examples of commercial ethers. The major uses of potato starch esters and ethers are in the textile industry (warp sizing: lubrication and abrasion resistance, printing), paper (internal sizing, coating and surface sizing: strength, stiffness and ink resistance), adhesives (gummed tape, paper, paperboard and bag adhesives; wall covering adhesives), water treatment (flocculation), oil industry (fluid loss reducer) and scale inhibition.

2. Converted Starches (See Also Chapters 17 and 20)

Converted starches, also called thin-boiling starches, are produced by degradation of the starch chains into small segments. They can be cooked in water at higher concentrations than native starches. Low-viscosity starches are needed in applications where a high solid starch paste with a pumpable and workable viscosity is required. There are four classes of commercial converted starches: dextrins (hydrolysis in solid-state); acid-modified starches (hydrolysis in a slurry); oxidized starches; and enzymically depolymerized starches.

Dextrination is the heating of powdered starch, mostly in the presence of small amounts of acids, at different temperatures and with different reaction times. The products are white dextrins, yellow dextrins and British gums (converted without acids). The yellow dextrins are the most highly depolymerized. Due to their DP of 20–50, it is possible to make low- to medium-viscosity solutions which contain up to 80% of solids. Dextrins are used as adhesives (gummed paper, bag adhesives, bottle labeling), in textiles (finishing textile fabrics, printing) and in the foundry industry. Dextrins are preferably made from potato starch or tapioca starch, because the amount of proteins, lipids and other impurities are lower in potato starch than in maize starch. Potato starch and tapioca starch convert easily to dextrins which give dispersions of excellent clarity, stability and adhesiveness.

Acid-modified starches are made by heating and stirring concentrated starch slurry with an acid at a temperature which is below the gelatinization temperature. When the desired viscosity or degree of conversion is reached, the acid is neutralized. Acid-modified starches are used in the food industry (candy).

Starch can be oxidized by a number of oxidizing agents (mainly sodium hypochlorite) in an aqueous starch suspension or in a paste. Depending on the oxidant and the conditions of the reaction, carboxyl groups and carbonyl groups are introduced, which in an alkaline environment result in depolymerization. The carboxyl groups reduce retrogradation and gelation. Oxidized starch is used in the paper industry (surface sizing, coating) and in the textile industry (fabric finishing, warp sizing).

Enzymes such as α-amylase and β-amylase can break the starch chains into maltodextrins and starch sugars such as glucose and maltose. The heat-stable enzymes convert the starch after dissolving the starch by jet-cooking. Enzymatically converted starches are used in the food industry (confectionery, baking products, sweeteners), paper industry (surface sizing) and the fermentation industry.

3. Crosslinked Starches (See Also Chapters 17 and 20)

When a starch is treated with bi- or multi-functional reagents, crosslinking occurs. The reaction takes place in a slurry. With increasing degrees of crosslinking, starch granules become more and more resistant to gelatinization. There is no change in appearance of the intact granules. Some crosslinked starches are distarch phosphate (from reaction with phosphoryl chloride or sodium trimetaphosphate) and distarchglycerol (from reaction with epichlorohydrin). Crosslinked starches are used when a stable high viscosity is needed. They are used in the food industry (desserts, bakery products, soups, sauces), the textile industry (printing), adhesives and pharmaceuticals.

4. The Preference for Potato Starch in Applications

Potato starch and its derivatives have special properties, such as a low gelatinization temperature and a high paste consistency. Potato starch is preferred in the food industry,[29] because its pastes have a good clarity (due to a small amount of lipids and protein) and a neutral flavor. In the paper industry, there is also a preference for potato starch. The reason for this is the high molecular weight of its amylose and its good solubility. Cationic potato starches are preferred for internal sizing due to the concurrent presence of phosphate groups.[30] Potato starch dextrins also have an advantage over other starches as an adhesive, because of the good remoistenability[31] and a desirable rheology resulting in a perfect direct tack. Textiles are manufactured better with potato starch due to its film properties,[32] paste penetration depth[33] and adhesive power.[34] Potato starch, tapioca starch and waxy maize starch perform better than other starches in oil drilling, due to their excellent fluid loss properties. The high phosphate content of potato starch is responsible for special flocculation effects in mining[35] and water treatment.[36] Due to its large granule size, potato starch is preferred as a precoat on filters. Extruded carboxymethylated potato starch is an excellent scale inhibitor in aqueous systems, being superior to other carboxymethylated starches.[37]

An overview of the average performance and preference of commercially available starches and their derivatives is given in Table 11.7.[38] The suitability of the starches for these uses is indicated by +++ (particularly suitable), ++ (suitable) and + (less suitable).

As can be seen, there are differences in performance between the starches and derivatives of different biological origin. Two groups of factors have a significant influence on the choice of the kind of starch used in certain applications: economics and performance factors. Starches produced from different sources are not interchangeable in every use. In some applications, a particular kind of starch is preferred, irrespective of price concerns. However, for most applications, there is a certain degree of interchangeability between different kinds of starch.

Table 11.7 Performance of starches in some applications

Application	Potato	Maize	Wheat	Tapioca	Waxy maize
Food	+++	+	+	++	++
Paper	+++	++	+	+++	+++
Adhesives	+++	++	+	++	++
Textile	++	++	++	+++	++

VII. Future Aspects of Potato Starch Processing

Although potato starch manufacturing is a mature technology, new developments will occur. The market is increasingly demanding a more defined starch quality. Starch production will move from a bulk production industry to a more commodity-oriented industry. This will result in different starch products from the potato process. For example, the Brabender or RVA viscosity or particle size distribution will have narrow specifications.

Potato starch processing will be improved by the use of new or improved technology and techniques. Examples of these improvements might be the use of enzymes, chemicals or organic solvents during processing, and the use of improved separation and classification techniques, such as other hydrocyclone geometries and three-phase centrifugal separation.

Processes and products can be maximized by optimizing the potato by means of bio-technological improvements, such as the amylopectin potato, increasing the yield of starch per hectare of the potato crop by reducing vulnerability to disease (or decreasing the use of herbicides and pesticides), and increasing the amount and quality of protein in the potato.

A third major field of developments will be in co-products. These streams will be treated as product streams rather than as waste streams. Much attention will be paid to product specification and GMP-like production. Also the recovery of better (more suited to specific applications) co-products will be investigated. Examples of this might include the recovery of asparagine, citric acid, low-potassium deproteinized concentrated potato juice, potassium nitrate, food-grade fiber and food-grade protein.

VIII. References

1. Van Houten EJ. *Anderhalve eeuw aardappelzetmeel industrie [One and a half centuries of potato starch industry]* [AVEBE brochure]. The Netherlands, (Dutch): Veendam; 1994.
2. Willard M. *Am Potato J.* 1993;70:405.
3. Brown CR. *Am Potato J.* 1993;70:363.
4. *Starches & Derivatives.* Global Supply/Demand Patterns 2003–2010 GIRACT, 24 Pré-Colomb, CH-1290 Versoix/Geneva Switzerland. www.giract.com. 2004.
5. Zijlstra K. *Bijdragen tot de kennis van het aardappelzetmeel (Contributions to the knowledge of potato starch).* The Netherlands: Rijksuitgeverij, The Hague; 1941 [Dutch].
6. *Starches & Derivatives.* Global Supply/Demand Patterns 2003–2010. GIRACT, 24 Pré-Colomb, CH-1290 Versoix/Geneva Switzerland; www.giract.com. 2004.
7. van Es A, Hartmans KJ. *Structure and Chemical Composition of the Potato.* The Netherlands: Pudoc, Wageningen; 1981.
8. de Willigen HA. Een en ander over de anatomie en chemische samenstelling van den aardappel, (More information about the anatomy and chemical structure of the potato). In: *Symposium Samenstelling en Voedingswaarde van aardappelen en aardappelproducten (Composition and Nutritional value of potatoes and potato products).* Utrecht, The Netherlands: 1943 [Dutch].

9. Kram A. University of Groningen, The Netherlands, personal communication.
10. Brunt K. NIKO TNO, Groningen, The Netherlands, personal communication.
11. NIKO TNO, Groningen, The Netherlands, personal communication.
12. Swinkels JJM. *Starch/Stärke*. 1985;37:1.
13. Potze J. Commercial Separation of Amylose and Amylopectin from Starch. In: Radley JA, ed. *Starch Production Technology*. London, UK: Applied Science Publishers; 1976.
14. de Vries JA. *Voedingsmiddelen Technol.* 1995;23:26.
15. Matheis G. *Chem Mikrobio. Techno. Lebensm.* 1987;11:5.
16. Iyengar R, McEvily AJ. *Trends Food Sci Technol.* 1992;3:60.
17. Bradley D, Svarosky L. *The Hydrocyclone. Hydrocyclones*. Oxford, UK. London, UK: Pergamon Press: Technomic Publishing Co.; 1965; 1984.
18. Bednarski S. *Stärke*. 1964;16:6.
19. Zeevalkink HA. AVEBE, Eur. Patent 0517965 B1. 1996.
20. Verberne P. Koninklijke Scholten-Honing NV. Dutch patent NL 157350. 1978.
21. Verberne P. *Stärke*. 1977;29:303.
22. Grommers HE, Krikken J, Bos BH, van der Krogt DA. *Minerals Engineering*. 2004;17:5.
23. Schwalbach WW. *Filtration Separation*. 1988:264.
24. van der Krogt DA. Eiwitwinning en zuivering op industriele schaal. (Recovery and purifying protein on an industrial scale). Lecture during the IOP course Industrial Proteins. Utrecht, The Netherlands. 1994 [in Dutch].
25. de Noord KG. *PT Procestechnologie*. 1975:11.
26. Wurzburg OB, ed. *Modified Starches: Properties and Uses*. Boca Raton, FL: CRC Press; 1986.
27. Rutenberg MW, Solarek D. Starch derivatives: production and uses. In: Whistler RL, BeMiller JN, Paschall EF, eds. *Starch Chemistry and Technology*. 2nd edn. Orlando, FL: Academic Press; 1984:312.
28. Swinkels JJM. *Industrial Starch Chemistry*. Veendam, The Netherlands: AVEBE.
29. Glicksman M. *Gum Technology in the Food Industry*. New York, NY: Academic Press; 1969.
30. Caldwell CG, Jarowenko W, Hodgkin ID. US Patent 3 459 632. 1969.
31. Whistler RL, Paschall EF, eds. *Starch: Chemistry and Technology*. Vol. II, *Industrial Aspects*. New York, NY: Academic Press; 1967.
32. Kerr RW, ed. *Chemistry and Industry of Starch*. New York, NY: Academic Press; 1950.
33. Treadway RH. *Stärke*. 1962;14:30.
34. Broell W. *Melliand Textilber Int.* 1971;3:270.
35. La Mer VK. *J Coll Sci.* 1956;11:710.
36. Vogh RP, Warrington JE, Black AP. *J Am Waterworks Assn.* 1969;61:267.
37. Johnston DP, Stone PJ, Magnon JL. US Patent 3 596 766. 1971.
38. Swinkels JJM. *Performance of potato starch products in various applications*. The Netherlands: AVEBE, Veendam.
39. Swinkels JJM. *Differences between commercial native starches*. The Netherlands: AVEBE, Veendam.

Tapioca/Cassava Starch: Production and Use

William F. Breuninger[1], Kuakoon Piyachomkwan[2] and Klanarong Sriroth[3]

[1]*National Starch and Chemical Company, Bridgewater, New Jersey, USA*
[2]*National Center for Genetic Engineering and Biotechnology, Pathumthani, Thailand*
[3]*Department of Biotechnology, Kasetsart University, Bangkok, Thailand*

I. Background	541
II. Processing	545
III. Tapioca Starch	550
IV. Modification	555
V. Food Applications	556
VI. Industrial Applications	563
VII. Outlook	564
VIII. References	564

I. Background

Tapioca starch is obtained from the roots of the cassava plant, which is found in equatorial regions between the Tropic of Cancer and the Tropic of Capricorn. Names for the cassava plant vary depending on the region: yucca (Central America), mandioca or manioca (Brazil), tapioca (India and Malaysia) and cassada or cassava (Africa and Southeast Asia). In North America and Europe, the name cassava is generally applied to the roots of the plant, whereas tapioca is the name given to starch and other processed products. The plant belongs to the spurge family (Euphoriaceae). Previously, cassava was described as two edible species of the genus *Manihot*, *Manihot ultissima* Phol and *Manihot palmata*, based on the presence of high and low cyanide contents in roots (or called 'bitter' and 'sweet' cassava), respectively. Recently, both bitter and sweet cassava classes were classified as being the same species of *Manihot esculenta*.

Cassava is a shrubby perennial crop which is well-recognized for the ease of plantation and low input requirement. The plant can grow in all soil types, but root formation is better in loose-structured soils, such as light sandy loams and/or loamy sands. It can grow even in infertile soil or acid soil (pH < 4.4), but not in alkaline soil (pH > 8). Cassava has excellent drought tolerance, so it can be planted in areas having less than 1000 mm/year of rain. Planting usually starts at the beginning of a rainy period. The plants are propagated readily by removing a piece of stem of ~20 centimeters in length and having at least four nodules (called a stake) and placing it horizontally, vertically or inclined into the soil. Vertical planting is preferred as it gives higher root yields and better plant survival rates, and is easy for plant cultivation and root harvest. To propagate effectively, good-quality stakes obtained from mature plants (9–12 months old) are planted vertically approximately 10 cm in depth below the soil surface and with all round spacing of 100 cm (10 000 plants/hectare). Less spacing (100 × 80 cm or 80 × 80 cm) is recommended for infertile sandy soil, whereas greater spacing (100 × 120 cm or 120 × 120 cm) is used in fertile soil. By vertical planting (or 'standing'), the mature plants have at least two large branches (dichotomous branching) or three large branches (trichotomous branching). At maturity, which is reached in 8–18 months, plants may be 1–5 meters high, with feeder roots extending 1 meter into the soil. Storage roots of commercial interest are just below the soil surface, extending radially from the plant. The roots are typically harvested on average at 10–12 months after planting. Too early or too late harvesting may result in lower root yields and root starch content. Optimal productivity of the plant for harvesting can be indicated by its harvest index, i.e. the ratio of root weight to total plant weight, which ranges from 0.5 to 0.7. Root harvesting can be accomplished manually by cutting the stem at a height of 40–60 centimeters above the ground. Roots are then pulled from the ground using an iron or wooden tool with a fulcrum point in between the branches of the plant. Plant tops are cut into pieces for replanting, and roots are delivered to the processing factory.

Yields of cassava roots vary with variety, plant growth conditions such as soil, climate, rainfall, and agronomic practices including fertilizer application, weed control, irrigation and multiple cropping.[1] Although cassava cultivation requires only low input, the crop provides high yields if it is well-managed by planting with good-quality stakes at the optimal time, weed control during the first few months of planting, application of chemical fertilizers or manures, and root harvesting at the appropriate time and climate. Cassava is often grown on lands that are inadequate for higher value crops, and crops are seldom rotated. The farming practices in some major producing areas are such that plants are often short of nutrients. Fertilizer demand is, therefore, high and necessary for highest yield. Root yields are usually in the range 5–20 metric tons/acre (12–48 T/hectare).

Typical mature roots have different shapes (conical, conical-cylindrical, cylindrical, fusiform) and different sizes (3 to 15 centimeters in diameter), depending on variety, age and growth conditions. The color of the outer peel varies from white to dark brown. The cross-section of cassava roots shows the two major components which are the peel and the central pith (Figure 12.1). The peel is composed of the outer layer (called the periderm) and the inner layer (called the cortical region or cortex), which contains sclerenchyma, cortical parenchyma and phloem tissue. The large central pith of the roots is the starch-reserve flesh, comprised of cambium and parenchyma tissue and xylem vessels.

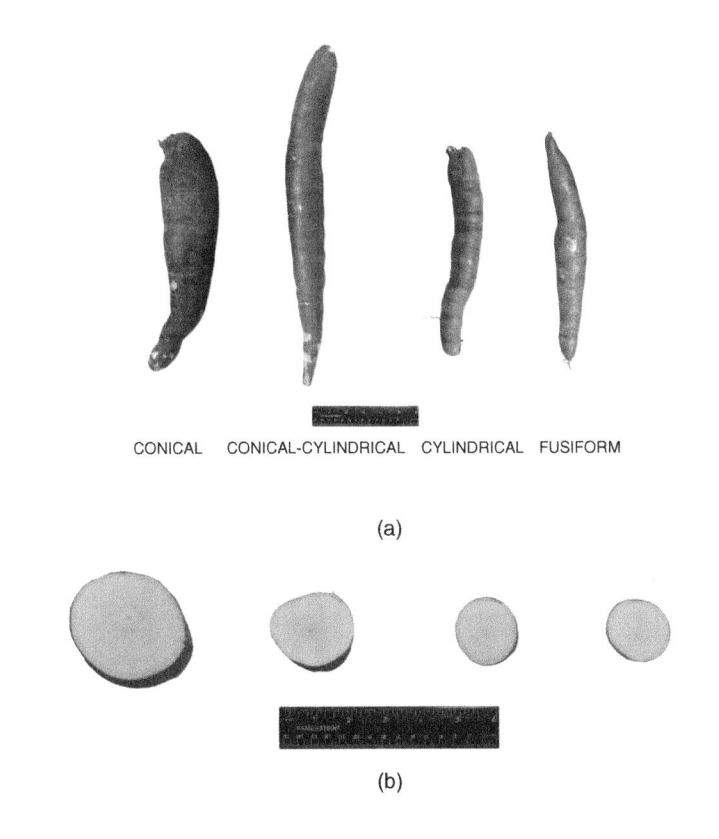

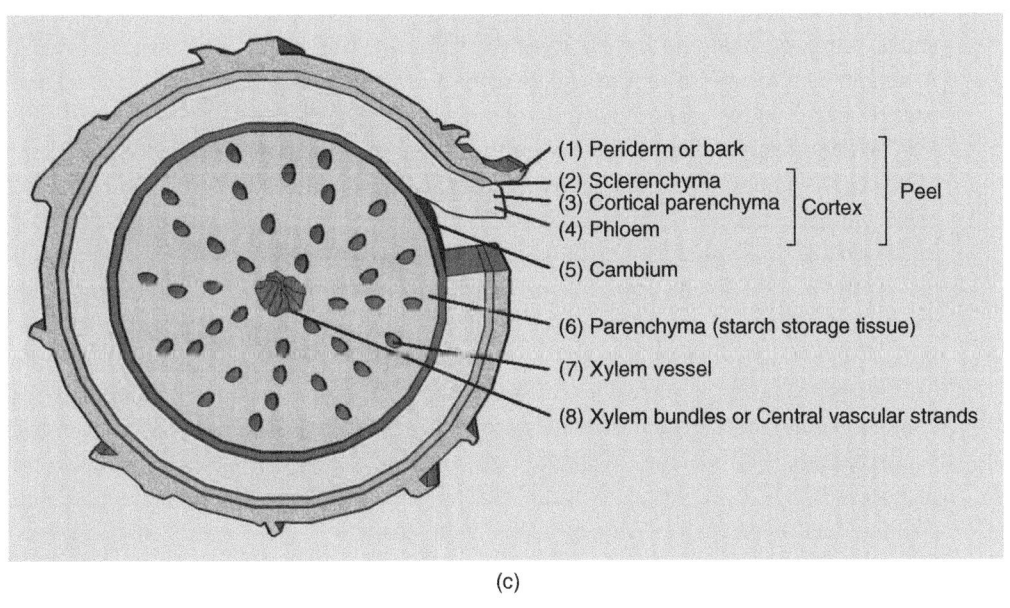

Figure 12.1 (a) Cassava roots with conical, conical-cylindrical, cylindrical and fusiform shapes; (b) cross-section of cassava roots; and (c) drawing of root cross-section containing different components including (1) periderm or bark; (2) schlerenchyma; (3) cortical parenchyma; (4) phloem (1 to 4 = peel); (5) cambium; (6) parenchyma (starch reserves); (7) xylem vessels; and (8) xylem bundles and fibers.

Table 12.1 Proximate analysis of cassava roots					
	Source[a]				
	1	2	3	4	5
Moisture (%, wet basis)	63.28	59.40	66.00	70.25	61.92
Carbohydrate (%, wet basis)	29.73	38.10	26.00	26.58	35.7
Protein (N × 6.25,%, wet basis)	1.18	0.70	1.00	1.12	0.70
Fat (%, wet basis)	0.08	0.20	0.30	0.41	0.10
Ash (%, wet basis)	0.85	1.00	n.a[b]	0.54	0.92
Crude fiber (%, wet basis)	0.99	0.60	1.00	1.11	0.68
Potassium (mg/kg)	0.26	n.a.	n.a	n.a	n.a
Phosphorus (mg/kg)	0.04	400	n.a	n.a	n.a
Iron (mg/kg)	n.a	n.a	n.a	n.a	n.a
Vitamin C (mg/kg)	n.a	252	n.a	n.a	n.a

[a] 1 = reference 2; 2 = reference 3; 3 = reference 4; 4 = reference 5; 5 = experimental data of authors Sriroth and Piyachomkwan
[b] n.a., not available

Typical mature roots have an average composition of 60–70% water, 30–35% carbohydrate, 1–2% fat, 1–2% fiber and 1–2% protein, with trace quantities of vitamins and minerals (Table 12.1). Mature roots can range in starch content from as low as 15% to as high as 33%, depending on the climate and harvest time. Starch content reaches a maximum at the end of the rainy season. Less mature roots will be lower in starch content and higher in water, while overly mature roots will be lower in recoverable starch content and have a woody texture, making starch processing difficult. Although cassava root is a poor source of protein, vitamins and minerals, it is a good energy source and is used as a staple food in many regions.

When cassava roots are to be used in the human diet or as animal feed, the toxicity of cassava is important. Cyanogenic glucosides are present in all parts of the cassava plant, including the edible roots. The two major types of cyanogenic glucosides, namely linamarin [2-(β-D-glucopyranosyloxy)isobutyronitrile] and lotaustralin [methyl-linamarin or 2-(β-D-glucopyranosyloxy) methylbutyronitrile] are synthesized from valine and isoleucine, respectively, and are found in cassava in an approximate ratio of 20:1.[6] The toxicity of cassava-based products is due to the release of hydrogen cyanide from these cyanogenic glucosides.[7] When plant tissues are damaged, intentionally or unintentionally, cyanogenic glucosides are hydrolyzed by linamarase, an endogenous, compartmentalized enzyme, to liberate glucose and acid-stable acetone cyanohydrin compounds.[8] The cyanohydrin is further decomposed spontaneously under neutral and alkaline conditions (pH > 5) or by enzymes to liberate volatile hydrogen cyanide or free cyanide.[5,6] The content of hydrogen cyanide in cassava roots varies, depending on variety, harvest time, environmental conditions of growth and agronomic practices.[11–13] The cyanide content is highly distributed in the peel portion (i.e. the periderm and cortex regions) rather than in the starchy parenchyma tissue (Table 12.2). Based on the cyanide contents in edible roots, cassava is classified into three classes: low toxic (or sweet) type; medium toxic type; and high toxic (or bitter) type, with cyanide contents

Table 12.2 Contents of total cyanogenic compounds in parenchyma tissue and peel of fresh cassava roots (harvested at 11 months after planting at the early and late rainy periods)

Varieties	Total cyanide content (mg HCN/kg dry weight)[a]			
	Planting at early rainy period		Planting at late rainy period	
	Parenchyma	Peel	Parenchyma	Peel
Rayong 5	194	2160	338	3211
Kasetsart 50	588	2447	671	3069
Huay Bong 60	291	2100	311	3080
KMUC 34-114-106	547	2360	477	4022
KMUC 34-114-235	479	3862	231	2787
KM99-1	204	2184	141	2215
KM98-5	590	2213	875	2284
KM140	245	2649	300	2643
Rayong 7	229	2015	373	3257

[a] As measured by an enzymic method according to reference 14

of <50, 50–100 and >100 mg HCN equivalent/kg fresh weight, respectively.[15] Sweet cassava is preferred for direct consumption. The bitter type is mostly processed to industrial products such as chips and starch. During processing, a portion of the cyanogenic compounds is removed, but some residues remain in finished products, the content depends on processing conditions. Proper processes such as cooking, boiling, drying, frying and fermentation are employed to effectively detoxify cassava and produce products with a safe level of hydrogen cyanide.[14,16,17]

II. Processing

Roots are sold to processing factories, usually within 24 hours of harvest. Delays in reaching the processing facility can result in lower starch yield and increased microflora content. The latter can be observed as brown and black discolorations in a freshly broken root a few days after harvest. As a rule, roots delivered to starch factories are in areas where transportation distances (and costs) are low; this ensures raw material freshness. When arriving at the factory, roots are sampled and tested for starch content by specific gravity using a beam balance called a Rieman balance (Figure 12.2) which has been adopted from the potato starch industry.[19] The analytical procedure is simple: 5 kilograms of clean roots are weighed into a sample basket, then dumped into a second basket suspended in water and the specific gravity is determined to estimate the dry matter and starch content of the fresh roots. Rebalancing of the scale gives a direct reading as percentage of starch. Calibration of this method is required for cassava grown under different growth conditions.[20] Shipman[21] has provided an example of specific gravity data on which the direct reading balance is based. A good correlation of starch content in fresh cassava roots as determined by a simple beam balance and a specific enzymic method was found (Figure 12.2).[22]

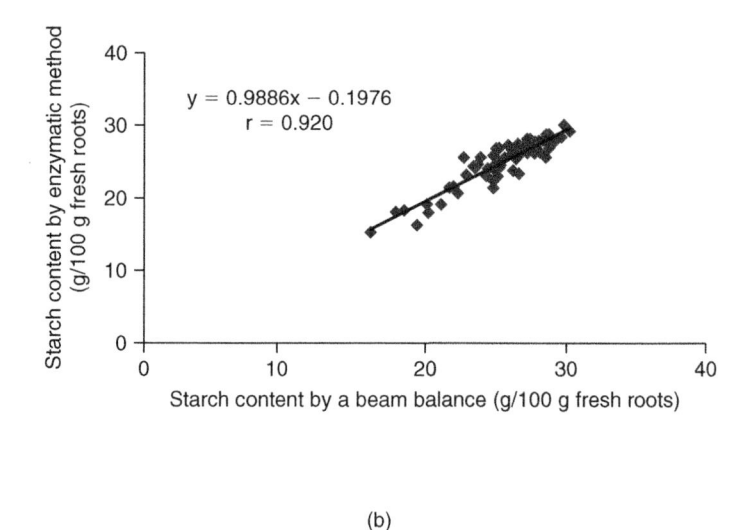

Figure 12.2 (a) A beam balance or Rieman balance for determining cassava root starch content by specific gravity analysis; and (b) a correlation between starch contents of fresh cassava roots as determined by a beam balance and by the AACC enzymic method.[18]

The machinery of tapioca processing is highly varied. In both Thailand and Brazil there are well-equipped factories that utilize local, custom-built devices for processing roots, product streams, by-products and effluent. In addition, some manufacturers have successfully utilized equipment that is more common to potato and corn processing. The basic process of screening and density separation remains common to all machine variations employed in the production of high-quality tapioca starch. Individual manufacturers have arrived at machines and configurations based on their own experiences, with raw material supply and finished product requirements influencing the selection process.

The basic process for isolating a high-quality tapioca starch is represented in Figure 12.3. At the factory, roots are transferred to a root hopper (Figure 12.4a) and are cleaned of dirt and sticks by a rotating slotted drum (Figure 12.4b). Cleaned roots are fed to a washer where recycled water is used (Figure 12.4b). In the early stage of the washer, stones sink in the water, while roots are lifted by the action of the washer paddles. The latter stages remove the peel from the roots. Roots are chopped into 1–2 centimeter chunks by a cutting blade and fed to a saw-tooth rasper for intense attrition into a pulpy slurry (Figure 12.4c). Liquid recycled from the process is fed along with chopped root into the rasper. Use of a decanter for fruit water separation, as applied in potato starch processing, is optional as the protein and other impurities of cassava are very low. Fresh rasped root slurry from the rasper is then pumped through a series of coarse and fine extractors, either vertical or horizontal (Figure 12.4d), where fiber is removed by screens arranged conically in continuous centrifugal perforated baskets. Recently, an extractor equipped with curved screens was introduced to improve filtering efficiency. Starch slurry exiting the coarse extractor equipped with a filter cloth and a screen with an aperture of 150 microns (100 mesh) to 125 microns (120 mesh) still contains a large amount of fine fiber which must be removed in a fine extractor equipped with a finer screen (140–200 mesh). Starch slurry during transport

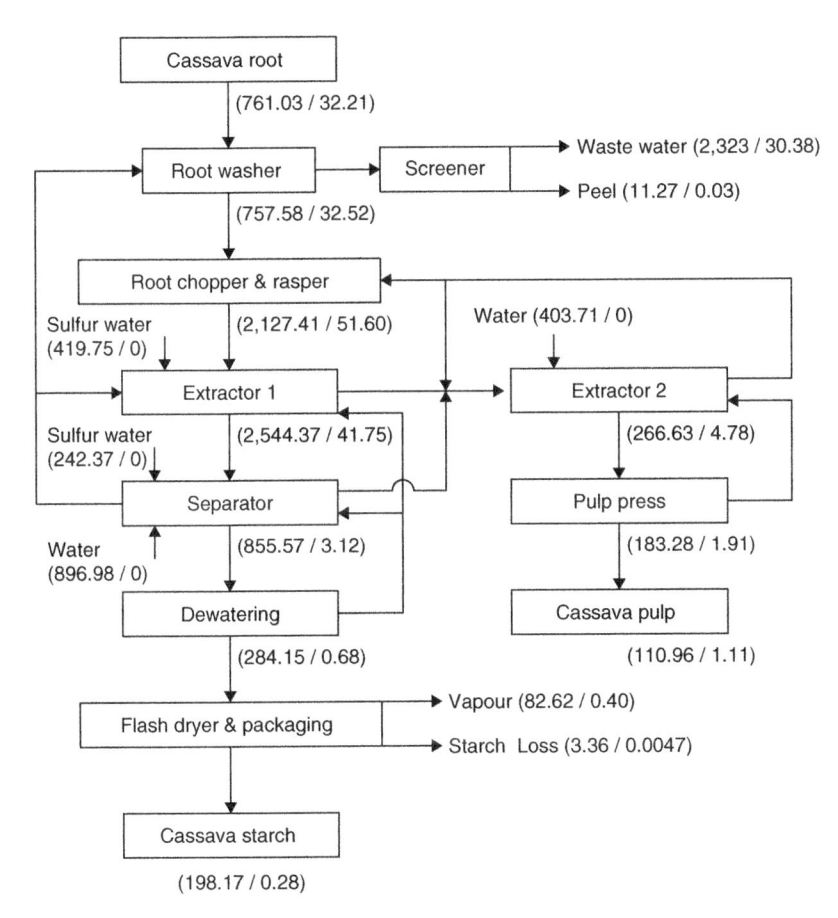

Figure 12.3 Basic process for isolating a high-quality tapioca/cassava starch. The numbers in parenthesis represent the mass balance (tons/day) and cyanide balance (kg/day) of a factory with the production of capacity ~200 tons of tapioca/cassava starch per day. (Modified from Sriroth et al.[23] and Piyachomkwan et al.[24])

to each extraction stage is passed through a sand-removing hydrocyclone to ensure complete removal of sand. Extra filtration, such as a rotary brush strainer, is installed to ensure against passage of a starch clump. Pulp from the coarse extractor is repeatedly re-extracted to achieve minimal loss of starch trapped in the moist pulp (60–70% moisture content and 45–55%, dry basis, starch content).[25] Starch slurry received from the fine extraction has a concentration of 20–35% (10–17° Baume). The starch slurry is further concentrated to 40% (20° Baume) using a two-phase or three-phase separator (Figure 12.4e) or a series of hydrocyclones (Figure 12.4f). Starch may be recovered from the final starch stream and dried, or the stream may be used to supply a chemical modification process. To produce native starch, the concentrated starch slurry is dewatered in a horizontal centrifuge (Figure 12.4g) to produce starch cake with a moisture content of 35–40%. Alternative processes of dewatering, e.g. high-pressure filtration[26] or use of a filter press, are used in some starch factories to reduce starch loss in the liquid stream. High-moisture starch cake is then subjected to a pneumatic conveying dryer, known as flash dryer, to lower the moisture content to <13% prior to packing (Figure 12.4h). Starch dust lost through the two cyclones can be recovered by being trapped with venturi scrubbers or filter bags.

548 Tapioca/Cassava Starch: Production and Use

(a) Root hopper

(b) Sand removal drum and root washer

(c) Rasper

(d) Horizontal extractor

(e) Separator

(f) Hydrocyclone

(g) Dewatering centrifugal

(h) Packaging

Figure 12.4 Machinery of tapioca starch processing.

In the cassava starch process, the consumption of water and energy is critical, since it affects both production cost and starch quality. During the extraction process, the starch is purified by multiple stages of fiber screens and starch washers to remove solubles and finely divided fiber. Water utilization is optimized by reuse prior to discharge. Most commonly, water flow is countercurrent to the starch flow. All process water streams contain a small amount of sulfur dioxide for control of microbes and as an aid in processing. Modern factories also practice pretreatment of water, which consists of flocculation, chlorination and filtration, ahead of process consumption. The final water discharge from the process is treated on-site to meet environmental requirements prior to release to public channels. By-products are consumed in useful applications, so there is no waste from the process. Fiber is used in animal feed formulations. Peels and waste-treatment sludge are good compost for the original cassava growing land or other agricultural purposes.

Starch recovery, as presently practiced, depends on a continuous supply of fresh root. Hence, factories are located close to the root producing areas. In Brazil, the processing season can be as short as six months. In Thailand, continuous processing is possible, although some starch producers are known not to operate during part of the rainy season when the starch content of cassava root is low and operating economics are less favorable.

Some creative approaches to try to match the regular demand of a tapioca production process with the irregular supply of feedstock have been taken. Meuser et al.[27] demonstrated that starch could be recovered from either cassava chips or pellets, although the starch is obtained at some sacrifice of quality. Nauta[28] has proposed large silos for the storage of tapioca starch, similar to those used in the potato processing industry. At present, the primary supply of tapioca starch remains fresh roots.

The quality of tapioca starches produced can be affected by fresh root quality, as well as the production practices of each factory. Nevertheless, commercial tapioca starches always comply well with industrial specifications which differ to some extent, depending on the manufacturers and end users. General specifications of cassava starch are summarized in Table 12.3.

Table 12.3 General specification of native cassava starch

Attribute	Specification
Moisture content (% maximum)	13
Starch content (% minimum)	85
Ash (% maximum)	0.2
pH	5.0 to 7.0
Whiteness (Kett scale, minimum)	90
Viscosity Barbender unit, BU, minimum at 6% dry weight concentration	600
Sulfur dioxide content (ppm, maximum)	30
Cyanide content	nil
Appearance	White, no speck, fresh odor

III. Tapioca Starch

Swinkels[29] collected published characterization data for tapioca starch and compared it to that for other starches of commercial significance (Table 12.4). Tapioca starch is differentiated from other starches by its low level of residual materials (fat, protein, ash), lower amylose content than for other amylose-containing starches, and high molecular weights of amylose and amylopectin. The small amount of phosphorus in tapioca starch is partially removable[30] and, therefore, not bound as the phosphate ester as in potato starch. It is also common to find protein and lipid values of zero, as reported by Hicks.[31] The very low protein and lipid content is an important factor which differentiates tapioca starch from the cereal starches.

Typically, cassava starch contains 17–20% amylose. Unlike corn (0–70% amylose content) and rice (0–40% amylose content), no significant variation of amylose content has been found in cassava starch. Similar to other starches, the amylose molecules of cassava starch are not completely unbranched as indicated by their beta-amylolysis limit, a value lower than that for corn, potato, rice and wheat starches. In addition, cassava amylose has a higher molecular weight than other starches (Table 12.5). The amylopectin structure of cassava starch is compared to that of other commercial starches in Table 12.6. Figure 12.5 demonstrates the chain length distribution in amylopectin fractionated from cassava starch. Each fraction can be classified in terms of its chain length (reported as degree of polymerization, DP) and position in amylopectin molecules. Cassava amylopectin consists mostly of short chains in Fraction III, which are A and B_1 chains. A small portion of extra long chains (Fraction I, less than 1% by weight) have also been observed in cassava amylopectin.[34] The variation in chain length distributions of amylopectin fractionated from different cassava cultivars seems to be less affected by genetics (Figure 12.5), which is in agreement with a previous report by Charoenkul et al.[34]

Microscopic examination of tapioca starch granules reveals smooth, spherical granules 4–35 μm in diameter, commonly with one or more spherical truncations.

Table 12.4 Tapioca starch in comparison to some other commercial starches[29]

Source	Granule diameter (μm)	Average diameter (μm)	Amylose content (%)	Amylose[a] average DP	$Tg^{a,b}$ range (°C)	Phosphorus (%)	Ash (%)	Protein (%)	Lipid (%)
Tapioca	4–35	15	17	3000	65–70	0.01	0.2	0.10	0.1
Potato	5–100	27	21	3000	60–65	0.08	0.4	0.06	0.05
Normal corn	2–30	10	28	800	75–80	0.02	0.1	0.35	0.7
Waxy corn	2–30	10	0	n.a.[c]	65–70	0.01	0.1	0.25	0.15
Wheat	1–45	bimodal	28	800	80–85	0.06	0.2	0.40	0.8

[a]Now known to be approximate values
[b]Gelatinization temperature
[c]not applicable

Table 12.5 Properties of amylose from different botanical sources[32]

Property	Chestnut	Kuzu	Corn	Potato	Rice	Sweet potato	Tapioca	Water chestnut	Wheat
Iodine binding capacity (g/100g)	19.9	20.0	20.0	20.5	20.0–21.1	20.2	20.0	20.2	19.9
$[\eta]$ at 22.5°C in M KOH (mL/g)	242	228	169	384	180–216	324	384	n.a.[a]	n.a.
Beta-amylolysis limit (%)	86	76	84	80	73–84	73	75	95	82
Degree of polymerization by weight, DP_w (range)	440–14900	480–12300	390–13100	840–21800	210–12900	840–19100	580–22400	160–8090	n.a.
Degree of polymerization by weight, DP_w (mean)	4020	3220	2550	6360	2750–3320	5430	6680	4210	n.a
Degree of polymerization by number, DP_n (mean)	1690	1540	960	4920	920–1110	4100	2660	800	1290
DP_w/DP_n	2.38	2.09	2.66	1.29	2.64–3.39	1.31	2.51	5.76	n.a.
Chain length, CL	375	320	305	670	250–370	380	340	420	270
Chain number	4.6	4.8	3.1	7.3	2.5–4.3	11	7.8	1.9	4.8
Unbranched amylose (mol%)	66	47	52		69	30	58	89	73
Phosphorus (µg/g)	10	10	n.a.	3	None	6	7	n.a.	n.a.

[a] n.a., not available

Table 12.6 Chain length distribution of components from amylopectin debranching[33]

Starch types		Chain					
		A1	A2	B1	B2	B3	B4
Wheat	Number-average DP[a]	10	15	24	49	84	110
	Weight-average DP	11	15	25	51	84	111
	Weight%	18.5	21.4	35.6	20.9	2.3	0.3
	Molar%	35.4	27.4	28.3	8.2	0.6	0.1
Waxy rice	Number-average DP[a]	10	15	24	44	69	86
	Weight-average DP	11	16	24	46	69	87
	Weight%	20.2	26.2	27.2	21.9	3.9	0.5
	Molar%	36.9	32.0	20.7	9.1	1.1	0.1
Potato	Number-average DP[a]	10	14	24	44	69	86
	Weight-average DP	10	14	24	46	69	87
	Weight%	6.2	18.5	33.7	29.5	11.6	0.5
	Molar%	15.0	31.8	35.2	14.5	3.4	0.1
Cassava	Number-average DP[a]	10	13	21	48	74	90
	Weight-average DP	10	13	22	49	74	91
	Weight%	9.4	17.0	45.6	24.7	2.7	0.6
	Molar%	18.9	26.3	43.6	10.4	0.7	0.1
Sweet potato	Number-average DP[a]	10	14	23	48	77	99
	Weight-average DP	10	14	24	50	78	100
	Weight%	9.4	17.0	45.6	24.7	2.7	0.6
	Molar%	15.9	37.2	34.3	11.2	1.3	0.1

[a] DP, degree of polymerization

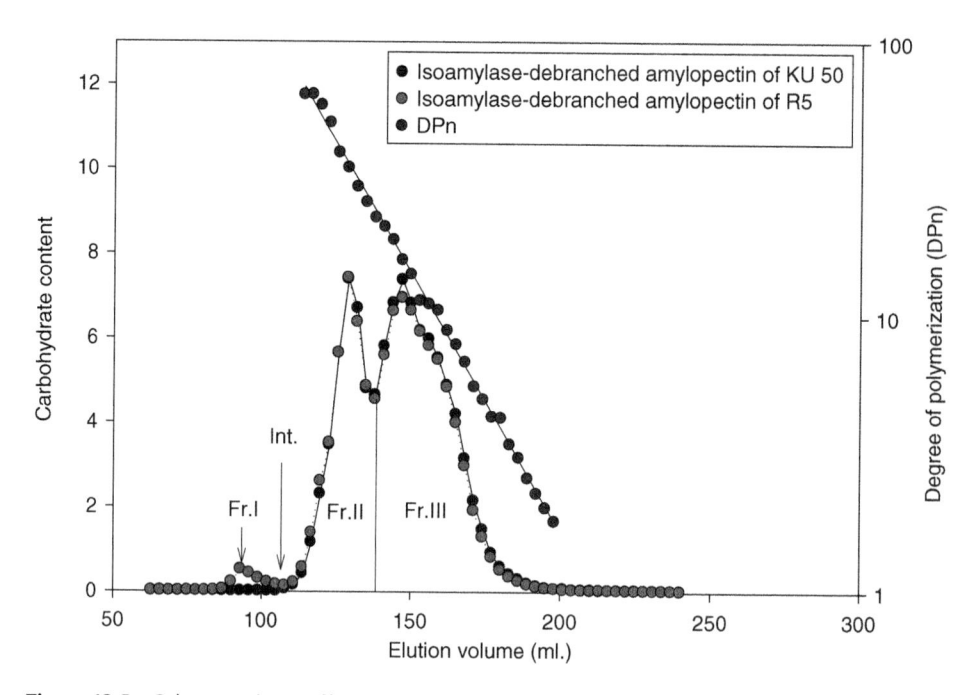

Figure 12.5 Gel permeation profiles of *Pseudomonas* isoamylase debranched amylopectin fractionated from cassava starches of two varieties (Kasetsart 50 or KU50 and Ranyong 5 or R5). (Isoamylase-debranched amylopectin was fractionated by gel permeation chromatography on Toyopearl HW55S and Toyopearl HW50S columns (300 × 20 mm). Each fraction was divided according to the wavelength of maximum absorption (λmax) of the glucan–iodine complex: fraction I (Fr. I), λmax $\geq$ 620 nm; intermediate fraction (Int. Fr.), 620 nm > λmax $\geq$ 600 nm; fraction II (Fr. II), 600 nm > λmax $\geq$ 525 nm and fraction III (Fr. III), λmax < 525 nm and the straight line represents the degree of polymerization (DPn from 2 to 85) of glucans.

The hilum is off-center, and there are usually noticeable fissures that cross the hilum. Using a light scattering technique, Finkelstein and Sarko[35] found that the layers in tapioca starch granules could be as thin as 0.2 μm and that a single granule could have at least 40 layers. By way of comparison, potato starch granules were found to have a much coarser organization of only a few structural layers that are several micrometers thick.

When heated in excess water, starch undergoes an irreversible structural transition. As a result, starch granules heated in excess water lose birefringence and crystallinity. The granules then swell and absorb more water, resulting in changes in the rheological properties of the starch–water mixture. These phenomena are well-known as starch gelatinization and pasting. A comparison of the processes for cassava and other starches is given in Table 12.7. There is more resistance to gelatinization in tapioca starch than would be expected based on its amylose content (compared to that of potato starch). This small inconsistency is likely due to a tighter organization of the tapioca granule.

Allen et al.[37] studied the gelatinization of tapioca starch by taking samples from a Brabender ViscoAmylograph. Water penetrates to the middle of the granule from the truncated end. At 50°C, well below the starch gelatinization temperature, there is

Table 12.7 Gelatinization, pasting and paste properties of starches from different botanical origins[36]

Type[a]	Gelatinization[b]			Retrogradation[c] (%)	Pasting and paste properties[d]			
	To (°C)	Range (°C)	Enthalpy (ΔH, J/g)		Pasting temperature (°C)	Viscosity (RVU)		
						Peak	Hot paste	Final
A-type starch								
Normal maize	64.1 ± 0.2	10.8	12.3 + 0.0	47.6	82.0	152	95	169
Waxy maize	64.2 ± 0.2	10.4	15.4 + 0.0	47.0	69.5	205	84	100
du Waxy maize	66.1 ± 0.5	14.4	15.6 + 0.2	71.2	75.7	109	77	99
Normal rice	70.3 ± 0.2	9.9	13.2 + 0.6	40.5	79.9	113	96	160
Waxy rice	56.9 ± 0.3	13.4	15.4 + 0.2	5.0	64.1	205	84	100
Sweet rice	58.6 ± 0.2	12.8	13.4 + 0.6	4.3	64.6	219	100	128
Wheat	57.1 ± 0.3	9.1	10.7 + 0.2	33.7	88.6	104	75	154
Cattail millet	67.1 ± 0.0	8.5	14.4 + 0.3	53.8	74.2	201	80	208
Mung bean	60.0 ± 0.4	11.5	11.4 + 0.5	58.9	73.8	186	161	363
Chinese taro	67.3 ± 0.1	12.5	15.0 + 0.5	32.0	73.1	171	88	161
Cassava	64.3 ± 0.1	10.1	14.7 + 0.7	25.3	67.6	173	61	107
B-type starch								
ae Waxy maize	71.5 ± 0.2	25.7	22.0 + 0.3	61.6	83.2	162	150	190
Potato	58.2 ± 0.1	9.5	15.8 + 1.2	43.4	63.5	702	165	231
C-type starch								
Lotus root	60.6 ± 0.0	10.5	13.5 + 0.1	43.2	67.4	307	84	138
Green banana	68.6 ± 0.2	7.5	17.2 + 0.1	47.7	74.0	250	194	272
Water chestnut	58.7 ± 0.5	24.1	13.6 + 0.5	47.9	74.3	61	16	27

[a]A, B and C types are classified based on the polymorphism of starch by x-ray diffraction patterns

[b]Starch gelatinization parameters were obtained by a differential scanning calorimeter. Range of gelatinization is the difference between conclusion temperature and onset temperatures

[c]% retrogradation is the percentage of the enthalpy of the first run (i.e. native samples) and the second run (i.e. samples stored at 4°C for 7 days)

[d]As determined by a Rapid ViscoAnalyzer, using 8% (w/w, dsb) starch in water (28 g of total weight)

disruption in the center of the granule. As the temperature increases further, the granule increases in diameter until, at the point of peak viscosity, outer layers of the granule are disrupted. Continued heating beyond the occurrence of peak viscosity gives continued imbibing of water, with some of the starch dispersing into solution and some swollen granular fragments remaining. Cooling of a freshly-prepared tapioca starch paste produces an increase in viscosity, some loss of clarity (but the pastes are still clearer than those of cereal starches, except for the all-amylopectin, i.e. waxy, starches), and a cohesive body. A tapioca starch paste may slowly form a weak gel. These changes are due to reassociation of amylose molecules in the aqueous system. While this property of gelation (also called setback or retrogradation) is common to all native starches containing amylose (see Chapter 8), the property is less noticeable in tapioca starch because of a lower amylose content and higher molecular weight amylose than is found in other starches. At 17% amylose, there is sufficient functionality that, by modification, it is possible to enhance gelation and/or suppress cohesiveness.

Similar to other starches, the physicochemical properties of starches are affected by many factors. While gelatinization, pasting and gelation of starches are a function of their botanical sources, starch preparations from the same basic origin may vary in their properties. The effect of cultivars, growing seasons and root ages on cassava starch properties have been repeatedly reported.[38–42] Figure 12.6 is an example of variation in gelatinization process of cassava starches with different cultivation practices. In addition, cassava starch properties can be affected by processing conditions. As previously described, sulfur dioxide is often used in the cassava starch isolation process at the centrifugal or extraction stage to increase extraction efficiency and to improve product whiteness. However, sulfur dioxide residues in finished products alter starch properties by lowering paste viscosity and increasing the gelatinization temperature.[23] The difference in granule stability of cassava starch indicates existing variation in granule architecture and can influence cassava starch uses, in particular, the modification process.

The low amylose, low lipid and low protein contents, combined with the high molecular weight of its amylose, make tapioca a unique native starch for direct use

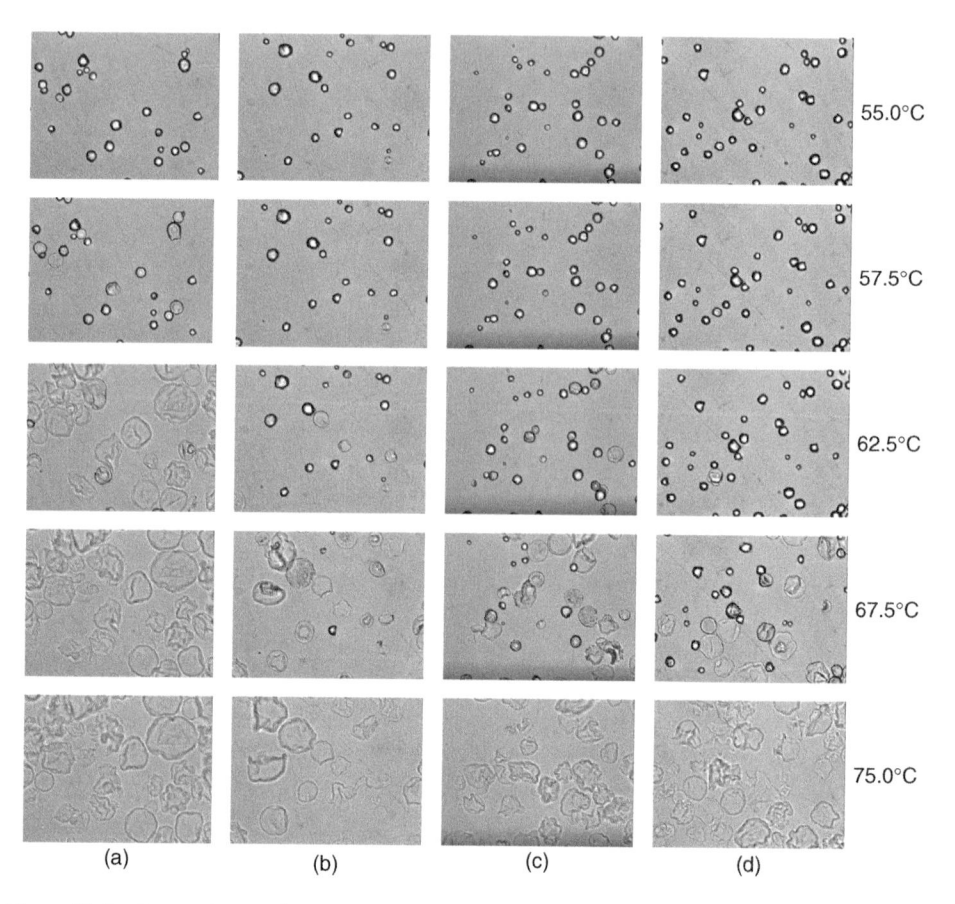

Figure 12.6 Changes in granular structure of cassava starch extracted from roots: (a) planted during the rainy period and harvested at 6 months; (b) planted during the rainy period and harvested at 12 months; (c) planted during the dry period and harvested at 6 months; (d) planted during the dry period and harvested at 12 months, when observed by hot-stage microscopy at different temperatures.

in food and industrial applications and an excellent starting material for modification into specialty products. This unique position of tapioca starch was recognized as early as 1944, when Kiesselbach[43] reported on the progress of waxy corn breeding for replacement of tapioca starch. While both waxy maize and tapioca starches have competed in some applications, tapioca remains differentiated by its gel-forming ability and bland taste. Tapioca starch utilization in specialty applications has continued to grow. In 1974, with a shortage of gum arabic, tapioca starch received new attention as a replacement for it.[44] Tapioca starch consumption continues to increase, as new applications where its properties are of unique value are found.

IV. Modification

Starch modifications can be classified as physical modifications, chemical modifications and genetic modifications.[45] Physical modification of cassava starch involves application of shear force, blending and thermal treatment. A combination of thermal treatment and shear force has been widely used to produce many extruded products and snacks. Well-known physically modified cassava starch products are alpha starch or pregelatinized starch and heat–moisture treated starch.

Tapioca starches, both native and chemically modified, are easily converted into an 'instant' (pregelatinized) form, also known as cold-water-soluble starch. This physical modification is brought about by pasting of the starch slurry (30–40% dry solids with or without some additives) and subsequent drying. The conventional means of preparation is the continuous feed of starch slurry onto a hot drum at 160–170°C.[46] The starch is cooked and dried, then scraped from the drum in a single process and milled to desired particle size. Pregelatinized starches have also been prepared by spray drying. The gelatinization can be performed by a cooking system ahead of the spray dryer or by steam injection to the slurry ahead of atomization.

Heat–moisture treated starch, sometimes called *Tao* starch in Thailand, is prepared from cassava starch by heating moistened starch (≈50% moisture content) at various temperatures and times. Heat–moisture treatment provides a modified starch that produces a less cohesive, shorter-textured paste with improved shear resistance and gel properties, as compared to the long, stringy, cohesive paste of native tapioca starch.

For chemical modification, tapioca starch is easily modified to all current commercial derivatives. There are no special precautions or equipment required beyond what already might be practiced for a particular derivative or reagent applied to other starches. Recovery of modified products is facilitated in conventional washing and drying equipment. The reader is referred to Wurzburg[47] and Chapter 17 for details of starch modifications, all of which may be practiced with tapioca starch. In the preparation and evaluation of some derivatives of tapioca starch, some of its unique characteristics have been revealed.

Tapioca starch granules have multiple layers and a tightly organized structure, as previously mentioned. Penetration of reagents might be more difficult than with other commercial starches. Patel et al.[48] studied graft copolymers of polyacrylonitrile and granular starches (Chapter 19). Their work showed that grafting took place on the surface of tapioca granules, while similar work with potato starch showed grafting within the granule. A study of the gel microstructure of a hydroxypropylated, crosslinked tapioca starch by Hood et al.[49] suggested that crosslinking occurs only in the outer regions of the granule. Later work by Mercier et al.[50] established that crosslinking with epichlorohydrin is also in the outer portion of the granule and located on the amylopectin molecules.

Franco et al.[51] and Piyachomkwan et al.[52] treated both corn and tapioca starches with α-amylase and/or amyloglucosidase. Corn starch was found to be more susceptible to hydrolysis than tapioca starch, which is consistent with the results of Hood and Mercier,[53] who studied molecular structures of both unmodified and modified tapioca starch and found that hydroxypropylation of tapioca starch took place only in the amorphous areas of the granule, and that parts of the amylopectin molecules were likely underivatized because of a more dense (crystalline packing) structure. However, the details of reagent penetration and derivatization sites within the granule are not fully understood, leaving considerable opportunities for future research. Evidence presented here is for the purpose of differentiating tapioca from the other commercial starches. From what is known, it can be concluded that tapioca starch granules are highly layered and have a tight organization. Both amylose and amylopectin may have higher molecular weights than found in other common starches. Tapioca starch is modifiable by common chemical and biochemical reactions, but the products are more heterogeneous than those of other modified starches.

Genetic modification of starches (see Chapters 3 and 4) has received attention, as it is a cost-effective tool for modifying starch structure and functionalities. Recently, genetic engineering of the cassava plant has been done by antisense inhibition of granule-bound starch synthase (GBSS), the enzyme responsible for the synthesis of amylose.[54] This cassava genotype produces amylose-free starch (all-amylopectin or waxy starch) with improved paste clarity and stability, which is more desirable in many applications, so an environmentally unfriendly chemical process is no longer required for modification after starch extraction. This can be another significant driving force to accelerate the growth of the cassava starch industry.

As a native starch, lipids and protein residuals are significantly lower than they are in many other commercial starches. These properties of tapioca starch have been utilized in many industries and further enhanced by means of physical and/or chemical modifications which give close control of its properties to fit the needs of customers in process and product applications. However, tapioca starch is regarded as a specialty starch outside of its local production area.

V. Food Applications

The greatest diversity of uses of tapioca starch is in the food industry (see Chapter 20). As an ingredient in foods, native and modified tapioca starch has been widely utilized. Tapioca pearls (previously called sago pearls since they were made from sago [*Metroxylon* spp.] starch) are also familiar to many. The pearls formed in spherical shape are a mixture of gelatinized and ungelatinized starch produced by heat–moisture treatment. To produce tapioca pearls the starch is wetted to equilibrium of 50%

moisture content. The moist starch is disintegrated and formed into spherical particles by continuous mechanical shaking. The particles are then subjected to dry heat processing or roasting at 250–300°C. The pearls are cooled before being subjected to another drying process at a lower temperature (50–80°C) to reduce the moisture content of finished products (Figure 12.7). When cooked, the pearls have unique

(a) Preparation of moistened starch

(b) Moistened starch

(c) Sphere/pearl forming machine

(d) Starting of shaking

(e) Formation of sphere-like pearls after continuously shaking

(f) Sizing by sieving

(g) Roasting

(h) Tapioca pearls after drying

Figure 12.7 The process of making tapioca pearls.

product characteristics, i.e. they produce very transparent and chewy pastes, making them popular in many dishes.

Other food applications have generally made use of tapioca starch as a thickener and stabilizer, with special emphasis on its lack of flavor contribution to food systems, allowing full and immediate detection of the flavor of the food itself.[55,56] Chemical modifications, in keeping with the broadly accepted limitations of the United States Food and Drug Administration,[57] have resulted in highly specialized tapioca starch products that give added value and functionality to foods. Thailand, as one of the world's leading manufacturers and exporters of cassava-based modified starches, also has regulations for chemically modified starches approved for food applications (Table 12.8).

Wood[59] reports that, of the seven most common food allergies, two are associated with significant sources of food starch, namely, wheat and corn. Some food processors, sensitive to consumer perceptions of labeling, have sought to reformulate using tapioca starch or to specify tapioca starch in the development of new products that address the concern of allergic reactions.

Tapioca starch has long been the starch of choice in baby food for its physical properties of texture and stability, as well as its low flavor contribution. The starch used may be modified in order to survive the rigors of processing and enhance the physical properties of the final product. Starches containing amylose were found by Wurzburg and Kruger[60] to take on some of the physical characteristics of modified starches when heated under certain conditions with salts and moisture. Smalligan et al.[61] used steam injection as the heat and moisture source to prepare a tapioca starch for baby food having the enhanced properties of modified tapioca starch, but without modification. It is the opinion of the authors that, in the drive toward physical modifications matching the properties of chemical modifications, tapioca starch will have an advantage in taste, molecular weight and non-starch residuals over cereal starches.

Tapioca starch is used broadly in the manufacture of Asian-style noodles in combination with other ingredients. Kasemsuwan et al.[62] used tapioca starch as an extender in mung bean noodles. The use of tapioca starch was based on paste clarity; the textural properties imparted by the starch were adjusted by crosslinking. The resulting noodle was acceptable, with improved economics. Ishigaki et al.[63] included tapioca starch in noodle dough of durum semolina in order to obtain an instant cooking noodle, instant cooking also being aided by grooving from the extruder die to aid in water uptake. Starches from other tuber and grain sources, substituted in the invention, did not give noodles of acceptable texture and mouthfeel compared to those from tapioca starch. In this application, tapioca starch moderates water uptake during noodle rehydration. Starches of higher amylose content have the disadvantage of more difficult rehydration, due to retrogradation in the noodle.

Dextrinization of tapioca starch is well-known. The difficulty in dextrinizing corn starch,[64] and the relative ease of tapioca starch conversion,[65] have been described. The near-absence of lipids, which interfere with the dextrinization process, has given tapioca dextrins an advantage in stability and color because of their ease of manufacture and control. However, the economics of base starch supply have resulted in a significant shift to waxy corn starch as a base for industrial dextrins, even though

Table 12.8 Specification of chemically modified starch for food applications according to Thai regulations

Type of modified starch	INS-number[a]	E-number[b]	Chemical reagents[c]	Standard/specification
Pregelatinized starch			—	—
Dextrin[d]	1400		Hydrochloric acid <0.15% or orthophosphoric acid <0.17%	pH 2.0 to 9.0
Thin-boiling starch[d]	1401		Hydrochloric acid or orthophosphoric acid <7.0% or sulfuric acid <2.0%	pH 3.0 to 7.0
Alkali-treated starch[d]	1402		Sodium hydroxide or potassium hydroxide <1.0%	pH 5.0 to 7.5
Bleached starch[d]	1403		Hydrogen peroxide and/or peracetic acid with active oxygen <0.45	Carboxyl group increase <0.1%
			Ammonium persulfate <0.075% and sulfur dioxide <0.05%	Sulfur dioxide <50 mg/ka
			Sodium chlorite <0.5%	No chlorite
			Potassium permanganate <0.2%	Manganese <50 mg/kg
Oxidized starch	1404	E1404	Chlorine (as sodium hypochlorite) <5.5%	Carboxyl groups <1.1%
Distarch phosphate	1412	E1412	Sodium trimetaphosphate Phosphorus oxychloride <0.1%	Phosphate (reported as phosphorus) <0.04%
Starch octenylsuccinate	1450	E1450	Octenylsuccinic anhydride <3%	Octenylsuccinate groups <0.03%
Hydroxypropylated starch	1440	E1440	Propylene oxide <10% (or 25%[e])	Propylene chlorohydrin <1 mg/kg Hydroxypropyl groups <7.0%
Acetylated starch	1420	E1420	Acetic anhydride Vinyl acetate <7.5% Adipic anhydride and acetic anhydride <0.12%	Acetyl groups <2.5%
Monostarch phosphate	1410	E1410	Orthophosphoric acid or sodium or potassium orthophosphate or sodium tripolyphosphate	Phosphate (reported as phosphorus) <0.4%
Hydroxypropyl distarch phosphate	1442	E1442	Sodium trimetaphosphate or phosphorus oxychloride and propylene oxide <10%	Propylene chlorohydrin <1 mg/kg (or <5 mg/kg[e]) Hydroxypropyl groups <4.0%
Acetylated distarch adipate	1422	E1422	Acetic anhydride and adipic anhydride < 0.12%	Acetyl groups <2.5% and adipate groups <0.135%
Acetylated distarch phosphate	1414	E1414	Phosphorus oxychloride followed by acetic anhydride < 10% (or 8%[e] or vinyl acetate <7.5%[e])	Acetyl groups <2.5% and phosphate (reported as phosphorus) <0.04%

[a]CCFAC International Numbering System ('INS', 1989 ftp://ftp.fao.org/codex/standard/volume1a/en/CXxot04e.pdf
[b]Modified starches, Annex 1 of EEC Directive No. 95/2/EC (an ingredient category)
[c]According to Thai Industrial Standard Institute (TISI) unless specified by others[58]
[d]Dextrin, bleached starch, starches modified by acid, alkali and enzyme or by physical treatment are not considered as food additives in the context of the EEC Directive 95/2/EC
[e]According to US Food and Drug Administration (FDA) 21 CFR Chapter 1[57]

production of tapioca food dextrins has experienced growth. Continued growth of tapioca starch dextrins is related to new uses, the properties of which cannot be matched by cereal starch dextrins. Tapioca starch dextrins have a long-established use as a high solids coating former in pan coating of confections. With current consumer interests in low-fat and no-fat foods, there has been a renewed interest in some old products. Tapioca starch dextrins have enabled food producers to develop low- or no-fat products having the taste, mouthfeel and texture expected by the consumer in the original food product. Current dextrin use for low- and no-fat products includes meats, sauces, soups, confections, baked goods and dairy products.[66,67]

Tapioca starch and chemically modified tapioca starches are easily converted to 'instant' (pregelatinized) forms (also known as cold-water-soluble starch). This physical modification is brought about by pasting of the starch and subsequent drying as described in Section 12.4. Control of particle size is critical to texture and rehydration rate when the product is redispersed in water.[68] Fine particle size results in a smooth texture on redispersion, e.g. in pudding preparation.[69] As a comparison, coarse texture could be more desirable in fruit- or vegetable-based foods.

Pudding products have been prepared in the form of cook-up, instant and frozen types, using tapioca starch in a wide variety of chemical and physical modifications. Tapioca starch is favored in such applications because of lack of flavor contribution and fast meltaway, allowing the flavors of the pudding to be released.[55] The water-binding capacity of gelatinized tapioca starch is good. Contributing to this property is its low amylose content and a good degree of dispersion on gelatinization. Hydroxypropylation increases its water-binding capacity and improves its ability to withstand freeze–thaw cycling. Hydroxypropylation also aids in redispersion if the starch is in a pregelatinized form. Control of dispersion on initial cooking has also been utilized.[69] Acetylated tapioca starch is also a functional pudding starch.[70] One of the more creative inventions utilizing the unique properties of tapioca starch is that of Glicksman et al.,[71] who used unmodified tapioca starch, along with fat and emulsifier, to produce a freeze–thaw stable pudding. When other unmodified commercial starches were used in the same formulation, syneresis (water separation) occurred. This invention made use of the formation of complexes between amylose molecules and fats, moderating both gelatinization and retrogradation.

Formation of complexes between tapioca amylose and lipids was studied in detail by Mercier et al.[50] who found that, as the chain length of the fatty acid or monoglyceride increased, complexing observed by x-ray crystallography, iodine absorbance and solubility also increased. Complexes were only possible when the lipid employed was capable of fitting inside the amylose helix. Diglycerides and triglycerides did not form complexes. However, such complexing is not unique to tapioca amylose. It does, however, give the user of tapioca starch a more complete understanding of interactions in a processed food system where the previously mentioned unique properties of tapioca are utilized.

Tapioca starch has a demonstrated ability to contribute to puffing or popping when heated in a microwave oven. Shachat and Raphael[72] prepared a puffable snack product in which a flour-free dough, containing tapioca starch and other ingredients, gave puffed, crisp pieces on microwaving. The use of tapioca starch is described as 'critical to the invention.' Birch and Goddard[73] prepared an expandable snack product with tapioca starch by sealing a tapioca starch containing core dough with an outer dough not containing tapioca starch, to produce a coated product that pops on microwaving. Exactly how the popping takes place is not explained in either invention. Both inventions require some, if not all, of the tapioca starch granules to be ungelatinized. A later study of tapioca extrusion ahead of cooking expansion was done by Seibel and Hu,[74] who found that die temperature and residence time were the primary influencing factors. A barrel temperature of 70–80°C was critical for a puffable cracker preparation which, under the conditions studied, did not fully gelatinize the tapioca starch. The puffing effect was thought to come from microcells of air in the extrudate. This leaves the probability that the presence of granular tapioca starch could simply be an indication of correct conditions to produce a puffable snack.

Sour cassava starch (known as *Polvilho azedo* in Brazil and *Almidon agrio* in Colombia), a naturally 30-day fermented and sun-dried starch, is used extensively to prepare snacks and biscuits. The unique characteristics of products made from this modified cassava starch are high specific volumes, baking expansion and crispness, i.e. it produces products with characteristics similar to those of extruded snacks.[75] The improved rheological properties are believed to be an effect of degradative oxidation by organic acid and ultraviolet (UV) irradiation, and can be observed only in the baking of acidified-irradiated cassava starch, but not corn starch.[76,77] Further investigation demonstrates that certain UV wavelengths, i.e. the short wavelengths of UVB and UVC, are essential for improving baking expansion of acidified cassava starch dough.[78]

Combined use of natural gums and starch, including tapioca starch, has been widely practiced in the food industry. Exact reasons for the synergistic effects that have been observed and reported are not fully understood at this time. While tapioca starch has been widely used in combination with gums in food systems, it does not likely contribute unique properties to the system beyond the properties already presented. Fujii[79] reported on the inhibition of hydroxypropylated tapioca starch pasting by low-methoxyl pectin. He also demonstrated that the combination of gelatin and tapioca starch can have variable results, depending on concentration. Kleemann[80] reported variability of viscosity based on the molecular weight of gelatin used, with high molecular weight gelatin being more effective and higher concentrations of gelatin increasing the observed viscosity change. Kuhn et al.[81] demonstrated an interaction between starch and xanthan under conditions of extrusion. Bahnassey and Breene[82] suggest that the interaction of starch and gum could be as simple as a competition for water in the use system. Furthermore, the electrostatic interaction between positively or negatively charged chemical groups introduced to tapioca starch and gum can occur and significantly change its gelatinization temperature and rheological behavior.[83,84] The properties of the starch can also be markedly altered by the addition of other ingredients. The effects of sucrose and sodium chloride, widely-used as ingredients in a variety of processed foods, on starch properties have been widely reported.[85–90] Like other starches, the addition of sucrose and sodium chloride shifts gelatinization of tapioca starch to higher temperatures, but does not affect the gelatinization enthalpy. The effect of salt on gelatinizing tapioca starch is, however, concentration-dependent. Salt tends to retard retrogradation of tapioca starch gels, whereas sucrose either has no effect or increases retrogradation of aged gels, depending on the aging temperature.[91] As evidenced by light microscopy, the addition of sucrose and sodium chloride have similar effects on cooking behaviors of various chemically modified tapioca starches used in food products (Figure 12.8). Increasing the gelatinization temperatures of starches in aqueous systems by sucrose is considered to be due to an antiplasticizing effect[92] and competition for available water,[93] whereas the ionic nature of salts presumably causes an alteration in the weakly-anionic hydroxyl groups of starch molecules,[94,95] as well as altering water structure.[96] Recently, confocal laser scanning microscopy has been used to reveal the presence of granule-associated protein in various starches,[97,98] including tapioca starch. The protein-containing envelope of tapioca starch can be modified by ionic salts (Ca^{2+}), resulting in modification of its granular structure, as well as its hydration properties.[99] Some additives such as glycerol, the most commonly used plasticizer, when blended with tapioca starch, increase the molecular mobility of glucan chains, thereby providing improved starch functionalities for promising production of biopolymer-based edible film.[100–102]

The effects of food ingredients on thermal transitions and physicochemical properties are of importance to the processing and properties of starch-based products, and can play major roles in the quality and storage stability of tapioca starch-containing products. For the food scientist, it remains vital to understand the role of ingredients and their properties, with the understanding that results can vary with the conditions of use.

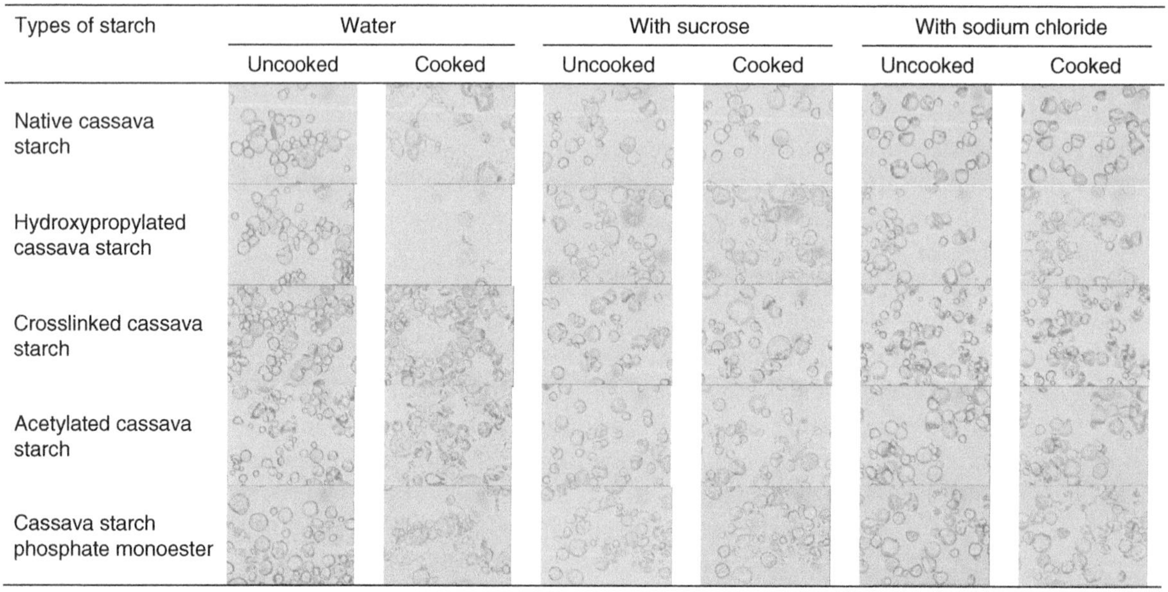

Types of starch	Water		With sucrose		With sodium chloride	
	Uncooked	Cooked	Uncooked	Cooked	Uncooked	Cooked
Native cassava starch						
Hydroxypropylated cassava starch						
Crosslinked cassava starch						
Acetylated cassava starch						
Cassava starch phosphate monoester						

Figure 12.8 Hot-stage microscope pictures of uncooked (at 30°C) and cooked (at 70°C) native and various chemically modified tapioca starches in water containing sucrose (20% by weight) and sodium chloride (5% by weight), indicating different ease of cooking as a result of food ingredients.

VI. Industrial Applications

Tapioca starch consumption in industrial applications has been more related to economics than to any unique functionality. While some performance characteristics of tapioca starch are advantageous in several applications, it is generally used in non-food applications close to the supply points, such as in Brazil, India and Southeast Asia. Non-food consumption of starch in North America and Europe is primarily from readily available sources in these areas, i.e. corn, wheat and potatoes.

Paper manufacturing is a significant consumer of starch, in both modified and unmodified forms (see Chapter 18). The role of starch is to bind fibers, retain additives and increase strength. Further, if the granules are not fully dispersed, strength gains are considerable, while the drainage rate on the forming wire is not impeded.[103] Chen[104] studied the role of amylose, amylopectin and tapioca starch in papermaking and the retention of titanium dioxide pigment. Protective colloid effects were present with all three additives, but as the molecular weight of amylose increased, titanium dioxide retention increased, demonstrating that protective colloid effects were overcome. Even though unmodified starch can perform well in papermaking, the process is enhanced by functional modification of the starch. Addition of positively charged chemical groups, usually tertiary or quaternary amines, establishes ionic binding with negatively charged cellulose fibers and pigment in the paper. Tapioca starch is suitable for such modifications. Additionally, modified tapioca starches and derivatives have been used in sizing and coating preparations. In corrugating, unmodified tapioca starch has demonstrated technical advantage over corn starch as the adhesive (ungelatinized) portion of the corrugating formulation. Leake[105] showed improvement in tack and green bond strength that could be translated to increased machine speed by virtue of the higher water-holding capacity when tapioca starch was used, compared to conventional formulations. It has also been possible to control uniformity and speed of gelatinization if the adhesive portion of the corrugating formulation is esterified tapioca starch.[106] Tapioca starch has also been used as the carrier portion of corrugating formulations by partial swelling with alkali, terminated by the addition of acid.[107] The carrier portion of the formulation may also be tapioca starch, as demonstrated by Sakakibara and Iwase.[108] Low-amylose starches, when cooked, provide viscous pastes with a cohesive texture and a low tendency to retrograde, which is preferable for use as adhesives. In addition, cassava starch is odorless and tasteless and its dried film can reabsorb water rapidly. It is, therefore, an excellent, remoistenable adhesive for use on postage stamps, envelope flaps, labels and gummed tape.

The textile industry is another significant user of starch for sizing, finishing and printing purposes. Starch, typically as a modified form, is primarily used as a warp size to increase yarn strength and abrasion resistance during the weaving process. Properties required include the formation of strong films to provide a protective coating (partly on the surface and partly inside the yarn), good affinity to the yarn, good film flexibility, stable viscosity during application and ease of size removal. Modified tapioca starch possesses desirable functionalities for this application.

A large volume of tapioca starch is also used to make sweeteners and related products. This application primarily involves enzyme-catalyzed hydrolysis. Similar to chemically modified starches, diverse products are produced by varying enzyme types, degree of starch hydrolysis and derivatization. Important products obtained by direct hydrolysis of starch are sugar syrups (glucose syrups and maltose syrups made using α-amylase plus glucoamylase or β-amylase, respectively, with dextrose equivalence (DE) values of >20) (see Chapters 7 and 21), maltodextrins (DE <20) and cyclodextrins (see Chapter 22). The advantages of using tapioca starch over the cereal starches for syrup production are the bland taste, clean flavor, high purity and ease of cooking due to a lower gelatinization temperature. Furthermore, many derivatives can be made from starch hydrolysis products by chemical or bioprocesses. The reduction of sugars and starch hydrolyzates, readily accomplished by hydrogenation, changes the carbonyl group to an additional hydroxyl group. These products called sugar alcohols (such as sorbitol, mannitol, maltitol and hydrogenated starch hydrolyzates) are reduced-calorie sweeteners that are useful in many food and clinical applications. Fructose is produced from glucose by isomerization with glucose isomerase (see Chapter 21). Fermentation of starch hydrolyzates with different microorganisms provides a wide range of chemicals, such as monosodium glutamate (MSG), lysine, organic acids and alcohol. Advances in biotechnology and environmental concerns are the major driving forces that accelerate starch demand for the production of eco-friendly chemicals. The demand for cassava starch for this purpose, especially for the production of bioethanol, has also increased.

VII. Outlook

There are bright prospects for continued growth of the cassava starch industry. The low residuals in the starch and favorable flavor properties are being widely used in the food industry with opportunities for growth as the consuming public shifts toward lower-fat foods, reduced dependence on chemical modifications and avoidance of cereal starches, without sacrifice of eating quality. Breeding holds promise of both increased and decreased amylose contents. Properties of these starches can be further modified physically and/or chemically. Root production will improve as yield and maturity factors are made a part of hybrid selection, and with improvements in crop and soil management techniques.

VIII. References

1. Howeler RH. Agronomica practices for sustainable cassava production in Asia. In: Centro Internacional de Agricultura Tropical (CIAT). Cassava research and development in Asia : Exploring new opportunities for an ancient crop. *Proceedings of the 7th Regional Workshop, held in Bangkok, Thailand, Oct 28–Nov 1, 2002*; 2007.
2. Rojanaridpiched C. *Cassava: Cultivation, industrial processing and uses*. Bangkok, Thailand: Kasetsart University; 1989.
3. Balagopalan C, Padmaja G, Nanda SK, Moorthy SN. *Cassava in Food, Feed, and Industry*. Boca Raton, FL: CRC Press; 1988.

4. Beynum G, Van MA, Roels JA. *Starch Conversion Technology*. New York, NY: Marcel Dekker; 1985.
5. Grace MR. *Cassava Processing*. Rome, Italy: Food and Agriculture Organization of the United Nations; 1977.
6. Essers SAJA, Bosveld M, van der Grift RM, Voragen AGJ. *J. Sci. Food Agr.* 1993;63:287.
7. Bokanga M. *Food Technol.* 1995;86.
8. Padmaja G. *Crit. Rev. Food Sci. Nutr.* 1995;35:299.
9. Jones DM, Trim DS, Bainbridge ZA, French L. *J. Sci. Food Agr.* 1994;66:535.
10. Mlingi NLV, Bainbridge ZA, Poulter NH, Rosling H. *Food Chem.* 1995;53:29.
11. Santisopasri V, Kurotjanawong K, Chotineeranat S, Piyachomkwan K, Sriroth K, Oates CG. *Industrial Crops Products*. 2001;13(2):115.
12. Sriroth K, Santisopasri V, Petchalanuwat C, Kurotjanawong K, Piyachomkwan K, Oates CG. *Carbohydr. Polym.* 1999;38:161.
13. Howerler RH. In: *Potassium in Agriculture, International Symposium*, Atlanta, GA, July 7–10, 1985; Madison, WI: ASA, CSSA, SSA; 1985.
14. O'Brien GM, Taylor AJ, Pouler NH. *J. Sci Food Agr.* 1991;56:277.
15. Jansz ER, Uluwaduge DI. *J. Natl. Sci. Council Sri Lanka*. 1997;25(1):1.
16. Cereda MP. In: Bokanga M, Essers AJA, Poulter N, Rosling H, Tewe O, eds. *Acta Horticulturae: International Workshop on Cassava Safety*. Working Group on Cassava Safety (WOCAS). Published by WOCAS, Printed by Drukkerij PJ, Jauzen BV, The Netherlands; 1994:225–226.
17. Oke OL. In: Bokanga M, Essers AJA, Poulter N, Rosling H, Tewe O, eds. *Acta Horticulturae: International Workshop on Cassava Safety*. Working Group on Cassava Safety (WOCAS). Published by WOCAS, printed by Drukkerij PJ, Jauzen BV, The Netherlands; 1994:163–174.
18. American Association of Cereal Chemists (AACC). *Approved Methods of the American Association of Cereal Chemists*. 8th edn. Vol. 2. St. Paul, MN: The Approved Methods Committee; 1990.
19. Brautlecht CA. *Starch: Its Sources, Productions and Uses*. New York, NY: Reinhold; 1953.
20. Bainbridge Z, Tomlins K, Wellings K, Westby A. *Methods for Assessing Quality Characteristics of Non-Grain Starch Staples. (Part 2. Field Methods)*. Chatman, UK: Natural Resources Institute; 1996.
21. Shipman L. In: Whistler RL, Paschall EF, eds. *Starch: Chemistry and Technology*. Vol. II. New York, NY: Academic Press; 1967:103–119.
22. Neamfug L. Comparison of Analytical Methods of Starch Content and Determination of Non-starch Saccharides Content in Cassava and Sweet Potato [M.S. Thesis]. Bangkok, Thailand: Kasetsart University; 2001.
23. Sriroth K, Wanlapatit S, Piyachomkwan K, Oates CG. *Starch/Stärke*. 1998;50:466.
24. Piyachomkwan K, Wanlapatit S, Chotineeranat S, Sriroth K. *Starch/Stärke*. 2005;57:71.
25. Sriroth K, Chollakup R, Chotineeranat S, Piyachomkwan K, Oates CG. *Bioresource Technol.* 2000;71:63.
26. Sriroth K, Wanlapatit S, Chollakup R, Chotineeranat S, Piyachomkwan K, Oates CG. *Starch/Stärke*. 1999;51:383.

27. Meuser F, Smolnik HD, Rajani Ch, Giesemann HG. *Starch/Stärke*. 1978;30:299.
28. Nauta DJ. *Muehle Mischfuttertechnik*. 1981;118:37.
29. Swinkels JJM. *Starch/Stärke*. 1985;37:1.
30. Singh V, Ali SZ. *Starch/Stärke*. 1987;39:277.
31. Hicks CP. *Process Biochem*. 1970;5:38.
32. Tester RF, Karkalas J, Qi X. *J. Cereal Sci*. 2004;39:151.
33. Ong MH, Jumel K, Tokarczuk PF, Blanshard JMV, Harding SE. *Carbohydr. Res*. 1994;260:99.
34. Charoenkul N, Uttapap D, Pathipanawat W, Takeda Y. *Carbohydr. Polym*. 2006;65:102.
35. Finkelstein RS, Sarko A. *Biopolym*. 1972;11:881.
36. Jane J, Chen YY, Lee LF, McPherson AE, Wong KS, Radosavljevic M, Kasemsuwan T. *Cereal Chem*. 1999;76:629.
37. Allen JE, Hood LF, Chalbot JF. *Cereal Chem*. 1977;54:783.
38. Asaoka M, Blanshard JMV, Rickard JE. *J Sci. Food Agr*. 1992;59:53.
39. Asaoka M, Blanshard JMV, Rickard JE. *Starch/Stärke*. 1991;43:455.
40. Defloor I, Swennen R, Bokanga M, Delcour JA. *J. Sci. Food Agri*. 1998;76:233.
41. Moorthy SN, Ramanujam T. *Starch/Stärke*. 1986;38:58.
42. Sriroth K, Santisopasri V, Petchalanuwat C, Kurotjanawong K, Piyachomkwan K, Oates CG. *Carbohydr. Polym*. 1999;38:161.
43. Kiesselbach TA. *J. Am. Soc. Agron*. 1944;36:668.
44. Anon. *Chem. Marketing Reptr*. 1974;Aug 19:5.
45. BeMiller JN. Starch modification. *Starch/Stärke*. 1997;49:127.
46. Powell EL. In: Whistler RL, Paschall EF, eds. *Starch: Chemistry and Technology*. Vol. II. New York, NY: Academic Press; 1967:523–536.
47. Wurzburg OB, ed. *Modified Starches: Properties and Uses*. Boca Raton, FL: CRC Press; 1986.
48. Patel AR, Patel MR, Patel NR, Suthar JN, Patel KG, Patel RD. *Starch/Stärke*. 1986;38:160.
49. Hood LF, Seifried AS, Meyer R. *J. Food Sci*. 1974;39:117.
50. Mercier C, Charbonniere R, Grebaut J, de la Gueriviere JF. *Cereal Chem*. 1980;57:4.
51. Franco CML, Preto SJD, Ciacco CF. *Starch/Stärke*. 1987;39:432.
52. Piyachomkwan K, Wansuksri R, Wanlapatit S, Chatakanonda P, Sriroth K. In: Tomasik P, Yuryev VP, Bertoft E, eds. *Starch: Progress in Basic and Applied Science*. Published by Polish Society of Food Technologists, Malopolska Branch. Printed in Poland; 2007:183–190.
53. Hood LF, Mercier C. *Carbohydr. Res*. 1978;61:53.
54. Raemakers K, Schreuder M, Suurs L, Furrer-Verhorst H, Vincken J, Vetten N, Jacobsen E, Visser RGF. *Molecular Breeding*. 2005;16:163.
55. Anon. *Dairy Ind. Intern*. 1991;56(11):19.
56. Anon. *Food Proc*. 1972;33(5):34.
57. US Code of Federal Regulations, Title 21, Chapter I, Part 172.892, Food Starch-Modified. Washington, DC: US Government Printing Office.
58. Thai Industrial Standard Institute: Standard for modified starch for food industry. UDC 547.458.61:591.133.1. Bangkok, Thailand: Ministry of Industry; 1992.
59. Wood R. *East West Natural Health*. 1992;22:50.

60. Wurzburg OB, Kruger LH. US Patent 3 977 897. 1976.
61. Smalligan WJ, Kelly VJ, Estela EG. US Patent 4 013 799. (1977).
62. Kasemsuwan T, Jane J-L, Bailey T. *Cereal Foods World*. 1995;40:670.
63. Ishigaki T, Saito H, Fujita A. US Patent 5 332 592. 1994.
64. Evans RB, Wurzburg OB. In: Whistler RL, Paschall EF, eds. *Starch: Chemistry and Technology*. Vol. II. New York, NY: Academic Press; 1967:253–278.
65. Kennedy HM, Fischer AC. In: Whistler RL, BeMiller JN, Paschall EF, eds. *Starch: Chemistry and Technology*. 2nd edn. New York, NY: Academic Press; 1984:593–610.
66. McAuley C, Mawson R. *Food Australia*. 1994;46:283.
67. Seyfried RJ. Intern. Zeit. Lebensmitt-Technol. Verfahrenstechnik, **41**, 103, 104, 106 1990.
68. Light JM. *Cereal Foods World*. 1990;35:1081.
69. O'Rourke JD, Katcher JH. US Patent 4 438 148. 1984.
70. Katcher JH, Ackilli JA. US Patent 4 238 604. 1981.
71. Glicksman M, Wankler BN, Silverman JE. US Patent 3 754 935. 1973.
72. Shachat MA, Raphael, SJ. US Patent 4 950 492. 1990.
73. Birch MR, Goddard MR. US Patent 5 080 914. 1992.
74. Seibel W, Hu N. *Starch/Stärke*. 1994;46:217.
75. Plata-Oviedo M, Camargo C. *J. Sci. Food Agr.* 1998;77:103.
76. Bertolini C, Mestres C, Colonna P. *Starch/Stärke*. 2000;52:340.
77. Demiate M, Dupuy N, Huvenne JP, Cereda MP, Wosiacki G. *Carbohydr. Polym.* 2000; 42:149.
78. Vatanasuchart N, Naivikul O, Charoenrein S, Sriroth K. *Carbohydr. Polym.* 2005; 61:80.
79. Fujii GS. *J. Jpn. Soc. Food Sci. Technol.* 1995;42:843.
80. Kleemann G. *Food Marketing Technol.* 1994;4–6:8.
81. Kuhn M, Elsner G, Gräber S. *Starch/Stärke*. 1989;41:467.
82. Bahnassey YA, Breene WM. *Starch/Stärke*. 1994;48:134.
83. Chaisawang M, Suphantharika M. *Carbohydr. Polym.* 2005;61:228.
84. Chaisawang M, Suphantharika M. *Food Hydrocoll.* 2006;20:641.
85. Chungcharoen A, Lund DB. *Cereal Chem.* 1987;64:240.
86. Wootton M, Bamunuarachchi A. *Starch/Stärke*. 1980;32:126.
87. Evans D, Haisman DR. *Starch/Stärke*. 1982;34:224.
88. Eliasson C. *Carbohydr. Polym.* 1992;18:131.
89. Baker LA, Rayas-Duarte R. *Cereal Chem.* 1998;75:308.
90. Chang SM, Lui LC. *J. Food Sci.* 1991;56:564–570.
91. Chatakanonda P, Wongprayoon S, Sriroth K. In: Tomasik P, Yuryev VP, Bertoft E, eds. *Starch: Progress in Basic and Applied Science*. Published by Polish Society of Food Technologists, Malopolska Branch. Printed in Poland; 2007:289–295.
92. Slade L, Levine H. *Pure Appl. Chem.* 1988;60:1841.
93. Spies RD, Hoseney RC. *Cereal Chem.* 1982;59:128.
94. Oosten BJ. *Starch/Stärke*. 1982;34:233.
95. Oosten BJ. *Starch/Stärke*. 1990;42:327.
96. Jane JL. *Starch/Stärke*. 1993;45:161.
97. Han XZ, Hamaker BR. *J. Cereal Sci.* 2002;35:109.

98. Han XZ, Benmoussa M, Gray JA, BeMiller JN, Hamaker BR. *Cereal Chem.* 2005;82:351.
99. Israkarn K, Hongsprabhas Pranithi, Hongsprabhas Parichat. *Carbohydr. Polym.* 2007;68:314.
100. Mali S, Sakanaka LS, Yamashita F, Grossmann MVE. *Carbohydr. Polym.* 2005;60:283.
101. Fama L, Flores SK, Gerschenson L, Goyanes S. *Carbohydr. Polym.* 2006;66:8.
102. Chang YP, Abd Karim A, Seow CC. *Food Hydrocoll.* 2006;20:1.
103. Mentzer MJ. In: Whistler RL, BeMiller JN, Paschall EF, eds. *Starch: Chemistry and Technology.* 2nd edn. New York, NY: Academic Press; 1984:543–574.
104. Chen D. The Effects of Tapioca Starch and its Components on the Retention of Titanium Dioxide in Paper, Ph.D. Thesis, North Carolina State University, Raleigh, NC, 1988.
105. Leake C.H. US Patent 4 787 937. 1988.
106. Nippon Shokuhin, Japan. Kokai Tokkyo Koho 8,256,030. 1983; *Chem Abstr* 1984;100:141120.
107. Sakakibara K, Iwase F, Nozaki Y. Japan. Kokai Tokkyo Koho 01,278,588. 1989; *Chem. Abstr.* 1989;112:141695.
108. Sakakibara K, Iwase F. Japan. Kokai Tokkyo Koho 63,251,486. 1988; *Chem. Abstr.* 1989;110:25654.

13

Rice Starches: Production and Properties

Cheryl R. Mitchell
Creative Research Management, Stockton, California, USA

I. Rice Production and Composition 569
   1. Rice Production. 569
   2. Rice Milling and Composition 570
II. Uses of Milled Rice and Rice By-products 571
   1. Milled Rice 571
   2. By-products. 572
III. Preparation of Rice Starch 573
   1. Traditional Method 573
   2. Mechanical Method 574
IV. Properties of Rice Starch 574
   1. General Properties Unique to Rice Starch 574
   2. Pasting Properties 575
V. Factors Affecting Rice Starch Properties. 575
   1. Rice Variety: Common Versus Waxy 575
   2. Protein Content. 576
   3. Method of Preparation 576
   4. Modification 577
VI. Rice Starch Applications 577
VII. References 578

I. Rice Production and Composition

1. Rice Production

Cultivated rice (*Oryza sativa* L.) is a major world food crop. Only wheat surpasses rice with regard to production and food use. About 90% of the world's rice is produced and consumed in Asia.[1] The production and consumption of rice in the United States, while not a staple food, increased more than 40% during the period from 1980 to 1995[2] and an additional 25% from 1995 to 2005.[3] The price of rice in

Table 13.1 Typical amylose content (%) by grain length of commercial US rice varieties[9]			
Long	Medium	Short	Waxy short
20–25	14–18	15–23	0–2

the United States showed an extraordinary increase during 2007, increasing by over 75% of its price during 2003. This increase was due to the increased demand for rice in the US and international markets, without an equivalent increase in production.

Worldwide usage of milled rice in 2005 was reported to be 406 million metric tons (mmt), with the United States reportedly producing 10.2 mmt and using 5.7 mmt.[4] Major rice producing areas include China, India, Indonesia, Bangladesh, Thailand, Burma, Japan, Korea, Vietnam and the Philippines. Other than Asia, major areas of rice production include Brazil, the United States, the Malagasy Republic, Egypt, Columbia, Nigeria and Italy.[5]

Major rice producing areas in the United States, in descending order of production, are Arkansas, California, Florida, Louisiana, Mississippi, Missouri and Texas.[6] The US focus is on three grain types: short; medium; and long. Long grain represents approximately 70% of the total US rice production. Approximately 93% of the California rice crop is medium grain, while in the south about 84% of the rice grown is long grain and 16% is medium grain.[7] In commercial crops grown in the US, the amylose content is to some extent associated with grain length, with long grain varieties having a higher amylose content than medium or short grain types (Table 13.1). Waxy rice currently cultivated in the US is a short grain type; long grain waxy rice varieties are available in the orient. Experimental rices having amylose contents of up to 40% (in 2% increments) have been reported, but are not commercially available.

Rice varieties are distinguished by their amylose content. There are two basic types: common and waxy. Waxy rice has an amylose content of 0–2%. Common or regular rice varieties are referred to as having amylose contents that are low (9–20%), medium (20–25%) or high (>25%) amylose. The availability of waxy rice is usually limited and large requirements for waxy rice are contracted in advance to ensure a supply.

A typical variety of waxy rice produced in the US is Calmochi-101.[9] Waxy rices are also referred to as 'glutinous' rices, due to their sticky cooking property, or as 'sweet' rices because of their use in Japanese 'mochi' cakes. Examples of low-amylose, medium-grain varieties are Calrose, Nato and Vista. Low-amylose, short-grain varieties are Caloro, Colusa and Nortai. Intermediate-amylose, long-grain varieties are Belle Patna, Bluebelle, Labelle and Starbonnet.[10]

2. Rice Milling and Composition

Harvested rice grain is composed of the outer hull, the bran layer and the inner white starchy endosperm. Rough rice, which is rice with the hull intact, is the form in which rice is stored for use throughout the year. Removal of the hull produces what is known as brown rice. Once the hull is removed, inadvertent scratching of the inner

Table 13.2 Typical mill yields and composition of products and by-products from rough rice[13]

Fraction	Yield from rough rice, %	Protein, N × 5.95, %	Crude fat, %	Crude fiber, % d.b.	Crude ash, %	Nitrogen-free extract, %	Free sugars, %
Hull	18–28	2–4	0.4–0.8	48–53	15–20	26–34	
Brown rice	72–82	7–15	2–4	1	1–12	79–90	1.3–1.5
Bran	4–5	12–17	15–22	9–16	9–16	40–49	6.4–6.5
Polish	3	13–16	9–15	2–5	5–9	54–71	
Milled rice	64–74	6–13	0.3–0.6	0.1–0.6	0.3–0.7	84–93	0.2–0.5
Head	56						
Seconds	9						
Brewers	3						

bran layer releases lipases which effect lipolytic hydrolysis and oxidation of the kernel oil. The subsequent rancidity results in bitter, soapy, off-tastes. Normally, once the hull is removed the rice is sold immediately for use as brown rice or continues to the next step of milling, which is bran removal and polishing.

Abrasive milling removes the outer bran layer to produce partially polished rice or, after polishing to remove the entire bran layer, white rice. Rice bran or polish may be subsequently stabilized by heat treatment to inactivate lipases. Stabilized rice bran has found use as an ingredient in human-grade processed foods.

After removal of the bran layer and polishing to the desired degree, the residual milled rice is divided into several fractions depending on size. Total milled rice contains head rice (whole unbroken kernels and kernels that are at least three-fourths of an unbroken kernel) and broken rice. Further screening of the total milled rice separates the head rice from second heads (largest of the broken kernels, usually between one-half and three-fourths of an unbroken kernel) and brewers rice (broken kernels that are about one-fourth of an unbroken kernel). The relative amounts in each category are a function of rice variety, growing conditions, harvesting, drying, handling and the type of milling operation. Typical mill yields and their compositions are given in Table 13.2.

II. Uses of Milled Rice and Rice By-products

1. Milled Rice

Milled rice is comprised of head rice, second heads and brewers' rice. Unlike most grains, the primary use of rice is its direct consumption as a table food. The two other major uses are in processed foods and to make beer. Head rice is the only rice that is used for table consumption and represents ~82% of the milled rice. Broken rice, which is comprised of second heads and brewers' rice, is considered to be a by-product of milling and represents ~18% of the milled rice. Broken rice generally has a lower market value than head rice. Due to the relatively high cost of head rice compared with other grains, lower cost by-products of head rice production are preferred for use in processed foods and beer production. In 2004–2005, the USDA estimated that the principle outlets for milled rice in the US included 52% for direct food use,

29% for use in processed foods and 17% for use as a beer adjunct.[13] While use in beer has remained rather constant, use in processed foods grew 38% during 1985–1995, with the fastest growing product categories being pet foods and rice ingredients.[14] Obviously, an imbalance exists between the availability of milling by-products (18%) and use in processed foods (29%).

Increase in demand for second heads and brewers rice by processed food manufacturers has resulted in a shortage of these by-products. Additionally, more and more rough and brown rice is being exported, thereby limiting even more the supply of milling by-products in the US. It is obvious that, as more and more processed food products are developed, that mill run rice (milled rice that has not been size separated) will have to be used. The cost of mill run rice is typically 2–3 times that of corn. Because of the economic disadvantage of rice over other grains, future use of rice or rice ingredients (including starch) in processed foods will be dependent on unique characteristics and nutritional aspects that rice products or ingredients contribute to processed food products. In 2006, the USDA advocated the consumption of at least six servings of grains per day, of which three should be wholegrain. The latter prompted an increase in the use of wholegrains in food products. Development of novel wholegrain rice products was catalyzed by the development of wholegrain brown rice ingredients, such as RiceLife®, which made it possible to make wholegrain beverages, frozen desserts and even wholegrain non-dairy yogurts. Other brown rice ingredients, such as brown rice or brown ricemilk syrup, and their resulting products have greatly expanded the use of brown rice throughout the food and beverage industry.

Many processed foods, such as cereals, packaged mixes, pet foods, rice cakes and baby foods, were developed and incorporate broken rice, meal or flour. However, in recent years ingredients derived from milled rice, and now brown rice, have not only greatly extended the use of rice in processed foods, but have also been the basis for the development of novel food products. Commercially available rice ingredients mimic those products available from the corn wet-milling industry. Rice starch, rice proteins, rice syrups, rice dextrins and rice hydrolyzates are finding increased use. All these ingredients are uniquely different from their corn-derived counterparts. Products based entirely on rice starch-based ingredients, such as vinegars, oral rehydration solutions, non-dairy beverages, frozen desserts and puddings, are stable commodities in the marketplace, often taking the lead in sales over their corn or soy counterparts in natural foods and conventional marketing segments. Whole brown rice, while still in its developmental infancy, has great potential for growth as a processed food ingredient, due to its unique characteristics. Rice starch from white rice or brown rice exhibits unique properties as functional ingredients in many food categories.

2. By-products

Hulls

A major fraction of paddy rice is the hull. The hull is non-digestible, fibrous and abrasive in character, and has a low bulk density and a high ash content. It is used only in limited, low-value applications, such as in animal feed, in chicken litter, as a juice pressing aid and as fuel.[15]

Bran

Unstabilized bran and polish have been used almost exclusively for animal feed, due to the bitter flavor that develops from the lipolytic action of enzymes on the oil found in them. However, development of a thermal process that inactivates the lipases has resulted in a stabilized rice bran product that is suitable for the food industry. The impressive nutritional qualities of the oil, fiber, carbohydrate and proteins of rice bran have made it a valuable food material. Removal of fiber from the bran by physical[16,17] or enzymic[18,19] processes produces a milk-like product having desirable nutritional and functional properties. The nutritional composition of the rice bran milk product described by California Natural Products has been shown to match the nutritional requirements of an infant formula. Originally, the anti-nutritional factor of the residual phytates was of concern. However, as of 2005, phytase enzymes are suitable for use to break down these phytates.

Rice bran oil can be extracted from either stabilized or unstabilized bran. The by-product resulting from stabilized bran extraction is suitable for human food use. Stabilized rice bran is currently being used in baked goods, energy bars and protein fortification of powdered drink formulations.

III. Preparation of Rice Starch

Rice starch is preferably prepared from broken rice for economic reasons, as discussed above. There are currently two commercial methods of rice starch isolation: traditional and mechanical. The traditional method involves alkali solubilization of rice protein, while the mechanical method releases starch via a wet-milling process.

Worldwide production of rice starch amounts to about 25 000 metric tons.[20] Approximately 75% of this was manufactured by the Belgium company Remy Industries, which has been manufacturing rice starch by the traditional alkali method for more than 100 years. Until 1990, rice starch prepared by the alkali process was the only commercially-available rice starch.

1. Traditional Method

The traditional method of starch production involves alkali solubilization of the glutelin which constitutes approximately 80% of the protein in rice. This method has been described by Hogan,[21] and is utilized in some form by almost every rice starch manufacturer (with the exception described below). It produces a starch containing <1% protein. The protein by-product of this process, while good-quality protein, has a distinct aftertaste (alkali, salt and amino acid), making it not easily acceptable as a food ingredient.

In the alkali process, broken milled rice is steeped in 0.3–0.5% sodium hydroxide solution for up to 24 hours at temperatures that may vary from room temperature to 50°C. This steeping process softens the grain and effects solubilization of the proteins. Wet-milling of the steeped grain, in the presence of sodium hydroxide solution, releases the starch, producing a starch slurry. The starch is kept in suspension and stored for

10–24 hours to further dissolve the proteins. The cell wall material is then removed by filtration, and the starch slurry is washed with water (to remove the protein), neutralized and dried. The commercial drying process, which necessarily involves initially low air temperatures to prevent gelatinization, has the potential for allowing high bacterial growth and, consequently, total plate counts must be carefully monitored. An advantage to the alkali process is the ease with which the alkali solution allows for modification of the starch, since most modification reactions are done at high pH values. A disadvantage is the alkali waste water that is produced and its ecological consequences. These two problems, along with the limited economic potential of rice starch, are the primary reasons why the alkali process has not been used in the US since 1943.[22]

2. Mechanical Method

The mechanical method of rice starch production is a wet-milling process first used on milled rice that permits optional protein removal. Protein is not solubilized in this process, but rather is removed by physical separation after the starch granules and protein have been mechanically liberated from the starchy endosperm. The released starch granules exist in clumps or small aggregates of 10–20 μm in diameter. Products containing from 0.25% to 7% protein may be produced by this commercial method. If the protein is removed, the resulting starch is similar in appearance to that produced via the traditional process. However, differences in pasting and functional characteristics (described below) exist. Unlike alkali-solubilized protein, the protein resulting from the mechanical process is a valuable by-product, having excellent taste qualities suitable for the food industry. A current patent pending variation of the mechanical method, as developed by this author and practiced by Creative Research Management, utilizes whole brown rice. The resulting whole brown rice starches are unique and contain both protein and rice bran oil. Both of these constituents of the whole brown rice impart novel physical properties to the starch-containing fraction. The advantage of the use of these types of starches is relatively new and is just now becoming available to the industry as an ingredient.

The primary advantage of the mechanical process is the variety of starches of different protein and fat content that may be produced, and their respective unique functional properties. Also, the waste water from the mechanical process has no negative environmental impact and is quite suitable for land application.

IV. Properties of Rice Starch

1. General Properties Unique to Rice Starch

Traditionally, there have been basic properties associated with rice starch that have given it advantages over other starches. These characteristics include hypoallergenicity, digestibility, consumer acceptance, bland flavor, small granules (2–10 μm), white color, greater freeze–thaw stability of pastes, greater acid resistance and a wide range of amylase:amylopectin ratios.

Rice is considered to be hypoallergenic, because it does not possess gliadins or parts of proteins that are normally associated with the allergenic responses, such as those that may be caused by wheat, barley or rye. Because of its extensive table use, consumers perceive rice as a balanced food, as indeed it is. Its better digestibility, as compared to other cereal grains, has led to its preferred use in infant and geriatric foods, to medical recommendations that rice be the first cereal grain given to infants, and to it being the preferred grain for recuperating patients. There is no specific scientific evidence to support why rice has this digestive advantage, only hundreds of years of anecdotal experience. However, research on the utilization of rice in oral rehydration solutions and in infant formulas is generating some understanding of the beneficial qualities unique to rice.[23]

The bland taste of rice starch, its whiteness and its small granule size have provided it with the advantages necessary in the manufacture of smooth gravies, sauces and puddings, having excellent mouthfeel and flavor profiles. Non-food uses of rice starch are also based on its small granule size; they include textile size, cosmetic and printing ink applications.

A spray dried form of rice starch in which the individual starch granules are dried in aggregates has been developed by CNP. These porous starch spheres are advantageous with regard to their greatly improved dispersibility over other starches. The aggregates also have special properties associated with absorption of other solutes within the sphere structure and slow release for an improved distribution.

Depending on the relative amylopectin:amylose ratio found in the rice, the starch can exhibit a variety of gelatinized textures and strengths, as well as resistance to acid. Overall, there are several basic factors that affect starch performance. These factors, which include rice variety, protein content, method of starch production and modification, are described below.

2. Pasting Properties

The Rapid ViscoAmylograph (RVA) has been the preferred instrument for determining the pasting properties of rice flour and rice starches[24,25] (see Chapter 8).

V. Factors Affecting Rice Starch Properties

1. Rice Variety: Common Versus Waxy

The functionality of rice starch depends on the amylose:amylopectin ratio. Differences among rice starches made from long-, medium- or short-grain rice are insignificant relative to the amylose content. Rice starches made from common rice tend toward higher peak, cooked and cooled viscosities, as well as paste textures that are short and pasty. The texture of waxy rice starch pastes tends to be long and stringy (Table 13.3).

Rice starch from common rice contains 0.3%–0.4% lipids, while waxy rice contains considerably less (0.03%). Complexes of these lipids are not easily removed from the starch, and are presumably responsible for the lack of paste clarity and the difficulty in clarification of starch hydrolyzates.

Table 13.3 Ranges of physicochemical properties of common and waxy rice starches[a,16]

	Common	Waxy
Amylose:amylopectin	20:80	2:98
Bound lipid, %	0.2–0.4	0.02–0.03
Initial paste temperature, °C	65–74	61–64
Peak temperature, °C	91–95	69–71
Brabender ViscoAmylograph, BU		
Peak	560–1020	670–685
Cooked	260–480	280–425
Cooled	490–785	320–675
Texture	short	long

[a] Prepared from US rice varieties by the mechanical method

2. Protein Content

Residual protein can greatly affect the pasting and functional properties of rice starch. The higher protein content of alkali-processed rice starches appears to reduce the pasting curve viscosity. However, among mechanically processed rice starches, this is not the case. Mechanically produced rice starches with a relatively high protein content, in most cases, give higher pasting curve viscosities than those starches with a lower protein content. This relationship is particularly dramatized in the case of a waxy rice starch having a protein content of 4.1% versus one containing 0.7% protein.

Increasing protein content among mechanically produced rice starches also significantly improves functional properties such as dispersibility, gel smoothness, stability during ultra-high-temperature (UHT) processing, acid resistance and freeze–thaw stability. It was found that freeze–thaw stability in common rice starch containing at least 2% protein is equivalent to that normally associated with waxy rice starch. Some of these functional advantages may be attributed to the unique sphere aggregates that are formed in the mechanical process on spray drying of the rice starch in the presence of the protein.

After spray drying, rice starch containing <0.5% protein is present as clusters of 10–20 μm. At 1.5% and again at 6.0% protein, increased formation of spheres of 30–70 μm is observed. The presence of these spheres is responsible for improved dispersibility and gel smoothness.[26] It has also been suggested that the unique absorption properties of the sphere aggregates may have application in holding and dispersement of flavor material or pharmaceuticals.[27]

3. Method of Preparation

RVA curves compare commercial rice starch prepared by traditional alkali methods and rice starches prepared by the mechanical method. In the case of common rice starches, the onset of pasting appears to occur earlier (at lower temperature) and the final cooled viscosity appears to be lower for the starches prepared using alkaline conditions. The latter differences may be due to either rice type or the effect of alkali on granules. In the case of waxy rice starches, mechanically produced starches, in general, have both higher paste and cooled viscosities as compared to those prepared by the alkali method.

4. Modification

Properties of rice starches are changed by chemical modification in the same way as the properties of other starches (see Chapter 18). Starches prepared via the alkali method have been modified to provide additional pH and shear stability. In general, hypochlorite-oxidized rice starch has a lower gelatinization temperature and lower maximum paste viscosity producing a softer, clearer gel. Hydroxypropylated rice starches have lower gelatinization temperatures, whereas crosslinked rice starch has an increased gelatinization temperature, increased shear resistance and acid stability.

VI. Rice Starch Applications

Rice starch applications are normally discussed in terms of common and waxy types and are specified that way in Table 13.4. A 10% solution of a common rice starch, when sheered and gelatinized simultaneously, produces a product that resembles a solid shortening in texture. The waxy rice starches do not produce this same texture; however, they too have been used very effectively for fat replacement due to a fat-like mouthfeel when blended with other food products. Waxy rice starches also tend to resist oil uptake when used in batters for fried foods. Most applications of rice starch may be attributed to one or more of the characteristics already discussed that are unique to rice starch.

Table 13.4 Rice starch applications[16]

Common	Waxy
Binder	**Fat-mimetic**
Confectioneries	Coatings for deep-fat frying
Dairy products	Dairy products
Infant foods	Pastries
Processed meats	Processed meats
Puddings/custards	Sauces/soups
Sauces/soups	
Dusting agent	**Binder**
Confectioneries	Canned foods
Pharmaceuticals	Infant foods
	Rapid cook (microwaveable) foods
	Ready-to-eat meals
Miscellaneous	**Crispness agent**
Cosmetics	Breakfast cereals
Laundry products	Extruded snacks
Pet foods	
Photographic	
	Frozen foods
	Low-fat ice cream
	Non-fat sauces

Source: reference 29

VII. References

1. Marshall WE, Wadsworth JI. In: Marshall WE, Wadsworth JI, eds. *Rice Science and Technology*. New York, NY: Marcel Dekker; 1993:5.
2. Newman R. Rice Forecast: What's in Store for 1995 and Beyond. *Rice Utilization Workshop*. Houston, TX: USA Rice Federation; 1995:33.
3. USA Rice Federation Report. www.USARice.com; June 2007.
4. USDA. Economic Research Services. www.ers.usda.gov. Rice Outlook/RCS-05F/ June 15.
5. Juliano BO. In: Whistler RL, BeMiller JN, Paschall EF, eds. *Starch: Chemistry and Technology*. Orlando, FL: Academic Press; 1984:507.
6. USDA. *Rice Market News*. 1992;73:8.
7. Marshall WE, Wadsworth JI. In: Marshall WE, Wadsworth JI, eds. *Rice Science and Technology*. New York, NY: Marcel Dekker; 1993:10–11.
8. McKenzie KS. In: Marshall WE, Wadsworth JI, eds. *Rice Science and Technology*. New York, NY: Marcel Dekker; 1993:91.
9. McKenzie KS. In: Marshall WE, Wadsworth JI, eds. *Rice Science and Technology*. New York, NY: Marcel Dekker; 1993:98.
10. Webb BD. *Texas Agr. Expt. Sta. Res. Monograph*. 1975;4:97 [Beaumont, TX].
11. Houston, TX: USA Rice Federation.
12. Juliano BO. In: Whistler RL, BeMiller JN, Paschall EF, eds. *Starch: Chemistry and Technology*. Orlando, FL: Academic Press; 1984:511.
13. US Rice Federation, Domestic Usage Report, MY 2004–2005.
14. Meyers R. Current Domestic Markets for Rice. In: *Rice Utilization Workshop*. Houston, TX: USA Rice Federation; 1995:44.
15. Luh BS. In: Luh BS, ed. *Rice Utilization*. Vol. II. New York, NY: Van Nostrand Reinhold; 1991:269.
16. Lathrop, CA: Courtesy of California Natural Products.
17. Patty Mayhew, Food Extrusion Inc., El Dorado Hills, California, Personal Communication.
18. Hammond NA. US Patent 5 292 537; 1995.
19. Ribus, Inc., St. Louis, MO, personal communication.
20. Orthoefer F. Riceland Foods, Inc., Stuttgart, Arkansas; 1995, personal communication.
21. Hogan JT. In: Whistler RL, Paschall EF, eds. *Starch: Chemistry and Technology*, vol. II, New York: Academic Press; 65.
22. Juliano BO. In: Whistler RL, BeMiller JN, Paschall EF, eds. *Starch: Chemistry and Technology*. Orlando, FL: Academic Press, Inc. 1984:513.
23. Khin-Maung-U, Greenough WB. *J. Pediatrics*. 1991:118.
24. Welsh LA, Blakeney AB, Bannon DR. *Modified R.V.A. for Rice Flour Viscometry* [unpublished method]. Australia: Yanco Agricultural Institute, NSW; 1983.
25. Deffenbaugh LB, Walker CE. *Cereal Chem*. 1989;66:493.
26. Courtesy of Hall J. California Natural Products, Lathrop, California.
27. Zhao J, Whistler RL. *Food Technol*. July 1994;104.

14

Rye Starch

Karin Autio[1] and Ann-Charlotte Eliasson[2]
[1]VTT Biotechnology and Food Research, VTT, Finland
[2]Department of Food Technology, Lund University, Lund, Sweden

I. Introduction	579
II. Isolation	580
1. Industrial	580
2. Laboratory	580
III. Modification	582
IV. Applications	582
V. Properties	582
1. Microscopy	582
2. Composition	583
3. X-Ray Diffraction Patterns	584
4. Gelatinization Behavior	584
5. Retrogradation	584
6. Amylose–Lipid Complex	584
7. Swelling Power and Amylose Leaching	584
8. Rheology	585
9. Falling Number	586
VI. References	586

I. Introduction

Rye is grown mainly in the northeastern countries of Europe and in some locations in North America and Argentina.[1] Rye is used mainly for human consumption and is the principal flour used in sour dough breads and in crisp bread.

The industrial use of rye starch is limited. Industrial production of rye starch was carried out during World War II, but as soon as the supply of wheat, maize and potato starches returned to normal production ceased.[2] Rye flour is a difficult raw material from which to isolate starch, because of its high pentosan content and the poor gluten-forming ability of the proteins. It has, therefore, been suggested that the whole rye kernel, and not the isolated starch, should be used in industrial applications (e.g. production of insulating and plastic material).[3] Rye is also a problematic raw material because of the variation in the climatic conditions in northeastern Europe, which results in variations in the properties of the starch. These circumstances seem to discourage the use of rye for starch production, but there still might be reasons for rye starch production. Schierbaum et al.[2] suggest that rye might be grown for its starch because of climates and acreages unsuitable for cultivation of wheat and maize, because potato cultivation and processing may not be expanded due to environmental or economic reasons, because of rising average yields of rye, and because of the need to utilize part of the harvest that is not used for human consumption, animal feed or fermentation.

II. Isolation

1. Industrial

Industrial production of rye starch received special attention in the former German Democratic Republic, and several patents were issued.[4,5] The flow sheet in Figure 14.1 illustrates a process that has been used in pilot plant production of rye starch.[2] Impact-milled rye flour is used as the starting material. The flour is suspended in water, followed by intense agitation after the addition of 0.5–0.15% sulfur dioxide (SO_2). The most important step in the production of rye starch is to reduce the viscosity of the flour suspension, and this can be achieved by heating it, preferably to 30–48°C, and incubating it with *lactobacilli, streptococci*, sour dough extract or sour dough culture for 5–24 hours. If the activity of the enzymes present in the native flour is utilized, the incubation time might vary between 4 and 24 hours, due to variation in enzyme activity of the raw material. In modifications of the process, in order to make it continuous, microbial enzyme preparations are used. After the incubation step, starch is separated by centrifugation and sieving. In the continuous process, the centrifugate is recycled into the suspension-preparation stage. Alkali treatment is used to reduce protein content from 0.3–0.4% to 0.12–0.15%. The starch-rich sediment obtained in the continuous process is collected on sieves, purified, dewatered and dried. The average yield is about 50%. The preparation contains mostly large granules (A-starch).

2. Laboratory

For the isolation of rye starch in the laboratory, a method more or less similar to the one outlined in Figure 14.2 is used. The laboratory procedure involves steeping and wet-milling,[6] and in a modified procedure, defatting by refluxing overnight with 80% methanol.[7] In a study of nine different rye samples,[8] the starch was isolated according to a method described by Meredith et al.[9] In another laboratory method, rye flour was mixed with water at low speed in a blender, the suspension was centrifuged and the starch was air dried.[10] The starch recovery in this process was 42%. The prime starch obtained as described can then be further purified by repeated washings with distilled water.

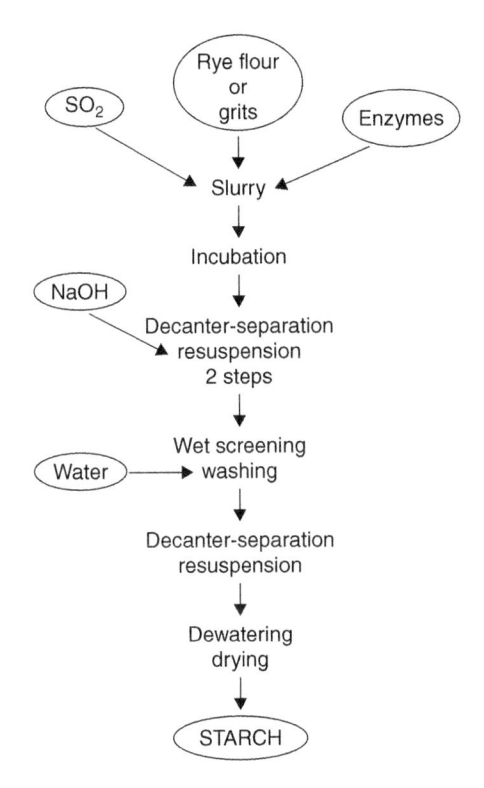

Figure 14.1 Pilot plant process for isolation of rye starch. (From reference 2, modified)

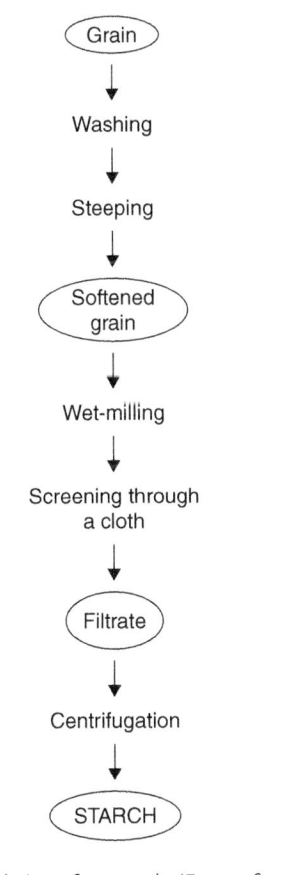

Figure 14.2 Laboratory process for isolation of rye starch. (From reference 6, modified)

III. Modification

Rye starch has been modified by succinylation (0.7–2.1% succinyl groups), acetylation (1.9–2.5% acetyl groups) and oxidation.[11,12] Succinylation resulted in more 'potato-like' behavior of rye starch.[13] Syrups have been produced from rye starch by hydrolysis.[12]

Physical modification has been done by annealing, steaming, extraction by aqueous solvents and drying.[11] Hydrothermal treatment has included gelatinization and autoclaving. Treatment with sodium hydroxide has also been used, as has association with soluble pentosans and gliadin.[14] It has been observed that the conditions during preparation, e.g. different drying methods or the incubation of flour – water slurries (35–40°C for 12–24 hours), affect the properties of the isolated starch.[11,13] When rye starch slurries were stirred for 1 hour at 50°C in a ViscoAmylograph bowl before the ordinary heating program, an increase of 2–8°C in onset temperature of consistency development was noticed.[11] When rye starch (25% moisture) was heated for 4 hours at 100°C, the treated starch produced lower viscosity at 96°C and at 50°C (after cooling). Annealing effects were also observed when rye flour was heat – moisture treated.[15]

IV. Applications

Rye starch has been tried as a gelling and thickening agent in the food industry and also in non-food applications.[12] Rye starch can be substituted for potato starch. A combination of rye starch and lipids has been used in dessert mixes.[16]

V. Properties

1. Microscopy

Rye starches, like those of wheat, triticale and barley, are composed of large (A-type) and small (B-type) granules with diameters of 23–40 μm and less than 10 μm, respectively.[2,17] The maximum in the distribution curve of the large granules occurs at about 24 μm,[18] and there are rye starch granules even larger than 40 μm in diameter.[10] This means that the large granules in rye starch are larger than those in wheat or barley starch. The small granules seem to be less frequent in rye starch than in wheat and barley starches.[19] The size distribution of a commercial rye starch depends on the isolation procedure.[2] Fissures are frequently observed on the surface of rye starch granules.[8] The fissures are not correlated with starch damage as determined with Congo red.[8] However, it is probable that the fissures are due to α-amylase action, since the falling number of rye flours used for starch preparation correlated with the percentage of fissures in the granules and with amylose leaching.[8] Pores are found in the equatorial groove of large granules.[20]

The swelling behaviour of rye starch is very similar to that of wheat starch and occurs at two temperatures. At temperatures of 50–55°C, the swelling of large granules is in one plane and radial in nature, whereas the small granules swell to a spherical shape.[13,21] At higher temperatures, the swelling of the large granules is tangential in the same plane. Amylose, beginning with the smaller molecules, is preferentially leached from rye starch granules, the equatorial groove of the A granules facilitating transport of amylose from the interior out of the granules.[13,21] Rye starch may contain low activities of α-amylase, which has an affect on granule structure and leaching of amylose.[17] Slight chemical modification, such as succinylation, changes the swelling behavior of A granules, in that their natural disk shape becomes more spherical as swelling proceeds.[13] Increase in succinyl content decreases the onset swelling temperatures and increases swelling power and solubility.

2. Composition

The gross composition of rye starch isolated by different procedures is given in Table 14.1. Values of amylose content as high as 30.1% have been reported,[22] although values around 25% seem to be more common (Table 14.1). The average molecular weight of amylose was determined to be 218000 and the intrinsic viscosity 2.60.[10] These values are lower than the corresponding values for wheat starch. The average degree of polymerization (DP) for amylose was reported to be in the range 223–242,[2] which corresponds to much lower molecular weights than reported earlier.[10] The λ_{max} of aqueous solutions of amylose–iodine complexes is 623–645 nm, which is higher than that for potato starch.[2]

There are two populations of rye starch amylopectin branch chains with average DPs of 11–25 and 52–60, respectively.[23] The overall average DP was found to be 26, which is greater than found for amylopectins of other starches examined in the same study.[23] An average DP of 21 has also been reported, which is lower than that of studied wheat samples.[10] The ratio of A-chains to B-chains was reported to be 1.7–1.8.[23] The degree of branching was determined to be 4.8%, which is greater than that of wheat starch, and the intrinsic viscosity was 1.33 dL/g, which is lower than that of wheat starch.[10]

The lipid content of rye starch has been determined to be 0.62 g/100 g starch,[24] about half the amount found in wheat starch. Lysolecithin constituted 45% of the lipids. The values given in Table 14.1 for the lipid content are higher, and it might again be concluded that the purity of the preparation is a function of the isolation procedure.

Table 14.1 Composition of laboratory-isolated rye starch[2,10]

Component	Amount, %
Protein	0.04;[a] 0.19–1.30
Ash	0.37
Total lipids	0.98; 1.05–1.74
Phosphorus	0.025
Pentoses	0.3–0.7
Amylose	24; 22.1–26.4[b]

[a]Nitrogen
[b]After lipid extraction

3. X-Ray Diffraction Patterns

Rye starch gives an A-type x-ray diffraction pattern,[25] as would be expected for a cereal starch. A content of 6–9% B-type also has been suggested for rye starch.[2] The degree of crystallinity was estimated at 15–17%.[2]

4. Gelatinization Behavior

Loss of birefringence occurs in the temperature interval 50–58°C,[23] but a lower range (47–56°C) has also been reported.[7] Eberstein et al.[25] determined the range to be 51–58°C using a hot-stage microscope, a value similar to the range determined using DSC.[25] The gelatinization temperature, measured as the peak maximum (T_m) of the gelatinization endotherm obtained by DSC, was found to differ between 54.8° and 60.3°C for nine different rye starch samples,[8] temperatures which are lower than those for wheat starch. The corresponding gelatinization enthalpy (ΔH) was determined to be in the range 10.6–12.0 J/g starch, which was also somewhat lower than that for a wheat starch. The gelatinization onset temperature, measured in the DSC, was found to be lower than for many other starches, such as those from wheat, barley and potato.[27] The starch preparation method influences the DSC parameters.[11]

5. Retrogradation

Retrogradation of rye starch was studied in gels composed of 40% starch and 60% water.[8] The starch suspensions were first pasted, then stored at ambient temperature (23°C) for up to 14 days, before they were analyzed (DSC). Retrogradation of amylopectin was measured as the melting enthalpy (ΔH_c) obtained for the endotherm obtained during reheating of the stored samples. The rye starch with the lowest tendency for retrogradation and the one with the highest tendency are compared with maize and wheat starches in Figure 14.3. It is evident that rye starch retrogrades to a much lower degree than maize and wheat starches. The low retrogradation tendency cannot be explained by the lipid content and formation of an amylose–lipid complex, as this was lowest for rye starch.

6. Amylose–Lipid Complex

In the presence of excess water, the transition enthalpy (ΔH_{cx}) of the rye starch–lipid complex was determined to be 0.84 J/g.[25] For the nine different rye starches described above, the ΔH_{cx} range was 0.3–0.6 J/g.[8] These values are lower than those for wheat starch, thus supporting the lower lipid content of rye starch compared to wheat starch.[24] Also, there were small differences in transition temperature; at a water content of 60%, the range was 100.0–106.6°C.

7. Swelling Power and Amylose Leaching

Water-binding capacity (WBC) was reported to be 141% for starch prepared from mature rye,[7] whereas it was somewhat higher for starch prepared from immature rye.

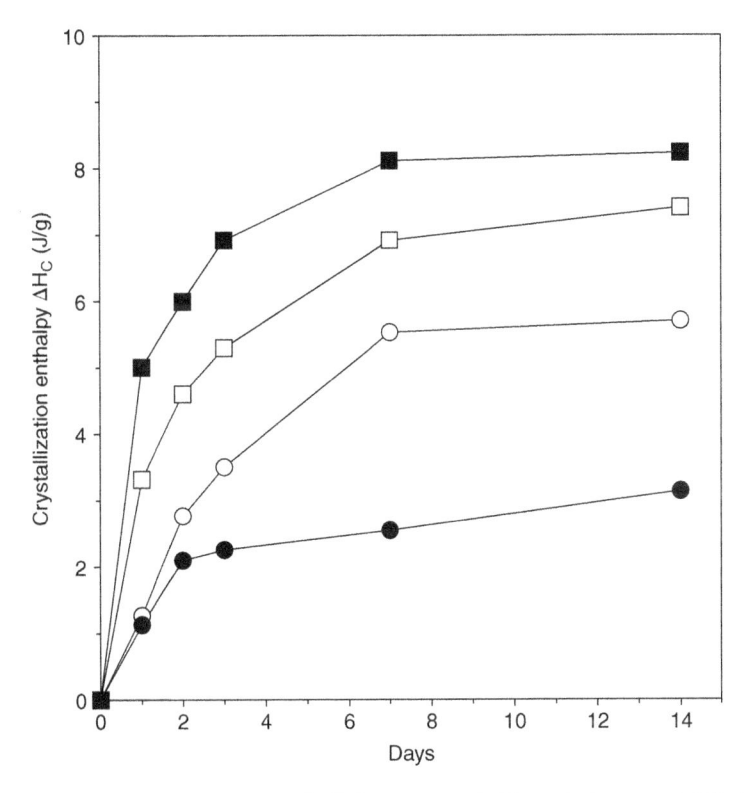

Figure 14.3 Retrogradation of: rye starch (○, ●, l); maize starch (■); and wheat starch (□). (From reference 8, used with permission)

Table 14.2 Swelling power and solubility determined for rye starch[7]		
Temperature, °C	Swelling power	Solubility, %
60	6.20	1.42
70	6.77	2.52
80	8.82	3.12
90	13.85	6.70

A WBC value of 86.5% has also been reported.[10] Examples of the swelling power and solubility are given in Table 14.2. These values were obtained for starch from mature rye; values for starch prepared from immature rye were higher. Values for rye starch were lower than those for wheat starch.[7] In a study of nine different rye starches, the gel volume at 90°C was determined to be in the range 14.8–19.7 mL/g, and the leached amylose to be 120–172 mg/g starch.[8]

8. Rheology

Heat-induced changes in rye starch dispersions have been studied using the Brabender ViscoAmylograph.[11] The problem with this method is that the changes monitored are also dependent on shear, and some starches, such as potato starch, are very sensitive to shear.[26] Dynamic viscoelastic measurements, which allow the measurement of rheological properties without shear, have not been used for rye starch pastes.

The pasting behavior of rye starch in relation to maize, potato, wheat and triticale starches has been determined.[11] The behavior is similar to that of wheat starch. Holding rye starch at the maximum temperature (96°C) effects no viscosity changes, an unusual behavior for an unmodified starch. There are, however, great variations in viscosities at this temperature for different rye starches. Values from 165 to 525 Brabender units (BU) have been reported. During cooling viscosity increases, and again great variations in viscosity values at 50°C (varying from 390 to 1100 BU), have been reported.

Succinylation of rye starch (0.7–2.1%) changes the pasting behavior towards that of potato starch, decreasing the onset temperature, increasing the maximum viscosity and decreasing stability during holding at 96°C. Acetylation, on the other hand, makes the behavior of rye starch more like that of maize starch, with increased maximum viscosity and viscosity at 50°C.

9. Falling Number

Falling number, being the most used parameter for rye quality, was determined to be 95–280 for nine different rye starch samples.[8] The falling number was correlated to the properties measured, and a strong correlation was found to percentage of fissures in the granules (−1.84***) and to amylose leaching (0.78***). The falling number also correlated to ΔH (0.64*) and to T_m (0.61*) (DSC).

VI. References

1. Lampinen R. In: Poutanen K, Autio K, eds. *International Rye Symposium: Technology and Products*. Espoo, Finland: VTT Technical Research Centre of Finland; 1995:9.
2. Schierbaum F, Radosta S, Richter M, Kettlitz B, Gernat C. *Starch/Stärke*. 1991;43:3313.
3. Kretschmer P, Gebhardt E. In: Poutanen K, Autio K, eds. *International Rye Symposium: Technology and Products*. Espoo, Finland: VTT Symposium; 161; 1995:137.
4. Schirner R, Bernhardt T, Kraetz H, Schulz P, Schmidts R, Kunz B, German Democratic Republic Patent 19860129; 1986.
5. Bernhardt T, Goermer L, Schmidt H, Nindel H, German Democratic Republic Patent 19860514; 1986.
6. Kulp K. *Cereal Chem*. 1972;49:697.
7. Park A, Lorenz K. *Lebensm. Wiss. Technol*. 1977;10:73.
8. Gudmundsson M, Eliasson A-C. *Cereal Chem*. 1991;68:172.
9. Meredith P, Dengate HN, Morrison WR. *Starch/Stärke*. 1978;30:119.
10. Berry CP, D'Appolonia BL, Gilles KA. *Cereal Chem*. 1971;48:415.
11. Schierbaum F, Kettlitz B. *Starch/Stärke*. 1994;46:2.
12. Kraetz H. *Nahrung*. 1985;29:847.

13. Radosta S, Kettlitz B, Schierbaum F, Gernat C. *Starch/Stärke*. 1992;44:8.
14. Michniewicz J, Jankiewicz M. *Z. Lebensm. -Unters. Forsch.* 1988;187:102.
15. Münzing K. *Thermochim. Acta.* 1989;151:57.
16. Decnop C. West Ger. Patent Appl. 1567354; 1970.
17. Forssell P, Härkönen H, Valkiainen J, Pessa E, Autio K. In: Poutanan K, Autio K, eds. *International Rye Symposium: Technology and Products*. 161. Espoo, Finland: VTT Symposium; 1995:210.
18. Karlsson R, Olered R, Eliasson A-C. *Starch/Stärke*. 1983;35:335.
19. Simmonds DH, Campbell WP. In: Bushuk W, ed. *Rye: Production, Chemistry, and Technology*. St. Paul, MN: American Association of Cereal Chemists; 1976:63.
20. Fannon JE, Hauber RJ, BeMiller JN. *Cereal Chem.* 1992;69:284.
21. Williams MR, Bowler P. *Starch/Stärke*. 1982;34:211.
22. Klassen AJ, Hill RD. *Cereal Chem.* 1971;48:647.
23. Lii C-Y, Lineback DR. *Cereal Chem.* 1977;54:138.
24. Acker L, Schmitz HJ. *Starch/Stärke*. 1967;19:275.
25. Eberstein K, Höpcke R, Konieczny-Janda G, Stute R. *Starch/Stärke*. 1980;12:397.
26. Svegmark K, Hermansson A-M. *Carbohydr. Polym.* 1991;15:151.

15

Oat Starch

Karin Autio[1] and Ann-Charlotte Eliasson[2]
[1]VTT Biotechnology and Food Research, VTT, Finland
[2]Department of Food Technology, Lund University, Lund, Sweden

I. Introduction	589
II. Isolation	589
1. Industrial	590
2. Laboratory	590
III. Modification	591
IV. Applications	591
V. Properties of Oat Starch	591
1. Microscopy	591
2. Chemical Composition	592
3. X-Ray Diffraction	594
4. Gelatinization	594
5. Retrogradation	595
6. Swelling Power and Amylose Leaching	596
7. Rheological Properties	597
VI. References	598

I. Introduction

Oat is primarily a cool season crop.[1] It is grown mainly in Russia, the United States, Canada and Europe. The principal use of oat is as an animal feed. However, the popularity of oat as a part of the human diet has increased because of reports describing the beneficial nutritional properties of oat β-glucans.[2–4] Oat also has other desirable nutritional properties. The protein content of oat is much higher than that of other cereal grains, and oat oil has a favorable ratio of polyunsaturated to saturated lipid.

Oat starch has properties that have captured industrial interest, but its use has been limited, largely because of difficulties in separation of the grain into starch, fiber and protein fractions.[5]

II. Isolation

Dry-milling of oat groats yields approximately 35% oat bran and 65% oat flour.[6] The starch content of the flour is 67.0–73.5%.[7] Separation of oat flour into starch, protein and β-glucan fractions is difficult, because of a strong bonding between starch and protein, and because of the presence of β-glucans.[6] Oat flour with a high starch content is preferred, not only because of the higher yield, but also because as a result, the content of impurities, such as lipid and β-glucan, in the starch fraction will be low.[8]

1. Industrial

A commercial fractionation process in which groats are dry-milled and the comminuted groats are soaked in a solution of cellulases and hemicellulases has been developed.[9] In addition to the starch the process produces fiber and protein fractions. Unlike wheat starch, oat starch cannot be separated from the grain by selective hydration and centrifugation, because of hydrated bran and protein layers.

2. Laboratory

Several different methods have been used for the isolation of oat starch in the laboratory in yields of 42.7–61.0%, depending on the variety.[7,10] Lowest yields are obtained from cultivars with high protein content. A much better yield (60–80%) was reported by Sowa and White.[8]

In earlier methods, sodium carbonate or sodium hydroxide solutions and a low shear rate or water and a high shear rate were used.[6] The alkali isolation procedure is shown in Figure 15.1a. A modification involving use of enzymes (protease and/or cellulase) (Figure 15.1b) gave a starch yield of 86%.[6] With another method employing a protease, 92–98% of the starch was recovered.[11]

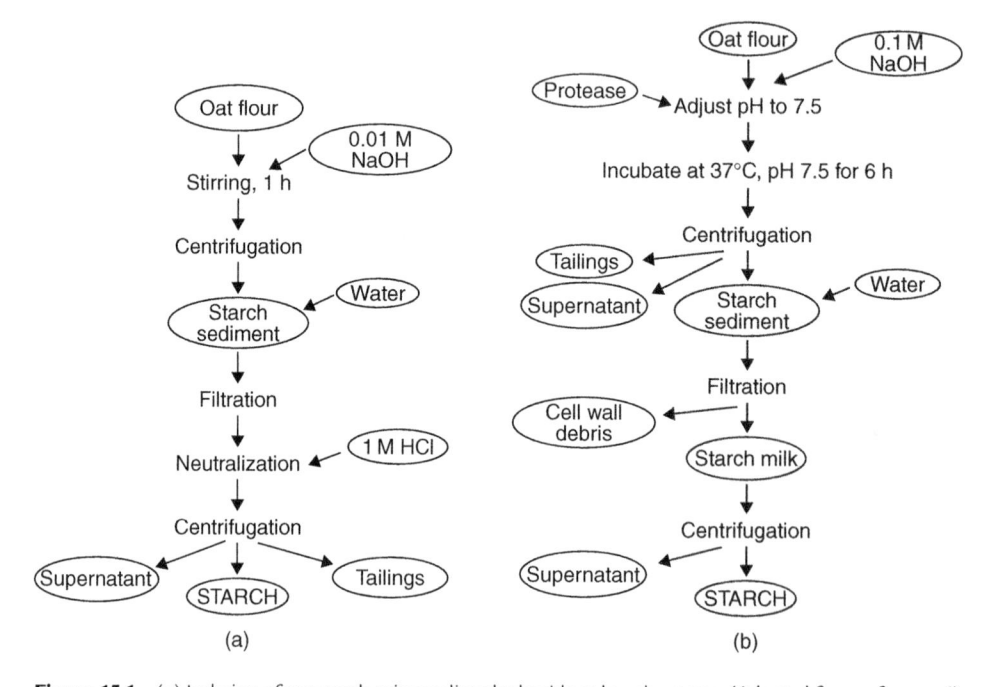

Figure 15.1 (a) Isolation of oat starch using sodium hydroxide at low shear rate. (Adapted from reference 6) (b) Isolation of oat starch using protease. (Adapted from reference 6)

III. Modification

Oatrim, a USDA-patented product, is made by conversion of oat starch in the bran or flour to maltodextrins using α-amylase.[12] Some such products, which are combinations of maltodextrins and β-glucan, can be used as fat mimetics. Modification of oat starch by a mild treatment with acid changes the pasting behavior of oat starch towards that obtained by acid thinning of wheat starch.[13] Cationic ethers of DS 0.014–0.042 have been prepared by reaction of oat starch with 2-chloro-3-hydroxypropyltrimethylammonium chloride at alkaline pH.[14] Oat starch has been modified by annealing, which changes its gelatinization behavior.[15] Heat–moisture treatment reduced the aggregation of oat starch granules and changed their gelatinization behavior.[16] Starch mutants of oat have recently been described where one type of mutant lacks amylose, and the other contains a phytoglycogen-like polysaccharide.[17]

IV. Applications

Oatrim can be used in cheeses as a fat replacer.[12] Other possible applications are in ice cream and frozen desserts, milk shakes, hot chocolate, instant-type breakfast drinks, cereals, salad dressings, soups, sauces and gravies. Oat starch in combination with oat hydrolysate or with xanthan gum has been used for thickening of sweet and sour sauces.[18]

Cationic oat starch was used to improve the dry strength of paper handsheets and found to be comparable to wheat starch in functionality.[14] Oat starch might possibly be a replacement for rice starch in pharmaceutical applications.[19] Oat flour has been used in adhesive preparations.[19]

V. Properties of Oat Starch

1. Microscopy

In oat grain, starch granules are clustered, the clusters being similar to those found in rice.[20] The size of the clusters varies from 20 to 150 μm in diameter, whereas individual, irregularly-shaped polygonal granules are typically 2–15 μm across.[8,10,21,22] These granules have surface indentations. Oat starch granules are weakly birefringent.[10] Non-contact atomic force microscopy of oat starch granules revealed a rough surface with depressions or pores with regular shapes with diameters below 40 nm.[23] Smooth areas without any depressions were also observed.

Morphological changes taking place during the heating of oat starch suspensions have been examined by light and scanning electron microscopy.[24,25] During the first phase of swelling (between 60°C and 80°C), the expansion of oat starch occurs along the three axes (in contrast to wheat, rye and triticale starches, which exhibit radial swelling), resulting in the formation of flattened discs.[24] During the second phase the swelling, which also occurs along the same three axes, is more pronounced.

In contrast to other cereal starches, great changes occur in the granule structure of oat starches below 100°C.[13,25] At 90°C, amylose forms a network structure around the granules and part of the granule structure is broken down. Heating to 95°C induces considerable changes in the granule structure of oat starch, and a network composed of both amylose and amylopectin is formed. This observation is in agreement with solubility studies which have shown that amylose and amylopectin are co-leached from oat starch granules at 95°C.[26] The microstructure also partly explains the special rheological properties of oat starch dispersions during cooling.

Even a very mild acid-catalyzed hydrolysis (0.5 hour with 1 M HCl at 40°C), followed by heating at 95°C, induced considerable changes in the oat starch granule, viz., a breaking down of the granule structure and formation of a continuous phase of amylopectin.[13] The tendency for aggregation decreased when oat starch was annealed.[15]

2. Chemical Composition

The chemical composition of isolated oat starch is given in Table 15.1. When an oat-flour slurry is centrifuged, the bottom layer is the prime starch. This starch can be purified with repeated washings, and a starch with a protein content of 0.44–0.6% and a lipid content of 0.67–1.11% (dry basis) can be obtained.[10,27] The moisture, ash and nitrogen contents are similar to those of wheat, whereas the lipid content is higher than that of wheat, and the amylose content is lower.[15]

Oat starch contains up to 2.5% lipid,[8,28,29] a very high lipid content that parallels the oil content of the oat grain used in the isolation process.[8,30] Oat is the cereal of highest lipid content; a lipid content as high as 15.5% has been reported.[8] Because of the range of lipid contents of the grain, several investigators have compared properties of starch isolated from groats differing in lipid content and found that many of the properties of oat starch are related to its lipid content.[8,29–33]

Lysophosphatidylcholine makes up about 70% of the lipids present in oat starch.[11] For AC Hill oat starch, free fatty acids dominated the neutral lipids (NL), lysophosphatidylcholine (a phospholipid, PL) and digalactosyldiglyceride (a galactolipid, GL).[20] The predominant fatty acid was C18:2 in NL and C16:0 in PL and GL. Most of the phosphorus present in oat starch is present in lipids.[8] There is a significant

Table 15.1 Proximate composition of native oat starches

Protein, %	Lipid, %	Amylose, %	Reference
	1.08–1.18	22.1–26.6	31
0.05[a]	1.13	19.4	15
0.85–0.95	2.1–2.5	30.3–33.6	8
1.1	1.7	17.3–20.7	14
	1.8–2.3	27.3–29.4	29
	1.38–1.52	25.2–29.4	11
0.44–0.60	0.67–1.11	25.5–27.9	7
0.02–0.09[a]	0.85–1.29	19.6–24.5	33

[a]Nitrogen

relationship between total amylose content and lysolecithin; the higher the amylose content, the higher the lipid content.[8,11,30] According to Table 15.1, the amylose content of oat starch is in the range of 17.3% to 33.6%. However, care must be employed when amylose values are interpreted, as the value probably depends on how the starch was defatted before determination of amylose content.

White et al.[8,30,31] have, in several papers, studied the chemical composition of oat starch amylose and amylopectin (Table 15.2). Gel permeation chromatography (GPC) of the total starch gave rise to two peaks, but some of the material had an unusual iodine affinity and was, therefore, described as an intermediate fraction.[8]

Fractionation of an oat starch after debranching resulted in three fractions denoted F1, F2 and F3.[30] F1 had a chain length (i.e. degree of polymerization, DP) of 593–703 and was believed to be amylose. F2 had a chain length of 42–44 and was reported as being long B-chains of amylopectin. F3 had a chain length of 17–22 and was reported to be A- and short B-chains of amylopectin. The chain length of the F3 fraction was positively correlated with the lipid content.

The amylose fraction from three different oat starches was characterized by high-performance size-exclusion chromatography (HP-SEC) and shown to be free of amylopectin.[31] The degree of polymerization is given in Table 15.2. By treating amylose with isoamylase, it was found to be branched.[31] The major fraction had a DP of 700, whereas a second fraction had a DP of 72. The amylose chain length was reported to decrease with increased amylose and starch-lipid content.[30] The intrinsic viscosity of amylose has been reported to be 246–299 mL/g.[7]

Fractionation of amylopectin after debranching with isoamylase resulted in three fractions, F1 (high molecular weight), F2 (intermediate molecular weight) and F3 (low molecular weight).[31] Of these fractions, F1 was a minor fraction. The chain length at the peak for each fraction is given in Table 15.2.

There were differences between oat varieties in chain lengths and chain length distributions of amylopectin. The degree of multiple branching decreased when the starch–lipid and amylose contents increased.[31] The intrinsic viscosity of amylopectin has also been determined (Table 15.2). In another investigation, values ranging from 170 to 207 mL/g were reported.[7]

Table 15.2 Characterization of amylose, amylopectin and intermediate material in three oat starches[31]

Parameter	Amylose fraction	Amylopectin fraction	Intermediate fraction
Iodine affinity (g/100 g starch)	18.4–18.9	0.30–0.58	0.62–1.26
λ_{max} (nm)	659–662	557–560	567–575
$[\eta]$ (mL/g)	167–173	124–146	145–148
DP[a]	392–568	181.7–204.2	280–310.9
	2149–2920	30.7–31.8	34.5–79.9
		16.6–20.1	21.8–22.6

[a]After debranching for amylopectin and intermediate material. For amylose, the range in the HP-SEC profile is given

White et al.[31] suggest the presence of an intermediate fraction (Table 15.2). It was suggested that this intermediate fraction was structurally close to amylopectin, but had longer branches. When debranched, three fractions were obtained; the chain lengths of these fractions are given in Table 15.2. The existence of intermediate material in oat starch was suggested by Banks and Greenwood in 1967.[34] They found an anomalous amylopectin, amounting to 4.5% of the starch, that was less branched than the ordinary amylopectin. An equal amount of anomalous amylose was also found.

Paton[19] also found three fractions by size-exclusion chromatography; 56% was amylopectin, 26% intermediate material and 18% amylose. It has been recommended that the presence of intermediate material should be interpreted with caution, as the intermediate fraction might be an artifact.[35] Mua and Jackson found the intermediate fraction after various harsh treatments of the starch, including dissolution and sonication in DMSO, heating at 100°C, sonication for 30–40 seconds, autoclaving and then sonication again, and suggested that the intermediate fraction is not inherent to the native granule.[35]

3. X-Ray Diffraction

The x-ray diffraction pattern of oat starch is the A-type, as expected for a cereal starch.[22,33,36] The crystallinity of oat starch was reported to be higher than that of wheat starch when determined at the same moisture content.[22] A relative crystallinity in the range of 28.0–36.5% was reported for starch from six different Canadian oat varieties.[33] Heat–moisture treatment caused the intensity of d-spacings in the x-ray diffraction pattern to increase.[16]

4. Gelatinization

DSC parameters (T_o = onset of gelatinization endotherm; T_p = temperature at the peak; T_c = conclusion; and ΔH = enthalpy) for the gelatinization of oat starch in excess water are given in Table 15.3. It has been observed that the melting temperature of the most perfect crystallite is higher for oat starch than for other starches.[28] The T_o values of oat starch were found to increase with increased amylose ($r = 0.97$) and starch–lipid ($r = 0.92$) content.[32] Defatting decreased T_o, T_p, and T_c about 2°C, and increased ΔH (Table 15.3).[37] Annealing caused an increase in gelatinization temperature and in gelatinization enthalpy.[15] The annealing temperature was found to be important, but not the moisture content of the starch undergoing annealing.

The high lipid content of oat starch is reflected in the high value of the transition enthalpy (ΔH_{cx}) measured for the amylose–lipid complex (Table 15.4). This transition is reversible, and on a rerun greater ΔH_{cx}, as well as T_{cx} (endotherm peak temperature), was found.[26] When the oat starch was defatted, the endotherm assigned to the amylose–lipid complex disappeared from the DSC thermogram.[26] Extraction at room temperature with 1-propanol:water (3:1 v/v) did not influence the thermogram of the starch, but if the starch was refluxed in the same solvent, a decrease in ΔH and a complete elimination of the endotherm ascribed to the amylose–lipid complex occurred.[28] For oat starches with a lipid content of 1.1–1.7%, it was possible to

Table 15.3 DSC parameters for gelatinization of oat starch in excess water

T_o, °C	T_p, °C	T_m, °C	ΔH, J/g starch	Reference
60.4	64.1	70.0	10.1	15
	66.8±0.2		9.13±0.14	28
55.5–62.4			8.6–9.2	8
56.1–69.5			10.1–12.9	32
	66.8		9.13	26
	57.8–61.6		9.4–10.6	29
52	58.3	64	9.2	36
61	66	73	10.4	22[a]
(58)	(64)	(71)	(11.1)	
51.1	58.1	65.2		35
56.0–63.5	59.5–66.0	65.5–74.0	12.4–14.6[b]	33

[a]Values in parentheses were obtained for defatted starch
[b]Expressed on amylopectin content

Table 15.4 DSC parameters for the transition of the amylose–lipid complex of oat starch in excess water

T_o, °C	$T_{c\infty}$, °C	$\Delta H_{c\infty}$, J/g starch	Reference
90.3–91.1		2.56–3.52	8
91.9–92.4		1.13–3.02	32
	102.3	3.57	26
	(104.3)[a]	(4.34)[a]	
	94.1–97.1	2.4–3.7	29
79	96	4.2	36
	106	0.84	22

[a]The values in parentheses were obtained on a rerun

reduce the lipid content to 0.7–0.9% with cold extraction (3:2:1 v/v chloroform:methanol:water) and to 0.1% by extraction with hot 3:1 v/v 1-propanol:water.[38] The proportion of amylose that is complexed with lipids has been estimated from DSC measurements and from iodine binding to be in the range 28–80%.[19,26,28,39] That the amylose and lipid exist in an inclusion complex in the native oat starch granule has been shown using ^{13}C CP/MAS-NMR.[40]

5. Retrogradation

When pastes of three oat starches isolated from different oat varieties were compared after storage (up to 28 days after cooking), no differences were found in their retrogradation behavior, even though the starches differed in gelatinization properties (Table 15.3).[8] When stored at the same conditions, the melting enthalpy of retrograded oat starch was considerably lower than that of retrograded maize, wheat, lentil or potato starch.[32,37] Oat starch retrogrades to a lesser extent than does high-amylose maize starch, i.e. a starch that has a lower content of amylopectin than oat starch.[29] The high level of lipids in oat starch have been suggested as an explanation.[32] When the oat starch was defatted, retrogradation increased, but it was still less than that of other starches.[29,37] It was also found that the lowest degree of retrogradation was obtained with the oat starch of highest lipid content.[29] The transition temperature for the amylose–lipid complex increased during storage, whereas the transition enthalpy decreased.[8]

6. Swelling Power and Amylose Leaching

Swelling power and solubility are given for some oat starches in Table 15.5. A negative correlation between lipid content and swelling power at 85°C has been found.[30] A correlation was found between swelling at 70°C and the amount of lipid-complexed amylose; the variety with highest amount of complexed amylose showed the lowest swelling, and *vice versa*.[33] Based on the absorption maximum of the iodine complex of the leached material, it was concluded that amylose and amylopectin co-leached from oat starch granules, perhaps also together with some intermediate material.[22] When the oat starch was defatted, swelling power decreased (from 46.3 g/g to 17.2 g/g at 95°C) and solubility increased (from 33.3% to 50.7%).[26] At the same time, λ_{max} of the iodine complex of the solubilized material increased from 598 nm to 626 nm, indicating that, after removal of lipids, the amylose leached out more easily. Similar results have been obtained in other investigations.[22,37–39]

Solubilization of oat starch with dilute, room temperature solutions of an acid and its digestibility by enzymes have been investigated.[22] The extent of acid-catalyzed hydrolysis during the first 12 days indicated that the amorphous regions of oat starch are more accessible to acids than the corresponding regions in wheat starch. However, the second stage of hydrolysis, i.e. when the crystalline regions are involved, was slower in oat starch than in wheat starch. On the twenty-fifth day, 83.1% of the oat starch was hydrolyzed.[22] The removal of lipids decreased the hydrolysis slightly. The swelling factor was found to decrease as a result of annealing, whereas leaching of amylose was unaffected.[15] Pancreatic α-amylase digested oat starch to a lesser extent than it did wheat starch (31.6% compared with 42.0%). The removal of lipids increased digestibility.[22] Freeze–thaw stability, measured as percentage synersis

Table 15.5 Swelling power and solubility of oat starches

Swelling power		Solubility		Reference
at 85°C,%	at 95°C,%	at 85°C,%	at 95°C,%	
8.7–9.6	27.8–34.8	4.1–6.0	33.5–43.3	30
	23.7[a]		30.8[a,c]	26
	17.5[b]		21.7[b,c]	
8.3–10.6[d]	28.0–28.5	3.3–13.3[d]	51.6–63.5	29

[a]High heating rate
[b]Low heating rate
[c]Sediment in g/g
[d]80°C

after up to four freeze–thaw cycles, was better for oat starch pastes than for wheat starch pastes, and it was further improved after defatting the oat starch.[22]

7. Rheological Properties

The method of preparation of starch dispersions for rheological measurements has a great influence on the behavior of pastes. Important parameters are initial temperature, rate of temperature increase, highest cooking temperature, stirring rate, cooking time, pH and presence of salt.[41] Measurements on instruments such as the Ottawa Starch Viscometer,[10,19] rotational viscometers[13,26,31] and dynamic rheometers[13,25,29] have been used. The Ottawa Starch Viscometer provides graphs similar to those obtained with a Brabender ViscoAmylograph, allowing measurements of viscosity during heating, cooking and cooling cycles, but the heating rate is much faster. The viscosity measurements performed by rotational viscometers must be taken at elevated temperatures, preferably above 60°C, to avoid interference of amylose gel formation. Viscoelastic measurements with dynamic rheometers allow the monitoring of structure formation during heating and cooling without destroying the structure.

Flow Properties

The viscosity of oat starch subjected to a controlled pasting procedure was studied with a rotational viscometer as a function of increasing and thereafter decreasing shear rate.[26,42] The area of up and down curves is related to the degree of thixotropy. In contrast to other starch dispersions, oat starch dispersions showed marked thixotropic behavior at 70°C,[26] and the degree of thixotropy was dependent on starch concentration; the higher the starch concentration, the higher was the area between the up and down curves. The structural basis for the behavior is related to gel formation and is discussed in more detail below. These results are in agreement with those reported by Paton[10,19] in which the Ottawa Starch Viscometer was used. Dilute acetic acid had a great affect on the pasting behavior of an oat starch suspension.[43] In the presence of acid, the unusual cooking peak at 80°C disappeared. Salts (0.1 M) had a similar effect.

Viscoelastic Behavior

The viscoelastic transitions occurring during cooling of oat starch pastes are dependent on heating temperatures.[13,25] The storage modulus (G′) increased and the phase angle (δ) decreased during cooling of 10% oat starch dispersions preheated at 95°C. In contrast to other cereal starches, two transitions have been observed: one below 90°C and the other below 30°C. No rheological changes occurred below 90°C when oat starch dispersions were preheated at only 90°C. This occurs because, at 90°C, the granule structure is primarily intact and most of the amylose is located inside the granule. At 95°C, the granule structure breaks down and amylose and amylopectin are co-leached from the granule.[25,26] It is highly probable that the transition below 90°C is related to an amylose–lipid complex in the continuous phase. The increase in G′ and the decrease in δ can thus be attributed to the formation of highly-ordered junction zones of helical inclusion complexes. It has been found that the oat variety with the highest oil content had the highest G′.[29] The slight increase in G′ below 30°C is related to gelation of amylose, the increase being small in comparison to that of wheat and barley starch pastes of the same concentration.

Mild acid modification of oat starch had a great affect on the viscoelastic behavior of pastes during cooling. Acid-modified oat starch underwent one transition in viscoelastic behavior; below 40°C, G′ increased and δ decreased, due to gelation of amylose. The transition below 90°C typical for native oat starch was not observed after acid modification. This finding is in agreement with that of Paton,[43] who found that treatment with acid almost eliminated the exceptionally high viscosity measured at ~80°C for native oat starch.

VI. References

1. Schrickel DJ. In: Webster FH, ed. *Oats, Chemistry and Technology*. St. Paul, MN: American Association of Cereal Chemists; 1986:1.
2. Anderson JW, Chen W-JL. In: Webster FH, ed. *Oats, Chemistry and Technology*. St. Paul, MN: American Association of Cereal Chemists; 1986:309.
3. Anderson JW, Seisel AE. In: Furda I, Brine CJ, eds. *New Developments in Dietary Fiber: Physiological, Physicochemical and Analytical Aspects*. New York, NY: Plenum Press; 1990:17.
4. Anderson JW, Deakins DA, Bridges SR. In: Kritchevsky D, Bonfield C, Anderson JW, eds. *Dietary Fiber: Chemistry, Physiology and Health Effects*. New York, NY: Plenum Press; 1990:339.
5. Lapveteläinen. Barley and Oat Protein Products from Wet Processes: Food Use Potential [Ph.D. Thesis] Finland: University of Turku; 1994.
6. Lim WJ, Liang YT, Seib PA, Rao CS. *Cereal Chem*. 1992;69:233.
7. MacArthur LA, D'Appolonia BL. *Cereal Chem*. 1979;56:458.
8. Sowa SM, White PJ. *Cereal Chem*. 1992;69:521.
9. Wilhelm E, Kempf W, Lehmussaari A, Caransa A. *Starch/Stärke*. 1989;41:372.
10. Paton D. *Stärke*. 1977;29:149.
11. Morrison WR, Milligan TP, Azudin MN. *J. Cereal Sci*. 1984;2:257.
12. Inglett GE. *Food Technol*. 1990;44:100.
13. Virtanen T, Autio K, Suortti T, Poutanen K. *J. Cereal Sci*. 1993;17:137.
14. Lim WJ, Liang YT, Seib PA. *Cereal Chem*. 1992;69:237.
15. Hoover R, Vasanthan T. *J. Food Biochem*. 1994;17:303.
16. Hoover R, Vasanthan T. *Carbohydr. Res*. 1994;252:33.
17. Verhoeven T, Fahy B, Leggett M, Moates G, Denyer K. *J. Cereal Sci*. 2004;40:69.
18. Gibinski M, Kowalski S, Sady M, Krawontka J, Tomasik P, Sikora M. *J. Food Eng*. 2006;75:407.
19. Paton D. *Starch/Stärke*. 1979;31:184.
20. Bechtel DB, Pomeranz Y. *Cereal Chem*. 1981;58:61.
21. Jane J-L, Kasemsuwan T, Leas S, Ia A, Zobel H, Il D, Robyt JF. *Starch/Stärke*. 1994;46:121.
22. Hoover R, Vasanthan T. *Carbohydr. Polym*. 1992;19:285.

23. Juszczak L, Fortuna T, Krok F. *Starch/Stärke*. 2003;55:8.
24. Williams MR, Bowler P. *Starch/Stärke*. 1982;34:221.
25. Autio K. *Food Structure*. 1990;9:297.
26. Doublier J-L, Paton D, Llamas G. *Cereal Chem*. 1987;64:21.
27. Youngs VL. *J. Food Sci*. 1974;39:1045.
28. Paton D. *Cereal Chem*. 1987;64:394.
29. Gudmundsson M, Eliasson A-C. *Acta. Agric. Scand.* 1989;39:101.
30. Wang LZ, White PJ. *Cereal Chem*. 1994;71:443.
31. Wang LZ, White PJ. *Cereal Chem*. 1994;71:263.
32. Wang LZ, White PJ. *Cereal Chem*. 1994;71:451.
33. Hoover R, Smith C, Zhou Y, Ratnayake RMWS. *Carbohydr. Polym*. 2003;52:253.
34. Banks W, Greenwood CT. *Stärke*. 1967;19:394.
35. Mua J-P, Jackson DS. *Starch/Stärke*. 1995;47:2.
36. Eberstein K, Höpcke R, Konieczny-Janda G, Stute R. *Starch/Stärke*. 1980;32:397.
37. Hoover R, Vasanthan T, Senanayake NJ, Martin AM. *Carbohydr. Res*. 1994;261:13.
38. Gibinski M, Palasinski M, Tomasik P. *Starch/Stärke*. 1993;45:354.
39. Shamekh S, Forssell P, Poutanen K. *Starch/Stärke*. 1994;46:129.
40. Morrison WR, Law RV, Snape CE. *J. Cereal Sci*. 1993;18:107.
41. Radley JA. In: Radley JA, ed. *Examination and Analysis of Starch and Starch Products*. London, UK: Elsevier; 1976:91.
42. Hoover R, Vasanthan T. *J. Food Biochem*. 1994;18:67.
43. Paton D. *Cereal Chem*. 1981;58:35.

Barley Starch: Production, Properties, Modification and Uses

16

Thava Vasanthan[1] and Ratnajothi Hoover[2]

[1]*Department of Agricultural, Food and Nutritional Sciences, University of Alberta, Edmonton, Canada*
[2]*Department of Biochemistry, Memorial University of Newfoundland, St. John's, Canada*

I. Introduction	601
II. Barley Grain Structure and Composition	602
III. Barley Starch	604
1. Isolation and Purification	604
2. Chemical Composition of Barley Starch	605
3. Granule Morphology	607
4. X-Ray Diffraction and Relative Crystallinity	607
5. Gelatinization	607
6. Swelling Factor and Amylose Leaching	610
7. Enzyme Susceptibility	612
8. Acid Hydrolysis	613
9. Pasting Characteristics	615
10. Retrogradation	618
11. Freeze-Thaw Stability	619
12. Chemical Modification	619
13. Physical Modification	621
IV. Resistant Barley Starch	621
V. Production and Uses of Barley Starch	623
VI. Conclusion	625
VII. References	625

I. Introduction

Barley *Hordeum vulgare* is a grass belonging to the family *Poaceae*, the tribe *Triticeae* and the genus *Hordeum*.[1] It is now the fourth most important cereal crop of the world after wheat, corn and rice. Historically, barley was the major food grain in many parts of the world. This grain was used as a staple food in the Near East several thousand years ago and was the chief form of nourishment of Greeks in Homeric times (800 BC). *Rieska*, an unleavened barley bread, was the earliest of all breads made in Finland. In medieval England, bread made from barley formed the staple diet of the peasant and poor people; the nobles ate wheaten bread. It is only since the beginning of this century that barley has been systematically replaced by wheat, rice and corn in human foods. Barley production in colonial America was predominantly for malt to produce beer. However, at present, most barley produced in the American continent is used for livestock feeding; very little is used in human foods and for further value-added processing. Pot or pearled barley, prepared by gradual removal of hull, bran and germ by abrasive action in a stone mill, is used in breakfast cereals, soups, stews, bakery products and baby foods. Barley malt extracts and syrups are included in fermented bakery products to enhance soluble sugar, protein and α-amylase in the dough; this improves texture, loaf volume, flavor and color of the final baked product. Other products of barley milling such as bran, grits and flour (native and malted) are now commercially available.

The world production of barley was 132 million metric tons in 2003.[2] Some of the major barley producing countries are Canada, the United States, France, Spain, Finland, Denmark, Germany, Turkey, China, England and the Russian Federation. Canada is the largest barley producer on the American continent, with production of 13.5 million metric tons, accounting for 10.2% of the total world production.[2] In Canada the prairie provinces, Alberta and Saskatchewan, contribute up to 88% of the national barley production.[3] The average Canadian yield is about 4 tons/hectare.

II. Barley Grain Structure and Composition

A longitudinal section of a barley grain[6] is illustrated in Figure 16.1. Barley grain structure and most of the basic components are relatively similar to that of other cereals. The hull or husk constitutes up to 10–13% of the dry weight of barley grain. The hull is rich in cellulose, insoluble arabinoxylans, lignin, polyphenols and minerals.[5] The lemma and palea of the hull covers the caryopsis (kernel) which includes pericarp, seed coat and endosperm. The pericarp contains cellulose, lignin and arabinoxylan. The endosperm, which contributes to 75–80% of the total kernel weight, is rich in starch which is embedded in a protein matrix.[4] The cell walls in the endosperm are mainly composed of β-glucan (70–75%), arabinoxylans (20–25%) and protein (5–6%) with minor amounts of glucomannans, cellulose and phenolic compounds.[3,5] The aleurone, the outer most layer of the endosperm comprising of arabinoxylans (60%), β-glucan (22%) and proteins (16%) contributes to about 5–10% of the total kernel weight. The embryo (2–4% kernel weight) is rich in lipids (13–17%), protein and amino acids (34%), sucrose and raffinose (5–10%), arabinoxylan, cellulose and pectin (8–10%) and ash (2–10%).[4] Barley grains have traditionally been covered with hulls or husks, where the lemma and palea adhere to the caryopsis and do not thresh freely. Recently, hull-less barleys (HB) have been developed through genetic engineering and are popular as food or feed grain, because they require no pearling and thus retain nutrients that would otherwise be lost during processing.[6–8] An average

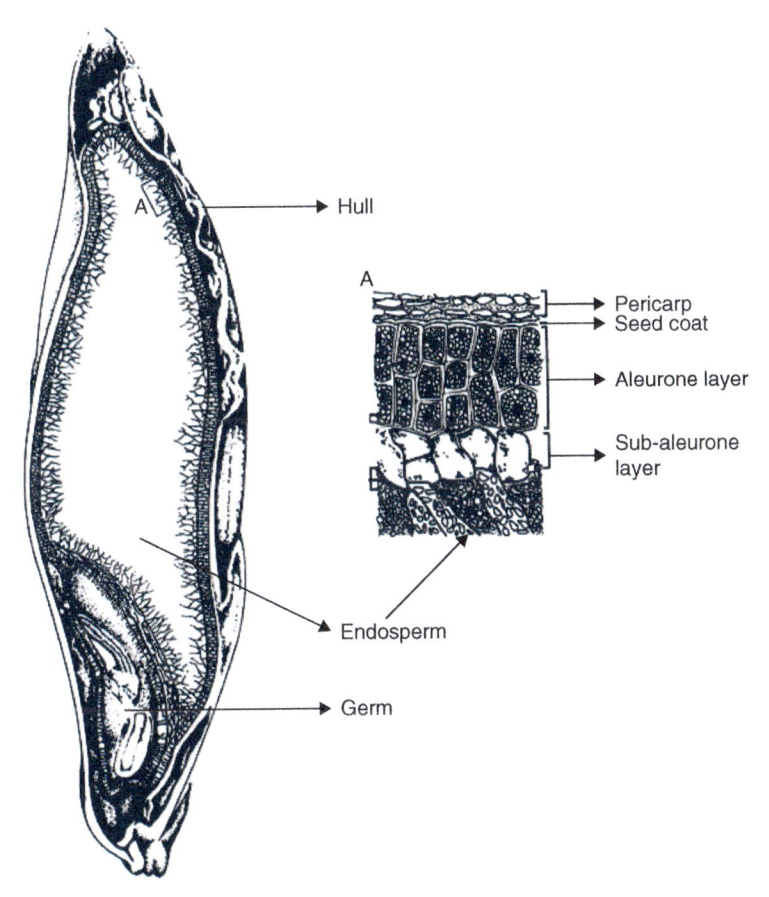

Figure 16.1 A longitudinal section of a barley grain. (Modified from Newman and Newman[6] with permission)

Table 16.1 Average chemical composition of hulled and hull-less barley grains[10–13,115]

Component	Hulled (%, db)	Hull-less (%, db)
Starch	53.7	59.7
Cellulose	4.1	2.0
Arabinoxylans	6.5	4.5
Lignin	2.0	9.0
β-glucan	5.2	5.6
Dietary fiber	18.6	13.8
Lipid	2.4	2.7
Protein	15.9	16.5
Ash	2.8	2.1

composition of hulled and hull-less barley grains[10–13,115] is presented in Table 16.1. Barley grains contain an average 14.5% moisture. Hulled barley has lower protein but higher fiber and ash content than hull-less. Hull-less barley contains more starch and lignin than hulled. The lipid content is nearly similar in both types.

III. Barley Starch

1. Isolation and Purification

Various laboratory methods[14–35] involving grain steeping, blending followed by screening, deproteinization and recovery of starch by centrifugation, have been used for isolation of starch from barley grains. The methodologies differ with respect to the media in which the grains are steeped (water, acid, mercuric chloride) prior to milling, isolation technique (dry-milling, wet-milling, dry-milling followed by wet-milling), enzyme source used for increasing starch yield (protease, cellulose, xylanase, lichenase, glucanase), screen size, centrifugal speed and the chemicals used for protein removal (sodium hydroxide, caesium chloride). Compared to maize starches, barley starch isolation by wet extraction presents a few difficulties. Beta-glucan, (1→3 and 1→4) a major cell wall polysaccharide of barley, produces high viscosities in aqueous solutions and impairs separation of starch by screening and subsequent centrifugation.[14]

One of the earliest methods of barley starch isolation was proposed by Greenwood and coworkers.[15–17] They used aqueous mercuric chloride solution (0.01 M) for steeping, nylon mesh (75 µm) for screening and repeated toluene washing deproteinization. McDonald and Stark[18] outlined a process for barley starch extraction that was based on a method developed by Morrison et al.[19] The scheme involved acid (pH = 2) steeping of cracked grains followed by neutralization, gentle grinding and screening through a nylon mesh (75 µm). The crude starch was recovered by centrifugation of the filtrate and by scraping of the brown proteinaceous layer. Small starch granules in the brown proteinaceous layer were recovered by protease treatment followed by toluene shaking, and then added back to the main stock starch. Tester and Morrison[20] isolated starch from waxy and non-waxy barley grains by steeping in water, maceration and then purification by centrifugation in the presence of caesium chloride (CsCl), CsCl was used to remove cellular material, storage proteins and some starch granule surface proteins. They obtained yields of 36% and 44%, respectively, for waxy and non-waxy barley starches. Zheng and Bhatty[14] isolated starch from hull-less barley grains (varying widely in amylose content) using an enzyme-assisted wet separation process. The enzyme cocktail consisting of cellulase, endo (1→3), (1→4)-β-D-glucanase and xylanase separated starch, protein, β-glucan, bran and tailings resulting in starch yields in the range 44–54% of total dry matter. The purity of the starch was 98.0%. Li et al.[115] isolated starches from 10 cultivars of hull-less (HB) barley grains following the wet-milling procedure of Wu et al.[30] They reported average yields and extraction efficiencies of 44.4% and 70.9%, respectively. Waxy HB barley genotypes with high β-glucan content gave low starch yields (42.1%) and extraction efficiencies (68.0%). Isolated starches contained 95–98% starch with β-glucan contents in the range 0.01 to 0.06%. Vasanthan and Bhatty[21] reported a procedure for pin-milling and air-classification of hull-less barley to obtain fractions rich in starch, protein and β-glucan. The starch-rich fraction, containing 77.78% starch, was subjected to a wet-processing method to obtain pure starch that contained mainly large granules. The starch yield from this dry- and wet-processing was ~34%, which is similar to that of the conventional wet-milling procedure. Barley starch has been shown to exist in two

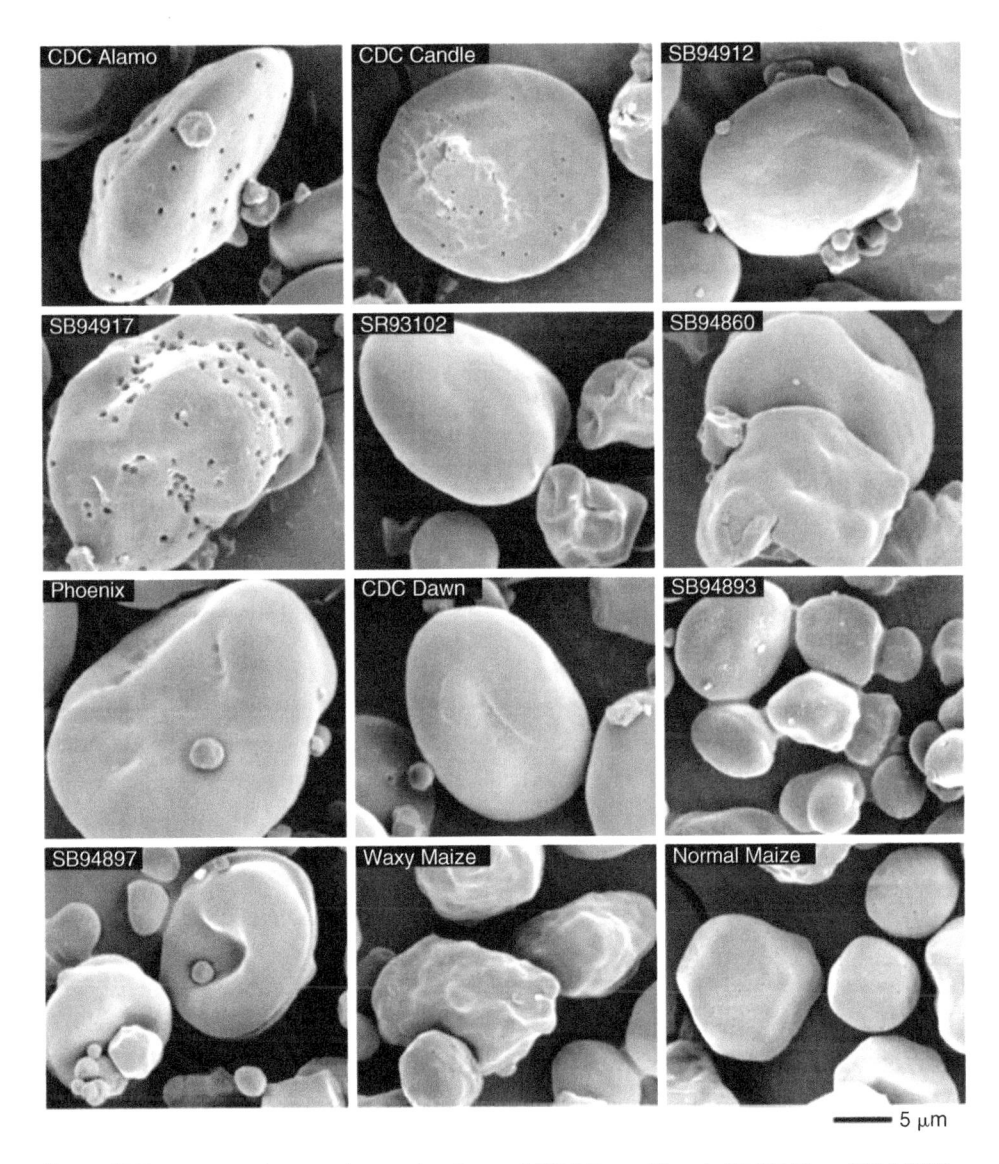

Figure 16.2 Scanning electron micrographs of normal (CDC Dawn, Phoenix, SR 93102, and SB 94860); waxy (CDC Alamo, CDC Candle, SB 94912, and SB 94917); and high-amylose (SB 94893 and SB 94897) barley starches.[115] (Reproduced with permission)

clearly defined populations[21] large, lenticular (A-type) and small, irregular shaped (B-type) granules (Figure 16.2). The A- and B-type granules have been fractionated by decantation[22] and by pin-milling and air-classification.[21]

2. Chemical Composition of Barley Starch

Carbohydrate Component

Amylose and amylopectin are the main carbohydrate components of the starch granule. Amylose is an essentially linear α-1, 4-D-glucan chain. About 25–55% of the total amylose (mole basis) in starch granules is branched with 4–18 branch points per molecule and branch chain length of 4 to over 100.[36] The molecular weight and degree of polymerization (DP) of amylose are usually in the range of 10^5–10^6 Da and 700–5000 anhydro-glucose units, respectively. Amylopectin is highly-branched and composed of thousands of linear α-1, 4-D glucan unit chains and 4–5.5% of α-1, 6-glucosidic bonds.[36] The average chain length and average molecular weight of amylopectin is 17–31 anhydro-glucose units[37] and 7.0×10^7–5.7×10^9 Da,[48] respectively. The amylose content of barley starches ranges from 0 to 46%.[14,21,33,38,39] The length of the amylopectin B-chains,[40] size of the crystalline lamella[41] and formation of amylose–lipid complexes[42] increases with increase in barley amylose content. The number average degree of polymerization (DP_n), average chain length ($\overline{CL}$), average number of chains per molecule (NC), weight average molecular weight (M_w) and radii of gyration (R_g) of barley amylose are in the range 940–1000,[15,33,43] 115–530,[44–45] 6–10,[45] 2.73×10^6–5.67×10^6 [32] and 64–148[32] respectively, whereas the β-amylolysis limit,[15,17,46] iodine affinity[15,17,33] and limiting viscosity number[15,17,46] are in the range 70–95%, 17.4–20% and 240–391 g/ml, respectively.

Morrison et al.[46,57] have shown that amylose in barley starches exists partially as lipid complexed amylose (L·AM) with a lysophosphatidyl choline (LPL) to L·AM ratio of 1:7 and partially as lipid-free amylose (F·AM). These authors have shown that L·AM and F·AM in waxy barley starches range from 0.8–4.0% and 0.9–6.4%, respectively, whereas the corresponding values for non-waxy barley starches are 5.1–7.2% and 23.1–25.9% respectively. In high- and low-amylose barley starches, L·AM is 12.1% and 7.1%, respectively in the small granules, and 4.1% and 3.2%, respectively, in the large granules.[42] Debranched amylopectins of normal, waxy and high-amylose barley starches exhibit similar chain length distribution profiles, all showing trimodal distributions of short chains (mostly A-chains), intermediate chains (mostly B_1- and some A-chains) and long chains (mostly B_2 and B_3 chains).[27,32,47,49,50,115] There is conflicting information in the literature with respect to barley amylopectin structure due to different methods (HPSEC-MALL-RI,[32] MALDI-MS,[115] HPAEC-ENZ-PAD,[27] gel permeation-HPLC[50]) being used for detection of debranched chains. The average chain length ($\overline{CL}$) and the degree of branching is in the range 17.6–22.6[27,33,45,50,115] and 4.2–5.7,[40,115] respectively. Song and Jane[27] and Suh et al.[51] have reported that in barley starches, the proportion of DP 6–9, DP 6–12, DP 13–24, DP 25–36 and DP >37 are in the range 3.1–5.3, 16.5–21.6, 40.9–47.5, 14.6–17.9 and 17.8–23.7%, respectively. The highest detectable DP has been shown to be in the range 67–133.[27,32,115] You and Izydorczyk[32] reported that the highest DP value was observed for high-amylose starch followed by normal, waxy and zero amylose. However, Song and Jane[27] reported that normal starch contained the highest DP followed by high-amylose and waxy starches.

Non-carbohydrate Components
Proteins, lipids and phosphorus are the minor non-carbohydrate components of barley starch. These components have been shown to influence hydration rates, thermal properties, retrogradation characteristics, rhelogical characteristics, susceptibility towards α-amylolysis and processing, and product qualities of starch hydrolysates.

The protein content of purified barley starches ranges from 0.1–0.4% db,[18,52] part of which cannot be removed even with stringent starch purification methods, and thus is considered an integral part of the starch granule interior.[52,53] Integral proteins have been shown to be present in the central and peripheral regions of the granule.[52] Starch lipids are present on the granule surface and interior.[52,54,55] The surface and internal lipid contents range from 0.1–0.2% and 0.3–1.7%, respectively.[115] The internal lipid (mainly lysophospholipids) content of small barley starch granules is higher than that of large granules,[42] and is proportional to the amylose content.[56] Phosphorus in barley starch occurs mainly in the form of phospholipids, and ranges from 9–37, 47–66, 75–106 mg/100 g in waxy, normal and high-amylose barley starches, respectively.[58]

3. Granule Morphology

Barley starch granules[115] consist mainly of a mixture of large lenticular granules (10–30 μm) and smaller irregularly shaped (<6 μm) granules (Figure 16.2). The large granules constitute 10–20% of the total number of starch granules and 85–90% of the weight of total starch mass; the small granules constitute 80–90% by number and 10–15% by weight.[4] The particle size distribution of barley starch granules has been reported in many investigations.[22,32,44,56,59,115] The wide variation observed between the results of these studies is probably due to genotypic and environmental effects, and to differences in starch extraction methodologies and size measurement techniques. Scanning electron microscopy[115] shows the presence of pin holes on the granule surface of high-amylose (up to 0.9 μm in diameter) and normal (<0.9 μm) barley starches (Figure 16.2). Transmission electron microscopy[52] shows the presence of internal channels only in waxy and normal starch granules (Figure 16.3).

4. X-Ray Diffraction and Relative Crystallinity

Barley starches generally exhibit an A-type x-ray pattern (Figure 16.4) which is characteristic of cereal starches.[27,39,42,44,45] However, certain barley cultivars exhibit a mixed A + B-type pattern which is characteristic of legume starches.[39] The relative crystallinity (RC) of normal, high-amylose and waxy barley starches are in the range 20–36%, 37–42% and 33–44%, respectively.[27,39,42,44,45] The wide difference in RC among normal, waxy and high-amylose starches can be attributed to varietal differences and/or to starch moisture content. In normal, waxy and high-amylose maize starches, RC ranges[61,62] are 17.6–28.0%, 17.2–21.8% and 25–42%, respectively. Large granules of barley starch generally exhibit a higher RC than small granules (Figure 16.4). Irrespective of amylose content, barley starches exhibit a V-amylose–lipid complex peak centered at 20 °2θ[39,42] representing crystalline V-amylose–lipid complexes.[39] However, this peak has been shown to be absent in waxy maize starch.[61]

5. Gelatinization

Starch, when heated in the presence of excess water, undergoes an order–disorder phase transition called gelatinization over a temperature range characteristic of the

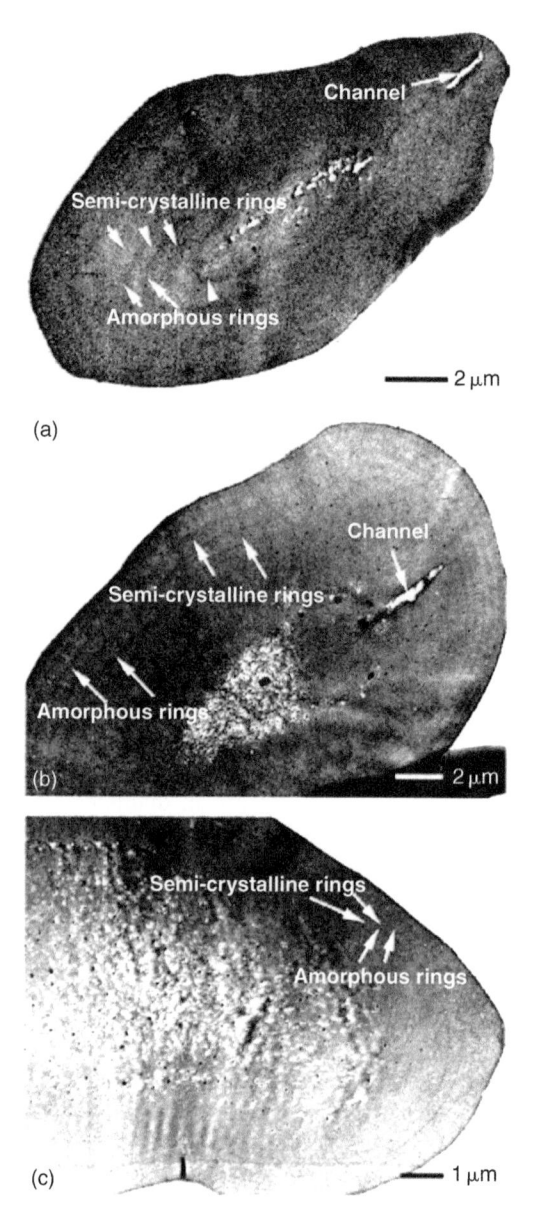

Figure 16.3 Transmission electron microscopy of (a) waxy; (b) normal; and (c) high-amylose barley starch granules.[52] (Reproduced with permission)

starch source. This phase transition is associated with the diffusion of water into the granule, water uptake by the amorphous background region, hydration and radial swelling of the starch granules, loss of optical birefringence, uptake of heat, amylose leaching and loss of molecular and crystalline order.[63–66] The gelatinization temperature of barley starches has been studied by polarized light microscopy and differential scanning calorimetry (DSC).[14,28,39,42,45,47,49,67–71] The DSC thermal characteristics of barley starch and those of its isolated granules are presented in Tables 16.2 and 16.3,

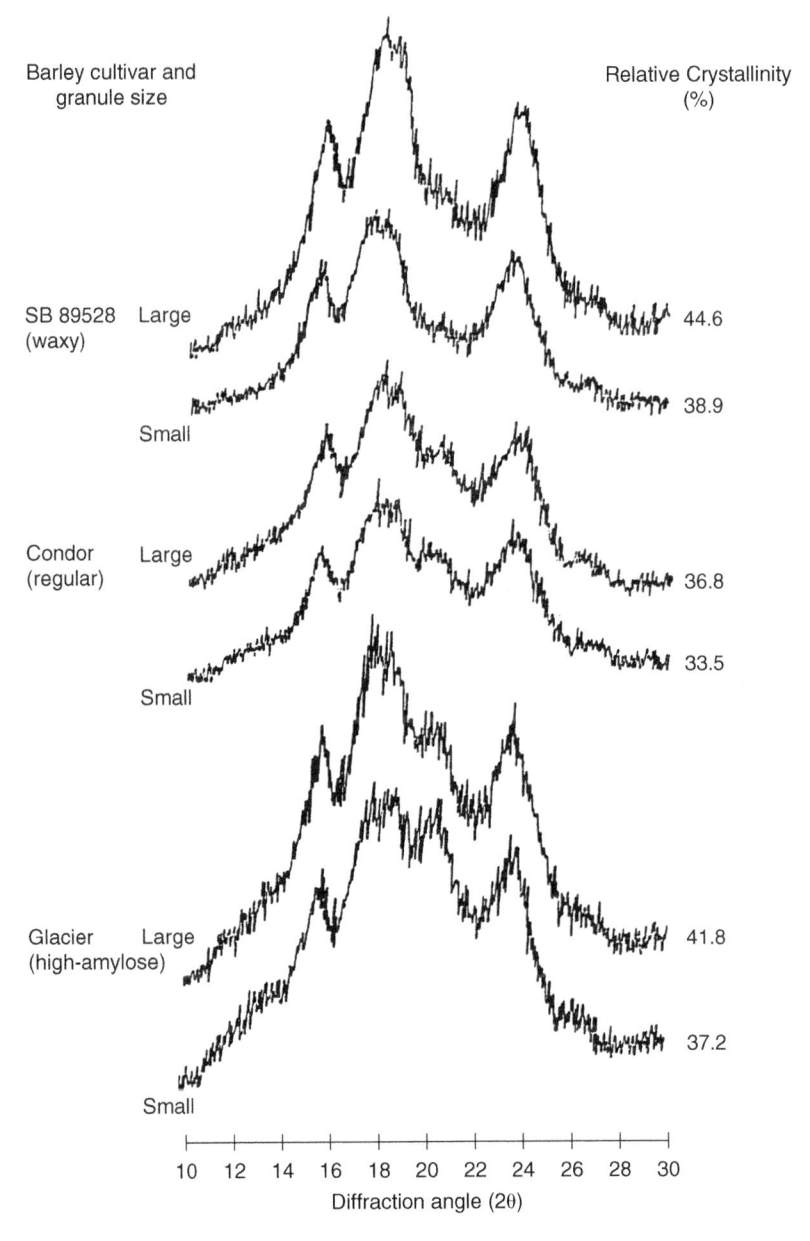

Figure 16.4 X-ray diffraction patterns and relative crystallinities of normal, waxy and high-amylose starches. (Adapted from reference 42 with permission)

respectively. The onset temperature (T_o) of gelatinization differs only marginally among the barley genotypes (Table 16.2). The magnitude of the peak temperature of high-amylose starch is higher than normal and waxy (waxy > normal). However, the conclusion temperature (T_c) is higher in waxy and high-amylose (waxy > high-amylose) than in normal starch (Table 16.2). The enthalpy of gelatinization follows the order: normal ~ waxy > high-amylose (Table 16.2). In isolated granules the

Table 16.2 Gelatinization parameters of barley starches[28,39,60]

Barley starch	Gelatinization transition temperature[d] (°C)			Enthalpy(J/g)
	Onset (T_o)	Peak (T_p)	End (T_c)	
Normal[a]	52.0–61.4	58.1–65.3	62.7–74.4	11.4–14.2
Waxy[b]	54.5–61.3	61.8–65.5	73.8–81.8	11.4–13.1
High-amylose[c]	53.0–61.0	62.0–68.0	74.3–76.6	6.6–12.2

[a]Amylose content 22.9–33.3%
[b]Amylose content 0–7.8%
[c]Amylose content 41–55%
[d]Starch:water 1:3

Table 16.3 Gelatinization parameters of isolated small and large granules of normal, waxy and high-amylose barley starches

Barley starch	Granule size	Gelatinization transition temperatures[a] (°C)			Range (°C) $T_c - T_o$	Enthalpy (ΔH) (J/g)
		T_o	T_p	T_c		
Normal	Small	54	61	70	16	6.8
	Large	54	58	67.5	13.5	7.9
Waxy	Small	60	64	76	16	9.9
	Large	59	62.6	72	13	11.4
High-amylose	Small	58	68.4	79.5	21.5	6.5
	Large	59.3	65.7	77.3	18	7.8

[a]Starch:water 1:3
From reference 42, reproduced with permission

difference in T_o between small and large granules of each genotype is only marginal (Table 16.3). Furthermore, small granules of each genotype exhibit higher T_p, T_c and $T_c - T_o$ and smaller ΔH than the large granules (Table 16.3). The higher T_p, T_c and the lower ΔH of small granules have been attributed to their higher content of lipid-complexed chains.[42] The wider $T_c - T_o$ of small granules probably reflected the presence of crystallites of varying stability.

The wide range in T_o, T_p, T_c and ΔH among genotypes (Table 16.2) could be attributed to differences in cultivars,[39,49,52] growth conditions,[49,58] and thermal and mechanical conditions employed during starch isolation.[66] DSC parameters of barley starches are generally lower than those reported for normal, waxy and high-amylose maize starches examined under identical conditions.[28]

6. Swelling Factor and Amylose Leaching

When starch is heated in excess water, the crystalline structure is disrupted (due to breakage of hydrogen bonds) and water molecules become linked by hydrogen bonding to the exposed hydroxyl groups of amylose and amylopectin. This causes an

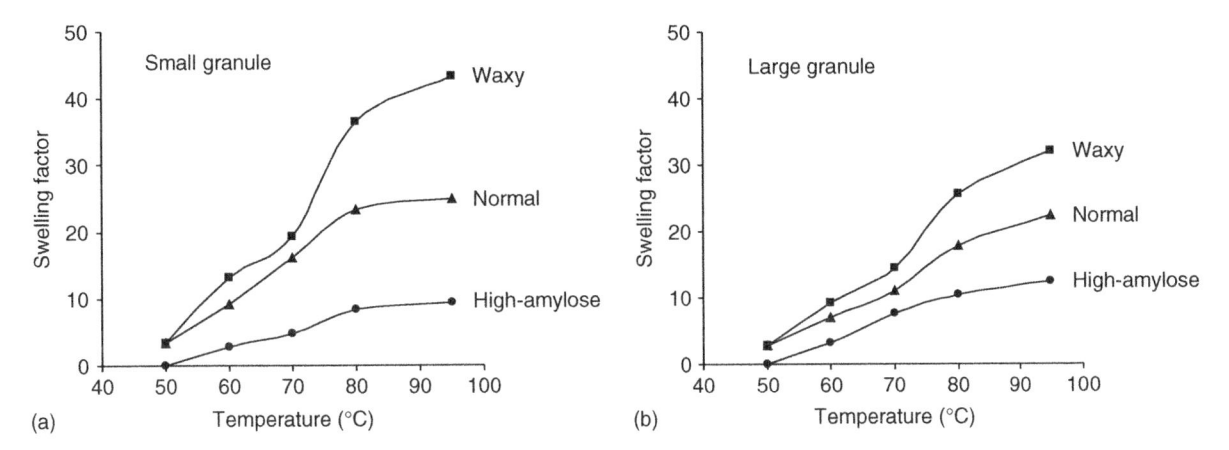

Figure 16.5 Swelling factor of small and large granules of normal, waxy and high-amylose starches in the temperature range 50–95°C: (a) small granules; (b) large granules. (Adapted from reference 42 with permission)

Table 16.4 Extent of amylose leaching (AML) in normal, waxy and high-amylose barley starches at 90°C

Barley source	Total amylose content (%)	AML (at 90°C) (%)
Normal	32.3	14.5
Waxy	7.8	3.2
High-amylose	43.7	27.4
	55.3	31.5

From reference 39, reproduced with permission

increase in granule swelling and amylose leaching. The extent of granule swelling is determined by measuring the swelling factor (SF), which is reported as the ratio of the volume of swollen granules to the volume of the dry starch.[72] SF has been shown to be influenced by: (a) granule size;[42] (b) amylose–lipid complexes;[39,42,72] (c) amylopectin structure;[73–75] and (d) extent of interaction between starch chains in the native granule.[76] The SF of barley starches has been shown to follow the order: waxy > normal > high-amylose.[39,42,77,78] This order is similar to that observed for maize starches.[42,79] In waxy and normal barley starches, small granules exhibit a higher SF than large granules (Figure 16.5).[42,44] However, in high-amylose barley starch, SF is lower in the small granules (Figure 16.5). The difference in SF between the small and large granules in the barley genotypes has been attributed to an interplay of the amount of lipid-complexed amylose chains and granule size.[42] When barley starches are heated in water at temperatures exceeding 55°C, AML follows the order high-amylose > normal > waxy (Table 16.4). A similar trend has also been reported for maize starches.[78] However, in isolated small and large barley starch granules, AML follows the order: high-amylose > normal > waxy and normal > high-amylose > waxy, respectively (Figure 16.6). The extent of AML has been shown

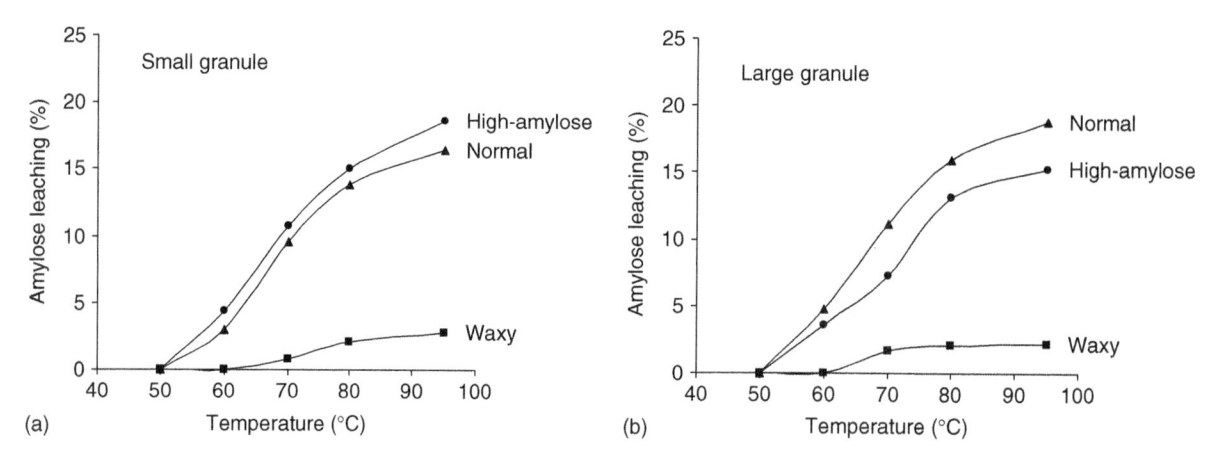

Figure 16.6 Amylose leaching (%) in the temperature range 50–90 °C: (a) small granules; (b) large granules. (Adapted from reference 42 with permission)

to be influenced by: (a) total amylose content; (b) extent to which starch chains are associated with each other and/or with the outer branches of amylopectin; and (c) amount of lipid-complexed amylose chains.[39]

Vasanthan and Bhatty[42] have shown that small granules of high-amylose barley starch contain more amylose–lipid complexes than the larger granules. The order of AML in small granules of barley genotypes (Figure 16.6a) suggests the total amylose content in high-amylose starch negates the influence of amylose–lipid complex content on AML. However, the order of AML in the large granules (Figure 16.6b), suggests that interaction between amylose–amylose and/or amylose–amylopectin are probably of a high order of magnitude in high-amylose starch such that it negates the effect of total amylose and lipid-complexed amylose chains on AML. Myllarinen et al.[80] have reported that at temperatures lower than at 90°C, amylose is preferentially leached out from large granules of normal barley starch, whereas more amylopectin is released from small granules under these conditions. This is probably due to the lower crystallinity and higher content of lipid amylose complexes in small granules.[42]

7. Enzyme Susceptibility

The susceptibility of barley starches to various amylases is important in syrup, feed and malting and brewing industries. Digestibility of barley starches by various amylases has been attributed to the interplay of many factors, such as surface area to volume ratio,[9,42,81] amylose–lipid complexes,[43] amylose content,[42] crystallinity,[42] presence of pin holes and channels in the native starch granule,[42] cultivar,[9,42,56,81,82] starch granule size,[42,44,56,81] amylase source[9,47] and proportion of small granules.[9]

Barley starches are hydrolyzed by porcine pancreatic α-amylase (PPA) in the order: waxy > normal > high-amylose.[9] The corresponding order for hydrolysis by α-amylase[9] from *Bacillus* species (BAA) and amyloglucosidase from *Aspergillus niger* (AAG) is waxy > high-amylose > normal (Table 16.5). The difference in hydrolysis[9] between normal and high-amylose barley starches is much smaller than

Table 16.5 Hydrolysis of barley starches by α-amylases and amyloglucosidase		
Enzymen source	Barley starch[5]	Hydrolysis[1] (%)
PPA[2]	Normal	94
	Waxy	97
	High-amylose	91
BAA[3]	Normal	29
	Waxy	78
	High-amylose	37
AAG[4]	Normal	30
	Waxy	67
	High-amylose	33

[1] After 72 hours' hydrolysis
[2] Porcine pancreatic α-amylase
[3] α-amylase from *Bacillus species*
[4] Amyloglucosidase from *Aspergillus niger*
[5] The total amylose content of normal, waxy and high-amylose used in this study was 23.2, 0 and 44.2%, respectively.[9] Reproduced with permission

that reported for normal and high-amylose maize starches.[82,83] Action of PPA[9] and BAA[9] on waxy barley starches results in the formation of large erosion areas (0.3 to 3 μm in diameter) which penetrate through several layers of the granule into the interior (Figure 16.7a,b). However, erosion areas are less pronounced in normal (Figure 16.7c,d) and high-amylose (Figure 16.7e,f) starches. Transmission electron microscopy[9] of PPA hydrolyzed residues shows the presence of large cavities (in the central region) and channels (in the peripheral regions) of all three barley genotypes (Figure 16.8). Channels in the periphery provided a pathway for the entry of enzymes into the granule interior. These channels may be derived from inherent surface pores/internal channels in starch granules and enlarged by PPA hydrolysis.[84] In isolated small and large granules from normal, waxy and high-amylose barley starches, the reactivity towards PPA[42] follows the order: normal > waxy > high-amylose (Figure 16.9). Isolated small granules from waxy and high-amylose starches are hydrolyzed to a greater extent than large granules (Figure 16.9).[42] However, the difference in the extent of hydrolysis between small and large granules of normal barley starch is less pronounced (Figure 16.9).[42] In all three genotypes, small granules are degraded more rapidly than large granules. Isolated small granules of normal, waxy and high-amylose starches exhibit surface erosion when treated with PPA. Whereas large granules of normal, waxy and high-amylose starches exhibit surface erosion and pin holes, and surface erosion, respectively.

8. Acid Hydrolysis

Barley starches exhibit a relatively high rate of acid hydrolysis during the first 10 days, followed by a slower rate thereafter.[28,39] The extent of degradation follows the order: waxy > normal > high-amylose (Table 16.6). This order is similar to that

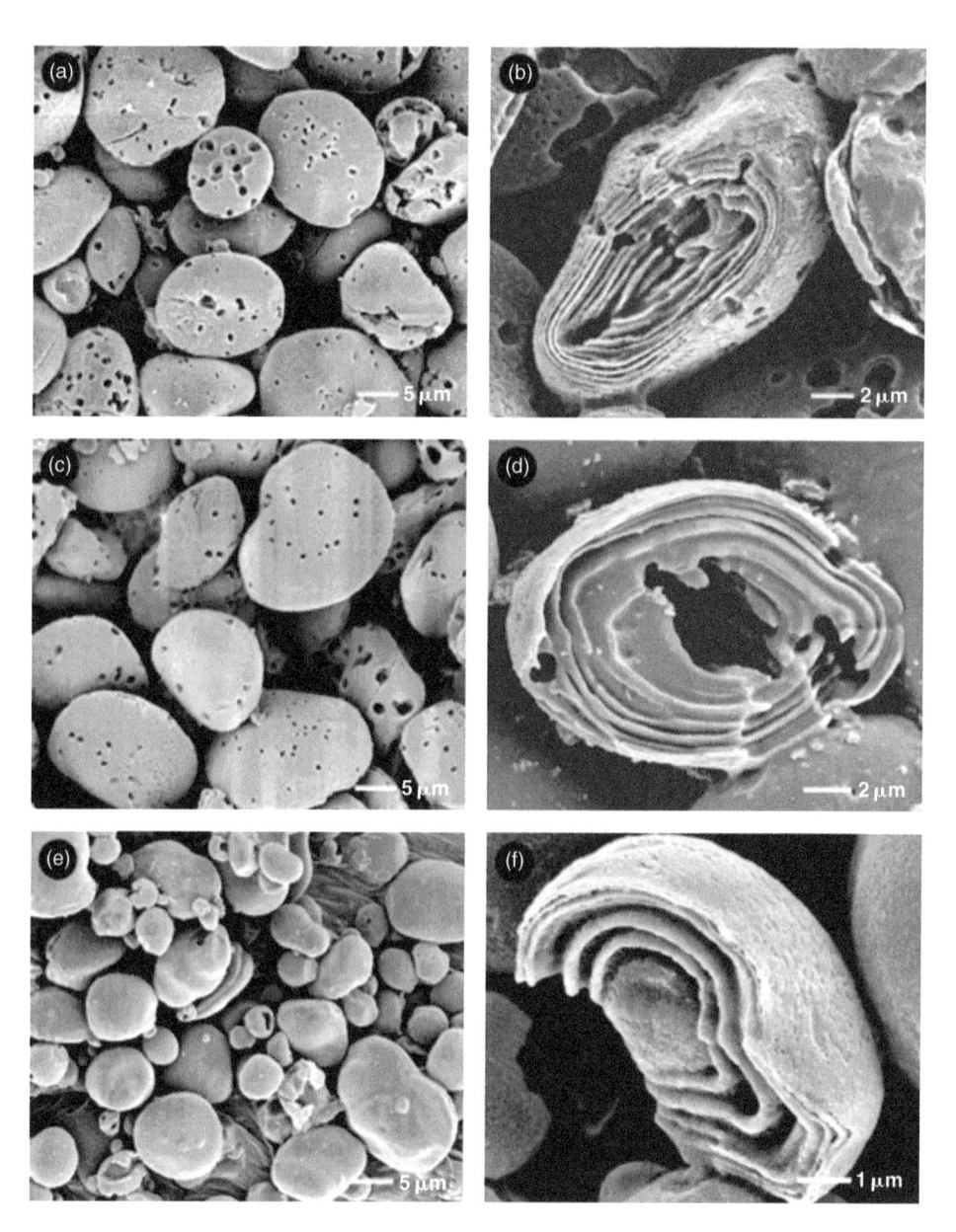

Figure 16.7 Scanning electron micrographs of waxy (a, b), normal (c, d) and high-amylose (e, f) barley starches, hydrolyzed by PPA at 37°C for 1 hour.[9] (Reproduced with permission)

reported for maize starches.[28,79] Among barley genotypes, small granules are hydrolyzed to a greater extent than large granules (Figure 16.10).[42] The difference in the extent of hydrolysis among genotypes and between small and large granules of each genotype have been attributed to the interplay of the following factors: (a) granule size;[42] (b) amount of lipid complexed amylose chains;[85] (c) double helical content;[85] (d) extent of interaction between amylose–amylose and amylose–amylopectin chains within the native granule;[28] and (e) extent of granular swelling.[27]

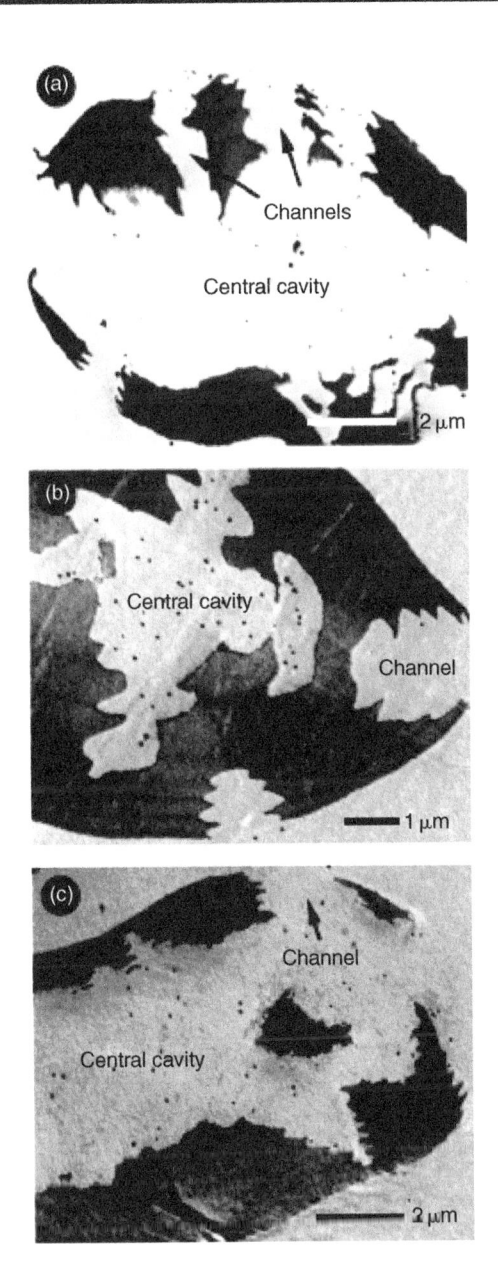

Figure 16.8 Transmission electron micrographs of waxy (a), normal (b) and high-amylose (c) barley starches, hydrolyzed by PPA at 37°C for 3 hours.[9] (Reproduced with permission)

9. Pasting Characteristics

The viscoamylography of barley starches provides information on their pasting characteristics and textural changes during cooking and cooling cycles. Marginal differences in the pasting characteristics have been reported among different barley varieties from the same type[86] and between small and large granule barley starches from the same variety.[42] However, greater differences have been reported[27,42,50,81,82]

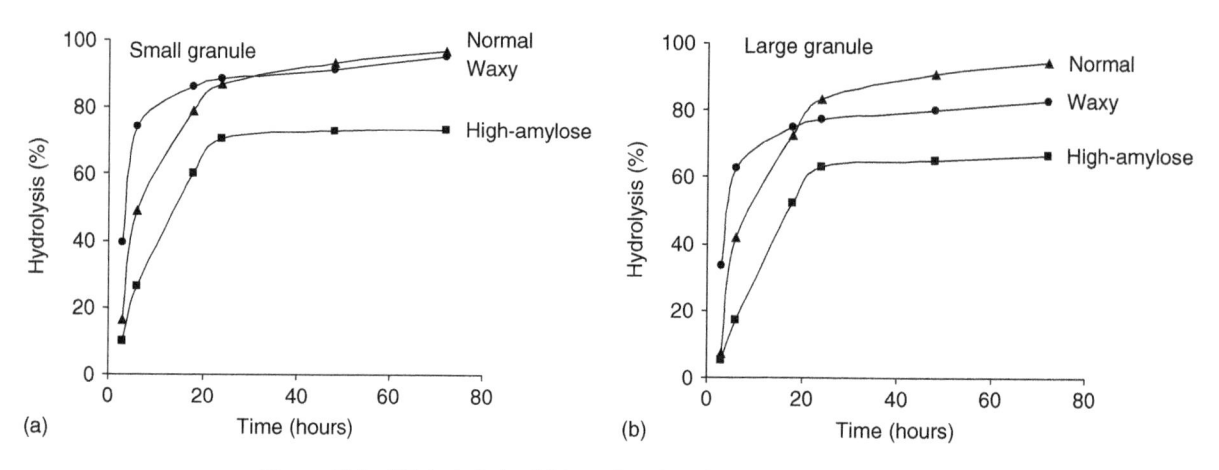

Figure 16.9 PPA hydrolysis of: (a) small; and (b) large granules of normal, waxy and high-amylose barley starches. (Adapted from reference 42 with permission)

Table 16.6 Acid hydrolysis of native barley starches with 2.2 N HCl at 35°C

Barley starch	Hydrolysis (%)[a]	
	10th day	18th day
Normal	28–35	42–44
Waxy	32–40	48–52
High-amylose	23–31	37–39

[a]Represents data from three cultivars of each genotype[28,39]

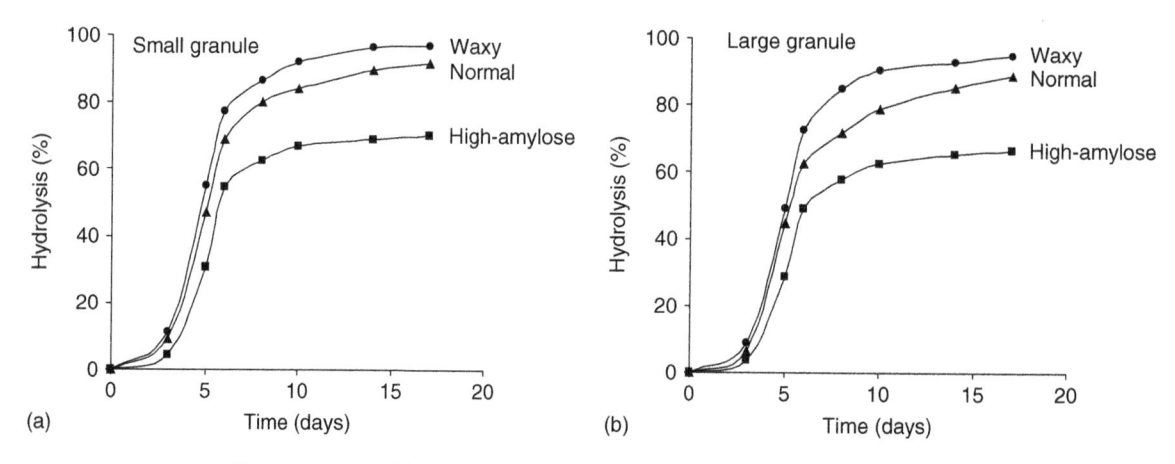

Figure 16.10 Acid hydrolysis of: (a) small; and (b) large granules of waxy, normal and high-amylose barley starches. (Adapted from reference 42 with permission)

among waxy, regular and high-amylose starches (Figure 16.11).[27] Substantially lower pasting temperature and higher peak viscosity have been recorded in waxy barley starch compared to normal or high-amylose types (Figure 16.11). The thermal and shear stabilities are highest in high-amylose followed by normal and waxy types (Figure 16.11). The pasting characteristics of isolated small and large granules from normal, waxy and high-amylose barley starches[42] are presented in Figure 16.12. Small granules of the three genotypes exhibit higher peak viscosities (Figure 16.12a) than the large granules (Figure 16.12b). In normal barley starch, setback is higher in the large granules than in the small granules, whereas setback is higher in the small granules of waxy and high-amylose starches than in the large granules. The difference in pasting characteristics among the genotypes and between small and large granules of a particular genotype have been attributed to the interplay of difference in: (a) granule size; (b) granule swelling; (c) extent of amylose leaching; (d) V-amylose–lipid complexes; (e) phospholipid content; (f) amylose content; and (g) granule crystallinity.[27,72,87,88] The pasting curves of the barley starches (Figure 16.12) are nearly similar to those reported for maize starches.[89]

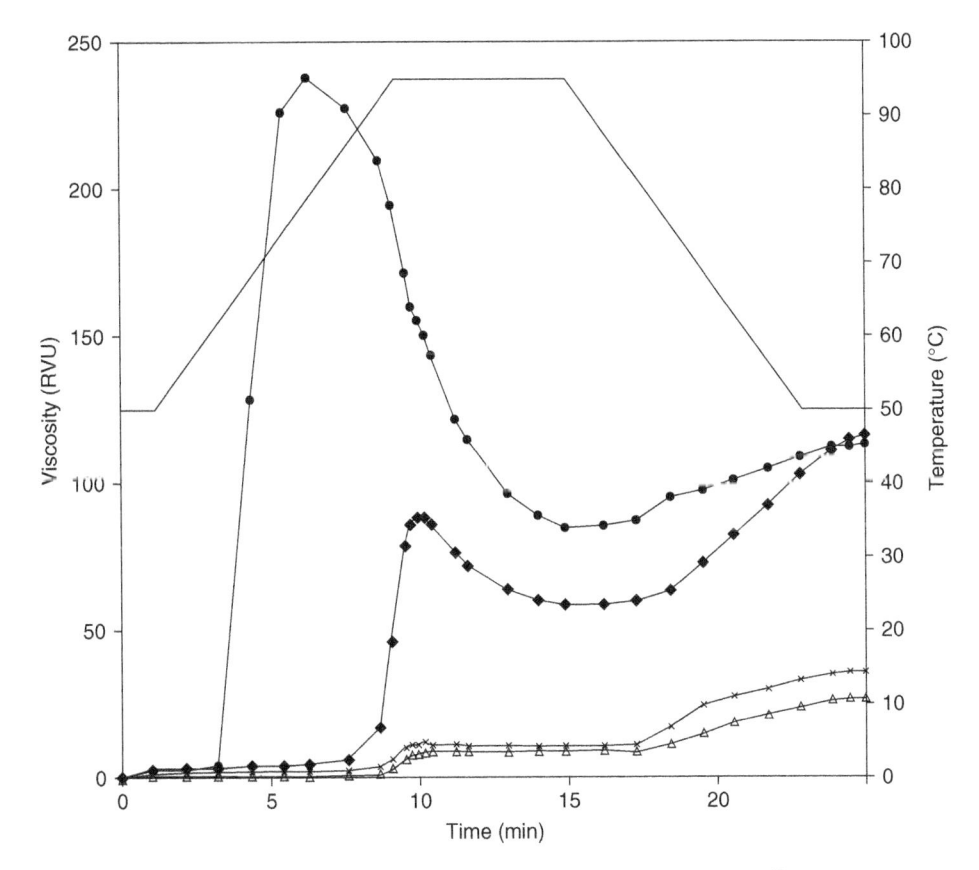

Figure 16.11 Pasting profiles of barley starches measured by a Rapid ViscoAnalyzer.[27] Waxy (●); normal (◆); (×) high-amylose; and (△) high-amylose hull-less. (Reproduced with permission)

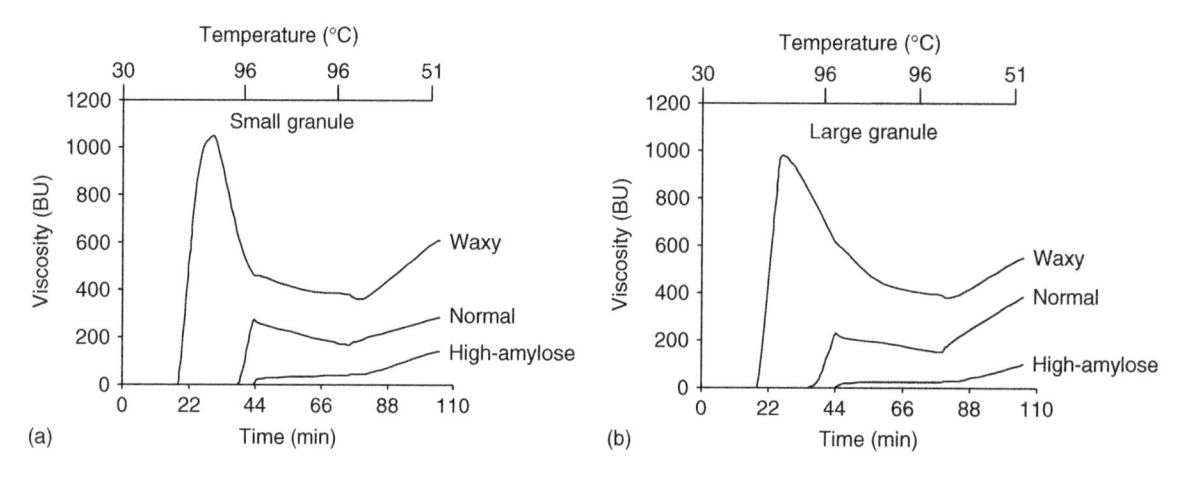

Figure 16.12 Pasting profiles of: (a) small; and (b) large granules of waxy, normal and high-amylose barley starches measured by Brabender Amylography. (Adapted from reference 42 with permission)

Table 16.7 Enthalpy of retrogradation (J/g amylopectin) of barley starches

Barley starch	ΔH_R^a	ΔH_R^b
Normal	10.3	10.6
Waxy	10.6	11.1
High-amylose	12.5	13.1

[a]Stored for 2 days at 6°C
[b]Stored for 4 days at 6°C
From reference 92, reproduced with permission

10. Retrogradation

Starch granules, when heated in excess water above their gelatinization temperature, undergo irreversible swelling resulting in amylose leaching into the solution. In the presence of high starch concentration this suspension will form an elastic gel on cooling. The molecular interactions (mainly hydrogen bonding between starch chains) that occur after cooling have been called retrogradation. These interactions are found to be time and temperature dependent. During retrogradation, amylose forms double helical associations of 40–70 glucose units,[90] whereas amylopectin crystallization occurs by association of the outermost short branches (DP = 15).[91] There is limited information on the retrogradation properties of barley starches. DSC measurement[92] has shown that the extent of retrogradation of barley genotypes follows the order: high-amylose > waxy > normal (Table 16.7) and that the rate of retrogradation of isolated small granules from normal barley starch is slower than that of larger granules (Figure 16.13).[93] Turbidity measurements have shown that small granules were the easiest to retrograde both in normal and waxy barley starches.[44]

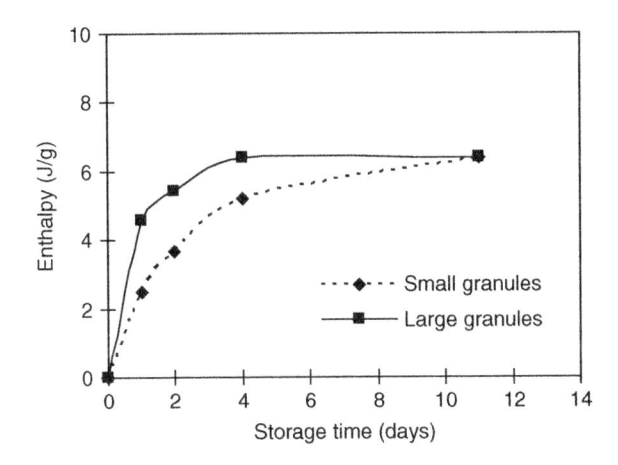

Figure 16.13 Retrogradation enthalpy of small and large granules of normal barley starch. (Adapted from reference 93 with permission)

11. Freeze–Thaw Stability

Freeze–thaw stability is an important characteristic of starches used in foods that are subjected to freezing and thawing. When a cooked starch paste is stored at low temperatures amylose and amylopectin chains reassociate and water is released (synerisis). Pastes of waxy starches remain stable for several weeks under refrigerated conditions, but the stability breaks down quickly after freeze–thaw.[94] The extent of synerisis is used as an index of freeze–thaw stability.[75,95,96] The freeze–thaw stability of barley starches of different genotypes has not been subjected to detailed study. Zheng et al.[95] showed by studies on waxy barley and waxy maize starches, that freeze–thaw (two cycles at 5% concentration) followed the order: waxy barley (0% amylose) > waxy barley (5% amylose) > waxy maize (1.1% amylose). None of the starches showed any stability beyond two cycles at 5% concentration. Tang et al.[77] have shown that gels from isolated large, medium and small granules of normal barley starch exhibit synerisis (after 12 freeze–thaw cycles) in the range 50–55%, 58–60% and 66–69%, respectively. The corresponding values for large and medium granules from waxy barley starch are 52–61% and 63–66% respectively. The difference in freeze–thaw stability between barley genotypes and among granule sizes has been attributed to differences in amylose content, amylopectin size and amylopectin chain length distribution.[98]

12. Chemical Modification

Cationization

Cationic starches are widely used as wet-end additives in the pulp and paper industry to enhance starch and filler retention during papermaking. Use of cationic starches increases paper strength and decreases biological oxygen demand (BOD) of paper mill effluent. Presently only cationic corn and potato starches are used by Canadian paper mills. The degree of substitution (DS) of normal, waxy, high-amylose barley

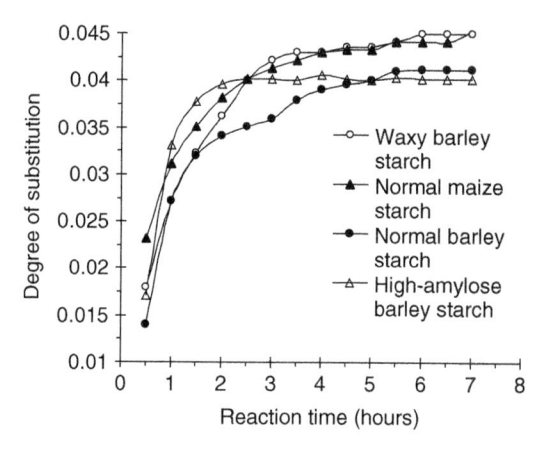

Figure 16.14 Increase in degree of substitution with reaction time during cationization of barley and normal maize starches in an alkaline-alcoholic semiaqueous medium. (Adapted from reference 97 with permission)

and normal maize starches with 3-chloro-2-hydroxypropyl-trimethylammonium chloride (cationizing reagent) in an alkaline alcoholic semiaqueous media for 7 hours, was shown to be 0.041, 0.044, 0.039 and 0.044, respectively.[97] The average rates of reaction (DS/hour) for the above starches calculated from the initial slopes of the reaction curves (Figure 16.14) are 0.016, 0.019, 0.021 and 0.019, respectively.[97] Laboratory evaluation of the strength characteristics of paper sheets containing cationic barley starches were found to be superior to that of paper sheets containing laboratory modified or commercial grade (Cato-15) cationic normal maize starch, with respect to breaking length and burst index.[97]

Crosslinking Followed by Hydroxypropylation or Acetylation

Native starches are sensitive to high temperature, shear and acid treatment when cooked in water. Therefore, starches used in the food industry are modified by crosslinking in order to improve their functionality in foods that are subjected to high temperature processing and shear. However, crosslinking does not impart freeze–thaw stability. Thus, starches are further modified by esterification (acetylation) or etherification (hydroxypropylation). Presently, waxy maize is the major chemically-modified starch in North America. Only a few studies have reported on acetylated[91] and hydroxypropylated[98,99] distarch phosphates from barley starches. Furthermore, the above studies have been mainly on waxy genotypes.[98,99] The optimum reaction conditions for crosslinking of waxy barley starch were shown to be pH 11.1–11.5, phosphorous oxychloride ($POCl_3$), 0.005–0.007% and a reaction time of 20–60 minutes. At the same pH and reaction time, the optimum level of $POCl_3$ to crosslink waxy maize starch is 0010–0.015% $POCl_3$.[98] This study showed that the thickening power, hot paste stability and clarity of optimally crosslinked waxy barley starch is similar to that of crosslinked waxy maize starch. Hydroxypropylation (molar substitution 0.10–0.13) of crosslinked waxy barley starch was found to impart higher freeze–thaw stability than commercial samples of similarly modified waxy maize starch.[98,99] In addition, hydroxypropylation was more efficient than acetylation in imparting freeze–thaw stability to crosslinked waxy barley starch.[98]

13. Physical Modification

Annealing

Annealing is a process whereby a material is held at a temperature somewhat lower than its melting temperature, which permits modest molecular reorganization to occur and a more organized structure of lower free energy to form.[100] Several studies[39,101–105] have shown that changes to starch structure and properties occur on annealing. Recently, Waduge et al.[39] showed that the relative crystallinity (RC) of waxy (zero-amylose, 7.8% amylose), normal (32 and 34% amylose) and high-amylose (44 and 55% amylose) barley starches increased on annealing (at 50°C for 72 hours at a moisture content of 75%). However, RC of the high-amylose barley starches remained unchanged. The x-ray pattern remained unchanged on annealing in normal (32% amylose) and waxy starches, but changed from A- to an A + B-pattern in normal (34% amylose) and high-amylose starches. In all starches the x-ray intensity of the V-amylose–lipid complex peak increased on annealing. Annealing increased the gelatinization transition temperature and decreased the gelatinization temperature range in all starches. The enthalpy of high-amylose (55% amylose) starches increase on annealing, whereas it remained unchanged in the other starches. In all starches, granular swelling decreased on annealing. Annealing decreased amylose leaching in normal (34% amylose) and high-amylose starches in the temperature range of 50–90°C, but increased amylose leaching in waxy (7.8% amylose) and normal (32% amylose) starches at higher temperatures. The different responses shown by the barley genotypes towards annealing were attributed mainly to the influence of amylose/amylopectin ratio and the packing arrangement of the starch chains within the amorphous and crystalline regions of the native granule.

Kiseleva et al.[106] showed that starches extracted from normal, waxy and high-amylose barley grains grown at relatively low temperatures contained a greater proportion of crystalline defects and were hence less 'perfect' than starches grown at high temperatures. Annealing of these starches caused a significant increase in gelatinization temperatures, a small increase in gelatinization enthalpies and little change with respect to crystal size. They postulated that annealing diminishes amylopectin crystalline defects by improving double helix registration and optimizing the length of the double helices within crystallites.

IV. Resistant Barley Starch

Most foods are heat processed prior to consumption. Although heat processing increases the availability of starch to enzyme, a fraction of starch remains resistant to amylase hydrolysis in the human gastrointestinal tract. This fraction is called resistant starch (RS). RS has been classified[107] into three groups: (a) RS1: starch that is physically inaccessible to digestive enzymes due to enclosure in structures such as

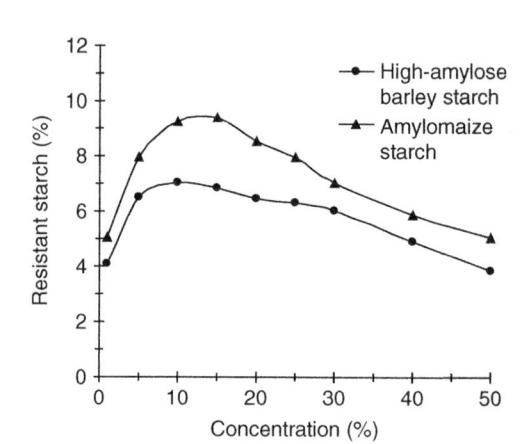

Figure 16.15 The effect of starch concentration on the resistant starch (RS3) content of gels from high-amylose barley and amylomaize starches. (Adapted from reference 111 with permission)

intact cells in legumes; (b) RS2: crystalline regions of native starch granules and retrograded amylopectin; (c) RS3: retrograded amylose. An additional type of RS, type RS4 is considered to result from chemical modification.[108] RS1 and RS2 are slowly but completely digested with appropriate preprocessing of foods, but RS3 has been shown to escape digestion in the small intestine and is slowly fermented in the large intestine.[107,109] RS3 *in vitro* consists of semicrystalline, mostly linear resistant starch that is present in two main molecular size subfractions: (a) DP >100 (arising from semicrystalline material present in the retrograded part of the gel); and (b) DP 20–30 (composed of recrystallized amylose fragments released during degradation by α-amylase). A third minor subfraction consisting of oligosaccharides (DP ≤5) is also present.[110] Vasanthan and Bhatty[111] have shown that the RS3 content of native high-amylose (40.2%) barley starch (HABS) is only 3.8%, which is lower than that of amylomaize (54% amylose) starch. However, the RS3 content of HABS gels increases significantly (3.8% to 7.0%) on storage at 20°C (Figure 16.15). The corresponding value for amylomaize is 9.2% (Figure 16.15). Annealing[110] of dried retrograded HABS starch gels in the temperature range 100–140°C (three heating and cooling cycles) was shown to increase RS3 content to 13.4, 13.1 and 11.5% at 100, 130°C and 140°C, respectively (Table 16.8) after the third cycle. The corresponding values for amylomaize starch gel were 19.1, 15.7 and 12.9 at 100, 130 and 140°C, respectively (Table 16.8). Pullalanase (debranching enzyme) treatment of the retrograded starch gels of HABS and amylomaize[111] increased RS3 from 7.0 to 8.2% and from 9.2 to 10.2%, respectively (Table 16.9). However, treatment of the pullalanase hydrolyzed HABS and amylomaize retrograded gels by annealing (3 cycles)[111] increased RS3 from 8.2% to 20.2% and from 10.2% to 23.8%, respectively (Table 16.9). Acid hydrolysis (for 30 minutes) of dried retrograded starch gels of HABS and amylomaize followed by annealing (at 100°C for 3 heating/cooling cycles) was found to increase RS3 content to 21.5% and 24.3%, respectively.[111] The authors[111] concluded from this study that acid hydrolysis may be more economical that pullalanase hydrolysis for RS3 production. However, no comparative evaluation of the

Table 16.8 Effect of annealing on the resistant starch (RS3) content of dried retrograded high-amylose barley and amylomaize starch gels

Starch source	Resistant starch (RS3) (% dry starch basis)		
	Annealing temperature °C	Heating/cooling cycles	
		0	3
High-amylose barley	100	7.0	13.4
	130	7.0	13.1
	140	7.0	11.5
Amylomaize	100	9.2	19.1
	130	9.2	15.7
	140	9.2	12.9

From reference 97, reproduced with permission

Table 16.9 Effect of pullalanase hydrolysis and annealing on the resistant starch (RS3 content) of dried retrograded high-amylose barley and amylomaize starch gels

Treatment	% resistant starch (RS3)	
	High-amylose barley	Amylomaize
Dried retrograded starch gel (DRSG)	7.0	9.2
DRSG (after pullalanase hydrolysis)	8.2	10.2
DRSG (after pullalanase hydrolysis and 3 cycles of annealing)	20.2	23.8

From reference 92, reproduced with permission

nutritional and functional properties of RS3 produced by pullalanase or acid pretreatment annealing techniques has been carried out.

V. Production and Uses of Barley Starch

The quantity of barley starch produced around the world is very little when compared to starch production from corn, wheat, rice, potato or tapioca. There are only a few processing plants around the world that produce starch concentrate or purified starch from barley. Barley starch concentrates containing up to ~78% (dry basis) starch are now produced in North America by milling and air-classification of barley grains. Milling disintegrates the grain into fine particles and air-classification separates them on the basis of differences in density, mass and projected area in the direction of air flow; generally, a protein-rich low yield fraction and a starch-rich high yield fraction are obtained. The starch concentrates thus produced are used in feed formulations, feed pelleting, as an adhesive in paper board and for clay removal in the potash mining industry. The presence of impurities such as protein (~7–10%), β-glucan (~2.5%) and bran material in the starch concentrate limit its applications in food.

One of the few countries in the world which produces barley starch is Finland. The wet-milling processes used by two Finnish companies, Alko Limited and Finnsugar Company, are shown in Figure 16.16. Alko Limited processes barley into prime starch (A-granule starch), a lower grade starch (B-granule starch) and animal feed (Figure 16.16a). In their operation, the hulled barley is initially dry-milled and sifted to removed most of the hull material; the flour produced is steeped in water in the presence of celluloytic enzymes so that the starch granules are liberated from other constituents. Multistage screens are then used to remove remaining fiber and cell wall materials. The slurry passing through the screens contains mainly starch granules, proteins and solubles. The starch is recovered by centrifugation and then backwashed in banks of very small wet cyclones known as Dorr-clones; this step separates the A- and B-starches. Highly pure A-starch is mainly used by the Finnish paper industries in coating and sizing applications. The relatively low gelatinization temperature and excellent water binding capacity of A-starch can lead to potential applications in the meat processing industry.[112] The B-starch contains some impurities (proteins) and is used for the production of potable alcohol. The by-products of the process are

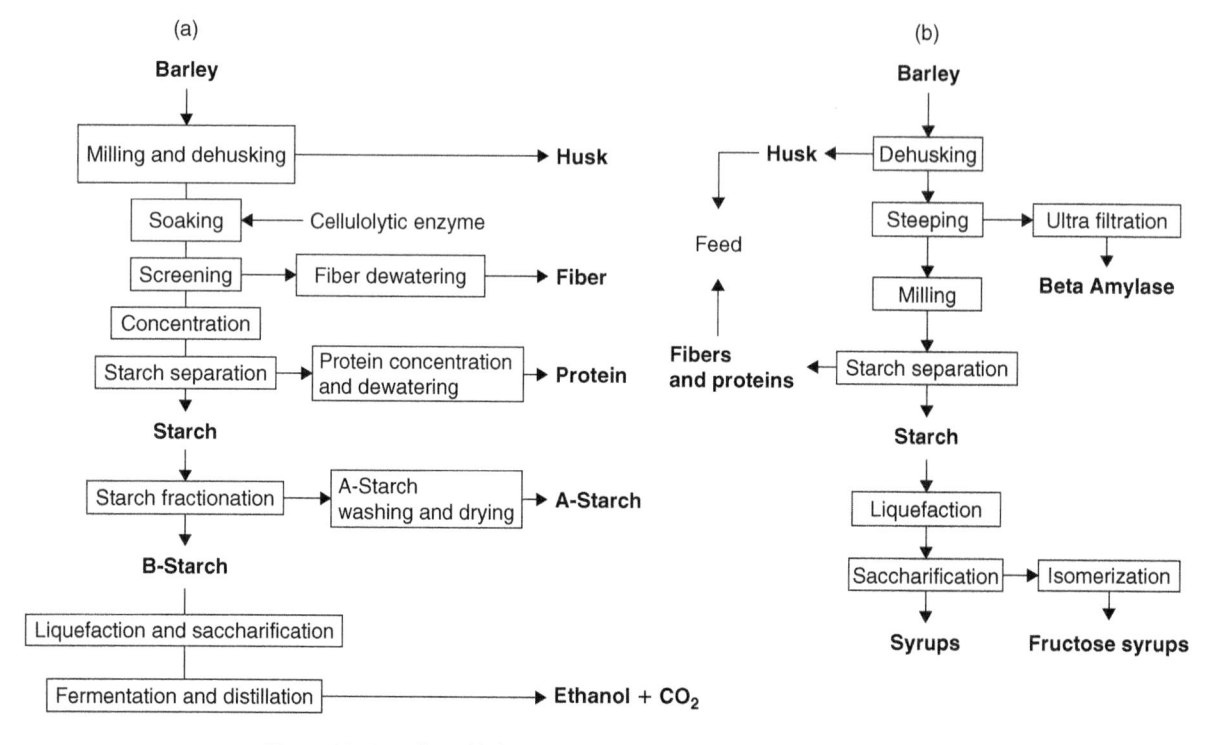

Figure 16.16 Value-added processing of barley grains by two Finnish companies: (a) the flow sheet of the integrated starch-ethanol-feed process of Alko Ltd; (b) the flow sheet of Finnsugar process.[12,113] (Reproduced with permission)

used for livestock feeds. The Finnsugar Company employs a patented wet-milling process[113] for the fractionation of hulled barley grains through unit operations such as dehusking, steeping, milling and starch separation (Figure 16.16b). The starch is subsequently processed to various nutritive sweeteners and maltodextrins by liquefaction, saccharification and isomerization. Presently β-cyclodextrin is used in the microencapsulation of essential oils or flavors. Recently[114] it has been shown that succinylated normal barley starch exhibits better flavor retention capabilities than β-cyclodextrin.

The abundance of corn in North America at a relatively low and steady price, and its higher starch content, make it a much more attractive raw material for starch production than barley. Therefore, the potential success of a barley wet-milling operation which produces starch as a main product would rely on the value of other grain components such as β-glucan, bran and protein.

VI. Conclusion

This review has shown that most studies on barley starches have been focused on understanding the fine structures, particle size distribution, chemical composition, gelatinization properties and susceptibility towards enzyme hydrolysis. However, there is a dearth of information on the rheological and retrogradation characteristics of barley starches from different cultivars. Furthermore, the response of small and large barley starch granules towards physical and chemical modification needs investigation. Research in the above areas is underway in our laboratories. It is hoped that this study may improve the utilization of different types of barley starches for specific products within the food and paper industry.

VII. References

1. Nilan RA, Ullrich SE. In: MacGregor AW, Bhatty RS, eds. *Barley: Chemistry and Technology*. St. Paul, MN: American Association of Cereal Chemists; 1993:88.
2. FAO 2003, http://apps.fao.org/default.htm
3. Jadhav SJ, Lutz SE, Ghorpade VM, Salunkhe DK. *Crit. Rev. Food. Sci. Nut.* 1998;38:123.
4. MacGregor AW, Fincher GB. In: MacGregor AW, Bhatty RS, eds. *Barley: Chemistry and Technology*. St. Paul, MN: American Association of Cereal Chemists; 1993:80.
5. MacGregor S. Composition of barley related to food uses. Presented at the International Food Barley Program. Manitoba, Canada: Canadian International Grains Institute in Winnipeg; 1988.
6. Newman RK, Newman CW. *Cereal Foods World.* 1991;36:800.
7. Bhatty RS. *Cereal Chem.* 1999;76:589.
8. Bhatty RS, Rossnagel BG. In: *Proceedings of the 4th International Barley Genetics Symposium*. Edinburgh, UK; 1981:341.
9. Li JH, Vasanthan T, Hoover R, Rossnagel BG. *Food Chem.* 2004;84:621.

10. Oscarsson M, Andersson R, Salomonsson AC, Aman P. *J. Cereal. Sci.* 1996;24:161.
11. Xue Q, Wang L, Newman RK, Newman CW, Graham H. *J. Cereal. Sci.* 1997;26:251.
12. Andersson AAM, Elfverson C, Andersson R, Regnér S, Aman P. *J. Sci. Food. Agric.* 1999;79:979.
13. Bhatty RS, Rossnagel BG. *Cereal Chem.* 1998;75:15.
14. Zheng GH, Bhatty RS. *Cereal Chem.* 1991;68:589.
15. Greenwood CT, Thomson J. *J. Inst. Brew.* 1959;65:346.
16. Adkins GK, Greenwood CT. *Starch/Stärke.* 1966;18:213.
17. Banks W, Greenwood CT, Muir DD. *Starch/Stärke.* 1973;25:225.
18. McDonald ML, Stark JR. *J. Inst. Brew.* 1988;94:125.
19. Morrison WR, Milligan TP, Azudin MN. *J. Cereal. Sci.* 1984;2:257.
20. Tester RF, Morrison WR. *Cereal Chem.* 1992;68:589.
21. Vasanthan T, Bhatty RS. *Cereal Chem.* 1995;72:379.
22. Tang H, Ando H, Watanabe K, Takeda Y, Mitsunaga T. *Cereal Chem.* 2000;77:27.
23. Anderson AAM, Andersson R, Aman P. *Cereal Chem.* 2001;78:507.
24. Anderson AAM, Andersson R, Autio K, Aman P. *J. Cereal Sci.* 1999;30:183.
25. Inagaki T, Seib PA. *Cereal Chem.* 1992;69:321.
26. Szczodrak J, Pomeranz Y. *Cereal Chem.* 1991;68:589.
27. Song Y, Jane J. *Carbohydr. Polm.* 2000;41:365.
28. Li JH, Vasanthan T, Rossnagel B, Hoover R. *Food Chem.* 2001;74:407.
29. Chmelik J, Krumlová A, Budinska M. *J. Inst. Brew.* 2003;107:11.
30. Wu YV, Sexson KR, Sanderson JE. *J. Food Sci.* 1979;44:1580.
31. Sulaiman BD, Morrison WR. *J. Cereal Sci.* 1990;12:53.
32. You S, Izydorczyk MS. *Carbohydr. Polym.* 2002;49:33.
33. Yoshimoto Y, Tashiro J, Takenouchi T, Takeda Y. *Cereal Chem.* 2000;77:279.
34. Lorenz K, Collins F. *Starch/Stärke.* 1995;47:14.
35. Myllarinen P, Schulman AH, Salovaava H, Poutanen K. *Acta. Agric. Scand.* 1998;48:85.
36. Hizukuri S, Takeda Y, Yasuda M, Suzuki A. *Carbohydr. Res.* 1981;94:205.
37. Hizukuri S, Kaneko T, Takeda Y. *Biochem. Biophys. Acta.* 1983;760:188.
38. Takenouchi T, Takeda Y. *Carbohydr. Polym.* 2002;47:159.
39. Waduge RN, Hoover R, Vasanthan T, Gao J, Li J. *Food Res. Intl.* 2006;39:59.
40. Salomonsson AC, Sundberg B. *Starch/Stärke.* 1994;46:325.
41. Jenkins PJ, Donald AM. *Int. J. Biol. Macromol.* 1995;17:315.
42. Vasanthan T, Bhatty RS. *Cereal Chem.* 1996;73:199.
43. Takeda Y, Takeda C, Mizukami H, Hanasuiro I. *Carbohydr. Polym.* 1999;38:109.
44. Tang H, Watnabe K, Mitsunaga T. *Carbohydr Polym.* 2002;49:217.
45. Tang H, Ando H, Watnabe K, Takeda Y, Mitsunaga T. *Carbohydr. Res.* 2001;330:241.
46. Morrison WR, Tester RF, Snape CE, Law R, Gidley MJ. *Cereal Chem.* 1993;70:385.
47. MacGregor AW, Morgan JE. *Cereal Chem.* 1984;61:222.
48. Yoo S-H, Jane J-L. *Carbohydr. Polym.* 2002;49:307.
49. Tester RF, South JB, Morrison WR, Ellis RP. *J. Cereal Sci.* 1991;13:113.
50. Yoshimoto Y, Takenouchi T, Takeda Y. *Carbohydr. Polym.* 2002;47:159.
51. Suh DS, Verhoeven T, Denyer K, Jane JL. *Carbohydr. Polym.* 2004;56:85.
52. Li JH, Vasanthan T, Hoover R, Rossnagel BG. *Food Res. Intl.* 2004;37:417.

53. Goldner WR, Boyer CD. *Starch/Stärke*. 1989;41:250.
54. Morrison WR. *Starch/Stärke*. 1981;33:408.
55. Vasanthan T, Hoover R. *Food Chem.* 1992;43:19.
56. Morrison WR, Scott DC, Karkalas J. *Starch/Stärke*. 1986;38:374.
57. Tester RF, Morrison WR. *J. Cereal Sci.* 1993;17:11.
58. Tester RF. *Int. J. Biol. Macromol.* 1997;21:37.
59. MacGregor AW, LaBerge DE, Meredith WOS. *Cereal Chem.* 1971;48:255.
60. Qi X, Tester RF, Snape CE, Yurev V, Wasserman LA, Ausell R. *J. Cereal Sci.* 2004;39:57.
61. Cheetham NWH, Tao L. *Carbohydr. Polym.* 1997;33:251.
62. Morsi MKS, Sterling C. *Carbohydr. Res.* 1960;3:97.
63. Cameron RF, Donald AM. *Carbohydr. Res.* 1993;244:225.
64. Hoover R, Hadziyev D. *Starch/Stärke*. 1981;33:290.
65. Jenkins PJ, Donald AM. *Carbohydr. Res.* 1998;308:133.
66. Biliaderis CG. In: Harwalker VR, Ma CY, eds. *Thermal Analysis of Foods*. London, UK: Elsevier Applied Science; 1990:168–220.
67. Kang MY, Sugimoto Y, Kato I, Sakamoto S, Fuwa H. *Agric. Biol. Chem.* 1985;49:1291.
68. Naka M, Sugimoto Y, Sakamoto S, Fuwa H. *J. Nutr. Sci. Vitaminol.* 1985;31:423.
69. MacGregor AW, Bazin SL, Izydorczyk MS. *J. Inst. Brew.* 2002;108:43.
70. Czuchajowska Z, Klamczynski A, Paszczynska B, Baik BK. *Cereal Chem.* 1998;75:747.
71. Yasui T, Seguchi M, Ishikawa N, Fujita M. *Starch/Stärke*. 2002;54:179.
72. Tester RF, Morrison WR. *Cereal Chem.* 1990;75:625.
73. Sasaki T, Matsuki S. *Cereal Chem.* 1998;75:525.
74. Shi YC, Seib PA. *Carbohydr. Res.* 1992;227:131.
75. Tester RF, Morrison WR, Schulman AH. *J. Cereal Sci.* 1993;17:1.
76. Hoover R, Manuel H. In: Fenwick CR, Hedley G, Richards RL, Khokhar S, eds. *Agri-food quality: An interdisciplinary approach*. Cambridge, UK: Royal Society of Chemistry; 1996:157–161.
77. Tang H, Mitsunaga T, Kawamura Y. *Carbohydr. Polym.* 2004;57:145.
78. Yasui T, Seguchi M, Ishikawa N, Fujita M. *Starch/Stärke*. 2002;54:179.
79. Jayakody L, Hoover R. *Food Res. Intl.* 2002;35:615.
80. Myllarinen P, Autio K, Schulman AH, Poutanen K. *J. Inst. Brew.* 1998;104:343.
81. MacGregor AW, Ballance DL. *Cereal Chem.* 1980;57:397.
82. Fujita S, Glover DV, Okuno K, Fuwa H. *Starch/Stärke*. 1989;41:221.
83. Fuwa H, Nakajima M, Hamada A, Glover DV. *Cereal Chem.* 1971;54:230.
84. Kimura A, Robyt JF. *Carbohydr. Res.* 1995;277:87.
85. Morrison WR, Tester RF, Snape CE, Law R, Gidley MJ. *Cereal Chem.* 1993;70:385.
86. Goering KJ, Slick RE, DeHass BW. *Cereal Chem.* 1970;48:592.
87. Shibanuma Y, Takeda Y, Hizukuri S. *Carbohydr. Polym.* 1996;29:253.
88. Takeda Y, Takeda C, Suzuki A, Hizukuri S. *J. Food Sci.* 1989;54:177.
89. Hoover R, Manuel H. *J. Cereal Sci.* 1996;23:153.
90. Jane JL, Robyt JF. *Carbohydr. Res.* 1984;132:105.

91. Ring SG, Colonna P, Ianson P, Kalicheversky KJ, Miles MT, Morris VJ, Orford PD. *Carbohydr. Res.* 1987;163:277.
92. Fredriksson H, Silverio J, Anderson R, Eliasson AC, Aman P. *Carbohydr. Polym.* 1998;35:119.
93. Myllarinen P, Autio K, Schulman AH, Poutanen K. *J. Inst. Brew.* 1998;104:343.
94. Schoch TJ. In: Tressler DK, van Arsdel WB, Copley MJ, eds. *The Freezing Preservation of Foods*. Vol. 4. Westport, CT: Avi; 45–46.
95. Zheng GH, Han HL, Bhatty RS. *Cereal Chem.* 1998;75:520.
96. Zheng GH, Sosulski FW. *J. Food Sci.* 1998;63:134.
97. Vasanthan T, Bhatty RS, Tyler RT, Chang P. *Cereal Chem.* 1997;74:25.
98. Wu Y, Seib PA. *Cereal Chem.* 1990;67:202.
99. Zheng GH, Han HL, Bhatty RS. *Cereal Chem.* 1999;76:182.
100. Blanshard JMV. In: Galliard T, ed. *Starch properties and potential*. New York, NY: Wiley; 1987:16–54.
101. Cameron E, Donald AM. *Polymer*. 1992;32:2628.
102. Genkina NK, Wasserman LA, Noda T, Tester R, Yurev V. *Carbohydr. Res.* 2004;339:1093.
103. Hoover R, Vasanthan T. *J. Food Biochem.* 1994;17:303.
104. Jacobs H, Delcour JA. *J. Agri. Food. Chem.* 1998;46:2895.
105. Gomez AMM, Mendes da Silva GE, Ricardo MPNS, Sasaki JM, Germani R. *Starch/Stärke*. 2004;56:419.
106. Kiseleva VI, Genkina NK, Tester R, Wasserman LA, Popov AA, Yurev VP. *Carbohydr. Polym.* 2004;56:157.
107. Englyst HN, Cummings JH. Cereals in a European Context. In: Morton ID, ed. *First European Conference on Food Science and Technology*. Chichester, UK/Weinheim Federal Republic of Germany: Ellis Horwood Ltd/VCH; 1987:221–233.
108. Asp NG, Van Amelsvoort JMM, Hautvast JGAJ. *Nutr. Res. Rev.* 1996;9:1.
109. Björck IM, Nyman M, Pedersen B, Siljestrom M, Asp NG, Eggum BO. *J. Cereal Sci.* 1987;6:159.
110. Cairns P, Morris VJ, Botham RL, Ring SG. *J. Cereal Sci.* 1990;12:203.
111. Vasanthan T, Bhatty RS. *Starch/Stärke*. 1998;50:286.
112. Linko Y. In: Sparrow DHB, Lance RCM, Henry RJ, eds. *Alternative End Uses of Barley*. Victoria, Australia: Royal Australian Chemical Institute; 1988:87–92.
113. Hakonen T, Lehmusaari A, Nevalainen P, Visuri K. Finnish Patent 772260. 1980.
114. Jeon YJ, Vasanthan T, Temelli F, Song BK. *Food Res. Int.* 2003;36:349.
115. Li JH, Vasanthan T, Rossnagel B, Hoover R. *Food Chem.* 2001;74:395.

Modification of Starches

Chung-wai Chiu[1] and Daniel Solarek[2]
[1]*National Starch and Chemical Co., Bridgewater, New Jersey (Retired)*
[2]*Akzo Nobel Surface Chemistry LLC*

I.	Introduction	629
II.	Cationic Starches	632
	1. Dry or Solvent Cationization	633
	2. Polycationic Starches	634
	3. Amphoteric Starch or Starch-containing Systems	635
	4. Cationic Starches with Covalently-reactive Groups	636
III.	Starch Graft Polymers (See Also Chapter 19)	637
IV.	Oxidation of Starch	638
V.	Starch-based Plastics (See Also Chapter 19)	640
VI.	Encapsulation/Controlled Release	642
VII.	Physically Modified Starch	644
	1. Granular Cold-Water-Swellable (CWS) and Cold-Water-Soluble Starch (Pregelatinized Granular Starch)	644
	2. Starch Granule Disruption by Mechanical Force	646
VIII.	Thermal Treatments	646
IX.	Enzyme-catalyzed Modifications	647
X.	References	648

I. Introduction

Starches are inherently unsuitable for most applications and, therefore, must be modified chemically and/or physically to enhance their positive attributes and/or to minimize their defects. Starch derivatives are used in food products as thickeners, gelling agents and encapsulating agents, in papermaking as wet-end additives for dry strength, surface sizes and coating binders, as adhesives (corrugating, bag, bottle labeling, laminating, cigarettes [tipping, side-seam], envelopes, tube-winding and wallpaper pastes), for warp sizing of textiles, and for glass fiber sizing. Various starch products are used to control fluid loss in subterranean drilling, workover and completion fluids (for oil, gas or water production). Modified starches are also used in tableting and cosmetic formulations. Some starch is incorporated into plastics to enhance environmental fragmentation and degradation. Thermoplastic starch and starch–polymer composites can replace petroleum-based plastics in some applications. Newer applications include use of nondigestible starch as neutraceuticals. The future of starch may include a role in detergents.

The properties required for a particular application, availability of the starch and economics play a role in selecting a particular native starch for subsequent chemical and/ or physical modification. Normal maize, waxy maize, high-amylose maize, tapioca, potato and wheat starch are the most available and accessible starches, but varieties of rice, including waxy rice, pea (smooth and wrinkled), sago, oat, barley, rye, amaranth, sweet potato and certain other exotic starches indigenous to the areas in which they are produced can be used as localized commercial sources. Conventional hybrid breeding and genetic engineering has the potential to provide even more options.[1–4]

Chemical modification of starch generally involves esterification, etherification or oxidation of the available hydroxyl groups on the α-D-glucopyranosyl units that make up the starch polymers.[i] Reactions used to produce most commercially-modified starches have been reviewed by others.[5,6] Many commercial derivatives are produced by the addition of reactive, organic reagents to aqueous starch slurries while controlling alkalinity (pH 7–9 for esterification and pH 11–12 for etherification) and temperature (typically <60°C). Sodium sulfate or sodium chloride is often added to restrict swelling of the starch granules during reaction. Neutralization of the reaction slurry, typically by hydrochloric or sulfuric acid, followed by water washing of the filter cake and drying, yields a powder. Generally, the degree of substitution (DS) of commercial starches is less than 0.2. Dry or semi-dry reactions and reactions in ethanol or isopropanol slurries are known. While these methods allow higher substitution, salts and modifying reagent by-products remain in the final product. A continuous method of hydroxypropylation in a static mixer reactor has been described.[7,8] Another process involves the use of a cylindrical turbo-reactor for etherification, esterification and acid modification.[9] A stirred, vibrating, fluidized-bed reactor for

---

[i]Editor's note: Starch modification was reviewed and discussed in the first edition of this work, viz., H.J. Roberts, *Nondegradative Reactions of Starch*, in Vol. I, 1st edn, 1965, pp. 439–493; P. Shildneck and C.E. Smith, *Production and Use of Acid-modified Starch*, in Vol. II, 1st edn, 1967, pp. 217–235; B.L. Scallet and E.A. Sowell, *Production and Use of Hypochlorite – Oxidized Starches*, in Vol. II, 1st edn, 1967, pp. 237–251; R.B. Evans and O.B. Wurzburg, *Production and Use of Starch Dextrins*, in Vol. II, 1st edn, 1967, pp. 254–278; J.W. Knight, in *Modification and Uses of Wheat Starch*, in Vol. II, 1st edn, 1967, pp. 279–291; H.J. Roberts, *Starch Derivatives*, in Vol. II, 1st edn, 1967, pp. 293–350; R.M. Hamilton and E.F. Paschall, *Production and Use of Starch Phosphates*, in Vol. II, 1st edn, 1967, pp. 351–368; L.H. Kruger and M.W. Rutenberg, *Production and Uses of Starch Acetates*, in Vol. II, 1st edn, 1967, pp. 369–401; E.F. Paschall, *Production and Use of Cationic Starches*, in Vol. II, 1st edn, 1967, pp. 403–422; E.T. Hjermstad, *Production and Use of Hydroxyethylstarch*, in Vol. II, 1st edn, 1967, pp. 423–432; C.H. Hullinger, *Production and Use of Cross-linked Starch*, in Vol. II, 1st edn, 1967, pp. 445–450; and in other chapters in Volumes I and II such as those on specific starches and starch applications in the food, paper, and textile industries. The subject was updated in the 2nd edition, M.W. Rutenberg and D. Solarek, *Starch Derivatives: Production and Uses*, 2nd edn, 1984, pp. 311–388; R.G. Rohwer and R.E. Klem, *Acid-modified Starch: Production and Uses*, in 2nd edn, 1984, pp. 529–541; and again in chapters on specific starches and starch applications. This chapter covers primarily the patent literature from the early 1980s through part of 1998. An additional update can be found in K.C. Huber and J.N. BeMiller, *Modified Starch*, in 'Starch and Other Biopolymers', A. Bertolini, ed., Taylor and Francis/CRC Press, in press.

modifying starch with gaseous ethylene oxide has been described.[10] Reactive extrusion to prepare starch succinates has been studied.[11] An extrusion process for preparing crosslinked, carboxymethylstarches as water absorbents has been developed.[12]

Unless some insolubilizing, crosslinking treatment or a hydrophobic substituent is added, increasing substitution will eventually make the starch cold-water-soluble. Steric consequences of substituent groups bring about the disruption of hydrogen bonding and weakening of the granular structure.[13] Anything that breaks glycosidic linkages, e.g. thinning under acidic conditions or oxidation under alkaline conditions, also weakens the granular structure.

Monofunctional reagents provide nonionic, cationic, anionic and hydrophobic or covalently reactive substituent groups that dramatically affect the properties of the particular starch being modified. The type of modification alters the gelatinization temperature and pasting characteristics of the starch and stabilizes the paste resulting from cooking a suspension by controlling or blocking associations between dissolved amylose and amylopectin molecules. Such so-called stabilizing modifications result in improved freeze–thaw and refrigerated storage stability, an important property for food systems. Differential scanning calorimetry (DSC), which measures the energy needed to disrupt recrystallized or retrograded starch after low temperature storage or repeated freeze–thaw cycles, has provided evidence that chemical modification reduces or eliminates aggregation and/or association of starch molecules during cold storage of pastes.[14]

Hydroxypropylstarches prepared by etherification with propylene oxide and starch acetates prepared by esterification with acetic anhydride are commonly used in food applications. Enhanced stability is achieved by using an all-amylopectin starch, e.g. waxy maize starch. Hydroxypropylation is more effective than acetylation in imparting low temperature stability. Similar reduced-retrogradation properties are provided by hydroxyethylation with ethylene oxide, but this modification is not permitted in food applications. Chemical modification provides improved stability and film-forming properties to partially-degraded starches used in paper surface sizing or coating, textile warp sizing and adhesives. In food applications, these modifications can be combined with crosslinking treatments to provide a range of products with a range of properties. Starch and modified starch for food applications has been reviewed.[15]

Difunctional reagents are capable of crosslinking starch polymers by reacting with more than one hydroxyl group and, thereby, reinforcing granules. The most common crosslinking agents for food applications are phosphoryl chloride (phosphorus oxychloride), adipic-acetic mixed anhydride and sodium trimetaphosphate. Epichlorohydrin may be used for industrial applications. The stability of the crosslinks provided by these respective reagents varies. The adipate diester crosslink is the most labile, particularly at higher pH. Crosslinking restricts swelling of starch granules and the solubility and mobility of the polymer molecules. Pastes of lightly crosslinked starches (1×10^{-3} to 5×10^{-2} percent of crosslinking reagent based on the weight of starch) have shorter textures, higher viscosity, greater resistance to shear-thinning and low pH, and overall greater stability than the native starch from which they are made. Covalent crosslinking partially compensates for the hydrolysis of the starch molecules that may occur at low pH. More highly crosslinked starches (>0.5% reagent) will not gelatinize in boiling water or under sterilization conditions and are typically used in dusting powder applications. Crosslinking of dispersed or swollen starch can be used to improve the water resistance of starch films; for example, acetone-formaldehyde condensates are used in starch-based corrugating adhesives. It has been reported that crosslinking of amylose-containing corn and potato starch granules joins amylose molecules to amylopectin molecules.[16]

Starch polymers are often partially depolymerized to produce products that generate less viscosity on cooking a unit weight of starch; such products are known as fluidity or thinned starches. Depolymerization may be effected by an acid or an oxidant. Such treatment is generally carried out on granular starch. Fluidity (thinned) starches result from treatment of a slurry of granular starch with dilute hydrochloric or sulfuric acid at 40–60°C. Dextrins are more highly degraded and are produced by heating dry acidified starch at 100–200°C. Some transglycosylation also occurs in this process, resulting in more highly branched polymer molecules. Acid-catalyzed hydrolysis of potato, high-amylose maize and waxy maize starches in aqueous methanol, ethanol, isopropanol, butanol and blends of alcohols has been examined.[17] A wide range of limit dextrins with specific DP values were produced using 0.36–5.0% hydrochloric acid at 5–65°C. It was proposed that crystalline regions in the starch granules were converted to amorphous regions during this hydrolysis.[18] Acid conversion of non-crosslinked starch esters or ethers in aqueous ethanol (6 weight percent water) at 50–150°C and under pressure produced degraded, cold-water-soluble starches.[19] Suggested applications for the products, which formed clear, transparent films, were for wall covering and other remoistenable adhesives, protective colloids for emulsion polymerization and encapsulation. Starch polymers can also be depolymerized using various enzymes (amylases). These conversions are typically done on cooked (pasted) starch and typically to a much greater degree, i.e. to produce D-glucose and/or malto-oligosaccharides.

II. Cationic Starches

Cationic starches have significant use in papermaking as wet-end additives for dry strength, as emulsion stabilizers for internal, synthetic sizing agents, such as alkyl ketene dimer and alkenyl succinic anhydride, and as surface sizing agents[20,21] (see also Chapter 18). In recent years, much work has been done to develop synergistic, wet-end additive systems with inorganic microparticles (colloidal silica, bentonite) and/or synthetic polymers or even other starches. The goal is improved retention of cellulose fines and filler, better sheet formation, enhanced drainage and greater strength. Overall, cationic starches provide both wet and dry strength in the final paper.

Quaternary ammonium cationic starch prepared by treatment with 2,3-epoxypropyltrimethylammonium chloride or the more stable chlorohydrin form (3-chloro-2-hydroxypropyltrimethylammonium chloride, which is converted to the reactive epoxide under the highly alkaline starch reaction conditions) is the major commercial cationic starch type. Quaternary ammonium cationizing reagents where one of the methyl groups is replaced with a hydrophobic group (e.g. dodecyl, cocoalkyl or octadecyldecyl) have been examined. Polysaccharides modified with them have enhanced thickening properties.[22]

Many products contain 0.1–0.4% nitrogen (<0.05 DS), but more highly substituted products accessible from dry cationization are available. Germany allows the use of starch products with up to 1.6% nitrogen from reaction with 2,3-epoxypropyltrimethylammonium chloride. A petition has been filed with the US Food and Drug Administration that proposes that 2,3-epoxypropyltrimethylammonium chloride be allowed in food-contact articles.[23] The current allowable level is 5%.

Tertiary amino starches made by etherification with diethylaminoethylchloride are also available. Protonation of the tertiary amine under acidic pH conditions produces the cationic charge, which diminishes as pH increases. The nature of the alkyl groups influences the pK_a of the tertiary amino group. Starches modified with 2-chloroethylmorpholine are useful additives for immobilizing paper coating compositions.[24] At pH 8.0–8.5, typical of many coating formulations, the morpholinoethyl substituent is not protonated, but during the process of applying the coating heat can drive off ammonia, causing a drop in pH, protonation (cationization) of the morpholinoethyl group, and flocculation of the pigment, which immobilizes the coating on the paper surface, improving surface properties.

An improvement to slurry cationization involved inline mixing of a solution of commercial 65% 3-chloro-2-hydroxypropyltrimethylammonium chloride with a 21% sodium hydroxide solution to convert the reagent rapidly to the reactive epoxide form just prior to addition to the starch slurry. This process results in less dilution and increased reaction kettle capacity.[25] In another process for which greater reaction efficiency was claimed, the chlorohydrin is added to a dilute aqueous caustic solution followed by the addition of starch, small amounts of sodium sulfate and finally calcium oxide to maintain the pH at 11.5–11.9.[26] Use of potassium rather than sodium hydroxide to catalyze reactions with the epoxide is claimed to provide higher nitrogen contents.[27] Reaction with the usual cationizing reagents on starch dispersions cooked at a high temperature to make cationic[28] or amphoteric[29] (via combined treatment with sodium trimetaphosphate) starches may be used for on-site derivatization. Countercurrent washing in hydrocyclones can be used to replace the chloride counterions of quaternary ammonium starches with other anions.[30]

Although cationic corn, tapioca, wheat and potato starches are the most common commercial products, preparations, properties and performance of cationic oat[31] and pea[32] starches have been reported. Improved retention performance via the use of a blend of cationic cereal (wheat, corn) starch and cationic potato starch has been reported.[33] Products for papermaking using all-amylopectin potato starch have been proposed.[34–39]

1. Dry or Solvent Cationization

Cationization of starch by dry reaction with 2,3-epoxypropyltrimethylammonium chloride is a commercially significant process. The key to a dry reaction is an intimate, homogenous mixture of the reagent and the catalyst. One process[38,39] describes an 'activator' consisting of spray dried, precipitated silica with a surface area of 190 m^2/g (BET) that contains an alkaline agent such as calcium oxide or calcium hydroxide and/or silicates. Different ratios of silica to alkali and 1–3% catalyst (based on dry starch) were used. The viscosity of the final starch product varied based on the nature and the amount of the catalyst. Use of 0.5% 3-chloro-2-hydroxypropyldimethylethanolamine based on the weight of epoxide or halohydrin was claimed to provide high viscosity in the final starch products. More importantly, reaction efficiencies of 90–95% (versus 70–85% for aqueous reactions) and higher DS values of 0.2–0.5 (versus 0.05–0.06) are possible.[40] With intensive mixing of reactants, complete cationization can occur in storage hoppers, silos or bags at ambient temperature. This feature depends on the reactivity of the particular reagent used rather than the process; other reagents may require a higher temperature for reaction. While the cationic modification is usually complete in 2–3 days, up to 7 days may be required to ensure residual epoxide is <100 ppm.[40] An extension of this process incorporates up to 1% of a mixture of sodium peroxodisulfate and sodium peroxocarbonate (1:2 to 1:4 w/w) to prepare 'adjustable viscosity' cationic starches.[41] Equipment was described for continuously preheating a mixture of starch and the cationizing reagent while maintaining the moisture content to achieve faster reaction.[42]

Aluminosilicate clays (kaolinite) with a cation exchange capacity of 2.2 meq/100 g were blended with calcium oxide and starch prior to spray addition of the epoxide. The reaction proceeded at ambient temperature without mixing. Greater reaction efficiencies are claimed.[43]

Salts and organic by-products, mostly the diol resulting from hydrolysis of the epoxide, from dry cationization are left in the starch. Trimethylamine, if formed, can be detected by its odor. It can be neutralized by subsequent addition of acid. Addition of a slightly soluble organic acid, such as fumaric or adipic acid, during the cationization both eliminates the odor and aids scale control in starch cooking equipment.[44]

Cationization and carboxymethylation of starch in an extruder has been reported.[45–47] Cationization of potato starch in a twin-screw extruder had an optimum reaction efficiency of 71%. Further work yielded 80% efficiency and products with 0.03–0.10 DS.[44] Additional heat treatment of extruded products (made via reaction with quaternary ammonium reagents) with sodium trimetaphosphate or citric acid has improved reaction efficiencies and/or viscosities.[48] Dry cationization in the presence of methanol and isopropanol[49] or combined with microwave irradiation has also been done.[50]

Cationization of waxy maize, corn and barley starches in aqueous alcohol slurries is most effective at 35–65% ethanol for all starch types; a 1:1 starch to water ratio gave highest DS values.[51] A process for making cationic or amphoteric starches in aqueous, alkaline alcoholic solvents has also been described.[52]

2. Polycationic Starches

It is possible to etherify starch with reagents containing two or more cationic groups (perhaps containing combinations of quaternary, tertiary and/or secondary amines).[53,54] A typical reagent is 1,3-bis(dimethylamino)-2-chloropropane.[53] Such derivatives gave improved drainage, retention and paper strength. A diquaternary cationizing reagent was prepared by the reaction of 3-chloro-2-hydroxypropyltrimethylammonium chloride with dimethylethanolamine to form a dicationic alcohol, which was then reacted with epichlorohydrin.[54]

Products that provide high dry strength in paper and paperboard can be prepared by co-cooking or heating solubilized potato starch with cationic polymers consisting of co-polymerized units of diallyldimethylammonium chloride, N-vinylamine (from hydrolyzed N-vinylformamide) or N-vinylimidazoline.[55] Usually, 8–12% of the cationic polymer on the weight of starch is used. Complete disruption of the starch granules is required for optimum dry breaking length and burst strength.[56] Enzymically degraded potato, wheat, corn, rice or tapioca starch cooked or heated with 5–15% of the polymers mentioned above increased retention onto the fibers and dry strength.[57] Finally, cationic starch graft polymers can be prepared by co-polymerizing degraded and/or modified starches with N-vinylformamide and vinyl acetate followed by hydrolysis.[58] The graft polymers, which contain amino and vinyl alcohol functionalities, are useful as dry and wet strength agents for paper and paperboard. Complexes of oxidized, carboxymethylated starch and poly(dimethyldiallylammonium chloride) have been proposed as binders for papermaking.[59]

3. Amphoteric Starch or Starch-containing Systems

Control of electrostatic interactions between the various components (cellulose fines and fibers, fillers, pigments, inorganic and natural or synthetic polymeric additives) is essential to papermaking (see Chapter 18). Amphoteric starches containing cationic tertiary amino or quaternary ammonium groups and anionic phosphate groups can interact with both anionic and cationic furnish components. Starches containing at least 0.12% phosphorus, prepared under conditions that preserve high molecular weight, give improved drainage efficiency while providing retention and dry strength.[60] Amphoteric starches can also be prepared via reaction of cationic starch with 2-chloroethylaminodipropionic acid.[61] The amino-dicarboxylate substituent is zwitterionic. Amphoteric potato starch made in this way gave significantly greater retention of fines and filler ($CaCO_3$) than the base cationic potato starch control in a microparticle-containing, alkaline papermaking system. Starches modified only with these groups also functioned in a paper furnish. This potato starch derivative has found commercial application as a thickener/emulsion stabilizer in cosmetic formulations, particularly in low pH systems.[62]

Use of multi-component systems can have synergistic effects. This usually involves oppositely charge polyelectrolytes that interact.[63] Sequential addition of cationic and a non-phosphorylated anionic starch provided improved retention and drainage. Typically, cationic potato starch and potato starch sulfosuccinate (0.05 DS) are used.[64] A further process uses anionic (e.g. phosphorylated, oxidized or carboxymethylated) starch and a high molecular weight cationic (usually synthetic) polymer to neutralize the pulp slurry and insolubilize the starch. Swollen starch can also be a component in the system. The separately added polymers improve concorra, ring crush and burst strength.[65] An amphoteric combination was created by partially swelling cationic starch in the presence of 2–3% (based on the weight of starch) 0.7 DS carboxymethylcellulose (CMC). This mixture was then added to a filler slurry (e.g. $CaCO_3$), followed by an inorganic microparticle which has a positive effect on the flocculated mixture, magnesium polyaluminum citrate complex being preferred.[66]

The treated filler slurry is finally mixed with a pulp slurry. Improved sheet properties are claimed. Yet another process involves separate addition of anionic gums and cationic starch to a papermaking furnish to achieve higher dry strength.[67] Amphoteric starch complexes prepared by co-cooking blends of cationic and anionic starch can also be used for paper and paperboard manufacture.[68,69] Tapioca starch modified with 3-chloro-2-hydroxypropyldimethylbenzylammonium chloride, propylene oxide and sodium hypochlorite is useful in stabilizing emulsions used in paper sizing.[70] Similarly, a blend of cationic corn starch and tapioca starch modified with an alkenylsuccinic anhydride,[71] complexes of tertiary amino cationic corn starch and sodium poly(acrylate), complexes of quaternary ammonium tapioca starch and poly(styrene sulfonate)[72,73] and degraded, high DS (0.5–1.0) cationic starches[74] can be used in the preparation of stable alkylketene dimer emulsions for paper sizing.

Combinations of cationic starch and anionic microparticles are useful commercial systems. Shear-sensitive flocculation occurs, allowing microscale reflocculation in the formed paper sheet, which improves dewatering and retention.[63,75,76] The microparticles can be colloidal silica, aluminum silicate, poly(silicic acid) or bentonite of specific size and surface area.[77–79] Cationic, anionic or polymeric aluminum-containing compounds can be additional components. A three-part coacervate system uses a high molecular weight anionic polyacrylamide, cationic starch and silica.[80] Cooking cationic starch in the presence of an anionic silica hydrosol was reported to improve drainage and retention.[81]

A different approach[82] uses sodium polysilicate microgels to prepare 'silicated cationic starch.'[82] The polysilicate is mixed with an aqueous cationic starch slurry at pH 10.4–10.8. Sodium sulfate may be present to control swelling of the starch granules and deposition of the polysilicate. The reaction is completed by acidification to less than pH 6.5. Products with 5–20% microgels are obtained. Analysis indicated some orthosilicate bonding. Improved drainage and fines retention over the typical two-component system (cationic potato starch and silica) was claimed. The polysilicate microgels and cationic starch can also be added separately.[82,83]

Controlled crosslinking of cationic starches improves performance in microparticle-containing papermaking systems.[84–86] Superior performance over cationic potato starch was achieved with crosslinked cationic or amphoteric waxy maize, tapioca or potato starch in microparticle systems when the starch cooking was optimized to produce the proper colloidal dispersions.[86]

4. Cationic Starches with Covalently-reactive Groups

Temporary wet strength is desirable in toweling or tissue that may be disposed of in septic systems and for applications where water resistance is needed only for short periods. Dialdehyde starch (DAS) is a good temporary wet strength agent.[87,88] Reaction of the aldehydo groups of DAS with the hydroxyl groups of cellulose creates a reversible, hemiacetal-bonded network that provides initial wet strength. DAS is prepared by treatment of starch with periodate (IO_4) which selectively oxidizes the adjacent hydroxyl groups on C-2 and C-3 of the α-D-glucopyranosyl units. Intra- or intermolecular reactions of the aldehydo groups produce a highly crosslinked structure within the starch itself.[89] Controlled degradation of DAS by heating it in mildly alkaline or acidic water results in complete dispersion of the starch and regeneration of the aldehyde groups.[88] Due to its degraded structure, a large number of reactive groups are needed for efficient crosslinking. Generally, products with 50–90% oxidation are best.[90] The process for making DAS requires electrolytic oxidation of the spent oxidant, iodate (IO_3), back to periodate.[91] An improved process for electrochemical regeneration of the periodate has been suggested.[92,93] Products with 40–65% oxidation appear to give the best balance of material consumption, reaction time and properties.

Reacting cationic starch with N-(2,2-dimethoxyethy1)-N-methylchloracetamide introduces acetal substituent groups. Cationic starch aldehydes can be generated from this product by cooking the starch at low pH just prior to use in the papermaking system.[94] Commercial products with less than 1% aldehyde functionality yielded superior wet-strength performance in tissue paper compared to cationic DAS, and higher dry strength than either DAS or conventional cationic starches.[95] As when DAS is used, the wet strength is temporary. Improved paper machine runability and sheet properties were obtained when the cationic starch aldehyde was used in line paper systems.[96] Cationic starch or maltodextrin can be esterified with cis-1,2,3,6-tetrahydrophthalic acid by heating a dried mixture of the carbohydrate and reagent. Subsequent oxidation of the ester substituent with ozone generates aldehydes (preferred products have 1.0 DS).[97]

Starches containing reactive silanol substituents were prepared by treatment with glycidoxypropyltrimethoxysilane. Under alkaline etherification reactions, the methoxyl groups are removed forming the silanol.[98] The starch products have good adhesion to glass. High amylose corn and potato starch derivatives are useful binders for glass filaments and facilitate their movement through the various processing steps. Cationic starches modified with silanol groups yielded higher dry strength than did typical cationic starches. The anionic nature of the silanol groups combined with the cationic modification provides amphoteric characteristics. The covalent reactivity of starch silanols with cellulose or starch hydroxyl groups creates networks that provide wet strength.

2-Nitroalkyl ethers are formed by reaction of starch with nitroalkenes generated in situ from α-nitro-β-acyloxy (or halogeno) alkanes, such as (2-nitropropyl)acetate, during alkaline slurry reactions at pH 10.[99] There is some evidence that the nitro group can be reduced with sodium dithionite ($Na_2S_2O_4$) to the primary amine. This treatment can be combined with cationic or anionic modification to give products claimed to be useful in papermaking, adhesives and oil-well drilling.

III. Starch Graft Polymers (See Also Chapter 19)

Starches (particularly hydroxyethylated starches) are commonly used as pigment binders in paper coating formulations (see Chapter 18). Starch binds pigment particles together and to the paper surface, and contributes to water retention in the coating. Starch graft polymers with 1,3-butadiene and styrene appear to produce paper with high gloss and smoothness while maintaining high porosity and ink receptivity.[100,101] The preferred compositions require enzyme-converted, lightly oxidized hydroxyethylstarch. Thinning is claimed to improve grafting efficiency. Reaction with a blend of styrene and 1,3-butadiene monomers is carried out under pressure. Ratios by weight of monomer to starch of between 6:10 and 8:10 are suggested. Persulfate or Fe^{++}/H_2O_2 are preferred initiators. Surfactants may be used. Films of the starch graft polymers or blends of degraded starches (or other appropriate water-soluble polymers) and latex dispersions have improved coating properties.[102,103] Films consist of a continuous, water-soluble polymer phase reinforced with non-coalesced, submicron-sized latex particles that provide toughness and mechanical strength. The water-insoluble polymer particles help absorb shrinkage stresses and ensure a smooth, uniform surface. Film continuity translates to more uniform surface wetability in paper coatings. Coalescence of latex particles in conventional coatings can result in surface variations leading to binder migration and the printing defect called mottle.[103] The improved compositions can be spray dried and subsequently redispersed to yield coatings with good performance. Starch graft copolymers with 1,3-butadiene styrene and optionally acrylonitrile and acrylic acid (as well as other monomers) require use of a dextrin, either a modified or an unmodified one.[104] Use of cationic and/or hydrophobically modified dextrins in emulsion copolymerization has been evaluated.[105–107] High solids, small particle size and water resistance are claimed properties. Hydrogen peroxide-oxidized or enzyme-degraded potato starch has been graft-copolymerized with styrene-butadiene-acrylic acid or vinyl acetate-butylacrylate and used in moisture-barrier coatings for paper.[108] Paper sacks made with the starch graft polymers had improved repulpability.

Enzyme-thinned cationic starch grafted with vinyl acetate can replace soy protein or casein in board coatings.[109] Enhanced strength and glueability are proposed benefits. Adhesives can be prepared by extruding starch and polymer dispersions, e.g. hydroxypropylstarch and poly(vinyl acetate).[110] Starch-based, non-formaldehyde, self-crosslinking binders for nonwovens have also been prepared.[111] Latexes prepared from maltodextrins (DE <10) polymerized with various acrylate monomers provide high surface tension so that the emulsion stays on the surface of the substrate and inhibits wicking or rewetting. Such starch grafts are subsequently mixed with additional granular starch derivatives and crosslinked by cooking the blend with a cyclic urea-glyoxal condensate. The granular starch, typically a derivative such as hydroxyethylated potato starch, provides film strength. The crosslinker provides water resistance and improved emulsion stability.

IV. Oxidation of Starch

Development or improvement of catalyzed oxidation processes for carbohydrates has resulted in better control, faster reaction rates and higher selectivity. For example, use of small amounts of hydrogen peroxide with 50 ppm of potassium permanganate provides a mild, reproducible starch degradation system.[112] The oxidation is done on an aqueous starch slurry at pH 11.4–12.0 at ambient or slightly elevated temperatures. Reaction times are much shorter than those with uncatalyzed hydrogen peroxide. Successive additions of hydrogen peroxide can be made to achieve specific amounts of degradation.

Interest in detergent products derived from renewable resources and with better biodegradability has driven evaluation of oxidized sugars and starches as builders or co-builders in detergents.[113] Builders and co-builders complex calcium and magnesium ions in hard water to prevent sealing or deposits due to precipitation of insoluble carbonate salts. In current powder detergents, the builders are usually zeolites used in combination with polycarboxylate polymers derived from synthetic acrylic-maleic acid copolymers.[114]

Oxidation of carbohydrates to carboxyl-containing materials is a logical approach to more biodegradable materials.[115–117] Maltodextrins oxidized under alkaline conditions with oxygen followed by bleaching with hydrogen peroxide yield poly-(hydroxycarboxylic acid) mixtures useful as builders, paper binders and thickening agents.[118]

Dinitrogen tetraoxide (N_2O_4) selectively oxidizes some of the primary hydroxyl groups. One solvent-based process involves selective oxidation of starch with N_2O_4 in the presence of oxygen[119] to give a product with 75% uronic acid units. Co-builders for use with zeolites have been produced by oxidation and hydrolysis of starch with a gas phase of NO_2/N_2O_4 in a fluidized bed.[120]

A two step process for producing a dicarboxyl starch derivative uses periodate oxidation to generate dialdehyde starch (DAS) followed by further oxidation with sodium chlorite and hydrogen peroxide.[121,122] Hydrogen peroxide destroys the by-product sodium hypochlorite. Without hydrogen peroxide, the hypochlorite reacts with chlorite, forming toxic chlorine dioxide and generating a requirement for higher amounts of sodium chlorite.[115,116]

$$DAS + 2\ NaClO_2 \rightarrow Dicarboxyl\ starch + 2\ NaOCl\ H_2O_2\ NaCl + H_2O + O_2$$

Dicarboxyl starch can also be made by oxidation with sodium hypochlorite or hypobromite (see reference 116). Reaction with hypochlorite is accelerated by use of catalytic amounts of sodium bromide.[123,124]

Incorporation of catalytic amounts of TEMPO (2,2,6,6-tetramethyl-1-piperidinyloxy) into the sodium hypochlorite–sodium bromide oxidation system creates a nitrosonium ion that is highly selective as an oxidant for primary hydroxyl groups of carbohydrates.[125–128] The hydroxymethyl group is selectively oxidized to a carboxyl group via the aldehyde (hydrate) intermediate. Optimum reaction pH is 9.2–9.7 and lower temperatures (0–5°C) are preferred. Although the reaction is faster at pH 10–11,[127] β-elimination, presumably due to a steady-state aldehyde concentration, is also favored at higher pH values, resulting in degradation of the polysaccharide chain.[128] At lower pH values the reaction is slower and less selective, and oxidative degradation occurs.[129] *In situ* formation of sodium hypobromite oxidizes the hydroxylamine back to the active nitrosonium ion.

The TEMPO – sodium hypochlorite – sodium bromide system has been applied to starch ether derivatives, particularly hydroxyethyl starch, which has a primary hydroxyl group on the hydroxyethyl ether group that can also be oxidized to a carboxyl group and carboxymethyl starch. The apparent goal was improved sequestering agents via higher carboxyl content and the proper multidentate conformations (in which ether oxygen atoms can participate).[130] A conventional use of sodium hypochlorite is to remove residual protein in starch slurries for absorbable dusting powder applications.[131]

V. Starch-based Plastics (See Also Chapter 19)

Problems associated with handling and disposal of solid waste and interest in environmentally-friendly products has created a significant market opportunity for starch.[132–135] The goal has been to increase the amount of starch in thermoplastic composites designed for various packaging materials, containers and one-time-use, shaped articles prepared by injection molding, blow molding, extrusion, co-extrusion or compression molding.[136–140]

Use of granular starch as a filler to enhance biodegradation of commodity plastics, such as conventional and linear low density polyethylene and high density polyethylene, polypropylene and polystyrene has been established.[141,142] Typically, corn starch granules are surface treated with silanes to improve compatibility with the hydrophobic, plastic matrix. An improved process involves the use of starch with a thermoplastic elastomer that functions as a compatibilizer and pro-oxidant.[143] The starch must be dried to <1% moisture to retard steam formation during extrusion processing. Normal corn starch processed in this way has a density of $1.28\,g/cm^3$ and a particle size of $15\,\mu m$ and is stable to 230°C. The usual starch content of the product is 6–20%. In addition to increasing biodegradability, starch filler has other claimed benefits: antiblocking, improved printability, improved water vapor permeability, low gloss finish in films; increased dimensional stability in injection molding; and increased stiffness in blow molding.[132,141] Smaller granules, such as those of rice starch, may be required for very thin films. Starch octenylsuccinate ionically crosslinked with aluminum sulfate can be used as a filler for linear, low density polyethylene (LLDPE). LLDPE-starch octenylsuccinate films had higher strength than did LLDPE-unmodified corn starch films, but a lower degree of biodegradation.[144] Granular corn starch modified with N-methylolstearamide proved to be useful for blending with LDPE.[145]

Granular starch and copolymers of ethylene and acrylic esters and alkyl(meth)acrylates or vinylacetate are produced as master batches for the production of mulch films, geotextiles and molded articles.[146] The polar copolymers act as compatibilizers by lowering the interfacial energy between starch and the polyolefin and eliminate the need to coat granules. Processing via a vented twin-screw extruder also eliminates the need for anhydrous starch.

The drive to use starch at higher addition levels requires it to contribute to the expected strength properties. For this to happen, the starch must be disrupted or 'destructured' so that it can form a continuous phase in an extruded matrix. This can be done by extrusion of starch under low moisture conditions, which effects granular fragmentation, melting of hydrogen-bonded crystallites and partial depolymerization. Thermoplastic blends of up to 50% starch and poly(ethylene-co-acrylic acid) (EAA) were produced in the presence of aqueous base, which solubilized EAA and increased its compatibility with starch and urea, which aids in starch gelatinization.[147,148]

V. Starch-based Plastics 641

Extrusion blown films containing 40% starch are uniform, flexible and transparent, and have good physical properties. Polyethylene can partially replace EAA to reduce raw material costs and provide improved properties. EAA forms V-type inclusion complexes with both amylose and amylopectin.[149,150] Jet-cooking EAA-normal starch slurry blends produced higher paste viscosities and different gel strengths (compared to starch pastes prepared without EAA).[151] Gel strengths of high-amylose corn starches were decreased with EAA, while waxy maize starch produced uncharacteristically firm gels in the presence of EAA.

Starch composites are formed with poly(ethylene-co-vinyl alcohol) and/or EAA and poly(vinyl alcohol) or another plasticizer, all components being hydrogen bond formers. Three phases, viz., destructured starch with a particle size of <1 micron, synthetic polymer and starch physically or chemically interacted with the polymer, are uniformly dispersed in an interpenetrating polymer network.[152–157] The mechanical properties of molded articles or films produced from these composites are between those of low density and high density polyethylene.[132] Breathable, water-impermeable, biodegradable, flexible films useful as backsheets for diapers, protective garments and other articles have been prepared using blends of polycaprolactone and a starch-ethylene copolymer blend made by the above process.[158]

Another proposed process employed injection molding in which starch and limited amounts of plasticizing water are heated under pressure to temperatures above the T_g and T_m to transform the native starch into a homogenous, destructured, thermoplastic melt. The process melt is then cooled to below the T_g of the system before pressure release to maintain the moisture content. Additives include natural and synthetic polymers, plasticizers and lubricants.[136–139,159,160] The technology has been used to prepare pharmaceutical capsules and shaped objects, such as disposable cutlery, straws and pens.

A large portion of the disposable plastics market consists of products made from expanded polystyrene (EPS). A starch foamed extrudate prepared as loose fill from a hydroxypropylated, high-amylose (70%) corn starch provided very acceptable resilience and compressibility as compared to EPS loose fill. Chemically and physically modified high-amylose starches are more resistant to molecular degradation during high temperature–high shear processing than unmodified starch, and they also provide excellent foam cell structure and cushioning properties. The starch-based loose fill is stable over a range of humidity and temperature conditions and dissolves only on direct contact with water. In a soil environment, biodegradation is essentially complete.[161] Resistance to humidity can be improved through the use of hydroxypropylated, high-amylose flours and hydrophobically modified high-amylose starches.[162,163] A process for making biodegradable packing materials from non-high-amylose starch has been claimed.[164] Foamed starch compositions for packing, insulation, filler and cat litter can be prepared by extrusion at 150–250°F (65–120°C) and 30–70 bar.[165]

Strengths of films made from a starch-poly(vinyl alcohol) blend containing glycerol and poly(ethylene-co-acrylic acid) have been examined.[166] High-amylose starches produced films with the most consistent properties. In a process for extruded blown film, blending of high-amylose starch with starches with more typical amylose contents and plasticizers or gelatinization aids improved properties.[167] Extruded blends of starch and starch hydrolysis products (particularly maltodextrins, oxidized starches and pyrodextrins) are claimed to be useful molding materials.[168]

Starch esters are useful in biodegradable applications; that application has been reviewed.[169,170] In particular, high DS starch acetates provide thermoplasticity, hydrophobicity and compatibility with other additives. Starch acetates with a DS of >2.4 are not readily biodegradable.[169] Intermediate DS starch acetates are very biodegradable and have interesting properties. An aqueous slurry process has been developed for producing starch acetates or propionates with 0.5–1.8 DS.[150] The hydrophobicity provided by the ester groups in a product with 1.5 DS or about 30% by weight acetate ester groups makes it water insoluble. Partially depolymerized high-amylose (70%) corn starch acetate of intermediate DS formulated with plasticizer and wax to produce a hot melt adhesive for paper-to-paper bonding applications is water dispersible.[172]

Other methods have been suggested for making high DS starch esters. Clear elastic films were produced from DS 2.5 starch esters made via reaction of starch in acetic anhydride with palmitic acid.[173] Starch acetates with >0.5 DS were prepared by reacting starch under nitrogen with equimolar amounts of acetic anhydride in N-methylpyrrolidone at the reflux temperature with 4-dimethylaminopyridine as a catalyst.[174] High DS (at least 2.8) starch acetates were prepared by reacting starch in excess acetic anhydride in the presence of sodium carbonate at the reflux temperature.[175] Uniform substitution was attained by swelling high-amylose corn starch in aqueous sodium hydroxide followed by precipitation and washing with methanol, then reacting the 'activated starch' with acetic anhydride.[176] Strength properties and water resistance are improved by extruding blends of starch acetate and low molecular weight esters (such as triethyl citrate).[177] Extruded blends of high DS starch esters and linear polyesters provide resistance to moisture-induced changes in molded articles.[178] A starch propionate (DS 1.5) plus other high DS (1.0–3.0) C_2–C_{18} carboxylic acid esters can be used as paper surface sizing agents to provide increased tensile and burst strengths.[179]

VI. Encapsulation/Controlled Release

Spherical porous aggregates of small starch granules (such as amaranth, rice and small wheat) have been prepared by spray drying a starch slurry containing a water-soluble gum as a bonding agent.[180] It was also suggested that calcium chloride could crosslink pectin or alginate used to coat the sphere to improve retention of the ingredient and sphere integrity.

Microporous starch granules are created by amylase treatment of native starch granules. This creates pinholes or pores leading from the surface to the granule interior. The sponge-like granules can absorb a variety of liquids. Chemical modification can be used to provide mechanical strength or alter receptivity. A coating can be applied to the starch granules to improve retention of the active or core material. Release is by mechanical compression, diffusion or degradation.[181]

Adherent starch particles were prepared by partially swelling pregelatinized starch in 30% 2-propanol and/or solutions of inorganic salts. A pest control agent was stirred into the gelled mass, which was dried, broken up and ground. The material could be applied to plants after watering, allowing the active ingredient-containing particles to become sticky and adhere. An alternative method involved partially swelling pregelatinized starch in water containing sugary material, adding the active ingredient, and spraying the mixture directly onto plants.[182] Swollen intact starch granules are prepared by controlled heating in a kneading extruder with 30–50% of a swelling agent, such as glycerol or triethanolarnine, an emulsifier, such as a polyoxyethylene derivative of a sorbitan ester, and an oily material, such as a triglyceride. The emulsifier facilitates mixing of an active component into the swollen starch granules; the oil prevents the granules from fusing.[183] Encapsulation of vitamins, fragrance oil and concrete additives was accomplished in this way.

Core materials can be encapsulated by injecting them into freshly-cooked, temperature-stabilized, amylose-containing starch dispersions. Retrogradation of the starch on rapid cooling encloses the core material in a protective gel that can be dried and ground.[184,185] The release rate of the core material decreases with increasing amylose content.[186] Incorporating an active agent into gelatinized native starch is conveniently accomplished by twin-screw extrusion.[187] Moisture content and particle size are important parameters. Use of highly crosslinked, swollen amylose matrices for controlled release of solid, orally delivered pharmaceutical products has also been described.[188] Extruded, biodegradable, controlled-release matrices for agricultural materials were prepared using high-amylose corn starch and synthetic polymers. Slight modification of the high-amylose starch with acetyl or hydroxypropyl groups improves processing of the hot melt. Crosslinking agents, fillers and plasticizers may also be used.[189] Stable starch–lipid composites are prepared by jet-cooking native starch slurries containing 20–50% of an oil, such as soybean oil.[190,191] The jet-cooked starch–oil composites are stable and can be converted into dry powders by drum drying. The powders redisperse easily and can be used to improve the properties of starch–lipid formulations. Complete solubilization of the starch during cooking allows intimate contact with the oil, which is microencapsulated as small droplets ($<10\mu m$) in the starch matrix.[192] Suggested applications for the composites are to thicken, suspend, stabilize and replace fat. They may be useful as encapsulating agents for flavor oils, antioxidants, medicinal agents and agricultural materials for seed coating. A starch–oil composite was blended with a polyol polyester and the mixture was reacted with an isocyanate to produce a polyurethane foam.[193] Amylose complexation as a method for molecular encapsulation has been investigated.[194,195] Chemical modification, e.g. by hydroxypropylation, improves the solubility of amylose complexes.[196,197]

Starch octenylsuccinates are important commercial products for stabilizing a wide range of oil-in-water emulsions.[198,199] Because they are good film formers, they provide an effective matrix for encapsulating volatile flavors and fragrances. They are approved for use as food ingredients. For spray drying applications, the starch is usually partially depolymerized to reduce the solution viscosity and allow a higher solids content to be used. Modified waxy starches are particularly effective in spray drying applications, due to enhanced solution stability. In a typical spray drying process, a starch octenylsuccinate is dissolved in water. Homogenization of the oil in the starch solution gives an effective emulsion with low particle size, which can now be spray dried to form hollow spheres. The oil is entrapped as droplets in the starch matrix that forms the spheres. When the spray dried powders are added to water, the emulsion re-forms.

Improved stability towards oxidation of spray dried flavor oils was achieved by using a combination of a high-maltose syrup, maltodextrin and a high molecular weight, film-forming polysaccharide, such as starch octenylsuccinate or gum arabic.[200,201] Emulsification performance of maltodextrins is improved by treatment with octenylsuccinic anhydride and aluminum sulfate.

A dual spray drying–extrusion process was used to produce glassy matrices for volatile or labile food components.[202] The spray dried, encapsulated food component was mixed with starch octenylsuccinate, a 10 DE maltodextrin, corn syrup solids (24 DE) and maltose, and the blend was extruded to produce an amorphous melt. Another dual spray drying–extrusion process used starch octenylsuccinates to encapsulate a bleach catalyst for incorporation into a non-aqueous detergent formulation.[202] A process for producing readily dispersible, pregelatinized starch octenylsuccinates by extrusion has been described.[204]

VII. Physically Modified Starch

1. Granular Cold-Water-Swellable (CWS) and Cold-Water-Soluble Starch (Pregelatinized Granular Starch)

Cold-water-swellable (CWS) starch with the desirable characteristics of a starch paste prepared by cooking granular starch is required for instant foods. Traditional pregelatinized starches are prepared by roll drying, spray drying or extrusion (see Chapter 20). In spite of their wide use, these starch products are not able to match all the attributes provided by a starch paste prepared by cooking a slurry of granular starch. However, attributes such as dispersibility in hot or cold water, high viscosity and smooth texture are generally desirable for food products that require minimal home preparation. The general approach to overcome some of the shortcomings of traditional pregelatinized starch is to maintain starch granule integrity while providing cold water thickening. Two major classes of technology have been developed. One controls the swelling of starch granules in a mixture of water and an organic solvent. The other involves spray drying an aqueous starch slurry under carefully controlled conditions.

Pregelatinization of Starch in a Water-nonsolvent Solution

Aqueous alcohol: controlled granule swelling in aqueous alcohol was first reported in 1971,[205] but was not put into practice until an improved procedure for preparing CWS was described.[206] In this method, a slurry of granular starch in a solution of water in a water-miscible organic solvent, characteristically 70–80% alcohol, is heated to 157–177°C (315–350°F) for 2–5 minutes. Only slight degradation occurs. A slurry of such CWS common corn starch in a sugar or corn syrup sets to a sliceable gel without cooking or chilling and is useful in pie fillings, confectioneries, demoldable desserts and instant puddings. Blends of waxy and normal starches treated this way produce CWS products that have the properties of chemically modified starch and disperse in hot or cold water without producing lumps, making them useful for instant and convenience foods.[207]

Alkaline alcohols: the aqueous alcohol process described above is not universally successful for all starches. Some products hydrate only in cold to warm sugar solutions without lumping, while others hydrate in cold water without lumping, but have reduced capacity to thicken. An improved process was described in 1991.[208] In it, starch is treated with aqueous ethanol containing a strong base, such as sodium or potassium hydroxide, at 20–40°C for 20–40 minutes. The process provides a non-degraded, non-lumping product. The alcoholic alkaline process was used to produce a CWS hydroxypropylated granular starch useful for low calorie formulations such as instant desserts.[209]

Aqueous polyols: the aqueous alcohol process described above requires a pressure reactor which makes it undesirable from a manufacturing standpoint. In 1991, an atmospheric pressure process for production of granular cold-water-soluble starches was disclosed.[210] In it, a slurry of starch granules in water and a polyhydric alcohol, such as 1,2-propanediol, is heated at 145–155°C for ~15 minutes. Heating the slurry converts the starch crystalline structure to a V-type single helix crystalline arrangement or to an amorphous structure. The process is effective on cereal, tuber, root and legume starches, and on many of their crosslinked and substituted forms to yield cold-water-soluble starch granules having cold-water-solubilities of 70–95%.

Spray drying pregelatinization of starch: in 1981, a dual nozzle spray drying system to manufacture granular CWS starch was disclosed.[211] A starch slurry is injected through an atomization aperture in a nozzle assembly to form a fine spray. Steam is injected through another aperture in the nozzle assembly into the atomized starch spray to gelatinize the starch, the entire operation taking place in an enclosed chamber. The time for passage of the material through the chamber, i.e. from the atomization aperture through the vent aperture, defines the cooking or gelatinization time. The gelatinized starch is recovered essentially as granules. The technology is broadly applied to both native and chemically-modified starches. For example, using this process, Schara and Katcher[212] developed a pregelatinized, modified normal maize starch that is essentially flavor-free and which has a viscosity building capacity equivalent to spray dried and pregelatinized tapioca starch. In 1992, an improved process was reported.[213] In it, starch is uniformly and simultaneously atomized and cooked in an aqueous medium by means of a single atomization step in an apparatus comprising a two-fluid, internal mix, spray drying nozzle, coupled to a means for drying the cooked, atomized starch to produce a uniformly pregelatinized CWS starch with desirable textural, visual and organoleptic properties.

Other methods: alternative methods of producing lump-free CWS starch have been described. One employs heat–moisture treatment of a mixture of granular starch, a surfactant containing a fatty acid moiety and (optionally) a gum.[214] A process for making a corn starch product giving a uniform viscous dispersion when added to boiling water employs heating a mixture of starch, surfactant and water, followed by microwave radiation.[215] Compositions that gel at low solids concentrations were prepared by complexation of starches of moderate (20–30%) amylose content with emulsifiers.[216]

2. Starch Granule Disruption by Mechanical Force

The size of fat globules or lipid micelles (<3 μm) has triggered an examination of small starch granules as fat mimetics. Some starches such as amaranth, cow cockle, quinoa and rice, and fractions of other starches such as wheat, have small granules that have shown promise as fat replacers. However, their high production cost, because of the difficulty in isolating and purifying them, presently prohibits them from large-scale commercialization.

A process has been developed to produce a starch fat replacer by controlled acid-catalyzed hydrolysis of corn starch, followed by high pressure shearing of the starch slurry to a stable, cream-like gel consisting of aggregated starch crystallites.[217] Amorphous areas joining crystallites within granules are preferentially hydrolyzed, thus making the granule more fragile; shearing frees the crystallites. The molecular weight range of the hydrolyzate is generally between 3000 and 12 000, preferably between 4500 and 6500. The preferred commercial starch is waxy maize starch. Typically, the product is supplied to users as hydrated granules, and a product with the consistency of a cream is prepared on-site by high pressure homogenization. A pregelatinized product has also been manufactured. This product has found wide applications as a fat substitute in products such as margarine, pourable and spoonable salad dressings, frostings and frozen novelties.[218,219]

Jane et al.[220] reported that mechanical attrition of the starch hydrolyzate in dry form also provides a small particle size and that the particles so produced also form a salve-like, fat-like paste when dispersed in water. Whistler[221] disclosed the preparation of starch-based fat substitutes from microporous starch prepared by amylase- or acid-catalyzed hydrolysis and then crosslinked before mechanical disintegration. The microporous starch can be modified by adsorption of a surface modifying agent or by reacting with an etherifying or esterifying reagent before the disintegration step.

VIII. Thermal Treatments

Effects produced by holding starch granules at various moisture contents and elevated temperatures have been the subject of extensive investigations.[222] Annealing or heat–moisture treatments alter the crystallinity of starch granules, especially those containing amylose. Typically, the starch gelatinization temperature is higher, the gelatinization endotherm is more defined and the energy value is increased (as determined by DSC) after the treatment. This scientific observation has been reduced to practice by Shi and Trzasko[223] who generated a high-amylose starch with a higher dietary fiber content. Under preferred conditions, linear chains within granules realign themselves in a more orderly manner, thus making it more difficult for amylase attack. The resulting granular, high-amylose starch has a peak gelatinization temperature greater than 110°C. Starch with high dietary fiber content, i.e. starch that is more resistant to digestion, is more desirable than traditional cereal dietary fiber which has higher water adsorbtivity and gives a gritty texture. In addition to its functional benefits, resistant starch has been associated with physiological benefits, i.e. lowering blood glucose and cholesterol concentrations and reducing the incidence of colon cancer.[224]

Non-chemically-modified starch that functions like chemically-modified starch is of great interest. Chemical modification is used to provide functionalities such as temperature, acid and shear tolerance during processing. Limited success has been achieved in simulating chemically crosslinked tapioca or potato starch by subjecting native amylose-containing starch to an annealing process.[225] Several new, functional, native starches which have functional performance equivalent to traditional chemically-modified starches have been prepared. Their viscosity and process-tolerance profiles are similar to those of chemically crosslinked starches. In addition, the starches have a unique sensory property and a beneficial flavor release profile in various food systems. One such product is prepared by adjusting the pH to an alkaline value and then drying it.[226] In another process, starch was treated with activated chlorine to produce a product that is temperature resistant like a conventially crosslinked starch.[227]

IX. Enzyme-catalyzed Modifications

Currently, only hydrolases (amylases) are used to modify starch. The use of amylases to produce products derived from hydrolysis of starch is described in Chapters 7, 20, 21 and 22. Starch hydrolyzates with good adhesion property that can be applied at high solids to minimize the energy required to remove moisture after application are very desirable for coating food items with seasonings, flavors and colorants. This property can be achieved by treating starch with an amylase or amylases to a dextrose equivalency (DE) (see Chapter 21) of 2–40.[228] Waxy maize is the preferred starch.

A low DE starch hydrolyzate with improved sweetness and browning capacity is prepared by treating starch with a combination of α- and β-amylase or α-amylase and glucoamylase (amyloglucosidase).[229] A further improved process employs a heat-stable α-amylase to convert the starch, with recovery of the products at high temperature.[230]

A starch hydrolyzate with a peak average molecular weight of <10 000 that is capable of forming a gel was prepared by treating high-amylose starch with an α-amylase to a DE between 5 and 15.[231] The hydrolyzate is useful for preparing a foodstuff with reduced fat. Hydroxypropyl high-amylose starch hydrolyzate functions as a fat replacer.[232]

A hydroxypropylated starch hydrolyzate with a DP 2–6 (DE 20–45) functions as a bulking agent. When combined with a high-intensity sweetener, it is useful as a reduced calorie replacement for sucrose.[233]

A low-viscosity granular starch can be produced by contacting raw corn starch granules with an α-amylase in water.[234] The preferred degree of hydrolysis is 0.1 to 1.0%.

Consumer healthy eating trends have generated the demand for functional fibers. Beta-glucan has been found to have desirable cholesterol-lowering benefits. Herwood[235] disclosed a process for degrading cereal flour. New processing and/or physical modification techniques continue to ensure a constant stream of functional starch products into the marketplace. Conventional hybrid breeding and genetic engineering offer the possibility of new base starches with varying properties that may show better performance in some applications than traditional native starches; chemical and/or physical modification of these new base starches should offer even more enhanced properties.

Concurrent with developments in starch technology, work is continuing with regulatory agencies to allow changes or additions to the list of approved starch derivatives. Enhanced characterization methods will aid this work and the development of new generations of starch products. A growing knowledge of the interactions of starch with other natural and synthetic polymers and various inorganic materials will allow the development of high performance, environmentally-friendly systems.

X. References

1. BeMiller JN. *Starch/Stärke*. 1997;49:127.
2. Gerritsen WJ. *Voeding*. 1995;56:32.
3. Friedman RB, Hauber RJ, Katz PR. *J. Carbohydr. Chem.* 1993;12:611.
4. Bruinenberg PM, Jacobson E, Visser RGF. *Chem. Ind (London)*. 1995:881.
5. Wurzburg OB, ed. *Modified Starches, Properties and Uses*. Boca Raton, FL: CRC Press; 1986.
6. Blanchard PH Industrial Chemistry Library. *Technology of Corn Wet Milling and Associated Processes*. Vol. 4. Amsterdam, The Netherlands: Elsevier Science Publishers B.V.; 1992.
7. Lammers G, Balt L, Stamhuis EJ, Beenackers AACM, PCT International Patent Appl. WO 96/06,866; 1996.
8. Lammers G, Stamhuis EJ, Beenackers AACM. *Starch/Stärke*. 1993;45:227.
9. Vezzani C. European Patent Appl. EP 710 670; 1996.
10. Kuipers N. *Carbohydr. Eur.* November 1995:28–29.
11. Wang L, Shogren RL, Willett JL. *Starch/Stärke*. 1997;49:116.
12. Franke GT, Venema B. Neth. Patent Appl. NL 9100249; 1992.
13. Wootton M, Bamunuarachchi A. *Starch/Stärke*. 1979;31:201.
14. White PJ, Abbas IR, Johnson LA. *Starch/Stärke*. 1989;41:176.
15. BeMiller JN. *Carbohydrate Chemistry for Food Scientists*. St Paul, MN: AACC International; 2007:173–223.
16. Jane J, Xu A, Radosavljevic M, Seib PA. *Cereal Chem*. 1992;69:405.
17. Robyt JF, Choe J-y, Hahn RS, Fuchs EB. *Carbohydr. Res.* 1996;281:203.
18. Robyt JF, Choe J-y, Fox JD, Hahn RS, Fuchs EB. *Carbohydr. Res.* 1996;283:141.
19. Eastman JE. US Patent 4 827 314; 1989.
20. Roberts JC. In: Roberts JC, ed. *Paper Chemistry*. Glasgow and London, UK: Blackie; 1991:114–131.
21. Lindstrom MJ, Savolainen RM. *J. Dispersion Sci. Technol*. 1996;17:281.
22. Yeh MH. US Patent 5 387 675; 1995.
23. *Federal Register*, October 3, 1996.
24. Fernandez J, Solarek D, Koval J. US Patent 5 093 159; 1992.
25. Buffa V, Johnson S. European Patent Appl. 542 236; 1993.

26. Tasset EL. US Patent 4 464 528; 1984.
27. Noguchi Y, Kadota T. Jpn. Patent 03 229 701; 1991.
28. Harvey RD, Hubbard D, Meintrup RA. US Patent 4 554 021; 1985.
29. Hubbard ED, Harvey RD, Hogen ML. US Patent 4 566 910; 1986.
30. Fuertes P, Druex J-L. US Patent 5 300 150; 1994.
31. Lim WJ, Lang YT, Seib PA. *Cereal Chem.* 1992;69:237.
32. Yook C, Sosulski F, Bhird PR. *Starch/Stärke.* 1994;46:393.
33. Huchette M, Fleche G, Gosset S. US Patent 4 613 407; 1986.
34. Hendricks J, Terpsta J. European Patent Appl. EP 703 314; 1996.
35. Faber A, Hulst AC. European Patent Appl. EP 737 777; 1996.
36. Wikstroem O. PCT International Patent Appl. WO 97 04 168; 1997.
37. Wikstroem O. PCT International Patent Appl. WO 97 04 167; 1997.
38. Stober R, Fischer W, Huss M, Udluft K. US Patents 4 785 087; 1988, 4 812 257; 1989.
39. Stober R, Fischer W, Huss M, Udluft K. US Patent 4 812 257; 1989.
40. Hellwig G, Bischoff D, Rubo A. *Starch/Stärke.* 1992;44:69.
41. Fischer W, Bischoff D, Huss M, Stober R, Roessler G. US Patent 5 116 967; 1992.
42. Ralvert K. US Patent 5 492 567; 1996.
43. Roerden DL, Wessels CD. US Patent 5 241 061; 1993.
44. Gunns J, Gielen JW. US Patent 4 906 745; 1990.
45. Meuser F, Gimmler N. *Starch/Stärke.* 1990;42:330.
46. Esan M, Brummer TM, Meuser F. *Starch/Stärke.* 1996;48:131.
47. Della Valle G, Colonna P, Tayeh J. *Starch/Stärke.* 1991;43:3.
48. Narkrugsa W, Berghofer E, Camargo LCA. *Starch/Stärke.* 1992;44:81.
49. Tetsuo S. Jpn. Patent 06 100 603 (94 100 603); 1994.
50. Shitai Y, Liu G, Zhang L. PRC Patent CN 1 050 024; 1991.
51. Kweon MR, Sosulski FW, Bhirud PR. *Starch/Stärke.* 1997;49:59.
52. Sosulski RW, Bhirud PR. PCT International Patent Appl. WO 96 09 327; 1996.
53. Tsai JT, Trzasko PT, Philbin MT, Billmers RL, Tessler MM, Van Gomple JA, Rutenberg MW. US Patent 5 227 481; 1993.
54. Roerdon DL, Frank RK. US Patent 5 616 800; 1997.
55. Juergen HD, Pfohl S, Weberndoerfer V, Rehmer G, Kroener M, Stange A. US Patent 4 818 341; 1989.
56. Linhart F, Stange A, Schuhmacher R, Hartmann H, Denzinger W, Niessner M, Nilz C, Reuther W, Meixner H. German Offen. DE 4 438 708 A1; 1996.
57. Stange A, Degen HJ, Auhorn W, Weherndoerfer V, Kroener M, Hartmann H. US Patent 4 940 514; 1990.
58. Hartmann H, Denzinger W, Kroener M, Nilz C, Linhart F, Stange A. US Patent 5 543 459; 1996.
59. Dikler YE, Oleinik AT, Burovikhin A. Russian Patent RU 2 026 913 C1 (1995); *Chem. Abstr.*, 1995; 124: 32360.
60. Solarek DB, Dirscherl TA, Hernandez HR, Jarowenko W. US Patents 4 876 336; 1989. 4 964 953; 1990.
61. Bernhard K, Tsai J, Billmers RL, Sweger RW. US Patent 5 455 340; 1995.
62. Sweger R, Tsai J, Pasapane J, Bernhard K. US Patent 5 482 704; 1996.

63. Eklund D, Lindstrom T. *Paper Chemistry: An Introduction*. Grankulla, Finland: DT Paper Science Publications; 1991:160–164.
64. Gosset S, Lefer P, Fleche G, Schneider J. US Patent 5 129 9899; 1992.
65. Owen D. PCT International Patent Appl. WO 96 05 373; 1996.
66. Sunden O, Sunden A. US Patent 4 710 270; 1987.
67. Taggart TE, Schuster MA, Schellhamer AJ. US Patent 5 104 487; 1992.
68. Kato M, Takasaki K. Japan Patent 6173193 (JP 941731963); 1994.
69. Nippon Shokuhin Kako KK (Niso). Japan Patent 08296193; 1996.
70. Ueda T. Japan Patent 04 363 301 (92 363 301); 1992.
71. Kondo J, Ikeda Y, Moriwaki H. Japan Patent 05 195 488 (93 195 488); 1993.
72. Kondo J, Inoe M, Urushibata H. Japan Patent 04 289 293 (92 289 293); 1992.
73. Kondo J, Inoe M, Moriwaki H. Japan Patent 04 289 295 (92 289 295); 1992.
74. Vihervaara T, Paakkanen M. FI Patent 90 679; 1992.
75. Kettunen K. *Pap. Puu.* 1996;78:265.
76. Hoffman J. *Wochenbl. Papierfabr.* 1994:785.
77. Johansson HE, Johansson K, Klofvers, Kloefver S. US Patent 277 764; 1994.
78. Batelson PG, Johnasson HE, Larsson HM, Svending PJ. US Patent 4 385 961; 1983.
79. Sunden O, Batelson PG, Johansson HE, Larsson HM, Svending PJ. US Patent 4, 388 150; 1983.
80. Johnson KA. US Patents 4 643 801; 1987, 4 750 974; 1988.
81. Nissan Chemical Industries Ltd., Japan Patent 02 12 096; 1990.
82. Rushmere JD. US Patent 5 185 206; 1993.
83. Rushmere JD. US Patent 4 954 220; 1990.
84. Maeda K. Japan Patent 02 l33 695; 1990.
85. Anderson KR. US Patent 5 122 231; 1992.
86. Solarek D, Peek L, Henley MJ, Trksak RM, Philbin MT. US Patent 5 368 690; 1994.
87. Britt KW. In: Casey JP, ed. *Pulp and Paper Chemistry and Chemical Technology.* 3rd edn. Vol. III. New York, NY: John Wiley; 1981:1618.
88. Hofreiter BT. In: *Wet-Strength in Paper and Paperboard, TAPPI Monograph Series,* No. 29. New York, NY: TAPPI; 1965.
89. Borchert PJ, Mirza J. *TAPPI.* 1964;47:208.
90. Neal CW. A Review of the Chemistry of Wet-Strength Development. In: *TAPPI Wet and Dry Strength Short Course Notes;* 1988: 1–24.
91. McGuire TA, Mehltretter CL. *Stärke.* 1971;23:42.
92. Veelaert S, DeWit D, Tourniois H. PCT Patent Appl. WO 9512619; 1995.
93. Veelaert S. *Dialdehyde Starch: Production, Properties and Applications.* Belgium: Ghent University; August 1996. [Thesis].
94. Solarek D, Jobe P, Tessler MM. US Patent 4 675 394; 1987.
95. Solarek D, Tessler MM, Jobe P, Peek L. *Cationic Starch Aldehydes – Wet End Additives for Temporary Wet Strength and Improved Dry Strength.* Paper Chemistry Symposium Stockholm, Sweden: STFI and SPCI organizers; 1988 Sept.: 27–290
96. Laleg M, Pikulik II. *J. Pulp Paper Sci.* 1991;17:206.
97. Smith DJ, Headlam MM. US Patent 5 656 746; 1997.
98. Billmers RL. US Patents 4 973 680; 1990, 5 004 791; 1991.

99. Gotlieb KF, Bleeker IP, Van Doren HA, Heers A. European Patent Appl. 710 671; 1996.
100. Nguyen CC, Martin VJ, Pauly EP. US Patents 5 003 022; 1991, 5 130 394; 1992.
101. Nguyen CC, Martin VJ, Luebke GR, Pauly EP, Tupper DE. US Patent 5 130 395; 1992.
102. Nguyen CC, Martin VJ, Pauly EP, Buccigross HL, Rudolph S. US Patents 5 416 181; 1995, 527 544; 1996.
103. Nguyen CC, Martin VJ, Pauly EP. US Patent 5 565 509; 1996.
104. Rinck G, Moller K, Fullert S, Krause F, Koch H. US Patent 5 147 907; 1992.
105. Bodiger M, Demharter S, Mulhaupt R. In *Carbohydrates as Organic Raw Materials III*. New York, NY: VCH publishers; 1996:141–154.
106. Demharter S, Riechtering R, Mulhaupt R. *Polym. Bull.* 1995;34:271.
107. Demharter S, Frey H, Drechsler M, Mulhaupt R. *Colloid Polym. Sci.* 1995;273:661.
108. Hietanummi M, Sievers V, Nurmi K, Hamunen A. European Patent Appl. 545 228; 1993.
109. Nguyen CC, Tupper DE. US Patent 5 536 764; 1996.
110. Ritter W, Gardenier KJ, Kempf W. German Offen. DE 4 038 700; 1992.
111. Floyd WC, Dragner LR. US Patent 5 116 890; 1992.
112. Kruger L. US Patent 4 838 944; 1989.
113. van der Wiele K. *Carbohydr. Eur.* 1995;13:3.
114. Koch H, Beck R, Roper H. *Starch/Stärke.* 1993;45:2.
115. van Bekkum H, Besemer AC. *Carbohydr. Eur.* 1995;13:16.
116. Besemer AC, van Bekkum H. In: van Bekkum H, Roper H, Voragen F, eds. *Carbohydrates as Organic Raw Materials III*. New York, NY: VCH Publishers; 1996:273–293.
117. Beck R. In: Finch CA, ed. *Industrial Water Soluble Polymers*. Cambridge, UK: The Royal Society of Chemistry; 1996:76–91.
118. Beck RHF, Lemmens HOJ. Eur. Patent Appl. 755 944; 1997.
119. Engelskirchen K, Fischer H, Verholt HW. German Offen. DE 4 203 923; 1993.
120. Engeiskirchen K, Fischer H, Kottwitz B, Upadek H, Nitsch C. US Patent 5 501 814; 1996.
121. Floor M. NL Patent 8 802 907; 1990.
122. Floor M, Hofsteede LPM, Groenland WPT, Verharr LATh, Kieboom APG, van Bekkum H. *Rec. Trav. Chim. Pays-Bas.* 1989;108:384.
123. Besemer AC, van Bekkum H. *Starch/Stärke.* 1996;46:95 101.
124. Besemer AC. European Patent Appl. EP 427 349; 1991.
125. Besemer AC. PCT International Patent Appl. WO 95/07303; 1995.
126. van Bekkum H, de Nooy AEJ, Besemer AC. *Tetrahedron.* 1995;51:8023.
127. van Bekkum H, de Nooy AEJ, Besemer AC. *Carbohydr. Res.* 1995;269:89.
128. de Nooy AEJ, Besemer AC, van Bekkum H, van Dijk JAPP, Smit JAM. *Macromol.* 1996;29:6541.
129. Floor M, Kiebom APG, van Bekkum H. *Starch/Stärke.* 1989;41:348.
130. Heeres A, Bleeker I, Gotlieb K, Van Doren HA. PCT International Patent Appl. WO 96/38484; 1996.
131. Fitt LE, McNary HT. US Patent 5 385 608; 1995.

132. Doane WM. *Cereal Foods World*. 1994;39:556.
133. Roper H, Koch H. *Starch/Stärke*. 1990;42:123.
134. Shogren RL, Fanta GF, Doane WM. *Starch/Stärke*. 1993;45:276.
135. Albertson AC, Karlsson S. *Acta Polym.* 1995;46:114.
136. Poutanen K, Forssell P. *Trends Polym. Sci.* 1996;4:128.
137. Stepto RFT, Dobler B. Eur. Patent Appl. 326 517; 1989.
138. Tomka, PCT Int. Patent Appl. WO 91/16375; 1991.
139. Wittwer F, Tomka I, Bodenmann T, Raible J, Louis SG. US Patent 4 738 724; 1998.
140. Witter F, Tomka I. US Patent 4 673 438; 1987.
141. Maddever WJ, Chapman GM. *Plastics Eng.* July 1989:31.
142. Griffin GJL. US Patents 4 016 177; 1977, 4 021 388; 1978, 4 125 495; 1978.
143. Griffin GJL. PCT International Patent Appl. WO 88/9354; 1988.
144. Evangelista RL, Sung W, Jane JL, Gelina RJ, Nikolov ZL. *Ind. Eng. Chem. Res.* 1991;30:1841.
145. Nakajima T, Takagi S, Ueda T. Japan Patent 04 202 543 (92 202 543); 1992.
146. Willet JL. US Patent 5 087 650; 1992.
147. Otey FH, Westhoff RP. US Patents 4 133784; 1979, 4 337 181; 1982.
148. Otey FG, Westhoff RP, Doane WM. *Ind. Eng. Chem. Res.* 1987;26:1659.
149. Fanta GF, Swanson CL, Doane WM. *J. Appl. Polym. Sci.* 1990;40:811.
150. Shogren RL, Greene RV, Wu YV. *J. Appl. Polym. Sci.* 1991;42:1701.
151. Fanta GF. Christianson DE, US Patent 5 550 177; 1996.
152. Bastiotti C, Lombi R, Del Tredicii G, Guanella I. European Patent Appl. 400 531; 1990.
153. Bastiotti C, Belloti V, Del Giudice L, Del Tredici G, Lombi R, Rallis A. PCT International Patent Appl.WO 90/1067; 1990.
154. Bastiotti C, Del Giudice L, Lombi R. PCT International Patent Appl. WO 91/02024; 1991.
155. Bastiotti C, Del Giudice L, Lombi R. PCT International Patent Appl. WO 91/02025; 1991.
156. Bastiotti C, Del Tredici G. PCT International Patent Appl. WO 91/02025; 1991.
157. Bastiotti C, Belloti V, Del Tredici G, Lombi R, Montino A, Ponti R. PCT International Patent Appl. WO 92/19680; 1992.
158. Wu PC, Palmer GW, High WR. PCT International Patent Appl. WO 93/03098; 1993.
159. Sachetto J-P, Silberger J, Lentz DJ. European Patent Appl. 408 501; 408 502; 408 503; 409 781; 409 782; 409 783; 1991.
160. Sachetto, J-P, Rehm J. European Patent Appl. 407 350; 1991.
161. Lacourse NL, Altieri PA. US Patents 4 863 655; 1989, 5 035 196; 1991, 5 043 196; 1991.
162. Altieri PA. US Patent 5 153 037; 1992.
163. Altieri PA, Tessler MM. US Patent 5 554 660; 1996.
164. Kazemzadeh M. PCT International Patent Appl. WO 92 13 004; 1992.
165. Borghuis J, Dijksterhuis JF. German Offen. DE 4 424 946; 1996.
166. Lawton JW. *Carbohydr. Polym.* 1996;29:203.

167. Weigel P, Schwarz W, Frigge K, Vorwerg W. European Patent Appl. EP 735 080; 1996.
168. De Bock ILHA, Van Den Broecke PMR, Bahr KH. European Patent Appl. EP 599 535; 1994.
169. Tessler MM, Billmers RL. *J. Environ. Polym. Degrad.* 1996;4:85.
170. Lower ES. *Iv. Ital. Sostanze Grasse.* 1996;73:159.
171. Billmers RL, Tessler MM. US Patent 5 321 132; 1994.
172. Billimers RL, Paul CW, Hatfield, SF, Kauffman TF. US Patent 5 360 845; 1994.
173. Frische R, Gross-Lannert R, Wollmann K, Schmid E, Buehler F, Best B. PCT International Patent Appl. WO 92 19 675; 1992.
174. Reinisch G, Radics U. German Offen. DE 4 423 681; 1996.
175. Reinisch G, Roatsch B. German Offen. DE 4 425 688; 1996.
176. Frische R, Best B. European Patent Appl. EP 579 197; 1994.
177. Borches G, Dake I, Dinkelaker A, Sacheto JP, Zdrahala R, Rimsa SB, Loomis G, Tatarka PD, Mauzac O. PCT International Patent Appl. WO 93 20 141; 1993.
178. Rimsa S, Tatarka P. PCT International Patent Appl. WO 94 07 953; 1994.
179. Ishikawa H, Tanaka H, Kuno T, Utsue I. PCT International Patent Appl. WO 95 15 414; 1995.
180. Zhao J, Whistler RL. *Food Technol.* 1994;48:104.
181. Whistler RL. US Patent 4 985 082; 1991.
182. Shasha RS, McGuire MR. US Patent 5 061 697; 1991.
183. Tomka I, Sala R. PCT International Patent Appl. WO 91/11178; 1991.
184. Eden J, Trksak R, Williams R. US Patent 4 812 445; 1989.
185. Eden J, Trksak R, Williams R. US Patent 4 755 397; 1988.
186. Doane WM, Maiti S, Wing RE. US Patent 4 911 952; 1990.
187. Carr ME, Doane WM, Wing RE, Bagley EB. US Patent 5 183 690; 1993.
188. Mateescu MA, Lenaerts V, Dumoulin Y. US Patent 5 456 921; 1995.
189. Knight AT. Canadian Patent Appl. CA 2 025 367; 1991.
190. Eskins K, Fanta GF. US Patent Appl. 233 173; 1994.
191. Fanta GF, Eskins K. *Carbohydr. Polym.* 1995;28:171.
192. Eskins K, Fanta GF, Felker FC, Baker FL. *Carbohydr. Polym.* 1996;29:233.
193. Cunningham RL, Gordon SH, Felker FC, Eskins K. *J. Appl. Polym. Sci.* 1997;64:1355.
194. Kubik S, Wulff G. *Starch/Stärke.* 1993;45:220.
195. Kubik S, Holler O, Steinert A, Tolksdorf M, Van der Leek Y, Wulff G. In: van Bekkum H, Roper H, Voragen AGJ, eds. *Carbohydrates as Organic Raw Materials.* New York, NY: VCH Publishers; 1996:169–187.
196. Wulff G, Kubik S. *Macromol. Chem.* 1992;193:1071.
197. Wulff G, Kubik S. *Carbohydr. Res.* 1992;237:1.
198. Trubiano PC. In: Wurzburg OB, ed. *Modified Starches: Properties and Uses.* Boca Raton, FL: CRC Press; 1986:131–147.
199. Trubiano PC, Hussain Z. National Starch & Chemical Company, unpublished results.
200. Boskovic MA, Vidal SM, Saleeb FZ. US Patent 5 124 162; 1992.
201. Morehouse AL. Can. Patent Appl. 2 034 639; 1990.

202. Levine H, Slade L, Van Lengerich B, Pickup JG. US Patent 5 087 461; 1992.
203. Altieri P, Eden J, Gribnau MC, Hoogendijk L, Krijnen LB, Solarek DB, Swarthoff T. US Patent 5 433 884; 1996.
204. Fitton MG. US Patent. 5 505 783; 1996.
205. Thurston RA, McConiga RE. US Patent 3 617 383; 1971.
206. Eastman JE, Moore O. US Patent 4 465 702; 1984.
207. Eastman JE. US Patent 4 634 596; 1987.
208. Jane J-L, Seib PA. US Patent 5 057 157; 1991.
209. Katt JL, Moore CO, Eastman JE. US Patent 4 623 549; 1986.
210. Rajagopalan S, Seib PA. US Patent 5 037 929; 1991.
211. Pitchon E, O'Rourke JD, Joseph JH. US Patent 4 280 851; 1981.
212. Schara RE, Katcher JH. US Patent 4 847 371; 1989.
213. Rubens RW. US Patent 5 149 799; 1992.
214. Dudacek WE, Kochan DA, Zobel HF. US Patent 4 491 483; 1985.
215. Mudde JP. US Patent 4 508 576; 1985.
216. Conde-Petit B, Escher F. US Patent 5 291 877; 1994.
217. Chiou RG, Brown CC, Little JA, Young AH, Schanefelt RV, Harris DW, Stanley KD, Coontz HD, Hamdan CJ, Wolf-Rueff JA, Slowinski LA, Anderson KR, Lehnhardt WF, Witczak ZJ. EP Patent 443 844 A1; 1991.
218. Chiou RG, Brown CC, Little JA, Young AH, Schanefelt RV, Harris DW, Stanley KD, Coontz HD, Hamdan CJ, Wolf-Rueff JA, Slowinski LA, Anderson KR, Lehnhardt WF, Witczak ZJ. US Patent 5 378 286; 1995.
219. Stanley KD, Harris DW, Little JA, Schanefelt RV. US Patent 5 409 726; 1995.
220. Jane J-L, Shen L, Maningat CC. *Cereal Chem.* 1990;69:280.
221. Whistler RL. US Patents 5 445 678; 1995, 5 580 390; 1996.
222. Hoover RH, Vasanthan T. *Carbohydr. Res.* 1994;252:33.
223. Shi YC, Trzasko PT. US Patent 5 593 503; 1997.
224. Björck I, Asp NG. *Trends Food Sci. Technol.* 1994;5:213.
225. Wurzburg OB, Kruger LH. US Patent 3 977 897; 1976.
226. Chiu C-W, Schiermeyer E, Thomas DJ, Shah MB. US Patent 5 725 676 1998; *Chem. Abstr.* 1998; 128: 193,928.
227. Kettliz BW, Coppin JV. US Patent 6 235 894; 2001.
228. Chiu C-W, Huang DP, Kasica JJ, Xu ZF. US Patent 5 599 569; 1997.
229. Morehouse AL, Sander PA. US Patents 4 699 670; 1987, 4 699 669; 1987, 4 684 410; 1987.
230. Morehouse AL, Sander PA. US Patent 4 782 143; 1988.
231. Harris DW, Harris JA. US Patent 5 435 019; 1995.
232. Furcski SL, Mauro DJ, Kornacki J, Faron EJ, Turnak FL, Owen R. US Patent 5 094 872; 1992.
233. Quarles JM, Alexander RJ. US Patent 5 110 612; 1992.
234. Kobayashi S, Miwa S, Tsuzuki W. US Patent 5 445 950; 1996.
235. Senkeleski J, Chiu CW, Xu Z-F, Mason WR, Chicalo-Kaighn KL. US Patent 5 562 937; 1996.
236. Chiu CW. US Patent 5 185 176; 1993.
237. Chiu CW. US Patent 4 977 252; 1990.

238. Ammerall R, Friedman R. US Patent 5 482 560; 1996.
239. Kaper FS, Aten J, Reinders MA, Dijdstra P, Suvee AJ. US Patent 4 780 149; 1988.
240. Yoshida M, Hirao M. US Patent 3 879 212; 1975.
241. Hathaway RJ. US Patent 3 556 942; 1971.
242. Chiu CW. US Patent 4 971 723; 1990.
243. Chui CW, Zallie JP. US Patent 4 886 678; 1989.
244. Zallie JP, Chiu CW. US Patent 4 937 091; 1990.
245. Chiu CW. US Patent 5 089 171; 1992.
246. Chiu CW, Henly M. US Patent 5 194 284; 1993.
247. Little JA, Scobell HD. US Patent 5 372 835; 1994.
248. Harris DW, Little JA. European Patent Appl. WO 93/03629 A1; 1993.
249. Harris DW, Little JA. US Patent 5 395 640; 1995.
250. Harris DW, Little JA. US Patent 5 374 442; 1994.
251. Stanley KD, Harris DW, Little JA, Schanefelt RV. US Patent 5 409 726; 1995.
252. Martino G, Chiu CW. US Patent 5 496 861; 1996.
253. Chiu CW, Kasica JJ. US Patent 5 468 286; 1995.
254. Besemer AC, van der Lugt JP. PCT Int. Patent Appl. WO 94/0191; 1994.
255. Besemer AC, Lerk CF. PCT Int. Patent Appl. WO 94/01092 A1; 1994.
256. Chiu CW, Henley M. US Patent 5 281 276; 1994.
257. Henley M, Chiu CW. US Patent 5 409 542; 1995.
258. Lyengar R, Zaks A, Gross A. US Patent 5 051 271; 1991.
259. Whistler RL. US Patent 4 985 082; 1991.

Starch in the Paper Industry

Hans W. Maurer
Highland, Maryland 20777

I.	Introduction to the Paper Industry	658
II.	The Papermaking Process	660
III.	Starch Consumption by the Paper Industry	662
IV.	Starches for Use in Papermaking	663
	1. Current Use	663
	2. Recent Trends	665
V.	Application Requirements for Starch	666
	1. Viscosity Specifications	666
	2. Charge Specifications	668
	3. Retrogradation Control	669
	4. Purity Requirements	671
VI.	Dispersion of Starch	672
	1. Delivery to the Paper Mill	672
	2. Suspension in Water	673
	3. Dispersion Under Atmospheric Pressure	674
	4. Dispersion Under Elevated Pressure	674
	5. Chemical Conversion	676
	6. Enzymic Conversion	677
VII.	Use of Starch in the Papermaking Furnish	681
	1. The Wet End of the Paper Machine	681
	2. Flocculation of Cellulose Fibers and Fines	681
	3. Adsorption of Starch on Cellulose and Pigments	682
	4. Retention of Pigments and Cellulose Fines	683
	5. Sheet Bonding by Starch	684
	6. Wet-end Sizing	685
	7. Starch Selection for Wet-end Use	687
VIII.	Use of Starch for Surface Sizing of Paper	688
	1. The Size Press in the Paper Machine	688
	2. The Water Box at the Calender	693
	3. Spray Application of Starch	693
	4. Starch Selection for Surface Sizing	693
IX.	Use of Starch as a Coating Binder	695
	1. The Coater in the Paper Machine	695
	2. Starch Selection for Paper Coating	698

X.	Use of Starch as Adhesive in Paper Conversion	700
	1. Lamination of Paper	700
	2. The Corrugator for Paperboard	700
	3. Starch Selection for Use in Corrugation and Lamination	702
XI.	Use of Starch in Newer Specialty Papers	703
XII.	Environmental Aspects of Starch Use in the Paper Industry	703
XIII.	Starch Analysis in Paper	705
XIV.	References	706

I. Introduction to the Paper Industry

The forest and paper products industry accounts for more than 6% of the total manufacturing output of the United States. It employs about 1.2 million people and ranks among the top 10 manufacturing employers in 42 states. Sales of forest and paper products exceed $230 billion annually, both domestically and abroad.[1]

The paper industry is capital and energy intensive. A modern paper machine costs about $400 million, requires three to five highly-trained people per shift, and uses 8 to 10 million BTU per metric ton of product, depending on machine layout and the manufacturing process applied. The industry's resources are, however, renewable, which is further highlighted by the facts that it plants more trees than it cuts, grows more wood than it harvests and generates 60% of the energy that it consumes.

Total US production of paper in 2005 was 37.6 million metric tons, and production of paperboard was 45.7 million tons at operating rates of 92% and 97% respectively. Production data and projections are analyzed by RISI,[1] and periodic updates on industry issues, grade profiles and production statistics are published in *Pulp and Paper, Paper Age, Pulp and Paper International* and other trade journals. The US paper industry reuses more than 51% of the product it produces with an ultimate goal of 55%.[2]

Paper grades are specified as wood-containing or wood-free, bleached or unbleached and coated or uncoated. Printing papers, which are generally bleached, are further separated by appearance ranking (numbers 1 through 5), which groups paper grades by brightness and opacity. Printing papers are finished to specific levels of smoothness and gloss with classifications of glossy, dull or matte. Packaging papers are mostly unbleached and graded by weight, fiber source, caliper and strength properties. Higher quality board grades are bleached (SBS) or coated to generate a white surface.

Paper products are sold into many market segments. Major grades are printing and writing papers, SBS board, linerboard, corrugating medium, saturating Kraft, container board, wallpaper and release liner. Tissue making refers to the production of soft absorbent grades with basis weights of 8 to 30 lb per 3000 ft^2 (13–32 g/m^2). Medium weight sheets for printing and writing are produced in the basic weight range of 30 to 100 lb/3300 ft^2 (45–150 g/m^2) and have smoothness, ink holdout and strength properties as needed for specific end uses. Board making refers to the production of heavyweight sheets in the basic weight range of 26 to 400 lb/1000 ft^2 (125–2000 g/m^2) with properties designed for packaging. The US demand for paper and paperboard grades by market segment is illustrated in Table 18.1. Growth predictions, derived from production data in 2005, range from −0.3% for total paper production to +1.4% for paperboard.[3,4]

There are numerous handbooks and monographs that describe the paper industry, its processes, and products.[5–7] The American Forest and Paper Association[2] is the US national trade association of the forest, pulp, paper, paperboard and wood products industry, and represents most US paper producing companies. TAPPI, the Technical Association of the Pulp and Paper Industry,[8] is an international professional society serving the pulp and paper, packaging, converting and allied industries. TAPPI collects and disseminates technical information for engineers, scientists and students who work in these industries, and promotes their professional development. Reports on scientific and technical progress in pulp and paper manufacturing are published in two monthly periodicals: *TAPPI Journal* and the *Journal for Pulp and Paper Science*, the latter in cooperation with the Canadian Pulp and Paper Association. A monograph 'Starch and Starch Products in Surface Sizing and Paper Coating' was published in 2001 by TAPPI PRESS;[9] a complementary 'Starch and Starch Products for Wet End Application' is scheduled for publication in 2007. Another recent monograph[10] and a research report[11] describe synthesis, properties and uses of modified starch. Previous editions of 'Starch: Chemistry and Technology' appeared in 1967[12] and 1984.[13] Leading international journals which report on progress in paper manufacturing and the use of starch are *Das Papier*, *Wochenblatt für Papierfabrikation*, *Nordic Pulp and Paper Research Journal*, *Paperi ja Puu*, *Japan Tappi Journal*, *Paper Technology*, *Appita Journal*, and *Starch/Stärke*.

Table 18.1 US production of paper and paperboard grades by market segment

Market segments	Production[a]	
	1998	2005
Printing and writing, total	26 501	24 495
• Uncoated groundwood paper	2062	1859
• Coated groundwood paper	4370	4704
• Uncoated free sheet	13 605	12 016
• Coated free sheet	4932	4626
Newsprint	7229	5392
Packaging papers	4425	4107
Linerboard	20 911	21 218
Corrugating medium	9369	10 212
Boxboard	14 994	14 813

[a]Thousand short tons

Reprinted by permission of TAPPI

II. The Papermaking Process

Paper is a hydrogen-bonded mat of cellulose fibers. The product may consist of softwood fibers (3–4 mm length, 30–45 μm width), hardwood fibers (1.0–1.5 mm length, 20 to 30 μm width), deinked fibers from paper recycling, and broke (a blend of fibers that is recycled within the papermaking process). Paper often contains pigments as filler and a surface coating to enhance brightness and opacity, dyes to give it a specific color, sizing agents to decrease its sensitivity to water after drying, and bonding agents to raise its internal and surface strength.

The papermaking process consists of several major steps: stock preparation, sheet forming, pressing, drying and surface finishing. A modern paper machine (Figure 18.1), although huge in size, is a precision device. Recent installations have a width of more than nine meters and are operated at speeds up to 1800 meters/minute. The production rate for a 75 g/m² (50 lb/ream) sheet on an eight meter wide machine at 1000 m/min is about 600 kg/min (864 tons/day).

Figure 18.1 (a) A modern high-speed paper machine (Courtesy of Voith Paper). (b) An on-machine coater installation (Beloit Corporation).

An aqueous dispersion of cellulose fibers, drawn from an integrated pulping process or obtained by dispersing purchased dry or wet pulp, is screened, refined, blended, cleaned and deaerated prior to use as furnish on the paper machine. In the refining or 'beating' process, fibers are passed through narrow gaps between fast-turning profiled disks, which fibrillate their outer layers, thus increasing fiber surface area through swelling and the propensity for hydrogen bonding between fibers during drying. The process is energy-intensive and generates some cut fibers and fiber fines in the product.

Refined fibers are combined with filler pigments, dyes, sizing agents and strength additives to form the papermaking furnish. At various stages during the process, flocculants are added to improve retention of the additives. Major flocculation agents are cationic starch, poly(acrylamides), poly(ethylene imine) and various aluminum compounds. Most flocculants have a cationic charge. New retention systems utilize amphoteric starch, dual systems of cationic and anionic polymers and small mineral particles, including nanoparticles. The papermaking furnish is diluted to a concentration of 0.3% to 0.8%, deaerated and delivered to the paper machine.

The stock is pumped through a manifold into the headbox of the paper machine, where the stock flow is decelerated and distributed over the width of the machine. Various baffles and step diffusors are used to avoid vortex flow and stagnation zones. The furnish leaves the headbox through the slice, a narrow gap with controlled profile, and impacts on one or two endless screens, the so-called papermakers wire. Water is removed from the fiber mat by the action of foils and vacuum.

The sheet leaves the wire with a water content of about 80%. The sheet is tender and must be supported by felts or fabrics for subsequent passage through the press section of the machine. By pressing, the sheet is dewatered to a solids concentration of 40% to 50%. Water removed from the web is collected in felts or fabrics and removed by vacuum foils. After pressing, the sheet passes through a system of steam-heated cylinders. During drying, the sheet is restrained by endless fabrics, which inhibit crossdirection shrinkage. Finally, the paper is collected on a reel.

Finishing processes are utilized to improve the surface of paper and to control its thickness (caliper) and density. In the calendering process, previously dried paper is compressed by one or more nips between hydraulically loaded cylinders, which improves paper smoothness and decreases its caliper. Machine calendering prepares the sheet for a subsequent coating application. Supercalendering, a multiple nips operation with alternating hard and soft rolls, is used to render uncoated paper suitable for color printing. When applied to coated paper, it will generate high levels of smoothness and gloss. In new installations, plastic (soft)-covered rolls and high temperature are used to achieve further gains in appearance and printability.

Surface sizing is practiced within the paper machine after about two-thirds of the drier section. A dispersion of starch or other binder, with or without pigments and sizing agents, is applied to paper while it passes through the nip between two rolls, the so-called size press. Surface sizing is primarily used to increase the strength of the paper surface, especially its resistance to picking and linting in offset printing. In alternative systems, starch is added to paper by spraying onto the wire of the machine or by feeding from a water box at a calender.

High-quality paper grades for printing are produced, on the paper machine or in a separate process, by coating the sheet surface with a concentrated dispersion of pigments, binders and additives. Starch is a major coating binder. In one system roll coaters, a train of hydraulically loaded rolls, occasionally with a cell (gravure) pattern, are used. The coating color is metered by film splitting through a sequence of two or more nips prior to application to the paper. Current practice relies primarily on the use of various forms of blade coaters. The coating color is applied to paper by a pickup roll, an overflow device (SDTA) or a fountain (jet). A stiff (scraping) or a bent (gliding) blade is used to remove the excess. This process levels the coating on the sheet and generates a substrate for quality printing. In board coating, a pressurized air curtain (air knife) is often used to remove excess coating fluid from the sheet. In the newest technical development, a free-falling curtain of a coating is applied to the surface of paper or paperboard. Corrugating and laminating are subsequent converting processes for paper and board that require large quantities of starch.

III. Starch Consumption by the Paper Industry

Starch is an important component of many paper grades. Starch consumption by weight in papermaking and paper conversion processes ranks third after cellulose fiber and mineral pigments. Starch is used as a flocculant and retention aid, as a bonding agent, as a surface size, as a binder for coatings and as an adhesive in corrugated board, laminated grades and other products.[14] Current consumption of industrial corn starch for paper and paperboard production in the US exceeds 2.5 billion pounds (1.1 million metric tons) of which >40% is chemically modified. Another 750 million pounds are used for corrugated and laminated paper products.[15] Data for starch use are summarized in Table 18.2. The shipment reports of CRA, the Corn Refiners Association,[15] are the main source for starch consumption data.

Table 18.2 North American demand for starch in the manufacture of paper products

Application	Starch grade	Actual use 1995[a]	Projected for 2000[a]
Wet end[b]	Corn starch	309	424
	Potato starch	287	349
Size press[c]	Unmodified starch	819	839
	Oxidized starch	718	703
	Hydroxyethylated starch	735	1034
	Cationic starch	118	137
Coating[c]	Unmodified starch	212	203
	Oxidized starch	55	81
	Hydroxyethylated starch	275	340
Corrugating and laminating[d]	Unmodified starch	899	
	Modified starch	134	

[a]Million pounds
[b]Cationic, anionic or amphoteric
[c]All corn starch
[d]1994 demand, members of CRA only
Reprinted by permission of TAPPI

IV. Starches for Use in Papermaking

1. Current Use

The major starch sources are corn, potato, waxy maize, wheat and tapioca. Refined starches are supplied in powder form or as slightly aggregated pearl starch.[16] Unmodified (native) starch is rarely used in the paper industry, except as a binder for laminates and in the corrugating process. Most starches for use in papermaking are specialty products that have been modified by controlled hydrolysis, oxidation or derivatization.[17]

Acid-hydrolyzed (thinned) starches are produced through depolymerization by hydrochloric or other acids.[18,19] Their low viscosity stability restricts their utilization in paper mills. They are used primarily for surface sizing at the size press or at the calender stack.

Oxidized starches are obtained by reaction with sodium hypochlorite or a peroxide.[20,21] They are available in a broad range of viscosity specifications and widely used as low-cost surface sizing agents or coating binders. Bleached starches, a subgroup with a carboxyl content below 0.3%, are partially crosslinked due to their carbonyl content[22] and resulting acetal bonding in the starch granule. Bleached starches require thermal conversion under pressure or high shear during cooking for complete dispersion. They are exclusively utilized in surface sizing. In recent years, use of oxidized starches has declined due to their dispersing properties, which will affect pigment retention and raise the TTS (total suspended solids) content of paper mill discharges. Oxidized corn starch may contain a small quantity of AOX compounds from the reaction of associated lipids with chlorine. The oxidation and depolymerization of starch can also be conducted in the paper mill prior to use of the product on the paper machine.[23]

Starch ethers are produced by a nucleophilic substitution reaction with an ethylenically unsaturated monomer, followed by acid-catalyzed hydrolysis for viscosity adjustment.[24] Widely-used substituents are ethylene oxide, epichlorohydrin, propylene oxide or reagents that contain a reactive halogen. The original reaction in aqueous suspension at elevated temperature has been supplemented by a gas/solid process in order to avoid excessive swelling of starch granules, to suppress side reactions of the monomer and to meet environmental restrictions on plant discharges.[25,26] Starch ethers are available in ranges of viscosity specifications (Figure 18.2). They are the products of choice for most coating and surface sizing applications. Starch esters are obtained by reaction with acetic acid anhydride or acid chlorides.[27] Their use in papermaking is limited, since they will saponify at alkaline pH.

Cationic starches are obtained by nucleophilic substitution reaction with tertiary or quaternary amines, using wet or dry production processes.[28,29] They are primarily utilized as retention or drainage aids in the paper-forming process. Another rapidly growing application is the use of cationic starch in surface sizing, primarily by mills that have strict waste discharge limits. Development work is under way to formulate and utilize cationic coatings, but so far their application has been limited by compatibility problems, difficulties with cationic pigment dispersion and an insufficient

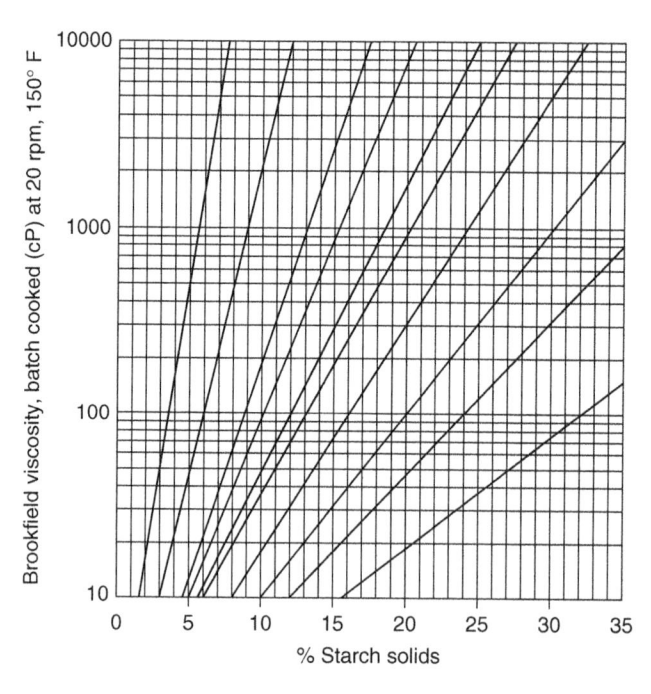

Figure 18.2 Relationship between solids content and viscosity for a series of commercial starch ethers. (Reprinted by permission of TAPPI)

supply of cationic dispersants. Cationic starch can be produced in the paper mill for use as a wet-end additive.

Anionic starches are obtained by reaction with phosphoric acid and alkali metal phosphates or by derivatization with carboxymethyl groups.[30,31] This modification is primarily used to introduce amphoteric properties into cationic corn starch for application on the wet end of the paper machine. Anionic starches with carboxymethyl substitution are used as thickeners in coating colors or as binders in coatings for specialty paper grades. Oxidized starches are inherently anionic but without thickening action. Potato starch already carries sufficient natural anionic charge to provide amphoteric properties after cationization.[32]

Grafted starches are produced by free radical copolymerization with ethylenically unsaturated monomers.[33] The products are mixtures of copolymer, homopolymer and unreacted starch.[34–36] A starch–latex binder can be obtained by grafting starch with styrene and butadiene.[37,38] There has been some use of grafted starch as surface size, particularly on the metered size press. The use of grafted starch copolymers as sole coating binder has not yet been commercially successful, since many paper mills prefer blending starch and synthetic binder to meet specific viscosity and product property targets. The requirement that the product be free of unreacted monomer sets a further restriction on the use of grafted starch.

Hydrophobic starches are generated by a variety of reactions, for example by esterification with octenyl-substituted succinic acid anhydride.[39] They can have the properties of a polymeric surfactant, which will generate a weak network with dispersed latex in a coating formulation. Their action is comparable to that of associative thickeners and is intended to achieve 'rheological engineering' by separately affecting the high and low shear rheological characteristics of coating colors. Hydrophobic starches are not yet widely used in papermaking.

Aldehyde starches are prepared by treatment with periodic acid/periodate ions, which selectively oxidize the adjacent hydroxyl groups on carbon atoms 2 and 3 to aldehyde groups. Dialdehyde starch can react with cellulose by forming covalent hemiacetal and acetal bonds.[40] It is primarily used as a wet strength agent in the production of tissue and other sanitary grades.

Waxy maize starch is obtained from a genetic variety of corn that contains only amylopectin. Its granules swell more easily and its polymer molecules have high molecular weight and are naturally resistant to retrogradation. Cationic waxy starches are used as an additive in papermaking furnish, where they enhance pigment retention and paper bonding.

High-amylose corn starch is obtained directly from high-amylose varieties of corn. Amylose can be isolated from regular starch using butanol or higher aliphatic alcohols for crystallization or specific salt solutions for selective precipitation. High product costs and difficulty in achieving and maintaining good dispersion are severe limits to the use of amylose in the papermaking process.

Dextrins are produced by dry heating (roasting) starch in the presence of an acid catalyst. They are produced in a range of viscosity and color specifications. Dextrins are primarily used as adhesives in paper conversion, such as laminating and envelope production. A low-viscosity dextrin is used in Europe as a total chlorine free (TCF) coating binder for application on TCF paper.

Starch fibrids are produced by spraying alkaline starch dispersion into an agitated, concentrated solution of ammonium sulfate.[41,42] Amylose fibers or films are likewise produced by extrusion of a hot paste into a coagulating bath. Starch fibrids (starch pulp) can be utilized as a bonding additive in paper, but its use is rare.

Pregelatinized starches are formed when a dispersion of starch is dried on a heated roll or in a hot air (flash) drier. Added surfactants will aid in rewetting. The use of a pregelatinized starch simplifies the preparation of coating colors, but this practice is restricted due to higher cost and difficulties in obtaining a homogeneous product.

Mechanically-modified starches are obtained by extrusion.[43-45] The high temperature and high shear forces experienced by the starch offers further options for starch modification, but applications in paper mills are rare.

2. Recent Trends

Recent trends in the paper industry have had an impact on starch supplies. The implementation of statistical process control has led to more awareness that the production of high-quality paper grades depends on the use of high-quality materials. As a consequence, there is increased emphasis on acceptance specifications for starch supplies according to ISO guidelines and more rigorous testing in order to ensure consistently uniform quality.

A previous trend in the paper industry of limiting starch purchases to unmodified grades and effecting modification on-site in the paper mill has changed. The variance in products thus obtained was frequently wider than in products supplied by the starch manufacturer. As a result, there is now more preference to utilize modified starches with specific application properties. Growth in paper recycling should lead to an increased use of starch as a coating binder in place of synthetic materials.

New starch products might be derived from emulsion copolymerization with synthetic monomers and the replacement of all-synthetic polymers. Potential applications could be in flocculation, sizing, modified rheological characteristics, bonding to a wide range of substrates, film formation and in effluent treatment. A critical requirement will be the removal of hazardous residuals and Food and Drug Administration (FDA) approval for use in specific paper grades.

Introduction of new starch products will require extensive technical services, especially for adaptation to closed paper machine wet-end systems, for use with deinked pulp and for the high shear conditions of high-speed paper coating.[46]

V. Application Requirements for Starch

Dispersions of starch have found wide use in papermaking and paper conversion due to their unique properties, viz., low-cost renewable adhesive, controlled viscosity, specific rheological characteristics, water-holding properties, electrostatic charge, film formation and bonding after drying.

Starches are chemically or physically modified to obtain specific properties of viscosity, charge, bonding to fibers and pigments, and bond strength. The viscosity of a dispersion of starch depends on concentration, chemical substitution on the starch molecule and molecular weight.[47] Natural starch has a slight anionic electrostatic charge. The charge can be modified by chemical substitution that introduces anionic and/or cationic ionizing moieties and generates a specific charge or amphoteric property. Film-forming and bonding properties depend on molecular weight, the state of the starch dispersion and its water-holding properties. Improvements are obtained by chemical substitution.

Modified starches, however, are only moderately different in their abilities to provide bonding strength and elongation. Humidity (moisture content) affects the strength and elongation of starch films and is often a dominating factor. Increasing humidity from 35% to 65% may decrease film strength by more than 40%.[48] Various starch products are, therefore, distinguished more by rheological and charge characteristics than by bonding strength. Trade associations of the starch, paper, and agriculture industries have defined standard analytical methods for starch characterization.

1. Viscosity Specifications

Starch is a natural product and as such is not uniform. Type, genetic variety and environmental factors of soil quality and weather during the growing season for the starch source may influence the rheological characteristics of the product. Additional effects result from differences in associated fat (lipids) and phosphate content. As a consequence, viscosity control is needed to adjust starch to a viscosity target and to minimize differences that result from property differences in the feedstock.

For applications of starch in the wet stage of papermaking, a high molecular weight and a corresponding high viscosity are desirable. For applications to the dried paper surface, starch is partially depolymerized and substituted to attain the specific rheological properties needed for surface sizing and coating. It is important that these rheological characteristics be maintained for a reasonable time prior to application. There is a strong natural trend for reassociation of dispersed starch (retrogradation), especially when the starch has not been chemically modified.

The viscosity of a starch paste is often characterized differently by the starch manufacturer and by users in the paper industry. In the manufacturing plant, viscosity tests serve to control the starch modification process. As an example, the test is used to determine when to terminate a starch oxidation reaction for viscosity reduction. In the paper mill, viscosity testing is required to characterize the flow properties of starch-based coating formulations.

The Brabender AmyloViscograph and the Rapid ViscoAnalyzer[49–51] are tools to monitor the starch dispersion process as a function of temperature, time and shear intensity. A typical viscosity trace is depicted in Figure 18.3. An initial ascending

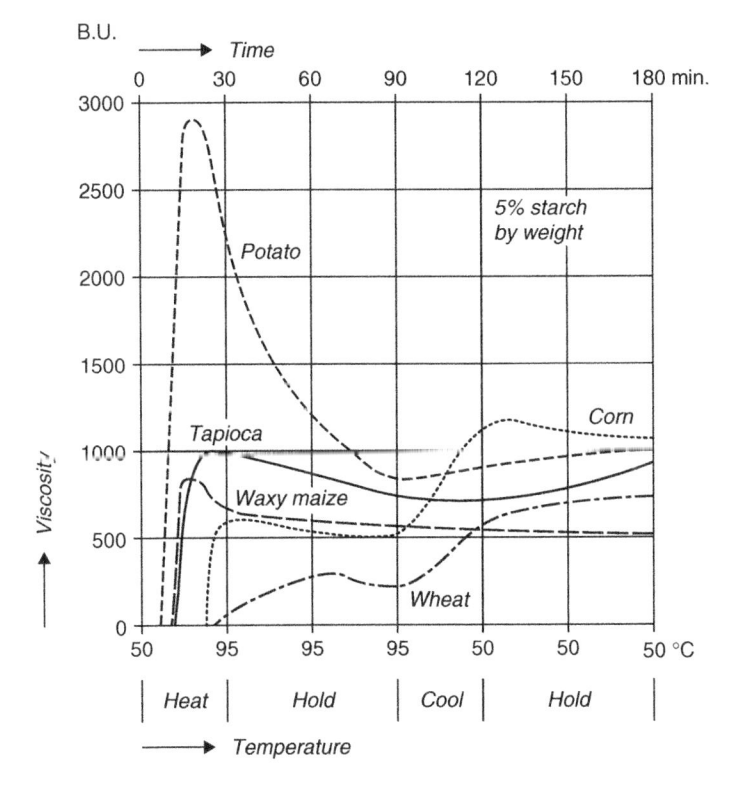

Figure 18.3 Brabender ViscoAmylograph traces for commercial starches (5% starch by weight in water). (Reprinted by permission of TAPPI)

viscosity slope indicates granule swelling. Termination of heating and cooling will reduce the size of the swollen granules, but some weakening of the structure will have occurred. Further heating will increase the slope. The temperature at which the slope magnifies indicates the beginning of irreversible starch swelling and the onset of gelatinization (the pasting temperature). After reaching a peak, swollen granules begin to disintegrate, as indicated by a descending viscosity slope. Prolonged heating close to the boiling temperature will disperse the starch, but a complete molecular dispersion is rarely achieved under atmospheric conditions, except for a highly depolymerized starch. The viscosity increases again when the temperature is lowered due to retrogradation (gelling). Starches are characterized by the temperature at which an increased slope appears during cooling of the paste, and by the magnitude of the slope. Unmodified or slightly modified starches will exhibit continued build up of viscosity, even at constant temperature, due to strong retrogradation. Steps in the ascending or descending portions of the viscosity trace indicate the presence of mixtures of starches. Starch granule disintegration and dispersion are also affected by mechanical energy. As a consequence, recording viscometers have to be operated at a constant rate of shear.[52]

The process of granule swelling and disintegration can be observed microscopically with a Kofler heating stage.[53] At a critical stage of expansion, optical anisotropy (birefringence) and x-ray diffraction disappear, indicating loss of crystallinity.[54]

Starch pastes obtained by heating in a recording viscometer are seldom completely dispersed, but are a mixture of granular fractions and dispersed molecules. The kinetics of phase transition of native and modified starches are analyzed by DSC.[55]

Frequently used single-point viscosity tests in the starch plant are orifice pipettes,[56] orifice funnels,[57] the Hot Scott viscometer, and various methods to determine alkaline fluidity.[58] For absolute measurements of the rheological properties, rotating viscometers with coaxial cylinders are used.[59] The paper industry uses mainly the Brookfield viscometer and the Hercules viscometer for determining shear-dependent viscosity, pseudoplasticity, and thixotropy. Oscillatory and capillary viscometers are used for more detailed viscosity characterization, such as yield value, elastic properties, and viscoelasticity.[60]

In a more fundamental test, the alkaline inherent viscosity, which is related to molecular weight, is determined.[61] This test uses the Cannon-Ubbelohde capillary viscometer. A recent modification is based on the use of a cylinder viscometer.[62]

A chemically stabilized and viscosity adjusted starch will have predictable rheological properties. For modified starches, the logarithm of viscosity correlates directly with starch concentration, as shown for hydroxyethylated starches in Figure 18.2. In contrast, unmodified starch might produce very noticeable changes in rheological behavior with only a slight variation in concentration.

2. Charge Specifications

Natural products carry a slight anionic charge due to oxidized sites and the presence of acidic groups.[63,64] The extent of charge depends on the properties of attached functional groups, the pH, and the electrolyte content of the suspending water. Some starches have attached ionizable phosphate groups. Potato starch contains 0.06–0.09% phosphorus, while corn starch has only 0.015–0.02%. Carboxylic and phosphate groups ionize, imparting a negative (anionic) charge at pH values above the specific isoionic (isoelectric) point, which prevails at a pH of ~6.5.

Recent years have seen a large increase in the use of starch products with a cationic charge. Such starches are used to control fiber flocculation and sheet formation during the papermaking process, to improve the retention of mineral pigments in the papermaking furnish, and to enhance alkaline sizing. They are also used to improve the bonding of cellulose fibers and pigments in the sheet, to impart picking resistance for offset printing, and to treat waste water streams from the papermaking and surface sizing operations. A new development is the use of cationic starch to produce cationic coatings.[65,66]

Further improvements in the retention of pigment in the paper and enhanced sheet formation are obtained when amphoteric starches are used. Phosphate groups have to be added to cationic cornstarch to produce amphoteric properties. Cationic potato starch has a natural source of anionic charge due to its phosphate content. In recent years, waxy corn starches, which consist entirely of amylopectin, have been modified for use as cationic or amphoteric agents for papermaking and surface sizing.

Electrokinetic effects that produce flocculation and retention are controlled by the zeta potential of the suspended phase.[67] In recent practice, the streaming current (potential) is used for characterizing charged particles and for monitoring wet-end charge.[68,69] The charge can be assessed in the measuring cell by adding a solution of material with opposite charge, which allows quantification of charge and the determination of the amount of required polymer additive for charge neutralization.[70] The instrument has been adapted to on-line use.[71] The wet-end system is normally adjusted to near the isoionic point, preferably slightly anionic, which produces best retention and paper strength. Charges on dispersed colloidal material can also be quantified by polyelectrolyte titration with the use of a suitable indicator.

3. Retrogradation Control

An essential requirement for the use of starch in papermaking is the prevention of starch retrogradation. Starch dispersion is transient. During the cooking process, crystallites in starch granules are melted, hydrogen bonds are broken, and starch molecules are hydrated and dispersed. However, on cooling, amylose and linear segments of amylopectin can reassociate, thus reversing the dispersion process.

The conformation of amylose in aqueous dispersion depends on its molecular weight. When the molecular weight of amylose is outside the so-called dissolving gap ($6500 < MW < 160\,000$), it behaves as a random coil; whereas when its molecular weight is within this gap, it easily aggregates, forming a rigid coil of retrograded amylose.[72] During cooling of the dispersion, hydrogen bonds are reformed between amylose molecules and chain segments of amylopectin. The retrogradation process leads to the formation of amylose particles when a dilute starch dispersion is held in the temperature range of 67 to 89°C, with a maximum at 77°C. An alternate route of retrogradation leads to the formation of a gel (network) at temperatures below 55°C.

On storage, the gel may shrink due to the loss of water (syneresis). Amylose undergoes more rapid retrogradation than does amylopectin.[73]

Kinetics of the retrogradation process have been studied by fast neutralization (stop gap) of an alkaline starch dispersion and recording the onset of turbidity by light scattering techniques. The retrogradation process can be monitored by differential scanning calorimetry (DSC),[74] Fourier-transform infrared spectroscopy,[75,76] nuclear magnetic resonance spectroscopy[77] and Raman spectroscopy.[78] The formation of amylose complexes and their decomposition at high temperature is indicated by changes in gelatinization enthalpy, as shown by differential scanning calorimetry (DSC).[79–81] The extent of retrogradation is quantified by digestion with α-amylase. Retrograded starch is not accessible to enzyme-catalyzed hydrolysis and remains as a residue.[82]

Retrogradation is enhanced when the starch has been thoroughly dispersed and subjected to limited hydrolysis. Particularly sensitive are the products of high temperature cooking with excess steam, and of high temperature oxidation with ammonium persulfate or hydrogen peroxide. A slightly hydrolyzed corn starch will retrograde faster than the original starch or the product of more extensive hydrolysis, because of an optimum in the relationship between molecular weight (size), radius of gyration and ease of association. Retrogradation is enhanced by slight acidity at a pH of about 6.5. Low-cost starches for use in the paper mill are, therefore, buffered to a pH of 7.5 or higher.

Starches with high amylose content, such as corn starch (28% amylose) retrograde more than starches with lower amylose content, such as potato starch (20% amylose). Redispersion of retrograded starch is energy-intensive due to extensive bonding. Temperatures of 115° to 120°C are required to solubilize amylose gels or crystals. Amylopectin gels can be redispersed at temperatures above 55°C.

Multivalent ions in starch dispersions, particularly those of aluminum, calcium, sulfate and oxalate, will induce retrogradation due to complexation or competition for water of hydration. The ions can be introduced by hard process water or accumulate by leaching from paper during surface sizing or coating. The destabilizing effect of ions follows the Schulze–Hardy rule.

Retrogradation of starch can be delayed by holding the dispersion in the temperature range of 60° to 65°C or above 91°C. Any contamination that might seed amylose formation has to be avoided. Seeding can occur when cellulose and hemicellulose fines are present in the aqueous starch dispersion, or when lipids, other fatty materials, or other compounds which are aliphatic or have aliphatic end groups are present or have been added.[83–85] Corn starch has a residual fat content of 0.4 to 0.6% which induces retrogradation by forming inclusion compounds with amylose. Similar reactions can occur with wheat and potato starch.[86–88] Any aliphatic compounds with four or more methylene groups are potential complexing and seeding agents. Other sources for such agents can be coating defoamers or lubricants. The dissociation temperature of complexes of amylose with fatty acids increases with increasing chain length of the fatty acid, but the dissociation enthalpy is independent of chain length.[89]

Particles of papermaker's amylose (reformed amylose particles, RAPS) appear in two shapes, Types I and II. The onset of retrogradation is indicated when a clear and transparent starch dispersion becomes opaque (turbid). Type I amylose particles are small. They are shaped like small bow ties or wedges of pie. Starch solutions with Type

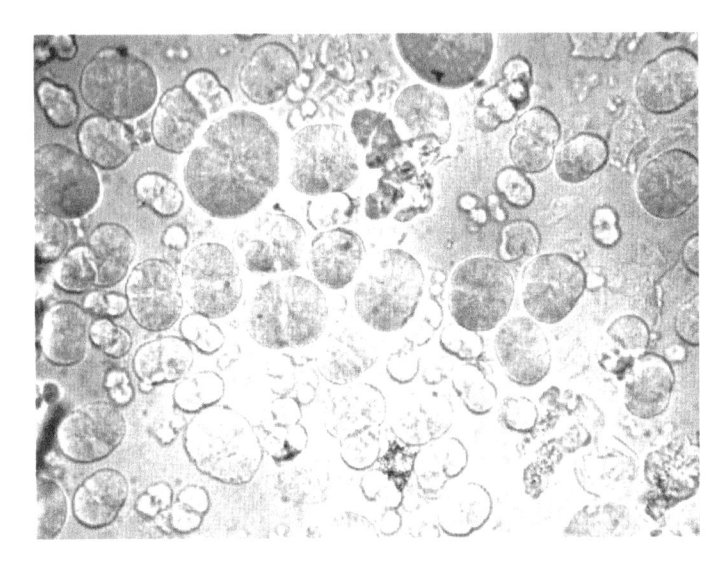

Figure 18.4 Type I and Type II amylose particles (RAPS) at ×300.

I retrogradation have lowered viscosity and are impaired in binding strength. The particles are not rejected at the coater blade or in the size press nip. Amylose particles of Type II are large (30 to 100 μm) and balloon-shaped. The presence of Type II particles raises the viscosity of a coating formulation and lowers binding strength. The particles are rejected by size press nips and at the coater blade, which can induce scratching and render the coated surface defective. The organic solids content of the coating color increases with time during recirculation, since the retrograded particles are not taken up by the paper and are retained in the color. An illustration of Type I and Type II amylose particle is shown in Figure 18.4. Most starches prepared for use in papermaking are modified by the starch supplier to retard, delay or eliminate retrogradation. The most frequently used modifications are oxidation, nucleophilic substitution and esterification.

4. Purity Requirements

Starch for use in papermaking has to meet specific purity requirements in residual oil, protein, bran and ash content. Industrial starches have a protein content ($N \times 6.25$), ranging from about 0.05% for potato starch to 0.3–0.6% for corn starch, depending on separation efficiency during production. Excess protein content will induce foaming in dispersions of starch and affect the quality and strength of the coated surface. Starch for use in the paper industry should not contain more than 0.4% protein. Oxidized starches tend to have the lowest protein content. Residual oil will cause retrogradation due to complex formation with amylose.

Starches often contain a substantial amount of salt (sodium chloride or sodium sulfate) as a residue from the modification process. Small quantities of salt in a hot starch dispersion will attack low-grade stainless steel and can cause severe corrosion of tanks, pipelines and coating application equipment. The salt content should not exceed 0.2% or 2000 ppm chloride.

VI. Dispersion of Starch

1. Delivery to the Paper Mill

Industrial starches are supplied to the paper mill in packages or bulk containers.[9] The shipment of starch as a suspension in water is possible, but it is costly due to the need to transport about 65% water. Starch settles easily, is difficult to resuspend, and can easily spoil. Bags are shipped stacked in layers on pallets. Super sacks, which hold about 1000 kg starch, have lifting straps for handling them with forklift trucks or overhead hoists. The sacks are fitted with an outlet sleeve.

Bulk transport uses trailer trucks (about 29 000 kg) or rail cars (up to 90 000 kg). Trucks are equipped with equipment for pressurizing and fluidizing the starch. At the paper mill, the starch is blown from the trailer into a silo. Various types of rail cars are also used for starch shipment. They are distinguished by their mode of unloading, such as air slide, pressure differential or pressure flow. Hopper cars have a sloping trench bottom that is covered with a permeable fabric. In one design, low-pressure air (3 psig) is used to aerate and fluidize the starch. The airborne suspension flows through a discharge gate into either mechanical or pneumatic conveying equipment. Complete discharge requires 4–6 hours. Other cars are designed for gravity discharge. They require the use of portable vibrators or impactors to prevent bridging of the starch during discharge. An air lance is often used to break any bridges that might form in the car. The starch flows into mechanical conveying equipment located under the railroad track or into vacuum pans for pneumatic conveying. Discharge time can take up to 24 hours. Tank cars for starch transport have 2–4 hoppers within the tank structure. The car requires a high air pressure of $1.03 \, \text{kg/cm}^2$ (14.7 psig) for fluidizing the starch. The airborne suspension leaves the car under pressure and is pneumatically conveyed to a storage silo. Some cars are equipped with vacuum unloading equipment. Unloading requires 2–3 hours.

The dry bulk and suspension viscosity of starch will be affected by the mode of drying; flash-dried starch tends to be more bulky, while tunnel-dried starch may contain aggregates of partially gelatinized starch. Mechanical or pneumatic conveying systems are used to transport dry starch to and from the storage silo. The use of screw or *en masse* conveyors is declining in favor of pneumatic systems that are easier to install and maintain. The material is made to flow by a positive or negative pressure air stream in dense-phase or dilute-phase flow, according to the air-to-starch ratio. Starch silos vary in size (up to 6.4 meters in diameter and 21.9 meters in height) and have a storage capacity of up to 360 000 kg. The required size depends on daily starch usage and possible in-transport delays of shipments. Large users of starch require two silos, which allow keeping products from different suppliers separate.

Starch dust can produce an extremely explosive mixture with air. All motors in confined areas must have explosion-proof ratings. The silo and all conveying equipment must be grounded to eliminate static electricity. Furthermore, starch silos must be equipped with an explosion relief deck to provide an escape route in case of a dust explosion. Another requirement is a vacuum relief valve to prevent the silo from caving in, in case a starch bridge in the silo suddenly collapses. The starch content in the silo is monitored by cable and reel, ultrasonic vibration or capacitance cables. Starch is discharged through a cone. Air pads or bin activators are used to prevent starch bridging.

2. Suspension in Water

Starch slurry make-down systems are designed to prepare a uniform suspension, using batch or continuous systems. In modern paper mills, automated batch or continuous systems are used. Water content and bulk density will affect the flow of dry starch.[90] In the batch slurry system, weighed increments of starch are drawn from the silo and periodically added to a measured quantity of water in the make-up tank. The suspension is screened and automatically pumped to a larger storage tank on demand by a level transmitter. Continuous make-down of starch slurry is accomplished by simultaneously feeding starch and water at controlled rates into the slurry make-down tank. The dry starch is metered with a volumetric screw feeder. The slurry is screened before dispersion by batch or continuous cooking.

In calculating starch and water quantities, adjustments need to be made for the water content of starch, which ranges from 10% to 12% for corn starch to 15–18% for potato starch. Starch solids in the slurry may fluctuate if the delivered product is not uniform. The bulk density of the starch will vary depending on the method of drying. Flash dried starch has a bulk density of about $608\,kg/m^3$, while belt dried starch has a bulk density of about $672\,kg/m^3$.

All starch systems have to be designed to prevent settling of starch granules, which have a density of about $1.5\,g/cm^3$. Piping has to be sized to maintain a velocity of at least 45 cm/s and must be equipped for self-draining, particularly in vertical sections.[91] In case of a power loss exceeding 30 minutes all slurry pipes should be flushed and drained. An air lance will be required for resuspension of settled starch.

When starch is added to cold water (below 29°C, 85°F), only negligible swelling will occur. However, the suspension volume expands, since the insoluble starch replaces water. Addition of starch to water at a concentration of 10% will increase the volume by 13%. The maximum in suspension solids is 40–45%. Various methods are used to determine the solids content of the starch slurry: aerometer,[92] density cells, densitometer, attenuation of vibration (Dynatrol) or a radiation-type density meter. Concentrated starch slurries have high viscosity and shear thickening (dilatent) rheology. Settling of starch from the slurry produces densely packed sediments that are difficult to disperse.

Modified starches tend to swell more after contact with water due to a weakened granular structure. Their suspensions have higher viscosity, lower maximum suspension solids and a slower settling rate. Suspension solids and viscosity are also affected by mechanical starch damage during drying.

Uncooked starch suspensions can easily spoil. If suspensions are to be held for more than a few hours, a preservative (biocide) needs to be added. Some commonly used reagents, such as chlorine or sodium hypochlorite, will attack (oxidize) starch. This reaction has to be considered when choosing a preservative.

3. Dispersion Under Atmospheric Pressure

Heating an aqueous suspension of starch will cause the granules to become hydrated and to swell to many times their original volume. Continued heating produces a paste, which is a mixture of swollen granules, granular fragments and starch molecules leached from the granules. Starch granules and their fragments will be increasingly disintegrated, generating a colloidal dispersion of amylose and amylopectin molecules. Complete dispersion is generally not obtained by batch cooking. Nevertheless, the final dispersion must meet specific application requirements for use at the wet end of the paper machine, in surface sizing or in coating. There is no discrete gelatinization or pasting temperature for specific types of starch. Rather, they occur over a temperature range, due to heterogeneity in the population of granules. Starch modification for paper industry use generally lowers gelatinization and pasting temperatures.

Starch is dispersed in the paper mill in large stainless steel tanks by injection of steam or by heat transfer from a steam-heated jacket. The tanks are stirred and equipped with baffles to prevent formation of a single vortex at the agitator shaft. A minimum heating time of 20 minutes at 95°C is normally required. Steam injection dilutes the starch paste by condensate, which must be considered for concentration control. Pastes that are prone to retrogradation are held at a temperature above 91°C or quickly cooled to 66°C to prevent amylose formation. Attention to storage temperature and water balance is an essential requirement for the effective use of starch in a paper mill.

When starch is cooked under atmospheric pressure in the paper mill, residues of granular fractions persist in the starch paste. Dispersion of most residues requires higher temperature (120–140°C), jet cooking with a high rate of shear (see below), autoclaving or the use of alkaline solutions. Complete colloidal dispersion of starch under atmospheric pressure can be obtained by adding a strong base to the suspension.[93] Dispersion will be affected in the presence of various salts[94] or by the action of urea.[95] Dispersion techniques range from the use of small amounts of base in conjunction with heat to cold dispersion by a strong base. Alkaline dispersions of starch are widely utilized in the production of the adhesive used to make corrugated board. Starch cooking is also practiced at temperatures below 100°C to obtain a partially-dispersed product for use as retention or bonding agent in papermaking furnish.

4. Dispersion Under Elevated Pressure

The constituent molecules of starch need to be completely dispersed in order to become an efficient binder. Heat transfer under pressure and elevated temperature can be achieved by injecting steam into a slurry of starch while it passes through a mixing device, such as a venturi nozzle, against a back pressure.[96] The stream of high-velocity steam mixes intimately and smoothly with the slurry and raises its temperature rapidly. Operation temperatures in excess of 120°C are normally used for this method of starch dispersion (jet cooking). The equipment requires a positive displacement pump for the starch slurry, a back-pressure valve, and means to control temperature and pressure. Jet cooking of starch is a continuous process and can be activated or terminated by level controllers in the receiving tank for starch paste.

Efficiency of starch dispersion can be further improved when an excess quantity of steam is used[97] or when the starch is held under pressure in a retention device (tank or coil) before discharge to the atmosphere. Both processes require a flash chamber (cyclone) to remove excess steam and steam relieved by the pressure drop. Excess steam refers to a multiple of the quantity required to reach target temperature and pressure. The quantity of saturated steam to reach a target temperature can be calculated from its heat content, the slurry temperature, the specific heats of starch and water, and the enthalpy of starch gelatinization. The action of excess steam in the dispersion zone between the jet and back-pressure valve causes a shear effect that enhances disintegration of the starch granules. In many applications, a three-fold excess of steam is used.

The products of starch cooking with retention are similar to those obtained by dispersion with excess steam. Retention of starch paste for several minutes under pressure requires that all steam is condensed; otherwise 'false body' of uncondensed steam will shorten the retention. Jet cooking with retention is practiced for on-site chemical starch modification.

In most cases of cooking starch at high temperature (thermal conversion), a slurry with higher solids content is processed than is needed for its application. The starch cooker is pressurized and preheated by initial passage of steam-heated water. After a set time (~5 minutes), starch slurry is injected. The cooked paste leaves by way of a flash chamber (cyclone). Excess steam is vented into a second flash chamber (cyclone), condensed by a spray of water and utilized as warm water in the paper mill. Dilution water to reach application concentration is injected into the cooked starch flow prior to and after the main flash chamber. A specific flow split is required to obtain the target discharge temperature concurrent with the target discharge concentration. The dilution water has to be preheated to about 38°C, which prevents 'snow-flaking' in the starch paste due to local chilling (thermal shock) that will generate nuclei for starch retrogradation. Steam condensation adds about 10% water to the cooked paste. Prior to shut-off, the cooker has to be purged with water, which is discharged to a sewer.

Excess steam cooking is required to disperse bleached starches for use on the size press of the paper machine. These starches can have a substantial content of carbonyl groups with internal crosslinking, which makes high shear mandatory to achieve complete disintegration of all granules. Slurries of bleached starch are dispersed at high concentration (~30% solids) and diluted to a discharge concentration in the range of 2 to 14% at a discharge temperature of 60–70°C. Jet cooking is also used to obtain complete dispersion of certain modified starches, especially starch ethers, which might contain granules with a 'hornified' (crosslinked) surface, and for enzyme inactivation after enzymatic conversion of starch.

Pigment and heat-stable coating adjuncts may be added to the starch slurry for processing through the cooker jet or added to the dilution water. It is thus possible to obtain complete size-press formulations or low solids pigmented coatings in a continuous and on-demand fashion. The specific heat of the additives needs to be considered in the thermal balance for the system.

5. Chemical Conversion

Starch dispersion, extensive viscosity reduction and chemical modification can be combined. Some processes require a retention vessel or coil after initial dispersion of the starch by jet cooking. In a widely-used process for the preparation of a starch paste for use as a surface size or coating binder, 0.05–0.3% ammonium persulfate (APS) (based on the weight of starch) is used to reach a viscosity target. The process is initiated in a jet cooker where high-pressure steam is injected into starch slurry with added oxidant.[98] Dispersion temperature is about 152°C and the retention time is about 2 minutes. Protons released by the decomposition of the persulfate or peroxide facilitate additional hydrolytic degradation for viscosity reduction. Besides APS, potassium persulfate (KPS), hydrogen peroxide and sodium hypochlorite are used. Occasionally, a catalyst such as copper sulfate or another transition metal salt is added to promote free-radical reactions and to exhaust the oxidant. Products obtained by oxidation with APS or KPS differ from those obtained with hydrogen peroxide.

The colloidal dispersion resulting from thermal/chemical oxidative conversion can rapidly retrograde, which makes immediate neutralization of the acid product to a pH 8.0 to 8.5 mandatory. Discoloration of the paste is lessened when sodium bisulfite is added to the starch slurry and/or sodium sulfite added to the paste. Multivalent ions in the paste can induce colloidal destabilization, and may require the addition of a sequestrant.[99]

Starch slurries for chemical/thermal conversion are prepared by dispersing dry starch in cold water at concentrations up to 35%. The slurry is maintained in suspension by agitation and pumping through a loop. The process requires pressurizing the converter with hot water prior to the introduction of starch. Water from a bypass is heated by steam injection and pumped into the coil against the back-pressure valve. This operation is continued until all air is expelled and the equipment has reached conversion temperature. Water leaves the coil through an atmospheric flash chamber, where steam is vented. After completing the pressurizing step, starch slurry with added oxidant is directed into the converter. The starch paste replaces the water in the retention coil, passes the back-pressure valve and is separated from steam in the flash chamber. It is rendered slightly alkaline (pH 8.0 to 8.5) by injecting a solution of base into the flash chamber or shortly afterwards (Figure 18.5). The cooking process is initiated and terminated using the signal from a level controller in the receiving tank for converted paste. On shutdown, the converter is automatically switched back to operation with water, which expels remaining starch paste from the coil. Flush water is directed to the sewer. On completion of the cycle, the system reverts to a stand-by position. A computer or a programer is used to control the cycling sequence.

All steam that is injected into the starch slurry has to be condensed in order to achieve the desired retention time under pressure for the reaction. Air released from the starch slurry will affect retention. Steam used for the reaction is superheated.

A minimum temperature of 88°C is required for paste storage of thermally oxidized starch in order to prevent retrogradation. Various retrogradation control agents have been recommended for stabilizing. The addition of 0.5 to 1.0% calcium stearate prevents the build up of viscosity, but could actually lead to the precipitation of amylose due to gradual stearate dissolution (ionization) and complex formation with

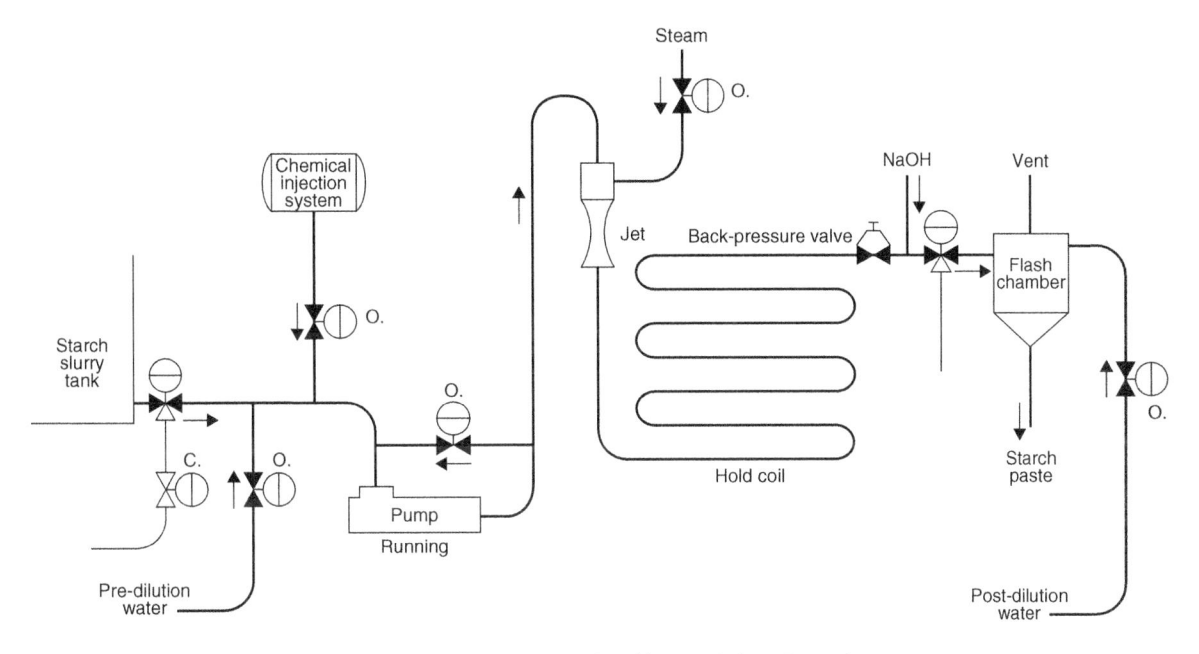

Figure 18.5 Thermal–chemical starch conversion system. (Reprinted by permission of TAPPI)

stearic acid. Better preservation of adhesive properties is obtained with the use of fatty acid derivatives.[100,104]

Cationic starch can be prepared in the paper mill.[100,101] Jet cooked starch (with or without APS addition) is cooled by injection of dilution water or by heat exchange. The resulting paste is blended with an alkali, e.g. sodium hydroxide, and a cationic monomer, e.g. (3–chloro-2–hydroxypropyl)trimethylammonium chloride, and fed through a second reactor with controlled temperature and retention. In the alkaline medium, the chlorinated cationizing agent converts into an epoxide, which reacts with the starch. The degree of substitution (DS) can be controlled in the range of 0.005 to 0.04, depending on reagent input and time in the reactor. The cooling step after initial jet cooking prevents unwanted hydrolysis of the cationic monomer at high temperature.

Some paper mills have tried to produce a cationic product on-site by complexing starch with a cationic polymer.[101,102] Starch has been cooked together with a quaternized fatty amine to render it cationic. The product has cationic functionality, but may contain unreacted reagent which is either adsorbed by the stock or adds to the chemical oxygen demand (COD) of the process water. Lower molecular weight amines in the reagent may also act as a dispersant and cause foaming. Recently, poly(vinylamine) and chitosan have been introduced as reagents for starch cationization.[102,103,104]

6. Enzymic Conversion

A widely-used method to adjust the viscosity of a starch dispersion to a target level uses amylases. Alpha-amylase catalyzes hydrolysis of starch molecules by breaking α-D-(1,4)-glucosidic bonds at random, generating depolymerization products, α-(1,4;

1,6)-glucosyl oligosaccharides and, ultimately, maltose.[105,106] The enzyme cannot effect hydrolysis of α-D-(1,6) bonds in amylopectin, thus leaving starch fragments (limit dextrins) in the product. Commercial α-amylases generally contain other enzymes, including other amylases. However, α-amylase for use in starch conversion should have minimal β-amylase and glucoamylase activities. Commercial amylases are distinguished by their potency, which can be determined by simple laboratory tests.[107] The amount of enzyme used for starch depolymerization depends on its potency and the desired viscosity reduction. The dosage may range from 0.01% to 0.1% (based on starch dry solids). Most applications for the production of coating binder require about 0.05% enzyme.

Alpha-amylase is most active at its pH optimum of 6.3 to 6.8.[108,109] It is inactive at pH values below 4 and above 9. Enzymic starch conversion is terminated by raising the temperature until enzyme denaturation occurs or by the addition of enzyme poisons, such as the ions of copper, mercury or zinc. Inactivation can also be achieved by moving the pH outside the enzyme's active limits or by the addition of oxidizing agents, such as sodium hypochlorite, hydrogen peroxide or barium peroxide.

Corn starch for use in an enzymic conversion process should not contain more than 0.4% protein. In some cases modified starches, such as starch ethers, are used as feedstock. Higher product cost is balanced by the substantial reduction of retrogradation in products thus obtained. The feedstock has to be buffered in order to reach the required pH level for optimum enzyme activity in the process water of the paper mill. Additional adjustments may be required at the mill site when filtered surface water, which varies with the seasons, is used. Calcium salts for improved heat stability of the enzyme are added by the starch supplier or the paper mill. Further addition of sodium chloride will promote enzyme activity, and urea will broaden the critical pH range. Starch preservatives have to be added after the enzyme has been inactivated.

Enzymic conversion is conducted in a manual batch process, as an automated batch process or in a continuous process. In some cases, starch is depolymerized in the presence of pigment. In the manual batch conversion process, starch is suspended in an agitated tank filled with a specified volume of water. Enzyme is added and steam is injected, either directly into the tank or into a jacket. A time–temperature cycle, which produces the desired viscosity of the product, is applied. A typical basic cycle with one plateau has the following stages: heat to 170°F (76.7°C) in 20 minutes; hold at this temperature for 30 minutes (conversion plateau); heat to 205°F (96.1°C) in 12 minutes; hold at this temperature for 30 minutes (inactivation stage). Enzyme dosage and the length of the conversion plateau determine the viscosity of the final product. Thinner starch pastes are obtained by increased use of enzyme or extended reaction time. Cycles with reaction times of 20 to 40 minutes are being used. A more refined cycle utilizes two plateaus for starch conversion at high solids. At the first conversion plateau, dispersed starch is partially depolymerized, which decreases the peak viscosity and improves agitation, heat transfer and temperature control. Peak Brookfield viscosity can reach 35 000 cP (35 Pa s) at 15% starch solids and 170 000 cP (170 Pa s) at 30% solids. In some cases, the peak viscosity will be so high that the agitator has to be turned off until the enzyme has sufficiently liquefied the starch paste. In the second conversion plateau, the reduction of viscosity to target is completed, followed by thermal or chemical termination of the process.

In the automated batch process, starch is drawn from silo storage. Two tanks are required. In the preparation tank, the starch suspension is prepared by proportional blending of dry starch and water. Slurry concentrations between 25% and 35% are used. The slurried batch is pumped (or dropped by gravity) into the conversion tank and enzyme is added. Steam is supplied by way of an automatic valve. A temperature controller is used in conjunction with a recording viscometer to regulate the flow of steam. The viscometer is activated near the end of the hold period, when a substantial decrease in viscosity has occurred. The temperature controller continues to hold the set-point temperature until the viscometer reaches its target level, which overrides the temperature controller and opens the steam valve to initiate deactivation. Deactivation time depends on the concentration of the starch paste. Generally, 15 minutes plus one additional minute for each percent of solids over 15% are required. Commercial deactivation times range from 15 to 40 minutes. Deactivation time can be shortened by passing the paste through a high-temperature jet cooker or retaining it for a short time in a retention device under pressure.

Most mills control the cooking cycle by automatic time–temperature controllers and recorders. The rate of temperature rise to the conversion plateau must be slow to prevent hot pockets or cold areas. The rate of temperature increase to the inactivation plateau must be rapid to prevent excessive depolymerization in the intermediate temperature range. The viscometers operate according to different mechanisms: time to expel paste from a sample device (Norcross); vibration of a probe in the paste (Dynatrol); torque readings (Brookfield); or pressure drop on passage through an orifice (Escher Wyss). Potential errors in viscosity can result from variations in starch solids due to differences in moisture content of the starch, errors in slurry preparation and the quantity of condensate added by the steam. The process yields a maximum paste concentration of about 32%.

In the continuous process, both the conversion temperature and the deactivation temperature are obtained instantaneously by steam injection in jet cookers.[110–112] Retention time is determined by the rate of plug flow through a retention vessel. An enzyme with higher heat tolerance is needed, since a higher temperature (85°C) is used. As a rule, continuous conversion requires more enzyme than does batch conversion. The deactivation temperature is raised to 132°C. The combination of higher reaction temperature and higher enzyme dosage shortens the reaction time by about 30 minutes, compared to the batch process. The starch slurry concentration can be increased to 36–38%, yielding a product with about 35% solids.

Various types of equipment used for continuous enzymic starch conversion are distinguished by differences in the design of the retention unit and in the control strategy.[113] The operational principle of one converter is illustrated in Figure 18.6. Starch suspension with added enzyme is passed through the first jet cooker at about 85°C and the product is distributed by a sweeping agitator over a perforated baffle near the top of the retention column. The paste moves in plug flow through the vessel, which takes 25–30 minutes. The thinned product passes through another baffle into the lower section of the column, which is agitated to ensure uniformity. The converted paste is then pumped through a second high-temperature (~120°C) jet cooker to deactivate the enzyme. Finally, the product passes a flash chamber, which decreases

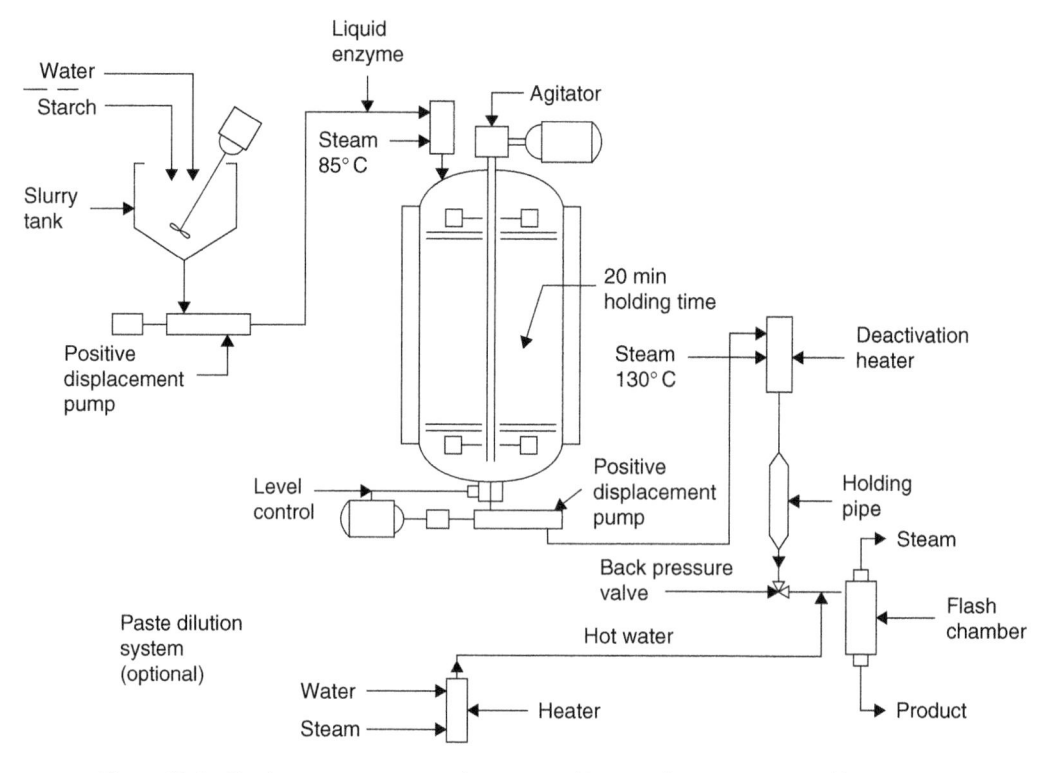

Figure 18.6 Continuous enzyme conversion process. (Corn Products Co., Reprinted by permission of TAPPI)

the temperature to about 100°C and vents steam before the starch is released into a storage tank. A viscometer mounted in the product exit line can be used to adjust enzyme flow into the slurry to adjust viscosity. However, at constant flow rate, about 30 minutes are required to correct viscosity.

The converter will operate most efficiently when it supplies coating binder at the same rate as it is consumed. Problems may occur when the converter has to be shut off and flushed with water. Channel flow and vortices during the purge with water affect cleaning time and may make it necessary to sewer starch product. In another converter, a single pump is used to transport the starch through the retention vessel. Jet cookers of the venturi type are placed at the entrance to the vessel and at its exit.[114] In an alternative design, a retention coil is used instead of a retention vessel.[115]

Starch can be enzymically converted in the presence of pigment. The conversion follows a similar time–temperature cycle as in neat starch conversion. The pigment will adsorb a portion of the enzyme; adsorption can be minimized by the addition of sodium silicate to the mixture prior to the addition of the enzyme (Vanderbilt process). Even with silicate treatment, a higher quantity of enzyme will be required to reach a specific viscosity target. Other coating components, such as latex and lubricants, have to be added after the conversion. The Vanderbilt process is now rarely used for the preparation of coating binder.

Since the action of α-amylase is random some reducing sugars will accumulate during conversion. The quantity of sugars depends on the amount and purity of the enzyme used. The sugar content in the converted paste should not exceed 3%. The product composition can be analyzed by titration and chromatographic separation.

Products of enzymic starch conversion are used as a surface size or coating binder. The presence of sugar provides an internal lubricant, which improves the plasticity of the coating during subsequent finishing. However, high sugar content lowers the surface strength of a coating, which may cause problems in offset printing. Enzymically converted starch, particularly in more viscous pastes, will retrograde during extended storage. As a rule, the product should be consumed shortly after preparation. Retrogradation control agents, such as fatty acid derivatives, can be used to extend shelf life.

VII. Use of Starch in the Papermaking Furnish

1. The Wet End of the Paper Machine

The wet-end operation is physically and chemically complex. The suspension of cellulose fiber in water has non-Newtonian behavior. Papermaking furnish is generally diluted to less than 1% consistency before sheet forming. Fibers are mutually attracted, leading to the generation, dispersion and regeneration of flocs by hydraulic shear forces.[116,117] During dewatering and consolidation, the suspension has to be deflocculated in order to obtain good and uniform paper formation. Charged and hydrophobic additives and various molecular and colloidal interactions[118,119] are required to achieve sufficient retention of filler pigments and cellulose fines in the sheet, enhance wet and dry paper strength, and attain paper sizing. Starch is an important wet-end additive for flocculation and retention control, sheet bonding and in support of paper sizing.

2. Flocculation of Cellulose Fibers and Fines

Stock composition, kinetics of adsorption and hydrodynamic shear dictate the point at which a cationic polymer is added to a papermaking furnish in order to induce flocculation. Flocculation of cellulose fibers in turbulent flow proceeds very rapidly and is completed in less than two seconds.[120-123] Flocs form due to charge interactions through a patch-type or a bridging-type mechanism. However, these flocs will be sensitive to shear force and deflocculation and reflocculation might occur.

Within the papermaking furnish, surface charge decays with time in a three-step process: adsorption of polymer segments onto fiber surfaces; reconformation of polymer molecules into a flat configuration; and diffusion of polymer molecules into fiber pores. High molecular weight polymers undergo the greatest reconformation and thus have the largest charge decay. Low molecular weight polymers (<10 000) are adsorbed in a very flat conformation and immediately begin to diffuse into the fiber pores. High molecular mass polymers cannot penetrate into the small pores of cellulose fibers. Their interaction is limited to charges on the external fiber surfaces. Consequently, it is important that the polymer addition points be selected to produce the desired effect.[123,125] Optimal paper formation with high retention is achieved when the papermaking furnish is effectively flocked prior to the headbox and deflocked when it exits onto the papermaker's wire. The rate of flocculation and flock size distribution can be measured by frequency analysis of backscattered laser light.[122,126]

Cationic starch is used during papermaking to balance flocculation and retention, to promote sheet dewatering and to aid in attaining good sheet formation. The charge density and amount of cationic starch added to cellulose fibers, the contact time with the fibers and the concentration of inorganic ions in the papermaking system all influence flocculation in the system. At low levels of cationic starch addition, the degree of flocculation is low and is unaffected by the degree of substitution (DS) of the starch. At higher levels of starch addition, a more highly-charged starch induces more flocculation due to higher electrostatic interaction. At high levels of addition of a highly-charged cationic starch, flocculation again decreases.[124,127]

Great improvement in formation can be achieved by using a microparticulate flocculation and retention system.[125,126,128–131] Small tight fiber flocks are produced by the addition of cationic potato starch and anionic colloidal silica to the papermaking furnish, resulting in an open sheet structure. Wet-end retention is improved. Previous formation and first pass problems due to over-flocculation are overcome by this combination of effects. The micro-flocks release water more readily, which improves sheet dewatering and paper production speed. Colloidal silica assists in the retention of cationic starch in the fiber mat; greater paper strength results. Cationic starch is also used in combination with bentonite (smectite kaolin), nano particles and aluminum hydroxide.[132,133]

The evenness of mass distribution in paper, the 'formation,' will affect paper strength, opacity, visual appearance, printability and convertibility. Over-flocculation results in poor formation and a decrease in paper strength due to weak regions between flocks, while drainage is increased. Formation is not a standardized quantity. Various methods that describe grammage distribution have been proposed.[120,134] Most frequently, variance of scanned light transmission through paper is used. Recently, image analysis techniques have been developed to evaluate the frequency and size distribution of flocks in paper.[121,127]

3. Adsorption of Starch on Cellulose and Pigments

Cellulose fibers have a slight negative (anionic) charge. The ionizing (acidic) groups on cellulose originate from cell wall constituents or are introduced during pulping and bleaching. The number of charged groups, primarily carboxyl groups, varies between 2 and 30 meq/100 g pulp for different types of papermaking fibers. Glucuronic acid residues on xylan remaining in the pulp and carboxyl groups in residual lignin are an additional source of acidic groups in alkaline delignified (Kraft) pulp.[123,133] The charged groups on fibers affect their swelling properties and electrochemical interactions with chemical additives, including starch.

Cationic starch is mainly adsorbed by the small fiber fractions and filler particles, according to their high specific surface. At low cationic starch levels (0.5%), only the small fiber fraction is saturated and rendered cationic; a mosaic system is established which results in small flocks and good retention, but poor drainage. Increasing the cationic starch level (1%) will satisfy the negative charges on the larger cellulose fibers and lead to the formation of larger flocks. The higher the DS, the sooner an overall saturation level is reached.

Cellulose fines have a much larger specific surface area than fibers, which accounts for their greater absorption of starch.[128,135] Fines from a never-dried bleached Kraft pulp can adsorb 120–150 mg of cationic starch (quaternized potato starch, DS 0.015–0.03) per gram of fines, while regular fibers absorb about 7.5 mg per gram of fibers. Fines from a dried pulp can adsorb more cationic starch (200–250 mg), due to a change in the structure of the fines during drying. The adsorption isotherms are of the high-affinity type. An increase in electrolyte concentration decreases adsorption of cationic potato starch. Excess cationic starch in solution can be precipitated on cellulose fibers by aluminum hydroxide formed *in situ*.[132,136] Adsorption of cationic amylose and amylopectin decreases at high salt concentration, due to shielding of the interaction between the charged sites.[134,137]

Cationic potato starch is actually an amphoteric starch due to the presence of phosphate groups. An increase in cationic charge results when phosphate groups protonate or react with aluminum ions in an acid papermaking system. Anionic phosphate starch will adsorb on cellulose, depending on pH and concentration in solution.[135,138]

In the papermaking furnish, there is a competition between cellulose fibers and pigments for the cationic polymer. The amount of polymer transferred from fibers to pigments increases with decreasing particle size and correspondingly larger specific surface area. At equilibrium, the fraction transferred approaches the ratio of charges on the pigment and on the outer surfaces of the fibers.[129,130,136,137]

Cationic starch adsorbs readily on kaolin particles due to the anionic charge on their faces. Anionic starches are adsorbed at low pH due to cationic charges on the edges of kaolin particles. Adsorption on calcium carbonate depends on the adsorbent concentration and is affected by the presence of electrolytes. Finely ground $CaCO_3$ is negatively charged, owing to the presence of impurities (SiO_3^{2-}, PO_4^{3-}), while precipitated $CaCO_3$ is positively charged with a point of zero charge at pH 9.5. Ca^{2+} ions increase the density of cationic surface sites, thereby decreasing the interaction between the cationic starch and the negatively-charged calcium carbonate surface, whereas SO_4^{2-} ions increase the amount of anionic surface sites, thereby increasing cationic starch adsorption. Excess cationic starch can be deposited on $CaCO_3$ when colloidal aluminum hydroxide formed *in situ* is present.[131,138]

4. Retention of Pigments and Cellulose Fines

Most paper grades contain pigments (kaolin, calcium carbonate, titanium dioxide, aluminum trihydrate, etc.) that add brightness, opacity and density to the sheet, and improve printability. The filler content in paper ranges from 5% to 15% for most grades, but can be as high as 35%. Pigments have a higher specific weight ($\sim$2.5 g/cm^3) than cellulose fibers ($\sim$1.5 g/cm^3). Successful retention of dispersed pigment and of cellulose fines in the sheet requires controlled flocculation of the papermaking furnish.

Only particles larger than the openings in the paper web and the papermaker's wire can be entrapped mechanically. Kaolin particles, for example, have a hydrodynamic diameter of 0.5 to 2 μm, while cellulose fibers have a diameter of about 50 μm. The average pore size of paper is commensurate with the average size of its compounds. Cellulose fibers and most pigments used for papermaking are negatively charged and repel each other.

Small particles must be retained in the sheet, despite dewatering at high speed on the paper machine. Retention is customarily described as first-pass and total retention. First-pass retention is defined as the amount of headbox solids retained on the wire. Retention ranges from >90% for fibers to 30–40% for filler pigments and fines. Consequently, returning pigments and fines enriches the headbox solids and a steady-state concentration is obtained. Total retention is calculated by considering the mass of paper leaving the machine and the amount of stock supplied.[139,140]

In order to improve retention, an electrostatic attraction between fibers and particles has to be facilitated. This can be achieved by manipulating the pH, when surface charge is pH-dependent, or by increasing the ionic strength of the furnish and thus decreasing repulsion. The most effective approach involves introduction of a cationic polyelectrolyte, which by adsorbing on the components can cause their heterocoagulation, due to either charge modification or a bridging mechanism.

Cationic starch is a frequently-utilized retention aid, especially when paper strength improvement is a second objective.[141,142] Low pH activates tertiary amino starch, but at pH 6 an amphoteric starch, and at pH above 7 a quartenary ammonium starch, will be needed.

A recent development is the use of dual polymer systems for pigments and fines retention. Cationic potato starch is used together with an anionic polymer or an anionic silica sol.[143] Such systems produce a smaller flock size at approximately the same overall formation as the other systems. For maximum effectiveness, the cationic polymer is added before the anionic polymer.[144] The molar ratio of the charges on the two polymers is of critical importance. A combination of cationic starch with hydrolyzed polyacrylamide generates increased dry strength, and improves drainage and retention.[145] In another method to improve retention, filler clay is preflocculated by cationic starch.[146] Paper grades with more than 30% filler content can be produced on modern, high-speed paper machines.

5. Sheet Bonding by Starch

During drying of the initial fiber mat on the paper machine, water will recede from the fiber surfaces, generating contact angles that draw adjoining fibers together. In close proximity, hydrogen bonds are formed between contacting fibers. However, since the fibers have different shapes, bonding will only occur in patches that are intimately drawn together. Starch readily adsorbs on cellulose, especially when it has cationic charge, and reinforces fiber-to-fiber linkages in regions that are in close physical contact. The relative bonded area in the paper sheet will be increased through hydrogen bonding of starch to fibers and retained pigment.

In past practice, native (unmodified) starch was added to the papermaking furnish in order to increase internal and surface strength of the product. The presence of starch in the papermaking furnish allows lower levels of pulp refining for improvements in sheet bulk while maintaining strength. The affinity of starch to cellulose in the absence of charge interaction is, however, low. Retention of starch in the papermaking furnish is improved by adding a dispersion with a high content of swollen granules, obtained through low-intensity cooking. However, in the absence of significant electrochemical attraction, dispersed starch can accumulate in the process water, which will cause production problems due to viscosity increase, press-felt filling, and enhanced microbiological activity due to high a biological oxygen demand (BOD) load in process and effluent water. In more recent practice, granular or dispersed unmodified starch is sprayed onto the stock on the papermaker's wire, but retention is still low and can lead to problems with press-felt filling.

In current practice, cationic starch products are widely used to improve the strength of paper. Cellulose fibers are capable of adsorbing up to 4–5% cationic starch. The addition of 0.5%–2% is often sufficient; further starch addition can adversely affect sheet formation due to charge reversion on fibers and fines. Starch added at the wet end for strength improvement should have a high molecular weight and contain a sufficient quantity of amylose. Use of cationic starch as a wet-end additive allows reducing refining energy while maintaining tensile strength.[147–150] An alternative use of starch to improve paper strength is the incorporation of a starch pulp into paper or the use of starch-treated cellulose fibers.[151,152]

The composition of process water will affect cationic starch performance.[153,154] A high amount of electrolytes will interfere in the retention process. Cationic starch can be inactivated by anionic trash in the papermaking stock. In wet-end systems that contain high-yield pulp, such as groundwood or TMP, or utilize deinked stock, electrolytes and colloidally dispersed substances (anionic trash) can build up to high levels, especially when the system is partially closed. Hydrophobic colloidal material can form undesirable deposits on the paper machine and cause paper breaks.[148,155] Cationic starches and other cationic polymers are used as scavengers to aggregate and precipitate these substances. The aggregates are subsequently removed by filtration retention in the paper.[151,154] Highly-charged cationic starches are used in closed water systems of paper machines.[149,153] In some paper mills, a synthetic low molecular weight polyelectrolyte is used as a scavenger to sweep the system before cationic starch is added.[150,155]

6. Wet-end Sizing

During pulping and bleaching of wood, resinous substances are dissolved. Paper made from such pulp will be hydrophilic, readily accept water, swell and lose strength due to the breaking of hydrogen bonds. Interactions of paper with water occur in pen-and-ink writing, offset printing, water-based gravure printing, ink-jet printing, glue application and in many other processes. Wetting by writing ink can cause feathering. Fountain water application in offset printing can change the paper dimensions and cause misregister. Vessel segments can be lifted off the surface and contaminate the printing press. The sheet can expand, cockle and curl when water-based inks or glues are applied, and many other problems might occur. The paper has to be sized in order to decrease water sensitivity and to maintain the dimensions of paper after exposure to high relative humidity or contact with water. This process returns resinous material to paper or adds synthetic agents to impart a controlled measure of hydrophobicity to paper. Sizing primarily affects the free surface energy and the acid–base character of paper.[156]

The process of wet-end (internal) sizing is different from surface sizing, although the same terminology is commonly used in the industry. In wet-end sizing, paper strength is decreased, since the hydrophobic size attached to the fibers interferes with hydrogen bonding between fibers. Starch frequently must be used as a wet-end additive in order to improve the strength of sized paper. Surface sizing relates primarily to the enhancement of paper strength and stiffness by the application of dispersed starch to dried paper.

Internal sizing agents are amphipatic; they have a hydrophobic (non-polar) portion to increase the contact angle with water and a polar (or reactive) portion to attach them to cellulose fibers and filler pigments. Sizing agents are, generally, added to the papermaking furnish prior to sheet formation. Sizing agents decrease the moisture sensitivity of cellulose by interacting with its hydroxyl groups. An acidic,[157] neutral[158] or alkaline[159,160] environment, or a microparticle system,[161] can be utilized to achieve the transfer of size to fibers and fillers. Sizing agents can be added to surface size for surface-specific application. In some operations, a combination of wet-end sizing and modified surface sizing is used.

The most commonly used acid sizing system consists of rosin and alum. At pH values of 4–5, a positively charged precipitate in this system is deposited on negatively charged fibers and kaolin particles. Sizing is induced during drying of the paper, when the precipitate spreads and orientates on the paper components. In acid sizing, alum competes with starch for cellulose surfaces; at high alum usage (2% Al^{3+}) cationic starch will not be effective. The activity of cationic starch is further reduced when a large quantity of dissolved anionic material is present in the stock.[162] Starch with very high cationicity is required to maintain a high level of sizing at a furnish conductivity as high as 500 microSiemens/cm.[163]

Neutral or alkaline sizing gained prominence when calcium carbonate was introduced as paper filler. The minimum level of pH that can be tolerated is 6.5. Neutral sizing is based on the use of polynuclear systems of aluminum that are reactive with dispersed rosin size. Cationic starch is used as a retention aid for the size. Most alkaline sizing is conducted at a pH of 7.5 to 8.2, but new developments have shifted the target pH increasingly toward neutral. Commonly used sizing agents are alkenyl succinic anhydride (ASA) and alkyl ketene dimer (AKD). Both systems require cationic starch as a protective colloid and retention aid. Sizing is achieved by precipitation, spreading and, partially, by the formation of covalent ester bonds with cellulose hydroxyl groups.[164] ASA is an unsaturated fatty acid anhydride that is very reactive and must be emulsified with cationic starch at the mill and used shortly thereafter.[165] Best results are obtained with a cationic starch of high DS and low molecular weight, which will aid in generating an optimal drop size for efficient sizing. Hydrolysis competes with sizing, decreasing sizing efficiency and producing a tacky by-product that can cause problems with press roll picking. AKD is less reactive and develops sizing during storage of hot paper off the paper machine. Cationic potato starch is utilized as a protective colloid to generate particles in the required size range. The reactivity with cellulose is catalyzed by pH in the range 6.5 to 8.5. AKD, however, can also form hydrolysis products which decrease friction and can cause problems in feeding mechanisms for paper sheets.

A novel sizing strategy utilizes a microparticle system that enhances retention.[166] A representative example is a combination of cationic potato starch (DS = 0.04), a metallosilicate hydrosol and AKD. In an approach to combine elevated filler retention, effective sizing and high paper strength, calcium carbonate is reacted with a starch–soap complex and combined with AKD.[167,168]

7. Starch Selection for Wet-end Use

Starch is added to the papermaking furnish for a variety of applications: paper formation control; furnish drainage improvement; filler and cellulose fines retention; size retention; internal paper strength improvement; surface strength enhancement; and reduction of waste water pollution.

Most frequently, corn or potato starches are used, but there are also applications of wheat starch, rice starch, tapioca starch and others. Recently, waxy maize starch has found commercial application in the manufacture of paper. Thermally dispersed or pregelatinized unmodified starches are used for paper strength improvement by addition to the furnish or by spraying onto the papermaker's wire.

Starches in current use at the wet end are, generally, modified by chemical substituents: tertiary amino groups; quartenary ammonium groups; phosphate groups; zwitterion groups; aldehyde groups; acetal groups; and graft copolymers. The degree of substitution (DS) describes the number of substituents per glucopyranosyl unit (molecular weight = 162). Since there are, on average, three hydroxyl groups in a monomer unit, the DS can range from 0 to 3. The DS of cationic starches ranges from about 0.02 to 0.04, or even higher for wet-end systems with a high electrolyte content. Wet or dry procedures are used to obtain cationic starches. Cationization occurs more easily in the amorphous material of the inner parts of the granule than in the crystalline amylopectin layer of the surface.[169] The degree of cationicity can be dertermined by Kjeldahl analysis of nitrogen content or solid-state ^{13}C-NMR spectroscopy.[170] Cationic starches are, generally, not depolymerized in order to maintain a high hydrated molecular volume for bridging flocculation.[171] Cationic waxy maize starch has superior resistance to retrogradation. Dry cationized starch can have a high chloride content.[172]

Starches that contain tertiary amino groups can only be used in an acid papermaking system (pH <5.5), since they require protonation in order to become charged. Quartenary ammonium starches are inherently charged. They can be used at pH levels above 7, but their efficiency may be reduced at high pH due to screening by hydroxyl ions in the dispersion.

Potato starch is inherently anionic, since it contains phosphate groups. The presence of both anionic and cationic groups in cationized potato starch enhances its flocculation power and renders it active over a wider pH range. Corn starch has to be rendered amphoteric by the introduction of phosphate groups. Charge modification has also been applied to waxy maize starch.

Cationic starch must be thermally dispersed before it is added to the papermaking furnish. In batch cooking, a slurry with up to 4% starch content is heated to a temperature of 93–96°C (200–205°F) for 20–30 minutes. In jet cooking, the solids can be raised to 6% and the temperature to 102–116°C (215–240°F) with short retention. For waxy maize starch, the cooking temperature should not exceed 107°C (225°F). The starch cook must be diluted to <1% solids before addition to the paper stock.

Point of addition will vary depending on the desired results. If strength development is the major concern, addition should be at an early stage of the wet-end system, ranging from the discharge of the refiners to the various blend chests and screening sites. If high retention is the major concern, addition should be at the headbox manifold. A preferred point of addition for balance between strength, good formation and retention is prior to the primary screen before the headbox. Often a split addition at two different locations is practiced. On occasions, cationic starch is added to the filler pigment or the broke in order to enhance retention.[173] Addition levels range from 2.3 to 23 kg/ton (5 to 50 lb/ton), with an average of 7–9 kg/ton (15–20 lb/ton).

Adding a dispersion of cationic starch to dispersed unmodified starch produces a polyelectrolyte complex, which extends the hydrodynamic radius for bridging flocculation between fibers and pigments. A blend of unmodified starch with a starch graft has been recommended for flocculation control.[174] Polyelectrolyte complexes of starch can become insoluble in water at a specific mixing ratio.[175] However, these complexes may separate again in the papermaking furnish and cause an undesirable accumulation of unreacted starch in the process water, leading to viscosity increase (reduced drainage rate), felt filling, lower production speed and increased incidence of web breaks.

In an alternative route, starch is complexed with a polyamine, poly(vinyl amine), poly(ethylene imine) or chitosan to render it cationic. Although stronger, these complexes may also separate with time and lead to an undesirable accumulation of a low molecular weight, high-charge polyamine fraction and interference in the retention process.

Dialdehyde starch,[176–178] block reactive starch[179] and cationic aldehyde starch[180,181] comprise a special class of starches that can react with cellulose through the formation of covalent hemiacetal or acetal bonds. The zwitterion group is another functional group that has been utilized in conjunction with cationic and acetal groups.

VIII. Use of Starch for Surface Sizing of Paper

1. The Size Press in the Paper Machine

The Flooded Size Press

On the paper machine, the size press is used to apply surface size to dried paper.[182,183] Starch is the most frequently used binder in surface sizing. Besides raising surface strength, starch also imparts stiffness, lowers water sensitivity, reduces dimensional changes and raises air leak density of the sheet. In conventional practice, the sheet passes through a pond of starch dispersion held above the nip between two large rotating cylinders. In the nip a high, transient, hydrostatic pressure is developed.

Excess starch dispersion is drained from the ends of the nip. The surface size is transferred to paper by capillary penetration, pressure penetration and by hydrodynamic force during nip passage.

The quantity of starch transferred to paper by a size press depends on several factors: concentration of dispersed starch in the surface size; viscosity of the starch dispersion; diameter of the size press rolls; size press pond height; cover hardness of the size press rolls; size press nip loading pressure; paper machine speeds; wet-end sizing of the sheet; and water content of the sheet. The concentration of starch in the surface size liquid can range from 2% to ~15%, depending on product requirements. Frequently, pigments and other materials are added, which further increases total dispersed and suspended solids content. The viscosity ranges from water thin to several hundred cP (mPa·s).

Viscosity of the starch dispersion is the primary rate-determining parameter for dynamic sorption of starch into paper during surface sizing. Surface size penetration into the capillaries of paper proceeds in lateral and normal directions. Lateral flow takes the shape of an ellipse, according to the bias of fiber orientation in machine direction.[184] Contributions by wetting and capillary penetration decrease with increasing paper machine speed, while the contribution by hydrodynamic force increases with speed. As a consequence, starch pick-up will pass through a minimum at a specific speed. The hydrodynamic force depends on the angle of convergence (which is determined by the diameter of the rolls), by the nip length (which is influenced by the hardness of the roll covers), by the paper machine speed and by the opposing loading force between the two rolls. High liquid viscosity, large roll diameter, soft roll covers and high paper machine speed increase starch transfer, while high nip pressure counteracts these drivers. Starch cationization has no affect on pressure-driven penetration, provided the hydrostatic pressure is high and the viscosity of the dispersion is low.

The transferred liquid penetrates into the sheet according to the void space between fibers and pigment particles. During drying, starch attaches to the fibers and pigment, and reinforces the sheet by 'spot welding' and bridging between paper constituents. The ultimate location of the starch in the sheet can be affected by chromatographic partitioning behind a front of water that advances into the sheet. This effect will primarily occur in heavyweight paper and board and may lead to a gradient in starch concentration in the sheet from the surface to the interior and a weakening of internal bond at the ultimate location of free water. Starch application to the sheet induces some desizing due to coverage of hydrophobic patches by hydrophilic starch.

Application of surface size to paper carries with it the transfer of a substantial quantity of water. As an example, surface sizing of a 75 g/m^2 ($50 \text{ lb}/3300 \text{ ft}^2$) sheet (with 1% residual water content) by a 5% starch solution for a coat weight of 1.5 g/m^2 ($1 \text{ lb}/3300 \text{ ft}^2$/side) will raise the water content of the sheet to 43%. This large quantity of water will weaken the paper. Web breaks at the size press can occur, particularly when the sheet is also weakened as a result of edge cracks or holes.

Surface sizing can induce structural changes in the paper sheet[185] due to the interaction of water sorption (which causes a relaxation of internal stresses) and machine direction tension (which increases anisotropy and creates additional stresses). Anisotropy can be lowered by reducing tension on the web during sheet passage through the size press and subsequent dryers, and by raising the moisture content prior to the size press.

When surface-sized paper leaves the size press, it will cling to a roll and has to be pulled off. The separation force due to film splitting depends on the free film thickness, its cohesiveness, and the rheological properties of the surface size, especially its viscoelasticity. Transfer defects, such as ribbing, orange peel, spatter or misting may result. It is important to control the starch viscosity, to use the correct take-off angle and to apply appropriate web tension. Surface-size splashing can occur due to the converging motion of paper sheet and roll surfaces in the pond and fluid rejection at the nip. Best pond stability is obtained at high or low viscosity, while intermediate viscosity is most prone to induce pond instability.

The Metered Size Press

Operational problems with high water transfer to paper during surface sizing and pond instability have led to the development of pre-metered size presses. Two design options are being used in modern high-speed paper machines: pre-metering rolls or short-dwell coater heads attached to the size press.[186,187]

The gate roll size press is a modified roll coater.[188] Surface size dispersion is fed into the nip between metering (gate) and transfer rolls with overflow of excess at the edges. The liquid film that passes the nip is split, and the amount left on the transfer roll is split again in a second nip between transfer roll and applicator roll. The film remaining on the applicator roll is shared with paper in a third nip between the applicator roll and a backing roll for the sheet. The same action proceeds from the other side in a simultaneous gate roll size press or in the second station of a tandem of gate roll size presses. Roll speeds in the size press train are scaled; the metering roll is operated at low speed. There is less pond agitation than in a flooded size press, which permits running higher starch solids and higher viscosity. As a consequence, water transfer at equal starch application is decreased, leading to less web breaks and increased paper machine speed. Higher moisture content is possible in the paper prior to surface size application.

In an alternative method, a liquid film is metered onto a size press roll by way of an attached short-dwell coater head.[189,190] A modern installation of a film press is illustrated in Figure 18.7a. Surface size is pumped against the size press roll and metered with a small profiled rod, a larger smooth rod or a bent blade. Excess surface size leaves the chamber by flowing back over a weir. The back flow effectively 'washes' the roll surface, removing any debris that might have been pulled from the sheet during nip passage. The metered film is transferred to paper in the size press nip. The fluid film thickness on the size press rolls can be monitored by scraping or by infrared absorption.[191]

Volumetric metering of surface size is obtained by using a profiled (grooved), counter-rotating rod. The quantity of fluid that passes between roll and rod is determined by the rod surface (gaps), film split at the rod nip and film split at the paper surface. Since metering takes place between the surfaces of rod and roll, it is important that the roll surface be smooth and level, and not contaminated by low-energy

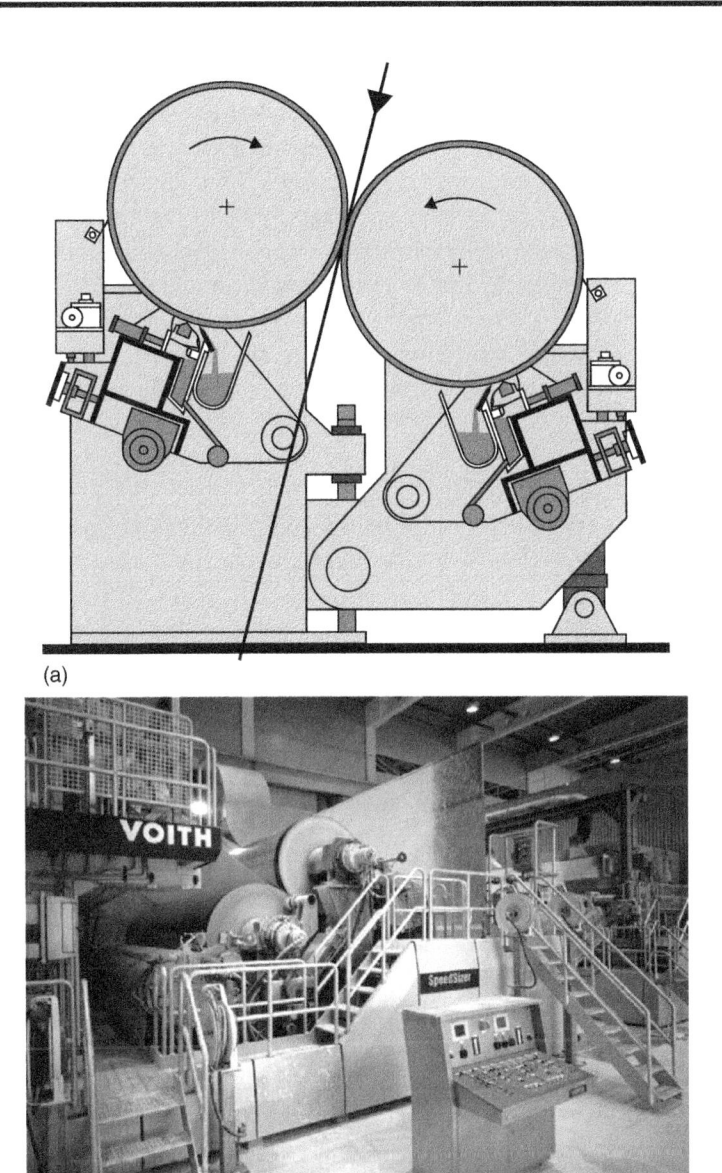

Figure 18.7 (a) Schematic illustration of a metered size press (Film Press); (b) a film press installation in a paper machine. (Courtesy of Voith Paper)

materials that interfere with wetting. Fluid metering by a profiled rod is largely independent of paper machine speed, surface size viscosity and solids content. In case of very low viscosity, fluid spraying at the rod can occur. Soft roll covers will decrease fluid passage through the metering gap due to indentation by the loaded rod.

Fluid passage through a bent blade nip is driven by hydrodynamic force and thus depends on paper machine speed and viscosity. Non-uniform fluid pressure in the feed chamber and fluctuations of viscosity can lead to variations in coat weight.

The blade metering system allows metering a thin liquid film for fluids with low viscosity. At the nip exit, misting can occur as a result of film splitting of coating fluids with low solids content at high operation speed. Filaments are formed that fracture into small droplets which are carried by air drafts. Film splitting and filamentation can also become the cause of an orange peel pattern on the coating surface.

Since metered size presses apply coating fluid simultaneously to both sides of paper, contactless air turning rolls and dryers are required to prevent scaling during the initial stages of coating drying.[192] Pre-metering the surface size allows control of the depth of starch penetration into the sheet and supports the design of paper products with specific properties of surface strength, stiffness, compressibility and fold strength.

Pigmented Surface Sizing
Adding pigment to starch-based surface size will facilitate various improvements in the properties of paper[193,194] viz., increased surface smoothness and ink holdout, improved optical properties, increased sheet density and stiffness, and improved printability. Pigmenting the surface size permits fiber replacement and an increase in paper machine speed, since a lower rawstock weight is required for a target product weight. The new application system will accommodate coating formulations with a wide range of compositions[195–197] viz., solids content up to 60%, use of high brightness pigments for enhanced visual quality, use of delaminated clay for paper density and stiffness, use of fluorescent whitening agents for paper brightness, replacement of internal size applications, use of synthetic co-binders and thickeners for improved paper printability. Additional benefits derive from changes in opacity, sizing and conversion properties.[198–200]

Metered size presses are operated at speeds exceeding 1800 meters per minute. Advantages of metered size press operations result from simultaneous two-side application, reduction in drying demand and web breaks, contour coating and holdout near the surface, higher surface strength and improved printability. Since coating fluid is applied at the same time to both sides of the paper sheet, contactless sheet guidance after the size press is required to prevent scaling on the drier cans. Various air turns, floating driers and contactless drying systems are being used.

Starch remains the dominant binder in size press coating.[201] Starch use has shifted from in-house depolymerized starch to modified products, such as those obtained by hydroxyethylation, acetylation or cationization.[202] Synthetic latices [styrene-butadiene, polyacrylics, poly(vinyl acetate)] are used as co-binders in high coat weight applications. Sometimes thickeners are added for adjustments in the rheological characteristics of the coating fluid. Starch-based coatings for high-speed application have to be designed for rapid fluid release, which forms a filter cake on the paper and moves the film split plane toward the applicator roll surface, thus decreasing the propensity for misting at the nip exit.[203]

In current practice, many paper grades are produced on-machine with a combination of precoating by metered size press and top coating by blade coater, followed by finishing with a soft calender.[204,205] Some of the new applications are: online precoating of wood-free or wood-containing papers, surface pigmentizing or coating of machine finished (MFP or MFC) papers, manufacture of ultra-lightweight coated (ULWC) papers, surface pigmenting of secondary fiber-based papers of low grammage (SC, newsprint), precoating and backside coating of folding boxboard, and pigmentizing of preprinted top liner for the corrugated board industry.

2. The Water Box at the Calender

An alternative method to apply starch to paper, especially to paperboard grades, is to use a water box at the calender. Water boxes are used for wet-finishing of paperboard prior to coating application. Use of starch in the water box will increase surface strength and lessen fiber reswelling after coating application. Dilute dispersions of oxidized starch are used, frequently in conjunction with a lubricant to prevent calender roll picking.

3. Spray Application of Starch

In the production of corrugating medium, linerboard and other unbleached (Kraft) paperboard grades, starch is added as a slurry spray to the fiber mat on the papermaker's wire.[205] This procedure provides better starch retention in the sheet than the addition of dispersed starch to the stock. According to the position of the spray boom, starch can be forced into the sheet for strength gain or held on the surface for improved surface quality.

Starch for use in spraying applications should have a low gelatinization temperature. Unmodified wheat starch gelatinizes at a lower temperature than does unmodified corn starch. On occasion, a cationic starch is used. The starch is sprayed as slurry, preferably after heating to a temperature just below onset on pasting. Dispersion occurs in the paper sheet during drying. Strength improvement by starch spraying ranges from 10% to 25% for 2–4% starch application.

Successful operation of a spraying system requires careful attention to starch suspension temperature, spray pressure, flow rate, drop size, nozzle alignment, spraying pattern and spray overlap.[206] Starch spraying is beneficial in the production of paperboard grades based on the use of waste paper and recycled fiber. High molecular weight starches are required to impart internal strength, while lower molecular weight starch is applied when high surface strength is the goal, such as in supercalendered (SC) groundwood grades. A major application of starch spraying is for the bonding of plies in the production of multiply packaging grades. Disadvantages of starch spraying are inefficiency at high paper machine speed, gradual plugging of wire and felts, and the accumulation of starch in water discharges due to poor retention.

4. Starch Selection for Surface Sizing

Many kinds of native and modified starches are being used on the size press.[207] The governing factors are dispersion viscosity, film formation and resistance to retrogradation. For low-cost applications, native corn starch is depolymerized in the paper mill by enzymic or thermal–chemical conversion. In-plant converted starches are widely used for surface sizing of commodity paper grades.[208,209] Variations in product viscosity have led to a decrease in their use on pre-metered size presses.

Acid-depolymerized starches are suitable for use at both the size press and the water box for improving surface sizing and internal bond with applications to SBS board, Kraft linerboard, cartonboard, bag stock and commodity grades. They are available from starch suppliers in a wide range of viscosity characteristics from thick boiling to thin boiling. These low-cost starches are sensitive to retrogradation and have to be consumed shortly after dispersion. The surface size has to be kept at elevated temperature and at a pH of 7.3 or higher. Various retrogradation control agents are offered, but strength loss due to amylose precipitation can occur even though viscosity is apparently stabilized.

Oxidized corn starches have improved water-holding properties, better film-forming characteristics and reduced gelling tendencies. These properties make oxidized starches particularly useful when high surface strength and ink holdout is desired. Oxidized starches are offered by starch suppliers in a wide range of viscosities, especially for use as surface size on uncoated wood-free sheets, such as bond, forms bond, offset, book and text grades. They are also used as a precoat to improve coating holdout on coated board and carrier board. Starches with a low degree of oxidation (bleached starches) require thermal conversion for complete dispersion, and pastes must be kept at pH levels above 7.3. Oxidized starches act as dispersants and can affect pigment retention in the sheet and waste water solids when added to the papermaking furnish as part of recycled broke. They are more easily discolored at high pH than other starch products. Recently it has been noticed that oxidized starch may contain AOX products, which were formed by the reaction of residual lipids with sodium hypochlorite. The presence of AOX products in starch can affect their use in consumer products.

Starch ethers and esters are offered in a wide range of viscosities. They have superior film-forming and water-holding properties. They are acid depolymerized, do not affect pigment retention and their retrogradation tendencies are greatly reduced. For these reasons, they are premium products for use at the size press, calender stacks and in coatings. They are used in high-quality printing and writing papers, as well as in specialty grades such as greaseproof paper and medical grades. Starch esters cannot be used at high alkalinity, since they will saponify, especially at the elevated temperature normally used at the size press.

The use of cationic starches is rapidly increasing, primarily in response to environmental concerns to reduce suspended solids and oxygen demand in waste streams from the paper mill.[210,211] Cationic starches are used in virtually all grades of paper and paperboard. Cationic starches impart strength to paper, reduce BOD and enhance film clarity. Tertiary starches require an acid pH and protonation to render them cationic. Quaternary starches are effective over the entire pH range normally encountered in papermaking, from acid to alkaline. When returned to the papermaking furnish as part of broke, they will aggregate anionic materials and improve their retention in paper. Hydroxyethylated cationic potato starch is used for embossed art paper, saturating grades, cotton content writing paper and as a precoat for specialty release papers.

Anionic starches, obtained by carboxymethylation of corn starch, provide improvements in film strength, pick resistance, opacity and air-leak density. They are used in the production of paperboard, polycoated papers, greeting card stock, barrier film and specialty papers with a high content of recycled fiber. They are compatible with poly(vinyl alcohol) and used in clear and pigmented size press applications.

Starch grafts are a new class of products for use on the size press.[212,213] They have high water retention and form flexible films. They are offered as a liquid, ready-to-use binder on conventional and metered size presses and on calender water boxes. Starch grafts are used in pigmented formulations for the production of film-coated offset, recycled offset, textbook and LWC (lightweight coated) grades, and as precoat for double-coated, sheet-fed offset grades.

Hydrophobic starches are another new class of starches for use on the size press. Some grades are based on waxy starches and are esterified. Hydrophobic starches provide improvements in water, ink and solvent holdout, as well as decreased sheet porosity.

In the surface sizing of paper with a high content of filler pigments and recycled fibers, addition of starch insolubilizers, such as formaldehyde-free glyoxal based crosslinker or zirconium salt, improves surface strength, wet pick resistance during offset printing and resistance to dusting.[214,215] When poly(vinyl alcohol) is used as a co-binder with starch, the starch can precipitate from solution, unless a carboxymethylated or hydroxyethylated and oxidized product with low viscosity is used.

Synthetic sizes, such as styrene maleic anhydride (SMA) copolymer or anionic polyurethane (PU) emulsion, are often added to starch-based surface size to develop water resistance, particularly in the production of label grades. This procedure may permit reducing or eliminating the use of internal size.[216]

In the production of copy paper grades, such as laser copy and ink-jet papers, hydrophobic polymer dispersions are used as an adjunct to the dispersion of hydrophilic starch. The hydrophobic/hydrophilic balance of the treated paper is controlled by the amount of applied material, the average particle size of the polymer particles, and the film-forming and water-absorption capacity of the starch. The surface energy of starch films has been studied by dynamic contact angle analysis. High-amylose, normal and waxy corn starch have similar critical surface energy of wetting in the range of 38 to 40 dynes/cm.[217]

The use of starch for surface sizing adds to the biological and chemical oxygen demand of water discharges from the paper mill, requiring appropriate effluent treatment. Cationic starches are held more tenaciously on cellulose and pigment surfaces. They are required for use in paper mills that have problems meeting environmental discharge standards.[218]

IX. Use of Starch as a Coating Binder

1. The Coater in the Paper Machine

The quality of a printed image is greatly enhanced when a coating is applied to the paper surface. Coating the paper lowers the roughness of the surface, generates gloss, provides an ink receptive surface with controlled porosity and improves ink holdout.[219] Paper coating is either practiced on-machine (in a unified operation immediately following sheet formation and drying) or off-machine on separate equipment. Off-machine coating is preferred when frequent grade changes are required. The coating is applied at speeds of up to 2500 meters per minute to paper webs up to 8 meters or more in width.[220] Early coater designs were derived from painting, such as the brush coater, or from printing, which has led to many versions of roll coaters. Most frequently used now are blade coaters, which first apply and then remove excess coating fluid from paper by a scraping action. Brush coating is limited to narrow specialty applications.

Roll coaters use multiple nips to split and meter the coating fluid. One parameter affecting coat weight is hydrodynamic force in the nip, which is provided by coater speed and the viscosity of the coating fluid.[221] Starch is a preferred binder for roll coatings. For high coat weights, a gravure roll coater is used. The equipment consists of a pick-up roll, the gravure roll with specific cell parameters and an offset roll in series. Roll coaters print a metered film of coating fluid onto the paper surface, resulting in a contour coating with a surface that follows the topography of the substrate. Gravure coating requires rheological characteristics that promote leveling of the coating into a uniform film.

In blade coating, an excess of coating color is transferred to the paper by an applicator roll[222] or a nozzle (jet). The subsequent scraping operation removes 90–95% of the excess color and levels the surface. The remaining coating fills the valleys of the sheet and covers the hills, resulting in a relatively smooth coating with a thickness principally dependent on blade load and roughness of the basestock. A modern blade coater installation is depicted in Figure 18.8a. In an alternative method, a small rotating rod is used instead of a blade to remove excess coating. In the coating of paperboard, the metering and leveling functions are frequently accomplished by an air knife.[223]

Roll coaters can apply a coating simultaneously to both sides of the sheet, while blade coaters generally treat one side at a time and require intermediate drying. Roll and blade systems have been combined in the Billblade coater, which has found use in the production of carbonless copy papers and for paper precoating.[224]

The coating fluid or 'color' consists of a dispersion of pigments in water at high solids, with binders and functional adjuncts added. The most commonly used inorganic pigments are kaolin clay, calcium carbonate, titanium dioxide and talc. The binders are either of natural origin, such as starch and protein, or synthetic products, such as emulsions of styrene-butadiene, poly(vinyl acetate), polyacrylics or various terpolymers. Polystyrene is frequently used as an organic pigment and poly(vinyl alcohol) as a water-soluble polymer. Coating adjuncts provide dispersing, thickening, lubricating, leveling and preservative functions.

Starch is used as a major coating binder on blade coaters, particularly for the production of LWC papers. New automated coating kitchens are equipped with jet cookers, which supply a uniform product to the proportioning system of coating preparation. The range of coating formulations for a specific grade line is illustrated in Table 18.3. Premodified starches, particularly starch ethers, are preferred for semi-continuous or continuous coating blending. Some paper mills use two starches of different viscosity.

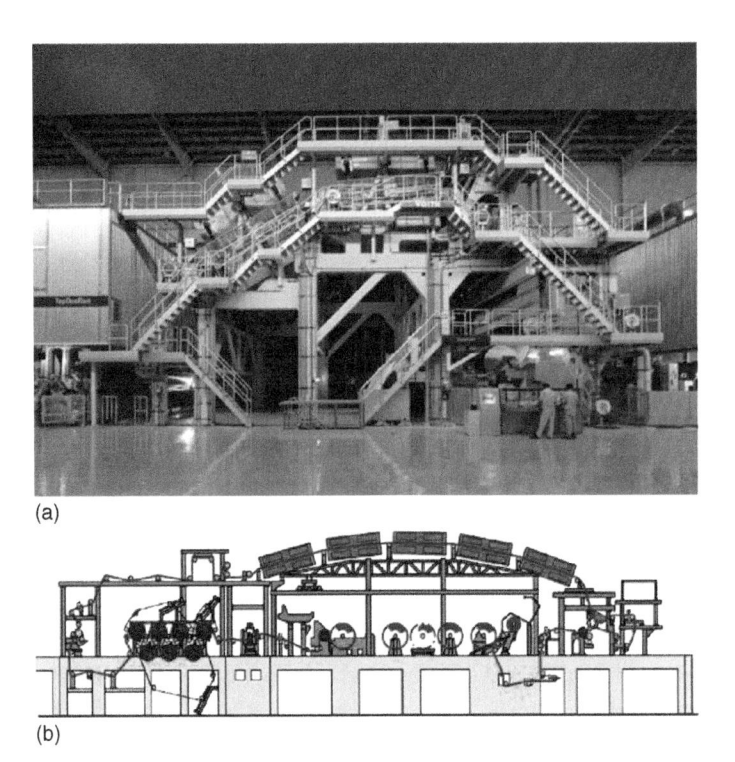

Figure 18.8 (a) A blade coater installation in a paper machine; (b) schematic illustration of paper coating equipment in a paper machine. (Courtesy of Voith Paper)

Table 18.3 Range of component amounts in coating formulations for publication web offset paper grades (lightweight coated paper)

Coating ingredients (parts)	High starch	Low starch
Clay (delaminated)	20	100
Clay, No. 2	80	0
Clay (calcined, structured)	0	10
Titanium dioxide	0	5
Synthetic pigment	0	10
Binder level	10	15
% starch	75	25
% latex (SBR or PVAc)	25	75
Crosslinker (% on binder)	2.5	10
Lubricant	0	2
% solids	Up to 57	>57
Coat weight (#/3300 ft²/side)	4	8

Reprinted by permission of TAPPI

The quantity of coating applied by a blade coater is determined by the roughness of the paper surface, the solids content and viscosity of the coating color, the speed of the coating operation, the geometry of the system, blade mechanical characteristics and the pressure applied at the blade.[225,226] Calendering prior to coating application improves the smoothness of the paper surface, but some of this effect is lost when the water-based coating is applied.[227]

After application to the paper, the coating color will release water and form a filter cake, which facilitates holdout on the surface. Water release from the coating is promoted by the pressure pulse in the application nip, followed by the action of capillary pressure on the paper and a second pressure pulse when the coating passes the metering blade nip.[228,229] The increased moisture content in the sheet will induce swelling of fibers and the release of stress, causing roughening of the paper surface due to raised fibers.[230] Water release from the coating color is controlled by the state of pigment dispersion, the association of binders and thickeners with the pigment fraction, and by the viscosity and water-holding properties of the liquid phase.[231]

Fast dewatering of the coating on the paper sheet results when the pigment dispersion is destabilized by a cationic polymer or a depeptizing salt and by large platey pigments in the coating.[232] Cationic starch has been used as an additive to lower the immobilization concentration of a paper coating. Rapid discharge of coating vehicle leads to a bulky cardhouse structure with good fiber coverage, high pore volume and opacity, low air-leak density and good printability, but diminished surface strength.

Starch is an effective water-retention aid in coating colors. With starch as the coating binder, the dewatering process will be slowed, leading to a consolidated coating structure. The dried coating normally has a slight excess of binder at the surface and at the interface with the substrate.[233] Water retention of coating colors can be tested by various static or dynamic methods. Useful data from static tests can be derived by any of the following methods: monitoring the time-dependent water loss into a porous ceramic plate or a sheet of paper by recording the time until disappearance of gloss from the surface;[234] determining the onset of conductivity through the coating and an underlying sheet of paper;[235] testing for coloration on the backside of paper after application of dyed coating color[236] or monitoring the sonic velocity through coating and paper.[237] Water retention can also be appraised by recording capillary penetration into a porous substrate[238] or by pressure filtration, with or without shear near the paper surface.[239]

Dynamic methods involve scraping off coating from paper at specific time intervals after the coater blade, recording the change in water content[240] or by monitoring the distance (time) when gloss disappears. Liquid migration through paper is normally not uniform but follows the capillary structure.[241] The mass distribution of coating on the paper surface is affected by paper formation.[242]

Coated paper is dried by infrared irradiation, hot air streams and by contact with hot drier cans. Heating and evaporation will cause water flow through the coating, which transports dispersed and emulsified material by viscous drag.[243] Depending on the drying rate and forces, binders are deposited uniformly within the coating structure or are enriched at the surface or at the interface with the paper.[244,245] Starch is especially sensitive to migration during drying. Non-uniform accumulation of starch at the surface is a frequent cause for print mottle.[246,247]

2. Starch Selection for Paper Coating

Starch is a major binder for paper coatings. It is used as a sole binder or combined with various synthetic binders. Blends with aqueous dispersions of poly(vinyl alcohol), alginates or cellulose derivatives are used also. Relatively low strength, water sensitivity and high surface energy[248] have limited the utilization of starch as a coating binder. On the other hand, starch is a low-cost, natural binder and facilitates redispersion of coated paper for recycling. In contrast to synthetic binders, starch raises the stiffness of coated paper, but when used in excess fold cracking might occur after web offset printing.[249]

Nearly all classes of starch are utilized as coating binders. They must be adjusted in molecular weight in order to obtain the rheological characteristics needed for uniform coating distribution and for controlled water retention at high application speed.[250–252] Oxidized starches are utilized as a coating binder,[253] but they can act as pigment dispersants when returned with broke to the paper machine, leading to an increase in suspended solids in paper mill discharges. Oxidized corn starch can contain AOX products from chlorination of residual lipids. Starch ethers[254] and esters,[255] which are resistant to retrogradation, are now most widely used as coating binders. Starch grafts combine the properties of starch with those of a synthetic binder.[256,257]

Excessive reaction of the substituting reagent at the surface of starch granules can impair their dispersion and may require high-temperature cooking to obtain a uniform paste. Starch esters are sensitive to alkalinity, which will break the ester bond. Recently, total chlorine free (TCF) starches have been introduced for use with TCF pulps. Dextrins, which are prepared by a dry roasting process, have improved paste stability due to induced branching and are used in Europe as a TCF starch for use in TCF paper grades.

Pregelatinized starch is sometimes used as a coating binder,[258] but these products are difficult to wet out uniformly and may cause problems due to clumping. Small granules of starch are used in specialty coatings.[259]

Special starch products have been developed for use as thickeners in coatings that are based on synthetic latex or resin binders. Starch-based thickeners can have associative properties when substituted with a hydrophobic reagent. They provide an alternative for cellulose-based thickeners, but their use is limited.

Starch use in the coating renders it sensitive to water. Starch can cause dimensional instability of the paper when the relative humidity is changed. In offset printing, where fountain water is used, it can lower the wet pick resistance of the surface. Various reagents are used to impart water resistance to starch,[9,260,261] Previously used formaldehyde- and melamine-based products have been replaced by new polymers, primarily based on glyoxal.[262] Zirconium salts are frequently applied as crosslinkers and insolubilizers.[263] The use of ammonia as a fugitive base to raise water resistance after drying is gradually being discontinued due to OSHA regulations.

Various attempts using a cationic binder have been made to render the coating color cationic.[264,265] Cationic starch, cationic poly(vinyl alcohol) and cationic latex have been used. Despite apparent product advantages in fiber coverage there has not been widespread use of this process, since it requires conducting the entire papermaking process under cationic conditions. A coating containing starch with tertiary amino substituents, adjusted to high pH with ammonium hydroxide, can be rendered cationic by evaporating the base.[266] All coating colors that contain starch as binder must also contain a biocide to prevent spoilage.

X. Use of Starch as Adhesive in Paper Conversion

Large quantities of starch are consumed as a component of adhesives for paper and paperboard, especially in the production of corrugated board. Dextrins, derived from corn or potato starch, are used as adhesives for laminated products and as gums for envelope construction and sealing.

1. Lamination of Paper

Many paper grades are bonded to one another or to a different substrate, primarily for use in packaging applications and as labels. The main component of the glue is often starch or modified starch. The glues are used in the building of various boxes for packaging applications. In laminating, the substrate can be a low-quality recycled paperboard to which a fine paper with good printability is attached. Label papers are coated on one side for printing, while the other side is treated with an adhesive. For papers with high internal sizing, an overcoat with starch may be required for good bonding by a dextrin-based glue. Glue is applied to paper by various roller systems that deliver the glue from a glue pan. A pressure nip is required for good contact. Rollers can be smooth or engraved with a pattern of small cells.

Various theories have been proposed to explain adhesion.[267,268] Requirements for good bonding are substrate wetting by the adhesive and formation of a boundary layer. These conditions will be met if the surface energy of the substrate is higher than the surface tension of the adhesive dispersion. Coated paper with a high amount of latex in the coating has low surface energy, which requires lowering the surface tension of the adhesive by the addition of a surfactant or an alcohol. Intermolecular bonds have to be established through the boundary layer. Improved bonding is obtained when acid/base interaction occurs due to the presence of polar groups.[269,270] The peeling force required for separation is used to determine the strength of a glue bond.[271,272]

2. The Corrugator for Paperboard

Corrugated board is obtained by combining corrugating medium and linerboard, two unbleached Kraft products. The medium is the fluted or corrugated portion; the liner is the surface sheet onto which the fluted tips are bonded. The corrugating process is illustrated in Figure 18.9. On the single facer, the fluted medium is attached to the first liner sheet. On the double backer, the second liner sheet is added. Various types of corrugated board can be constructed, depending on the fluting frequency (110 to 315 per meter), the flute height (0.47 to 0.11 cm), and the number of liners.

In the corrugating process, the medium is conditioned (softened) by heat and steam, and fluted between two corrugating cylinders. Adhesive is applied to the tips of the flutes by a roll applicator. The bond with the preheated liner is formed in a nip between a pressure roll and a corrugator roll and set by pressure and heat. The single-face board is transferred by a bridge to the double backer glue machine, where it is combined with the second liner, thus forming double-face corrugated board.

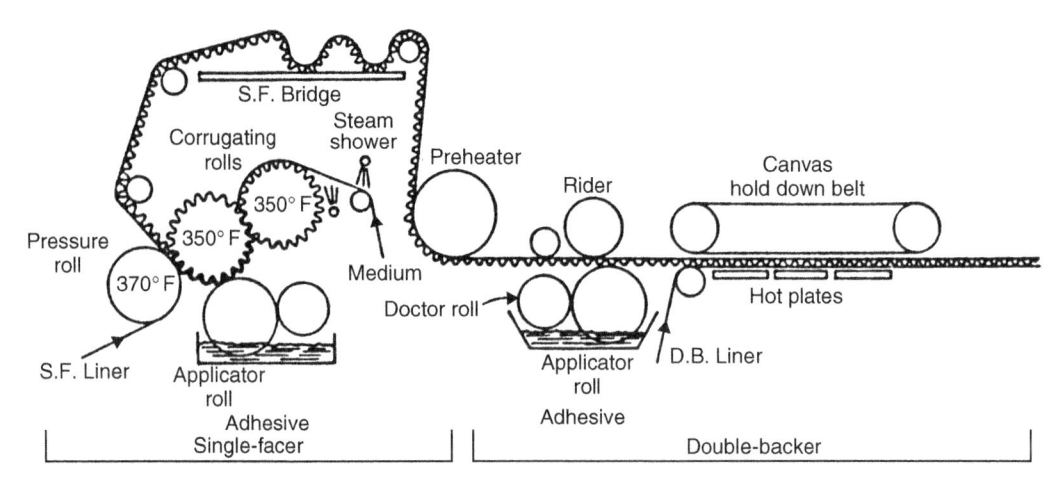

Figure 18.9 Schematic illustration of equipment for the manufacture of corrugated board. (Reprinted by permission of TAPPI)

Table 18.4 Corrugating adhesive formulation[a]

Primary mixer (carrier starch)
Step 1	Add water	378.5 L	(100 gal)
Step 2	Add corn starch	90.9 kg	(200 lb)
Step 3	Add caustic soda, dissolved in about 37.9 liters (10 gal) of water	13.6 kg	(30 lb)
Step 4	Heat to	66–71 °C	
Step 5	Hold under agitation	15 minutes	
Step 6	Add cooling water	227.1 L	(60 gal)

Secondary mixer (raw starch)
Step 7	Add water	1514.0 L	(400 gal)
Step 8	Heat to	27–32 °C	
Step 9	Add corn starch	454.5 kg	(1000 lb)
Step 10	Add borax (10 mole)	13.6 kg	(30 lb)
Step 11	Transfer primary mix into secondary mix in about 30 minutes		

[a]Stein-Hall

Reprinted by permission of TAPPI

The equipment is operated at 162–182°C, which is obtained using high-pressure steam of 965–1240 kPa (140–180 psi).

The major ingredient of corrugating adhesive is corn starch.[273] Most formulations contain a partially gelatinized carrier starch and ungelatinized raw starch.[274–276] Carrier starch is prepared in the primary mixer by suspension in water, addition of sodium hydroxide and heating to 66–71°C. The presence of base decreases the gelatinization temperature of corn starch, inducing swelling of the starch granules at operating conditions. In the secondary mixer, a suspension of raw starch and borax is prepared and heated to 27–32°C. Finally, the contents of the primary mixer are added to the secondary mixer and the product is pumped to the starch application pan. A corrugating starch adhesive formulation is illustrated in Table 18.4.

The carrier portion of the corrugator adhesive provides the viscosity needed to hold the raw starch in suspension and to retain water during application. Heating and stirring of the carrier and of the final product need to be carefully controlled in order to maintain the correct viscosity. In the majority of corrugator formulations, starch solids range from 18% to 21%, but may increase to 30% for specialty applications. About 15–17% of the total starch is used as carrier.

The addition of caustic, such as NaOH, aids in wetting the paperboard surface, thus enabling good contact and penetration into the paper in the short time available for the bond to form. Borax performs several functions in the adhesive. It provides tack to the adhesive, improves water holding and increases viscosity stability. Borax reacts with raw starch when it gelatinizes on the glue line, providing high viscosity and aiding in bond formation.

Adhesive is continuously circulated through the glue pan with overflow to a return pump. Constant flow is necessary to avoid dead spots (hot spots) which could cause premature gelling of the adhesive. The amount of glue on the applicator roll is determined by the clearance to a metering doctor roll. The surface speed of the applicator roll is slightly lower than that of the flutes, resulting in a wiping action that transfers more of the adhesive.

After starch application to the flutes, it must be gelatinized/pasted and dehydrated to form a minimum green-bond strength, which is necessary to resist delamination during subsequent mechanical operations. The bond between medium and liner is formed *in situ* by heat and pressure. Penetration of the carrier into the liner occurs as soon as contact is made. The depth of penetration is usually not more than 25% of the thickness of the sheet. Penetration must occur very quickly, since the contact time for the fluid under the pressure roll is only about 0.03 seconds at normal operating speed. It is unlikely that the starch is completely cooked after exiting from the pressure nip. Green-bond formation is completed on the bridge using residual heat in the paper. The gelatinization process requires about 190 kJ/kg, while much more energy (1500 kJ/kg) is needed for dehydration. Final cured bond strength is only obtained after the board is removed from the corrugating machine.

Besides the conventional corrugating adhesive, various modifications are in use. In a no-carrier formulation, all starch granules are slightly swelled in a controlled manner. When the desired viscosity is reached, boric acid metered into the adhesive neutralizes some of the caustic soda and stabilizes the adhesive viscosity. In some cases, continuous cookers are used for adhesive preparation. A cold corrugating process has been developed, which utilizes fully cooked starch obtained by pressure cooking with ammonium persulfate. The bond is formed by setback, brought on by cooling instead of heating. The adhesive is applied at a temperature of 87–95°C, which is maintained by means of a jacketed pan.[277]

3. Starch Selection for Use in Corrugation and Lamination

For most corrugating applications, unmodified corn starch is used. Specialty starches are being offered for increased corrugator speed or better warp control through minimizing the amount of water added to the board. Hydroxyethylated corn starch will lower the gelatinization temperature and improve dispersion stability.[278] Chemically-modified high-amylose or waxy maize starch improves bond strength or water retention respectively. The setting rate can be improved by addition of partially-hydrolyzed poly(vinyl alcohol). In the production of corrugated board, which involves the application of a blend of partially-cooked and uncooked starch, a specific depth of glue penetration is required in order to set the initial bond.[279,280] Wheat starch has found acceptance in areas where economy and availability are favorable. Water-resistant starch formulations are required for refrigerated boxes or outside box storage. Water resistance is obtained by addition of an alkaline curing resin, based on urea-formaldehyde, ketone-formaldehyde or resorcinol. New crosslinking resins that meet current environmental regulations are being developed. Starch-based hot melts have been developed for use as adhesive in packaging applications, such as in case and carton sealing, bag ending and roll wrap.[281] Adhesive viscosity is measured with a brass cup viscometer. The cup is filled with adhesive, and the time it takes for flow through a capillary is measured and compared to the flow time for water.

XI. Use of Starch in Newer Specialty Papers

New non-contact printing technologies have found wide application in office copying, computer printing and in print-on-demand systems. The image is processed digitally and transferred to paper by xerography, laser, indigo, xeikon, ink-jet and other computer-to-print methods. Paper for xerography, laser and related printing is generally surface sized with starch, which imparts stiffness, sizing and dimensional stability.[282] Various cationic and hydrophobic agents are used to enhance toner bonding and image fidelity.

Paper for ink-jet printing requires good dye or pigment holdout, image sharpness, absence of bleeding and the ability to carry a large quantity of water without distortion, especially in multi-color printing. The coating must meet critical requirements in porosity and surface energy.[283,284] Coatings for ink-jet printing are based on the use of fine particle size silica or calcium carbonate as pigment, cationic and hydrophobic additives, and starch as primary binder. Ink-jet papers must accept ink in a subsecond timeframe, requiring an open surface aided by electrostatic interaction between coating and dye. Fiber swelling in contact with water will degrade print sharpness and has to be depressed by a hydrophobic additive. Oxidized or cationic starch is applied to the paper surface by a flooded or metered size press. Internal alkaline sizing is enhanced by the use of cationic starch. In carbonless copy paper, wheat starch is used as a stilt material (spacer) to protect microcapsules from rupturing during paper rewinding and other manipulations prior to use.[285]

XII. Environmental Aspects of Starch Use in the Paper Industry

Starch is a rich nutrient for microbiological action and is easily degraded. Starch may spoil when exposed to high humidity. The rate of bacterial growth at 85% relative humidity increases with temperatures up to a maximum at 65.5°C (150°F). Bacteria grow in the pH range 4–9, with highest activity in the pH range 6–8. In the paper mill, starch degradation may already begin in the slurry tank and will continue throughout starch dispersion, holding and application. Spoilage originates from the starch supply (when it contains a heavy load of spores of starch degrading bacteria), the use of poorly treated process water and by infection from ambient air. The temperature required for batch cooking and holding will not suffice to preserve starch paste, since thermophiles can still be activated. Complete spore kill is only achieved by thermal starch conversion (300°F), but the potential for secondary infection may still require the use of a biocide.

Microbiological action in starch dispersions results in a drop in pH, loss of viscosity and the development of odor. Retrogradation may be accelerated by the drop in pH or especially if butanol, which complexes with amylose, is generated via starch fermentation. Sulfate-reducing bacteria will cause black deposits due to reaction with iron in the process water. For quality control, preservatives are added to starch slurry, cooked starch, surface size and coating color.

Starch dispersions have a high biological oxygen demand (BOD) and chemical oxygen demand (COD) which requires reduction before effluents from a starch operation can be discharged into the environment. BOD and COD are means to assess the potential contribution to pollution by the discharge of waste water. The 5-day BOD of size press starch is about 0.5–0.75 lb per lb of starch or about 10–15 lb/ton effluent, depending on the starch type. According to the 20-day test, native starch generates an approximately equal weight of BOD, similar to the COD. BOD is reduced to about 7 lb/ton discharge when cationic starch is used as surface size. Paper mills, which are subject to environmental regulation, must therefore use cationic starch as surface size.

During papermaking and coating, a substantial portion of surface-sized or coated paper is returned to the papermaking furnish as broke. Finished paper may contain 15–35% broke. This recycled portion has to be finely dispersed in order to avoid the incorporation of fiber clumps and broke chips into new paper. About 50–60% of the attached starch is released into the water system of the paper machine and, ultimately, into the paper mill effluent. The highest loss, about 15 lb starch per ton paper, will result when an anionic starch is used as surface size or coating binder.[210,286] Dissolution of starch is aided by elevated temperature, alkalinity, dispersing agents and, occasionally, the use of amylase. Oxidized starch in waste streams can act as a powerful dispersant, which hampers clarification of the effluent.[287,288] This disturbance has to be controlled by the use of cationic reagents or flocculants.[289] Cationic starches remain more strongly bonded to cellulose during broke processing and, if redispersed, associate with anionic materials to form larger aggregates that are retained in the paper, thus reducing BOD in the effluent.[290,291]

As part of a coating operation, a large quantity of coating material and filtered process water is lost to the sewer; this loss results from grade changes in coating preparation, screen dumps and wash-ups after paper breaks. Membrane filtration is a maturing technology that appears suited to remedy coating material recovery for reuse, effluent clarification, recovery of process water and landfill cost reduction in a single operation. The efficiency of membrane filtration of starch depends on molecular weight and weight distribution. Only molecules larger than 0.001 μm in diameter or with a molecular weight above 10 kD can be sufficiently retained by ultrafiltration. For smaller molecules, nanofiltration is required. Membrane passage may cause some degradation of starch, as indicated by decreased molecular weight.[292]

Paper is increasingly recycled. In 2005, the recycling rate reached 51.5%. Current AF&PA guidelines target the return of 55% consumer waste paper into new paper production by 2012. Various processes have been developed to de-ink waste papers that carry a wide range of inks and coatings.[293,294] Most de-inking is accomplished by the use of surfactants, dispersion and flotation.

Since starch is biodegradable, it has found use as filler in polymers, such as extruded films applied to paperboard for use in packaging.[295,296] The starch is incorporated in granular form,[297] as an octenylsuccinate derivative[298] and in other forms. Starch-based blown films are prepared from starch-poly(ethylene-co-acrylic acid) formulations.[299]

Oxidized corn starch may contain AOX materials from the chlorination of associated lipids. Avoidance of AOX discharges in paper mill effluents is an environmental priority.[300]

XIII. Starch Analysis in Paper

The starch content in paper is determined by a TAPPI test method.[301] Special methods are needed to assay starch in the presence of other adhesives.[302] Starch on the paper surface is determined by ATR (attenuated total reflection) Fourier transform IR spectroscopy.[303] New methods are being developed by the International Standards Organization (ISO).[304]

Various refinements of starch content analysis have been reported. The methods are based on starch hydrolysis, followed by polarimetry,[305] high-pressure liquid chromatography[306] or reaction with glucose oxidase/peroxidase.[307,308] An iodine reaction can be used to determine the botanical origin of starch.[309] The molecular weight distribution is determined by size-exclusion chromatography.[310,311]

Lateral variance in surface chemistry and surface porosity will interfere with ink receptivity during printing.[312,313] Non-uniform distribution of starch on the surface of a coating is revealed by staining with iodine vapor[315] or by ammonium chloride burnout.[316,317] IR reflectance microscopy, ESCA (electron spectroscopy for chemical analysis), EDXA (electron dispersive x-ray analysis), electron beam irradiation, Raman microscopy or TOF SIMS (time-of-flight secondary image mass spectroscopy) have been used to characterize the surface chemistry of a coating.[314] The distribution of binder throughout the coating and the paper is analyzed in cross-sections by staining with iodine and image analysis.[318,319] In coatings containing starch and styrene butadiene latex, staining with osmium tetroxide will differentiate between the two adhesives. For a comprehensive analysis of the paper surface, various optical, physical, and chemical procedures have been developed.[320]

XIV. References

1. Resource Information Systems, Inc. (RISI), Pulp and Paper Review.
2. American Forest and Paper Association, 1111 19th Street NW, Washington, DC 20036 USA. Internet: www://afandpa.org
3. Cutler PH. *Paper Technol.* 1995;36(9):36.
4. Cody HM. *Proc. TAPPI Coating Conf.* 1993:115.
5. Casey JP, ed. *Pulp and Paper, Chemistry and Chemical Technology.* 3rd edn. New York, NY: John Wiley; 1981.
6. Kocurek MJ, ed. *Pulp and Paper Manufacture.* 3rd edn. Atlanta, GA: Joint Textbook Committee TAPPI/CPPA; 1989.
7. Smook GA. *Handbook for Pulp and Paper Technologists.* 3rd edn. Atlanta, GA: TAPPI PRESS; 2002.
8. Technical Association of the Pulp and Paper Industry (TAPPI), Atlanta, GA; Internet: www.tappi.org
9. Maurer HW, ed. *Starch and Starch Products in Surface Sizing and Paper Coating*, Atlanta, GA: TAPPI PRESS; 2001.
10. Wurzburg OB. *Modified Starches, Properties and Uses.* Boca Raton, FL: CRC Press; 1988.
11. BeMiller JN. *Starch/Stärke.* 1997;49:127.
12. Whistler RL, Paschall EF, eds. *Starch: Chemistry and Technology.* New York, NY: Academic Press; 1967.
13. Whistler RL, BeMiller JN, Paschall EF, eds. *Starch: Chemistry and Technology.* 2nd edn. Orlando, FL: Academic Press; 1984.
14. Glittenberg D. *Wochenbl. Papierfabr.* 2001;129(1413):1508.
15. Corn Refiners Association (CRA), Shipment Reports; Internet: 1995. www.corn.org.
16. Swinkels JJM. *Starch/Stärke.* 1985;37:1.
17. Glittenberg D, Hemmes JL, Bergh NO. *Paper Technol.* 1995;36(11):18.
18. Singh V, Ali SZ. *Starch/Stärke.* 1987;39:492.
19. Pessa E, Suortti T, Autio K, Poutanen K. *Starch/Stärke.* 1992;44:64.
20. Hebeish A, El-Thalbuth IA, Refai R, Ragheb A. *Starch/Stärke.* 1989;41:293.
21. Forssell P, Hamunen A, Autio K, Suortti T, Poutanen K. *Starch/Stärke.* 1995;47:371.
22. Parovuori P, Hamunen A, Forssell P, Autio K, Poutanen K. *Starch/Stärke.* 1995;47:19.
23. Wing RE. *Starch/Stärke.* 1994;46:414.
24. van Warners A, Stamhuis EJ, Beenackers AC. *Ind. Eng. Chem. Res.* 1994;33:981.
25. Kuipers NJM, Stamhuis EJ, Beenackers AC. *Starch/Stärke.* 1996;48:22.
26. van Warners A, Lammers G, Stamhuis EJ, Beenackers AACM. *Starch/Stärke.* 1990;42:427.
27. Khalil MI, Hashem A, Hebeish A. *Starch/Stärke.* 1995;47:394.
28. El-Alfy EA, Samaha SH, Tera FM. *Starch/Stärke.* 1991;43:235.
29. Hellwig G, Bischoff D, Rubo A. *Starch/Stärke.* 1992;44:69.
30. Khalil MI, Hashem A, Hebeish A. *Starch/Stärke.* 1990;42:60.
31. Lapasin R, Pricl S, Tracanelli P. *J. Appl. Polym. Sci.* 1992;46:1713.
32. Muhrbeck P, Svensson E, Eliasson AC. *Starch/Stärke.* 1991;43:466.

33. Mark H. *TAPPI J.* 1963;46:653.
34. Hebeish A, El-Rafie MH, Higazy A, Ramadan M. *Starch/Stärke.* 1996;48:175.
35. Hebeish A, El-Alfy E, Bayazeed A, Hebeish A, Bayazeed A, El-Alfy E, Khalil MI. *Starch/Stärke.* 1988;40;(191):223.
36. Touzinski GF, Maurer HW. US Patent 3 640 925. 1972.
37. Carr ME. *Starch/Stärke.* 1992;44:219.
38. Abell S. *Proc. TAPPI Coating Conf.* 1992:95.
39. Järnström L, Lason L, Rigdahl M. *Nordic Pulp Pap. Res. J.* 1995;10(3):183.
40. Veelaert S, Polling M, de Witt D. *Starch/Stärke.* 1994;46:263.
41. Hernandez HR, Grief DS, Barna AN, Thorton DS. US Patent 4 243 480. 1981.
42. Hart JR, Juergens SG, McCormack WE. US Patent 4 340 442. 1982.
43. Tomasik P, Wang YJ, Jane J. *Starch/Stärke.* 1995;47:96.
44. Chinnaswamy R, Hanna MA. *Starch/Stärke.* 1991;43:396.
45. Narkrugsa W, Berghofer E, Camargo LCA. *Starch/Stärke.* 1992;44:81.
46. Maurer HW, Kearney RL. *Starch/Stärke.* 1998;50:396.
47. Zobel HF. *Starch/Stärke.* 1988;40:44.
48. Lloyd NE, Kirst LC. *Cereal Chem.* 1963;40:154.
49. Deffenbaugh LB, Walker CE. *Starch/Stärke.* 1989;41:461.
50. Deffenbaugh LB, Walker CE. *Cereal Chem.* 1989;66:493.
51. Haase NU, Mintus T, Weipert D. *Starch/Stärke.* 1995;47:123.
52. Viscosity of Starch and Starch Products, TAPPI Test Method T 676 cm-97. 1997.
53. Leszczynski W. *Starch/Stärke.* 1987;39:375.
54. Hari PK, Garg S, Garg SK. *Starch/Stärke.* 1989;41:88.
55. Yeh AI, Li JY. *Starch/Stärke.* 1996;48:17.
56. Ewing FG, DeGroot HS. *Paper Trade J.* 1960;(12):48.
57. Fetzer WR, Kirst LC. *Cereal Chem.* 1959;36:108.
58. DeGroot HS, Ewing FG. *Paper Trade J.* 1966;(36):32.
59. Janas P. *Starch/Stärke.* 1991;43(168):172.
60. Bohlin L, Eliasson AC, Mitta T. *Starch/Stärke.* 1986;38:120.
61. Cannon MR, Manning RE, Bell JD. *Anal.Chem.* 1960;32:355.
62. Heitmann T, Mersmann A. *Starch/Stärke.* 1995;47:426.
63. Sjöström E. *Nordic Pulp Pap. Res. J.* 1989;4(2):90.
64. Diniz JM. *Langmuir.* 1995;3617.
65. von Raven A. *TAPPI J.* 1988;71:141.
66. Kogler W, Spielmann D, Huggenberger L. *Proceedings of the TAPPI Coating Conference.* 1992: 313.
67. Arno JN, Frankle WE, Sherdan JL. *TAPPI J.* 1974;57(12):97.
68. Pruszinski P. *Pulp Paper Can.* 1995;96(3):73.
69. Kaunonen A. *Pap. Puu.* 1989;71:46.
70. Richter R, et al. *Wochenbl. Papierfabr.* 1989;117:682.
71. Baumgartner H, Bley L. *Wochenbl. Papierfabr.* 1994;122:894.
72. Kodama M, Noda H. *Biopolym.* 1978;17:985.
73. Kalichevski MT, Orford PD, Ring SG. *Carbohydr. Res.* 1990;198:49.
74. Roulet P, MacInnes WM, Gumy D, Wuersch P. *Starch/Stärke.* 1990;42:99.
75. van Soest JJG, deWit D, Tournois H, Vliegenthart JFG. *Starch/Stärke.* 1994;46:453.

76. Wilson RH, Kalichevsky MT, Ring SG, Belton PS. *Carbohydr. Res.* 1987;166:162.
77. Teo CH, Seow CC. *Starch/Stärke.* 1992;44:288.
78. Bulkin BJ, Kwak Y. *Carbohydr. Res.* 1987;160:95.
79. Kubik S, Wulff G. *Starch/Stärke.* 1993;45:220.
80. Fujita S, Iida T, Fujiyama G. *Starch/Stärke.* 1992;44:456.
81. Islam MN, Mohd BMN. *Starch/Stärke.* 1994;46:388.
82. Tsuge H, Tatsumi E, Ohtani N, Nakazima A. *Starch/Stärke.* 1992;44:29.
83. Davies T, Miller DC, Procter AA. *Starch/Stärke.* 1980;32:149.
84. Eliasson AC, Finstad H, Ljunger G. *Starch/Stärke.* 1988;40:95.
85. Bulpin PV, Welsh EJ, Morris ER. *Starch/Stärke.* 1982;34:335.
86. Nierle W, El Baya AW. *Starch/Stärke.* 1990;42:268.
87. Nierle W, El Baya AW, Kersting HJ, Meyer D. *Starch/Stärke.* 1990;42:471.
88. Morrison WR, Coventry AM. *Starch/Stärke.* 1989;41:24.
89. Raphaelidis S, Karkalas J. *Carbohydr. Res.* 1988;172:65.
90. Ollett AL, Kirby AR, Clark SA, Parker R, Smith AC. *Starch/Stärke.* 1993;45:51.
91. Roco MC, Shook CA. *AIChE J.* 1985;31:1401.
92. Potze J. *Starch/Stärke.* 1992;44:283.
93. Ragheb AA, El-Thalouth IA, Tawfik S. *Starch/Stärke.* 1995;47:338.
94. Jane J. *Starch/Stärke.* 1993;45:161.
95. Hebeish A, El-Thalouth IA, Kashouti ME. *Starch/Stärke.* 1981;33:84.
96. Goos H, Maurer HW, Kurz AW. US Patent 3 219 483. 1965.
97. Windfrey V, Black W. US Patent 3 133 836. 1962.
98. Harvey RD. *Proc. TAPPI Coating Conf.* 1995:491.
99. Maurer HW. US Patent 3 475 215. 1979.
100. Harvey RD. US Patent 4 544 021. 1985.
101. Brouwer PH. *Wochenbl. Papierfabr.* 1993;121:1032.
102. Lorenzak P, Stange A, Niessner M, Esser A. *Wochenbl. Papierfabr.* 2000;128:14.
103. Borchers B. *Wochenbl. Papierfabr.* 2005;133:1159.
104. Harvey RD, Welling LJ. *TAPPI J.* 1976;59(12):192.
105. Antrim RL, Solheim BA, Solheim L, Auterinen AL, Cunefare J, Karppelin S. *Starch/Stärke.* 1991;43:355.
106. Koch R, Antranikian G. *Starch/Stärke.* 1990;42:397.
107. Marciniak GP, Kuyla MR. *Starch/Stärke.* 1982;34:422.
108. Danilenko AN, Bogomolov AA, Yuryev VP, Dianova VT, Bogatyrev AN. *Starch/Stärke.* 1993;45:63.
109. Gorinstein S. *Starch/Stärke.* 1993;45:91.
110. Cave LE, Adam FR. *TAPPI J.* 1968;51(11):109A.
111. Ewing FG, Harrington HR. *Chem. Eng. Progr.* 1957;63(3):65.
112. Goos H, Maurer HW. US Patent 3 308 037. 1967.
113. Schink NF, McConiga RE. Canadian Patent 723 142. 1965.
114. Goos H, Maurer HW. US Patent 3 404 071. 1968.
115. Gallaher TL, Small TL. US Patent 4 957 563. 1990.
116. Hughes J. *Paper Technol.* 1992;33(5):26.
117. Dauplaise DL. *PIMA J.* 1985(10):28.
118. Roberts JC. *Paper Technol.* 1993;34(4):27.

119. Nazir BA, Carnegie-Jones J. *Paper Technol.* 1991;32(12):37.
120. Kajuanto IM, Komppa A, Ritala RK. *Nordic Pulp Pap. Res. J.* 1989;4(3):219.
121. Laleg M, Nguyen N. *J. Pulp Pap. Sci.* 1995;21:J356.
122. Wågberg L, Lindström T. *Colloids Surf.* 1987;27:29.
123. Lindström T. *Nordic Pulp Pap. Res. J.* 1992;7(4):181.
124. Björklund M, Wågberg L. *Colloids Surf. A.* 1995;105:199.
125. Koethe JL, Scott WE. *TAPPI J.* 1993;76(12):123.
126. Wågberg L. *Svensk Papperstidn.* 1985(6):R48.
127. Järnström L, Lason L, Rigdahl M, Eriksson U. *Colloids Surf. A: Physicochem. Eng. Aspects.* 1995;104:207.
128. Wågberg L, Björklund M. *Nordic Pulp Pap. Res. J.* 1993;8(4):399.
129. Andersson K, Lindgren E. *Nordic Pulp Pap. Res. J.* 1996;11(1):15 [57].
130. Carre B. *Nordic Pulp Pap. Res. J.* 1993;8(1):21.
131. Swerin A, Sjödin U, Ödberg L. *Nordic Pulp Pap. Res. J.* 1993;8(4):389.
132. Hedborg F. *Nordic Pulp Pap. Res. J.* 1993;8(2):258.
133. Bobu E, Parpalea RR. *Professional Papermaking.* 2005(1):30.
134. van der Steeg HGM, de Ketzer A, Bjisterbosch BH. *Nordic Pulp Pap. Res. J.* 1989;4(2):173.
135. Nedelcheva MP, Sokolova JS. *Starch/Stärke.* 1988;40:151.
136. Tanaka H, Swerin A, Ödberg L. *Nordic Pulp Pap. Res. J.* 1995;10(4):261.
137. Tanaka H, Swerin A, Ödberg L. *TAPPI J.* 1993;76(5):157.
138. Hedborg F. *Nordic Pulp Pap. Res. J.* 1993;8(3):319.
139. Jaycock MJ, Swales DK. *Paper Technol.* 1994;35(10):26.
140. Gill RIS. *Paper Technol.* 1991;32(8):34.
141. Laleg ML, et al. *Paper Technol.* 1991;32(5):24.
142. Howard RC, Jowsey CJ. *J. Pulp Pap. Sci.* 1989;15:J225.
143. Swerin A, Glad-Norsmark G, Sjödin U. *Pap. Puu.* 1995;77(4):215.
144. Wågberg L, Lindström T. *Nordic Pulp Pap. Res. J.* 1987;2(2):49.
145. Gaiolas C, Silva MS, Costa AP, Belgacem MN. *TAPPI J.* 2006;5(6):3.
146. Gerischer G, Murray LJ, Van Wyk WJ. *Pap. Puu.* 1996;78:51.
147. Helle TM, Hintermaier J. *Professional papermaking.* 2004;1:6.
148. Bobacka V, Eklund D. *Wochenbl. Papierfabr.* 1996;124;106.
149. Formento JC, Maximino MG, Mina LR, Srayh MI, Martinez MJ. *Appita J.* 1994;47:305.
150. Glittenberg D. *TAPPI J.* 1993;76(11):215.
151. Wågberg L, Ödberg L. *Nordic Pulp Pap. Res. J.* 1991;6(3):127.
152. Okomori K, Isogai A, Yoshizawa J, Onabe F. *Nordic Pulp Pap. Res. J.* 1994;9(4):237.
153. Hemmes JL, Glittenberg D, Bergh NO. *Wochenbl. Papierfabr.* 1993;121:162.
154. Webb LJ. *Paper Technol.* 1991;32(6).
155. Glittenberg D, Hemmes JL, Bergh NO. *Paper Technol.* 1994;35(9):18.
156. Huang Y, Gardner DJ, Chen M, Biermann CJ. *J. Adhesive Sci. Technol.* 1995;9:1403.
157. Fishman D. *Amer. Ink Maker.* 1993;(5):32.
158. Rogers C. *Paper Technol.* 1990;31(3):19.
159. Marton J. *TAPPI J.* 1990;73(11):139.

160. Taniguchi R, Isogai A, Onabe F, Usuda M. *Nordic Pulp Pap. Res. J.* 1993;8(4):352.
161. Hedborg F, Lindström T. *Nordic Pulp Pap. Res. J.* 1993;8(3):331.
162. Nachtergaele W. *Starch/Stärke.* 1989;41:27.
163. Beaudoin R, Gratton R, Turcotte R. *J. Pulp Pap. Sci.* 1991;21:J238.
164. Isogai A, Onabe F, Usuda M. *Nordic Pulp Pap. Res. J.* 1992;7(4):193.
165. Lindström MJ, Savolainen RM. *J. Dispersion Sci. Technol.* 1996;17:281.
166. Swerin A. *Pap. Puu.* 1995;77:215.
167. Kurrle FL. US Patent 5 514 212. 1996.
168. Yoon SY, Deng Y. *TAPPI J.* 2006;5(9):3.
169. Hamunen A. *Starch/Stärke.* 1995;47:215.
170. Kilinger WE, Murray D, Hatfield GR, Hassler T. *Starch/Stärke.* 1995;47:311.
171. Khalil MI, Farag S, Hashem A. *Starch/Stärke.* 1993;45:226.
172. Backman R, Bruun HH. *Starch/Stärke.* 1993;45:396.
173. Alince B, Lebreton R, St-Amour S. *TAPPI J.* 1990;73(3):191.
174. Willett JL. *Starch/Stärke.* 1995;47:29.
175. Heath HD, Ernst AJ, Hofreiter BT, Phillips BS, Russell CR. *TAPPI J.* 1974;57(11):109.
176. Mehltretter CL, Yeates TE, Hamerstrand GE, Hofreiter BT, Rist CE. *TAPPI J.* 1962;45(9):750.
177. Kaser WL, Mizra J, Curtis JH, Borchet PJ. *TAPPI J.* 1965;48(10):583.
178. Hofreiter BT, Hamerstrand GE, Kay DJ, Rist CE. *TAPPI J.* 1962;45(3):177.
179. Peek LR. *Paper Technol.* 1989;30(12):20.
180. Laleg M, Pikulik II. *Nordic Pulp Pap. Res. J.* 1993;8(1):41.
181. Laleg M, Pikulik II. *J. Pulp Pap. Sci.* 1993;19:J248.
182. Klass CP. *TAPPI J.* 1990;73(12):69.
183. Dobbe D. *Paper Technol.* 1993;34(3):16.
184. Danino D, Marmur A. *J. Coll. Interf. Sci.* 1994;166:245.
185. Osterovo MA, Ostorova IY. *Bum. Promst.* 1990;(8):6.
186. TAPPI. *Proceedings of the Metered Size Press Forum.* Atlanta, GA: TAPPI PRESS; I (1996), II (1998), III (2000).
187. Frei HP. *Wochenbl. Papierfabr.* 1993;121:390.
188. Hirakawa M, Twase H. *TAPPI J.* 1988;71(5):53.
189. Trefz M. *TAPPI J.* 1996;79(1):223.
190. Rantanen R. In: Brander J, Thorn I, eds. *Surf. Appl. Pap. Chem.* 21. London, UK: Blackie; 1997.
191. Knop R, Fathke V. US Patent 5 296 257. 1994.
192. Knop R, Lang F. *Wochenbl. Papierfabr.* 1994;122:671.
193. Rantanen R. *Wochenbl. Papierfabr.* 1992;120:193.
194. Weigl J, Laber A, Bergh NO, Ruf F. *Wochenbl. Papierfabr.* 1995;123:634.
195. TAPPI. *Proceedings of the Coating Conference.* Atlanta, GA.: TAPPI PRESS; 1996–2000 [CD-ROM].
196. Baily DF, Brown R. *TAPPI J.* 1990;73(9):131.
197. Kustermann M. *Das Papier.* 1989;43(10A):V139.
198. Knop R, Sommer H. *Wochenbl. Papierfabr.* 1995;123:436.
199. Bergh NO, Glittenberg D, Weinbach H. *Das Papier.* 1988;42(10A):V40.

XIV. References

200. Balzereit B, Drechsel J, Burri P, Naydowski C. *TAPPI J.* 1995;78(5):182.
201. Glittenberg D, Bergh NO. *Wochenbl. Papierfabr.* 1992;120:507.
202. Oja ME, Frederick CH, van Nuffel J, Lambrechts P. *TAPPI J.* 1991;74(8):115.
203. Dunlop J. *Paper Technol.* 1994;35(12):26.
204. Glittenberg D, Bergh NO. *Paper Technol.* 1993;34(5):30.
205. Turunen R. *Paper Technol.* 1994;35(3):33.
206. Sieniawski R. *Pulp & Paper.* 2004; 78(2):38.
207. Sirois RF. *Wochenbl. Papierfabr.* 1993;121:402.
208. Brogly DA. *TAPPI J.* 1978;61(4):43.
209. Brogly DA, Harvey RD. *Proc. TAPPI Coating Conf.* 1993:145.
210. Bristol K. *Pulp Pap.* 1974;48(10).
211. Sirois RF. *Paper Technol.* 1992;33(11):31.
212. Abell S. *Pulp Pap.* 1995(5):99.
213. Abell S. *Proc. TAPPI Coating Conf.* 1992:95.
214. Trouve C, Takala K. *Pap. Puu.* 1993;75:586.
215. Pandian VE. US Patent 5 362 573. 1994.
216. Maxwell CS. *TAPPI J.* 1970;53:1464.
217. Lawton JW. *Starch/Stärke.* 1995;47:62.
218. Wiseman N, Ogden G. *Paper Technol.* 1996;37(1/2):31.
219. Booth GL. *Coating Equipment and Processes.* New York, NY: Lockwood Publishing; 1970.
220. Teirfolk JE, Laaja V. *TAPPI J.* 1996;79(2):206.
221. Carvalho MC, Scriven LE. *TAPPI J.* 1994;77(5):201.
222. Sommer H, Rückert H, Knop R. *Wochenbl. Papierfabr.* 1987;115:1056.
223. Thomin WH, Bergh NO, Kogler WL. *TAPPI J.* 1973;56(1):66.
224. Funderburk K, Keirstaad JM, McGuire JH. *TAPPI J.* 1980;63(2):167.
225. Prankh R, Scriven LE. *AIChE J.* 1990;36:587.
226. Schachtl M, Weigl J, Baumgarten HL. *Wochenbl. Papierfabr.* 1993;121:661.
227. Engström G, Morin V. *Nordic Pulp Pap. Res. J.* 1994;9(2):106.
228. Letzelter P, Eklund D. *TAPPI J.* 1993;76(5):63 [No. 6, 93].
229. Chen KSA, Sciven LE. *TAPPI J.* 1990;73:151.
230. Skowronski J, Lepoutre P. *TAPPI J.* 1985;68(11):98.
231. Eriksson U, Rigdahl M. *J. Pulp Pap. Sci.* 1994;20:J333.
232. Ström G, Märdin AM, Salminen P. *Nordic Pulp Pap. J.* 1995;10(4):227.
233. Enomae T, Onabe F, Usuda M. *Japan Tappi J.* 1993;47:1002.
234. Herbet AJ, Gautam N, Whalen-Shaw MJ. *TAPPI J.* 1990;73(11):171.
235. Mark WR. *TAPPI J.* 1969;52(1):70.
236. Beazley KM, Climpson M. *TAPPI J.* 1970;53:2227.
237. Taylor DL, Dill DR. *TAPPI J.* 1967;50:536.
238. Lee YK, Kuga S, Onabe F, Usuda M. *Japan Tappi J.* 1992;46:72.
239. Sandas SE, Salminen PJ, Eklund DE. *TAPPI J.* 1989;72(12):207.
240. Kobayashi T, Okuyama T, Koike T. *Japan Tappi J.* 1995;49:324.
241. Windle W, Beazley KM, Climpson M. *TAPPI J.* 1970;53:2232.
242. Engström G, Rigdahl M, Kline J, Ahlroos J. *TAPPI J.* 1991;74(5):171.
243. Ranger AE. *Paper Technol.* 1994;35(12):40.

244. Dappen JW. *TAPPI J.* 1951;34:321.
245. Baumeister M. *Coating.* 1980;13(2):32.
246. Stanislawska A, Lepoutre P. *Polym. Mater. Sci. Eng.* 1993;70:297.
247. Eklund D, Norrdahl P, Heikkinen ML. *Drying Technol.* 1995;13:919.
248. Al-Turaif H, Unertl WN, Lepoutre P. *J. Adhes. Sci. Technol.* 1995;9:801.
249. Lepoutre P. *Trends Polym. Sci.* 1995;3(4):112.
250. Jones A. *Paper Technol.* 1992;33(12):24.
251. Hemmes JL, Bergh NO, Glittenberg D. *Wochenbl. Papierfabr.* 1994;122:259.
252. Bergmann W. *Starch/Stärke.* 1986;38:73.
253. Autio K, Suortti T, Hamunen A, Poutanen K. *Starch/Stärke.* 1992;44:393.
254. Glittenberg D, Hemmes JL, Bergh NO. *Paper Technol.* 1995;36(9):18.
255. Nachtergaele W, van der Meeren J. *Starch/Stärke.* 1987;39:135.
256. Carr ME. *Starch/Stärke.* 1992;44:219.
257. Möller K, Glittenberg D. *Proc. TAPPI Coating Conf.* 1990:85.
258. Croce A, Brizzi F, Caobianco J, Busetti M, Portico M. *Ind Carta.* 1993;31:289.
259. Wilhelm EC. *Spec. Publ., Roy. Soc. Chem.* 1993;134:180.
260. Anon. *Pulp Pap.* 1987;5:49.
261. Ramp JW, ed. *Binder Insolubilization: How to Achieve It.* Atlanta, GA: TAPPI PRESS; 1988.
262. Eldred NR, Spicer JC. *TAPPI J.* 1968;46:608.
263. McAlpine I. *Proc. TAPPI Coating Conf.* 1982:165.
264. Husband JC, Brown R, Drage PG. US Patent 5 384 013. 1995.
265. Kogler W, Spielmann D, Huggenberger L. *Wochenbl. Papierfabr.* 1993;121:549.
266. Fernandez J, Solarek D, Koval J. US Patent 5 093 159. 1992.
267. Bristow JA. *Svensk Papperstidn.* 1964;64:775.
268. Chung FH. *J. Appl. Polym. Sci.* 1991;42:1319.
269. Lepoutre P. *Svensk Papperstidn.* 1986;89:20.
270. Lepoutre P. *Wochenbl. Papierfabr.* 1993;121:752.
271. Yamauchi T, Cho T, Imamura R, Murakami K. *Nordic Pulp Pap. Res. J.* 1988;3(3):128.
272. Daub E, Göttschink L. *Das Papier.* 1990;44(5):45.
273. Wolf M. *Wochenbl. Papierfabr.* 1993;121:213.
274. Kroeschell WO, ed. *Preparation of Corrugating Adhesive.* Atlanta, GA: TAPPI PRESS; 1977.
275. Williams RH, Leake CH, Silano MA. *TAPPI J.* 1977;60(4):86.
276. Glittenberg D. *Paper Technol.* 1992;33(1):34.
277. Sprague CH. *TAPPI J.* 1979;62(6):45.
278. Chung KM, Seib PA. *Starch/Stärke.* 1991;43:441.
279. Inoue M, Lepoutre P. *Nordic Pulp Pap. Res. J.* 1989;4(3):206 [213].
280. Inoue M, Lepoutre P. *Starch/Stärke.* 1989;41:287.
281. Blumenthal M, Paul CW. *TAPPI J.* 1994;77(9):193.
282. Glittenberg D, Bergh NO. *Wochenbl. Papierfabr.* 1995;123:182.
283. Watson M. In: Brander J, Thorn I, eds. *Surf. Appl. Pap. Chem.* 192. London, UK: Blackie; 1997.

284. J. Borch. ICPS' 94: Phys. Chem. Imaging Syst., IS&T's 47th Annual Conference, No. 2. 1994;820; *Chem. Abstr.* 1995;123:317303n.
285. Nachtergaele W, van Nuffel J. *Starch/Stärke*. 1989;41:386.
286. Herrick R. *TAPPI J.* 1966;49(1):79A.
287. Gillespie WJ, Mazzola CO, Marshall DW. *Paper Trade J.* 1970;March 2.
288. Davies Jr WS. *TAPPI J.* 1964;47(8):129A.
289. Glittenberg D, Hemmes JL, Bergh NO. *Paper Technol.* 1994;35(7):18 [22, 25].
290. Friend W. *TAPPI J.* 1957;40(1):79A.
291. Rebhuhn M, Sperber H, Saliternik Ch. *TAPPI J.* 1967;50(12):62A.
292. Bayazeed A, Trauter J. *Starch/Stärke*. 1991;43:18.
293. Karneth AM, et al. *Wochenbl. Papierfabr.* 1995;123:8.
294. Heimonen J, Stenius P. *Wochenbl. Papierfabr.* 1996;124:181.
295. Röper H, Koch H. *Starch/Stärke*. 1990;42:123.
296. Fritz HG, Aicholzer W, Seidenstücker T, Widmann B. *Starch/Stärke*. 1995;47:475.
297. Endres HJ, Pries A. *Starch/Stärke*. 1995;47:384.
298. Evangelista RL, Sung W, Jane JL, Gelina RJ, Nikolov ZL. *Ind Eng. Chem. Res.* 1991;30:1841.
299. Otey FH, Westhoff RP, Doane WM. *Ind Eng. Chem. Res.* 1987;26:1659.
300. Demel I, Möbius CH, Cordes-Tolle M. *Paper Technol.* 1990;31(10):27.
301. Starch in Paper, TAPPI Test Method T 419 om-91; 1991.
302. Boast WH, Trosset SW. *TAPPI J.* 1962;45(11):873.
303. Reif L, Seelemann R, Wallpott G. *Wochenbl. Papierfabr.* 1991;119:666.
304. Mitchell G, Wijnberg AC. *Starch/Stärke*. 1995;47:46.
305. Mitchell GA. *Starch/Stärke*. 1990;42:131.
306. Birosel-Boettcher N. *TAPPI J.* 1993;76(3):207.
307. Holm J, Bjoerck I, Drews A, Asp NG. *Starch/Stärke*. 1986;38:224.
308. Kennedy JF, Cabalda VM. *Starch/Stärke*. 1993;45:44.
309. Richter M. *Starch/Stärke*. 1995;46:350.
310. Papantonakis M. *TAPPI J.* 1980;63(5):65.
311. Stone RG, Krasowski JA. *Anal. Chem.* 1981;53:736.
312. Arai T, Yamasaki T, Suzuki K, Ogura T, Sakai Y. *TAPPI J.* 1988;71(5):47.
313. Osada T, Katsu Y, Miyagawa T. *Japan Tappi J.* 1995;49(11):73.
314. Whalen-Shaw JM, ed. *Binder Migration in Paper and Paperboard Coating.* Atlanta, GA: TAPPI PRESS; 1993.
315. Hirabayashi T, Suzuki H, Fukui T, Osada T. *Proc. TAPPI Coating Conf.* 1995:247.
316. Dobson RL. *Proc. TAPPI Coating Conf.* 1975:123.
317. Engström G, Righahl M, Kline J, Ahlroos J. *TAPPI J.* 1991;74(5):171.
318. Le PC, Zeilinger H, Weigl J. *Wochenbl. Papierfabr.* 1993;121:284.
319. Lipponen J, Lappalainen T, Astola J, Grön J. *Nordic Pulp Pap. Res. J.* 2004;19(3):300.
320. Conners TE, Banerjee S, eds. *Surface Analysis of Paper.* Boca Raton, FL: CRC Press; 1995.

Starch in Polymer Compositions[1]

J.L. Willett

Plant Polymer Research, National Center for Agricultural Utilization Research, Agricultural Research Service, US Department of Agriculture, Peoria, Illinois, USA

I.	Introduction	715
II.	Starch Esters	717
III.	Granular Starch Composites	719
IV.	Starch in Rubber	724
V.	Starch Graft Copolymers	726
VI.	Thermoplastic Starch Blends	731
VII.	Starch Foams	735
VIII.	References	737

I. Introduction

Starch has been considered an attractive raw material for polymer applications for almost 200 years. Kirchoff's discovery in 1811 that treatment of starch with an acid yields a sweet substance was an unexpected result of the search for a low-cost substitute for natural rubber.[1] Considerable research in the development of starch-based polymer materials has been stimulated by the facts that starch is produced from wide variety of sources, is an annually renewable resource and is inherently biodegradable.

Motivation for interest in starch as a polymeric material is the result, in part, of the perceived negative impact of plastics in solid waste disposal. This perception is primarily due to the extent of use of plastics in packaging of consumer products, where they have displaced traditional materials such as metals, glass and paper. The estimated annual growth rate of plastics such as low- and high-density polyethylene (LDPE and HDPE), polypropylene (PP) and polystyrene (PS) averaged 8.9% in the United States during the period 1990–1995; current consumption of plastics in packaging and disposable applications is approximately 18×10^9 lb/yr (8×10^9 kg/yr)[2] Figure 19.1 shows the relative amounts of various commercial plastics used in packaging or other disposable applications in the United States.

---

[1]Names are necessary to report factually on available data; however, the USDA neither guarantees nor warrants the standard of the product, and the use of the name USDA implies no approval of the product to the exclusion of others that may also be available.

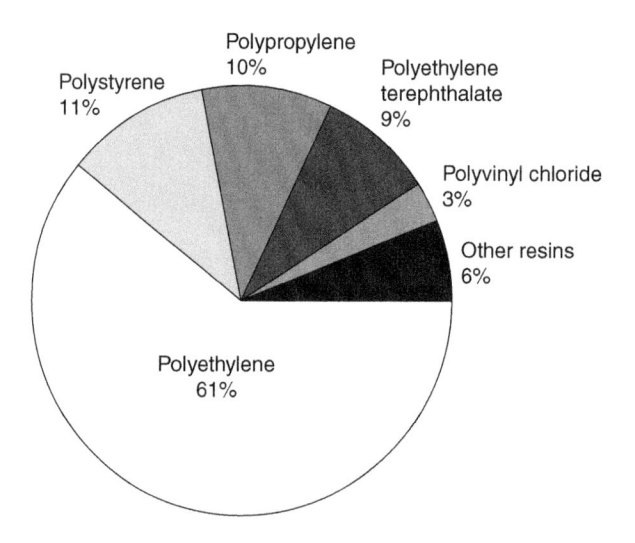

Figure 19.1 Plastics use in packaging and disposable applications. (Data adapted from reference 1)

Because of their low density, plastics comprise approximately 20% by volume of landfilled solid waste, although they contribute only about 7% of the total weight. Public opposition limits the disposal of plastics via incineration, despite their comparatively high energy content. These factors have stimulated considerable interest in the development of biodegradable plastics as replacements for conventional plastics, particularly in packaging and disposable applications.

Starch offers several potential advantages as a raw material for plastics applications. It is annually renewable, obtained from a variety of plant sources and is a low-cost material. Interest in its use in biodegradable plastics is also driven by the inherent biodegradability of starch and the ubiquity of microorganisms capable of utilizing starch as a carbon source.

These advantages are tempered by challenges presented by the structure of starch. The three hydroxyl groups per D-glucosyl unit impart a high degree of hydrophilicity to starch. (The equilibrium moisture content of starch under ambient conditions is 10–12%.) Since water is an effective plasticizer for starch, the gain or loss of moisture to maintain equilibrium can result in significant changes in physical and mechanical properties of starch-based materials, unlike the hydrophobic synthetic polymers currently in use. Its water sorption properties also impact its use in plastics for packaging, since the sorption and permeation of significant quantities of water can compromise package utility. In addition, the branched nature of amylopectin yields a polymer with poor properties relative to linear polymers. The extensive branching reduces the ability of amylopectin to form entanglements in the melt and solid states that are necessary to achieve useful properties in high polymers.[3] This low entanglement density is reflected in the brittle behavior of amylopectin films or extrudates. Much of the research in the area of starch-based plastics has focused on mitigating either or both of these properties.

While starch has been utilized for hundreds of years as a paper additive (see Chapter 18), as an adhesive and as a sizing for textiles, its use as a structural polymer is a relatively recent development. Considerable effort has been given to the use of starch in water- or solvent-based coatings and lacquers. These materials generally form crosslinked thermosets on curing. This chapter presents the use of starch in polymer composites and blends prepared using conventional plastics processing techniques, such as extrusion and injection molding. These processes generally utilize thermoplastic formulations, which can be reprocessed multiple times.

Several approaches have been utilized for incorporating starch into plastics. They can be broadly categorized in the following manner: chemical modification via reaction of the hydroxyl groups; granular starch composites; starch graft copolymers; thermoplastic starch; and expanded starch foams. It is clear that some overlap may exist in these distinctions; for example, a graft material may be extruded into foam. However, these categories provide a useful basis for distinguishing between general types of materials developed using starch.

II. Starch Esters

Historically, chemical modification of starch was the first widely-investigated method for producing starch-based plastics. Attempts at forming cast films from starch or amylose showed that the properties were highly dependent on relative humidity; the films became brittle at low humidities.[4] Figure 19.2 shows data for films of various amylose/amylopectin ratios conditioned at 50% relative humidity and 23°C. Efforts to overcome

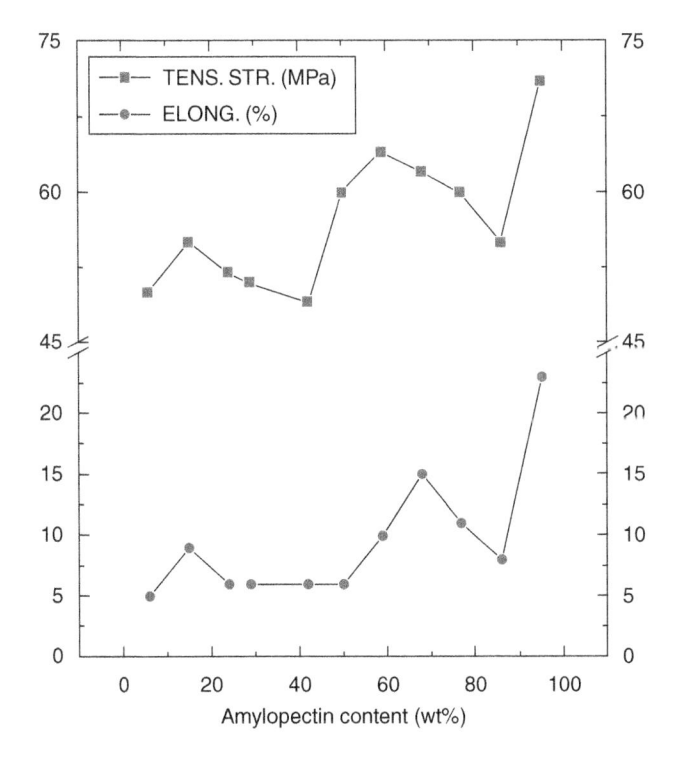

Figure 19.2 Tensile strength and elongation of films with various amylose/amylopectin ratios. (Data adapted from reference 4)

Table 19.1 Properties of plasticized amylose triesters[a]

Ester type	Dibutyl phthalate, %	Elongation, %	Tensile strength, MPa
Acetate	0	22	62
Acetate	10	60	50
Acetate	25	93	35
Acetate	40	124	23
Propionate	0	19	42
Propionate	10	76	28
Propionate	25	104	18
Propionate	40	140	11
Butyrate	0	17	26
Butyrate	10	114	15
Butyrate	25	146	8
Butyrate	40	170	4
Caproate	0	122	12
Caproate	10	228	7
Palmitate	0	24	8
Palmitate	10	50	7

[a]Data adapted from reference 10

this problem focused on the preparation of esters of starch or isolated amylose.[5–10] Plasticized starch triacetate films possessed tensile properties comparable to cellulose triacetate.[9,10] Tensile data for amylose triesters is shown in Table 19.1. Amylopectin esters or esters of whole starch, on the other hand, were weak and brittle. This decline in properties was attributed to the highly-branched nature of amylopectin. It was concluded that starch esters (and other derivatives) did not achieve significant market penetration because 'as a class, their cost of production is not significantly lower than competitive products to be attractive, particularly in view of the fact that the products appear to possess no outstanding merits to warrant a preference for their use.'[11]

Mixed esters of amylose were prepared with formate, acetate, propionate, butyrate or benzoate radicals.[12] Clear disks with tensile strengths of the order of 7000 psi (4.9×10^6 kg/m^2) and elongations of 12% to 20% were molded. Formate esters were brittle, aged rapidly and had smaller thermal processing windows than esters with longer substituents. Amylose triacetate fibers had properties similar to, but generally lower than, cellulose triacetate.[13] It was stated that any advantages of using amylose acetates would rely on the relative economics of separating and processing amylose.

Sagar and Merrill[14] prepared butyrate, valerate and hexanoate esters of high-amylose starch with DS values of approximately 2.8.[14] As the ester chain length increased, the melting points of neat and plasticized esters decreased smoothly, as did the glass transition temperature as measured by dynamic mechanical analysis. At low strain rates butyrate and hexanoate esters formed exhibited stable necking during tensile testing; the valerate ester formed a neck which did not propagate. At higher strain rates, the butyrate ester displayed almost brittle behavior, whereas the hexanoate ester displayed a stable neck during extension. Blends of these hydrophobically modified starches with unmodified starch are claimed.[15]

Rivard et al.[16] prepared starch esters of various DS values and ester chain lengths of C-2 through C-6. Anaerobic biodegradation tests indicated that increasing ester chain length or DS reduced the time required to achieve substantial degradation. Bioconversion curves were sigmoidal when plotted against DS; the DS at which a substantial decrease in bioconversion occurred shifted to lower DS as chain length increased.

Blends of hydrophobic, water-repellant starch esters and biodegradable polyesters have been claimed.[17-21] Starch esters prepared using high-amylose starches with DS values of at least 1.5[17] and blends of these with various plasticizers[18,19] can be thermoplastically processed. Starch propionate (DS 2.4) prepared from high-amylose starch was blended with either polycaprolactone or poly(hydroxy butrate valerate) (PHBV) to give materials with good stability at 90% humidity.[20] Starch propionate (DS 1.7) was compounded with PHBV and extruded into translucent, flexible films.[21] Similar materials were obtained when polycaprolactone was substituted for the PHBV. Appropriate selection of polyester, plasticizer and talc yielded injection molded products with properties comparable to general purpose polystyrene.[21]

III. Granular Starch Composites

Fillers are widely used in the plastics industry to modify the properties of polymer resins.[22] Properties which can be modified by fillers include stiffness, strength, toughness, heat distortion, damping, permeability, electrical characteristics, density and cost. In general, the positive attributes of filler are offset somewhat by a decline in other useful properties, such as tensile strength, impact strength or elongation. The properties of filled polymer materials are highly dependent on the size and shape of the filler, and the degree of adhesion between the matrix polymer and the filler surface.

The decline in tensile and impact properties often observed when starch is incorporated into a polymer follows the general trend observed in particulate-filled composites in the absence of specific interactions, such as hydrogen or covalent bonding between the matrix and the particle surface.[22] A high interface energy exists between the surface of the hydrophilic starch granule and hydrophobic polymer matrix, resulting in poor adhesion between the two. This poor adhesion essentially reduces the effective load-bearing cross-section of the material, leading to reduced tensile strength. A model developed using this principle[22] predicts the reduction in tensile strength:

$$\sigma = \sigma_0 \left(1 - 1.21\phi^{2/3}\right) \quad (19.1)$$

where σ and σ_0 are the tensile strengths of the filled and unfilled composites, respectively, and ϕ is the filler volume fraction. Equation 19.1 assumes no bonding between the filler and matrix, and a uniform distribution of particles of constant size. Equation 19.1 can be generalized to the form:

$$\sigma = \sigma_0 \left(1 - (\phi/\phi_m)^{2/3}\right) \quad (19.2)$$

where the symbols are the same as Equation 19.1, and ϕ_m is the maximum filler volume fraction.[22] Equation 19.2 recognizes that there is a distribution of particle sizes and that the volume fraction cannot exceed the bulk density of the filler. It has been shown that Equations 19.1 and 19.2 can be used to describe the tensile strength of starch-PE composites.[23]

Particulate-filled composites generally have greater elastic moduli compared to the unfilled resin, and starch-filled composites are no exception. Various models have been developed to describe this effect, and are generally of the form:

$$\frac{E}{E_0} = \frac{1 + AB\phi}{1 - B\phi} \qquad (19.3)$$

where E and E_0 are the unfilled resin and filled composite moduli, respectively, A is a geometric factor, and B is a function of A and the ratio of the filler and matrix moduli.[22] Several studies of the effect of starch content on modulus have been reported.[23-25] A simple additive model yielded starch granule modulus values of 2.7 GPa[24] to 4.2 Gpa,[25] while the use of Equation 19.3 yielded a value of 15 Gpa.[23] The additive model strictly holds only for long fibers and is obtained from Equation 19.3 by taking the limit $A = \infty$.[22] It has been shown[23] that Equation 19.3 describes the modulus data of Schroeter and Hobelsberger[24] using the 15 GPa value for the granule modulus.

Equations 19.1–19.3 do not explicitly include terms to account for particle size. It is well-known, however, that average particle size may significantly affect mechanical properties.[22,26] Typically, the modulus and tensile strength increase as the particle size decreases. Lim et al.[27] have shown that tensile strength, yield strength and elongation of blown films of starch and polyethylene decrease as granule size increases. Starch contents were 20% by weight or less. Fritz and Widmann[28] observed no particle size effect in polyethylene filled with 40% starch. In both studies, tensile strength and elongation decreased with increasing starch content. Other properties of starch-filled polymers that have been characterized include permeability of various gases,[29] water vapor transmission[30] and the kinetics of water sorption.[31]

Interest in the use of starch as a filler in plastics and rubber increased significantly in the late 1960s and early 1970s. Independent efforts by Griffin in England and the US Department of Agriculture's Northern Regional Research Center investigated the use of granular starch in various plastics. Griffin's work[32] was stimulated by the desire to find a particulate filler for low-density polyethylene (LDPE) which would impart paper-like characteristics to films. As part of a general research program in agricultural product utilization, USDA researchers investigated the use starch as a filler in polyurethanes[33-35] and poly(vinyl chloride) (PVC).[36]

Bennett et al.[34] found that the incorporation of granular starch or dextrin into rigid urethane foams reduced the compressive strength, but gave self-extinguishing foams on ignition. Granular starch could be incorporated into rigid urethane plastics at levels up to 60% by weight with good strength and hardness, but reduced elongation and impact strength.[33,35]

In the early 1970s, environmental concerns about the disposal of plastics generated considerable interest in the development of degradable materials.[37] This concern led to consideration of the use of starch in biodegradable replacements for non-biodegradable resins. Initial attempts to incorporate starch were based on the concept of simply blending in granular starch with synthetic resin. Westhoff et al.[36] blended granular starch with PVC at levels up to 40% by weight to produce materials susceptible to microbial attack. The biodegradability was offset somewhat by the reduction in tensile strength and elongation as the starch content increased. While it was recognized that the PVC matrix was resistant to microbial attack, it was conjectured that 'the extensive degradation of starch filler would lead to erosion of the plastics through natural forces.'

At the same time, Griffin[38–41] independently explored incorporating granular starch into a variety of polymers. It was recognized that while the starch was readily assimilated by microorganisms in soil, the synthetic polymer was not. The incorporation of autoxidizable substances, typically unsaturated fatty acid salts or esters, was used to produce peroxides, which led to chain scission of the polyolefin.[38,39] The resulting degraded plastic might then be more easily assimilated. Examples containing up to 50% by weight of starch were demonstrated using polyethylene, polystyrene and PVC. In general, the mechanical properties relative to the base polymer decreased as starch content increased.

Griffin[32] demonstrated that the decline in properties of starch-filled composites could be mitigated somewhat by treating the surface of the starch granules to make them more hydrophobic. This treatment improved the adhesion and stress transfer across the particle/matrix interface and resulted in improved properties relative to no treatment, although properties were still generally reduced compared to the unfilled polymer.

Resurgence in interest in starch-based biodegradable plastics in the mid-1980s stimulated further development of granular starch composites. These efforts reflected the desire to enhance biodegradability by incorporating starch, while improving the adhesion between the granule and the matrix to minimize property loss.

Jane et al.[42,43] showed that modifying starch with *n*-octenyl succinate (NOS) improved properties relative to unmodified starch at constant starch level in linear low density PE films. The NOS starch-containing film properties were typically 15% greater than films with untreated starch. With both starches, there was a steady decline in properties as starch content increased. In enzyme digestion tests, the NOS-treated starch was hydrolyzed at a slower rate relative to the untreated starch. It was also reported that the use of high molecular weight oxidized PE (OPE) improved the tensile strength of starch–PE films when the OPE was used at a ratio of 0.25 to 0.50 based on the starch, for starch contents up to 50%.[44] This improvement in properties was attributed to interactions between the carboxyl groups of the OPE and the hydroxyl groups of the starch granules.

Another approach to improving the properties of starch-filled polyolefin materials involves the use of ethylene–acrylate copolymers in blends with PE.[45] Addition of copolymers of ethylene with methyl acrylate, ethyl acrylate or butyl acrylate were shown to improve the properties of PE films, allowing for higher starch contents. Coextrusion of starch-containing films with outer layers incorporating oxidative pro-degradants has also been utilized.[46] The inner layer can contain up to 40% starch; the outer layers without starch provide the desired mechanical properties. On disposal, the outer layers degrade oxidatively, allowing access to the inner starch-containing layer.

Weil[47] disclosed the use of maleated polypropylene (MPP) to produce 'biodeteriable' starch–PP composites. Tensile strengths of composites made with copolymers of 0.2% to 0.4% maleic anhydride were significantly improved at any starch content, compared to unmodified PP. For instance, unmodified PP lost approximately 40% of its original tensile strength with the addition of 40% starch, while a PP–MA copolymer (0.2% MA) lost only 22% of its original tensile strength at the same starch loading. The improvement was attributed to the reaction between the anhydride functionality of the MPP and hydroxyl groups on the starch granules. The covalent bonding between the matrix and filler improved the stress transfer across the interface. Representative data showing the effect of maleation on properties is shown in Figure 19.3.

Similar results were obtained when corn starch was melt blended with either maleated ethylene–propylene copolymer (EPMA) or styrene–maleic anhydride copolymer (SMA).[48] Starch contents were in the range 50% to 80% by weight. The torque generated during melt blending was higher for the functionalized copolymers than for the corresponding unmodified polymers. Tensile strengths were also greater with the maleated polymers, being approximately double the values for unmodified polymers at 30% to 40% starch. Dynamic mechanical analysis indicated two glass transitions

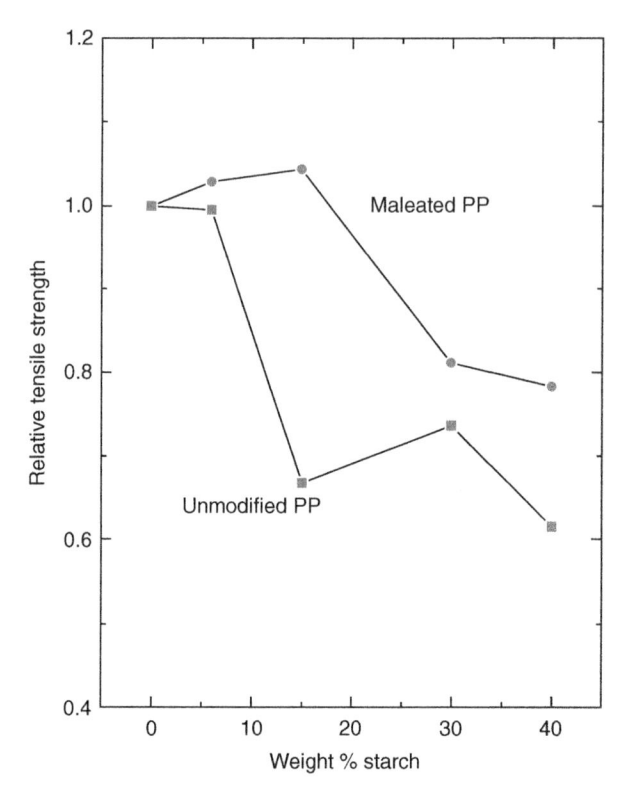

Figure 19.3 Relative tensile strength of starch-filled polypropylenes showing the effect of maleation. (Data adapted from reference 47)

for the starch–EPMA blends and one broad transition for the starch–SMA blends.[49] Increasing the starch content, mixing time or mixing speed generally increased both the storage and loss moduli for both blends. Blends with 60% to 70% starch exhibited shear-thinning behavior and considerable reduction in melt viscosity upon re-extrusion due to degradation.[50] Interpolymer compositions produced using this process with starch,[51] protein[52] and/or flours[52] have been patented.

Reactive extrusion of low molecular weight coupling agents in starch–polyethylene blends has been reported.[53] Coupling agents included maleic anhydride and methacrylic anhydride. Polyethylene pellets were coated with a solution of coupling agent, a free radical initiator and other additives, and then extruded with up to 60% starch. It is claimed that the coupling between the starch and the matrix minimized the loss in properties generally encountered when adding starch to thermoplastics.

Pretreatment of starch granules by esterification with ethylene–acrylic acid ionomers has been reported to improve the properties of compression molded PE–starch materials compared to simple melt mixing of the three components.[54] Improvements over no treatment with ionomer were limited to starch contents less than about 20%.

While the starch in hydrophobic polyolefins is accessible for biodegradation, the resin matrix comprising the bulk of the material is resistant to biodegradation by microorganisms. Although the starch portion rapidly degrades, a residue of synthetic polymer remains. This was recognized by Griffin in the early 1970s, and led to the use of transition metal catalysts to enhance the oxidative degradation of the polyolefin matrix. Controversy over labeling such materials as 'biodegradable' has limited market acceptance. Because of this, considerable attention has been given to blends of starch with biodegradable polymers.

Composites of granular starch with poly(hydroxy alkanoates) (PHAs) have been extensively studied. PHAs are produced as energy storage polymers by a wide range of microorganisms. Poly(hydroxy butyrate-co-valerate) (PHBV) copolymers have been commercialized. These copolymers are completely biodegradable, but currently are several times more costly than commodity polymers such as PE and PS. Because of its low cost, starch is an attractive filler for these materials.

Yasin et al.[55] investigated the effects of incorporating various polysaccharides in the hydrolytic degradation of PHBV. The presence of polysaccharide fillers significantly increased the hydrolysis rate over wide ranges of pH and temperature. The increased porosity due to removal of the filler led to eventual matrix collapse. Ramsay et al.[56] showed that incorporating wheat starch at levels up to 50% increased the rate of biodegradation in mixed microbial cultures. While the microbial culture needed more than 20 days to degrade pure PHBV, the presence of starch reduced the time to eight days. At a starch content of 50%, the tensile strength was reduced by approximately half, while the modulus increased by ~65%.

Mechanical properties of starch–PHBV composites was improved by coating the starch with poly(ethylene oxide) (PEO).[57] PEO was chosen because of its partial compatibility with PHBV. When the starch was precoated with PEO (~9% by weight) before extrusion, tensile strength and elongation were roughly doubled compared to uncoated starch. Exposure of starch–PHBV materials to activated municipal sewage sludge showed that coating the starch reduced the rate of weight loss.[58]

Table 19.2 Mechanical properties of PHBV-starch composites[a]

Formula[b]	Tensible strength, MPa	Elongation, %	Modulus, GPa
80/15/0/5	22.0	24.20	1.60
65/25/0/10	14.9	28.10	1.21
70/15/0/15	14.1	26.40	0.88
60/25/10/5	15.0	15.2	1.90
65/15/10/10	14.5	17.0	1.23
50/25/10/15	8.9	15.2	0.98
65/15/15/5	16.9	11.6	1.77
50/25/15/10	10.8	11.4	1.38
55/15/15/15	10.6	13.7	0.91
45/25/15/15	8.1	11.4	0.95
100/0/0/0	31.80	13.20	2.10

[a]Data adapted from reference 59
[b]Numbers correspond to PHBV, starch, calcium carbonate and plasticizer (wt%)

Kotnis et al.[59] demonstrated that statistical experimental design methods could be used to generate predictive equations for composites of plasticized PHBV with starch and calcium carbonate fillers. Tensile strength, elongation and tensile modulus could be predicted with coefficients of determination of 0.95 or greater. By changing the relative amounts of plasticizer and filler, formulations ranging from rigid to flexible could be made. Table 19.2 shows the tensile properties of materials made using this approach.

Copolymers of polyethylene terephthalate (PET) with nonaromatic acids, poly(ethylene ethers) or hydroxy acids have been blended with starch to produce compostable products such as fibers and films.[60] Starch contents up to 80% by weight are claimed.

Mayer et al.[61,62] developed composites of starch with cellulose esters. Starch contents ranged from 30% to 70% by weight. Formulations of cellulose acetate (DS 2.5), 25% starch and 19% propylene glycol had mechanical properties similar to general purpose polystyrene.[62] The addition of approximately 5% calcium carbonate significantly improved the mechanical properties by neutralizing acetic acid liberated during processing. Biodegradable composites of starch with poly(vinyl alcohol),[63] poly(lactic acid)[64,65] and polycaprolactone[66] have also been reported.

In an application unrelated to biodegradability, it is reported that the addition of up to 1% by weight of starch to polycarbonate resins is useful in the preparation of films with low static coefficient of friction, high light transmission and low haze.[67]

IV. Starch in Rubber

Rubber compounds generally contain significant amounts of carbon black as a reinforcing agent.[22] To provide reinforcement, the filler particle size must be small, on the order of 2 μm or less. Since most native starch granules are much larger than 2 μm, methods of reducing starch particle size have been investigated.

Buchanan et al.[68,69] demonstrated that starch xanthate (SX) could act as a reinforcing agent in various rubber compounds. Master batches were prepared by coprecipitating SX (DS 0.07) with elastomer latexes by the addition of zinc sulfate. A curd was formed in which the SX was the continuous phase. After drying, the crumb was milled, whereupon phase inversion occurred, yielding a rubber compound containing a fine particle dispersion of SX. The reinforcing effects were dependent on rubber type. Natural rubber showed little change, even after the addition of 30 parts per hundred rubber (phr) zinc SX, while a carboxylic-modified rubber was steadily reinforced up to a level of 40 phr; a styrene-butadiene rubber showed similar results. The addition of greater than 10 phr zinc SX accelerated the curing of the rubber compounds.

Stephens et al.[70–72] examined the commercial potential of SX-reinforced rubber. SX reinforced styrene–butadiene and nitrile–butadiene rubber compounds, but displayed no reinforcement of natural rubber. The addition of resorcinol–formaldehyde resin at 8% of the starch weight significantly improved the mechanical properties for all three rubber compounds. The preparation of starch xanthide–rubber compounds by crosslinking the SX with sodium nitrite produced results similar to those obtained by adding resorcinol–formaldehyde resin.

Powdered compounds with starch and flour xanthates that could be milled or extruded were prepared by Buchanan et al.[73,74] The formulations could accommodate high loadings of extender pigments. Milling or extruding the compounds before molding improved the properties of the finished goods. Starch xanthide–rubber compounds prepared with xanthides of DS >0.1 could be vulcanized without additives such as sulfur or accelerators.[75] Starch xanthide–encased rubber was also prepared using alcohol dehydration to prepare a dry crumb suitable for milling or extrusion.[76,77] Xanthide reinforced ethylene–propylene terpolymer compounds were prepared by emulsifying hexane solutions of EPDM in starch xanthate, followed by coprecipitation using an acid and sodium nitrite.[78]

Starch poly(ethyleniminothiourethanes) prepared by the reaction of SX with poly(ethylenimine) were shown to be suitable reinforcing agents for rubber compounds.[79,80] Master batches were prepared using SX with DS values ranging from 0.08 to 0.58 and starch contents of 15 to 50 phr. Best results were obtained at a DS of ~0.22 and a loading of 25 phr starch, with a PEI stoichiometric ratio of 3.5. Hardness generally increased with DS and starch content, while compression set and abrasion resistance decreased.

In an extensive study of parameters controlling the properties of starch–elastomer compounds, Buchanan et al.[81] utilized experimental design methods to determine which factors determined compound properties. Xanthate derivatives were prepared using pearl corn starch, acid-modified corn starch, waxy maize starch, high-amylose corn starch, and corn and wheat flours. Variables included in the design were starch loading, xanthate DS, crosslinking, co-reaction with resorcinol–formaldehyde (RF) resin, filler or oil extenders, and processing conditions. In general, high-amylose starch xanthates had poorer physical properties and reduced water absorption. The flour-based xanthates gave materials with similar properties to starch xanthates, but with a darker color. A xanthate DS of 0.06 was optimal. Co-reaction with RF resin increased the reinforcing effect of SX. Extrusion drying was better than hot wet-milling or dry-milling. Optimal rubber compounds contained 30 phr SX or 45 phr starch xanthide.

Starch xanthide-reinforced rubber compounds swell more than conventionally reinforced compounds.[82] The swelling ratio after 90 days immersion was less than that calculated using the volume fraction of starch xanthide and its swelling behavior. Swelling effects were reversible on drying the swollen samples, both in sample dimensions and in mechanical properties. The use of RF resin or aminosilane coupling agents reduced swelling in water.[82–85] Although their swelling behavior was superior to normal starch xanthides, high-amylose starch xanthides generally exhibited poorer mechanical properties.[83,86] Xanthide-reinforced compounds with water resistance needed for most applications could be produced by appropriate formulation and compounding procedures,[87] and exhibited little difference during outdoor weathering when compared to conventional rubber compounds.[88] Formulations suitable for injection molding applications were demonstrated.[89,90] Manipulation of the coprecipitation variables allowed for control of xanthide particle size and structure,[91] which significantly affect the properties of the reinforced compounds.[92]

Carboxylic elastomers have been reinforced with cationic starch products.[93] Compounds were prepared either by solution mixing followed by drying, or by using a torque rheometer. The addition of cationic starch significantly reduced the toluene solubility of the carboxylated rubber. Physical properties of these compounds were greater than compounds prepared with unmodified pearl starch.

V. Starch Graft Copolymers

Another method of combining starch and synthetic polymers is through graft copolymerization. This approach has the advantage that the synthetic polymer and the starch are coupled together by covalent bonds formed during the grafting reaction. Methods such as the reaction of starch alkoxides with chloroformylated poly(ethylene oxide)[94] or ethylene oxide[95,96] have been used to prepare starch graft copolymers. Starch-g-polystyrene (PS) copolymers have been prepared by reacting anionically polymerized PS with nucleophilic end groups and modified starch.[97,98] This method allows the grafting of polymers with well-controlled molecular weight distributions onto starch.

Starch graft copolymers can also be synthesized initiating free radicals on the starch molecule in the presence of unsaturated monomers, such as vinyl or acrylic compounds. This approach has considerable potential for producing a wide range of products, given the different methods of free radical initiation and the number of monomers, which can be used alone or in combinations. Various cationic or anionic monomers have been used to produce copolymers for use as flocculants, dispersing agents, paper additives or superabsorbents. The use of starch graft copolymers in these applications has been reviewed.[99–101]

An early application of starch graft copolymers claims that incorporation of graft copolymers of starch or other polyols with ethyl or butyl acrylate into polystyrene improves the impact resistance if the polyol content is less than 20% of the graft.[102]

Generation of free radicals in starch by mastication[103] was used to prepare block copolymers of starch with polystyrene[104] and polyacrylamide.[105] Thewlis[106] prepared graft copolymers by masticating wheat starch or flour with glycerol and monomer in a bowl mixer for 30 minutes at 10°C. Styrene, methyl methacrylate, ethyl acrylate, acrylonitrile and methacrylic acid were used as monomers. Styrene and methyl methacrylate yielded graft contents of 8% and 6%, respectively. The addition of pyrogallol, a free radical scavenger, substantially reduced the amounts of grafted monomer, indicating that free radicals generated in the starch by mastication were responsible for the polymerization. The range of products obtained by this procedure was described as 'tough and rubbery to hard and rock-like,' depending on the selection of monomer and mastication time.

A wide range of starch graft copolymers have been prepared and characterized, using different methods of free radical initiation, many different monomers and comonomer combinations, and granular or gelatinized starches from various sources. Free radical initiation mechanisms have included ionizing radiation, such as that from a Co^{60} source, ferrous sulfate/hydrogen peroxide and ceric ammonium nitrate. Copolymers with potential plastics applications have generally utilized monomers such as methyl methacrylate (MMA), styrene (S), vinyl acetate (VA) and methyl acrylate (MA).[107–115]

Bagley et al.[107,108] prepared starch–g–PS, starch–g–PMMA, starch–g–PMA and starch–g–butyl acrylate (PBA) copolymers with copolymer add-ons of 40–50%. The monomers were selected to give a broad range of T_g values. Starch–g–PS copolymers, containing approximately 9% PS homopolymer, gave continuous, well-formed extrudates at 175°C, but poorer quality extrudates at 150°C or 190°C. The copolymer extruded at 175°C had tensile strengths of approximately 7500 psi, comparable to that of PS homopolymer. Starch–g–PMMA (47% add-on) extrudates of good quality could only be obtained by the addition of plasticizer. Copolymers with MA or BA at comparable add-ons required less torque for extrusion, and yielded soft, flexible extrudates with tensile strengths of approximately 3000 psi (2.1×10^6 kg/m^2). The extrudates exhibited little die swell; the extrusion mechanism was suggested to be powder flow followed by sintering under the high-pressure conditions experienced in the die.

Increasing the number of grafted chains improved the properties of starch–g–PMA copolymers (~60% add-on) prepared using ceric ammonium nitrate (CAN) initiation.[109] Increasing the CAN concentration also reduced the molecular weight of the grafted PMA chains from 936 000 (3.3 mmoles CAN) to 252 000 (71.2 mmole CAN). The tensile strength and elongation reached approximately constant values of 22 MPa and 230%, respectively, at PMA molecular weights of 465 000 and less. Exposure of the copolymers to various inocula for three weeks resulted in significant weight loss and reduction in tensile properties. The copolymers were suggested to have potential applications as agricultural mulch films.

Swanson et al.[116] investigated the role of copolymer content on starch–g–PMA extrudates. As the graft content increased from 42% to 77%, the ultimate tensile strength decreased from 30 MPa to 20 MPa. Elongation at break increased from 65% to 320%. The copolymers with higher PMA contents were therefore tougher materials, since the area under the stress–strain curve increased with increasing PMA content.

Extraction of the PMA homopolymer had little effect on the tensile strength, regardless of the PMA content. The homopolymer content was ~16% of the PMA present.

The effects of polymerization conditions on the structures and properties of CAN initiated starch–g–PMA copolymers has been extensively studied by Patil and Fanta.[113,115,117] It was shown that approximately 1 ceric ion per 100–200 D-glucosyl units (AGU) was sufficient to give high conversions to grafted copolymer.[113] High-amylose corn starch gave grafts with lower molecular weights compared to regular corn starch. Gelatinizing the starch before polymerization led to higher molecular weights and less frequent grafting compared to granular materials, although the difference was not as marked as with starch–polyacrylonitrile copolymers. Increasing the solids content of the reaction mixture by up to ~50% led to monomer conversions greater than 90% in 30 minutes at 25°C.

The effects of starch type and portion-wise addition of CAN initiator were also investigated.[115] Under identical polymerization conditions, copolymers with waxy maize starch yielded higher extractables, lower graft add-on and higher graft molecular weights. These effects were more pronounced at lower starch/MA ratios. Starch type had little effect on tensile strength, but elongation decreased and tear strength increased with increasing amylose content, as shown in Table 19.3. Addition of the CAN initiator in several aliquots rather than a single addition yielded copolymers with slightly reduced graft content and greater graft molecular weights. In addition, the tensile strength was slightly reduced by stepwise CAN addition, but substantial increases were observed in elongation and tear strength.

Substitution of cornflour for cornstarch yielded more flexible copolymers, with lower tensile strengths but greater elongation and tear strength, compared to starch copolymers.[117] Cornflour–g–PMA could be extruded into blown films, in contrast with starch–g–PMA prepared from granular starch. The effects of sequential CAN addition were similar to those observed for starch–g–PMA.[115]

Henderson and Rudin[111] examined the effects of water on starch–PMA and starch–g–PS extrudates of similar add-on. When immersed in water, starch–g–PS gained approximately 9% in weight and 7% in cross-section area, with essentially no water extractables. Starch–g–PMA gained 25% in weight and 50% in cross-section under the same immersion conditions. Extractables for the starch–g–PMA were 12.4%. Water absorption significantly reduced the tensile strengths of the copolymers; the

Table 19.3 Properties of starch–g–MA copolymers[a]

Starch type	Acetone soluble, %	Graft content	Graft $M_w \times 10^{-5}$	Tensile strength, Mpa	Elongation, %	Tear strength, N/mm
Waxy	31	40	11.3	13.6	158	7.4
Normal	18	50	9.0	14.8	111	7.5
High-amylose	20	59	7.6	13.3	98	11.0

[a]Data adapted from reference 115. Starch/MA weight ratio = 60/40. CAN initiator added in 9 portions. Tensile data for extruded ribbons

reduction was more drastic with starch–g–PMA: 85% compared to 45% for starch–g–PS. Elongation of the starch–g–PMA increased from 70% to ~300% on water absorption, although no correlation was observed between weight gain and elongation. No significant increase in elongation was observed between samples immersed for one hour and those immersed for 10 days. On drying, the starch–g–PS regained its original cross-section and tensile strength. Starch–g–PMA exhibited residual increases in cross-section after drying; the tensile strength returned to its original value, while the elongation remained at 250%, significantly greater than the original value. Dynamic mechanical analysis of the graft copolymers indicated that the starch was plasticized by water absorbed during immersion.

Graft content and extrusion condition effects on starch–g–PMA were examined by Trimnell et al.[114] Copolymers with PMA contents of 10%, 30%, 46% and 58% were prepared using CAN as an initiator and extruded with either 10% or 30% moisture content at 140°C or 180°C. At 10% PMA, the copolymers could not be extruded into testable ribbons. In general, extrudates with 10% moisture were smooth, while those with 30% moisture were rough. Increasing the moisture content yielded extrudates with lower tensile strengths and higher elongations, regardless of PMA content or extrusion temperature. At constant moisture content, extrusion at 180°C gave improved tensile strength and elongation. Water absorption was enhanced by increasing the moisture content or temperature during extrusion and decreased with increasing PMA content. Table 19.4 lists the effects of extrusion content and PMA content on the properties of these materials. Evidence of crosslinking of the starch was indicated by differential scanning calorimetry measurements of starch melting endotherms.

While starch copolymers are typically prepared by batch polymerization methods, reactive extrusion of starch has been studied. Mixtures of starch, polystyrene, styrene, sodium hydrogen carbonate, citric acid and water were extruded at temperatures between 100°C and 200°C.[118] It was reported that expanded graft copolymers

Table 19.4 Properties of extruded starch–g–PMA copolymers[a]

% PMA in starch–g–PMA	Extrusion		Tensile properties[b] (28 days)		Water sorption, %
	H$_2$O content, %	Maximum temperature, %	UTS	% E[c]	
58	10	180	16	53	34
58	30	180	14	102	48
58	10	140	13	24	28
58	30	140	–	–	47
46	10	180	15	10	60
46	30	180	11	108	67
43	10	140	12	8	34
46	30	140	9	63	67
30	10	180	11	2.7	95
30	30	180	24	32	123

[a]Data adapted from reference 114

[b]Aged at 23°C and 50% relative humidity. UTS = ultimate strength (MN/mm^2)

[c]% E = percentage elongation at break

are produced by this process, with starch contents of approximately 60% by weight. A combination of free radical and carbocationic mechanisms was speculated to be active during extrusion. Evidence of grafting between starch and polystyrene was indicated by β-amylase treatment of water and DMSO extracts of the extrudates. Benzene ring absorption at 262 nm, due to the PS, was observed to shift to lower molecular weights (as measured chromatographically) when the extracts were treated with the enzyme.

Carr et al.[119] investigated grafting via reactive extrusion of starch with cationic methacrylate, acrylamide and acrylonitrile monomers. Starch, monomer and CAN initiator were metered into a twin-screw extruder at starch contents of approximately 35% solids. The cationic methacrylate monomer showed poor reactivity during extrusion, with essentially no add-on. Acrylamide–starch systems (1:1 w/w) gave conversions of approximately 20% and add-ons of 16% to 18%. Acrylonitrile displayed the greatest reactivity during extrusion, with conversions of 74% and 63% for 1:1 and 1:2 w/w acrylonitrile/starch ratios, respectively. The corresponding add-ons were 27% and 42%.

Starch graft copolymers exhibit unusual extrusion characteristics, such as reduced die swell.[107,112,114,120] The granule structure of starch–PMA has been shown to persist through the extrusion process at low moisture contents.[112,114] These effects have been attributed to sintering of the deformed granules in the high pressure region of the die.[107,112] Granule deformation during extrusion was observed to be dependent on PMA content.[114] As the PMA decreased, granule disruption during extrusion increased. At a grafting level of 30%, complete phase inversion was observed at a moisture content of 30%, with the starch becoming the continuous matrix phase and the PMA the dispersed phase.

Starch–g–PS and starch–g–PMA exhibit shear thinning behavior, characteristic of most thermoplastic polymer melts.[112,120] Satisfactory extrudates of starch–g–PS required a screw with a compression ratio greater than 3:1.[120] Starch–g–MA disintegrates, but does not dissolve when immersed in solvents for PMA, perhaps because starch graft copolymers flow by superparticle mechanisms, wherein the flow of deformable particles is indistinguishable from the flow of polymer melts or solutions.[101] This behavior might be dependent on graft content, given the phase inversion behavior demonstrated by Trimnell et al.[114]

Grafted starch acylates have been reported.[121] Various starches and amylose were grafted with either ethyl acrylate (EA) or butyl acrylate (BA) and then acylated. The grafted starch acylates were brittle, but appeared to have good electrical insulating properties. Grafted amylose acetates were flexible and could be molded without added plasticizer, which was not the case for ungrafted amylose acetate. Amylose acetate with 49% graft add-on had a tensile strength of 4150 psi (4.2×10^6 kg/m^2) and elongation of 27%, compared to 6000 psi and 7.9% for ungrafted amylose acetate.

Graft copolymers of starch and vinyl acetate (VA) have been prepared using Co^{60} irradiation.[110] Near-quantitative conversions of monomer to polymer were reported when the radiation dose was 1.0 Mrad, although the grafting efficiency was <50%. Lower radiation doses gave lower conversions and add-ons. When 10% methacrylate was added to the monomer mixture, grafting efficiency was improved to 70%. Selected copolymers were treated with methanolic sodium hydroxide to yield starch-poly(vinyl alcohol) copolymers. Films cast from hydrolyzed graft copolymer had greater tensile strengths than those of comparable physical mixtures of starch and poly(vinyl alcohol).

An interesting feature of starch–g–PMA copolymers is their ability to shrink under high humidity conditions.[122] When the PMA content is between approximately 40% and 70%, blown films exhibit shrinkages of more than 60% at 100% relative humidity. The shrunken films were reported to be stable for more than a year after returning to ambient temperature and relative humidity. They were easily removed from objects by immersion in water after shrinking.

Starch graft copolymer latexes were prepared by sonication of reaction mixtures.[123–127] Clear adhesive films were prepared by grafting polyacrylonitrile onto cationic starches.[123,125] Films of cationic starch–polychloroprene graft copolymers were softer and more pliable if prepared from sonicated latexes.[124] Copolymers of starch with isoprene and acrylonitrile (2:1) with 30% starch could be masticated to give tough, pliable films suitable for vulcanization.[126] Particle diameters in latexes prepared by sonication of starch graft reaction mixtures were typically in the range of 300Å to 1500Å.

Graft copolymers of starch with acrylonitrile (AN), methyl methacrylate (MMA) or mixtures of the two monomers were studied as fillers for poly(vinyl chloride).[128] Grafted starch gave blends with significantly greater tensile strengths compared to ungrafted starch. At starch levels above ~30%, tensile strengths were greater than for the PVC resin. AN–MMA grafts displayed superior properties to grafts of either PAN or MMA alone. Opacity was reduced by gelatinizing the starch before grafting, and then wet blending the graft copolymer with PVC resin. Plastics with most clarity were obtained using a PVC latex and gelatinized graft copolymer. The use of dialdehyde starch gave blends with the greatest strength.

VI. Thermoplastic Starch Blends

Due to the loss in tensile strength described by Equations 19.1 and 19.2, and because of the difficulties of processing highly filled polymer materials, granular starch composites are generally limited to starch contents of 40% or less. Starch content can be increased significantly by converting granular starch into a thermoplastic by addition of water or other plasticizers. Extrusion processing converts starch into a thermoplastic material at lower moisture contents than are required for pasting. Transforming starch into a thermoplastic material offers a route by which blends with synthetic hydrophilic polymers can be processed to take advantage of the useful properties of each component. Early efforts to blend starch and other polymers this way used conventional gelatinization techniques, but led directly to the use of extrusion in order to eliminate the need for water removal. This work is presented together with later work which deals primarily with extrusion to produce thermoplastic or destructurized starch blends.

One of the first synthetic polymers to be blended with starch was poly(vinyl alcohol) (PVA). Otey et al.[129,130] prepared cast films of starch and PVA from aqueous solutions containing a plasticizer (glycerol). Films were cast onto glass plates and air-dried at 130°C. Small amounts of crosslinking agent, such as formaldehyde, generally increased elongation while decreasing tensile strength. At higher levels of glycerol, crosslinking increased tensile strength. Coating the starch–PVA films with a hydrophobic polymer greatly increased wet strength.

Starch–PVA films plasticized with glycerol embrittled with age. A study of various polyols was undertaken to identify plasticizers which would reduce this tendency.[131] It was demonstrated that sorbitol and glycol glycoside performed well. When sorbitol alone was used, it migrated to the surface and crystallized. When sorbitol and glycerol were used in a 3:1 w/w ratio, the films were stable and showed no evidence of sorbitol crystallization. Substituting plasticizer with PVA did not adversely affect the films; when starch was substituted for plasticizer, the films became more susceptible to aging.

Effects of small amounts of ethylene–acrylic acid copolymer (EAA) on starch–PVA–glycerol films have been reported.[132] Addition of 6% or less EAA improved elongation significantly. At high PVA and glycerol concentrations, EAA prevented phase separation. On the other hand, too much EAA made the films brittle. Statistical analysis indicated that elongation depended positively on interactions between EAA, PVA and glycerol. It was suggested that the EAA formed helical inclusion complexes with both starch and PVA.

Films of starch–EAA blends were prepared by either casting aqueous dispersions or by dry fluxing both the polymers.[133] Cast films were relatively transparent, flexible and stable on immersion in water. In contrast, the dry blended materials were opaque at starch levels greater than 30% and had poorer tensile properties at high starch contents. No plasticizers were needed to prepare flexible films. Starch–EAA films have also been prepared by extrusion.[134]

Polyethylene (PE) was also incorporated into starch–EAA compositions.[135,136] Starch, water and EAA were blended at 95–100°C in a heated mixer. Aqueous ammonia was then added, which produced an immediate increase in the viscosity of the mixture. (Optionally, PE could be added during this mixing step.) The mixture was then extruded through a strand die at approximately 130°C; multiple passes were sometimes required to reduce the moisture content to a desirable level. The extrudate was transparent and flexible, and could easily be blown into film. The addition of ammonia was important to the process, as streaky, inhomogeneous films were produced when it was omitted. Films with best appearance were produced when the moisture content was 5% to 8%. Tensile strengths of the starch–EAA–PE films varied little with starch or PE content, up to 30% PE, while elongation decreased with increasing starch content (Table 19.5). Urea, polyols and PVA were also incorporated into film compositions.[136,137] Formulations with urea as a plasticizer were successfully demonstrated on a pilot scale.[137] Statistical analysis methods have been used to study the component effects in these formulations for both film[138] and injection molding.[139] Various improvements of the technology described in reference 136 have been claimed.[140–142]

The use of a strong alkali other than ammonia in starch–EAA formulations yielded water-stable films with greater transparency and permeability.[143,144] Permeabilities of the films to solutes such as urea, sodium chloride and sugars was reported.[143] Urea exhibited the greatest diffusivity. Increasing the starch content significantly increased diffusion rates of all solutes investigated.

Table 19.5 Mechanical properties of starch–EAA–PE films[a]

Formula (starch–EAA–PE)	Tensile strength, psi[b]	Elongation to break, %
10/90/0	3470	260
20/80/0	4140	120
30/70/0	3225	150
40/60/0	3870	920
50/50/0	3940	61
40/50/10	3570	80
40/40/20	3477	66
40/30/30	3150	85
40/40/10	2920	34
40/10/50	1840	10

[a] Data adapted from reference 135
[b] 1.0 psi = 703.1 kg/m^2

The nature of the interaction between starch and EAA has been extensively investigated. Fanta et al.[145,146] demonstrated that precipitates obtained on mixing either amylose or amylopectin solutions with EAA dispersions contained both components, which could not be separated by solvent extraction. Low molecular weight starch and high molecular weight dextran did not form precipitates when mixed with EAA, suggesting that a helical inclusion complex was formed between the starch and EAA. Further characterization of starch–EAA complexes in solution[147] and solid-state[148] by chiroptical methods, x-ray diffraction and NMR supported this conclusion. Complex formation in extruded starch–EAA films was greater when water was replaced with solutions of ammonia.[149]

The starch–EAA complex exhibits interesting effects on the morphology of extruded materials. Shogren et al.[150] found that PE was immiscible with the starch–EAA complex, with each forming sheet-like structures during extrusion. Blown films with hydrophobic skins and water sensitive cores were produced by replacing intact starch with a dextrin.[151] Recycling the dextrin-containing films reduced their water sensitivity; and thin, semipermeable PE membranes could be prepared by water-soaking the films and floating off the outer PE skin.

Stepto and Tomka[152–155] reported that starch could be injection molded at a moisture content of less than 20%. Micrographs taken of starch from various points along the screw of the injection molder showed a progressive homogenization and loss of granule structure.[152] The molded articles prepared from starch and water were transparent, brittle and sensitive to atmospheric moisture.[153,154] They were claimed to be useful as capsules or controlled release matrices. The use of non-aqueous plasticizers for starch has been reported for injection molding and extrusion processing.[156]

These discoveries provided the technology base for commercialization efforts of starch-based plastics by the Novon division of the Warner-Lambert Company.[157,158] The thermoplastic compositions were based on blends of 'destructurized' starch and various hydrophilic polymers, such as ethylene–vinyl alcohol copolymers (EVOH). Hydrophobic polymers and additives, such as plasticizers and lubricants, could also be included. Several grades of product were offered for injection molding and extrusion processing. George et al.[159] have reported on the processing and properties of blends of starch and EVOH. Blends with high-amylose starch exhibited higher injection pressures during molding than starches with lower amylose content. Waxy maize starch blends were stiffer, while blends with high-amylose starch gave greater elongation. Tensile strengths of the starch–EVOH compounds were on the order of 3000 psi ($2.1 \times 10^6 \text{kg/m}^2$), with elongation increasing from about 100% to about 250% as the starch content decreased from 75% to 25%.

The term 'destructurized,' apparently first referred to by Stepto and Tomka,[152] describes the process of heating the starch under pressure to temperatures above the glass transition and melting points of its components.[153,158] Melting and disordering of the molecular and granular structure take place, yielding a thermoplastic material. It has been argued that the term destructurized describes starch-based materials developed prior to its appearance in the literature.[160] A comparison of extrusion processing of thermoplastic or destructurized starch with extrusion cooking has been given.[161]

During the same period, commercialization of thermoplastic starch polymer blends was pursued by Novamont, a division of the Ferruzzi Group of Italy.[162–172] Their products, marketed under the trade name Mater-Bi, are typically comprised of at least 60% starch or natural additive and hydrophilic, biodegradable synthetic polymers.[64,165] It is stated that these blends form interpenetrated or semi-interpenetrated structures at the molecular level. Properties of typical commercial formulations have properties similar to those in the range of low- and high-density PE. Blends of Mater-Bi products with biodegradable polyesters have been claimed for use as water impervious films.[173]

A wide range of thermoplastic starch compounds have been claimed in recent years. Formulations of thermoplastic starch with linear, biodegradable polyesters, including polycaprolactone and PHBV,[174–176] and with polyamides[175] have been reported. Laminated structures have been claimed using thermoplastic starch or starch blends as one or more of the layers.[175,177,178] The use of polymers latexes as components of thermoplastic starch blends has also been claimed.[179–181] Blends with natural polymers are also claimed, including cellulose esters[182,183] and pectin.[184] A crosslinked thermoplastic material of dialdehyde starch and protein has been reported.[185]

The rheology and melt-spinning of thermoplastic starch compounds was reported by Simmons et al.[186] Melts of Novon and Mater-Bi materials exhibited shear thinning behavior and could be melt-spun into fibers, which could be cold-drawn. Rheological studies of starch–EVOH[187] and starch–PVA blends[188,189] have been reported. Shear-thinning behavior was seen for all the blends studied. Non-Newtonian behavior increased with increasing starch content or increasing amylose content, as indicated by changes in the power law index.[187,189] The power law index decreased with increasing moisture content in starch–EVOH blends with glycerol.[188] Other factors which influence the rheology of thermoplastic starch include moisture content during prior extrusion and low molecular weight additives.[190,191] It has been suggested that glycerol monostearate forms stable inclusion complexes with starch during extrusion at 15% moisture content.[190,191] Endres et al.[192] determined the energy inputs required to plastify starch during extrusion, including thermal and mechanical terms. Values range from 380 kJ/kg for corn starch to 651 kJ/kg for potato starch. By comparison, a high-density PE requires 585 kJ/kg.

Deformation and fracture properties of thermoplastic starch have been reported. Fracture stress and fracture strain of extruded wheat starch ribbons exhibited a maximum near 10% to 12% moisture.[193] Similar results were observed in starch–xylitol and starch–glycerol materials, while the modulus exhibited a steady decrease with increasing moisture content.[194] Microscopy showed increasing plastic deformation in the starch ribbons as the moisture content approached 10%.[194] Warburton et al.[195] showed that the strain required for deformation zones to appear in thin starch films was dependent on extrusion temperature. A maximum in strain was seen for samples extruded at temperatures of about 140°–150°C. At lower extrusion temperatures, deformation zones initiated at granule remnants; the remnants disappeared as the extrusion temperature increased. Above ~150°C, however, greater degradation of the starch reduced the strain needed for initiation. The role of sub-T_g annealing (physical aging) to deformation properties of extruded starches has been reported.[196,197] High-amylose starch showed reduced rate of aging compared to normal starch, which was attributed to the increased number of chain ends from amylopectin in normal starch.[197]

Graft copolymers of thinned, gelatinized starch, including hydroxyethyl starch, with 1,3-butadiene-styrene latexes and other polymers that are claimed to have useful properties as paper coating materials are presented in Chapter 17.

VII. Starch Foams

Starch has received attention as a component of polyurethane foams, due to the reactivity of its hydroxyl groups with isocyanates. Dosmann and Steel[198] claimed flexible, shock-absorbing polyurethane foams that incorporated granular starch; the starch acted as an extender and significantly increased the stiffness of the foams. Bennett et al.[34] investigated starch as an extender in rigid polyurethane foams. The addition of 10–40% hammer-milled starch reduced compressive strength and imparted self-extinguishing behavior on ignition.

Cunningham et al.[199–203] investigated the use of starches and dextrins as extenders for polyurethane foams. Starch generally had little effect on the foam density, while compressive strength moderately decreased with increasing starch content. Hydrophilic foams prepared using starch were suggested to have useful horticultural properties.[202,203]

Extrusion of starch and water at temperatures above 100°C leads to the formation of voids, due to the expansion of steam as the extrudate exits the high-pressure conditions of the die. The food industry has long recognized the utility of starch extrusion to produce expanded snack foods[204] and various models have been proposed to describe this process.[205] The properties of extruded starch foams display power law behavior with respect to density.[206] Environmental concerns and the high visibility of polystyrene in foam packaging and food serving applications has sparked the development of various technologies for producing starch-based foams as polystyrene replacements. Consumption of expanded polystyrene products in disposable packaging applications has been estimated at approximately 1 billion lbs/yr (4.5×10^8 kg/yr) for 1995; the expanded PS foam loose-fill market was estimated at 90 million lbs/yr (41×10^6 kg/yr)^2 Typical properties of expanded PS loose-fill materials are a density of 0.25 lb/ft^3 (4.0 kg/m^3), 65% resiliency and compressibilities of 200 Newtons. Tatarka[207] has reviewed the commercial aspects of starch-based loose-fill products.

Lacourse and Altieri[208] claimed expanded starch foam products with low density, good resilience and compressibility, and a closed cell structure. It was demonstrated that the properties required starches with amylose contents of greater than ~45%. Normal corn starch, with approximately 25% amylose, gave foams with poor, open cell structures, which were brittle, easily crushed and displayed no resilience. Starch with about 70% amylose yielded foams with densities and resiliences approaching that of commercial PS foams, but with significantly lower compressibility. Hydroxypropylation of the starch, at levels up to ~10%, improved compressibility by a factor of four. It was shown that optimal properties with lightly hydroxypropylated high-amylose starches were obtained when the starch was extruded with moisture contents less than 21%, preferably between 13% and 19%. Up to 10% poly(vinyl alcohol) could also be added without diminishing the properties. Incorporation of inorganic salts, such as sodium sulfate, at levels up to 10% gave expanded products with improved resilience and compressibility, higher conductivities and more uniform closed cell structures.[209]

Expanded products from compositions including starches with amylose contents up to 35% have been claimed.[210] Compositions are based on mixtures of corn grits, corn meal or corn starch, with minor proportions of cellulosic components, gums, surfactants, boron compounds and/or blowing agents. A mixture containing corn grits (98%), carboxymethylcellulose (CMC) (1%), cellulose fiber (0.5%) and sodium aluminum acid pyrophosphate (0.5%) were extruded at 15% moisture content. The product showed improved resilience, crush strength and dusting compared to corn grits with CMC (0.2%), sodium bicarbonate (0.4%) and 0.05% of a poly(sorbate oleate). It is stated that the CMC functions to enhance the expansion and improve the resilience of the foams. Other additives impart strength through crosslinking or improve the cellular structure.

The manufacture of foams using unmodified starches in combination with poly-(alkylene glycol) derivatives and nucleating agents has been claimed.[211,212] The addition of up to ~10% of poly(ethylene glycol) or poly(propylene glycol) generally increased the volume expansion, regardless of the starch type. It is suggested that the polyglycols may form helical inclusion complexes with the amylose in the starch, resulting in increased molecular alignment in the melt. Once the polyglycol concentration exceeds that needed to fully complex the amylose, volume expansion increases are due to the plasticizing effect on the starch melt, making bubble formation and growth less difficult. Silicon dioxide or amorphous silica with surface areas in the range 190–450 m^2/g is added to increase bubble nucleation.

Starcevich[213] claimed packaging materials comprised of a grain and a binding agent at moisture contents of 14–16%. Guar gum at a level 0.5% by weight was incorporated as a binding agent. It was stated that maintaining the die temperature in the range 345–351°F (174–177°C) is critical to the extrusion of acceptable products.

Use of unmodified starches in combination with mild acids and carbonates has been claimed.[214] Starch moisture content should be no greater than 25%. Acids such as tartaric, citric or malic acids are added at levels between 0.2% and 7%, while the carbonate level is 0.1% to 2%, both based on total starch composition. The acid is stated to serve a two-fold function: it depolymerizes the starch during extrusion via hydrolysis, improving expansion; at the same time, it liberates carbon dioxide, which acts as a blowing agent, from the carbonate. A typical formula extruded in a twin-screw extruder at 170–195°C yielded a product with a density of approximately 1 lb/ft^3 (16 kg/m^3) and resilience of 60–85%. The foam had a continuous skin with a closed cell structure.

Bastioli et al.[172] claimed expanded articles from extruded compositions of starch, mixtures of EAA and EVOH copolymers, and an inorganic carbonate. Typically, the synthetic copolymers comprised 20–40% of the composition. Preferred extrusion temperatures were in the range 180°C to 210°C. Extrusion of a typical blend in a single-screw extruder at a temperature of 180°C produced a closed cell foam with a density of approximately 1.2 lb/ft^3 (19 kg/m^3).

A batch process similar to baking has been reported for the manufacture of starch foam trays.[215] A mixture of starch, water, fibers and other additives are introduced into a mold. The heated top of the mold is closed and foaming takes place *in situ* by the generation of steam. Vent cycles reduce the moisture content of the material by steam evaporation, thereby raising the glass transition temperature of the starch. Porous articles such as food trays and protective packaging with densities on the order of 0.1 g/cm^3 are produced. For many applications, water sensitivity of the starch is a limiting factor. Various materials have been used to provide waterproof coatings, particularly for tray applications.[216] Expanded foams have been made using starch graft copolymers.[217] Graft copolymers with at least 15% add-on are combined with unmodified starch, and extruded with 15–25% moisture. Additives such as CMC and talc are included to improve strength and control cell size.

Microcellular foams using starch as a raw material have been reported.[218] Aerogels are produced by displacing the water in aqueous gels with air. This cannot be done by simply air-drying starch gels as surface tension compresses the gel into a dense material. Starch aerogels were prepared by displacing the water from aqueous gels with ethanol to make an alcogel. The alcogels were air dried, extracted with liquid CO_2 and dried in CO_2 vapor, or critical point dried. High-amylose starch microcellular foams could be made only by liquid CO_2 extraction and critical point drying. Foam densities were generally 0.1–0.2 g/cm^3. The microcellular foams were comprised of pores with diameters of less than 2 μm. Thermal conductivities were comparable to commercial polystyrene foams. Foams were produced in either cylindrical or slab forms. Shrinkage during ethanol dehydration appeared to be the critical factor in determining aerogel density.

VIII. References

1. Morawetz H. *Polymers: The Origin and Growth of a Science*. New York, NY: Wiley; 1985:37.

2. *Modern Plastics Encyclopedia '96.* New York, NY: McGraw-Hill. 1995.
3. Sauer JA, Hara M. In: Kasch HH, ed. *Crazing in Polymers.* Vol. 2. Berlin, Germany: Springer-Verlag; 1990:69–118.
4. Wolff IA, Davis HA, Cluskey JE, Gundrum LJ, Rist CE. *Ind. Eng. Chem. Res.* 1951;43:915.
5. Mack DE, Shreve RN. *Ind. Eng. Chem.* 1942;34:304.
6. Mullen JW, Pacsu E. *Ind. Eng. Chem.* 1942;34:1209.
7. Mullen JW, Pacsu E. *Ind. Eng. Chem.* 1943;35:381.
8. Whistler RL, Schieltz NC. *J. Am. Chem. Soc.* 1943;65:1436.
9. Whistler RL, Hilbert GE. *Ind. Eng. Chem.* 1944;36:796.
10. Wolff IA, Olds DW, Hilbert GE. *Ind. Eng. Chem.* 1951;43:911.
11. Kerr RW. In: Kerr RW, ed. *Chemistry and Industry of Starch.* 2nd edn. New York, NY: Academic Press; 1950:652.
12. Wolff IA, Olds DW, Hilbert GE. *Ind. Eng. Chem.* 1957;49:1247.
13. Wolff IA. *Ind. Eng. Chem.* 1958;50:1552.
14. Sagar AD, Merrill EW. *J. Appl. Polym. Sci.* 1995;58:1647.
15. Merrill EW, Sagar AD. US Patent 5 459 258. 1995.
16. Rivard C, Moens L, Roberts K, Brigham J, Kelley S. *Enz. Micro. Technol.* 1995;17:848.
17. Borchers G, Dake I, Dinkelaker A, Sachetto J-P, Zdrahala R, Rimsa S, Loomis G, Tatarka PD, Mauzac O. International Patent Appl. WO 93/20110. 1993.
18. Borchers G, Dake I, Sachetto nd J-P, Mauzac O. International Patent Appl. WO 93/20140. 1993.
19. Borchers G, Dake I, Dinkelaker A, Sachetto J-P, Zdrahala R, Rimsa S, Loomis G, Tatarka PD, Mauzac O. Int. Patent Appl. WO 93/20141. 1993.
20. Rimsa S, Tatarka PD. International Patent Appl. WO 94/07953. 1994.
21. Bloembergen S, Narayan R. US Patent 5 462 983. 1995.
22. Nielsen LE. *Mechanical Properties of Polymers and Composites.* Vol. 2. New York, NY: Marcel Dekker; 1974:386–414.
23. Willett JL. *J. Appl. Polym. Sci.* 1994;54:1685.
24. Schroeter J, Hobelsberger M. *Starch/Stärke.* 1992;44:247.
25. Endres H-J, Pries A. *Starch/Stärke.* 1995;47:384.
26. Alter H. *J. Appl. Polym. Sci.* 1965;9:1525.
27. Lim S, Jane J-L, Rajagopalan S, Seib PA. *Biotechnol. Prog.* 1992;8:51.
28. Fritz H-G, Widmann B. *Starch/Stärke.* 1993;45:314.
29. Kshiragar NJ, Rangaprasad R, Naik VG, Kale DD. *J. Polym. Mater.* 1992;9:17.
30. Holton EE, Asp EH, Zottola EA. *Cereal Foods World.* 1994;39:237.
31. Willett JL. *Polym. Eng. Sci.* 1995;35:1184.
32. Griffin GJL. In: Griffin GJL, ed. *Chemistry and Technology of Biodegradable Polymers.* London, UK: Blackie Academic & Professional; 1994:18–47.
33. Otey FH, Bennett FL, Mehltretter CL. US Patent 3 405 080. 1968.
34. Bennett FL, Otey FH, Mehltretter CL. *J. Cell. Plast.* 1967;3:369.
35. Otey FH, Westhoff RP, Kwolek WF, Mehltretter CL, Rist CE. *Ind. Eng. Chem., Prod. Res. Dev.* 1969;8:267.
36. Westhoff RP, Otey FH, Mehltretter CL, Russell CR. *Ind. Eng. Chem., Prod. Res. Dev.* 1974;13:123.

37. Polymer Science and Technology. In: Giullet J, ed. *Polymers and Ecological Problems*. Vol. 3, New York, NY: Plenum Press; 1973.
38. Griffin GJL. US Patent 4 016 117. 1977.
39. Griffin GJL. US Patent 4 021 388. 1977.
40. Griffin GJL. US Patent 4 125 495. 1978.
41. Griffin GJL. US Patent 4 218 350. 1980.
42. Evangelista RL, Nikolov ZL, Sung W, Jane J-L, Gelina RJ. *Ind. Eng. Chem. Res.* 1991;30:1841.
43. Jane JL, Gelina RJ, Nikolov Z, Evangelista RL. US Patent 5 059 642. 1991.
44. Jane J-L, Schwabacher AW, Ramrattan SN, Moore JA. US Patent 5 115 000. 1992.
45. Willett JL. US Patent 5 087 650. 1992.
46. Knott JE, Gage PD. US Patent 5 091 262. 1992.
47. Weil RC. US Patent 5 026 754. 1991.
48. Vaidya UR, Bhattacharya M. *J. Appl. Polym. Sci.* 1994;52:617.
49. Vaidya UR, Bhattacharya M, Zhang D. *Polym.* 1995;36:1179.
50. Seethamraju K, Bhattacharya M, Vaidya UR, Fulcher RG. *Rheol. Acta.* 1994;33:553.
51. Vaidya UR, Bhattacharya M. US Patent 5 321 064. 1994.
52. Vaidya UR, Bhattacharya M. US Patent 5 446 078. 1995.
53. Yoo Y-D, Kim Y-W, Cho W-Y. US Patent 5 461 093. 1995.
54. Kim YJ, Lee HM, Park OO. *Polym. Eng. Sci.* 1995;35:1652.
55. Yasin M, Holland SJ, Jolly AM, Tighe BJ. *Biomater.* 1989;10:400.
56. Ramsay BA, Langlade V, Carreau PJ, Ramsay JA. *Appl. Environ. Microbiol.* 1993;59:1242.
57. Shogren RL. *J. Environ. Polym. Degrad.* 1995;3:75.
58. Imam SH, Gordon SH, Shogren RL, Greene RV. *J. Environ. Polym. Degrad.* 1995;3:205.
59. Kotnis MA, O'Brien GS, Willett JL. *J. Environ. Polym. Degrad.* 1995;3:97.
60. Gallagher FG, Shin H, Tietz RF. US Patent 5 219 646. 1993.
61. Mayer JM, Elion GR. US Patent 5 288 318. 1994.
62. Mayer JM, Elion GR, Buchanan CM, Sullivan BK, Pratt SD, Kaplan DL. *J. Macro. Sci., Pure Appl. Chem.* 1995;A32:775.
63. Nwufo BT, Griffin GJL. *J. Polym. Sci., Polym. Chem. Ed.* 1985;23:2023.
64. Ajioka M, Enomoto K, Yamaguchi A, Shinoda H. US Patent 5 444 107. 1995.
65. Ajioka M, Enomoto K, Yamaguchi A, Shinoda H. Eur. Patent Appl. 530 987 A1. 1993.
66. Koenig MF, Huang SJ. *Polym.* 1995;36:1877.
67. Carter RP. US Patent 4 405 731. 1983.
68. Buchanan RA, Weislogel OE, Russell CR, Rist CE. *Ind. Eng. Chem., Prod. Res. Dev.* 1968;7:155.
69. Buchanan RA, Russell CR, Weislogel OE. US Patent 3 442 832. 1969.
70. Stephens HL, Murphy RJ, Reed TF. *Rubber World.* November 1969.
71. Stephens HL, Roberts RW, Reed TF, Murphy RJ. *Ind. Eng. Chem. Prod. Res. Dev.* 1971;10:84.
72. Stephens HL, Reed TF. US Patent 3 645 940. 1972.
73. Buchanan RA, Katz HC, Russell CR, Rist CE. *Rubber J.* 1971;153:28.
74. Buchanan RA, Russell CR. US Patent 3 673 136. 1972.
75. Bagley EB, Dennenberg RJ. *Rubber Age.* June 1973.

76. Abbott TP, Doane WM, Russell CR. *Rubber Age.* August 1973.
77. Abbott TP. US Patent 3 830 762. 1974.
78. Dixon RE, Bagley EB. *J. Appl. Polym. Sci.* 1975;19:1491.
79. Douglas JA, Maher GG. US Patent 3 542 708. 1970.
80. Douglas JA, Maher GG, Russell CR, Rist CE. *J. Appl. Polym. Sci.* 1972;16:1937.
81. Buchanan RA, Kwolek WF, Katz HC, Russell CR. *Stärke.* 1971;23:350.
82. Buchanan RA. *Stärke.* 1974;26:165.
83. Katz HC, Kwolek WF, Buchanan RA, Doane WM, Russell CR. *Stärke.* 1976;28:211.
84. Buchanan RA, Russell CR. US Patent 3 480 572. 1969.
85. Buchanan RA, Russell CR. US Patent 3 714 087. 1973.
86. Katz HC, Kwolek WF, Buchanan RA, Doane WM, Russell CR. *Stärke.* 1974;26:201.
87. Buchanan RA, Doane WM, Russell CR, Kwolek WF. *J. Elast. Plast.* 1975;7:95.
88. Buchanan RA, McBrien J, Otey FH, Russell CR. *Starch/Stärke.* 1978;30:91.
89. Abbott TP, James C, Doane WM, Russell CR. *J Elast. Plast.* 1975;7:114.
90. Abbott TP. US Patent 3 941 767. 1976.
91. Buchanan RA, Seckinger HL, Kwolek WF, Doane WM, Russell CR. *J. Elast. Plast.* 1976;8:82.
92. Dennenberg RJ, Buchanan RA, Bagley EB. *J. Appl. Polym. Sci.* 1977;21:141.
93. Abbott TP, James C, Otey FH. *J. Appl. Polym. Sci.* 1979;23:1223.
94. Ezra G, Zilkha A. *J. Appl. Polym. Sci.* 1969;13:1493.
95. Tahan M, Zilkha A. *J. Polym. Sci. A–1.* 1969;7:1815.
96. Tahan M, Zilkha A. *J. Polym. Sci. A–1.* 1969;7:1825.
97. Narayan R, Stacy N, Lu Z. *Polym. Preprints.* 1989;30:105.
98. Narayan R, Tsao GT, Biermann CJ. US Patent 4 891 404. 1990.
99. Bagley EB, Fanta GF. In: Mark HF, Bikales NM, eds. *Encyclopedia of Polymer Science and Technology.* New York, NY: John Wiley and Sons; 1977:665–699.
100. Stannett VT, Fanta GF, Doane WM. In: Chatterjee PK, ed. *Absorbency.* Amsterdam, The Netherlands: Elsevier Science Publishers B.V.; 1985:257–279.
101. Fanta GF, Doane WM. In: Wurzburg OB, ed. *Modified Starches: Properties and Uses.* West Palm Beach, FL: CRC Press, Inc.; 1986:149–178.
102. Segro NR, Hodes W. US Patent 3 044 972. 1962.
103. Augustat S, Grohn H. *Stärke.* 1962;14:39.
104. Berlin AA, Penskaya EA, Volkova GI. *J. Polym. Sci.* 1962;56:477.
105. Whistler RL, Goatley JL. *J. Polym. Sci.* 1962;62:S123.
106. Thewlis BH. *Stärke.* 1964;16:279.
107. Bagley EB, Fanta GF, Burr RC, Doane WM, Russell CR. *Polym. Eng. Sci.* 1977;17:311.
108. Bagley EB, Fanta GF, Doane WM, Gugliemelli LA, Russell CR. US Patent 4 026 849. 1977.
109. Dennenberg RJ, Bothast RJ, Abbott TP. *J. Appl. Polym. Sci.* 1978;22:459.
110. Fanta GF, Burr RC, Doane WM, Russell CR. *J. Appl. Polym. Sci.* 1979;23:229.
111. Henderson AM, Rudin A. *J. Appl. Polym. Sci.* 1982;27:4115.
112. Swanson CL, Fanta GF, Bagley EB. *Polym. Composites.* 1984;5:52.
113. Patil DR, Fanta GF. *J. Appl. Polym. Sci.* 1993;47:1765.

114. Trimnell D, Swanson CL, Shogren RL, Fanta GF. *J. Appl. Polym. Sci.* 1993;48:1665.
115. Patil DR, Fanta GF. *Starch/Stärke.* 1994;46:142.
116. Swanson CL, Fanta GF, Fecht RG, Burr RC. In: Carraher CE, Sperling LH, eds. *Polymer Applications of Renewable Resource Materials.* New York, NY: Plenum; 1983:59–71.
117. Patil DR, Fanta GF. *Starch/Stärke.* 1995;47:110.
118. Chinnaswamy R, Hanna MA. *Starch/Stärke.* 1991;43:396.
119. Carr ME, Kim S, Yoon KJ, Stanley KD. *Cereal Chem.* 1992;69:70.
120. Henderson AM, Rudin A. *Makromol. Chem.* 1992;194:23.
121. Ray-Chaudhuri DK. *Stärke.* 1969;21:47.
122. Fanta GF, Otey FH. US Patent 4 839 450. 1989.
123. Gugliemelli LA, Swanson CL, Baker FL, Doane WM, Russell CR. *J. Polym. Sci., Polym. Chem. Ed.* 1974;12:2683.
124. Gugliemelli LA, Swanson CL, Doane WM, Russell CR. *J. Polym. Sci., Polym. Lett. Ed.* 1976;14:215.
125. Gugliemelli LA, Swanson CL, Doane WM, Russell CR. *J. Appl. Polym. Sci.* 1976;20:3175.
126. Gugliemelli LA, Doane WM, Russell CR. *J. Appl. Polym. Sci.* 1979;23:635.
127. Gugliemelli LA, Swanson CL, Russell CR. US Patent 3 984 361. 1976.
128. Otey FH, Westhoff RP, Russell CR. *Ind. Eng. Chem., Prod. Res. Dev.* 1976;15:139.
129. Otey FH, Mark AM, Mehltretter CL, Russell CR. *Ind. Eng. Chem., Prod. Res. Dev.* 1974;13:90.
130. Otey FH, Mark AM. US Patent 3 949145. 1976.
131. Westhoff RP, Kwolek WF, Otey FH. *Starch/Stärke.* 1979;31:163.
132. Lawton JW, Fanta GF. *Carbohydr. Polym.* 1994;23:275.
133. Otey FH, Westhoff RP, Russell CR. *Ind. Eng. Chem., Prod. Res. Dev.* 1977;16:305.
134. Otey FH, Westhoff RP. US Patent 4 133 784. 1979.
135. Otey FH, Westhoff RP, Doane WM. *Ind. Eng. Chem., Prod. Res. Dev.* 1980;19:592.
136. Otey FH, Westhoff RP. US Patent 4 337 181. 1982.
137. Otey FH, Westhoff RP, Doane WM. *Ind. Eng. Chem. Res.* 1987;26:1659.
138. Jasberg BK, Swanson CL, Nelsen T, Doane WM. *J. Polym. Mater.* 1992;9:153.
139. Jasberg BK, Swanson CL, Shogren RL, Doane WM. *J. Polym. Mater.* 1992;9:163.
140. Wool RP, Oelschlaeger P, Willett JL. International Patent Appl. WO 90/14388. 1990.
141. Schiltz DC. US Patent 5 449 708. 1995.
142. Wool RP, Schiltz DC, Steiner D. US Patent 5 162 392. 1992.
143. Otey FH, Westhoff RP. *Ind. Eng. Chem., Prod. Res. Dev.* 1984;23:284.
144. Otey FH, Westhoff RP. US Patent 4 454 268. 1984.
145. Fanta GF, Swanson CL, Doane WM. *J. Appl. Polym. Sci.* 1990;40:811.
146. Fanta GF, Swanson CL, Doane WM. *Carbohydr. Polym.* 1992;17:51.
147. Shogren RL, Greene RV, Wu YV. *J. Appl. Polym. Sci.* 1991;42:1701.
148. Shogren RL, Thompson AR, Greene RV, Gordon SH, Cote G. *J. Appl. Polym. Sci.* 1991;47:2279.
149. Fanta GF, Swanson CL, Shogren RL. *J. Appl. Polym. Sci.* 1992;44:2037.
150. Shogren RL, Thompson AR, Felker FC, Harry-O'Kuru RE, Gordon SH, Greene RV, Gould JM. *J. Appl. Polym. Sci.* 1992;44:1971.

151. Swanson CL, Fanta GF, Salch JH. *J. Appl. Polym. Sci.* 1993;49:168.
152. Stepto RFT, Tomka I. *Chimia.* 1987;41:76.
153. Wittwer F, Tomka I. US Patent 4 673 438. 1987.
154. Wittwer F, European Patent Appl. 118 240 B1. 1989.
155. Sachetto JP, Stepto RFT, Zeller H. US Patent 4 900 361. 1990.
156. Tomka I. In: Levine H, Slade L, eds. *Water Relationships in Food.* New York, NY: Plenum Press; 1991:627–637.
157. Warner-Lambert. European Patent Appl. 0 326 517; 0 327 505; 0 391 853; 0 404 723; 0 404 727; 0 404 728; 0 407 350; 0 408 501; 0 408 502; 0 408 503; 0 409 781; 0 409 782; 0 409 783; 0 409 788; 0 409 789; 1989.
158. Lay G, Rehm J, Stepto RF, Thome M, Sachetto JP, Lentz DJ, Silbiger J. US Patent 5 095 054. 1992.
159. George ER, Sullivan TM, Park EH. *Polym. Eng. Sci.* 1994;34:17.
160. Shogren RL, Fanta GF, Doane WM. *Starch/Stärke.* 1993;45:276.
161. Wiedmann W, Strobel E. *Starch/Stärke.* 1991;43:138.
162. Butterfly SRL. European Patent Appl. 0 400 531; 0 400 532. 1990.
163. Butterfly SRL. International Patent Appl. WO 91/02024; WO 91/02025. 1991.
164. Bastioli C, Tinello E. In: *Proceedings of Corn Utilization Conference IV.* St. Louis, MO; June 24–26, 1992.
165. Bastioli C, Bellotti V, Del Guidice L, Gilli G. In: Vert M, Feijen J, Albertsson A, Scott G, Chiellini E, eds. *Biodegradable Polymers and Plastics.* Cambridge, UK: The Royal Society of Chemistry; 1992:101–110.
166. Bastioli C, Bellotti V, Montino A, del Tredici G, Lombi R. International Patent Appl. WO 9201743 A1. 1992.
167. Bastioli C, Bellotti V, Montino A, del Tredici G, Lombi R. US Patent 5 234 977. 1993.
168. Bastioli C, Bellotti V, Montino A. US Patent 5 292 782. 1994.
169. Bastioli C, Bellotti V, Montino A, Tredici GD, Lombi R, Ponti R. US Patent 5 412 005. 1995.
170. Bastioli C, Bellotti V, del Tredici G, Ponti R. US Patent 5 462 981. 1995.
171. Bastioli C, Lombi R, del Tredici G, Guanella I, US Patent 5 462 982. 1995.
172. Bastioli C, Bellotti V, del Giudici L, Lombi R, Rallis A. US Patent 5 360 830. 1994.
173. Toms D, Wnuk, AJ. US Patent 5 422 387. 1995.
174. Tokiwa Y, Takagi S, Koyama M. US Patent 5 256 711. 1993.
175. Buehler FS, Schmid E, Schultze H-J. US Patent 5 346 936. 1994.
176. Verhoogt H, St-Pierre N, Truchon FS, Ramsay BA, Favis BD, Ramsay JA. *Can. J. Microbiol.* 1995;41(Suppl. 1):323.
177. Uemura T, Akamatsu Y, Yoshida Y, Moriwaki Y. US Patent 5 384 187. 1995.
178. Wnuk AJ, Koger TJ, Young TA. US Patent 5 391 423. 1995.
179. Bortnick NM, Graham RK, LaFleur EE, Work WJ, Wu J-C. US Patent 5 403 875. 1995.
180. Ritter W, Bergner R, Kempf W. US Patent 5 439 953. 1995.
181. Ritter W, Bergner R, Kempf W. International Patent Appl. 92/10539. 1992.
182. Tomka I. US Patent 5 280 055. 1994.
183. Tomka I. European Patent Appl. 542 155 A2. 1993.

184. Fishman ML, Coffin DR. US Patent 5 451 673. 1995.
185. Jane J, Spence KE. US Patent 5 397 834. 1995.
186. Simmons S, Weigand CE, Albalak RJ, Armstrong RC, Thomas EL. In: Ching C, Kaplan DL, Thomas EL, eds. *Biodegradable Polymers and Packaging*. Lancaster, PA: Technomic Publ. Co.; 1993:171–207.
187. Villar MA, Thomas EL, Armstrong RC. *Polym.* 1995;36:1869.
188. Dell PA, Kohlman WG. *J. Appl. Polym. Sci.* 1994;52:353.
189. Wang XJ, Gross RA, McCarthy SP. *J. Environ. Polym. Degrad.* 1995;3:161.
190. Willett JL, Jasberg BK, Swanson CL. In: Fishman ML, Friedman RB, Huang eds. *Polymers from Agricultural Coproducts, ACS Symposium Series* Vol. 575 Washington, DC: American Chemical Society; 1994:5068.
191. Willett JL, Jasberg BK, Swanson CL. *Polym. Eng. Sci.* 1995;35:202.
192. Endres H-J, Kammerstetter H, Hobelsberger M. *Starch/Stärke*. 1994;46:474.
193. Attenburrow GE, Davies AP, Goodband RM, Ingman SJ. *J. Cereal Sci.* 1992;16:1.
194. Kirby AR, Clark SA, Parker R, Smith AC. *J. Mater. Sci.* 1993;28:5937.
195. Warburton SC, Donald AM, Smith AC. *Carbohydr. Polym.* 1993;21:17.
196. Shogren RL, Swanson CL, Thompson AR. *Starch/Stärke*. 1992;44:335.
197. Shogren RL, Jasberg BK. *J. Environ. Polym. Degrad.* 1994;2:99.
198. Dosmann LP, Steel RN. US Patent 3 004 934. 1961.
199. Cunningham RL, Carr ME, Bagley EB. *Cereal Chem.* 1991;68:258.
200. Cunningham RL, Carr ME, Bagley EB. *J. Appl. Polym. Sci.* 1992;44:1477.
201. Cunningham RL, Carr ME, Bagley EB, Nelsen TC. *Starch/Stärke*. 1992;44:141.
202. Cunningham RL, Carr ME, Bagley EB, Gordon SH, Greene RV. *J. Appl. Polym. Sci.* 1994;51:1311.
203. Cunningham RL, Carr ME, Bagley EB, Gordon SH, Greene RV. In: Fishman ML, Friedman RB, Huang SJ, eds. *Polymers From Agricultural Coproducts*. Washington, DC: American Chemical Society; 1994:101–110.
204. Mercier C, Linko P, Harper J, eds. *Extrusion Cooking*. St. Paul, MN: American Association of Cereal Chemists; 1989.
205. Alvarez-Martin L, Kondury KP, Harper JM. *J. Food Sci.* 1988;53:609.
206. Hutchinson RJ, Siodlak GDE, Smith AC. *J. Mater. Sci.* 1987;22:3956.
207. Tatarka PD. *SPE ANTEC Proc.* 1995;53:2225.
208. Lacourse NL, Altieri PA. US Patent 4 863 655. 1989.
209. Lacourse NL, Altieri PA. US Patent 5 043 196. 1991.
210. Anfinsen JR, Garrison RR. International Patent Appl. WO 92/08759. 1992.
211. Neumann PE, Seib PA. US Patent 5 185 382. 1993.
212. Neumann PE, Seib PA. US Patent 5 208 267. 1993.
213. Starcevich BK. US Patent 5 186 990. 1993.
214. Jeffs HJ. US Patent 5 252 271. 1993.
215. Tiefenbacher KF. *J. Macro. Sci., Pure Appl. Chem.* 1993;A30:727.
216. Haas K, Haas J, Tiefenbacher KF. International Patent Appl. WO 94/13734. 1994.
217. Boehmer EW, Hanlon DL. US Patent 5 272 181. 1993.
218. Glenn GM, Irving DW. *Cereal Chem.* 1995;72:155.

Starch Use in Foods

20

William R. Mason
Formerly of National Starch and Chemical Co., Bridgewater, New Jersey, USA

I. Introduction.. 746
    1. First Enhancement of Starch for Foods........................ 747
    2. Modern Use of Starch in Foods................................ 747
    3. Development of Crosslinking 747
    4. Development of Monosubstitution.............................. 747
    5. 'Instant' Starches 748
    6. Improvement of Starch Sources (See Also Chapter 3)............ 748
II. Functions of Starch in Food Applications........................ 748
    1. Starch Structures Relevant to Foods 749
    2. Gelatinization and Pasting 749
    3. Changes During Cooking 750
III. Impact of Processing and Storage on Foods Containing Cooked Starch .. 751
    1. Concentration During Cooking................................. 751
    2. Effects of Time and Temperature 751
    3. Effects of Shear... 752
    4. Comparison of Food Processing Equipment...................... 753
    5. Impact of Processing and Storage............................. 754
    6. Changes that Occur During Cooling, Storage and Distribution .. 754
    7. Recommended Processing 755
IV. Modified Food Starches (See Also Chapter 17) 756
    1. Why Starch is Modified....................................... 756
    2. Derivatizations 756
    3. Conversions 760
    4. Oxidation 761
    5. Physical Modifications 762
    6. Native Starch Thickeners..................................... 767
V. Starch Sources (See Also Chapters 9–16)......................... 767
    1. Dent Corn.. 768
    2. Waxy Corn 768
    3. High-amylose Corn.. 769
    4. Tapioca.. 770
    5. Potato... 770
    6. Wheat.. 770
    7. Sorghum.. 771
    8. Rice... 771
    9. Sago 772
    10. Arrowroot... 772

	11. Barley	772
	12. Pea	772
	13. Amaranth	773
VI.	Applications	773
	1. Canned Foods	774
	2. Hot-filled Foods	775
	3. Frozen Foods	775
	4. Salad Dressings	776
	5. Baby Foods	777
	6. Beverage Emulsions	777
	7. Encapsulation	777
	8. Baked Foods	778
	9. Dry Mix Foods	778
	10. Confections	778
	11. Snacks and Breakfast Cereals	779
	12. Meats	780
	13. Surimi	781
	14. Pet Food	781
	15. Dairy Products	781
	16. Fat Replacers	782
VII.	Interactions with Other Ingredients	783
	1. pH	783
	2. Salts	783
	3. Sugars	784
	4. Fats and Surfactants	784
	5. Proteins	785
	6. Gums/hydrocolloids	786
	7. Volatiles	786
	8. Amylolytic Enzymes	786
VIII.	Resistant Starch	787
IX.	References	788

I. Introduction

Starch is used in a wide range of foods for a variety of purposes including thickening, gelling, adding stability and replacing or extending more costly ingredients. Starches are favored for their availability, comparatively low cost and unique properties. The history of the starch industry, with an emphasis on the US industry, is presented in Chapter 1. With reference to food uses, in the first century AD, Celsus, a Greek physician, described starch as a wholesome food.[1] Starch was added to rye and wheat breads during the 1890s in Germany and to beer in 1918 in England.[2] Moffett, writing in 1928,[3] describes the use of corn starch in baking powders, pie fillings, sauces, jellies and puddings. During the 1930s, Kraft began extending mayonnaise with a paste containing dent corn and arrowroot starch to produce salad dressings.

Subsequently, combinations of dent corn and tapioca starches or dent corn starch and locust bean gum were used by salad dressing manufacturers.[4]

1. First Enhancement of Starch for Foods

Sweeteners produced by acid-catalyzed hydrolysis of starch were used in the improvement of wines in what is now Germany after large-scale potato starch mills opened in about 1830.[5] Naegeli wrote in 1874 that acid-catalyzed hydrolysis of starch granules produced a residue containing short linear molecules.[6] His products and those of Lintner,[7] who published ten years later, are the so-called fluidity or thin-boiling starches of today. Bellmas[8] and Duryea[9] filed patents for acid-thinned starches at the turn of the century.

2. Modern Use of Starch in Foods

In 1950, Kerr[10] described the use of starch by the US food industry in the late 1940s. Roughly 100 000 metric tons of starch were used annually in foods, but only about 30 000 metric tons were used in processed food, minor by today's standards. Kerr's discussion of food starches ends with a brief but auspicious reference to crosslinking, for which he held a patent. The leading users of starch were the brewing, baking powder and confectionery industries. By 1995, 2.6 million metric tons of corn starch, modified corn starch and corn dextrins were produced in the US.[11] Of that amount, 950 000 metric tons were used in foods. Since corn starch represents roughly 95% of US food starch, this corresponds to a ten-fold increase in starch use in 50 years. In 1989, 5.1 million metric tons of starch were produced in Europe, including ~60% maize, ~20% potato and ~20% wheat starch. As reported in 1992, of this amount, 55% was used in food.[12]

3. Development of Crosslinking

The observation that bleached tapioca flour with a high protein content improved process stability in fruit pies led to a 1943 patent for starch oxidized in the presence of protein.[13] Crosslinking with bifunctional reagents, as we know it, was first practiced in the late 1940s using epichlorohydrin[14] to make distarch ethers and phosphoryl chloride[15] and sodium trimetaphosphate[16] to make distarch phosphates. These starches were widely used in salad dressings, commercial baking of fresh pies, canned pie fillings and canned vegetables, particularly cream-style corn. Crosslinked waxy maize starch improved the quality of canned soups, sauces and gravies, although some manufacturers continued to use native corn starch. The US corn wet-milling industry discontinued use of epichlorohydrin in food starches in 1978 via self-regulation. This was due to concern for the safety of starch plant workers and possible residual reagent in food ingredients.

4. Development of Monosubstitution

Acetylation was the first form of monosubstitution used in food starches. Waxy maize acetylated distarch adipates came into use in the mid-1950s, driven by demand for pie fillings with improved stability to winter distribution. Acetylation was achieved by reaction with acetic anhydride[17,18] or vinyl acetate.[19] The improved smoothness and sheen attainable led canned soup and sauce processors to switch from using only native corn or potato starch, and popularized the more general use of modified waxy maize starch. The improved freeze–thaw stability of acetylated crosslinked waxy maize starch led to the marketing of frozen sauces, initially on vegetables, but also in entrees and pies by the 1960s.[20] The use of hydroxypropylation[21] was commercialized in the early 1970s with epichlorohydrin or phosphoryl chloride as the crosslinker. Hydroxypropylation provided an increased level of stability so that the quality attainable in puddings and frozen sauces was dramatically improved. Encapsulation with emulsion-stabilizing dextrins began in the mid-1960s. Corresponding modified food starches for use in beverage emulsions came into use in the mid-1970s.[22]

5. 'Instant' Starches

Drum-dried, crosslinked and stabilized starches became available in about 1960, making instant dessert mixes and in-plant cold processing of starch-containing foods possible. Granular, cold-water-swelling starches appeared in the 1980s, affording much higher quality in instant desserts and cold processing of dressings and other foods. Cold-water-swelling starches were prepared by spray drying[23] and hot aqueous ethanol treatment.[24]

6. Improvement of Starch Sources (See Also Chapter 3)

Amylopectin-rich corn was first found in China in 1908. It was called waxy corn due to the appearance of the kernels. At first it was only maintained as a genetic curiosity, although its pasting properties were known to be similar to that of tapioca starch. Interest in waxy varieties of maize, barley, rice and grain sorghum increased when access to Southeast Asian tapioca supplies were interrupted during World War II. Waxy sorghum hybrids were developed for production in Kansas, Texas and Nebraska, and waxy maizes were developed for Iowa. Starch production was commercialized in the mid-1940s.[25] Maize double mutants involving the waxy gene and other loci have unique starch properties; e.g. *wx su2* has shorter outer branches and improved freeze–thaw stability. Commercial high-amylose starch arose from the Bear Hybrid Corn Company's incorporation of the *ae* gene into adapted hybrids during the 1950s. Starch from these high-amylose corn hybrids became available in 1958.[20]

II. Functions of Starch in Food Applications

Starch has a range of roles in a variety of foods, as shown in Table 20.1. An understanding of the mechanism underlying each effect is necessary to make the best use of starch in these functions. To gain that understanding, it is helpful to track the changes that starch undergoes during pasting and cooling, and the impact these have on the structures of foods. Moreover, it is important to recognize how the cooked

II. Functions of Starch in Food Applications

Table 20.1 Roles starches play in various food systems

Function	Foods
Adhesion	Battered and breaded foods
Binding	Formed meat, snack seasonings
Clouding	Beverages
Crisping	Fried and baked foods, snacks
Dusting	Chewing gum, bakery products
Emulsion stabilization	Beverages, creamers
Encapsulation	Flavors, beverage clouds
Expansion	Snacks, cereals
Fat replacement	Ice cream, salad dressings, spreads
Foam stabilization	Marshmallows
Gelling	Gum drops, jelly gum centers
Glazing	Bakery, snacks
Moisture retention	Cakes, meats
Thickening	Gravies, pie fillings, soups

starch paste changes during storage and the resulting affects on the texture and appearance of the foods. The selection of a starch for a given use depends on the desired food properties, as well as the processing and distribution stresses involved.

1. Starch Structures Relevant to Foods

Prior to heating in water, starch granules are insoluble and will absorb only a limited amount of water. During processing, starch granules swell and are fragmented and solubilized to varying degrees, according to the severity of heating and mechanical shear.[26,27] The distribution of starch between swollen granules, fragmented granules and solubilized polymers determines the texture, appearance and stability of the paste, and the food.[28] Modifications to be discussed later enhance a starch's performance by favoring the desired form and the desired placement in the food structure, such as at the interface for emulsification. The following describes changes occurring during cooking and the form that starch must take to perform its desired function.

2. Gelatinization and Pasting

Both gelatinization and pasting are used to describe changes starch undergoes when heated in water; clarification of these is helpful for a meaningful discussion. According to Atwell et al.,[29] gelatinization is the collapse (disruption) of molecular order within the starch granule manifested in irreversible changes in properties such as granular swelling, native crystallite melting, loss of birefringence and starch solubilization. The point of initial gelatinization and the range over which it occurs is governed by starch concentration, method of observation, granule type and heterogeneities within the granule population under observation. Gelatinization on a macroscopic scale causes thickening and loss of opacity. (Note that gelatinization is very different from gel formation [gelation], which creates a semi-solid from a liquid or sol and may be accompanied by increased turbidity.) Pasting, according to Atwell et al.,[29] is the phenomenon following gelatinization in the dissolution of starch. It involves granular swelling, exudation of molecular components from the granule and, eventually, total disruption of granules. By this definition, pasting goes beyond optimal cooking for thickening by starch, but describes what is needed where starch is in the form of dissolved polymer molecules as required for emulsification, gelation and mouthfeel. Both gelatinization and pasting are general terms which do not precisely define the condition of processed starch, but include a number of phenomena which occur during starch processing. To characterize a process or product, it is helpful to reference specific changes in the structure, texture or appearance of the starch paste or food. Depending on the conditions used, the quality of the cook can vary across a broad range. In swelling or fragmenting granules and in solubilizing polymers, gelatinization and pasting affect the contribution of starch to texture in several ways. Each of these is necessary for starch to perform one or more of the functions outlined above.

3. Changes During Cooking

When heated in water, the starch granule begins to swell as thermal energy breaks hydrogen bonds between adjoining starch polymers. Amorphous regions are disrupted first.[30] Bonds in the crystalline areas hold the granule intact until a point is reached where they are also broken. With continued heating, the granule swells to many times its original volume. Intact swollen starch granules have a major impact on the rheological properties of a starch slurry. The physical force or friction between these highly-swollen granules and interactions with large solubilized polymers causes the paste to thicken.[31–33]

During granular swelling amylose, if present, leaches from the granule and, if allowed to reassociate in the surrounding solution, makes the paste cloudy or opaque.[34] Amylose can initiate gelation (setback or retrogradation) (see Section 20.3.6).[34] The gel may shrink and express water and become rubbery. The swollen granule is susceptible to disruption via additional heating or shear. If granules swell excessively and rupture, a dispersion of amylose, amylopectin and granule fragments is formed and the paste loses viscosity and becomes long.[35] Because it is highly-branched amylopectin does not gel or form films as strong as those of amylose, but amylopectin films remain pliable and are useful in emulsification and encapsulation.

Where starch is used to thicken, the desired properties are obtained when the starch granule is optimally swollen, but not disrupted or fragmented. On cooling, the properties of the sols reflect the composition and degree of swelling and solubilization achieved. Overcooking usually results in rubbery gels with amylose-containing starch like normal corn starch, or gummy pastes in the case of waxy maize starch. Conversely, an undercooked gel is opaque and watery. The optimum degree of swelling is not always easy to obtain because of the susceptibility of swollen granules to fragmentation due to excessive heat, acidity and/or shear. This is especially true for unmodified starch, where the range of tolerance between undercooking and overprocessing is narrow. In some food starch applications, it is necessary to cook the product until all granules are disrupted and a molecular dispersion is obtained.

These applications exploit the polymeric nature of starch in gelling, emulsifying and mouthfeel enhancement.

III. Impact of Processing and Storage on Foods Containing Cooked Starch

Optimum processing of starch establishes the desired distribution of the appropriately swollen granules, fragments and solubilized polymers for the intended starch function. Suboptimal processing, as often happens during product scale-up, fails to achieve the desired texture and stability. Various processing factors influence the physical structures and distribution of a starch in a food system. They include starch concentration, cooking time and temperature, rate of heating and cooling, and shear during cooking, cooling and filling. Environmental factors affecting changes during storage include temperature, water activity and the presence of protective or antagonistic ingredients. The effects of botanical source and formulation (e.g. pH and competition for water) are discussed in Sections 20.7 and 20.5, respectively.

1. Concentration During Cooking

The concentration at which a starch is cooked influences its behavior, including the extent of swelling and solubilization, which in turn affects viscosity and overall texture. At temperatures above the gelatinization temperature, there is a minimum moisture content required for complete gelatinization.[36] When there is insufficient water of hydration present, swelling of granules will be retarded and the texture will be affected. The critical starch:water ratio, that at which loose packing occurs, is characteristic of each starch source, as is the impact of starch concentration on flow properties. These properties vary with raw granule size and with diameter increases during cooking.[37] Cooking above the critical starch concentration prohibits amylose solubilization.[38] Also above the critical starch concentration, friction between granules increases and they become more susceptible to rupture by shear. Likewise, the behavior of starch sols on cooling is influenced by the starch:water ratio. In general, the higher the starch concentration is, the greater the viscosity and the greater the tendency to thicken and gel on cooling.

2. Effects of Time and Temperature

Time and temperature interact with regard to granular swelling. At higher temperatures swelling is faster, and below a certain temperature cooked starch will swell no further, because a lower percentage of crystallites can melt at the lower temperature.[36] Christianson and Bagley[26] found that, if cooked at 67°C, 12% corn starch was required to achieve the viscosity of 8% corn starch cooked at 80°C. Eliasson et al.[39] observed a reduction in the storage modulus of waxy maize starch sols cooked at 90°C compared to those cooked at 80°C, possibly reflecting increased deformability of more highly swollen granules at the higher temperature. Doublier et al.[27,34]

found that more rapid heating gave higher viscosities in wheat starch sols, possibly due to minimized overcooking of individual granules within the sample.

3. Effects of Shear

Swollen starch granules are susceptible to disintegration if subjected to physical impact or a severe pressure drop. This is an issue where starch is used to develop viscosity and granular integrity is required. Raw, unswollen starch granules are generally not damaged by shear in the slurry before cooking and can be safely dispersed with high speed mixing or homogenization. But once pasted (cooked) starch granules can be disrupted by shear, resulting in lost viscosity, shortness and textural stability. Problematic shear occurs in heat exchange and size reduction equipment (e.g. homogenizers or colloid mills), due to their close clearances, high product velocities and high pressure drops. Table 20.2 presents a qualitative rating of processing equipment with regard to impacts on swollen starch granules.

Positive or centrifugal transfer pumps can cause viscosity reductions in cooked starch formulations, especially where recirculation is used. Homogenizers, colloid mills and high shear mixers are even more severe, achieving shear rates as high as 150 000 sec^{-1}[40]. The effects of shear are exacerbated by high temperatures and acidity, which also favor disruption of the starch granule. Crosslinked starch, which is less subject to degradation due to shear and gives a shorter, more viscous paste after severe shear, is required for applications involving high shear after cooking. For products such as salad dressings, some starch granule destruction is deliberately introduced to achieve the desired, slightly cohesive texture.

Agitation during cooking and cooling may affect the texture of the final product. High shear rates during heating will accelerate heat transfer, and shear itself can enhance swelling of highly crosslinked granules, as evidenced by increased paste viscosities after shearing.[40] If shear is excessive, it can reduce the final viscosity by fragmenting granules.[34] Shear applied to products during cooling can reduce ultimate gel

Table 20.2 Relative shear of processing equipment

Cooking equipment	Shear intensity
Kettle cooker	Low
Tubular	Low
Steam infusion	Low to moderate
Scraped surface	Moderate to high
Plate exchanger	High
Steam injection	Moderate to very high
Cooling equipment	
Tubular	Low
Scraped surface	High shear
Plate exchanger	High shear
Vacuum cooling	High shear
Milling equipment	
Colloid mill	High shear
Homogenizer	Very high shear

strength by interfering with the retrogradation of amylose, if it is present. Similarly, rapid cooling produces softer gels, since the amylose has less time to become oriented and form significant junction zones.

4. Comparison of Food Processing Equipment

The following brief overview of starch issues pertaining to the use of processing is only intended to illustrate the stresses that starch is subjected to during processing. Equipment manufacturers can provide detailed information about heat exchanger design and performance.[41] Since much of the relevant shear in food processing arises during cooking and cooling, a brief discussion of heat exchanger design and application is warranted. The type of cooker used is fundamental to starch choice and can have an affect on final product texture. Cookers employing steam can be classified as direct and indirect. In direct steam heaters, the steam comes into contact with the food stream, while in indirect steam heaters the steam heats a surface that heats the food stream. Direct steam heaters include steam injection and steam infusion equipment. Indirect steam heaters include scraped surface, tubular and plate exchangers.

Steam injection may be done batch-wise in a kettle or continuously as a fluid passes through a mixing chamber in a pipe. The latter is called jet cooking, and at high steam flow rates and chamber temperatures it introduces considerable turbulence and a large pressure drop – both of these generating high shear stresses. For processing of high-amylose starch, steam injection at 165°C with accompanying high turbulence is required, especially at the high solids of candy formulations. Jet cooking is more commonly used where solubilization of starch is needed, as in the confectionery industry, but it is also common in the European dairy and baby food industries. Steam injection in a kettle is comparatively gentle and is used in making cream-style corn and processed cheese. Steam infusion systems are used for ultra-high temperature treatment of gelatin and dairy products. In steam-infusion systems, the food stream flows through a chamber of steam so the shear stresses achieved are much less than those in jet cooking.

Indirect heat exchangers include plate heat exchangers, scraped-surface heat exchangers and tubular heat exchangers. Indirect heat exchangers can also be used for cooling. Plate heat exchangers are indirect steam heaters involving a pack of parallel steel plates through which a food stream and a heat exchange medium flow in adjacent streams. The heat exchange medium could be steam itself, but water heated by steam is often preferred for greater control. Plate heat exchangers provide the highest levels of heat exchange area per unit floor space and can process at high product throughput. They are favored unless high product viscosity or particulates make them inappropriate. Tubular heat exchangers take product through stainless steel tubes surrounded by steam or other heat exchange media. High product flow velocities promote rapid heat transfer and offer the greatest efficiency per unit heat exchange surface area. Tubular exchangers can handle higher viscosities than plates and can heat and cool fluids more quickly. Starches suitable for plate heat exchangers are suitable for use in tubular cookers. Scraped-surface heat exchangers can process viscous fluids like puddings and sauces. They are basically a jacketed cylinder with a rotating shaft in the center. As the fluid passes through the cylinder, the shaft drives scrapers that mix the product and continually renew the heat exchange surface. Scraped surface units can process fluids containing particulates up to 15 mm in size. Larger particles require the use of ohmic heaters.

Particle size reduction equipment is used for homogenizing emulsions and preventing grittiness in various foods. Piston homogenizers and colloid mills can introduce severe shear, depending on operating conditions. Typically, a homogenizer running at pressures greater than 70 bar or a colloid mill with a gap clearance of 1.5 mm or less is destructive to starch. One approach to avoiding starch damage during processing is to homogenize prior to the high temperature hold when the starch granules are less swollen and less susceptible to mechanical damage.

5. Impact of Processing and Storage

Laboratory cooking procedures seldom simulate production residence times, temperatures and shear stresses. A promising laboratory formulation may perform less well when scaled-up, due to dissimilar processing and adverse, unanticipated changes during abusive storage. For that reason, special care is required in scale-up to maintain product viscosity and texture. Final starch selection is often based on performance in production equipment.

6. Changes that Occur During Cooling, Storage and Distribution

From the viewpoint of the consumer, storage and distribution stresses are part of the packaged food process. Convenience foods continue to change after packaging and processed starch within the food also changes. These changes can be minimized for better control of product quality.

During distribution and storage, starch chains within and between molecules can associate or retrograde. Atwell et al.[29] defined retrogradation as a process which occurs when the molecules comprising gelatinized starch begin to reassociate in an ordered structure. In its initial phases, two or more starch chains may form a simple juncture point, which then may develop into more extensively ordered regions. This phase separation also causes syneresis and appearance changes like graininess and opacity and texture changes, including gelling and loss of smoothness. Ultimately, under favorable conditions, a crystalline order appears.

The nature and extent of changes in appearance and texture depend on the type of starch, the type of modification and the degree of substitution.[42] Formulation factors such as the concentration of the starch and other ingredients and processing history, including the rate of freezing, are also important.[43] An understanding of these changes is helpful in achieving maximum product stability. As a cooked paste cools, there is less motion separating the adjacent polymers and increasing hydrogen bonding holds the polymers together. The initial development of firmness during gelation of a starch paste is caused by the relatively rapid formation of an amylose gel, the amylose apparently having fully retrograded by the time the paste has cooled to room temperature. The slow increase in firmness thereafter is due to crystallization of amylopectin within the swollen starch granules.[44] Although it is much slower, amylopectin retrogradation affects paste properties more profoundly.[45]

In starches containing only amylopectin, retrogradation will cause synersis, graininess and loss of clarity, but minimal gelling. Syneresis occurs when water is expelled from within the swollen starch granules or by other ingredients. The ability of starch to entrain water and provide a sink for water expressed by changes in the food can be determined by measuring the swelling power.[46] In most foods, the starch concentration exceeds the critical concentration relative to the water available. Obviously, where syneresis occurs, the swelling power has decreased below the critical concentration where close packing occurs, either because water has been released elsewhere or because the starch gel has contracted.[47] All else being equal, starches with reduced amylose content are generally more freeze–thaw stable. The relative stability of amylopectin sols accounts for its effectiveness as a thickener where solution stability and freedom from loss of water-holding properties are desired.

Retrogradation occurs at room temperature, in the refrigerator and in frozen storage, and is accelerated by freeze–thaw cycling.[48] These changes have been monitored via water separation,[42] differential scanning calorimetry[49] and small strain rheology.[43] Defects resulting from cold storage are only partially reversible by heating.[43]

7. Recommended Processing

Maximum viscosity in a starch suspension is achieved with uniform swelling and minimum granular disruption. This requires good dispersion for hydration and continued agitation for efficient heat exchange. High tip speeds or pressure drops after the starch is swollen should be avoided. In open kettles, it is important to replace high speed propeller or turbine agitators with slowly turning scrapers that scrape against the kettle wall or cooking surface.[50] Inadequate stirring can prevent thorough cooking, as thickened starch will form an insulating layer on the cooking surface if scraping is not sufficient. Large kettles should have baffles that generate currents, to work the center slurry outward to the cooking surface, since a thickened starch slurry has poor heat transfer properties.[51] Because of differences in residence times, shear rates and heat transfer rates, it is not possible to say what temperature is best for a given starch, modified or otherwise. Since unswollen granules are shear-resistant, homogenization before cooking is preferable for maximum starch viscosity. Where homogenization follows pasteurization, pressures below 1000 psi are preferred. Cooling after cooking ensures that overcooking will not occur during a protracted cool-down. Freezing is less destructive to texture and appearance if it is done quickly.[52]

The effect of acid and dissolved solids is discussed under ingredient interaction. Since the combined effects of heat, acid and shear are additive, delay in the addition of acid until after cooking will allow a less-crosslinked starch to suffice. Similarly, delay in the addition of sugar will allow a more-crosslinked starch to be adequately swollen or to be cooked out.

IV. Modified Food Starches (See Also Chapter 17)

Unmodified starches are used in many foods. Native corn, sorghum and wheat starches are useful as thickeners and gelling agents where their retrogradation and ensuing opacity and texture changes are not limiting. They are used as dusting powders and as flow agents. Native starches are also used in dry mixes for foods eaten shortly after preparation, such as gravies and puddings, or in starch blends for products where some gelling is desired, such as salad dressings. Native waxy maize and tapioca starches offer improved stability over corn, sorghum and wheat starches, and are used in some packaged foods such as salad dressings and buttermilk, where a slightly long texture is acceptable. Native waxy maize and potato starches are also used to provide fill-viscosity for metering of particulates during filling of cans prior to retorting. Native high-amylose starches are used in jelly gum candies and batters. However, in general, foods made with native starches have inadequate process tolerance for commercial manufacturing and inadequate shelf stability for retail distribution. Typical processing will lead to a loss in viscosity and a short texture. Prolonged holding will lead to deterioration, especially with repeated freezing and thawing. Therefore, the specialty starch industry has developed techniques to enhance specific properties of starch and improve its usefulness in food. For instance, where granular integrity is important, crosslinking will prevent disruption during processing and ensure uniformly high product viscosity.

1. Why Starch is Modified

Modification allows starch to maintain desirable appearance and texture despite stresses during food processing and distribution, and to expand its range of utility in foods. Modified starches will retain fresh appearance, e.g. clarity and smoothness, and eating properties longer. Starches are also modified to change processability, e.g. to reduce gelatinization temperature or reduce hot paste viscosity. Modification will also improve performance in a wide variety of functions including film formation, mouthfeel enhancement and emulsification (see Sections 20.4.2–20.4.4). The following is an overview of the improved utility of chemically and physically modified starches. Table 20.3 summarizes the effects of chemical and physical modification.

2. Derivatizations

Crosslinking

The acid, heat and shear stability of starches is improved by crosslinking. Most food processes involve a combination of these beyond the tolerance of unmodified starches. Pastes of crosslinked starches are shorter, more heavy-bodied and less likely to break down with excessive cooking, acid or shear, resulting in improved food product quality. Because unmodified granules are only held together by hydrogen bonds, the granules are weakened by high temperatures, shear and acid. Introduction of bifunctional

Table 20.3 Functional improvements via modification

Chemical modification	Effect
Crosslinking	Retains viscosity despite heat, acid and shear[12,15,53,54]
Monosubstitution	Improved stability to cold storage[51,59]
	Reduced gelation[61]
	Reduced gel temperature for low moisture foods[59]
	Improved emulsification properties[25,66,67]
Acid-thinning	Reduced hot-paste viscosity, improved gelling[59]
Oxidation	Improved batter adhesion[61]
Dextrinization	Emulsification[59]
	Increased stability
	Reduced viscosity
	Improved film forming
Redrying	Improved flow and moisture binding[80]
Oil addition	Holds shape for molding
Pregelatinization	Dissolves and thickens without heat[82,83,90,92,95]
Annealing	Increased gel temperature[97,98]
	Increased enthalpy
	Narrowed gel temperature range
Heat-moisture	Increased gel temperature[101]
	Reduced swelling[101–103]
	Increased stability, broadened gel temperature range[105]
	Reduced enthalpy[104]
Dry heat	Increased viscosity[109]
	Increased process tolerance[110]
	Increased gel temperature range
Shear	Higher gel strength[107]
	Improved stability[111]
	Cold gelling[115]
Radiation	Increased gel temperature[119]
	Improved stability
	Reduced viscosity
Solvents	Pregelatinization, thin-thick properties[24,84,88,89]

reagents reinforce the granule by linking hydroxyl groups on adjacent polymers with covalent bonds,[15,53] thus retarding the rate of granule swelling and reducing the tendency to rupture. Reactions with as little as 0.0005% added phosphoryl chloride (based on starch weight, DS approximately 5×10^{-6}) can affect starch rheology, but starch for food is generally crosslinked with 0.001 to 0.002 degree of substitution (DS), providing a crosslink for every 1000 to 2000 D-glucosyl units.[54] With more frequent crosslinks, the granule becomes more tolerant to chemical and physical abuse and will continue to thicken with additional heat, acid or shear.[12] Low levels of crosslinking have minimal affect on gelatinization temperature, but significant affects on pasting properties. High levels of crosslinking increase the starch's gelatinization temperature and, with extreme crosslinking, starches can be prepared that do not gelatinize when they are boiled in water or sterilized in an autoclave. Cooking requirements for optimum viscosity also increase with increased crosslinking.

Effect of crosslinking on granule swelling and pasting: when similarly cooked, crosslinked starch granules swell less than native granules, but with the correct level of crosslinking, the desired viscosity and texture can be achieved for each process condition.[53] Crosslinked starches generate less solubles during cooking than the native starch from which they are made, providing a shorter texture. Because it reduces rupture and solubilization, increased crosslinking reduces the gel strength of a cooked amylose-containing starch.

Crosslinking typically improves freeze–thaw performance, in part because native starches are easily overcooked and the effects of retrogradation are amplified. But the impact of crosslinking on stability is unclear. Wu and Seib[55] found that crosslinking of unstabilized starch increased syneresis of waxy maize starch pastes after freeze–thaw cycling. It is noteworthy that they cooked the samples very carefully to achieve equal swelling. In a separate study, crosslinked and stabilized (Section 20.4.2b) rice starch had a lower retrogradation enthalpy after 5 days at 4°C than stabilized rice starch (see Section 20.4.2d).[56] Crosslinking delayed retrogradation and therefore the time when the starch gel became opaque.

Crosslinking reagents: food starches are crosslinked with phosphoryl chloride (phosphorus oxychloride),[15] sodium trimetaphosphate (STMP)[57] and adipic acetic mixed anhydride.[17] Food and Drug Administration (FDA) regulations stipulate that, by weight of starch, no more than 0.1% phosphoryl chloride, 1% STMP or 0.12% adipic acetic mixed anhydride should be used.[58] The first two reagents form phosphate diester linkages, which are more acid resistant than the adipate diester linkage. Phosphoryl chloride is favored in starches subjected to severe heat, acid and shear. Formerly, epichlorohydrin was used in the US to form diester linkages, which are very acid resistant; it is still used in some countries.

Monosubstitutions (stabilizations)
Monosubstitution or stabilization is the addition of blocking substituents to the hydroxyl groups of starch. These substituents sterically interrupt the association of starch polymers in a starch paste or food and thus deter retrogradatation. There are four major types of monosubstitution: starch acetates; starch hydroxypropyl ethers; starch monophosphate esters; and starch sodium octenylsuccinates.[58] The first three interfere with retrogradation and improve cold temperature stability. They are used most often in combination with crosslinking, but may be used alone or with converted (see Section 20.4.3) starches. Substitution with octenylsuccinate provides a product with emulsifying capability. It is more commonly used with hydrolyzates and is discussed separately in Section 20.4.2c. The following describes generically the utility of the three other monosubstituents.

Monofunctional substitution improves freeze–thaw and water-holding properties, lowers the swelling (pasting) temperature, increases paste clarity and reduces gel formation.[59] In cold storage, even waxy maize and tapioca starch pastes become cloudy and chunky and synerese as adjacent polymers associate. The addition of blocking substituents on hydroxyl groups interrupts this association sterically.[59] The net effect is a reduction in opacity, syneresis, gelling and graininess, despite storage at reduced temperatures. After freezing, the water produced during melting can rehydrate the starch chains, and the starch gel or food retains many of its original properties.

A further effect of monosubstitution is to counteract the forces holding the granule intact. With substitution, the granule tends to swell and the polymers dissolve more easily during gelatinization, giving a starch paste with a higher viscosity, but with poorer resistance to granular disruption. Monosubstitution lowers gelatinization temperatures,[59] which is important in systems with high solids or where avoidance of high temperatures is preferred. With a high level of substitution (e.g. a DS of 0.7 monophosphorylation), starches will gelatinize when added to room temperature water. Stabilization reduces the tendency of corn starch pastes to gel and become opaque.[60] The monofunctional substituents most often used for food starch are acetate esters and hydroxypropyl ethers.

Acetylation: acetylation is the addition of acetyl mono-ester groups to the starch polymers. Acetylation is primarily used to reduce syneresis and texture changes in waxy maize, tapioca and potato starches used in refrigerated and frozen foods. It is also used to reduce gelation in normal corn starch pastes and to increase the ease of cooking in high-amylose corn starch.[61] The impact of acetylation on the clarity of normal corn starch pastes is greater than it is on tuber starches or waxy maize starch.[54] Acetic anhydride[18] or vinyl acetate[62] are used as derivatizing reagents to produce starch acetates. FDA regulations stipulate that no more acetic anhydride or vinyl acetate can be used than that required to achieve 2.5% acetate substitution.[58] This corresponds to a degree of substitution (DS) of 0.1. Corn starch paste gelling is reduced at 0.0275 DS and almost eliminated at 0.05 DS; commercial products typically have DS values of much less than 0.05.[54]

Hydroxypropylation: hydroxypropylation is the formation of hydroxypropyl ether groups from hydroxyl groups on the starch polymers. Hydroxypropylation has several advantages over acetylation.[21] Hydroxypropyl substitution is more stable to prolonged high temperatures as occurs in retorting, especially above pH 6.[63] Since a higher level of substitution is allowable than with acetylation (e.g. DS 0.2 versus DS 0.1), greater freeze–thaw stability can be achieved. Moreover, hydroxypropyl groups appear to create more steric hindrance. Wu and Seib[55] found that a measured 0.01 DS hydroxypropylation provided greater freeze–thaw stability than an equal amount of acetylation on waxy barley distarch phosphate. Hydroxypropylation has been more popular with the dairy industry than acetylation, apparently because hydrolytic liberation of acetyl groups causes protein destabilization and graininess.

Monophosphorylation: pastes made from starch monophosphates also have greater clarity, viscosity and stability than unmodified starches,[64] but are sensitive to salts, especially polyvalent cations.[65] Variability in residual ash can lead to variability in the viscosity of monophosphorylated starches. Monophosphate substitution also lowers the gelatinization temperature; at 0.07 DS, a value much greater than is found in food starches, the gelatinization temperature is below room temperature. Native potato starch contains 0.07 to 0.09% bound phosphorus and wheat starch contains 0.055% phosphorus, primarily as phosphoglycerides in the latter case. The FDA allows up to 0.4% phosphate as phosphorus.[58] Monophosphates were used commercially in the US until about 1970.

Octenylsuccinylation

Starches esterified with octenylsuccinic anhydride are also monosubstituted, but are primarily produced not for stabilization, but because they have hydrophobic substituents, have interfacial properties and can be used in emulsification and encapsulation,[22] i.e. in many of the same applications in which gum arabic is used. Sodium octenylsuccinate derivatives of starches with a broad range of viscosities are used to stabilize oil-in-water food emulsions, such as beverage concentrates containing flavor and clouding oils.[66] Flavor oils emulsified in this way and spray dried are protected against oxidation in storage.[67] Similarly, low-viscosity substituted hydrolyzates are used to encapsulate flavor and clouding oils and vitamins.[68] For reduced viscosity and improved performance, hydrolyzates (from either enzyme- or acid-catalyzed hydrolysis) and dextrinized products are employed.

The allowed level of treatment is 3% octenylsuccinic anhydride.[58] This corresponds to a DS of 0.02 or 1 substituent for every 50 D-glucosyl units.[69] Depending on pH, starch octenylsuccinate half-esters may be present as the free acid or a salt. Octenylsuccinylation can also be used to render granular starch free-flowing and water repellent.[70] Such a starch remains free-flowing even at 16% moisture, where corn starch will clump, making the modified starch more effective as a dusting compound.

3. Conversions

Conversion (partial cleavage; Section 20.4.3) of the glycosidic linkages reduces hot viscosity, allowing the starch to be cooked and used at higher concentrations than would otherwise be possible. The use of increased starch solids allows the starch to provide greater gelling, adhesion and oil encapsulation. Hydrolysis also affects gelatinization temperature (either increased or decreased), cold water solubility (increased), rate of gelation (either increased or decreased), gel strength (either increased or decreased) and adhesive properties (improved). Depending on the method used for conversion, stability may also be improved through molecular rearrangement and formation of carboxyl substituents. Hydrolyzed or converted starches include thin-boiling or fluidity starches, oxidized starches, white dextrins, British gums and yellow dextrins. Depending on the process and desired product, conversions are performed both on raw granular and cooked, dispersed starch.

Fluidity (thin-boiling) Starches

Acid modification is the oldest modification, having first been described by Lintner in 1874[7] and Naegeli in 1886,[6] and applied around the turn of the century.[8,9] Thin-boiling (water fluidity) starches are prepared by mild acid treatment of starch slurries at non-gelatinizing temperatures. Acid treatment reduces the molecular weight of the starch, while leaving the crystalline structure of the granule intact. The gelatinization temperature of the starch may increase,[71] but the starch granule doesn't swell; rather, it dissolves on heating in water. The term fluidity came into use because the low hot viscosity of the cooks of these products allowed greater starch solids to be used, and because the gel strength and other desired properties were less affected by acid-catalyzed hydrolysis than was the hot viscosity.[72] Glycosidic linkages of the amylopectin are cleaved preferentially, lowering the paste viscosity.[10,73] On cooling, pastes of fluidity corn, wheat and sorghum starches form rigid, opaque gels. Fluidity waxy maize starch pastes cool to a stable sol useful in coatings and adhesives; lightly converted tapioca starch has similar properties. Heavily converted tapioca and acid converted potato starches cooked at high concentrations form opaque gels on cooling.[54]

Thin-boiling starches are usually made from corn starch. Acid treatment is conducted on intact granules either in a slurry or under dry or semi-dry conditions. The most common use of thin-boiling starches is in jelly gum candies where a hot fluid material is deposited into molds and allowed to gel to a specific texture and shape.[74] Fluidity starches form strong gels, even after cooking with high sugar.

Dextrins

Dextrins (pyrodextrins) are made by heating dry starch with or without acid. Since it is a dry process, recovery of water-soluble materials is simpler than with aqueous fluidity and oxidized starches. Depending on reaction conditions, greater or lesser amounts of three reactions will occur: (a) hydrolysis; (b) transglycosidation; and (c) repolymerization. According to which predominates, the product is a white dextrin, a yellow dextrin or a British gum. Like other converted materials, these products offer a way to use higher solids to increase performance.

Dextrins differ from fluidity starches in that their cold-water-solubility increases, while their gel strength and their mean molecular weight are reduced. Dextrins are formed through acid modification of a dry powder. Dextrins are more completely hydrolyzed products than fluidity starches. Hydrochloric acid is favored, but sulfuric and orthophosphoric acids are also used. Dextrins are used where dispersions or sols having high solids are desired. The choice of a dextrin is a function of application requirements (concentration of sol, color, film strength, ability to be moistened, tack, etc.). A typical application is the pan coating of confections, where the clear dextrin film prevents separation of the sugar shell from the base center material. Dextrins are also used to provide gloss to bakery goods as fat replacers. Highly soluble British gums and yellow dextrins are used as carriers for active food flavorings, spices and colorants, where rapid dissolution in water is desired. Yellow corn dextrin is also used in the encapsulation of water-insoluble flavorings and oils, replacing gum arabic. A white dextrin is marketed as a fat replacer.[75]

4. Oxidation

Starches treated with oxidants fall into two broad classes: oxidized and bleached. Oxidized starches are treated with hypochlorite. Oxidation is typically carried out on intact granules and causes the granule to dissolve, rather than swell and thicken. Oxidation by aqueous hypochlorite is a random reaction converting hydroxyl groups to aldehydo, keto and carboxyl groups. Some depolymerization also occurs. Higher pH favors formation of carboxylate groups over aldehydes and ketones. Introduction of carboxylate groups provides both steric hindrance and electrostatic repulsion. The reaction takes place throughout the granule and can introduce up to 1.1% of carboxyl groups.[54] Thus, pastes of oxidized starches have a reduced tendency to gel as compared to those of thin-boiling starches of comparable viscosity. Other oxidants, e.g. chlorine, hydrogen peroxide and potassium permanganate, are less commonly used.

Oxidation with chlorine or sodium hypochlorite reduces the tendency of amylose to associate or retrograde. Oxidized starches are used where intermediate viscosity and soft gels are desired, and where the instability of acid-converted starches is unacceptable.[76] Oxidized starches are reported to give batters improved adhesion to meat products and are widely used in breaded foods.[76] Fuller[77] reported that candies made with hypochlorite-oxidized starches gel and dry faster and have increased clarity, longer shelf life and better taste than those made with acid-thinned counterparts.

For bleaching starch, low levels of hydrogen peroxide, potassium permanganate, sodium hypochlorite or other oxidants are used to remove color from naturally occurring pigments. Bleaching is done to improve whiteness and/or destroy microorganisms. Reagent levels of about 0.5% are usually employed, inducing some loss of starch viscosity due to hydrolysis.[78]

5. Physical Modifications

A variety of physical treatments are used to alter food starches, including heat with or without moisture, radiation and mechanical processing. These treatments provide improved processability or improved texture and stability.[79] Moreover, physical modification (e.g. pregelatinization) may be used, as well as chemical modification for maximum overall performance. The following sections review the types of physical modification used or described for use in foods. A discussion of control of flow will be followed by a review of pregelatinization and physical modifications intended to otherwise alter starch performance.

Altered Flow Properties

The flow of a starch powder can be enhanced by drying or reduced with moisture or other liquids. This range in flow properties, along with starch's edibility, make it useful as a diluent, bulking agent or moisture-absorbing agent in various food powders, and a medium for casting gels. Redrying starch to a level below its equilibrium moisture content allows it to draw moisture from systems having a lower affinity for moisture. Baking powder is an example of where starch is used, at 15% to 40%, to protect the active ingredients from becoming moist, dissolving and reacting.[80] Starch flow is improved via the addition of agents (such as tricalcium phosphate) to coat the starch granules.

Confectioners use starch in molding beds to cast gum candies of specific shapes. The tendency of powdered starch to receive such an impression is improved through the addition of tenths of a percent of mineral oil. This liquid promotes bridging between granules and reduces powder flow even after the starch is dried to 6% moisture. The hygroscopic molding starch also aids in drying the cast candy fluid. Pregelatinization is another form of physical modification.

Pregelatinization

Starch can be pregelatinized (precooked) so that it swells and thickens on contact with water. Pregelatinized starches are cooked then dried in one of several ways, including drum drying, spray drying, cooking in aqueous ethanol and extrusion cooking. These so-called instant starches are used for convenient, in-home preparation (e.g. instant puddings) and in food production to reduce energy costs, where limited moisture prevents cooking, and to avoid flavor loss associated with high temperatures. They are also used in processes that traditionally don't involve heat, such as those used to produce pourable salad dressings.

Drum drying: traditionally starch has been pregelatinized by pouring a starch slurry onto a hot drum and scraping off the cooked sheet with a knife. Powell[81] reviewed drum drying technology. The rate of rehydration and texture of the finished pastes can be controlled through chemical modification, drum operation and grinding of the dried sheets. The feed starch can be a chemically modified product to further extend the range of finished properties. Solids concentration, time and temperature can be varied to produce different products. The size of the particles can be varied. Drum-dried products have slightly less viscosity than counterpart cook-up starches (as non-pregelatinized products are called) and produce less glossy and less smooth pastes when dispersed in water. Drum drying typically destroys starch granule integrity, resulting in loss of viscosity and shortness of texture. Newer technologies have addressed this shortcoming. Still, drum drying is favored for economy and for its ability to make large particles used in texturizing starches, which are products that are slow to hydrate. Texturizing starches are used in cookie fillings and toppings to control boil-out when subjected to high oven temperatures, as well as to provide a pulpy fruit texture.[82] They are typically made from highly crosslinked starches that are subsequently drum dried.

Extrusion: extrusion gelatinizes (cooks) starch[83] but damages granules more than drum drying, thus reducing viscosity. Extrusion may even fragment amylopectin molecules.[84] Colonna et al.[85] compared drum-dried and extruded wheat starch. Both processes created a continuous phase of melted starch. In comparison to drum-dried starch, extruded starch had lower average molecular weights in the amylose and amylopectin fractions. Davidson et al.[86] showed that this molecular fragmentation was due both to temperature and shear stresses. Doublier et al.[87] compared the rheology of drum-dried and extruded starches, and found that extruded starches had lower cold-water-swelling power and greater solubility. Starches extruded at lower temperatures and at higher moisture contents were more like the drum-dried starches in these properties. Extrusion has been recommended as a means of reducing molecular weight and increasing solubility for improved emulsification by octenylsuccinylated starches.[88]

Use of pregelatinized starches: pregelatinized (instant) starch products are used in low-fat salad dressings, high-solids fillings, bakery fillings and dry mixes. Problems may occur in cold processes if flour, fruit, spices or other ingredients contain amylases. In a hot process, thermal inactivation of the enzymes typically precedes swelling of the starch and their ensuing vulnerability to enzymes. Instant starches should first be blended with other dry ingredients before addition to the aqueous phase, to slow hydration and minimize lumping. For formulations with little sugar or other dry ingredients, starches are agglomerated to facilitate dispersion. A light coating of oil will aid dispersion.

Agglomeration: coarse ground starches hydrate slowly and disperse easily, especially in hot water. (Since hot water wets starch more quickly, its use exacerbates lumping and generally requires coarser or agglomerated products.) Coarse ground starches may, however, produce a grainy or pulpy texture, which is sometimes but not always desirable. Starches with a finer particle size produce smoother solutions, but hydrate rapidly and are prone to lumping. Accordingly, where dispersion is a problem because agitation is inadequate, less finely ground drum-dried starch or agglomerated spray-dried starch is used. Agglomeration is typically done by adding water and then drying starch on a fluidized bed.[89]

Cold-water-swelling Starches

Cold-water-swelling (CWS) is a term used for granular starches that swell upon wetting.[90] Cold-water-swelling starches, unlike drum-dried starches that are effectively flakes of cooked, congealed paste, form pastes more like those of cook-up starches. They typically retain a high percentage of granular integrity with minimum solubilization of amylose or amylopectin. CWS starches are made by spray drying[23] or by heating in aqueous ethanol[24] or other solvents. Hence, they produce pastes that are generally shiny and have a short, heavy texture. Unmodified CWS starches have greater process tolerance than unmodified drum-dried starches of the same source and comparable viscosity. Crosslinked and stabilized CWS starches can have the same acid, shear and cold temperature stability as cook-up starches. Because they consist of very small particles, CWS starches may need to be agglomerated for adequate dispersibility in certain food processes.

Spray drying: spray-dried starches are cooked in water and then atomized into droplets which yield moisture as they fall through dry air. The challenge with spray drying starch is to delay starch swelling to allow a high rate of throughput. Pitchon et al.[23] described spray drying with a nozzle which first atomized a slurry via pressure and then cooked the starch granules with steam before releasing them into a drying tower. Rubens[91] described spray drying with a simultaneous cooking and atomization step after mixing steam with the starch slurry in a two-fluid nozzle. These methods provide starch granules that reswell in water to form a product with the high viscosity and short texture of a cooked paste.

Hot aqueous ethanol treatment: CWS starches can also be formed by heating starch in hot aqueous ethanol[24,92,93] or other solvents,[94] including at atmospheric pressures.[95] In this process starch crystallites are melted, but the water is insufficient to paste the granules. The solvent is then removed. The dried starch is stable.

Other Physical Modifications

Heat treatments at various moisture levels have been shown to affect starch properties, including paste stability to heat, shear and storage. These treatments can be classified as annealing with moist heat and dry heat, and are best characterized by their impact on moisture absorption, differential scanning calorimetry (DSC), x-ray diffraction patterns and pasting properties. There is some similarity in the affects of the three on starch properties; only relatively recently have the differences been recognized.

Annealing: annealing, i.e. heating in excess water at a temperature above the glass transition temperature (T_g) but below the gelatinization temperature[96] increases the DSC onset temperature, narrows the gelatinization temperature range and may increase gelatinization enthalpy.[97,98] Typically, starch is heated in four times its weight of water at 40°–55°C for 12 hours or longer. Annealing does not change the x-ray diffraction pattern[99,100] or the sorption isotherms.[99] Annealing also affects pasting properties, increasing pasting onset temperature, lowering peak viscosity and increasing viscosity on cooling (potato starch). These changes were explained as reduced swelling power and enhanced shear stability.[99] Possible mechanisms for annealing include crystallite growth and perfection, and altered couplings between crystallites and the amorphous matrix. Starches from different botanical sources differ in the impact of annealing on their pasting properties, possibly due to differences in shear sensitivity. The temperature profile and shear rate used for generating a pasting curve has been shown to influence observed differences between botanical sources.[98]

Moist heat: moist heat has come to mean heating starch at 25% moisture at ~100°C for several hours. During the 1940s, Sair[101] studied the effect of heat–moisture treatment on the physical properties of corn and potato starches. He found that heating between 95°C and 100°C at 100% relative humidity caused an increase in gelatinization temperature from 61°C to 80°C for potato starch and from 67°C to 75°C for corn starch. Moist heat-treated starches were observed to have reduced swelling power and reduced solubility, and to produce gels with less strength and greater opacity. These observations suggest increased intermolecular associations. Leach et al.[30] showed that heat–moisture treatment drastically reduced loss of amylose on gelatinization. Osman[51] noted that moist heat effects might occur in food processes.

Lorenz and Kulp[102] saw that cereal and tuber starches reacted differently to moist heat treatment. While barley, triticale, arrowroot and cassava (tapioca) starches were all reduced in swelling power, the solubility of cereal starches increased, while that of tuber starches decreased. By x-ray diffraction, cereal starches show reduced crystallinity while tuber starches changed from the B-type to the A-type pattern. They found that the performance of potato starch in breads and cakes improved by heat treatment, although it failed to perform as well as untreated wheat starch. Heat treatment reduced thickening properties as tested with pie fillings.[103] Using DSC, Donovan et al.[104] also found that heat–moisture treated wheat and potato starches gelatinized over temperature ranges that were broader and higher than those of the native starches. Moreover, the enthalpy was reduced and the water-binding capacity, enzyme susceptibility and paste stability were all increased. Abraham[105] also observed increased paste stability and freeze–thaw stability in cassava (tapioca) starch after heating.

Moist heat has practical implications. Smalligan et al.[106] have a patent on moist heat treatment of starch for improving storage stability in canned baby food. They claim heating at 95–100°C at 25% moisture for 16 hours produced moist, heated starch with an increased gelatinization temperature, reduced peak viscosity, and reduced breakdown, similar to a crosslinked starch. Similar effects may be responsible for the impact of parboiling on grain properties. Schierbaum and Kettlitz[107] suggested that moist heat might be used for upgrading low quality wheat starch and could be achieved during belt drying.

Dry heat: dry heat is claimed to improve texture and provide a more heat- and shear-tolerant starch. Krake[108] claimed drying starch under pressure with continuous rotation from 14–22% moisture to 10–12% moisture at 60–75°C then heating to 90–128°C for 0.75 to 1.5 hours provides a starch which dissolves easily and has high viscosity and a creamy consistency for use in instant soups and sauces and quick-cooking products. Dry heating corn starch of original moisture content of 11.4% above 190°C increased the enthalpy of gelatinization while decreasing peak viscosity.[109]

Mechanical energy: besides extrusion, various types of mechanical energy have been shown to pregelatinize starch, increase gel strength and improve the stability of starch pastes. These effects are achieved through various degrees of disrupting starch granules and degrading starch polymers. Carter et al.[110] claimed the use of cooked, sheared, waxy maize starch as a protectant for native amylose-containing starch thickeners. They patented the use of granular amylose as a dispersed phase and sheared waxy maize starch, maltodextrin or a gum as the continuous phase. This combination yielded improved storage stability and retention of smoothness relative to native thickeners for gravies, soups, sauces, dressings and spreads.

Ajinomoto[111] has a patent on using 20-MHz ultrasound in alkaline conditions to achieve higher gel strength and better low-temperature paste stability. Such starches show a synergistic effect with meat protein in gel formation. Redding[112] reported that an abrupt pressure change through a liquid medium alters the melting point, decreases the solubility rate and increases clarity, viscosity, thermal stability and resistance to shear. Kudta and Tomasik[113] formed dextrins via mechanical damage through ball milling. Morrison et al.[114] showed that cold gelling behavior can also be achieved via ball milling. Pressure alone has been shown to enhance pasting and sol properties of starch. Meuser et al.[115] used a pressure treatment to achieve cold-water-swelling starches that form stable pastes with reduced retrogradation. These pastes also display improved shear stability and better binding of aromas.

Solvents: in addition to making CWS starches, hot aqueous ethanol has been used for preparing so-called thin–thick starches. Tuschoff et al.[116] claimed that flour heated to 140°C in 80–95% ethanol provides thin–thick properties useful in retorting, especially for baby foods. Schierbaum and Kettlitz[107] found that 3 hours at 80°C in 75% 1-propanol reduced the swelling power of starch. Hot solvent extraction reduced the enthalpy and changed the pasting curve, perhaps because it is similar to heat–moisture treatment.[108] Solvent extractions typically remove surface lipids. While removal of lipids may account for some of the differences observed, Donovan et al.[104] reported that defatting had no effect on temperature or enthalpy unless coupled with heat treatment. Radosta et al.[117] reported that lipid extraction of rye starch did not affect swelling power. It did, however, increase amylose leaching and overall solubility. Room temperature extraction of rye starch changed neither pasting properties nor enthalpy of gelatinization.

Radiation: radiation, depending on wavelength and dose, has been shown to have a variety of effects, including increased gelatinization temperature, improved stability and reduced viscosity, swelling power and enzyme susceptibility. Radiation has also been used to promote chemical modification. Marquette et al.[118] have a patent for microwaving starch to improve rheological stability and hot water dispersibility. They report treating 10% to 40% moisture starch with 2450 MHz for 20 to 300 seconds; this treatment also provides higher swelling temperature, lower peak viscosity and increased enzyme susceptibility. Muzimbaranda and Tomasik[119] used microwaves for dextrinization of cassava, maize and potato starches. They reported that starch microwaved in the presence of titanium chloride produced gels of low viscosity, while starch irradiated in presence of acetylene or formaldehyde had increased viscosity.

6. Native Starch Thickeners

Since the 1980s, 'natural' foods have become more popular both in North America and in Europe, as consumers have become increasingly mindful of diet and its role in their vitality and freedom from disease. At the same time, changes in European food labeling laws allowed certain texturizing agents to be viewed as 'clean label' or 'healthy' ingredients as opposed to 'chemical' additives. In 1988, European Economic Community Council Directive 89/107/EEC (as amended by Directive 94/34/EC) stipulated that, along with many other ingredients, chemically modified starches had to be labeled with an 'E number.'[120] For example, hydroxypropyl distarch phosphate would be labeled as E1442, a name intended to help regulate the additive and inform consumers. Meanwhile, unmodified (or native) starches could simply be labeled with their class name: starch. E numbers have since been termed INS numbers, referring to the International Numbering System outlined in the Codex Alimentarius General Standard for the Labeling of Packaged Foods.[120] However, consumers and food manufacturers alike remained committed to convenience foods containing retorted or precooked and frozen or chilled sauces. Thus, there arose a demand for thickening ingredients that did not require 'chemical' or 'additive' labels. The 'native' starches, nonetheless, needed to be process-tolerant and capable of providing the desired short texture, even after freeze–thaw and other distribution stresses.

The approaches that were offered are illustrated in the patent literature of the time. Chiu et al.[121] claim that starch heated after being adjusted to an alkaline pH value and then dried would behave as though it was crosslinked. In a patent awarded in 1999, Mahr and Trueck[122] claim the use of amylose-containing starches complexed with lipid to enhance tolerance to thermal processing. Kettlitz and Coppin[123] treated starch with activated chlorine to create a native product that was temperature-tolerant like a conventionally crosslinked starch.

V. Starch Sources (See Also Chapters 9–16)

Corn, tapioca, potato and wheat starches are the most commonly used starches in the US and Europe. In the US, 95% of the starch is made from corn, an amount representing 3.4% of the total corn crop, excluding that wet-milled to make sweeteners (see Chapter 22) and alcohol[11,124] (see Chapter 2). In Europe, about 60% of the starch produced is made from corn and about 20% each from potato and wheat.[12] In select regions, rice, sorghum, arrowroot, sago and other starches are also used. The relative utility of these starches in foods is a function of differences in viscosity, stability to processing and distribution, and gel strength. Differences in percentage of amylose and amylopectin in part account for differences in pasting properties. Root and tuber starches and those containing only amylopectin (waxy starches) swell to a greater extent, demonstrate a greater viscosity drop on prolonged cooking,[31,51] are more apt to develop a cohesive or rubbery texture on overcooking, and do not become opaque or gel on cooling. Because of the unique properties of waxy maize, tapioca and high-amylose maize starches, they command premium prices.

1. Dent Corn

Dent (normal or common) corn starch (see Chapter 9) is available as its native powder and in modified forms. Many unmodified or powdered corn starch variants have evolved, including pH adjusted, bleached, oiled, agglomerated and redried, all for specific purposes.[50] Powdered corn starch is often used as a dusting powder, molding starch, filler and bulking agent, and as a cooked stabilizer in certain cook-and-serve and shelf-stable, canned goods.[51,59] Native dent corn readily cooks at atmospheric pressure in formulations with at least 50% moisture. Cooked pastes develop a thick, heavy viscosity and translucent appearance and produce an opaque, resilient, short-textured gel on cooling. Native corn starch is used to thicken retail or institutional cook-and-serve products like gravies, sauces, puddings and pie fillings where shelf life expectancy is hours or a few days. The gelling character of native dent corn starch makes it useful in spoonable salad dressings where it provides a stiff, short texture when used with a crosslinked and stabilized waxy maize starch. However, the texture of some dressings change after prolonged storage, due to aging of the gel made with unmodified starch. The most common change is that the gel shrinks until cracks appear and free liquid is released from the matrix. Generally, unmodified corn starch is avoided in frozen foods. It is used in canned goods like soups, stews and chilli which are disturbed, diluted and reheated by the consumer, allowing the product to recover enough consistency to be acceptable when served. Such products generally have a poor appearance in the can and, even after reheating, appear marginal alongside products containing modified starch. Modified (Section 20.4) dent corn starch is used in applications where clarity and resistance to setback and gelling are not required. These include puddings, canned cheese sauce and salad dressings, where it is used with a modified waxy maize starch. Dent corn starch is also used as an opacifier for gravies, providing a home-style, flour-like appearance. Fluidity (Section 20.4.3) corn starches are used as gelling agents in gum candies.

2. Waxy Corn

Waxy maize (see Chapters 3 and 9) today accounts for only 1% of the total corn acreage in the US,[124] but accounts for roughly one-third of corn starch produced and a large fraction of modified starch used in food. Sufficient waxy maize to produce about 800 000 metric tons of starch is grown each year in the US.[125] It is primarily grown under contract to wet-milling companies, although some is grown for feed due to apparent feed efficiency advantages over dent corn.[126] The high amylopectin content of waxy maize starch

(~99%) results in paste properties which are more like those of root and tuber starches than those of other cereal starches. Cooked pastes of waxy maize starches develop heavy viscosity and good clarity and, being essentially free of amylose, have improved resistance to gel formation and syneresis during cold storage.[127] Within waxy maize starches, there are differences in degrees of branching and chain lengths[25] and a double mutant with shorter branches has been claimed as a freeze–thaw stable thickener.[128] Crosslinked (Section 20.4) waxy maize starches are used where excellent thickening ability, clarity and processing tolerance are required. They do demonstrate a somewhat stringy consistency when cooled if inadequately crosslinked or otherwise overcooked or damaged by shear. By comparison, normal corn starch gels are short and resilient. Waxy maize starch is generally thought to have a more bland taste than dent corn starch. Dried waxy maize films yield a translucent, water-soluble coating on drying.

3. High-amylose Corn

High-amylose corn (see Chapters 3 and 9) has been bred to provide class 5 and class 7 cultivars whose starches contain 50–60% and 70–80% amylose, respectively.[124] More recent breeding efforts have resulted in still higher levels of amylose and dramatically reduced amounts of amylopectin.[129] The higher concentration of linear polymer in high-amylose corn starch effects more rapid gelation and stronger gels than occur with dent corn starch. These gels produce strong, tough films which are brittle when dry. Applications of high-amylose starches exploit these properties. In jelly gum candies high-amylose starch gels faster than fluidity corn starches (Section 20.4.3), reducing the time required in the starch mold. Fluidity high-amylose starches allow a reduction in hot viscosity so that candies can be deposited at higher solids, further reducing drying time and increasing plant capacity.[130] Dried high-amylose starch films are especially brittle, a property exploited in fried, batter starches, including those for frozen products to be reheated in the microwave.[131] High-amylose starches have been used together with an instant starch (Section 20.4.3) or food gum as binder to provide a crisp coating for French fries, which also reduces oil absorption,[132] and as part of a binder-texturizing blend for fabricated French fries from potato granules or flakes.[133] Chinnaswamy and Hanna[134] observed that increased amylose content improved expansion and increased shear strength in extruded starch. Accordingly, high-amylose starches are used for increased crispness and bowl-life in cereals and snacks.[135]

Several claims which exploit the gelling character of high-amylose starches, including their use in imitation cheese where some or all of the caseinate has been replaced,[136] have been made. Other patented applications which use the gel character are pasta and tortillas for retorting,[137] extruded shrimp analogs[138] and aerated confections like nougats, where the starch is used with protein hydrolyzates in a whipping agent[139] and a fat replacer.[140] Other novel uses of high-amylose starches include their use in sausage casings and food wrappers,[141] incorporation into bread crusts and pasta for more uniform heating in the microwave[142] and as a thickener with delayed swelling for retorted gravies and puddings.[143] Native high-amylose starch added to unleavened pastry dough, e.g. pie crusts, reduces stickiness to improve handling and avoid sogging during baking.[144]

4. Tapioca

Root and tuber starches show a sharper rise in viscosity during pasting than do cereal starches. They also rise to a higher viscosity and show greater relative breakdown on continued cooking. Tapioca (also known as cassava and manioc, see Chapter 12) and potato (see Chapter 11) starches with amylose contents of about 18% and 23%, respectively, give relatively stable, clear sols on cooling, in spite of their amylose content. This is because the amylose is higher in molecular weight than corn amylose and may be more branched – both factors interfering with retrogradation and gelation. Traditionally, delicately flavored puddings and pastry fillings have been made with tapioca starch, due to its bland flavor. It has also been the choice for baby foods, because of its wholesome image. While cassava is grown in other regions, most used in wet-milling comes from Brazil and Thailand. It is imported in powdered form or as pearls. Tapioca pearls are agglomerates of uncooked and partially gelatinized tapioca starch granules made by stirring or rolling damp starch in a container to the desired particle size before drying. Native tapioca starch forms a clear but stringy and cohesive paste when cooked and, like unmodified waxy maize starch, finds limited usage. Because some food processors favor tapioca starch for its bland flavor, it commands a premium price when modified to overcome its textural shortcomings. Modified tapioca starch also contributes a slight gel to puddings and pie fillings. However, the use of tapioca starch as a general thickener in the US has given way to waxy maize and waxy maize–dent starch blends for economic reasons.

5. Potato

Potato starch (see Chapter 11) is more commonly used in Europe, where it represents 20% of the starch produced, than in North America. As reported in 1984, about 30% of the potato starch made in the US is used in food.[145] Potato starch granules are large and swell and solubilize more readily than those of cereal starches. Potato starch produces pastes with high viscosity and a subtle grainy appearance. Cooked potato starch is more sensitive to shear than cereal starches.[50] On cooling, overcooked, unmodified potato starch forms a gummy but clear and pliable gel. Potato starch pastes have good clarity, but synerese like corn starch, especially if frozen. Potato starch typically has 20% amylase, though amylose-free potatoes have been developed.[146] In the US, potato starch is primarily used in canned soups and in blends where its thickening power is exploited, especially for fill viscosity. It is also used as a base for gelling agents in confections, for thickeners in products like pastry and pie fillings, and in instant puddings. The native mono-phosphate substitution on potato starch causes its viscosity to vary with solution ionic strength.

6. Wheat

Wheat starch (see Chapter 10) is a by-product of vital gluten manufacture, but is also isolated from wheat flour in its own right. It is a major starch in Australia and New Zealand; in Europe, it represents 20% of the total starch production.[12] Residual protein in the starch gives it a flour-like odor, flavor and appearance. Wheat starch granules are bimodal in size distribution with large lenticular granules, ~25 micrometers, and small spherical ones <10 micrometers in diameter (see Chapter 10). Rheological properties of pastes of wheat starch are similar to those of corn starch, although viscosity and gel strength are lower.[147] Takahashi et al.[148] compared modified wheat, tapioca and waxy maize starches and found that modified wheat starch has greater freeze–thaw stability than similarly modified tapioca starch and less than similarly modified waxy maize starch. Its principal use is in baking, where it replaces portions of wheat flour. Modified wheat starches can have superior emulsifying properties in certain food products.[148] Wheat starch is also used in confectionery products such as Turkish delight and has been found to improve head retention in beer.[81] Wheat starch with a high proportion of damaged starch can be used for gluten-free bread for celiac patients[149] and for sausage and meat rolls.[81]

7. Sorghum

Adapted white waxy sorghum cultivars were developed in the US during the 1940s when cassava/tapioca was unavailable.[25] In 1967, Osman wrote that sorghum and corn starches were similar and used interchangeably.[51] Waxy sorghum starch fell out of use around 1970 because of its increased gelatinization/pasting temperature relative to waxy maize starch, but the higher temperature provides greater process tolerance and may revitalize its food use.

8. Rice

Champagne[150] has reviewed the composition and properties of rice starch (see Chapter 13). From the sticky, non-gelling waxy types to the intermediate amylose varieties which produce dry, firm gels, rice starch pastes demonstrate a wider range in texture than those of other common crop species.[151] This range is illustrated by the resiliency of rice noodles and the low gumminess of rice crackers. Texture is largely controlled by percentage of amylose, and amylose content is controlled by one or two major genes with modifiers.[152] The amylose content of milled rice is classified as waxy, 1–2%; low, 7–20%; intermediate, 20–25%; and high, >25%. No varieties with amylose contents greater than 40% have been reported. The soft set of low-amylose and waxy rice starches have given those starches a reputation for smooth and creamy mouthfeel. Bost et al.[153] have a patent for rice starch in ice cream, and Bakal et al.[154] have a patent for using raw rice starch as a fat replacer in dressings. Rice starch hydrolyzates are marketed as fat replacers. Rice starch granules are typically 3 to 9 micrometers in diameter and occur in compound granules up to 39 micrometers in diameter.[151] Certain applications take advantage of the small granule size. These include dusting powders for cosmetics and on bakery products and certain candies.[155] Similarly, the small granules may allow rice starch to be used as an opacifier, and they may have special mouthfeel properties. Rice flour and starch produce films with tender crispness. This may explain their use in batters and ice cream cones and as coatings and glazing agents for nut meats and candies. Rice is favored in bland systems because of its neutral flavor,[156] explaining its use as a beer adjunct.[155]

Schoch[157] reported that waxy rice starch had superior freeze–thaw stability, with a 5% starch paste showing syneresis only after 20 cycles. Similarly treated waxy corn and waxy sorghum starches had syneresed after three freeze–thaw cycles. This native stability has allowed waxy rice starch to be used in thickening pie-fillings, sauces and soups that are frozen, and as a moisture retention agent in toppings and icings for cakes. The mechanism of this stability is not clear. Rice amylopectins of both non-waxy and waxy rice starch have mean chain lengths similar to those of other cereal amylopectins, 20–28 D-glucosyl units. The fine structures of amylopectins of waxy and non-waxy rices are almost identical with A-chain:B-chain ratios of 1.1 to 1.5, even for samples differing in birefringent end point temperature (BEPT).[151]

9. Sago

Sago starch comes from the stem pith of several palm trees native to the East Indies, including palms in the genera *Sagus*, *Cycus* and *Areca*. Radley[80] reports that sago was used in breakfast foods, puddings, pastries and breads. Fluidity (Section 20.4.3) sago starches have the high gel strength desirable in gum candy, but lose clarity on standing.[10] Sago starch production and consumption is giving way to that of rice.[158]

10. Arrowroot

Arrowroot starch is obtained from the rhizomes of the tropical plant *Maranta arundinacea* L. and related species cultivated in the West and East Indies, Australia and elsewhere. Arrowroot starch has been a home remedy for gastrointestinal disorders, especially in Britain, since it is reputedly the most digestible starch. It is made into puddings by heating it in milk and is blended with eight parts of wheat flour in arrowroot biscuits. It is also used in jellies, cakes and various infant and invalid food mixtures. What is marketed as arrowroot starch from some locations may be tapioca starch. If arrowroot starch is required, its identity should be verified.

11. Barley

Barley starch (see Chapter 16) has a viscosity profile similar to that of potato starch and also has a similar range of applications.[159] Seib and Wu[160] claimed excellent freeze–thaw stability in a hydroxypropylated waxy barley starch.

12. Pea

Relative to more commonly used starches, smooth pea starches show a restricted swelling and increased shear and acid stability.[161] These properties led Stute[162] to propose that pea starches be used: (a) to make gelled foods where pea starch would allow less starch to be used; (b) for extruded and pregelatinized products which can be produced without the significant loss in viscosity that occurs with other starches; (c) to produce fruit and vegetable flakes having a pulpy texture (using roll-dried products); and (d) for the production of gelling, roll-dried, instant (Section 20.4.3) starches for use in flans and other gelled desserts. Chevalier[163] noted that the native process tolerance of pea starch makes it suitable where unmodified starch is preferred. Wrinkled pea starch has a comparatively higher amylose content than smooth pea starch and forms strong gels.

13. Amaranth

Amaranth starch has very small and very uniform granules, the majority being less than 1 micrometer in diameter. Starch isolated from two *Amaranthus* species was compared and found to contain approximately 90% amylopectin and 10% amylose.[164] Those authors prepared distarch phosphates and found that *A. hypochondriacus* starch responded more to crosslinking, as evidenced by reduced swelling power at 85°C and an increased gelatinization temperature range than did *A. cruentus* starch.

VI. Applications

Starch can perform many functions, including thickening, gelling, emulsification and mouthfeel enhancement. Before describing starches favored in primary applications, a general discussion of selection criteria and trends is appropriate. The optimum starch for a particular application provides the desired product properties while accommodating the effects of other ingredients, processing and distribution. Accordingly, the food product developer must consider desired texture, appearance, flavor and labeling, as well as the rigors of heat, acid and shear during processing. Stresses of prolonged storage, possibly involving low temperatures, must also be addressed.

Starch is used to thicken many foods; soups, sauces and pie fillings being good examples. Thickening generally involves granular starches swollen by thermal processing. Modified waxy maize, potato and tapioca starches are generally preferred because of their comparative stability to textural changes during distribution. However, granular starch will not thicken effectively at low concentrations unless other thickeners are present, because it will settle out; so soluble, stabilized hydrolyzates are used instead in thin fluids. In beverages and light syrups, these have the advantage of minimal opacity.

A variety of starches are used for gelling. Dent corn starch is used in puddings, lemon pie and cheese products for shorter texture and shape retention or set. Blends of gelling and non-gelling starches can be used to provide intermediate properties, such as a slight cuttability in products like salad dressings. Tapioca starch can also be used, but will give a softer set. Fluidity (Section 20.4.3) corn and potato starches are used for gel formation in high solids candy formulations, such as jelly gum centers. High-amylose corn is also used in candies where higher cook temperatures are possible. Because brittle food films are dried gels, fluidity and high-amylose starches are used to make battered and fried foods crisper.

The choice of starch can affect mouthfeel. Tapioca starch, highly stabilized products, and unmodified waxy maize starch are thought to contribute more mouth coating and prolonged residence on the palate. Starch hydrolyzates are used for mouthfeel enhancement, especially where cohesiveness or pastiness is to be avoided. Both flavor and flavor release varies between starches. However, the blandness of a starch in a product may be a function of the familiarity of its flavor contribution combined with the taster's expectation. Waxy maize starch is typically blander than dent (normal) corn starch, a difference evident in bland foods. Tapioca starch is thought to be blander than waxy maize starch. Consumer tastes often reflect regional preferences for familiar starch sources. For example, potato starches may be preferred in potato dishes or snacks. Labeling may also be a factor in starch choice, since tradition dictates the use of certain starches in certain foods. Moreover, processors may have a preference for a specific profile, e.g. native labeling, depending on the targeted market.

The available processing equipment will affect starch choice. Without cooking equipment, instant starches will be used instead of cook-up starches. Similarly, where heat treatment or shear after cooking is severe, more crosslinking is required. Tapioca and potato starches are more shear-sensitive and to be avoided where processing is abusive.

Starch choice is affected by formulation. Increased competition for water (e.g. by sugars) will require a less crosslinked or more stabilized starch to be used. Acidic formulations will require more crosslinking. The manner in which the food is to be distributed is also important regarding the base choice and amount of stabilization. Frozen or refrigerated distribution requires greater monosubstitution for texture and appearance stabilization. The following overview of starch use in specific applications illustrates how food starches are chosen.

1. Canned Foods

Canning preserves food for up to several years by achieving a temperature sufficient to destroy or inactivate food poisoning or spoilage microbes. Starch is most commonly used to thicken, stabilize and enhance the mouthfeel of canned foods such as puddings, pie fillings, soups, sauces and gravies. It is also used to provide viscosity at the time of filling for controlled metering of suspended particulates and for control of splashing. Canning involves preheating products, filling and sealing the container, and providing an adequate temperature treatment for sterilization. Times and temperatures required vary with the food, container and equipment, but involve holding the coldest part of the food for at least 20 minutes above 120°C and at least 145°C for several seconds.[165] Minimization of ramp-up time reduces quality loss during retorting. Delaying thickening improves convective heat penetration, especially where the cans are rolled or tumbled during retorting. Foods with particulates, however, require viscosity during filling to control distribution of suspended pieces to containers. The need for high viscosity at filling and low viscosity during ramp-up can be met by using fill-viscosity starches. A fill-viscosity starch thickens in the make-up kettle, but breaks down quickly in the retort. They are most typically used where no residual viscosity is needed, such as in chicken noodle soup. Light to moderately crosslinked starches are used to thicken many canned foods, but when delayed thickening is needed a highly crosslinked starch, usually waxy maize starch, is favored. Canning starches will contain various levels of monosubstitution depending on the stability and desired texture of the food. Fill-viscosity starches are typically waxy maize or potato starches with or without monosubstitution or acid thinning, both of which enhance breakdown. These give high initial viscosity for suspension of particulates during filling, but then break down to provide good heat penetration. Other canned foods have other specific requirements. Vegetable soups often contain only a lightly crosslinked fill-viscosity starch which breaks down to form a light broth in the final product. Cream-style canned corn is improved by the use of a highly-stabilized, lightly crosslinked waxy maize starch. This starch prevents curdling by coating soluble protein which is denatured early in retorting.[50] More highly crosslinked products are required for short texture in acidic foods like lemon or cherry pie fillings. Retorted cheese sauces are often thickened with a highly crosslinked and stabilized dent corn starch for the short cuttable texture it provides. Unmodified dent corn starch is sometimes added to gravies to enhance the opacity for a homemade appearance. Unmodified waxy maize starch is added to provide gloss and slight cohesiveness in products such as dips.

Aseptic processing is a rapid heating and packaging process used to preserve the quality of certain foods, such as dairy products, where prolonged heating causes denaturation and loss of flavor.[166] It involves addition of sterilized food to sterilized packaging in a sterile environment. The rapid heating and cooling allows a brief exposure to very high temperatures for an effective microbial kill with minimum damage to food quality. Aseptically processed cheese sauce is often thickened with a highly crosslinked and stabilized waxy maize starch for a smooth and flowable texture or blends of highly crosslinked and stabilized dent and waxy maize starches to provide a smooth and short cuttable texture.

2. Hot-filled Foods

Microbial stability can be achieved via the cumulative effects of several barriers to growth, e.g. initial heating, low water activity, antimicrobials, low pH, etc.[165] Low pH, e.g. below 4, effectively obviates sterilization. Acidic foods can be packed and sealed while hot enough to kill organisms that would survive that pH at lower temperatures during brief storage. A pH of 4.6 or lower and temperatures greater than 155°F (68°C) are required. Because of the low pH and because the hot package is allowed to cool slowly, high levels of crosslinking are necessary. Hot-filled products include tomato sauces, salad dressings and fruit pie fillings. White sauces and cream pie fillings cannot be hot-filled because of their higher pH.

3. Frozen Foods

Starch is used in frozen foods for the same reason it is used in fresh, refrigerated or canned foods, i.e. for thickening, low-temperature stability and control of the flow character of the food. Freezing exacerbates syneresis in other components of the food system, such as fruit tissue, increasing the demand for water entrainment by the starch. Frozen dinners are cooked, cooled, packaged and flash frozen. Fruit pie fillings are mixed, heated, cooled, filled into shells and flash frozen. Freezing introduces stability demands beyond the other forms of distribution. The stability of the starch is critical to the stability of the frozen food.[167] Freezing and especially freeze–thaw cycling accelerates retrogradation. Frozen foods are subjected to repeated temperature cycling en route to the end user. This includes cycling in transit to distribution warehouses, to retail outlets, and within the store and home freezers during defrost cycles. Manufacturers typically allow for six cycles. Retrogradation in frozen starch-based foods is evident as gelling, graininess, opacity and syneresis.[127] Reheating partially reverses these defects, e.g. it may remove graininess, but may not restore the original viscosity or flow character. The most demanding case, thaw and serve foods, require the greatest freeze–thaw stability. In these products, even crosslinked waxy maize starch becomes opaque and gels, so a higher level of monosubstitution is used. Because of the cost of heating and cooling, pregelatinized starches are sometimes used.

Starch is used in frozen battered foods to provide better adhesion and reduced oil pick-up during frying, and also for improved crispness after oven or microwave reconstitution. Oxidized starches are used for improved adhesion, reputedly because of covalent bonding with protein on the substrate surface. High-amylose starches and flours are favored for crisping. Preprocessed high-amylose starch has also been employed to increase the crispness of breadings.[168] Cochran et al.[169] claim a microwave frozen bakery product containing monosubstituted starch. Waxy maize has been used in frozen tofu to control textural changes.[170]

4. Salad Dressings

Hot-process spoonable salad dressings are typically made by cooking starch, vinegar, salt, sugar and water to a paste, cooling the paste, and adding the remaining ingredients. The oil is added last, typically with egg yolks, whites or another emulsifier. A finer, more stable emulsion is achieved by sending the complete formula through a colloid mill, a very high shear process. Cold-process dressings follow a similar flow, but without heating. Pourable dressings are more often made via a cold process. Starch is used in dressings to thicken and stabilize the dressing and to provide the desired cuttability and flow character. Because of the low pH, high temperatures and high shear involved, use in cook-up dressings requires the starch to be highly crosslinked. The amount of monosubstitution present will vary with the formulation and desired texture. Blends containing modified waxy maize and dent corn starches are commonly used in higher fat, spoonable dressings; the modified dent corn starch is included to shorten the texture and provide cuttability. Sometimes unmodified dent corn starch is used for additional cuttability. Modified waxy maize starch products alone are more typical in pourable dressings. Potato and tapioca starches have not been successful in this application, because of their greater susceptibility to acid and shear. Fat-reduced and fat-replaced dressings generally use a more highly stabilized, crosslinked waxy maize starch. No-fat dressings employ fat replacers. These are typically dextrins, maltodextrins or highly monosubstituted hydrolyzates (Section 20.4.3).

The texture of fruit pulp or tomato paste can be achieved in nonstandardized products, such as pizza sauces, through the introduction of texturizer starches. These are typically highly crosslinked starches that are subsequently drum dried.

Brueckner et al.[171] have claimed use of hydrophobic starches with modified protein in sterilizable dressing emulsions and canned fish products.

5. Baby Foods

Starch is used in retorted baby foods to thicken and provide short texture. In addition to high temperature stability, fruit-based, dessert-type products require resistance to low pH and stability in the jar, including (though it is secondary) freeze–thaw stability. Accordingly, crosslinked acetylated waxy maize and tapioca starches are favored. Tapioca starch products are favored for flavor and because they are thought to be more wholesome for babies.

6. Beverage Emulsions

A beverage emulsion is a concentrate added to sugar and carbonated water to make soda and fruit drinks. The oil-in-water emulsion provides flavor as well as opacity in products such as orange soda. Traditionally, gum arabic has been used to stabilize these emulsions. Interfacial starch derivatives (Section 20.4.2) are used to prevent creaming (phase separation), sedimentation, and loss in flavor and opacity, where desired, both in the concentrate and in the finished beverage. The concentrate is made by homogenizing the oils with an equal amount of the solubilized lipophillic starch, citric acid, sodium benzoate and color. A fine emulsion, typically 1 micrometer or less, is required for stability and for opacity, where desired.

7. Encapsulation

Starch derivatives are used to encapsulate flavors, beverage clouds, creamers and vitamins.[69,172] These are typically made by dispersing the oil in a solution containing roughly four times as much encapsulant, followed by homogenization and spray drying. Various other processes can be used, including fluidized-bed addition, extrusion and drum drying. The starch derivative stabilizes the emulsion, which forms films and ultimately provides a dry matrix occluding the oil.[173] The encapsulant minimizes loss of volatiles during drying and storage, and oxidation of the oil during storage. Sustained or delayed release can also be achieved. Treatment of starches with lipophillic reagents such as octenylsuccinic anhydride (OSA) renders the starch sufficiently interfacial to function as a stabilizer for beverage emulsions or as a encapsulant for oil-soluble flavor and vitamins. Lipophilic starches provide greater stability to oxidation during storage than gum arabic or maltodextrins.[66] Octenylsuccinylated dextrins and other starch hydrolyzates are used, as are both cook-up and pregelatinized starch derivatives. The starch is hydrolyzed to achieve low viscosity for more efficient spray drying. Related applications include liquid and powdered non-dairy creamers, where the starch provides a functional alternative to caseinate as the emulsion stabilizer. Other approaches have been used: Zhao and Whistler[174] have demonstrated encapsulation with porous spheres made by spray drying starch with a bonding agent. Maier et al.[175] report encapsulated freeze dried emulsions stabilized by various starches. Similarly, modified food starch can improve encapsulation by extrusion.[176]

8. Baked Foods

Baked foods containing added starch as a functional ingredient include extra-moist cake formulas, glazes and fillings. Pregelatinized starch is used in cake mixes to soften the cake crumb and retain moisture in the baked product. They are also used to provide convenience. Takashima[177] claims the use of a pregelatinized starch and a thermosetting protein for the baking of sponge cakes in a microwave oven. Pregelatinized starches are also used in cookies to control spread.[51] Film-forming starches are added to glaze mixes to provide an attractive coating, moisture barrier and decreased tackiness. Starches used include dextrins, stabilized hydrolyzates and maltodextrins.[178] Starches are used to stabilize and thicken fruit and cream fillings in pies and tarts. They also provide resistance to boil-out during baking, and resistance to weeping during distribution. Highly stabilized waxy maize starch products are used for fruit filling; modified tapioca starch is favored in cream fillings. Moderately high levels of crosslinking are needed for low-pH fillings and where shear is encountered in kettles with scraper agitation. Pregelatinized derivatized starches are sometimes used to avoid the costs of heating and cooling fillings. It is important that dusting flour be unmalted and that the fillings be free of enzymes if pregelatinized products are used.

9. Dry Mix Foods

Instant dry mixes are used for home and institutional preparation of a variety of foods; examples include puddings, desserts, soups and gravy and sauce bases. Upon hydration, the starch provides thickening, a smooth and creamy texture, and fast meltaway. Coarsely ground drum-dried starches and agglomerated spray-dried starches improve dispersibility if dilution of starch with sugar or other non-hygroscopic materials is not possible. In instant puddings and desserts, instant, derivatized tapioca starches are favored for their bland flavor. In soup mixes, instant, derivatized waxy maize starches are used. A cook-up, highly-stabilized starch can be used in an instant soup if hot water is to be added.

10. Confections

In confections, starch is used for gelling centers and to provide attractive coatings. It is also used as a dusting powder and as an impressionable bed where candies are cast. To make jelly bean centers, starch is cooked with water, corn syrup and sugar. Flavors may then be added, and the fluid is cast in impressions formed in trays of molding starch. Stacks of trays are placed in heated rooms until the candy is dry enough to be shaken from the mold, coated and packaged.

Cooking is traditionally done in kettles with acid-thinned corn starch chosen for its shorter texture and reduced hot viscosity, allowing more starch to be used for greater gel strength. With jet or tubular cookers, high-amylose starches are favored because they gel more quickly, allowing more rapid removal from starch molds and greater factory throughput. However, while thin-boiling starches (Section 20.4.3) will be cooked at 140°C, unmodified high-amylose starches require 170°C. Acid-thinned high-amylose starches are also used, as are monosubstituted, i.e. stabilized, starches. By using fluidity high-amylose starches which gel faster and allow higher solids, the traditional method of drying thin-boiling starch gum drops for two days in molding starch has been reduced to a curing time of as little as eight hours, allowing greater outputs.[50] Monosubstitution is used to reduce gelatinization temperature, shorten texture and improve shelf life. Hard gum candies are generally made with tapioca dextrins for structure, stability and clarity. The starch is used at a higher level, along with an equal amount of corn syrup and a lesser amount of sugar. Examples include wine gums, lozenges and chewy cough drops. Oxidized waxy maize starches are also used for hard gums because they provide clarity.[12]

Candy coatings are applied via repeated, successive application of dextrin and sugar solutions, followed by drying. Dextrins and monosubstituted starch hydrolyzates are used in candy coatings because of their clear, appetizing films. In this application, clarity, sheen and reduced cracking are desired. Oxidized waxy maize starches are also used.[12] Starch used as an impressionable bed for casting candies is typically dent corn starch with several tenths of a percent of mineral oil. The oil causes the starch to retain the imprinted shape for a more precisely formed candy piece.

11. Snacks and Breakfast Cereals

Starches have been used by the cereal and snack industries to achieve specific textures, e.g. increased crispness, especially in high-temperature, short-time processes.[135,179] Snack market trends emphasize lower fat, including fried textures from baked products, high fiber and simpler processes. Starch can contribute significantly to the achievement of each of these targets.

Ready-to-eat Breakfast Cereals

The thermal and mechanical stresses of cooking, forming and expanding cereal pieces may cause native starches of corn, oat and other commonly used flours and meals to break down. If the stresses are excessive, optimum expansion of dough will not result in the desired bulk density and texture. An example is the severe shear of direct expanding cereal pieces from an extruder. Addition of a high-amylose or crosslinked corn starch improves shape definition and process control, as well as texture and bowl-life. Conversely, where milder processing conditions are used, expansion is enhanced with starches with lower cooking requirements, i.e. with waxy maize starches, including those that have been monosubstituted or pregelatinized. Bindzus and Altieri[180] claim use of a phosphate ester as an extrusion expansion aid.

Cereal brans and other ingredients added for health benefits such as fecal bulk and colonic fermentation typically impair expansion and mouthfeel of puffed cereal products. Resistant starches can be used in the manufacture of snacks and breakfast cereals without compromising product appearance, texture or palatability.[135] Such resistant starches, offered for use in expanded cereals and snacks, use high-amylose maize with biochemical and physical processing to optimize retrogradation and nutritive value.

Fried or Baked Snacks

Starches have been used in fried snacks to add dough machinability and to improve the texture of the fried product. As with cereals, increased amylose levels will provide a firmer texture. High-amylose starches can be used to control oil absorption. French fries are commonly battered with high-amylose starches or dextrins to provide increased crispness and stability to holding under heat lamps. Woerman and Wu[181] claim use of crosslinked tapioca starch, and Rogols and Woerman[182] use of low-DE starch hydrolyzates in this application. Modified dent corn and potato starches are also used.[183] High-amylose starches also give fried products greater crispness, due to the brittleness of the films they form when cooked and dried. Starch has been used with sodium alginate to form fabricated onion rings, which are extruded and fried.[184]

Following the trend towards lower fat levels, expanding and crisping starches are used in baked doughs to provide the appearance and eating properties of fried products. The starches used are modified to allow expansion under the more mild heating of an oven. A pregelatinized or a highly monosubstituted starch is favored. As another means of reducing fat levels, the oil used to provide sheen is replaced with a dextrin or other soluble starch. Starch-based tackifiers, which exploit the film-forming and adhesion properties of converted starches, have been introduced.[135] Again, to further reduce fat levels, these starches replace the oils used to attach seasonings and particulates to the surface of snacks or to agglomerate cereals like granola.

12. Meats

Starch is used in meats, including those reduced in fat, to increase moisture retention for succulence and purge (free package moisture) control, to reduce shrinkage and for firmness of bite. Moisture retention is improved with a highly-stabilized, moderately crosslinked waxy maize or tapioca starch.[185] A firmer bite is generally achieved via use of a modified corn starch. Meats containing starches hold up better on the steam table than meats with other texture modifiers.[186] Starch is added to meats via injection of a raw slurry, tumbling or addition to formed meats. Choice of the base starch affects texture. Starches that break down may become cohesive, adversely affecting palatability. The internal temperature of smokehouse meats does not exceed 70°C, so the starch must swell below that temperature and at the low water activity of the product. Starches are monosubstituted so that they cook-out under these conditions. Highly-stabilized, moderately crosslinked waxy and tapioca starches are typically used.

Modified food starch is approved for use in standardized frankfurters and bologna at 3.5% usage. In these products, it increases the water-holding capacity of the meat and reduces purge during storage. Cook-up starches are preferred to avoid increasing batter viscosity and to survive shear on high-speed emulsion mixers. They must still swell under smokehouse conditions. The starch is added at the same time as the seasonings; in fact it may be part of a seasoning binder blend. Modified food starch is approved for use in 'Ham Water Added' and 'Ham and Water – X%' at the approved usage of 2.0%, where it reduces purge. The modified food starch is added to the brine tank (requiring agitation). The starch gelatinizes after the ham is in the smokehouse. Poultry loaf was among the first applications where the use of 'food starch – modified' was permitted by the USDA. Here, starch is used to firm texture, increase juiciness and sliceability, and to control purge. Modified food starch can be used in poultry meats at levels up to 3.0%. In meats like poultry franks, 3.5% is permitted. A derivatized pregelatinized waxy maize starch has been used in a chicken coating to produce a fried appearance after baking.[187] Shake-on coatings use dextrins, octenylsuccinylated starches and some pregelatinized products.

13. Surimi

Surimi is fish paste from deboned fish used to make simulated crab legs and other seafood. For preservation the paste is blended with cryoprotectants, such as sucrose, sorbitol and phosphates, and frozen. To make the final product, the frozen paste is thawed, blended with starch and extruded as a film onto a belt. The belt takes the film into an oven that heat-denatures the fish protein and cooks the starch. The film is then rolled to form striations, shaped, colored and cut. Depending on the required distribution, the product is frozen or refrigerated. Potato and tapioca starch were used in surimi products 400 years ago, since they provided a cohesive, elastic matrix consistent with seafood. Frozen distribution has made the use of highly-stabilized, moderately crosslinked tapioca starch popular, alone or with native tapioca starch. Modified waxy maize products are used, as is unmodified corn starch, for increased cuttability. Kim[188] reported that the gel strengthening ability of starch correlates with starch paste viscosity.

14. Pet Food

Starches are used primarily in canned pet food, not in dry or semi-moist products. Canned pet foods with gravy require a stable, non-gelling thickener, so a highly-stabilized waxy maize starch is used. Fluidity corn starches are used as gel formers for firm, cuttable loaf products.[189]

15. Dairy Products

Modified food starches are used in a wide variety of dairy products to provide a variety of effects, including enhanced viscosity, cuttability, mouthfeel and stability. In puddings, starch is used to impart viscosity and a smooth, short texture. Starches are used in yogurts and sour cream to control syneresis, thicken, replace milk solids and improve mouthfeel. In cottage cheese, dressing starch improves cling on the curd. Other dairy products containing starch include buttermilk, cheese, dips and ice cream. Resistance to shear is particularly important in those dairy foods that are homogenized. Swollen granular starches that are homogenized after pasteurization must be very highly crosslinked to resist viscosity loss. One approach is to homogenize prior to pasteurization, thus avoiding mechanical stress on the swollen granule.

In that case, the starch needs sufficient crosslinking to swell little during preheating prior to homogenization. Undercooked products are equally undesirable. They are thin and grainy with poor stability to storage stress. Starch selection depends on the process and equipment used, as well as on the intended function and mode of distribution. Highly crosslinked, moderate to highly stabilized waxy maize and tapioca starches are generally used in products homogenized during or after pasteurization.

In yogurt, starch is used to control syneresis, thicken, improved mouthfeel and replace non-fat milk solids. Typically, highly-crosslinked, stabilized waxy maize and tapioca starches are favored. In imitation cheese, e.g. pizza cheese, starch is used as a casein replacer for cost reduction. The starch products are typically enzyme hydrolyzates with special gelling and melting properties. In chilled desserts and ultra-high temperature (UHT) puddings, highly stabilized, crosslinked waxy maize and tapioca starch products provide stability, viscosity, mouthfeel and a short, smooth texture. Unmodified dent corn starch is commonly used for a gelled structure in pudding and other dairy desserts, but the products undergo syneresis on prolonged storage. Modified corn starch provides a short, firm structure desired in certain desserts, such as flans.

Batch pasteurization following preheating and homogenization requires only moderate crosslinking. More highly crosslinked starches are required for direct UHT processing than for indirect heating. Aseptic UHT processing at high pressure is the most severe condition as far as starch is concerned, since the temperature and shear rate are additive in disrupting starch granules. Aseptic processing of dairy products, such as puddings, is increasingly popular.[12] It involves a brief (3 to 10 seconds) exposure to very high (e.g. 140°C) temperature followed by packing in sterile containers under sterile conditions. While providing long product life, aseptic processing has advantages over traditional heat treatment (i.e. canning), viz., better flavor; lower energy costs; light, less expensive containers; and lower shipping and storage costs. The heat treatment may be achieved via plate, tubular, or swept surface cookers or via steam injection.

16. Fat Replacers

Starch-based fat replacers are commonly used in sauces, salad dressings and dairy products to enhance mouthfeel and to give products the flow character of the full-fat product.[190] The original approach was to replace fat with an equivalent weight of a 25% paste of the fat replacer. Ingredient suppliers now make a product-specific recommendation, including other types of ingredients. Preparing fat replacers from starch began in East Germany at the Academy for Science where reversible, salve-like gels from hydrolyzed potato starch were proposed as a way of simulating fatty textures.[191,192] Starch hydrolysis products (SHP) were suggested as replacements for fat in liquid and semi-solid foods such as ice cream and salad dressings. Since that time, starch derivatives representing diverse technologies have been offered for fat replacement. The SHP patents claimed gelling alpha-amylase digests with a DE range of 5 to 25. A subsequent patent by Lenchin et al.[74] claimed a gelling hydrolyzate with a DE range of 0 to 5 achieved via enzyme- or acid-catalyzed hydrolysis, including dextrinization. Morehouse and Lewis[193] described a corn-based maltodextrin of 5 to 10 DE for use in margarine spreads. Inglett[194,195] developed a fat replacer based on alpha-amylase hydrolyzed oat bran or oat flour. The beta-glucan present was thought to complement the fat sparing maltodextrin by lowering serum cholesterol.[196] Furcsik et al.[140] claimed a high-amylose starch with a high level of hydroxypropylation which would form a translucent paste. Chiou et al.[197] claimed a microcrystalline starch for use as a fat replacer, basically as a particle gel former. The product is made by treating starch granules with acid to weaken the amorphous regions, and then shearing the granule to form micrometer-sized particles. Whistler[198] patented a microcrystalline starch fat replacer made by mechanically or chemically disintegrating granular starch rendered microporous by enzymes or mineral acid. Capitani et al.[199] claimed starch hydrolyzates monosubstituted to prevent gelling. Chiu and Mason[200] claimed short chain amylose formed by debranching starch as a gelling fat replacer with unique shear-thinning properties. Mallee et al.[201] patented the use of a sheared, high-amylose starch paste for use as a fat replacer, especially in opaque, gelled foods like no-fat sour cream. Fanta and Eskins[202] have a process for making a fat replacer based on a jet cooked blend of starch, fats and water. Baensch et al.[203] claim a fat replacer and opacifying agent based on high-amylose starch that has been cooked with minimal shear.

VII. Interactions with Other Ingredients

Discussions of starch and its performance in foods too often overlook the other ingredients present and the role that interactions play in the starch's functionality. The following sections describe the effects of common food ingredients on starch function.

1. pH

The pH of most foods is between 4 and 7; between these values variations in acidity produce little effect on starch viscosity during heating.[204] Salad dressings, fruit fillings and other foods may have a pH of 3 or less. Such a pH will accelerate starch swelling and breakdown and even effect hydrolysis of glycosidic bonds at elevated temperatures, as occurs when fruit pies are baked. A high level of crosslinking can compensate for the effects of acid, allowing the hydrolysis during baking to actually increase the viscosity of the final product.[205] Acidic breakdown of starch is exploited in canning where a fill-viscosity starch is used, but minimum final viscosity is desired. The incorporation of less than 0.2% ascorbic acid, araboascorbic acid or dihydroxymaleic acid allows lower final viscosities in products such as retorted meat-in-broth soups. Potato starch pastes achieve a maximum dynamic viscosity at pH 8.5, while those of tapioca starch are not affected by pH. Very high pH values, i.e. greater than pH 11, will effect gelatinization under mild heating.[206]

2. Salts

Potato starch contains 0.07 to 0.09% phosphate mono-ester groups.[53] This native monosubstitution makes the starch anionic and sensitive to the presence of other ions. Similarly, the US Code of Federal Regulations allows up to 0.4% phosphate substitution. Muhrbeck and Eliasson[207] found that the dynamic viscosity of potato starch decreases with increased ionic strength, while that of tapioca starch was unaffected. Divalent cations had a more pronounced effect than monovalent cations. They concluded that cations may interrupt electrostatic bonding within the granule. These cation exchange properties can cause potato and other monophosphorylated starches to be affected by residual minerals present during wet-milling and modification. The viscosity of a paste of a corn starch phosphate was little affected by heating with sodium chloride at concentrations up to 0.1 M.[50] Ruggeberg[208] showed that the sodium salt of potato starch gave a higher maximum viscosity than the untreated base starch, while the calcium salt provided less viscosity than the control. Corn starch similarly treated was unaffected. Sandstedt et al.[209] found that 2 M sodium chloride increased the gelatinization temperature of wheat starch by 13°C, while 5 M salt lowered it. The overriding effect of salt on non-ionic starches is competition for water, but salt does have complex and subtle effects, such as increasing then decreasing gelatinization enthalpy and onset temperature with increasing sodium chloride concentration.[210,211] The effects of other salts are interesting, even if not commonly encountered in foods.[50] Calcium chloride at a concentration of 3 M will gelatinize starch[212] and sodium sulfate increases the gelatinization temperature of potato starch from 62° to 80°C, while sodium bromide reduces it to 44°C.[213] Salt was found to lower the flow behavior index of corn starch and combinations of xanthan and corn starch.[214] Salt reduces the tendency of starch to retrograde.[215]

3. Sugars

Hydrophilic solutes such as sucrose, glucose and glucose syrups compete for water, and can delay and inhibit starch swelling if present in adequate amounts.[216] Sugars and oligosaccharides influence gelatinization onset temperature in relation to their impact on water activity and water volume fraction.[210] Large concentrations of sugar decrease the rate of thickening and the enthalpy of gelatinization. Freke[217] found that the gelatinization temperature of starch increased 3.5°C for each 10% increase in sucrose concentration above 20% weight per volume solution. If the concentration of sugar is very high, e.g. with high brix fillings, it can prevent proper cooking. Addition of the sugar after starch cooking is complete may be necessary to develop viscosity. Sugars also make swollen granules less sensitive to mechanical disruption. By reducing the degree of swelling, they also minimize the tendency of the granules to rupture or overcook.[211,218] Sugars and other hydroxyl-containing compounds also stabilize starch pastes.[214,219] Addition of sugar can reduce opacity and syneresis, presumably by minimizing interaction between hydroxyl groups on adjacent polymer molecules.[50]

4. Fats and Surfactants

Fats and emulsifiers affect paste viscosity and can modify mouthfeel by smoothing and shortening starch pastes. The effects are mostly due to physical blending, but may involve specific complexes. Interactions between starch and lipids are complicated and vary with the ingredients involved and the conditions under which they come in contact. Fat can interfere with granule swelling, presumably by occluding the starch and preventing hydration. But in the laboratory, with vigorous mixing, native and hydrogenated fats have minor effects on the pasting of starch. They accelerate the cooking and lower the temperature at which the starch develops its maximum viscosity.[220] It was found that the temperature corresponding to the viscosity maximum decreased as fat was added until fat was 8–12% of the mixture. The source of the fat and its iodine number had no affect.[220]

Certain surfactants, e.g. long chain fatty acids and glycerol mono-esters, will complex with starch by forming a clathrate with the helical amylose wrapped around the fatty moiety. Clathrate formation will occur with acyl chains of 4 carbon atoms, but 12 to 18 carbon atoms is optimal. Unsaturation is advantageous, depending on the process.[221] This complexation, which requires elevated temperatures for formation, can have several affects on the starch's contribution to texture. If complexing occurs with amylose while it is still in the granule, swelling and solubilization are delayed and the gelatinization temperature is increased.[222] It may also cause marked increases in hot viscosity, possibly because of reduced deformability of the granules.[223] If the starch has been cooked before the complexing lipid is introduced, then gel strength and other physical properties tied to retrogradation are mitigated. Melting of crystals containing complexed amylose and lipid can be measured as a DSC transition at about 100–120°C.[224] While this complexation is more pronounced with amylose-containing starches, it occurs with waxy starches,[59] although there is no DSC endotherm on melting as occurs with amylose–lipid complexes.[225] Surfactants such as glyceryl monostearate can increase the opacity of cooled starch pastes.

Starch–lipid complexing is exploited in dehydrated potatoes and retorted pasta where the integrity of the product is improved by adding monoglycerides to minimize solubilization during processing.[205] Similarly, mono- and diglycerides are used to prevent stickiness in rice kernels,[35] and sodium stearoyl-2-lactylate is used to control firming of bread. Aseptic puddings often include sodium-2-stearoyl lactylate to avoid stringiness and textural graininess.

5. Proteins

Proteins can compete for water, delaying cooking and increasing the pasting temperature, but proteins and starch interact in ways that are more specific and profound. When starch and milk protein are cooked together, the resulting viscosity may be greater than if they are cooked separately and mixed.[50] Lelievre and Husbands[226] found that blends of starch and caseinate were synergistic in terms of viscosity formation. Similarly, starch increases the rigidity of surimi at elevated temperatures; the presence of the fish protein increased the gelatinization temperature of the starch.[227] The improved gelling may be a function of particulate starch filling the protein network to make it more rigid, in which case the tightness of binding at the interface would be critical. Proteins vary with regard to how tightly they absorb onto the surface of gelatinized granules. Starches also vary with regard to protein binding. Eliasson and Tjernald[228] found that bovine serum albumin bound more tightly to potato starch granules than to wheat or corn starch granules. Brownsey et al.[229] found that amylose and egg white blends formed stronger gels than either did alone. Phase separations were evident in microscopic analysis. Muhrbeck and Eliasson[207] found that the respective transition temperatures and rates of gelation of the components were critical for a variety of mixed gels.

6. Gums/hydrocolloids

The texture-enhancing affects of starch and gums are typically additive, so the two can be used to achieve specific flow properties or reduce cost. An example is the use of carrageenan with starch in a refrigerated dairy dessert. Dairy desserts are characteristically thixotropic, while appropriately cooked modified starches display minimal thixotropic behavior. Selective use of gelling gums such as kappa-carrageenan allows manipulation of thixotropy and viscosity after shear.[230] A pregelatinized modified waxy maize starch may be used with xanthan in a pourable dressing; the xanthan provides shear thinning flow and the starch provides cost-effective viscosity.

There are many cases where gums and starches interact.[231] Coffey[232] found that adding modified starch to derivatized celluloses produced unexpectedly high viscosities at temperatures above 50°C. Kohyama and Nishinari[233] reported that methylcellulose decreased retrogradation of sweet potato starch, while carboxymethylcellulose, fibrous cellulose and microcrystalline cellulose increased retrogradation. Alloncle and Doublier[234] found that, while G' increased, elasticity of pastes decreased when guar gum, xanthan or locust bean gum were added to corn starch. Adding rice starch to gellan caused G' to increase with temperature.[235] The greatest increase occurred as the starch gelatinized, suggesting that the gelatinized starch reinforced the gel by binding water (increasing the effective concentration) or by an improved adhesion between gel and filler. Stress and strain at failure were reduced, but increased with increased amounts of gelatinized starch. Coffin and Fishman[236] cast films from mixtures of pectin and various maize starches. Films containing waxy maize starch were too brittle, but addition of high-amylose starch made good films. More starch reduced both storage and loss moduli. It was shown that the viscosity could be varied by adding different amounts of starch, although tan delta remained constant and could be varied independently by varying the kappa-carrageenan concentration. An efficient synergism was observed for corn starch, as seen from Brabender amylograph peak viscosity values. Freeze–thaw stability, as well as stability under acidic conditions, retort conditions of 15 psig/30 minutes, and mechanical shear improved admirably on interactions with CMC.

7. Volatiles

Polysaccharides, including starches and dextrins, have been shown to reduce the volatility of aroma compounds, such as limonene, isoamyl acetate, ethyl hexanoate and beta-ionones.[237]

8. Amylolytic Enzymes

Many food ingredients contain amylases which will catalyze hydrolysis of cooked or solubilized starch and impair its ability to thicken, gel or otherwise provide the desired texture. Examples include blue cheese in salad dressings, blueberries in pies and flour used for dusting baked goods. Where possible, it is desirable to cook the amylase-containing ingredient along with the starch; typically the enzyme is denatured before the starch is rendered susceptible to attack. For cold processes involving pregelatinized starches, the problem is less easily overcome, so amylase-free ingredients are required. Where ingredients are added to cooked starch, low levels of enzyme may have a delayed affect. This is especially true if high temperatures are encountered during product distribution or if the ingredient contains microbes producing amylase activity. In the case of dusting flours, unmalted flour should be used.

VIII. Resistant Starch

Recognition that some starch is neither completely digested nor absorbed in the small intestine has aroused interest in non-digestible starch fractions and their potential to perform functions similar to dietary fiber in the large intestine.[238,239] Nutritionists now agree that starch plays an important role in colonic physiology and can provide protective affects against colorectal cancer.[240,241] Resistant starch is not digested by pancreatic amylases in the small intestine, but reaches the colon where it provides benefits, including the growth of favorable bacteria.[242] Fermentation of resistant carbohydrates by anaerobic bacteria produces acetic, propionic and butyric acids which are known to be the preferred respiratory fuel of cells lining the colon.[243,244,245] Production of short chain fatty acids is accompanied by a decrease in pH in the lumen, increased colonic blood flow and reduced development of abnormal colonic cell populations.[246] The health benefits of resistant starch have been reviewed by Cummings and Englyst[247] and Nugent.[246]

Englyst et al.[248] classified starch into three types: RS1, RS2 and RS3. RS1 is a starch protected from digestion in the small intestine by a food matrix, such as the intact endosperm of a wheat kernel. RS2 is simply unswollen granular starch. RS3 is starch that was gelatinized, but then retrograded to become enzyme-resistant. Chemically modified (e.g. crosslinked) starch (RS4) has been added to the list.[242] The four forms of resistant starch and their measurement are described by Nugent[246] and Sajilata et al.[249]

There is no food labeling convention for resistant starch. Resistant starch is included within the definitions of dietary fiber by the American Association of Cereal Chemists[250] and is measured as fiber via the AOAC method.[251] Where the Prosky fiber method is used (US, UK, Australia and Japan) commercially-manufactured sources of resistant starch can be used as vehicles to increase the total dietary fiber content of foods and food products. In addition to being enriched in resistant starch, they may also have to be more process tolerant; to remain 'resistant' a starch must survive food processing, distribution and preparation for consumption as an indigestible entity.[249] Different approaches have been used to manufacture resistant starch ingredients for various food applications. These reflect efforts to provide cost-effective ingredients with physiological and labeling benefits and the ability to survive food processes presenting a range of thermal severity. McNaught et al.[252] claim the use of starches with 80% or more amylose as a source of resistant fiber (RS2). Shi and Jeffcoat[253] describe a method for annealing a high-amylose starch to increase both dietary fiber content and process tolerance. Haynes et al.[254] claim a process for gelatinizing, then annealing, starch to make an RS3 product for use in baked food products. In such a process, granular swelling, nucleation and crystal propagation are optimized to maximize effective total dietary fiber (TDF). DSC may be used to confirm the stability of the retrograded starch to processing. Several patents describe preparation of an RS3 by debranching amylose-containing starches with isoamylase or pullulanase followed by allowing the short chain amylose thus created to retrograde into amylase-resistant complexes.[255,256,257,258] These are intended for use in nutritional supplements, as well as edible products.

Seib and Woo[259] describe a process for making a highly-crosslinked distarch phosphate for use as an RS4 starch. Thompson and Brumovsky[260] claim another approach to making RS4. Their method involves annealing the linear portions of acid-hydrolyzed starches to produced TDF values as high as 60%.

IX. References

1. Knight JW. *The Starch Industry*. London, UK: Pergamon Press; 1969.
2. Walton RP. In: Walton RP, ed. *A Comprehensive Survey of Starch Chemistry*. Vol. 1. New York, NY: Chemical Catalog Co.; 1928:235.
3. Moffett GM. In: Walton RP, ed. *A Comprehensive Survey of Starch Chemistry*. Vol. 1. New York, NY: Chemical Catalog Co; 1928:130.
4. Lipschultz M, personal communication.
5. Preuss E. In: Walton RP, ed. *A Comprehensive Survey of Starch Chemistry*. Vol. 1. New York, NY: Chemical Catalog Co.; 1928:139.
6. Naegeli CW. *Ann. Chem.* 1874;173:218.
7. Lintner CJ. *J. Prakt. Chem.* 1886;34:378.
8. Bellmas B. German Patent 110 957. 1897.
9. Duryea CB. US Patent 675 822. 1901.
10. Kerr RW. *Chemistry and Industry of Starch*. 2nd edn. New York, NY: Academic Press; 1950:535.
11. Corn Refiner's Association. *Annual Report*. DC: Washington; 1996.
12. Rapaille A, Vanhemelrijck J. In: Imeson A, ed. *Thickening and Gelling Agents for Food*. London, UK: Blackie Academic and Professional; 1992:171.
13. Fuller AD. US Patent 2 317 752. 1943.
14. Koenigsberg M. US Patent 2 50 950. 1950.
15. Felton GE, Schopmeyer HH. US Patent 2 328 537. 1943.
16. Kerr RW, Cleveland FC. US Patent 2 801 242. 1957.
17. Caldwell CG. US Patent 2 461 139. 1949.
18. Wurzburg OB. US Patent 2 935 510. 1960.
19. Tuschoff JV. US Patent 3 022 289. 1962.
20. Wurzburg OB. *Cereal Foods World*. 1986;31:897.
21. Kessler CC, Hjermstad ET. US Patent 2 516 633. 1950.

22. Caldwell CG, Wurzberg OB. US Patent 2 661 349. 1953.
23. Pitchon E, O'Rourke JD, Joseph TH. US Patent 4 280 851. 1981.
24. Eastman JE. US Patent 4 452 978. 1984.
25. White PJ. In: Hallauer AR, ed. *Specialty Corns*. Boca Raton, FL: Chemical Rubber Co; 1994:29.
26. Christianson DD, Bagley EB. *Cereal Chem*. 1983;60:116.
27. Doublier JL, Paton D, Llamas G. *Cereal Chem*. 1987;64:21.
28. Eliasson AC, Bohlin L. *Starch/Stärke*. 1982;34:267.
29. Atwell WA, Hood LF, Lineback DR, Varriano-Marston E, Zobel HF. *Cereal Foods World*. 1988;33:306.
30. Leach HW, McCowen LD, Schoch TJ. *Cereal Chem*. 1959;36:534.
31. von Hofstee J. *Starch/Stärke*. 1962;9:318.
32. Miller BS, Derby RI, Trimbo HB. *Cereal Chem*. 1973;50:271.
33. Morris VJ. *Trends Food Sci. Technol*. 1990;1:2.
34. Doublier JL. *Starch/Stärke*. 1981;33:415.
35. Steenken PA, Woortman AJJ. *Roy. Soc. Chem., Spec. Publ*. 1993;113:147 [Food Colloids and Polymers: Stability and Mechanical Properties].
36. Lund D. *CRC Critical Rev. Food Sci. Nutr*. 1984;20:249.
37. Evans ID, Lips A. *Roy Soc. Chem., Spec. Publ*. 1992;113:214 [Food Colloids and Polymers: Stability and Mechanical Properties].
38. Eliasson AC, Gudmundsson M. In: Eliasson AC, ed. *Carbohydrates in Food*. New York, NY: Marcel Dekker; 1996:431.
39. Eliasson AC, Finstad H, Ljunger G. *Starch/Stärke*. 1988;40:95.
40. Kuhn K, Schlauch S. *Starch/Stärke*. 1994;46:208.
41. Parrott D. APV Co., personal communication.
42. Schoch TJ. In: Tressler DK, Van Arsdel WB, Copley MJ, eds. *The Freezing Preservation of Foods*. Vol. 4. New York, NY: Van Nostrand Reinhold/AVI; 1968:44.
43. Eliasson AC, Kim HR. *J. Text. Stud*. 1992;23:279.
44. Schoch TJ. *Bakers' Dig*. 1965;39:48.
45. Ring SG, Colona P, Anson KJ, Kalichevsky MT, Miles MJ, Morris VJ, Orford PD. *Carbohydr. Res*. 1987;162:277.
46. Schoch TJ. *Meth. Carbohydr. Chem*. 1965;4:106.
47. Ellis HS, Ring SG, Whittam MA. *Food Hydrocoll*. 1988;2:321.
48. Albrecht JJ, Nelson AI, Steinberg MP. *Food Technol*. 1960;14:57.
49. White PJ, Abbas IR, Johnson LA. *Starch/Stärke*. 1989;41:176.
50. Moore CO, Tuschoff JV, Hastings CW, Shanefelt RV. In: Whistler RL, BeMiller JN, Paschall EF, eds. *Starch: Chemistry and Technology*. 2nd edn. New York, NY: Academic Press; 1984:575.
51. Osman EM. In: Whistler RL, Paschall EF, eds. *Starch: Chemistry and Technology, Vol. II, Industrial Aspects*. New York, NY: Academic Press; 1967:163.
52. Navarro AS, Martino MN, Zaritzky NE. *J. Food Eng*. 1995;26:481.
53. Evans JW. US Patent 2 806 026. 1957.
54. Wurzburg OB. In: Stephen AM, ed. *Food Polysaccharides and their Applications*. New York, NY: Marcel Dekker; 1995:67.
55. Wu Y, Seib PA. *Cereal Chem*. 1990;67:202.

56. Yook C, Pek UH, Park K-H. *J. Food Sci.* 1993;58:405.
57. Kerr RW, Cleveland FC. US Patent 3 021 222. 1962.
58. Food and Drug Administration, 21 Code of Federal Regulations 172.892; 1995.
59. Wurzburg OB. In: Furia TE, ed. *Handbook of Food Additives*. 2nd edn. Cleveland, OH: CRC Press; 1972:361.
60. Schoch TJ. *Cereal Chem.* 1964;18:121.
61. Rutenberg MW, Jarowenko W, Ross LJ. US Patent 3 038 895. 1962.
62. Smith CE, Tuschoff JV. US Patent 2 928 828. 1960.
63. de Conninck V, Vanhemelrijck J, Peremans J. *Food Market. Technol.* 1995;9:5.
64. Neukom H. US Patent 2 884 727. 1959.
65. Kerr RW, Cleveland FC. US Patent 2 961 440. 1960.
66. Trubiano P. In: Wurzburg OB, ed. *Modified Starches – Properties and Uses*. Boca Raton, FL: CRC Press; 1968:131.
67. Wurzburg OB, Herbst W, Cole HM. US Patent 3 091 567. 1963.
68. Marotta NG, Boettger RM, Nappen BH, Szymanski CD. US Patent 3 455 838. 1969.
69. Kenyon MM. In: Risck SJ, Reineccius GA, eds. *Encapsulation and Controlled Release of Food Ingredients*. 590: ACS Symposium Series; 1995:42.
70. Caldwell CG. US Patent 2 613 206. 1952.
71. Leach HW, Schoch TJ. *Cereal Chem.* 1962;39:318.
72. Bechtel WG. *J. Colloid Sci.* 1950;5:260.
73. Kruger LH, Rutenberg MW. In: Whistler RL, Paschall EF, eds. *Starch: Chemistry and Technology, Vol. II. Industrial Aspects*. New York, NY: Academic Press; 1967:369.
74. Lenchin JM, Trubiano P, Hoffman S. US Patent 4 510 166. 1985.
75. Rutenberg MW, Solarek D. In: Whistler RL, BeMiller JN, Paschall EF, eds. *Starch: Chemistry and Technology*. 2nd edn. New York, NY: Academic Press; 1984:312.
76. Langan RE. In: Wurzberg OB, ed. *Modified Starches: Properties and Uses*. Boca Raton, FL: CRC Press; 1986:199.
77. Fuller AD. US Patent 2 173 878. 1939.
78. Hullinger CH. In: Whistler RL, Paschall EF, eds. *Starch: Chemistry and Technology, Vol. II. Industrial Aspects*. New York, NY: Academic Press; 1967:445.
79. Colonna P, Buleon A, Mercier C. In: Galliard T, ed. *Starch: Properties and Potential*. New York, NY: John Wiley; 1987.
80. Radley JA. *Starch Production Technology*. London, UK: Chapman and Hall; 1976.
81. Powell EL. In: Whistler RL, Paschall EF, eds. *Starch: Chemistry and Technology, Vol. II. Industrial Aspects*. New York, NY: Academic Press; 1967:523.
82. Marotta NG, Trubiano P, Ronai KS. US Patent 3 443 964. 1969.
83. Millauer C, Wiedmann M. *Starch/Stärke*. 1984;36:228.
84. Mercier C. *Stärke*. 1977;29:48.
85. Colonna P, Doublier JL, Melcion JP, de Monredon F, Mercier C. *Cereal Chem.* 1984;61:538.
86. Davidson VJ, Paton D, Diosady LL, Larocque GJ. *J. Food Sci.* 1984;49:453.
87. Doublier JL, Colonna P, Mercier C. *Cereal Chem.* 1986;63:240.
88. de Conninck V. *Food Market. Technol.* 1996;10:5.
89. Boersen AC. *Stork Agglomeration Processes*. Gorredijk, The Netherlands: Stork Friesland BV.; 1993.

90. Jane JL. *Trends Food Sci. Technol.* 1992;3:145.
91. Rubens RW. US Patent 5 149 799. 1992.
92. Eastman JE, Moore LD. US Patent 4 465 702. 1984.
93. Eastman JE. US Patent 4 634 596. 1987.
94. Jane JL, Seib PA. US Patent 5 057 157. 1991.
95. Rajagopalan S, Seib PA. US Patent 5 037 929. 1991.
96. Larsson I, Eliasson AC. *Starch/Stärke.* 1991;43:227.
97. Krueger BR, Knutson CA, Inglett GE, Walker CE. *J. Food Sci.* 1987;52:715.
98. Jacobs H, Eerlingen RC, Delcour JA. *Starch/Stärke.* 1996;48:266.
99. Stute R. *Starch/Stärke.* 1992;44:205.
100. Gough BM, Pybus JN. *Stärke.* 1971;23:210.
101. Sair L. *Cereal Chem.* 1967;44:8.
102. Lorenz K, Kulp K. *Starch/Stärke.* 1982;34:50.
103. Lorenz K, Kulp K. *Starch/Stärke.* 1982;34:76.
104. Donovan JW, Lorenz K, Kulp K. *Cereal Chem.* 1983;60:381.
105. Abraham TC. *Starch/Stärke.* 1993;45:131.
106. Smalligan WJ, Kelly VJ, Enad EG. US Patent 4 013 799. 1977.
107. Schierbaum F, Kettlitz B. *Starch/Stärke.* 1994;46:2.
108. Krake B, Peer R, Mocka E, Pfeifer K. WPI 91-000179/01; 1991.
109. Suh C-S, Kim S-K. *Agric. Chem. Biotechnol.* 1995;38:353 [*Biol. Abstr.* 1995; **100**:183466].
110. Carter JB, Brown JM, Hodges RC, Appelqvist IA, Brown CR, Norton IT. US Patent 5 538 751. 1994.
111. Ajinomoto, Japan Patent 58/071900. 1983.
112. Redding BK. WPI 93-351489/44. 1993.
113. Kudta E, Tomasik P. *Starch/Stärke.* 1992;44:167.
114. Morrison WR, Tester RF, Gidley MJ. *J. Cereal Sci.* 1994;19:209.
115. Meuser F, Klingler RW, Niedick EA. *Stärke.* 1978;30:376.
116. Tuschoff JV, Eastman JE, Schanefelt RV. US Patent 4 256 509. 1981.
117. Radosta S, Kettlitz B, Schierbaum F, Gernat C. *Starch/Stärke.* 1992;44:8.
118. Marquette GHA, Gonze M, Lane C. WPI 82-74938E/36; 1982.
119. Muzimbaranda C, Tomasik P. *Starch/Stärke.* 1994;46:469.
120. FAO Codex Alimentarius Commission. *Food Labelling.* 2005;4.2.3.1:5.
121. Chiu CW, Schiermeyer E, Thomas DJ, Shah MB. US Patent 5 725 676. 1998; *Chem. Abstr.* 1998;128:193,928.
122. Mahr B, Trueck HU. US Patent 5 928 707. 1999.
123. Kettlitz BW, Coppin JV. US Patent 6 235 894. 2001.
124. Fergason V. In: Hallauer AR, ed. *Specialty Corns*. Boca Raton, FL: Chemical Rubber Co.; 1994:55.
125. Edith Monroe, Feed Grain Institute, personal communication.
126. McDonald TA. *Proceedings of the 28th Corn Sorghum Research Conference*. Washington, DC: ASTA; 1973 [98–107; cited in reference 122].
127. Palmer HH. In: Tressler DK, Van Arsdell WB, Copley MJ, eds. *The Freezing Preservation of Foods*. Vol. 4. Wesport, CT: AVI Publ. Co.; 1968:314.
128. Wurzburg OB, Fergason VL. US Patent 4 428 972. 1981.

129. Fergason V, Jeffcoat R, Fannon J, Capitani T. US Patent 5 300 145. 1994.
130. Lacourse NL, Zallie JP. US Patent 4 726 957. 1988.
131. Lenchin JM, Bell H. US Patent 4 529 607. 1985.
132. Van Patten EM, Freck JA. US Patent 3 751 268. 1973.
133. Cremer CW. US Patent 3 987 210. 1976.
134. Chinnaswamy R, Hanna MA. *Cereal Chem.* 1988;65:138.
135. Huang DP. *Cereal Foods World.* 1995;40:528.
136. Zwiercan GA, Lacourse NL, Lenchin JM. US Patent 4 608 265. 1986.
137. Miller BJ, Bell H, Wojcak E. US Patent 4 590 084. 1986.
138. Morimoto K. US Patent 4 562 082. 1985.
139. Moore CO. US Patent 4 120 987. 1978.
140. Furcsik SL, Mauro DJ, DeBoer E, Yahl K, Delgado G. US Patent 4 981 709. 1991.
141. Bridgeford DG. US Patent 4 592 795. 1986.
142. Wolke M, Zakin R. US Patent 5 104 669. 1992.
143. Tessler MM, Jarowenko W. US Patent 3 970 767. 1976.
144. Van Patten EM. O'Rell DH, Kondrot LB. US Patent 3 792 176. 1972.
145. Mitch EL. In: Whistler RL, BeMiller JN, Paschall EF, eds. *Starch: Chemistry and Technology.* 2nd edn. New York, NY: Academic Press; 1984:479.
146. Hofvander P, Persson P, Tallberg A, Wikstrom O. WO 92/11376 A1. 1992.
147. Shi YC, Seib PA. In: Pomeranz Y, ed. *Wheat is Unique.* St. Paul, MN: American Association of Cereal Chemists; 1989.
148. Takahashi S, Maningat CC, Seib PA. *Cereal Chem.* 1989;66:499.
149. Russo JW, Doe CA. *J. Food Technol.* 1970;5:365.
150. Champagne ET. *Cereal Foods World.* 1996;41:833.
151. Juliano BO. In: Whistler RL, BeMiller JN, Paschall EF, eds. *Starch: Chemistry and Technology.* 2nd edn. New York, NY: Academic Press; 1984:507.
152. McKenzie KS. In: Marshall WE, Wadsworth JI, eds. *Rice Science and Technology.* New York, NY: Marcel Dekker; 1994:83.
153. Bost R, Chigurupati SR, Scherph D. WP 92/13465. 1992.
154. Bakal AI, Cash PA, Galbreath T. US Patent 5 137 742. 1992.
155. Sharp RN, Sharp CQ. In: Marshall WE, Wadsworth JI, eds. *Rice Science and Technology.* New York, NY: Marcel Dekker; 1994:405.
156. Orthoefer FT, McCaskill DR, Dooper DS. *Food Technol. Eur.* 1995;3(2):36.
157. Schoch TJ. In: Whistler RL, Paschall EF, eds. *Starch: Chemistry and Technology, Vol. II. Industrial Aspects.* New York, NY: Academic Press; 1967:79.
158. Corbishley DA, Miller W. In: Whistler RL, BeMiller JN, Paschall EF, eds. *Starch: Chemistry and Technology.* 2nd edn. New York, NY: Academic Press; 1984:469.
159. Petersen PB, Munck L. In: MacGregor AW, Bhatty RS, eds. *Barley: Chemistry and Technology.* St. Paul, MN: American Association of Cereal Chemists; 1993:437.
160. Seib PA, Wu Y. US Patent 4 973 447. 1990.
161. Blenford D. *Int. Food Ingred.* 1994;27.
162. Stute R. *Starch/Stärke.* 1990;42:207.
163. Chevalier O. *Food Technol. Eur.* 1994;1:126.
164. Perez E, Bahnassey YA, Breene WM. *Starch/Stärke.* 1993;45:215.
165. Potter NN. *Food Science.* Westport, CT: Avi Publishing Co.; 1968.

166. Holdsworth SD. *Aseptic Processing and Packaging of Food Products*. New York, NY: Elsevier Applied Science; 1992.
167. Hanson HL, Campbell A, Lineweaver H. *Food Technol.* 1951;5:256.
168. Zallie J, Eden J, Kasica J, Chiu CW, Zwiercan GA, Plutchok G. US Patent 5 281 432. 1994.
169. Cochran SA, Benjamin EJ, Crocker ME, Seidel WC, Cipriano VL. European Patent 305 105 A2. 1989.
170. Nakao Y, Yamaguchi T, Taguchi T. *J. Jpn. Soc. Food Sci. Technol.* 1994;41:141.
171. Brueckner J, Muschiolik G, Mieth G, Ackermann K. DDR Patent 250 048. 1987.
172. Trubiano PC, Lacourse NL. In: Risch J, Reineccius G, eds. *Flavor Encapsulation*. 370: ACS Symp. Ser.; 1986:45.
173. Reineccius GA. *Food Technol.* 1991;45:144.
174. Zhao J, Whistler RL. *Food Technol.* 1994;48:104.
175. Maier HG, Moritz K, Ruemmler U. *Starch/Stärke*. 1987;39:126.
176. Barnes JM, Steinke JA. US Patent 4 689 235. 1987.
177. Takashima H. US Patent 6 884 448. 2005.
178. Shi YC, Chiu CW, Huang DP, Janik D. US Patent 6 410 073. 2005.
179. Howling D. *Eur. Food Drink Rev.* 1991;69(3).
180. Bindzus W, Altieri PA. US Patent 6 461 656. 2002.
181. Woerman JH; Wu Y. US Patent 5 750 168. 1998.
182. Rogols S, Woerman JH. US Patent 5 897 898. 1998.
183. Sloan JL, Middaugh KF, Jacobsen GB. US Patent 5 059 435. 1991.
184. Bretch EE. US Patent 3 703 378. 1972.
185. Demos BP, Forrest JC, Grant AL, Judge MD, Chen LF. *J. Muscle Foods*. 1994;5:407.
186. Rust RE. *Meat Process.* 1992;20:42.
187. Fazzina T, Gilmore G, McSweeney D. US Patent 3 852 501. 1974.
188. Kim JM. *Diss. Abstra. Int.* 1987;B48:1562.
189. King W. *Cereal Sci. Today*. 1974;19:385.
190. Yackel WC, Cox C. *Food Technol.* 1992;46:16.
191. Richter M, Schierbaum F, Agustat S, Knoch K-D. US Patent 3 962 465. 1976.
192. Richter M, Schierbaum F, Agustat S, Knoch K-D. US Patent 3 986 890. 1976.
193. Morehouse AL, Lewis CJ. US Patent 4 536 408. 1985.
194. Inglett GE. US Patent Appl. 373 978. 1990; US Patent 5 225 219. 1993.
195. Inglett GE. *Abstr. Papers, Am. Chem. Soc.* 1990;199 [AGFD 15].
196. Duxbury D. *Food Process.* 1990(8):48.
197. Chiou RG, Brown CC, Little JA, Young AH, Schanefelt RV, Harris DW, Stanley KD, Coontz HD, Hamdan CJ, Wolf-Rueff JA, Slowinski LA, Anderson KA, Lehnhardt WF, Witczak ZJ. European Patent 443 844 A1. 1991.
198. Whistler RL. US Patent 5 580 390. 1996.
199. Capitani TA, Trzasko P, Zallie JP, Mason WR. US Patent 5 512 311. 1995.
200. Chiu CW, Mason WR. European Patent 486 936. 1994.
201. Mallee FM, Finoccharo ET, Stone JA. US Patent 5 470 391. 1994.
202. Fanta GF, Eskins K. *Carbohydr. Polym.* 1995;28:171.
203. Baensch J, Gumy D, Seivert D, Würsch P. PCT Int. Appl. WO 96/03,057 A1. 1996.

204. Jongh G. *Cereal Chem.* 1961;38:140.
205. Hoseney RC. *Principles of Cereal Science and Technology.* 2nd edn. St. Paul, MN: American Association of Cereal Chemists; 1984.
206. Wootton M, Ho P. *Starch/Stärke.* 1989;41:261.
207. Muhrbeck D, Eliasson AC. *Carbohydr. Polym.* 1987;7:291.
208. Ruggeberg H. *Stärke.* 1953;5:109.
209. Sandstedt RM, Kempf W, Abbott RC. *Stärke.* 1960;12:333.
210. Evans ID, Haisman DR. *Starch/Stärke.* 1982;34:224.
211. Wootton M, Bamunuarachchi A. *Starch/Stärke.* 1980;32:126.
212. Takahashi K, Wada K. *J. Food Sci.* 1992;5:1140.
213. Sudhakar V, Singhal RS, Kulkarni PR. *Starch/Stärke.* 1992;44:369.
214. Bello-Perez LA, Paredes-Lopez O. *Starch/Stärke.* 1995;47:83.
215. Woodruff S, Nicolai L. *Cereal Chem.* 1931;8:243.
216. Beleia A, Miller RA, Hoseney RC. *Starch/Stärke.* 1996;48:259.
217. Freke CD. *J. Food Technol.* 1971;6:281.
218. Buck JS, Walker CE. *Stärke.* 1977;40:353.
219. Germani R, Ciacco CF, Rodriguez-Amaya DB. *Stärke.* 1983;35:377.
220. Osman EM, Dix MR. *Cereal Chem.* 1960;37:464.
221. Krog N. *Stärke.* 1971;23:206.
222. Ghiasi K, Hoseney RC, Varriano-Marston E. *Cereal Chem.* 1982;59:81.
223. Eliasson AC. *J. Texture Stud.* 1987;17:357.
224. Villwock VK, Eliasson A-C, Silverio J, BeMiller JN. *Cereal Chem.* 1999;66:292.
225. Evans ID. *Starch/Stärke.* 1986;38:227.
226. Lelievre J, Husbands J. *Starch/Stärke.* 1989;41:236.
227. Marzin C, Doublier JL, Lefebvre J. In: Dickinson E, Lovient D, eds. *Food Macromolecules and Colloids.* 156: Roy. Soc. Chem. Spec. Publ.; 1995.
228. Eliasson AC, Tjernald E. *Cereal Chem.* 1990;67:366.
229. Brownsey GJ, Orford PD, Ridout MJ, Ring SG. *Food Hydrocoll.* 1989;3:7.
230. Nadison J, Doreau A. In: Phillips GO, Williams PA, Wedlock DJ, eds. *Gums and Stabilisers for the Food Industry 6.* New York, NY: IRL Press; 1992:287.
231. Shi X, BeMiller JN. *Carbohydr. Polym.* 2002;50:7.
232. Coffey DG. US Patent 4 915 970. 1990.
233. Kohyama K, Nishinari K. *J. Food Sci.* 1992;57:128.
234. Alloncle M, Doublier JL. In: Phillips GO, Williams PA, Wedlock DJ, eds. *Gums and Stabilisers for the Food Industry 5.* Oxford, UK: IRL Press; 1990:111.
235. Liu H, Lelievre J. *Cereal Chem.* 1992;69:597.
236. Coffin DR, Fishman ML. *J. Agric. Food Chem.* 1993;41:1192.
237. Langoreieux S, Crouzet J. In: Charalambous G, ed. *Food Flavors: Generation, Analysis and Process Influence, Proceedings of the 8th International Flavor Conference.* Amsterdam, The Netherlands: Elsevier; 1995 [Cos, Greece, July 1994].
238. Cummins JH, Englyst HN. *Can. J. Physiol. Pharmacol.* 1991;69:121.
239. Asp NG. *Am. J. Clin. Nutr.* 1994;59:5679.
240. Cassidy A, Bingham SA, Gummings JH. *Brit. J. Cancer.* 1994;69:119.
241. Silvi S, Rumney CJ, Cresci A, Rowland IR. *J. Appl. Microbiol.* 1999;86:521.
242. Thompson DB. *Trends Food Sci. Technol.* 2000;11:245.

243. Macfarlane GT, Gummings JH. In: Phillips SF, Pemberton JH, Shorter RG, eds. *The Large Intestine: Physiology, Pathophysiology, and Disease*. New York, NY: Raven Press; 1991:51.
244. Kleessen B, Stoof G, Proll J, Schmiedl D, Noack J, Blaut M. *J. Animal Sci.* 1997;75:2453.
245. Le Blay G, Michel C, Blottiere HM, Cherbut C. *Brit. J. Nutr.* 1999;82:419.
246. Nugent AP. *Nutr. Bull.* 2005;30:27.
247. Cummings JH, Englyst HN. *Am. J. Clin. Nutr.* 1995;61(Suppl.):938S.
248. Englyst HN, Kingman SM, Cummings JH. *Eur. J. Clin. Nutr.* 1992;46(Suppl. 2):S33.
249. Sajilata MG, Singhal RS, Kulkarni PR. *Comp. Rev. Food Sci. Food Safety.* 2006;5:1.
250. Jones J. *Cereal Foods World.* 2000;45:219.
251. Prosky L, Asp NG, Furda I, DeVries JW, Schweizer TR, Harland B. *J. Assoc. Off. Anal. Chem.* 1985;68:677.
252. McNaught KJ, Moloney E, Brown IL, Knight AT. US Patent 5 714 600. 1998.
253. Shi YC, Jeffcoat R. US Patent 6 664 389. 2003.
254. Haynes L, Gimmler N, Locke JP, III, Kweon MR, Slade L, Levine H. US Patent 6 013 299. 2000.
255. Haralampu SG, Gross A. US Patent 5 849 090. 1998.
256. Kettlitz BW, Moorthamer ET, Petrus BM, Anger H, Stoof GR. US Patent 6 090 594. 2000.
257. Shi YC, Cui X, Burkitt AM, Thatcher MG. US Patent 6 890 571. 2005.
258. Shi YC, Cui X, Burkitt AM, Thatcher MG. US Patent 6 929 817. 2005.
259. Seib PA, Woo K. US Patent 5 855 946. 1999.
260. Thompson DB, Brumovsky J. US Patent 6 468 355. 2002.

21

Sweeteners from Starch: Production, Properties and Uses

Larry Hobbs

I. Introduction	797
1. History	797
2. Definitions	799
3. Regulatory Status	800
II. Production Methods	800
1. Maltodextrins	800
2. Glucose/corn Syrups	802
3. High-fructose Syrups	808
4. Crystalline Fructose	813
5. Crystalline Dextrose and Dextrose Syrups	813
6. Oligosaccharide Syrups	816
III. Composition and Properties of Sweeteners from Starch	817
1. Carbohydrate Profiles	817
2. Solids	818
3. Viscosity	819
4. Browning Reaction and Color	821
5. Fermentability	822
6. Foam Stabilization and Gel Strength	823
7. Freezing Point Depression	824
8. Boiling Point Elevation	824
9. Gelatinization Temperature	824
10. Humectancy and Hygroscopicity	825
11. Crystallization	826
12. Sweetness	827
13. Selection of Sweeteners	828
IV. References	829

I. Introduction

1. History

Commercial production of sugar in the Indus valley was reported during Alexander the Great's invasion in the period around 325 BCE, but cane sugar did not reach Europe until the crusades.[1] Received at first as a novelty, Europeans developed a taste for sugar and demand for this new ingredient developed rapidly. During the sixteenth century, the Caribbean islands became major producers of this crop until the supply was disrupted during the Napoleonic Wars. With cane sugar supplies cut off by the British blockade, France turned to sugar beets. In 1811, G.S.C. Kirchoff, a Russian chemist, discovered that acid-catalyzed hydrolysis of starch produced a sweet substance.[2] By 1831, an American syrup plant capable of producing 30 gallons (115 liters) of syrup per day utilizing this new technology had been built; 150 years later, 140 American plants were producing starch from corn, wheat, potatoes and rice.[3]

Figure 21.1 provides a dramatic presentation of the growth of the industry since 1910.[4] The use of corn by the United States' corn refining industry increased to 1.4 billion bushels (39×10^6 tons, 36×10^9 kg) in 1999, which was about 15% of the total crop harvested. From that production came 33 billion pounds (15×10^9 kg) of sweeteners, more than 2.5 times the amount produced in 1984 when the last edition of this book was published.[5]

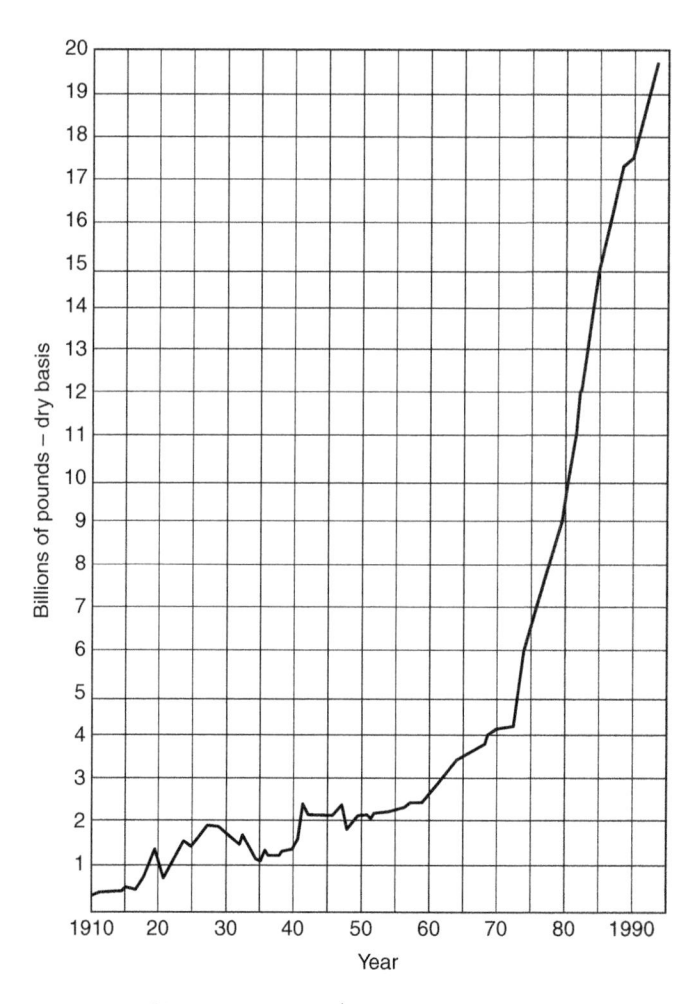

Figure 21.1 Corn sweetener shipments since 1910.[4]

In order to discuss the changes that have happened in the industry and the forces driving them, it is necessary to first define the types of sweeteners covered.

2. Definitions

Dextrose equivalence (DE) is a measure of the total reducing sugars calculated as D-glucose on a dry weight basis. The approved method for determining DE is the Lane–Eynon titration, which measures reduction of a copper sulfate solution. Unhydrolyzed starch has a DE value of zero, while the DE value of anhydrous D-glucose is 100. Glucose/corn syrups range from 20 to 95 DE.

Maltodextrins are the dried products or purified aqueous solutions of saccharides obtained from edible starch having a dextrose equivalency of less than 20. Outside the US, the products may be known as dextrins; only the US has an official definition of maltodextrins.

Glucose syrups, also know as *corn syrups* in the US, are purified aqueous solutions of nutritive saccharides obtained from edible starch having a dextrose equivalency of 20 or more.

Dried corn syrups or *corn syrup solids* are glucose/corn syrups from which most of the water has been removed.

High fructose syrups are purified aqueous solutions of nutritive saccharides obtained from edible starch in which a portion (at least 42%) of the dextrose (D-glucose) has been isomerized to fructose.

Crystalline fructose is crystalline product containing not less than 98.0% fructose and not more than 0.5% glucose.

Dextrose monohydrate is purified, crystalline D-glucose containing one molecule of water of crystallization per molecule of D-glucose.

Anhydrous dextrose is purified, crystalline D-glucose without water of crystallization.

Baume (Be) units arise from an arbitrary system of graduating hydrometers in degrees for determining the specific gravity of a solution. Within the corn refining industry, Baume is related to specific gravity by the following equation:[6]

$$\text{Baume } (60°F/60°F) = 145 - \frac{145}{\text{true sp.gr.}} (60°F/60°F)$$

The modulus of 145 is the ratio of the total volume displaced in water by the hydrometer and the volume displaced by the unit scale length of the hydrometer stem. Corn syrups are commercially available with Baume values of 42, 43 and 44.

Degree of polymerization (DP) is the number of glucosyl (saccharide) units in an oligo- or polysaccharide. DP_1 refers to a monosaccharide, DP_2 refers to disaccharides and so on.

Refractive index (RI) is a measure of the refraction of light rays as they pass obliquely from one solution to another of different density. Refractive index is commonly used to measure the solids level of sweeteners. The refractive index of a sweetener is a function of the carbohydrate profiles, ash level, solids level and temperature of the solution.

Retrogradation is the reassociation of solubilized starch polymers in their native state or those in dextrins or in low-DE hydrolyzates resulting in an insoluble precipitate. Dextrins are depolymerized starches produced by heating a starch moistened with dilute hydrochloric acid or heating a moist starch in the presence of gaseous hydrogen chloride until a cold-water-soluble product is formed.

Reversion is the condensation reaction of reducing sugars to form di- and higher oligosaccharides.

3. Regulatory Status

Corn syrups, maltodextrins and D-glucose are affirmed as 'generally recognized as safe' (GRAS) in the US Code of Federal Regulations (CFR) 21, Section 184. High fructose corn syrup was affirmed as GRAS in 21CFR, Section 184.1866 on August 23, 1996.

II. Production Methods

The pathways for production of the various sweeteners share many common steps. A generalized sweetener process is shown in Figure 21.2.[7] Production of each of the sweeteners discussed will utilize one or more steps in this process.

1. Maltodextrins

The GRAS affirmation contained in 21 CFR, Section 184.1444, defines maltodextrins as non-sweet, nutritive saccharide polymers consisting of D-glucosyl units linked primarily with alpha-1,4 bonds and having a DE less than 20. The document has been modified to include maltodextrins derived from potato starch as GRAS.[8] In 1992, more than 328 million pounds (149×10^6 kg) of maltodextrins and corn syrup solids were produced in the United States from various starch sources.[9]

Maltodextrins may be manufactured either by acid or by acid–enzyme processes. Maltodextrins produced by acid conversion of starch from dent corn contain a high percentage of linear fragments, which may slowly reassociate into insoluble compounds causing haze in certain applications.

Haze formation, which results from retrogradation, can be overcome by use of alpha-amylases. Alpha-amylases preferentially cleave the alpha-1,4-D-glucosidic bonds of amylose and amylopectin (see Chapter 7), leaving a higher proportion of branched fragments, decreasing the ability of the fragments to reassociate. Maltodextrins made from waxy corn starch also have a lower tendency to haze, because such starch is composed almost entirely of the highly branched molecule, amylopectin.

In a maltodextrin process using enzyme-catalyzed conversion, the starch slurry (30% to 40% dry solids) is first pasted at a temperature of 80–90°C, and is then treated with a 'heat-stable' bacterial alpha-amylase for liquefication. When stabilized with calcium ions, alpha-amylases from *B. licheniformis* or *B. stearothermophilus* can withstand temperatures of 90–105°C for at least 30 minutes,[10] allowing sufficient process time to split the 1,4 bonds and form maltose and limit dextrins (see Chapter 7).

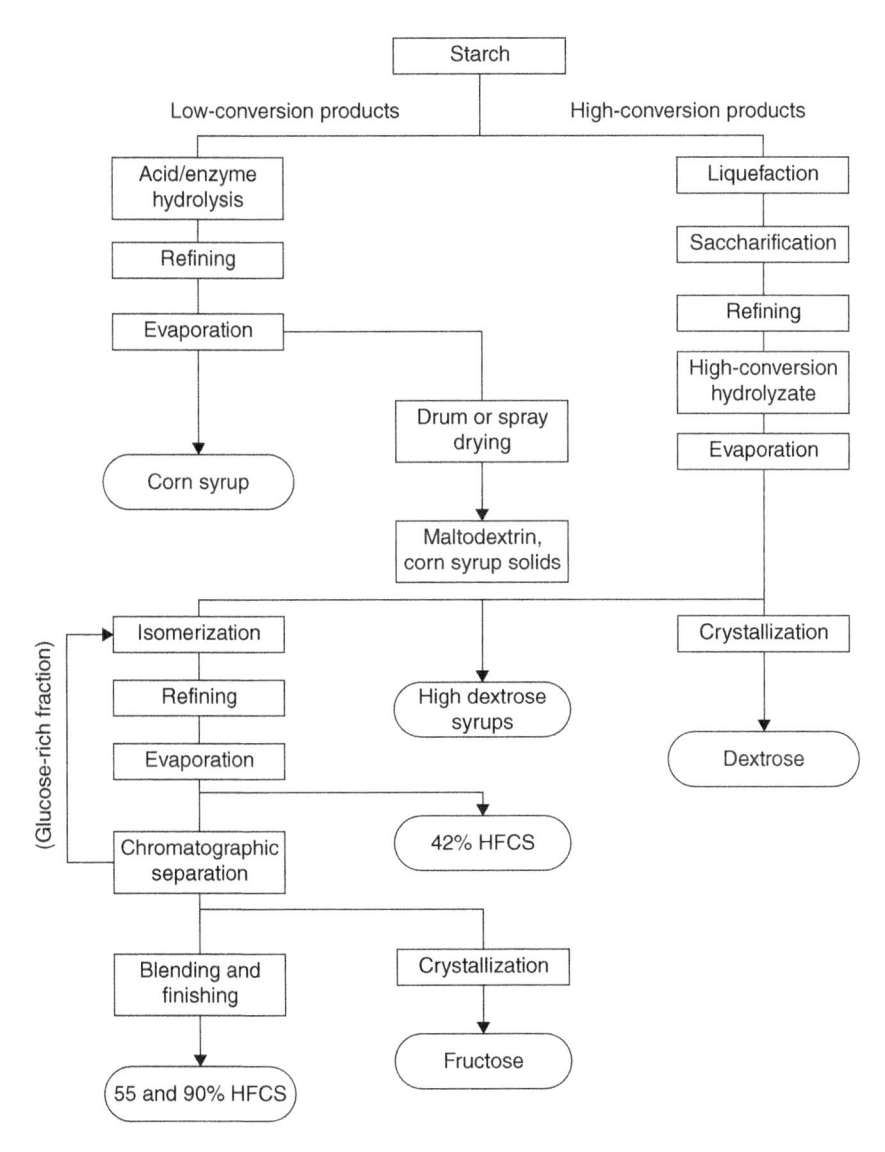

Figure 21.2 General process flow for starch-derived sweeteners (corn/glucose syrups, high fructose syrups, dextrose, fructose, maltodextrins and syrup solids).[7]

The fragmentation reaction proceeds until there is a preponderance of maltohexoses and maltoheptoses and the liquor has a DE of 12–15.[11]

A number of maltodextrin production methods using multiple enzyme treatments have been described.[12,13] In one such process, a starch paste is first treated with acid and/or an alpha-amylase at 95–105°C for liquefication, then cooled to 90–102°C, at which time addition of a second enzyme takes place. If desired, the slurry may be further cooled to 85°C for addition of a third enzyme.[14]

After conversion, the pH of the crude slurry is adjusted to about 4.5 and the solution is filtered to remove protein and fats. The clarified liquor is then refined in a

Table 21.1 Typical carbohydrate profile of commercial maltodextrins[15]

DP	5 DE	10 DE	15 DE	20 DE
1	<1	<1	<1	<1
2	1	3	6	8
3	2	4	7	9
4	2	4	5	7
5	2	4	5	8
6	3	7	11	14
7+	90	78	66	53

Maltodextrin composition (% dry basis)

process similar to that for glucose/corn syrups discussed in the next section. After refining and decolorization, the liquor is evaporated to a solids level of approximately 77% or dried to about 5% moisture.

Maltodextrin solutions are not evaporated to as low a solids level as is typical of most glucose syrups because the viscosity of the latter is extremely high (see Table 21.9). At the higher solids level of 43 or 44 Baume typical for corn syrups, maltodextrin solutions would be extremely difficult to pump. It should also be noted that, since the water activity of maltodextrins at a given solids level is so much higher than that of other syrups, some care must be exercised in the handling of these products to prevent microbial fermentation. Commercial maltodextrins, as shown in Table 21.1,[15] are used in applications where high viscosity coupled with a bland, neutral taste is desirable.

2. Glucose/corn Syrups

Corn syrups are affirmed as GRAS in 21 CFR, Section 184.1865 and meet the further standards of identity in Sections 168.120 and 168.121. In 1999, more than 7 billion pounds (3×10^9 kg) of syrups were produced from starch in the US.[16] Figure 21.3 outlines a conventional glucose/corn syrup manufacturing process.[17]

Glucose/corn syrups may be manufactured by either an acid or an acid–enzyme process. Acid-catalyzed hydrolysis was the traditional method of corn syrup production and is still the most common method for producing sweeteners up to about 42 DE. Since acid-catalyzed hydrolysis of sweeteners to 55 DE or above creates products of reversion, such as gentiobiose, isomaltose and trehalose, which give unacceptable flavors to the syrup, these syrups are usually made by acid–enzyme processes.[18]

Acid-catalyzed Hydrolysis

In an acid process the slurry, containing about 35–45% starch solids, is pumped into a pressure vessel called a 'converter' and acidified to a pH of about 2.0 with dilute hydrochloric acid at 140–160°C and a pressure of 80 psi (5.4 atm). Although acid-catalyzed hydrolysis is a rather (but not completely) random process,[19,20] carefully controlled hydrolysis produces syrups in the 25 to 45 DE range with very predictable carbohydrate profiles as shown in Table 21.2.[21]

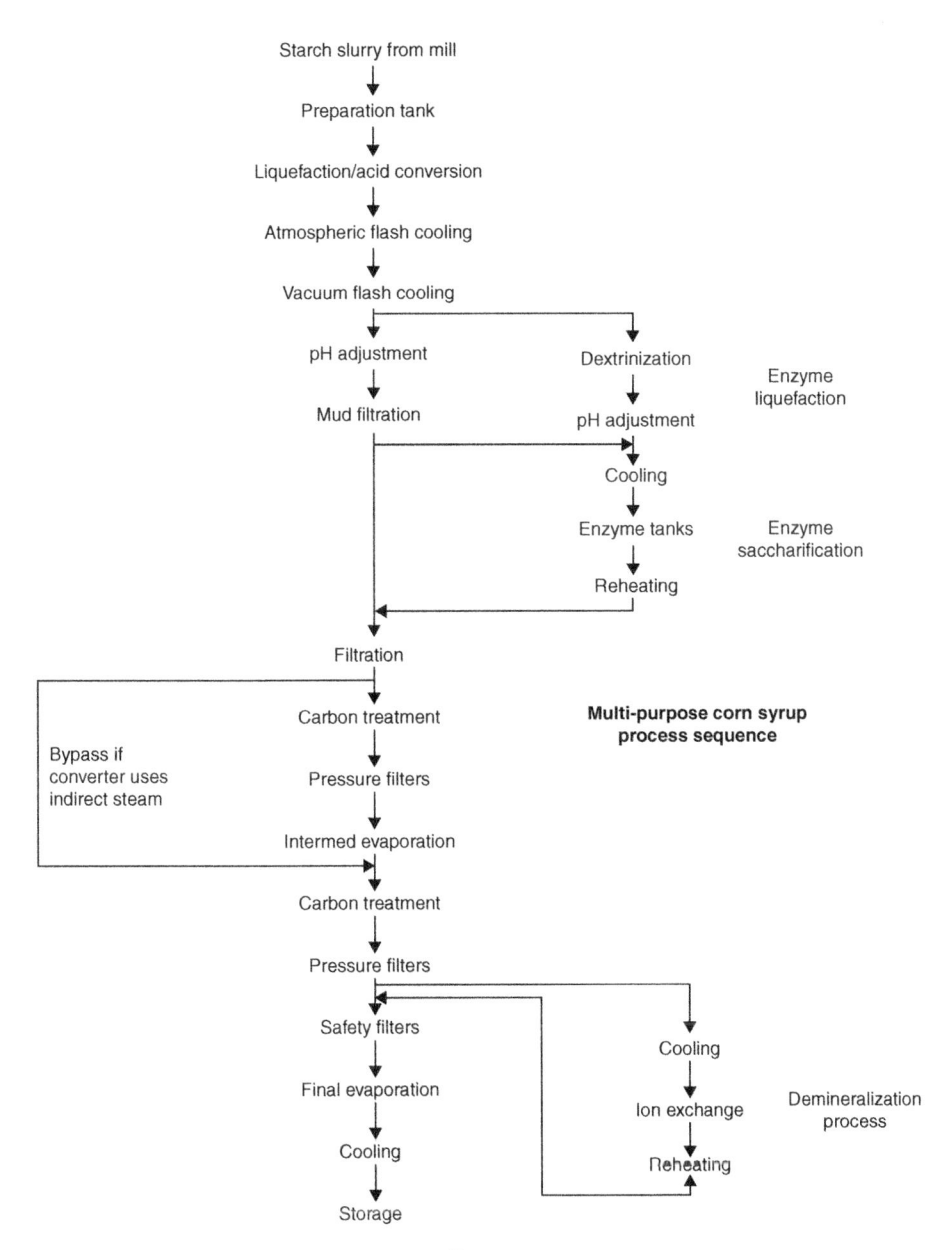

Figure 21.3 Typical corn/glucose syrup process.[17]

During hydrolysis, both 1,4 and 1,6 linkages are cleaved, converting the starch molecule to increasingly lower molecular weight products. A typical residence time in the converter is 5–10 minutes for low DE syrups; high DE syrups may require 15–20 minutes at the necessary temperature. It is important to keep the conversion time short to prevent unnecessary color development. A typical commercial acid converter system is shown in Figure 21.4.[22]

Table 21.2 Composition of typical starch-derived sweeteners[21]

Designation	Ash	Saccharides, carbohydrate basis			
		DP_1	DP_2	DP_3	DP_{4+}
28 DE	0.3	8	8	11	73
36 DE	0.3	14	11	10	65
34 HM	0.3	9	34	24	33
43 HM	0.3	9	43	18	30
43 DE	0.3	19	14	12	55
43 DE (IE)	0.03	19	14	12	55
53 DE	0.3	28	18	13	41
63 DE	0.3	36	31	13	20
63 DE (IE)	0.03	36	31	13	20
66 DE	0.3	40	35	8	17
95 DE	0.3	95	3	0.5	1.5
95 DE (IE)	0.03	95	3	0.5	1.5
HFCS 42	0.03	95	3	0.7	1.3
HFCS 55	0.05	95.7	3	0.4	0.9
Crystalline fructose	0.05	100			

DP_1 = Monosaccharides (dextrose, dextrose + fructose in HFS, fructose in crystalline fructose)

DP_2 = Disaccharides, primarily maltose

DP_3 = Trisaccharides, primarily maltotriose

DP_{4+} = Oligosaccharides, maltotetraose and higher saccharides

HM = High maltose

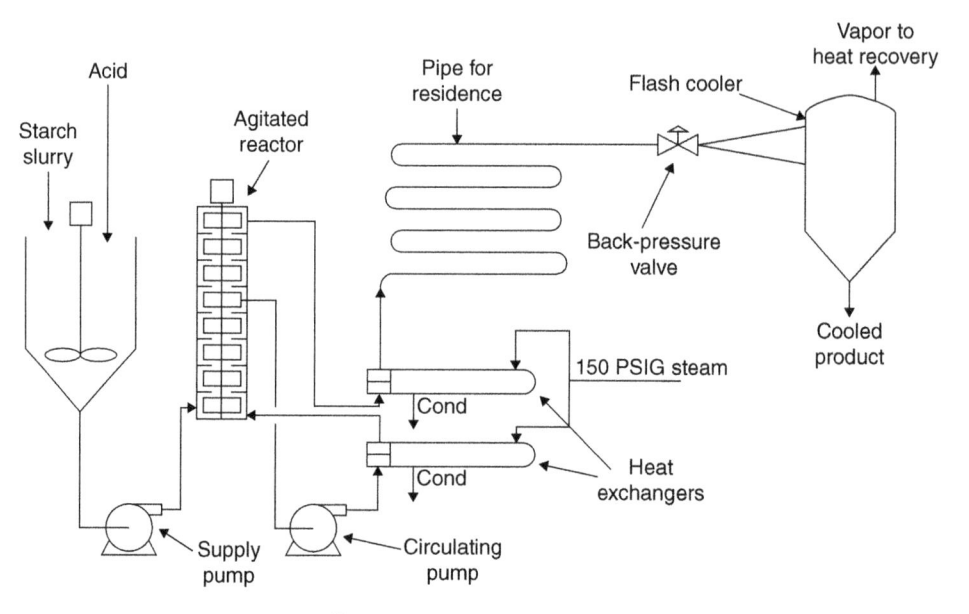

(IE) = Ion-exchanged

Figure 21.4 Typical acid converter.[22]

After conversion to the proper DE, the reaction is stopped in the neutralizer tank by raising the pH with soda ash (sodium carbonate) to 4.5–5.0. This pH is critical not only to optimize the conditions under which the proteins and fats can be removed, but also to reduce the risk of unnecessary color development. At this point, the liquor may be pumped to an enzyme tank for further enzyme-catalyzed conversion, or clarified, bleached and evaporated.

If the liquor is not to be enzyme-converted, it is pumped to 'mud' centrifuges and rotary drum filters which remove the suspended fats and insoluble impurities from the filtrate. Amino acids and peptides which may react with carbohydrates are also removed. Then the filtrate is passed through pulsed beds of activated carbon for clarification and bleaching. The temperature in the carbon column is maintained at 150–170°F (69–77°C) with a typical contact time of 90–120 minutes for optimum removal of impurities. Usually these columns contain packed granular carbon, although powdered carbon may also be used.

It has long been known that activated carbon removes color precursors and off flavors and is particularly effective in removing 5-(hydroxymethyl)-2-furaldehyde (HMF), a glucose decomposition product created during acid-catalyzed hydrolysis.[23] In typical systems, as shown in Figure 21.5, the carbon beds are used in a counterflow fashion in which the spent carbon is removed, regenerated in a furnace and repacked at the top of the column.[24] After the carbon beds, the liquor is passed through 'check' filters designed to remove escaping carbon fines.

Some syrups are ion-exchanged at this point in the process. Ion exchange is essential in the production of certain types of sweeteners, such as high fructose syrups. Not only does ion exchange improve the color and color stability of the syrup by

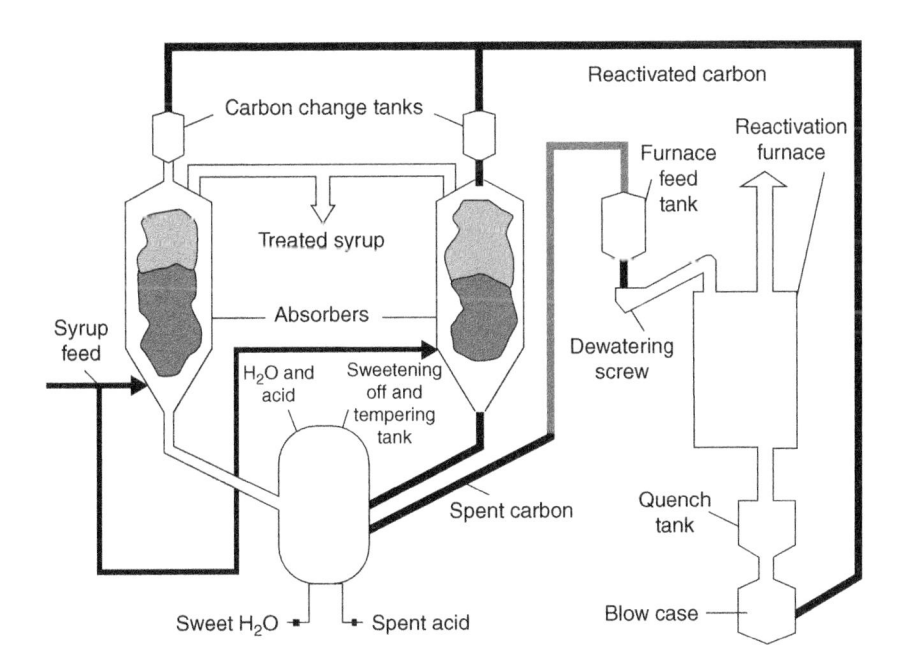

Figure 21.5 Carbon treatment and regeneration system.[24]

removing components that could otherwise participate in a Maillard reaction with the reducing sugars, it also substantially reduces the ash level and improves the flavor.

Ion exchange resins are synthetic organic polymers containing functional groups that exchange mobile ions in a reversible reaction based on affinities. A cation exchange resin in the hydrogen ion form will exchange hydrogen ions for equally charged cations and become converted into a salt. Cation exchange resins in corn syrup manufacturing are typically strong acid exchangers with a sulfonic acid functional group. The anion exchange resins employed contain tertiary amino groups and act as weak bases.[25]

Typical ion exchange processes consist of 'trains' of three cation and three anion beds arranged in pairs. The first pair takes the heaviest load, while the second pair acts as the polishing unit and the third pair undergoes regeneration. As the resins become exhausted, ions start to leak through the primary units. At a predetermined level of exhaustion, the primary units are taken offline and the secondary units are moved to the primary position. The units that were in regeneration are put online as the secondary units and the exhausted primary units are regenerated (Figure 21.6).[26]

Following the carbon columns or the demineralizers, the pH of the filtrate is adjusted and the liquor is evaporated. The solids level of the filtrate prior to the evaporators is about 30% dry solids (DS). Typical evaporators are multiple-effect, falling-film evaporators in which the temperature is increased under precisely controlled conditions that prevent formation of unwanted flavors or color in the syrup. As shown in Figure 21.7, the flow is generally countercurrent, i.e. the hottest portion of the evaporation contains the syrup of lowest solids.[11] After evaporation, the syrup is pumped to large storage tanks where it is held under agitation and analyzed prior to shipment.

Acid–Enzyme Processes

As in the case of acid-catalyzed hydrolysis, the starch molecule is hydrolyzed to the desired starting DE in a converter, but further conversion is carried out with enzymes until the final DE or carbohydrate profile is reached. This is done by adding the appropriate enzymes to the acid-converted slurry and allowing them to react in a holding vessel called an 'enzyme tank.' Several enzymes may be used to achieve the desired carbohydrate profile.

The *alpha-amylases* (EC3.2.1.1) used are bacterial or fungal enzymes that hydrolyze alpha-1,4 linkages in both amylose and amylopectin, eventually producing dextrose and maltose. The reaction is initially rapid, then relatively slow[27] (see Chapter 7).

The *beta-amylases* (EC3.2.1.2) used are enzymes of barley and yeast that act on the non-reducing ends of starch molecules and produce maltose in the beta form from the starch polymers. These enzymes are used to produce high-maltose syrups. Although beta-amylase converts linear chains completely to maltose, the enzyme cannot cleave branch points and the yield of maltose from amylopectin[11] (see Chapter 7) is only 55% of the molecule.

Glucoamylases (EC3.2.1.3) are fungal enzymes which hydrolyze maltose to produce glucose (dextrose). These enzymes catalyze hydrolysis of alpha-1,3, alpha-1,6 and beta-1,6 linkages. Their primary reaction is on the 1,4-linked α-D-glucopyranosyl units of

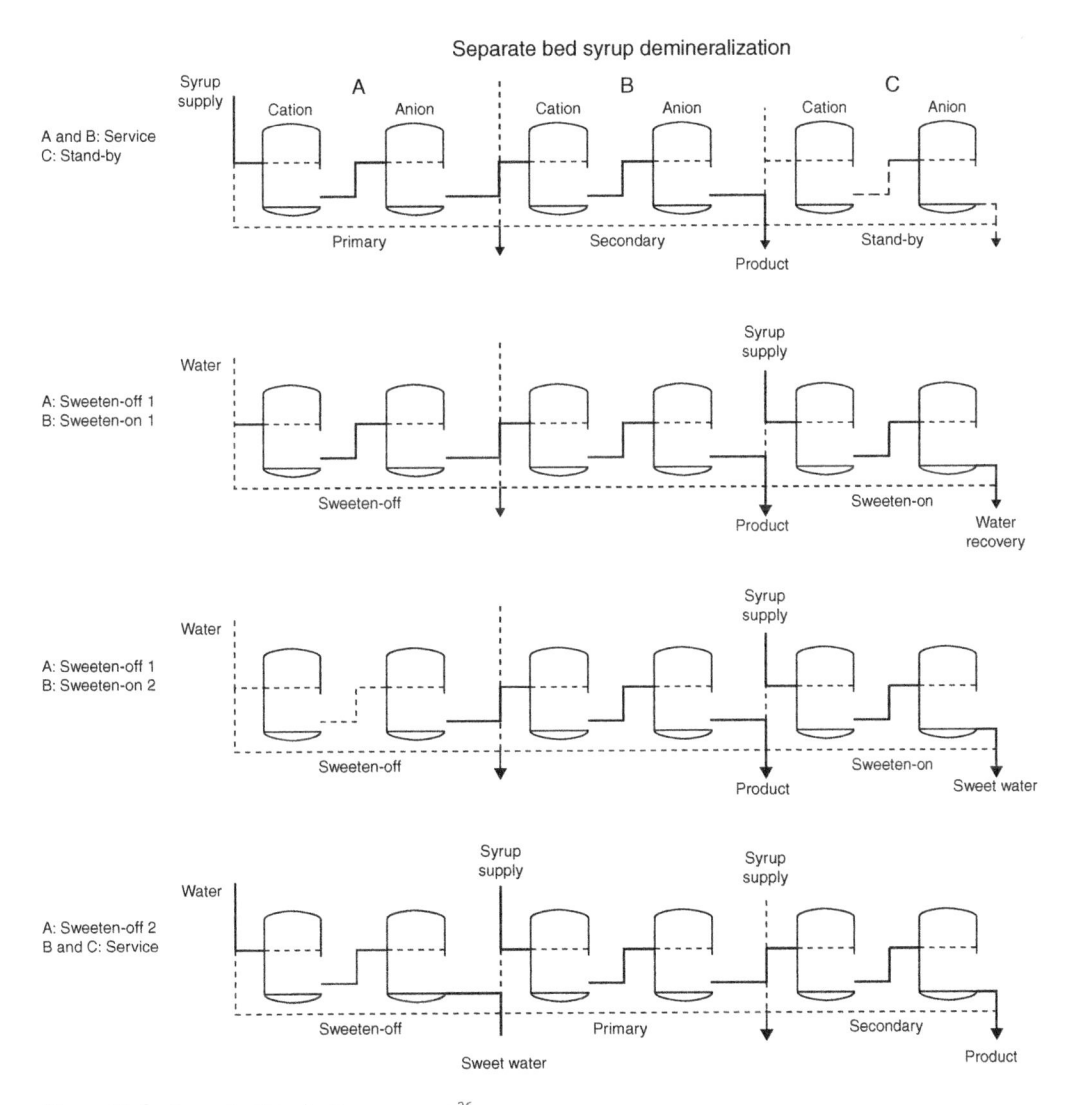

Figure 21.6 Syrup demineralization sequence.[26]

non-reducing ends, releasing β-D-glucopyranose (see Chapter 7). When branch points are reached, the enzyme will cleave the 1,6 bonds, but at a much slower rate.[28]

Pullulanase (EC3.2.1.41) and *isoamylase* (EC3.2.1.68) are so-called debranching enzymes because they catalyze the hydrolysis of the 1,6 linkages without effect on the 1,4 linkages[29] (see Chapter 7). These enzymes are particularly useful in the production of extremely high maltose syrups with maltose levels of 50 to 90%.[30] Table 21.3 provides an overview of commercial enzymes used in the corn refining process today and typical operating requirements.

Proper pH and temperature control is critical during batch enzyme conversion processes, which usually last about 48 hours. In such processes, a number of enzyme tanks are filled sequentially from the converter at the adjusted temperature for treatment (140–150°F) and then dosed with the necessary enzymes. Progress of the reaction

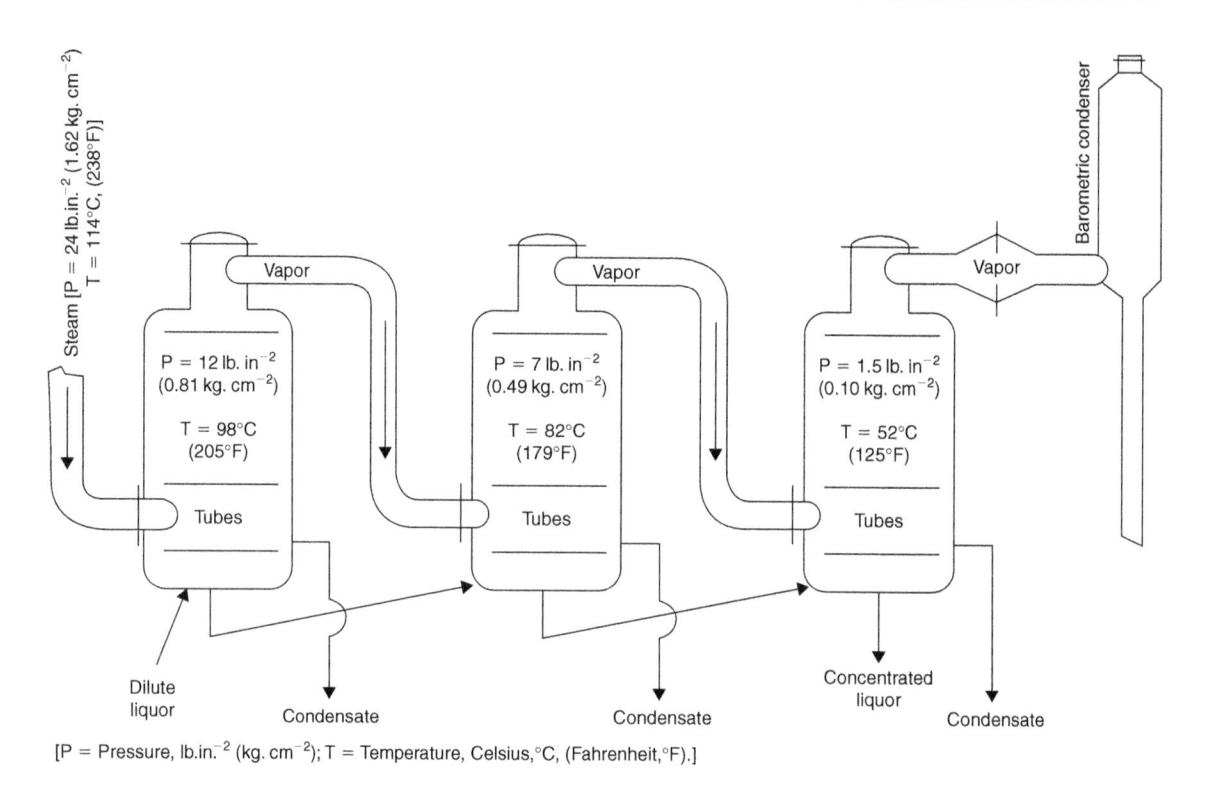

Figure 21.7 Multiple-effect evaporator[11]

is monitored by occasional checks of the DE or the carbohydrate profile until the desired conversion is reached.

At completion of the enzyme conversion, the tanks are emptied in succession and the liquor is processed through filtration and carbon bleaching, as previously described, and evaporated to the proper solids level. The advent of enzyme-converted syrups lessens the importance of traditional methods of measurement, such as determination of DE. It is possible to have two syrups with the same DE and completely different carbohydrate profiles and performance characteristics, as shown in Table 21.4.[31]

3. High-fructose Syrups

Using immobilized enzyme technology, it is possible to produce high-fructose syrups containing 42%, 55% or 90% fructose. In 1999, US shipments of high-fructose syrups exceeded 24 billion pounds (11×10^9 kg) (dry basis).[5]

The typical process for producing a 42% high-fructose syrup is shown in Figure 21.8.[32] A starch solution at about 35% solids and a pH of about 6.5 is drawn into a steam jet at 180°F (82°C) in the presence of a calcium-stabilized, thermostable alpha-amylase. The slurry is maintained at this temperature through a series of loops for 3–5 minutes and then cooled to 95°C (200°F) in a secondary reactor, where further alpha-amylase additions occur. A holding time of up to 120 minutes in the secondary reactor produces a solution of approximately 12 DE. The pH is adjusted to about 4.3

II. Production Methods

Table 21.3 Enzymes used in corn milling processes

Enzyme type	Enzyme activity	Enzyme dose, % on DS	Dry substance,%	Temperature °F	Typical enzyme reaction conditions pH	Time, hours	Start DE	Final DE	Activators	Inactivated by	Notes
High-temperature bacterial alpha-amylase	60 K-NOVO units/g	0.03	30-35	221	6.0-6.5	0.1	0	2	Calcium 50-70 ppm	pH < 4.5	Starch liquefaction
High-temp. bacterial alpha-amylase	60 K-NOVO units/g	0.12	30-35	203	6.0-6.5	1.5	2	12	Calcium 50-70 ppm	pH < 4.5	Starch liquefaction
Bacterial alpha-amylase	120 K-NOVO units/g	0.15	30-40	185	6.0-6.5	1.5	0	15	Calcium 50-70 ppm	pH < 4.5	Starch liquefaction
Bacterial alpha-amylase	120 K-NOVO units/g	0.04	35-45	170	5.8-6.2	6	15	22	Calcium 50-70 ppm	pH < 4.5	Destarching Acid conversion CSU
Fungal gh maltose alpha-amylase	40 000 SKB units/g	0.01	35-45	130-133	4.9-5.3	48	22	48	–	Heating to 175°F	High 42-DE syrup
Fungal maltose alpha-amylase	40 000 SKB units/g	0.02	35-45	130-133	4.9-5.3	48	22	48	–	Heating to 175°F	High 48-DE syrup
Gluco-amylase	150 NOVO AG units/mL	0.175	30-35	136-140	4.0-5.0	48	15	95	–	Heating to 175°F	Dextrose syrup from acid liquid
Gluco-amylase	150 NOVO AG units/mL	0.175	30-35	136-140	4.0-5.0	48	15	98	–	Heating to 175°F	Dextrose syrup from acid liquid
Malt beta-maltose amylase	1500°F Lintner	0.01	35-45	130-133	4.9-5.3	48	22	48	–	Heating to 175°F	High 42-DE syrup
Malt beta-maltose amylase	1500°F Lintner	0.02	35-45	130-133	4.9-5.3	48	22	48	–	Heating to 175°F	High 48-DE syrup
Fungal alpha-amylase + gluco-amylase syrup	40 000 SKB units/g 150 NOVO AG units/g	0.0075 0.01	35-45	130-133	4.9-5.3	48	22	48	–	Heating to 175°F	62-DE acid-enzyme
Glucose isomerase	150 IGIC units/g	–	35-45	140-150	7.5-8.5	–	95% dextrose	–	Magnesium Calcium 0.0004 M	Deaerated <1 ppm	Demineralized

Table 21.4 Saccharides, total carbohydrate basis[31]

Designation	DE	DP_1, %	DP_2, %	DP_3, %	DP_{4+}, %
43 DE	43	19	14	12	55
43 HM (high-maltose)	43	9	43	18	30

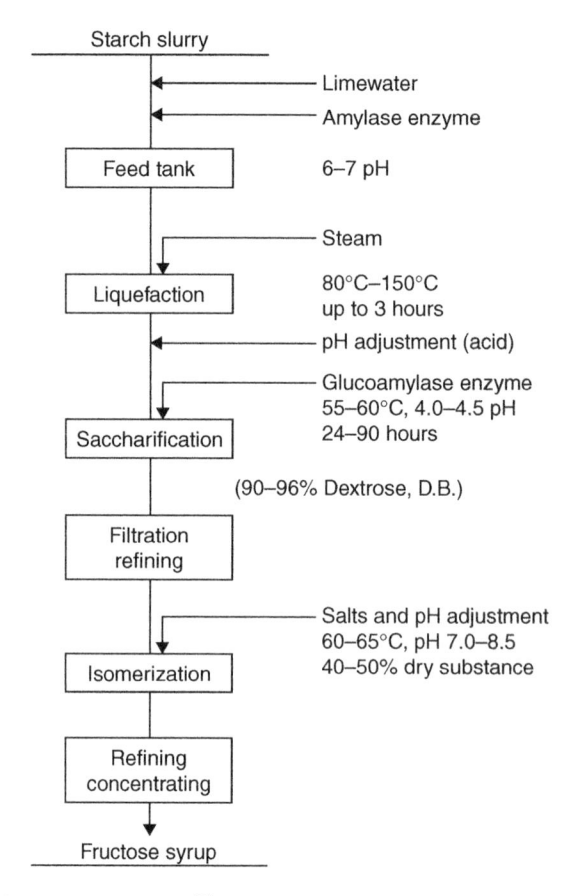

DP_n = degree of polymerization

Figure 21.8 High-fructose syrup process.[32]

and glucoamylase is added. Then the product is pumped to saccharification tanks where the enzyme reacts for 24–90 hours. The glucoamylase reaction produces liquor containing 94% dextrose, which is then filtered to remove residual protein and fats before being passed through beds of activated carbon, as was described for the corn syrup process. Following carbon purification, the hydrolyzate is demineralized through anion and cation exchange resins prior to being isomerized.

The conversion step to high-fructose syrup takes place in a reactor containing immobilized glucose isomerase. Although nonenzymic processes to isomerize glucose to fructose have been developed,[11] these processes result in undesirable by-products of ash, color, and flavor, and other faults.

$$S + E \underset{k_2}{\overset{k_1}{\rightleftharpoons}} ES \underset{k_4}{\overset{k_3}{\rightleftharpoons}} E + P$$

Where S, E, ES and P are substrate, enzyme, enzyme-substrate complex, and product respectively. K_1, K_2, K_3, and K_4 are the rate constants. The mechanism can be described by the rate equation:

$$V = \frac{E[(K_3S)/KS - (K_2P)/KP]}{1 + (S/KS) + (P/KP)}$$

Where V is the rate of product formation and KS and KP are the Michaelis constants for substrate and product, respectively.

Figure 21.9 Michalis–Menten kinetics.

Considerable effort based on research work initiated in the 1950s resulted in enzyme technology able to convert glucose to fructose on a commercial scale.[32–34] Current production of high-fructose syrups generally uses immobilized, rather than soluble, enzymes. Sources of the enzyme include *Streptomyces*, *Bacillus*, *Actinoplanes* and *Arthrobacter* species.

The advantage of fixed bed systems is that the relatively high activity per unit weight allows manufacturers to process large quantities of product through relatively small reactors in short times. The short residence time in these reactors also reduces development of undesirable color and flavor compounds.

The isomerization of glucose to fructose has been extensively studied and the mechanism is well-documented.[35–37] The reaction is essentially first order and reversible, following Michalis–Menten characteristics shown in Figure 21.9.

Glucose isomerase requires divalent cations, such as Co^{++}, Mg^{++} or Mn^{++} for catalytic activity and is inhibited by the presence of Cu^{++}, Ni^{++} Ag^{++}, Hg^{++}, Ca^{++} and Zn^{++} ions. Therefore, proper demineralization of the liquor prior to isomerization is essential. Depending on the source of the enzyme, optimum operating conditions include a pH range of 6.5 to 8.5 and a temperature of 40–80°C. Residence time in the reactor is usually less than four hours. Enzyme decay is exponential; therefore the typical system will contain a number of reactors containing enzyme in varying stages of output. A typical half life of such a column may be as long as 200 days.

The fructose level of the output of each of these columns can be controlled by varying the reaction time (flow rate), temperature and pH. Once conversion is complete the liquor is pumped through beds of activated carbon and then evaporated to the proper solids level, generally 71% or 80% dry solids as previously described.

Forty-two percent high-fructose syrup produced by this method is used in many applications as a replacement for liquid sucrose. However, in some applications, 42% high-fructose syrup is not sweet enough and a higher level of fructose is necessary. Although it is possible to create enriched fructose syrups by forming complexes with borate compounds during isomerization[38] and by liquid–liquid extraction,[39] commercial production of such products generally involves adsorption–separation technology. This technology employs the relative differences in the affinity of dextrose and fructose for strong acid

Figure 21.10 Adsorption separation columns.[41]

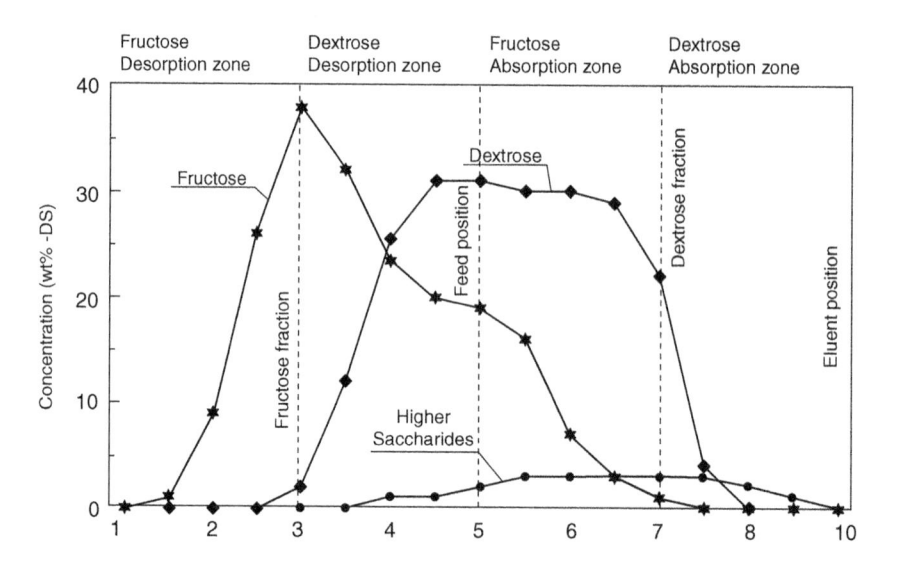

Figure 21.11 Separation of glucose and fructose on adsorption columns.[41]

ion exchange resins in the Ca^{++} salt form. By controlling the feed and elution rate of the components or by monitoring elution of certain components in the direction of flow[40] it is possible to separate the sugars in a continuous process in a large column.

The 42% fructose syrup from the isomerization column is first demineralized to remove trace components picked up during isomerization, and is then pumped into the separator at 36–60% solids. The relative difference in affinity of the resin for fructose and dextrose allows separation of the carbohydrates into two enriched streams. A typical system, shown in Figures 21.10 and 21.11, is based on the concept of a simulated moving bed.[41] As the 42% fructose feedstock is pumped through the bed, the fructose portion is selectively absorbed relative to dextrose, resulting in separation of the two carbohydrates.

Through a series of automatic valves, it is possible to draw off a stream of enriched dextrose, as well as one of enriched fructose. Some consideration must be given to the fact that the higher is the purity of this stream, the lower are the solids and the greater is the evaporation cost. Typical purity of the separated streams of dextrose and fructose is 85% to 90%. The dextrose fraction ('raffinate') is returned to the front of the system to be reisomerized, while the enriched fructose fraction is blended with a 42% fructose stream to produce 55% high-fructose syrup. The enhanced fructose phase may also be isolated as a separate product stream to make 90% high-fructose syrup or crystalline fructose.

4. Crystalline Fructose

If the ratio of fructose to non-fructose materials is high enough, it is possible to crystallize the fructose. In one such process, the fructose must have a purity of at least 90% to achieve sufficient saturation.[42] The liquor is seeded with crystals of dry fructose and cooled to crystallize the pure fructose. The resulting crystals are washed and separated from the mother liquor by centrifugation, then dried, producing crystalline fructose material that is 99.5% pure. Since fructose is only moderately soluble in alcohol, processes improving the crystallization of fructose by adding ethanol or methanol to the solution have also been described.[43] Crystalline fructose may exist as anhydrous β-D-fructopyranose. Crystals of the dihydrate of fructose require careful handling, because of their ability to dissolve in their own water of hydration. A typical crystalline fructose process is outlined in Figure 21.12.[44] Because of its extremely hygroscopic nature crystalline fructose should be stored under conditions below 50% relative humidity. Figure 21.13[45] shows the equilibrium water content of fructose and sucrose.

5. Crystalline Dextrose and Dextrose Syrups

Several processes of separating and crystallizing dextrose by repeated seeding, washing and crystallization were developed during the 1800s and early 1900s,[46] but these methods were time consuming and costly. Today, commercial processes begin with the solution produced by liquefaction of the starch in a jet cooker, as previously described.

Saccharification produces a 94% dextrose liquor, which can be processed in several ways. To make a dextrose syrup, the fats and protein are removed from this liquor, as in the high fructose process. The syrup is then carbon bleached, demineralized and evaporated to 71% solids. The 94% dextrose liquid may also be further refined to 99% dextrose by adsorption-separation chromatography prior to being bleached, demineralized and evaporated.

Either anhydrous dextrose or dextrose monohydrate can be obtained by crystallization. Monohydrate crystallizers are large horizontal, cylindrical batch tanks or

```
                Concentrated fructose feed
      Seeding ─────────▶ │
                         ▼
                   Crystallization
                         │
                         ▼
             (Alternate crystallization step)
                         │
                         ▼
                    Centrifugation
                         │
                         ▼ ◀───────── Washing
                       Drying
                         │
                         ▼
                   Conditioning
                         │
                         ▼
                     Screening
                         │
                         ▼
                    Packaging
```

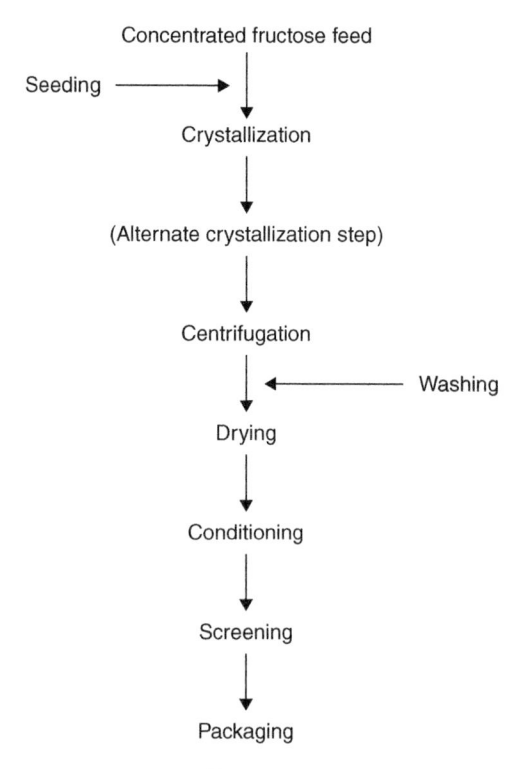

Figure 21.12 Fructose crystallization process.[44]

continuous systems in which the crystal mass is continuously removed, leaving about 20–25% of the batch to seed the next. During crystallization, the syrup (75% solids, 95% dextrose) is cooled carefully and in a controlled manner below 50°C. Since crystallization of dextrose is an exothermic reaction, constant cooling is essential to maintain the proper level of supersaturation. When the magma of crystallized dextrose (α-D-glucopyranose) monohydrate is formed, the material is washed and centrifuged in basket centrifuges to remove the mother liquor ('first greens'). The first greens may be reprocessed to yield a second crop of crystals. The mother liquor from this step is known as 'second greens' or hydrol. Both hydrol streams are combined to improve the yield. The remaining monohydrate crystals are dried in a stream of hot air and packaged. A typical batch crystallizer and basket centrifuge are shown in Figures 21.14 and 21.15.[47]

Anhydrous dextrose is produced by dissolving the monohydrate in hot purified water and refining it again. During the crystallization step, proper temperature control is essential to ensure formation of nuclei in the anhydrous dextrose (α-D-glucopyranose).[48] These nuclei are grown under controlled conditions, separated from the mother liquor, and washed and screened as before. The mixture is then evaporated under reduced pressure with heat and agitation and dried to a moisture level of 0.1%, resulting in a free-flowing powder. These crystals are separated and packaged as in the case of the monohydrate. A typical process for both anhydrous and monohydrate production is shown in Figure 21.16.[48]

II. Production Methods 815

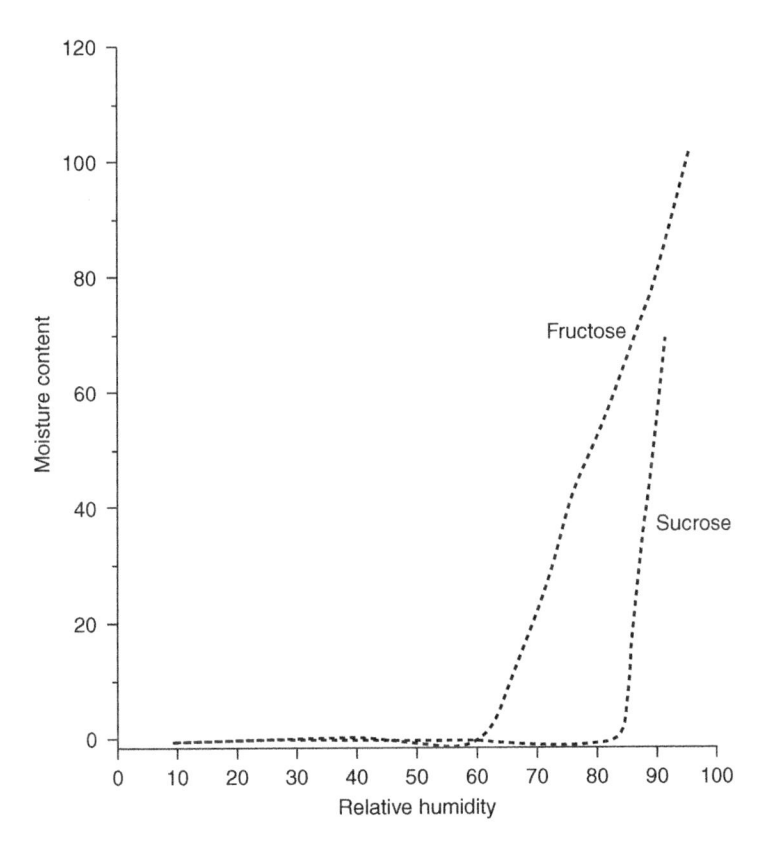

Figure 21.13 Equilibrium water content of fructose and sucrose.[45]

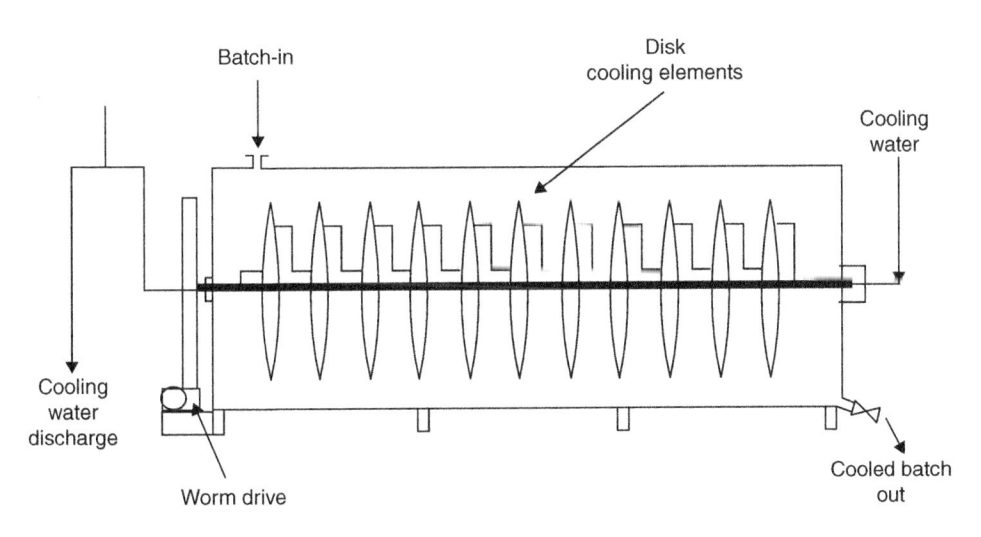

Figure 21.14 Batch crystallizer.[47]

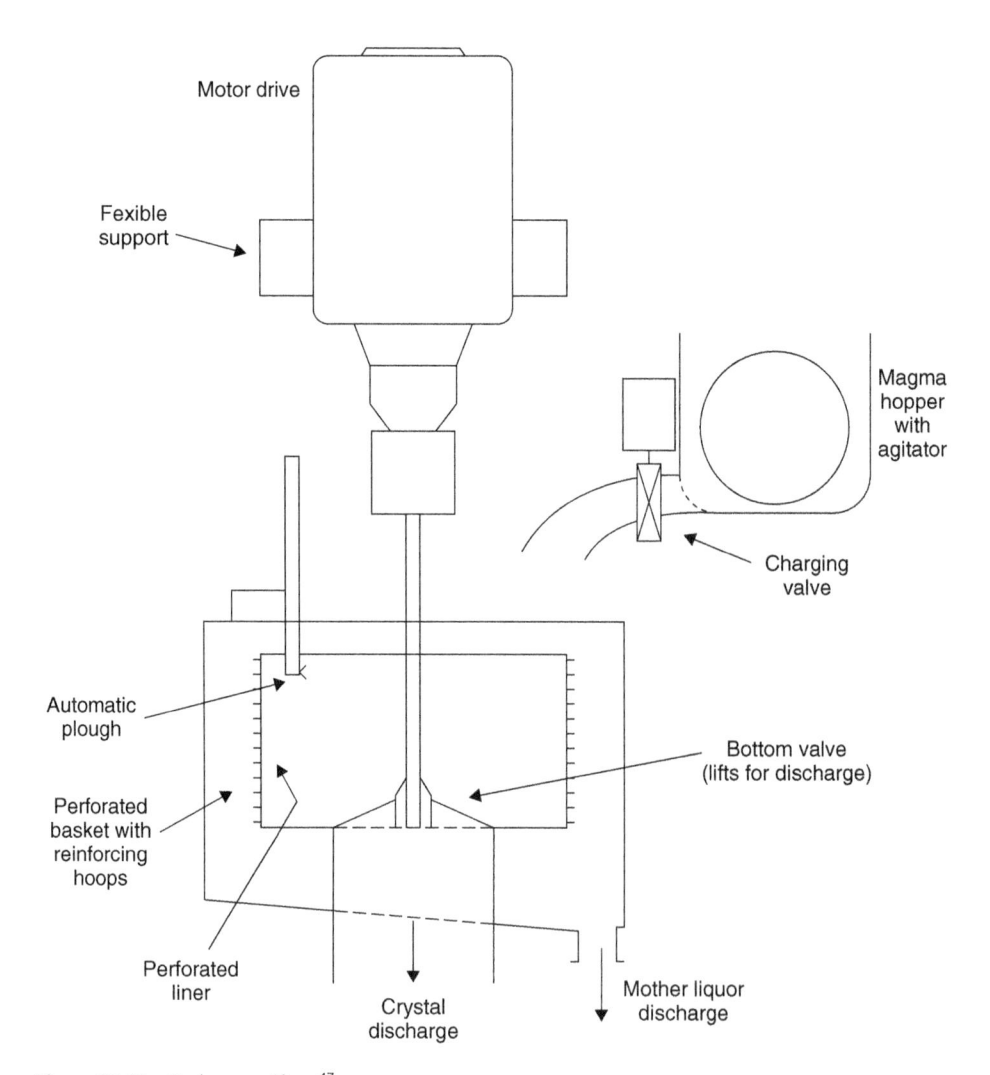

Figure 21.15 Basket centrifuge.[47]

6. Oligosaccharide Syrups

Beginning in the 1990s, oligosaccharide sweeteners produced from sucrose, soy flour or corn starch gained increasing attention. Malto-oligosaccharides and isomalto-oligosaccharides from starch contain 1,6 and/or 1,4 linkages. Glucose/corn syrups may be described as a concentrated solution of glucose and varying amounts of a mixture of malto-oligosaccharides, including isomalto-oligosaccharides. In the production of oligosaccharide syrups, oligosaccharides are emphasized (over glucose/dextrose) through the use of certain enzymes, a transglucosidase in particular[49] (see Chapter 7).

Starting with a feedstock of 65–70% maltose, made by one of the methods previously described, the liquor is passed through an adsorption separation column. This results in a product with a maltose level of approximately 98% which may be crystallized, leaving a predominately maltotriose fraction which is purified as a syrup.

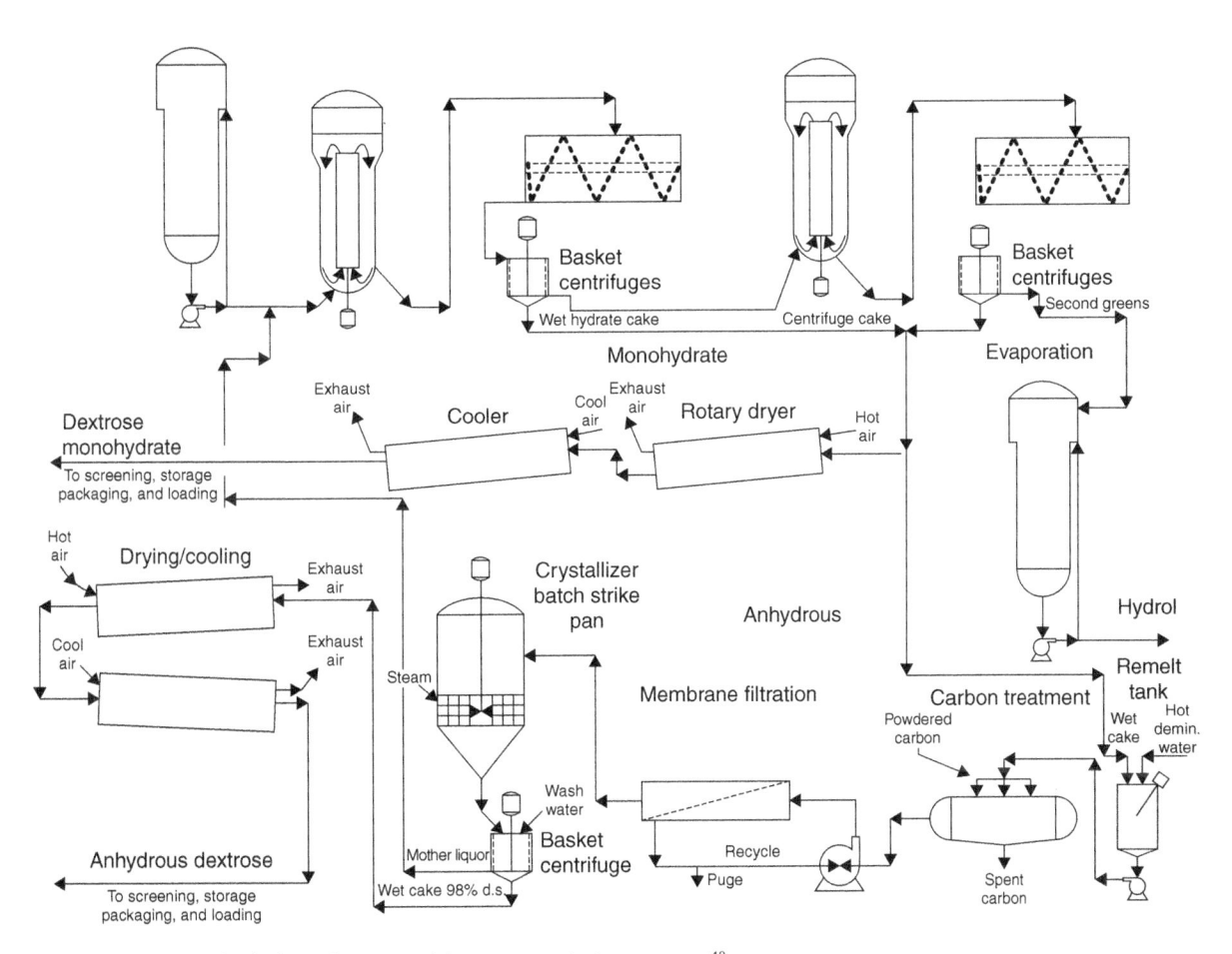

Figure 21.16 Typical anhydrous dextrose and dextrose monohydrate process.[48]

Maltotriose syrups may also be prepared by the use of maltotriose-forming amylases combined with pullanase or isoamylase enzymes (see Chapter 7). In one such process, waxy maize starch is treated with alpha-amylase, followed by pullulanase. The syrup resulting from this procedure contains 60–85% low molecular weight oligosaccharides.[50] Maltose syrups may also be treated with transglucosidases from *Aspergillus* sp. to produce high isomalto-oligosaccharide levels. Typical profiles of some commercial syrups are given in Table 21.5.[51]

III. Composition and Properties of Sweeteners from Starch

1. Carbohydrate Profiles

Early work by Hoover[52] provides a framework for determining how physical properties would change on the basis of the degree of conversion of the sweetener being considered (Table 21.6). At that time, most syrups were divided into loose classifications

Table 21.5 Saccharide compositions of several syrups[51]

		High maltotetraose syrup		Acid-conversion syrup	Enzyme-conversion syrup
		Type I	Type II		
DE:		30.8	36.2	32.6	47.0
Glucose		1.0	2.0	10.3	3.0
Maltose		7.8	8.5	8.3	53.5
Maltotriose		10.2	11.0	8.4	15.8
Maltotetraose		50.5	72.0	8.1	3.2
Maltopentaose		2.5	1.0	6.9	2.0
Others (>DP5)		28.0	5.5	58.0	22.5

Table 21.6 Relationship between degree of conversion and functional property

Property or functionality that increases with an increasing degree of conversion	Property or functionality that decreases with an increasing degree of conversion
Browning	Bodying
Fermentability	Cohesiveness
Flavor enhancement	Foam stabilization
Flavor transfer	Prevention of sucrose crystallization
Freezing point depression	Prevention of coarse ice crystal formation during freezing
Hygroscopicity	Viscosity
Osmotic pressure	
Sweetness	

based on DE as a measure of the relative degree of conversion. As already mentioned, subsequent advances in enzyme technology and the proliferation of syrups based on the carbohydrate profile diminished the importance of DE in describing the nature of a syrup. Rapid and inexpensive methods of analysis, such as liquid chromatography, have allowed producers to focus on the carbohydrates present in sweeteners and how they impact the physical properties of the syrup.

Physical properties of a syrup depend heavily on its carbohydrate profile. The carbohydrate profile, in turn, is determined by the type of conversion and the nature of the enzyme treatment (previously discussed). Table 21.2 gives typical DE and carbohydrate profiles of syrups in common production today. Because enzyme treatments can provide sweeteners with different carbohydrate profiles but the same DE value, it is usual to refer to a product using more than one descriptor, e.g. a 43 DE, high-maltose syrup. This issue becomes particularly important when addressing functional differences and applications of starch-derived sweeteners.

2. Solids

In addition to DE values and carbohydrate profiles, syrups are usually identified by their solids level. The traditional means of expressing the solids of a glucose/syrup is the Baume number. Extensive work comparing the Baume number to refractive index

Table 21.7 Refractive index-dry substance tables for typical glucose syrups[58]

Syrup	% DS	RI 20°C	RI 45°C	Be Comm 140°/60°F +1	Sp. G. AIR/AIR 100°/60°F	Total lbs/ gal, 100°F	Dry sub lbs/ gal, 100°F
28 DE	78.00	1.4943	1.4892	42.00	1.4064	11.73	9.15
36 DE	80.40	1.4993	1.4941	43.02	1.4202	11.84	9.52
34 HM	80.60	1.4988	1.4936	42.99	1.4197	11.84	9.54
43 HM	80.90	1.4988	1.4937	43.01	1.4199	11.84	9.58
43 DE	80.70	1.4988	1.4936	43.02	1.4201	11.84	9.55
43 DE IX	80.80	1.4990	1.4938	43.00	1.4198	11.84	9.56
53 DE	83.50	1.5044	1.4992	44.13	1.4354	11.97	9.99
63 DE	82.00	1.4982	1.4931	43.02	1.4201	11.84	9.71
63 DE IX	84.00	1.5064	1.4985	44.01	1.4337	11.95	10.04
66 DE	84.30	1.5044	1.4993	44.09	1.4347	11.96	10.04
95DE	71.00	1.4644	1.4595	36.39	1.3348	11.13	7.90
HFCS 42	71.00	1.4643	1.4589	NA	1.3372	11.15	7.92
HFCS 55	77.00	1.4789	1.4728	NA	1.3809	11.51	8.98
Liquid fructose	77.00	1.4780	1.4715	N/A	1.3763	11.47	8.84

Table 21.8 Refractometer corrections for HFS 55, 0.05% ash (dry basis)[59]

Dry substance, % by weight	Refractive index, 20°C	Brix (1936)	Refractometer correction	Brix (1966)	Refractometer correction
10.00	1.34771	9.92	0.08	9.93	0.07
20.00	1.36352	19.81	0.19	19.82	0.18
30.00	1.38054	29.69	0.31	29.68	0.32
40.00	1.39881	39.53	0.47	39.48	0.52
50.00	1.41841	49.24	0.76	49.22	0.78
60.00	1.43944	58.93	1.07	58.92	1.08
70.00	1.46197	68.70	1.30	68.59	1.41
77.00	1.47870	75.52	1.48	75.35	1.65
80.00	1.48611	78.43	1.57	78.24	1.76

has produced easily-used tables to convert the various measures of solids measurement.[53–56] Unlike glucose syrups the solids content of high-fructose syrups is stated as the dry substance.[57] Table 21.7 shows the comparison of dry solids to degrees Baume for several typical sweeteners.[58]

When the solids content of glucose syrups is measured by refractometers calibrated in degrees Brix, some correction must be applied to obtain the true solids level. Brix measurements are commonly used in the sucrose industry and refer to the percentage of sucrose in solution. The Brix tables were modified in 1936, 1966 and 1974, resulting in minor changes as shown in Table 21.8.[59] These corrections have been incorporated into high-fructose syrup tables commonly used in the beverage industry (Table 21.9).[60]

3. Viscosity

Glucose syrups exhibit the viscosity characteristics of Newtonian fluids. The coefficient of viscosity is constant and is measured in poises (dyne-seconds/cm^2), which is

Table 21.9 Relationship of refractive index to Brix for high-fructose syrups[60]

42% High-fructose syrup

Solids, %	Refractive Index		Ref. Brix (20°C)		Density (vac) 20°C	Sp. Grav (air) 20°C	Lbs/US gal. (20°C)		Hydr. Brix 20°C
	20°C	45°C	1936	1966			Total	Solids	
70.500	1.463	1.458	9.270	69.170	1.344	1.347	11.211	7.903	69.540
70.600	1.464	1.458	9.370	69.270	1.345	1.348	11.216	7.918	69.630
70.700	1.464	1.458	9.460	69.370	1.346	1.348	11.221	7.933	69.730
70.800	1.464	1.459	9.560	69.460	1.346	1.349	11.226	7.948	69.830
70.900	1.464	1.459	9.660	69.560	1.347	1.350	11.231	7.963	69.920
71.000	1.465	1.459	9.760	69.660	1.347	1.350	11.236	7.977	70.020
71.100	1.465	1.459	9.860	69.750	1.348	1.351	11.241	7.992	70.120
71.200	1.465	1.460	9.950	69.850	1.349	1.351	11.246	8.007	70.210
71.300	1.465	1.460	0.050	69.950	1.349	1.352	11.251	8.022	70.310
71.400	1.466	1.460	0.150	70.040	1.350	1.353	11.256	8.037	70.410
71.500	1.466	1.460	0.250	70.140	1.350	1.353	11.261	8.052	70.500

55% High-fructose syrup

Solids, %	Refractive Index		Ref. Brix (20°C)		Density (vac) 20°C	Sp. grav (air) 20°C	Labs/US gal. (20°C)		Hydr. Brix 20°C
	20°C	45°C	1936	1966			Total	Solids	
76.5	1.477	1.472	75.000	74.830	1.382	1.385	11.522	8.814	75.380
76.6	1.478	1.472	75.090	74.930	1.382	1.385	11.527	8.829	75.480
76.7	1.478	1.472	75.190	75.020	1.383	1.386	11.532	8.845	75.580
76.8	1.478	1.472	75.290	75.120	1.384	1.386	11.537	8.861	75.670
76.9	1.478	1.473	75.390	75.220	1.384	1.387	11.543	8.876	75.770
77.0	1.479	1.473	75.480	75.310	1.385	1.388	11.548	8.892	75.870
77.1	1.479	1.473	75.580	75.410	1.385	1.388	11.553	8.907	75.960
77.2	1.479	1.473	75.680	75.510	1.386	1.389	11.558	8.923	76.060
77.3	1.479	1.473	75.770	75.600	1.387	1.390	11.564	8.939	76.160
77.4	1.480	1.474	75.870	75.700	1.387	1.390	11.569	8.954	76.250
77.5	1.480	1.474	75.970	75.800	1.388	1.391	11.574	8.970	76.350

the force per unit area necessary to maintain unit difference in velocity between two parallel layers of fluids that are one unit distance apart. The poise is the absolute viscosity of the solution. Viscosity in starch-derived sweeteners is due primarily to the solids level and the percentage of higher saccharides present. As might be expected, at the same solids level, sweeteners with relatively long chains of branched molecules have higher viscosity than sweeteners with high concentrations of monosaccharides, an effect compounded by the presence of weak hydrogen bonds.

The viscosity of syrups is usually reported in either centipoises (mPa s) or SSU (Saybolt Seconds Universal) units. SSU units are related to centipoises by the following equation:[61]

$$SSU = \frac{centipoises \times 4.55}{specific\ gravity}$$

Table 21.10 Viscosity (centipoises) of syrups of different DS and temperatures[62]

DS	Temperature °F	Dextrose equivalent			
		35.4	42.9	53.7	75.4
85	60				457 000
	80	7 080 000	1 410 000	537 000	83 200
	100	1 000 000	227 000	85 200	17 000
	120	188 000	50 100	20 000	4270
	140	44 900	13 000	6310	1660
	160	13 000	5190	2290	589
	180	4420	1760	944	275
80	60		266 000	89 100	24 000
	80	126 000	59 600	17 800	4570
	100	29 900	15 000	5010	1550
	120	8910	4840	1800	603
	140	3350	1860	785	282
	160	1410	851	367	141
	180	687	386	196	75.9
75	60	39 800	18 200	7590	6030
	80	10 000	5390	2140	741
	100	3020	1880	807	331
	120	1260	817	372	159
	140	620	389	191	83.2
	160	325	197	103	47.9
	180	180	110	62.0	28.8

Table 21.11 Color designations of glucose syrups[63]

Optical density	Visual color
0.025	Water white
0.035	Very light straw
0.050	Light straw
0.075	Straw to very light yellow
0.100	Medium light yellow
0.125	Light yellow
0.150	Medium yellow
0.200	Yellow

Typical values for the viscosity of a number of syrups produced by acid conversion measured at various temperatures and solids levels are given in Table 21.10.[62]

4. Browning Reaction and Color

The color of starch-derived sweeteners is often referred to as 'water white,' but it is more meaningful to express color of syrup in terms of absorbance (optical density, Table 21.11).[63] Typically, the color of commercial corn sweeteners, particularly high-fructose and dextrose syrups, is expressed in absorbance measured against a reference standard of water at 450 nm and 600 nm, as shown in Figure 21.17.[64]

Several reactions can cause color development in starch-derived sweeteners. Because they contain reducing sugars, they will react with proteins and amino acids

$$\text{Color} = \frac{(A_{450} - A_{600})}{N} \times 100$$

Where N is the path length in centimeters

Figure 21.17 Equation for calculating color of syrups.[64]

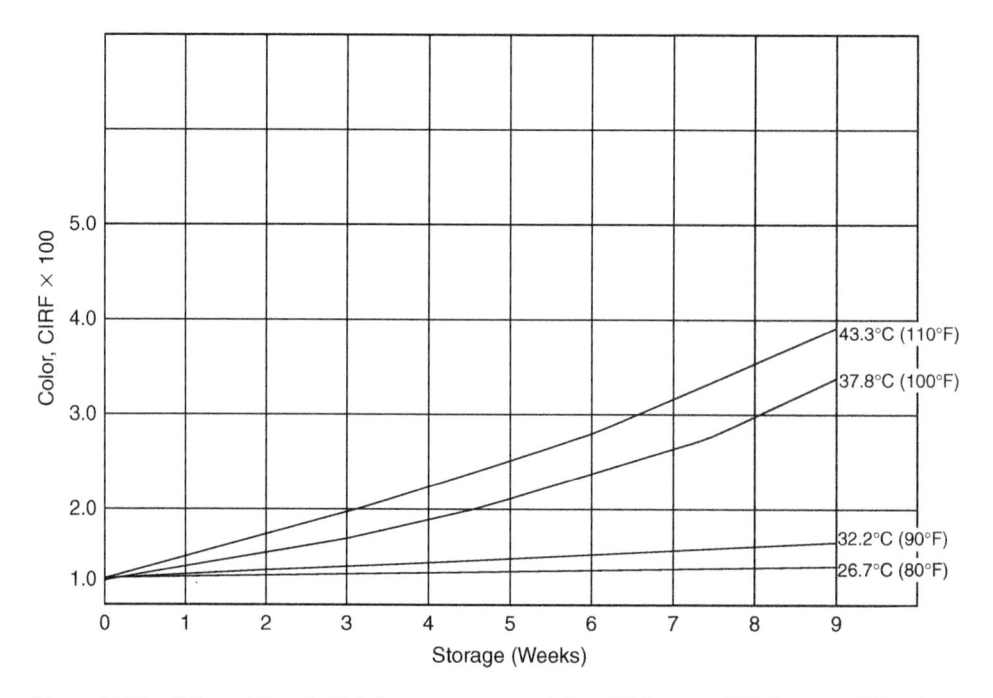

Figure 21.18 Color stability of a high-fructose syrup containing 42% fructose, 50% glucose and 8% other saccharides (71% solids).[67]

via the non-enzymic browning reaction between sugars and primary or secondary amines (Malliard reaction) to form what are referred to as 'color bodies.'[65,66] Color development that occurs in the absence of nitrogenous compounds with the application of heat or acids is the result of caramelization. Excessive heating of starch-derived sweeteners will result in partial caramelization and development of undesirable flavors. The degree and rapidity of color development will be in direct proportion to the amount of reducing sugars present. This is used to advantage in certain food applications, such as in baking. The rate of color development may be retarded by demineralization of the sweetener. Selection of a sweetener with a lesser amount of monosaccharides will also slow color development. Figure 21.18 shows the rate of color development of a 42% high-fructose syrup under different conditions of storage.[67]

5. Fermentability

Starch-derived sweeteners provide a highly fermentable substrate for many industrial applications. The ability of yeast to ferment starch-derived syrups is directly

Table 21.12 Fermentable extract values of corn syrups compared with dextrose[68]

Glucose syrup or dextrose	Fermentable extract, % db[a]
36 DE	24.8
42 DE	41.5
54 DE	53.4
62 DE	70.0
64 DE	76.1
68 DE	79.0
HFCS	95.0
Dextrose	100.0

[a] db = dry basis

Table 21.13 Theoretical versus actual fermentability[69]

Adjunct	Weight %				Fermentability	
	DP$_1$	DP$_2$	DP$_3$	DP$_{4+}$	Calc.	Det.
Dextrose	99.8	0.2	0.0	0.0	100.0	99.8
Maltose syrup	7.1	58.8	19.7	14.4	85.6	85.5
High-conversion syrup	32.7	28.8	14.3	24.2	75.8	72.8
Acid syrup	18.8	15.3	14.7	51.2	48.8	47.1
Maltodextrin	0.9	3.2	2.8	93.1	6.9	11.4

proportional to the monosaccharide and disaccharide content of the sweetener. Table 21.12 gives the fermentable extract value for a number of sweeteners.[68] The fermentable extract of a sweetener is generally calculated as the sum of the DP$_1$, DP$_2$ and DP$_3$ components, although there may be some slight variation between theoretical and actual results, as shown in Table 21.13.[69] The fermentation pathway is different for glucose and fructose, and high levels of glucose have been shown to impair the mechanism for maltose and maltotriose utilization.[70]

6. Foam Stabilization and Gel Strength

In whipped and aerated products, low DE sweeteners have the ability to stabilize foam structures, because of the relatively high polysaccharide level and hydrogen bonding that enhances the effectiveness of the albumin or gelatin. The concentration and type of sweetener affects the gel–sol transition process, in part by regulating water availability.[71] Syrups may be used to achieve the high soluble solids concentrations (>55%) required to effect gelation of high methoxyl pectin (HM pectin) solutions used to make jams, jellies and preserves. When glucose syrups are used, the gels may be firmer than when high-fructose syrups are used.[72]

Table 21.14 Estimating the freezing point depression of various sweeteners[73]

Sweetener	Sucrose equivalence value (SE)[a]	Freezing point equivalence factor[a]
Sucrose	100	1.00
Lactose	100	1.00
Maltose	100	1.00
Dextrose	180	0.55
Fructose	180	0.55
90%	187	0.53
55%	185	0.54
42%	180	0.55
Glucose syrups		
62 DE	114	0.58
52 DE	95	1.05
42 DE	79	1.27
36 DE	72	1.39
32 DE	63	1.59
28 DE	58	1.67
Dextrin (C18-C40)	67	1.47

[a]The sucrose equivalence value is based on the molecular weight of sugar. The freezing point equivalence factor for freezing point depression of a sweetener is based on the molecular weight relative to that of sucrose. The percentage of sweetener is multiplied by the appropriate freezing point equivalence factor to give the freezing point depression

7. Freezing Point Depression

The freezing point of solutions containing sweeteners can be calculated based on their concentration in that solution and their average molecular weight. For a given condition, the effect on the freezing point is inversely proportional to the average molecular weight of the sweetener. Lower conversion syrups in solution will not depress the freezing point as much as higher conversion syrups. The relative effect of various sweeteners on the freezing point of a solution compared to sucrose has been reported by Arbuckle[73] and is shown in Table 21.14 and Figure 21.19. The relative degrees to which these sweeteners affect the freezing point also depends on the solids level.[74] Sweeteners also act in a manner similar to stabilizers, determining the time and degree of ice crystal formation by controlling the migration of free water.[75]

8. Boiling Point Elevation

As in the case of freezing point depression, boiling point elevation depends on the carbohydrate profile. In general, the boiling point is increased as the level of conversion increases. Table 21.15 shows the relationship of boiling point to solids content for various sweeteners.[76]

9. Gelatinization Temperature

This relationship is important to aspects of baking where sweetener concentration and replacement can impact the gelatinization temperatures of starch granules. Substitution of fructose for sucrose lowers the gelatinization temperature (Figure 21.20).[77]

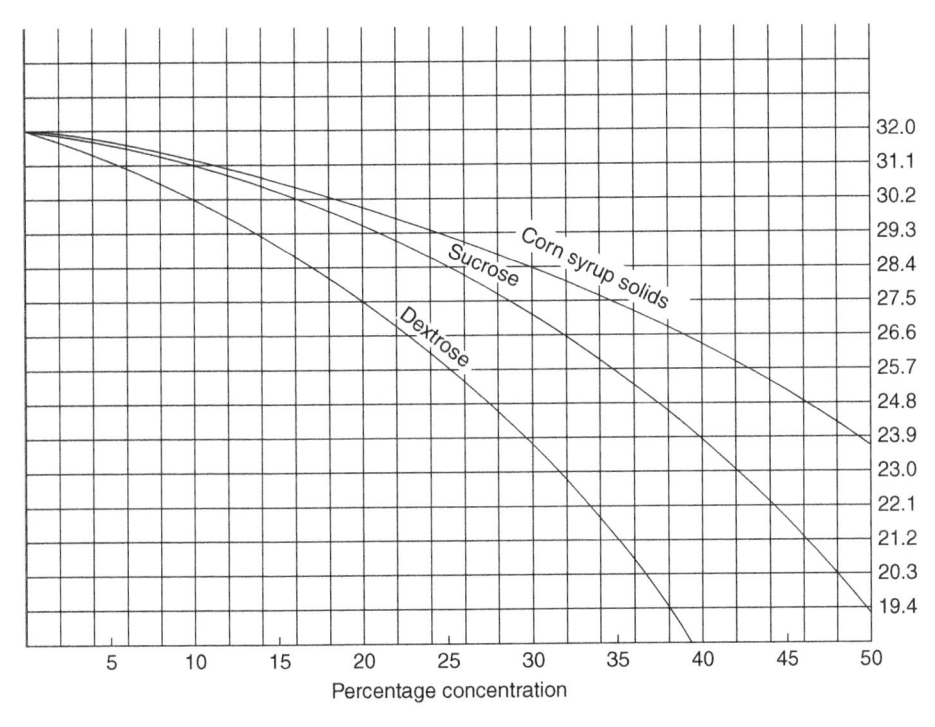

Figure 21.19 Freezing point relationships of solutions of dextrose, sucrose and syrup solids.[73]

Table 21.15 Boiling point elevation of glucose syrups (760 mm Hg)[76]

Syrup DE	% dry substance					
	20	40	60	70	80	85
30	0.39	1.18	3.55	6.3	11.1	15.1
42	0.56	1.60	4.81	8.3	14.3	19.2
55	0.76	2.09	5.98	10.1	16.9	22.3
65	0.91	2.51	6.86	11.3	18.5	23.8
80	1.17	3.16	8.15	12.9	20.3	25.3
95	1.38	3.62	8.95	13.9	21.1	26.1

In addition to lowering the water activity and forming sugar bridges between starch chains restricting flexibility, there is also an antiplasticizing effect of saccharide solutions relative to water.[78]

10. Humectancy and Hygroscopicity

A hygroscopic material absorbs moisture from its surrounding atmosphere, while a humectant material is one that resists changes in relative moisture content. The gain or loss of moisture in a corn syrup is dependent on the relative humidity of the atmosphere surrounding the syrup. Moisture absorption values for several sweeteners are shown in Table 21.16.[79]

11. Crystallization

Starch-derived syrups are able to crystallize, depending on the type of carbohydrates present, the solids level and the temperature. This property can be used to advantage, as in the manufacture of hard candy, or can be one to be avoided, as in the case of

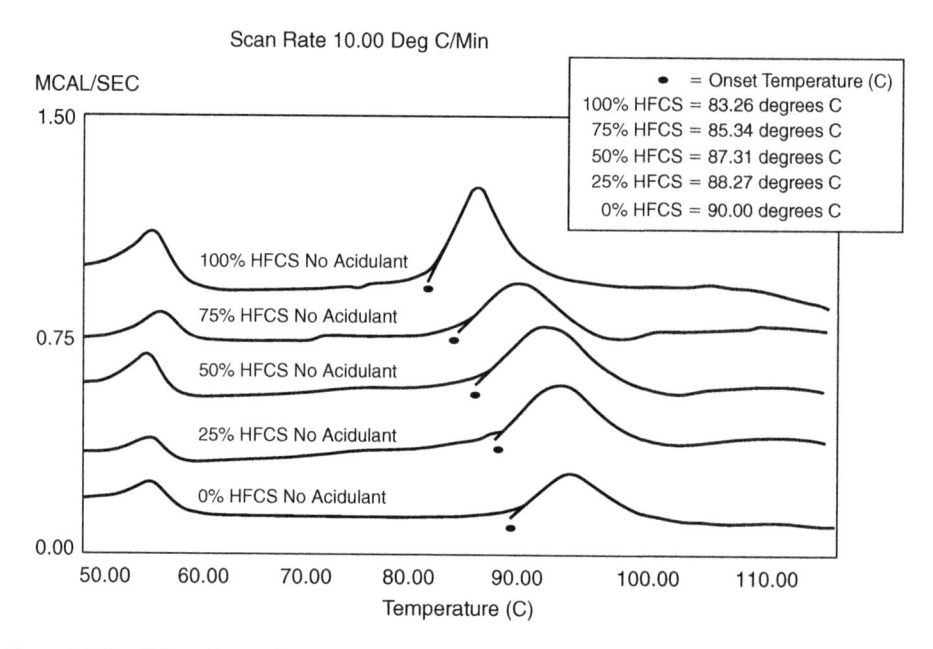

Scan Rate 10.00 Deg C/Min

• = Onset Temperature (C)
100% HFCS = 83.26 degrees C
75% HFCS = 85.34 degrees C
50% HFCS = 87.31 degrees C
25% HFCS = 88.27 degrees C
0% HFCS = 90.00 degrees C

Figure 21.20 Effect of increasing amounts of high-fructose syrups on the gelatinization temperature of starch in a cake batter.[77]

Table 21.16 Moisture absorption by nutritive sweeteners[79]

Sugar	Relative humidity, %	Percent moisture absorption from 1 to 76 days (25°C)											
		1	3	7	11	17	20	26	30	40	50	60	76
Sucrose[a]	62.7	0.06	0.05	0.05	–	0.05	–	–	0.05	–	0.05	0.05	0.05
	81.8	0.05	0.05	0.05	–	0.05	–	–	0.05	–	0.05	0.05	0.05
	98.8	1.31	4.85	13.53	20.81	–	33.01	38.53	–	45.62	–	–	–
Dextrose[b]	62.7	0.04	0.04	0.04	–	0.38	–	–	0.43	–	0.79	1.07	1.74
	81.8	0.62	2.04	5.15	–	9.70	–	–	9.62	–	9.77	9.60	9.60
	98.8	4.68	8.61	15.02	20.78	–	28.43	33.95	–	42.82	–	–	–
Fructose[c]	62.7	0.65	1.41	2.61	–	7.09	–	–	13.01	–	18.35	21.85	21.40
	81.8	4.18	10.22	18.58	–	29.16	–	–	35.05	–	36.32	35.30	35.50
	98.8	11.09	18.43	30.74	37.61	–	45.95	49.41	–	54.99	–	59.14	–
Lactose	62.7	0.03	0.03	0.03	–	0.05	–	–	0.05	–	0.05	0.05	0.08
	81.8	0.07	0.07	0.07	–	0.07	–	–	0.11	–	0.11	0.11	0.07
	98.8	0.05	0.05	0.09	0.12	–	0.13	0.26	–	0.33	–	–	–

[a]Sucrose crystals liquefied after absorption of 16–18% moisture
[b]On reaching 15–18% moisture, crystals began to dissolve; when absorption reached 42%, the dextrose was completely liquefied
[c]All fructose crystals liquefied

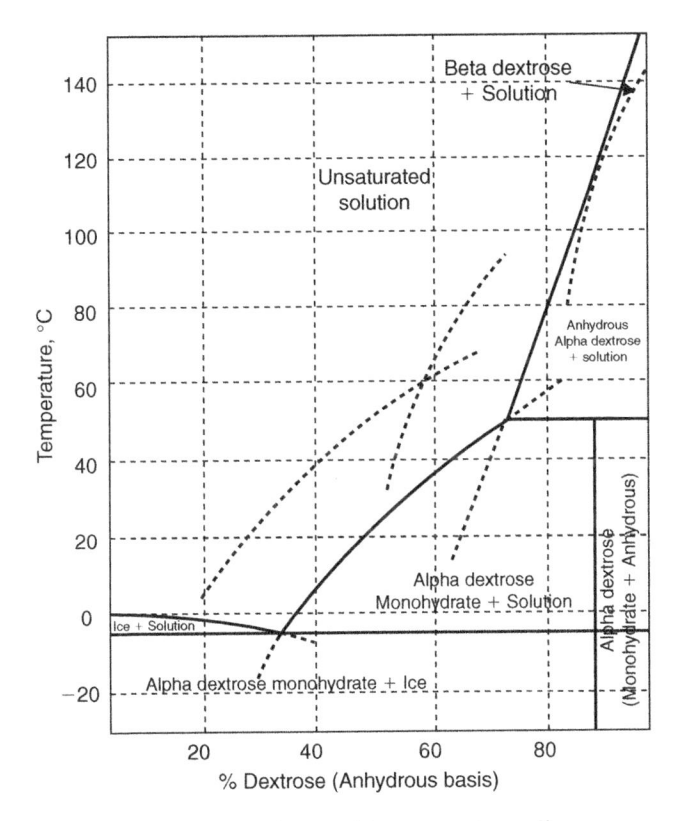

Figure 21.21 Phase diagram for aqueous solutions of dextrose/D-glucose.[80]

jam and jelly manufacture. In jelly manufacturing, one role of the syrup is to increase the solids level and the osmotic strength to the point that it is difficult for the system to support microbial growth, while maintaining a smooth, clear, non-crystallized structure. As previously discussed, the sweetener also binds most of the water present and induces gelation of the HM pectin solution. This process must be carefully controlled because too high a level of crystallizing sugars, such as dextrose, may induce crystallization on storage. The phase diagram for glucose is shown in Figure 21.21.[74]

12. Sweetness

Sweetness is an important and easily identifiable characteristic of glucose- and fructose-containing sweeteners. The sensation of sweetness has been extensively studied.[80–82] Shallenberger[83] defines sweetness as a primary taste. He furthermore asserts that no two substances can have the same taste. Thus, when compared to sucrose, no other sweetener will have the unique properties of sweetness onset, duration and intensity of sucrose. It is possible to compare the relative sweetness values of various sweeteners, as shown in Table 21.17,[84] but it must be kept in mind that these are relative values. There will be variations in onset, which is a function of the chirality of the sweetener,[85] variations in duration, which is a function of the molecular weight profile and is impacted by the viscosity, and changes in intensity, which is affected by

Table 21.17 Relative sweetness values of various sweetener[84]

Type of sweetener	Sweetness relative to sucrose
30 DE acid-converted syrup	30–35
36 DE acid-converted syrup	35–40
42 DE acid-converted syrup	45–50
54 DE acid-converted syrup	50–55
62 DE acid-converted syrup	60–70
HFS (42% fructose)	100
HFS (55% fructose)	100–110
HFS (90% fructose)	120–160
Lactose	40
Dextrose	70–80
Fructose	150–170
Sucrose	100

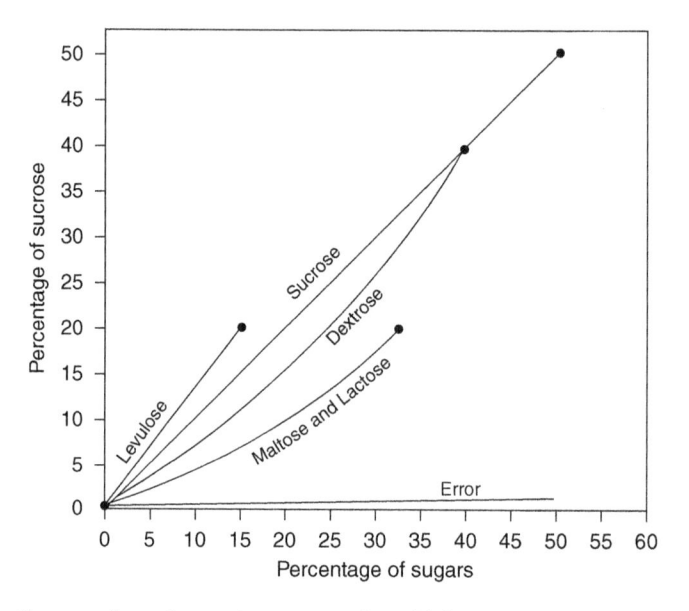

Figure 21.22 Concentrations of sugars (on a percent dry solids basis) needed to give the same sweetness as sucrose solutions (levulose = fructose).[85]

the solids level and the particular isomers present (Figure 21.22).[85] Such variables are demonstrated by the performance of fructose in solution. The fructose molecule may exist in any of several forms. The exact concentration of any of these isomers depends on the temperature of the solution. At cold temperatures the sweetest form, β-D-fructopyranose, predominates, but at hot temperatures, fructofuranose forms predominate and the perceived sweetness lessens (Figure 21.23).[86]

13. Selection of Sweeteners

Selection of sweeteners for food applications is driven by cost, availability and consideration of the functional properties. In many cases, the desired properties may be

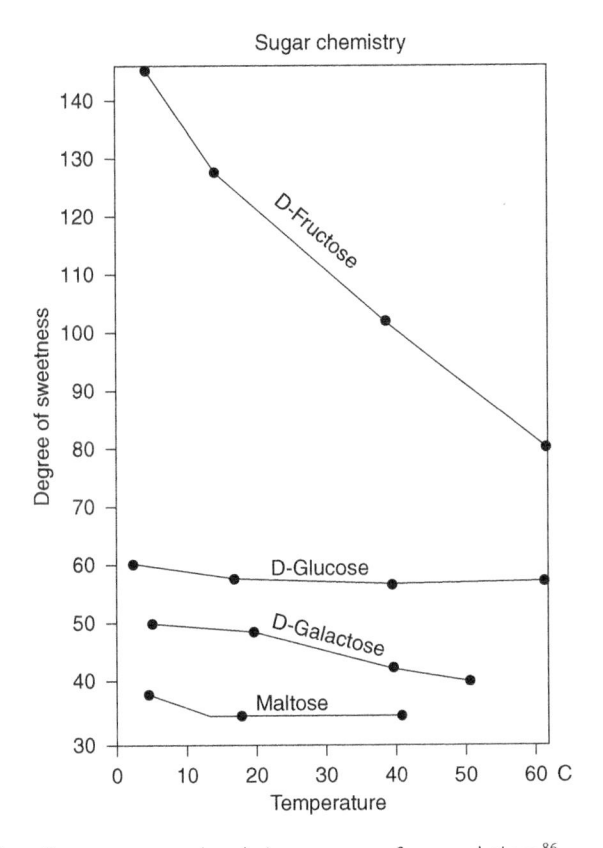

Figure 21.23 Effect of temperature on the relative sweetness of sugar solutions.[86]

mutually exclusive in a given sweetener. This would be the case, for example, if a high degree of sweetness combined with a high viscosity were required in the sweetener. In such cases, blends of sweeteners or the use of other food ingredients would be required. Desirable physical properties of starch sweeteners as they apply to many food applications can be found in reference 87.

IV. References

1. Kretchmer N. In: Kretchmer N, Hollenbeck CB, eds. *Sugars and Sweeteners*. Orlando, FL: CRC Press; 1991:8–9.
2. Kooi ER, Armbruster FC. In: Whistler RL, Paschall EF, eds. *Starch: Chemistry and Technology*. Vol. II. New York, NY: Academic Press; 1967:553–568.
3. Whistler RL. In: Whistler RL, BeMiller JN, Paschall EF, eds. *Starch Chemistry and Technology*. Orlando, FL: Academic Press; 1984:4–6.
4. Anon. *Nutritive Sweeteners From Corn*. Washington, DC: Corn Refiners Association; 1993:6.
5. Anon. *Corn Annual 2000*. Washington, DC: Corn Refiners Association; 2000:3.
6. Pancoast H, Junk WR. *Handbook of Sugars* 1980. Westport, CT: AVI Publishing; 1980 180.

7. Lloyd NE, Nelson WJ. In: Whistler RL, BeMiller JN, Paschall EF, eds. *Starch Chemistry and Technology*. Orlando, FL: Academic Press; 1984:611–660.
8. Anon. *Federal Register*. 1995 September 21;60(No. 183):48939.
9. Anon. *Census of Products, Grain Mill Products*. Washington, DC: U.S. Department of Commerce; 1992:20D-22.
10. Teague WM, Brumm PJ. In: Schenck FW, Hebeda RE, eds. *Starch Hydrolysis Products*. New York, NY: VCH Publishers; 1992:45–77.
11. MacAllister RV. *Adv. Carbohydr. Chem. Biochem.* 1979;36:15.
12. Katz F; *1995 Regulatory Seminar*. Washington, DC: Corn Refiners Association; 1995:78.
13. Alexander R. In: Schenck FW, Hebeda R, eds. *Starch Hydrolysis Products*. New York, NY: VCH Publishers; 1992:233–275.
14. Vance RV, Rock AO, Carr PW. US Patent 3 654 081. 1972.
15. Hebeda RE. In: Watson SA, Ramstad PE, eds. *Corn: Chemistry and Technology*. St. Paul, MN: American Association of Cereal Chemists; 1987:501–534.
16. Vuilleumeir SW. *The Outlook for Sweeteners*: Mckeany-Flavell Company; June 2000.
17. Blanchard PH. *Technology of Corn Wet Milling and Associated Processes*. Amsterdam, The Netherlands: Elsevier Science Publishers; 1992:232.
18. Scallett B, Ehrenthal I. US Patent 3 305 395. 1967.
19. BeMiller JN. In: Whistler RL, Paschall EF, eds. *Starch: Chemistry and Technology*. Vol. I. New York, NY: Academic Press; 1965:495–520.
20. BeMiller JN. *Advan. Carbohydr. Chem.* 1967;22:25.
21. Watson SA. In: Whistler RL, Paschall EF, eds. *Starch: Chemistry and Technology*. Vol. II. New York, NY: Academic Press; 1967:1–51.
22. Wright KN. In: Watson SA, Ramstad PE, eds. *Corn: Chemistry and Technology*. St. Paul, MN: American Association of Cereal Chemists; 1987:447–478.
23. Hassler JW. *Activated Carbon*. New York, NY: Chemical Publishing Company; 1963:115.
24. Conlee J. *Stärke*. 1971;23:366.
25. Schenck FW, Cotilion M. In: Schenck FW, Hebeda RE, eds. *Starch Hydrolysis Products*. New York, NY: VCH Publishing; 1992:531–554.
26. Rooney LW, Serna-Saldivar SO. In: Watson SA, Ramstad P, eds. *Corn: Chemistry and Technology*. St. Paul, MN: American Association of Cereal Chemists; 1987:399–429.
27. Komaki T. *Agri. Biol. Chem.* 1968;32:123.
28. Pazur JH, Kleppe K. *J. Biol. Chem.* 1962;237:1002.
29. Linko P. In: Kruger JE, Lineback D, Stauffer CF, eds. *Enzymes and their Role in Cereal Technology*. St. Paul, MN: American Association of Cereal Chemists; 1987:364.
30. Norman BE. *Starch/Stärke*. 1982;34:340.
31. Anon. *Nutritive Sweeteners from Corn*. Washington, DC: Corn Refiners Association; 1993:17.
32. Antrim RL, Colilla W, Schnyder BJ. *Appl. Biochem. Bioeng.* 1979;2:97.
33. Casey JP. *Research Management*. 1976;28.
34. Pedersen S. In: Tanaka A, Tosa T, Kobayashi T, eds. *Industrial Application of Immobilized Biocatalysts*. New York, NY: Marcel Dekker; 1993:185–207.

35. Havewala Jr NB, Pitcher WH. In: Pye EK, Wingard LB, eds. *Enzyme Engineering*. Vol. 2. New York, NY: Plenum Press; 1974:315–328.
36. MacAllister Jr RV. In: Pitcher WH, ed. Boca Raton, FL: CRC Press; 1980:82–112.
37. Takasaki Y, Kosugi Y, Kanbayashi A. In: Penman D, ed. *Fermentation Advances*. New York, NY: Academic Press; 1969:561–589.
38. Lloyd NE, Khaleeluddin K. *Cereal Chem.* 1976;53:270.
39. Takasaki Y. *Agr. Biol. Chem.* 1967;31:309.
40. Takasaki Y. *Agri. Biol. Chem.* 1971;35:1371.
41. Shioda K. In: Schenck FW, Hebeda RE, eds. *Starch Hydrolysis Products*. New York, NY: VCH Publishers; 1992:565–566.
42. Kusch T, Gosewinkel W, Stoeck G. US Patent 3 513 023. 1970.
43. Hara K, Samoto M, Sawai M, Nakamura S. US Patent 3 704 168. 1972.
44. Hanover LM, White JS. *Am. J. Clin. Nutr.* 1993;58(5, Suppl):7245.
45. Hanover LM. In: Schenck FW, Hebeda RE, eds. *Starch Hydrolysis Products*. New York, NY: VCH Publishers; 1992:201–231.
46. Dean GR, Gottfried JB. *Adv. Carbohydr. Chem.* 1950;5:127.
47. Weber EJ. In: Watson SA, Ramstad PE, eds. *Corn: Chemistry and Technology*. St. Paul, MN: American Association of Cereal Chemists; 1987:311–349.
48. Mulvihill PJ. In: Schenck FW, Hebeda RE, eds. *Starch Hydrolysis Products*. New York, NY: VCH Publishers; 1992:121–176 [Figure 5.5].
49. Nakakuki T. Japan Patent JP 8916599 A2. 1989.
50. Heady RE. US Patent 3 535 123. 1970.
51. Okada M, Nakakuki Teruo. In: Schenck FW, Hebeda RE, eds. *Starch Hydrolysis Products*. New York, NY: VCH Publishers; 1992:335–367.
52. Hoover WJ. In: Food Processing Catalogue, 1964–65 Edition. 1964.
53. Wartman AM, Hagberg C, Eliason MA. *J. Chem. Eng. Data.* 1976;21:459.
54. Kurtz F, Eliason MA. *J. Chem. Eng. Data.* 1979;24:44.
55. Wartman AM, Bridges AJ, Eliason MA. *J. Chem. Eng. Data.* 1980;25:277.
56. Wartman AM, Spawn TD, Eliason MA. *J. Agr. Food Chem.* 1984;32:971.
57. Anon. *Quality Guidelines and Analytical Procedures*: International Society of Beverage Technologists; 1994:2.
58. Anon *Nutritive Sweeteners From Corn*. Washington, DC: Corn Refiners Association; 1993:30–31.
59. Marov GJ. *Proceedings of the Society of Soft Drink Technologists*. 1982:91–98.
60. Anon. *Quality Guidelines and Analytical Procedures*. NY: International Society of Beverage Technologists; 1994 [Table 1].
61. Pancoast H, Junk WR. *Handbook of Sugars* 1980. Westport, CT: AVI Publishing; 1980:223.
62. Erickson ER, Berntsen RA, Eliason MA. *J. Chem. Eng. Data.* 1966;11:485.
63. Pancoast H, Junk WR. *Handbook of Sugars* 1980. Westport, CT: AVI Publishing; 1980:186.
64. Anon. *Standard Analytical Methods*. 6th edn. Washington, DC: Corn Refiners Association; 1991 [Method E-16].
65. Shallenberger RS, Birch GG. *Sugar Chemistry*. Westport, CT: AVI Publishing Company; 1975.

66. Waller GR, Feather MS. *The Maillard Reaction in Foods and Nutrition*. Washington, DC: American Chemical Society; 1983.
67. Pancoast H, Junk WR. *Handbook of Sugars* 1980. Westport, CT: AVI Publishing; 1980:246.
68. Pancoast H, Junk WR. *Handbook of Sugars* 1980. Westport, CT: AVI Publishing; 1980:181.
69. Hebeda RE, Styrlund CR. *Cereal Foods World*. 1986;31:685.
70. Phaweni M, O'Conner-Cox ESC, Pickerell ATW, Axcell BC. *J. Am. Soc. Brewing Chem.* 1993;51:10.
71. Keeney P. *Food Technol.* 1982;65.
72. da Silva JAL, Rao MA. *Food Technol.* 1995;49:70.
73. Arbuckle WS. *Ice Cream*. 4th edn. Westport, CT: AVI Publishing Company; 1986:74.
74. Critical Data Tables, Washington, DC: Corn Refiners Association; 1975.
75. Wittinger SA, Smith DE. *J. Food Sci.* 1986;51:1463.
76. Smith ER, Torgesen JL. *Phys. Chem. Sect. Rept. 5*. Washington, DC: National Bureau of Standards; 1950.
77. Johnson JM, Harris CH, Barbeau WE. *Cereal Chem.* 1989;66:155.
78. Kim CS, Walker CE. *Cereal Chem.* 1992;69:212.
79. Pancoast H, Junk WR. *Handbook of Sugars* 1980. Westport, CT: AVI Publishing; 1980:391.
80. Shallenberger RS, Birch GG. *Sugar Chemistry*. Westport, CT: AVI Publishing; 1975:180.
81. Birch GG, Green LF, Coulson CB, eds. *Sweetness and Sweeteners*. London, UK: Applied Science Publishers; 1971.
82. Shallenberger RS. *Advanced Sugar Chemistry*. Westport, CT: AVI Publishing; 1982.
83. Shallenberger RS. *Taste Chemistry*. London, UK: Blackie Academic&Professional; 1993.
84. Pancoast H, Junk WR. *Handbook of Sugars* 1980. Westport, CT: AVI Publishing; 1980:388.
85. Shallenberger RS, Birch GG. *Sugar Chemistry*. Westport, CT: AVI Publishing; 1975:154.
86. Shallenberger RS, Birch GG. *Sugar Chemistry*. Westport, CT: AVI Publishing; 1975:156.
87. Pancoast H, Junk WR. *Handbook of Sugars* 1980. Westport, CT: AVI Publishing; 1980:403.

22 Cyclodextrins: Properties and Applications

Allan Hedges
Consultant, Crown Point, Indiana USA

I.	Introduction	833
II.	Production	835
III.	Properties	837
IV.	Toxicity and Metabolism	838
V.	Modified Cyclodextrins	840
	1. Hydroxyalkylcyclodextrins	840
VI.	Complex Formation	842
VII.	Applications	845
VIII.	References	848

I. Introduction

Cyclodextrins are produced from starch by the action of an enzyme, cyclodextrin glycosyltransferase (CDTGase; ED 2.4.1.19). A chemical synthesis has been reported,[1] but it is too tedious for commercial production of cyclodextrins. Alpha-, beta- and gamma-cyclodextrins contain 6, 7 and 8 (1→4)-linked α-D-glucopyranosyl units, respectively, and are the predominate cyclodextrins formed by the enzyme. Delta- and nu-cyclodextrins, which respectively contain 9 and 12 glucosyl units in the ring, have been isolated in small quantities and characterized,[2,3] and evidence has been reported for cyclodextrins containing up to 16 glucosyl units in the ring.[4] Because of the predominance of the α-, β- and γ-cyclodextrins, they are the cyclodextrins of current commercial interest.

Enzymes for the production of cyclodextrins (CD) are produced by various bacteria (Table 22.1). No fungi or other organisms have been reported to produce CDTGase. Enzymes obtained from these bacteria differ in the relative amounts of α-, β- and γ-cyclodextrins produced. Typically, a mixture of two or three cyclodextrins

Table 22.1 Bacteria which produce CDTGase

Bacillus macerans	(6, 7)
Bacillus megaterium	(8)
Bacillus circulans	(9)
Klebsiella pneumoniae	(10)
Alkolophilic *Bacillus sp.*	(11)
Bacillus stearothermophilus	(12)
Bacillus amyloliquefaciens	(13)
Thermoanaerobacter sp.	(14)
Clostridium thermoamylolyticum	(15)
Clostridium thermohydrosulfuricum	(15)
Bacillu (all sp.)	(16)
Bacillus lentus	(17)

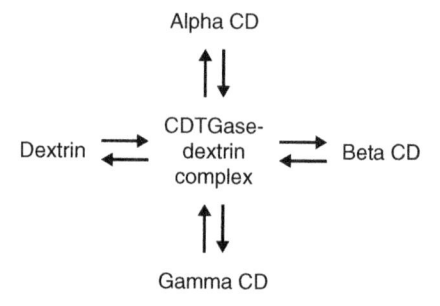

```
                    Alpha CD
                       ⇅
             CDTGase-
Dextrin  ⇌   dextrin   ⇌   Beta CD
             complex
                       ⇅
                    Gamma CD
```

Figure 22.1 'Dextrin' represents the initial starch hydrolyzate used for the reaction. It is a hydrolyzate resulting from alpha-amylase and the disproportionation activity of the CDTGase, or a combination of these activities. This 'dextrin' and cyclodextrins are both substrates and products of the CDTGase.

is produced. As the reaction proceeds, a second or third cyclodextrin also begins accumulating. While α-cyclodextrin might be the cyclodextrin initially produced, β-cyclodextrin might accumulate in larger quantities by the time the reaction is complete. In some cases, some of the α-cyclodextrin might be converted to β-cyclodextrin. The final ratio of the α-, β- and γ-cyclodextrins depends on the enzyme used, the conditions of the reaction and the duration of the reaction.

CDTGase will catalyze three different enzyme reactions: cyclization; transglycosylation; and hydrolysis[18–20] (see Chapter 7). Figure 22.1 shows a reaction scheme for the reaction. Starch or a starch hydrolyzate can be used as the substrate. The starch or starch hydrolyzate forms a complex with the enzyme; cleavage and joining of the ends results in cyclization to form the cyclodextrin. The reaction is reversible, and a cyclodextrin can serve as the substrate for the enzyme, resulting in ring opening. If the reaction is allowed to go to equilibrium, either with starch or any one of the cyclodextrins as the substrate, a mixture of cyclodextrins and linear malto-oligosaccharides result.

In the transglycosylation reaction, a cyclodextrin or starch chain is cleaved and another starch chain is added to the cleaved chain, resulting in one longer linear chain and one shorter linear chain of glucopyranosyl units. This allows for the formation of a mixture of cyclodextrins from a single reaction.

CDTGase displays α-amylase activity. In this case, water is the acceptor and starch or the cyclodextrin is cleaved, resulting in a shorter starch chain or a linear malto-oligosaccharide when a cyclodextrin is the substrate.

By using water-soluble organic solvents, usually referred to as precipitants, the reaction can be directed to produce only one cyclodextrin. In the presence of toluene, the enzyme from *B. macerans* produces only β-cyclodextrin and linear starch chains. Beta-cyclodextrin can be separated from the soluble starch chains and recovered as a precipitated toluene complex. In the presence of decanol, α-cyclodextrin is the only cyclodextrin produced using the enzyme from *B. macerans*.[21,22] Precipitants such as large cyclic compounds similar to musk oil[23] or α-naphthol and methyl ethyl ketone (butanone)[24] can be used to produce δ-cyclodextrin.

II. Production

Two processes are used for the commercial production of cyclodextrins. One is referred to as the non-solvent process; in it, the enzyme is allowed to react with starch in a completely aqueous environment. In the other, referred to as the solvent process, an organic precipitant is added to direct the reaction to produce only one cyclodextrin.

Both processes use starch (corn, potato, rice, wheat, etc.) as the substrate. Corn and potato starches are most commonly used. In both processes, the starch is hydrolyzed to a dextrose equivalent (DE) of 3 to 8 (see Chapter 21) prior to use.[25] If liquifaction/hydrolysis is not sufficient, retrogradation occurs. This limits the availability of substrate, resulting in low yields of cyclodextrins, and interferes with later recovery steps. If the starch is over-hydrolyzed, the disproportionation reaction dominates and yields of cyclodextrins are low. If an α-amylase is used to hydrolyze the starch, it must be inactivated by acidification, raising the temperature, otherwise, the yield will be greatly reduced.

For the non-solvent process,[26–28] the pH of the starch hydrolyzate is adjusted to the optimum pH before adding the enzyme. Starch concentration is usually ~15%. At higher concentrations, the yield of cyclodextrins decreases. At lower concentrations, the amount of cyclodextrins per unit weight of the starch increases, but the yield per unit volume of the reaction decreases. It is economically beneficial to utilize more starch than to recover a smaller amount of cyclodextrin per unit volume.

When the reaction is complete, a mixture of α-, β- and γ-cyclodextrins and linear malto-oligosaccharides are present. The relative amounts of these components are dependent on the source of CDTGase, which is inactivated, usually by heating, to prevent the loss of cyclodextrins when α-amylase is used to hydrolyze starch to reduce the viscosity. After hydrolyzing the starch with an α-amylase that will not attack the cyclodextrins, the reaction mixture is deionized with a mixed-bed ion exchange resin and treated with activated charcoal. Then, the reaction mixture is concentrated to a solids concentration of about 45% using vacuum evaporation. This concentration is selected so that β-cyclodextrin, the least soluble of the cyclodextrins, will crystallize on cooling, but the other cyclodextrins will remain in solution. The concentration can vary, depending on the enzyme used and the proportion of cyclodextrins in the mixture. The concentrated reaction mixture is then cooled to crystallize β-cyclodextrin. If the starch is not sufficiently hydrolyzed, the viscosity will be too high to allow recovery of β-cyclodextrin crystals. The yield is ~15%.

Crystals of β-cyclodextrin are then recovered by filtration or centrifugation and washed with a small amount of water. They are then dried or dissolved in water to be recrystallized to make a purer product. The β-cyclodextrin produced from this process has a purity of 98% or greater. If it is recrystallized, the purity can be 99%. The major impurities in both cases are starch hydrolyzate and α- and γ-cyclodextrins.

The mother liquor contains linear starch hydrolyzate and a mixture of cyclodextrins. This can be further concentrated to about 70% solids and sold or further treated to recover the α- and γ-cyclodextrins.[29] To recover the latter, the residual starch hydrolyzate is converted to D-glucose using glucoamylase/amyloglucosidase. The hydrolyzate is then passed through an ion exchange resin to separate the glucose from the cyclodextrins. The resulting mixture of cyclodextrins is then passed through a size-exclusion column to separate the cyclodextrins.

In the solvent process,[30,31] a solvent is used to direct the reaction to produce only one cyclodextrin. Higher starch concentrations can be used in this process than in the non-solvent process. (Concentrations of 30% are typically used.) Maximum yields per unit volume are achieved at 30% solids in the presence of solvents, compared to a maximum yield at 15% solids using the non-solvent process. The pH of the hydrolyzate is adjusted to the optimum for the enzyme, and the enzyme and solvent are added to the starch hydrolyzate. Solvents typically used for β-cyclodextrin production include toluene, trichloroethylene and cyclohexane. As the reaction proceeds, β-cyclodextrin precipitates as an insoluble complex. Yields per unit weight of starch are typically 30–45%, much higher than the yield obtained by the non-solvent process.

On completion of the reaction, the complex of cyclodextrin and precipitant is recovered by filtration or centrifugation. The precipitate is slurried in water, and the mixture is heated to distill off the precipitant. At the completion of the distillation an aqueous solution of cyclodextrin, containing <1 ppm of precipitant, results. The solution of β-cyclodextrin is treated with activated carbon and then cooled for crystallization. Crystals of β-cyclodextrin are collected by filtration or centrifugation and then dried. The purity of the product is at least 98%, the major impurity being starch hydrolyzate. Since the reaction is directed to produce only β-cyclodextrin, α- and γ-cyclodextrins are not present in detectable quantities.

Alpha-cyclodextrin can also be produced using an organic precipitant.[21,22] In the presence of decanol, the enzyme from B. macerans produces predominantly α-cyclodextrin. In the early stages of the reaction, a mixture of α- and β-cyclodextrins are present, so the reaction must be allowed to proceed long enough for the β-cyclodextrin to disappear. Choice of enzyme and precipitant is important. Some precipitants, such as cyclohexane, will produce a mixture of α- and β-cyclodextrins; by controlling the temperature, relative proportions of each can be controlled.[32] As the temperature is increased, the relative amount of β-cyclodextrin increases.

Decanol can be removed either by steam distillation[21,22] or by solvent extraction. For steam distillation, the precipitated complex with decanol is slurried in water, and the slurry is heated. Alternatively, a slurry can be extracted by mixing it with a solvent of lower boiling temperature than decanol. Alpha-cyclodextrin forms a complex with the extraction solvent and precipitates. The precipitate is recovered by centrifugation or filtration and slurried in water, and the extraction solvent is then distilled and recovered for further use. Alpha-cyclodextrin solution, free of decanol and extraction solvent, is treated with activated carbon and filtered, and the product is crystallized.

Gamma-cyclodextrin can also be formed using a precipitant. Macrocyclic compounds, such as cyclotridecanone, which do not fit into the cavity of α- and β-cyclodextrins are used.[23,33] Such compounds do not dissolve in water or melt at temperatures compatible with the enzyme and require the presence of a solvent, such as methyl isobutyl ketone, to dissolve the precipitant and make it available to form a complex with γ-cyclodextrin. The precipitated complex of γ-cyclodextrin is collected by centrifugation or filtration. Solvent extraction is used for purification of γ-cyclodextrin in the same manner as described for α-cyclodextrin.

III. Properties

Cyclodextrins are soluble in water and insoluble in most organic solvents (Table 22.2).[34,35] Solubility increases as the temperature increases. Beta-cyclodextrin is the least soluble cyclodextrin in water. In β-cyclodextrin, the C2 and C3 hydroxyl groups of adjacent α-D-glucopyranosyl units orientate in such a way as to have maximum interaction with each other, compared to the other cyclodextrins.[36,37] As a result, they are not as much available to be hydrated compared to the other cyclodextrins. The α-cyclodextrin ring is more strained, resulting in these hydroxyl groups being further apart and more able to interact with water, providing greater solubility. The γ-cyclodextrin ring is less strained than that of α- or β-cyclodextrins. Water solubility results from the hydroxyl groups interacting much less with each other and much more with water.

Interference with the hydrogen bonding of the hydroxyl groups increases the water solubility of the cyclodextrins. In the presence of urea, the solubility of β-cyclodextrin increases to 250 g/L.[38] At high pH, ionization of hydroxyl groups increases water solubility; at pH 12.5 the solubility of β-cyclodextrin is 750 g/L.[38,39]

Cyclodextrins are insoluble in most organic solvents (Table 22.3). Beta-cyclodextrin is soluble in some polar aprotic solvents and is more soluble in some of these solvents than in water. Cyclodextrins are somewhat soluble in mixtures of water and

Table 22.2 Solubility of cyclodextrins in water[a]

Temperature, °C	Alpha	Beta	Gamma
25°	12.8	1.8	25.6
45°	29.0	4.5	58.5
60°	66.2	9.1	129.2

[a]g/100 mL

Table 22.3 Solubility of cyclodextrins in organic solvents

Solvent	Solubility, %
Dimethylsulfoxide	>41
Dimethylformamide	28.3
N-Methylpyrrolidone	14.8
Ethylene glycol	7.0
Pyridine	3.5
Propylene glycol	0.5
Tetrahydrofuran	0
Methyl isobutyl ketone	0
Methyl isopropyl ketone	0
Acetone	0
Alcohols	0

water-miscible organic solvents. In general, as the concentration of the organic solvent increases, the solubility of the cyclodextrin decreases.

Cyclodextrins are hydrolyzed by strong acids at a rate dependent on the temperature and the acid concentration.[40–42] The first step in hydrolysis of the cyclodextrins is ring opening, a slow step compared to the hydrolysis of starch by acids. Once the ring is opened, hydrolysis proceeds at the same rate as the hydrolysis of malto-oligosaccharides. Depending on the conditions, the products are a mixture of D-glucose and chains of (1→4)-linked α-D-glucopyranosyl units up to a degree of polymerization equal to that of the cyclodextrin being hydrolyzed. Weak organic acids do not readily hydrolyze cyclodextrins. Cyclodextrins are stable in the presence of bases.

The differential scanning calorimetry (DSC) thermograms for α-, β- and γ-cyclodextrins are identical. Two heat absorption peaks are present: the first occurs at 100°C as water is evaporated from the crystals; the second, occurring at ~250°C, is a result of crystal melting and thermal decomposition.

Hydrolysis of cyclodextrins can be affected by enzymes[43–45] but not by glucoamylases or β-amylases, because both require a non-reducing end group to initiate hydrolysis. Alpha-amylases are endoenzymes and can open the ring of the cyclodextrin molecule (see Chapter 7 for a description of various amylases). Fungal α-amylases are better catalysts for the hydrolysis of cyclodextrins than bacterial α-amylases. The rate of hydrolysis is dependent upon the cyclodextrin. Alpha-cyclodextrin is hydrolyzed very slowly compared to β-cyclodextrin. Gamma-cyclodextrin is hydrolyzed much more rapidly than β-cyclodextrin. Hydrolysis products consist of a mixture of D-glucose and short malto-oligosaccharides.

IV. Toxicity and Metabolism

Beta-cyclodextrin has been the most studied of the cyclodextrins. Results from numerous investigations have demonstrated its safety and metabolism. Table 22.4 shows the results of acute toxicity studies from various routes of administration. At the maximum dose of β-cyclodextrin presented orally, no mortality related to it was

Table 22.4 Acute toxicity of beta-cyclodextrin

Mode of administration	LD$_{50}$ (mg/Kg)	Animal	Reference
Oral	<12 500	Mouse	46
	<12 500	Rat	46
Intravenous	788	Rat	47
Intraperitoneal	<900	Mouse	46
	<1200	Rat	46
	700	Rat	48
Subcutaneous	<900	Mouse	46
	3700	Rat	48

observed. Injection of β-cyclodextrin did result in the death of animals, because it crystallized in the kidneys and caused renal damage.[47]

Various long-term feeding studies have been carried out with β-cyclodextrin. Dogs and rats were fed diets containing up to 5% β-cyclodextrin for one year.[49] Some changes in levels of plasma liver enzymes and biochemical parameters were found at high dose levels, but these were not considered to be of toxicological significance. The results indicated a non-toxic level of 650 mg/Kg/day (1.25% of diet) of β-cyclodextrin consumed for rats and ~1800 mg/Kg/day (5.0% of diet) of β-cyclodextrin consumed for dogs. No evidence of oncogenicity was found in studies where mice and rats were fed up to 675 mg/Kg/day of β-cyclodextrin during a lifetime study of about 2 years for mice and 2.5 years for rats.[50] In reproductive and developmental toxicity tests, no adverse effect level of 1.25% β-cyclodextrin in the diet of rats was determined.[51] At the next higher dietary level (5% β-cyclodextrin in the diet), some retardation of the growth of nursing pups was observed.

Beta-cyclodextrin is not attacked by salivary or pancreatic α-amylases.[52] As a result, it passes through the digestive tract to the colon intact.[53] Very little is absorbed intact. Only at very high concentrations (5 and 10% in the diet), much higher concentrations than would be used in actual use of the cyclodextrin, was any β-cyclodextrin detected in the urine and serum of test animals.[54,55] When labeled β-cyclodextrin was fed, no radioactivity appeared in the serum or exhaled carbon dioxide until about 12 hours after administration of the cyclodextrin, which corresponds to the transit time to the colon.[56] In the colon, the cyclodextrin ring is opened by enzymes from the intestinal flora; enzymes from *Bacteroides* species have been reported to be capable of doing this.[57] The resulting maltoheptaose is catabolized as any starch fragment reaching the colon.

The toxicity of γ-cyclodextrin has also been studied. The acute oral LD$_{50}$ was >16 000 mg/Kg for mice and >8000 mg/Kg for rats.[58] Rats were fed diets containing up to 20 grams of δ-cyclodextrin for 90 days; no toxicological effects were observed.[61] The LD$_{50}$ for intraperitoneal administration was >4600 mg/Kg in rats.[61] For subcutaneous administration, the LD$_{50}$ was >2400 mg/Kg in rats.[58] Intravaneous administration indicated an LD$_{50}$ of 10 000 mg/Kg in mice and >3750 mg/Kg in rats.[58]

Mice were fed diets containing α-cyclodextrin.[59,60] Alpha-cyclodextrin was neither digested nor absorbed by the animals. At the highest dose (24% α-cyclodextrin in the diet), weight loss, poor appetite and gas accumulation in the colon occurred. Late in the study, the animals adapted to the diet containing the large amount of cyclodextrin.

V. Modified Cyclodextrins

Modification of cyclodextrin allows them to be used in some applications where the unmodified cyclodextrin does not give the desired performance. Modification can increase the solubility of cyclodextrins and their complexes, allowing the modified cyclodextrin to be used in applications where increased solubility in water or organic solvents is needed. Derivatization of oxygen or ozone disrupts the network of hydrogen bonding around the rim of the cyclodextrin molecule, resulting in more interaction of the hydroxyl groups with water and increased solubility of the modified cyclodextrin. Due to the random nature of the substitution on the available hydroxyl groups, concentrations of a single species do not reach high enough concentrations to crystallize. Both the modified cyclodextrin and the complexes formed can be more soluble than the unmodified cyclodextrin.

The nature of the group added to the cyclodextrin and the amount of substitution can affect the solubility of the cyclodextrin. If a nonpolar group is added, solubility of the modified cyclodextrin depends on the size of the group and the degree of substitution (DS). To make an acetate ester of β-cyclodextrin that is insoluble in water, at least 17 of the 21 hydroxyl groups must be acetylated. If hexanoate ester is made, only three hydroxyl groups need to be esterified to create insolubility. These derivatives are more soluble in organic solvents than are the unmodified cyclodextrins.

Binding and release characteristics are also altered by modification. Some modifications can provide additional surface area for binding of some guests, resulting in stronger binding or binding of some guests that do not complex with the unmodified cyclodextrins. The addition of ionic or nonpolar groups can be used to achieve stronger binding of ionic guests. Substitution can also result in decreased binding of some guests to allow them to be released more easily.

1. Hydroxyalkylcyclodextrins

Hydroxyalkyl derivatives are made by reacting cyclodextrins with propylene or ethylene oxide. The average number of hydroxyalkyl groups per cyclodextrin molecule ranges from ~3 to ~10 in commercial products. Lower amounts of substitution result in excessive amounts of unmodified cyclodextrin in the product. Substitution is random on the hydroxyl groups so that a mixture of molecular species is present in the product, both with respect to the number of substitutions on a molecule and the position of the substitution on the hydroxyl groups. Because the substitution disrupts the hydrogen bonding of the secondary hydroxyl groups around the rim of the cyclodextrin, and because a mixture of molecular species is created by the random substitution, crystallization of the modified cyclodextrins does not occur and the derivatives are highly soluble in water. Solubilities of 60% are typical; above this concentration, the solution becomes very viscous. Complexes made with hydroxyalkyl ethers are much more soluble than complexes made with unmodified cyclodextrins.

Differential scanning calorimetry thermograms of hydroxyalkylated cyclodextrins are consistent with the noncrystalline nature of the derivatives, showing no sharp melting peak that would be found for a crystalline substance. Instead, a change of slope attributed to glass transitions occurs. The temperature at which the glass transition occurs is dependent on the degree of substitution (DS) of the derivative and increases as the DS increases.[62] The specific rotation is also dependent on the DS and increases as it increases.

No toxicological effects of hydroxypropyl β-cyclodextrins were observed on oral administration at doses up to 5000 mg/Kg in rats and 1000 mg/Kg in rabbits.[63] Intravenous administration was tolerated at doses of 200 mg/Kg in rats and monkeys.[64] Due to its high solubility, crystallization causing renal damage does not occur in the kidneys; clearance occurs in about four hours.[63]

Methylated derivatives are also highly soluble in water. The major methylated derivatives are the 2,6-dimethyl, 2,3,6-trimethyl and randomly methylated derivatives. The dimethyl and trimethyl derivatives are the derivatives most frequently described in the literature. The main advantage of the randomly methylated derivatives is lower cost. Production of the dimethyl and trimethyl derivatives involves several purification steps. Due to the random substitution and mixture of constituents, the randomly substituted derivatives are more soluble than the dimethyl and trimethyl derivatives. All methyl ethers have a decreased solubility at higher temperatures, similar to the behavior of other methylated carbohydrates.

Toxicology of the methylated derivatives has not been studied extensively. The LD_{50} for oral administration of dimethyl β-cyclodextrin is >300 mg/Kg and >200 mg/Kg for intravenous administration in mice.[65] Methylated derivatives interact with cellular membranes, extracting cholesterol and disrupting the membranes. This has limited interest in the derivatives for intravenous delivery of pharmaceuticals, but there has been some interest in use of methylated derivatives to enhance transdermal administration of drugs.[66]

Ionic derivatives of cyclodextrins have also been made. These derivatives exhibit high solubility in aqueous systems. Complexes made with ionic derivatives are also very soluble. A sulfated derivative has been found to be useful for solubilization of drugs. It also has been found to mimic heparin and act as an inhibitor of infection of cells by viruses. A sulfobutyl ether derivative is also capable of solubilizing various drugs.[67,68]

Branched cyclodextrins are also used to increase the solubility of complexes. Two methods are used to make branched cyclodextrins, an enzymic method and a pyrolytic method. In the enzymic method, a starch debranching enzyme, such as pullulanase, is added to a solution of cyclodextrin and a large excess of D-glucose or maltose to force the reaction to proceed in the reverse direction, i.e. to add rather than remove a branch.[69] Since the equilibrium favors the debranching reaction, yields are low and the product typically contains ~15% branched cyclodextrin and ~85% glucose or maltose. Purification is difficult because of the high solubility of both the glucose or maltose and the branched cyclodextrin, but much of the unreacted cyclodextrin can be removed by crystallization.

Branched cyclodextrins can also be formed by heating dry cyclodextrin in the presence of a small amount of hydrogen chloride.[70] The pyrolysis product is dissolved in a small amount of water to dissolve the branched cyclodextrin, leaving behind most of the β-cyclodextrin which has limited solubility.

Derivatives with decreased solubility in water have also been made. By crosslinking cyclodextrins with epichlorohydrin, an insoluble polymer is formed. The polymeric cyclodextrin is able to form complexes with soluble compounds and remove them from solution.

VI. Complex Formation

Several forces are involved in the formation of complexes (Table 22.5). All of them are weak forces, but they are strong enough to allow for formation of a stable complex that can be isolated and dried, if desired. The relative importance of these forces varies from guest to guest and, in most cases, more than one force is involved in formation and stabilization of the complex. While the forces are strong enough for formation and stabilization of the complex, they are also weak enough to allow release of the guest. An equilibrium, such as that described in Figure 22.2, is involved in the formation and release of guests from complexes.

Complexation is a molecular phenomenon involving bringing one molecule of cyclodextrin into contact with one molecule of the guest. Several factors are involved, some of which can be controlled and some of which involve properties of the guest and cannot be controlled. A solvent is needed to dissolve both the cyclodextrin and the guest. Generally this solvent is water. In addition to solvating the cyclodextrin and guest, and allowing the mixing to speed the rate of dissolution and complexation, water provides a driving force for complexation. The hydrophobic cavity of the cyclodextrin molecule provides a more thermodynamically favorable environment for the hydrophobic guest or hydrophobic moiety of the guest than the aqueous environment. While, for most applications, water is the solvent of choice, there are examples of complexation taking place in organic solvents.[71,72] Due to the lipophilic nature of organic solvents, the hydrophobic driving force for the complexation is decreased,

Table 22.5 Possible driving forces for formation and stabilization of complexes

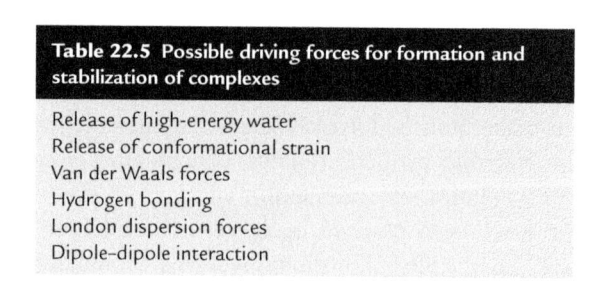

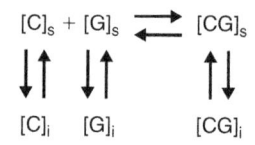

Release of high-energy water
Release of conformational strain
Van der Waals forces
Hydrogen bonding
London dispersion forces
Dipole–dipole interaction

$$[C]_s + [G]_s \rightleftarrows [CG]_s$$
$$\updownarrow \quad \updownarrow \quad \quad \updownarrow$$
$$[C]_i \quad [G]_i \quad \quad [CG]_i$$

Figure 22.2 The complexation reaction is an equilibrium reaction. The magnitude of the arrows is dependent on the guest. Subscript i denotes insoluble materials, and subscript s denotes soluble materials.

because of the similar or greater lipophilicity of the solvent compared to that of the cyclodextrin cavity, resulting in weak binding compared to binding of guests in water.

Temperature also affects the rate of complexation. As the temperature increases, the solubility of the cyclodextrin and guest increases, the number of molecules in solution increases, and the probability of a molecule of cyclodextrin and a molecule of guest coming into contact to form a complex increases. However, heat can also destabilize complexes, so the positive effect of increased solubility must be balanced against the negative effect of decomplexation. The optimum temperature for formation of complexes varies from guest to guest and must be determined for each guest.

Some guests are not sufficiently soluble or dispersible in water to form complexes rapidly. A water-miscible solvent, such as ethanol, acetone or methanol, can be used to increase the solubility or dispersibility of the guest. A small amount of solvent, usually <5%, can be added to the water. This can increase the solubility of the guest sufficiently to allow complexation without altering the hydrophobic nature of the water–solvent solution enough to lose the hydrophobic driving force. Another approach is to dissolve the guest in a small amount of solvent and add the dissolved guest to the water. The guest might precipitate out of solution as it is added, but it is generally dispersed as much smaller particles than the previously undissolved guest, resulting in a faster rate of complexation.

Coprecipitation is commonly used for forming complexes in the laboratory. The cyclodextrin is dissolved in water and the guest is added. Conditions are selected so that the solubility of the complex in water will be exceeded and a precipitate will form as the guest is complexed. The amount of β-cyclodextrin in solution and the rate of complexation can be increased by heating. The guest does not need to be completely dissolved. As the guest in solution is complexed, the insoluble guest will dissolve to be complexed. As the solubility of the complex is exceeded, the complex precipitates from solution. This allows more complex to be formed. The rate of dissolution of the complex is generally slower than the rate of formation of the complex, so that complex accumulates. The precipitate is collected by filtration or centrifugation and dried if desired.

Coprecipitation is used for initial studies of the complex or when only small quantities of complex are needed. Because of the large amounts of water used, this method is not suitable for large-scale production of complexes. Various properties of the complex can be characterized to assist in development of other methods for making complexes which are more suitable for large-scale production.

Dissolving the cyclodextrin is the preferred means of making complexes with soluble derivatives of cyclodextrins. When the guest is added to a solution of cyclodextrin, the insoluble guest will disappear as it is complexed. Because of the high solubility of complexes made with the soluble derivatives, the solubility of the complex is generally not exceeded; therefore a precipitate is not formed. If the solubility of the complex is exceeded and a precipitate is formed, most of the complex is soluble and the precipitate frequently has a tendency to remain in suspension. If a solid complex is desired, the solution is dried by spray drying or other suitable means.

Complexes can also be formed using a slurry of cyclodextrin. The guest is added to the slurry and stirred. Since there are no visual indications of formation of the complex, as in the coprecipitation method, the length of time of stirring must be determined experimentally. The amount of time required varies from guest to guest. Samples are removed and characterized to determine if a complex having properties the same as the complex formed using the coprecipitation method has been formed. The slurry method does have the advantage of using less water than the coprecipitation method and good yields of complex can be achieved without heating, which can be detrimental to some heat labile and reactive guests. Generally slurry concentrations of up to 50% can be achieved. Some guests affect the slurry, making it very viscous and difficult to mix, requiring use of lower concentrations. While the slurry is being mixed, it is mixable and pourable. If agitation is stopped, the slurry becomes firm, requiring the input of much energy to resume mixing. Slurry complexations can be carried out in a tank equipped with an agitator or, in the case of more concentrated slurries, a roller mill or ball mill can be used. The complex can be collected by filtration or centrifugation.

The amount of water can be further reduced to form a paste. Solids concentrations can be as high as 80%. The amount of time required for mixing varies with the guest and the type of mixer. As with the slurry method, samples must be removed to determine the required mixing time to complete complex formation. A variety of mixing devices can be used. The rate of complexation is faster using mixing devices providing the most shear. Because a minimum of water is used in this method, the complex can be removed from the mixer and dried without an isolation step. Heat can also be used to increase the rate of complexation.

The paste method can also be used with soluble derivatives. Since drying can be problematic with complexes of soluble derivatives, a minimum amount of water (20% maximum) is generally used. In order to reduce viscosity and associated mixing problems, a higher temperature is employed. A glass forms as the complex is removed from the mixer and cools. The glass can be milled and dried if necessary.

Dry mixing has also been used to form complexes. This method can be rather slow, taking days or hours to form a complex, compared to the other methods which generally require less than an hour. Both addition of a small amount of water and heating in a sealed vessel increase the rate and efficiency of complexation.[73] For some applications it is not necessary to have a complex in the dry state. Rather, thorough mixing in the dry state can result in thorough distribution of the guest in the cyclodextrin, so that a complex is formed immediately on contact with water.

The means of drying a complex is dependent on the nature of the complex. Complexes which are not highly soluble, such as those that can be readily collected using the coprecipitation method, can be readily dried in a hot air oven. Small amounts dry to a fine pourable powder in the laboratory. On a larger scale, some hard masses of complex requiring milling to form a pourable powder can result. The hard masses can be milled or prevented from forming by selecting a dryer which keeps the complex constantly turning over to allow uniform exposure of the complex to the heat.

Soluble complexes are typically dried using a spray dryer, vacuum dryer or freeze dryer, because if they are dried in a conventional oven, a glass is formed. Spray drying, vacuum drying and/or freeze drying are also used for heat-labile and reactive compounds such as complexes of some acids. With prolonged heating, some compounds decompose or react with the cyclodextrin. However, the time of exposure to heat in a spray dryer is sufficiently short so that the compound does not decompose or react with the cyclodextrin.

When drying complexes, water should be removed as quickly as possible. If a solid complex is collected, filtration or centrifugation is used to remove as much water as possible before drying. Temperatures of 100–110°C are typically used. Lower temperatures are used for vacuum drying and freeze drying. For guest compounds that have boiling temperatures <100°C, a lower temperature (3–5 degrees below the boiling temperature of the guest) is used.

VII. Applications

Applications of cyclodextrins have been reviewed extensively,[74] so emphasis here is placed on more recent products. The inclusion of a guest molecule into the cavity of a cyclodextrin results in it essentially surrounding the molecule or a portion of the molecule. Isolation of the molecule inside the cyclodextrin cavity and interaction of the hydroxyl groups of the cyclodextrin with water results in an increased solubility of the guest molecule. Virilan contains a complex of α-cyclodextrin and prostaglandin E1 (PGE1).[75] A significant increase in solubility is achieved by this complexation. At 20°C, the solubility of the PGE1 is 0.06 mg/mL, while the complexed PGE1 has a solubility of 12.0 mg/mL. Unglut and Lommiel contain complexes of benexate in β-cyclodextrin.[76] The low solubility of benexate in water does not provide enough available compound to be an effective anti-ulcer drug, but the greater solubility of the drug when complexed with β-cyclodextrin allows it to be an effective agent.

Glycolic acid is released slowly from a β-cyclodextrin complex, increasing its efficacy in cosmetics as a hydrating, plasticizing and exfoliating agent.[77] The slow release results in less uncomplexed or free glycolic acid being present at any one time, resulting in a reduction of irritation.

The solubility of a complex is generally greater than the solubility of the complexed compound, but less than the solubility of the cyclodextrin. If insufficient solubility has been achieved using unmodified cyclodextrins, greater solubility may be achieved with derivatives of cyclodextrins. Soluble derivatives, such as hydroxypropyl-β-cyclodextrin, can increase the solubility of insoluble drugs sufficiently to allow them to be administered parentally. Flunarizine, used to treat stroke victims, is very insoluble and, after oral administration, reaches therapeutic concentrations very slowly. When injected as a complex with hydroxypropyl-β-cyclodextrin, a therapeutic concentration is achieved in the brain in minutes.[78]

Cyclodextrins can also be used to decrease the solubility of compounds. Insoluble derivatives of cyclodextrins decrease the rate of release of drugs, resulting in maintaining a level of drug over a prolonged period, because two steps are involved in the release of the drug: dissolution of the complex and release of the drug from the solubilized complex.

Cyclodextrins can also be used as process aids to remove specific compounds from a mixture of materials.[79–81] A precipitate forms as the compound or compounds of interest are complexed. The complex is collected by centrifugation or filtration.

In treating some aqueous products, such as milk or eggs, some residual cyclodextrin might remain in the product. The residual cyclodextrin can be removed, if desired, by treating the product with a cyclodextrin transglycosylase or an α-amylase.[82] If an immobilized enzyme is used, the enzyme can be removed from the product. When oils are treated with cyclodextrin, undetectable levels, if any, of cyclodextrin remain in the oil phase if the proper proportions of oil and cyclodextrin solution are used.

A complexed compound can be removed from the complex and isolated as a product, and the cyclodextrin can be reused to treat further batches of material.[83] To recover the cyclodextrin, the complex can be suspended in water at a concentration of ~30% and the suspension heated to 90–95°C. At this temperature the complex dissociates. The dissociated complex is passed through a centrifuge to separate the organic layer or guest that was formerly complexed from the aqueous solution of cyclodextrin. Depending on the material being treated, the cyclodextrin can be reused after cooling and dilution, or it might be treated by filtration, activated carbon, ion exchange or other means to remove any impurities that might have accumulated before reuse.

Cholesterol can be removed from eggs, dairy products and animal fats. Milk and egg yolk can be mixed with β-cyclodextrin to complex the cholesterol, forming a precipitate that can be removed by centrifugation or filtration. Using cholesterol as a process aid, 80% of the cholesterol has been removed from cheeses[84,85] and liquid eggs.[86] Similar results have been obtained by treating lard and tallow.[79]

Limonin and naringen, which can give citrus juices an undesirable bitter flavor, can be removed selectively by passing juice through a crosslinked polymer of β-cyclodextrin.[87,88] Only the bitter components are removed. The polymer can be regenerated for reuse by washing the polymer with a 1–2% solution of sodium hydroxide.

Many fruit and vegetable juices darken in color with time due to enzymic browning. The phenolic compounds involved in the reaction complex with cyclodextrin. Thus, the browning reaction can be prevented, either by passing the juice through a crosslinked cyclodextrin polymer to remove the phenol, or by adding cyclodextrin to the juice, since polyphenoloxidase, the enzyme responsible for oxidizing the phenolic compounds, is unable to act on the phenolic compounds while they are complexed with the cyclodextrin.[98,99]

Garlic cloves were homogenized in water and the garlic oil complexed using β-cyclodextrin. The dried complex had very little odor, but had a strong garlic flavor, more like the natural garlic than many garlic oils obtained by conventional means. The complex could be used as isolated or further processed to release the garlic oil.

Cyclodextrins can also be used to recover oil from tar sands and shale.[92] The oil bearing material was mixed with a 2% solution of cyclodextrin at 60°C. When mixing was stopped, the oil floated to the surface and the sand or shale settled to the bottom. About 90% of the oil in the tar sand was recovered. The solution of cyclodextrin could be reused repeatedly.

Isomers of xylene,[93] ethyl benzene[94] and nitrotoluene[95] have been separated using cyclodextrins. Individual components of petroleum fractions, such as polynuclear aromatic hydrocarbons, have been identified and quantified using cyclodextrin.[96,97]

Complexation with cyclodextrins greatly reduces the rate of evaporation of volatile compounds in both the wet and dry states, creating a longer shelf life for many products. There is also a decreased loss of volatile material during processing, resulting in less material being released into the atmosphere. Complexes of flavors, such as apple, cinnamon, banana and mango, have been added to tea.[98] The use of flavor complexes for tea eliminated the need for the use of hermetically sealed boxes for storage of the tea bags. Dry complexes retain guest compounds very well, with little or no odor or aroma being released from the complex while the complex is dry. When the tea bag is added to water, the fruit flavors are released immediately.

Seasoning salts have been made by adding complexes of onion, tarragon, laurel, caraway, smoke, dill and garlic oils to the salt.[99] The flavor oil is prevented from evaporating because it is complexed with the cyclodextrin, but is readily released when the complex is moistened in the mouth. Complexation of the oils with cyclodextrin also converts them to a solid powder which can be easily mixed with the salt without caking.

Use of cinnamon complexed with cyclodextrin in the manufacture of dried cinnamon-flavored apple slices prevents loss of cinnamon due to evaporation and protects the cinnamon from oxidation.[100] Good flavor is released on consumption of the apple slices. Mustard oils are very irritating and care must be taken while working with them to prevent their release into the atmosphere. Complexing mustard oils with cyclodextrin reduces their volatility, making them easier to work with. While the volatility of the mustard oil is greatly reduced when it is in the complexed form, the mustard oil is readily released in the mouth when consumed. A complex of mustard oil has been used to prepare steak sauce.[101]

Perfumes and fragrances can be complexed with cyclodextrins. On application to the skin, the perfume is released over a longer time than perfume applied to the skin in a non-complexed form. Some perfumes are irritating to the skin. Because the perfume is contained within the cavity of the cyclodextrin, contact with the skin is minimized, resulting reduction or elimination of irritation. Moisture and oils in the skin release the perfume slowly, so that the minimum concentration of perfume needed to elicit an irritable response is not reached.

Fragrance complexes can also be used in laundry products.[102-104] Sheets for fabric softening with a complex of cyclodextrin and fragrance incorporated into them are placed into the clothes dryer, where the fragrance is released during the drying process.

Cyclodextrins can stabilize some unstable molecules against the effects of light, heat and oxidation. Association of the molecule or a portion of the guest molecule with the walls of the cavity of the cyclodextrin or hydroxyl groups on the rims of the cyclodextrin can result in increased activation energy required in order to cause a chemical reaction to occur. The cavity of the cyclodextrin is a finite space. If the space is filled, other molecules cannot enter the cavity to react with the included molecule. Some steric hindrance can also be provided to included molecules to prevent reactive molecules from approaching the reactive sites of the guest molecule.

Cyclodextrins have been used to protect vitamins and pharmaceuticals containing easily oxidizable double bonds, such as prostaglandins,[75] against oxidation. Activity of biologically active peptides, such as thymopentin, which undergo hydrolysis and lose biological activity rapidly in aqueous solution, is greatly increased when complexed with hydroxypropyl-β-cyclodextrin.[105] Cyclodextrins have been used in skin tanning compositions to provide stability against physical and chemical degradation over time and prevent development of an unpleasant odor during tanning.[106] The enzymic activity of superoxide dismutase is also protected by cyclodextrins.[107] Cyclodextrins have been investigated for their ability to complex explosives to reduce sensitivity to impact and friction.[108]

Cyclodextrins can catalyze certain chemical reactions. Styrenesulfonate polymerization was accelerated in the presence of cyclodextrin, and the polymers had a higher molecular weight.[109] Platinum-catalyzed hydroxylation reactions were accelerated in the presence of cyclodextrin.[110]

VIII. References

1. Ogata T, Takahashi Y. *Carbohydr. Res.* 1995;138:C5.
2. Miyazawa I, Endo T, Ueda H, Kobayashi S, Nagai T. In: Osa T, ed. *Proceedings of the 7th International Cyclodextrins Symposium*. Tokyo: Business Center for Academic Societies of Japan; 1994:214.
3. Endo T, Ueda H, Kobayashi S, Natai T. In: Osa T, ed. *Proceedings of the 7th International Cyclodextrins Symposium*. Tokyo: Business Center for Academic Societies Japan; 1994:66.
4. French D, Pulley AO, Effenberger JA, Rougvie MA, Abdullah M. *Arch. Biochem. Biophys.* 1965;118:153.
5. Handbook of Cyclodextrin Technology. Hammond, IN: American Maize-Products Company; 1993.
6. Schardinger F. *Woschr.* 1904;17:204.
7. Tilden EB, Hudson CS. *J. Am. Chem. Soc.* 1939;63:2900.
8. Okada S, Tsuyama N, Kitahata S. *Amylase Symp. Japan*. 1973;7:61.
9. Okada S, Kitahata S. *Amylase Symp. Jpn*. 1973;8:21.
10. Bender H. *Arch. Microbiol.* 1977;111:271.
11. Horikoshi K, Akiba T. *Alkalophilic Microrganisms*. Tokyo, Japan: Japan Scientific Societies Press; 1982:101–110.
12. Kitahata S, Okada S. *J. Jpn. Soc. Starch Sci.* 1982;29:7.
13. Yu EKC, Aoki H, Misawa M. *Appl. Microbiol. Biotechnol.* 1988;28:377.
14. Starnes RL, Flint VM, Kalkoren DM. In: Duchene D, ed. *Minutes of the 5th International Symposium on Cyclodextrins*. Paris, France: Editions de Sante; 1990:55.
15. Starnes RL, Flint VM. In: Hedges AR, ed. *Minutes of the 6th International Symposium on Cyclodextrins*. Paris, France: Editions de Sante; 1992:39.
16. Ruchtorn U, Mongholkin P, Limpaseni T, Kamelsicipchalporn S, Pongsawasdi P. In: Hedges AR, ed. *Minutes of the 6th International Symposium on Cyclodextrins*. Paris, France: Editions de Sante; 1992:29.
17. Park YK, Sabioni JG, Ferrarezo EM, Pastore GM. In: Hedges AR, ed. *Minutes of the 6th International Symposium on Cyclodextrins*. Paris, France: Editions de Sante; 1992:23.
18. Bender H. In: Szejtli J, ed. *Proceedings of the 1st Inernational Symposium on Cyclodextrins*. Dordrecht, The Netherlands: Reidel; 1982:77.

19. Horikoshi K, Akiba T. *Alkalophilic Microorganisms*. Tokyo, Japan: Japan Scientific Society Press; 1982.
20. Kitahata S, Okada S, Fukui T. *Agric. Biol. Chem.* 1978;42:2369.
21. Flaschel E, Landert JP, Renton A. In: Szejtli J, ed. *Proceedings of the 1st International Symposium on Cyclodextrins*. Dordrecht, The Netherlands: Reidel; 1982:41.
22. Flaschel E, Landert JP, Spieser D, Renton A. *Ann. N.Y. Acad. Sci.* 1984;434:70.
23. Schmid G, Huber O, Eberle HJ. In: Huber O, Szjetli J, eds. *Proceedings of the 4th International Symposium on Cyclodextrins*. Dordrecht, The Netherlands: Kluwer Academic Publishers; 1988:82.
24. Seres G, Sarai M, Piiukovich S, Gebanyi MS, Szejtli J. US Patent 4 835 109. 1989.
25. Armbruster FC, Kooi ER. US Patent 3 425 910. 1969.
26. Horikoshi K, Nakamura N, Matsuzawa N, Yamomoto M. In: Szejtli J, ed. *Proceedings of the 1st International Symposium on Cyclodextrins*. Budapest, Hungary: Akademiai Haido; 1982:25.
27. Horikoshi K. US Patent 3 923 598. 1975.
28. Horikoshi K, Nakamura N. US Patent 4 135 977. 1979.
29. Okada M, Matsuzawa M, Uezima O, Nakakuki T, Horikoshi K. US Patent 4 418 144. 1983.
30. Armbruster FC. In: Huber O, Szejtli J, eds. *Proceedings of the 4th International Symposium on Cyclodextrins*. Dordrecht, The Netherlands: Kluwer Academic Publishers;1988:33.
31. Vakaliu H, Miskolczi-Torak M, Szejtli J, Jarai M, Seres G. Hungarian Patent 16 098. 1979.
32. Shieh WJ, Hedges AR. US Patent 5 326 701. 1994.
33. Schmid G. In: Duchene D, ed. *New Trends in Cyclodextrins and Derivatives*. Paris, France: Editions de Sante; 1991:25.
34. Wiedenhof N, Lammers JNJJ. *Carbohydr. Res.* 1968;7:1.
35. Jozwiakowski MJ, Connors KA. *Carbohydr. Res.* 1985;143:51.
36. Manor PC, Saenger W. *Nature*. 1972;237:392.
37. Manor PC, Saenger W. *J. Am. Chem. Soc.* 1974;96:3630.
38. Pharr DY, Fu ZS, Smith TK, Hinze WL. *Anal. Chem.* 1989;61:275.
39. Coleman AW, Nicolis I, Keller N, Dalbiez JP. *J. Inclusion Phenom. Mol. Recognit. Chem.* 1992;13:139.
40. Swanson MA, Cori CF. *J. Biol. Chem.* 1948;172:797.
41. Myrback K, Jarnestrom T. *Kemi*. 1949;1:129.
42. French D, Knapp DW, Pazur JH. *J. Am. Chem. Soc.* 1950;72:5150.
43. French D. *Advan. Carbohydr. Chem.* 1957;12:189.
44. Schenk W, Sand D. *J. Am. Chem. Soc.* 1961;83:2312.
45. Szejtli J. *Cyclodextrins and Their Inclusion Complexes*. Budapest, Hungary: Academiai Kiado; 1982: 43.
46. Mifune A. *Yugaka*. 1974;32:889.
47. Frank DW, Gray JE, Weaver RN. *Am. J. Pathol.* 1976;83:367.
48. Makita T, Oshima N, Hashimoto H, Ide H, Tsuji M, Fujisaka Y. *Oyo Yakuri*. 1975;10:449.

49. Bellringer ME, Smith TG, Read R, Capinath C, Olivier P. *Food Chem. Toxicol.* 1995;33:367.
50. Waner T, Borelli G, Cadel S, Privman I, Nysha A. *Arch. Toxicol.* 1995;69:631.
51. Barrow P, Olivier P, Marzin D. *Reproductive Toxicol.* 1995;9:389.
52. Marshall JJ, Miwa I. *Biochem. Biophys. Acta.* 1981;661:142.
53. Gergely V, Sebestyen G, Virag S. In: Szejtli J, ed. *Proceedings of the 1st International Symposium on Cyclodextrins.* Budapest, Hungary: Academiai Kiado; 1982:109.
54. Gerloczy A, Fornagy A, Heresztes P, Pulaky L, Szejtli J. *Arzneim. Forsch./Drug Res.* 1985;35:1042.
55. Olivier P, Verwaerde F, Hedges A. *J. Am. Coll. Toxicol.* 1991;10:407.
56. Szejtli J, Gerloczsy A, Fonagy A. *Arzneim. Forsch./Drug Res.* 1980;30:808.
57. Antenucci RN, Palmer J. *J. Agric. Food Chem.* 1984;32:1316.
58. Matsuda K, Mora Y, Sagaws Y, Uchida I, Yokomino A, Tagaki K. *Oyo Yokura.* 1983;26:287.
59. Suzuki M, Satoh A. *J. Jpn Soc. Starch Sci.* 1983;30:240.
60. Suzuki M, Satoh A. *J. Nutr. Sci. Vitaminol.* 1985;31:209.
61. Schmid G. In: Duchene D, ed. *New Trends in Cyclodextrins and Derivatives.* Paris, France: Editions de Sante; 1991:25.
62. Szeman J, Szente J, Szabo T. In: Huber O, Szejtli J, eds. *Proceedings of the 4th International Symposium on Cyclodextrins.* Kluwer Dordrecht, The Netherlands: Academic Publishers; 1988:275.
63. Coussenent W, Van Cauleren H, Vandenbergh J, Vanparys P, Teuns G, Lampo A, Moreboom R. In: Duchene D, ed. *Minutes of the 5th International Symposium on Cyclodextrins.* Paris: Editions de Sante; 1990:522.
64. Brewster M, Estes K, Bodor N. *Int. J. Pharm.* 1990;59:231.
65. Szejtli J. *J. Inclusion Phenom.* 1983;11:135.
66. Hirai S, Okada H, Yashiki T. US Patent 4 659 696. 1987.
67. Stella VJ, Lee HY, Thompson DO. In: Osa T, ed. *Proceedings of the 7th International Symposium on Cyclodextrins.* Tokyo, Japan: Business Center for Academic Societies of Japan; 1994:365.
68. Stella VJ, Lee HY, Thompson DO. In: Osa T, ed. *Proceedings of the 7th International Symposium on Cyclodextrins.* Tokyo, Japan: Business Center for Academic Societies of Japan; 1994:369.
69. Kobayashi S, Shibuya N, Young BM, French D. *Carbohydr. Res.* 1984;126:215.
70. Ammeraal RN, Benko L, DeBoer ED, Kozlowski RJ. In: Hedges AR, ed. *Minutes of the 6th International Symposium on Cyclodextrins.* Paris, France: Editions de Sante; 1992:69.
71. Gerasimowicz WV, Wojcik JF. *Bioorg. Chem.* 1982;11:420.
72. Siegel B, Breslow R. *J. Am. Chem. Soc.* 1975;97:6869.
73. Nakai Y, Yamamoto K, Terada K, Watanabe D. *Chem. Pharm. Bull.* 1987;35:4609.
74. Szejtli J. *Cyclodextrin Technology.* Dordrecht, The Netherlands: Kluwer Academic Publishers; 1988.
75. Schwarz Pharma, Product Literature.
76. *Cyclodextrin News.* 1994;8:180.
77. Instituto Recherche Appliate IRA. *Cosmet. Toilet.* 1995;110:59.

VIII. References

78. Janssen Biotech NV, Encapsin HPB, Product Literature.
79. Couregelongue J, Maffrand JP. US Patent 4 880 573. 1989.
80. Rouderbourg H, Dalemans D, Boubon R. US Patent 5 232 725. 1993.
81. Cully J, Vollbrecht RR. US Patent 3 342 633. 1994.
82. Cully J, Vollbrecht RR. US Patent 4 980 180. 1990.
83. Shieh WJ, Hedges AR. US Patent 5 371 209. 1994.
84. *Cyclodextrin News* 1994;8:72.
85. *Cyclodextrin News* 1994;8:89.
86. *Cyclodextrin News* 1994;8:54.
87. Shaw PE, Tatum JH, Wilson CW. *J. Agric. Food Chem.* 1989;32:832.
88. Shaw PE, Wilson CW. *J. Food Sci.* 1985;50:1205.
89. Irwin PL, Pfeffer PE, Doner LW, Sapers GM, Brewster JD, Nagahashi G, Hicks KB. *Carbohydr. Res.* 1994;256:13.
90. Hicks KB, Sapers GM. US Patent 4 975 293. 1990.
91. Sapers GM, Hicks KB, Philips JG, Garzarells L, Pondrick DL, Matalaites RM, McCormack TJ, Sondey SM, Seib PA, El-Atawy YS. *J. Food Sci.* 1989;54:997.
92. Horikoshi K, Shibanai I. US Patent 4 444 647. 1984.
93. Gerhold CG, Broughton D. US Patent 3 456 028. 1969.
94. Glein WKT, Walker RC, Ramquist FC. US Patent 3 456 055. 1969.
95. Uemase I, Takahashi H, Hara K, Hashimoto H. In: Hedges AR, ed. *Minutes of the 6th International Symposium on Cyclodextrins*. Paris, France: Editions de Sante; 1992:579.
96. Elliot NB, Prenni AJ, Ndou TT, Warner IM. *J. Coll. Interface Sci.* 1993;156:359.
97. Tachibana M, Furusawa M. *Analyst.* 1995;120:437.
98. *Cyclodextrin News* 1994;8:108.
99. *Cyclodextrin News* 1993;7:295.
100. *Cyclodextrin News* 1994;8:126.
101. *Cyclodextrin News* 1993;7:276.
102. Gardlick JM, Trinh T, Barks TJ, Benvegnu F. US Patent 5 234 610. 1993.
103. Trinh T, Bacon DR, Benvegnu F. US Patent 5 234 611. 1993.
104. Trinh T, Bacon DR. US Patent 5 236 615. 1993.
105. Brown ND, Butler DL, Chiang PK. *J. Pharm. Pharmacol.* 1993;45:666.
106. Lentini PJ, Zecchino JR. Int. Patent WO 95/22960. 1995.
107. Meyer R. US Patent 5 464 614. 1995.
108. Cahill S, Bulusu S. *Mag. Res. Chem.* 1993;31:731.
109. Yamamoto Y, Shiraki S, Gao D. *J. Chem. Soc., Perkin Trans.* 1993;2:1119.
110. Lewis LN, Sumpter CA. *J. Mol. Cat. A: Chem.* 1996;104:293.

Index

A

A allomorph, 157, 159, 164
A-starch, 445, 447, 448, 451, 624, 447, 451, 624
A-type pattern, 321, 461
   of cereal starches, 151, 153
A-type starch crystals, 153, 154, 156–8, 331
Absolute amylose content, 201
Acarbose, 259–60, 272–3
Acarviosine, 272
Acetic anhydride, 457, 479, 480, 631, 632, 748, 759
Acetyl/hydroxypropyl groups:
   wheat starch, monosubstitution of, 478–9
Acetylated tapioca starch, 560
Acetylation, 457, 586, 620–1, 747–8, 759
Acid-catalyzed hydrolysis, 153, 220, 239, 269, 596, 632, 747, 802–6
Acid converter, 804
Acid-depolymerized starches, 694
Acid detergent residue (ADR), 381
Acid–enzyme processes, 806–8
Acid hydrolysis, 220–1, 613–14, 616, 622
Acid-hydrolyzed (thinned) starches, 663
Acid-modified starches, 536
Acid oil, see Hydrolyzed vegetable oil (HVO)
Acid-thinning, of wheat starch, 481
Actinomyces sp. 272
Actinoplanes, 811
Adenylate translocator, 42, 43
Adipic acetic mixed anhydride, 225, 480, 631, 758
ADM Arkady, 10
ADM Corn Processing, 9
ADP-glucose pyrophosphorylases (ADPGlc PPase), 36
   activation, by thioredoxin, 107–8

amino acid residues identification within, 111–13
   domain characterization, 113
catalysis, in large subunit, 91–5
characterization, from different sources, 108–11
interaction between 3PGA and Pi, differences in, 105–6
kinetic properties and quaternary structure, 87–91
large and small subunits, phylogenetic analysis of, 95
potato tuber, crystal structure of, 95
   ADP-glucose binding, 101–2
   allosteric regulation, 103–4
   ATP binding, 101
   catalysis, implication for, 102–3
   homotetramer structure, 97–8
   sulfate binding mimics phosphate inhibition, 98–100
regulation of, 104–5
synthesis, 85, 105, 108
AE gene, 132
A.E. Staley Manufacturing Company, 4
Aerobacter aerogenes, 247, 248
Agrobacterium tumefaciens, 113, 128
Albumins, 405
Aldehyde starches, 665
Alfa-Laval/Raisio process, 447
Alkali, 429
Alkenyl succinic anhydride (ASA), 632, 686
Alko Limited, 624
Alkyl ketene dimer (AKD), 632, 686, 687
All-amylopectin potato, 521–2
'Allosteric switch', 113
Almidon agrio, 561
Alpha-amylase, see Amylases
Alpha-amylolysis, see Amylolysis

Amaranth starch, 195, 196, 773
Amaranthus, 45
Amaranthus cruentus starch, 773
Amaranthus hypochondriacus starch, 773
Amaranthus retroflexus, 33
American Maize-Products Company, 4
Amino acids, 8–9, 104, 113, 125, 258, 406, 805
    involved in substrate binding and catalysis, 127–8, 261–2
    residues, within ADPGlc Ppases, 111
Amino acids at active-site:
    involved in catalysis and substrate binding, 261–2
Amorphous structural domains, glass transitions of, 311–19
Amphoteric starch, 635–6, 683
Amstar Corporation, 4
Amylase action inhibitors, 272–6
Amylases, 786
    α-amylase, 39–40, 209, 238–44, 247, 253–7, 261–2, 536, 591, 612, 800, 806
        in amylose-V complexes and retrograde amylose, 267–9
        hydrolysis of barley starches by, 613
    β-amylase, 39–40, 60, 115, 205, 220, 239, 244, 265, 280–1, 282–4, 458, 536, 730, 806, 838
    archaebacterial amylases, 248–50
    characterization, 593
    cyclomaltodextrin glucanosyltransferase, 250–3
    endo-acting α-amylases, 238–44
    exo-acting β-amylases, 244–6
    isoamylases, 247–8
    maltodextrin-producing amylase, 246–7
    native starch granules, action with, 269–72
Amyloglucosidase, 556, 612, 836
    hydrolysis of barley starches by, 613
    see also Glucoamylases
Amylolytic enzymes, 134, 135, 212, 786–7
    domains structure and function, 262–4
Amylopectin, 28, 61, 85, 86, 125, 151–3, 161, 164, 194, 212–13, 244, 316, 451, 521, 606, 800
    β-amylase limit dextrins formation, 282–4
    and amylose:
        helical interactions, 184–5
        interactions, 340–2
    branching points, 162, 163
    chain length distribution, 551
    characterization, 593
    chemical structure, 212–18
    cluster models, 218–24
    crystallinity, 297
    fine structures, 457–64
        determination, 282–4
    gels, 337–40
    growing temperature and kernel maturity, effects of, 224–5
    molecular compositions, 201
    orientation, 159
    in potato starch, 168
    structure, 151, 152
    in wheat, 460
Amylopectin-rich corn, 748
Amyloplasts, 26, 33, 34, 40, 85, 107
Amylose, 5–6, 28, 51, 123, 151–2, 194, 225–7, 244, 451, 583, 605–6, 779, 806
    and amylopectin:
        helical interactions, 184–5
        interactions, 340–2
    blocklet organization, 184
    chemical structure, 205–8
    cooking, 750
    double helical structures, 211–12
    fine structures, 457–64
    gelation, 334–6
    growing temperature and kernel maturity, effects of, 224–5
    iodometric assay, of, 455
    molecular compositions, 201–2
    molecular properties, in wheat, 458
    properties, 551
    single helical structures, 208–11
    structure, 151
    of wheat starch, 451–7
    within granules, location and state of, 184–6
Amylose-extender (*ae*), 44, 50–3, 132, 133
Amylose-extender dull (*ae du*), 61
Amylose-extender dull sugary (*ae du su*), 64
Amylose-extender dull sugary-2 (*ae du su2*), 65
Amylose-extender dull sugary waxy (*ae du su wx*), 68–9

Amylose-extender dull waxy (*ae du wx*), 65–6
Amylose-extender sugary (*ae su*), 59–60
Amylose-extender sugary-2 (*ae su2*), 60
Amylose-extender sugary-2 waxy (*ae su2 wx*), 67
Amylose-extender sugary sugary-2 (*ae su su2*), 66
Amylose-extender sugary waxy (*ae su wx*), 66
Amylose-extender waxy (*ae wx*), 58–9
Amylose leaching and swelling powder:
    in barley starch, 610–12
    in oat starch, 596–7
    in rye starch, 584–5
Amylose–lipid complex, 186, 210, 298, 299, 357
    of oat starch, 595
    of rye starch, 584
Amylose retrogradation, 309, 346, 521
Amylose-V complexes and retrograde amylose:
    α-amylases action, 267–9
Amylolysis:
    α-amylolysis, 165, 181, 606
    β-amylolysis, 30, 31, 47, 67, 206, 283, 458
Anabaena, 112
Anheuser-Busch, 4
Anhydrous dextrose, 799, 813, 814
Anionic starches, 664, 694–5
Annealing, 320–3, 582, 621, 765
Anti-plasticization, 330
ApL1, 90
ApL2, 90
ApL3, 90, 91
APL4, 90, 91
APS1, 90
Arabidopsis thaliana, 84, 90, 111, 120, 126, 137
Arabinoxylans, 352, 456, 488, 602
Archaebacterial amylases, 248–50
Archer Daniels Midland Co, 4, 9, 17
Arrowroot starch, 490, 772
Arthrobacter species, 811
Ascorbic acid, 275, 783
Aseptic processing, 775, 782
Asp142, 94, 112
Asp145, 93, 94, 102, 112
Asp160, 94, 112
Asp253, 113
Aspergillus awamori, 244, 245, 246
Aspergillus awamori var. *kawachi*, 244, 258
Aspergillus flavus, 387, 388
Aspergillus niger, 244, 246, 261, 489, 612
Aspergillus oryzae, 238, 240, 243, 244, 253, 262, 276
Atomic force microscope (AFM), 170, 180, 182
    granular starch imaging, sample preparation for, 172–3
    principles, 171–2
    in starch research:
        high-resolution AFM (and LVSEM) studies, 173–5
        interpretation, with respect to granule structure, 175–7
        real-time AFM studies, 173
ATP binding, 101
Avena sativa L., 30

B

B allomorph, 151, 159, 165
B-nanocrystals, 335
B-starch, 154, 445, 447, 448, 449, 451, 624–5
B-type pattern, 151, 153
B-type starch crystals, 153, 154, 156–8
Baby foods, 572, 777
Bacillus, 248, 263, 811
Bacillus α-amylase, 261
Bacillus amyloliquefaciens, 238, 240, 241, 267, 268, 271
Bacillus caldovelox, 247
Bacillus cereus, 244
Bacillus circulans, 247, 248, 251, 261, 272
Bacillus coagulans, 238
Bacillus licheniformis, 247, 800
Bacillus macerans, 250, 251, 253, 265, 275, 835
Bacillus megaterium, 244, 251
Bacillus polymyxa, 244
Bacillus species, 272, 612
Bacillus stearothermophilus, 251, 272, 800
Bacillus subtilis var. *saccharitikus*, 238
Bacillus subtilis α-amylase, 209, 211
Baked products, 347, 778
    wheat starch role in, 481–5

Banana starch, 9
Barley, 1, 2, 27, 34, 108–9, 131, 210
  longitudinal section, 603
Barley kernels, 33
Barley starch, 482, 582, 601, 772
  acid hydrolysis, 613–14, 616
  chemical composition:
    carbohydrate component, 605–6
    non-carbohydrate components, 606–7
  chemical modification:
    cationization, 619–20
    hydroxypropylation/acetylation, crosslinking followed by, 620–1
  composition, 602–3
  enzyme susceptibility, 612–13, 614, 615, 616
  freeze–thaw stability, 619
  gelatinization, 607–10
  granule morphology, 607, 608
  isolation, 604–5
  pasting characteristics, 615, 617–18
  physical modification:
    annealing, 621
  production and uses, 623–5
  purification, 604–5
  resistant starch (RS), 621–3
  retrogradation, 618–19
  structure, 602–3
  swelling factor and amylose leaching, 610–12
  X-ray diffraction and relative crystallinity, 607, 609
  see also High-amylose barley starch
Batch crystallizer, 814, 815
Batter process, 446–7
Bauer attrition mill, 411
Baume number, 799, 818
Beccari, Jacob, 442
Belle Patna, 570
Beta-amylase, see Amylases
Beta-amylolysis, see Amylolysis
Beverage emulsions, 777
Biochemistry and molecular biology, of starch biosynthesis, 83
  1,4-α-glucan-synthesizing enzymes, properties of:
    α-1,4-glucanotransferase, 136–8

ADPGlc PPase, see ADP-glucose pyrophosphorylases (ADPGlc PPase)
  branching enzyme (BE), 129–36
  isoamylase, 136
  starch synthase, 114–28
enzyme-catalyzed reactions, in plants and algae, 85–6
glycogen synthesis, in cyanobacteria, 86
in plants:
  leaf starch, 84
  storage tissues, 85
Biological oxygen demand (BOD), 619, 704
Birefringence loss, measurement of, 301
Birefringence patterns, 296
Biscuits, 483, 561
Bleaching, of large granule wheat starch, 480–1
'blocklets concept' of starch granule structure, 180–4
'Blökchen Strüktur', 180
Bluebelle, 570
Boiling point elevation, 824, 825
Brabender AmyloViscograph, 667–8
Brabender units (BU), 586
Brabender ViscoAmylograph, 303, 552, 585, 597
  for potato starch, 522
Bran, uses of, 573
Branch-linkage assay, 129
Branched cyclodextrins, 841
Branching enzyme (BE), 129
  and α-amylase family, 134–5
  branching enzyme-deficient mutants, genetic studies on, 132–3
  catalysis, functional amino acid residues in, 135–6
  cDNA clones encoding isozyme genes, isolation of, 133–4
  isozymes, characterization of, 129–32
  localization in plastid, 134
  purification and characterization, 129
Bread and wheat starch, 481–3
Breakfast cereals, 779–80
Brevibacterium sp., 251
Brewers rice, 571, 572
Brittle-1 (*bt1*), 42, 43
Brittle-2 (*bt2*), 43

Brix, 819, 820
Broken corn, foreign material (BCFM), 389
Broken kernels and foreign material (BKFM), 390
Brookfield viscometer, 668
Brown rice, 570, 572, 574
Brown ricemilk syrup, 572
Browning reaction:
    and color, 821–2
    prevention, 846
Building industry, 491
By-products, of rice:
    bran, 573
    hulls, 572

C

C allomorph, 159
C-type pattern, 151, 461
C-type polymorphism, 159
Cakes, 483
Calcium chloride, 784
California Natural Products, 573
Caloro, 570
Calrose, 570
Calvin cycle, 38
Canned foods, 774–5
Carbohydrate profiles, 817–18
    of maltodextrins, 802
    of sweeteners, 817–18
Cargill Inc., 4, 9, 17
Cassava starch, *see Tapioca starch*
Cationic potato starch, 537, 669, 683, 684, 686–7
Cationic starches, 619–20, 632, 663–4, 677, 682–3, 684, 685, 688, 694, 698
    amphoteric starch/starch-containing systems, 635–6
    with covalently-reactive groups, 636–7
    dry/solvent cationization, 633–4
    polycationic starches, 634–5
Cationic waxy maize starch, 687
Cationic waxy starches, 665
Cellular developmental gradients, 26–7
Centrifugal sieves, 527
Cereals, 471
Charles Pope Glucose Company, 4
Chemical conversion, of starch, 676–7
Chemical oxygen demand (COD), 695, 704

Chemically modified starches, 6
Chenmopodium quinoa, 33
Chlamydomonas reinhardtii, 26, 46, 86, 105, 108, 126, 136, 137
Chloroplasts, 34
Citric acid, 8
Clean Air Acts:
    1977 Clean Air Act, 15
    1990 Clean Air Act Amendments, 15
Clinton Corn Syrup Refining Company, 4
Clinton Sugar Refining Company, 4
Clostridium butyricum, 26, 272
Clostridium spp., 406
Clostridium thermocellum, 259
Cluster models, of amylopectin, 85, 163, 213, 218–24
Coating binder, starch as:
    paper coating, starch selection for, 698–9
    paper machine, coater in, 695–8
Coix lachryma-jobi, 44
Cold-process dressings, 776
Cold-water-soluble starch, 555, 560, 645
Cold-water-swellable starch, 644–5, 748, 764
Colocasia esculenta, 33
Colusa, 570
Commercial esters, 536
Commercial ethers, 536
Commercial starches, differences between, 519, 521
Common rice starches, 570
    versus waxy rice starches, 575–7
Concanavalin-A procedure, 468
Condensed corn fermentation extractives, 426
Confectioners, starch in, 762
Confections, 778–9
Confocal laser-light scanning micrographs (CLSM), 178, 200, 562
Consumption of starch, by paper industry, 662
Conventional processes:
    for wheat starch production, 446–8
Conversions, 760–1
    acid, 220–1, 481, 536, 613–14, 616, 622, 663
    acid–enzyme, 806–8
    dextrins, 761

Conversions (*Continued*)
  enzymic, 677–81
  fluidity (thin-boiling) starches, 760–1
Converted starches, 536
Cook-up starches, 764, 780
Cooked starch, processing and storage impact on, 751
  changes during cooling, storage and distribution, 754–5
  concentration:
    during cooking, 751
    role, 751
  food processing equipment, comparison of, 753–4
  processing and storage, impact of, 754
  recommended processing, 755
  shear, effects of, 752–3
  time and temperature, effects of, 751–2
  see also Foods, starch use in
Coprecipitation, 843, 844
Corn and sorghum starches production, 373
  alternative fractionation procedures, 427–9
  composition, 375, 381–5
  future directions:
    biosolids as animal food, 429–30
    biotechnology and genetic engineering techniques, 430
    continued expansion into fermentation products, 429
    corn starch industry, segregation of, 430
    hybrid-specific processing, 430
  grain quality, 375, 385–91
  products:
    corn oil, 425
    ethanol, 424–5
    feed products, 426
    starch, 423
    sweeteners, 423–4
  structure, 375, 376–81
  wet-milling, 391, 393
    automation, 423
    energy use, 421–2
    and fraction separation, 408–21
    grain cleaning, 392
    pollution control, 422, 423
    product drying, 421–2, 423
    starch processing, 421

  steeping, 394–408
Corn bran, 426
Corn germ meal, 426
Corn gluten feed, 426
Corn gluten meal, 426
Corn milling processes, enzymes used in, 809
Corn price variability, effects of, 18–19
Corn production, in United States, 12
Corn Products Company, 4
Corn Products International, Inc, 4, 10
Corn Products Refining Company, 4
Corn Refiners Association, Inc, 5, 662
Corn Starch & Syrup Company, 4
Corn starch, 556
  applications:
    in baked products, 483
    in foods, 747, 765, 767, 768–9, 779, 780, 782
    in paper industry, 663, 664, 670, 678, 693, 694, 701, 702–3
    starch-based plastics, 640
  behaviors, 301, 311, 318, 320, 328, 329
  cationization, 634
  consumption, 662
  genetics, 6, 329
  modifications, 6, 320, 321
  oxidized corn starches, 663, 694, 699, 705
  powdered corn starch, 768
  producers, 9–10
  properties, 302, 304, 305, 320, 329
  US corn starch industry, economic growth and organization of, 11
  corn price variability, effects of, 18–19
  extent and direction, 11–13
  fuel alcohol, 15–16
  future industry prospects, 20
  high-fructose syrup (HFS) consumption, 13–15
  industry organization, 16–18
  international involvement, 19–20
  plant location, 16
  technical progress, 16
  see also Corn and sorghum starches production; High-amylose corn starch; High-amylose maize starch; Maize starch; Waxy maize starch

Corn starch gel, 352
Corn starch molasses, 426
Corn starch producers, 9–10
Corn sweeteners, 5, 12
Corn syrup solids, 799
Corn syrups, 799
Corrugated board, 490, 700, 701, 703
Corrugating process, 700–2
Corrugating starch adhesive formulation, 701
Cosmetics, starch in, 491
Cotton finishing, wheat starch in, 490
CPC International, 17
Creative Research Management, 574
Crosslinked starches, 536–7
Crosslinked waxy maize starch, 747, 748
Crosslinking:
    in food processes, 756–8
    of wheat starch, 475–7
Crosslinks, 332
Crystalline dextrose:
    and dextrose syrups, 813–16, 817
Crystalline fructose, 799, 813, 814, 815
Crystalline ultrastructural features, of starch, 158–60
Crystallites in granular starch, melting transitions of, 323–32
Crystallization, of sweeteners, 826–7
Cyanogenic glucosides, 544
Cyclization reaction, 834
Cyclodextrin glycosyltransferase, 220, 833
Cyclodextrin (CD), 251, 258, 833
    applications, 845–8
    complex formation, 842–5
    α-cyclodextrins, 250, 834, 835, 836, 837, 838, 839
    β-cyclodextrins, 250, 625, 834, 835, 836, 837, 838, 839
    γ-cyclodextrins, 250, 834, 835, 836, 837, 838, 839
    hydrolysis of, 838
    hydroxyalkyl, 840–2, 845
    modification, 840–2
    production, 835–7
    properties, 837–8
    toxicity and metabolism, 838–9
Cyclohexane, 381
Cyclohexitol, 272
β-Cyclomaltodextrin, *see Cyclodextrin*

Cyclomaltodextrin glucanosyltransferase (CGTase), 250–3, 262–3, 265
Cyclomaltoheptaose, *see Cyclodextrin*
Cyclomaltohexaose, *see Cyclodextrin*
Cyclomaltooctaose, *see Cyclodextrin*
Cyclotridecanone, 837
Cys377 mutant, 128
Cytophaga sp., 248
Cytosolic fructose 1,6-bisphosphatase, 38
Cytosolic starch formation, 25–6

D

D-enzyme, 136, 137
D145N mutant, 112
Dairy products, starch in, 781–2
Debranching enzymes, 36, 55, 69, 86, 807
Decanol, 835, 836
Decanter®, 451
Degree of polymerization (DP), definition of, 799
Degree of substitution (DS), 619–20
Dent corn starch, 768, 773, 779
Deoxynivalenol, 388
Derivatizations, 756–9
    crosslinking, 756–8
    monosubstitutions, 758–9
    octenylsuccinylation, 759–60
Development of starch, genetics and physiology of, *see Genetics and physiology, of starch development*
Dextrin, 3, 536, 665, 761, 778, 779, 799, 834
Dextrose, *see Glucose, D-*
Dextrose equivalence (DE), 799
Dextrose equivalency starch hydrolyzate, 647
Dextrose monohydrate, 799, 817
Dextrose syrups:
    and crystalline dextrose, 813–16, 817
Dialdehyde starch (DAS), 636, 637, 639, 688
Dicarboxyl starch, 639
Dietary fiber, 787, 788
Diffenbachia starch, 195
Differential scanning calorimetry (DSC), 208, 210, 296, 608, 631, 788, 838, 840
Dihydronojirimycin, 273
Dihydroxyphenylalanine, 525

Dinitrogen tetraoxide (N$_2$O$_4$), 639
Diquaternary cationizing reagent, 634
Dispersed starch, 29
Dissolving gap, 669
Donnan potential, 331
Dorr-clones, 624
Double helical structures, of amylose, 211–12
Douglas & Company, 4
Dpe1 mutant, 126
Dried corn syrups, 799
Dry cationization, 633–4
Dry heat, 766
Dry mix foods, 778
DSM (Dutch State Mines) screens, 415
DTDP-glucose pyrophosphorylase (dTDPGlc PPase), 94, 95
  see also ADP-glucose pyrophosphorylases
Dual derivatization, of wheat starch, 479–80
Dual spray drying–extrusion process, 644
Dull (*du*), 57–8
Dull soft starch amylose extender waxy, 6
Dull sugary (*du su*), 61–2
Dull sugary sugary-2 (*du su su2*), 67
Dull sugary waxy (*du su wx*), 67–8
Dull sugary-2 (*du su2*), 6, 62
Dull sugary-2 waxy (*du su2 wx*), 68
Dull waxy (*du wx*), 6, 62–3
Dull1 (*du1*) gene, 116, 117
Dynamic mechanical thermal analysis (DMTA), 296, 315, 316
Dynamic rheometers, 597

E

E number, 767
Encapsulation, 748, 777–8
Encapsulation/controlled release, 642–4
Endo-acting α-amylases, 238–44
  relation of structure with action of, 253–7
Entoleter mill, 414
Environmental aspects:
  of potato starch, 534–5
  of starch use, in paper industry, 703–5
Environmental scanning electron microscopes (ESEM), 170–1
Enzyme-catalyzed modifications, of starch, 647–8

Enzyme tank, 806
Enzymes' action, on starch:
  amylases:
    archaebacterial amylases, 248–50
    cyclomaltodextrin glucanosyltransferase's action, 250–3
    endo-acting α-amylases, 238–44
    exo-acting β-amylases, 244–6
  inhibitors, 272–6
    isoamylases' action, 247–8
    maltodextrin-producing amylase, 246–7
  glycosidic bond, enzymatic hydrolysis of, 264–7
  insoluble starch substrates:
    action of amylases on, 267–72
  plant phosphorylase, 276–7
  starch lyase, 277–8
  starch molecules, enzymic characterization of, 278–84
  structure, relation of:
    with action of endo-acting α-amylases, 253–7
    amino acids at active-site:
      involved in catalysis and substrate binding, 261–2
    domains in amylolytic enzymes, structure and function of, 262–4
    of glucoamylases, 257–61
    of soybean β-amylase, 257
  see also Starch molecules, enzymic characterization of
Enzymic conversion, of starch, 677–81
Enzymically depolymerized starches, 536
Epichlorohydrin, 477, 556, 631, 747, 758
Ethanol, 7–8, 424–5
Ethylene–propylene copolymer (EPMA), 722
Ethylene–vinyl alcohol copolymers (EVOH), 733
Europe, 448, 490, 770
  potato starch production in, 514–15
Exo-acting β-amylases, 244–6
Expanded polystyrene (EPS), 641, 735–6
Extruded starch–g–PMA copolymers, properties of, 729
Extrusion, 644

F

Fat replacers, 782–3
Fats and surfactants, 784–5
Feed products:
   of corn and sorghum starches, 426
Fermentablilty, 822–3
Ferredoxin–thioredoxin system, 107
Fesca processes, 447
Filipendula vulgaris, 34
Finishing processes, 661
Finnsugar Company, 624, 625
Fixed bed systems, 811
Flavobacterium sp., 248
Flory–Huggins theory, 324–5
Floury endosperm, 197, 376, 378
Floury-1 (fl1), 43, 383
Floury-2 (fl2), 43
Floury-3 (fl3), 43
Fluidity starches, 632, 760–1
Flunarizine, 845
Fluorophore-assisted carbohydrate electrophoresis (FACE), 459
Foam stabilization and gel strength, 823
Food products, wheat starch in, 488–9
Foods, starch use in, 746
   applications, 748, 773
      baby foods, 777
      baked foods, 778
      beverage emulsions, 777
      canned foods, 774–5
      changes during cooking, 750–1
      confections, 778–9
      dairy products, 781–2
      dry mix foods, 778
      encapsulation, 777–8
      fat replacers, 782–3
      frozen foods, 775–6
      gelatinization and pasting, 749–50
      hot-filled foods, 775
      meats, 780–1
      pet food, 781
      salad dressings, 776–7
      snacks and breakfast cereals, 779–80
      starch structures relevant to foods, 749
      surimi, 781
   cooked starch, processing and storage impact on, 751
      changes during cooling, storage and distribution, 754–5
      concentration during cooking, 751
      food processing equipment, comparison of, 753–4
      processing and storage, impact of, 754
      recommended processing, 755
      shear effects, 752–3
      time and temperature, effects of, 751–2
   crosslinking, development of, 747
   ingredients, interactions with, 783
      amylolytic enzymes, 786–7
      fats and surfactants, 784–5
      gums/hydrocolloids, 786
      pH, 783
      proteins, 785–6
      salts, 783–4
      sugars, 784
      volatiles, 786
   instant starches, 748
   modern use, 747
   modification, 756
      conversions, 760–1
      derivatizations, 756–9
      native starch thickeners, 767
      oxidation, 761–2
      physical modifications, 762–7
   monosubstitution, development of, 747–8
   resistant starch, 787–8
   starch sources, 767
      amaranth, 773
      arrowroot, 772
      barley, 772
      dent corn, 768
      high-amylose corn, 769
   improvement of, 748
      pea, 772–3
      potato, 770
      rice, 771–2
      sago, 772
      sorghum, 771
      tapioca, 770
      waxy corn, 768–9
      wheat, 770–1
   see also Functions of starch, in food applications; Modified food starch
Freeze–thaw stability, 480, 596–7, 619
Freezing point depression, 824, 825

Friabilin, 472
Fried/baked snacks, 780
Frozen foods, 775–6
Fructose, D-:
　crystalline, 799, 813–15
　equilibrium water content of, 815
　fructopyranose, β-D-, 813, 828
Fructose 1,6-bisphosphatase (F1,6BPase), 38
Fuel alcohol, 12, 15–16, 20
Fumonusin, 388
Functions of starch, in food applications, 748
　changes during cooking, 750–1
　gelatinization and pasting, 749–50
　starch structures relevant to foods, 749
　see also Foods, starch use in
Future, of starch:
　new starches for industry, 9
　present American companies, 9–10

G

Galactomannans, 353
Gasohol, 7, 15
Gel permeation chromatography (GPC), 206, 593
Gelatinization, 301
　of barley starch, 607–10
　electrolyte effects on, 331
　of oat starch, 594–5
　and pasting, 749–50
　and pasting properties of native starches, 305
　of rye starch, 584
　salts effect on, 331
Gelatinization temperature, of starch granules, 824–5
Gelation and retrogradation, of starch, 332
　amylopectin gels, 337–40
　amylose and amylopectin, interactions of, 340–2
　amylose gels, 332–6
　starch pastes and gels, 342–54
'Generally recognized as safe' (GRAS), 800
Genetic modification, of tapioca starch, 556
Genetics and physiology, of starch development, 24
　cellular developmental gradients, 26–7
　mutant effects, 43

amylose-extender (*ae*), 50–3
amylose-extender dull (*ae du*), 61
amylose-extender dull sugary (*ae du su*), 64
amylose-extender dull sugary-2 (*ae du su2*), 65
amylose-extender dull sugary waxy (*ae du su wx*), 68–9
amylose-extender dull waxy (*ae du wx*), 65–6
amylose-extender sugary (*ae su*), 59–60
amylose-extender sugary-2 (*ae su2*), 60
amylose-extender sugary-2 waxy (*ae su2 wx*), 67
amylose-extender sugary sugary-2 (*ae su su2*), 66
amylose-extender sugary waxy (*ae su wx*), 66
amylose-extender waxy (*ae wx*), 58–9
dull (*du*), 57–8
dull sugary (*du su*), 61–2
dull sugary-2 (*du su2*), 62
dull sugary-2 waxy (*du su2 wx*), 68
dull sugary sugary-2 (*du su su2*), 67
dull sugary waxy (*du su wx*), 67–8
dull waxy (*du wx*), 62–3
sugary (*su*), 53–6
sugary-2 (*su2*), 56–7
sugary-2 waxy (*su2 wx*), 63–4
sugary sugary-2 (*su su2*), 64
sugary sugary-2 waxy (*su su2 wx*), 68
sugary waxy (*su wx*), 63
Waxy (*wx*), 44–50
non-mutant starch granule and plastid morphology:
　average starch granule size, developmental changes in, 34
　description, 33
　formation and enlargement, 34–6
　species and cultivar effects, 33–4
non-mutant starch granule polysaccharide composition:
　developmental changes, 31–2
　environmental effects, 32
　polysaccharide components, 28–30
　species and cultivar effects, 30–1
occurrence, 25
　cytosolic starch formation, 25–6

general distribution, 25
plastids, starch formed in, 26
polysaccharide biosynthesis:
   compartmentation and regulation of starch synthesis:
      in amyloplasts, 40–3
      and degradation in chloroplasts, 37–40
   enzymology, 36–7
Germ, of corn, 376, 377, 379
Glass transitions:
   of amorphous structural domains, 311–19
Glassy state relaxation, *see Physical aging*
Glg B gene, 133
Globulin, 405
Glucan β-, 446, 590, 591, 604
Glucan components, α-D-, 293
Glucan-synthesizing enzymes, properties of:
   1,4-α-glucan-synthesizing enzymes, 87
   ADP-glucose pyrophosphorylases (ADPGlc PPase)
      activation, by thioredoxin, 107–8
      amino acid residues identification within, 111–13
      catalysis, in large subunit, 91–5
      characterization, from different sources, 108–11
      interaction between 3PGA and Pi, differences in, 105–6
      kinetic properties and quaternary structure, 87–91
      large and small subunits, phylogenetic analysis of, 95
      potato tuber ADPGlc PPase, crystal structure of, 95–104
      regulation of, 104–5
   branching enzyme (BE), 129–36
   starch synthase, 114–28
Glucanotransferase α-1,4, 136–8
Glucoamylases, 244, 245, 806, 810
   structure and action, 257–61
Gluconic acid, 8
Glucose, D-, 3, 7, 239, 241, 277, 424, 799, 800, 827
   β-D-glucose, 244
Glucose 1-phosphate thymidylyltransferase (Rffh), 102
Glucose/corn syrups, 802–8, 809, 810
   acid-catalyzed hydrolysis, 802–6

   acid–enzyme processes, 806–8
Glucose Sugar Refining Company, 4
Glucose syrups, 7, 799
   boiling point elevation, 825
   color designations, 821
   refractive index-dry substance for, 819
Glutelin, 405, 573
Gluten, 408, 409, 414, 418, 426
Glutinous rices, 44, 570
Glycerol monostearate (GMS), 350, 358, 734
Glycogen synthesis, in cyanobacteria, 85, 114, 128
Glycolic acid, 845
Glycosidic bond, enzymatic hydrolysis of, 264–7
Gordon–Taylor equation, 316
Grafted starch acylates, 730
Grafted starches, 664
Grain Processing Company, 4, 10
Grain seeds, 194
Grain sorghum, *see Corn and sorghum starches production*
Granular cold-water-swellable starch, 332
   and cold-water-soluble starch, 644–5
Granular starch, crystallites in:
   melting transitions of, 323–32
Granular starch composites, 719–24
Granule-bound enzyme, 122
Granule-bound starch synthase I (GBSSI), 123, 125, 465, 467, 472
Granule-bound starch synthase II (GBSSII), 123, 125, 465
Granule-bound starch synthases (GBSS), 114, 122, 123, 124, 125–6, 134, 465, 556
Granule swelling, 668
   and pasting:
      crosslinking effects on, 758
'Growth rings', 85, 153, 182, 454, 461
Gums/hydrocolloids, 786

H

H/M-treated starches, *see Heat–moisture treatments*
Halobacterium halobium, 247
Havea brasiliensis, 30
Head rice, 571
Heartland Wheat Growers, 10

Heat–moisture treatments, 158
  annealing and structural modifications by, 320–3
Heat treatments, 764
Hercules viscometer, 668
Hexose phosphates, 42
High-amylose barley starch (HABS), 52, 610, 611, 616, 618, 622, 623
High-amylose corn starch, 5–6, 358, 384, 665, 769, 773
High-amylose maize starches, 50, 195, 196
High-amylose wheat starch, 470–1
High DS starch acetates, 642
High-fructose syrups (HFS), 7, 12, 14, 19, 20, 799, 808, 810–13, 820
  color stability, 822
  consumption, 13–15
  on gelatinization temperature, 826
  price list, 13
  US production capacity for, 18
High Oil TC Blends™, 382
High performance size-exclusion chromatography (HP-SEC), 467, 593
High-pressure disintegration (HD) process:
  corn starch production, 428
  wheat starch production, 450–1
High pressure liquid chromatography (HPLC), 279, 705
History, of starch:
  during 1500–1900, 2–4
  1900–present, 4–5
  early history, 1–2
  products:
    amino acids, 8–9
    ethanol, 7–8
    organic acids, 8
    polyols, 8
    sweeteners, 6–7
  specialty starches, development of, 5
    chemically modified starches, 6
    high-amylose corn starch, 5–6
    naturally modified corn starches, 6
    waxy corn starch, 5
Hordeum vulgare, 601
Horny endosperm:
  of corn, 376, 378, 379
  of grain sorghum, 380
Hot-filled foods, 775

Hoya carnosa, 106
Hulled and hull-less barley grains, 602–3
Hulls, 377, 572, 602
Human salivary α-amylase, 209, 211, 242, 243, 267
Humectancy, 825
Huron Milling Company, 4
Hydrocyclone process, 448–9, 449
Hydrocyclone separator, 411, 412, 413
Hydrocyclones, 547, 548
Hydrolyzed vegetable oil (HVO), 426
Hydrophobic starches, 664–5, 695
Hydroxyalkylcyclodextrins, 840–2
Hydroxyethylated cationic potato starch, 694
Hydroxypropyl-β-cyclodextrin, 845
Hydroxypropylated starch, 478, 647
Hydroxypropylation, 457, 560, 620–1, 631, 736, 748, 759
Hydroxypropylstarches, 631
Hygroscopicity, 825

I

Illinois Sugar Refining Company, 4
Indirect heat exchangers, 753
Insoluble starch substrates, amylases action on, 267–72
'Instant' starches, 748, 763
Intermediate material, 116
  characterization, 593
  and phytoglycogen, 202, 204
Intermittent milling and dynamic steeping process (IMDS), 427
Iodine affinities and amylose contents, 201, 202
Ion exchange resins, 806
Ipomoea batatas L., 33
'IR8' rice, 31, 32
'IR28' rice, 31
Isoamylase hydrolysates, 206
Isoamylases, 136, 247–8, 807, 817
Isoascorbic acid, 275
Isomaltooligosaccharides, 816
Isomerases, 7

J

Japanese white salted noodle, 465, 487
J.C. Hubinger Brothers Company, 4
Jet cooking, 641, 643, 674, 675, 688, 753
Junction zones, 332

K

K198, 101, 103
Katahdin potatoes, 32
Keever Starch Company, 4
Kirchoff, G.S.C., 3, 6, 715, 798
Kneen's Inhibitor, 272
Kuzu amylose, 206

L

L-α-lysophosphatidylcholine (LPC), 358
Labelle, 570
Lactic acid, 8, 402, 403
    bacteria, 405–8
Lactobacilli, 406, 580
Lactobacillus bulgaricus, 406
Lactobacillus delbrukii, 407
Lactobacillus leichmanii, 407
Lamella single crystals, 153
LargeK44R/T54K, 91, 92
Laundry sizing, wheat starch in, 490
LD_{50}, 839, 841
Leaf starch, 84
Lily amylose, 206
Limonin, 846
Linamarin, 544
Linear, low density polyethylene (LLDPE), 640
Lipid complexed amylose (L·AM), 606
Lipids and phospholipids, 204
Liquefying amylases, 239
Lotaustralin, 544
Low-density polyethylene (LDPE), 720
Low-voltage scanning electron microscopes (LVSEM), 171, 173, 174, 177
Lysophosphatidylcholine, 358, 592
Lysophospholipids, 456, 472

M

Macerans amylase, 250
Maize, 32, 43, 196, 374
    see also Corn and sorghum starches production
Maize endosperm, 110, 129, 130
Maize endosperm mutant, 56, 105, 116, 132
Maize starch granules, 132, 181, 197
Malic acid, 8, 737
Malto-oligosaccharides, 125, 126, 816
Maltodextrins, 252, 489, 591, 639, 778, 799, 800–2
    carbohydrate profiles of, 802, 817–18
    products, 239, 246–7, 801
Maltododecaose (G12), 275
Maltoheptaose (G7), 239, 247
Maltohexaose (G6), 239, 241, 247, 275
Maltopentaose (G5), 239, 242
Maltose β-, 244
Maltose (G2), 239, 817
Maltotetraose (G4), 239, 241
Maltotriose (G3), 125, 162, 239, 240, 241, 817
Manihot, 541
Manihot esculenta, 541
Manihot palmate, 541
Manihot utilissima, 31, 541
Manildra Milling Corporation, 10
Manioc, *see Tapioca starch*
Martin process, 446
Meats, starch in, 489, 780–1
Mechanical energy, 766
Melanin, 525
Merco starch separation centrifuge, 417
Mesophyll cells, 38
Methyl isobutyl ketone, 837
Methylcellulose, 786
Metroxylon sp., 33, 556
Michalis–Menten kinetics, 811
Micrococcus halobius, 247
Microporous starch granules, 642
Microscopy, of oat starch, 591–2
Midwest Grain Products, Inc, 10
Mill starch, 416–17, 418
Milled rice, uses of, 571–2
Minnesota Corn Processors, 5, 10
Modifications, of starches, 629
    cationic starches, 632
        amphoteric starch, 635–6
        with covalently-reactive groups, 636–7
        dry/solvent cationization, 633–4
        polycationic starches, 634–5
        starch-containing systems, 635–6
    encapsulation/controlled release, 642–4
    enzyme-catalyzed modifications, 647–8
    oxidation, 638–40
    physically modified starch:
        granular cold-water-swellable and cold-water-soluble starch, 644–5
        starch granule disruption, by mechanical force, 646

Modifications, of starches (*Continued*)
   starch-based plastics, 640–2
   starch graft polymers, 637–8
   thermal treatments, 646–7
Modified food starch, 756
   conversions, 760
      dextrins, 761
      fluidity (thin-boiling) starches, 760–1
   derivatizations, 756–9
      crosslinking, 756–8
      monosubstitutions, 758–9
      octenylsuccinylation, 759–60
   native starch thickeners, 767
   oxidation, 761–2
   physical modifications, 762–7
      altered flow properties, 762
      annealing, 765
      cold-water-swelling starches (CWS), 764
      dry heat, 766
      heat treatments, 764
      mechanical energy, 766
      moist heat, 765
      pregelatinization, 762–4
      radiation, 766–7
      solvents, 766
Modified wheat starch, 484, 485, 488
   amylograph viscosities and pasting temperatures, 479
Moist heat, 765
Molar distribution technique, 459
Monoglycerides, 210, 354
Monophosphorylation, 759
Monostarch phosphate esters, 205
Monosubstitutions, 758–9, 779
   development, 747–8
   of wheat starch, 480
Multiple melting transitions, for V-structures, 356
Mustard oils, 847
Mutant effects, 43
   amylose-extender (*ae*), 50–3
   amylose-extender dull (*ae du*), 61
   amylose-extender dull sugary (*ae du su*), 64
   amylose-extender dull sugary-2 (*ae du su2*), 65
   amylose-extender dull sugary waxy (*ae du su wx*), 68–9
   amylose-extender dull waxy (*ae du wx*), 65–6
   amylose-extender sugary (*ae su*), 59–60
   amylose-extender sugary-2 (*ae su2*), 60
   amylose-extender sugary-2 waxy (*ae su2 wx*), 67
   amylose-extender sugary sugary-2 (*ae su su2*), 66
   amylose-extender sugary waxy (*ae su wx*), 66
   amylose-extender waxy (*ae wx*), 58–9
   dull (*du*), 57–8
   dull sugary (*du su*), 61–2
   dull sugary-2 (*du su2*), 62
   dull sugary-2 waxy (*du su2 wx*), 68
   dull sugary sugary-2 (*du su su2*), 67
   dull sugary waxy (*du su wx*), 67–8
   dull waxy (*du wx*), 62–3
   sugary (*su*), 53–6
   sugary-2 (*Su2*), 56–7
   sugary-2 waxy (*su2 wx*), 63–4
   sugary sugary-2 (*su su2*), 64
   sugary sugary-2 waxy (*su su2 wx*), 68
   sugary waxy (*su wx*), 63
   Waxy (*wx*), 44–50
Mutant S5, 124
Mutant S9, 124

N

N-acetylglucosamine 1-phosphate uridylyltransferase (GlmU), 102
N-alkyl glucosides, 352
Naringen, 846
National Adhesives Corporation, 4
National Candy Company, 4
National Starch and Chemical Company, 10
National Starch and Chemical Corporation, 4
National Starch Company of New Jersey, 4
National Starch Manufacturing Company, 4
National Starch Products, Inc., 4
Native corn starch, 350, 693, 756, 768
Native dent corn starch, 768
Native potato starch, 535, 756, 759
Native sorghum starch, 756
Native starch granule, 150, 173, 296, 724
   amylases action with, 269–72
Native starch thickeners, 767
Native starches, 328, 475, 547, 620, 756

pasting properties of, 304, 305
Native tapioca starch, 770
Native waxy maize starch, 756
Native wheat starch, 314, 455, 489, 756
Nato, 570
Naturally modified corn starches, 6
Netherlands, history of potato processing in, 512–14
Neutral detergent residue (NDR), 381
Newer specialty papers, starch in, 703
Nitroalkyl ethers, 637
NMR, 156, 295, 308
   ^{13}C-CP/MAS-NMR, 210, 211, 299–300
Non-mutant starch granule and plastid morphology:
   average starch granule size, developmental changes in, 34
   description, 33
   formation and enlargement, 34–6
   species and cultivar effects, 33–4
Non-mutant starch granule polysaccharide composition:
   developmental changes, 31–2
   environmental effects, 32
   polysaccharide components, 28–30
   species and cultivar effects, 30–1
Noodles, 469–70, 485–6, 487
Nortai, 570

O

Oat starch, 589
   applications, 591
   isolation, 589
      industrial, 590
      laboratory, 590
   modification, 591
   properties:
      chemical composition, 592–4
      flow properties, 597
      gelatinization, 594–5
      microscopy, 591–2
      retrogradation, 595–6
      rheological properties, 597
      swelling power and amylose leaching, 596–7
      viscoelastic behavior, 597–8
      X-ray diffraction, 594
Oatrim, 591
Occurrence, of starch, 25
   cytosolic starch formation, 25–6
   general distribution, 25
   plastids, starch formed in, 26
Octenylsuccinylated dextrins, 777
Octenylsuccinylation, 759–60
Oligosaccharide syrups, 816–17, 818
Opaque-2 (*o2*), 43
Opaque 2 corn, 383
Opaque-6 (*o6*), 43
Opaque-7 (*o7*), 43
Optimum™ High Oil Corn, 382
Ordered and amorphous structural domains, 296–301
Organic acids, 8
Oriental wheat noodles, 487
Oriented two-dimensional diffraction patterns, 158
Oryza sativa L., 569
Ottawa Starch Viscometer, 304, 473, 597
Oxidation, 761–2
   of starch, 638–40
   of wheat starch, 480–1
Oxidized starches, 536, 663, 664, 694, 699, 776

P

Paper, starch analysis in, 705
Paper industry, starch in, 657, 658
   as adhesive in paper conversion, 700
      corrugation and lamination, starch selection for use in, 702
      corrugator for paperboard, 700–2
      lamination of paper, 700
   as coating binder:
      coater in paper machine, 695–8
      paper coating, starch selection for, 698–9
   application requirements for, 666
      charge specifications, 668–9
      purity requirements, 671
      retrogradation control, 669–71
      viscosity specifications, 666–8
   current use, 663–5
   dispersion:
      chemical conversion, 676–7
      enzymatic conversion, 677–81
      paper mill, delivery to, 672–3
      under atmospheric pressure, 674
      under elevated pressure, 674–5

Paper industry, starch in (*Continued*)
   water, in suspension, 673
   environmental aspects, 703–5
   newer specialty papers, 703
   papermaking furnish:
      cellulose and pigments, starch adsorption on, 682–3
      cellulose fibers and fines, flocculation of, 681–2
      paper machine, wet end of, 681
      pigments and cellulose fines, retention of, 683–4
      sheet bonding, by starch, 684–5
      wet-end sizing, 685–7
      wet-end use starch, selection for, 687–8
   papermaking process, 660–2
   recent trends, 665–6
   starch analysis, in paper, 705
   starch consumption by, 662
   surface sizing, of paper:
      size press, in paper machine, 688–93
      spray application of starch, 693
      surface sizing, starch selection for, 693–5
      water box at calendar, 693
Partial waxy and waxy wheat starch, 465–9
Particle size distribution (PSD), 526
Pasta, 485–8
Pasting and gelatinization, 749–50
Pasting properties, of native starches, 304, 305
Pastries, 483
Pea embryos, 109, 117, 124
Pea starch, 160, 226, 346, 772
Penford Food Ingredients, 10
Penford Products Company, 4, 10
Penick & Ford, Ltd, 4
Pericarp, of corn, 376, 377
Periodic acid (thiosemicarbazide) silver (PATAg), 180
Pet food, 781
PGM deficient mutants, 127
pH, of food, 783
Phaseolamin, 272
Phosphate monoesters, 194, 205
Phosphoglycerate, 3- 37, 104
Phosphoglycerides, 169, 210
Phospholipids and lipids, 204
Phosphoryl chloride, 477, 747, 758

Phosphorylase-stimulation assay, 129
Physical aging, 317
Physical properties of starch, in water, 301–10
Physically modified starch:
   granular cold-water-swellable (CWS) and cold-water-soluble starch, 644–5
   starch granule disruption, by mechanical force, 646
Physiology and genetics, of starch development, *see* Genetics and physiology, of starch development
Phytoglycogen, 48, 53, 54, 55, 57, 58, 59, 63, 69, 70, 86, 136, 194, 204
Piel Brothers Starch Company, 4
Pisum sativum L., 31
Plant phosphorylase, 276–7
Plasticization of starch, 317–18, 330
Plasticized amylose triesters, properties of, 718
Plastics, 491, 715, 731
   starch-based plastics, 640–4
Plastids, starch formed in, 26
Plate heat exchangers, 753
P-nmr spectroscopy, 204, 205
Polarized light microscopy, 195, 608
Polvilho azedo, *see* Sour cassava starch
Polycationic starches, 634–5
Poly(ethylene-co-acrylic acid) (EAA), 640, 643
Polyethylene terephthalate (PET), 724
Poly(hydroxy butyrate-co-valerate) (PHBV) copolymers, 723
   mechanical properties, 724
Polymer compositions, starch in, 715
   granular starch composites, 719–24
   rubber, starch in, 724–6
   starch esters, 717–19
   starch foams, 735–7
   starch graft copolymers, 726–31
   thermoplastic starch blends, 731–5
Polyols, 8
Polyphenoloxidase, 525, 846
Polysaccharide biosynthesis:
   compartmentation and regulation of starch synthesis:
      in amyloplasts, 40–3
      and degradation in chloroplasts, 37–40
   enzymology, 36–7

Polysaccharide hydrocolloids, 331
Porcine pancreatic α-amylase (PPA), 209,
    211, 238, 243, 253, 274, 280, 612
  inhibitors and KI values, 276
  primary products, 242
Porous structures, of starch granules, 195,
    198–200
Potato, under water weight of, 518, 519
Potato amylose, 206, 281
Potato phosphorylase, 276, 277
Potato protein:
  environmental aspects, 534–5
  properties, 535
  protein recovery, 535
  uses, 535
Potato starch, 182, 195, 195, 328, 489, 537,
    770, 774, 783
  amylopectin arrangement in, 168
  cationic, 537, 669, 683, 684, 686–7
  granule surfaces, 174
  potato processing:
    future aspects, 538
    in Netherlands, history of, 512–14
  potato protein:
    environmental aspects, 534–5
    properties, 535
    protein recovery, 535
    uses, 535
  processing, 522
    fiber extraction, 524–7, 528
    grinding, 525
    juice extraction, 525–6
    removal of water, from starch, 532–3
    sideline extraction, 530–2
    starch classification, 527, 529
    starch drying and storage, 533–4
    starch refinery, 529–30
  production:
    in Europe, 514–15
    world starch production, 514
  structure and chemical composition:
    all-amylopectin potato, 521–2
    anatomy, of tuber, 516–18
    chemical composition, 518–19, 520
    commercial starches, differences
      between, 519, 521
    formation and morphology, of tuber, 515
  utilization, 535
    converted starches, 536
    crosslinked starches, 536–7
    preference, for potato starch, 537
    substitution, 535–6
Potato starch producers, 10
Potato tuber:
  anatomy of, 516–18
  formation and morphology of, 515
Potato tuber ADPGlc PPase, 89
  crystal structure, 95
    ADP-glucose binding, 101–2
    allosteric regulation, 103–4
    ATP binding, 101
    catalysis, implication for, 102–3
    homotetramer structure, 97–8
    sulfate binding mimics phosphate
      inhibition, 98–100
  random mutagenesis experiments on, 113
Pourable dressings, 776
Preamylopectin, 36, 136
Pregelatinized starch, 644–5, 665, 699,
    762–4, 778
Pressure fed screens, 415, 416
Prime mill starch, 412
Proplastids, 35, 54
Prostaglandin E1 (PGE1), 845
Pseudodisaccharide, 272
Pseudomonas amylodermosa, 248
Pseudomonas isoamylase, 248, 281
Pseudomonas saccharophila, 238, 246
Pseudomonas stutzeri, 246, 247
Ptyalin, 238
Pullulanase, 248, 281, 283, 622, 623, 807
Pulse-amperometric detection system
    (PAD), 279
Pyridoxal 5-phosphate (PLP), 112
Pyridoxal phosphate, 43
Pyrococcus furiosus, 247, 249
Pyrococcus woesei, 249, 250

Q

Quaternary ammonium cationic starch, 632
Quaternary starches, 694
'Quick Germ' process, 427, 428

R

R-enzyme, 248
Rapid ViscoAmylograph, 575
Rapid ViscoAnalyzer, 303, 456, 473, 488,
    576, 667–8

Ready-to-eat breakfast cereals, 779–80
Refining/beating process, 661
Refined fibers, 661
Refractive index (RI), 799–800
Regular rice, *see Common rice*
Rieman balance, 545, 546
Remy Industries, 573
Resistant starch (RS), 212, 309–10, 621–3, 787–8
Retrogradation, 308, 618–19, 754, 755, 800
   and gelation, of starch, *see Gelation and retrogradation, of starch*
   of oat starch, 595–6
   process, 669–70
   of rye starch, 584
Reversion, 800
Rheometers, 306, 307, 597
Rhizopus delemar, 244, 246
Rhizopus niveus, 244, 269, 271
Rhizopus sp., 260
Rhodopsedudomonas gelatinosa, 272
Rice, by-products of:
   bran, 573
   hulls, 572
Rice starches, 569, 771–2
   applications, 577
   by-products, uses of, 572–3
   composition, 570–1
   factors affecting properties of:
      common rice versus waxy rice, 575–7
      modification, 577
      preparation, 576–7
      protein content, 576
   milled rice, uses of, 571–2
   milling, 570–1
   preparation, 573
      mechanical method, 574
      traditional method, 573–4
   production, of rice, 569–70
   properties:
      general properties, 574–5
      pasting properties, 575
RiceLife®, 572
Rieska, 602
RNA interference (RNAi) technology, 133, 471
Roquette America, Inc, 4, 10
Rotational viscometers, 597

Rough rice, 570, 572
Rubber, starch in, 724–6
Rubisco, 38
Rug5 mutant, 117
Rye starch, 579
   applications, 582
   isolation:
      industrial production, 579, 580, 581
      laboratory process for, 580, 581
   modification, 582
   properties:
      amylose–lipid complex, 584
      composition, 583
      falling number, 586
      gelatinization behavior, 584
      microscopy, 582–3
      retrogradation, 584
      rheology, 585–6
      swelling power and amylose leaching, 584–5
      X-ray diffraction patterns, 584

S

Saccharomycopsis α-amylase, 261
Sago pearls, *see Tapioca pearls*
Sago starch, 772
Salad dressings, 776–7
Saybolt Seconds Universal (SUU) units, 820
SbeIc allele codes, 130
Scanning electron micrographs, 607
   of high-amylose barley starches, 605, 614
   of normal barley starches, 605, 614
   of starches, 196, 198, 199
   of waxy barley starches, 605, 614
Scanning electron microscopy, 180
   granule imaging by, 170–1
   of maize starch granules, 181
Scanning force microscopy (SFM), *see Atomic force microscope*
Scanning probe microscopy (SPM), 170
   starch granule surface imaging, 177–9
   future prospects, 179–80
Scanning tunneling microscope (STM), 170
Scholten, W.A., 512
Schulze–Hardy rule, 670
Scilla ovatifolia, 33
Scraped-surface heat exchangers, 753–4
Screen bend device, 412, 415

Second greens/hydrol, 814
Second heads, 571, 572
'Selkirk' wheat, 32
Sequential extraction (SE) process, 428
Sex6 mutants, 118
Sheet bonding, by starch, 684–5
Short flow milling process, 448
Short grain, 570, 575
Shrunken-1 (sh), 43
Shrunken-2 (sh2), 43
Shrunken-4 (sh4), 43
Simple granule, 33
Single helical structures, of amylose, 208–11
Single nucleotide polymorphisms (SNPs), 468
Site-directed mutagenesis, 98–9, 111, 112, 253
Size exclusion column chromatography (SEC), 29, 459
Small angle x-ray scattering (SAXS), 225, 295, 456
SmallD145N, 91–2
SmallD145NLargeK44R, 92
SmallD145NLargeK44R/T54K, 92, 93
SmallD145NLargeK44R/T54K, 92, 93, 94
SmallD145NLargeT54K, 92
SmallD145NLargeWT, 92
SmallWTLargeWT, 92
Snacks industries, 779–80
Sodium bromide, 331
Sodium chloride, 630, 636, 671, 784
Sodium stearoyl lactylate (SSL), 350
Sodium sulfate, 331, 630
Sodium trimetaphosphate (STMP), 477, 747, 758
Solanum tuberosum L., 91
Solvent cationization, 633–4
Sorghum, 771
  see also Corn and sorghum starches production
Sorghum bicolor Moench, 375
Sorghum starch, 195, 196, 198, 373
Sour cassava starch, 561
Sources, of starches, 767–73
  amaranth, 773
  arrowroot, 772
  barley, 772
  dent corn, 768
  high-amylose corn, 769
  improvement, 748
  pea, 772–3
  potato, 770
  rice, 771–2
  sago, 772
  sorghum, 771
  tapioca, 770
  waxy corn, 768–9
  wheat, 770–1
Soybean β-amylase:
  exo-mechanism, 244–6
  structure and action, 257
Soybean cells, 41
Spaghetti, 485, 486, 487
Specialty starches, development of, 5
  chemically modified starches, 6
  high-amylose corn starch, 5–6
  naturally modified corn starches, 6
  waxy corn starch, 5
Spherulites, 153
Spinach (*Spinacia oleracea*), 37, 107, 108, 112
Spray application, of starch, 693
Spray dried starches, 764
Spray drying pregelatinization, of starch, 644–5
Squeegee fraction, 408, 409
Squeegee starch, *see* B-starch
Sta2 mutants, 126
Stabilizations, *see* Monosubstitutions
Stable starch–lipid composites, 643
Starch acetates, 631, 642
Starch-based plastics, 640–2
Starch-branching enzyme (SBE), 36, 471
Starch-containing systems, 635–6
Starch-derived sweetener, 6–7
  composition of, 804
  consumption, 14
  general process flow for, 801
Starch dust, 547, 672–3
Starch–EAA–PE films, mechanical properties, 733
Starch esters, 663, 717–19
Starch fibrids, 665
Starch foams, 735–7
Starch–g–MA copolymers, properties of, 728

Starch–gluten separation, 416–20
Starch graft copolymerization, 638, 726–31
Starch graft copolymers, 638, 726–31
Starch graft polymers, 637–8
Starch grafts, 695
Starch granule-bound proteins (SGP), 452, 472
Starch granule disruption, by mechanical force, 646
Starch granules, 149, 193
   amylopectin, 212
      chemical structure, 212–18
      cluster models, 218–23
      growing temperature and kernel maturity, effects of, 223–4
   amylose:
      chemical structure, 205–8
      double helical structures, 211–12
      growing temperature and kernel maturity, effects of, 223–4
      single helical structures, 208–11
   amylose within granules, location and state of, 184–6
   architecture:
      crystalline structures, molecular organization of, 153–8
      crystalline ultrastructural features, 158–60
      granule structure, 153
      supramolecular organization, 160–7
   blocklets concept, 180–4
   characteristics:
      gelatinized starch granules, shapes of, 200–1
      granule shapes, sizes and distributions, 194–5
      porous structures, 195, 198–200
   molecular components in granule, locations of, 225–7
   molecular compositions:
      amylopectin and amylose, 201–2
      intermediate material, 202, 204
      lipids and phospholipids, 204
      phosphate monoesters, 205
      phytoglycogen, 204
   storage tissues, 85
   surface, 167
      surface chemistry and composition, 168–9
      surface-specific chemical analysis, 169
   surface imaging, 170
      atomic force microscope (AFM), 170, 171–7
      by Scanning Probe Microscopy (SPM), 177–80
      by SEM methods, 170–1
   surface pores and interior channels, 186–7
   swelling, 302
Starch–gum interactions, 331
Starch hydrolysis products (SHP), 782
Starch hydrolyzates, 647, 773–4
Starch hydroxypropyl ethers, 758
Starch–latex binder, 664
Starch lyase, 277–8
Starch molecules, enzymic characterization of, 278–84
   amylopectin, β-amylase limit dextrins formation of:
      fine structure determination, 282–4
   branch linkage nature, determination of, 279–80
   slightly branched amyloses, identification and structure determination of, 280–2
   see also Enzymes' action, on starch
Starch octenylsuccinates, 640, 643–4
Starch pastes, 667, 668, 675, 676, 801
   and gels, 342–54
Starch synthase, 36, 86, 114
   bound to starch granule, 122–3
   characterization of, 114
   granule-bound starch synthases (GBSS) and isoforms, 124–7
   soluble starch synthases, 114
      double mutants of, 121–2
   starch synthase I (SSI), 114–16
   starch synthase II (SSII), 116–19
   starch synthase III (SSIII), 119
   starch synthase IV (SSIV), 119–21
   substrate binding and catalysis, amino acids involved in, 127–8
   waxy protein structural gene, isolation of, 123–4
Starch xanthate (SX), 725
State and phase transitions, of starch, 310, 311
   crystallites in granular starch, melting transitions of, 323–32

gelation and retrogradation, 332
    amylopectin gels, 337–40
    amylose gels, 332–6
    starch pastes and gels, 342–54
glass transitions, of amorphous structural
        domains, 311–19
heat–moisture treatments, annealing and
        structural modifications by, 320–3
V-structures, phase transitions and
        properties of, 354–9
Steam injection, 558, 753
Storage modulus (G) profiles, 307, 334, 337,
        345
Storage tissues, starch in, 85
Streptococci, 406, 580
Streptomyces, 248, 811
Streptomyces griseus, 246
Structural transitions and physical
        properties, of starch, 293, 295,
        310
    ordered and amorphous structural
            domains, 296–301
    state and phase transitions, 310
        crystallites in granular starch, melting
                transitions of, 323–32
        gelation and retrogradation, 332–54
        glass transitions, of amorphous
                structural domains, 311–19
        heat–moisture treatments, annealing
                and structural modifications by,
                320–3
        V-structures, phase transitions and
                properties of, 354–9
    in water, 301–10
Styrene maleic anhydride (SMA)
        copolymer, 695, 722, 723
Styrenesulfonate polymerization, 848
Su-am (*amylaceous*), 55, 57
Su-Bn2 (*Brawn-2*), 55
Su-Ref, 55, 56
Su-st (*starchy*), 55, 56
Su1 mutation, 136
Substrate-binding residues, 127
Sucrose, equilibrium water content of, 815
Sucrose export defective-1 mutant, 38
Sucrose-P synthase (SPS), 38–9
Sugar alcohols, 8, 270, 329, 564
Sugars, effect of, 784
Sugary (*su*) mutants, 53–6

Sugary-2 (*Su2*), 6, 56–7
Sugary-2 maize starch, 195, 196
Sugary-2 waxy (*su2 wx*), 63–4
Sugary enhancer (*se*), 56
Sugary sugary-2 (*su su2*), 64
Sugary sugary-2 waxy (*su su2 wx*), 68
Sugary waxy (*su wx*), 63
Sulfolobus, 249
Sulfolobus acidocaldarius, 249
Sulfolobus solfataricus, 249
Sulfur dioxide effect, in corn steeping,
        403–5
Supercalendering, 661
Supramolecular organization, of starch
        granules, 160–7
Surface pores and interior channels, of
        starch granules, 186–7
Surface sizing of paper, use of starch for,
        661, 689–90
    size press, in paper machine:
        flooded size press, 688–90
        metered size press, 690–2
        pigmented surface sizing, 692–3
    spray application, of starch, 693
    starch selection, for surface sizing, 693–5
    water box at calendar, 693
Surface-specific analysis, of starch granule,
        169
Surfactants, effects of, 785
Surimi, 781
Sweeteners, from starch, 6–7, 13, 14, 423–4,
        797
    composition and properties:
        boiling point elevation, 824, 825
        browning reaction and color, 821–2
        carbohydrate profiles, 817–18
        crystallization, 826–7
        fermentablilty, 822–3
        foam stabilization and gel strength, 823
        freezing point depression, 824, 825
        gelatinization temperature, 824–5
        humectancy and hygroscopicity, 825,
                826
        selection, 828–9
        solids, 818–19
        sweetness, 827–8, 829
        viscosity, 819–21
    definitions, 799–800
    history, 797–9

Sweeteners, from starch (*Continued*)
  production methods, 800, 801
    crystalline dextrose, and dextrose syrups, 813–16, 817
    crystalline fructose, 813, 814, 815
    glucose/corn syrups, 802–8, 809, 810
    high-fructose syrups, 808, 810–13
    maltodextrins, 800–2
    oligosaccharide syrups, 816–17, 818
  regulatory status, 800
Swelling power and amylose leaching:
  of barley starch, 610–12
  of oat starch, 596–7
  of rye starch, 584–5
Sycamore cells, 41
Syrup demineralization sequence, 807

T

Tailings starch, *see B-starch*
Tao starch, 555
Tapioca amylose, 206, 560
Tapioca pearls, 556, 770
  making process, 557
Tapioca starch, 302, 536, 537, 541, 550–5, 636, 770, 773, 774, 779, 781, 782
  background, 541–5
  food applications, 556–62
  industrial applications, 563–4
  modification, 555–6
  outlook, 564
  physical modification, 555
  processing, 545–9
TAPPI (Technical Association of the Pulp and Paper Industry), 659
Tate & Lyle North America, 10, 17
TEMPO (2,2,6,6-tetramethyl-1-piperidinyloxy), 639
TEMPO–sodium hypochlorite–sodium bromide system, 639–40
Tertiary amino starches, 633, 687
Texturizing starches, 763
Thermal-chemical starch conversion system, 677
Thermal treatments, of starch, 646–7
Thermococcale, 249
Thermococcus aggregans, 249
Thermococcus celer, 249
Thermococcus guaymagensis, 249

Thermococcus hydrothermalis, 249
Thermococcus litoralis, 249
Thermococcus profundus, 249
Thermococcus zilligii, 249
Thermo-mechanical analysis (TMA), 296, 319
Thermophilic archaebacterial amylolytic enzymes, 249
Thermoplastic starch blends, 731–5
Thin-boiling starches, *see Converted starches; Fluidity starches*
Thymopentin, 847
TILLING (targeted induced local lesions in genomes) method, 467–8
Time-of-flight secondary-ion mass spectrometry (TOF-SIMS), 169, 705
TL25 mutant, 105
TL46 mutant, 105
Tomato fruit, 87, 110
Total chlorine free (TCF) starches, 665, 699
Total dietary fiber (TDF), 787, 788
Transglucosidase, 816, 817
Transmission electron microscopy (TEM), 165, 166, 180, 181, 299–300, 463, 607
  of high amylose barley starches, 608, 615
  of normal barley starches, 608, 615
  of waxy barley starches, 608, 615
  of waxy maize starch, 166
Tricanter®, 450
Triose phosphates, 37, 38
Triticum aestivum, 119, 131
Trp120, 261, 262
TSS (total suspended solids), 663
Tubular heat exchangers, 753
Tyr116, 262

U

UDP-Glc:protein transglucosylase (UPTG), 35, 36
UDP-glucose (UDPGlc), 86, 122, 123, 227
UDP-N-acetyl-glucosamine pyrophosphorylase (UDPGlcNAc PPase), 94, 95
Ultra-high temperature (UHT), 576, 782
Under water weight (UWW), 518, 519
Union Starch, 4

United Starch Company, 4
Unmodified starches, 484, 488, 688, 736, 737, 756
US corn starch industry, economic growth and organization of, 11
  corn price variability, effects of, 18–19
  extent and direction, 11–13
  fuel alcohol, 15–16
  future industry prospects, 20
  high-fructose syrup (HFS) consumption, 13–15
  industry organization, 16–18
  international involvement, 19–20
  plant location, 16
  technical progress, 16
US Department of Agriculture's Northern Regional Research Center, 720
US Grade Standards:
  for corn, 386
  for grain sorghum, 389, 390
USDA, 571, 572, 720, 781

V

V-amylose, 297–8, 299
V-complexes, of amylose, 208–11
V-structures, phase transitions and properties of, 354–9
Vacuum drum filter, 533
Vanderbilt process, 680
Vertical planting, 542
Vicia faba, 111
Vinyl acetate (VA), 635, 638, 730, 748, 759
Vista, 570
Vomitoxin, see Deoxynivalenol

W

Water-binding capacity (WBC), 584, 585, 765
Water-soluble polysaccharide (WSP), 25, 44, 47, 53, 61, 62, 63, 65, 66, 67, 68, 136, 204, 352
Waxy corn starch, 5, 669, 768–9
  see also Waxy maize starch; Waxy rice starchWaxy wheat starch
Waxy maize starch, 327–8, 521, 522, 537, 620, 665, 774, 817
  cationic, 687
  crosslinked, 747–8

Waxy mutants, 44, 45, 86
Waxy rice starch, 195, 198, 205, 214, 551, 553, 570, 575, 576, 577, 771, 772
  DSC traces of, 325
  physicochemical properties of, 576
  versus common rice starches, 575–6, 577
Waxy wheat starch, 204, 465–9, 470, 473
Western Glucose Company, 4
Western Polymer Corporation, 10
Westfalia HD Process, 428
Wet-end sizing, 685–7
Wet-end use, starch selection for, 687–8
Wet-milling, of corn and sorghum starches, 391, 393
  automation, 423
  corn starch manufacturing process, flow diagram of, 393
  grain cleaning, 392
  milling and fraction separation:
    component yields, 408–10
    germ separation, 410–13
    grain sorghum processing innovations, 420–1
    second milling and fiber separation, 413–16
    starch–gluten separation, 416–20
  product drying, energy use and pollution control, 421–3
  starch processing, 421
  steeping:
    commercial steeping, mechanics of, 395–401
    lactic acid bacteria, role of, 405–8
    principles, 394–5
    sulfur dioxide, effect of, 403–5
    water absorption and solubles removal, 401–3
Wheat gluten, 443, 445, 448, 451, 481
Wheat starch, 2, 204, 282, 441, 443, 770–1
  color, 471
  dual derivatization, 479–80
  flavor, 472–3
  in food products, 488–9
  gel strength, 474
  and gluten separation techniques, 446
  granule surfaces, 174–5
  industrial processes, for production, 444
    conventional processes, 446–8

Wheat starch (*Continued*)
   high-pressure disintegration process, 450–1
   hydrocyclone process, 448–9
  laundry sizing, 490
  modification, 475
   acid-thinning, 481
   bleaching, 480–1
   crosslinking, 475–7
   dual derivatization, 479–80
   oxidation, 481
   substitution, 478–9
  modified and unmodified starches:
   baked products, role in, 481–5
   food uses, 488–9
   industrial uses, 489–91
   noodles and pasta, functionality in, 485–8
  paste clarity, 473
  paste texture, 473
  paste viscosity, 473
  production, 442–4
  properties, 451, 471–4
   amylase and amylopectin, fine structures of, 457–64
   high-amylose wheat starch, 470–1
   large versus small granules, 452–7
  purity, 472
  see also Modified wheat starch; Partial waxy and waxy wheat starchWaxy wheat starch

Wheat starch gel, 339, 344, 351, 352
Wheat starch producers, 10
Whole brown rice, 572, 574
Wide-angle x-ray scattering and diffraction (WAXD), 295, 296, 297, 354, 356
World starch production, 514
Wx protein, 134

X

X-ray diffraction, 46, 127, 220, 222, 262, 321, 350
  of barley starch, 607
  of oat starch, 594
  of rye starch, 584
X-ray powder diffractogram, 153, 154
Xanthosoma sagittifolium, 34
Xerosicyos danguyi, 106

Y

Yellow grain sorghum, 391
Zea diploperennis, 374
Zea mays, 25, 124, 374
Zea mays L., 374
Zearalenone, 388
Zein, 43, 405, 428

Food Science and Technology

Series List

Maynard A. Amerine, Rose Marie Pangborn, and Edward B. Roessler, *Principles of Sensory Evaluation of Food*. 1965.
Martin Glicksman, *Gum Techology in the Food Industry*. 1970.
Maynard A. Joslyn, *Methods in Food Analysis*, second edition. 1970.
C. R. Stumbo, *Thermobacteriology in Food Processing*, second edition. 1973.
Aaron M. Altschul (ed.), *New Protein Foods*: Volume 1, *Technology, Part A*—1974. Volume 2, *Technology, Part B*—1976. Volume 3, *Animal Protein Supplies, Part A*—1978. Volume 4, *Animal Protein Supplies, Part B*—1981. Volume 5, *Seed Storage Proteins*—1985.
S.A. Goldblith, L. Rey, and W. W. Rothmayr, *Freeze Drying and Advanced Food Technology*. 1975.
R.B. Duckworth (ed.), *Water Relations of Food*. 1975.
John A. Troller and J. H. B. Christian, *Water Activity and Food*. 1978.
A. E. Bender, *Food Processing and Nutrition*. 1978.
D.R. Osborne and P. Voogt, *The Analysis of Nutrients in Foods*. 1978.
Marcel Loncin and R.L. Merson, *Food Engineering: Principles and Selected Applications*. 1979.
J.G. Vaughan (ed.), *Food Microscopy*. 1979.
J.R.A. Pollock (ed.), *Brewing Science*, Volume 1—1979. Volume 2—1980. Volume 3—1987.
J. Christopher Bauernfeind (ed.), *Carotenoids as Colorants and Vitamin A Precursors: Technological and Nutritional Applications*. 1981.
Pericles Markakis (ed.), *Anthocyanins as Food Colors*. 1982.
George F. Stewart and Maynard A. Amerine (eds), *Introduction to Food Science and Technology*, second edition. 1982.
Malcolm C. Bourne, *Food Texture and Viscosity: Concept and Measurement*. 1982.
Hector A. Iglesias and Jorge Chirife, *Handbook of Food Isotherms: Water Sorption Parameters for Food and Food Components*. 1982.
Colin Dennis (ed.), *Post-Harvest Pathology of Fruits and Vegetables*. 1983.
P.J. Barnes (ed.), *Lipids in Cereal Technology*. 1983.
David Pimentel and Carl W. Hall (eds), *Food and Energy Resources*. 1984.

Joe M. Regenstein and Carrie E. Regenstein, *Food Protein Chemistry: An Introduction for Food Scientists*. 1984.

Maximo C. Gacula, Jr. and Jagbir Singh, *Statistical Methods in Food and Consumer Research*. 1984.

Fergus M. Clydesdale and Kathryn L. Wiemer (eds), *Iron Fortification of Foods*. 1985.

Robert V Decareau, *Microwaves in the Food processing Industry*. 1985.

S. M. Herschdoerfer (ed.), *Quality Control in the Food Industry*, second edition. Volume 1—1985. Volume 2—1985. Volume 3—1986. Volume 4—1987.

F.E. Cunningham and N.A. Cox (eds), *Microbiology of Poultry Meat Products*. 1987.

Walter M. Urbain, *Food Irradiation*. 1986.

Peter J. Bechtel, *Muscle as Food*. 1986. H.W.-S. Chan, *Autoxidation of Unsaturated Lipids*. 1986.

Chester O. McCrokle, Jr., *Economics of Food Processing in the United States*. 1987.

Jethro Japtiani, Harvey T. Chan, Jr., and William S. Sakai, *Tropical Fruit Processing*. 1987.

J. Solms, D. A. Booth, R. M. Dangborn, and O. Raunhardt, *Food Acceptance and Nutrition*. 1987.

R. Macrae, *HPLC in Food Analysis,* second edition. 1988.

A.M. Pearson and R.B. Young, *Muscle and Meat Biochemistry.* 1989.

Marjorie P. Penfield and Ada Marie Campbell, *Experimental Food Science,* third edition. 1990.

Leroy C. Blankenship, *Colonization Control of Human Bacterial Enteropathogens in Poultry*. 1991.

Yeshajahu Pomeranz, *Functional Properties of Food Components*, second edition. 1991.

Reginald H. Water, *The Chemistry and Technology of Pectin*. 1991.

Herbert Stone and Joel L. Sidel, *Sensory Evaluation Practices*, second edition. 1993.

Tilak Nagodawithana and Gerald Reed, *Enzymes in Food Processing,* third edition. 1993.

Dallas G. Hoover and Larry R. Steenson, *Bacteriocins*. 1993.

Takayaki Shibamoto and Leonard Bjeldanes, *Introduction to Food Toxicology*. 1993.

John A. Troller, *Sanitation in Food Processing,* second edition. 1993.

Harold D. Hafs and Robert G. Zimbelman, *Low-fat Meats*. 1994.

Lance G. Phillips, Dana M. Whitehead, and John Kinsella, *Structure-Function Properties of Food Proteins*. 1994.

Robert G. Jensen, *Handbook of Milk Composition*. 1995.

Yrjö H. Roos, *Phase Transitions in Foods*. 1995.

Reginald H. Walter, *Polysaccharide Dispersions*. 1997.

Gustavo V Barbosa-Cànovas, M. Marcela Góngora-Nieto, Usha R. Pothakamury, and Barry G. Swanson, *Preservation of Foods with Pulsed electric Fields*. 1999.

Ronald S. Jackson, *Wine Tasting: A Professional Handbook*. 2002.

Malcolm C. Bourne, *Food Texture and Viscosity: Concept and Measurement*, second edition. 2002.

Benjamin Caballero and Barry M. Popkin (eds), *The Nutrition Transition: Diet and Disease in the Developing. World*. 2002.

Dean O. Cliver and Hans P. Riemann (eds), *Foodborne Diseases,* second edition. 2002. Martin

Kohlmeier, *Nutrient Metabolism*, 2003.
Herbert Stone and Joel L. Sidel, *Sensory Evaluation Practices,* third edition. 2004.
Jung H. Han, *Innovations is Food Packaging*. 2005.
Da-Wen Sun, *Emerging Technologies for Food Processing*. 2005.
Hans Riemann and Dean Cliver (eds) *Foodborne Infections and Intoxications,* third edition. 2006.
Ioannis S. Arvanitoyannis, *Waste Management for the Food Industries*. 2008.
Ronald S. Jackson, *Wine Science: Principles and Applications*, third edition. 2008.
Da-Wen Sun, *Computer Vision Technology for Food Quality Evaluation*. 2008.
Kenneth David and Paul B. Thompson, *What Can Nanotechnology Learn From Biotechnology?* 2008.
Elke K. Arendt and Fabio Dal Bello, *Gluten-Free Cereal Products and Beaverages*. 2008.
Da-Wen Sun, *Modern Techniques for Food Authentication*. 2008.
Debasis Bagchi, *Nutraceutical and Functional Food Regulations in the United States and Around the World*, 2008.
R. Pual Singh and Dennis R. Heldman, *Introduction to Food Engineering,* fourth edition. 2008.
Peter J. Clark, *Practical Design, Construction and Operation of Food Facilities*. 2008.
Zeki Berk, *Food Process Engineering and Technology*. 2009.
Abby Thompson, Mike Boland and Harjinder Singh, *Milk Proteins: From Expression to Food*. 2009.
Wojciech J. Florkowski, Stanley E. Prussia, Robert L. Shewfelt and Bernhard Brueckner (eds) *Postharvest Handling,* second edition. 2009.
J. BeMiller and R. Whistler, *Starch*. 2009.
Maximo Gacula, Jr., Jaqbir Singh, Jian Bi and Stan Altan, *Statistical Methods in Food and Consumer Research,* 2009.
Ronald Jackson, *Wine Tasting*. 2009.
Takayuki Shibamoto and Leonard Bieldanes, *Introduction to Food Toxicology*. 2009.